Reactions
of
Coordinated
Ligands

Volume 1

Reactions of COORDINATED LIGANDS

Volume 1

Edited by

Paul S. Braterman

University of Glasgow
Glasgow, Scotland

PLENUM PRESS • NEW YORK AND LONDON

Library of Congress Cataloging in Publication Data

Main entry under title:

Reactions of coordinated ligands.

Includes bibliographies and index.
1. Coordination compounds. 2. Ligands. 3. Reactivity (Chemistry) I. Braterman,
Paul S.
QD474.R38   1985   541.2′242   85-24443
ISBN 0-306-42201-8

© 1986 Plenum Press, New York
A Division of Plenum Publishing Corporation
233 Spring Street, New York, N.Y. 10013

Printed in the United States of America

PREFACE

This book is aimed at graduate students and research workers in all branches of chemistry, who wish to gain insight into what continues to be one of the fastest growing areas of the subject.

Bonding to a metal center may stabilize a ligand towards some reagents, activate it towards others, or modify its chemical behavior in more subtle ways. All these effects have their uses, and all invite understanding in terms of mechanism. Thus mechanistic insight is linked to control of reaction pathways. The detailed working out of this relationship provides the central theme of the book.

The effect of the metal may be electronic or steric, and may involve the energy or the entropy of activation. It may depend on changes induced in the initial state of the ligand, or on those that only arise further along the reaction pathway. It may involve one coordination site or several, and the effects may be more, or less, specific to the metal involved and more, or less, amenable to control through the other ligands. These remarks apply equally strongly to the carbon-bound ligands which occupy the major part of this work, and to those attached by other atoms. Thus the reactions discussed here are relevant in such diverse areas as bulk homogeneous catalysis, stereoselective stoichiometric synthesis, and bioinorganic chemistry.

Some arbitrary boundaries are inevitable. I have chosen to concentrate on ligands attached to transition metals, and on those cases where the metal and/or other ligands strongly affect reactivity. Volume 1 deals with the chemistry of carbon-bound ligands in ascending order of complexity, while

Volume 2 covers carbon dioxide and nitrogen, and other ligands bound through nitrogen, oxygen, phosphorus and sulfur. The aim throughout is to present a unified perspective on this central area of present-day Chemistry.

I would like to thank the many friends and collaborators who have made this work possible; my colleagues at Glasgow for their understanding and support while it was being assembled; Jill Shiel and Angela Hamilton, who prepared the camera-ready text; and Ken Derham and his colleagues in Plenum Publishing for converting this into the finished article.

Paul S. Braterman

CONTENTS

ONE-CARBON, TWO-CARBON AND THREE-CARBON LIGANDS

P.S. Braterman

Department of Chemistry

University of Glasgow

1. GENERAL

In this Chapter, the chosen emphasis is on mechanism, reflecting the author's own interests, catalytic and synthetic applications of organo-transition metal reagents being considered in terms of the component reaction steps. The processes described are grouped according to reaction type and, where knowledge permits, by mechanism. Within each subsection, a general discussion of issues is followed by examples grouped to illustrate the points that emerge, and then further loosely ordered, when the argument allows, according to the Periodic Table. The choice of material is inevitably subjective, and much excellent work has been excluded or treated all too briefly. The General Bibliography[1] to this Chapter, as well as the more specialized reviews cited in the relevant Sections, may go some way towards rectifying these defects. Some individual reviews of general interest cover free energy relationships in organometallic chemistry,[2] the role of free radicals in coordination chemistry,[3] electron transfer in organometallic catalysis,[4] fluxional and non-rigid behavior[5] and σ-π rearrangements,[6] transition metal alkyls and aryls,[7-8] methylene bridged[9] and hydrocarbon and hydrocarbyl bridged species in general,[10] species with M...H-C interactions,[11] cyclometalation[12] and platinacyclobutane chemistry,[13] isotope crossover in investigating C-C bond formation,[14] the activation of hydrocarbons,[15] hydrogenation[16] and the heterolytic activation of dihydrogen,[17] hydrosilation,[18] transition metal catalyzed rearrangements of small rings,[19] metal basicity,[20] the organic chemistry of titanium,[21] osmium alkyls,[22] cobalt[23] and $HCo(CO)_4$,[24] nickel (synthetic aspects),[25] palladium in natural product synthesis,[26] in allylic alkylation,[27] and related reactions,[28] platinum (with reference to metallo-carbonium ions),[29] copper[30] and gold,[31] Ziegler-Natta catalysis,[32] nickel-catalyzed olefin oligomerization,[33] the chemistry of $Rh(PPh_3)_3Cl$,[34] polymerization and co-polymerization reactions of olefins and dienes,[35] the organic and hydride chemistry of transition metals,[36] asymmetric catalysis of allylic alkylation,[37] and the photogeneration of reactive organometallic species.[38] The plenary lectures of the First and Second[39] International Symposia on Organometallic Chemistry directed towards Organic Synthesis have been published; the former includes a most interesting discussion of the fundamentals of catalysis in terms of the energies and symmetries of coupled wave-functions.[40] The stereochemistry of reactions of metal-carbon σ-bonds has been authoritatively reviewed.[41]

2.   THERMOCHEMICAL STUDIES

The thermodynamics of metal-carbon bonds have been reviewed,[42] as have those of oxidation addition.[43] The mean titanium-carbon bond dissociation energy in Cp_2TiPh has been estimated as 310 kJ mol^{-1},[44] and dissociation energies for compounds $Cp*_2ThR_2$ have been given.[45] Perhaps because of handling and purification problems, relevant studies are all too scarce. Mean dissociation energies of 149.5 and 197.8 kJ mol^{-1} have been reported for the metal-methyl bonds in Cp_2MoMe_2 and Cp_2WMe_2 respectively.[46] This should be compared with 159 kJ mol^{-1} for WMe_6.[47] The thermochemistry of pentacarbonylmanganese hydride, acyls and alkyls have been examined in relation to Fischer-Tropsch reactions.[48] In cis-[$(Ph_3P)_2Pt(CH_3)I$], the platinum-methyl bond dissociation energy is only slightly (6 ± 5 kJ mol^{-1}) greater than D(Pt - I).[49] The oxidative addition of acyl halides to a range of species of type trans-[$Ir(Cl)(CO)L_2$] has been studied, and D(Ir-CO.Me) found to be 205 ± 50 kJ mol^{-1}.[50]

In alkylcobalamines and related complexes, dissociation energy increases with the basicity of the ligand trans to carbon, consistent with formal reduction on cobalt-carbon bond cleavage.[51]

The binding energies of organic ligands to a range of transition metal cations in the vapor phase have been studied by ion cyclotron resonance.[52]

3.   ONE-CARBON LIGANDS

Carbonylation and the chemistry of acyls, isonitriles, and carbenes are covered in other Chapters. This Section is therefore confined to alkyls, aryls, and closely related species.

3.1   Formation of metal-carbon sigma bonds

3.1.1   General considerations

It was at one time thought that σ-bonds between transition metals and carbon were inherently unstable, and that if they were formed at all they would collapse unless special factors were present. This view is now recognized as too simple. It is more correct to say that these bonds are often highly reactive, and indeed the enormous usefulness of organometallic reagents stems from this very fact. The fundamental problem of synthesis then becomes the discovery of reagents that will produce the desired bond without side-reactions (based, for instance, on those same reagents' reducing power, electrophilicity, or nucleophilicity), and under such conditions that the desired product will not itself react further.

A number of strategies are employed to reduce the lability of the new metal-carbon bond, by impeding the decomposition processes of Section 3.2 below, so that the organometallic species can be isolated or, at the least, conveniently studied in solution. These include the addition to the reaction mixture of phosphines and other 'soft' ligands (bipyridyl is often effective), the choice of very simple or very bulky organic groups to become ligands, an absence of β-hydrogens on these groups, the use of aromatic or other electron-withdrawing substituents, and the incorporation of the new metal-carbon bond into a chelate ring. Despite such measures, however, any novel combination of reaction type and conditions must be treated as an exercise in exploratory chemistry; this is part of the fun. There remain important examples where the σ-bonded species reacts further in situ; for instance, where that species forms part of a catalytic cycle.

3.1.2 Methods available

The main methods available for forming a new transition metal-carbon bond include transmetalation (eq 1,2), attack by a metal nucleophile on an organic halide or similar species (eq 3), oxidative addition to an organic halide (eq 4), insertion of a metal into a carbon-hydrogen or carbon-carbon bond (eqs 5,6), addition of a metal hydride to an alkene or alkyne (eq 7) or of a nucleophile to an alkene already coordinated (eq 8), the reductive coupling of alkenes to form metallacycles (e.g. eq 9), the interaction of metal complexes with ylids (e.g. eq 10), and other miscellaneous reactions. Each of these is discussed in one of the following subsections, together with the role and changing numbers of the 'supporting' ligands. These are omitted for clarity, as are substituents on the organic groups, in the highly schematic equations 1 - 10. In these, M represents the metal of interest, M' some other (usually main group) metal, R may be alkyl, aryl, benzyl etc., X is halogen or other leaving group precursor, and nothing is implied about stereochemistry.

$$MX + M'R \longrightarrow MR + M'X \tag{1}$$

$$2MX + 2M'R \longrightarrow M\overset{R}{\underset{R}{\diamond}}M + 2M'X \tag{2}$$

$$M^- + RX \longrightarrow MR + X^- \tag{3}$$

$$M + RX \longrightarrow R\text{-}M\text{-}X \text{ or } RM^+ + X^- \tag{4}$$

$$M + R\text{-}H \longrightarrow R\text{-}M\text{-}H \tag{5a}$$

$$MX + R\text{-}H \longrightarrow M\text{-}R + HX \tag{5b}$$

$$M + R\text{-}R' \longrightarrow R\text{-}M\text{-}R' \tag{6}$$

$$MH + C=C \longrightarrow M\text{-}C\text{-}C\text{-}H \tag{7}$$

$$M(\text{alkene}) + Nu \longrightarrow M\text{-}C\text{-}C\text{-}Nu \tag{8}$$

$$M + 2C=C \longrightarrow \overline{M\text{-}C\text{-}C\text{-}C\text{-}C} \tag{9}$$

$$MX + 2R_2Q(CH_3){:}CH_2 \longrightarrow M(CH_2)_2QR_2 + R_2QMe_2^+X^- \ (Q=R,As) \tag{10}$$

3.1.3 Transmetalation leading to normal products

The most familiar reaction of this kind involves exchange of halogen attached to a transition metal for an organic group attached to a more electropositive metal, which plays the role of M' in the alkylating agent M'R of eq 1. Such reactions are, formally, nucleophilic displacements at the transition metal, but the true mechanisms are not always known, and it is plausible that bond formation between M' and halogen simultaneously assists the removal of X. Alkylating agents derived from lithium, magnesium and aluminum, which are the most popular class, are themselves of course more complex species than their conventionalized formulae indicate, and are in principle capable of halogen or organocarbon bridging to the transition metal. There is also a second class of alkylating agent, typified by organomercury and organosilicon compounds. In these the new M'X bond is clearly covalent, and electrophilic attack by M' on the MX bond is at least as important as nucleophilic attack by R on M. In the limit, the mechanism presumably becomes oxidative addition of M'R to M followed by reductive elimination of M'X:

$$M\text{-}X + M'R \longrightarrow XM(R)(M') \longrightarrow MR + M'X \tag{11}$$

The choice of alkylating agent presents interesting problems. It must be remembered that reagents in the first class are nucleophiles and reducing agents as well as alkylating agents, while those of the second kind may give rise to stable M-M' bonds rather than the desired product. The solvent is important, not only directly (as a ligand, or in stabilizing more or less polar transition states), but indirectly in terms of eventual product separation; thus alkyllithium reagents are often preferred to Grignard reagents, not only because they are more effective but because reactions can then with advantage be performed in solvents in which lithium halides are insoluble. In many cases, product lability makes it desirable to carry out the synthesis at reduced temperature and a polar solvent may then be preferred in order to speed polar bond formation; thus at dry ice temperature lithium alkyls are commonly used in diethyl ether, rather than in hydrocarbons. Some typical successful reactions involving organolithium reagents are:

$$MCl_3 \ (M = Er, Lu) + 6LiMe + 3tmen \longrightarrow Li(tmen)_3MMe_6 \qquad (12)$$

where tmen stabilizes Li^+, increasing the nucleophilicity of LiMe, and helps form a large cation, thus reducing product solubility:[53]

$$MCl_3 + 3Me_3SiCH_2Li \longrightarrow (Me_3SiCH_2)_3M + 3LiCl \qquad (13)$$
(M = Cr, V, TiCl, ZrCl, HfCl)[54] and

$$MX_n + nLi(adme) \longrightarrow M(adme)_n + nLiX \qquad (14)$$

(adme = 1-adamantylmethyl; $M(adme)_n$ = $Ti(adme)_4$, $Zr(adme)_4$, $Cr(adme)_4$, $Nb(OBu-t)_3(adme)_2$, $TaCl_2(adme)_3$, $Mn(adme)_2$ - an interesting species with three unpaired spins - $Ni(bpy)(adme)_2$ or $Pd(bpy)(adme)_2$[55] in which the large bulk of the organic group and the absence of β-hydrogen contribute to product stability, and the use of large leaving groups X (OPr-i, OBu-t) facilitates the original reaction:

$$CpNi(Cl)P(OR)_3 + MeLi \longrightarrow CpNi(Me)P(OR)_3 + LiCl \qquad (15)^{[56]}$$
$$IrCl(CO)L_2 + ArLi \longrightarrow Ir(Ar)(CO)L_2 + LiCl \qquad (16)^{[57]}$$
$$MX + LiR \longrightarrow MR + LiX \qquad (17)$$

where MX = $CpFe(CO)_2I$, $W(CO)_5Br^-$; R = 2,4,6-cycloheptatrienyl,[58] and

$$CoCl(PMe_4)_3 + MeLi + PMe_3 \longrightarrow (Me_3P)_4CoMe + LiCl \qquad (18)$$

(note the change from 4- to 5-coordinate cobalt(I) with decreasing electronegativity).[59] With gold(III), the reaction proceeds with retention at metal:[60]

$$\underline{cis}\text{-}Me_2Au(PPh_3)I + CD_3Li \longrightarrow \underline{cis}\text{-}Me_2Au(PPh_3)(CD_3) + LiI \qquad (19)$$

Lithium alkyls BuLi, Me_3SiCH_2Li formally add across the V=O bond in $V(O)(OCMe_3)_3$ to give products $Li[V(O)(OCMe_3)_3R]$ without reduction of the metal.[61]   However, $MoCl_5$ with LiC_6H_4Me-4 in ether gives $(Li.OEt_2)_3Mo(C_6H_4Me$-$4)_6$,[62] and products $(Li.Et_2O)_2Fe(II)(naph)_4$[63] and $(Li.Et_2O)_4Fe(O)Ph_4$[64] are formed from $FeCl_3$ and α-naphthyllithium and phenyllithium respectively.

Two other examples illustrate the limitations of organolithium reagents. In the reactions[65]

$$Cp_2ZrPh_2 + 2MeLi \longrightarrow \text{'MeZrPh.3Et}_2O\text{'} + CpLi + CH_4 + C_6H_6 + ... \quad (20)$$
$$\text{or}$$
$$Cp_2ZrMe_2 + 2PhLi$$

the nucleophilic group attached to lithium has displaced the less nucleophilic cyclopentadienyl; this is not generally observed with the later transition metals (cf. eq 15), presumably because of π-bonding [the formation[66] of species $CpNiR(\eta^2$-$C_3H_6)$ from nickelocene, RMgX and propene is an interesting special case, in which an 18-electron product is generated from a 20-electron precursor]. There has also been decomposition with reduction of the metal. The attempted reaction

$$MCH_2O.CO.CMe_3 + MeLi \longrightarrow MCH_2OLi + MeCO.CMe_3 \quad (21)$$

(in which the initial metal-carbon bond has been prepared by other routes) succeeds for M = $CpMo(CO)_3$, but for M = $CpFe(CO)_2$ or $Mn(CO)_5$, the result is decomposition, presumably by electron transfer, and the implication[67] is that organolithium or Grignard alkylating reagents are liable to cause problems with these metals because of attack by unreacted organolithium on the product as it is generated.

Exchange between a transition metal and aluminum is of course important in generating Ziegler-Natta alkene polymerization catalysts, which depend for their functioning on bridging, and repeated exchange, between the two different metals. For the preparation of metal alkyls,

however, what is required is a simple, irreversible exchange. Some examples
are:

$$2Cp_2MoCl_2 + (EtAlCl_2)_2 \longrightarrow 2Cp_2Mo(Et)Cl + Al_2Cl_6 \qquad (22)$$

(note incomplete exchange),[68]

$$Mn(acac)_3 + 3AlPh_3 \cdot Et_2O + PCy_3 \longrightarrow Ph_2MnPCy_3 + Al(acac)_3 + ... \quad (23)$$

(note the use of acac to solubilize the manganese in aprotic solvents, and the
reduction of the manganese during the reaction),[69] and

$$Fe(acac)_2 + Et_2AlOEt + 2bpy \longrightarrow Et_2Fe(bpy)_2 + ... \qquad (24)^{[70]}$$

With $Co(acac)_3$, the products obtained depend on the ratio of reagents:

$$Co(acac)_3 + bpy + AlR_3/Et_2O \longrightarrow CoR_2(acac)(bpy) + ... \qquad (25)$$
(R = Me, Et) when Al:Co :: 1.5 - 2.0:1, but

$$Co(acac)_3 + 2bpy + 2AlR_3/Et_2O \longrightarrow [CoR_2(bpy)_2][AlR_4] + ... \qquad (26)$$

when the aluminum reagent is in seven-fold excess.[71] The effect of initial
product structure on overall stoichiometry is nicely illustrated by the
sequence[72]

$$Ni(PCy_3)_2 + AcOD \longrightarrow (Cy_3P)_2Ni(D)(OAc)$$
$$(Cy_3P)_2Ni(D)(OAc) + Me_3Al \longrightarrow (Cy_3P)_2Ni(D)Me + Me_2AlOAc$$
but
$$(Cy_3P)_2Ni(D)(OAc) + Et_3Al \longrightarrow (Cy_3P)_2Ni(C_2H_4) + HD + ... \qquad (27)$$

the latter reaction presumably proceeding by β-elimination from an initial
ethylnickel species, followed by reductive elimination (Section 3.2 below).
$Ni(acac)_2$ reacts with dialkylaluminum ethoxides below $-30°C$ in the

presence of phosphines to give products NiL_2R_2, $NiL(acac)R$; but with phosphites, the products are Ni(0) phosphites or Ni-Al complexes.[73]

The reaction of $(Cp*_2IrCl_2)_2$ with trimethylaluminum gave, surprisingly, $Cp*IrMe_4$ in 11% yield as part of a complicated redox process.[74] The rhodium analogue, however, reacts with methyllithium in a rather more straightforward manner to give, ultimately, trans-$[Cp*Rh(Me)]_2(\mu-CH_2)_2$, via a cis species in which the methylene groups show some Rh...H-C bonding.[75]

Highly polar metal alkyls and aryls tend to reduce such metals as palladium, thereby (as mentioned above), limiting their usefulness. To overcome this difficulty, and gain access to catalytic cycles involving palladium, organomercurials have been employed:[76]

$$PhHgOAc + Pd(OAc)_2 \longrightarrow PhPdOAc + Hg(OAc)_2 \qquad (28)$$

An elegant variation is the simultaneous replacement and reduction of Hg(II):

$$ArHgCl + ML_n \longrightarrow ArMCl + Hg + nL \qquad (29)$$

where ML_n is a phosphine complex of zerovalent palladium or platinum.[77] In a closely related reaction, species $HgAr_2$ react with $Pt(PPh_3)_3$ to give $(Ph_3P)_2Pt(Ar)HgAr$, hydrolyzed by trifluoroacetic acid to $(Ph_3P)_2Pt(Ar)O.COCF_3$, HAr, and free mercury.[78] Organosilicon and organotin complexes have also been widely used, to avoid the problem of reduction of product, in the formation of metal-alkyl groups, as well as proving useful transporters of heavily substituted cyclopentadienyl groups.[79-80]

The chemistry of molybdenum halides shows clearly the effects of conditions and of choice of alkylating agent. For instance, the reaction

$$MoCl_3(PR_3)(thf)_2 + Me_2Mg + arene \longrightarrow (\eta^6\text{-arene})MoMe_2(PR_3)_2 + ... \quad (30)$$

involves reduction of Mo(III) to Mo(II), and the uptake of phosphine and arene ligands.[81] Reaction between $MoCl_4$ and $ZnMe_2$ in ether gives $MeMoCl_3$ as a mono-etherate, but the same reagents do not interact in thf or pyridine at low temperature, while at room temperature the main reaction is reduction to $MoCl_3$. $MoCl_5$ in ether reacts with $ZnMe_2$ at $-60°C$ to give $MoCl_4$ and CH_3Cl, while with $ZnEt_2$ the main organic products are ethane and ethylene, (strongly suggesting, given present knowledge, the decomposition of an ethylmolybdenum intermediate).[82] A Mo(V) alkyl has, however, been successfully prepared by the deliberate choice of low-polarity solvent and alkylating agent, in the reaction of $MoCl_5$ with $HgMe_2$ in dichloromethane at $-48°C$, followed by addition of dppe, to give $MeMoCl_4 \cdot dppe$ and $HgCl_2$.[83] Mo(VI) alkyls, like Mo(VI) species in general, are also stabilized by oxide as a ligand. For instance, cis-(bpy)Mo(O)$_2$Me$_2$ can be prepared directly from the corresponding bromide and MeMgBr, and is stable at room temperature to air and water.[84]

The exchange reaction is consistent with the formation or preservation of multiple metal-metal bonds. Thus $Li_4Cr_2Me_8 \cdot 4thf$, containing a quadruple bond between eclipsed $CrMe_4$ moieties,[85] is formed from $CrCl_2$ and methyllithium in thf.[86] Reaction between 1,4-dilithiobutane and $CrCl_2$ gives rise[87] to the very similar[88] $Li_4Cr_2(C_4H_8)_4 \cdot 4thf$. $Cr_2[(CH_2)_2PMe_2]_4$, from chromium(II) acetate and $Me_2P(CH_2)_2Li$, has the structure I with bridging organic ligands.[89] $Re_2Cl_2(OAc)_4$ reacts with species R_2Mg in ether to give products II,[90] where R is a group (CH_2SiMe_3, CH_2CMe_3, CH_2Ph) which cannot undergo β-hydride elimination.

I   II   III

As some of the above examples show, the exchange reaction may be used in the formation of metallacycles. Since these are discussed further in Chapter 10, a few examples here must suffice. $Li_2C_4H_8$ reacts with dppePdCl$_2$ to give the palladacyclopentane, dppePd(C$_4$H$_8$),[91] while chelated diphosphine complexes (L-L)NiBr$_2$ with LiCPh:CPh.CPh:CPhLi give the metallacyclopentadienes (L-L)Ni(CPh)$_4$ (**III**).[92] Metallaheterocycles can also be obtained by a combination of metal-metal exchange and simple ligand replacement:

$$M(CO)Br + ClMg(CH_2)_3PPh_2 \longrightarrow \overline{M.PPh_2.CH_2.CH_2CH_2} + ClMgBr \quad (31)$$

where M = Mn(CO)$_4$, Re(CO)$_4$, CpFe(CO).[93] An interesting new class of metallacycles, accessible by trans-metalation, are the ortho-phenylene bis(methyl) or bis(methylene) derivatives, which contain a ligand possibly mesomeric between **IVa** and **IVb**. Thus WOCl$_4$ reacts with o-C$_6$H$_4$(CH$_2$MgCl)$_2$ to give the thermally robust, 1-electron reducible species **V**,[94] while RuCl$_2$L$_4$ (L = PMe$_2$Ph, PMePh$_2$) reacts with o-CH$_3$C$_6$H$_4$CH$_2$MgBr, eliminating HCl as well as magnesium halide, to give products formulated as **VI**.[95]

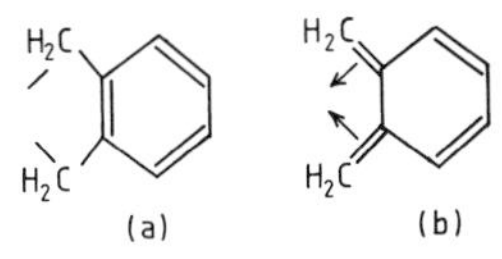

| IV | V | VI |

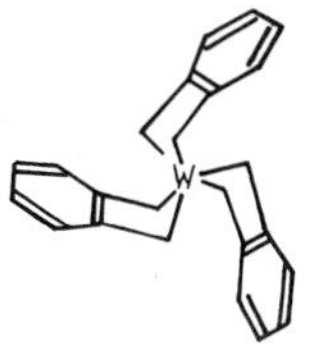

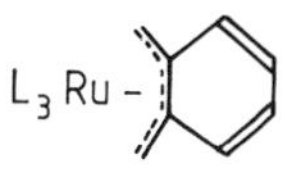

3.1.4 Transmetalation leading to alkyl-bridged and related products

Alkyl bridging between main group elements is well established for lithium, beryllium, magnesium and aluminum,[96] and has long been suspected in materials formulated (too simply, according to more recent work),[97] as manganese alkyls. Recently, more examples have begun to emerge:[98-9]

$$(Cp_2MCl)_2 + LiAlR_4 \longrightarrow Cp_2M(\mu\text{-}R)_2AlR_2 \quad (32)$$
(M = Ti,Sc,Y,Ln; R = e.g. Me),
$$2Cp_2M(\mu\text{-}Me)_2AlR_2 + 2py \longrightarrow Cp_2M(\mu\text{-}Me)_2MCp_2 + ... \quad (33)$$

Manganese(II) chloride reacts with $ClMg(CH_2SiMe_3)$ to give a polymer, $[Mn(CH_2SiMe_3)_2]_n$, in which Mn is tetrahedral and all ligands bridge. Long-established 'Me$_2$Mn' and 'Ph$_2$Mn' prepared similarly contained magnesium and halide. Organolithium reagents give rise to anions $[MnR_4]^{2-}$, which for $R = CH_2SiMe_3$ can be oxidized to MnR_4.[97]

Treatment of chromium(II) acetate with $Mg(CH_2SiMe_3)_2$ (chosen, rather than the lithium reagent, to avoid formation of anions; cf. Reference 97) gave, in the presence of trimethylphosphine, a produce shown crystallographically to be $[(Me_3SiCH_2)Cr(PMe_3)_2](\mu\text{-}CH_2SiMe_3)_2$, with alkyl bridges and a quadruple metal-metal bond.[100-101] Re_3Cl_9, with sufficient $MgMe_2$, gives Re_3Me_9,[102] while $MeMgCl$, followed by PEt_2Ph, led to $Re_3Me_6(\text{-}Me)_3(PEt_2Ph)_2$, **VII**.[103]

3.1.5 Displacement of a leaving group by anionic metal

Reaction (3) is an important special case of oxidative addition (3.1.6. below), and has been used with metal carbonyl anions for many years (for an early review, see Reference 104). The reaction succeeds most readily with activated (allylic or benzylic) derivatives. The nucleophilicity of M^- correlates with the reducing power of the $M_2/2M^-$ couple. Nucleophilicities have been estimated from the rate of attack of metal on methyl iodide (or methyl tosylate, which is generally far less reactive to metals), and the order of increasing nucleophilicity for anions and for neutral species which take part in directly related reactions is[105-6]

$$
\begin{aligned}
&RhCl(CO)(PPh_3)_2 < PPh_3 < CpCo(CO)PPh_3 < IrCl(CO)(PPh_3)_2 < \\
&CpRh(CO)PPh_3 < Pt(PPh_3)_2 < CpIr(CO)PPh_3 < Co(CN)_3^{3-} < Pd(PPh_3)_3 < \\
&Co(CO)_4^- < CpCr(CO)_3^- < LiAuMe_2 < Li_2PtMe_4 < (LiCuMe_2)_2 < \\
&CpMo(CO)_3^- < CpW(CO)_3^- < Mn(CO)_5^- < Pt(PEt_3)_3 < Ni(PPh_3)_3 < \\
&Fe(CO)_4^{2-} < Re(CO)_5^- < Rh(CN)_4^- < Pd(PEt_3)_3 < Co(dmgH)_2py^- < \\
&CpNiCO^- < CpRu(CO)_2^- < Ni(PEt_3)_4 < CpFe(CO)_2^-.
\end{aligned}
\qquad (34)
$$

Equation (3) idealizes the displacement reaction by ignoring the opposed effects of ion pairing (reducing anion nucleophilicity) and countercation assistance in the departure of the leaving group. The relative

importance of these is finely balanced; thus extensively ion-paired Na^+M^-
[M = $Mn(CO)_5$,[107] $CpMo(CO)_3$[108]] reacts more rapidly in thf with benzyl
chloride than does ion-separated $Na(HMPA)_x^+M^-$, but for alkyl halides, at
least where M = $CpMo(CO)_3$, the reverse is true.[108]

Some typical examples of the reaction are

$Ph_3PMn(CO)_4^- + ClCH_2COMe \longrightarrow \underline{cis}\text{-}Ph_3PMn(CO)_4CH_2COCH_3 + Cl^-$ (35)[109]

$CpFe(CO)_2^- + ClCH_2SMe \longrightarrow CpFe(CO)_2CH_2SMe + Cl^-$ (36)[110]

$2CpFe(CO)_2^- + \textbf{VIII} \longrightarrow \textbf{IX} + 2Br^-$ (37)[111]

$Co(CO)_4^- + BrCH_2COOEt \longrightarrow (OC)_4Co.CH_2COOEt + Br^-$ (38)[112]

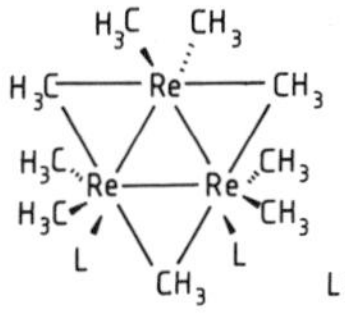

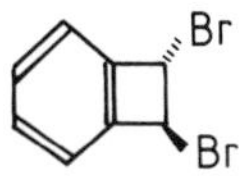

The dianion $[Fe_2(CO)_8]^{2-}$ reacts with diiodomethane to give $(OC)_4\overline{Fe\text{-}}$
$\overline{CH_2}\text{-}Fe(CO)_4$, one of the growing group[9] of methylene-bridged species.[113]

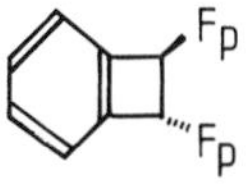

The reaction proceeds, as expected, with inversion at carbon, as
shown by Schemes 1[114] and 2.[115]

Scheme 1

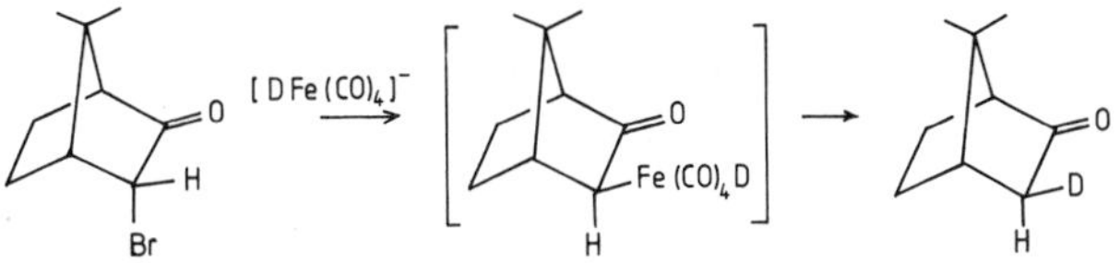

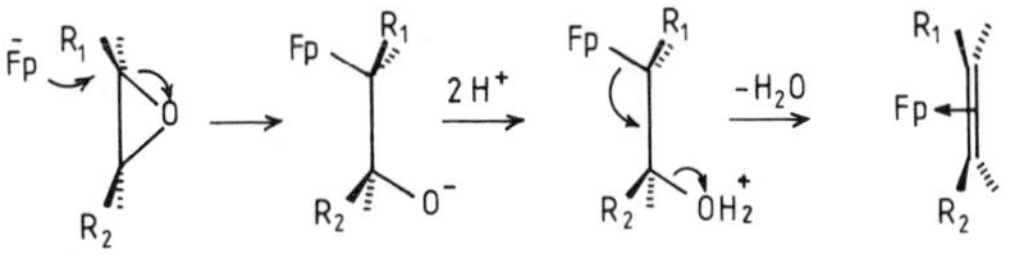

Scheme 2

Closely related are such reactions as

$\frac{1}{4}(Cp_2WHLi)_4 + 2CH_2{:}CR.CH_2Cl \longrightarrow Cp_2W(CH_2CR{:}CH_2)_2 + \ldots$ (39)[116]

$Cp_2ReLi + MeOTs \longrightarrow Cp_2ReMe + LiOTs$ (40)[117]

in which the metal nucleophile is covalently attached to a highly electropositive metal. The base-promoted reaction of $CpRe(CO)_2H_2$ with methyl iodide gives $CpRe(CO)_2Me_2$ via intermediates $[CpRe(CO)_2H]^-$, $CpRe(CO)_2(H)(Me)$ and $[CpRe(CO)_2Me]^-$.[118]

Scheme 3

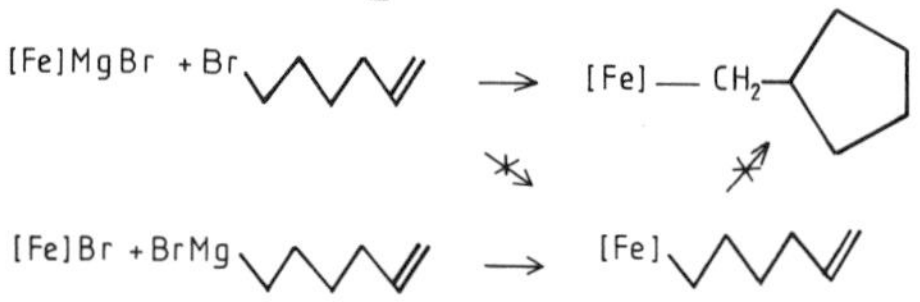

Some reactions that might be expected to conform to the stoichiometry of eq 3 in fact proceed by a free radical path. Thus the organometallic Grignard reagent $CpFe(dppe)MgBr$ reacts with alkyl bromides RBr to give $CpFe(dppe)Br$, R-R, RH and R(-H), as well as the expected $CpFe(dppe)R$. That the free radical process is a feature of the reagents rather than the products is shown by Scheme 3, cyclization being characteristic of the 5-hexenyl radical.[119] A free radical mechanism is also involved in the process

$$1\text{-adI} + 2Co(CN)_5^{3-} \longrightarrow 1\text{-adCo(CN)}_5^{3-} + Co(CN)_5I^{3-} \tag{41}$$

(1-ad = 1-adamantyl), for which kinetics, and radical scavenger effects, imply a mechanism

$$1\text{-adI} + Co(CN)_5^{3-} \longrightarrow 1\text{-ad}^{\cdot} + ICo(CN)_5^{3-} \tag{41a}$$
$$1\text{-ad}^{\cdot} + Co(CN)_5^{3-} \longrightarrow 1\text{-adCo(CN)}_5^{3-} \tag{41b}$$

and presumably in other related reactions less amenable to kinetic study.[120]

The dinuclear free radical $[Cp_2Co_2(CO)_2]^-$ reacts with halides RX to give cobalt-cobalt bonded species $(CpCoR)_2(\mu\text{-CO})_2$; CH_2I_2 and 1,3-diiodopropane give stable products with methylene and trimethylene bridges.[121]

3.1.6 Oxidative addition

This term is used for any process in which the coordination number and oxidation number of a metal both increase. The ease of such a process depends on many factors, such as the oxidizability of the metal in its actual environment, and can be enhanced by a judicious use of ligands (for example, replacing aryl by alkyl phosphines). Cases in which the metal is initially anionic were discussed in 3.1.5. above, while those in which a new metal-carbon bond is formed by cleavage of existing C-H or C-C bonds are discussed in 3.1.7. and 3.1.8. below.

These exclusions still leave a variety of reaction types and mechanisms. The process may or may not involve overall loss of an original ligand (oxidative elimination):

$$ML_n + RX \longrightarrow ML_{n-1}(R)X + L \tag{42}$$

The reaction may leave two new groups attached to the metal, or only one (the alternatives of eq 4). The oxidation state of the metal may increase by two (as in the above examples), or by one:

$$M + RX \longrightarrow MR + ... \tag{43}$$

A similar complexity attaches itself to each of the reaction steps. Overall addition may involve direct insertion of an unsaturated metal into the R-X bond; or nucleophilic displacement followed by electrophilic recombination of MR^+ and X^-; or homolysis, e.g.

$$M + RX \longrightarrow MX + R \tag{44}$$

followed by radical trapping at the same metal center

$$MX + R \longrightarrow R\text{-}M\text{-}X \tag{45}$$

or a different one:

$$M + R \longrightarrow MR \tag{46}$$

which can lead, by reaction between MR or MX and RX, to a radical chain reaction.

Since mechanism is sensitive to solvent and to small structural changes, it is naive to hope for too high a degree of generality, and each particular reaction of interest requires separate study. One aid to such study is product stereochemistry, both at the metal and at the R groups involved. Direct insertion of 2-coordinate d^{10} or 4-coordinate d^8 metals takes place with cis-addition of R and X, as required by frontier orbital considerations.[122-3] Any observed trans products must arise, either by rearrangement of an initial complex, or by some more indirect addition pathway. Moreover, direct insertion, which cleaves the R-X bond by populating the initial R-X σ^* orbital, should proceed with retention at R, while nucleophilic and radical attack should cause inversion and non-specificity respectively. The possible reversibility of some individual steps, even when the overall reaction is irreversible, and the existence of competing pathways, can lead to more complex results, such as incomplete specificity, and partial racemization of starting material.

Work on the mechanism of oxidative addition to Group VIII complexes has been reviewed,[124] as has the related topic of reaction between zerovalent palladium and platinum, and main group organometallics.[125] The reaction

$$M + MeSO_3F \longrightarrow MMe^+ + SO_3F^- \qquad (47)$$

is shown by M = $Mo(phen)(PPh_3)_2(CO)_2$, Cp_2MoH_2, $Rh(ttp)Cl$, $Ir(CO)(PPh_3)_2Cl$, or $Pt(PPh_3)_3$; trimethyloxonium salts are alternative methylating agents for these metals.[126-7] $Ru[P(OMe)_3]_5$ with methyl iodide gives $\{MeRu[P(OMe)_3]_5\}^+$,[128] while $[MeFe(CO)_3(PMe_3)_2]^+$ is probably an intermediate in the formation of $Fe(CO)_2(PMe_3)_3(COMe)I$ from $Fe(CO)_3(PMe_3)_2$ and methyl iodide.[129] $Mo(CO)_2(dppe)_2$ reacts according to a variety of mechanisms, giving initial $[RMo(CO)_2(dppe)_2]^+$ decomposing to $[HMo(CO)_2(dppe)_2]^+$ with methyl, ethyl, or t-butyl iodides, but abstracting halogen from benzyl bromide to give $Mo(CO)_2(dppe)_2Br^+Br^-$ and bibenzyl.[130]

Apparently straightforward oxidative addition, with R and X mutually cis in the product, is shown by methyl iodide and "$Fe(PMe_3)_4$"[131] (for the isomerism of this compound, see 3.1.7. below), by $(Me_3P)_4CoMe$ and MeBr (with elimination of one PMe_3),[59] and by platinum(II) alkyl complexes with a wide range of organic halides (for examples, see References 132-3). Complexes of type trans-$Pt(X)(CCR)L_2$ react similarly with methyl iodide, but more slowly, as might be expected for an oxidation.[134]

Oxidative addition of **X** to $Pd(PPh_3)_4$ gives **XI**, presumably by nucleophilic displacement followed by recombination.[135] Such a two-step process is clearly indicated for the addition (in methanol) of $IrCl(CO)(PMe_2Ph)_2$ to allyl halides; the allyl and halide groups are mutually trans in the product, and ionic halide in solution can be incorporated in the product in place of halide originally bound to carbon.[136]

The addition of methyl iodide to species $RhCl(CO)(PR_3)_2$ is kinetically highly complex. Dissociation of a phosphine ligand leads to a series of interlocking processes including reversible quaternisation of the free phosphine by methyl iodide, and addition of the iodide ions so formed to the initial complex, as well as direct reaction between methyl iodide and the complexes present.[137]

Free radical processes in oxidative addition occur across the periodic table. Complexes Cp_2ZrL_2 react with alkyl halides to give mixtures of Cp_2ZrX_2 and $Cp_2Zr(R)X$;[138] Zr(III) intermediates are directly observable, and the intensity of their esr signals has been ascribed to the balance between ease of halide abstraction by unsaturated Cp_2ZrL from RX, and the reactivity of R' (and RX) towards the Zr(III) species.[139] Chromium(II) reacts with active organic halides to give equimolar mixtures of Cr(III) halide and Cr(III) organic derivatives, according to reactions (44) (slow, first order in each component), and (46) (fast).[140] The photochemical reaction between $(Ph_3P)_2Pt(C_2H_4)$ and dichloromethane is retarded by duroquinone and gives $(Ph_3P)_2PtCl_2$ as well as $(Ph_3P)_2Pt(Cl)(CH_2Cl)$, strongly indicating a free radical pathway;[141] in this field, as elsewhere, light should be avoided in studies directed towards thermal processes, especially where colored, highly reactive species are involved.

The situation with iridium(I), palladium(0), and platinum(0) is so complicated that the concept of a single dominant mechanism seems out of place. The rate of oxidative addition of aryl iodides to $IrCl(CO)(PAr_3)_2$ (in 1-methylnaphthalene at $150^{\circ}C$) contains two terms, one first order in complex only while the other is first order both in complex and in substrate. The latter term is decreased by added phosphine, suggesting a two-step mechanism with reversible dissociation of phosphine as the first step.[142] The oxidative addition to related trialkylphosphine complexes of optically active ethyl 2-bromopropionate, and of cis- and trans-2-fluorobromocyclohexane (in which the stable CF bond acts as a marker), occur with complete lack of stereospecificity.[143] Such evidence is not conclusive, however. Addition of active $PhCHClCF_3$ to palladium(0) gives racemic products; but under CO, organic ligands can be trapped by carbonyl insertion (cf. Chapter 6), as optically active acyl groups.[144] So unless there is a difference in mechanism between the reactions of Pd(0) phosphines and Pd(0) carbonyl phosphines, it would seem that the organometallic first formed does retain stereochemical information, and that this is only lost in a later stage (such as Pd - Pd exchange) which can be pre-empted by carbonylation. With iridium(I),[145] mechanism is highly sensitive to the nature of the substrate, with saturated alkyl halides (other than methyl halides), vinyl and aryl halides, and α-haloesters said to add by a radical chain path. A broadly similar duality exists for Pt(0);[146] and solvent effects and catalysis by ethyl bromide both support a radical chain mechanism for addition of bromobenzene to Pt(0) [and, incidentally, to Sn(II)].[147] The addition of iodobenzene to $Pd(PPh_3)_3$, however, is quite different, with addition to an intermediate $Pd(PPh_3)_2$, perhaps by initial coordination of iodine.[148] Evidence for radical pathways comes from CIDNP studies of the reactions of $Pd(PEt_3)_3$ and $Pt(PEt_3)_3$ with isopropyl iodide or benzyl bromide. Such pathways tend to generate metal dihalides as side-products, and part of the resulting mechanistic complexity is summarized in Scheme 4.[149] Incomplete inversion occurs in the reaction between $Pd(PEt_3)_3$ and PhCHDCl; this is not due to exchange nor to radical pathways (since standing with excess Pd(0) has no effect, and no bibenzyl is formed) and may reflect pseudorotation at 5-coordinate carbon in the transition state during

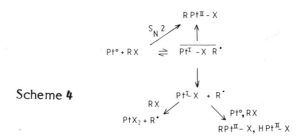

Scheme 4

nucleophilic displacement of halide.[150] Free radical intermediates may be detected in favorable cases by using spin traps, such as t-butylnitrosyl, although great care is necessary in interpretation since such species can themselves initiate chains.[151] An obvious related caveat is that traces of oxygen may affect the mechanism in unforeseen ways.

For nickel complexes, behavior is further complicated by the existence of stable odd-electron oxidation states. The reaction between $(Et_3P)_4Ni$ and aryl halides is thought to proceed through a charge transfer radical pair $[(Et_3P)_3Ni]^+[ArX]^-$, which gives eventually both <u>trans</u>-$(Et_3P)_2Ni(Ar)X$ and nickel(I) halide complexes.[152] However, the odd-electron species could be generated by homolysis or electron transfer <u>after</u> oxidative addition; reaction of $(Cy_3P)_2Ni$ with chlorobenzene gives the same products (mainly biphenyl) as does the thermolysis of $(Cy_3P)_2Ni(Ph)Cl$, and the reaction of $(Cy_3P)_2Ni$ with ethyl bromide is thought to involve oxidative addition, a combination of homolysis and β-hydride elimination, and radical attack on nickel hydride (Scheme 5).[153]

$$L_2Ni + EtBr \longrightarrow L_2Ni(Et)Br$$
$$L_2Ni(Et)Br \longrightarrow L_2Ni(H)Br + C_2H_4$$
$$L_2Ni(Et)Br \longrightarrow \tfrac{1}{2}(L_2NiBr)_2 + Et^{\cdot}$$
$$Et^{\cdot} + L_2Ni(H)Br \longrightarrow C_2H_6 + \tfrac{1}{2}(L_2NiBr)_2$$

Scheme 5

In the presence of excess aryl halide, the reactions of "nickel(0)" become of great overall complexity, involving both nickel(I) and nickel(III) species;[154] as in other transition metal catalyzed coupling reactions (3.3.2. below), the predominant mechanisms may differ from those governing related stoichiometric reactions. $Ni(PEt_3)_4$ in toluene reacts more rapidly with methyl bromide and iodide, and with ethyl bromide, than does the corresponding Ni(I) species $Ni(PEt_3)_3X$,[155] so that alkyl and aryl halides may differ in their coupling mechanisms.

More recently, free radicals have been implicated in the oxidative addition of alkyl iodides RI to $Pt(bpy)Me_2$, which in the presence of alkenes $R'CH:CH_2$ gives products $Pt(bpy)Me_2I(CHR'.CH_2R)$ by initial attack of R on alkene.[156]

Bridged Au(I) ylid complexes $Au_2(\mu\text{-}CH_2PR_2CH_2)_2$ react with methyl iodide to give $MeAu(\mu\text{-}CH_2PR_2CH_2)_2AuI$; these are formal Au(II) complexes with a gold-gold bond.[157]

Nucleophilic attack by Ir(I) on styrene epoxide, followed by β-elimination from CH-O, gives products of type $PhCOCH_2Ir(H)(Cl)L_3$.[158]

3.1.7 Insertion into carbon–hydrogen bonds

Reactions of this type are also known as metalations, and include cyclometalations (where the new metal-carbon bond completes a chelate ring), and other processes which include protolysis of existing metal-ligand bonds by carbon acids, electrophilic substitution by metal at unsaturated carbon, and oxidative addition of C - H to highly reducing or unsaturated metals. Cyclometalations themselves often occur with allylic or benzylic CH, with aromatic CH, or in cases where the ring formed contains bulky substituents.   Each of these overlapping classes is illustrated in what follows; other examples can be found in Section 3.2.7. of this Chapter and in Chapter 10 below.

3.1.7.1 Cyclometalation at benzylic carbon:- This reaction is shown by such electrophilic metals as palladium(II) and phosphite-bound, rather than phosphine-bound, iridium(I). For instance, when complexes such as $Ir(C_6H_4\text{-}o\text{-}CH_3)(CO)L_2$ (L = phosphine) are treated with phosphites, Ir(III) species XII (L = phosphite) are formed.[57]   Palladium(II) chloride reacts with 8-methylquinoline to give XIII. Substitution at C(2) impedes the reaction, implying steric constraints on a transition state in which nitrogen is already coordinated to the metal.[159-60]

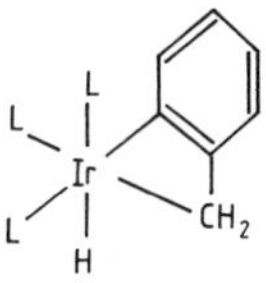

XIV   XV

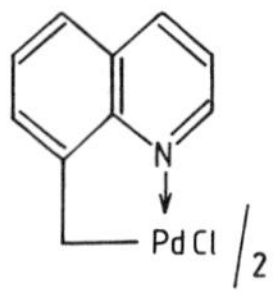

XII   XIII

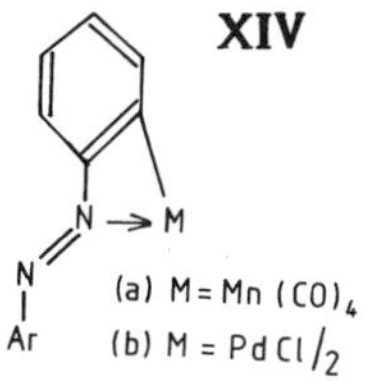

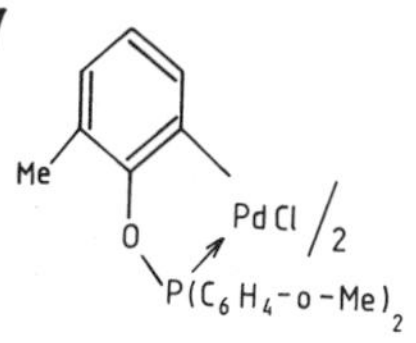

3.1.7.2 Cyclometalation at aromatic and vinylic carbon:- Azobenzenes are cyclometalated by a variety of reagents. Methylmanganese pentacarbonyl and palladium(II) chloride give **XIVa** and **XIVb** respectively. The effects of ring substituents on reactivity and regiospecificity confirm that $MeMn(CO)_5$ acts as a nucleophile (methyl presumably attacking ring hydrogen), while $PdCl_2$ is an electrophile.[161] The same is true no doubt for the reaction between $PdCl_2$ and tris(o-tolyl)phosphite, with metalation of the ring, rather than the side-chain, to give **XV**,[162] while electrophilic substitution by Ir(III) is presumably responsible for the formation of carbon-bonded[163] $[Ir(bpy)_2(C_{10}H_7N_2)]^{2+}$ as a major contaminant in attempted thermal preparations of $[Ir(bpy)_3]^{3+}$. The aromatic C - H bond in phosphorus ligands can undergo oxidative insertion as well as substitution; thus the substance of composition "$(dppe)_2Fe$" actually has the structure **XVI**,[164] while heating $[Ir(PMe_2Ph)_4]^+$ in solution to 80°C gives **XVII**.[165]

$R_2PCH_2C_6H_5$ and related ligands undergo ortho-insertion into ring C-H by Rh(I) and Ir(I) chlorides, to give complexes containing the grouping $(R_2P-C)M(III)(H)(Cl)$, where R_2P-C is the bidentate ligand $R_2PCH_2C_6H_4$-o-. Palladium and platinum(II) chlorides, on the other hand, react with overall substitution to give products $(R_2P-C)_2M(II)$. Substituent effects show Rh(I) to be a nucleophile, and Pd(II) an electrophile, with Ir(I) and Pt(II) apparently intermediate.[166] Bis(cyclopentadienyl)titanium(III) alkyls react with pyridines to give 1-azabenzyne complexes **XVIII**.[167]

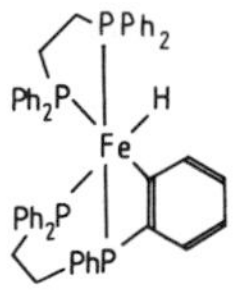

XVI 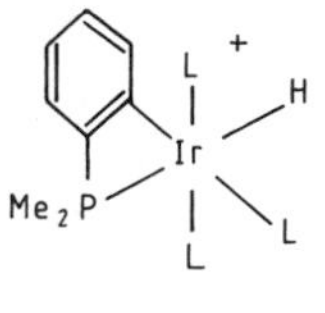   **XVII** 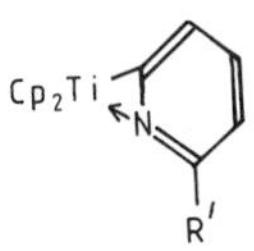   **XVIII**

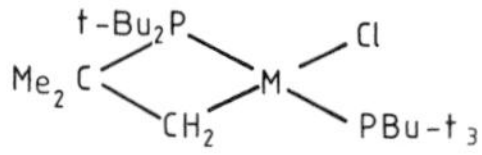

XIX (a) M = Pd
        (b) M = Pt

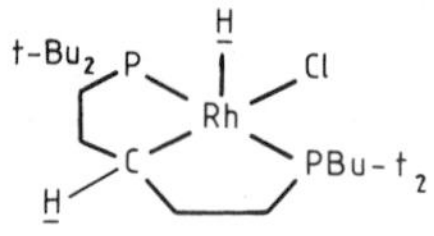

XX

3.1.7.3 Cyclometalation of ligands with bulky substituents:- The general tendency of bulky substituents to favor cyclization[168] is shown clearly in ligands with heavily substituted alkyl groups.[169] For instance, both (t-Bu$_3$P)$_2$PdHCl[170] and its platinum analogue[171] lose hydrogen to give **XIXa** and **XIXb** respectively. Rhodium(III) chloride trihydrate reacts with t-Bu$_2$P(CH$_2$)$_5$PBu-t$_2$ (L-L) to give the bimetallic macrocycle RhHCl$_2$(μ-L-L)$_2$RhHCl$_2$, which with base (o-lutidine) gives **XX**, in which the hydrogen atoms underlined exchange rapidly on the nmr timescale at room temperature.[172] The substituent effect is by no means confined to phosphine ligands, as shown by the reaction sequence of Scheme **6**, where labeling proves that the hydrogen transferred to the eliminated neopentane originated on the cyclizing neopentyl group.[173] The reaction has been extended to 5- and 6-membered ring formation.[174]

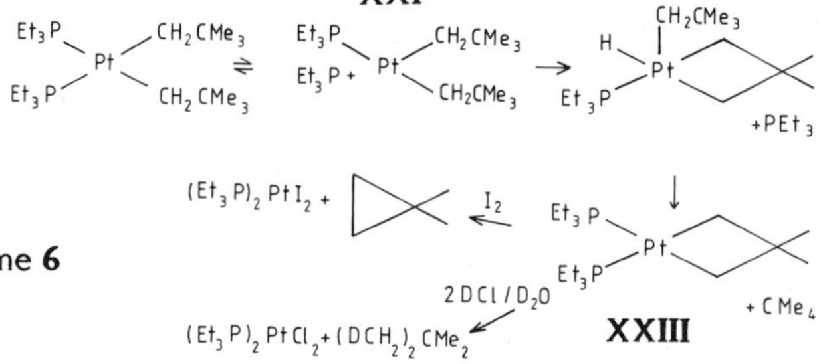

Scheme **6**

3.1.7.4 Reduced phosphine complexes:- Formally zerovalent, highly nucleophilic Group VIII metal complexes can be generated by reduction of chlorides in the presence of basic phosphines. In these species, the metal is capable of oxidative addition to C-H bonds that would normally be considered inert, either within the complex itself or in a (preferably unsaturated) substrate. Thus "Fe(PMe$_3$)$_4$", generated from iron(II) chloride, trimethylphosphine, and sodium amalgam or magnesium, undergoes the equilibria of Scheme **7** and may react in either the Fe(0) or Fe(II) form. For

Scheme 7

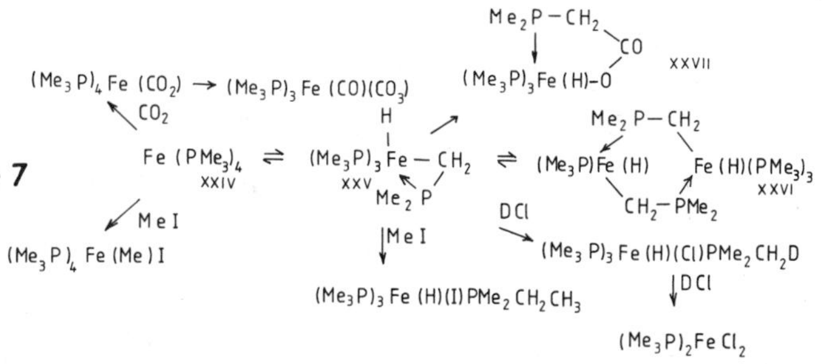

$$Me_2P-CH_2$$
$$(Me_3P)_4Fe(CO_2) \rightarrow (Me_3P)_3Fe(CO)(CO_3) \qquad (Me_3P)_3Fe(H)-O \overset{CO}{\diagdown} \quad \textbf{XXVII}$$
$$\searrow CO_2$$

$$Fe(PMe_3)_4 \rightleftharpoons (Me_3P)_3Fe-CH_2 \rightleftharpoons (Me_3P)Fe(H)\diagup Fe(H)(PMe_3)_3$$
$$\textbf{XXIV} \qquad \textbf{XXV} \underset{Me_2P}{\diagup} \qquad \overset{DCl}{\searrow} \quad CH_2-PMe_2 \quad \textbf{XXVI}$$

$$\swarrow MeI \qquad \downarrow MeI \qquad \rightarrow (Me_3P)_3Fe(H)(Cl)PMe_2CH_2D$$
$$(Me_3P)_4Fe(Me)I \qquad (Me_3P)_3Fe(H)(I)PMe_2CH_2CH_3 \qquad \downarrow DCl$$
$$(Me_3P)_2FeCl_2$$

instance, the reaction with methyl iodide gives two products, $(Me_3P)_4Fe(Me)I$ (3.1.6. above) and $(Me_3P)_3Fe(H)(I)PMe_2CH_2CH_3$, while carbon dioxide can either coordinate to the iron in **XXIV**, with activation of the C-O bond, or insert into the iron-carbon bond of **XXV** (or **XXVI**) to give **XXVII**.[131,175-6] The equilibria have been further demonstrated by following H-D interchange in $[(CD_3)_3P]_4Fe(D)CH_2PMe_2$, prepared from $[(CD_3)_3P]_4Fe(D)I$ and $LiCH_2PMe_2$.[177] The ruthenium analogue may be prepared from sodium amalgam and $Ru(PMe_3)_4Cl_2$ in benzene. It adopts predominantly the monomeric Ru(II) hydride form, but unlike the iron complex it is rigid on the nmr timescale at room temperature, and does not take up CO. Attempts to prepare the osmium analogue led instead to cis-$(Me_3P)_4Os(H)(C_6H_5)$.[178-9]

Related behavior in chelating diphosphine complexes is well established. Reduction of $(dmpe)_2MCl_2$ (M = Fe,Ru) by sodium naphthenide, $NaC_{10}H_8$, gives $(dmpe)_2M(H)(\beta-C_{10}H_7)$.[180-1] However, $(dppe)_2Rh$, generated electrochemically from the stable Rh(I) cation, is a free radical, and attacks toluene by methyl hydrogen atom abstraction.[182] The material of composition $(dmpe)_2Ru$ has been characterized crystallographically as a dimeric hydride with RuPCRu bridges.[183] The reaction of $Ir(PMe_3)_4Cl$ with a range of organolithium or -magnesium derivatives leads by γ- (or δ-) elimination to cyclic products $(Me_3P)\overline{Ir(H)-C_nC}$, where the C_nC group may be $CH_2CMe_2CH_2$, CH_2COCH_2, $CH_2C_6H_4$-o- or $CH_2CMe_2C_6H_4$-o-.[184]

3.1.7.5 Other reactions involving activated CH groups:- The reaction

$$MR + HR' \longrightarrow MR' + HR \qquad (48)$$

generally proceeds from left to right if HR' is a sufficiently strong proton acid. The strength of CH acids generally increases with unsaturation at carbon; under the influence of electron-withdrawing groups; and at positions α to carbon-carbon and, more especially, carbon-nitrogen and carbon-oxygen multiple bonds. There are many cases where Reaction 48 proceeds but where it is unlikely that HR' ever actually ionizes, and if the initial complex is a hydride then the reaction may involve elimination of HR, followed by insertion of metal into the H-R' bond, rather than any direct interaction between entering and departing ligands. For example, $(dmpe)_2Fe(H)(2$-naphthyl) reacts with acetonitrile, benzene, or methyl methacrylate to give products $(dmpe)_2Fe(H)(R)$ $(R = CH_2CN, C_6H_5,$ or CH:CMeCOOEt). C_6D_6 gives $(dmpe)_2Fe(D)(C_6D_5)$, showing that in this case at least the reaction is oxidative insertion of the $Fe(dmpe)_2$ fragment.[180-85]

Hydrogen transfer between R and R' is, however, probably involved in the reactions of tetramethylphosphonium chloride with organochromium compounds. Li_3CrPh_6 reacts to give **XXVIII** and thence **XXIX**, which reacts in turn with the stronger acid phenylacetylene to give **XXX**,[186] while $Li_4Cr_2Me_8$ provides an alternative route (cf. Reference 89) to **I**.[187] Deprotonation equilibria are presumably involved in the catalysis by tris(triethylphosphine)platinum (present in equilibrium with $HPt(PEt_3)_3{}^+OH^-$) of H_2O/D_2O exchange with the α-hydrogens of ketones. Aldehydic hydrogen is also exchanged but perhaps by a slightly different mechanism.[188] The decarbonylation of aldehydes by Rh(I) proceeds via metal insertion into the CO-H bond to give an acylmetal hydride.[189] Ir(I) inserts similarly, especially where the acyl is potentially chelated to phosphine.[190]

XXVIII **XXIX** **XXX**

3.1.7.6 Palladation of aromatic CH:- Palladium(II) chloride complexes, or, more conveniently,[191] the more soluble palladium(II) acetate, attack aromatic CH. Methyl groups sterically hinder,[192] but the activating and directing effects of methoxy groups confirm the electrophilic nature of the reaction, which can be applied to furans and thiophenes as well as to benzenoids. In the presence of CO, the reaction leads to carboxylation:[193]

$$ArH + CO + Pd(OAc)_2 \longrightarrow ArCOOH + Ac_2O + Pd \tag{49}$$

while styrenes react with the palladium aryl to give stilbenes[192] (cf. Section 3.3.3. below).

3.1.7.7 Metalation of allylic CH:- This important reaction is discussed in reviews cited above,[26-8] and, since it generally gives rise to η^3-allylic ligands, in Section 5. below.

3.1.7.8 Insertion into completely inactivated C-H bonds:- This can occur if the metal is sufficiently electron-rich and unsaturated. For instance, photolysis of $Cp^*Ir(PMe_3)H_2$ in R-H as solvent gives rise to products $Cp^*Ir(PMe_3)(H)(R)$, presumably by insertion of the photolysis fragment Cp^*IrPMe_3, where R is phenyl, cyclohexyl, or even neopentyl (the PMe_3 ligand was chosen as being itself a poor substrate for insertion).[194-5] The reaction has been extended to insertion by $Cp^*Rh(PMe_3)$[196] and $Cp^*Ir(PMe_3)$[195] generated thermally from $Cp^*M(PMe_3)(R)(H)$, and to $CpIr(CO)$ and $Cp^*Ir(CO)$ generated photochemically in C_6F_{12} from the corresponding dicarbonyl; these last were shown to insert into the C-H bond of methane.[197] The insertion of rhodium and iridium into C-H bonds has been reviewed.[198] Methane actually undergoes thermal exchange with species Cp_2LuMe and Cp_2YMe, possibly via species of type $[Cp_2M]_2\mu$-CH_3.[199]

Molybdenocene and tungstocene, which have been detected by matrix photolysis, are key intermediates in a number of C-H insertions including self-insertion to give $\eta^1:\eta^5$-C_5H_4 ligands.[200]

The vapor phase insertion of transition metal cations[52,201-2] and hydride cations,[203] and the subsequent fragmentation patterns, have been studied by such physical techniques as ion cyclotron resonance and Fourier transform mass spectroscopy.

3.1.8 Insertion into carbon–carbon bonds

Zirconium atoms react with neopentane on vapor deposition to give methyl(t-butyl)zirconium among other products,[204] and both Fe^+ and Co^+ are known to insert into carbon–carbon bonds under vapor phase or molecular beam conditions.[202]

Insertion of a metal into a strained carbon–carbon bond is a common route to metallacycles, which if not trapped may function as intermediates in transition metal catalyzed rearrangements (see Chapter 10; also Reference 123 and references therein). The insertion of platinum(II) into the strained CH_2-CH_2 bond of **XXXI** gives **XXXII** with retention at both carbons, as shown by deuterium labeling, confirming that the reaction is concerted.[205] However, the insertion into bicyclobutane, to give **XXXIII**, is thought to involve an initial delocalized electrophilic attack rather than simple insertion.[206] On the other hand, electrophilic metals such as silver(I), which cannot readily increase their oxidation number, attach themselves to one carbon only of strained hydrocarbons giving metallocarbonium ions which rearrange further.[207]

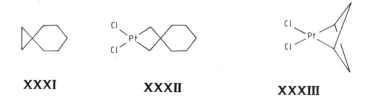

XXXI   **XXXII**   **XXXIII**

Iron(0) derivatives attack the ring-sidegroup bond of cyclopentenes $(CH)_4CHR$ to give cyclopentadienyliron alkyl complexes by an intramolecular process without crossover,[208] while molybdenum atoms react with C_5Me_6 to give $Cp^*_2MoMe_2$.[209] $Ni(PCy_3)_2$ oxidatively inserts into the carbon–carbon bond of acetonitrile to give an unstable product; one mole of free phosphine inhibits, presumably by forming a nickel(0) trisphosphine complex.[210]

3.1.9   Addition of metal hydride to alkenes and alkynes

As discussed in Chapter 11, transition metal promoted attack on an unsaturated ligand can take place by either of two distinct paths. In most cases, the entering nucleophile is external to the complex, and the elements of metal and nucleophile are added across the unsaturated species in mutually trans positions. Hydride, however, is usually coordinated to the metal before attack (as are alkyls; see Section 3.3.1. below) and the addition is at least initially cis. Metal alkyls generated in this way are key intermediates in the hydrogenation of alkenes, including the asymmetric hydrogenation of prochiral alkenes (Chapter 12).

The orbital energetics of addition of M - H to an alkene have been explored. In systems $Cp_2M(alkene)H$, the lowest metal-centered d-orbital is transformed from metal-alkene π-bonding to metal-alkyl σ-antibonding, thus presenting a higher barrier for Mo^+ or Nb than for Ti.[211] With Pt(II), the reaction is calculated to be easiest when alkene and hydride are mutually cis, and, less obviously, in a 4- rather than a 5-coordinate intermediate.[212] In the model system $RhCl(H)_2(PH_3)_2(C_2H_4)$, calculations show the formation of $RhClH(PH_3)_2Et$ to take place by movement of alkene, rather than of hydride, from its initial coordination position.[213] All these calculations relate to a concerted process, but it should not be assumed that metal hydride addition is invariably concerted or heterolytic, and there is evidence, with both $HMn(CO)_5$ and $HCo(CO)_4$, for free radical metal hydride addition.[214-5] Combination in the $\overline{M^{\cdot}\ R^{\cdot}}$ pair can lead on to hydroformylation, while separation gives ultimately M_2 and RH.

Hydrozirconation has been reviewed, along with related reactions of niobium hydrides.[216] Cp_2ZrClH, accessible from Cp_2ZrCl_2 and aluminum hydride complexes, is a convenient reagent, adding to alkenes or alkynes. The addition to alkynes, at least, is cis, and the zirconium attaches itself to the less hindered carbon.[216-9] The unsaturated species Cp_2NbH, which exists in equilibrium with dimeric "niobocene" and hydrogen, takes up ethylene to form an isolable adduct, from which the ethylene can be converted to ethane in the catalytic hydrogenation sequence of Scheme

Scheme 8

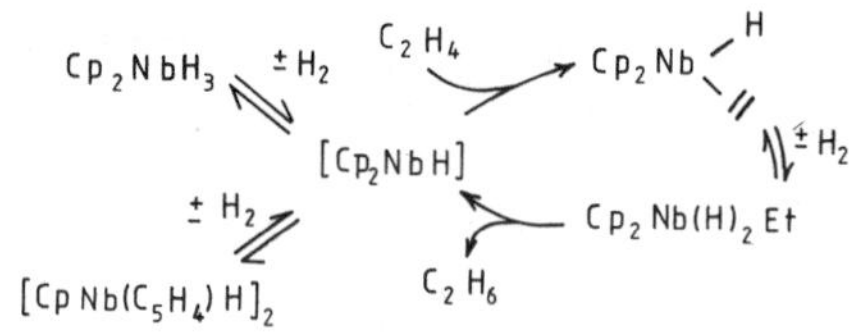

$$Cp_2NbH_3 \xrightarrow{\pm H_2} \xrightarrow{C_2H_4} [Cp_2NbH] \rightarrow Cp_2Nb\begin{matrix}H\\ \\ \end{matrix} \xrightarrow{\pm H_2} Cp_2Nb(H)_2Et$$

$$[CpNb(C_5H_4)H]_2 \xleftarrow{\pm H_2} \quad C_2H_6$$

8.[220] Relatedly, carbon monoxide or isonitriles promote Reaction 50.[221]

endo-$Cp_2Ta(H)(CHR:CH_2) + L \longrightarrow Cp_2TaL(CH_2CH_2R)$ (50)

Since the starting material is an eighteen-electron species, the reaction presumably proceeds by reversible hydride addition to alkene, followed by addition of the incoming ligand to the unsaturated center so generated. Closely related is the CO-induced rearrangement of $Cp_2Nb(H)$(alkyne) to give products $Cp_2Nb(CO).CR:CHR$, with cis-addition across the triple bond.[222]

The addition of molybdenum hydride in Cp_2MoHI across the double bond of 2-vinylpyridine can occur in either sense to give **XXXIV** or **XXXV**.[223] Addition of Cp_2MoH-H to CF_3CCH gives $Cp_2MoH.C(CF_3):CH_2$ by cis-addition, but the (isolated) product with hexafluorobut-2-yne indicates trans addition,[224] as does the product from $CpW(CO)_3H$ and dimethyl acetylenedicarboxylate.[225] These unexpected trans-additions may reflect either mechanistic complexity, or rotation about a weakened carbon-carbon bond in an initial cis adduct, mesomeric between structures m.C:CH and $M^+:C.C^-H$. The reaction of $\{CpMo[P(OMe_3)]_2(PhCCPh)\}^+$ with hydride presumably proceeds by hydride displacement of phosphite and metal hydride addition to give an intermediate $CpMo[P(OMe_3)]_2.CPh:CHPh$, but this latter isomerizes before isolation to give **XXXVI**.[226] The addition of H-$Mn(CO)_5$ to dimethyl acetylenedicarboxylate gives the CO-eliminated trans-adduct **XXXVII**.[227]

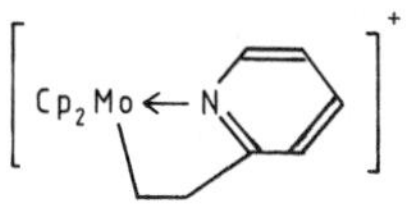

$$\left[Cp_2Mo\leftarrow N \right]^+ \qquad \left[Cp_2Mo\leftarrow N \right]^+ \qquad \underset{L}{\overset{L}{Mo}}\begin{matrix}Ph\\ \\H\\Ph\end{matrix} \qquad MeO\begin{matrix}CO.OMe\\ \\Mn(CO)_4\\O\end{matrix}$$

XXXIV 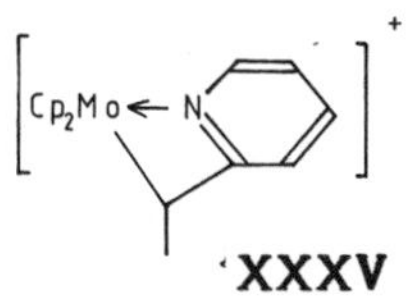   **XXXV** 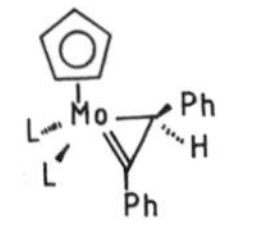   **XXXVI** 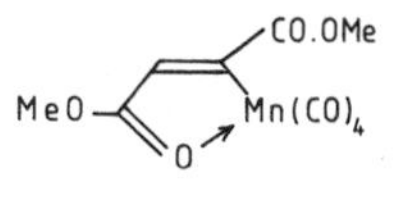   **XXXVII**

The dinitrogen hydride complex $Fe(H)_2(N_2)(PEtPh_2)_3$ reacts reversibly with cyclohexene in thf to give solvated $Fe(H)(Cy)(PEtPh_2)_2$ and free dinitrogen and phosphine.[228] The anion $HFe(CO)_4^-$ adds to acrylic esters to give synthetically useful intermediates $[RCH_2.CH(COOEt).Fe(CO)_4]^-$, which may be converted by alkyl halides to β-ketoesters.[229] CpFe(dppe)H adds across alkynes RCCR; for R = CF_3, the addition is cis.[230]

The reaction between stilbene or fluorene and cobalt tetracarbonyl hydride gives an anomalous isotope effect, showing that in the transition state the new C-H bond is largely formed.[231] $HCo(C_2H_4)(PMe_3)_3$ undergoes averaging of the position of a deuterium label, through an intermediate ethyl complex $(Me_3P)_3CoC_2H_5$.[232]

Rhodium alkyls, formed from Rh(I), alkenes, and dihydrogen, are key intermediates in the catalytic hydrogenation of alkenes. The detailed mechanism is sensitive to coordinating substituents on the alkene (cf. Chapter 12), but rhodium(III) hydride intermediates have been directly observed.[233] The addition of trimethylsilane to t-butylacetylene, catalyzed by $(Ph_3P)_3RhCl$, gives initially trans-t-BuCH:CHSiMe$_3$ which in the presence of excess silane isomerizes to the cis isomer. The reaction is thought to involve oxidative addition of $H-SiMe_3$, attack on coordinated alkyne, reductive elimination of trans-alkene, and, in the isomerization step, reversible addition-elimination of Rh-H across the alkene double bond with free rotation in the saturated intermediate.[234] An aqueous solution of $[Rh(NH_3)_5H]^{2+}$, formed by zinc reduction of the chloride, adds to alkenes, tetrafluoroethylene, or hexafluorobut-2-yne. Excess ammonia inhibits, as expected if the reaction proceeds by ammonia ligand loss followed by coordination of the unsaturated species.[235]

Hydride transfer via metal is presumably responsible for Reaction (51):

$$cod_2Ni + PCy_3 + H_2C:C(Me)CONH_2 \longrightarrow Cy_3P\overline{NiCH_2CHMeCON}H + 2cod \quad (51)$$

(the Ni(II) species formed is probably square planar, with oligomerization by N - Ni bonds.)[236]

Palladium hydrides $PdH(NO_3)(PCy_3)_2$ or $[PdH(NCMe)(PCy_3)_2]^+$ add cis across alkynes with an electron-withdrawing group; the hydride adds to the more electronegative carbon.[237] For platinum hydrides, stereochemistry at the metal presents an apparent problem, as in eq 52 (R = CF_3,[238] CH_2CN[239]):

$$\text{trans-}(Ph_3P)_2Pt(R)H + C_2H_4 \longrightarrow \text{cis-}(Ph_3P)_2Pt(R)Et \tag{52}$$

For R = CH_2CN, added phosphine inhibits the reaction, which is not shown at all by $(dppe)Pt(CH_2CN)H$, supporting the view that alkene insertion requires the prior dissociation of a ligand, followed by isomerization at platinum.

3.1.10 Attack of miscellaneous species on coordinated alkenes

Nucleophilic attack on coordinated alkenes is the subject of Chapter 11 and will be discussed only briefly here. Complexes of type $[CpFe(CO)(L)(un)]^+$ can be reduced by hydrides to σ-alkyls or σ-vinyls.[240] $CpFe(CO)(PPh_3)H$ in acetonitrile can act as a hydride source to $[CpFe(CO)_2(\text{isobutene})]^+$, giving $CpFe(CO)_2Bu\text{-}t$ by addition of hydride to the less substituted carbon.[241] Many other nucleophiles attack alkenes coordinated to the Fp^+ grouping, including the nucleophilic allyl group of $Fp.CH_2CH{:}CH_2$ (Scheme 9).[242]

Metal alkyls can also be formed by electrophilic attack on alkene complexes of oxidizable metals, as in the attack of proton acids on cod_2Pt to give **XXXVIII**.[243] Norbornene, carbon dioxide and $Ni(cod)_2$ react together in the presence of bipyridyl to give the 5-membered ring species $(bpy)\overline{Ni(1,2\text{-}C_7H_{10})\text{-}CO.O}$.[244]

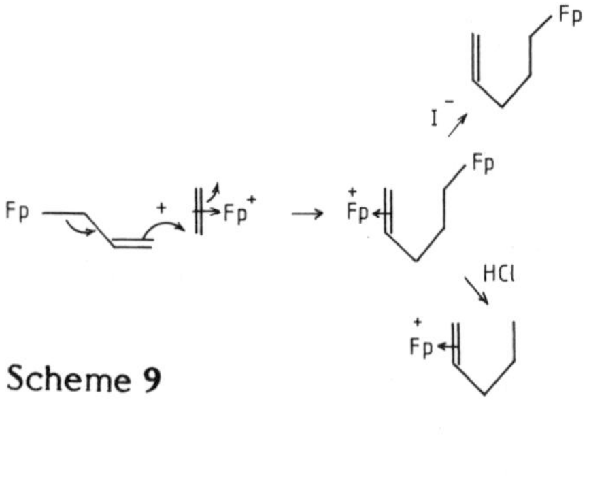

Scheme 9

3.1.11 Reductive coupling of unsaturated species

This reaction leads to metallacycles - see Chapter 10 and, for acetylenic species, Chapter 13.

3.1.12 Attack by main group ylids

Main group methyl ylids attack halides of rhodium,[245-6] iridium,[247] and gold[248] according to eq 10 to give monomeric or in some cases[246,248] dimeric products. Ylids can also coordinate to metals by simple displacement of other ligands, as in the formation of $(Me_3PCH_2)_2PtMe_2$ by ylid displacement of cod,[247] and of species $[Ph_3P.CHR.AuL]^+$ by displacement of chloride.[249] In a closely related reaction, $Me_3P:N.PMe_2:CH_2$ reacts with trimethylphosphine complexes of nickel(II) or platinum(II) chlorides to give products of type **XXXIX**.[250] Nickelocene reacts with appropriate ylids to give complexes of type $[CpNi(CR_2PR'_3)_2]^+$, among other species.[251]

XXXVIII $\left[\begin{array}{c} \end{array} \rightarrow Pt(cod) \right]^+$

$$\begin{array}{c} Me_2P\!\!-\!\!\diagdown \quad \diagup\!\!-\!\!PMe_2 \\ N\diagup\!\!\diagdown \!\!(\quad)\!\!\diagdown\!\!\diagup N \\ Me_2P\!\!-\!\!\diagup M \diagdown\!\!-\!\!PMe_2 \end{array}$$ **XXXIX**

Ylid complexes can also be formed indirectly by phosphine attack on α-halomethyl compounds, MCH_2Cl, (M = Fp, $CpW(CO)_3$) to give products $[MCH_2PR_3]^+$.[252] A process of this kind is thought to be responsible for the overall reaction[253]

$$Pt(Ph_3)_4 + CH_2ClI \longrightarrow \underline{cis}\text{-}[(Ph_3P)_2Pt(CH_2PPh_3)Cl]^+I^- \quad (53)$$

3.1.13 Miscellaneous preparations

The phosphite complex $Ru[P(OMe)_3]_5$ isomerizes on heating to give $\underline{cis}$-$[(MeO)_3P]_4Ru(Me)P(O)(OMe)_2$, which like its parent attacks methyl iodide to give $[(MeO)_3P]_5Ru(Me)^+I^-$.[128] With strong bases, $\underline{mer}$-$IrCl_3(PMe_2Ph)_3$ eliminates one unit of HCl to form a three-membered $IrPC(H_2)$ ring; the reaction is reversed by added HCl.[254] The pentacarbonylrhenium anion attacks $\underline{trans}$-$[(Ph_3P)_2Ir(CO)(Cl)(H)(C_2H_4)]^+$ to give $Re(CO)_5Et$ and $(Ph_3P)_2Ir(CO)Cl$; presumably the rhenium attacks the coordinated alkene (cf. 3.1.10 above) to give an ethylene-bridged species, which undergoes reductive elimination at Ir (3.2.6. below).[255] Pentacarbonylmanganese hydride adds across the C=O double bond of o-diphenylphosphinobenzaldehyde to give **XL**, an α-hydroxy derivative stabilized by chelation against β(M-H) elimination.[256] Metal alkyls can be formed by nucleophilic attack on carbenes:[257]

$$[CpRe(PPh_3)(NO):CHPh]^+ \underset{Ph_3C^+}{\overset{MeO^-}{\rightleftharpoons}} CpRe(PPh_3)(NO).CH(OMe)Ph \quad (54)$$

The direct addition of radicals to copper(II) gives organo-copper(III) intermediates in copper-moderated free radical processes.[258-60] Free methyl radicals also add directly to Cr(II) (as hydrate) and low-spin Co(II) (as pentacyano complex), interfering with the use of these metals to generate free radicals from the DMSO-H_2O_2 reagent.[261]

Diazonium cations add to palladium(0) triphenylphosphine, to give palladium(II) aryls after elimination of dinitrogen.[262] A methylene group from diazomethane inserts into the metal-metal bond of $(dppm)_2(Pt-Pt)$ framework complexes.[263]

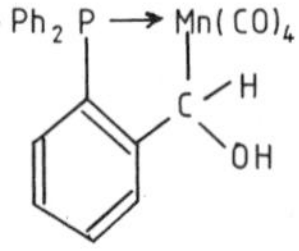

XL

3.2   Breaking of metal-carbon sigma bonds

3.2.1   General

For more than a decade, it has been recognized that the behavior of metal-carbon bonds must be understood in terms of mechanism, and several reviews have been written from this standpoint (see e.g. References 122-3, 264). In reaction against earlier assumptions, one-electron mechanisms were for a time unduly ignored, but more recent work[1a] has helped redress this balance. It is now clear that the choice of mechanism in each case depends, not only on the various pathways in principle available, but also on the energy of the species to which each of these would give rise.[265] Thus the actual pathway followed depends on many factors. As a result, theorizing about bond cleavage is a risky business, and a route predicted by sound reasoning to be facile may nonetheless not be the one actually followed. There is also a special technical problem[123] due to the existence of non-bonding orbitals in the unsaturated fragments that could result from cleavage; unlike the cases usually discussed in organic chemistry,[266] a change in ground state symmetry may indicate a process less energy-demanding than an "allowed" process in which symmetry is conserved.

The arrangement of this Section reflects these limitations. We begin with a review of reactions in which one particular pathway is well-characterized and dominant (3.2.2. to 3.2.15.), and continue (3.2.16.) with a discussion of more complicated examples, in which more than one route may be involved. 3.2.16. is arranged according to the periodic table, and for convenience gives references back to the earlier parts for each group. Finally, some generalizations that emerge are collected in 3.2.17., in the awareness that they present hostages to fortune.

3.2.2 Free radical pathways

3.2.2.1 General aspects:- Free radical mechanisms can be inferred by the usual techniques of kinetics and of isotopic labeling. In addition, CIDNP can show that particular atoms in the products (or in regenerated starting material) have formed part of paramagnetic species during the reaction, and such species may be directly detectable by esr. Radical traps can be used to intercept short-lived species, which can then be identified by esr, although care is then necessary to separate out the effects of any reaction between the substrate and the traps themselves. Further clues can be obtained from irregular kinetics and induction periods, initiation by radical sources or by light, inhibition by radical scavengers such as galvanoxyl, and initiation by the decomposing organometallic of free radical polymerization. Obvious precautions for mechanistic studies then include the exclusion of light, and eliminating or allowing for any effects of dissolved dioxygen, which can act either as a precursor or as a scavenger for free radical intermediates.

Free radical decompositions, unlike any of the other pathways discussed here, involve bond-breaking without any concomitant bond-making; they also involve a formal reduction by one unit of the oxidation state of the metal. For these reasons, they are of most significance in photochemical processes, in cases where other mechanisms have been precluded (by ligand design, for example), and when the central metal is reducible by one unit. Free radical <u>substitutions</u> involve bond formation as well as bond-breaking, and occur in at least some species that do not show free radical decomposition; the scope of such reactions in organometallic chemistry has been reviewed.[3]

3.2.2.2 Photolytic homolysis:- Photolytic homolysis can cleave the metal-methyl bond in cyclopentadienylmetal alkyls; this occurs in Cp_2TiMe_2,[267] species Cp_2ZrR_2,[268] Cp_2VMe_2 and Cp_2NbMe_2,[269] and $[Cp_2WMe_2]^+$.[270] For the species $CpM(CO)_3R$ (M = Cr, Mo, W), a possible primary photolytic step is CO loss, followed by alkyl radical loss from an unsaturated fragment;[271-3] but there is also evidence for direct loss of methyl both under matrix isolation conditions and at room temperature in pvc films.[274]

Photolysis of either tetra(neopentyl)chromium, or of (mesityl)$_3$ Cr(thf)$_3$, gives free radicals, which can initiate polymerization, and free chromium, which can either form a mirror or be trapped by CO as the hexacarbonyl.[275] Cr(norbornyl)$_4$ reacts photochemically with carbon tetrachloride to give norbornyl chloride. This process requires uv irradiation; the d-d region is without effect.[276] Similarly, Ti(norbornyl)$_4$ undergoes radical cleavage on irradiation in what must be assigned as the ligand-metal charge transfer band.[277]

The uv photolysis of L_2PtMe_2 (L = ligand phosphine) in $CDCl_3$ gives $L_2PtMeCl$. There is CIDNP evidence for methyl radicals and also for Me + $CDCl_2$ triplet pairs. In benzene, there is no net reaction but enhanced signals may be due to the formation of radical pairs which then recombine. The mechanism of thermolysis (3.2.13. below) is completely different.[278] Photolysis of (bpy)PtIMe(Ph)$_2$[279] or of $CpPtMe_3$[280] gives methyl radicals which abstract hydrogen either from solvent or from other ligands, leading to a range of products. Free radicals have been trapped by nitrosodurene in the photolysis of species $Mn(CO)_5$-R, Fp-R, Ph_3PAu-Me, and trimethylsilylplatinum species, and both alkyl- and metal-centered radicals have been identified.   Neopentylplatinum species gave both t-butyl and neopentyl radicals, while metal acyls gave free alkyl groups, probably by direct cleavage of the CO-R bond.[281]

3.2.2.3 Homolysis coupled to oxidation:- The reaction of isopropylpenta-aquochromium(III) with oxygen is thought to be initiated by homolysis of the alkylchromium bond, and propagated by a radical chain mechanism, to give eventually hydrated chromium(III), acetone, and some isopropanol.[282] The oxidative cleavage of benzylpentaaquochromium(III) by iron(III), oxygen, or cobalt(III) is first order in substrate and independent of the nature and concentration of the oxidizing agent. In the absence of an oxidizing agent, the chromium-carbon bond reversibly homolyzes, with some decomposition as benzyl radicals combine to give bibenzyl. In the presence of an oxidizing agent (which effectively traps the chromium(II) by oxidation) the initial homolysis becomes rate-determining.[283]   Likewise, in the absence of an oxidizing agent to trap Cr(II) or neopentyl radicals, the equilibrium between $[Cr(III)(H_2O)_5$-cyclo-$C_5H_9]^{2+}$ and its homolysis products favors the organocobalt species, which is however destroyed by heterolytic protonolysis.[284]   A related reaction is that between chromium(III) α-hydroxyalkyl complexes and Fe(III), Cu(II) or Hg(II) to give as primary products Cr(II), ketones, and Fe(II), Cu(I), or Hg_2^{2+}.[285] Oxygenolysis of $[(H_2O)_5Cr(III)Pr^i]$ proceeds by a radical chain mechanism, with Pr^iOO replacing Pr^i in an S_H2 reaction step.[286]   The homolyses of Cr-C and Co(III)-C bonds have been reviewed;[287] see also Chapter 2.

The cleavage of species FpR by copper(II) halides has been compared with cleavage following electrochemical oxidation. The reactions were found to involve electron transfer, halogen atom transfer, and either homo- or heterolytic cleavage of the oxidized substrate. When R was methyl or benzyl, the initial electron transfer pair $FpR^+CuX_2^-$ rearranged to give RX (with inversion), CuX, and Fp radicals, which latter in turn accepted a halogen atom from CuX_2. When R was n-butyl, diffusion and homolysis led to products Fp^+ and $R^{\cdot}$, while the 2-phenethyl compound gave $Fp^{\cdot}$ radicals and the phenonium-bridged $[CH_2CH_2Ph]^+$ cation.[288]

3.2.2.4 Other cases of homolysis:- Thermolysis of tetra-t-butylchromium at 70°C gives mainly isobutane and isobutene in a 6:1 ratio, consistent with an initial homolysis.[289] Direct homolysis of Cr(III)-R bonds (R = e.g. CMe_2OH) is favored by steric congestion, and by resonance stabilization of the resulting radical. In addition, that radical can itself attack a fresh substrate unit to give R-R and Cr(II).[290] (This process will give more R-R than expected from the behavior of R˙ alone, but without the intervention of a dinuclear intermediate.)

The reaction

$$PhCH_2Mn(CO)_5 + HMn(CO)_5 \longrightarrow PhCH_3 + Mn_2(CO)_{10} \tag{55}$$

is thought to proceed by initial homolysis of the benzyl-metal bond, followed by hydrogen atom transfer to the benzyl radical.[3]

Heating cyclopentadienyltrimethylplatinum at 165°C gives methane and free metal. The slow step is homolysis of a metal-methyl bond, for which the dissociation energy is found to be ca. 160 kJ mol^{-1}.[291] Neophylcopper, $Ph.CMe_2.CH_2Cu$, decomposes at 30°C to a mixture of t-butyl benzene and isobutylbenzene, characteristic of hydrogen abstraction by the partly rearranged neophyl radical (contrast t-butylcopper, unstable at -78°C). Results for neophylsilver and dineophylmercury were similar.[292]

Dicyclopentadienyldibenzylzirconium reacts with $t-Bu_2NO$ radicals to give $Cp_2Zr(CH_2Ph).O.NBu-t_2$ and benzyl radicals, which in turn form $PhCH_2ONBu-t_2$.[293] The formation of the radicals (bpy)Ni(Et).NR.O˙ and Et.NR.O˙ from $(bpy)NiEt_2$ and nitroso compounds is due, not to direct homolysis, but to the initial formation of an unstable adduct $(bpy)Ni(Et)_2.N(O).R$.[294] This may be relevant to the observation of species MeCH(COOEt).CH(COOEt).NBu-t.O˙ in the diethylfumarate-promoted decomposition of $(bpy)PtMe_2$,[295] but when species cis-L_2PtR_2 (R = Me, Et, $PhCH_2$) react with t-butoxy or thiophenoxy radicals in the presence of t-BuNO, simple R˙ radicals can be observed as well as R.N(Bu-t).O˙.[296] Free radical effects have been detected by CIDNP in the reaction of benzyl peroxide with platinum(II) and gold(I) methyls,[297] while the reactions of these methyls with thiophenol is also thought, on kinetic grounds, to involve attack by PhS˙ radicals. Gold(I) is more amenable to radical attack than platinum(II), but gold(III) reacts by a non-radical pathway.[298]

3.2.3 Cleavage by proton acids

The innocent-seeming reaction

$$MR + HX \longrightarrow MX + RH \tag{56}$$

is a special case of electrophilic attack at the M-R bond, and could proceed by any of four different limiting mechanisms. The proton may attack the metal-bound carbon atom directly, in which case the transition state could either be open (leading to inversion at carbon) (eq 57) or closed (giving retention) (58). The closed transition state, in which the proton attacks the electron density of the M-R bond itself, merges with oxidative protonation of the metal followed by reductive elimination of hydrocarbon (59), and a further possible variant (60) involves oxidative addition to the metal of the elements of undissociated acid.

$$M\text{-}C + H^+ \longrightarrow M...C...H^+ \longrightarrow M^+ + C\text{-}H \tag{57}$$

$$M\text{-}C + H^+ \longrightarrow M\overset{\displaystyle H}{\underset{..... }{.+.}}C \longrightarrow M^+ + H\text{-}C \tag{58}$$

$$M\text{-}C + H^+ \longrightarrow H\text{-}M^+\text{-}C \longrightarrow M^+ + H\text{-}C \tag{59}$$

$$M\text{-}C + HX \longrightarrow X\text{-}M(H)\text{-}C \longrightarrow MX + H\text{-}C \tag{60}$$

The choice of mechanism depends on an exceptionally fine balance of considerations, as illustrated by the reaction between methylpenta-carbonylmanganese and proton donors in the vapor phase, which has been studied by ion cyclotron resonance. This may proceed according to (61a) or (61b):-

$$Mn(CO)_5CH_3 + BH^+ \longrightarrow Mn(CO)_5(CH_3)H^+ \text{ (XLI)} + B \tag{61a}$$

$$Mn(CO)_5CH_3 + BH^+ \longrightarrow Mn(CO)_5^+ + CH_4 + B \tag{61b}$$

It would be natural to assume that **XLI** is an intermediate in (61b), undergoing reductive elimination of methane. In that case, we would expect to identify this intermediate most readily in reactions of poor proton donors, while the reactions of good proton donors would tend to generate enough surplus energy to trigger the second, elimination, step. The reverse is found. **XLI** is formed preferentially from good proton donors. It seems then that Reactions 61a and 61b are not connected. Direct protonolysis gives (61b) by a relatively low-energy pathway, while the formation of **XLI**, presumed to occur by protonation of the metal, is an independent, more energy-demanding, reaction.[299]

Regardless of such subtleties, acidolysis is a standard way of characterizing metal-carbon bonds. When the metal is electropositive, the reaction is often facile, presumably proceeding by direct protonolysis, and requires only weak acids such as water or an alcohol. When the metal-carbon bond is less polar, the reaction may be sluggish, and the identity of the acid becomes important. However, $[W(O)(CH_2CMe_3)_3]_2O$ is stable in water over the pH range 4-10, and reacts reversibly with hydrochloric acid to give $W(O)(CH_2CMe_3)_3Cl$.[300]

The compounds $Li(tmen)_3MMe_6$ (M = Er, Lu) are quantitatively destroyed by water; D_2O gives CH_3D.[53] Tetrabenzyltitanium reacts with two moles of ethanol to give $Ti(CH_2Ph)_2(OEt)_2$ and toluene, and with hydrogen chloride at $-20^{\circ}C$ gives rise to a range of species $TiCl_n(CH_2Ph)_{4-n}$.[301] The complexes $VR(OR')_3$ and $VR_2(OR')_2$ (R = Me, Ph, CH_2Ph) react with acids to liberate HR; at $90^{\circ}C$, even t-butanol is effective.[302] Protonolysis of cis-$Cp_2Nb(CO)-CR:CHR'$ by D_2SO_4 gives cis-$CDR:CHR'$, consistent with oxidative addition of proton to metal.[222] The protonolysis of benzylchromium(III) complexes gives toluene and inorganic Cr(III). Unlike their oxidation,[283] this may sometimes be a heterolytic reaction, and is assisted by certain anions which possibly ligate to chromium.[303] Hydrolysis of 4-pyridinomethylpentaaquochromium(III) is heterolytic.[304]

Alkyls in the triangulotrirhenium series show considerable hydrolytic stability. $Re_3Cl_3(CH_2SiMe_3)_6$ forms a stable hydrate, while Re_3Me_9 undergoes only partial protonolysis with β-diketones. Bridging groups are more resistant then those in terminal positions.[102,305]

Several studies[306-8] support a mechanism for acidolysis of cyclopentadienyldicarbonyliron alkyls in which initial protonation at the metal is followed by reductive elimination. Aryl derivatives, however, show higher rates which are very sensitive to ring substitution, in accord with direct proton attack at the iron-bound carbon.[308] $\underline{Cis}$-$(Me_3P)_4FeMe_2$ reacts with trimethylphosphine hydrochloride, or with ammonium hexafluorophosphate in the presence of trimethylphosphine, to give $[(Me_3P)_5FeMe]^+$ and one mole of methane. $(Me_3P)_3CoMe_3$ reacts similarly to give $[(Me_3P)_4CoMe_2]^+$.[309] The bridged ruthenium complex **XLII**, formed from the 'basic acetate' $[Ru_3O(OAc)_6(H_2O)_3]^+OAc^-$ and the methyl Grignard reagent in the presence of trimethylphosphine, is protonolyzed in stages according to Scheme 10; note the methyl-bridged intermediate.[310]

$$H^+$$
$$(Me_3P)_3Ru(\mu\text{-}CH_2)_3Ru(PMe_3)_3 \longrightarrow [(Me_3P)_3Ru(\mu\text{-}CH_2)_2(\mu\text{-}CH_3)Ru(PMe_3)_3]^+$$
$$\textbf{XLII}$$

$$H^+$$
$$\longrightarrow [(Me_3P)_3Ru(\mu\text{-}CH_2)_2Ru(PMe_3)_3]^{2+} + CH_4$$

Scheme 10

Treatment of the metallacycle **XLIII** with hydrogen chloride in thf gives 60% butane and 40% butene. This is consistent with protonolysis of one rhodium-carbon bond, followed by competition between β-elimination (3.2.11 below) and a second protonolysis.[311]

XLIII

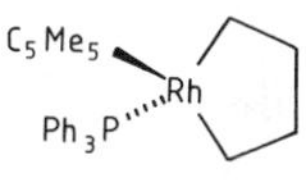

The conversion of **XLIV** to **XLV** by hexafluorophosphoric acid proceeds in two successive steps, each first order in protons. However, only one proton is added in the overall reaction, and the formation of intermediates **XLVIa** and **XLVIb** is inferred, with subsequent non-rate-determining deprotonation of the latter at carbon to give the product (Scheme 11). The important inference is drawn that the initial proton addition is at metal-bound carbon; if it had been at the metal, **XLIV** would have been converted to **XLV** directly.[312]

Scheme 11

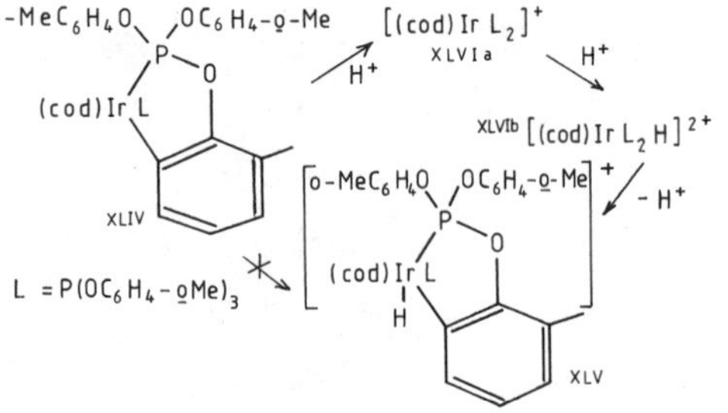

Reaction of <u>trans</u>-$(Me_3P)_2NiMe_2$ with water gives $(Me_3P)(Me)Ni(\mu\text{-}OH)_2Ni(Me)(PMe_3)$. Removal of the remaining methyl groups requires a stronger acid than water.[313] Protonolysis of <u>trans</u>-$(Et_3P)_2PtAr_2$ and of <u>trans</u>-$(Et_3P)_2PtMeX$ in protic solvents takes place through a pre-equilibrium proton addition, followed by reductive elimination. There is a large isotope effect, at least in the aryl species, showing hydrogen transfer to be part of the rate-controlling step.   The rate is increased by electron-donating substituents on the rings, but is far slower than for related Pt(II) methyls, indicating that even in the aryl series the initial protonation is at platinum rather than carbon (contrast the iron aryls[308] discussed above). One aryl group is lost much more readily than the second, and in the presence of chloride the reaction proceeds with retention at platinum, but in the absence of any nucleophile more effective than the solvent (MeOH-H_2O), isomerization at metal occurs, presumably in a T-shaped intermediate $[(Et_3P)_2PtAr]^+$.[314]   For the protonolysis of <u>trans</u>-$(Et_3P)_2PtMeX$ in methanolic HCl, the rate law contains terms in $[Pt][H^+]$ and also in $[Pt][H^+][Cl^-]$, indicating that oxidative addition either of a proton, or of HCl, can precede the elimination of methane.[315]   However, in the protonolysis of <u>trans</u>-$(Et_3P)_2Pt(Ar)Me$ to <u>trans</u>-$(Et_3P)_2Pt(Ar)Cl$, added chloride does not affect the rate.[316] Electron-withdrawing substituents can stabilize platinum(II) alkyls to protonolysis, and in species $L_2Pt(H)(CH_2CN)$ either hydride or cyanoalkyl can be lost depending on stereochemistry and conditions.[317-8]

The gold(I) lithium "ate" complexes LiAu(Me)R (R = Et, Bu-t) both react with hydrogen chloride to give mixtures of HR and methane. This suggests addition of a proton to the gold atom, followed by a non-specific reductive elimination of alkane from the T-shaped intermediate so formed.[319] Controlled protonolysis of gold(III) complexes $LAuR_3$, where L is a phosphine ligand, gives cis-$LAuR_2Cl$, as expected on trans-effect arguments.[320]

All the above reactions involve strong or moderate proton sources. There are also reactions where the proton source is a weak C-H acid. Some such cases, leading to new metal-carbon bonds, were included in (3.1.7). Other cases include reaction of Cy_3PMnPh_2 with benzophenone to give benzene and $(Cy_3P)_nMn(CH_2OPh)_2$,[69] and of bis(triphenylphosphine)copper(I) methyl with vinyl or isopropenyl acetate,[321] nitroalkanes, isonitriles, or β-diketones[322] to give methane.

3.2.4 Hydrogenolysis
Reactions of the type

$$M\text{-}R + H_2 \longrightarrow M\text{-}H + HR \qquad\qquad (62)$$

are extremely important in connection with catalytic hydrogenation and hydrosilation, and have been extensively reviewed (see e.g. References 1 (particularly 1c,d,j,k,l), 16, 18, 36, 40; also Chapter 12 and references therein). Attention is also drawn to the variety of products resulting from partial hydrogenolysis and condensation in triangulo-trirhenium systems,[323] and to the formation of hydride bridges in the hydrogenolysis of $(C_5Me_4Et)Ta(CH_2CMe_3)_2Cl_2$ to $[(C_5Me_4Et)TaCl_2]_2[\mu\text{-}H]_2$.[324]

The hydrogenolysis of $Cp_2Zr(R)(X)$ (X = Cl, H, D) gives $Cp_2Zr(H)X$, and is accompanied, for X = D, by H - D exchange. The process is described as a heterolytic cleavage of the hydrogen molecule, after the electrons originally holding it together have become involved in 3-center bonding with the low-lying vacant orbital of the metal.[325]

3.2.5 Reaction with metal hydrides
The reaction

$$M\text{-}R + M'H \longrightarrow M\text{-}M' + R\text{-}H \qquad\qquad (63)$$

provides an elegant but under-used route to metal-metal bonded compounds, is of possible relevance to catalytic hydrogenation and hydroformylation, and plays a role, in conjunction with β-elimination, in the decomposition of some metal alkyls. The driving force of the reaction is no doubt the formation of a strong carbon-hydrogen bond, but the detailed mechanisms are in general unknown. One example is the reaction[326]

$$Cp_2ZrMe_2 + CpMo(CO)_3H \longrightarrow CH_4 + Cp_2MeZr\text{-}MoCp(CO)_3 \qquad (64)$$

of which the product contains an $\eta^1;\eta^2$-bridging CO group, and a bridging -O.C(Me)- group, both attached to zirconium through oxygen. The reaction involves intermediates $Cp_2Mo(CO)_3.Zr(Me)Cp_2$ and (in the presence of labeled CO) $Cp_2Mo(CO)_3.Zr[\eta^2\text{-}*C(O)Me]Cp_2$, both formulated with zwitterionic $Mo^-\text{-}CO\rightarrow Zr^+$ bridges.[327] The same molybdenum hydride reacts with complexes $CpMo(CO)_3R$ to give aldehydes and a mixture of $Cp_2Mo_2(CO)_4$ and $Cp_2Mo_2(CO)_6$ by a second order process which occurs without scrambling, and is probably elimination between Mo-H and Mo-CO.R functionalities.[328] A related reaction is the formation of acetaldehyde from this hydride and methylpentacarbonylmanganese. Thus a contrast can be drawn between $CpMo(CO)_3H$, which reacts with metal alkyls but not simple carbonyls by a hydride transfer mechanism, and isoelectronic $[CpV(CO)_3H]^-$, which exchanges hydride with a range of carbonyls by a sequence suggested to involve initial electron transfer.[329]

The reaction between the substrate p-methoxybenzylpentacarbonyl-manganese and pentacarbonylmanganese hydride is influenced by conditions. In hydrocarbons, reversible carbon monoxide loss from substrate is followed by insertion of the unsaturated tetracarbonyl derivative into a manganese-hydrogen bond, and reductive elimination of p-methoxytoluene. In donor solvents such as acetone or acetonitrile, however, p-methoxyphenylacetaldehyde is the main organic product, being formed by attack of hydride on the solvated acyltetracarbonyl complex produced by benzyl migration to CO. Replacement of one CO in the substrate by a phosphine renders it more susceptible to homolytic cleavage; under carbon monoxide there is loss of acyl as well as of methoxybenzyl radicals.[330]

The catalytic hydrogenation of acrylamide by $(Ph_3P)_2RuHCl$ is thought, on kinetic evidence, to involve steps

$$(Ph_3P)_2RuHCl + (Ph_3P)_2Ru(CH_2CH_2CONH_2)Cl \rightleftharpoons \qquad (65)$$
$$[(Ph_3P)_2RuCl]_2 + CH_3CH_2CONH_2$$
$$[(Ph_3P)_2RuCl]_2 + H_2 \longrightarrow 2(Ph_3P)_2RuHCl \qquad (66)$$

with k_2/k_{-1} around 2000.[331] Cobalt-catalyzed hydroformylation reactions (Chapters 6, 7 below) may involve cleavage by cobalt carbonyl hydride as well as, or instead of, hydrogenolysis of organocobalt intermediates.[332]

The unusual reducing agent tri-n-butylphospinecopper(I) deuteride converts a variety of copper(I) derivatives RCu to specifically labeled RD and copper metal (R = 1-endo-norbornyl, cis- or trans-but-2-enyl). In each case, copper was replaced by deuterium with conservation of stereochemistry, showing the non-radical nature of the reactions.[333]

Reduction by metal hydride may play quite a widespread role in apparently simple reactions, such as hydrogenolysis. An example is the process

$$CpCoLMe_2 + H_2 + L \longrightarrow CpCoL_2 + 2CH_4 \qquad (67)$$

$(L = PPh_3)$, for which an addition-elimination sequence is implausible since it would require a Co(V) intermediate. The reaction is promoted by product but inhibited by free L, in accord with a mechanism

$$CpCoL_2 \rightleftharpoons CpCoL + L \qquad (68)$$
$$CpCoL + H_2 \rightleftharpoons CpCoLH_2 \qquad (69)$$
$$CpCoLMe_2 + CpCoLH_2 \longrightarrow 2CpCoL + 2CH_4 \qquad (70)$$

The mechanism of the last step is obscure, and may involve exchange of hydride and methyl ligands, followed by reductive elimination (3.2.6 below) from a hydridoalkyl intermediate.[334]

3.2.6 Reductive elimination of hydrogen and carbon ligands

Reductive elimination of alkane from a hydriodoalkyl is often facile, and indeed such species are scarcer than dihydrides or dialkyls. The driving force of the reaction is, as in those of the immediately preceding Sections, the tendency to carbon-hydrogen bond formation (review:Reference 335; theoretical study:Reference 336). The facility of the reaction may be connected with a lack of the steric constraints that affect reductive elimination of two alkyl groups (3.2.13. below). The reaction has already been discussed as a possible step in hydrogenolysis or protonolysis, and this Section is concerned with those cases where the hydridoalkyl may be considered as a reagent rather than an intermediate, though the distinction is one of convenience only.

The ease of the elimination is limited by the stability of the fragment left behind, and as a result the reaction is often promoted by addition of potential ligands, or is complementary to oxidative addition, while the unassisted reaction may sometimes proceed by surprising and indirect pathways. All these features are well displayed in the chemistry of dicyclopentadienylzirconium alkyl hydrides.[325,337-40] Hydridic reduction of Cp_2ZrRCl (R = CH_2Cy) gives the hydride-bridged dimer $(Cp_2ZrRH)_2$, which decomposes only slowly to give methylcyclohexane. More rapid elimination (Scheme 12) ensues under the influence of added pent-3-yne, basic phosphines (which give Cp_2ZrL_2 + RH), or hydrogen (compare Ref. 325); the suggestion that H_2 coordinates to Zr(IV) during these reactions gains plausibility from the dimeric nature of the hydridoalkyls, which demonstrates both the Lewis acidity of these species and their capacity for 'electron-deficient' H-bonding. For pentamethylcyclopentadienyl complexes,

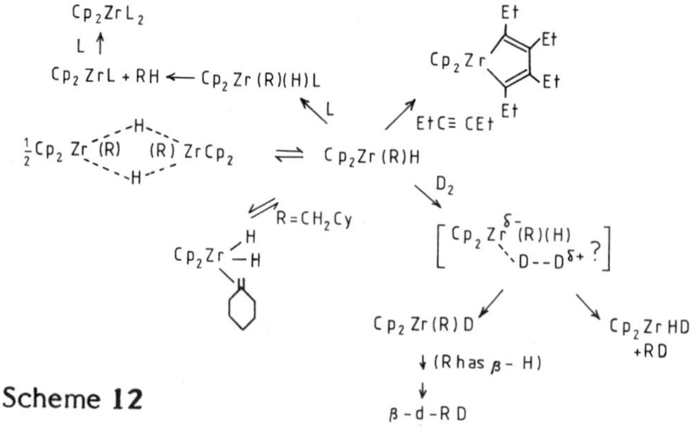

Scheme 12

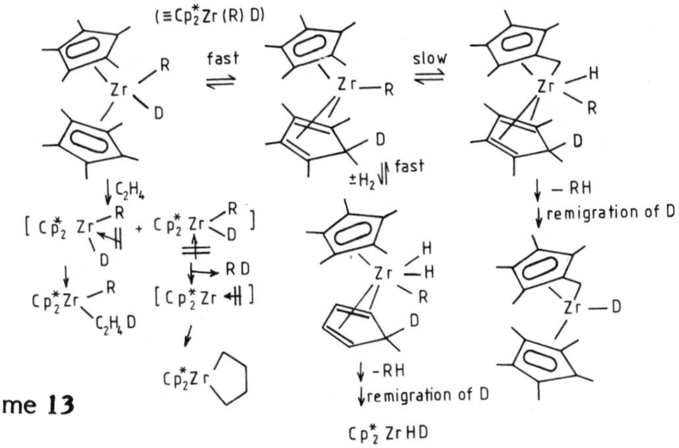

$(\equiv Cp_2^* Zr(R)D)$

fast   slow

Zr⟨R,D   Zr—R   Zr⟨H,R

$\pm H_2$ ⇃ fast

↓ C_2H_4

$[Cp_2^* Zr⟨^R_H + Cp_2^* Zr⟨^R_D]$ ⇌   RD

$Cp_2^* Zr⟨^R_{C_2H_4D}$ $[Cp_2^* Zr⟨H]$

$Cp_2^* Zr$ ⬠

Zr⟨H,R   ↓ −RH ↓ remigration of D   Zr—D

↓ −RH ↓ remigration of D

$Cp_2^* Zr HD$

Scheme 13

the possibility of methylene bridging between metal and ring leads to further complexities which have been elucidated by labeling and kinetic studies, and are shown, together with the effects of added alkene, in Scheme 13. Warming Cp_2WHMe gives methane and, presumably, a tungstenocene fragment which in benzene or toluene is trapped as Cp_2WHR (R = Ph, CH_2Ph), but in cyclohexane attacks starting material to form dinuclear species.[341]

Methane is evolved from species cis-L_2PtHMe (L = PPh_3) at $-25°C$, leaving PtL_3 and free metal. Added phosphine is without effect on the rate of the reaction, which is intramolecular, but prevents free metal formation, no doubt by trapping the PtL_2 fragments first formed. Electron-withdrawing groups in the phosphines promote the reaction, presumably by stabilizing Pt(0). Other species cis-L_2PtHR show the reaction, at rates depending on R in the order $CH_2CH{:}CH_2 <$ Me $<$ Et $<$ Ph.[342] The related A-frame cation **XLVII** has high thermal stability, since H and Me are constrained trans,[343] and species trans-L_2PtHMe with large L (e.g. PCy_3, $PPr-i_3$) are likewise stable.[344] Electron-withdrawing substituents on R, as in e.g. $L_2PtH(CH_2CN)$, inhibit the reaction although even here it does occur on addition of L or of phenylacetylene, presumably by an associative mechanism.[317-8]

XLVII

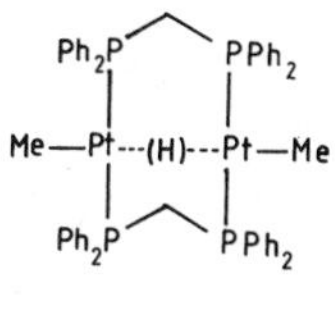

$$Ph_2P\diagdown \quad \diagup PPh_2$$
$$Me—Pt{\cdots}(H){\cdots}Pt—Me$$
$$Ph_2P\diagup \quad \diagdown PPh_2$$

The reduction of alkyl halides to alkanes by $[HFe(CO)_4]^-$, as in the conversion of 1-endo-bromocamphor to 1-exo-deuteriocamphor by $[DFe(CO)_4]^-$, involves reductive elimination from iron with retention at C (Scheme 1).[114]

3.2.7 Hydrogen abstraction from a remaining by a departing ligand

In reactions of this type, the hydrogen attached to the departing organic ligand originates entirely within the complex, often from specific sites, in contrast to the loss of free radicals, which abstract hydrogen atoms from all possible sources, including the solvent and each other. The process may sometimes be accompanied by formation of a new metal-ligand bond within the remaining complex (eqs 71-3) or by more complex reorganizations involving perhaps dimerization and metal-metal bond formation.

$$M\overset{\displaystyle R}{\underset{\displaystyle CH}{\diagup}} \quad \longrightarrow \quad M=C\diagup \quad + \quad RH \qquad (71)$$

$$M\overset{\displaystyle R}{\underset{\displaystyle C-C}{\diagup}}H \quad \longrightarrow \quad M\overset{\displaystyle C}{\underset{\displaystyle C}{-\!\!|}} \quad + \quad RH \qquad (72)$$

$$M\overset{\displaystyle R}{\diagup}H \quad \longrightarrow \quad M\bigcirc \quad + \quad RH \qquad (73)$$

Some reactions of this kind have already been presented in Section 3.1.7, under the heading of cyclometalation.[161,173-4,186-7] Other examples include the formation of R-H (note, not H-H) from $RuHR(PPh_3)_2$ in the presence of triphenylphosphine,[345] and of methane from $CpRuMe(PPh_3)_2$,[346] in both cases with loss of an ortho-hydrogen from one triphenylphosphine ligand to give complexes of the chelating ligand $Ph_2PC_6H_4$-o (P-C); while formation of an ethyl group, followed by elimination of ethane, is suggested for the overall reaction[347]

$$RuH_2(PPh_3)_4 + 2C_2H_4 \longrightarrow RuH(C_2H_4)(PPh_3)_2(P\text{-}C) + C_2H_6 \qquad (74)$$

Bulky R groups facilitate the process: thus the reaction

$$(Ph_3P)_3RhR \longrightarrow (Ph_3P)_2Rh(P-C) + RH \qquad\qquad (75)$$

is more facile when R is trimethylsilylmethyl or neopentyl[348] than when R is methyl.[349]   Bulk is also important in facilitating the α-abstraction reaction (eq 71), which provides a standard route to alkylidene complexes (Chapter 3).

The thermolysis of titanocene and zirconocene diaryls in the presence of alkynes produces free arene and metalla-indenes;[350-3] labeling experiments[352] show that the reaction proceeds through a titanocene benzyne, and the steric effects of substitution at the second (alkyne insertion) stage have been elegantly demonstrated (Scheme 14).[354]   In contrast, the thermolysis of niobocene diphenyl involves loss of benzene by a process of hydrogen abstraction from a cyclopentadienyl ring rather than from the other phenyl ring or from solvent. The process is first order, and the activation energy is increased by deuteration of the phenyl ring but not of the cyclopentadienyl ring, showing that the C_5H_4-H bond cleavage step occurs _after_ the rate-determining transition state.[355] The thermolysis of $Cp^*_2TiMe_2$ in toluene is first order and intramolecular. The products are $Cp^*Ti(C_5Me_4CH_2)(Me)$ and methane, but $Cp^*_2Ti(CD_3)_2$ gave CD_4.[356] Closely related, perhaps, is the thermolysis of thorium complexes Cp_3ThR to give **XLVIII** and alkane. The hydrogen abstraction by R is from ring, not solvent, and alkenyl groups R eliminate with retention of stereochemistry, confirming the non-radical nature of the reaction.[357]

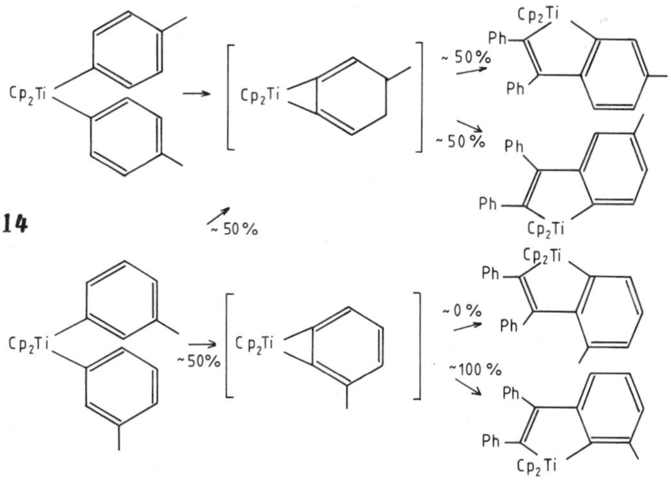

Scheme 14

XLVIII

3.2.8 Cleavage by electrophiles

The important special case of protonolysis is discussed in Section 3.2.3 above. The remarks made there about possible mechanistic variation are of general application, and indeed varying the electrophile adds one more dimension of diversity.

Titanium tetramethyl and aluminum trimethyl exchange methyl groups, presumably through bridged, electron-deficient intermediates. Added ether slows down the exchange by coordination to titanium, with formation of salts such as $Me_3TiOEt^+AlMe_4^-$.[358] Presumably nucleophilic attack by ether at titanium complements electrophilic attack by aluminum at carbon. This complementarity is also important in the electrophilic cleavage of species Cp_2ZrRCl. With bromine, for example, reaction proceeds with retention in a closed transition state, in which a developing bromide ion acts as nucleophile to the metal. Carbon electrophiles are unreactive, since they lack a good nucleophilic partner, but reaction with Al_2Cl_6 gives Cp_2ZrCl_2 and $(AlRCl_2)_n$. The reaction succeeds for alkenyls as well as for alkyls.[218]

In complete contrast, the attack by halogens, interhalogens, or mercuric compounds on hydrated chromium(III) benzyls proceeds by an open S_E2 process, without halide attachment to Cr(III). Substituent effects show some C-X or C-Hg bond formation in the transition state of the electrophilic attack.[359-61] Tetramesitylmolybdenum is oxidized rather than cleaved by bromine or iodine, giving salts $[Mo(mes)_4]^+X_3^-$.[362] For mercuric halide cleavage of W-C bonds see below (reference 363).

The stereochemistry of cleavage of <u>threo</u>-Ph.CHD.CHD-Mn(CO)$_4$-<u>cis</u>-PEt$_3$ depends greatly on the choice of reagent. Mercuric chloride gives predominantly inversion by an outer S_E2 mechanism, while iodine reacts with retention by a closed (or oxidative addition - reductive elimination) process. Iodine monochloride gives mainly <u>erythro</u> (inverted) PhCHD.CDH.I, with a trace of <u>threo</u> PhCHD.CHD.Cl. In contrast with FpCH$_2$CH$_2$Ph (see below), C(1) and C(2) do not exchange.[363-5]

The electrophilic cleavage of species FpR and Fp'R [Fp = CpFe(CO)$_2$; Fp' = CpFe(CO)PPh$_3$] has been very thoroughly investigated. Questions of interest are overall mechanism, the direction of attack with unsymmetrical electrophiles, stereochemistry at carbon and at iron, and in some cases rearrangement of the fragment R. As in the case of the manganese derivatives discussed above, stereochemistry at carbon can be established by nmr studies on species FpCHD.CHD.R', where R' is a bulky group so that in a substantial excess of the molecules Fp and R' are mutually <u>trans.</u>

Species Fp.CHD.CHD.Bu-t react with bromine to give FpBr and Br.CDH.CHD.Bu-t with inversion at C, but mercuric chloride gives FpCl and ClHg.CHD.CHD.Bu-t with retention.[366] However, halogenolysis of PhCHD.CHD.Fp occurs with retention at carbon.[367] Moreover, treatment of PhCH$_2$CD$_2$Fp with HgCl$_2$ or HBr gives FpX and PhCH$_2$CD$_2$HgCl and PhCH$_2$CD$_2$H respectively, (X = Cl or Br), but halogenolysis of PhCH$_2$CD$_2$Fp gives equal amounts of PhCH$_2$CD$_2$X and PhCD$_2$CH$_2$X, together with FpX.[367-8] Unless a change in β-substituent has brought about a total change in mechanism (which seems most unlikely), it follows that the brominolysis of Fp.CHD.CHD.Bu-t is not a classical open S$_E$2 process but that the observed inversion occurs at a later stage. A mechanism that

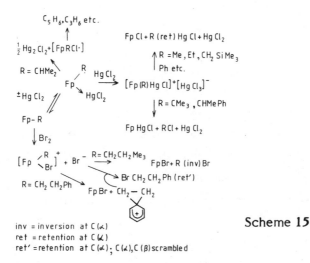

Scheme 15

inv = inversion at C(α)
ret = retention at C(α)
ret' = retention at C(α); C(α),C(β) scrambled

unites these disparate observations, together with others described below, is given in Scheme 15. Initial attack is at iron, whatever the reagent. A mercuric chloride adduct rearranges to eliminate RCl in a closed system with retention at R. Bromine, however, gives rise to the formally iron(IV) intermediate $[BrFpCH_2CH_2R']^+$ and free bromide. For R' = Bu-t, this is susceptible to classical <u>nucleophilic</u> attack at carbon by bromide, causing the observed inversion. For R = Ph, however, the phenyl group itself acts as an internal nucleophile, displacing FpBr to give the symmetrical phenethyl cation, which in turn undergoes nucleophilic attack by bromine at either CH_2 group. Thus C(1) and C(2) are mixed, but the overall process proceeds with two inversions, and thus with overall retention of <u>threo</u> vs. <u>erythro</u> character.

It would then seem that cleavage by $HgCl_2$ with retention requires a basic, or at the very least an oxidizable, metal center. This is what is found; cleavage of <u>threo</u>-PhCHD.CHD.M proceeds with retention when M is Fp or <u>trans</u>-$W(CO)_2PEt_3Cp$, but with inversion when M is <u>cis</u>-$Mn(CO)_4PEt_3$.[363-5]

Mercuric chloride cleavage of Fp-R bonds has been further investigated using both kinetic and product analysis.[369-70] R groups can be divided into three classes. Class (a) includes methyl, ethyl, phenyl and (by reference to the work cited above) 3,3-dimethylbutyl and phenethyl. These give RHgCl and FpCl by a reaction which is first order in FpR and second order in $HgCl_2$. Class (b) (R = Bu-t, CHMePh) react according to similar kinetics, but give RCl and FpHgCl, the CHMePh group racemizing during the reaction. Class (c), which includes isopropyl and neopentyl, show first order kinetics in both reagents, but lead to a complicated mixture of products including Hg_2Cl_2. The proposed mechanisms are included in Scheme 15.

So far we have only considered stereochemistry at carbon. It is also possible to study stereochemistry at iron, using resolved Fp'R (prepared via Fp'CH$_2$O.menthyl, for example), or by studying epimerization in $CpFe(CO)(PPh_2R*)R$, where R* is optically active, or in 1-methyl,3-phenylcyclopentadienyl derivatives. There is always predominant retention at iron, consistent with the general mechanism proposed above. Addition of an electrophile XY to iron, followed by reductive elimination of RX, gives

RX and Fp'Y with retention both at carbon and at iron, while formation of $(Fp'RX)^+$ and Y^-, followed by nucleophilic attack of Y^- on R, gives Fp'X with retention at iron and RY with inversion at carbon.[371-3]

Oxidative cleavage of RCHDCHDPdCl complexes by copper(II) chloride - lithium chloride generally proceeds with inversion to give RCHDCDHCl, but 2-phenethyl palladium(II) chloride reacts with retention and with scrambling of C(1), C(2). This is consistent with _nucleophilic_ attack at metal-bound carbon, either by halide or, for 2-phenethyl, by the phenyl group, followed by oxidation of Pd(0).[374-5]

Organoplatinum derivatives show duality of mechanism. Thus _cis_-PtMe(Ar)(PMe$_2$Ph)$_2$ loses methyl rather than aryl. This is attributed to an oxidative addition - reductive elimination process, where the basic phosphine has favored formation of a Pt(IV) intermediate, and support for this suggestion comes from the isolation of (bpy)PtMe$_2$(HgCl)Cl as an intermediate in the conversion by mercuric chloride of (bpy)PtMe$_2$ to (bpy)PtMeCl and MeHgCl. On the other hand, it is the Pt-Ar bond that is preferentially cleaved in (cod)PtMe(Ar), and in the non-oxidizable _cis_-AuMe$_2$(Ar)PPh$_3$, as expected for electrophilic attack at carbon.[376-7]

The cleavage of diphenylphosphineplatinum(II) alkyls by $IrCl_6^{2-}$ is accompanied by oxidation to platinum(IV) species, and kinetic and relative rate data have been interpreted in terms of an initial single electron transfer, followed by radical loss from a platinum(III) intermediate, and chlorine atom transfer to the radical.[378] Copper(I) also shows evidence for both one- and two-electron pathways, in systems protected from rapid decomposition by other means. Thus oligimeric $(CuCH_2SiMe_3)_n$ reacts with electrophiles EX (= Me$_3$SiCl, CH$_2$:CH$_2$Br, PhI) to give ECH$_2$SiMe$_3$ and copper(I) halide, but attack by benzyl bromide gives some bibenzyl as well as PhCH$_2$CH$_2$SiMe$_3$.[379]

3.2.9 Cleavage by non-hydridic nucleophiles

Reactions of this type provide the final stage for several of the "oxidative" cleavage reactions described in the last Section. Direct nucleophilic displacement from stable starting materials is less usual, but occurs (with the expected inversion) in the hydrolysis or chloridolysis of

Co(IV) species of type $[RCo(dmgH)_2H_2O]^+$,[380-1] and (with respect to the M-C-M bridges) in the depolymerization of $[Mn(CH_2SiMe_3)]_n$ by trimethylphosphine to give $Mn_2(PMe_3)_2(CH_2SiMe_3)_4$.[382]

3.2.10 α-Elimination

Once other, more facile, processes (and in particular β-elimination; 3.2.11 below) have been excluded by judicious ligand design, it becomes clear that α-elimination, the activation and loss of hydrogen and other groups on the carbon directly attached to metal, is an important and widespread phenomenon. For convenience, this Section includes cases of such α-elimination whether it leads to complete loss of ligands or merely to their modification. In addition to the cases discussed here, some examples of α-elimination from one ligand, leading to loss of another one, are described in Section 3.2.7 above, and also in Chapters 3 and 5. A distinction is sometimes made between elimination in the narrower sense (transfer of an atom or grouping to the metal) and abstraction (ultimate transfer to some other site). These types of process have much in common, however, and are both included in the present discussion. Metal-H(α) interaction is evident in the structure of $Ti(dmpe)MeCl_3$,[383] and of some alkylidene complexes (Chapter 3).

It has been argued that carbenes, formed by α-elimination, are the true catalysts in transition metal catalyzed olefin polymerization, rather than the metal alkyls usually assumed.[384] Given the tendency for mechanisms to diversify on inspection, it would be no surprise to find both mechanisms operating in different systems (see 3.3.1 below). However, the reaction sequence of eqs (76-8), in which the intermediate $[Cp_2W(D)CD_2PR_3]^+$ is isolable, gives clear evidence of reversible α-hydrogen migration to metal, while in this case at least insertion of alkene into the metal-carbon bond is conspicuously absent. Labeling experiments confirm that the elimination is intramolecular.[385-6]

$$[Cp_2W(CD_3)(C_2H_4)]^+ + PR_3 \rightleftharpoons CP_2W(CD_3)(CH_2CH_2PR_3)]^+ \qquad (76)$$

$$[Cp_2W(CD_3)(C_2H_4)]^+ + PR_3 \longrightarrow [CP_2W(D)(CD_2PR_3)]^+ + C_2H_4 \qquad (77)$$

$$[Cp_2W(D)(CD_2PR_3)]^+ \longrightarrow [Cp_2W(CD_3)PR_3]^+ \qquad (78)$$

XLIX

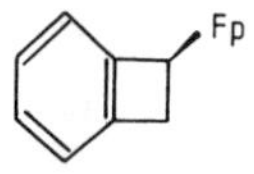

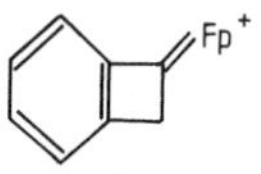

L

α-Hydrogen can also be abstracted by external reagents. For instance, the 16-electron cation $[Cp_2TaMe_2]^+$ can be deprotonated to the 18-electron carbene complex $Cp_2Ta(Me){:}CH_2$.[387] More common, however, is hydride abstraction from an electron-rich species, with no change in electron count:

$$M{-}CH + CPh_3^+ \longrightarrow M^+{:}C + HCPh_3 \qquad\qquad (79)$$

This reaction is shown by $[(OC)_5W{-}C_7H_7]^-$ and $Fp{-}C_7H_7$ (where C_7H_7 represents the η^1-cycloheptatrienyl group).[58] More surprisingly, the same reaction occurs for the species $CpRe(CO)(PPh_3)(CH_2R)$ where R = Me or Et as well as Ph,[388] and for **XLIX**, which is converted to **L**.[389] α-Elimination is occurring even though β-elimination[390] is available. Hydride abstraction from α-alkoxy derivatives gives alkoxycarbenes, which react further with iodide in the remarkable reaction sequence of Scheme 16, converting the initial $-CH_2OR$ ligand to CO.[391]

Facile α-elimination is not confined to hydrogen. The ready phosphine migration of eq 78 is itself a case of α-elimination, as is the acidolysis of $FpCH_2OMe$ to $[FpCH_2]^+$, which disproportionates to give $[Fp(C_2H_4)]^+$,[392] and the related formation of the more stable $[Fp{:}CHPh]^+$ and $[Fp'{:}CHPh]^+$.[393] Closely related is the conversion of species such as $Fp.CHMe.SPh$ to $[Fp.CHMe.SMePh]^+$, an S-methylated cation which in the presence of alkenes acts as a source of methyl carbene and thioether.[110,394]

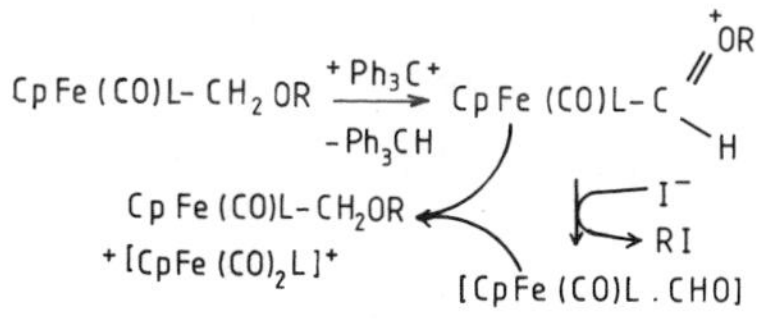

$$CpFe(CO)L{-}CH_2OR \xrightarrow[-Ph_3CH]{+Ph_3C^+} CpFe(CO)L{-}C\overset{+}{\underset{H}{\diagdown}}{\overset{OR}{\diagup}}$$

$$CpFe(CO)L{-}CH_2OR \longleftarrow$$
$$+[CpFe(CO)_2L]^+$$

$$\begin{array}{c} I^- \\ \longrightarrow RI \end{array}$$
$$[CpFe(CO)L.CHO]$$

Scheme 16

Halogen shift from α-carbon to metal is implicated in the reactions[395]

$$(dppe)PtMeCl + LiCHCl_2 \longrightarrow ?(dppe)Pt(Me)CHCl_2$$
$$\longrightarrow (dppe)Pt(Cl)CHCl.Me \qquad (80)$$

$$L_2PtClMe + LiCCl_3 \xrightarrow{-100^\circ C} L_2Pt(Me)CCl_3 \longrightarrow L_2Pt(Cl)CCl_2Me$$
$$\longrightarrow L_2Pt(Cl)CCl:CH_2 + HCl \qquad (81)$$

and in the decomposition of $[(H_2O)_5CrCCl_3]^{2+}$ to $[(H_2O)_5CrCl]^{2+}$, CO and HCl.[396] However, the rearrangement[397] of $[CpRh(PMe_3)_2 CH_2I]^+$ to $[CpRh(CH_2PMe_3)(PMe_3)I]^+$ is catalyzed by amines, and presumably involves nucleophilic displacement of iodide from carbon. Homolytic removal of iodine takes place[398] in the reaction sequence (water omitted for simplicity)

$$Cr^{2+} + [ICH_2Cr(III)]^{2+} \longrightarrow [Cr(III)I]^{2+} + [CH_2Cr]^{2+} \qquad (82)$$
$$[CH_2Cr]^{2+} + Cr^{2+} + H^+ \longrightarrow [CH_3Cr(III)]^{2+} + Cr^{3+} \qquad (83)$$

Added alkene can intercept the carbene ligand, so that by a combination of Reactions (44), (46) and (82), chromium(II) - gem-dihalide reagents can be used in the formation of ring-substituted cyclopropanes.[399] The apparently related reaction by which diiodomethane and copper powder convert alkenes to cyclopropanes is probably more mechanistically complex, since in the absence of added alkene the reaction of CH_2I_2 with Cu is slow, and the CH_2I_2/Cu reagent also fails to show carbene insertion into activated C-H bonds.[400]

3.2.11 β-Elimination of metal hydride

Transfer of β-hydride to metal,

$$M-C-C-H \rightleftharpoons MH(C=C) \rightleftharpoons MH + C=C \qquad (84)$$

is usually the dominant pathway in the decomposition of transition metal organometallics, unless specific steps are taken to prevent it. The reaction involves formal insertion of the metal into a β-CH bond in a (preferably

Scheme 17

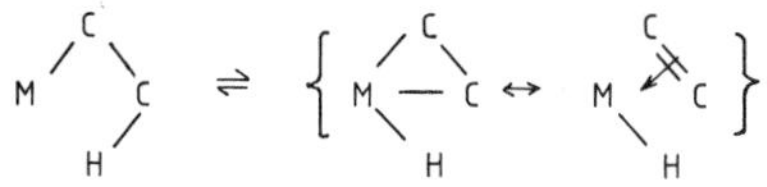

planar) cyclic transition state, as in Scheme 17, thus giving <u>cis</u>-elimination of M-H from an alkene which at this stage remains coordinated. Thus, in addition to the obvious requirement that a β-hydrogen should exist, the reaction requires a vacant coordination site and, preferably, an unused electron pair on the metal. Flexibility in the organic grouping will also aid the reaction. The process is inhibited by bulky and strongly held ligands, a fact that helps explain the importance of phosphines and similar supporting ligands in the older literature. The reaction is in general reversible, and the combination of forward and back reaction provides a pathway for isomerization, whether or not the alkene ligand formed is actually liberated; thus isomerization may or may not be accompanied by exchange. All these factors are of enormous practical importance in those processes (such as hydrogenation, olefin isomerization, hydroformylation and hydrocarboxylation) where Reaction (84) is or can be involved.

The metal hydride functionality generated by eq 84 can of course give rise to secondary reactions, and in particular to reductive elimination (3.2.6 above) if the starting material was a dialkyl, or to hydridic reduction (3.2.5) of a second molecule of starting material. These processes both give rise to the production of equal amounts of alkane RH and alkene R(- H) from species MR or MR_2, and convert metallacycles to 1-alkenes.

Reversible β-elimination was suggested many years ago[401] as a general route to transition metal catalyzed isomerization. As was pointed out, the metal hydride fragment can originate from many different possible sources including protic acids and solvents, and hydrogen itself. As required by the mechanism, H-D exchange between alkane and solvent CH_3OD is catalyzed more efficiently than isomerization itself, since addition in the sense of eq 85 is more facile than that of eq 86:

$$MD + CH_2{:}CH.CH_2R \rightleftharpoons M.CH_2.CHD.CH_2R \rightleftharpoons MH + CH_2{:}CD.CH_2R \quad (85)$$
$$MD + CH_2{:}CH.CH_2R \rightleftharpoons CH_2D.CH(M).CH_2R \rightleftharpoons MH + CH_2D.CH{:}CHR \quad (86)$$

The preferential addition of M to C(1) is as expected on steric and, in most cases, on electronic grounds. The relative rates of the different equilibria are crucial to the control of selectivity in metal-catalyzed additions to alkenes.

If we take seriously the formal oxidative nature of the metal - C(β)-H insertion, or (what amounts to the same thing) the desirability of back-donation in stabilizing the hydridometal-alkene bond produced, then the inertness of Ti(dmpe)Cl$_3$Et to β-elimination is not surprising, Of interest, however, is a crystallographically detectable Ti-H(β) interaction,[402] indicating an electron-deficient interaction such as invoked for Zr(IV) elsewhere in this Chapter. The high thermal stability of species Ta(NMe$_2$)$_4$Bu-t is also relevant.[403]

Configurational constraints on β-elimination are expected in small-ringed metallacycles. These seem to inhibit β-elimination effectively towards the end of the transition series, but not for the earlier members, even with five-membered rings. Thus β-elimination followed by reductive elimination is central to the ethylene dimerization sequence

$$M(C_2H_4) + C_2H_4 \longrightarrow \overline{M.CH_2CH_2CH_2CH_2}$$
$$\longrightarrow HM.CH_2CH_2CH:CH_2 \longrightarrow M + \text{but-1-ene} \qquad (87)$$

for which efficient catalysts or catalyst precursors are Ti(butadiene)(dmpe) and (Me$_3$P)$_2$Ta(C$_2$H$_4$)$_2$Et. (The latter species is itself formed, by a sequence involving β-elimination and reductive elimination, from (Me$_3$P)$_2$TaCl$_3$ and three moles of ethyl Grignard reagent.)[404] In a related reaction, complexes Cp*$_2$TaCl(alkene) react with diolefins CH$_2$:CH(CH$_2$)$_n$CH:CH$_2$ to give eventually cyclic products CH$_2$:$\overline{C(CH_2)_n}$CH.CH$_3$; for n = 5 the reaction is catalytic.[405] For platinum, however, the effect of ring size on stability is very marked. L$_2$Pt(CH$_2$)$_6$ decomposes at a rate comparable to the corresponding acyclic L$_2$PtEt$_2$, but cyclic species L$_2$Pt(CH$_2$)$_4$ and L$_2$Pt(CH$_2$)$_5$ decompose only slowly,[406] and the decomposition of platinacyclobutanes (3.2.16 below) is at least sometimes an α-, rather than a β-abstraction process.[407]

The β-elimination reaction shows an isotope effect, as expected for C - H bond loosening in the transition state. For species $RCHDCH_2Ir$ $(CO)(PPh_3)_2$, β-elimination gives $RCD:CH_2$ and $RCH:CH_2$ in the ratio 2.28:1.[408] Exo-addition of methoxide to $CpPt(PPh_3)(\underline{cis}\text{-}CHD:CHD)^+$ gives **LI**, which decomposes by β-elimination of Pt-H or Pt-D to give $(MeO)CD:CHD$ and $(MeO)CH:CDH$ respectively in 2:1 ratio.[409]

LI

Presumed intermediates Cp_2NbR and Cp_2TaR, formed in the reaction of metallocene halides with Grignard reagents, decompose rapidly to species $Cp_2M(H)(R - H)$; for tantalum, the reaction is highly stereospecific, with n- and s-propyls giving <u>endo</u> and <u>exo</u> isomers of $Cp_2Ta(H)(CH_2:CHMe)$. These decompose to give eventually RH, presumably by reversal of the β-elimination followed by H-abstraction from a ring.[410] In $(Bu\text{-}t)_4Cr$, the high degree of steric congestion interferes with β-elimination, and decomposition occurs with high activation energy (120 kJ mol^{-1} in heptane) to give isobutane, presumably by simple homolysis.[411] Protonation of <u>trans</u>-$Mo(dppe)_2(C_2H_4)_2$ leads to hydrogen exchange, detectable by nmr, between the added proton and one ethylene only. This indicates a reversible insertion-elimination sequence in a species $[(C_2H_4)Mo(dppe)_2(C_2H_4)H]^+$, in which the hydrogen and one ethylene ligand are trapped together on the same side of the $Mo(P)_4$ system.[412]

The need for coordinative unsaturation at the metal is well illustrated by the reactions following the photolysis of $CpW(CO)_3C_5H_{11}$-n at 77K. The fragment $CpW(CO)_2C_5H_{11}$-n undergoes β-elimination on warming to 175K, giving $CpW(CO)_2(pent\text{-}1\text{-}ene)(H)$, unless trapped as $CpW(CO)_2L.n\text{-}C_5H_{11}$ by added ligand L (phosphine or carbon monoxide).[413] Similarly, photolysis of $CpW(CO)_3Et$ gives in turn the fragment $CpW(CO)_2Et$, the species $CpW(CO)_2(C_2H_4)H$ (by β-elimination), $CpW(CO)_3H$ (by exchange of ethylene with CO liberated in the primary step of photolysis), and

$Cp_2W_2(CO)_6$ and ethane (by reaction of hydride with starting material).[414] There is spectroscopic evidence for intermediates $CpM(CO)_2CH_2CH_2$--H (M = Mo,W), distinct from $CpM(CO)_2(C_2H_4)H$ but less reactive to CO or PPh_3 than fully unsaturated $CpM(CO)_2Et$.[415] $CpFe(CO)_2Et$, $CpRu(CO)_2Et$ and related species likewise give $CpM(CO)_2H$ on photolysis, and the intermediate $CpRu(CO)(C_2H_4)H$ has been identified by matrix spectroscopy.[416]

Manganese(II) halides in thf react with Grignard reagents RMgX to give RH and R(- H), by β-elimination followed by reductive elimination of HR. The Mn-H intermediates can trap and ultimately hydrogenate added alkenes.[417]

β-Elimination from species Fp'R gives up to 90% free alkene; the unstable Fp'H has been isolated in 50% yield. The reaction is inhibited by excess phosphine, in accord with a dissociative mechanism that once again illustrates the importance of a vacant coordination site. Isomerization and migration of a deuterium label both occur, showing that loss of alkene is a slower step in the reaction than reversible elimination-addition.[418] This reversible reaction is of great importance in determining the products of cobalt-catalyzed hydroformylation. At high CO pressure, deuteriated alkene substrates do not exchange with reagent hydrogen, even though $HCo(CO)_4$ is known to do so. The implication is that, although an alkene may isomerize and show migration of label by reversible elimination-addition within the coordination sphere of cobalt, once it has been trapped by a cobalt atom it is not released until hydroformylated. At low CO pressure, deuterium exchange occurs between alkenes, suggesting the formation of bis(alkene) complexes from which alkene may be lost as such.[419] The decays of species $R_2Co(acac)(PMe_2Ph)_2$ are first order in complex and inhibited by added phosphine (the dissociative mechanism again). For R = CH_2CD_3, the products below room temperature or from the solid are CH_2CD_2 and CH_2DCD_3, as predicted for elimination without rearrangement, but above room temperature in benzene other products are also formed.[420] In contrast, $EtNi(acac)PPh_3$ shows positional scrambling on standing at room temperature, although not on the nmr timescale. Decomposition in pyridine gives equal amounts of ethane and $Ni(PPh_3)_2(C_2H_4)$.[421] Thus for nickel the

β-elimination equilibrium is faster than alkane loss, while for cobalt the reverse is true. This may be because for nickel the hydride intermediate does not itself carry an alkyl group.

In the absence of excess ligand, complexes $RCD_2CH_2Ir(CO)(PPh_3)_2$ decompose to give $RCD:CH_2$, RCD_2CH_2D, and dimeric Ir(0) complexes, but if excess phosphine is added the only products are $RCD:CH_2$ and $DIr(CO)(PPh_3)_3$. This shows that $HIr(CO)(PPh_3)_2$, formed by the initial elimination, can reduce the starting material but that $HIr(CO)(PPh_3)_3$ cannot.[422]

With triethylaluminum, $(Cy_3P)_2Ni(D)(OAc)$ gives $(Cy_3P)_2Ni(C_2H_4)$ and HD, by β-elimination from an ethylnickel intermediate followed by reductive elimination of HD. Trimethylaluminum gives rise to $(Cy_3P)_2Ni(D)Me$, a hydridoalkyl thermally stable at room temperature (but giving CH_3D on photolysis).[72]

The decomposition of cis-$(Ph_3P)_2Pt(Bu-n)_2$ in dichloromethane to butane and but-1-ene occurs by phosphine ligand dissociation, several elimination-insertion cycles leading to positional scrambling without exchange with added free butene, before eventual reductive elimination of butane. In the absence of added ligand, the initial dissociation is rate-determining, but in its presence there is competition between ligand recombination and alkane elimination, which becomes rate-determining.[423] A similar effect has been demonstrated for the related decomposition of cis-$(Et_3P)_2PtEt_2$, in which there is also evidence for alkane loss from 5-coordinate $(Et_3P)_2Pt(C_2H_4)(H)Et$ at high enough phosphine concentration. The $Pt(C_2D_5)_2$ complex decomposes at the same rate, at least in the

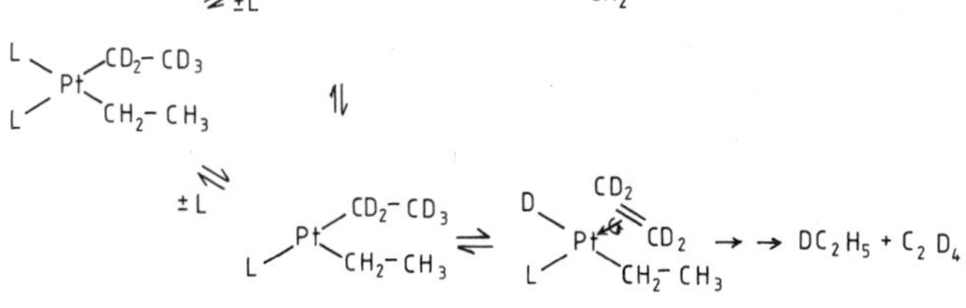

Scheme 18

absence of added phosphine, confirming that C-H bond breaking is not part of the rate-determining step. Despite this, the complex $(Et_3P)_2Pt(C_2H_5)(C_2D_5)$ shows isotopic selectivity, HC_2D_5 being favored over DC_2H_5 by a factor of 2.3. This implies isomerization in the coordinatively unsaturated intermediate (Scheme 18), since loss of either phosphine should occur at almost exactly the same rate. There is also evidence, from the products given by $(Et_3P)_2Pt(CH_2CD_3)_2$, for rotation and readdition reactions of coordinated ethylene.[424] However, β-elimination from trans-$(R_3P)_2PdEt_2$ is non-dissociative,[425] and reactions of this class can show both dissociative and non-dissociative components.[426]

The rearrangement of Ph_3PAuMe_2Bu-t to the corresponding s-butyl complex is another case of reversible addition-elimination, inhibited by added phosphine, and unaffected by added propene, which would have reacted to give propyl complexes had there been any dissociation of ligand isobutene from the metal.[427]

3.2.12 Other β-eliminations and abstractions

The reaction[390] of $FpCH_2CH_3$ with the trityl cation, to give $[Fp(C_2H_4)]^+$ and triphenylmethane, may be contrasted with closely related α-abstractions discussed above.[388-9] The Fp grouping assists the β - anti elimination of water in Scheme 2,[115] while the similar elimination of methanol on acidification of β-methoxyalkyls of palladium and platinum[428] is the reverse of the trans-addition of nucleophiles to alkenes, discussed in Chapter 11. However, the β-migration of bromide to metal in the Cr(II) reduction of vic-dibromides is a cyclic process with cis-elimination:[429]

$$Br-C-C-Br + 2Cr(II) \longrightarrow Br-C-C-Cr(III) + Cr(III)Br \qquad (88)$$

$$Br-C-C-Cr(III) \longrightarrow C=C + Cr(III)Br \qquad (89)$$

3.2.13 Intramolecular reductive elimination of two carbon ligands

The concept of conservation of orbital symmetry has been applied to reductive elimination, with apparently only partial success.[430] To the present author, the problem seems to be that where there is a manifold of non-bonding orbitals, incompletely occupied, a change in overall symmetry during reaction may simply represent a rearrangement of the non-bonding electrons, so as to lower electron-electron repulsion energy within the manifold. Such a rearrangement will be kinetically irrelevant if it occurs after the transition state, and may indeed facilitate the process if it occurs earlier. A treatment that ignores this possibility will inevitably reject as "forbidden" many perfectly feasible processes, such as reductive elimination from d^0 systems.

One must also consider the effects of changes in preferred geometry during an elimination reaction, due to changes in the interactions of d-electrons with ligands as the numbers of both of these alter. From such considerations, it is predicted that reductive elimination from octahedral d^6 to give square planar d^8 systems is only allowed when the ligands lost are originally mutually _cis_, and when the angle between the ligands _trans_ to those lost can increase from the original value of $90°$ towards $180°$ during the reaction. This result has been used[431] to explain the profound effect of supporting ligand chelation on the course of such reactions.[432] Reductive elimination from d^8 systems has been the subject of much more detailed theoretical study, which is summarized further below.

Reductive eliminations promoted by oxidation:- since reductive elimination is, by definition, reductive, it is hardly surprising that removal of one or more electrons from the central metal promotes the process. For instance, species of the type $Fe(bpy)_2R_2$ are thermolyzed to RH and R(- H) by β-elimination followed by reductive elimination of alkane,[70,433] but the addition of an oxidizing agent, such as bromine or Na_2IrCl_6, leads to reductive elimination of R-R. The process is intramolecular as shown by lack of crossover, and under electrochemical conditions is associated[434] with the second one-electron oxidation. Elimination also occurs on the oxidation of $CpCo(PPh_3)Me_2$,[435] $Co(acac)(PMe_2Ph)_2R_2$ (R = Me, Et) or $Ni(bpy)Et_2$, while _trans_-$Ni(Ar)_2(PEt_3)_2$ gives biaryl, presumably after isomerization. Oxidation of _trans_-$NiAr(PEt_3)_2X$ by Ir(IV) gives $ArPEt_3^+$, by what could be regarded as reductive elimination of aryl together with phosphine, and there is direct evidence for Ni(III) intermediates.[436] Oxidation of organocuprates $LiCuR_2$ (R = alkenyl, neophyl) by oxygen, copper(II), or nitrobenzene gives R-R without isomerization or rearrangement.[437]

Reductive elimination promoted by donors:- while reductive elimination reduces the oxidation number of the central metal, it also reduces the total electron count, as well as, of course, the coordination number and degree of steric crowding. As a result, added ligands sometimes facilitate the reaction. Thus addition of bipyridyl to $Fe(bpy)_2Et_2$ promotes reductive elimination at the expense of β-elimination, giving butane and neutral $Fe(bpy)_3$.[70] Electron-demanding alkenes promote reductive elimination from the compounds $Ni(bpy)R_2$, presumably by increasing the electron demand at the metal.[438] Aldehydes, diketones, and Schiff bases also promote elimination of R-R from complexes L_2NiR_2,[439] as do bidentate phosphines $Ph_2P(CH_2)_nPPh_2$, where n = 1 or 4; with n = 2 or 3 the principal reaction is simply displacement of monodentate L. Monodentate phosphines promote elimination of toluene derivatives from species $Ni(dmpe)(Me)(C_6H_4X)$.[440]

An important special case of donor-promoted reductive elimination arises when the donor is carbon monoxide. Here alkyl migration (Chapters 6-8) will commonly lead to (alkyl)(acyl) complexes, from which reductive elimination of ketones seems particularly facile, perhaps because it unites nucleophilic and electrophilic carbon, and may be further facilitated by a second CO as incoming ligand. Examples include the formation of acetone from species $Me_2Mo(arene)L_2$,[81] and diethyl ketone from $(bpy)NiEt_2$. However, $(dppe)NiEt_2$ gives ethylene and propionaldehyde, showing CO incorporation in this case to be slower than β-elimination although faster than alkane elimination,[441] while $(arene)Ni(C_6F_5)_2$ reacts with CO to give arene, tetracarbonylnickel, and uncarbonylated perfluorobiphenyl.[442] Carbonylation of a range of species L_2PdR_2 gives R.CO.R quantitatively.[443] Further examples of this reaction are discussed in Chapter 7.

Other cases:- reductive elimination as a step in transition metal modified coupling reactions is discussed in 3.3.3 below.

The orbital energies involved in reductive elimination from d^8 systems have been investigated by extended Hückel theory.[444-7] In square planar complexes, the groups eliminated must be mutually cis for a facile pathway to exist.[122] Better σ-donating leaving groups favor the reductive elimination. Other effects arise from the interaction between the antisymmetric d-orbital (or d-p hybrid), which becomes filled during the reaction, and the remaining ligands. It follows that stronger σ-donors trans to the leaving groups should inhibit the reaction, that the activation energy should be lower for nickel than for palladium and platinum, and that reductive elimination should be more facile from T-shaped intermediates LMR_2, formed by ligand dissociation, than from the parent complexes. If the R groups are originally trans, then elimination is preceded by isomerization within the T-shaped intermediate, and this process also will be more facile the poorer the donor properties of L.[444] For $[Me_4Au]^-$ in particular, elimination is further inhibited by what are described as antibonding interactions involving hyperconjugative π-orbitals on the methyl groups;[445] this presumably corresponds to the familiar but imprecise notion of "steric hindrance". For the metallacycles $L_nNi(CH_2)_4$, reductive elimination of cyclobutane is predicted to be facile for n = 2 (square planar) or 3 (square pyramidal; L apical), but for n = 2 (approx. tetrahedral) or 3 (square pyramidal; CH_2 apical) the preferred process is predicted to be ring cleavage to give two ethylene fragments.[446]

In accord with the above theory, replacement of triethylphosphine by triethylphosphite in complexes $L_2Ni(Ph)CN$ facilitates the reductive elimination of benzonitrile.[448] The thermolysis of cis-$(Et_3P)_2PdMe_2$ gives ethane[449] (no radicals being detected by CIDNP, in contrast to the photochemical reaction of the platinum analogue).[278] Such reductive elimination has been shown to be dissociative, again in accord with theory, and ligand loss precedes reductive elimination of gem-dimethylcyclopropane from species $L_2PtCH_2CMe_2CH_2$.[450] This reductive elimination is in contrast to the non-dissociative β-elimination shown by related trans species.[425] For trans-L_2PdMe_2, where L is a phosphine, isomerization to the cis species precedes reductive elimination. The elimination is facilitated by polar solvents (which perhaps facilitate loss of the strongly σ-donating L), in which there is no crossover between CH_3 and CD_3 derivatives. Reductive elimination is slow for (dppe)$PdMe_2$, either because of failure to lose L or because of unfavorable coordination geometry in the product. The constrained trans complex (TRANSPHOS)$PdMe_2$ does not itself show reductive elimination, but reacts with CD_3I to give CH_3CD_3, clearly via palladium(IV) intermediates.[451]   The trans – cis isomerism, in monodentate phosphine complexes, is assisted by added LiMe, or substrate (either isomer), and methyl exchange between palladium centers is proposed.[452]

The thermolysis of species cis-$L_2Pt(C_6H_4$-p-$Me)_2$ in toluene gives specifically 4,4'-dimethylbiphenyl by a first order reaction, which is if anything accelerated by added phosphine, as is the thermolysis of the neat complexes; substituent effects have been studied in some detail.[453]   Here, at least, the elimination does not require prior dissociation; this may, in the light of present theory, reflect smaller steric hindrance in the elimination of unsaturated carbon.

When species $L_2Pt(CH_2)_4$ are heated in dichloromethane, they undergo oxidative addition of $Cl-CH_2Cl$, followed by reductive elimination. If L_2 represents two monodentate ligands, cyclobutane is a major product, but where L_2 represents dmpe, cyclobutane formation is suppressed and five-carbon fragments predominate, implying reductive elimination between C_1 and C_4 ligands.[432]   This is because the low-energy pathway for reductive elimination from octahedral d^6 complexes requires the ligands initially trans to those lost to become trans to each other in a square planar product.[431]

Other examples of reductive elimination from platinum(IV) include the reaction of fac-$Me_3Pt(PPh_2Me)_2I$ to give ethane by an intramolecular, kinetically first order process, of activation energy around 130 kJ mol^{-1},[454] and the elimination of both $(C_3H_5)_2$ and C_3H_5Me (C_3H_5 = cyclopropyl) from $(PhMe_2P)_2Pt(C_3H_5)_2(Me)I$.[455]

Reductive elimination from Au(III) shows the effects predicted by theory. For instance, reductive elimination from cis-R_2AuXL is favored by a poor donor X, and strongly inhibited by excess L. In addition, bulky groups X and L both favor the reaction.[320]

3.2.14 Reductive elimination of carbon ligands on different metal centers

This Section covers reactions of the types

$$M-R + M'R' \longrightarrow R-R' + \dots \qquad (90)$$

and   R-MM'-R' $\longrightarrow$ R-R' + ...   (91)

where M,M' are transition metals, whether or not there is direct bonding between M and M' at any stage, and whether or not reductive elimination is preceded by migration of a leaving group from one metal center to the other. (In the limit where one M-R bond is broken before the other is affected, this class merges with others, such as simple homolysis followed by homolytic substitution; cf. Ref. 290.) Reactions of this kind are easily demonstrated when there is only one R group on each metal to start with, but may be much more common than so far realised among the apparently unimolecular processes of the previous Section. For example, cross-products CH_3CD_3 are obtained, in addition to ethane and ethane-d_6, from the thermolysis of mixtures of Me_3AuPPh_3 and $(CD_3)_3AuPPh_3$ in non-polar solvents, although exchange of labels in starting material did not occur.[447] Relatively small structural differences can affect whether a given overall reaction involves one or two centers. Thus the degree of crossover between $CpCo(L)Me_2$ and $CpCo(L)(CD_3)_2$ in the carbon monoxide-induced elimination of acetone is highly dependent on the nature of L,[14,456-8] and may involve phosphine dissociation and exchange of CH_3 and CD_3 in a preliminary step.[459] A binuclear process must be presumed for the reaction of $CH_3Co(PPh_3)_3$ with styrene to give a material of composition $Co(PPh_3)_3(styrene)$, together with 0.4 moles of ethane.[460]

Bi- and polynuclear processes are particularly well documented for the elements of the copper group, which are also known to form a wide range of cluster compounds. Thermolysis of CH_3AuPPh_3 gives gold, free phosphine, and ethane.[461] The decomposition of alkenylmetal complexes of copper(I) or silver(I), either as "simple" species or as tri-n-butylphosphine complexes, gives alkenyl dimer with retention of configuration, a finding of historical importance in helping to establish that the thermolysis of organometallics is not in general a free radical process.[462] Elimination from mixed cluster compounds has been studied, especially for aryls of copper and silver, and intermediates identified in which the average oxidation state of the metal is less than one.[463-5] Grignard reagents RMgX are coupled to give R-R by silver nitrate in a sequence in which nitrate re-oxidizes silver.[466]

3.2.15 Miscellaneous

Other reductive eliminations:- the rhodium(III) methyl complex $RhMeCl_2(CO)(PPh_3)_2$ reductively eliminates chloromethane by a first order, intramolecular process, of activation energy 95 kJ mol^{-1}, and undergoes attack at carbon by excess triphenylphosphine to give $Ph_3PMe^+Cl^-$.[467] The direction of reductive elimination at platinum is highly sensitive to structure:[468]

$$(Me_2PhP)_2PtMeCl_3 \longrightarrow MeCl + ...$$
$$(Me_2PhP)_2PtMe_2Cl_2 \longrightarrow MeCl + C_2H_6 + ...$$
$$(Me_2PhP)_2PtMe_2Br_2 \longrightarrow MeBr + ...$$
$$(Me_2PhP)_2PtMe_3X \longrightarrow C_2H_6 + ... (X = Cl, Br)$$
$$(Me_2PhP)_2PtMe_2(COMe)X \longrightarrow MeCOMe + \qquad (92)$$

(note again the preferential loss of ketone).

Exchange between metal centers can be followed using methyl and deuteriomethyl groups as labels for alkyl, and Cp and Mecp as labels for metal. In this way it was shown that Cp_2ZrMe_2, shows exchange while Cp_2MoMe_2, an eighteen-electron system, does not. Exchange in $CpCoLMe_2$, where L is a phosphine ligand, occurs by a bimolecular mechanism following on initial loss of L; for L = PMe_3, there is no prior dissociation and therefore no exchange.[459] All these facts are consistent with exchange through 3-center, 2-electron alkyl bridges, which can only form when at least one of the moieties to be bridged has available vacant orbitals.

Exchange has been observed between platinum(II) methyls and either platinum(II)[469] or platinum(IV)[470] halides and related species. For the Pt(II) - Pt(II) exchange at least, there is evidence for second order kinetics and for prior dissociation of L. Exchange also occurs between species $Ar_4M_2Li_2$ (M = Cu, Ag, Au) and $Rh_2(CO)_4Cl_2$, leading to ArM polymers and $ArRh(CO)_2$ when Ar is the chelating group C_6H_4-o-CH_2NMe_2, but giving only biaryl and ArCOAr for simple aryl groups.[471]

Abstraction of organic groups from phosphine ligands by insertion of zerovalent metal into phosphorus-carbon bonds, and subsequent decomposition by a variety of pathways, is a recognised secondary process following reductive elimination from such species as platinum(II) diaryl complexes (see e.g. Reference 453). Biaryls, and diarylmethylphosphines, are formed in the thermolysis of species $(Ar_3P)_3CoMe_3$.[472] Complexes (Ar)Cu(dppe) (Ar = C_6H_4-o-CH_2NMe_2) react with excess diphosphine according to Equation 93:[473]

$$ArCu(dppe) + dppe \longrightarrow \tfrac{1}{2}(dppeCuPPh)_2 + Ph_2PCH:CH_2 + ArH \qquad (93)$$

3.2.16 Further examples of metal-carbon bond cleavage

This Section surveys metal-carbon bond cleavage across the Periodic Table. Reactions referred to in Sections 3.2.1 to 3.2.15 are listed under each group, and a discussion is then presented of further cases, where the mechanism is unclear, where several mechanisms are involved, or where the interest centers on change of mechanism between one system studied and another rather than on any one clearly defined mechanism individually.

Lanthanides and actinides:- see Section 3.2.3, Ref. 53; 3.2.7, Ref. 357.

The uranium and thorium complexes $Cp*_2M(Cl)Me$ rapidly insert acetone to give $Cp*_2M(Cl).O.CMe_3$, while $Cp*_2ThMe_2$ is hydrogenolyzed, according to Equation 94, to a bridged hydride (which catalyzes the hydrogenation of ethylene:[474]

$$2Cp*_2ThMe_2 + 4H_2 \longrightarrow (Cp*_2ThH)_2(\mu\text{-}H)_2 + 4CH_4 \qquad (94)$$

Titanium, zirconium, hafnium:- see Section 3.2.2, Refs. 267-8,277,293; 3.2.3, Ref. 301; 3.2.4, Ref. 325; 3.2.5, Refs. 326-7; 3.2.6, Refs. 337-40; 3.2.7, Refs. 350-4,356; 3.2.8, Refs. 358,218; 3.2.10, Refs. 383-4; 3.2.11, Refs. 402,404,411; 3.2.15, Ref. 459.

The decomposition of tetramethyltitanium is a complicated process, leading to ethane and ethylene as well as methane, and being catalyzed by the decomposition products themselves.[475] In ether, at least, the residue is hydrolyzable and carbon-containing.[476] Tetramethylsilane was the only product characterizable when $(Me_3SiCH_2)_4Ti$ was decomposed in an nmr tube at 80°C.[477] The neopentyls Np_4M (M = Ti, Zr, Hf) gave neopentane as the only detectable product: stability increases down the group and in the order Me $\ll$ Np $\sim$ Me$_3$SiCH$_2$. Np_4Ti reacts with oxygen in benzene to give $(NpO)_4$Ti.[478] Tetrabenzyltitanium can abstract oxygen from refluxing ether, giving $(Bz_3Ti)_2$O (Bz = benzyl).[479] Thermolysis of tetrabenzylzirconium gives toluene, bibenzyl, and less expected products such as benzene and biphenyl;[480] it is tempting to postulate CH$_2$ - Ph bond cleavage giving a zirconium carbene fragment.

Thermolysis of $(C_5D_5)_2$TiMe$_2$ gives methane with some CH$_3$D and CH$_2$D$_2$, while $(C_5D_5)_2$TiPh$_2$ gives benzene with some C$_6$H$_5$D and C$_6$H$_4$D$_2$;[481] Cp$_2$Ti(C$_6$D$_5$)$_2$ gives benzene-d$_6$ with some C$_6$D$_5$H and C$_6$D$_4$H$_2$.[482] These results suggest hydrogen abstraction from one group by another, a process independently established (3.2.7 above) with the phenyl derivatives. In diethyl ether, Cp$_2$TiMe$_2$ decomposes with hydrogen abstraction from solvent rather than from the rings, but in hydrocarbon solvents the abstraction may be either from the rings or from a methyl group (not necessarily in the same molecule); the reaction is autocatalytic and reaction of starting material with an intermediate Cp$_2$Ti:CH$_2$ has been suggested.[483] Thermolysis of solid Cp$_2$Ti(CD$_3$)$_2$ has also been reported to give CD$_4$, CD$_3$H, C$_5$H$_6$, C$_2$H$_4$ and C$_5$H$_4$CD$_3$. Thus we have (at least!) abstraction from methyl or from ring by methyl, abstraction by one ring from another, ring fragmentation, and attachment of methyl to ring. The Zr and Hf analogues have also been studied.[484] In the photolysis, though not the thermolysis, of Cp$_2$TiMe$_2$, CIDNP shows the involvement of radicals. The starting material becomes polarized, showing that recombination is the usual fate of the singlet radical pair first formed.[485] The methyl radicals can initiate the polymerization of methyl methacrylate.[486]

The thermolysis of species Cp$_2$TiAr$_2$ is first order, with activation energy around 100 kJ mol^{-1}. Deuteriation of the phenyl rings increased activation energy but that of the cyclopentadienyl rings did not;[487] similar results were found for Cp$_2$TiBz$_2$.[488]

Photolysis of Cp$_2$TiPh$_2$ in C$_6$D$_6$ gave black TiC$_{10}$H$_{10}$, biphenyl, and biphenyl-d$_5$. Thus both reductive elimination and free radical loss occur, but the phenyl-phenyl abstraction so characteristic of the thermal reaction does not.[489] The species $[C_6H_4\text{-}o\text{-}(CH_2)_2]M$ (M = Ti, Zr, Hf) are stable to 190°C in vacuo, then decomposing to give o-xylene.[490] The Ti(III) derivatives Cp_2TiR (R = Ar, Bz) thermolyze to give RH,[491] presumably by H-abstraction from the ring.

Vanadium, niobium, tantalum:- see also Sections 3.2.2, Ref. 269; 3.2.3, Refs. 302,222; 3.2.4, Ref. 324; 3.2.7, Ref. 355; 3.2.10, Ref. 387; 3.2.11, Refs. 403,405,410.

The species $V(CH_2SiMe_3)_4$ and $VO(CH_2SiMe_3)_3$ are remarkably stable, being sterically restricted and lacking any obvious decomposition route; this illustrates the general point that alkyls of metals in high oxidation states are not necessarily unstable.[492] Likewise, species $RV(O)(OR')_2$ decompose below 0°C to give methane and some ethane when R is methyl, but are stable to 70°C if R is a sec-alkyl.[493] Thermolysis of tetrabenzylvanadium etherate gives more than two molecules of toluene, and a residue from which further toluene can be obtained by hydrolysis.[494] $LiV(mesityl)_4 \cdot 2Et_2O$ decomposes exothermically at 161°C to give mainly mesitylene.[495] Not only $Cp_2V(CH_2SiMe_3)_2$, but also Cp_2VEt_2, are stable enough to be handled at room temperature and stored at -40°C.[496] This may be because β-elimination, the most obvious decay route for the ethyl, would generate a 19-electron intermediate. Heating Cp_2VAr gives vanadocene, arene, and $CpV(C_5H_4Ar)$, while Cp_2NbAr_2 gives arene.[497] When Cp_2VPh is pyrolyzed alone at 150°C, or thermolyzed in benzene, biphenyl is formed from ligand phenyl, and benzene from ligand phenyl and cyclopentadienyl hydrogen, without uptake of material from solvent.[498] Photolysis of Cp_2TaMe_2 gives Cp_2TaMe_3, polymeric "tantalocene", and methane; the hydrogen required can come from solvent, ring, or the other methyl group.[499]

Chromium, molybdenum, tungsten:- see also Section 3.2.2, Refs. 270-6,282-7(a),289-90; 3.2.3, Refs. 300,303-4; 3.2.5, Ref. 328; 3.2.6, Ref. 341; 3.2.7, Refs. 186-7; 3.2.8, Refs. 359-63; 3.2.10, Refs. 385-6,58,396,398-9; 3.2.11, Refs. 411-5; 3.2.12, Ref. 429; 3.2.13, Ref. 81; 3.2.15, Ref. 459.

Like the related vanadium compounds, $[Cr(CH_2SiMe_3)_4]^-$, $Cr(CH_2SiMe_3)_4$, and $Mo_2(CH_2SiMe_3)_6$ show high thermal stability. The ease of oxidation of the Cr(III) anion is noteworthy; the alkyl group plays a role comparable to that of other bulky good σ-donors, such as t-butoxide, in stabilizing tetrahedral Cr(IV).[492]

Pyrolysis of $MeCrCl_2(pyridine-d_5)_3$ gives mainly unlabeled CH_4 and C_2H_6, and a residue which with D_2O gives mainly CH_3D and some CH_2D_2. Thus the decomposition is essentially intermolecular α-abstraction. The ethyl analogue gives ethane:ethylene:n-butane :: 1:6:5, suggesting a free radical pathway.[500] Mixtures of CD_3Li and WCl_6 decompose to give CD_4 and C_2D_6; high ratios of CD_3Li favor the former over the latter. Carbene intermediates are proposed, and addition of $Me_3SiCH:CH_2$ gives some metathesis to $CH_2:CD_2$.[501] Hexamethyltungsten itself, and its phosphine adducts, decompose and react according to Scheme 19.[502-4]

Scheme 19

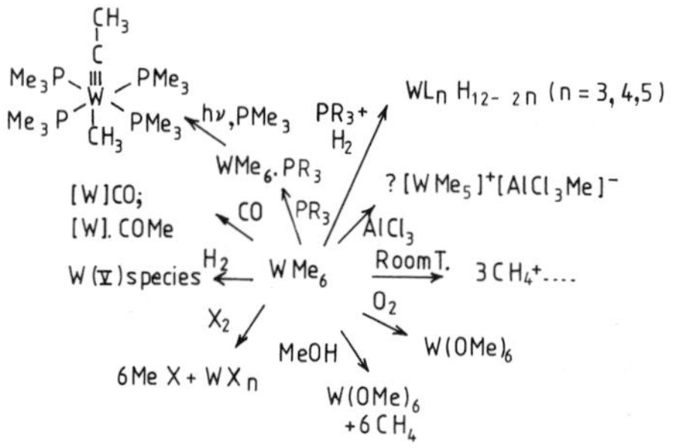

Manganese, technetium, rhenium:- see also Section 3.2.2, Refs. 3,281; 3.2.3, Refs. 299,102,305,69; 3.2.4, Ref. 323; 3.2.5, Refs. 329-30; 3.2.7, Ref. 161; 3.2.8, Refs. 363-5; 3.2.9, Ref. 382; 3.2.10, Ref. 388; 3.2.11, Ref. 417.

Manganese(II), like other metals, increases the yield of ethane relative to methane when methyl radicals are generated from t-butyl hydroperoxide.[505]

Iron, ruthenium, osmium:- see also Sections 3.2.2, Ref. 288; 3.2.3, Refs. 306-10; 3.2.5, Ref. 331; 3.2.6, Ref. 114; 3.2.7, Refs. 345-7; 3.2.8, Refs. 363-73; 3.2.10, Refs. 58,389-94,110; 3.2.11, Refs. 416,418; 3.2.12, Refs. 390,115; 3.2.13, Refs. 433-5,70.

Species $Fe(bpy)_2R_2$ are pyrolyzed mainly to methane for R = Me, and to roughly equimolar amounts of RH and R(- H) for R = Et, Pr-n. $Fe(bpy)_2Et_2$ reacts with D_2O to give C_2H_5D and with iodine to give $Fe(bpy)_2I_2$ and, mainly, n-butane; n-butane is also formed from $Fe(bpy)_2Et_2$ with methyl iodide, or (3.2.13 above) with bipyridyl.   Reaction with butadiene leads to evolution of ethane, ethylene, and C_6 species, and gives a dimerization catalyst, while electronegative olefins such as TCNE or maleic anhydride give solids $Fe(bpy)_2(olefin)$, with evolution of ethylene, ethane, and butane.[70,506] Electrochemical studies of this species illustrate the effects of oxidation state on reaction route. Oxidation to Fe(III) promotes loss of ethyl radicals, while a second oxidation is completely irreversible since the Fe(IV) species rapidly undergoes reductive elimination of butane.[434] $Fe(dppe)_2Me_2$ reacts with carbon monoxide by insertion, CO ligand incorporation, and reductive elimination to give acetone and a product formulated as $Fe(CO)_2(dppe)_2$; with D_2 to give $Fe(dppe)_2D_2$ and CH_3D; on thermolysis in CD_2Cl_2 to give mainly CH_4 and C_2H_6, with some C_2H_4 and $C_2H_2D_2$; and on recrystallization from toluene, with methane loss by hydrogen abstraction from a phenyl group of the ligand, to give $(Ph_2PCH_2CH_2PPh.C_6H_4\text{-}o\text{-})Fe(dppe)Me$.[507]

The decomposition of $Os(CO)_4(H)(Me)$ and related species has been studied in illuminating detail.[265] The decomposition is by a bimolecular process, as shown by crossover of isotopic labels, to give methane and $Me[Os(CO)_4]_2H$, which in turn reacts with starting material to give methane and $Me.[Os(CO)_4]_3.Me$. Nucleophiles such as triethylphosphine slow down the decomposition in accord with Scheme **20**.[508-9] Pyrolysis of $Os(CO)_4Me_2$ occurs slowly, at high temperature, to give methane by hydrogen abstraction from solvent.[510] Similarly, $Me[Os(CO)_4]_3Me$ gives $Os_3(CO)_{12}$ at 90°C in toluene, with methane again generated by abstraction from solvent.[511] The principles illustrated are (i) that reductive elimination

Scheme **20**

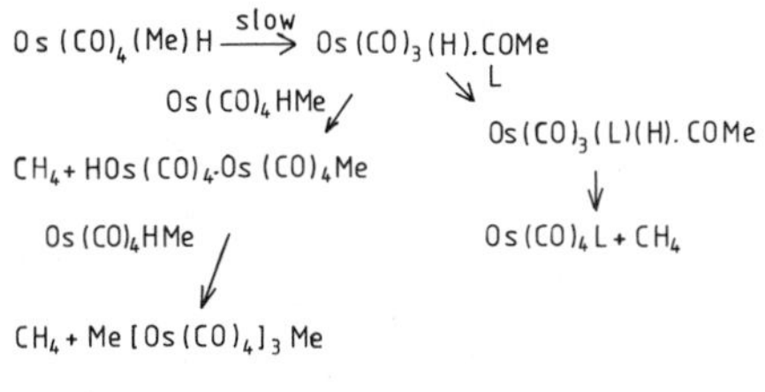

$Os(CO)_4(Me)H \xrightarrow{slow} Os(CO)_3(H).COMe$

$Os(CO)_4HMe \nearrow \qquad \searrow L$

$\qquad\qquad\qquad Os(CO)_3(L)(H).COMe$

$CH_4 + HOs(CO)_4.Os(CO)_4Me \qquad \downarrow$

$Os(CO)_4HMe \nearrow \qquad Os(CO)_4L + CH_4$

$CH_4 + Me[Os(CO)_4]_3Me$

is only possible when the remaining fragment is of not too high energy, (ii) that de-saturation (either by CO loss or by alkyl migration) makes possible the formation of binuclear species, (iii) that at least in this system acyl species are readily formed whereas formyl species are not (compare results for the decomposition of $H_2Os(CO)_4$[512]), and (iv) that hydrogen bridging or transfer is more facile than methyl bridging or transfer. These factors taken together explain the behavior of this family of compounds as well as the relative paucity of hydridoalkyls, especially hydridoalkyl carbonyls.

Cobalt, rhodium, iridium:- see also Sections 3.2.2, Ref. 287(b); 3.2.3, Refs. 309,311-2; 3.2.5, Refs. 332,334; 3.2.7, Refs. 348-9; 3.2.9, Refs. 380-1; 3.2.10, Ref. 397; 3.2.11, Refs. 408,419-20,422; 3.2.13, Refs. 435-6; 3.2.14, Refs. 14, 456-60; 3.2.15, Refs. 467,459,472, and Chapter 2.

Cobalt cations assist in the formation of ethane from methyl radicals.[505] Decomposition of $[CoL_4(CH_3)(CD_3)]^+$ $[L = P(OMe)_3]$, formed by attack of Me_3O^+ on CD_3CoL_4, gives CH_3D, CD_3H, and (surprisingly) CD_4, but no CH_2D_2; the hydrogens are not abstracted from solvent. CH_3CoL_4, treated with d-acid, gives CH_3D.[513]

Co(IV) species of type $[RCo(IV)(chel)]^+$, where chel^{2-} is a planar 4-coordinate ligand, are generally attacked by pyridine according to an S_N2 mechanism at C(α), to give pyR$^+$ and [Co(II)chel]; however, complexes where chel is a Schiff's base show what are presumably S_Ni processes, with R transfer to the chelate ligand oxygen. The p-methoxybenzyl cation is lost from Co(IV) by an S_N1 process while complexes $[(RCo(acacen)]^+$, $[MeCo(salen)]^+$ homolyze to free R$^\cdot$ and Co(III) species.[514] With $[EtCo(dmgH)_2]^+$, there is also a bimolecular decomposition path, thought to involve disproportionation.[515]

Nickel, palladium, platinum:- see also Sections 3.2.2, Refs. 278-81,291,294-8; 3.2.3, Refs. 313-8; 3.2.6, Refs. 342-4,317-8; 3.2.7, Refs. 173-4; 3.2.8, Refs. 374-8; 3.2.10, Ref. 395; 3.2.11, Refs. 406-7,409,421,72,423-6; 3.2.12, Ref. 428; 3.2.13, Refs. 431-2,436,438-43,445-8,278,425,449-53,432,454-5; 3.2.15, Refs. 468-70.

Nickel cations are even more effective than those of cobalt or manganese in increasing the yield of ethane from methyl radicals relative to that of methane.[505]

While the interaction of $Ni(PEt_3)_4$ with aryl halides gives arenes and nickel(I) species by a free radical process,[152] $Ni(PCy_3)_2$ reacts with chlorobenzene to give 95% biphenyl and traces only of benzene; $Ni(Cl)(Ph)(PCy_3)_2$ gives the same products on heating. With ethyl bromide, oxidative addition followed by β-elimination gives $Ni(H)(Br)(PCy_3)_2$ and ethylene, while atom transfer gives $[(Cy_3P)_2NiBr]_2$ and ethyl radicals, which abstract hydrogen from the hydride complex to give ethane.[153] Decomposition of bis(neophyl)nickel(II) complexes gives t-butyl benzene (neophane), dineophyl, $PhCMe_2CH{:}CHCMe_2Ph$ (presumably from carbenes formed by α-abstraction), and other species. Added phosphines increase the yield of dineophyl (presumably by favoring reductive elimination).[516]

Photolysis of $CpNi(PPh_3)Me$ gives methane, probably by homolysis. For the ethyl analogue, the products are ethane and ethylene (by β-hydride elimination etc. after loss of phosphine), while the phenyl or tolyl derivatives give biaryl.[517] Possibly related is the thermolysis of species CpNiR(propene), which after ligand loss gives $(CpNi)_3CH$ and methane (R = CH_3) or $[CpNi]_x$ and RH together with (CpNiH) and R - H (R = Pr-n).[66]

Oxidative addition of halides RX to $Pd(PPh_3)_4$ gives intermediates $(Ph_3P)_2PdRX$. For R = benzyl, these are isolable, and react with CO to give an acyl complex, with $ArCH_2Cl$ to give $ArCH_2CH_2Ph$, and with acyl chlorides R'COCl to give $R'COCH_2Ph$. (The latter two reactions are expected oxidative addition - reductive elimination processes). For R = $PhCH_2CH_2$, decomposition is by β-elimination; for R = Me_3SiCH_2, by radical loss; and for R = $PhCOCH_2$, by loss of R^-.[518]

Oxidative addition of organic halides to platinum(II) dialkyls, followed by thermolysis, is a convenient way to study the decomposition of platinum(IV) compounds. Complexes $L_2Pt(Me)_2(COMe)Cl$ give acetone, showing the expected alkyl-acyl reductive elimination, while $L_2Pt(Me)_2(Bz)Br$ gives a mixture of ethane and ethylbenzene, as expected for competing reactions. Elimination is accompanied by isomerization in the Pt(II) product, presumably by reversible loss of L. Reaction of L_2PtEt_2 with iodomethane gives an unstable Pt(IV) product, which decomposes, after loss of L, by β-elimination. Thus, when available, β-elimination in Pt(IV) appears to be more facile than either reductive elimination from Pt(IV), or reductive elimination or β-elimination from Pt(II). In the thermolysis of

$PtMe_2(CD_3)L_2X$, CH_3 and CD_3 species were statistically scrambled, consistent with prior dissociation of L. The activation energy for the reductive elimination of ethane was 69 kJ mol^{-1}, compared with a mean platinum-methyl bond energy of 144 kJ mol^{-1} (suggesting considerable C-C bond formation in the transition state), but there is a sizeable isotope effect, with $k(CH_3)/k(CD_3) = 1.07$.[132-3]

With excess methylpyridine, Pt(IV) metallacyclobutanes eliminate alkene. Surprisingly, the mechanism is α- rather than β-elimination, with a 1,2 shift in the carbene intermediate; $(Cl_2\overline{PtCD_2CHMeCHMe})_4$ gives CHD:CMeCHDMe, presumably via $Cl_2Pt(L):CDCHMeCHDMe$, but no $CD_2:CMeCH_2Me$.[407]

Copper, silver, gold:- see also Sections 3.2.2, Refs. 281,292,297-8; 3.2.3, Refs. 319-22; 3.2.5, Ref. 333; 3.2.8, Refs. 376,379; 3.2.10, Ref. 400; 3.2.11, Ref. 427; 3.2.13, Refs. 437,444-5,447,320; 3.2.14, Refs. 447,461-6; 3.2.15, Refs. 471,473.

Solid $Cu(CH_3)(PPh_3)_2.0.5Et_2O$ decomposes at $75^{\circ}C$ to give 90% ethane, while $Cu(CH_3)(PPh_3)_3$ gives 25% methane and 75% ethane; thus reductive elimination and hydrogen abstraction reactions are both occurring to different extents. At 56°, $Cu(Et)(PPh_3)_2$ gives mainly ethylene, with some hydrogen and ethane, as expected for β-elimination followed by decomposition. $Cu(CH_3)(PCy_3)$ gives 100% ethane. Solution processes were similar to those in the solid but faster, first order in complex, and inhibited by excess ligand. Activation energies are quite similar for methyl and ethyl compounds, strongly suggesting that metal-carbon bond loosening is important in β-elimination as well as in other decomposition pathways.[519] Strongly donating and bridging ligands, such as R_2P or R_2N, stabilize organocopper(I) species.[520]

The copper(II)-moderated decay of free radicals R˙, using copper(II) acetate in non-coordinating solvents, proceeds by oxidation of the radical to a carbonium ion, which may then undergo characteristic skeletal rearrangements. If a Cu(III) species is formed at all, it decomposes to R^+ and Cu(I). In polar solvents or in the presence of donor ligands (both of which favor dissociation of dimeric Cu(II) acetate) radicals with β-hydrogen atoms tend to lose them, giving alkenes and acetic acid. This process occurs without rearrangement, by hydrogen migration within an alkylcopper(III) acetate.[521]

While thermolysis of n-butylsilver (as the tri-n-butylphosphine complex) gives n-octane above $-50^{\circ}C$ by dinuclear elimination, photolysis gives butyl radicals. Thermolysis of the sec-butyl gives butenes and butane, indicating β-elimination.[522] Thermolysis of neat Ph_3PAuR gives R-R for R = methyl, ethyl; roughly equal amounts of RH and R(- H), with some R-R, for isopropyl, and 47% R(- H) and 13% RH for t-butyl. In solution, R-R is the main product for R = methyl, ethyl, or n-propyl. Excess ligand inhibits, indicating a dissociative mechanism,[523] possibly involving electron-deficient polymer.

3.2.17 The choice of decomposition pathway

It is clear from the examples cited throughout Sections 3.2 that the choice of decomposition pathway is often finely balanced. A number of generalizations, however, are possible. These are not put forward as absolute rules, but rather as guidelines to what may be expected, breaches of which may occur but will invite special explanation.

(i)   Decomposition pathways are only facile if they can lead to low-energy fragments by symmetry-allowed paths. However, in contrast to most organic reactions, the symmetry-allowed low energy fragment need not be the final, observable, ground state of the product.

(ii)   The achievement of a low energy fragment may require changes in bond angle between the remaining ligands, and this may interfere with the reactivity of chelate complexes.

(iii) Reductive elimination, in the case where both groups lost come from the same metal, requires the metal to accept electrons into an orbital initially antibonding with respect to the remaining ligands. This may require ligand dissociation, or major bond angle changes.

(iv)   The initial reaction coordinates of ultimately diverse processes may show common features, such as metal-carbon bond loosening.

(v)   The normal rules of inorganic chemistry are not suspended for the metal fragment merely because we are interested in the fate of the organic groups. For instance, strong field ligands will favor processes that convert Cr(II) to Cr(III), while high oxidation states, and ligands stabilizing metals in low valencies, will favor reductive elimination.

(vi)   Reductive elimination is more facile for sp^2 or sp than for sp^3 carbon ligands; this may in part be a steric effect.

(vii)   Reductive elimination is more favored for systems R-H and RCO-R than for systems R-R.

(viii)   Sec- and tert-alkyls are more likely than n-alkyls to favor β-elimination [but see (xii)].

(ix) β-Elimination requires the presence of a vacant coordination site at the metal, the minimum of steric constraints on the movement of the β-hydrogen to that site, and in general the presence of two or more d-electrons.

(x) α-Elimination is favored by oxidizing and highly unsaturated metals, expecially on the left hand side of the Periodic Table.

(xi)   Mechanisms involving the cooperation of two or more metal centers generally require a degree of coordinative unsaturation, and may be favored by bridge-forming ligands such as hydride.

(xii)   Constraints on the freedom of movement of bulky organic groups in sterically crowded complexes will affect the range of decomposition pathways available, and in extreme cases [such as the species M(1-norbornyl)$_4$, M = Ti, Zr, Hf, Cr, Mn, Fe, Co[524]] can severely inhibit every possible pathway.

(xiii)   Simple homolysis involves bond-breaking without any compensatory bond-making in the transition state, and for thermal reactions only occurs when other possibilities are excluded.

3.3   Further reactions of metal-carbon single bonds

3.3.1   Insertion of alkenes and alkynes

Two mechanisms are currently proposed for insertion of an unsaturated group into a metal-carbon bond. The traditional mechanism is cis-attack by an alkyl ligand on a coordinated unsaturated species, analogous to the well-established reaction of metal hydrides (3.1.9). A more recently

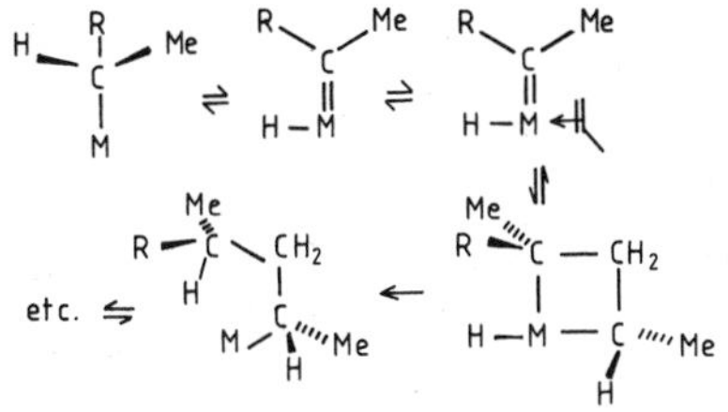

Scheme 21

proposed alternative, illustrated in Scheme 21 for the polymerization of propene, is α-migration from the alkyl group, forming a hydridometal carbene. The alkene then adds to the metal-carbon double bond to form a metallacyclobutane, and finally reductive elimination of C-H gives the insertion product.[384]

It is certainly true[384-6] that some saturated alkylmetal alkene complexes fail to show insertion, even though related hydridometal alkene complexes readily give alkyls. The difference is strongly reminiscent of that between hydridoalkyls and dialkyls in the relative ease with which they undergo reductive elimination, as may be seen if the alkene complex is written as a metallacyclopropane (Scheme 22). This may explain the failure of the cation $[Cp_2W(CH_3)(C_2H_4)]^+$ to exhibit insertion of alkene directly into the metal-methyl bond, while the lack of insertion by the α-elimination mechanism would then be due to the fact that this migration increases the electron count at the metal by two, and thus cannot take place while the ethylene remains coordinated. Indeed, all the relevant reactions[384-6] fit the equilibria

$$[Cp_2W(CH_3)L]^+ \rightleftharpoons [Cp_2W(CH_3)]^+ + L \rightleftharpoons [Cp_2W(CH_2)H]^+ + L \quad (95)$$
$$[Cp_2W(CH_2)H]^+ + L \rightleftharpoons [Cp_2W(CH_2L)H]^+ \quad (96)$$

Scheme 22

It is difficult to decide whether any given alkene insertion occurs by the direct mechanism or via α-migration, especially as the migrating hydrogen is returned to the carbon to which it was originally attached. However, some ingenious experiments on metal dialkyl complexes give clear evidence for direct insertion.[458,525] The overall reaction

$$CpCoL(CD_3)_2 + 2C_2H_4 \longrightarrow CpCoL(C_2H_4) + CD_3H + CH_2{:}CH.CD_3 \quad (97)$$

(L = PPh_3) can be understood as a sequence of steps

$$CpCoL(CD_3)_2 + C_2H_4 \longrightarrow CpCo(C_2H_4)(CD_3)_2 + L \qquad\qquad (98)$$
$$CpCo(C_2H_4)(CD_3)_2 \longrightarrow CpCo(CH_2CH_2CD_3)(CD_3) \qquad\qquad (99)$$
$$CpCo(CH_2CH_2CD_3)(CD_3) \longrightarrow CpCo(CH_2{:}CH.CD_3)(CD_3)H \qquad (100)$$

followed, not necessarily in this order, by reductive elimination of $H\text{-}CD_3$, exchange of labeled propene for ethylene, and recombination of lost phosphine. If α-migration had occurred in the species $CpCo(CD_3)_2$, which is an intermediate in Reaction (98), it would have given a species $CpCo(CD_2)(CD_3)D$. This would have been prone to reductive elimination of $D\text{-}CD_3$, in the same way as the product of Equation 100 is prone to elimination of $H\text{-}CD_3$; the possible replacement of alkene by phosphine in that product is irrelevant, since basic ligands do not promote reductive elimination.

There is no possibility of α-hydride migration in metal phenyl complexes, which are nonetheless known to add to alkenes, as in the sequence[76]

$$PhHgOAc + Pd(OAc)_2 \longrightarrow PhPdOAc + Hg(OAc)_2$$
$$PhPdOAc + C_2H_4 \longrightarrow PhCH_2CH_2PdOAc$$
$$PhCH_2CH_2PdOAc + Pb(OAc)_4 \longrightarrow PhCH_2CH_2Pb(OAc)_3 + Pd(OAc)_2 \quad (101)$$

by which palladium(II) catalyzes both insertion and _trans_-metalation reactions. The sequence of Scheme 23[526] involves two successive insertions of alkene into palladium-carbon bonds.

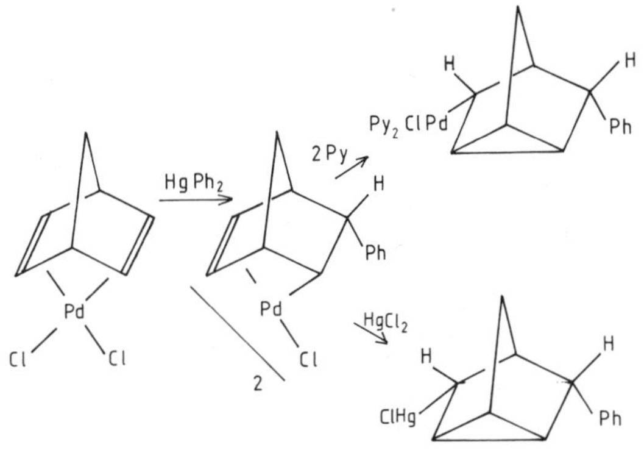

Scheme **23**

Irrespective of these mechanistic niceties, a very large number of important coupling reactions between alkenes and organotransition metal compounds give rise to substituted alkenes without the isolation of the insertion product intermediate. Examples are discussed in Section 3.3.2.

Alkyne insertion is generally more facile than alkene insertion; as in the case of alkene insertion into metal-aryl bonds, the analogy with reductive elimination may help explain why greater unsaturation at the carbons to be bonded facilitates the reaction. The metal and organic group are added <u>cis</u> across the triple bond to give a vinylic ligand. The first step is presumably alkyne coordination, which will commonly occur by dissociative ligand exchange, as in the insertion of alkyne into the nickel-methyl bond of $Ni(PPh_3)(acac)Me$, which is inhibited by excess phosphine. The methyl group migrates to the <u>less</u> sterically hindered carbon (in accord with the suggestion in this Section of a steric barrier to the carbon-carbon bond forming step. There is rapid <u>cis</u>-<u>trans</u> isomerization across the C:C bond of the vinylic ligand formed, suggesting delocalization and mesomerism between forms $Ni-C=C$ and $Ni^+=C-C^-$, or the formation of intermediates $Ni^-=C-C-P^+Ph_3$.[527]

Acetylene inserts into the copper-carbon bond of lithium organocuprates R^1_2CuLi to give <u>cis</u>-adducts $(R^1CH:CH)_2CuLi$. These react further with reagents $PhSO_2CH:CR^2R^3$ to give, after protonolysis, products $PhSO_2CH_2C(R^2R^3)CH:CHR^1$, reducible to $MeC(R^2R^3)CH:CHR^1$.[528] (The addition of stannyl cuprates to ethyl butynoate is actually reversible.[529]) The reaction of vinyl cuprates with alkynes, followed by treatment with carbon dioxide or other electrophiles, provides a route to dienecarboxylic acids and other functionalized dienes.[530]

Coordinated carbene has been shown to insert into a metal-alkyl bond in the very system used to demonstrate the reluctance of coordinated alkene to do so. The cation $[Cp_2W(:CH_2)CH_3]^+$, inferred as an intermediate in the reaction sequences initiated by hydrogen atom abstraction from $[Cp_2WMe_2]^+$, can be trapped by L(= PMe_2Ph) either directly, as $[Cp_2W(Me)CH_2L]^+$, or, after carbene insertion, as $[Cp_2WL(CH_2CH_3)]^+$.[531]

3.3.2 Coupling reactions

In this Section we include a variety of reactions that lead to the formation of organic products with new carbon-carbon bonds. Reactions of this kind generally fall into one of the following classes:-

a)   "Simple" coupling;

$$R-X + R'M \longrightarrow R-R' + MX \tag{102}$$

where M represents a transition metal with its attendant ligands. Reactions of this kind include catalytic couplings of organic halides with lithium, magnesium, or zirconium reagents, as well as organocuprate coupling

b)   Reductive coupling;

$$2\ R-M \longrightarrow R-R + ... \tag{103}$$

(including the transition metal induced reductive coupling of Grignard reagents)

c)   Conjugate addition, e.g.

$$R_2M^- + C=C-C=O \longrightarrow RM + R-C-C=C-O^- \tag{104}$$

d)   Allylic coupling

$$RM + C=C-C-X \longrightarrow R-C-C=C + MX \tag{105}$$

e)   Vinylic coupling

$$RM + HC=C \longrightarrow R-C=C + MH \tag{106}$$

Coupling clearly due to insertion of carbene ligands or of organic carbonyls is also discussed in 3.3.4 below.

Combinations of these processes with others that regenerate the organometallic starting material can lead to catalytic cycles. In such cases, the ratios of reagents may be very different from those present during reactions carried out stoichiometrically, and this may lead to changes in dominant mechanism.

The development of new organometallic reagents to exploit these processes, and their modification and control, now form a major branch of synthetic organic chemistry (see e.g. Reference lq). We give here only a few examples from across the Periodic Table to illustrate the range and sensitivity of these reagents. Nickel- and palladium-catalyzed cross coupling,[532] and palladium-catalyzed vinylation of organic halides,[533] have been reviewed.

Methyltitanium(IV) derivatives methylate tertiary chlorocarbons (in preference even to primary and secondary derivatives!), or tertiary alcohols, giving a route to asymmetrically substituted quaternary carbon. Ketones are converted to gem- dimethyls. The methyltitanium reagent may be pre-formed, or may be generated in situ from aluminum methyls and titanium tetrachloride, which in the case of attack on chlorocarbons can therefore act as a catalyst.[534] Zirconium reagents, such as $Cp_2Zr(Cl)CH:CHR$, show catalyzed coupling to aryl halides or conjugate (1 - 4) addition to enones, in the presence of copper(I) chloride or of catalytic amounts of nickel(II) acetylacetonate. Ni(0) derivatives are ineffectual, and a catalytic cycle involving Ni(I) and Ni(III) intermediates has been suggested.[217,219,535] The use of Ti- and Zr-centered coupling in organic synthesis has been reviewed.[21]

Benzylic chlorides $ArCH_2Cl$ add to styrene in the presence of iron(II) chloride or copper(I) chloride and bipyridyl, to give products $ArCH_2CH_2CH(Cl)Ph$ and $ArCH_2CH:CHPh$.[536] However, the attempted catalytic coupling of alkyl halides and Grignard reagents, using iron(II) or iron(III) catalysts, gave a mixture of alkanes and alkenes. The mechanism suggested included the formation of organoiron species by radical processes, and their decomposition by β-elimination and hydridic reduction.[537] Species

$[R^1R^2CH.CH(CO_2Et).Fe(CO)_4]^-$ are accessible either by addition of a carbon nucleophile to the $Fe(CO)_4$ complex of unsaturated esters $RCH:CHCO_2Et$, or by addition of $[HFe(CO)_4]^-$ to such esters, and react with alkyl halides by Fe-C bond formation, carbonyl insertion, and reductive elimination to give β-ketoesters $R^1R^2CH.CH(CO_2Et).CO.R^3$.[229,538]

The dimerization of ethylene to butenes by Rh(I) in acid may proceed by a mechanism involving insertion of ethylene into a rhodium-ethyl bond, followed by β-elimination,[539] while metalation of a phenyl ring, followed by a double insertion and 1-2 shift, are presumed to be involved in the $RhCl_3/PPh_3$-catalyzed conversion of aniline by ethylene to 2-methylquinoline.[540]

The complex $NiPh(acac)(PPh_3)$ couples with alkyl halides to give alkylbenzenes, and also undergoes insertion of alkenes into the metal-carbon bond, followed by β-elimination, to give styrenes.[541] In situ thermolysis of species $L_2Ni(Ar)Me$ generates ArMe and nickel(0) species, which can then undergo oxidative addition reactions. This is not, however, the true mechanism of the many coupling reactions of types (102), (103), induced by nickel reagents. The reaction of aryl halides with nickel(0) complexes contains a one-electron component, generating nickel(I) species which can then take part in oxidative addition - reductive elimination cycles between oxidation states I and III. These cycles, being more rapid than those between Ni(II) and Ni(0), dominate the catalytic reactions. Scrambling occurs between added aryl halides, and those aryl groups originally attached to the nickel, and this is explained by transfer of groups between Ni(II) and Ni(III) species.[154,542] Despite these complexities, the nickel-catalyzed cross-coupling of aryl Grignards and aryl halides has been investigated using a wide range of nickel complexes,[543] and Ni(0) complexes have been shown to catalyze the coupling of aryl halides and $BrZnCH_2CO_2Et$.[544] Optically active phosphine complexes of nickel have been used in attempts to induce chiral selectivity into the coupling of organic halides and Grignards, but the magnitude, and even the direction, of the chiral preference depended on the halogen.[545] It is tempting to suggest that this reflects the relative importance of one-electron and two-electron pathways. However, a 67%

optical yield has been claimed for the coupling of 1-phenethyl magnesium bromide to allyl bromide, catalyzed by $NiCl_2$(-)NORPHOS.[546] Chiral ferrocenylphosphines have been used as ligands in nickel- and palladium-based catalysts.[547]

Coupling reactions occur (not, perhaps, surprisingly) in the electrochemistry of organonickel compounds. Anodic oxidation of species $(Et_3P)_2Ni(Ar)R$ leads to reductive elimination of Ar-R, as well as to formation of R· radicals.[548] Reduction of nickel(II) phosphine halides in the presence of aryl halides leads to biaryl formation, via species $L_2Ni(Ar)X$. The overall reaction is then catalysis by nickel(II) of the electrochemical coupling of aryl halides to give biaryls.[549]

The coupling of organopalladium derivatives to alkenes proceeds by insertion followed by β-elimination of a palladium hydride, functionally equivalent to palladium(0) plus acid:[550]

$$RPd(II)X + C=C \longrightarrow R-C-C-PdX \longrightarrow R-C=C + H-Pd(II)X \qquad (107)$$

$$HPd(II)X \longrightarrow HX + Pd(0) \qquad (108)$$

A variant is the use of halide to expel palladium as Pd(0):[551]

$$R-C-C-PdX + Cl^- \longrightarrow R-C-C-Cl + Pd(0) + X^- \qquad (109)$$

Copper(II)[551] or oxygen[552] can re-oxidize, thus leading in suitable cases to a catalytic cycle. The organopalladium reagent may be generated from palladium(II) and organic derivatives of magnesium,[553] mercury, tin, lead,[551] or thallium;[554] by electrophilic substitution of palladium(II) into an aromatic ring,[192] by reaction of palladium(0) with a diazonium salt or an acyl halide followed by elimination of dinitrogen[262,555] or CO,[556] or by oxidative addition of aryl halide to palladium(0).[557] The reaction has been used to form styrenes,[550-2,555] stilbenes,[192,553,557] and even vinylferrocenes,[558] and to introduce aromatic substituents into quinones.[559] Anhydrous potassium carbonate may be used to mop up the acid generated in eq 108.[553] In the $PdCl_2$-catalyzed coupling of methyllithium and styrene to give methylstyrene, labeling experiments show that palladium methyl

Scheme 24

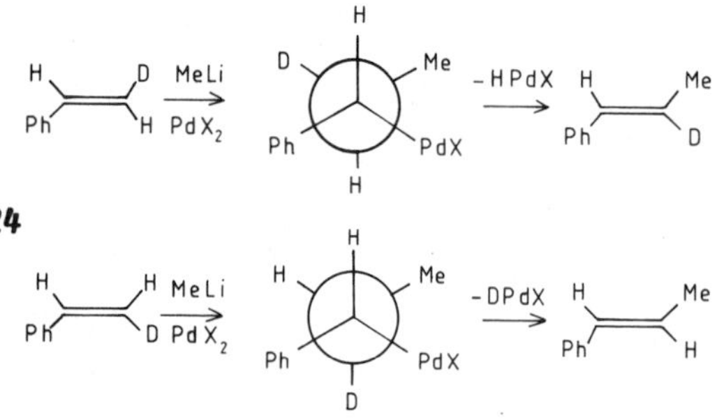

addition and palladium hydride elimination both have the same stereochemistry, which may safely be taken as <u>cis</u> (Scheme 24).[560] It is noteworthy that in all these reactions palladium adds specifically to $C(\alpha)$ of styrenes.

Vinyl halides couple to organozinc or to Grignard species under the influence of a palladium(0) catalyst. In the coupling of vinyl halides to $BrZnCH_2CO_2Et$, it has been shown that the slow step is formation of the vinylpalladium(II) intermediate. For β-bromostyrenes, one group reports considerable <u>cis</u> to <u>trans</u> isomerization during this reaction, although the 1-bromopropenes react with retention,[561] and other workers state that the coupling of α-bromostyrenes to methyllithium is stereospecific with retention. (The difference could be due to alterations to the lifetime of the styrylpalladium(II) bromide intermediate.) There is evidence that, in this case, under catalytic conditions the reaction involves a palladium(II) - (IV) cycle.[562]

Allylic alkylations using palladium have been reviewed;[27] see also Section 5 below.

$(Ph_3P)_2PtMe_2$ reacts with PhCHDBr to give PhCDHMe with inversion. The suggested mechanism is oxidative addition, with inversion at benzylic carbon, followed by reductive elimination with retention. $(Ph_3P)_2Pt(CH_2Ph)Cl$ reacts with a stoichiometric amount of tetramethylsilane to give toluene, by metathesis followed by α-abstraction:

$$L_2Pt(CH_2Ph)Cl + Me_4Si \longrightarrow L_2Pt(CH_2Ph)Me + Me_3SiCl \qquad (110)$$

$$L_2Pt(CH_2Ph)Me \longrightarrow H\text{-}CH_2Ph + \dots \qquad (111)$$

In the presence of excess benzyl chloride, the reaction takes a different course, with oxidative addition to the unstable intermediate, followed by reductive elimination, giving catalytic production of ethylbenzene:[563]

$$L_2Pt(CH_2Ph)Me + ClCH_2Ph \longrightarrow L_2Pt(CH_2Ph)_2MeCl$$
$$\longrightarrow L_2Pt(CH_2Ph)Cl + MeCH_2Ph \qquad (112)$$

Copper-induced coupling reactions have been reviewed.[30] Organocopper reagents, especially in the presence of halides, exist as aggregates in solution, as in general do lithium or magnesium organocuprates, and the overall mechanisms and stereo- and regioselectivities vary with the exact composition of the reagent, as well as with solvent. This important complexity is ignored in the idealized mechanisms presented here, but will be crucial in any attempt to advance control of these reactions beyond the trial and error stage (see e.g. References 564-6).

Copper "ate" reagents R_2CuLi are at once more stable and more nucleophilic than the corresponding RCu. Coupling with these reagents can proceed by a simple S_N2 mechanism:

$$LiCuPh_2 + \underline{R}-C(Et)(Me)(H)Br \longrightarrow LiBr + CuPh + \underline{S}-C(Et)(Me)(Ph)H \qquad (113)$$

or by trans-metalation, oxidative addition, and reductive elimination:[567]

$$PhI + Me_2CuLi \longrightarrow Ph(Me)CuLi + MeI \qquad (114)$$
$$Ph(Me)CuLi + MeI \longrightarrow PhMe + MeCu + LiI \qquad (115)$$

Copper(I) catalyzes the coupling of Grignard reagents and alkyl halides, even where the alkyl group contains β-hydrogen. For instance, ethyl Grignard and ethyl bromide couple to give butane. This cannot involve the interaction of two copper(I) ethyls, which is known (cf. 3.2.11 above) to give ethane and ethylene. A plausible reaction sequence is[568]

$$EtMgBr + CuBr \longrightarrow EtCu + MgBr_2 \qquad (116)$$
$$EtCu + EtBr \longrightarrow Et-Et + CuBr \qquad (117)$$

We may then ask whether the crucial carbon-carbon bond forming step (117) occurs by nucleophilic attack on bromine-bound carbon, which would give inversion at the latter, or by oxidative addition - reductive elimination or by reaction through a closed transition state, so as to give retention. It is known that the alkylation of cyclopropyl iodides by species R_2CuLi occurs with retention.[569] So does the more indirect reaction[570]

$$R\text{-}Br + n\text{-}Bu_2CuLi \rightleftharpoons R(n\text{-}Bu)CuLi + n\text{-}BuBr \qquad (118)$$

$$R(n\text{-}Bu)CuLi + R'X \longrightarrow R\text{-}R' + (n\text{-}BuCu) + LiBr \qquad (119)$$

where R is a cyclopropyl and R'X is allyl bromide or methyl iodide. The reagent Li_2CuMe_3 is a particularly effective coupling reagent, replacing halide by methyl even in the case of an aryl fluoride.[571]

Reagents Li_2CuR_2CN efficiently open up epoxides to give C(R)-C(OH) units. In this case the mechanism is nucleophilic attack, since $\underline{C}(R)$ is inverted.[572]

A useful characteristic reaction of organocuprate reagents is conjugate addition, e.g.

$$C=C\text{-}C=O + R_2CuLi \xrightarrow{\quad} R\text{-}C\text{-}C=C\text{-}OLi \overset{H^+}{\underset{^-Li^+}{\xrightarrow{\hspace{1cm}}}} R\text{-}C\text{-}C(H)\text{-}C=O \qquad (120)$$

This process has been reviewed,[30,573] and can be used to modify catalytically the effect of Grignard reagents on enones, since the copper-mediated 1,4-addition (120) is much faster than the uncatalyzed Grignard 1,2-addition.[574] The addition is non-radical, and proceeds with retention at carbon of the group added to the enone.[575] The conjugate addition can be impeded by steric effects. For example, lithium diallyl cuprate transfers an allyl group to C(3) of 2-cyclohexenone, but to C(1) of isophorone (**LII**) to give **LIII** after hydrolysis. However, the complexes $R_4MeCu_3(MgBr)_2$ are extremely selective for conjugate addition, and even when R is isobutyl and isophorone is the substrate, react with transfer of R, rather than Me, to give **LIV**.[576]

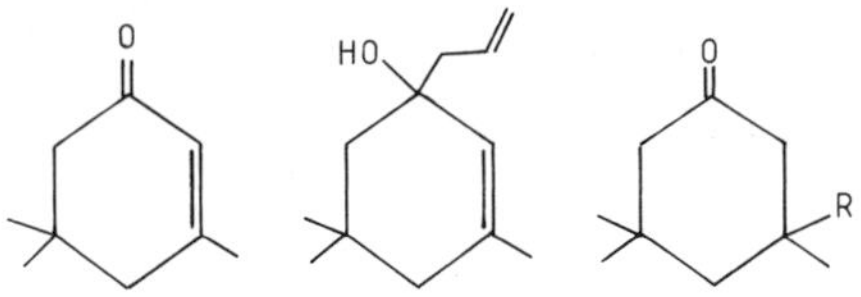

LII   LIII   LIV

Yet another class of organo-copper reagents is that formulated as $RCu.BF_3$. These again show strong preference for conjugate addition, and react even with such weakly activated molecules as unsaturated acids and esters, as well as alkylating allyl alcohols and acetates by a formal S_N2' process.[577]

The catalytic coupling by silver(I) of organic halides RX to Grignard reagents $R'X$ proceeds by a free radical pathway completely different from that of the superficially similar copper(I)-catalyzed reaction:

$$R'MgX + AgX \longrightarrow R'Ag + MgX_2 \tag{121}$$

$$2R'Ag \longrightarrow R'-R' + 2Ag^o \tag{122}$$

$$Ag^o + RX \longrightarrow AgX + R^· \tag{123}$$

$$R^· + Ag^o \longrightarrow RAg \tag{124}$$

$$RAg + R'Ag \longrightarrow R-R' + 2Ag^o \tag{125}$$

This mechanism requires the organic group added as halide to become at one stage a radical, while that added as Grignard reagent does not. In accord with this requirement, stereochemistry is retained in the coupling of vinyl Grignards to methyl bromide, but when methyl Grignards are coupled to vinyl halides, cis-trans isomerization occurs.[578] The Ag^o of the reaction sequence need not represent free silver atoms, but is plausibly part of a cluster of material of average oxidation state less than one.

3.3.3 Sulfur dioxide insertion

The sulfur dioxide insertion reaction

$$M-R + SO_2 \longrightarrow M.S(O)_2.R \tag{126}$$

was authoritatively reviewed some years ago.[579] Questions of importance were, and remain, the nature of the original attack of sulfur dioxide on the complexes, and of any intermediates formed, and the stereochemistry of the process both at metal and at the affected carbon atom. The stereochemistry at titanium has been studied in complexes (Cp)(Cp')Ti(Ph)Me, where Cp' is a cyclopentadienyl group with an optically active substituent, and it is found that SO_2 inserts into the metal-methyl bond with retention at metal.[580] For t-BuCHDCHDZrCp$_2$Cl, insertion occurs with retention of configuration at the metal, suggesting coordination of SO_2 followed by alkyl migration, as in CO insertion (Chapters 6-8).[581] In contrast, saturated species M-R [M = CpFe(CO)$_2$, CpMo(CO)$_3$, Mn(CO)$_5$, Re(CO)$_5$] are thought to react via an intermediate tight ion pair of type $M^+SO_2R^-$, which equilibrates with species M.O.SO.R and more slowly isomerizes to the final product M.SO$_2$.R. Among these saturated species, the relative rates for different metals vary with R, but more oxidizable groups are generally more reactive.[582] For different CpFe(CO)$_2$R, the rate decreases with increasing σ^* of R and with increasing CO stretching frequency, confirming the electrophilic behavior of SO_2. Despite steric effects, methylation of the Cp ring facilitates the reaction, which shows a large negative entropy of activation.[583] For M = CpW(CO)$_2$(PEt$_3$), Mn(CO)$_4$PEt$_3$, or CpFe(CO)$_2$, there is inversion at carbon, while for the iron species at least there is retention at metal.[584] All these facts are consistent with the proposed mechanism, the initial electrophilic attack by SO_2 taking place via an open transition state.

In organoplatinum and -gold species, SO_2 insertion occurs with retention of stereochemistry at metal, and with insertion into metal-methyl, rather than metal-phenyl, bonds, when both are available. In cis-Me$_2$Au(CD$_3$)PMe$_3$, the insertion occurs cis to phosphine, with equal amounts of insertion into Au-CH$_3$ and Au-CD$_3$ bonds. The corresponding reaction of cis-Me$_2$Au(Ph)PPh$_3$ gave unstable Au(III) products, which at room temperature decomposed to give MeSO$_2$AuPPh$_3$ and toluene.[585] The more electron-withdrawing nature of sulfinate as opposed to methyl, and of PPh$_3$ as opposed to PEt$_3$, is no doubt responsible for destabilizing Au(III) in this case.

3.3.4 Other insertion reactions

Insertion of alkenes and alkynes was discussed in Sections 3.3.1, for metal-carbon bonds, and 3.1.9, for metal-hydrogen bonds, while that of CO is covered in Chapters 6 - 8, and of isonitriles in Chapter 9.

Insertion of acetone, rather than mere transfer of Ph^- from Cr(III), is the proposed pathway for the production of $PhCMe_2OH$ and $[Cr(OH)(acac)_2]_2$ from acetone and $Cr(acac)_2Ph$.[586] Metal-bound carbene inserts into a metal-carbon bond in the formation of hydridometal alkene complexes from $Ru(C_6Me_6)(PPh_3)Me_2$,[587] Cp_2WMe_2 and $Cp_2W(Me)Et$ on attack by the triphenylmethyl cation. For the tungsten complexes, it has been shown that the attack by CPh_3^+ proceeds via initial electron rather than hydride removal, and leads to α-abstraction even when β-abstraction is available.[588]

The insertion of coordinated allene into the platinum-methyl bond of salts $[trans-(Ph_2MeP)_2Pt(Me)(\eta^2-CH_2:C:CH_2)]^+Z^-$ in dichloromethane gives π-allyl salts $[(\eta^3-CH_2CMeCH_2)Pt(PMe_2Ph)_2]^+Z^-$ at a rate dependent on the identify of Z, and the reactive species is therefore presumed to be a tight ion pair.[589] Carbon dioxide reversibly inserts into copper-carbon bonds in the tri-n-butylphosphine complexes of CuC_2Ph and $CuCH_2CN$.[590] Tetracyanoethylene inserts into iron-carbon bonds in systems FpR to give products $Fp-C(CN)_2.C(CN)_2.R$ and $Fp-N:C:C(CN).C(CN)_2.R$, where R is alkyl or benzyl. The rates of reaction fall from benzyl to methyl to phenyl, perhaps indicating a free radical mechanism. Species $CpMo(CO)_3R$ failed to react, but the more electron-rich $CpMo(CO)_2LR$ [L = PPh_3, $P(OPh)_3$] gave both types of insertion product.[591]

The insertion of germanium or tin dihalides into iron-alkyl bonds in species Fp-R occurs by a radical process, being inhibited by scavengers, promoted by sunlight, and showing induction periods and erratic progress.[592] For species of the type $Fp-(\eta^1-allyl)$, the reaction occurs with allylic rearrangement.[593]

Scheme 25

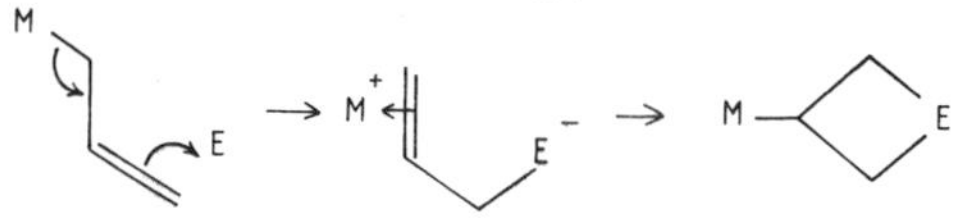

3.3.5 Shift cyclization on electrophilic attack at unsaturated ligands

Electrophilic attack on a σ-allylic organometallic can proceed with σ-π rearrangement to give an η^2-alkene complex. The original electrophile accepts an electron in the process, and can in suitable cases then attack the carbon originally σ-bonded to metal. The overall result (Scheme **25**) is 1,3-addition of the electrophile across the allylic group, with concomitant 1,2-shift of the metal. Similar reactions of σ-propargyl complexes give allene complexes which are converted to cyclic vinyl ligands. The study of this class of reactions requires the ready preparation of robust σ-bonded complexes of functionalized and unsaturated ligands, which may explain why most work in this area has involved the Fp or similar groupings. Reactions of this type using the Fp group have been reviewed.[594]

Protons attack C(3) of Fp-η^1-allyls, giving η^2-alkene complexes of Fp$^+$, as in the first stage of Scheme **25**,[595] thus giving a convenient route to isobutene complexes:[596]

$$Fp.CH_2.CH:CH_2 + D^+ \longrightarrow [Fp(\eta^2-CH_2:CH-CH_2D)]^+ \qquad (127)$$
$$Fp.CH_2.CMe:CH_2 + H^+ \longrightarrow [Fp(\eta^2-CH_2:CMe_2)]^+ \qquad (128)$$

The reaction can even be used to form vinyl alcohol complexes:[597]

$$Fp.CH_2COR + R^+ \longrightarrow [Fp(\eta^2-CH_2:C(OH)R)]^+ \qquad (129)$$

Mercuric chloride also adds to C(3), giving products $[Fp^+(\eta^2-CH_2:CH.CR_2HgCl_2^-)]$, which lose chloride to give isolable cations.[598] Among the electrophiles capable of inducing the η^1-η^2 shift are

Scheme 26

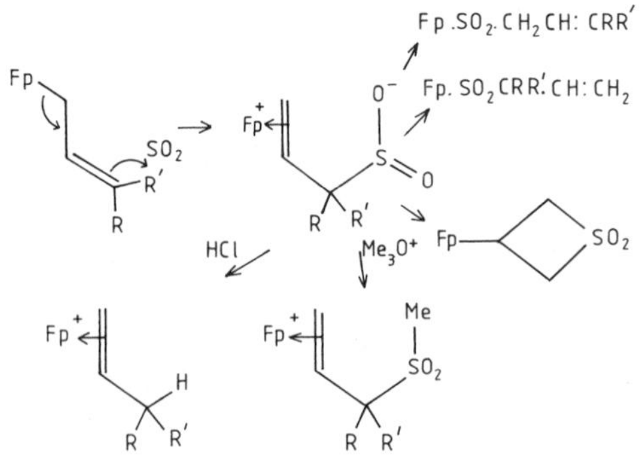

LV   **LVI**

[Fp(alkene)]$^+$ complexes themselves, giving binuclear products containing both η^1 and η^2 functionalities (Scheme **9**).[242]

Attack by sulfur dioxide on η^1-allylic complexes is thought to take place initially at C(3), causing an η^1-η^2 shift; the initial product can be trapped by methylation or decomposed by acid. Subsequent reaction may involve rearrangement to the C(1) insertion product, or to the allylically rearranged product, or cyclization to a 4-membered S-sulfinato ring (Scheme **26**). Propargyl derivatives, however, cyclize to 5-membered sultones such as LV (M = Fp or CpW(CO)$_3$).[599]

The electrophilic cyclization reaction is of wide application. The reaction

$$M.CH_2CH:CH_2 + TCNE \longrightarrow LVI \qquad\qquad (130)$$

is established for M = Fp, CpMo(CO)$_3$, CpW(CO)$_3$, Co(dmgH)$_2$py and CpCr(NO)$_2$, and some of the additions established for M = Fp are shown in Scheme **27**. Since the allylic starting materials are accessible by

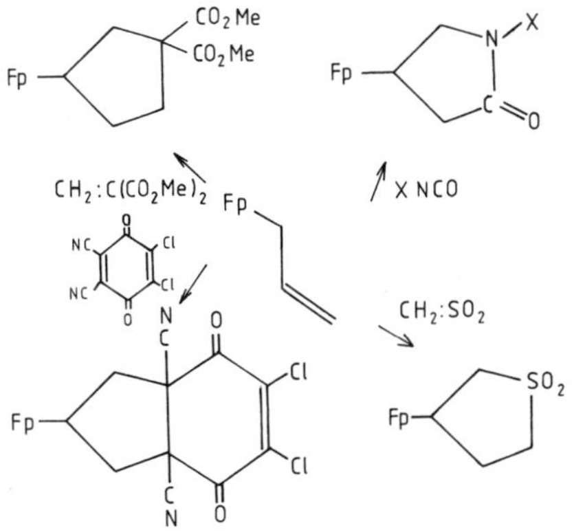

Scheme 27

deprotonation of $[Fp(unsat)]^+$, formed by a number of routes (eqs 131-3), the reaction provides a method of some generality for the synthesis of heterocycles:[600]

$$[Fp(\eta^2\text{-}CH_2\text{:}CMe_2)]^+ + unsat \longrightarrow [Fp(unsat)]^+ + CH_2\text{:}CMe_2 \qquad (131)$$

$$Fp\text{-}CH_2CHRR' + Ph_3C^+ \longrightarrow [Fp(\eta^2\text{-}CH_2\text{:}CRR')] + Ph_3CH \qquad (132)$$

$$Fp^- + \overline{CR_2.CR'_2O} \longrightarrow FpCR_2CR'_2\text{-}O^- \xrightarrow[-H_2O]{2H^+} [Fp(\eta^2\text{-}CR_2CR'_2)]^+ \quad (133)$$

Propargyl complexes show a set of reactions closely related to those of allyls, giving η^2-allene derivatives in the first step. These also cyclize in favorable cases, to give heterocyclic vinyl ligands, as in Scheme **28**.[601] An alternative reaction, shown by metal carbonyl derivatives in alcoholic solution, is carboxylation at C(2) to give η^3-allylic complexes **LVII** [M = $Mn(CO)_4$, $CpMo(CO)_2$, $CpW(CO)_2$].[602]

Alkynyl complexes can show clearly related reactions. FpCCPh is protonated to the carbene $[Fp\text{:}C\text{:}C(H)Ph]^+$, which adds TCNE to give **LVIII**.[603] More remote electrophilic attack is shown by acids HX on $CH_2\text{:}CMe(CH_2)_2Fp$, to give gem-dimethylcyclopropane and FpX; the metal is activating C(4) by a homoallylic process.[604] Cyclopropyl and cyclopropylmethyl derivatives also show related reactions, in these cases involving ring opening.[605]

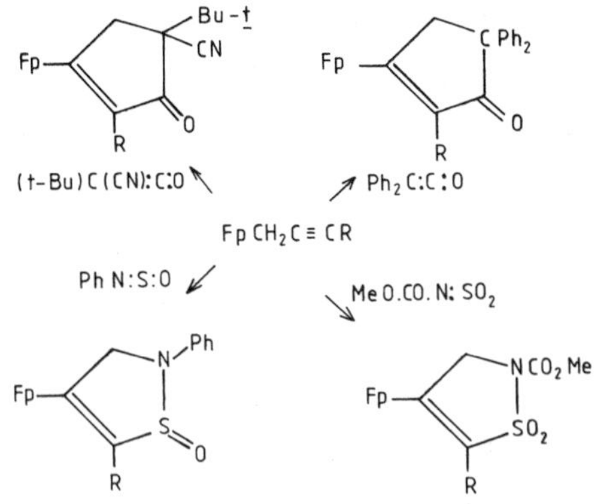

Scheme **28**

LVII   [structure: >—CO.OR with M below]

LVIII   [structure: Fp—[cyclobutane ring with Ph, (CN)₂, (CN)₂]]

4.   ALKENE COMPLEXES

4.1   General

Asymmetric hydrogenation, and nucleophilic attack at coordinated alkenes, are reviewed in other Chapters, while hydrogenation, hydrosilylation and the insertion of alkenes into MH and MC bonds, are discussed in Sections 3.1.9, and 3.3.1 and 3.3.2 above, and in reviews.[16-18,606] Reviews have also appeared on structural aspects of the coordination of unsaturated molecules to transition metals,[607] on complexes of ligands chelated by a one-carbon and a two-carbon site,[608] and on developments in transition metal alkene complexes including complexes in which the metal is in a very low or negative oxidation state.[609] The attack of metal on the allylic position of coordinated alkenes, and the co-dimerization of alkenes and dienes, are discussed in Section 5 below.

4.2   Theory of bonding and reactivity

The coordination of alkenes to transition metals in complexes is still interpreted in terms of the so-called Chatt-Dewar-Duncanson model, proposed more than thirty years ago.[610] In this model, ligand to metal σ-donation involves a suitable "vacant" metal orbital accepting electron density from the carbon-carbon π-bond, while π-back-bonding involves a "full" metal d-orbital (or hybrid of largely d character) so positioned that lobes of opposite phase point towards the two carbon atoms, overlapping the "vacant" ligand π^* orbital. There is continuity between the descriptions of these systems as metal-alkene complexes and as metallacyclopropanes, and even such superficially far-removed species as alkene oxides and aziridines can be described by the same formalism but with different coefficients.[611] Low-valent metals, at least, are overall electron-releasing, and a range of complexes of type $(OC)_4Fe$(cinnamic acid) have been found to be less acidic than the corresponding free ligands.[612] Despite this fact, and the related fact that the back-bonding raises the energy of the ligand LUMO, coordination of alkenes to metal activates them towards nucleophilic attack (see Chapter 11). This paradox can be resolved by suggesting a lowering of symmetry of the metal-alkene substrate during the attack, the LUMO becoming localized on the carbon atom more remote from the metal.[613] In substituted alkenes, steric effects can also be important.[614]

The preferred orientation of alkenes in complexes has been the subject of several theoretical investigations. For species such as $Ni(C_2H_4)_3$, the predicted orientation is with metal and all carbons co-planar, but for $Ni(C_2H_4)_2$ the calculated energy difference between coplanar and perpendicular ligand arrangements is predicted to be small. In square planar d^8 complexes, such as Zeise's anion, $[PtCl_3(C_2H_4)]^-$, it is mainly steric effects that hold the ligand perpendicular to the coordination plane.[615-6] In species of the type $[Fp(ene)]^+$, the best π-donor orbital of the Fp^+ fragment is odd under reflection in the mirror plane, so that alkene ligands should lie across (and carbene ligands lie in) this plane.[617]

4.3   Preparation and alkene ligand exchange

Among the earliest organometallic syntheses reported was the displacement of chloride from the $[PtCl_4]^-$ anion by ethylene, to give Zeise's salt, $KPtCl_3 \cdot C_2H_4$.[618] Such simple displacement remains one of the most useful routes to alkene complexes. Where the departing ligand is halide, the process may be assisted by Lewis acids:

$$MX + AlCl_3 + C_2H_4 \longrightarrow [M(C_2H_4)]^+ + AlCl_3X^- \qquad (134)$$

$[M = Fp, CpMo(CO)_3, CpW(CO)_3]$;[619] η^2-complexes of dienes can be prepared by this route.[620] Ligand exchange is sometimes reversible, as in the process (chx = cyclohexene, L = PCy_3)[621]

$$Ni(chx)L_2 + chx \rightleftharpoons Ni(chx)_2L + L \qquad (135)$$

Other routes to alkene complexes are described in Scheme 2 and in Sections 3.2.11, 3.2.12 and 3.3.5 above.

Displacement of alkene from complexes of type $[Fp(alkene)]^+$ by alkyne or allene[622-3] is an important route to interesting intermediates. The replacement reaction

$$[Fp(C_2H_4)]^+ + PR_3 \longrightarrow [Fp.PR_3]^+ + C_2H_4 \tag{136}$$

(R = Ph, Bu-n) is less straightforward than it seems, being first order in phosphine as well as in substrate, and showing a negative entropy of activation. These data suggest an associative mechanism, with alkene loss occurring from an intermediate such as $[Fp.CH_2CH_2PR_3]^+$,[624] and just such intermediates have been observed in the displacement of alkene by phosphine from ruthenium(II).[625] Closely related is the addition of the triphenylphosphine methylene ylid to $[Fp'(C_2H_4)]^+$, giving $[Fp'.CH_2CH_2CH_2PPh_3^+]$.[626]

4.4   Reactions of coordinated alkenes

For reviews of hydrogenation, see References 16, 606. Hydrogenation does sometimes occur without prior coordination of the alkene, as in the reactions of $[HCo(CN)_5]^{3-}$ and Cp_2MoH_2,[16,627] and also perhaps in the hydrogenation of dimethylmaleate or -fumarate by $IrH_3(CO)(PPh_3)_2$, where the alkene complexes, if generated, would be too stable kinetically to account for the catalysis.[628]

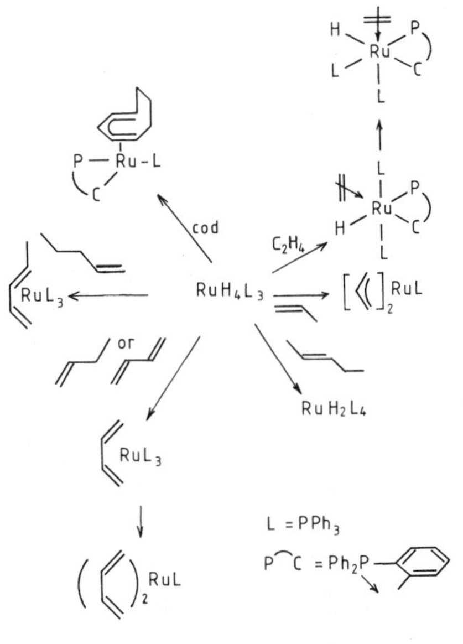

Scheme 29

The interaction of coordinated alkenes with metal hydrides can be of great complexity, as in the reactions of $RuH_4(PPh_3)_3$ (Scheme **29**).[629] Even in simple cases, the interpretation of labeling experiments requires some care. For instance, the conversion of allylbenzene, $PhCH_2CH:CH_2$, to methylstyrene, $PhCH:CH.CH_3$, is catalyzed on addition of $DCo(CO)_4$. The obvious inference, that a [1,3] shift is occurring without direct intervention by the deuteride ligand, is wrong. Addition followed by elimination without isomerization converts the "catalyst" via $PhCH_2CHD.CH_2Co(CO)_4$ to $HCo(CO)_4$, while isomerization is only initiated by that minority of additions that give $PhCH_2CH(CH_3)Co(CO)_4$. In agreement with this interpretation, $PhCD_2CH:CH_2$ is converted by $HCo(CO)_4$, in the presence of excess p-tol$CH_2CH:CH_2$, into $PhCD:CHCH_3$.[630]

Chlorination of metal and of ligand are combined in the reactions

$$[PtCl_3(C_2H_4)]^- + Cl_2 + Cl^- \longrightarrow [Pt(IV)Cl_5(CH_2CH_2Cl)]^{2-} \qquad (137)$$
$$[PtCl_5CH_2CH_2Cl]^{2-} + H_2O \rightleftharpoons [PtCl_5CH_2CH_2OH]^{2-} + HCl \qquad (138)$$
$$[PtCl_5CH_2CH_2OH]^{2-} \longrightarrow [Pt(II)Cl_4]^{2-} + ClCH_2CH_2OH \qquad (139)$$

The last step in this sequence formally resembles a reductive elimination, but occurs at a rate dependent on chloride concentration, suggesting a more complicated process (such as chloridolysis).[631]

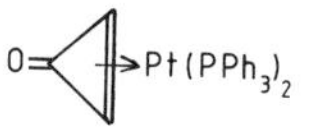

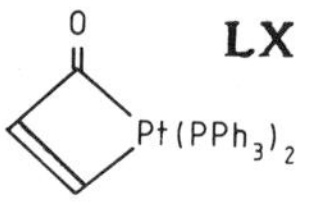

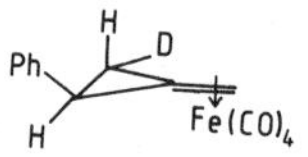

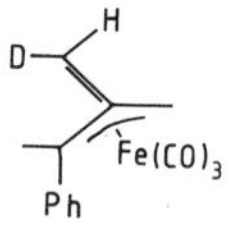

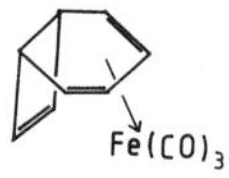

The coordination of small ring alkenes commonly leads to ring opening reactions, as in the reaction of $Pt(PPh_3)_2(C_2H_4)$ with cyclopropenone to give **LIX** at $-65°C$, and its conversion at $-30°C$ to **LX**.[632] Methylenecyclopropane also undergoes ring opening, reacting with platinum hydrides to give η^3-crotyl complexes.[633] With tetracarbonyliron(0)

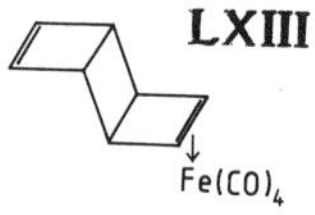

complexes, the ring opening occurs with loss of CO from the metal to give a trimethylenemethane complex, as in the conversion of **LXI** to **LXII.** The disrotatory opening process is symmetry-forbidden, but the outward rotation has a reasonably low activation energy. In contrast, the conversion of **LXIII** to **LXIV** is symmetry-allowed and disrotatory inwards, the broken metal-carbon σ-bond becoming part of the diene-iron bonding system.[634]

Electronegative vinyl halides form η^2 complexes with platinum(0), which rearrange to give insertion of metal into the carbon-halogen bond. There must be subtle differences of mechanism within this class of reactions, since the rate of rearrangement of $Pt(C_2Cl_4)(PPh_3)_2$ to $(Ph_3P)_2Pt(Cl).CCl:CCl_2$ is sensitive to solvent polarity, while that of $Pt(C_2F_3Br)(AsPh_3)_2$ to $(Ph_3As)_2Pt(Br).CF:CF_2$ is not.[635]

4.5 Some reactions of coordinated allenes

The role of allene and alkyne complexes in homogeneous catalysis has been reviewed,[636] as has the chemistry of coordinated allenes in general.[637] The formation of allenes from propargyl complexes is discussed in Section 3.3.6, and references therein. Allene complexes of Pt(II) are subject to nucleophilic attack in the same way as alkene complexes, as in the reaction

$$\underline{cis}\text{-}Cl_2PtL(\eta^2\text{-}CH_2:C:CMe_2) + NR_3 \longrightarrow \underline{cis}\text{-}Cl_2Pt^-L.C(:CMe_2).CH_2N^+R_3$$

$$(140)$$

(L = e.g. PPh_3, Me_2SO), of which the product can be cleaved by hydrochloric acid to give $Me_2C:CH.CH_2NR_3^+$. When NR_3 is an aniline, the initial product rearranges to give the enamine $CH(NHAr):CH.CHMe_2$ as a ligand, which can be released from the complex by cyanide.[638-9] Intramolecular nucleophilic attack (methyl migration) has also been described, and is facilitated when the allene is activated by a net positive charge on the complex.[589]

Allene undergoes oligomerization and cooligomerization at low valent metal centers; examples are trimerization at Ni(0) to give **LXV**,[640] dimerization by $Pt(cod)_2$ to form the metallacyclopentane **LXVI**,[641] and dimerization-insertion with $PdCl_2$ to give $\{ClPd[\eta^3\text{-}C_3H_4\text{-}2\text{-}C(:CH_2)CH_2Cl]\}_2$.[642]

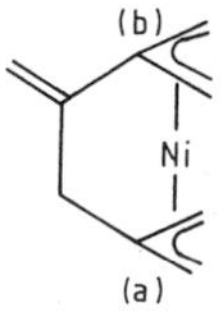

LXV 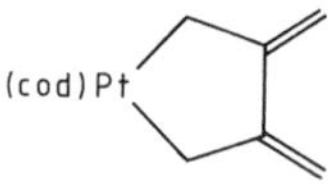   LXVI

5.   FORMATION AND REACTIONS OF η^3-ALLYL COMPLEXES

5.1   Reviews

In addition to the more general reports cited in Section 1 [especially Reference 1(f)], reviews have appeared on the formation of carbon-carbon bonds via η^3-allyl nickel complexes (including useful preparative hints on precursors),[643] π-allyl derivatives in organic synthesis,[644-5] and control of reactivity in nickel allyl systems,[646] while π-allyl complexes are key intermediates in the metal-catalyzed coupling of alkenes to dienes,[27,35] and in the palladium-catalyzed coupling of dienes in the presence of other reagents,[28] and are employed in elegant syntheses of a range of important and structurally complex organic molecules.[647]

5.2   Preparation

In this Section the distinction between η^1- and η^3-allyls will not be stressed; the two are readily interconverted by the process

$$(\eta^1\text{-allyl})ML \rightleftharpoons (\eta^3\text{-allyl})M + L \qquad\qquad (141)$$

considered explicitly in 5.3.1 below. The η^3-allyl, or π-allyl, ligand is formally a three-electron donor, and is usually regarded as taking up two coordination sites at the metal.   As long as all three carbons remain coordinated, rotation about both C-C bonds is restricted, and syn and anti sites (cis and trans respectively to the group bonded to the middle carbon atom) retain their separate identities.

Methods available for the formation of metal alkyls can usually be applied at least as well to metal allyl generation, since the double bond is a strongly activating group; the pioneer work was reviewed some years ago.[648] However, the problem of secondary reaction is sometimes more acute, as metal allyls can react (5.3.3 below) both with Grignard reagents and with allyl halides. The formation of allylic intermediates in the reactions of dienes with each other and with alkenes is discussed in 5.4 below, while the deprotonation of suitable η^2-alkene complexes to η^1-allyls[600] is mentioned in 3.3.5 above.

Allylmercuric halides react with halides of Ru, Rh, Os, Ir, Pt to give π-allyl complexes.[649] Nickel(0) and platinum(0) complexes, and palladium even as the bulk metal, react with allyl halides to give halide-bridged dimers $(\eta^3$-allylMX)$_2$[650] and Ni(cod)$_2$ with allyl acetate gives $(\eta^3$-C$_3$H$_5)_2$Ni and nickel(II) acetate, via an intermediate (C$_3$H$_5$)NiOAc which can be trapped by added phosphine.[651] Phosphine complexes of Pd(0) and Pt(0) react with allylic acetates to give products [$(\eta^3$-allyl)ML$_2$]$^+$ and, for M = Pd, $(\eta^3$-allyl)M(OAc)L.[652-3] Strong dehydrating acid (HPF$_6$/Ac$_2$O/Et$_2$O) converts CpMn(CO)$_2$(CH$_2$:CHCH$_2$OH), an η^2 complex, into the η^3-allylic cation [CpMn(CO)$_2$(C$_3$H$_5$)]$^+$.[654]

Protonation of a diene gives a π-allylic complex; for instance, CpMn(CO)$_2(\eta^2$-butadiene) is specifically protonated to [(exo- syn-η^3-crotyl)Mn(CO)$_2$Cp]$^+$.[655] If the diene is originally η^4-coordinated, the product will be coordinatively unsaturated, as in the protonation of η^4-(1,3-cyclooctadiene) iron tris(trimethylphosphite) to {$(\eta^3$-cyclooctenyl) Fe[P(OMe)$_3$]$_3$}$^+$.[656] Allyls can also be generated by formal addition of metal hydride to conjugated dienes, including cyclic dienes. Thus isopropylmagnesium bromide, which functions as a metal hydride precursor because of the ease of β-elimination of propene from transition metal isopropyls, has been used to generate η^3-2,3,4-pentenyl ligands from 1,3-pentadiene at rhodium(I)[657] and iridium(I),[658] while Ir(cod)(C$_6$H$_9$) exists in equilibrium with Ir(cod)(C$_6$H$_8$)H (C$_6$H$_9$ and C$_6$H$_8$ being η^3-cyclohexenyl and η^4-cyclohexadiene respectively).[659] Pre-formed [HPd(dppe)$_2$]$^+$ reacts with either isomer of 1,3-pentadiene to give the syn-syn-η^3-2,3,4-pentenyl complex [Pd(dppe)(C$_5$H$_9$)]$^+$.[660] The reaction of [H$_2$Ir(PPh$_3$)(acetone)$_2$]$^+$ with 2,3-dimethylbutadiene gives the product LXVII, in which one coordination site is blocked by a terminal methyl group. This methyl group is distorted by a metal-hydrogen bonding interaction, and shows reversible

LXVII

hydrogen migration to metal.[661] Dienes can also be converted to π-allyls by nucleophiles other than hydride; for instance, 1,3-pentadiene reacts with $NaSO_2Nph$ and palladium(II) chloride to give $[Pd(\eta^3\text{-CHMe.CH.}$ $CHCH_2SO_2Nph)Cl]_2$, the anion having attacked at the sterically less hindered terminal carbon of the diene grouping.[662] Complexes cis-$Mn(CO)_4$ $(CH_2Ph)(\eta^2\text{-diene})$, accessible photochemically, rearrange to give $Mn(CO)_4(\eta^3\text{-benzyldienyl})$ products.[663]

The replacement of allylic hydrogen by metal is an important and useful process, occurring especially with unsaturated or weakly coordinated derivatives of low-valent metals late in the transition series. Thus $Ru(PPh_3)_2(\text{styrene})_2$ reacts with propene or 1-hexene to give $HRu(PPh_3)_2(\eta^3\text{-}C_3H_4R)$ (R = H or Pr-n respectively),[664] while the ortho-metalated species $RuH(CH_3CN)(P\text{-}C)(PPh_3)_2$,[665] originally mis-formulated as $Ru(CH_3CN)_2(PPh_3)_4$, reacts with alkenes to give allylic hydrides $HRu(CH_3CN)(\eta^3\text{-allyl})(PPh_3)_2$. The hydride ligand originates from the alkene, since propene-d^6 gives $DRu(CH_3CN)(C_3D_5)(PPh_3)_2$. The reaction is presumably reversible, since the $\eta^3\text{-}C_3H_5$ species reacts with isobutene to give the 2-methallyl complex and free propene.[666] Since there is no need for the hydride ligand to be returned to the same end of the allyl group as that from which it originated, the reaction provides a pathway for alkene isomerization, and converts allylbenzene into a mixture of both isomers of methylstyrene.[667] If the allylic intermediate is too stable the reaction will not be catalytic, and allylbenzene, methylstyrene, and phenylcyclopropane all react with trans-$IrCl(N_2)(PPh_3)_2$ to give the same product, $IrCl(H)(\eta^3\text{-}CH_2CHCHPh)(PPh_3)_2$.[668]

The insertion of palladium(II) into the allylic C-H bond involves overall conversion of allylic hydrogen to H^+, and is often carried out in the presence of a proton sink, such as sodium acetate or anhydrous potassium carbonate. The process, followed by alkylation, is of great synthetic importance, and there is much interest in its detailed stereochemistry and mechanism (5.3.3 below). Studies of the palladation of unsaturated steroids by $PdCl_2$ show that the metal coordinates preferentially to the less hindered face.[669]   There is a normal H/D isotope effect, and loss of hydrogen is <u>syn</u> to the coordinated metal.[670] This is consistent with allylic hydrogen transfer to the metal complex after initial coordination to the double bond, but does not distinguish between hydrogen transfer to the metal itself (with subsequent reductive elimination of HCl) and direct transfer to chloride.   In this context, the reaction of alkenes with palladium(II) hexafluoroacetate   is   instructive.   Acyclic   alkenes   (including methylenecyclohexane)   give   products   of   the   expected   type $[(\eta^3-$ allyl)Pd(O_2C.CF_3)]_2$, but hexene is disproportionated to cyclohexane and benzene. In the presence of malonic anhydride, cyclohexene is converted to benzene. These data all point to initial hydrogen transfer to metal; this is followed by reductive elimination of acid, unless the hydride ligand is intercepted.[671] Keto groups on the substrate can, however, alter the mechanism and affect the stereochemistry.[672]

Methylenecyclopropanes react rapidly with palladium(II) chloride complexes to give π-allyl complexes, the single bond between the saturated carbon atoms of the ring being broken in the process (Scheme 30).[673]

The reaction of $[HFe(CO)_4]^-$ with acetylenes RCCR' gives **LXVIII** (surprisingly,   the   formal   <u>trans</u>-addition   product),[674]   while   iron pentacarbonyl itself reacts with diene monoxides to give allylic products, which can be converted into β - or δ -lactones, depending on structure (Scheme 31).[675]

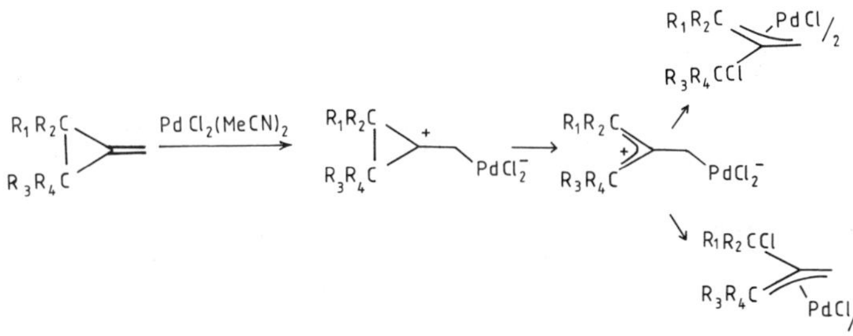

Scheme **30**

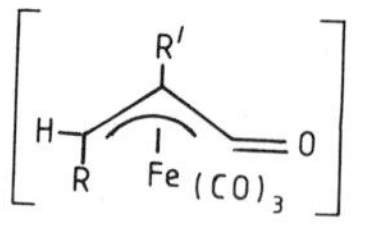

LXVIII

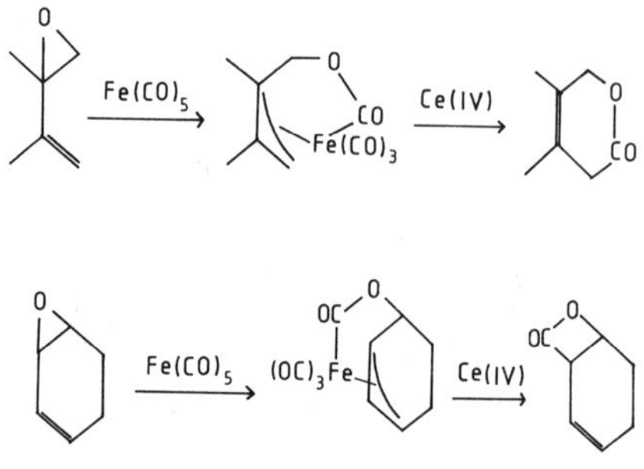

Scheme 31

5.3   Reactions of allylic ligands

5.3.1   Syn–anti and η^1-η^3 rearrangements

The nmr spectra of η^3-allyl complexes show independent syn–anti and left-right isomerisms, and the mechanisms of these processes have been reviewed.[5,6,676] Important among these is the so-called π-σ-π process, in which the η^3-allyl ligand becomes bound by one carbon only, rotates around the metal-carbon σ-bond, and then becomes η^3-bonded in its new position. This process interchanges left and right sides, and syn and anti positions at the carbon attached to metal during the rearrangement, while maintaining the syn- anti distinction at the end of the ligand that becomes detached. Left-right interchange can also occur without syn-anti interchange, by dissociation of one of the other ligands, followed by isomerization around the metal before recombination; an example is the left-right interchange in $(\eta^3$-crotyl)Pd(PPh$_3$)(C$_6$Cl$_5$).[677] For η^1-allyl complexes with labile ligands, a σ-π-σ rearrangement is possible, which can interchange C(1) and C(3). In $(\eta^1$-C$_3$H$_5$)Pt(Cl)(CNMe)(PPh$_3$), this interchange can involve any one of the supporting ligands.[678] An interesting example of η^1-η^3 rearrangement is the process[679]

$$(\eta^5\text{-}C_5H_5)Pd(PPr\text{-}i_3)(\eta^1\text{-}CH_2.CMe:CH_2) \rightleftharpoons$$
$$(\eta^1\text{-}C_5H_5)Pd(PPr\text{-}i_3)(\eta^3\text{-}CH_2CMeCH_2) \qquad (142)$$

The $\eta^1\text{-}\eta^3$ rearrangement is often a key step in the synthesis of η^3-allyls, as in the reaction sequence

$$[M(CO)]^- + \text{all}X \longrightarrow M(CO)(\eta^1\text{-allyl}) + X^-$$
$$M(CO)(\eta^1\text{-allyl}) \longrightarrow M(\eta^3\text{-allyl}) + CO \qquad (143)$$

The second step will occur more sluggishly when M is a good back-donating group, and for that reason the barrier to (143) was suggested many years ago[104] as an empirical measure of the nucleophilicity of $[M(CO)]^-$. Room temperature decarbonylation of cis- or trans-(σ-crotyl)Mn(CO)$_5$ by (dppe)$_2$IrCl gives anti- or syn-(η^3-crotyl)Mn(CO)$_4$, with retention about the initially double bond; thermal decarbonylation of either isomer at $90°C$ gave the more stable syn isomer only.[680] The equilibria between tricarbonylcobalt-η^3-allyls, tetracarbonylcobalt-η^1-allyls, and the corresponding acyls have been investigated.[681]

5.3.2 Reactions with proton acids, hydrogen, and hydrides

Hydrogen and hydridic reducing agents can convert η^3-allyl to η^2-propene ligands. The metal is formally reduced by one unit, thus compensating for the replacement of a 3- by a 2-electron donor. The allyl hydride complexes MoH(C$_3$H$_5$)(dppe)$_2$[412] and NiH(C$_3$H$_5$)PF$_3$[682] both take part in equilibria

$$MH(\eta^3\text{-allyl}) \rightleftharpoons M(\eta^2\text{-alkene}) \qquad (144)$$

The molybdenum species is stable to $110°C$, but the nickel complex decomposes above $-30°C$ to give free propene. The forward reaction (144) is the formal reverse of the formation of η^3-allyls by insertion of metal into allylic C-H bonds, and may be assisted by bulky or basic supporting ligands, as in the evolution of propene when $(\eta^3\text{-}C_3H_5)Pt(PR_3)H$ (R = Bu-t, Cy) formed by borohydride reduction of $[(\eta^3\text{-}C_3H_5)Pt(PR_3)(OMe_2)]^+$, are warmed above $-30°C$.[683]

Protonolysis of palladium allyls may conveniently be combined with removal of the metal by using dimethylglyoxime; an example is the isolation of $MeCH:CH.CH_2CH_2SO_2Nph$ from $[(\eta^3\text{-}MeCH.CH.CH\text{-}CH_2SO_2Nph) PdCl]_2$.[662]

Hydrogenation[16] and deuteriation[684] of alkenes and arenes by $(\eta^3\text{-}C_3H_5)Co[P(OMe)_3]_3$ proceeds with formation of hydrido- or deuteridocobalt phosphite complexes, which are probably the true catalysts. An elegant reduction-removal of π-allyl ligands has been used by Joshi, Thompson and colleagues[685] and more recently by Vahrenkamp,[686] in the formation of unsymmetric metal-metal bonds by the reduction of a π-allyl complex with coordinated diphenyl- or dimethylphosphine (Scheme 32; $M = Fe(CO)_4$, $M' = Mn(CO)_4$, $Co(CO)_3$, $\frac{1}{2}(PdCl)_2$;[685] $M = $ e.g. $CpCo(CO)$, $M' = Co(CO)_3$[686]).

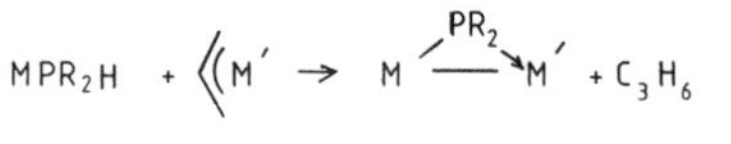

$$MPR_2H \; + \; \langle\!\langle M' \; \to \; M \overset{PR_2}{\diagdown\!\!\diagup} M' \; + C_3H_6 \qquad\qquad \text{Scheme 32}$$

5.3.3 Reactions of η^3-allyls with nucleophiles

It is predicted on general grounds (Chapter 14) that nucleophilic attack on an η^3-allylic ligand should take place at the central carbon, if electron demand at the metal is low, so as to generate a metallacyclobutane ring (Scheme **33a**). More electron-demanding metals, such as palladium(II), encourage attack at the terminal carbon (Scheme **33b**). The formation of an allyl complex by insertion of metal into an allylic C-H or C-X bond, followed by attack at terminal carbon, gives the useful procedures of allylic alkylation and allylic coupling.

Scheme **33**

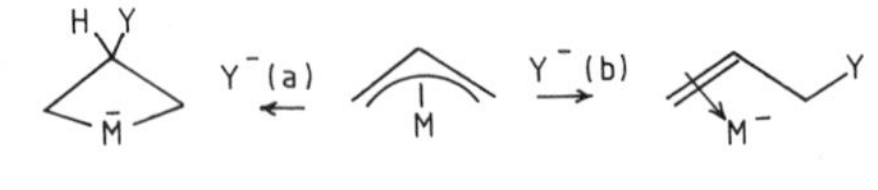

Nucleophilic attack, e.g. by $NaBD_4$, on species $[Cp_2W(\eta^3\text{-allyl})]^+$ occurs at C(2) to give $Cp_2\overline{W}.CH_2CHDCH_2$. Direct reaction between Cp_2MoCl_2 and excess allylmagnesium chloride gives a product $Cp_2\overline{Mo.CH_2CH(CH_2CH:CH_2)CH_2}$, presumably by nucleophilic attack on a π-allyl precursor.[687,116] The complex $C_3H_5Fe(CO)_2NO$ reacts reversibly with phosphines in thf to give the zwitterion $(\eta^2\text{-}R_3P^+CH_2CH=CH_2)Fe^-(CO)_2NO$.[688]

Species $[(\eta^3\text{-allyl})\text{PdCl}]_2$, formed from alkene and palladium(II) chloride, can be activated towards nucleophilic attack at terminal carbon, a reaction in which the original <u>syn</u> or <u>anti</u> orientation of substituents at C(1) and C(3) is retained. Activation can be by a coordinating solvent, such as DMSO,[689] which can cleave the dimer, or by the addition of two moles of phosphine ligand, which convert it to cationic $[(\eta^3\text{-allyl})\text{PdL}_2]^+$.[690] Calculations[691] support the expectation that the positive charge of this cation, and the electron-withdrawing power of the ligands, should help promote attack by nucleophiles. The reaction has been extensively studied by Trost and co-workers[690] as a method of regio- and stereospecific organic synthesis. Carbon nucleophiles, such as the anion of malonic ester, are highly effective, so that the reaction provides a method of carbon-carbon bond formation. Generally the nucleophile attacks preferentially at the less crowded end of the allylic group, but very sterically demanding ligands (such as tris-o-tolylphosphine) can reverse this preference. The overall alkylation reaction involves reduction of Pd(II) to Pd(0), and a possible technique is re-oxidation using Cu(II). This however changes the specificity of the initial allylation. For instance, 2-methylbut-1-ene with palladium chloride in acetic acid-sodium acetate gives the 1,2-dimethylallyl ligand, but in the presence of Cu(II), hydrogen abstraction occurs from terminal rather than internal carbon, to give 2-ethylallyl. It is suggested (Scheme 34) that the allylation involves reversible insertion of palladium into the allylic C-H bond (for independent evidence of the reality of the hydridic intermediate see Reference 603 in Section 5.2). There is eventual thermodynamic control of the distribution among the kinetically accessible products, but Cu(II) intercepts the hydride ligand and increases the importance of initial kinetic control.

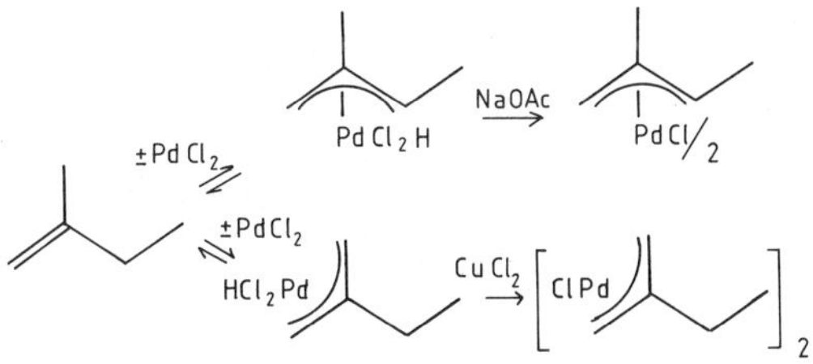

Scheme 34

The scope of the reaction may be extended by using as the initial step the attack of palladium(0) or platinum(0) on an allylic C-O, rather than C-H, bond. This reaction can be built into a catalytic cycle

$$allOAc + Pt(0) \longrightarrow (\eta^3\text{-all})Pt(II)OAc \qquad\qquad (145)$$
$$(\eta^3\text{-all})Pt(II)OAc + CH^-(COMe)_2 \longrightarrow all\text{-}CH(COMe)_2 + Pt(0) + OAc^- \quad (146)$$

(all = allylic grouping; phosphine ligands omitted for clarity).[652] Since the attack of metal on the starting material, and the attack by the external nucleophile on the allylic ligand, are both S_N2-type processes, occurring with inversion, the overall reaction is stereospecific. The reaction may be applied to allyl lactones as substrates, with relaying of specificity (Scheme 35),[692] and, since C- are favored over O-nucleophilies, to the isomerization of Scheme 36.[693] If the C-nucleophile is already linked to the allyl ligand, the reaction will lead to ring closure, directed by the preference for attack at the least substituted carbon (Scheme 37). So the reaction makes unusual ring sizes accessible, and is of use in macrolide synthesis.[694] A further refinement is the use of chiral ligands on palladium, causing asymmetric induction.[37,695]

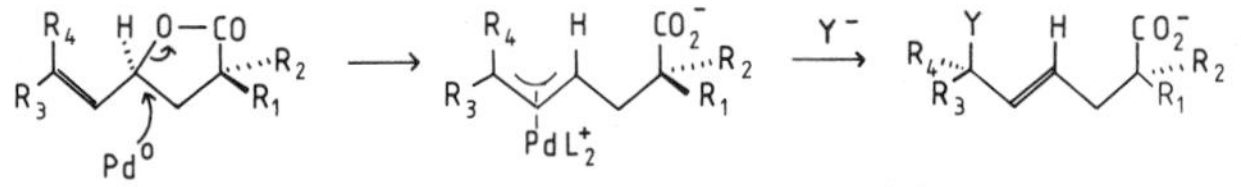

Scheme 35

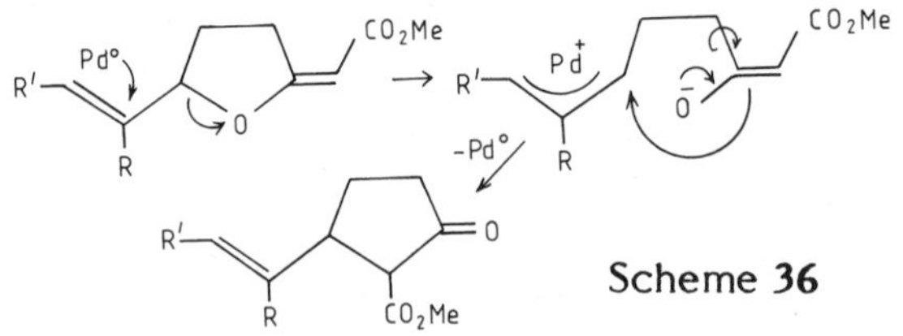

Scheme 36

A further variant on the theme of nucleophilic attack consists in the transfer of an alkyl group to the palladium atom, followed by intermolecular reaction (Scheme 38).[696] Several points about this Scheme deserve comment; the site of the initial palladation, the use of an alkylating agent which attacks at palladium rather than directly at the allyl ligand, the change in regiospecificity on cleaving the dimeric allyl complex with maleic

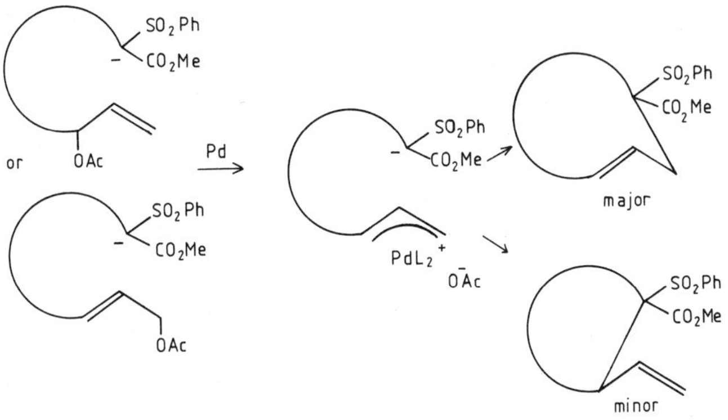

Scheme 37

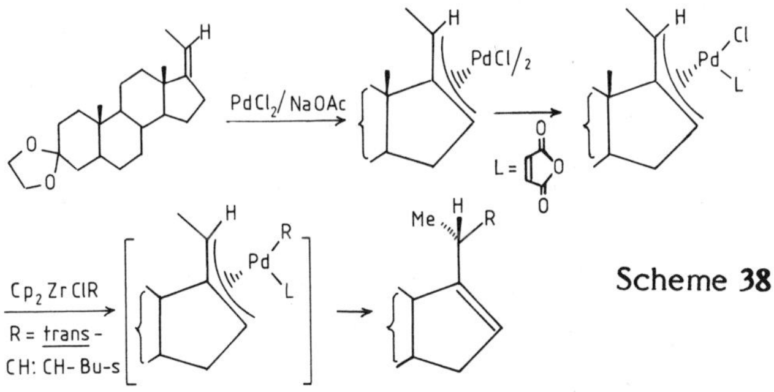

Scheme 38

anhydride (phosphines had diverse effects), and finally the cis-specificity of reductive elimination from palladium. This gives the opposite chirality at the new tertiary carbon atom from that which would have been afforded by intermolecular exo attack. Nucleophilic attack by a group already attached to metal is in fact a long-established process, and may be regarded as a ligand-assisted reductive elimination. An example is the conversion of $(\eta^3$-$C_3H_5)Pd(acac)$ by carbon monoxide to 3-allylacetylacetone; in accord with the general pattern, acac⁻ has behaved as a C-centered, rather than an O-centered, nucleophile to C(1) of the allyl group.[697]

The above discussion assumes a duality of detailed mechanism, similar to that described in Chapter 11 for nucleophilic attack on alkenes. Such a duality has been elegantly demonstrated for the reaction of acetate with the 4-methoxy-η^3-cyclohexenyl ligand (Scheme 39). In the absence of

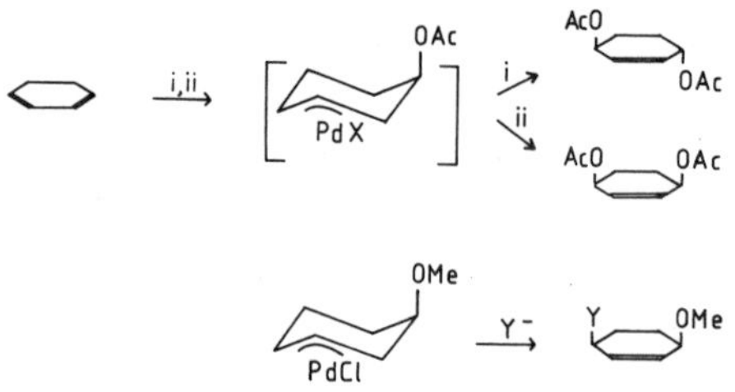

Scheme **39**

chloride, acetate coordinates to metal before attacking the ligand. Coordinated chloride blocks this process, and acetate then functions as an external nucleophile, attacking _trans_ to the metal.   In either case, eliminated Pd(0) can be re-oxidized by benzoquinone, completing the cycle.[698]   The reaction of potassium acetonate confirms the ability of chloride to block the metal, and the stereochemistry of external attack.[699]

Coordination of chloride can be strong enough to tip the balance between η^3- and η^1-allylic bonding (Scheme **40**). This can lead to a gross change in products. As discussed above, external nucleophiles attack π-allyl at the less crowded end. This is the end to which the metal itself migrates in the σ-allyl complex, so that attack now occurs preferentially, by the S_N2' mechanism, at the opposite end of the allylic group.[700] Thus techniques are available to control both the stereo- and regiospecificity of these reactions.

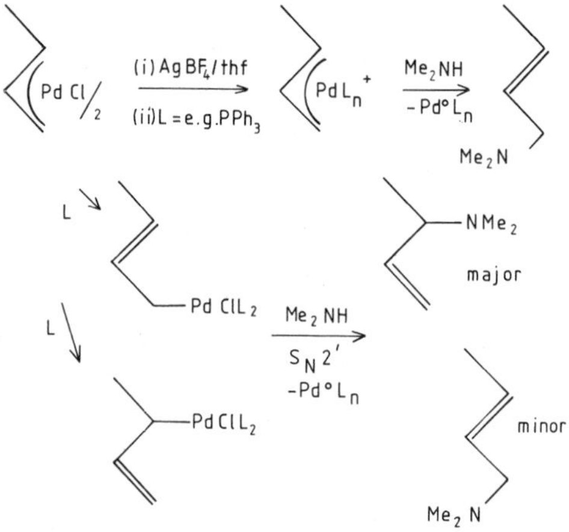

Scheme **40**

Differences in steric crowding may explain the striking difference between $NiCl_2$.dppf and $PdCl_2$.dppf [dppf = 1,1'-bis(diphenylphosphino)ferrocene] as catalysts for the coupling of allylic derivatives to phenylmagnesium bromide. 1- or 3-substituted compounds $RCH:CH_2CH_2X$ and $CH_2CH:CH(R)X$ (X = e.g. Cl, $OSiMe_3$) give the same products, indicating an η^3- or at least a rapidly isomerizing η^1-allylic intermediate, from which the nickel-catalyzed reaction forms $CH_2:CH.CH(R)Ph$, while the palladium catalyst gives $PhCH_2.CH:CHR$.[701] With palladium, the attack is taking place, as usual, at the less crowded end of the allyl group; but the smaller metal, tied to a fairly bulky ligand, itself tends to migrate to that less cluttered position, reversing the regiospecificity.

A useful extension of the allylic alkylation reaction is illustrated in Scheme **41**. A diene epoxide is converted by palladium(0) to an alkoxide-containing allylic ligand. The alkoxy group is basic enough to convert a CH-acid to a nucleophilic anion, which then attacks the allyl. The overall reaction is catalytic in Pd(0), is of wide application (since diene monoxides are readily available), and is compatible with the presence of either acid- or base-sensitive groups in the substrate.[702] The reaction should be compared with that of diene monoxides with $Fe(CO)_5$ (Scheme **31** above), where a similar alkoxide intermediate attacks metal-bound CO.[675]

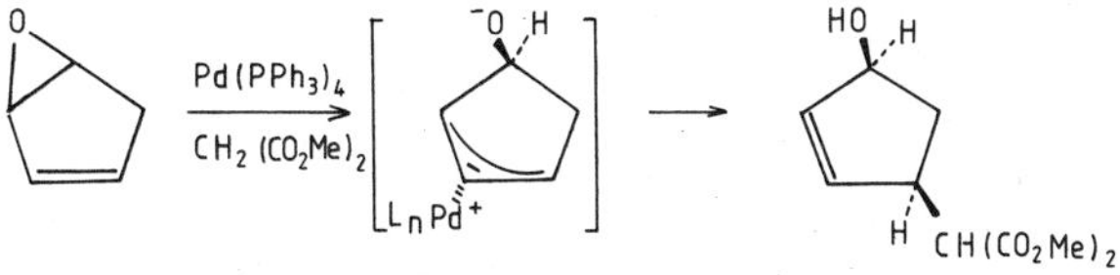

Scheme **41**

5.3.4 Reaction of metal allyls with electrophiles

While the π-allyl group is vulnerable to nucleophilic attack, the σ-allyl group is itself nucleophilic, so that especially in the presence of coordinating solvents the equilibrium of eq 141 can lead to the reaction of an initially π-bonded allyl with electrophiles, as in Section 3.3.5 above. This leads to an elegant selectivity in the reactions of **LXV**. Aldehydes add to functionality (a), which is more readily converted to η^1-allyl than is (b), giving ultimately $RCHOH.CH_2.C(:CH_2).CH_2.C(:CH_2).C(:CH_2)CH_3$, while nucleophile precursors YH such as amines and active methylenes add to functionality (b), to give $CH_3.C(:CH_2).CH_2.C(:CH_2).C(:CH_2)CH_2Y$.[703]

Allylnickel halides also couple with a wide range of organic halides, eliminating nickel(II) halide; there is evidence that the reaction proceeds by a free radical mechanism after initial electron transfer from complex to organic halide.[704] The coupling of (2-vinyl)η^3-allylnickel bromide dimer to allyl halides and to 3-methylbutanal has been used in terpene synthesis.[705]

The coupling of titanium allyls to carbonyl groups occurs with migration of metal to the <u>less</u> hindered of C(1) and C(3), together with coordination of oxygen to metal. As a result, it is the <u>more</u> hindered of C(1) and C(3) that is coupled to carbonyl carbon in the product.[706]

5.3.5 Reductive coupling of allyl groups

Reactions of the type

$$M(\eta^3\text{-allyl})_2 + nL \longrightarrow ML_n + \text{bis(allyl)} \tag{147}$$

are among the earliest known or surmised for such complexes,[707] and are a special case of ligand-assisted reductive elimination (Section 3.2.13). A striking example is the conversion of **LXV** by CO to 1,2,4-trimethylenecyclohexane, with elimination of nickel carbonyl.[640] Other examples are presented in Section 5.4 below.

Thermolysis of $Zr(C_3H_5)_4$ gives 1,5-hexadiene without any CIDNP effects, but photolysis of this species proceeds through trapped radical pairs, with polarization of starting material, either by recombination or by exchange of polarized free radicals with molecules of parent.[708] Photolysis of allylpalladium chloride dimers gives bis-allyls in high yield under anaerobic conditions, but in the presence of oxygen the isolated products are enones in which one terminal carbon has been converted to a keto group with loss of a hydrogen atom;[709] allyl radical intermediates have been trapped with nitroso-durene.[710]

5.3.6   Insertion into metal-allyl bonds

The insertion of alkenes is covered in 5.4 below, and that of carbon monoxide in other Chapters.

$Cp_2Ti(\eta^3\text{-}CHRCHCH_2)$ reacts with a large number of unsaturated electrophiles (CO_2, PhNCO, PhN:CHPh, Me_2CO) to give insertion products in which, in most cases, the insertion is between the metal and the substituted terminal carbon. Hydrolysis then gives products CH_2:CH.CHRR' [R' = COOH, CO.NH.Ph, CHPh.NHPh, $C(Me_2)OH$].[711] For CO_2 at least, chiral substituents on the Cp rings lead to some chiral selectivity in the insertion.[712] Bis(2-methallyl)nickel reacts quantitatively with one mole of CO_2 and one of phosphine to give the insertion product $(\eta^3\text{-}CH_2CMeCH_2)Ni(PR_3).O.CO.CH_2CMe:CH_2$.[713]

5.4   Coupling of unsaturated hydrocarbons and related reactions

The oligomerization of butadiene and its co-oligomerization with other unsaturated species at low-valent metal catalysts has been studied with great thoroughness; for reviews see References 1f, 28, 33, 707, and those cited in Section 5.1, as well as Chapter 15. The key reactions are reductive coupling of two butadienes to give a bis-allylic ligand (which in some reactions may display one η^3- and one η^1-allyl functionality), insertion of alkene, alkyne, or diene into metal-allyl groupings, reductive coupling of bis-allyl systems under the influence of added ligands (which can include excess of starting diene), electrophilic and nucleophilic attack on the allyl groups, β-hydrogen elimination, protonolysis of metal-carbon bonds, and reductive elimination of metal. The number of isomers that could be involved at each step is bewildering, but a surprising degree of control is possible through judicious choice of conditions and supporting ligands.

The reaction of η^3-allylnickel acetate with norbornadiene gives initially the insertion product **LXIX**, the conformation of which precludes β-elimination but allows alkene insertion, to give **LXX**. This can undergo β-elimination directly, to give **LXXI**, as well as prior insertion of metal into the strained C-C bond, to give **LXXII** and hence, by β-elimination, **LXXIII**.[714]

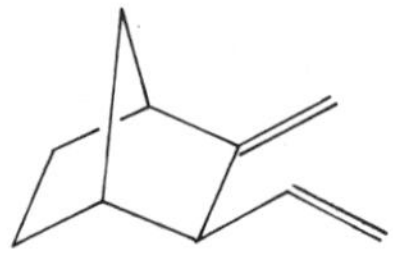

This reaction serves to confirm the mechanism accepted for the addition of ethylene to butadiene, as catalyzed under suitable conditions by complexes of iron, ruthenium, cobalt, rhodium, nickel or palladium:[715-6,35]

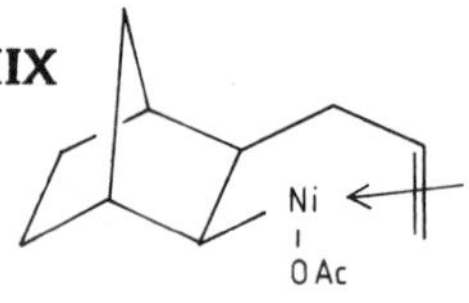

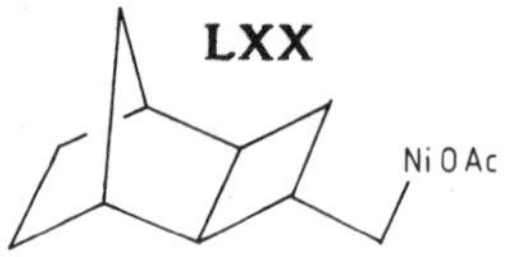

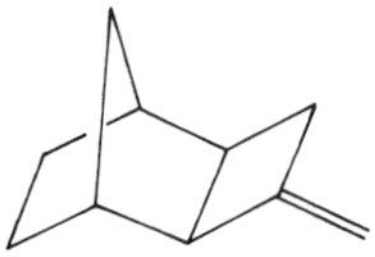

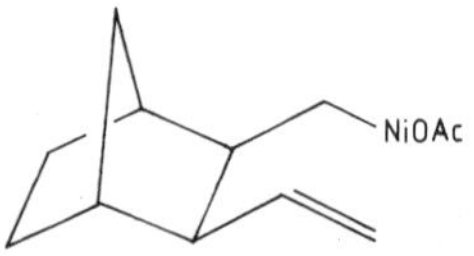

$$CH_2=CH-CH=CH_2 + MH \longrightarrow \eta^3\text{-}(MeCHCHCH_2)M \qquad (148)$$

$$\eta^3\text{-}(MeCHCHCH_2)M + C_2H_4 \longrightarrow MeCH=CH.CH_2.CH_2CH_2.M \qquad (149)$$

$$MeCH=CH.CH_2.CH_2CH_2.M \longrightarrow MeCH=CH.CH_2.CH=CH_2 + MH \qquad (150)$$

The reaction is not completely regiospecific (except at cobalt), generally giving some 3-methyl-1,4-pentadiene, presumably by the alternative insertion to (149). The hexadiene may be formed as the cis or the trans isomer. The yields of these depend on the metal, and may possibly reflect the conformation of the (coordinated) butadiene immediately prior to the hydride transfer step (148).

The much-discussed coupling of butadiene at nickel(0) has already been mentioned. It would be desirable to insert carbon monoxide between the two allylic ligand functions of $Ph_3PNi(C_8H_{12})$, where C_8H_{12} is the ligand formed by reductive dimerization of butadiene, but it is known that CO gives only C_8H_{12} isomers and nickel carbonyl phosphine complexes.[1f] However, the insertion of n-butylisonitrile is successful, and subsequent hydrolysis leads to the formation of cyclononadienone and vinylcycloheptenone, providing a route to the elusive C_9 rings.[717]

Butadiene inserts into the allyl-palladium system of π-2-methallyl-palladium acetylacetonate to give η^3-CH$_2$CHCH(CH$_2$CH$_2$.CMe:CH$_2$) Pd(acac). This reaction contrasts with the CO-induced coupling of ligand acac to allyl, and the position of the methyl label in the sidechain confirms that the new allylic group, and the first CH$_2$ group of the sidechain, originate from the added butadiene.[697] Substituted dienes CH$_2$:CH.CR:CH$_2$ couple so that the new allylic group has R at C(3). If R is t-butyl, the butenyl sidechain is forced into the <u>anti</u> position, and ring closure generally occurs, by insertion of the butenyl double bond into the Pd–C(1) bond. If the starting material is 2-chloroallyl palladium chloride, the double bond of the butenyl group is deactivated and the uncyclized intermediate can be isolated.[718]

Reductive coupling at palladium(0) (review: reference 28) is in one way more convenient than that at nickel, since palladium(II) acetate is reduced by butadiene itself in the presence of phosphines, with formation of AcO.CH$_2$CH:CHCH$_2$OAc.   Butadiene is converted into open chain dodecatriene (with presumably uptake of hydrogen from some source). It is thought that the intermediates involved are similar to those in the formation of cyclododecatriene at nickel(0), but that the larger size of palladium prevents ring closure. More widely employed are dimerization reactions, thought to proceed through the reductive coupling intermediate **LXXIV** of Scheme **42**. Nucleophilic attack at C(1) [or in side-reaction C(3); but note that the main reaction obeys the rule of attack at the less crowded carbon] leads to overall formation of <u>trans</u>-2,7-octadienyl-Y from the nucleophile precursor HY, and the site of deuteration in CH$_3$OD is consistent with this mechanism. Attack by the nucleophilic C(6) on unsaturated electrophiles such as acetaldehyde (cf. Reference 703) gives an alkoxide which, presumably because of steric constraints, attacks the η^3-allyl group at C(3) rather than C(1) to give the observed product. With formic acid, a source of both hydride and protons, **LXXIV** gives rise to linear dienes and CO$_2$. The detailed mechanism of this reduction has been studied using labeled formic acid, and with isoprene as substrate.[719]

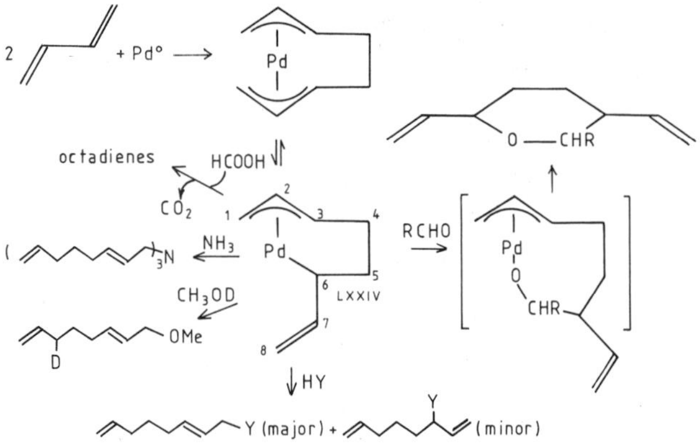

Scheme **42**

The related reactions of platinum(0) have been studied by Green, Stone and co-workers.[641] The results (Scheme **43**) show the greater ease of formation of η^1- as opposed to η^3-allylic systems at platinum, bonding to C(8) rather than C(6) in the η^1,η^3 intermediate, and <u>trans</u> configuration in the divinylmetallacyclobutane species, in contrast to the <u>cis</u> configuration that analogy with the nickel(0) butadiene system[1f] would suggest. These facts illustrate once more the large effects on organotransition metal reactions of subtle differences in reactivity at the metal, and how these effects can point the way to detailed mechanistic control of selectivity.

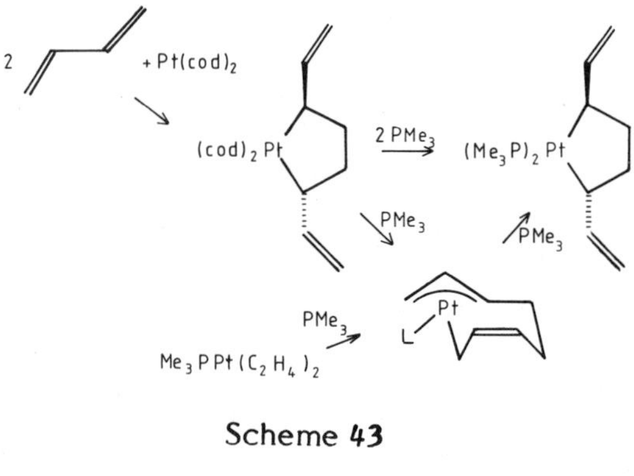

Scheme **43**

6.   MORE RECENT DEVELOPMENTS

Reviews have appeared on organometallics in the preparation of supported catalysts[720] and surface organometallic chemistry in relation to heterogeneous catalysis,[721] organolanthanide[722] and -actinide[723] chemistry, organocobalt carbonyls,[724] metal alkynyls,[725] proton transfer reactions in organometallic chemistry,[726] nucleophilic attack on organopalladium and related species,[727] and coupling of organotin, -mercury and -copper species catalyzed by palladium.[728] Also of interest is a book devoted to catalysis by phosphine complexes[729] and the papers from the first Texas A and M industry-university cooperative symposium, on organometallic compounds.[730]

Bond disruption energies have been determined for a range of species Cp_3ThR and $Cp*_2ThR_2$.[731]

Transmetalation of <u>trans</u>-$MnX_2(dmpe)_2$ with $MgMe_2$ gives <u>trans</u>-$MnMe_2(dmpe)_2$, with one unpaired spin. $MgEt_2$, however, gives diamagnetic <u>trans</u>-$Mn(H)(C_2H_4)(dmpe)_2$,[732] while $Mn(acac)_3$, methyllithium and dmpe give a mixture of $MnMe_2(dmpe)_2$ and $MnMe_4(dmpe)$.[733] Among the products from $RMgCl$ (R = trimethylsilylmethyl, neopentyl) and $Ru_2(OAc)_4Cl$ is Ru_2R_6, with an unbridged formal triple bond.[734] $Cp*RhCl_2PMe_3$ with triethylaluminium gives $Cp*RhEt_2PMe_3$, but in the absence of phosphine $(Cp*RhCl_2)_2$ gives $Cp*Rh(C_2H_4)_2$.[735]

The solid products $Li(dme)_3MMe_6$ and $Li(tmed)_3MMe_6$ (M=Er,Lu), formed from MCl_3 and methyllithium in etherial solvents, contain $M(\mu$-$Me)_2Li$ bridges.[736] Bridging methyl groups also occur in the asymmetric dimer $Cp*_2Lu(\mu$-Me$)LuCp*_2Me$, which exists in rapid equilibrium with monomer; the monomer exchanges its alkyl group with methane and other saturated hydrocarbons.[199,737] A zirconocene ketene complex with trimethylaluminum gave bridging, trigonal bipyramidal carbon with zirconium atoms apical.[738]

Electrophilic attack by platinum, and metal-metal transfer, are implicated in the oxidation of methane to methyl chloride or meth anol, via Pt(IV) methyls[739]

$$Pt(II) + CH_4 \longrightarrow Pt(II)\text{-}Me + H^+ \tag{151}$$

$$Pt(II)Me + Pt(IV) \longrightarrow Pt(II) + Pt(IV)Me \tag{152}$$

$$Pt(IV)Me \longrightarrow MeCl \text{ or } MeOH + Pt(II) \tag{153}$$

Reactions of phosphorus ylids with transition metals have been reviewed.[740] The suggested[453] direct insertion of metal into a phosphorus-aryl bond has been confirmed, and can give rise to aryl exchange, to the degradation of hydroformylation catalysts, and to the freeing of immobilized phosphine ligands from their supports.[741]

The photolysis of transition metal alkyls has been reviewed.[742] Homolysis of Cr(III) alkyls shows a large volume of activation, consistent with the need for cage disruption as Cr(II) and the organic radical diffuse apart.[743]

Cobalt tetracarbonyl hydride cleaves the cobalt-carbon bond in $EtO_2C.CH_2Co(CO)_4$[744] and $n\text{-}C_5H_{11}COCo(CO)_4$;[745] the kinetics have been investigated. $Cp_2W(H)CH_3$ in dilute solution undergoes intramolecular reductive elimination of methane; in more concentrated solutions, intermolecular exchange between metal- and carbon-bound hydrogen occurs.[746] Matrix photolysis of $CpCr(CO)_3Me$ gives rise to $CpCr(CO)_2(CH_2)(H)$,[747] while photolysis or thermolysis of $Cp^*Os(CO)_2.CH_2OH$ gives $Cp^*Os(CO)_2H$ by α-hydrogen elimination.[748]

Sequences involving α-elimination, carbene insertion and β-hydrogen elimination have been invoked to explain the reactivity of a methoxymethyl rhodium complex (L=PMe$_3$):[749]

$$RhL_3Br(CH_3)(CH_2OH) + SiMe_3Br \longrightarrow (Presumed)[Rhl_3Br(CH_3)(CH_2)]^+ \text{ etc.}$$
$$\longrightarrow [RhL_3Br(C_2H_5)]^+ \longrightarrow C_2H_4 \text{ etc.} \tag{154}$$
$$RhL_3Br(CH_3)(CH_2OMe) + Ag^+/CH_3CN \longrightarrow$$

$$(presumed)[RhL_3(CH_3)(CH_2OMe)(CH_3CN)]^+ \text{ etc.}$$
$$\longrightarrow [RhL_3(H)(CH_3)(=CHOMe)]^+ \longrightarrow [RhL_3(H)CH(Me)(OMe)]^+$$
$$\longrightarrow CH_2{:}CHOMe \text{ etc.} \tag{155}$$

The complex $Rh(H)(C_2H_4)(Pr\text{-}i_3)_2$ shows exchange between hydride and ethylene hydrogen by reversible β-insertion in the <u>cis</u> complex, as well as cis- trans isomerism.[750] Insertion and elimination are even more finely balanced in $[(C_5Me_4Et)Co(C_2H_4)(C_2H_5)]^+$ and $[Cp*Co(PPh_3)(C_2H_5)]^+$, in both of which there is strong interaction between metal and an ethyl group β-hydrogen; such bridging hydrogen has been termed 'agostic'.[751]

Calculations support the view that neutral ligands can promote reductive elimination from species L_2NiR_2; the experimental evidence for this effect has been summarised.[752] The elimination of ethyl cyanide from $LNi(CN)(C_2H_5)(C_2H_4)$ (L = tris-o-tolylphosphite) is promoted associatively by L,[753] and CO promotes the formation of both RCOR and R-R from $(dppe)PdR_2$ (R = neopentyl).[754]

Protonolysis of optically active $CpRe(NO)(PPh_3)Me$ gives $CpRe(NO)(PPh_3)X$ with retention, but the product from halogenolysis by X_2 is racemic.[755] The oxidative cleavage of species CpRuLL'R by halogen, mercuric chloride, HCl or $CuCl_2$ generally resembles the corresponding reactions at iron.[756] Reaction of $(bpy)NiMe_2$ or $(bpy)NiEt_2$ and organic halides proceeds by coordination of halogen, elimination of ethane or butane, and insertion of nickel into the carbon-halogen bond. Trans-$(Et_3P)_2NiMe_2$ with bromobenzene gives $(Et_3P)_2Ni(Ph)Br$, but $(dppe)NiMe_2$ gives mainly toluene and $(dppe)Ni(Me)Cl$.[757]

Reversible insertion of a coordinated double bond is thought to be responsible for the reaction sequence[758]

$$[(dmpe)\overset{\frown}{Pt\text{-}CH_2\text{-}CMe_2\text{-}CD_2\text{-}CH}=CH_2]^+ \longrightarrow \text{(presumed)} \qquad\qquad (156)$$

$$[(dmpe)Pt\text{-}CH_2\text{-}\overset{\frown}{CH\text{-}CD_2CMe_2CH_2}]^+ \longrightarrow [(dmpe)\overset{\frown}{PtCH_2CD_2CMe_2CH}=CH_2]^+$$

(It is worth noting that the α-elimination-carbene insertion cycle would imply highly strained intermediates in this case.) Aldehydes insert (stereospecifically) into the metal-carbon bond of CpZrCl(2-butenyl),[759] while formaldehyde inserts into the mesityl-copper bond.[760] Carbon dioxide inserts into the metal-alkyl bond of species $[M(CO)_5Me]^-$ (M=Cr,Mo,W); replacement of CO by more basic ligands accelerates the reaction,[761] which is also shown by $\overset{\frown}{Mn(CO)_4CH_2CH_2PPh_2}$.[762]

Attack by secondary amines on platinum(II)-bound ethylene leads to cyclic products of type $\overline{L\overset{\frown}{Pt}(Cl)CH_2CH_2NR_2}$;[763] the stereochemistry of the related reaction of 2-butene shows the initial attack to be <u>trans</u>.[764] Amphiphilic reagents can also attack both carbon and metal, as in the reaction[765]

$$[Pt(C_2H_4)(Cl)(tmen)]^+ + 2\ NCO^- \longrightarrow (tmen)Pt(Cl)(NCO)CH_2CH_2.NH.C=O$$
(157)

Uranium tetra-allyl reacts with bipyridyl to give a product involving allyl-bipyridyl coupling.[766] Conjugated allylic ligands show differences in regioselectivity of alkylation when coordinated to tungsten or to palladium; the synthetic usefulness of the reaction can be further extended by internal Diels-Alder reactions.[767] Zwitter-ionic allylic ligands on palladium couple to alkenes with electron-withdrawing substituents to give 5-membered ring products.[768] The stereochemistry of alkylation at palladium allyls is highly sensitive to the structure of the nucleophile, occurring at C(1) of [(-)-(1S,2R,3R)-MeCH.CH.CHPhPdCl]$_2$ with inversion using sodium malonic ester or dimethylamine, but with retention using phenyl or allyl Grignards.[769]

Species $Cp_2Ti(\eta^3$-allyl) couple regio- and stereoselectively to aldehydes.[770] Carbon dioxide inserts into chromium(III) allyls to give butenoate complexes.[771] Maleic anhydride, being a good π-acceptor, promotes reductive elimination between the less crowded carbon atoms of palladium bis-allyls. This reaction can be used to prepare asymmetric bis-allyls from allylpalladium chloride dimers and allyl Grignards, using dioxane to precipitate magnesium-containing residues and reduce allyl ligand metathesis.[772] Cyclo-oligomerization of butadiene to, e.g., vinylcyclohexene can be catalyzed by palladium, as well as by nickel, provided phosphine-free materials, such as palladium bis-allyl, are used as catalyst precursors.[773]

REFERENCES

1. (a) J.K. Kochi, "Organometallic Reactions and Catalysis", Academic Press, New York (1978); (b) J.P. Collman and L.S. Hegedus, "Principles and Applications of Organotransition Metal Chemistry", University Science Books, Mill Valley, California (1980); (c) C. Masters, "Homogeneous Transition Metal Catalysis - a Gentle Art", Chapman and Hall, London (1981); (d) R.P. Houghton, "Metal Complexes in Organic Chemistry", Cambridge University Press (1979); (e) R.F. Heck, "Organotransition Metal Chemistry - a Mechanistic Approach", Academic Press, New York (1974); (f) P.W. Jolly and G. Wilke, "The Organic Chemistry of Nickel", Academic Press, New York (1974-5); (g) P.M. Maitlis, "The Organic Chemistry of Palladium", Academic Press, New York (1971); (h) U. Belluco, "The Organometallic and Coordination Chemistry of Platinum", Academic Press, New York (1974); (i) "New Applications of Organometallic Reagents in Organic Synthesis", D. Seyferth, ed. (J. Organomet. Chem. Library 1), Elsevier Scientific, Amsterdam (1976); (j) "Homogeneous Catalysis", Adv. Chem. Ser. 70 (1968), 132 (1974); (k) G.W. Parshall, "Homogeneous Catalysis by Soluble Transition Metal Complexes", Wiley - Interscience, New York (1980); (l) A. Nakamura and M. Tsutsui, "Principles and Applications of Homogeneous Catalysis", Wiley - Interscience, New York (1980); (m) E.A. Koerner von Gustorf, F.W. Grevels and I. Fischler, "The Organic Chemistry of Iron", Academic Press, New York (Vol. 1, 1978; Vol. 2, 1981); (n) J. Tsuji, "Organic Synthesis with Palladium Compounds", Springer-Verlag, New York (1980); (o) R.S. Dickinson, "The Organometallic Chemistry of Rhodium and Iridium", Academic Press, New York (1983); (p) G. Wilkinson, F.G.A. Stone and E.W. Abel, ed., "Comprehensive Organometallic Chemistry" (9 volumes), Pergamon, Oxford and New York (1982); (q) S.G. Davies, "Organo-transition Metal Chemistry: Applications to Organic Synthesis", Pergamon, Oxford (1982).

2. T. Bartik, P. Heimbach and H. Schenkuln, Kontakte (Darmstadt) 16 (1983).

3. M.F. Lappert and P.W. Lednor, Adv. Organomet. Chem. 14:345 (1976); J. Halpern, Pure Appl. Chem. 51:2171 (1979).

4. J.K. Kochi, Acc. Chem. Res. 7:351 (1979).

5. J.W. Faller, Adv. Organomet. Chem. 16:211 (1977).

6. M. Tsutsui and A. Courtney, Adv. Organomet. Chem. 16:241 (1977).

7. (a) P.J. Davidson, M.F. Lappert and R. Pearce, Acc. Chem. Res. 7:209 (1974); (b) P.J. Davidson, M.F. Lappert and R. Pearce, Chem. Rev. 76:219 (1976).

8. R.R. Schrock and G.W. Parshall, Chem. Rev. 76:243 (1976).

9. W.A. Herrmann, Adv. Organomet. Chem. 20:160 (1982); idem, Pure Appl. Chem. 54:65 (1982); idem, J. Organomet. Chem. 250:319 (1983); P.B. Mackenzie, K.C. Ott and R.H. Grubbs, Pure Appl. Chem. 56:59 (1984).

10. J. Holton, M.F. Lappert, R. Pearce and P.I.W. Yarrow, Chem. Rev. 83:135 (1983).

11. M. Brookhart and M.L.H. Green, J. Organomet. Chem. 250:395 (1983).

12. J. Dehand and M. Pfeffer, Coord. Chem. Rev. 18:327 (1976); M.I. Bruce, Angew. Chem. Int. Ed. Engl. 16:73 (1977).

13. R.J. Puddephatt, Coord. Chem. Rev. 33:149 (1980); idem, ACS Symp. Ser. 211:353 (1983).

14. R.G. Bergman, Acc. Chem. Res. 13:113 (1980).

15. G.W. Parshall, Acc. Chem. Res. 8:113 (1975); D.E. Webster, Adv. Organomet. Chem. 15:147 (1977); A.E. Shilov and A.A. Shteinman, Coord. Chem. Rev. 24:97 (1977); G.W. Parshall, Catalysis (Chem. Soc. Spec. Period. Rep.) 1:335 (1977).

16. G. Dolcetti and N.W. Hoffman, Inorg. Chim. Acta 9:269 (1974); B.R. James, Adv. Organomet. Chem. 17:319 (1979); E.L. Muetterties and J.R. Bleeke, Acc. Chem. Res. 12:324 (1979); B.R. James, "Homogeneous Hydrogenation", Wiley - Interscience, New York (1973).

17. P.J. Brothers, Prog. Inorg. Chem. 28:1 (1981).

18. E. Lukevics, Z.V. Belyakova, M.G. Pomerantseva and M.G. Voronokov, J. Organomet. Chem. Library (Organomet. Chem. Rev.) 5:1 (1977); J.L. Speier, Adv. Organomet. Chem. 17:407 (1979).

19. K.C. Bishop III, Chem. Rev. 76:461 (1976).

20. H. Werner, Pure Appl. Chem. 54:177 (1982).

21. R.J.H. Clark, S. Moorhouse and J.A. Stockwell, J. Organomet. Chem. Library (Organomet. Chem. Rev.) 3:223 (1977); M.T. Reetz, Top. Curr. Chem. 106:1 (1982); B. Weidmann and D. Seebach, Angew. Chem. Int. Ed. Engl. 22:31 (1983).

22. D.S. Moore, Coord. Chem. Rev. 44:127 (1982).

23. D. Dodd and M.D. Johnson, J. Organomet. Chem. 52:1 (1973).

24. M. Orchin, Acc. Chem. Res. 14:259 (1981).

25. G.P. Chiusoli and G. Salerno, Adv. Organomet. Chem. 17:195 (1979).

26. J. Tsuji, Top. Curr. Chem. 91:29 (1980).

27. B.M. Trost, Tetrahedron 33:2615 (1977); idem, Acc. Chem. Res. 13:385 (1980).

28. J. Tsuji, Adv. Organomet. Chem. 17:141 (1979); idem, Pure App. Chem. 54:197 (1982).

29.   M.H. Chisholm and H.C. Clark, Acc. Chem. Res. 6:202 (1973).

30.   G.H. Posner, "An Introduction to Synthesis Using Organocopper Reagents", John Wiley - Interscience, New York (1980); A.E. Jukes, Adv. Organomet. Chem. 12:215 (1974); J.F. Normant in 1(i), p. 219; J.G. Noltes, Philos. Trans. R. Soc. London Ser. A 308:35 (1982).

31.   G.K. Anderson, Adv. Organomet. Chem. 20:40 (1982).

32.   H. Sinn and W. Kaminsky, Adv. Organomet. Chem. 18:99 (1980).

33.   B. Bogdanovic, Adv. Organomet. Chem. 17:105 (1979).

34.   F.H. Jardine, Prog. Inorg. Chem. 28:63 (1981).

35.   D.H. Richards, Chem. Soc. Rev. 6:235 (1977); A.C.L. Su, Adv. Organomet. Chem. 17:269 (1979).

36.   G.L. Geoffroy and J.R. Lehman, Adv. Inorg. Chem. Radiochem. 20:190 (1977); see also "Transition Metal Hydrides", Adv. Chem. Ser. 167 (1978).

37.   B. Bosnich and P.B. Mackenzie, Pure Appl. Chem. 54:189 (1982).

38.   M.S. Wrighton, J.L. Graff, R.J. Kazlauskas, J.C. Mitchener and C.L. Reichel, Pure Appl. Chem. 54:161 (1982).

39.   Pure Appl. Chem. 53:2307 (1981); ibid. 55:1669 (1983).

40.   P. Heimback, H. Schenkluhn and K. Wisseroth, Pure Appl. Chem. 53:2419 (1981).

41.   T.C. Flood, Top. Stereochem. 12:37 (1981).

42.   J.A. Connor, J. Organomet. Chem. 94:195 (1975); idem, Top. Curr. Chem. 71:71 (1977); J. Halpern, Acc. Chem. Res. 15:238 (1982).

43.   J.U. Mondale and D.M. Blake, Coord. Chem. Rev. 47:205 (1982).

44.   A.R. Dias, M.S. Salema and J.A. Martinho Simoes, Organometallics 1:971 (1982).

45.   J.E. Bruno, T.J. Marks and L.R. Morss, J. Am. Chem. Soc. 105:6824 (1983).

46.   J.C.G. Calado, A.R. Dias, J.A. Martinho Simoes and M.A.V. Ribeiro Da Silva, J.C.S. Chem. Commun. 737 (1978).

47.   F.A. Adedeji, J.A. Connor, H.A. Skinner, L. Galyer and G. Wilkinson, J.C.S. Chem. Commun. 159 (1976).

48.   J.A. Connor, M.T. Zafavani-Moattar, J. Bickerton, N.I. El Saied, S. Suradi, R. Carson, G. Al Takhin and H.A. Skinner, Organometallics 1:1166 (1982).

49.   C.J. Mortimer, M.P. Wilkinson and R.J. Puddephatt, J. Organomet. Chem. 165:265 (1979).

50.   G. Yoneda, S-M. Lin, L-P. Wang and D.M. Blake, J. Am. Chem. Soc. 103:5768 (1981).

51. F.T.T. Ng, G.L. Rempel and J. Halpern, J. Am. Chem. Soc. 104:621 (1982); T-T. Tsou, M. Loots and J. Halpern, ibid. 104:623 (1982).

52. R.R. Corderman and J.L. Beauchamp, J. Am. Chem. Soc. 98:3998 (1976); J.S. Uppal and R.H. Staley, ibid. 104:1238 (1982).

53. H. Schumann and J. Müller, Angew, Chem. Int. Ed. Engl. 17:276 (1978).

54. G.K. Barker, M.F. Lappert and J.A.K. Howard, J.C.S. Dalton Trans. 734 (1978).

55. M. Bochmann, G. Wilkinson and G.B. Young, J.C.S. Dalton Trans. 1879 (1980).

56. N. Kuhn and H. Werner, Synth. React. Inorg. Met.-Org. Chem. 8:249 (1978).

57. L. Dahlenburg, V. Sinnwell and D. Thoennes, Chem. Ber. 111:3367 (1978).

58. N.T. Allison, Y. Kawada and W.M. Jones, J. Am. Chem. Soc. 100:5224 (1978).

59. H-F. Klein and H.H. Karsch, Chem. Ber. 108:944 (1975).

60. C.F. Shaw and R.S. Tobias, Inorg. Chem. 12:965 (1973).

61. F. Preuss and L. Ogger, Z. Naturforsch. 37B:957 (1982).

62. B. Sarry and P. Velling, Z. Anorg. Allg. Chem. 500:199 (1983).

63. T.A. Bazhenova, R.M. Lobkovskaya, R.P. Shibaeva, A.K. Shilova, M. Grouselle and E. Deschamps, J. Organomet. Chem. 244:375 (1983).

64. T.A. Bazhenova, R.M. Lobkovskaya, R.P. Shivaeva, A.K. Shilova, A.E. Shilov, M. Grouselle, J. Leny and B. Choubard, Kinet. Katal. 23:246 (1982).

65. G.A. Razuvaev, L.I. Vyshinskaya, G.A. Vasil'eva, A.V. Malysheva and V.P. Mar'in, Inorg. Chim. Acta 44:L285 (1980).

66. H. Lehmkuhl, S. Pasynkiewicz, R. Benn and A. Rufinska, J. Organomet. Chem. 240:C27 (1982); H. Lehmkuhl, C. Naydowski and M. Bellenbaum, ibid. 246:C5 (1983).

67. J.A. Labinger, J. Organomet. Chem. 187:287 (1980).

68. F.W.S. Benfield and M.L.H. Green, J.C.S. Dalton Trans. 1324 (1974).

69. K. Maruyama, T. Ito and A. Yamamoto, Bull. Chem. Soc. Jpn. 52:849 (1979).

70. A. Yamamoto, K. Morifuji, S. Ikeda, T. Saito, Y. Uchida and A. Misono, J. Am. Chem. Soc. 90:1878 (1968).

71. S. Komiya, M. Bundo, T. Yamamoto and A. Yamamoto, J. Organomet. Chem. 174:343 (1979).

72. K. Jonas and G. Wilke, Angew. Chem. Int. Ed. Engl. 8:519 (1969).

73. T. Yamamoto, M. Takamatsu and A. Yamamoto, Bull. Chem. Soc. Jpn. 55:325 (1982).

74. K. Isobe, P.M. Bailey and P.M. Maitlis, J.C.S. Chem. Commun. 808 (1981).

75. K. Isobe, D.G. Andrews, B.E. Mann and P.M. Maitlis, J.C.S. Chem. Commun. 809 (1981).

76. R.F. Heck, J. Am. Chem. Soc. 90:5542 (1968).

77. V.I. Sokolov, V.V. Bashilov and O.A. Reutov, J. Organomet. Chem. 97:299 (1975).

78. O. Rossell, J. Sales and M. Seco, J. Organomet. Chem. 236:415 (1982).

79. C. Eaborn, K.J. Odell and A. Pidcock, J.C.S. Dalton Trans. 357, 1288 (1978); 134, 758 (1979); C. Eaborn, K. Kundu and A. Pidcock, ibid. 933 (1983); and references therein.

80. M.L.H. Green and R.B.A. Pardy, J.C.S. Dalton Trans. 355 (1979).

81. E. Carmona-Guzman and G. Wilkinson, J.C.S. Dalton Trans. 1139 (1978).

82. K-H. Thiele and V. Dieckmann, Z. Anorg. Allg. Chem. 394:293 (1972).

83. C. Santini-Scampucci and J.G. Riecs, Nouveau J. Chim. 4:75 (1980).

84. G.N. Schrauzer, L.A. Hughes, N. Strampach, P.R. Robinson and E.O. Schlemper, Organometallics 1:44 (1982).

85. J. Krausse, G. Marx and G. Schödl, J. Organomet. Chem. 21:159 (1970).

86. E. Kurras and J. Otto, J. Organomet. Chem. 4:114 (1965).

87. J. Otto, Dissertation, Jena, 1966.

88. J. Krausse and G. Schödl, J. Organomet. Chem. 27:59 (1971).

89. F.A. Cotton, B.E. Hanson, W.H. Ilsley and G.W. Rice, Inorg. Chem. 18:2713 (1979).

90. R.A. Jones and G. Wilkinson, J.C.S. Dalton Trans. 1063 (1978).

91. P. Diversi, G. Ingrosso and A. Lucherini, J.C.S. Chem. Commun. 735 (1978).

92. H. Hoberg and W. Richter, J. Organomet. Chem. 195:355 (1980).

93. E. Lindner, G. Funk and S. Hoehne, Angew. Chem. Int. Ed. Engl. 18:535 (1979).

94. M.F. Lappert, C.L. Raston, B.W. Skelton and A.H. White, J.C.S. Chem. Commun. 485 (1981).

95. S.D. Chappell and D.J. Cole-Hamilton, J.C.S. Chem. Commun. 319 (1981); S.D. Chappell, D.J. Cole-Hamilton, A.M.R. Galas and M.B. Hursthouse, J.C.S. Dalton Trans. 1867 (1982).

96. G.E. Coates, M.L.H. Green and K. Wade, "Organometallic Compounds", 4th edn., vol. 1 (Main Group Elements), Chapman and Hall, London (1979).

97. R.A. Andersen, E. Carmona-Guzman, J.F. Gibson and G. Wilkinson, J.C.S. Dalton Trans. 2204 (1976).

98. D.G.H. Ballard and R. Pearce, J.C.S. Chem. Commun. 621 (1975).

99. J. Holton, M.F. Lappert, D.G.H. Ballard, R. Pearce, J.L. Atwood and W.E. Hunter, J.C.S. Dalton Trans. 45, 54 (1979).

100. R.A. Andersen, R.A. Jones and G. Wilkinson, J.C.S. Dalton Trans. 446 (1978).

101. M.B. Hursthouse, K.M.A. Malik and K.D. Sales, J.C.S. Dalton Trans. 1314 (1978).

102. A.F. Masters, K. Mertis, J.F. Gibson and G. Wilkinson, Nouveau J. Chim. 1:389 (1977).

103. P. Edwards, K. Mertis, G. Wilkinson, M.B. Hursthouse and K.M.A. Malik, J.C.S. Dalton Trans. 334 (1980).

104. R.B. King, Adv. Organomet. Chem. 2:157 (1964).

105. R.E. Dessy, R.L. Pohl and R.B. King, J. Am. Chem. Soc. 88:5121 (1966).

106. R.G. Pearson and P.E. Figdore, J. Am. Chem. Soc. 102:1541 (1980).

107. M.Y. Darensbourg, D.J. Darensbourg, D. Burns and D.A. Drew, J. Am. Chem. Soc. 98:3127 (1976).

108. M.Y. Darensbourg, P. Jimenez and J.R. Sackett, J. Organomet. Chem. 202:C68 (1980).

109. J. Engelbrecht, T. Greiser and E. Weiss, J. Organomet. Chem. 204:79 (1981).

110. S. Brandt and P. Helquist, J. Am. Chem. Soc. 101:6473 (1979).

111. A. Sanders, C.V. Magatti and W.P. Giering, J. Am. Chem. Soc. 96:1610 (1974).

112. V. Galamb, G. Palyi, F. Cser, M.G. Furmanova and Yu. T. Struchkov, J. Organomet. Chem. 209:183 (1981).

113. C.E. Sumner, jr., J.A. Collier and R. Pettit, Organometallics 1:1350 (1982).

114. H. Alper, Tetrahedron Lett. 2257 (1975).

115. W.P. Giering, M. Rosenblum and J. Tancrede, J. Am. Chem. Soc. 94:7170 (1972).

116. M. Ephritkhine, B.R. Francis, M.L.H. Green, R.E. Mackenzie and M.J. Smith, J.C.S. Dalton Trans. 1131 (1977).

117. D. Baudry and M. Ephritkhine, J. Organomet. Chem. 195:213 (1980).

118. G.K. Yang and R.G. Bergman, J. Am. Chem. Soc. 105:6500 (1983).

119. H. Felkin and B. Meunier, Nouveau J. Chim. 1:281 (1977).

120. S.H. Goh and L.Y. Goh, J.C.S. Dalton Trans. 1641 (1980).

121. K.H. Theopold and R.G. Bergman, Organometallics 1:1971 (1982); and references therein.

122. P.S. Braterman and R.J. Cross, Chem. Soc. Rev. 2:271 (1973).

123. P.S. Braterman, Top. Curr. Chem. 92:149 (1980).

124. J.K. Stille and K.S,Y. Lau, Acc. Chem. Res. 10:434 (1977).

125. V.I. Sokolov and O.A. Reutov, Coord. Chem. Rev. 27:89 (1978).

126. D. Strope and D.F. Shriver, J. Am. Chem. Soc. 95:8197 (1973).

127. J.L. Peterson, T.E. Nappier, jr. and D.W. Meek, J. Am. Chem. Soc. 95:8195 (1973).

128. R.K. Pomeroy and R.F. Alex, J.C.S. Chem. Commun. 1114 (1980); R.F. Alex and R.K. Pomeroy, Organometallics 1:453 (1982).

129. G. Bellachioma, G. Cardaci and G. Reichenbach, J. Organomet. Chem. 221:291 (1981).

130. J.A. Connor and P.I. Riley, J.C.S. Chem. Commun. 149 (1976).

131. H.H. Karsch, Chem. Ber. 111:1650 (1978).

132. M.P. Brown, R.J. Puddephatt, C.E.E. Upton and S.W. Lavington, J.C.S. Dalton Trans. 1613 (1974).

133. M.P. Brown, R.J. Puddephatt and C.E.E. Upton, J.C.S. Dalton Trans. 2457 (1974).

134. M.H. Chisholm and L.A. Rankel, Inorg. Chem. 16:2177 (1977).

135. M. Onishi, H. Yamamoto and K. Hiraki, Bull. Chem. Soc. Jpn. 51:1856 (1978).

136. R.G. Pearson and A.T. Poulos, Inorg. Chim Acta 34:67 (1979).

137. S. Franks, F.R. Hartley and J.R. Chipperfield, Inorg. Chem. 20:3238 (1981).

138. G.M. Williams, K.I. Gell and J. Schwartz, J. Am. Chem. Soc. 102:3660 (1980).

139. G.M. Williams and J. Schwartz, J. Am. Chem. Soc. 104:1122 (1982).

140. J.K. Kochi and J.W. Powers, J. Am. Chem. Soc. 92:137 (1970).

141. O.J. Scherer and H. Jungmann, J. Organomet. Chem. 208:153 (1981).

142. R.J. Mureinik, M. Weitzberg and J. Blum, Inorg. Chem. 18:915 (1979).

143. J.A. Labinger and J.A. Osborn, Inorg. Chem. 19:3230 (1980).

144. K.S.Y. Lau, R.W. Fries and J.K. Stille, J. Am. Chem. Soc. 96:4983 (1974); K.S.Y. Lau, P.K. Wong and J.K. Stille, ibid. 98:5832 (1976).

145. J.A. Labinger, J.A. Osborn and N.J. Colville, Inorg. Chem. 19:3236 (1980).

146. A.V. Kramer, J.A. Labinger, J.S. Bradley and J.A. Osborn, J. Am. Chem. Soc. 96:7145 (1974).

147. M.J.S. Gynane, M.F. Lappert, S.J. Miles and P.P. Power, J.C.S. Chem. Commun. 192 (1978).

148. J-F. Fauvarque, F.P. Pflüger and M. Troupel, J. Organomet. Chem. 208:419 (1981).

149. A.V. Kramer and J.A. Osborn, J. Am. Chem. Soc. 96:7832 (1974).

150. Y. Becker and J.K. Stille, J. Am. Chem. Soc. 100:838 (1978).

151. T.L. Hall, M.F. Lappert and P.W. Lednor, J.C.S. Dalton Trans. 1448 (1980).

152. T.T. Tsou and J.K. Kochi, J. Am. Chem. Soc. 101:6319 (1979).

153. A. Morville and A. Turco, J. Organomet. Chem. 208:103 (1981).

154. T.T. Tsou and J.K. Kochi, J. Am. Chem. Soc. 101:7547 (1979).

155. A. Morvillo and A. Turco, J. Organomet. Chem. 224:387 (1982).

156. P.K. Monaghan and R.J. Puddephatt, Organometallics 2:1698 (1983).

157. J.P. Fackler, jr. and J.D. Basil, Organometallics 1:871 (1982).

158. D. Milstein and J.C. Calabrese, J. Am. Chem. Soc. 104:3773 (1982).

159. G.E. Hartwell, R.V. Lawrence and M.J. Smas, J.C.S. Chem. Commun. 912 (1970).

160. A.J. Deeming and I.P. Rothwell, J.C.S. Chem. Commun. 344 (1978); idem., J. Organomet. Chem. 205:117 (1981).

161. M.I. Bruce, B.L. Goodall and F.G.A. Stone, J.C.S. Dalton Trans. 687 (1978).

162. D.J. Tune and H. Werner, Helv. Chim. Acta 58:2240 (1975).

163. W.A. Wickramsinghe, P.H. Bird and N. Serpone, J.C.S. Chem. Commun. 1284 (1981); G. Nord, A.C. Hazell, R.G. Hazell and O. Farver, Inorg. Chem. 22:3429 (1983).

164. S.D. Ittel, C.A. Tolman, P.J. Krusic, A.D. English and J.P. Jesson, Inorg. Chem. 17:3432 (1978).

165. R. Crabtree, J.M. Quirk, H. Felkin, T. Fillebeen-Khan and C. Pascard, J. Organomet. Chem. 187:C32 (1980).

166. S. Hietkamp, D.J. Stufkens and K. Vrieze, J. Organomet. Chem. 168:351 (1979).

167. E. Klei and J.H. Teuben, J. Organomet. Chem. 214:53 (1981).

168. C.K. Ingold, J. Chem. Soc. 119:305, 951 (1921); W.H. Perkin, jr. and J.F. Thorpe, J. Chem. Soc. 75:61 (1899).

169. B.L. Shaw, J. Am. Chem. Soc. 97:3856 (1975).

170. H.C. Clark, A.B. Goel and S. Goel, Inorg. Chem. 18:2803 (1979).

171. H.C. Clark, A.B. Goel, R.G. Goel and W.O. Ogini, J. Organomet. Chem. 157:C16 (1978).

172. C. Crocker, R.J. Errington, W.S. McDonald, K.J. Odell, B.L. Shaw and R.J. Goodfellow, J.C.S. Chem. Commun. 498 (1979); C. Crocker, R.J. Errington, R. Markham, C.J. Moulton and B.L. Shaw, J.C.S. Dalton Trans. 387 (1982).

173. P. Foley and G.M. Whitesides, J. Am. Chem. Soc. 101:2732 (1979); P. Foley, R. DiCosimo and G.M. Whitesides, ibid. 102:6713 (1980); J.A. Ibers, R. DiCosimo and G.M. Whitesides, Organometallics 1:13 (1982).

174. R. DiCosimo, S.S. Moore, A.F. Sowinski and G.M. Whitesides, J. Am. Chem. Soc. 104:124 (1982).

175. H.H. Karsch, H-F. Klein and H. Schmidbaur, Chem. Ber. 110:2200 (1977); H.H. Karsch, ibid. 110:2213, 2222 (1977).

176. T.V. Harris, J.W. Rathke and E.L. Muetterties, J. Am. Chem. Soc. 100:6966 (1978).

177. H.H. Karsch, Angew. Chem. Int. Ed. Engl. 21:311 (1982).

178. H. Werner and R. Werner, J. Organomet. Chem. 209:C60 (1981).

179. R. Werner and H. Werner, Angew. Chem. Int. Ed. Engl. 20:793 (1981).

180. S.D. Ittel, C.A. Tolman, A.D. English and J.P. Jesson, J. Am. Chem. Soc. 98:6073 (1976).

181. J. Chatt and J.M. Davidson, J. Chem. Soc. 843 (1975).

182. J.A. Sofranko, R. Eisenberg and J.A. Kampmeier, J. Am. Chem. Soc. 102:1163 (1980).

183. F.A. Cotton, D.L. Hunter and B.A. Frenz, Inorg. Chim. Acta 15:155 (1965).

184. T.H. Tulip and D.L. Thorn, J. Am. Chem. Soc. 103:2448 (1981).

185. C.A. Tolman, S.D. Ittel, A.D. English and J.P. Jesson, J. Am. Chem. Soc. 101:1742 (1979).

186. E. Kurras and U. Rosenthal, J. Organomet. Chem. 160:35 (1978).

187. E. Kurras, U. Rosenthal, H. Mennenga, G. Oehme and G. Engelhardt, Z. Chem. 14:160 (1974).

188. T. Yoshida, T. Matsuda, T. Okano, T. Kitani and S. Otsuka, J. Am. Chem. Soc. 101:2027 (1979).

189. D. Milstein, Organometallics 1:1549 (1982).

190. E.F. Landvatter and T.B. Rauchfuss, Organometallics 1:506 (1982).

191. M.A. Gutierrez, G.R. Newkome and J. Selbin, J. Organomet. Chem. 202:341 (1980).

192.   M. Watanabe, M. Yamamura, I. Moritani, Y. Fujiwara and A. Sonoda,
        Bull. Chem. Soc. Jpn. 47:1035 (1974).

193.   Y. Fujiwara, T. Kawauchi and H. Taniguchi, J.C.S. Chem. Commun.
        220 (1980).

194.   A.H. Janowiez and R.G. Bergman, J. Am. Chem. Soc. 104:352 (1982).

195.   A.H. Janowicz and R.G. Bergman, J. Am. Chem. Soc. 105:3929 (1983).

196.   W.D. Jones and F.J. Feher, Organometallics 2:562 (1983).

197.   J.K. Hoyano, A.D. McMaster and W.A.G. Graham, J. Am. Chem. Soc.
        105:7190 (1983).

198.   A.H. Janowicz, R.A. Periana, J.M. Buchanan, C.A. Kovac, J.M.
        Stryker, M.J. Wax and R.G. Bergman, Pure Appl. Chem. 56:13 (1984).

199.   P.L. Watson, J.C.S. Chem. Commun. 276 (1983); idem, J. Am. Chem.
        Soc. 105:6941 (1983).

200.   J. Chetwynd-Talbot, B. Greberiik and R.N. Perutz, Inorg. Chem.
        21:3647 (1982).

201.   L.F. Halle, P.B. Armentrout and J.L. Beauchamp, Organometallics
        1:963 (1982); R. Houriet, L.F. Halle and J.L. Beauchamp, ibid. 2:1818
        (1983); G.D. Byrd, R.C. Burnier and B.S. Frieser, J. Am. Chem. Soc.
        104:3565 (1982); G.D. Byrd and B.S. Frieser, ibid. p. 5944; J.B.
        Jacobson and B.S. Frieser, ibid. 105:5197, 7484, 7492 (1983).

202.   J. Allison, R.B. Freas and D.P. Ridge, J. Am. Chem. Soc. 101:1332
        (1979); P.B. Armentrout and J.L. Beauchamp, J. Am. Chem. Soc.
        102:1736 (1980).

203.   T.J. Carlin, L. Sallans, C.J. Cassady, D.B. Jacobson and B.S. Frieser,
        J. Am. Chem. Soc. 105:6320 (1983).

204.   R.J. Remick, T.A. Asunta and P.S. Skell, J. Am. Chem. Soc. 101:1320
        (1979).

205.   N. Dominelli and A.C. Oehlschlager, Can. J. Chem. 55:364 (1977).

206.   A. Miyashita, M. Kakahashi and H. Takaya, J. Am. Chem. Soc.
        103:6257 (1981).

207.   P.E. Eaton and U.R. Chakraborty, J. Am. Chem. Soc. 100:3634 (1978),
        and references therein.

208.   P. Eilbracht and P. Dahler, J. Organomet. Chem. 135:C23 (1977).

209.   J.C. Green, M.L.H. Green and C.P. Morley, J. Organomet. Chem.
        233:C4 (1982).

210.   G. Favero, A. Morvillo and A. Turco, Gazz. Chim. Ital. 109:27 (1979).

211.   J.W. Lauher and R. Hoffman, J. Am. Chem. Soc. 98:1729 (1976).

212.   D.L. Thorn and R. Hoffman, J. Am. Chem. Soc. 100:2079 (1978).

213.   A. Dedieu, Inorg. Chem. 20:2803 (1981).

214. T.E. Nalesnik and M. Orchin, J. Organomet. Chem. 222:C5 (1981);
      T.E. Nalesnik, J.H. Freudenberger and M. Orchin, ibid. 236:95 (1982).

215. J.A. Roth, P. Wiseman and L. Ruszala, J. Organomet. Chem. 240:271
      (1982).

216. J. Schwartz and J.A. Labinger, Angew. Chem. Int. Ed. Engl. 15:333
      (1976).

217. M. Yoshifuji, M.J. Loots and J. Schwartz, Tetrahedron Lett. 1303
      (1977).

218. D.B. Carr and J. Schwartz, J. Am. Chem. Soc. 101:3521 (1979).

219. J. Schwartz, M.J. Loots and H. Kosugi, J. Am. Chem. Soc. 102:1333
      (1980).

220. F.N. Tebbe and G.W. Parshall, J. Am. Chem. Soc. 93:3793 (1971).

221. A.H. Klazinga and J.H. Teuben, J. Organomet. Chem. 192:75 (1980).

222. J.A. Labinger and J. Schwartz, J. Am. Chem. Soc. 97:1596 (1975).

223. M.J. Calhorda and A.R. Dias, J. Organomet. Chem. 198:41 (1980).

224. A. Nakamura and S. Otsuka, J. Mol. Catal. 1:285 (1976).

225. R.M. Laine and P.C. Ford, J. Organomet. Chem. 124:29 (1977).

226. M. Green, N.C. Norman and A.G. Orpen, J. Am. Chem. Soc. 103:1267
      (1981).

227. B.L. Booth and R.G. Hargreaves, J. Chem. Soc.(A) 2766 (1969).

228. V.D. Bianco, S. Doronzo and N. Gallo, J. Organomet. Chem. 124:C43
      (1977).

229. T. Mitsudo, Y. Watanabe, M. Yamashita and Y. Takegami, Chem.
      Lett. 1385 (1974).

230. P.M. Treichel and D.C. Molzahn, Inorg. Chim Acta 36:267 (1979).

231. J.A. Roth and M. Orchin, J. Organomet. Chem. 182:299 (1979); T.E.
      Nalesnik and M. Orchin, J. Organomet. Chem.199:265 (1980).

232. H-F. Klein, R. Hammer, J. Gross and U. Schubert, Angew. Chem. Int.
      Ed. Engl. 19:809 (1980).

233. A.S.C. Chan and J. Halpern, J. Am. Chem. Soc. 102:838 (1980).

234. H.M. Dickers, R.N. Haszeldine, A.P. Mather and R.V. Parish, J.
      Organomet. Chem. 161:91 (1978).

235. K. Thomas, J.A. Osborn, A.R. Powell and G. Wilkinson, J. Chem. Soc.
      (A) 1801 (1968).

236. T. Yamamoto, K. Igarashi, S. Komiya and A. Yamamoto, J. Am.
      Chem. Soc. 102:7448 (1980).

237. H.C. Clark and C.R. Milne, J. Organomet. Chem. 161:51 (1978).

238. R.A. Michelin, U. Velluco and R. Ros, Inorg. Chim. Acta 24:L33
      (1977).

239. R. Ros, R.A. Michelin, R. Battaillard and R. Roulet, J. Organomet.
      Chem. 165:107 (1979).

240. D.L. Reger and C.J. Coleman, Inorg. Chem. 18:3155 (1979).

241. T. Bodnar, S.J. LaCroce and A.R. Cutler, J. Am. Chem. Soc. 102:3292
      (1980).

242. A. Rosan, M. Rosenblum and J. Tancrede, J. Am. Chem. Soc. 95:3062
      (1973); P.J. Lennon, A. Rosan, M. Rosenblum, J. Tancrede and P.
      Waterman, ibid. 102:7033 (1980).

243. M. Green, D.M. Grove, J.L. Spencer and F.G.A. Stone, J.C.S. Dalton
      Trans. 2228 (1977).

244. H. Hoberg and D. Schaefer, J. Organomet. Chem. 236:C28 (1982).

245. H. Schmidbaur, G. Blaschke, H.J. Füller and H.P. Scherm, J.
      Organomet. Chem. 160:41 (1978).

246. R.L. Lapinski, H. Yue and R.A. Grey, J. Organomet. Chem. 174:213
      (1979).

247. G. Blaschke, H. Schmidbaur and W.C. Kaska, J. Organomet. Chem.
      182:251 (1979).

248. H. Schmidbaur, J.E. Mandl, W. Richter, V. Bejenke, A. Frank and G.
      Huttner, Chem. Ber. 110:2236 (1977).

249. Y. Yamamoto and Z. Kanda, Bull. Chem. Soc. Jpn. 52:2560 (1979).

250. H. Schmidbaur, H-J. Füller, V. Bejenke, A. Franck and G. Huttner,
      Chem. Ber. 110:3536 (1977).

251. B.L. Booth and R.G. Smith, J. Organomet. Chem. 178:361 (1979);
      idem, ibid. 220:219, 229 (1981).

252. C. Botha, J.R. Moss and S. Pelling, J. Organomet. Chem. 220:C21
      (1981); J.R. Moss and S. Pelling, J. Organomet. Chem. 236:221 (1982).

253. J.R. Moss and J.C. Spiers, J. Organomet. Chem. 182:C20 (1979).

254. S. Al-Jibori, C. Crocker, W.S. McDonald and B.L. Shaw, J.C.S. Dalton
      Trans. 1572 (1981).

255. B. Olgemüller and W. Beck, Angew. Chem. Int. Ed. Engl. 19:834
      (1980).

256. G.D. Vaughn and J.A. Gladysz, J. Am. Chem. Soc. 103:5608 (1981).

257. A.G. Constable and J.A. Gladysz, J. Organomet. Chem. 202:C21
      (1980).

258. C.L. Jenkins and J.K. Kochi, J. Am. Chem. Soc. 94:843 (1972).

259. T. Cohen, R.J. Lewarchik and J.Z. Tarino, J. Am. Chem. Soc.
      96:7753 (1974).

260. M. Freiberg and D. Meierstein, J.C.S. Chem. Commun. 127 (1977).

261. V. Gold and D.L. Wood, J.C.S. Dalton Trans. 2462 (1981).

262. R. Yamashita, K. Kikukawa, F. Wada and T. Matsuda, J. Organomet. Chem. 201:463 (1980).

263. K.A. Azam, A.A. Frew, B.R. Lloyd, Lj. Manojlovic-Muir, K.W. Muir and R.J. Puddephatt, J.C.S. Chem. Commun. 614 (1982).

264. M.C. Baird, J. Organomet Chem. 64:289 (1974).

265. J.R. Norton, Acc. Chem. Res. 12:139 (1979).

266. R.B. Woodward and R. Hoffmann, "The Conservation of Orbital Symmetry", Verlag Chemie, Weinheim (1970).

267. E. Samuel, P. Maillard and C. Giannotti, J. Organomet. Chem. 142:289 (1977).

268. A. Hudson, M.F. Lappert and R. Pichon, J.C.S. Chem. Commun. 374 (1983).

269. D.F. Foust, M.D. Rausch and E. Samuel, J. Organomet. Chem. 193:209 (1980).

270. S.M.B. Costa, A.R. Dias and F.J.S. Pina, J. Organomet. Chem. 175:193 (1979); idem, J.C.S. Dalton Trans. 314 (1981).

271. H.G. Alt, J. Organomet. Chem. 124:167 (1977).

272. R.G. Severson and A. Wojcicki, J. Organomet. Chem. 157:173 (1978).

273. E. Samuel, M.D. Rausch, T.E. Gismondi, E.A. Mintz and C. Giannotti, J. Organomet. Chem. 172:309 (1979).

274. R.B. Hitam, R.H. Hooker, K.A. Mahmoud, R. Narayanaswamy and A.J. Rest, J. Organomet. Chem. 222:C9 (1981).

275. E.A. Mintz and M.D. Rausch, J. Organomet. Chem. 171:345 (1979).

276. H.B. Abrahamson and E. Dennis, J. Organomet. Chem. 201:C19 (1980).

277. H.B. Abrahamson and M.E. Martin, J. Organomet. Chem. 238:C58 (1982).

278. P.W.N.M. van Leeuwen, C.F. Roobeek and R. Huis, J. Organomet. Chem. 142:233 (1977).

279. D.C.L. Perkins, R.J. Puddephatt and C.F.H. Tipper, J. Organomet. Chem. 166:261 (1979).

280. O. Hackelberg and A. Wojcicki, Inorg. Chim Acta 44:L63 (1980).

281. A. Hudson, M.F. Lappert, P.W. Lednor, J.J. MacQuitty and B.K. Nicholson, J.C.S. Dalton Trans. 2159 (1981).

282. D.A. Ryan and J.H. Espenson, J. Am. Chem. Soc. 101:2488 (1979).

283. R.S. Nohr and J.H. Espenson, J. Am. Chem. Soc. 97:3392 (1975).

284. J.H. Espenson, P. Connolly, D. Meyerstein and H. Cohen, Inorg. Chem. 22:1009 (1983).

285. A. Bakac and J.H. Espenson, J. Am. Chem. Soc. 103:2721 (1981); J.H. Espenson and A. Bakac, ibid. 103:2728 (1981); G.W. Kirker, A. Bakac and J.H. Espenson, ibid. 104:1249 (1982).

286. D.A. Ryan and J.H. Espenson, J. Am. Chem. Soc. 104:704 (1982).

287. (a) J.H. Espenson, Prog. Inorg. Chem. 30:189 (1983); (b) J. Halpern, Pure Appl. Chem. 55:1059 (1983).

288. W.N. Rogers, J.A. Page and M.C. Baird, Inorg Chem. 20:3521 (1981).

289. W. Kruse, J. Organomet Chem. 42:C39 (1971).

290. W.A. Mulac, H. Cohen and D. Meyerstein, Inorg. Chem. 21:4016 (1982).

291. K.W. Egger, J. Organomet. Chem. 24:501 (1970).

292. G.M. Whitesides, E.J. Panek and E.R. Stedronsky, J. Am. Chem. Soc. 94:232 (1972).

293. P.B. Brindley and M.J. Scotton, J. Organomet. Chem. 222:89 (1981).

294. E. Dinjus, D. Walther, R. Kirmse and J. Stach, J. Organomet. Chem. 198:215 (1980).

295. N.G. Hargreaves, R.J. Puddephatt, L.H. Sutcliffe and P.J. Thompson, J.C.S. Chem. Commun. 861 (1973).

296. D.J. Cardin, M.F. Lappert and P.W. Lednor, J.C.S. Chem. Commun. 350 (1973).

297. R. Kaptein, P.W.N.M. van Leeuwen and R. Huis, J.C.S. Chem. Commun. 568 (1975).

298. A. Johnson and R.J. Puddephatt, J.C.S. Dalton Trans. 115 (1975).

299. A.E. Stevens and J.L. Beauchamp, J. Am. Chem. Soc. 101:245 (1979).

300. I. Feinstein-Jaffe, S.F. Pedersen and R.R. Schrock, J. Am. Chem. Soc. 105:7176 (1983).

301. U. Zucchini, E. Albizzati and U. Giannini, J. Organomet. Chem. 26:357 (1971).

302. G.A. Razuvaev, V.N. Latyaeva, V.V. Drobotenko, A.N. Linyova, L.I. Vishinskaya and V.K. Cherkasov, J. Organomet. Chem. 131:43 (1977).

303. J.K. Kochi and D. Buchanan, J. Am. Chem. Soc. 87:853 (1965).

304. A.R. Schmidt and T.W. Swaddle, J. Chem. Soc. (A) 1927 (1970).

305. P.G. Edwards, F. Felix, K. Mertis and G. Wilkinson, J.C.S. Dalton Trans. 361 (1979).

306. W.N. Rogers and M.C. Baird, J. Organomet. Chem. 182:C65 (1979).

307. S.N. Anderson, C.J. Cooksey, S.G. Holton and M.D. Johnson, J. Am. Chem. Soc. 102:2312 (1980).

308. N. De Luca and A. Wojcicki, J. Organomet. Chem. 193:359 (1980).

309. H.H. Karsch, Chem. Ber. 110:2712 (1977).

310. M.B. Hursthouse, R.A. Jones, K.M.A. Malik and G. Wilkinson, J. Am. Chem. Soc. 101:4128 (1979).

311. P. Diversi, G. Ingrosso, A. Lucherini, P. Martinelli, M. Benedetti and S. Pucci, J. Organomet. Chem. 165:253 (1979).

312. D.J.A. de Waal, E. Singleton and E. van der Stok, J.C.S. Chem. Commun. 1007 (1978).

313. H-F. Klein, Angew. Chem. Int. Ed. Engl. 19:362 (1980).

314. R. Romeo, D. Minniti, S. Lanza, P. Uguagliati and U. Belluco, Inorg. Chim. Acta 19:L55 (1976); idem., Inorg. Chem. 17:2813 (1978).

315. U. Belluco, M. Giustiniani and M. Graziani, J. Am. Chem. Soc. 89:6494 (1967).

316. R. Romeo, D. Minniti and S. Lanza, J. Organomet. Chem. 165:C36 (1979).

317. R. Ros, R.A. Michelin, R. Bataillard and R. Roulet, J. Organomet. Chem. 161:75 (1978).

318. P. Uguagliati, R.A. Michelin, U. Belluco and R. Ros J. Organomet. Chem. 169:115 (1979).

319. A. Tamaki and J.K. Kochi, J.C.S. Dalton Trans. 2620 (1973).

320. S. Komiya and J.K. Kochi, J. Am. Chem. Soc. 98:7599 (1976).

321. M. Kubota, A. Miyashita, S. Komiya and A. Yamamoto, J. Organomet. Chem. 139:111 (1977).

322. T. Yamamoto, M. Kubota, A. Miyashita and A. Yamamoto, Bull. Chem. Soc. Jpn. 51:1835 (1978).

323. K. Mertis, P.G. Edwards, G. Wilkinson, K.M.A. Malik and M.B. Hursthouse, J.C.S. Chem. Commun. 654 (1980); idem, J.C.S. Dalton Trans. 705 (1981).

324. M.R. Churchill and H.J. Wasserman, J.C.S. Chem. Commun. 274 (1981).

325. K.I. Gell, B. Posin, J. Schwartz and G.M. Williams, J. Am. Chem. Soc. 104:1846 (1982).

326. P. Renaut, G. Tainturier and B. Gautheron, J. Organomet. Chem. 150:C9 (1978).

327. B. Longato, J.R. Norton, J.C. Huffman, J.A. Marsella and K.G. Caulton, J. Am. Chem. Soc 103:209 (1981); J.A. Marsella, J.C. Huffman, K.G. Caulton, B. Longato and J.R. Norton, J. Am. Chem. Soc. 104:6360 (1982).

328. W.D. Jones and R.G. Bergman, J. Am. Chem. Soc. 101:5447 (1979).

329. W.D. Jones, J.M. Huggins and R.G. Bergman, J. Am. Chem. Soc. 103:4415 (1981).

330. M.J. Nappa, R. Santi, S.P. Diefenbach and J. Halpern, J. Am. Chem. Soc. 104:619 (1982).

331. B.R. James and D.K.W. Wang, J.C.S. Chem. Commun. 550 (1977).

332. R.F. Heck in "Organic Syntheses via Metal Carbonyls", ed. I. Wender and P. Pino, Vol. 1, Wiley-Interscience, New York (1968), p. 373; P. Pino, Piacenti and Bianchi, idem. Vol. 2 (1977), p. 233; H. Adkins and G. Krsek, J. Am. Chem. Soc. 70:383 (1948); I. Wender, H.W. Sternberg and M. Orchin, J. Am. Chem. Soc. 75:3041 (1953); N.H. Aldemaroglu, J.C.M. Penninger and E. Oltay, Monatsh. Chem. 107:1153 (1976).

333. G.M. Whitesides, J. San Filippo, Jr., E.R. Stedronsky and C.P. Casey, J. Am. Chem. Soc. 91:6542 (1969).

334. A.H. Janowicz and R.G. Bergman, J. Am. Chem. Soc. 103:2488 (1981).

335. J. Halpern, Acc. Chem. Res. 15:332 (1982).

336. A.C. Balazs, K.H. Johnson and G.M. Whitesides, Inorg. Chem. 21:2162 (1982).

337. D.R. McAlister, D.K. Erwin and J.E. Bercaw, J. Am. Chem. Soc. 100:5966 (1978).

338. K.I. Gell and J. Schwartz, J. Am. Chem. Soc. 100:3246 (1978).

339. K.I. Gell and J. Schwartz, J.C.S. Chem. Commun. 244 (1979); idem., J. Am. Chem. Soc. 103:2687 (1981) and references therein.

340. M. Yoshifuji, K.I. Gell and J. Schwartz, J. Organomet. Chem. 153:C15 (1978).

341. N.J. Cooper, M.L.H. Green and R. Mahtab, J.C.S. Dalton Trans. 1557 (1979).

342. L. Abis, A. Sen and J. Halpern, J. Am. Chem. Soc. 100:2915 (1978).

343. M.P. Brown, S.J. Cooper, A.A. Frew, L. Manojlovic-Muir, K.W. Muir, R.J. Puddephatt and M.A. Thompson, J. Organomet. Chem. 198:C33 (1980); idem., J.C.S. Dalton Trans. 299 (1982).

344. L. Abis, R. Santi and J. Halpern, J. Organomet. Chem. 215:263 (1981).

345. D.J. Cole-Hamilton and G. Wilkinson, J.C.S. Dalton Trans. 797 (1977).

346. B.R. James, L.D. Markham and D.K.W. Wang, J.C.S. Chem. Commun. 439 (1974).

347. D.J. Cole-Hamilton and G. Wilkinson, Nouveau J. Chim. 1:141 (1977).

348. C.S. Cundy, M.F. Lappert and R. Pearce, J. Organomet. Chem. 59:161 (1973).

349. W. Keim, J. Organomet. Chem. 14:179 (1968).

350. H. Masai, K. Sonogashira and N. Hagihara, Bull. Chem. Soc. Jpn. 41:750 (1968).

351. J. Mattia, M.B. Humphrey, R.D. Rogers, J.L. Atwood and M.D. Rausch, Inorg. Chem. 17:3257 (1978).

352. J. Dvorak, R.J. O'Brien and W. Santo, J.C.S. Chem. Commun. 411 (1970).

353. G. Erber and K. Kropp, J. Am. Chem. Soc. 101:3659 (1979).

354. M.D. Rausch and E.A. Mintz, J. Organomet. Chem. 190:65 (1980).

355. C.P. Boekel, J.H. Teuben and H.J. de Liefde Meijer, J. Organomet. Chem. 128:375 (1977).

356. C. McDade, J.C. Green and J.E. Bercaw, Organometallics 1:1629 (1982).

357. T.J. Marks, Adv. Chem. Ser. 150:232 (1976).

358. L.S. Bresler, A.S. Khachaturov and I. Ya. Poddubnyi, J. Organomet. Chem. 64:335 (1974).

359. J.C. Chang and J.H. Espenson, J.C.S. Chem. Commun. 233 (1974).

360. J.H. Espenson and G.J. Samuels, J. Organomet. Chem. 113:143 (1976).

361. J.P. Leslie II and J.H. Espenson, J. Am. Chem. Soc. 98:4839 (1976).

362. W. Seidel and I. Bürger, J. Organomet. Chem. 177:C19 (1979).

363. D. Dong, D.A. Slack and M.C. Baird, Inorg. Chem. 18:188 (1979).

364. D. Dong, B.K. Hunter and M.C. Baird, J.C.S. Chem. Commun. 11 (1978).

365. D. Dong and M.C. Baird, J. Organomet. Chem. 172:467 (1979).

366. G.M. Whitesides and D.J. Boschetto, J. Am. Chem. Soc. 93:1529 (1971).

367. D.A. Slack and M.C. Baird, J. Am. Chem. Soc. 98:5539 (1976).

368. T.C. Flood and F.J. DiSanti, J.C.S. Chem. Commun. 18 (1975).

369. L.J. Dizikes and A. Wojcicki, J. Am. Chem. Soc. 97:2540 (1975).

370. L.J. Dizikes and A. Wojcicki, J. Am. Chem. Soc. 99:5295 (1977).

371. H. Brunner and G. Wallner, Chem. Ber. 109:1053 (1978).

372. T.C. Flood and D.L. Miles, J. Organomet. Chem. 127:33 (1977).

373. T.G. Attig, R.G. Teller, S-M. Wu, R. Bau and A. Wojcicki, J. Am. Chem. Soc. 101:619 (1979).

374. J.E. Bäckvall, B. Åkermark and S.O. Ljunggren, J. Am. Chem. Soc. 101:2411 (1979).

375. J.E. Bäckvall and R.E. Nordberg, J. Am. Chem. Soc. 102:393 (1980).

376. J.K. Jawad and R.J. Puddephatt, J.C.S. Chem. Commun. 892 (1977).

377. J.K. Jawad and R.J. Puddephatt, Inorg. Chim. Acta 31:L391 (1978).

378. J.Y. Chen and J.K. Kochi, J. Am. Chem. Soc. 99:1450 (1977).

379. M.F. Lappert and R. Pearce, J.C.S. Chem. Commun. 24 (1973).

380. J. Halpern, M.S. Chan, J. Hanson, T.S. Roche and J.A. Topich, J. Am. Chem. Soc. 97:1606 (1975).

381. R.H. Magnuson, J. Halpern, I. Ya. Levitin and M.E. Vol'pin, J.C.S. Chem. Commun. 44 (1978).

382. J.I. Davies, C.G. Howard, A.C. Skapski and G. Wilkinson, J.C.S. Chem. Commun. 1077 (1982).

383. Z. Dawoodi, M.L.H. Green, V.S.B. Mtetwa and K. Prout, J.C.S. Chem. Commun. 1410 (1982).

384. K.J. Ivin, J.J. Rooney, C.D. Stewart, M.L.H. Green and R. Mahtab, J.C.S. Chem. Commun. 604 (1978).

385. M.J. Cooper and M.L.H. Green, J.C.S. Chem. Commun. 208, 761 (1974); idem., J.C.S. Dalton Trans. 1121 (1979).

386. M. Canestrari and M.L.H. Green, J.C.S. Chem. Commun. 913 (1979); idem., J.C.S. Dalton Trans. 1789 (1982).

387. R.R. Schrock and P.R. Sharp, J. Am. Chem. Soc. 100:2389 (1978).

388. W.A. Kiel, G-Y. Lin and J.A. Gladysz, J. Am. Chem. Soc. 102:3299 (1980).

389. A. Sanders, L. Cohen, W.P. Giering, D. Kenedy and C.V. Magatti, J. Am. Chem. Soc. 95:5430 (1973).

390. M.L.H. Green and P.L.I. Nagy, Proc. Chem. Soc. 74 (1962).

391. A.R. Cutler, J. Am. Chem. Soc. 101:604 (1979).

392. P.W. Jolly and R. Pettit, J. Am. Chem. Soc. 88:5044 (1966).

393. M. Brookhart and G.O. Nelson, J. Am. Chem. Soc. 99:6099 (1977); M. Brookhart, J.R. Tucker and G.R. Husk, ibid. 105:258 (1983).

394. K.A.M. Kremer, P. Helquist and R.C. Kerber, J. Am. Chem. Soc. 103:1862 (1981).

395. P.W.N.M. van Leeuwen, C.F. Roobeek and R. Huis, J. Organomet. Chem. 142:243 (1977).

396. P. Sevcik, Inorg. Chim. Acta 32:L16 (1979).

397. R. Feser and H. Werner, Angew. Chem. Int. Ed. Engl. 19:940 (1980).

398. R.S. Nohr and L.O. Spreer, Inorg. Chem. 13:1239 (1974).

399. C.E. Castro and W.C. Kray, jr., J. Am. Chem. Soc. 88:4447 (1966).

400. N. Kawabata, I. Kamemura and M. Naka, J. Am. Chem. Soc. 101:2139 (1979).

401. R. Cramer and R.V. Lindsey, jr., J. Am. Chem. Soc. 88:3534 (1966).

402. Z. Dawoodi, M.L.H. Green, V.S.B. Mtetwa and K. Prout, J.C.S. Chem. Commun. 802 (1982).

403. M.H. Chisholm, L-S. Tan and J.C. Huffman, J. Am. Chem. Soc. 104:4879 (1982).

404. S. Datta, M.B. Fischer and S.S. Wreford, J. Organomet. Chem. 88:353 (1980).

405. J.D. Fellmann, G.A. Rupprecht and R.R. Schrock, J. Am. Chem. Soc. 101:5099 (1979); G. Smith, S.J. McLain and R.R. Schrock, J. Organomet. Chem. 202:269 (1980).

406. J.X. McDermott, J.F. White and G.M. Whitesides, J. Am. Chem. Soc. 98:6521 (1976).

407. S.S.M. Ling and R.J. Puddephatt, J.C.S. Chem. Commun. 412 (1982).

408. J. Evans, J. Schwartz and P.W. Urquhart, J. Organomet. Chem. 81:C37 (1974).

409. T. Majima and H. Kurosawa, J.C.S. Chem. Commun. 610 (1977).

410. A.H. Klazinga and J.H. Teuben, J. Organomet. Chem. 194:309 (1980).

411. J. Holton, M.F. Lappert and R. Pearce, cited in 7(b).

412. J.W. Byrne, H.U. Blaser and J.A. Osborn, J. Am. Chem. Soc. 97:3871 (1975).

413. R.J. Kazlauskas and M.S. Wrighton, J. Am. Chem. Soc. 102:1727 (1980).

414. H.G. Alt and M.E. Eichner, Angew. Chem. Int. Ed. Engl. 21:78, suppl. 121 (1982).

415. R.J. Kazlauskas and M.S. Wrighton, J. Am. Chem. Soc. 104:6005 (1982).

416. R.J. Kazlauskas and M.S. Wrighton, Organometallics 1:602 (1982).

417. M. Tamura and J. Kochi, J. Organomet. Chem. 29:111 (1971).

418. D.L. Reger and E.C. Culbertson, J. Am. Chem. Soc. 98:2789 (1976); idem., Inorg. Chem. 16:3104 (1977).

419. M. Bianchi, F. Piacenti, P. Frediani and U. Matteoli, J. Organomet. Chem. 135:387, 137:361 (1977).

420. T. Ikariya and A. Yamamoto, J. Organomet. Chem. 120:257 (1976).

421. T. Yamamoto, T. Saruyama, Y. Nakamura and A. Yamamoto, Bull. Chem. Soc. Jpn. 49:589 (1976).

422. J. Schwartz and J.B. Cannon, J. Am. Chem. Soc. 96:2276 (1974).

423. G.M. Whitesides, J.F. Gaasch and E.R. Stedronsky, J. Am. Chem. Soc. 94:5258 (1972).

424. T.J. McCarthy, R.G. Nuzzo and G.M. Whitesides, J. Am. Chem. Soc. 103:1676, 3396 (1981).

425. F. Ozawa, T. Ito and A. Yamamoto, J. Am. Chem. Soc. 102:6457 (1980).

426. S. Komiya, Y. Morimoto and A. Yamamoto, Organometallics 1:1528 (1982).

427. A. Tamaki and J.K. Kochi, J.C.S. Chem. Commun. 423 (1973).

428. R. Pietropaolo and F. Cusmano, J. Organomet. Chem. 168:363 (1979).

429. J.K. Kochi, D.M. Singleton and L.J. Andrews, Tetrahedron 24:3503 (1968).

430. B. Åkermark and A. Ljungquist, J. Organomet. Chem. 182:47 (1979).

431. P.S. Braterman, J.C.S. Chem. Commun. 70 (1979).

432. G.B. Young and G.M. Whitesides, J. Am. Chem. Soc. 78:5808 (1978).

433. A. Yamamoto, K. Morifuji, S. Ikeda, T. Saito, Y. Uchida and A. Misono, J. Am. Chem. Soc. 87:4652 (1965).

434. W. Lau, J.C. Huffman and J.K. Kochi, Organometallics 1:155 (1982).

435. B. Åkermark, M. Almemark and A. Jutand, Acta Chem. Scand. B36:451 (1982).

436. T.T. Tsou and J.K. Kochi, J. Am. Chem. Soc. 100:1634 (1978).

437. G.M. Whitesides, J. SanFilippo, jr., C.P. Casey and E.J. Panek, J. Am. Chem. Soc. 89:5302 (1967).

438. T. Yamamoto, A. Yamamoto and S. Ikeda, J. Am. Chem. Soc. 93:3350, 3360 (1971).

439. D. Walther, Z. Chem. 17:348 (1977).

440. T. Kohara, T. Yamamoto and A. Yamamoto, J. Organomet. Chem. 192:265 (1980); S. Komiya, Y. Abe, A. Yamamoto and T. Yamamoto, Organometallics 2:1466 (1983).

441. T. Yamamoto, T. Kohara and A. Yamamoto, Chem. Lett. 1217 (1976); idem., Bull. Chem. Soc. Jpn. 54:2161 (1981).

442. R.G. Gastinger, B.B. Anderson and K.J. Klabunde, J. Am. Chem. Soc. 102:4959 (1980).

443. T. Ito, H. Tsuchiya and A. Yamamoto, Bull Chem. Soc. Jpn. 50:1319 (1977).

444. S. Komiya, T.A. Albright, R. Hoffmann and J.K. Kochi, J. Am. Chem. Soc. 99:8440 (1977).

445. K. Tatsumi, R. Hoffmann, A. Yamamoto and J.K. Stille, Bull. Chem. Soc. Jpn. 54:1857 (1981).

446. R.J. McKinney, D.L. Thorn, R. Hoffmann and A. Stockis, J. Am. Chem. Soc. 103:2595 (1981).

447. S. Komiya, T.A. Albright, R. Hoffmann and J.K. Kochi, J. Am. Chem. Soc. 98:7255 (1976); S. Komiya and J.K. Kochi, ibid. 98:7599 (1976).

448. G. Favero, M. Gaddi, A. Morvillo and A. Turco, J. Organomet. Chem. 149:395 (1978); G. Favero, A. Morvillo and A. Turco, J. Organomet. Chem. 162:99 (1978).

449.   G. Calvin and G.E. Coates, J. Chem. Soc. 2008 (1960).

450.   R. DiCosimo and G.M. Whitesides, J. Am. Chem. Soc. 104:3601 (1982).

451.   A. Gillie and J.K. Stille, J. Am. Chem. Soc. 102:4933 (1980).

452.   F. Ozawa, T. Ito, Y. Nakamura and A. Yamamoto, Bull. Chem. Soc. Jpn. 54:1868 (1981).

453.   P.S. Braterman, R.J. Cross and G.B. Young, J.C.S. Dalton Trans. 1306, 1310 (1976), 1892 (1977); U. Bayer and H.A. Brune, Z. Naturforsch. 38B:621 (1983).

454.   M.P. Brown, R.J. Puddephatt and C.E.E. Upton, J. Organomet. Chem. 49:C61 (1973).

455.   R.L. Phillips and R.J. Puddephatt, J.C.S. Dalton Trans. 1732 (1978).

456.   H.E. Bryndza and R.G. Bergman, J. Am. Chem. Soc. 101:4766 (1979).

457.   N.E. Schore, C. Ilender and R.G. Bergman, J. Am. Chem. Soc. 98:7436 (1976).

458.   E.R. Evitt and R.G. Bergman, J. Am. Chem. Soc. 101:3973 (1979), 102:7003 (1980).

459.   H.E. Bryndza, E.R. Evitt and R.G. Bergman, J. Am. Chem. Soc. 102:4948 (1980).

460.   Y. Kubo, A. Yamamoto and S. Ikeda, J. Organomet. Chem. 59:353 (1973).

461.   G.E. Coates and C. Parkin, J. Chem. Soc. 421 (1963).

462.   G.M. Whitesides and C.P. Casey, J. Am. Chem. Soc. 88:4541 (1966).

463.   G. van Koten, J.T.B.H. Jastrzebski and J.G. Noltes, Inorg. Chem. 16:1782 (1977), and references therein.

464.   G. van Koten, R.W.M. ten Hoedt and J.G. Noltes, J. Organomet. Chem. 42:2705 (1977).

465.   H. Hofstee, J. Boersma and G.J.M. van der Kerk, J. Organomet. Chem. 168:241 (1979).

466.   M. Tamura and J.K. Kochi, Bull. Chem. Soc. Jpn. 45:1120 (1972).

467.   E.L. Weinberg and M.C. Baird, J. Organomet. Chem. 179:C61 (1979).

468.   J.D. Ruddick and B.L. Shaw, J. Chem. Soc (A) 2969 (1969).

469.   R.J. Puddephatt and P.J. Thompson, J.C.S. Dalton Trans. 1219 (1977).

470.   R.J. Puddephatt and P.J. Thompson, J. Organomet. Chem. 166:251 (1979).

471.   G. van Koten, J.T.B.H. Jastrzebski and J.G. Noltes, J. Organomet. Chem. 148:317 (1978).

472.   M. Michman, V.R. Kaufman and S. Nussbaum, J. Organomet. Chem. 182:547 (1979).

473. G. van Koten, J.G. Noltes and A.L. Spek, J. Organomet. Chem. 159:441 (1978).

474. P.J. Fagan, J.M. Manriquez, E.A. Maatta, A.M. Seyam and T.J. Marks, J. Am. Chem. Soc. 103:6650 (1981).

475. G.A. Razuvaev, V.N. Latyaeva and A.V. Malysheva, Dokl. Akad. Nauk SSSR 173:1353 (1967); CA 68:39766h (1968).

476. V.N. Latyaeva, A.P. Balatov, A.N. Malysheva and V.I. Kulemin, Zh. Obshch. Khim. 38:280 (1968); CA 69:52257h (1968).

477. M.R. Collier, M.F. Lappert and R. Pearce, J.C.S. Dalton Trans. 445 (1973).

478. P.J. Davidson, M.F. Lappert and R. Pearce, J. Organomet. Chem. 57:269 (1973).

479. A. Jacot-Guillarmod and D. Roulet, Chimia 28:15 (1974).

480. K-H. Thiele, E. Köhler and B. Adler, J. Organomet. Chem. 50:153 (1973).

481. G.A. Razuvaev, V.P. Mar'in and Yu. A. Andrianov, J. Organomet. Chem. 174:67 (1979).

482. C.P. Boekel, J.H. Teuben and H.J. de Liefde Meijer, J. Organomet. Chem. 81:371 (1974).

483. G.J. Erskine, D.A. Wilson and J.D. McCowan, J. Organomet. Chem. 114:119 (1976); G.J. Erskine, J. Hartgerink, E.L. Weinberg and J.D. McCowan, J. Organomet. Chem. 170:51 (1979).

484. H.G. Alt, F.P. di Sanzo, M.D. Rausch and P.C. Uden, J. Organomet. Chem. 107:257 (1976); G.A. Razuvaev, V.P. Mar'in, O.N. Drushkov and L.I. Vyshinskaya, J. Organomet. Chem. 231:125 (1982).

485. P.W.N.M. van Leeuwen, H von der Heijden, C.F. Roobeek and J.H.G. Frijns, J. Organomet. Chem. 209:169 (1981).

486. C.H. Bamford, R.J. Puddephatt and D.M. Slater, J. Organomet. Chem. 159:C31 (1978).

487. C.P. Boekel, J.H. Teuben and H.J. de Liefde Meijer, J. Organomet. Chem. 102:161 (1975).

488. C.P. Boekel, J.H. Teuben and H.J. de Liefde Meijer, J. Organomet. Chem. 102:317 (1975).

489. M.D. Rausch, W.H. Boon and E.A. Mintz, J. Organomet. Chem. 160:81 (1978).

490. M.F. Lappert, T.R. Martin, J.L. Atwood and W.E. Hunter, J.C.S. Chem. Commun. 476 (1980).

491. J.H. Teuben, J. Organomet. Chem. 69:241 (1974).

492. W. Mowat, A. Shortland, G. Yagupsky, N.J. Hill, M. Yagupsky and G. Wilkinson, J. Chem. Soc. (A) 533 (1972).

493. A. Lachowicz and K-H. Thiele, Z. Anorg. Allg. Chem. 431:88 (1977).

494. G.A. Razuvaev, V.N. Latyaeva, L.I. Vyshinskaya, A.N. Linyova, V.V. Drobotenko and V.K. Cherkasov, J. Organomet. Chem. 93:113 (1975).

495. W. Seidel and G. Kreisel, Z. Anorg. Allg. Chem. 426:150 (1976).

496. A.G. Evans, J.C. Evans, D.J.C. Epsley, P.H. Morgan and J. Mortimer, J.C.S. Dalton Trans. 57 (1978).

497. C.P. Boekel, A. Jelsma, J.H. Teuben and H.J. de Liefde Meijer, J. Organomet. Chem. 136:211 (1977).

498. G.A. Razuvaev, V.N. Latyaeva, A.N. Lineva and M.R. Leonov, Dokl. Akad. Nauk SSSR 208:1116 (1973); CA 78:159787u (1973).

499. D.F. Foust and M.D. Rausch, J. Organomet. Chem. 226:47 (1982).

500. A. Yamamoto, Y. Kano and T. Yamamoto, J. Organomet. Chem. 102:57 (1975).

501. S.A. Smirnov, I.A. Oreshkin and B.A. Dolgoplosk, Dokl. Acad. Nauk SSSR 239:1375 (1978); CA 89:24417e (1978).

502. A.J. Shortland and G. Wilkinson, J.C.S. Dalton Trans. 872 (1973).

503. K.W. Chiu, R.A. Jones, G. Wilkinson, A.M.R. Galas, M.B. Hursthouse and K.M.A. Malik, J.C.S. Dalton Trans. 1204 (1981).

504. D. Gregson, J.A.K. Howard, J.N. Nicholls, J.L. Spencer and D.G. Turner, J.C.S. Chem. Commun. 572 (1980).

505. J.K. Kochi and F.F. Rust, J. Am. Chem. Soc. 83:2017 (1961).

506. T. Yamamoto, A. Yamamoto and S. Ikeda, Bull. Chem. Soc. Jpn. 45:1104 (1972).

507. T. Ikariya and A. Yamamoto, J. Organomet. Chem. 118:65 (1976).

508. J. Evans, S.J. Okrasinski, A.J. Pribula and J.R. Norton, J. Am. Chem. Soc. 98:4000 (1976); W.J. Carter, J.W. Kelland, S.J. Okrasinski, K.E. Warner and J.R. Norton, Inorg. Chem. 21:3955 (1982).

509. S.J. Okrasinski and J.R. Norton, J. Am. Chem. Soc. 99:295 (1977).

510. J. Evans, S.J. Okrasinski, A.J. Pribula and J.R. Norton, J. Am. Chem. Soc. 99:5835 (1977).

511. J.W. Kelland and J.R. Norton, J. Organomet. Chem. 149:185 (1978).

512. J. Evans and J.R. Norton, J. Am. Chem. Soc. 96:7577 (1974).

513. E.L. Muetterties and P.L. Watson, J. Am. Chem. Soc. 98:4665 (1976).

514. M.E. Vol'pin, I. Ya. Levitin, A.L. Sigan, J. Halpern and G.M. Tom, Inorg. Chim Acta 41:271 (1980).

515. J. Halpern, M.S. Chan, R.S. Roche and G.M. Tom, Acta Chem. Scand. A33:141 (1979).

516. B. Åkermark and A. Ljundquist, J. Organomet. Chem. 149:97 (1978).

517. A. Emad and M.D. Rausch, J. Organomet. Chem. 191:313 (1980).

518. J.K. Stille and K.S.Y. Lau, J. Am. Chem. Soc. 98:5841 (1976).

519. A. Miyashita, T. Yamamoto and A. Yamamoto, Bull. Chem. Soc. Jpn. 50:1109 (1977).

520. S.H. Bertz and G. Dabbagh, J.C.S. Chem. Commun. 1030 (1982).

521. J.K. Kochi and A. Bemis, J. Am. Chem. Soc. 90:4038 (1968); J.K. Kochi, A. Bemis and C.L. Jenkins, ibid. 90:4616 (1968).

522. G.M. Whitesides, D.E. Bergbreiter and P.E. Kendall, J. Am. Chem. Soc. 96:2806 (1974).

523. A. Tamaki and J.K. Kochi, J. Organomet. Chem. 61:441 (1973).

524. B.K. Bower and H.G. Tennent, J. Am. Chem. Soc. 94:2512 (1972).

525. R.B.A. Pardy, J. Organomet. Chem. 216:C29 (1981).

526. A. Segnitz, E. Kelly, S.N. Taylor and P.M.M. Maitlis, J. Organomet. Chem. 124:113 (1977).

527. J.M. Huggins and R.G. Bergman, J. Am. Chem. Soc. 101:4410 (1979), 103:3002 (1981).

528. A. Alexakis, G. Cahiez and J.F. Normant, J. Organomet. Chem. 177:293 (1977); G. De Chirico, V. Fiandanese, G. Marchese, F. Naso and O. Sciacovelli, J.C.S. Chem. Commun. 523 (1981).

529. S.D. Cox and F. Wundl, Organometallics 2:184 (1983).

530. A. Alexakis and J.F. Normant, Tetrahedron Lett. 23:5151 (1982); A. Alexakis, C. Chuit, M. Commercon-Bougain, J.P.Foulon, N. Jabri, P. Mangeney and J.F. Normant, Pure Appl. Chem. 56:91 (1984).

531. J.C. Hayes, G.D.N. Pearson and N.J. Cooper, J. Am. Chem. Soc. 103:4648 (1981).

532. E. Negishi, Acc. Chem. Res. 15:340 (1982).

533. R.F. Heck, Org. React. (NY) 27:345 (1982).

534. M.T. Reetz, B. Wenderoth, R. Peter, R. Steinbach and J. Westermann, J.C.S. Chem. Commun. 1202 (1980); M.T. Reetz, J. Westermann and R. Steinbach, Angew. Chem. Int. Ed. Engl. 19:900 (1980); idem., J.C.S. Chem. Commun. 237 (1981).

535. F.M. Dayrit, D.E. Gladkowski and J. Schwartz, J. Am. Chem. Soc. 102:3976 (1980).

536. J.L. Fabre, M. Julia, B. Mansour and L. Saussine, J. Organomet. Chem. 177:221 (1979).

537. M. Tamura and J. Kochi, J. Organomet. Chem. 31:289 (1971).

538. B.W. Roberts, M. Ross and J. Wong, J.C.S. Chem. Commun. 428 (1980).

539. R. Cramer, J. Am. Chem. Soc. 87:4747 (1965).

540. S.E. Diamond, A. Szalkiewicz and F. Mares, J. Am. Chem. Soc. 101:490 (1979).

541. K. Maruyama, T. Ito and A. Yamamoto, J. Organomet. Chem. 155:359 (1978).

542. D.G. Morrell and J.K. Kochi, J. Am. Chem. Soc. 97:7262 (1975); G. Smith and J.K. Kochi, J. Organomet. Chem. 198:199 (1980).

543. E. Ibuki, S. Ozasa, Y. Fujioka, M. Okada and K. Terada, Bull. Chem. Soc. Jpn. 53:821 (1980).

544. J.F. Fauvarque and A. Jutand, J. Organomet. Chem. 177:273 (1979).

545. G. Consiglio, O. Piccolo and F. Morandini, J. Organomet. Chem. 177:C13 (1979).

546. H. Brunner and M. Pröbster, J. Organomet. Chem. 209:C1 (1981).

547. T. Hayashi, M. Konishi, M. Fukushira, T. Mise, M. Kagotani, M. Tajika and M. Kumada, J. Am. Chem. Soc. 104:180 (1982) and references therein.

548. M. Almemark and B. Åkermark, J.C.S. Chem. Commun. 66 (1978).

549. M. Troupel, Y. Rollin, S. Sibille, J. Perichon and J-F. Fauvarque, J. Organomet. Chem. 202:435 (1980).

550. R.F. Heck, Acc. Chem. Res. 12:146 (1979).

551. R.F. Heck, J. Am. Chem. Soc. 90:5538 (1968).

552. R.S. Shue, J. Catal. 26:112 (1972).

553. N-T. Luong-Thi and H. Riviere, J.C.S. Chem. Commun. 918 (1978).

554. T. Spencer and F.G. Thorpe, J. Organomet. Chem. 99:C8 (1975).

555. K. Kikukawa and T. Matsuda, Chem. Lett. 159 (1977); K. Kikukawa, K. Nagira, N. Terao, F. Wada and T. Matsuda, Bull. Chem. Soc. Jpn. 52:2609 (1979).

556. H-U. Blaser and A. Spencer, J. Organomet. Chem. 233:267 (1982); A. Spencer, ibid. 240:209 (1982).

557. H.A. Dieck and R.F. Heck, J. Am. Chem. Soc. 96:1133 (1974).

558. T. Izumi, K. Endo, O. Saito, I. Shimizu, M. Maemura and A. Kasahara, Bull. Chem. Soc. Jpn. 51:663 (1978).

559. T. Itahara, J.C.S. Chem. Commun. 859 (1981).

560. S.I. Murahashi, M. Yamamura and N. Mita, J. Organomet. Chem. 42:2870 (1977).

561. J-F. Fauvarque and A. Jutand, J. Organomet. Chem. 177:273 (1979), 209:109 (1981).

562. M.K. Loar and J.K. Stille, J. Am. Chem. Soc. 103:4174 (1981).

563. D. Milstein and J.K. Stille, J. Am. Chem. Soc. 101:4981, 4992 (1979).

564. H. Westmijze, A.V.E. George and P. Vermeer, Recl. J.R. Neth. Chem. Soc. 102:322 (1983).

565. E.C. Ashby and A.B. Goel, J. Org. Chem. 48:2125 (1983).

566. B.H. Lipshutz, J.A. Kozlowski and R.S. Wilhelm, J. Org. Chem. 48:546 (1983).

567. G.M. Whitesides, W.F. Fischer, jr., J. San Filippo, jr., R.W. Bashe and H.O. House, J. Am. Chem. Soc. 91:4871 (1969).

568. M. Tamura and J.K. Kochi, J. Organomet. Chem. 42:205 (1972).

569. R. Mathias and P. Weyerstahl, Chem. Ber. 112:3041 (1979).

570. T. Hiyama, H. Yamamoto, K. Nishio, K. Kitatani and H. Nozaki, Bull. Chem. Soc. Jpn. 52:3632 (1979).

571. E.C. Ashby and J.J. Lin, J. Org. Chem. 42:2805 (1977).

572. B.H. Lipschutz, R.S. Wilhelm and D.M. Floyd, J. Am. Chem. Soc. 103:7672 (1981); B.H. Lipschutz, J. Kozlowski and R.S. Wilhelm, ibid. 104:2305 (1982).

573. G.H. Posner, Org. React. 19:1 (1972); E. Singleton, J. Organomet. Chem. 158:413 (1978); L.S. Hegedus, ibid. 207:185 (1981); and references therein.

574. H.O. House, W.L. Respess and G.M. Whitesides, J. Org. Chem. 31:3128 (1966).

575. G.M. Whitesides and P.E. Kendall, J. Org. Chem. 37:3718 (1972).

576. J. Drouin, F. Leyendecker and J-M. Conia, Nouveau J. Chim. 2:267 (1978); F. Leyendecker, J. Drouin and J-M. Conia, ibid. 2:271 (1978).

577. Y. Yamamoto and K. Maruyama, J. Am. Chem. Soc. 100:3240 (1978); Y. Yamamoto, S. Yamamoto, H. Yatagai and K. Maruyama, ibid. 102:2318 (1980).

578. M. Tamura and J.K. Kochi, J. Am. Chem. Soc. 93:1483 (1971).

579. A. Wojcicki, Adv. Organomet. Chem. 12:31 (1974).

580. A. Dormond, C. Moïse, A. Dahchour and J. Tirouflet, J. Organomet. Chem. 177:181 (1979); A. Dormond, C. Moïse, A. Dahchour, J.C. Leblanc and J. Tirouflet, ibid. 177:191 (1979).

581. J.A. Labinger, D.W. Hart, W.E. Seibert III and J. Schwartz, J. Am. Chem. Soc. 97:3851 (1975).

582. S.E. Jacobson, P. Reich-Rohrwig and A. Wojcicki, Inorg. Chem. 12:717 (1973); S.E. Jacobson and A. Wojcicki, J. Organomet. Chem. 72:113 (1974).

583. S.E. Jacobson and A. Wojcicki, J. Am. Chem. Soc. 95:6962 (1973); idem., Inorg. Chim Acta 10:229 (1974).

584. D. Dong, D.A. Slack and M.C. Baird, J. Organomet. Chem. 153:219 (1978); P.L. Bock, D.J. Boschetto, J.R. Rasmussen, J.P. Demers and G.M. Whitesides, J. Am. Chem. Soc. 96:2814 (1974); T.C. Flood and D.L. Miles, ibid. 95:6460 (1973); S.L. Miles, D.L. Miles, R. Bau and T.C. Flood, ibid. 100:7278 (1978).

585. R.J. Puddephatt and M.A. Stalteri, J. Organomet. Chem. 193:C27 (1980).

586. T. Ito, T. Ono, K. Maruyama and A. Yamamoto, Bull. Chem. Soc. Jpn. 55:2212 (1982).

587. H. Kletzin, H. Werner, O. Serhadli and M.L. Ziegler, Angew. Chem. Int. Ed. Engl. 22:46(1983).

588. J.C. Hayes and N.J. Cooper, J. Am. Chem. Soc. 104:5570 (1982).

589. M.H. Chisholm and W.S. Johns, Inorg. Chem. 14:1189 (1975).

590. T. Tsuda, Y. Chujo and T. Saegusa, J.C.S. Chem Commun. 963 (1975); idem., J. Am. Chem. Soc. 100:630 (1978).

591. S.R. Su and A. Wojcicki, Inorg. Chem. 14:89 (1975).

592. J.D. Cotton and G.A. Morris, J. Organomet. Chem. 145:245 (1978).

593. J.D. Cotton, J. Organomet. Chem. 159:465 (1978).

594. M. Rosenblum, Acc. Chem. Res. 7:122 (1974).

595. M.L.H. Green and P.L.I. Nagy, J. Chem. Soc. 189 (1963).

596. W.P. Giering and M. Rosenblum, J. Organomet. Chem. 25:C71 (1970).

597. J.K.P. Ariyatne and M.L.H. Green, J. Chem. Soc. 1 (1964).

598. L.J. Dizikes and A. Wojcicki, Inorg. Chim Acta 20:L29 (1976), J. Organomet. Chem. 137:79 (1977).

599. L.S. Chen, S.R. Su and A. Wojcicki, Inorg. Chim. Acta 27:79 (1978); R.L. Downs and A. Wojcicki, ibid. 27:91 (1978); D.A. Ross and A. Wojcicki, ibid. 28:59 (1978); J.O. Kroll and A. Wojcicki, J. Organomet. Chem. 66:95 (1974).

600. A. Cutler, D. Ehntholt, W.P. Giering, P. Lennon, S. Raghu, A. Rosan, M. Rosenblum, J. Tancrede and D. Wells, J. Am. Chem. Soc. 98:3495 (1975); S. Raghu and M. Rosenblum, ibid. 95:3060 (1973).

601. J.P. Williams and A. Wojcicki, Inorg. Chem. 16:2506, 3116 (1977); L.S. Chen, D.W. Lichtenberg, P.W. Robinson, Y. Yamamoto and A. Wojcicki, Inorg. Chim. Acta 25:165 (1977).

602. C. Charrier, J. Collin, J.Y. Merour and J.L. Roustan, J. Organomet. Chem. 162:57(1978).

603. A. Davison and J.P. Solar, J. Organomet. Chem. 155:C8 (1978), 166:C13 (1979).

604. A. Bury, M.D. Johnson and M.J. Stewart, J.C.S. Chem. Commun. 622 (1980).

605. W.P. Giering and M. Rosenblum, J. Am. Chem. Soc. 93:5299 (1971); A. Cutler, R.W. Fish, W.P. Giering and M. Rosenblum, 94:4354 (1972).

606. F.J. McQuillin, "Homogeneous Hydrogenation in Organic Chemistry", D. Reidel, Dortrecht (1978); A.J. Birch and D.H. Williamson, Org. React. 24:1 (1976).

607. S.D. Ittel and J.A. Ibers, Adv. Organomet. Chem. 14:33 (1976).

608. I. Omae, Angew. Chem. Int. Ed. Engl. 21:889 (1982).

609. K. Jonas and C. Krüger, Angew. Chem. Int. Ed. Engl. 19:520 (1980).

610. F.A. Cotton and G. Wilkinson, "Advancced Inorganic Chemistry", 4th ed., Wiley-Interscience, New York (1980), Fig. 3.10; M.J.S. Dewar, Bull. Soc. Chim. Fr. C71 (1951); for an account of the history of the concept, see ref. 611.

611. M.J.S. Dewar and G.P. Ford, J. Am. Chem. Soc. 101:783 (1979).

612. G. Reichenbach, G. Cardaci and G.G. Aloisi, J. Organomet. Chem. 134:47 (1977).

613. O. Eisenstein and R. Hoffmann, J. Am. Chem. Soc. 102:6148 (1980), 103:4308 (1981).

614. T.C.T. Chang, B.M. Foxman, M. Rosenblum and C. Stockman, J. Am. Chem. Soc. 103:7361 (1981).

615. N. Rösch and R. Hoffmann, Inorg. Chem. 13:2656 (1974); T.A. Albright, R. Hoffmann, L.C. Thibeault and D.L. Thorn, J. Am. Chem. Soc. 101:3801 (1979).

616. R.M. Pitzer and H.F. Schaefer III, J. Am. Chem. Soc. 101:7176 (1979).

617. B.E.R. Schilling, R. Hoffmann and D.L. Lichtenberger, J. Am. Chem. Soc. 101:585 (1979).

618. W.C. Zeise, Ann. Phys. (Leipzig) [2]9:632 (1827).

619. E.O. Fischer and K. Fichtel, Chem. Ber. 94:1200 (1961).

620. E.O. Fischer and K. Fichtel, Chem. Ber. 95:2063 (1962).

621. A. Musca, I. Georgii and B.E. Mann, Inorg. Chim. Acta 45:L149 (1980).

622. D.J. Bates, M. Rosenblum and S.B. Samuels, J. Organomet. Chem. 209:C55 (1981); S.B. Samuels, S.R. Berryhill and M. Rosenblum, J. Organomet. Chem. 166:C9 (1979).

623. D.F. Marten, J.C.S. Chem. Commun. 341 (1980).

624. L. Cosslett and L.A.P. Kane-Maguire, J. Organomet. Chem. 178:C17 (1979).

625. M. Stephenson and R.J. Mawby, J.C.S. Dalton Trans. 2112 (1981).

626. D.L. Reger and E.C. Culbertson, J. Organomet. Chem. 131:297 (1977).

627. A. Nakamura and S. Otsuka, J. Am. Chem. Soc. 95:7262 (1973).

628. M.G. Burnett and C.J. Strugnell, J. Chem. Res. S250, M2935 (1977).

629.   D.J. Cole-Hamilton and G. Wilkinson, J.C.S. Chem. Commun. 59 (1977).

630.   W.E. McCormack and M. Orchin, J. Organomet. Chem. 129:127 (1977).

631.   J. Halpern and R.A. Jewsbury, J. Organomet. Chem. 181:223 (1979).

632.   W. Wong, S.J. Singer, W.D. Pitts, S.F. Watkins and W.H. Baddley, J.C.S. Chem. Commun. 677 (1972); J.P. Visser and J.E. Ramakers-Blom, J. Organomet. Chem. 44:C63 (1972); W.E. Carroll, M. Green, J.A.K. Howard, M. Pfeffer and F.G.A. Stone, J.C.S. Dalton Trans. 1472 (1978).

633.   R.L. Phillips and R.J. Puddephatt, J.C.S. Dalton Trans. 1736 (1978).

634.   A.R. Pinhas and B.K. Carpenter, J.C.S. Chem. Commun. 15, 17 (1980).

635.   J. Burgess, M.M. Hunt and R.D.W. Kemmitt, J. Organomet. Chem. 134:131 (1977).

636.   S. Otsuka and A. Nakamura, Adv. Organomet. Chem. 14:245 (1976).

637.   F.L. Bowden and R. Giles, Coord. Chem. Rev. 20:81 (1976).

638.   A. De Renzi, B. Di Blasio, A. Panunzi, C. Pedone and A. Vitagliano, J.C.S. Dalton Trans. 1392 (1978).

639.   A. De Renzi, P. Ganis, A. Panunzi, A. Vitagliano and G. Valle, J. Am. Chem. Soc. 102:1722 (1980).

640.   M. Englert, P.W. Jolly and G. Wilke, Angew. Chem. Int. Ed. Engl. 11:136 (1972).

641.   G.K. Barker, M. Green, J.A.K. Howard, J.L. Spencer and F.G.A. Stone, J.C.S. Dalton Trans. 1839 (1978).

642.   L.S. Hegedus, N. Kambe, R. Tamura and P.D. Woodgate, Organometallics 2:1658 (1983).

643.   M.F. Semmelhack, Org. React. 19:115 (1972).

644.   R. Baker, Chem. Rev. 73:487 (1973).

645.   L.S. Hegedus, J. Organomet. Chem. Library 1:329 (1976).

646.   P. Heimbach and H. Schenkluhn, Top. Curr. Chem. 92:45 (1980).

647.   E. Negishi, Pure Appl. Chem. 53:2333 (1981); B.M. Trost, ibid. 53:2357 (1981); J. Tsuji, ibid. 53:2371 (1981).

648.   M.L.H. Green and R.L.I. Nagy, Adv. Organomet. Chem. 2:325 (1964).

649.   A.N. Nesmeyanov and A.Z. Rubezhov, J. Organomet. Chem. 164:259 (1979).

650.   E.O. Fischer and G. Bürger, Z. Naturforsch. B. 16:702 (1961).

651.   T. Yamamoto, J. Ishizu and A. Yamamoto, J. Am. Chem. Soc. 103:6863 (1981).

652. H. Kurosawa, J.C.S. Dalton Trans. 939 (1979).

653. T. Yamamoto, O. Saito and A. Yamamoto, J. Am. Chem. Soc. 103:5600 (1981).

654. A.M. Rosan, J.C.S. Chem. Commun. 311 (1981).

655. M. Ziegler and R.K. Sheline, Inorg. Chem. 4:1230 (1965); J.W. Faller and A.M. Rosan, Ann. N.Y. Acad. Sci. 295:186 (1977).

656. S.D. Ittel, F.A. Van-Catledge, C.A. Tolman and J.P. Jesson, J. Am. Chem. Soc. 100:1317 (1978).

657. H-O. Stühler and J. Müller, Chem. Ber. 112:1359 (1979).

658. J. Müller, W. Hähnlein, H. Menig and J. Pickardt, J. Organomet. Chem. 197:95 (1980).

659. J. Müller, H. Menig and P.V. Rinze, J. Organomet. Chem. 181:387 (1979).

660. D.J. Mabbott and P.M. Maitlis, J. Organomet. Chem. 102:C34 (1975).

661. O.W. Howarth, C.H. McAteer, P. Moore and G.E. Morris, J.C.S. Chem. Commun. 506 (1981).

662. Y. Tamaru, M. Kagotani and Z. Yoshida, J.C.S. Chem. Commun. 367 (1978).

663. W. Lipps and C.G. Kreiter, J. Organomet. Chem. 241:185 (1983).

664. B.N. Chaudret, D.J. Cole-Hamilton and G. Wilkinson, J.C.S. Dalton Trans. 1739 (1978).

665. D.J. Cole-Hamilton and G. Wilkinson, J.C.S. Chem. Commun. 883 (1978).

666. E.O. Sherman, jr. and P.R. Schreiner, J.C.S. Chem. Commun. 223 (1978).

667. E.O. Sherman, jr. and M. Olson, J. Organomet. Chem. 172:C13 (1979).

668. T.H. Tulip and J.A. Ibers, J. Am. Chem. Soc. 100:3252 (1978), 101:4201 (1979).

669. J.Y. Satoh and C.A. Horiuchi, Bull. Chem. Soc. Jpn. 52:2653 (1979).

670. I.J. Harvie and F.J. McQuillin, J.C.S. Chem. Commun. 747 (1978).

671. B.M. Trost and P.J. Metzner, J. Am. Chem. Soc. 102:3572 (1980).

672. K.H. Henderson and F.J. McQuillin, J.C.S. Chem. Commun. 15 (1978).

673. R.P. Hughes, D.E. Hunton and K. Schumann, J. Organomet. Chem. 169:C37 (1979); B.K. Dallas and R.P. Hughes, ibid. 184:C67 (1980); T.A. Albright, P.R. Clemens, R.P. Hughes, D.E. Hunton and L.M. Margerum, J. Am. Chem. Soc. 104:5369 (1982).

674. T. Mitsudo, Y. Watanabe, H. Nakanishi, I. Morishima, I. Inubushi and Y. Takegami, J.C.S. Dalton Trans. 1298 (1978).

675. G.D. Annis and S.V. Ley, J.C.S. Chem. Commun. 581 (1977).

676. K. Vrieze in "Dynamic NMR Spectroscopy", ed. L.M. Jackman and F.A. Cotton, Academic Press, New York (1975), p. 441.

677. H. Kurosawa and S. Numata, J. Organomet. Chem. 175:143 (1979).

678. G. Carturan, A. Scrivanti, U. Belluco and F. Morandini, Inorg. Chim. Acta 26:1 (1978), 27:37 (1978).

679. H. Werner and A. Kühn, Angew. Chem. Int. Ed. Engl. 18:416 (1979).

680. N.N. Druz, V.I. Klepikova, M.I. Lobach and V.A. Kormer, J. Organomet. Chem. 162:343 (1978).

681. V. Galamb and G. Palyi, J.C.S. Chem. Commun. 487 (1982).

682. H. Bönnemann, Angew. Chem. Int. Ed. Engl. 9:736 (1970).

683. G. Carturan, A. Scrivanti and F. Morandini, Angew. Chem. Int. Ed. Engl. 20:112 (1981).

684. J.R. Bleeke and E.L. Muetterties, J. Am. Chem. Soc. 103:556 (1981).

685. B.C. Benson, R. Jackson, K.K. Joshi and D.T. Thompson, J.C.S. Chem. Commun. 1506 (1968).

686. E. Keller and H. Vahrenkamp, Z. Naturforsch. B. 33:537 (1978).

687. M. Ephritkhine, M.L.H. Green and R.E. MacKenzie, J.C.S. Chem. Commun. 619 (1976).

688. G. Cardaci, J. Organomet. Chem. 202:C81 (1981).

689. W.R. Jackson and J.U. Strauss, Austral. J. Chem. 30:553 (1977).

690. B.M. Trost and numerous others, J. Am. Chem. Soc. 100:3407, 3416, 3426, 3435 (1978).

691. S. Sasaki, M. Nishikawa and A. Ohyoshi, J. Am. Chem. Soc. 102:4062 (1980).

692. B.M. Trost and T.P. Klun, J. Am. Chem. Soc. 101:6756 (1979).

693. B.M. Trost, T.A. Runge and L.N. Jungheim, J. Am. Chem. Soc. 102:2840 (1980).

694. B.M. Trost and T.R. Verhoeven, J. Am. Chem. Soc. 101:1595 (1979), 102:4743 (1980).

695. T. Hayashi, K. Kanehira, H. Tsuchiya and M. Kumada, J.C.S. Chem. Commun. 1162 (1982).

696. J.S. Temple and J. Schwartz, J. Am. Chem. Soc. 102:7381 (1980); J.S. Temple, M. Riedeker and J. Schwartz, ibid. 104:1310 (1982).

697. Y. Takahashi, S. Sakai and Y. Ishii, J.C.S. Chem. Commun. 1092 (1967).

698. J-E. Bäckvall and R.E. Nordberg, J. Am. Chem. Soc. 103:4959 (1981).

699. B. Åkermark and A. Jutand, J. Organomet. Chem. 217:C41 (1981).

700.   B. Åkermark, G. Åkermark, L.S. Hegedus and K. Zetterberg, J. Am. Chem. Soc. 103:3037 (1981).

701.   T. Hayashi, M. Konishi, K. Yokota and M. Kumada, J.C.S. Chem. Commun. 313 (1981).

702.   B.M. Trost and G.A. Molander, J. Am. Chem. Soc. 103:5969 (1981).

703.   R. Baker, A.H. Cook and M.J. Crimmin, J.C.S. Chem. Commun. 727 (1975).

704.   L.S. Hegedus and L.L. Miller, J. Am. Chem. Soc. 97:459 (1975).

705.   L.S. Hegedus and S. Varaprath, Organometallics 1:259 (1982).

706.   D. Seebach, A.K. Beck, M. Scheiss, L. Widler and A. Wonnacott, Pure Appl. Chem. 55:1807 (1983).

707.   G. Wilke et al., Angew. Chem. Int. Ed. Engl. 2:105 (1963), 5:151 (1966); F. Brille, P. Heimbach, J. Kluth and H. Schenkluhn, ibid. 18:400 (1979); P. Heimbach, J. Kluth, H. Schenkluhn and B. Weimann, ibid. 19:569, 570 (1980); H. Buchholz, P. Heimbach, H-J. Hey, H. Selbeck and W. Wiese, Coord. Chem. Rev. 8:129 (1972).

708.   R. Benn and G. Wilke, J. Organomet. Chem. 174:C38 (1979).

709.   J. Muzart and J-P. Pete, J.C.S. Chem. Commun. 257 (1980); J. Muzart, P. Pale and J-P. Pete, ibid. 668 (1981).

710.   M.P. Crozet, J. Muzart, P. Pale and P. Tordo, J. Organomet. Chem. 244:191 (1983).

711.   B. Klei, J.H. Teuben and H.J. de Liefde Meijer, J.C.S. Chem. Commun. 342 (1981); E. Klei, J.H. Teuben, H.J. de Liefde Meijer, E.J. Kwak and A.P. Briuns, J. Organomet. Chem. 224:327 (1982).

712.   F. Sato, S. Iijima and M. Sato, J.C.S. Chem. Commun. 180 (1981).

713.   P.W. Jolly, S. Stobbe, G. Wilke, R. Goddard, C. Krüger, J.C. Sekutowski and Y-H. Tsay, Angew. Chem. Int. Ed. Engl. 17:124 (1978).

714.   M. Catellani, G.P. Chiusoli, E. Dradi and G. Salerno, J. Organomet. Chem. 177:C29 (1979).

715.   T. Anderson, E.L. Jenner and R.V. Lindsey, jr., J. Am. Chem. Soc. 87:5638 (1965).

716.   R. Cramer, J. Am. Chem. Soc. 89:1633 (1967).

717.   R. Baker and A.H. Copeland, Tetrahedron Lett. 4535 (1976).

718.   J. Kiji, Y. Miura and J. Furukawa, J. Organomet. Chem. 140:317 (1977).

719.   J.P. Neilan, R.M. Laine, N. Cortese and R.F. Heck, J. Org. Chem. 41:3455 (1966).

720.   Yu. I. Ermakov, J. Mol Catal. 21:35 (1983).

721.   J.M. Basset and A. Choplin, J. Mol Catal. 21:95 (1983).

722. W.J. Evans, J. Organomet Chem. 250:217 (1983); H. Schumann, Comments Inorg. Chem. 2:247 (1983); H. Schumann, Angew. Chem. Int. Ed. Engl. 23:474 (1984).

723. P.J. Fagan, E.A. Maatta, J.M. Manriquez, K.G. Moloy, A.M. Seyam and T.J. Marks, Actinides Perspect., Proc. Actinides Conf. 433 (1981).

724. V. Galamb and G. Palyi, Coord. Chem. Rev. 59:203 (1984).

725. R. Nast, Coord. Chem. Rev. 47:89 (1982).

726. R.F. Jordan and J.R. Norton, ACS Symp. Ser. 198:403 (1982).

727. B. Åkermark, J-E. Bäckvall and K. Zetterberg, Acta Chem. Scand. B36:577 (1982); U. Belluco, R.A. Michelin, P. Uguagliati and B. Crociani, J. Organomet. Chem. 250:565 (1983).

728. I.P. Beletskaya, J. Organomet. Chem. 250:551 (1983).

729. "Homogeneous Catalysis with Metal Phosphine Complexes", L.H. Pignolet, ed., Plenum Press, New York (1983).

730. "Organometallic Compounds Synthesis, Structure and Theory", B.L. Shapiro, ed., Texas A & M University Press, College Station, Texas (1983).

731. D.C. Sonnenberg, L.R. Morss and T.J. Marks, Organometallics 4:352 (1985).

732. G.S. Girolami, G. Wilkinson, M. Thornton-Pett and M.B. Hursthouse, J. Am. Chem. Soc. 105:6752 (1983).

733. C.G. Howard, G.S. Girolami, G. Wilkinson, M. Thornton-Pett and M.B. Hursthouse, J.C.S. Chem. Commun. 1163 (1983).

734. R.P. Tooze, M. Motevalli, M.B. Hursthouse and G. Wilkinson, J.C.S. Chem. Commun. 799 (1984).

735. A. Vasquez de Miguel and P.M. Maitlis, J. Organomet. Chem. 244:C35 (1983).

736. H. Schumann, H. Lauke, E. Hahn and J. Pickardt, J. Organomet. Chem. 263:29 (1984); H. Schumann, H. Lauke, E. Hahn, M.J. Heeg and D. van der Helm, Organometallics 4:321 (1985).

737. R.M. Waymouth, B.D. Santarsiero and R.H. Grubbs, J. Am. Chem. Soc. 106:4050 (1984).

738. P.L. Watson, J.C.S. Chem. Commun. 276 (1983); idem, J. Am. Chem. Soc. 105:6491 (1983).

739. L.A. Kushch, V.V. Lavrushko, Yu. S. Misharin, A.P. Moravskii and A.E. Shilov, Nouveau J. Chim. 7:729 (1983).

740. H. Schmidbaur, Angew. Chem. Int. Ed. Engl. 22:907 (1983).

741. A.G. Abatjoglou and D.R. Bryant, Organometallics 3:932 (1984); A.G. Abatjoglou, E. Billig and D.R. Bryant, ibid. 3:923 (1984); R.A. Dubois, P.E. Garriou, K.D. Lavin and H.R. Allcock, Organometallics 3:649 (1984).

742.   H.G. Alt, Angew. Chem. Int. Ed. Engl. 23:766 (1984).

743.   M.J. Sisley, W. Rindermann, R. van Eldik and T.W. Swaddle, J. Am. Chem. Soc. 106:7432 (1984).

744.   C.D. Hoff, F. Ungváry, R.B. King and L. Markó, J. Am. Chem. Soc. 107:666 (1985).

745.   J. Azran and M. Orchin, Organometallics 3:197 (1984).

746.   R.M. Bullock, C.E.L. Headford, S.E. Kegley and J.R. Norton, J. Am. Chem. Soc. 107:727 (1985).

747.   K.A. Mahmoud, A.J. Rest and H.G. Alt, J.C.S. Chem. Commun. 1011 (1983).

748.   C.J. May and W.A.G. Graham, J. Organomet. Chem. 234:C49 (1982).

749.   D.L. Thorn, Organometallics 4:192 (1985).

750.   D.C. Roe, J. Am. Chem. Soc. 105:7770 (1983).

751.   M. Brookhart, M.L.H. Green and R.B.A. Pardy, J.C.S. Chem. Commun. 691 (1983); M. Brookhart and M.L.H. Green, J. Organomet. Chem. 250:395 (1985) and references therein; R.B. Cracknell, A.G. Orpen and J.L. Spencer, J.C.S. Chem. Commun. 326 (1984).

752.   K. Tatsumi, A. Nakamura, S. Komiya, A. Yamamoto and T. Yamamoto, J. Am. Chem. Soc. 106:8181 (1984).

753.   R.J. McKinney and D.C. Roe, J. Am. Chem. Soc. 107:261 (1985).

754.   P. Diversi, D. Fasce and R. Santini, J. Organomet. Chem. 269:285 (1984).

755.   J.H. Merrifield, J.M. Fernandez, W.E. Buhro and J.A. Gladysz, Inorg. Chem. 23:4022 (1984).

756.   M.F. Joseph, J.A. Page and M.C. Baird, Organometallics 3:1749 (1984).

757.   T. Yamamoto, T. Kohara, K. Osakada and A. Yamamoto, Bull. Chem. Soc. Jpn 56:2147 (1983).

758.   T.C. Flood and S.P. Bitler, J. Am. Chem. Soc. 106:6076 (1984).

759.   K. Mashima, H. Yasuda, K. Asami and A. Nakamura, Chem. Lett. 219 (1983).

760.   P. Leoni and M. Pasquali, J. Organomet. Chem. 255:C31 (1983).

761.   D.J. Darensbourg and R. Kudaroski, J. Am. Chem. Soc. 106:3672 (1984).

762.   A. Behr, U. Kanne and G. Thelen, J. Organomet. Chem. 269:C1 (1984).

763.   J.K.K. Sarhan, M. Green and I.M. Al-Najjar, J.C.S. Dalton Trans. 771 (1984).

764.   B. Åkermark and K. Zetterberg, J. Am. Chem. Soc. 106:5560 (1984).

765. L. Maresca, G. Natile, A-M. Manotti-Lanfredi and A. Tiripicchio, J. Am. Chem. Soc. 104:7661 (1982).

766. J.C. Vanderhoof and R.D. Ernst, J. Organomet. Chem. 233:313 (1982).

767. B.M. Trost and M-H. Hung, J. Am. Chem. Soc. 106:6837 (1984); B.M. Trost, M. Lautens, M-H. Hung and C.S. Carmichael, ibid. 106:7641 (1984).

768. B.M. Trost and T.N. Nanninga, J. Am. Chem. Soc. 107:1075 (1985); B.M. Trost, T.N. Nanninga and T. Satoh, ibid. 107:721 (1985).

769. T. Hayashi, M. Konishi and M. Kumada, J.C.S. Chem. Commun. 107 (1984).

770. F. Sato, H. Uchiyama, K. Iida, Y. Kobayashi and M. Sato, J.C.S. Chem. Commun. 921 (1983).

771. W. Kramarz, S. Kurek, M. Nowak, M. Urban and A. Wlodarczyk, Pol. J. Chem. 56:1187 (1983).

772. A. Goliaszewski and J. Schwartz, J. Am. Chem. Soc. 106:5028 (1984).

773. A. Goliaszewski and J. Schwartz, Organometallics 4:415 (1985).

REACTIONS OF ONE-CARBON LIGANDS IN
COMPLEXES OF MACROCYCLES

Michael D. Johnson

University College London

1.   TYPES OF LIGAND ENCOUNTERED.

In the field of organo-metal macrocycle chemistry, two types of ligand have so far provided the majority of complexes encountered.   These are (i) the corrin ring as found in coenzyme B_{12} (**I**) and related cobalamins such as methylcobalmin (**Ia**) and (ii) the bis(dimethylglyoximato) ligands (**II**) from which a very wide range of organo-cobalt complexes have been prepared. Several tetradentate ligands such as SALEN and its derivatives (**III**), ACACEN and its derivatives (**IV**), and the dioximato ligands doenH and dotnH (**V**) are also fairly common, as are several porphyrins, especially tetraphenylporphyrin (**VI**), aetioporphyrin and polyalkylporphyrins (**VII**).   Less common, but being used with increasing frequency, are a variety of macrocyclic tetra-aza ligands, abbreviated to [14]aneN$_4$ (**VIII**), Me$_4$[14]tetraeneN$_4$ (also referred to in some work as TIM) (**IX**), Me$_6$[14]4,11-dieneN$_4$ (**X**), Me$_2$[14]4,11-dieneN$_4$ (**XI**), Me$_2$pyo[14]trieneN$_4$ (also referred to as CR) (**XII**), and the octaaza ligand (**XIII**).   The above list is by no means exhaustive.

Nearly of all the complexes to be considered are either octahedral or square pyramidal, with the macrocyclic ligand occupying the equatorial positions.   Even the bidentate dioximato ligands tend to occupy the equatorial positions in organometallic complexes.   The η^1-organic ligand

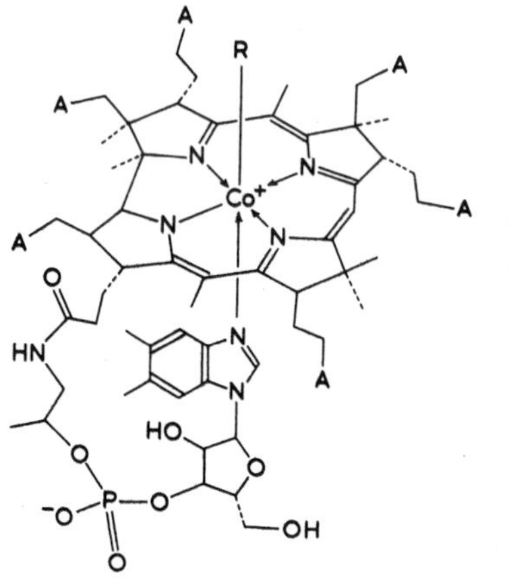

A = CONH$_2$

I   R =

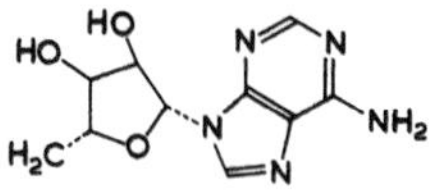

I a   R = Me

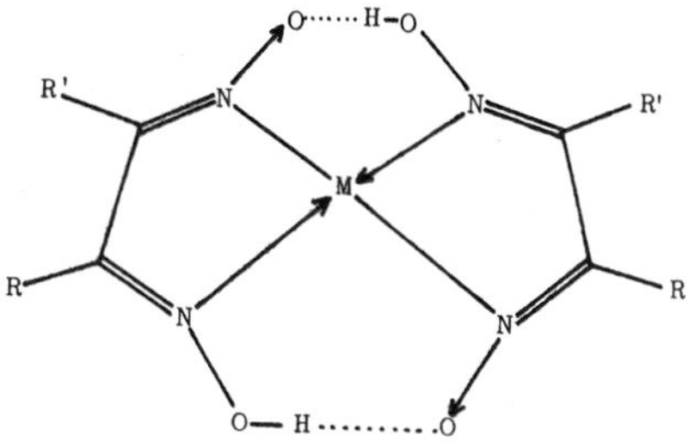

II   M(dmgH)$_2$;   R = R^1 = Me

IIa   M(chgH)$_2$;   RR^1 = (CH$_2$)$_4$

occupies the axial position, and there is usually (e.g. with the dioximato complexes) but not necessarily (e.g. with the porphinato complexes) a sixth ligand in the other axial position.   The reactions of the η^1-organic ligand are therefore strongly influenced both by the macrocyclic ligand and by the absence or nature of the sixth axial ligand.   The macrocyclic ligand is dominant, the extent of domination being dependent on its rigidity, which is most marked in the unsaturated macrocycles such as the porphyrins, but is less in the case of the more flexible partially saturated macrocyclic ligands. With the bidentate ligands, there is significant rigidity, reinforced by hydrogen bonding between the dioximato ligands, but this can be markedly reduced by protonation or alkylation of the dioximato oxygens. This rigidity has an influence on the interaction between the equatorial coordinating atoms and the metal, but also greatly reduces the chance of reaction through prior coordination of an incoming reagent adjacent to the η^1-organic ligand.   Thus carbonyl insertion reactions, for example, are uncommon in organometal macrocycle chemistry.

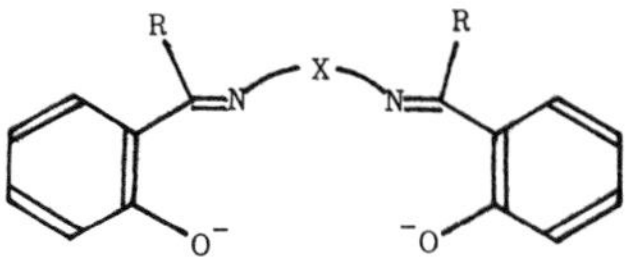

III SALEN R = H, X = $(CH_2)_2$
III a SALPN R = H, X = $(CH_2)_3$
III b 7,7'-Me_2SALEN R = Me, X = $(CH_2)_2$
III c SALOPHEN R = H, X = o-phenylene

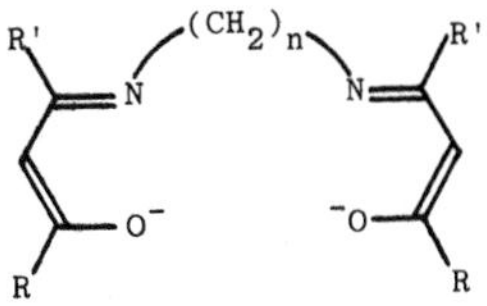

IV ACACEN (alias BAE)
R = R' = Me, n = 2
IV a TFACEN R = Me,
R' = CF_3, n = 2

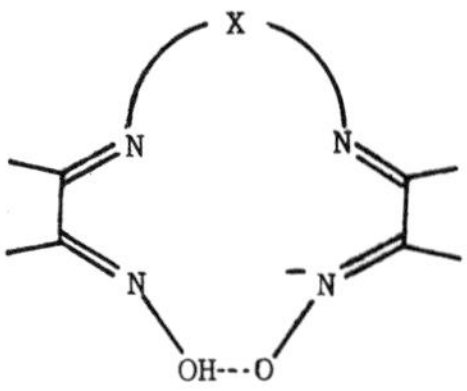

V dotnH X = $(CH_2)_3$
VI doenH X = $(CH_2)_2$

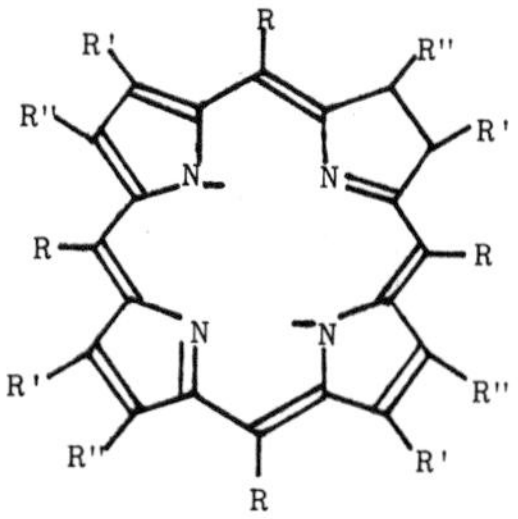

VII (TPP) R = Ph, R' = R'' = H
VII a (aetpor) R = H, R' = Et
R'' = Me
VII b (Octetpor) R = H,
R' = R'' = Et

Predissociation and preassociation involving both the equatorial ligand(s) and the sixth (axial) ligand also play an important, but often ill-defined role in the reactions of the η^1-organic ligand.

Though there have been a number of studies of the equilibria pertaining to reversible loss, exchange,[1-17] and modifications[18,19] of the sixth axial ligand, many mechanistic studies of reactions at the carbon-metal bond have been carried out without specific investigation of this problem, i.e. with the assumption that the sixth ligand remains intact, or is not present, in the reactive species. The η^1-organic ligand has a strong trans-influence and hence it is often not only the extent to which the sixth axial ligand dissociates, but also the rate at which it does so, that can influence the rate of any reaction occurring at the organic ligand.

Nowhere is this more apparent than in the reaction of coenzyme B_{12}, for which there are a number of reactions where the reagent (e.g. mercury

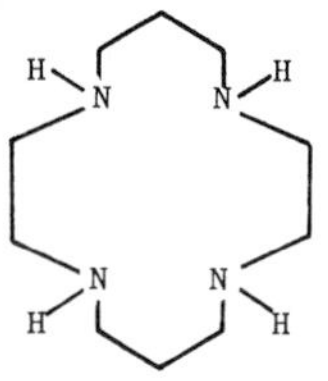

VIII [14]aneN$_4$

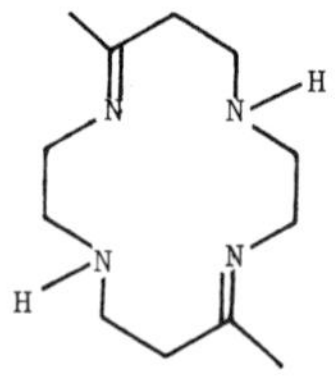

XI Me$_2$[14]4,11-dieneN$_4$

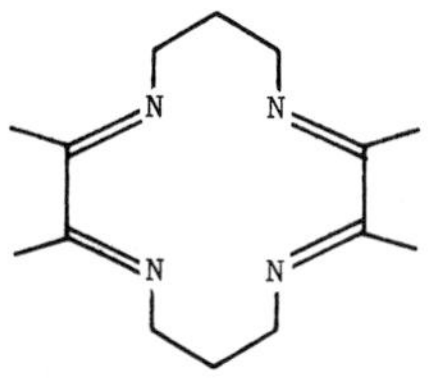

IX Me$_4$[14]tetraeneN$_4$
Also known as TIM

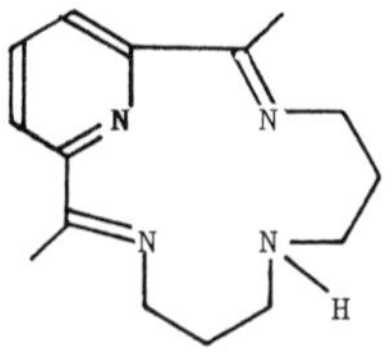

XII Me$_2$pyo[14]trieneN$_4$

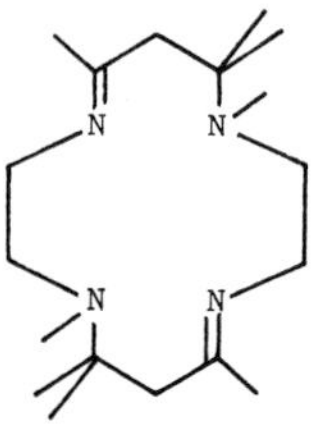

X Me$_6$[14]4,11-dieneN$_4$

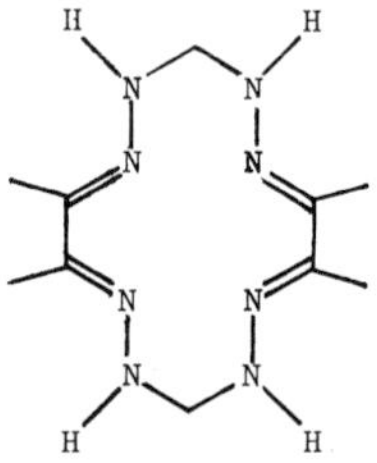

XIII

(II) species, Section 3.3.1) not only attacks the organic ligand at about the same rate as dissociation of the axial benziminazole ligand occurs, but also combines with that benziminazole as it becomes free. Since the equilibrium constant for the loss of an axial monodentate ligand has the units of concentration, the extent to which axial ligand dissociation occurs is concentration dependent; different results may therefore be obtained by different workers for the same reaction if widely different concentrations are used. Some of the discrepancies between results may be explained in this way, others by the slightly different solvents used. It should also be remembered that the equatorial ligands are themselves not inert chemically; they may undergo irreversible reaction prior to attack of a reagent at the organic ligand, they may undergo reversible protonation etc., or they may participate with the metal, as in some redox reactions.

The majority of reactions to be described in this chapter are of organocobalt (III) complexes.   This is because the determination of the structure of coenzyme B_{12} led to an interest not only in its chemistry but also in the chemistry of 'model' compounds and hence in the synthesis of an enormous range of relatively stable yet interestingly reactive organobis(dimethylglyoximato)(ligand)cobalt (III) complexes, henceforth described as organocobaloximes.   Though a number of complexes of other elements have been described,[20-31] the study of their reactions is only just beginning, and it is hoped that this chapter will service as a useful guide to their potential.   For simplicity, the words chelate and equatorial ligand will be used throughout this chapter to include all or any of those ligands described in the first paragraph.

2.   HOMOLYTIC CLEAVAGE OF THE CARBON-METAL BOND

2.1   Thermal decomposition

The stability of the carbon-metal bond in organometal macrocyclic complexes varies widely, from the barely detectable hydroxyalkyltetraglycinecopper (II) complexes[32] to the alkylcobaloximes and the alkylrhodium(TPP) complexes.   The isolation and characterisation of some thousands of different octahedral organocobalt (III) complexes with a variety of macrocyclic ligands is testament to the strength of the carbon-metal bond.   Few data are available for the actual strengths of carbon-cobalt bonds in these systems, but estimates have been made from studies of chemical equilibria and of thresholds for photochemical decomposition. Halpern[33] has estimated, from the temperature dependence of the equilibrium constant for reaction (1) (6.4 x $10^{-6}M^{3/2}$ at $18.8^{\circ}C$; $\Delta H = 92.8$ kJ mol^{-1}) and from the heats of formation of styrene and of the α-phenylethyl radical, that the carbon-cobalt bond dissociation energy is 83.6 kJ mol^{-1} for the complex (XIV).

Complex (XIV) is one of the less stable organocobaloximes by virtue of the high stability of the α-phenylethyl radical;   estimates of the bond dissociation energies of benzyl and alkylcobaloximes are therefore higher. Endicott has estimated from the exchange equilibria of various cobalt (II) and methylcobalt (III) complexes and photochemical threshold energies that bond dissociation energies of methylcobalt (III) complexes are as follows:[34] methylcobalamin 193, methyl(aquo)cobalamin ca. 185, methylCo(dmgH)$_2$ (aquo) 197, methylCo([14]aneN$_4$)(aquo)$^{2+}$ 168, methylCo(Me$_4$[14]4,11-dieneN$_4$)(aquo)$^{2+}$ 139, methylCo(Me$_2$[14]4,11-dieneN$_4$)(aquo)$^{2+}$ 185, methylCo(Me$_2$pyo[14]trieneN$_4$)(aquo)$^{2+}$ 172 kJ mol^{-1}. Though the above results suggest that the bond dissociation energies of the alkylcobalamins and the alkylcobaloximes are similar, it has also been suggested that because benzylcobaloxime can be readily prepared but benzylcobalamin is very much less stable, the bond dissociation energies of the organocobalamins must be some 40 kJ mol^{-1} smaller than those of the organocobaloximes. Further work in this area is clearly important, especially in view of the belief that much of the action of coenzyme B$_{12}$ in biological systems depends upon the homolysis of the carbon-cobalt bond and subsequent recombination of the fragments (Section 2.3.).

$$PhCHMeCo(dmgH)_2py \;\rightleftharpoons\; \tfrac{1}{2}H_2 \;+\; PhCH{:}CH_2 \;+\; Co^{II}(dmgH)_2py \qquad (1)$$

XIV

$$PhCH{:}CH_2 \;+\; \tfrac{1}{2}H_2 \;\longrightarrow\; PhCHMe \qquad (2)$$

$$PhCHMeCo(dmgH)_2py \;\rightleftharpoons\; PhCHMe \;+\; Co^{II}(dmgH)_2py \qquad (3)$$

$$PhCHMe \;+\; Co^{II}(dmgH)_2py \;\rightleftharpoons\; PhCH{:}CH_2 \;+\; HCo^{III}(dmgH)_2py \qquad (4)$$

$$PhCHMeCo(dmgH)_2py \;\rightleftharpoons\; PhCH{:}CH_2 \;+\; HCo^{III}(dmgH)_2py \qquad (5)$$

$$HCo^{III}(dmgH)_2py \;\rightleftharpoons\; \tfrac{1}{2}H_2 \;+\; Co^{II}(dmgH)_2py \qquad (6)$$

The mechanism of eq 1 does not have any bearing on the validity of the arguments above, but is of interest. It is not yet known whether it occurs by a homolytic process (eq 3 followed by eq 4), by a concerted molecular elimination of the hydridocobalt (III) species as in eq 5, followed by decomposition of the hydrido complex (eq 6), or by both mechanisms. There is little doubt that radicals can be formed in these thermal reactions, as will be evident from the reactions described below (Section 2.3.), but the large difference in susceptibility to thermal decomposition of adenosylcobalamin and ethylcobalamin does suggest that <u>both</u> mechanisms may occur.[35]   A study of the reverse reactions, i.e. the capture of an organic radical by cobalt (II) and the addition of hydridocobalt (III) complexes to unsaturated organic molecules, is instructive.   First, the capture of organic radicals by cobalt (II) complexes is very efficient, and approaches the encounter rate. For example, the rate of reaction of methyl radicals with $Co^{II}(Me_6[14]$-dieneN$_4$) $(H_2O)_2^{2+}$ is 7×10^8 $M^{-1}s^{-1}$,[36] and the rates of reaction of $HOCH_2$, OH, H, and e_{aq}^- with the same complex are 7×10^7, 2.7×10^9, 1.8×10^9, and 4.4×10^{10} $M^{-1}s^{-1}$, respectively.[37,38] Secondly, though the addition of hydridocobaloxime (III) to unsaturated organic molecules is apparently both regio- and stereospecific (e.g. eq 7) consistent with a concerted molecular addition process, [39,40] it is clear from studies of equilibration that the stereospecificity is less than complete.   Thus, the thermolysis of the deuteriated α-phenylethylcobaloxime (XV) (which has diastereotopic β-protons readily distinguishable by ^{1}H n.m.r.) at 40°C over 3 h gives the undeuteriated compound (XVI) expected from a concerted elimination of the metal deuteride together with the deuteriated compound (XVII) expected from a non-concerted elimination (eq 8).   Clearly, unless any <u>cis</u>-phenylpropene formed is very much more reactive than the <u>trans</u>-isomer, the elimination reactions do not proceed completely stereospecifically; this is consistent either with concerted or radical reactions taking place concurrently, or with a reversible homolytic process which takes place with a measure of specificity because of the high rate of hydrogen atom abstraction by the first-formed cobalt (II) complex, perhaps in a caged pair.

Studies of the kinetics of thermal decomposition of a series of primary and secondary alkylcobalamins support the duality of homolytic reactions. [41,42,35] Thus, the order of stability in aqueous solution measured spectrophotometrically under anaerobic conditions is 3-pentyl > 4-heptyl > 2-butyl > 2-pentyl ~ 2-octyl ~ 2-propyl > pentyl ~ cyclopentyl > cyclohexyl > neopentyl > iso-butyl > butyl > propyl > ethyl ~ methyl, the rate difference over the series 3-pentyl to cyclohexyl being rather small, i.e. approximately two orders of magnitude. The reactions are independent of pH above 6-7, i.e. under conditions where the axial base ligand is coordinated, but decrease in rate as the pH is lowered below 6. The duality of mechanism is illustrated by the fact that decomposition reactions of the secondary alkylcobalamins, which lead to olefinic products, are barely influenced by the presence of oxygen, whereas those of the primary alkylcobalamins, particularly neopentylcobalamin which cannot give olefinic products directly, are accelerated by oxygen, by added thiols, and by added isopropanol. In the latter two cases, the organic product is the alkane formed by hydrogen abstraction from the thiol or propanol by the alkyl radical. It seems very likely therefore, that the first formed radical pair will undergo hydrogen atom transfer where there is an appropriately placed hydrogen atom, as in secondary alkylcobalamins, but will recombine where no such hydrogen atom is available or in sufficient proximity to the metal centre. The unusual rate sequence (estimated in part on the assumption that some of the work described refers to ca.25°C) has been ascribed to the influence of the bulky organic ligands causing substantial distortion of the equatorial corrin ligand from the purely equatorial plane, with consequent weakening of the carbon-cobalt bond and alteration of the binding of the sixth axial ligand. [43,44] The thermal decomposition of alkylCo(SALEN) derivatives in methanol also gives alkenes where possible, together with traces of alkanes, except in the case of the methyl derivative which gives substantially methane with only traces of ethane.

The need for a suitable hydrogen available for abstraction may explain the surprising thermal stability of several tertiary organocobaloximes, such as the adamantyl(pyridine)cobaloxime, the preparation of which from the alkyl bromide and cobaloxime(I) cannot take place by a conventional bimolecular nucleophilic displacement process.[45]

2.2   Photochemical decomposition

There have been a large number of studies of the photochemical decomposition of organocobalt (III) complexes, many in the search for examples of rearrangement of organic ligands during or after separation from the metal, others in the presence of spin traps, and others with coenzyme B_{12} in in vivo enzyme systems.

The cleanest example of a radical rearrangement through homolysis without significant decomposition is in the photolysis, under very mild conditions, of the unsaturated tetramethylene organocobaloxime (XVIII) which is smoothly converted into the more stable vinylcobaloxime (XIX) via the organic radicals (XX) and (XXI) (Scheme 1).[45]   Most other attempts to observe such rearrangements have utilised the conversion of a primary alkylcobaloxime or -cobalamin into a less stable secondary or tertiary alkyl derivative (Section 2.3.).   When the above reaction is carried out in methylene chloride, or in pyridine containing potential halogen donors such as carbon tetrachloride or bromotrichloromethane, organic halides such as (XXII) are formed through abstraction of halogen atoms by the radical (XXI). In contrast, the corresponding trimethylene derivative neither rearranges nor gives cyclic vinyl halides on photolysis.

Good evidence for the formation of organic radicals and cobalt(II) complexes comes from the studies by Endicott on the photolysis of

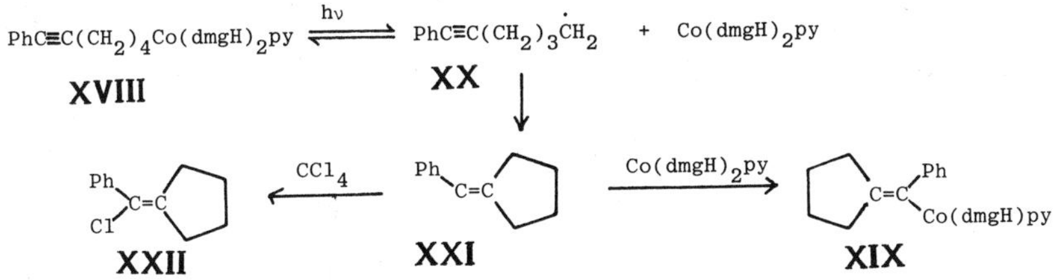

PhC≡C(CH$_2$)$_4$Co(dmgH)$_2$py $\underset{}{\overset{h\nu}{\rightleftarrows}}$ PhC≡C(CH$_2$)$_3\overset{\bullet}{C}H_2$ +   Co(dmgH)$_2$py

XVIII   **XX**

XXII   **XXI**   **XIX**

Scheme 1

methylcobalamin and other methylcobalt macrocyclic complexes.[47-49] In the photolysis of methylCo([14]aneN$_4$)$^{2+}$ the quantum yield is high and wavelength independent in the region 250-540 nm, and the threshold energy (corresponding to 610 nm) is very low. In the photolysis of methylCo([14]tetraeneN$_4$)$^{2+}$, the rate constant for recombination of methyl radicals and the cobalt(II) complex is 10^8 $M^{-1}s^{-1}$, and methane is the main product, formed by hydrogen abstraction from the equatorial ligand. Scavenging of the alkyl radical by oxygen is extremely efficient (k_2 = 4.7 x 10^9 $M^{-1}s^{-1}$) giving MeOȮ which may be recaptured by the cobalt(II) complex with a rate coefficient of 1.2 x 10^8 $M^{-1}s^{-1}$.

Flash photolysis generates high concentrations of radicals leading to dimer formation; this is not so with the less intense irradiation conditions of conventional photochemical studies. Similar rapid reactions have been detected in the flash photolysis of methylcobalamin at >480 nm and of methylcobaloxime at >380 nm, for which the recombination rate constants for methyl radicals and the cobalt(II) complexes are 1.5 x 10^9 and 5 x 10^7 $M^{-1}s^{-1}$, respectively.[47]

Organic radicals from the photolysis of organocobalt(III) have also been detected by the use of spin traps. For example, the photolysis of ethyl- and adenosyl-cobalamin in the presence of t-butylnitroxide or nitrosodurene at 50°C in water gives the appropriate ethyl- or adenosyl-nitroxyl radical, detected by esr spectroscopy,[50] and in the photolysis of several complexes RCo(chelate) (R = Me, Et, n-Pr; chelate = SALEN, SALOPHEN, bis(dimethylglyoximato),[51,52] not only has the radical R been trapped by phenyl-t-butylnitrone and identified by esr but some radical traps also capture a hydrogen atom in those cases where R = ethyl or n-propyl. It is almost certain that this hydrogen atom is abstracted from a hydridocobalt(III) species formed during an elimination reaction similar to those described above.

The anaerobic photolysis of coenzyme B$_{12}$ gives the 8,5'-cyclic adenosine (XXIII) by internal capture and subsequent loss of a hydrogen atom,[35] and in the photolysis of benzylcobaloximes, a nitroxyl radical formed by capture of a benzyl radical by a dimethylglyoximato ligand has been identified.[53]

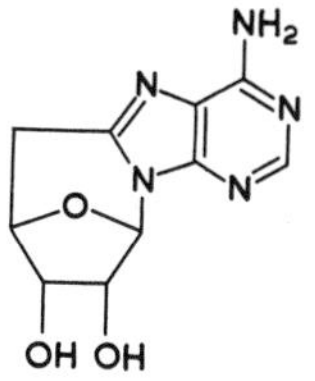

XXIII

The esr spectrum of the cobalt(II) fragment formed on photolysis has been identified at low temperature in several studies.[45,53,54] The cobalamin(II) radical shows coupling to [59]Co and ^{14}N, and at 77K in aqueous solution at pH 7.3, g = 2.006 ± 0.007.[55] In the laser-induced photolysis of solid PhCo(BAE), exchange-coupled radical pairs have been identified.[56]

The subtleties associated with radical formation and hydride elimination are revealed by the observation that the photolysis of alkylcobaloximes is markedly influenced by pH in aqeous solution.[57] At pH 2, the cobaloxime(II) product is unstable and the organic products are those derived from a clean homolysis of the carbon cobalt bond, i.e. alkanes, together with dimethylglyoxime and cobalt(II), but at pH 7 the photolysis is some 10 times less efficient and olefins are formed from C_2 and higher alkyl groups (Scheme **2**). Some hydrogen was also detected in those reactions carried out at higher pH. It was shown by deuterium labelling that the alkanes result from dimerisation of the organic radical and from abstraction

$$RCo(dmgH)_2OH_2 \xrightarrow{h\nu} olefin + HCo(dmgH)_2OH_2 \longrightarrow \tfrac{1}{2}H_2 + Co(dmgH)_2(OH_2)_2$$

$$\downarrow H^+$$

$$[RCo(dmgH)(dmgH_2)OH_2]^+ \xrightarrow{h\nu} \dot{R} + [Co(dmgH)(dmgH_2)OH_2]^+ \longrightarrow RH + RR + Co^{2+} + 2dmgH_2$$

$$\downarrow H^+$$

$$[RCo(dmgH_2)_2OH_2]^{2+} \xrightarrow{\Delta} \dot{R} + [Co(dmgH_2)_2OH_2]^{2+} \longrightarrow RR + Co^{2+} + 2dmgH_2$$

Scheme **2**

of a hydrogen atom from the dioxime OH groups but not from the dioximato-
methyl groups.   Since the equatorial ligands of organocobaloximes are
substantially protonated at pH 2, the increased quantum yield may be the
result either of a weakening of the carbon-cobalt bond or of a reduction in
the rate of recapture of the alkyl radical due to competing decomposition of
the cobalt(II) fragment.   The corresponding benzyl(pyridine) cobaloxime,
which is stable in the dark in aqueous solution, undergoes ready thermolysis
when doubly protonated in sulphuric acid,[58] the rate of this homolysis
remaining constant over the acid range 30 - 75% H_2SO_4, i.e. in that range
where the doubly protonated complex predominates.   Clearly, the carbon-
cobalt bond decreases in strength as the equatorial dioximato ligands
become more fully protonated.   Estimates of the pKa of the mono- and diprotonated species are 0.7 and -2.9 respectively, the latter being on the H_o
scale.

2.3   Reactions consequent on homolysis of the carbon-metal bond

2.3.1 Radical rearrangements

Having established the conditions under which homolysis of the carbon-
metal bond can occur, we are in a position to consider the consequences of
radical formation in the promotion of other reactions.   Such radical
reactions are of special interest because of proposals that the mode of
action of coenzyme B_{12} is through the homolysis of the carbon-cobalt bond,
hydrogen abstraction from the substrate by the adenosyl radical,

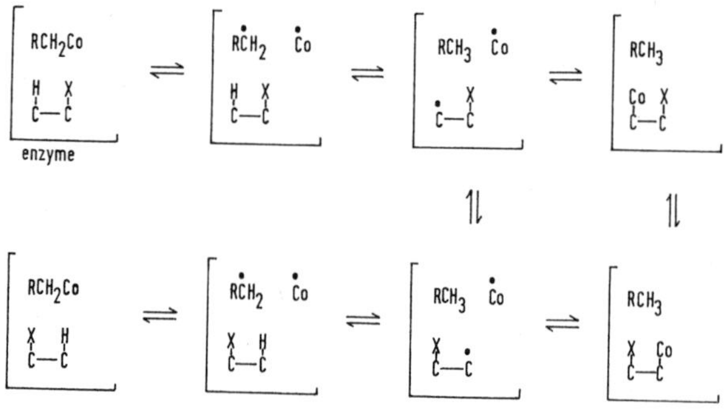

Scheme 3

rearrangement of the substrate radical either when free or when bound to cobalt, hydrogen transfer back to the substrate radical, and recapture of the adenosyl radical by the cobalamin(II) fragment as shown in Scheme **3.** [59,60] One approach to the investigation of such processes has been to study the homolysis of organocobalt complexes in which the organic ligand itself resembles a biologically important substrate, another approach is to study the homolysis of simple organocobalt complexes in the presence of potential substrates,[61-63] and a third, which we will not consider, has been to study model organocobalt complexes in biological systems.   The first approach, related to the biological transformation of vicinal diols into aldehydes, catalysed by coenzyme B_{12} and dioldehydrase, is illustrated by the photolysis of 4,5-dihydroxypentylcobaloximes **(XXIV)** and appropriately deuteriated analogues.[64]

Thus, **(XXIV)** on photolysis gives 30% pentanal through a 1,5-hydrogen shift in the intermediate 4,5-dihydroxypentyl radical.   This is confirmed by the photolysis of the 1,1,5,5-tetradeuterio-derivative which gives 1,2,5,5-tetradeuteriopentanal.   Other products, formed from the dihydroxypentyl radical, are shown in Scheme **4.**   2-Hydroxyethyl and 2-hydroxypropylCo(aetpor) give acetaldehyde and acetone, respectively, on photolysis,[65] and 2-hydroxypropylcobaloxime also gives acetone on photolysis.   However, the photolysis of 2-hydroxyisopropylcobaloxime gives allyl alcohol and propanal together with some acetone, formed following photochemical rearrangement to the 2-hydroxy-n-propyl isomer.[66] These

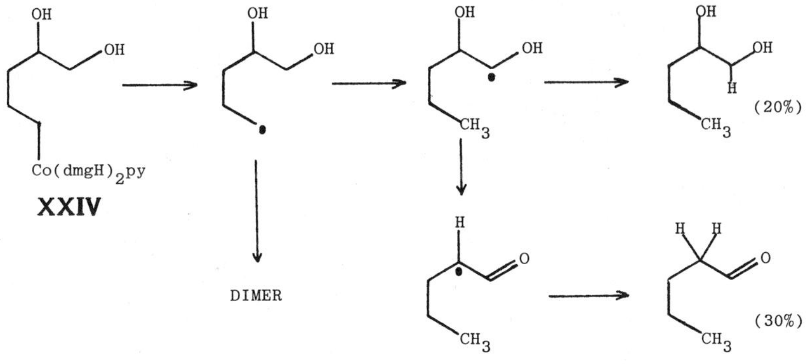

Scheme **4**

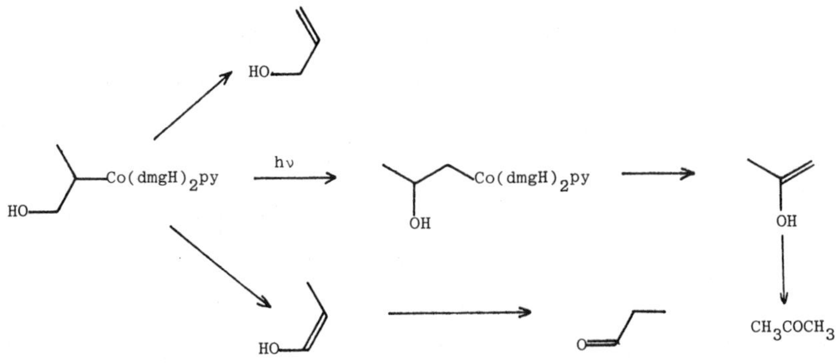

Scheme 5

reactions take place by an elimination of the hydridocobaloxime and subsequent rearrangement of the enol products (Scheme 5).

Several skeletal rearrangements, related to the interconversion of methylmalonylcoenzyme A and succinylcoenzyme A have also been studied by a similar method. Thus, the capped alkylcobaloxime (**XXV**) decomposes, on photolysis with a mercury lamp, to products which, on further hydrolysis, give both methylsuccinic acid and glutaric acid. In order to form the methylsuccinic acid, the free radical formed by homolysis of the carbon-cobalt bond must have undergone the skeletal rearrangement shown in Scheme 6. Since the rearrangement is from a primary to a tertiary organic radical, any recapture by the cobaloxime(II) would be short-lived.[67]

A similar result was obtained in the rearrangement of the diester

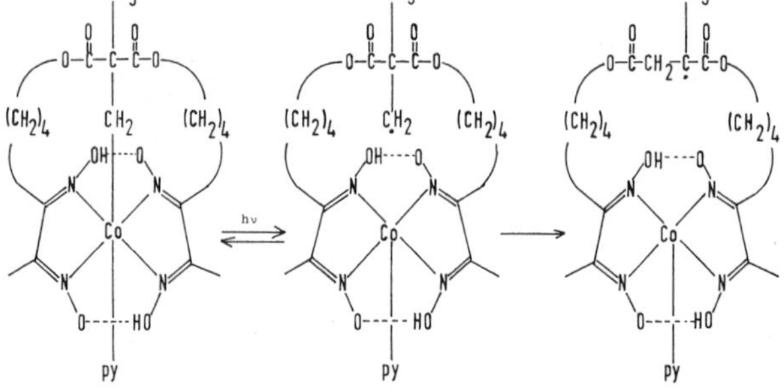

XXV

Scheme 6

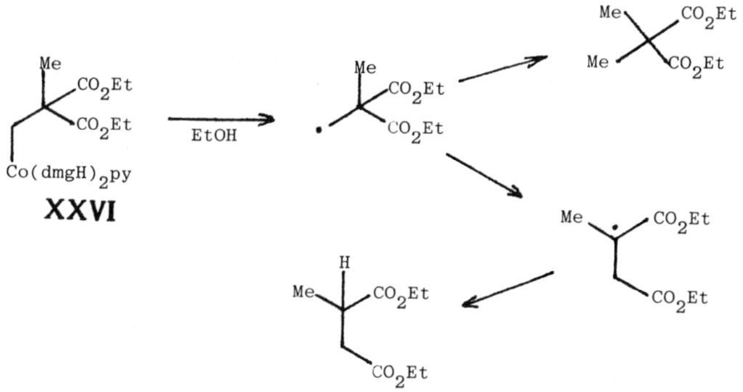

Scheme 7

(XXVI), which gives products expected from both rearranged and unrearranged substrate _via_ the intermediate radicals (Scheme 7).[68] Again, the rearrangement is from primary to tertiary radical and recapture cannot be observed. Some of the reported reactions of related organocobalamins are suspect because of possible formation of rearranged substrate during the preparation, and the formation of carbanionic intermediates through reaction with borohydride ion present in excess. (see Section 4.3.1.)[69-72]

2.3.2 Bimolecular reactions of radicals with organometallic substrates

Radicals, formed directly by the homolysis of the organometallic substrate (eq 9) or indirectly from an added precursor (eqs 10 and 11) may also react with the organic ligand of the organometallic substrate. Such a reaction may have a number of important consequences, but the formation of a new organic molecule with the displacement of a paramagnetic metal

$$RM(chel)L \rightleftharpoons \dot{R} + \dot{M}(chel)L \qquad (9)$$

$$\dot{R} + XY \longrightarrow RY + \dot{X} \qquad (10)$$

$$\dot{M}(chel)L + XY \longrightarrow YM(chel)L + \dot{X} \qquad (11)$$

$$\dot{X} + RM(chel)L \longrightarrow R'X + \dot{M}(chel)L \qquad (12)$$

$$XY + RM(chel)L \longrightarrow R'X + YM(chel)L \qquad (13)$$

complex is perhaps the most important (eq 12).   If the paramagnetic complex is capable of further reaction with the radical precursor, so regenerating the radical $\dot{X}$, then a chain reaction may ensue with the overall result shown in eq 13.   Moreover the organic ligand R may be retained in the same form in the product RX, or may be rearranged into an isomeric species R'X.

These chain reactions readily occur when an allylcobaloxime is mixed with a radical precursor such as bromotrichloromethane,[73] dimethylsulfamoyl chloride,[74] or a tosyl halide.[75]   Allylcobaloximes are known to undergo ready homolysis at ca. $40^{\circ}C$ or at lower temperatures under irradiation by tungsten light,[76] (e.g. eq 14) and the cobalt(II) complex reacts very rapidly with the precursor to generate trichloromethyl or tosyl

$$RCH:CH.CH_2Co(dmgH)_2py \rightleftharpoons RCH:CH.\dot{C}H_2 + \dot{C}o(dmgH)_2py \qquad (14)$$

$$\dot{C}o(dmgH)_2py + BrCCl_3 \longrightarrow BrCo(dmgH)_2py + Cl_3\dot{C} \qquad (15)$$

$$Cl_3\dot{C} + RCH:CH.CH_2Co(dmgH)_2py \longrightarrow Cl_3C.CHR.CH:CH_2 + \dot{C}o(dmgH)_2py \qquad (16)$$

$$BrCCl_3 + RCH:CH.CH_2Co(dmgH)_2py \longrightarrow Cl_3C.CHR.CH:CH_2 + BrCo(dmgH)_2py \qquad (17)$$

$$\dot{C}o(dmgH)_2py + ArSO_2X \longrightarrow XCo(dmgH)_2py + Ar\dot{S}O_2 \qquad (18)$$

$$Ar\dot{S}O_2 + Me_2C:CH.CH_2Co(dmgH)_2py \longrightarrow ArSO_2CMe_2CH:CH_2 + \dot{C}o(dmgH)_2py \qquad (19)$$

$$ArSO_2X + Me_2C:CH.CH_2Co(dmgH)_2py \longrightarrow ArSO_2CMe_2CH:CH_2 + XCo(dmgH)_2py \qquad (20)$$

radicals which then regiospecifically attack the γ-carbon of the allyl ligand to give the trichlorobutene or allyl sulfone (eqs 15-17 and 18-20) respectively.   These reactions provide a useful source of allylsulfones which are difficult to synthesis in conventional ways, and are useful intermediates for the introduction of allyl groups into more complicated organic molecules,   However, the attack of radicals at the γ-position is not confined to trichloromethyl and sulfonyl radicals.   Other carbon radicals such as the diethylmalonyl, diethyl bromomalonyl, and diethyl methylmalonyl radicals also react with allylcobaloximes providing a ·useful means of extending the carbon chain length regiospecifically (eq 21; R = H, Me, Br).[77]

$$(EtOCO)_2\overset{\cdot}{C}R \; + \; PhCH{:}CH.CH_2Co(dmgH)_2py \longrightarrow (EtOCO)_2CR.CHPh.CH{:}CH_2 \; + \; \overset{\cdot}{C}o(dmgH)_2py \qquad (21)$$

$$MeCH{:}C{:}CHCo(dmgH)_2imid \; + \; Ar\overset{\cdot}{S}O_2 \longrightarrow ArSO_2CHMe.C{\equiv}CH \; + \; \overset{\cdot}{C}o(dmgH)_2imid \qquad (22)$$

(23)

The same radicals can also attack the γ-carbon of propadienylcobaloximes as shown in eq 22, [74,77] and the δ-carbon of but-3-enylcobaloximes, leading to cyclopropylcarbinyl compounds in high yield, (e.g. Scheme 8 and eq 23) almost certainly through an intermediate substituted propyl radical such as (XXVII).[78-80]

Attack of the organic radical at the α-carbon of allyl groups is only found where the γ-carbon is highly hindered, but attack at the α-carbon of benzyl groups seems relatively facile.[81] Thus, benzylcobaloximes react thermally with bromotrichloromethane, or with tosyl iodide to give trichloroethyl benzene and benzyl bromide, or tolyl benzylsulfone and benzyl iodide, respectively (eqs 24-26). The yield of the trichloroethylarene is much lower with p-nitrobenzylcobaloxime, and higher with methyl substituted benzyl cobaloximes, but is increased in all cases when imidazole is the axial base, or is added in situ. It would appear that the chain length for attack at the α-benzylic carbon atom is much less than in reaction with allylcobaloxime, and that the halide is formed by attack of benzyl radicals on the radical precursor. Indeed, it seems likely that when the radical reagent reacts preferentially at the metal or the equatorial ligand, the carbon-metal bond is weakened and organic radicals are likely to be formed. In support of this hypothesis, the reaction of a five-coordinate complex such as n-propylCo(TPP) with bromotrichloromethane leads exclusively to n-propyl bromide, probably through attack of the trichloromethyl radical at

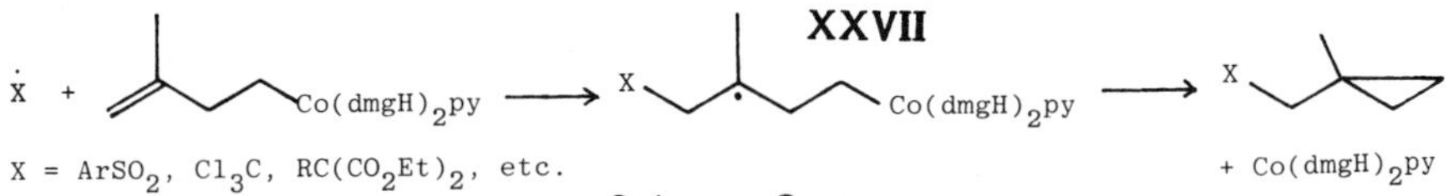

XXVII

X = ArSO$_2$, Cl$_3$C, RC(CO$_2$Et)$_2$, etc.

Scheme **8**

$$PhCH_2Co(dmgH)_2imid \rightleftharpoons Ph\dot{C}H_2 + Co(dmgH)_2imid \qquad (24)$$

$$Ph\dot{C}H_2 + XHal \longrightarrow PhCH_2Hal + \dot{X} \qquad (25)$$

$$\dot{X} + PhCH_2Co(dmgH)_2imid \longrightarrow PhCH_2X + Co(dmgH)_2imid \qquad (26)$$

$$X = Cl_3C, \; ArSO_2, \; etc.$$

$$RCo(chel) + Cl_3\dot{C} \rightleftharpoons Cl_3C.Co(chel) + \dot{R} \longrightarrow Products \qquad (27)$$

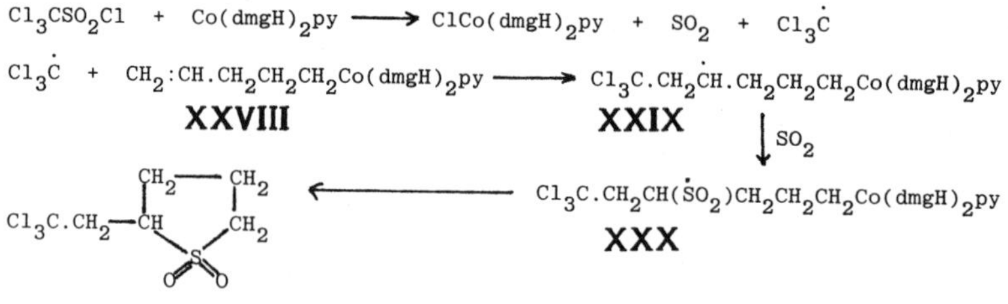

$$Cl_3CSO_2Cl + Co(dmgH)_2py \longrightarrow ClCo(dmgH)_2py + SO_2 + Cl_3\dot{C}$$

$$Cl_3\dot{C} + CH_2{:}CH.CH_2CH_2CH_2Co(dmgH)_2py \longrightarrow Cl_3C.CH_2\dot{C}H.CH_2CH_2CH_2Co(dmgH)_2py$$

XXVIII   **XXIX** $\downarrow SO_2$

$$Cl_3C.CH_2\overset{CH_2 \rule[0.5ex]{1em}{0.4pt} CH_2}{\underset{\underset{S}{\overset{|}{\rule{0pt}{1.5ex}}}}{\underbrace{CH}} \rule{0pt}{0pt}} \longleftarrow Cl_3C.CH_2CH(\dot{S}O_2)CH_2CH_2CH_2Co(dmgH)_2py$$

XXXI   **XXX**

Scheme **9**

the metal atom with formation of the n-propyl radical (eq 27).[82] However, as seen in Section 4.3.1., these trichloromethylmetal complexes are seldom stable and are difficult to isolate.

From the above results, it is apparent that homolytic attack at the α-saturated carbon of an alkyl ligand is rather rare, and will only occur if the metal and the equatorial ligand are well protected, or if other factors are particularly favourable. Such attack is believed to occur in the reaction of trichloromethanesulfonyl chloride with pent-5-enylcobaloxime.[83] Trichloromethanesulfonyl chloride is a useful source of trichloromethyl radicals and also gives sulfur dioxide as a by-product. In its photochemical reaction with pent-5-enylcobaloxime (**XXVIII**), the trichloromethyl radicals, formed on chlorine abstraction by cobaloxime(II) (Scheme **9**), react at the terminal unsaturated carbon of the substrate (**XXVIII**). The organometallic radical (**XXIX**) so formed is captured by sulfur dioxide to give a new sulfonyl radical (**XXX**) which chooses to attack the α-carbon intramolecularly to give the five-membered sulfolane derivative (**XXXI**) in high yield, at the same time displacing cobaloxime(II) for further reaction in the chain. The formation of sulfolane as a by-product in the reaction of sulfur dioxide with the dicobaloxime (**XXXII**) probably arises from a similar intramolecular

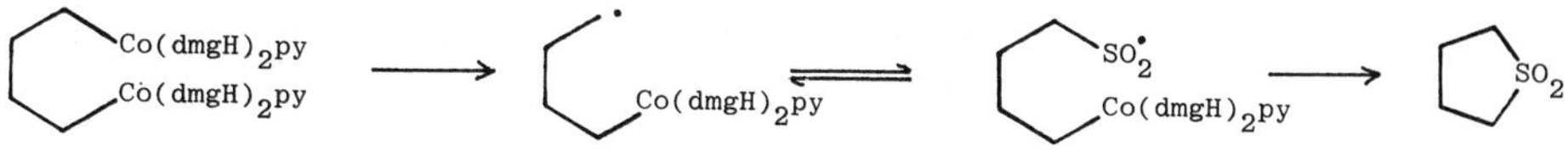

XXXII

Scheme **10**

homolytic displacement of cobaloxime(II) by attack at the α-carbon as in Scheme **10** (c.f. Section 2.3.4).

In the corresponding reaction with saturated alkylcobaloximes no such intramolecular reaction can occur and the main product is the alkanesulfonyl chloride, formed as in reactions 28-30. As expected for such a mechanism, the reaction of the monodeuteriated complex (**XXXIII**) gives a 1:1 mixture of the two diastereoisomeric sulfonyl chlorides (eq 31), which are readily distinguished, without decoupling, from the areas of the well separated methylene proton or deuterium resonances.

In the reactions of diaryldisulfides and of diaryldiselenides with organocobaloximes, both the thermal and photochemical reactions give reasonable yields of organoaryl selenides and sulfides; the reactions do not appear to proceed through homolytic attack of arylthiyl or aryl selenyl radicals at the organic ligand, but by the well known reaction of the organic radicals with the disulfide or diselenide (eqs 32-34).[84]

$$RCo(dmgH)_2L \rightleftharpoons \overset{.}{R} + Co(dmgH)_2L \tag{28}$$

$$\overset{.}{R} + SO_2 \rightleftharpoons R\overset{.}{S}O_2 \tag{29}$$

$$R\overset{.}{S}O_2 + Cl_3CSO_2Cl \longrightarrow Cl_3\overset{.}{C} + SO_2 + RSO_2Cl \tag{30}$$

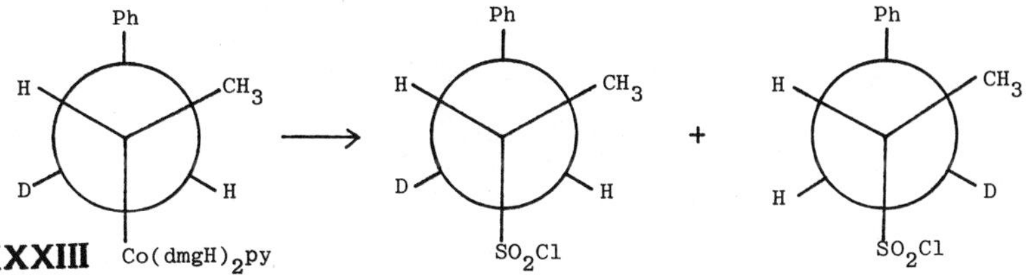

XXXIII

$$RCo(dmgH)_2py \rightleftharpoons \overset{.}{R} + Co(dmgH)_2py \tag{32}$$

$$\overset{.}{R} + PhXXPh \longrightarrow RXPh + Ph\overset{.}{X} \tag{33}$$

$$Ph\overset{.}{X} + RCo(dmgH)_2py \longrightarrow PhXCo(dmgH)_2py + \overset{.}{R} \tag{34}$$

X = S, Se

Table 1.

Kinetics of transfer of alkyl groups from organocobalt(III) complexes (RM(chel)L) to metal(II) complexes (M'(chel)L)

Solvent	RM(chel)L	M'(chel)L	T/°C	$k/M^{-1}s^{-1}$	K	Ref
0.1M aq NaClO$_4$	Mecobalamin	Co(Me$_4$[14]tetraeneN$_4$)$^{2+}$	25	1.07	4	34
0.1M aq NaClO$_4$	Mecobalamin	Co(Me$_2$[14]4,11-dieneN$_4$)$^{2+}$	25		0.25	34
0.1M aq NaClO$_4$	Mecobalamin	Co(dmgH)$_2$aq	25		59	34
MeOH	MeCo(dmgH)$_2$py	Co(chgH)$_2$py	0	44	1	89
0.1M NaOAc	MeCo(Me$_2$pyo[14]trieneN$_4$)$^{2+}$	Co(dmgH)$_2$aq	25	4.6	29	34
0.1M NaOAc	MeCo([14]aneN$_4$)$^{2+}$	Co(dmgH)$_2$aq	25	2.95	500	34
0.1M NaOAc	MeCo(Me$_2$[14]4,11-dieneN$_4$)$^{2+}$	Co(dmgH)$_2$aq	25	0.059	10^5	34
0.1M NaOAc	MeCo(Me$_2$pyo[14]trieneN$_4$)$^{2+}$	Co([14]aneN$_4$)$^{2+}$	25	0.32	6	34
0.1M CF$_3$SO$_2$H	MeCo(Me$_6$[14]4,11-dieneN$_4$)$^{2+}$	Co(Me$_4$[14]tetraeneN$_4$)$^{2+}$	25.6	0.01	7×10^{-3}	34
aq HClO$_4$	Mecobalamin	Cr(H$_2$O)$_6^{2+}$	25	3.6×10^{-2}		85
aq HClO$_4$	Etcobalamin	Cr(H$_2$O)$_6^{2+}$	25	4.4		85
aq HClO$_4$	MeCo(dmgH)$_2$aq	Cr(H$_2$O)$_6^{2+}$	25	23		86
aq HClO$_4$	iso-BuCo(dmgH)$_2$aq	Cr(H$_2$O)$_6^{2+}$	25	6.1×10^{-5}		86
MeOH	iso-BuCo(dmgH)$_2$py	Co(chgH)$_2$py	28	2.6×10^{-4}		89
DMSO	MeCo(ACACEN)	Co(TFACEN)	44	fast	0.9	88
DMSO	EtCo(ACACEN)	Co(TFACEN)	44	fast	0.27	88
DMSO	PrCo(ACACEN)	Co(TFACEN)	44	fast	0.25	88
DMSO	n-BuCo(ACACEN)	Co(TFACEN)	44	fast	0.22	88
DMSO	PrCo(TFACEN)	Co(dmgH)$_2$aq	44	fast	large	88

Solvent					
DMSO	MeCo(TFACEN)	Pt^{II}(SALEN)	44	no reaction	88
DMSO	MeCo(TFACEN)	Fe^{II}(SALEN)	44	no reaction	88
DMSO	MeCo(TFACEN)	Mn^{II}(SALEN)	44	no reaction	88
Benzene	MeCo(ACACEN)	Co(SALEN)		>1	90
Benzene	MeCo(ACACEN)	Co(SALOPHEN)		>1	90
Benzene	MeCo(Me$_2$SALEN)	Co(SALEN)		>1	90
Benzene	MeCo(SALEN)	Co(SALOPHEN)		>1	90
Benzene	MeCo(SALEN)	Co(dotnH)aq$^+$		>1	90
Benzene	MeCo(ACACEN)	Co(dotnH)aq$^+$		>1	90

2.3.3 **Homolytic Metal-for-Metal Displacement Reactions**

The homolytic attack of a <u>metal</u> radical on the -carbon of an alkylcobalt complex does occur much more readily than the attack of an organic radical because the alternative reactions at the equatorial ligand are much less facile.   A large number of examples of metal-for-cobalt displacement reactions have now been studied (eq 35), and the results are of value in deducing the relative dissociation energies of methyl-cobalt bonds in different macrocyclic complexes.   Thus, the rates of the near-symmetrical displacement reaction (eq 36) have been determined in methanol and in methylene chloride for a wide range of primary and secondary alkyl groups.[85]   The reactivity decreases down the series Me   Et

Pr   Bu   Oct   i-Pr   i-Bu   sec-Bu, the rate coefficient for the reaction with R = Me being some six orders of magnitude greater than with R = sec-butyl.   The symmetrical exchange between the diastereoisomeric <u>erythro</u>-cobaloxime (**XXXIVa**) and cobaloxime(II) occurs with racemisation at the α-carbon (eq 37), giving a 1:1 mixture with the <u>threo</u>-diastereoisomer (**XXXIVb**), distinguishable by the large difference in vicinal proton coupling constants with configuration at carbon.

Several rate coefficients and equilibrium constants have also been measured for unsymmetrical exchange reactions, a selection of which are shown in Table 1.[34,86-88]   A knowledge of the equilibrium constant allows us, after due consideration of reorganisational factors and solvation changes, to make the estimates of relative bond dissociation energies discussed in

$$M'(chel')L^{n+} + RM(chel)L^{m+} \rightleftharpoons RM'(chel')L^{n+} + M(chel)L^{m+} \quad (35)$$

$$py(dmgH)_2Co + RCo(chgH)_2py \rightleftharpoons py(dmgH)_2CoR + Co(chgH)_2py \quad (36)$$

chgH = conjugate base of cyclohexanedionedioxime

XXXIV a   **XXXIV b**

Section 2.1.   It is instructive and amusing to note that each of the above reactions can equally well be described as a homolytic displacement at saturated carbon in the language of the organic chemist, and an inner-sphere one-electron redox process through a saturated carbon bridge in the language of the inorganic chemist.   However, it is surprising that the Pt(II) (salen) does not react with the otherwise fairly reactive MeCo(tfacen).[88]

In view of the very ready reaction of organic radicals at the γ-carbon of allylcobaloximes, it is not surprising that metal radicals also react even more readily at that γ-carbon than at the α-carbon of methylcobaloximes. Thus, when allyl(pyridine)cobaloxime is freshly prepared and dissolved in chloroform, it gives a proton n.m.r. spectrum characteristic of a static η^1-allyl complex.[89]   After a few minutes, as decomposition begins and cobaloxime(II) is formed, the spectrum changes;   the α-pyridine proton resonance becomes very broad and four terminal allyl proton resonances begin to coalesce.   The spectrum of the η^1-allyl complex can be restored by the addition of a trace of carbon tetrachloride or bromotrichloromethane, but the coalescence reappears after a few minutes.   This behaviour can be ascribed to the symmetrical homolytic displacement of cobaloxime(II) from the α-carbon by attack of the cobaloxime(II), formed in the decomposition, at the γ-carbon of the allyl ligand.   The broadening of the allyl resonances is therefore due to the rapid interchange of C-1 and C-3 but the broadening of the α-pyridine resonances is due to the fact that the pyridine is spending a small part of its time on the paramagnetic cobaloxime(II) (eq 38).

$$py(dmgH)_2\overset{*}{Co} \; + \; CH_2\!:\!CH.\overset{*}{C}H_2\overset{*}{Co}(dmgH)_2py \; \rightleftharpoons \; py(dmgH)_2CoCH_2CH\!:\!\overset{*}{C}H_2 \; + \; \overset{*}{Co}(dmgH)_2py \qquad (38)$$

2.3.4 Insertion Reactions

Oxygen, sulfur dioxide and sulfur insertion reactions are known in organometal macrocyle chemistry.   There is little doubt that oxygen insertion is a homolytic process, though there may well be more than one homolytic mechanism that occurs.   Insertion can take place thermally with those organocoaloximes which have weak carbon-metal bonds, i.e. allyl and benzylcobaloximes,   or   photochemically   with   more   stable organocobaloximes,[91-94]   alkylcobalt(TPP),[95]   alkyltin(TPP),[96]   and alkylgermanium(TPP)[96] complexes.   In the case of the   dialkylM(TPP) complexes (M = Sn or Ge) selective insertion of oxygen into one of the carbon-metal bonds can be obtained by the use of appropriate temperatures and wavelengths (eq 39).

The thermal reaction of oxygen with α-phenylethylcobaloxime (eq 40) can readily be monitored by proton n.m.r. spectroscopy, because it occurs at room temperature.   The x-ray crystal structure of the 4-methyl derivative of the insertion product has been determined.[97,98] Thermal insertions of oxygen into the carbon cobalt bond of benzylcobaloximes are approximately first order in oxygen and in the substrate.   The rate coefficient is only slightly depressed by a 100-fold increase in the concentration of the axial ligand pyridine, and the rate coefficients for insertion in a series of substituted benzylcobaloximes differ only by a factor of two. The rates of photochemical insertion into a wide range of organocobaloximes also vary little with the nature of the organic ligand.[94]

These results have been interpreted as an initial homolysis of the carbon metal bond, capture of the alkyl organic radical by oxygen (for some rate coefficients see Section 2.2.) and capture of the organoperoxy radical by the cobalt (II)   fragment.   The formation of the organic radical is supported by the observation that optically active sec-octylcobaloxime gives racemic sec-octylhydroperoxide on photochemical reaction with oxygen followed by cleavage of the alkylperoxycobaloxime with trifluoroacetic acid.[93]   One interesting feature of the insertion into the alkylcobalt(TPP) complexes is the large change in the chemical shifts of the organic ligand protons consequent on insertion.[95]   In butylCo(TPP), for example, the chemical shifts of the α-, β-, γ-, and δ-protons are -3.56, -4.64, -1.35, and -0.73 ppm, the large negative shifts being a result of the strong ring current of the porphyrin ring.   In the insertion product the butyl group is much further from this ring and the shifts are -1.09, -0.71, -0.71, -0.04, and +0.16 ppm, respectively.

$$R_2M(chel) + 2O_2 \longrightarrow ROOM(chel)R + O_2 \longrightarrow (ROO)_2M(chel) \qquad (39)$$

$$PhCH(CH_3)Co(dmgH)_2py \; + \; O_2 \longrightarrow PhCH(CH_3)OOCo(dmgH)_2py \qquad (40)$$

The products of sulfur dioxide insertion reactions have been studied.[30,99-101]   Good evidence of radical character comes from the formation of the <u>four</u> insertion products (**XXXVII** to **XL**) in the reaction of a 1:1 mixture of benzylrhodoxime (**XXXV**) and 4-bromobenzylcobaloxime (**XXXVI**) in liquid sulfur dioxide at room temperature.[102]   These observations have been interpreted as a homolytic chain reaction in which the radical fragments are each captured by sulfur dioxide to give organosulfonyl or cobaltsulfonyl radicals (eqs 41-48).   Reaction of the former at the benzylic carbon would lead to dibenzyl sulfone; reaction of the latter at the benzylic carbon of either (**XXXV**) or (**XXXVI**) would lead to the observed insertion products and the displacement of rhodoxime (II) or cobaloxime(II) as a chain propagating agent.

The formation and further reaction of organic radicals is evident from the formation, as a by-product during the insertion of sulfur dioxide into 3-methylbut-2-enylcobaloxime, of a single diallyl sulfone (**XLII**) as shown in eqs 49-51.[75]   Reaction 51 is directly analogous to the homolytic displacement reaction of eq 19 and to the intramolecular displacement of Scheme **10**.   The photochemical insertion of sulfur into the carbon-cobalt bond of organocobaloximes has also been studied, but the products $RS_nCo(dmgH)_2$ py tend to be photolysed further under irradiation, and a number of subsequent products are also obtained.[103]

$$ArCH_2Co(dmgH)_2py \; \rightleftharpoons \; Ar\dot{C}H_2 \; + \; \dot{C}o(dmgH)_2py \tag{41}$$
XXXVI

$$Ar\dot{C}H_2 \; + \; SO_2 \longrightarrow ArCH_2\dot{S}O_2 \tag{42}$$

$$\dot{C}o(dmgH)_2py \; + \; SO_2 \longrightarrow py(dmgH)_2Co\dot{S}O_2 \tag{43}$$

$$py(dmgH)_2Co\dot{S}O_2 \; + \; PhCH_2Rh(dmgH)_2py \longrightarrow py(dmgH)_2CoSO_2CH_2Ph \; + \; \dot{R}h(dmgH)_2py \tag{44}$$
$\quad\quad\quad\quad$ **XXXV** $\quad\quad\quad\quad\quad\quad$ **XXXVII**

$$py(dmgH)_2\dot{R}h \; + \; SO_2 \longrightarrow py(dmgH)_2Rh\dot{S}O_2 \tag{45}$$

$$py(dmgH)_2Co\dot{S}O_2 \; + \; ArCH_2Co(dmgH)_2py \longrightarrow py(dmgH)_2CoSO_2CH_2Ar \; + \; \dot{C}o(dmgH)_2py \tag{46}$$
$\quad\quad\quad\quad\quad\quad\quad\quad\quad\quad\quad$ **XXXVIII**

$$py(dmgH)_2Rh\dot{S}O_2 \; + \; PhCH_2Rh(dmgH)_2py \longrightarrow py(dmgH)_2RhSO_2CH_2Ph \; + \; \dot{R}h(dmgH)_2py \tag{47}$$
$\quad\quad\quad\quad\quad\quad\quad\quad\quad\quad\quad$ **XXXIX**

$$py(dmgH)_2Rh\dot{S}O_2 \; + \; ArCH_2Co(dmgH)_2py \longrightarrow py(dmgH)_2RhSO_2CH_2Ar \; + \; \dot{C}o(dmgH)_2py \tag{48}$$
$\quad\quad\quad\quad\quad\quad\quad\quad\quad\quad\quad\quad\quad$ **XL**

$$Me_2C:CH.CH_2Co(dmgH)_2py \; \rightleftharpoons \; Me_2C:CH.\overset{\cdot}{C}H_2 \; + \; Co(dmgH)_2py \qquad (49)$$

XLI

$$Me_2C:CH.\overset{\cdot}{C}H_2 \; + \; SO_2 \; \rightleftharpoons \; Me_2C:CH.CH_2\overset{\cdot}{S}O_2 \qquad (50)$$

$$Me_2C:CH.CH_2\overset{\cdot}{S}O_2 \; + \; Me_2C:CH.CH_2Co(dmgH)_2py \longrightarrow Me_2C:CH.CH_2SO_2CMe_2CH:CH_2 \; + \; Co(dmgH)_2py \quad (51)$$

XLII

3.   REACTIONS WITH ELECTROPHILES

Electrophilic species are also oxidising agents and Lewis acids and we may therefore expect that the reaction of any particular electrophile with an organometal macrocycle may take place by a variety of mechanisms at a variety of sites in the complex.   The proton is the least readily reduced of the common electrophiles and, since many reactions of electrophiles are carried out in acidic solution, it is convenient to discuss reactions with the proton first.

3.1   Reactions in acidic solution

A major problem in the reactions of organometal macrocycles in acidic solution is the incidence of equatorial and axial ligand protonation.   For example, it was shown in Scheme 2 that the organocobaloximes undergo mono- and di-protonation in aqueous acid and that this influences the strength of the carbon-metal bond.   It also tends to decrease the rate of attack of electrophiles on the organic ligand, as will be seen below.   A similar example, relevant to the rearrangement described in Section 5., is the influence of trifluoroacetic acid on the character of organocobaloximes. Trifluoroacetic acid first protonates the dimethylglyoximato ligand and then, on addition of an excess of acid, removes the pyridine ligand as pyridinium trifluoroacetate and probably coordinates to the metal as trifluoroacetate (eq 52).[104]   Complexes such as (**XLIII**) and (**XLIV**) have been isolated and characterised.[105,106,80]   The reactions of coenzyme B_{12} and related organocobalamins are markedly dependent on the acidity of the solution because of the reversible protonation and hence de-ligation of the axial dimethylbenzimidazole ligand.   The pH at which there is 50% 'base-on' and 50% 'base-off' cobalamin shown schematically as (**XLV**) and (**XLVI**) in eq 53 has been measured for a very wide range of complexes.[44]

$$RCo(dmgH)_2py \underset{}{\overset{CF_3COOH}{\rightleftharpoons}} \underset{\textbf{XLIII}}{[RCo(dmgH_2)(dmgH)py]^+} \underset{}{\overset{CF_3COOH}{\rightleftharpoons}} RCo(dmgH_2)(dmgH)OCOCF_3 + pyH^+ \quad (52)$$

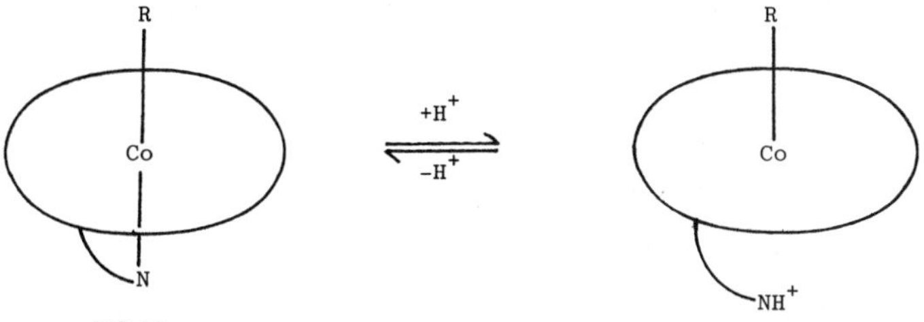

$$(53)$$

XLV   **XLVI**

Reactions of organic ligands may occur simultaneously with the above processes with and without cleavage of the carbon-metal bond. Thus, the ethanolysis of β-acetoxyethylcobaloxime occurs smoothly at a rate greater than expected for an unactivated primary alkyl acetate ($k_{obs}=4.4 \times 10^{-6}s^{-1}$ at 298 K).[107] Deuterium and carbon-13 labeling of the methylene groups has shown that scrambling of intact methylene groups takes place during ethanolysis, and it has been postulated that there is an intermediate η^2-ethylene complex on the reaction path (Scheme 11).[107-110] The kinetics of production of olefins from several substituted β-hydroxyethylcobaloximes have been measured in aqueous solution, the observed rate law being

$$k_{obs} = (k_o + k_a[H^+] + k'_o K_H[H^+])/(1 + K_H[H^+]) \quad (54).$$

where k'_o refers to the rate of reaction of the equatorial ligand-protonated complex with water, and K_H is the equilibrium constant for that protonation. The reactions are further complicated by the isomerisation of 2-hydroxy-iso-propyl- to 2-hydroxypropylcobaloxime during the decomposition, indicating that the formation of the η^2-olefin complex is reversible (Scheme 12).[109,66]

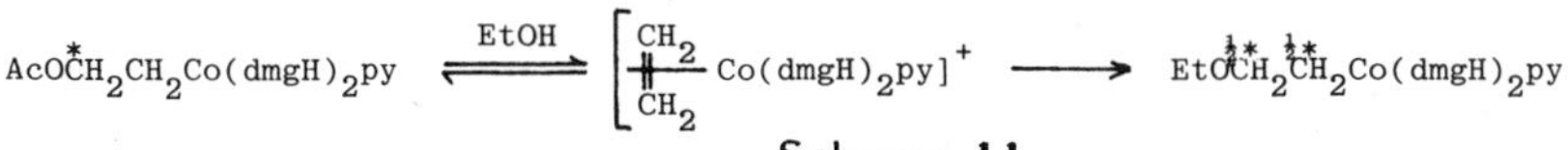

$$AcO\overset{*}{C}H_2CH_2Co(dmgH)_2py \rightleftharpoons \begin{bmatrix} CH_2 \\ \| \\ CH_2 \end{bmatrix} Co(dmgH)_2py]^+ \longrightarrow EtO\overset{**}{C}H_2\overset{**}{C}H_2Co(dmgH)_2py$$

Scheme 11

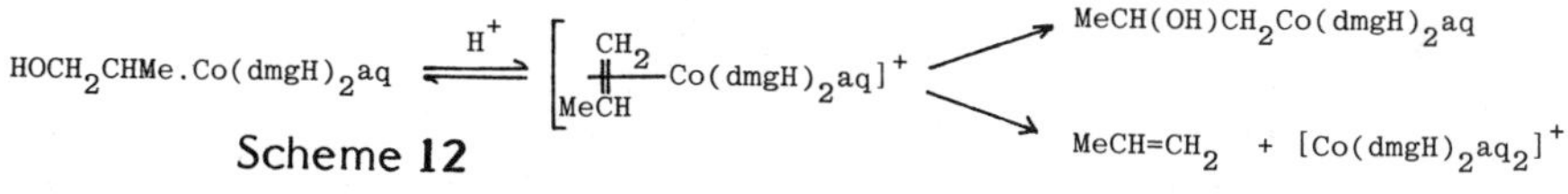

$$HOCH_2CHMe.Co(dmgH)_2aq \overset{H^+}{\rightleftharpoons} \begin{bmatrix} CH_2 \\ \| \\ MeCH \end{bmatrix} Co(dmgH)_2aq]^+ \begin{array}{c} \nearrow MeCH(OH)CH_2Co(dmgH)_2aq \\ \searrow MeCH=CH_2 + [Co(dmgH)_2aq_2]^+ \end{array}$$

Scheme 12

An interesting variation of this reaction is the use of substituted ethylcobalt(III) complexes as protecting groups for carboxylic acids. Thus, not only can esters of β-hydroxyethylcobalt(phthalocyanine) be prepared, characterised, and cleaved to give the acid, ethylene and cobalt(III)phthalocyanine[111] but aquocobalamins and dibromo(1-hydroxy-2,2,3,3,7,7,8,8,12,12,13,13,17,17-hexadecamethyl-10,20-diazaoctahydroporphinato)cobalt(III) can be used as catalysts (10^{-3} mol equiv), in the presence of an excess of zinc dust, for the reductive hydrolysis to bromide ion, ethylene, and the acid, of esters of β-bromoethyl alcohol, such as penicillinV β-bromoethyl ester (Scheme **13**).[112]

Not only is the formation of the η^2-olefin complex reversible, but exchange of olefins through η^2-complexes can also take place to a limited extent.[113] For example, 2-hydroxyethylcobaloxime reacts with the unsaturated alcohol (**XLVII**) in the presence of $BF_3.Et_2O$ or perchloric acid to give the acyclic and the cyclic complexes (**XLVIII**) and (**XLIX**), respectively (Scheme **14**). The latter can be synthesised by the

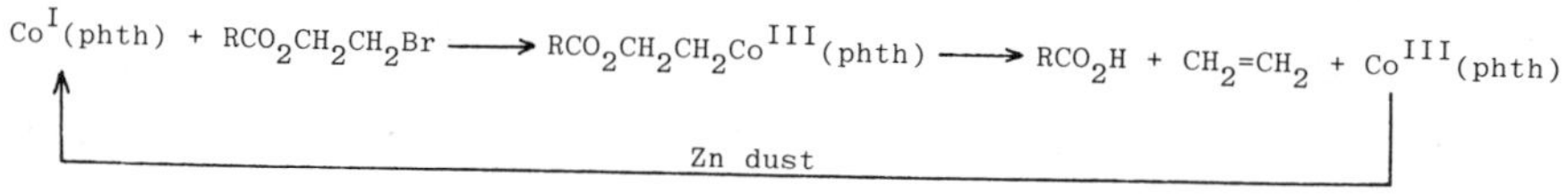
Co^I(phth) + RCO2CH2CH2Br → RCO2CH2CH2Co^III(phth) → RCO2H + CH2=CH2 + Co^III(phth)
Zn dust

Co(phth) = dibromo(1-hydroxy-2,2,3,3,7,7,8,8,12,12,13,13,17,17-hexadecamethyl-10,20-diaza-octahydroporphinato)cobalt

Scheme **13**

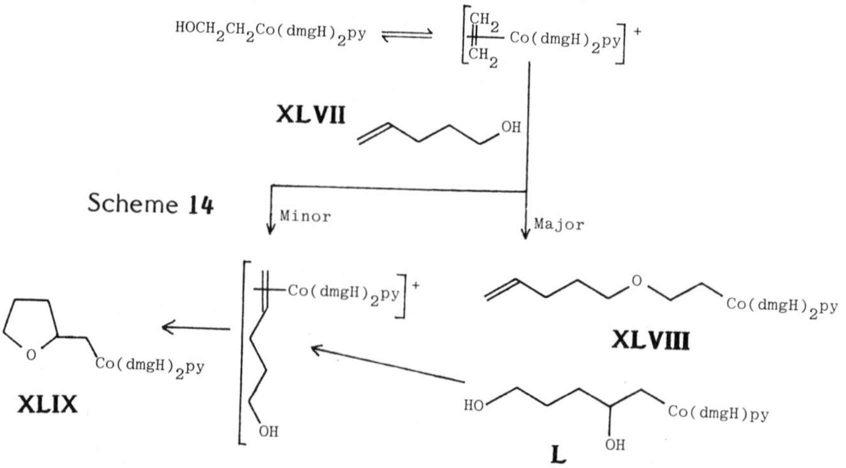
HOCH2CH2Co(dmgH)2py
Co(dmgH)2py
XLVII
OH
Minor
Major
Co(dmgH)2py
XLVIII
XLIX
L
Co(dmgH)py

Scheme **14**

intramolecular reaction of the dihydroxy complex (L) with silicic acid during chromatography.[114]   The configurational stability of the olefin complex is apparent from the formation of (+)-(S)-2-benzyloxyethylcobaloxime from (+)-(S)-2-acetoxyethylcobaloxime in benzyl alcohol (Scheme 15).[115]

Attack at a β-oxygen substituent is also responsible for the acid-catalysed decomposition of acetals derived from formylmethylcobalamin and cobaloxime.[116,117]   Thus, the cyclic acetal (LI), which can be prepared by the reaction of aquocobalamin(III) with the olefin (LII) in the presence of triethylamine, is hydrolysed under mildly acidic conditions to formylmethylcobalamin and subsequently to aquocobalamin and acetaldehyde as shown in Scheme 16.

3.2   Reaction with oxidising agents

Because of the possibility that a number of reactions with electrophiles occur by electron transfer processes, it is convenient next to discuss the consequences of such oxidation.   Halpern has studied the oxidation of several organocobalt(III) complexes by reagents such as Ce(IV) and $IrCl_6^{2-}$.[118,119]   The product, the organocobalt(IV) species such as $RCo(dmgH)_2py^+$, is short lived at 25°C, but can be characterised in solution at -50°C.   The esr parameters show the presence of a low spin $3d^5$ complex

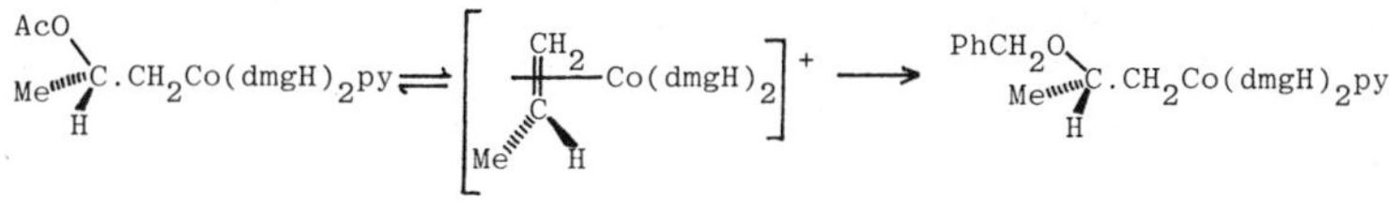

Scheme 15

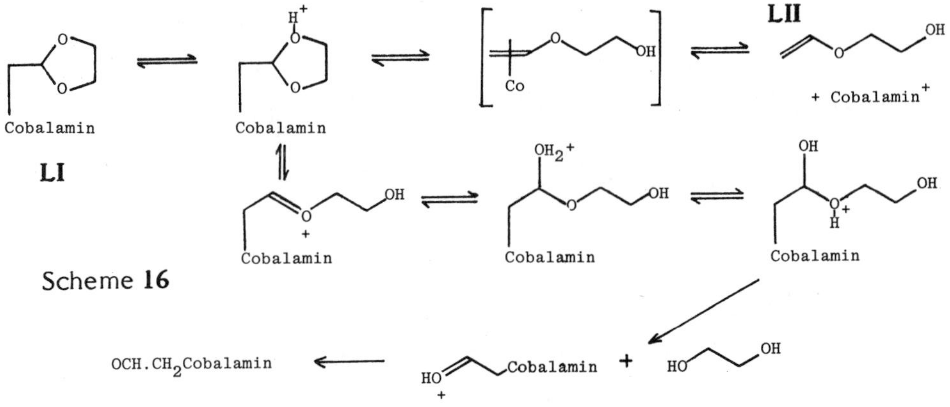

Scheme 16

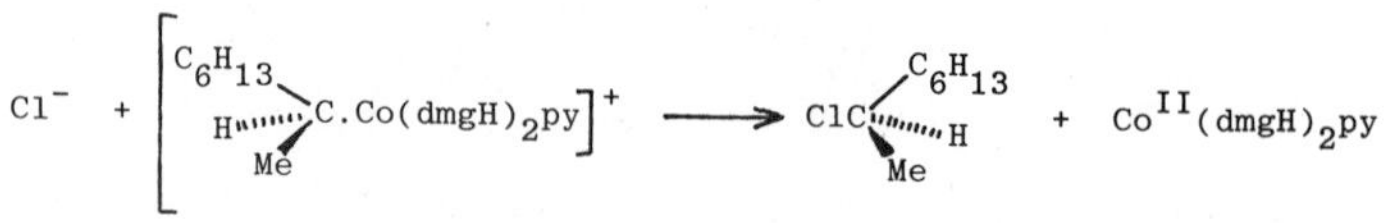

$$Cl^- + \left[\begin{array}{c} C_6H_{13} \\ H^{\text{\tiny\dots}}C.Co(dmgH)_2py \\ Me \end{array} \right]^+ \longrightarrow ClC^{\text{\tiny\dots}}\begin{array}{c} C_6H_{13} \\ H \\ Me \end{array} + Co^{II}(dmgH)_2py \quad (55)$$

with a $3d_{x^2-y^2}$ HOMO (x,y axes chosen between in-plane Co-N bonds) in which the unpaired electron has 30% $4p_z$ character but with very little interaction with the equatorial ligand orbitals.[120]

A key feature of the reactions of these organocobalt(IV) complexes is their susceptibility to nucleophilic attack (compare for example, Section 4.1.). Thus in eq 55 the optically active (R)-sec-octylcobaloxime, prepared at -50°C, reacts with chloride ion to give a mixture of (S)-sec-octyl chloride and octenes.[121] The relevance of this result to halogenation of organocobaloximes will be discussed in Section 3.5. The kinetics of oxidation of organocobaloximes by $IrCl_6^{2-}$ have been studied and the formation of $IrCl_6^{3-}$ and the organocobaloxime(IV) cations has been found to be reversible (Scheme 17) in aqueous methanol and influenced by the rate of nucleophilic attack of the solvent on the α-carbon.[122] The rate law

$$-d[RCo(III)]/dt = k_1k_2[RCo(III)][IrCl_6^{2-}]/(k_{-1}[IrCl_6^{2-}]+k_2) \quad (56)$$

allows the direct evaluation of k_1 and the ratio k_1k_2/k_{-1} from which k_2 can also be determined on consideration of the oxidation potential since k_1/k_{-1} = K (Table 2). The values of k_2 measured independently compared well with those calculated in the above manner.

For those substituents where the attack at the α-carbon is hindered, olefins are usually the main products of decomposition of the organocobalt(IV) intermediates (e.g. with $PhCH_2CH_2Co(dmgH)_2py^+$) and the rate of reaction is second order in the substrate. In these reactions some 50% of the substrate is recovered at the end of the reaction and a disproportionation mechanism (Scheme 18) has been proposed.

$$RCo(dmgH)_2OH_2 + IrCl_6^{2-} \underset{k_{-1}}{\overset{k_1}{\rightleftharpoons}} IrCl_6^{3-} + [RCo(dmgH)_2OH_2]^+ \overset{k_2}{\longrightarrow} ROH + Co(dmgH)_2py$$

Scheme 17

$$2[RCo(dmgH)_2L]^+ \overset{slow}{\longrightarrow} RCo(dmgH)_2L + [RCo(dmgH)_2L]^{2+} \overset{fast}{\longrightarrow} Olefin + Co(III)?$$

Scheme 18

Table 2.

Rate constants for reactions with $IrCl_6^{2-}$, and oxidation potentials, of organocobaloximes $RCo(dmgH)_2$ aq in 1 M perchloric acid.

R	$k_1/dm^3 mol^{-1} s^{-1}$	k_2/s^{-1}	$E_{\frac{1}{2}}$
4-MeO-benzyl	3.4×10^5	10^2	0.849
4-Me-benzyl	1.4×10^5		0.859
Benzyl	5×10^4	8.3	0.873
4-F-benzyl	5×10^4	9.1	0.876
4-Cl-benzyl	1×10^4	1.7	0.907
4-NO_2-benzyl	1×10^2	1.7×10^{-2}	0.902
Me		1.8×10^{-2}	0.878
Et		1.0×10^{-2}	0.867
n-Pr		7.3×10^{-3}	0.856
i-Pr		38	

Volpin and Costa [123-125] have obtained similar evidence for the formation of organocobalt(chelate)(IV) in their electrochemical studies. The oxidation potentials of a number of complexes have been measured, including: 4-nitrobenzylcobaloxime (0.907 v) > 4-chlorobenzylcobaloxime (0.876 v) > 4-fluorobenzylcobaloxime (0.873 v) > benzylcobaloxime (0.859 v) > 4-methylbenzylcobaloxime (0.849 v), and methyl(pyridine)cobaloxime (0.920 v) > MeRh(SALEN)py (0.838 v) > MeCo(SALEN) (0.739 v) > Pr^iRh(SALEN)py (0.772 v) > Pr^iCo(SALEN) (0.458 v) in acetonitrile. The potential for the oxidation seems to be markedly influenced by the nature of the alkyl group.

A novel oxidation is that of the carbeneiron(II) complex (**LIII**) which can be oxidised by Cu(II) to the corresponding carbeneiron(III) cation (Scheme **19**).[126]

3.3 Reaction with Metallic Electrophiles

3.3.1 Mercury(II)

It has been widely suggested that the reaction between methylcobalamin from sewage bacteria and mercury(II) compounds from chemical plants, is responsible for the formation of the highly toxic methylmercury(II) compounds in aquatic environments, particularly in the Great Lakes of North America.[127] As a result, the chemical reactions of mercury(II) compounds with organocobalt(III) complexes have received a disproportionate amount of attention, though still insufficient to resolve all the problems of mechanism.

There is little doubt that most primary organocobalt(III) complexes react with Hg^{2+} and with $MeHg^+$ to give RHg^+ and RMeHg respectively (eqs 57 and 58), but the reactions are complicated when coordinating ligands, particularly soft anions, are present. Mercury(II) coordinates to up to four halide ions or other ligands and also undergoes appreciable hydrolysis in other than rather acidic conditions. Unfortunately, these factors have not been taken into account in all the studies.

A large number of rate measurements have been made on the reactions of mercury(II) species with synthetic organocobalt complexes.[128-136] For each metal system, the order of reactivity is Me ~ $PhCH_2$ > Et > Pr, and the

$$Ar_2CHCCl_3 \quad + \quad Fe(TPP) \quad \longrightarrow \quad Ar_2C{=}Fe(TPP) \quad \xrightarrow{\quad Cu(II) \quad} \quad Ar_2C{=}Fe(TPP)^+$$

$$\textbf{LIII} \qquad \qquad \text{Scheme 19}$$

$$RM(chel)L \quad + \quad Hg^{2+} \longrightarrow RHg^+ \quad + \quad [M(chel)L]^+ \qquad (57)$$

$$RHg^+ \quad + \quad RM(chel)L \longrightarrow R_2Hg \quad + \quad [M(chel)L]^+ \qquad (58)$$

influence of the chelating ligand is such that the rate of attack of the Hg^{2+} ion is highest when the charge on the central metal ion is least.   Thus, for example, the dipositive cation $RCo([14]aneN_4)^{2+}$ is less reactive than the monopositive cation $RCo(dotnH)aq^+$ which is less reactive than most of the neutral complexes shown in Table 3.   That this is not simply an electrostatic effect is evident from a more detailed examination of the properties of the equatorial ligands and of their influence on other properties of the molecule, and also because the corresponding dipositive complex $MeCr([15]aneN_4)^{2+}$ is among the most reactive of those studied.[137]

The rates of many of the reactions, particularly of the cobaloximes and the dotnH complexes, are reduced in the presence of acid (necessary to prevent hydrolysis of the Hg^{2+} ion) through protonation of the equatorial ligand, and the rate coefficients shown in Table 3 have been corrected accordingly.   It has also been suggested that some equatorial ligands, such as dotnH, complex reversibly with mercury(II).[134]   Rate coefficients are not quoted in the table for reactions of organocobalt complexes with mercury(II) acetate, chloride, cyanide, etc., for, though the conditions of such reactions have usually been properly defined, the exact nature of the mercury(II) species present in solution has not usually been examined, making comparisons of less value.   Some reactions have been carried out in the presence of anionic surfactants which have a marked accelerating effect, through their action as micellar catalysts,[131] even when they do not complex directly with the mercury(II).

The above reactions have all been interpreted as bimolecular displacements at the α-carbon of the primary alkyl or benzyl ligand, through attack of the metal electrophile at the highest occupied orbital of the complex, namely the C-Co or C-Cr bonding orbital.   This is consistent with the observation that the diastereoisomeric dideuteriated cobaloxime (**LIV**)

Table 3.

Kinetics of reaction of Hg^{2+} with organometal chelate complexes RM(chel)L at 25°C.

Medium	R	M(chel)L	$k_2/M^{-1}s^{-1}$	Ref
aq acid μ=1.0	Me	$Co(dotnH)aq^+$	177	129
$HClO_4$ μ=2.0	Me	$Co(dotnH)aq^+$	5.9	135
aq acid μ=1.0	Me	$Co(dmgH)_2aq$	54	132
" "	Me	$Co(dmgH)_2aq$	65	129
" "	Me	$Co(Me_4[14]tetraeneN_4)^{2+}$	45×10^{-4}	129
" "	Me	$Co(Me_2SALEN)$	2.1×10^4	129
" "	Me	$Co(SALEN)$	2.6×10^4	129
" "	Me	$Cr([15]aneN_4)aq^{2+}$	3×10^6	137
" "	nPr	$Cr([15]aneN_4)aq^{2+}$	2.5×10^3	137
" "	$PhCH_2$	$Cr([15]aneN_4)aq^{2+}$	1.1×10^3	137
" "	cyclohexyl	$Cr([15]aneN_4)aq^{2+}$	1.6×10^{-3}	137
" "	adamantyl	$Cr([15]aneN_4)aq^{2+}$	3×10^{-3}	137
$HClO_4$ μ=2.0	Et	$Co(dotnH)aq^+$	1.8×10^{-2}	135
" "	Et	$Co(doenH)aq^+$	1.4×10^{-3}	134
" "	nPr	$Co(dotnH)aq^+$	5.7×10^{-3}	134
" "	$PhCH_2$	$Co(dotnH)aq^+$	7.7×10^{-3}	134
" "	$PhCH_2$	$Co(dmgH)_2aq$	7.5×10^{-2}	132
" "	$4-NO_2C_6H_4CH_2$	$Co(dmgH)_2aq$	6.5×10^{-3}	132
" "	$4-MeOC_6H_4CH_2$	$Co(dmgH)_2aq$	1.1×10^{-1}	132

HClO$_4$ μ=1.0	Et	Co(dmgH)$_2$aq	1.2×10^{-1}	132
" "	Ph	Co(dmgH)$_2$aq	4.4×10^{2}	132
" "	Me	Co(dmgBF$_2$)$_2$aq	3.5×10^{-4}	133
" "	Me [a]	Cr([14]aneN$_4$)$^{2+}$	1.6×10^{3}	137
" "	Et [a]	Cr([14]aneN$_4$)$^{2+}$	9.9	137
" "	PhCH$_2$ [a]	Cr([14]aneN$_4$)$^{2+}$	5.2	137
HClO$_4$ μ=0.10	Me [b]	Co([14]aneN$_4$)Me$^+$	1.3×10^{5}	136
" "	Me [b]	Co(Me$_2$pyo[14]trieneN$_4$)Me$^+$	2.1×10^{5}	136
" "	Me [b]	Co(dotnH)Me	2×10^{-6}	136

[a] MeHg$^+$ as reagent. [b] PhHg$^+$ as reagent.

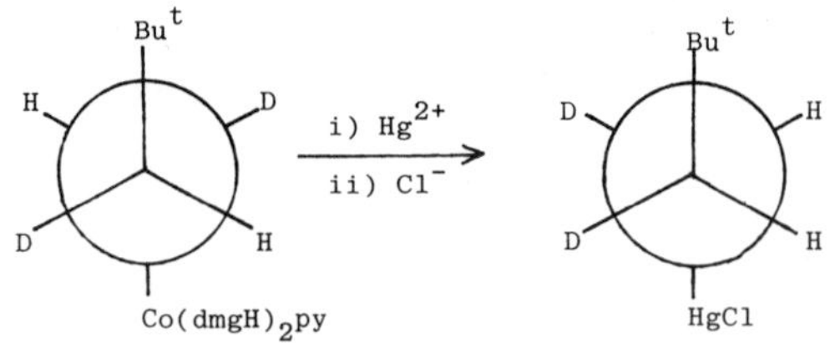

$$R_2Co(dotnH) \quad + \quad Hg^{2+} \longrightarrow RHg^+ \quad + \quad [RCo(dotnH)aq]^+ \tag{60}$$

$$[RCo(dotnH)aq]^+ \quad + \quad Hg^{2+} \longrightarrow RHg^+ \quad + \quad [Co(dotnH)aq_2]^{2+} \tag{61}$$

$$R_2Co(dotnH) \quad + \quad R'Hg^+ \longrightarrow RHgR' \quad + \quad [RCo(dotnH)aq]^+ \tag{62}$$

$$R = Me, PhCH_2, Ph \qquad R' = Me, Ph$$

(eq 59) and 4-Bu^{t}cyclohexylcobaloxime react with Hg^{2+} with _inversion_ of configuration at the α-carbon,[138,139] since the front side of C(α) is blocked by the equatorial ligands. However, with secondary alkylcobalt complexes the situation is more complicated for whereas _cis-2-_ methoxycyclohexylcobaloxime reacts with Hg^{2+} followed by chloride ion to give mainly _trans-2-methoxycyclohexylmercury(II)_ chloride, the corresponding _trans_-isomer gives mainly elimination products characteristic of a one-electron oxidation process.[139] Indeed, clean substitution is seldom found with secondary alkylcobalt(III) complexes and mercury(II) salts, probably because the alternative oxidation processes proceed by default.

There is a substantial difference between the rates of reaction of Hg^{2+} and MeHg$^+$, the former being generally about two or three orders of magnitude the more reactive. Similarly, the displacement of the first methyl group from the complex Me$_2$Co(dotnH)aq$^+$ is very much faster than the displacement of the second methyl group (eqs 60-62).[140]

Mercury(II) species also react with vinylcobaloximes, in some cases with retention of configuration, in others, with little stereospecificity. It would appear that the degree of coincidence between bond making and bond breaking may depend on the substituents present, being greater in the styrylcobaloximes[141,142] (eq 63) than in the octenylcobaloximes.[142,143] All the above reactions are irreversible, no organocobalt complex being detected in the reactions between alkylmercury compounds and cobalt(III) complexes. Organocobalt(III) complexes can however be prepared by the

$$\text{cis-PhCH=CHCo(dmgH)}_2\text{py} \xrightarrow[\text{ii) Cl}^-]{\text{i)Hg(OAc)}_2\text{/AcOH}} \text{cis-PhCH=CHHgCl} \qquad (63)$$

reaction between the corresponding cobalt(I) complexes and organomercury compounds.[140]

The reactions of the organocobalamins, especially of methylcobalamin, are very similar to those described above, except that there is one severe complicating factor.[144-150] The rate of loss of the benziminazole ligand from the metal, and hence of its capture by the mercury(II) species or by the proton, is comparable with the rate of attack of the mercury(II) species on the organic ligand (Scheme **20**). The kinetic analysis is therefore complicated, especially as these reactions are usually studied by ultraviolet spectroscopy, and the sometimes small spectral changes must both be interpreted correctly and used in the determination of the rate profile. Comparision of the results is difficult because of the different mercury(II) species used and the variations in solvent composition. In the case of methylcobalamin, the addition of a surface active agent causes a decrease in the rate of reaction with mercury(II) acetate probably because of its influence on the proportion of 'base-off' species in solution.[151]

3.3.2 Platinum, palladium, and gold

The reaction of $PdCl_4^{2-}$ with methylcobalamin is reported to give the unstable $MePdCl_3^{2-}$ and hence methyl chloride and palladium(0), but the corresponding $PtCl_4^{2-}$ merely acts as a catalyst in the reaction of $PtCl_6^{2-}$ (or vice versa).[152,153] It seems likely that some association between the $PtCl_4^{2-}$ and the equatorial ligand activates the complex to electrophilic

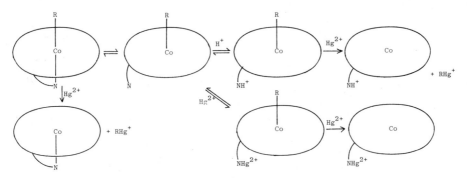

Scheme **20**

attack by the platinum(IV) species.[154-156]   Such interaction has been detected by proton nmr (K' = 0.46 at pH 1.22 and ionic strength 2.0 M at 25°C in HCl/LiCl), and the rate law

$$\text{Rate} = \frac{kK' [PtCl_4^{2-}][PtCl_6^{2-}][\text{Mecobalamin}]_{\text{Total}}}{1 + K' [PtCl_4^{2-}]} \qquad (64)$$

has been deduced.   Thus, at low $PtCl_4^{2-}$ the rate becomes first order in both $PtCl_4^{2-}$ and $PtCl_6^{2-}$.   These reactions are, of course, markedly dependent on the chloride ion concentration.   Similar cooperative effects have been noted in the reaction of methylcobalamin with Au(I) and Au(III) complexes.

Methyl transfer to palladium(II) probably also occurs in several useful alkylation reactions.   Thus, both quinones and activated olefins can be methylated by methylcobaloxime in the presence of palladium(II) salts (eqs 65 and 66).[157,158]

$$O{=}\langle\ \rangle{=}O \ +\ MeCo(dmgH)_2py \xrightarrow{\ Pd(II)\ } O{=}\langle\overset{Me}{\ }\rangle{=}O \qquad (65)$$

$$PhCH{=}CH_2 \ +\ MeCo(dmgH)_2py \xrightarrow{\ Pd(II)\ } PhCH{=}CH.CH_3 \qquad (66)$$

3.3.3 Cobalt

A series of transmethylation reactions have been described in which a cobalt(III) chelate reacts with a methylcobalt(III) chelate complex as in eqs 67 to 70.[90]   It is not certain whether all of these reactions proceed directly by an electrophilic displacement of one cobalt(III) by attack of another on the α-carbon, or through a faster catalytic process in which a trace of one cobalt(II) complex displaces the other cobalt(II) complex (cf eq 36 in Section 2.3.3.).   However, the equilibrium studies, which are independent of the choice of mechanism, show that the strength of the methylcobalt bond increases in the series: ACACEN < 7,7'-Me$_2$SALEN < SALEN < SALOPHEN < cobalamin < dotnH, with the weakest being that for the loss of the first methyl group from $Me_2Co(dotnH)$ and $Me_2Co(Me_4[14]tetraene\ N_4)$.[136]

$$Me_2Co(dotnH) \ +\ [Co(chel)OH_2]^+ \longrightarrow [MeCo(dotnH)aq]^+ \ +\ MeCo(chel)$$

$$chel = 7,7'\text{-}Me_2salen,\ salophen,\ salen,\ cobalamin \qquad (67)$$

$$Me_2Co(dotnH) + [PhCo(dotnH)]^+ \longrightarrow Ph(Me)Co(dotnH) + [MeCo(dotnH)]^+ \quad (68)$$

$$[MeCo(chel)aq]^+ + [Cobalamin]^+ \longrightarrow MeCobalamin + [Co(chel)aq]^+ \quad (69)$$

$$MeCo(chel)aq + [Co(dotnH)aq_2]^{2+} \longrightarrow [MeCo(dotnH)aq]^+ + [Co(chel)aq]^+ \quad (70)$$

3.3.4. Other metals

Thallium(III),[132,155] tin(II),[159] lead(IV),[161] lead(II),[160] cadmium(II),[160] zinc(II),[160] and arsenic[162] have all been postulated as capable of demethylating methylcobalamin through attack of the metal species at the α-carbon. However, iron(III) complexes react with the dioximato ligands of cobaloximes[163] and catalyse the reaction of tin(II) with methylcobalamin.[159] The latter has been explained as a prior oxidation of tin(II) to a tin(III) radical which then takes part in homolytic attack at the α-carbon.

Copper(II) complexes probably react through one-electron oxidation of the organocobalt(III) substrate.[164] Manganese(III) acetate in acetic acid also oxidises allylcobaloximes, probably through the formation of reactive carboxyalkyl radicals ($HO_2\dot{C}CH_2$), and the main product is the O-allyl derivative of dimethylglyoxime.[165]

3.4 Tetracyanoethylene and hexafluoroacetone

Tetracyanoethylene reacts with allylcobaloximes in methylene chloride to give cyclic addition products.[89] These reactions are believed to be sequential processes in which an initial attack of the olefin on the γ-carbon of the allyl group and synchronous migration of the metal gives a zwitterionic η^2-olefin complex intermediate (**LV**), the anionic end of which then attacks the terminal carbon of the η^2-olefin complex with migration of the metal to the β-carbon to give the cyclic complex (**LVI**) (Scheme 21). The crystal structure of the product (**LVI**, R = Ph) from trans-cinnamylcobaloxime shows that it has the trans-stereochemistry,[166] consistent with a trans-antarafacial migration of the cobalt. Hexafluoroacetone reacts similarly with propadienylcobaloximes via a η^2-alkyne complex to give the cyclic product (**LVII**) (Scheme 22).

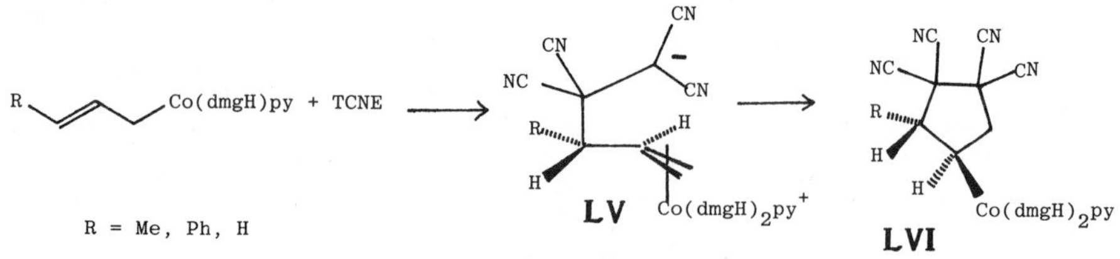

Scheme 21

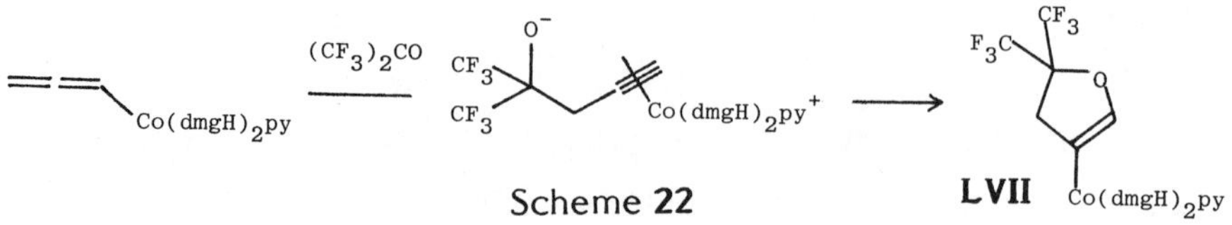

Scheme 22   LVII Co(dmgH)$_2$py

3.5 Nitrosation

The secondary aminomethylcobaloxime (**LVIII**) undergoes a smooth nitrosation in aqueous acidic solution, with concurrent exchange of the axial ligand (eq 71).[167]   However, nitrosonium sulphate in acetic acid oxidises benzylcobaloxime and the main product is the O-benzyl derivative of dimethylglyoxime[80] (see 3.3.4.).

3.6 Halogenation

Halogenation of organometal chelate complexes is perhaps the most difficult of the electrophilic reactions to interpret because of competition from one- and two-electron oxidation processes and the resulting confusion between organic halide formed directly by halogenation and indirectly by attack of halogen atoms or halide ions on intermediates. It is seldom that a single organic product is formed in these reactions, and the kinetics are rarely straightforward, and give little information about product forming steps when the first step is a rate determining oxidation of the substrate. It has been proposed that bromination of alkylcobaloximes proceeds predominantly through a one-electron oxidation followed by attack of bromide ion on the α-carbon of the intermediate organocobaloxime(IV)

$$C_6H_5NHCH_2Co(dmgH)_2py \xrightarrow{\text{H}^+/\text{NO}_2^-} [C_6H_5N(NO)CH_2Co(dmgH)_2NO_2]^- \qquad (71)$$

LVIII

$$RCo(dmgH)_2py + Br_2 \longrightarrow [RCo(dmgH)_2py]^+ + Br_2^{\cdot-} \qquad (72)$$

$$[RCo(dmgH)_2py]^+ + Br^- \longrightarrow RBr + Co(dmgH)_2py \qquad (73)$$

complex (eqs 72 and 73) i.e. a process similar to that described with cerium(IV) and halide ions (Section 3.2.). Indeed, the products of reaction of sec-octylcobaloxime with bromine are very similar to those of oxidation in the presence of bromide ion,[168] i.e. the organic bromide and octenes. As in the case of mercuration, the bromination of cis- and trans-2-methoxycyclohexylcobaloxime gives trans-2-methoxycyclohexyl bromide and elimination products, respectively.[139]

Several kinetic studies of halogenation have been carried out.[169-171] The reactions of organocobaloximes with iodine monochloride illustrate the problems well, because the organic products contain mixtures of organic chloride and organic iodide each formed with only partial inversion of configuration.[172,173] It seems likely that oxidation is the main process in this case also and that chloride and iodide ions and chlorine and iodine atoms are also intermediates, as well as some organic radicals from the organocobaloxime(IV) intermediate, as shown in Scheme 23.

The bromination of cyclodecylcobalamin has been used as a key reaction in determining the position of a deuterium label in the substrate. It has been proposed that hydrogen atom transfer occurs during the formation of cyclodecylcobalamin from cobalamin(I) and cyclodecyl bromide (Scheme 24).[174] The evidence for this is based on the fact that the final product formed on bromination of the decylcobalamin made from $1-[^2H_1]$-cyclodecyl bromide, has the deuterium label in positions 1,5, and 6. In view of the oxidative nature of the bromination reaction and the fact that this is a somewhat hindered secondary alkyl ligand, it seems possible that the hydrogen atom transfer occurs during the bromination step.

Vinylcobaloximes react more smoothly with halogens, probably because of the greater propensity of olefins to electrophilic attack. Thus, cis-styrylcobaloxime, conveniently prepared by the reaction of cobaloxime(I)

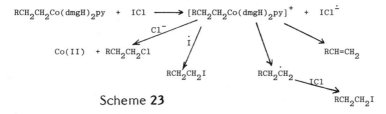

Scheme 23

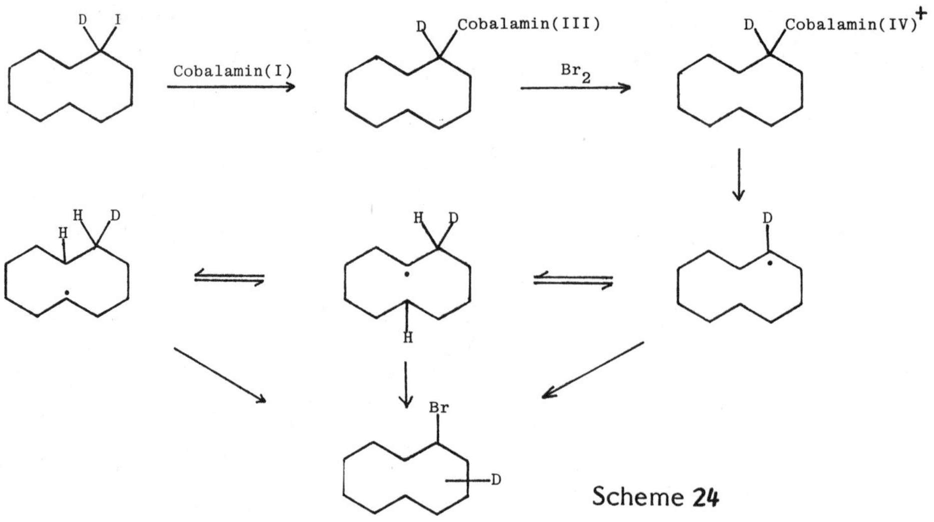

Scheme 24

with phenylacetylene under basic conditions (Scheme 25), reacts with chlorine, bromine, and iodine with complete retention of configuration and in high yield. This reaction provides a very quick and convenient synthesis of isomerically pure cis-styryl halides from phenylacetylene.[141] The bromination of 1-octenylcobaloximes is less stereospecific, the degree of stereospecificity being a function of the nature of the solvent.[142,143] Benzyl and substituted benzylcobaloximes react with bromine in acetic acid to give both cleavage and aromatic substitution products, depending on the nature of the substituents initially present in the benzene ring.[175]

3.7   Sulfenyl halides

Phenylsulfenyl halides do not react readily with organocobaloximes, but 2,4-dinitrophenylsulfenyl chloride reacts rapidly and regiospecifically with allylcobaloximes and with but-3-enylcobaloximes to give moderate yields (40-50%) of the corresponding rearranged allyl or cyclopropylcarbinyl phenyl sulfides, through attack of the sulfur electrophile at the γ- and δ-carbon, respectively, of the allyl or but-3-enyl ligand (eqs 74 and 75).[176]

$$PhC \equiv CH \;+\; \left[Co(dmgH)_2py\right]^- \longrightarrow \underline{cis}\text{-}PhCH=CH.Co(dmgH)_2py \xrightarrow{\;X_2\;} \underline{cis}\text{-}PhCH=CHX \;+\; XCo(dmgH)_2py$$

X = Cl, Br, I   Scheme 25

$$\text{(74)}$$

$$\text{(75)}$$

4.   REACTIONS WITH NUCLEOPHILES

4.1   Reactions at the α-carbon of the organic ligand

There are three main problems associated with the reactions of nucleophiles with organometal chelate complexes.   First, many good nucleophiles are also good bases and hence may take part in equilibria in which protons are removed from axial and equatorial ligands prior to or in competition with reaction elsewhere.   Secondly, many nucleophiles are also good ligands to transition metals and hence may replace the sixth axial ligand before, or in competition with, attack elsewhere on the complex. Complications often arise because the rate at which a particular axial ligand is replaced varies by many orders of magnitude depending on the nature of the organic ligand.   Thirdly, and more fundamentally, where a nucleophilic displacement takes place at the organic ligand, the displaced low-valent metal ion is itself likely to be a very nucleophilic species.   Such displacements are therefore reversible and to an extent that the equilibrium usually lies dramatically in favour of the substrate and reagent nucleophile (eq 76).   Indeed, the reverse of reaction 76 is one of the main means of preparation of organometal chelate complexes.[177]

We would therefore expect to find such nucleophilic displacement reactions where (i) the attacking nucleophile is of superior or comparable nucleophilicity to the displaced metal species, or (ii) where the nucleophile has moderate strength but the equilibrium is shifted to the right by

$$N^- \; + \; RM(chel) \; \rightleftharpoons \; RN \; + \; M(chel)^- \qquad (76)$$

$$RCo(dmgH)_2py \; + \; [Co(chgH)_2py]^- \; \rightleftharpoons \; RCo(chgH)_2py \; + \; [Co(chgH)_2py]^- \qquad (77)$$

$$[RCo(dotnH)]^- \; + \; [RCo(dotnH)]^+ \; \longrightarrow \; R_2Co(dotnH) \; + \; Co(dotnH) \qquad (78)$$

$$[MeCo(dotnH)PPh_3]^+ \; + \; [Co(salen)]^- \longrightarrow \; MeCo(salen) \; + \; Co(dotnH)PPh_3 \qquad (79)$$

irreversible reaction of the displaced low-valent metal complex with some reagent such as oxygen or a reactive organic halide. In the first category are the reactions of cobalt(I) nucleophiles with organocobalt(III) chelates. The rates of exchange for a number of the near-symmetrical exchange reactions (eq 77) have been measured,[85] and bear a surprising resemblance to those for the corresponding cobalt(II) exchange reactions. The rate variation from R = Me to R = sec-Bu is some six orders of magnitude and, since the erythro-2-phenylethyl complex (**XXXIII**) undergoes racemerisation at a rate comparable with that of the displacement reaction, it is assumed that the displacement takes place with inversion of configuration at the α-carbon, as expected for an S_N2 reaction. Similar exchange reactions have been observed with other metal chelates as in eqs 78 and 79.[90]

Some care is needed in the interpretation of reaction 78 because it may either be a nucleophilic displacement of cobalt(I) by attack of the alkylcobalt(I) complex on the α-carbon of the alkylcobalt(III) complex, or else an electrophilic displacement of cobalt(I) by attack of the alkylcobalt(III) complex on the α-carbon of the alkylcobalt(I) complex.

Nucleophilic displacements of the second type, with milder nucleophiles, are much more rare and are usually rather poorly defined, except in the case of the transient, reactive, organocobalt(IV) complexes described in Section 3.2. Other cases will be discussed under the headings of the individual nucleophiles in subsequent sections.

4.2   Nucleophiles as proton bases

A number of axial and equatorial ligands used in organometal chelate chemistry may undergo reversible loss of protons in alkaline media. The acid dissociation constants of several such complexes have been measured and have been shown to be a function of the nature of the organic ligand and of the axial base.[178,179] For example, the acid dissociation constants of the organo(pyridine)cobaloximes and organo(aquo)cobaloximes, respectively, decrease down the series (the numbers refer to the pKa for R =) Pr^i (13.93 and 13.28) > Et (13.88 and 12.97) >Me (13.61 and 12.68) >$PhCH_2CH_2$ (13.56 and 12.78) >$NCCH_2CH_2CH_2$ (13.37 and 12.22) >CF_3CH_2 (12.3 and 10.96)> $NCCH_2$ (11.77 and 10.56).[179] It is not known whether the lower values of the

$$RCo(dmgH)_2OH_2 \quad + \quad OH^- \rightleftharpoons [RCo(dmg)(dmgH)OH_2]^- \rightleftharpoons [RCo(dmgH)_2OH]^-$$

Scheme 26

pKa for the aquo complexes refer to the loss of a proton from the axial water molecule or from the equatorial ligand (Scheme **26**); but probably the latter.

The organic ligand may also take part in reversible and irreversible reactions without cleavage of the carbon-metal bond.   For example, carboxymethyl and carboxyethylcobaloximes have pKa's of 4.89 and 6.30 respectively, at $25°C$, indicative of a strong inductive electron donation by the cobaloxime moeity.[180]   The rates of alkaline hydrolysis of methyl esters of 3- and 4-carboxyphenyl(aquo)cobaloximes have also been measured ($k_2 = 3.45 \times 10^{-2}$ and 2.47×10^{-2} $M^{-1}s^{-1}$, respectively) (eq 80).   From these data, the corresponding rates of hydrolysis of the hydroxo-complexes were also estimated ($k_2 = 1.93 \times 10^{-2}$ and 1.54×10^{-2} $M^{-1}s^{-1}$, respectively). [180] Comparison with the rate of hydrolysis of methyl benzoate ($k_2 = 9.07 \times 10^{-2}$ $M^{-1}s^{-1}$) confirms that the cobaloxime moeity is electron donating, with a substituent constant $\sigma*$ of -2.16 for $Co(dmgH)_2OH_2$ and -2.51 for $Co(dmgH)_2OH^-$, and, from other work, -3.6 for cobalamin(III).

Direct reaction of the hydroxide ion with the β-hydrogen of some β-substituted ethylcobaloximes and cobalamins leads to elimination of the substituted olefin and the formation of cobaloxime(I) or cobalamin(I) (Scheme 27).[181,182]   The rates of reaction are first order in substrate and in hydroxide ion in that region of pH where dissociation of other ligands does not take place to a significant degree, and there is a primary kinetic isotope effect k_H/k_D = 1.6, suggesting that the proton is removed in the rate-determining step, probably with the formation of a transient η^2-olefin complex.   The reaction is reversible[183,184] unless carried out in the presence of oxygen or methyl iodide to remove the cobalt(I) product, and the

$$MeOCOC_6H_4Co(dmgH)_2OH_2 \quad + \quad 2OH^- \longrightarrow MeO^- \quad + \quad {}^-OCOC_6H_4Co(dmgH)_2OH_2 + H_2O \qquad (80)$$

$$OH^- \quad + \quad XCH_2CH_2Co(chel) \longrightarrow \left[XCH\begin{subarray}{c} \\ \mid \\ Co(chel) \end{subarray}CH_2\right]^- \longrightarrow H_2O + XCH=CH_2 \quad + \quad \left[Co(chel)\right]^-$$

X = CN, CO_2R, etc

Scheme 27

reverse reaction has been used as a method of preparation of certain cobaloximes. The apparent reaction of cyanide ion with these complexes is almost certainly a result of hydroxide ion formed by hydrolysis of the cyanide ion in aqueous solution.[182]   Hydroxide ion also promotes elimination (eq 81) of the unsaturated compounds (LIX) and (LX) in the reactions of coenzyme B_{12} and its analogue (LXI).[184]

In contrast, 2-trimethylammoniumethylcobalamin [185] and 2-methoxyethylcobaloximes do not undergo elimination of olefin even in strongly basic media, whereas 2-hydroxyethylcobaloximes and cobalamin not only react readily, but also undergo a hydrogen transfer from C2 to C1 to give aldehydes and ketones.[186]   Comparison with the 2-methoxyethyl complexes shows that the hydroxyl proton loss is crucial and the high primary kinetic isotope effect in the second order reaction of 2-hydroxypropylcobaloxime-2-[$^2H_2$] in 2M NaOH (k_H/k_D = 5.5) suggests that the migration of hyrogen from C_2 to C_1 occurs in the rate determining step. The mechanism shown in Scheme 28 has been proposed.   It is, however, surprising that no epoxide is formed in these reactions in view of the common reactions of bromohydrins with aqueous base.

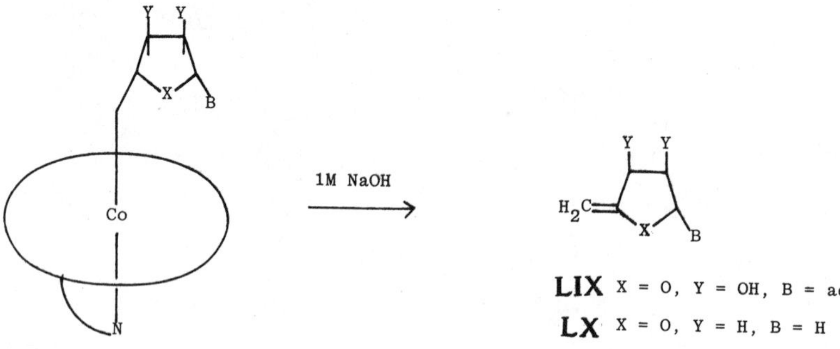

$$ \text{(81)} $$

LIX X = 0, Y = OH, B = adenine
LX X = 0, Y = H, B = H

I   B = adenine, X = 0, Y = OH
LXI B = H, X = 0, Y = H

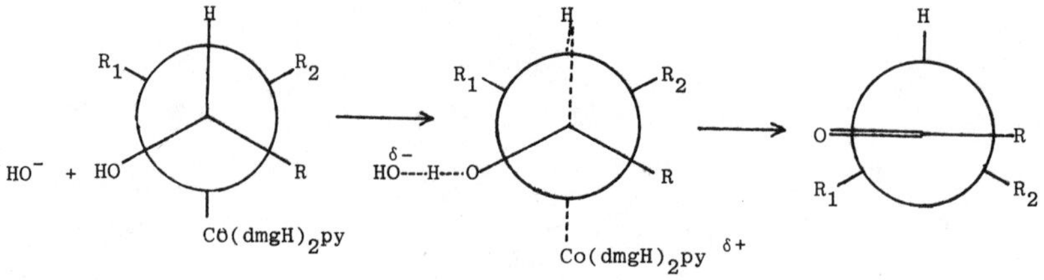

Scheme **28**

4.3   Other reactions in which the carbon-metal bond is cleaved

Many of the following reactions are not at all well-defined and some are even subject to dispute between different workers, in some cases probably because of the major influence of minor changes in reaction conditions.

4.3.1 Hydroxide and alkoxide ion

In strong alkaline solution, under conditions where deprotonation of equatorial ligands also occurs, methylcobaloxime decomposes to give methane in approximately 70% yield, and other products which include a hydrate of a dioximato ligand.[187]   In deuteriated protic solvents, the product is CH_3D and exchange of protons takes place both in the oxime and in the methyl protons of the equatorial ligands at a rate faster than that of the formation of methane.[188]   The mechanism of the formation of methane is not fully understood, but it is supposed that decomposition is initiated by attack of hydroxide ion on an unsaturated carbon of a dioximato ligand, which promotes the heterolysis, or possibly the homolysis, of the carbon-cobalt bond.

Methoxide ion probably attacks the carbonyl carbon of the carbonate ester (**LXII**) giving the mixture of products shown in eq 82.   Under the same conditions methoxide ion does not react with the n-butyl or methoxymethylcobalt(III) complexes, and it cleanly regenerates ethylene glycol from the carbonate ester of uncomplexed ethylene glycol.   The steps following the attack of methoxide ion are not clear, but it is supposed that the formation of radicals ($\dot{C}H_2CHO$ or possibly $\dot{C}H(OH)CH_2O^-$), is responsible for the production of acetaldehyde.[189]

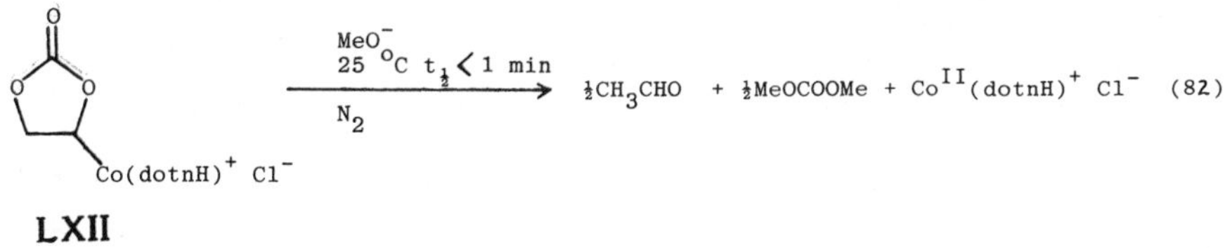

$$\xrightarrow[N_2]{\substack{MeO^- \\ 25\,^{\circ}C\ t_{\frac{1}{2}} < 1\ min}} \tfrac{1}{2}CH_3CHO\ +\ \tfrac{1}{2}MeOCOOMe\ +\ Co^{II}(dotnH)^+\ Cl^-\quad (82)$$

LXII

$$XCH_2Co(dmgH)_2py\ +\ OMe^- \longrightarrow XCH_2OMe\ +\ [Co(dmgH)_2py]^- \longrightarrow X^-\ +\ MeOCH_2Co(dmgH)_2py$$

Scheme 29

$$X_2CHCo(dmgH)_2py \quad + \quad 3OH^- \xrightarrow{\ H_2O\ } \quad CO \quad + \quad [Co(dmgH)_2py]^- \quad + \quad 2X^- \quad + \quad 2H_2O \quad (83)$$

$$Cl_3CCo(dmgH)_2py \quad + \quad OMe^- \xrightarrow{\ MeOH\ } \quad MeOCO.Co(dmgH)_2py \quad + \quad 3Cl^- \quad\quad\quad (84)$$

$$RCo(dmgH)_2L \quad + \quad CN^- \underset{-L}{\overset{}{\rightleftharpoons}} [RCo(dmgH)_2CN]^- \xrightarrow{\ RCo(dmgH)_2L\ } [RCo(dmgH)_2CNCoR(dmgH)_2py]^-$$

Scheme 30

Halogenomethylcobalt(III) complexes are also susceptible to attack by
hydroxide and alkoxide ion.   For example, monochloromethylcobaloxime
reacts with methoxide ion to give methoxymethylcobaloxime (Scheme 29),
dichloromethylcobaloxime reacts with hydroxide ion to give carbon
monoxide and cobaloxime(I) (eq 83), and trichloromethylcobaloxime, which is
rather unstable and difficult to characterise, reacts with methoxide ion to
give moderate yields of methoxycarbonylcobaloxime (eq 84).[190] In all the
above cases, halide ion is liberated.   It has been proposed that the formation
of methoxymethylcobaloxime (Scheme 29) proceeds through the formation of
the halomethyl methyl ether and cobaloxime(I) which subsequently react in
the absence of oxygen to give the observed product, but a direct
displacement of halide ion cannot be ruled out.   It has also been reported
that allyl chloride is formed in the reaction of alkoxide ion with
$ClCH_2CH(OAc)CH_2Co(dmgH)_2py$.[191]

4.3.2 Cyanide ion

Besides being a moderate base, significantly hydrolysed in protic
solvents, cyanide ion coordinates very strongly to many transition metals.
With organocobaloximes, for example, the first reaction to occur is usually
the replacement of the axial ligand by cyanide ion followed, in some cases,
by bridge formation (Scheme 30).[192,193]   It has also been reported that
cyanide ion reacts under very forcing conditions with methylcobalt(III)
complexes to give acetonitrile,[194] but any mechanistic conclusions from
such reactions may be suspect.   Cyanide ion does however, rapidly displace
and cleave the 5'-deoxyadenosyl group of coenzyme B_{12} and related
nucleosides in other cobalamin complexes (Scheme 31).[195,196] Indeed, one
of the early problems that arose in the quest to isolate coenzyme B_{12} was
the fact that a crystalline product could be obtained by addition of cyanide

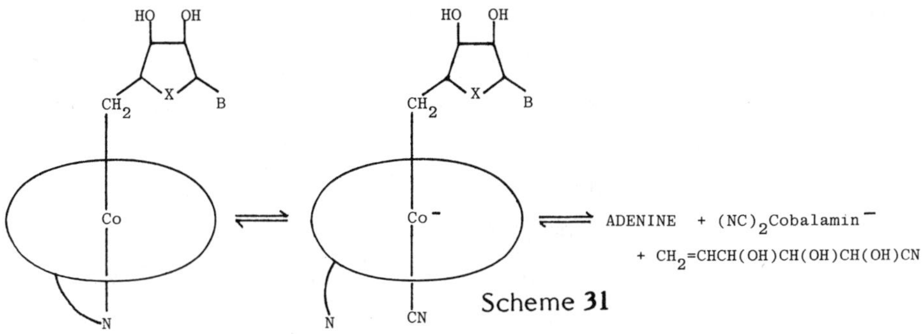

Scheme 31

I X = O   B = adenine

ion.   This crystalline product, cyanocobalamin, though correctly identified, was not of course the coenzyme, but was given the name vitamin B_{12} in the initial belief that it was the active constituent being sought.   Here again the nature of the cleavage reaction is uncertain.   The first step is, as expected, the rapid replacement of coordinated benzimidazole by cyanide ion, and the second slow step, the rate of which is a function of the nature of the nucleoside and the concentration of cyanide ion, gives the cyanohydrin of D- erythro-2,3-dihydroxypent-4-en-1-al.   This almost certainly takes place by attack of cyanide ion at the deoxyribose moeity, because the corresponding alicyclic complex (I, X = CH_2) does not react in this manner,[197] but the remaining details are still obscure.

4.3.3 Borohydride ion

Borohydride ion reacts with polyhalogenomethylcobaloximes causing successive reduction through other halogenomethylcobaloximes to methylcobaloxime (Scheme **32**).[198]   Since the protons of the product methylcobaloxime come from the solvent and not from the borohydride ion,

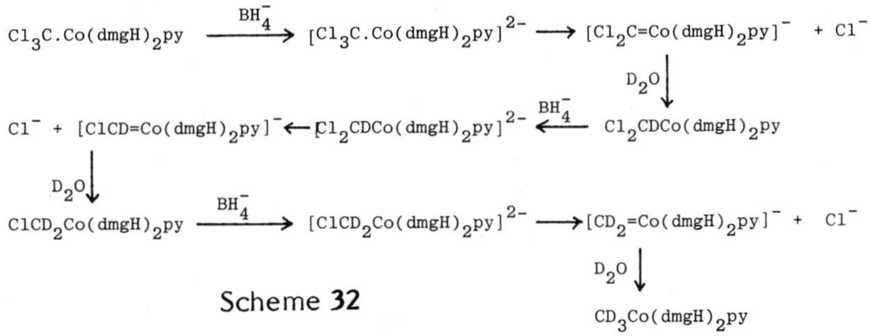

Scheme **32**

as shown by deuterium labeling of both solvent and borohydride ion, it is apparent that the reaction takes place by a reduction of the complex with loss of halide ion, and probably with the formation of carbene complexes as intermediates (Scheme 32). It is interesting that even fluoromethylcobalt(III) complexes are reduced in this way, and that, in biological systems, fluoromethylcobalamins can take the place of methylcobalamin as substrates for methane formation.[199] It is very likely that a variety of other reactions, such as those that occur when organocobalt(III) complexes are prepared in the presence of an excess of borohydride ion, also proceed through reductive activation of the organic ligand to electrophilic attack by protic solvents or even to carbanion formation. Such processes may include the formation of propene in the preparation of allylcobaloxime, and the formation of rearranged products from several organocobaloximes and organocobalamins during their preparation from organic halides.[69-72] Typical of the latter is the reaction (Scheme 33) of the organic bromide (LXIII) with cobalamin(I) in the presence of an excess of sodium borohydride in EtOD as solvent.[69,72] Not only are rearranged products formed, but they also contain substantial quantities of deuterium, indicative of carbanion formation followed by abstraction of deuterium from the solvent, rather than radical formation followed by H-atom abstraction from the solvent. An additional feature of the above reaction is that it seems likely that the two complexes (LXIV) and (LXV) are formed through an electron transfer from the cobalamin(I) to the organic

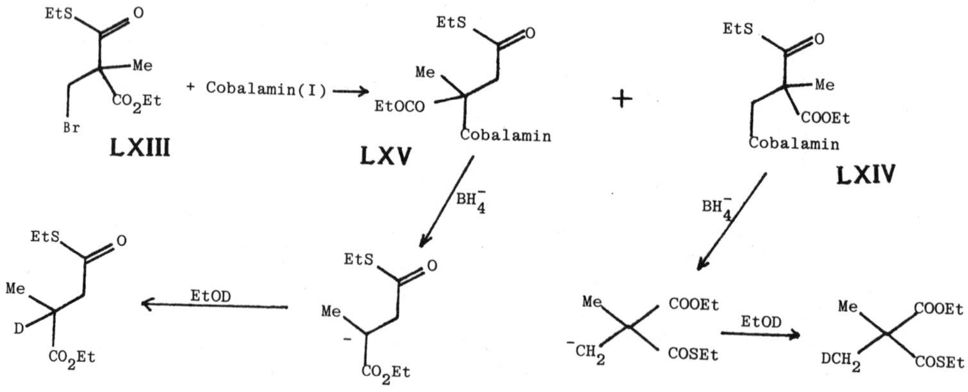

Scheme 33

halide, loss of bromide ion and partial rearrangement of the organic radical, and capture of rearranged and unrearranged radicals by cobalt(II).[72]

4.3.4 Thiols and thiolate ions

Like cyanide ions, thiolate ions are good ligands towards many transition metals. Fortunately they are not strong bases, but they are very susceptible to oxidation to thiyl radicals even when uncoordinated. Not surprisingly, therefore, the reactions of thiols and thiolate ions with organocobalt(III) complexes are among the messiest known, and it is difficult to come to many firm conclusions despite a deal of mechanistic investigation.

It has been reported that organic thioethers are formed in the reaction of thiolate ions with organocabalamins and organocobaloximes, and a bimolecular nucleophilic displacement mechanism has been proposed.[194,200] Such a reaction is important because of the biochemical formation of methionine from methylcobalamin in vivo. However, either the nature of the conditions is absolutely crucial or these reactions are not nucleophilic displacements. For example, in their study of the axial ligation of thiols and thiolate ions in aqueous solution, Brown and Kallen could find no evidence of thioether or alkane formation in the dark,[201] whereas Schrauzer has observed that methylcobaloximes and other methylcobalt(III) chelates react with a variety of thiols and thiolate ions at from pH 2 to pH 12 to give methane in good yield.[202] The rates of evolution of methane were a function of the pH, the thiol concentration, the nature of the buffer, and the axial ligand, and were greatly increased by irradiation. On the other hand, coenzyme M ($HSCH_2CH_2SO_3^-$) is reported to react with methylcobalamin in the pH range 6 to 14 to yield the S-methyl derivative $MeSCH_2CH_2SO_3^-$, methane being formed as well at lower pH.[203] However, two keys to these reactions come from the observation that some sulfur ligands, when coordinated to cobalt, increase the susceptibility of the carbon-metal bond to photolysis,[204] and the observation by Agnes et al., that there is an induction period in the reaction between thiolate ions and methylcobalamin.[205] The length of this induction period was inversely proportional to the thiolate ion concentration, but dependent on the oxygen concentration.   They proposed a mechanism in which the induction period depends on the formation of cobalt(II) complexes (eqs 85-90).   It is possible that the penultimate step might also take place not only by attack of RS˙ on the metal, as shown in eq 88, but also by attack of RS˙ at the α-carbon, giving the thioether directly and displacing more cobalamin(II) (eq 90).

The case for reductive processes is much stronger with dithiols such as dithioerythritol (DTE).[206-208]   It seems likely that when one of the two thiol groups becomes coordinated, the other can assist in the reduction with the formation of a cyclic disulfide as the oxidation product (Scheme 34). The reduced organocobalt product is then susceptible to attack by even the weakest of electrophiles such as carbon dioxide and water.   The former reaction has been put to excellent use in the synthesis of labelled amino acids, as shown in Scheme 35.[207]

4.3.5 Electrochemical Reduction

Electrochemical reduction is clearly, from the above discussion, of interest in the study of the cleavage of carbon-metal bonds.   So far much of the work has been devoted to the identification of the reduced species and

$$\text{Cobalamin(III)} \ + \ \text{RS}^- \longrightarrow \text{Cobalamin(II)} \ + \ \text{R}\overset{.}{\text{S}} \qquad (85)$$

$$\text{Cobalamin(II)} \ + \ \text{RSH} \longrightarrow \text{Cobalamin(II).RSH} \qquad (86)$$

$$\text{Cobalamin(II).RSH} \ + \ \text{O}_2 \longrightarrow \text{Cobalamin(III)} \ + \ \text{HO}_2^- \ + \ R\overset{.}{\text{S}} \qquad (87)$$

$$R\overset{.}{\text{S}} \ + \ \text{MeCobalamin(III)} \longrightarrow \text{Cobalamin(III)} \ + \ \overset{.}{\text{C}}\text{H}_3 \ + \ \text{RS}^- \qquad (88)$$

$$R\overset{.}{\text{S}} \ + \ \overset{.}{\text{C}}\text{H}_3 \longrightarrow \text{RSCH}_3 \qquad (89)$$

$$R\overset{.}{\text{S}} \ + \ \text{MeCobalamin} \longrightarrow \text{RSMe} \ + \ \text{Cobalamin(II)} \qquad (90)$$

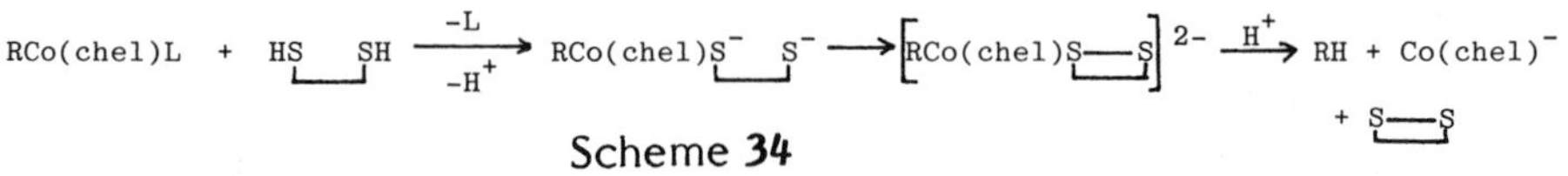

Scheme 34

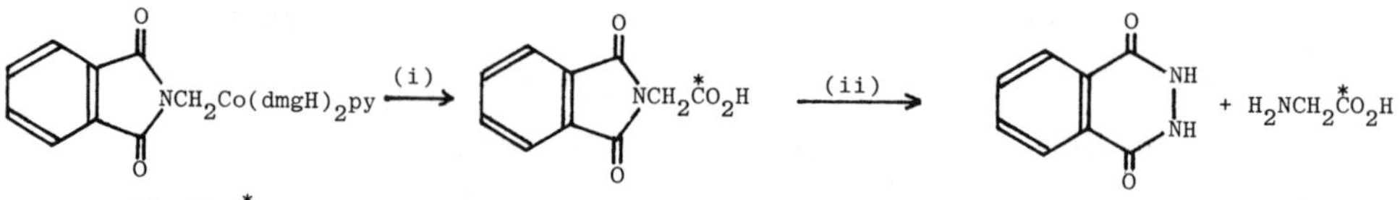

(i) DTE/C̊O$_2$; (ii) H$_2$NNH$_2$/reflux

Scheme 35

the determination of the reduction potentials under different conditions. Less work has been done on the determination of reaction products from decomposition of the reduced organometallic complexes, but the general outline below (Scheme **36**) gives an indication of the products normally expected on reduction of organocobalt(III) complexes.

In general, the carbanion route for decomposition of the organocobalt(II) complexes is only found where the carbanion is particularly stable, as in the case of perfluoroalkylcobalt complexes[209] and with acetonyl and related organocobalt complexes.[210] The free radical route is common with alkylcobalt complexes and, since the inorganic cobalt(II) products are themselves more readily reduced than the organocobalt substrates, they are not usually observed under the reaction conditions.

The half-wave potentials for reduction of the organometal chelate complexes are dependent on the nature of the equatorial and axial ligands, and on the character of the solvent system used. We have not therefore quoted specific values of these potentials, which can be found in the works of, for example, Costa[124] Hogenkamp,[211] and West.[209] In general most neutral organocobalt(III) complexes are reduced in the range -1.3 to -1.8v, but the cationic complexes such as $RCo(dotnH)^+$ are reduced more readily. The influence of the organic ligand on the reduction potential is clearly seen from the series RCo(SALEN) for which $E_{\frac{1}{2}}$ (Co(III) to Co(II)) varies from -0.989 to -0.908 v for from CF_3 to C_3F_7, and from -1.5 to -1.7 for from CH_3 to C_3H_7. The value of $E_{\frac{1}{2}}$ for the reduction from Co(II) to Co(I) is much less ligand dependent, varying from -2.0 to -2.2 over the same two series.[209] The influence of the equatorial ligand has been clearly demonstrated by

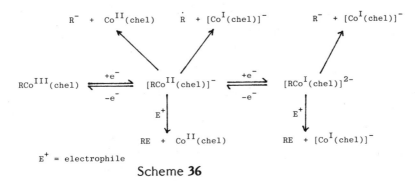

E^+ = electrophile

Scheme **36**

Costa[124] who has shown that the difference in reduction potentials $E_{\frac{1}{2}}(Co(III)-Co(II)) - E_{\frac{1}{2}}(Co(II)-Co(I))$ is -0.44 and the difference in reduction potentials $E_{\frac{1}{2}}(Co(IV)-Co(III)) - E_{\frac{1}{2}}(Co(III)-Co(II))$ is -1.84 for the wide series of complexes $EtCo(chel)^{n+}$ and that the absolute values of the reduction potentials all decrease in the series (chel) = dotnH >(dmgH)$_2$ >cobalamin SALOPHEN >SALEN >Me$_2$SALEN >ACACEN, the spread being nearly 1v over each series.

Electrochemical reduction has recently been put to very good use in the synthesis of bicyclic ketones using cobalamins and (1-hydroxy-2,2,3,3,7,7,8,8,12,12,13,13,17,17,18,18-hexadecamethyl-10,20-diazaoctahydro-porphinato)dibromocobalt(III) as catalysts.[212] Reaction of an organic halide RX, an olefin $CH_2{:}CHY$ (as part of the same or different molecules), the catalyst, a protic solvent, and an electrolyte, gives an excellent yield of the organic product RCH_2CH_2Y at a potential more negative than -1.4 v. It is presumed that the alkyl halide reacts with the electrochemically generated cobalt(I) species to give the organcobalt(III) complex which, when further reduced electrochemically to the organocobalt(I) complex, is susceptible to attack by the electrophilic olefin. On capture of a proton from the solvent, the observed product is formed (eq 91).

5.   UNIMOLECULAR REARRANGEMENTS

There are a number of reactions of organocobalt(III) chelates which proceed unimolecularly without the direct participation of external reagents. For example, allylcobaloximes, besides undergoing the cobalt(II)-promoted isomerisation outlined in Section 2, appear to undergo a unimolecular isomerisation via an η^3-allyl intermediate. This reaction is evident from the temperature dependence of the proton nmr spectrum of allyl(aquo)cobaloximes. Thus the concentration independent spectrum of 2-methylallylbis(cyclohexanedionedioxmato)aquocobalt(III) shows characteristics of an η^1-allylmetal complex at -50°C, but at higher

$$\text{(CH}_2\text{)}_4\text{Br} \xrightarrow[\text{Co(III)/H}^+]{2e^-} \text{(95\%)}$$

(91)

temperatures coalescence of the methylene proton resonances at $\delta 4.90$, 4.45 and 2.33 occur and the spectrum becomes characteristic of a pair of η^1-2-methylallyl complexes (LXVI) and (LXVII) in rapid equilibrium, with a four-proton singlet at $\delta 3.50$. It was proposed that the loss of the axial water ligand allows the formation of a transient symmentrical η^3-allyl complex as shown in Scheme 37.[213]

A closely related, but much slower rearrangement occurs with but-3-enylcobaloximes and cyclopropylcarbinylcobaloximes. Thus, it has been observed that cyclopropylcarbinylcobalamin and α-[^{13}C]-cyclopropylcarbinylcobaloxime spontaneously rearrange[214,215] in solution to open chain but-3-enylcobalt(III) complexes. In the latter case the label ends up exclusively as the terminal olefinic carbon of the but-3-enyl ligand (eq 92). It has also been shown that 1-methylbut-3-enyl(pyridine)cobaloxime (LXVIII) equilibrates spontaneously with 2-methylbut-3-enyl(pyridine)cobaloximes (LXIX) at 55°C in $CDCl_3$, and it was suggested that, under these conditions, a cobatl(II) promoted rearrangement via cyclopropylcarbinylcobaloxime complexes might be involved (Scheme 38), especially as 1,2-dimethylbut-3-enyl(pyridine)cobaloxime spontaneously cyclises at room temperature to isomers of 2,3-dimethylcyclopropylcarbinyl(pyridine)cobaloxime (eq 93).[216]

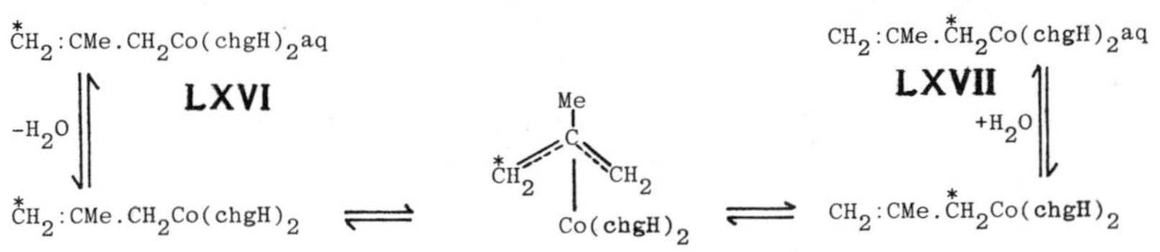

Scheme 37

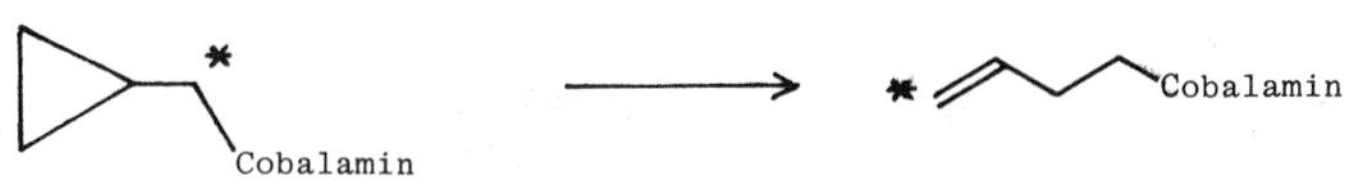

(92)

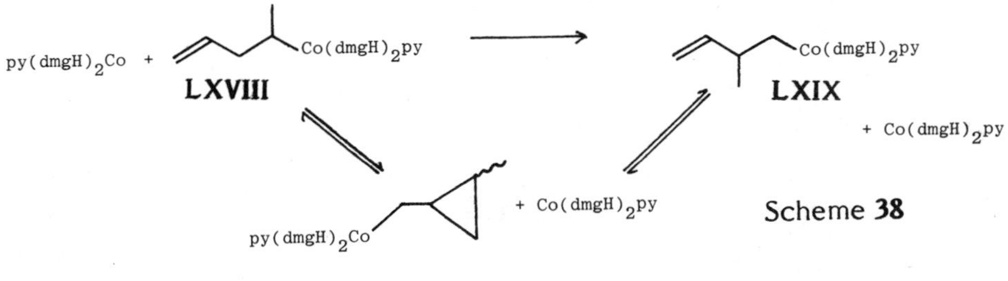

Scheme 38

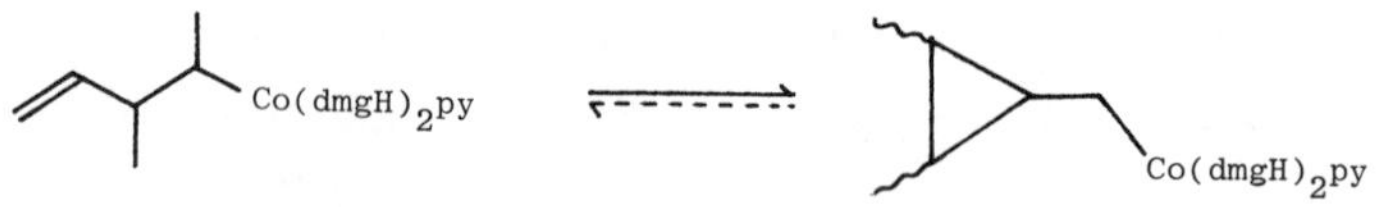

$$(93)$$

However, the isomeration of 1-methylbut-3-enyl(pyridine)cobaloxime is also catalysed by trifluoroacetic acid, the catalysis being most effective under conditions where the equatorial dioximato ligand pair is monoprotonated and the pyridine is removed as pyridinium trifluoroacetate (cf eq 52).   Under these conditions (S)- and (R)-1-methylbut-3-enylcobaloxime equilibrate stereospecifically with (R)- and (S)-2-methylbut-3-enylcobaloxime, respectively,[217] there being negligible loss of optical activity over the course of several exchanges.   It has therefore been proposed that the rearrangement arises via novel η^3-homoallylcobalt(III) complexes and cis-and trans-2-methylcyclopropylcarbinylcobalt(III) complexes as shown in Scheme 39.   The racemic cis- and trans-cyclopropylcarbinyl intermediates have been independently synthesised and they both spontaneously rearrange, at similar rates, directly to the equilibrium mixture (1:9) of 1- and 2-methylbut-3-enylcobaloxime.

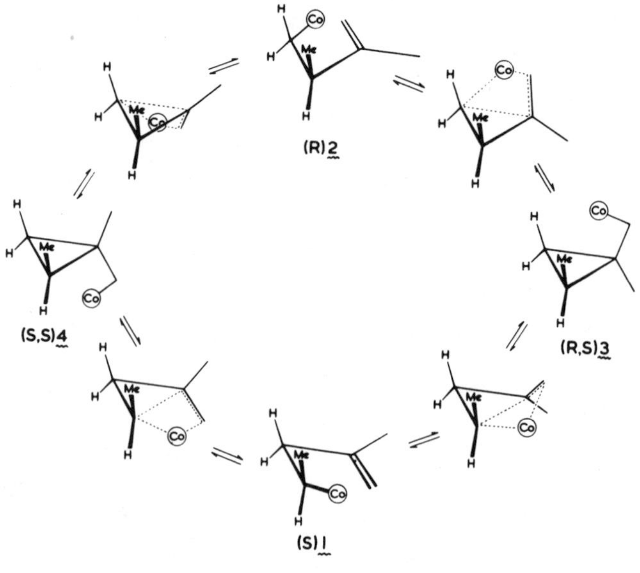

Scheme 39

6.   SUMMARY

Many of the earlier reviews of organocobalt chemistry have discussed the reactions with the premise that they could be based mainly on the three modes of fission of the carbon-cobalt bond, namely carbanion/cobalt(III), radical/cobalt(II), and carbocation/cobalt(I).   The above review shows clearly that a great many of the reactions of organometal chelate complexes are concerted reactions at the organic ligand, for which the above premise does not apply.   The exceptions are (i) homolysis, and (ii) reactions of the highly destabilised organocobalt(II) and organocobalt(I) complexes.   There is almost no evidence as yet for simple heterolysis of carbon-cobalt bonds in isolable organocobalt(III) complexes.

REFERENCES

1. P.A. Milton and T.L. Brown, J. Am. Chem. Soc. 99:1390 (1977).

2. G. Costa, G. Mestroni, G. Tauzher, D.M. Goodall, M. Green and H.A.O. Hill, J. C. S. Chem. Commun. 34 (1970).

3. K.L. Brown and R.G. Kallen, J. Am. Chem. Soc. 94:1894 (1972).

4. T. Sakurai, J.P. Fox and L.L. Ingraham, Inorg. Chem. 10:1105 (1971).

5. A.L. Crumbliss and W.K. Wilmarth, J. Am. Chem. Soc. 92:2593 (1970).

6. R.J. Guschl and T.L. Brown, Inorg. Chem. 12:2815 (1973).

7. R.J. Guschl, R.S. Stewart and T.L. Brown, Inorg. Chem. 13:417 (1974).

8. K.L. Brown, D. Lyles, M. Pencovici and R.G. Kallen, J. Am. Chem. Soc. 97:7338 (1975).

9. F.R. Jensen and R. Kiskis, J. Am. Chem. Soc. 97:5820 (1975).

10. R.C. Stewart and L.G. Marzilli, J. Am. Chem. Soc. 100:817 (1979).

11. J.A. Ewan and D.J. Darensbourg, J. Am. Chem. Soc. 98:4317 (1977).

12. D.P. Graddon and I.A. Siddiqi, J. Organomet. Chem. 133:87 (1977).

13. R.L. Courtright, R.S. Drago, J.A. Nusz and M.S. Nozari, Inorg. Chem. 12:2809 (1973).

14. K.L. Brown, D. Chernoff, D.J. Keljo and R.G. Kallen, J. Am. Chem. Soc. 94:6697 (1972).

15. L. Sables and E. Deutsch, Inorg. Chem. 16:2273 (1977).

16. K.L. Brown and A.W. Awtrey, Inorg. Chem. 17:111 (1978).

17. R.L. Gruschl and T.L. Brown, Inorg. Chem. 13:959 (1974).

18. J.M. Ciskowski and A.L. Crumbliss, Inorg. Chem. 18:638 (1979).

19. J.H. Espenson and R. Russell, Inorg. Chem. 13:7 (1979).

20. H. Ogoshi, E. Watanabe, N. Kuketsu and Z. Yoshida, Bull. Chem. Soc. Jpn. 49:2529 (1976).

21. H. Ogoshi, J. Setsune, Y. Nanto and Z. Yoshida, J. Organomet. Chem. 159:329 (1979).

22. H. Ogoshi, J. Setsune and Z. Yoshida, J. Organomet. Chem. 159:317 (1979).

23. C. Cloutour, D. Lafargne and J.C. Pommier, J. Organomet. Chem. 161:327 (1979).

24. R.M. McAllister and R.C. Kiskis, J. Organomet. Chem. 49:C46 (1973).

25. G.J. Samuels and J.H. Espenson, Inorg. Chem. 18:2587 (1979).

26. H. Ogoshi, J. Setsune, T. Onura and Z. Yoshida, J. Am. Chem. Soc. 97:6461 (1976).

27. M.C. Rakowski and D.H. Busch, J. Am. Chem. Soc. 97:2540 (1976).

28. H. Ogoshi, J.I. Setsune and Z.I. Yoshida, J. Am. Chem. Soc. 99:3869 (1977).

29. V.L. Goedken and S.-M. Peng, J. Am. Chem. Soc. 96:7826 (1976).

30. R. Guilard, P. Cocolios, P. Fournari, C. Lecompte and J. Protas, J. Organomet. Chem. 169:C49 (1979).

31. A.M. Abeysekera, R. Grigg, J.T. Grimshaw and V. Viswanatha, J. C. S. Chem. Commun. 227 (1976).

32. L.J. Kirschenbaum and D. Meyerstein, Inorg. Chem. 19:1373 (1980).

33. J. Halpern, F.T.T. Ng and G.L. Rempel, J. Am. Chem. Soc. 101:7124 (1979).

34. J.F. Endicott, J.P. Balakrishnan and C.L. Wong, J. Am. Chem. Soc. 102:5519 (1980).

35. H.P.C. Hogenkamp, P.J. Vergamini and N.A. Marwiyoff, J. C. S. Dalton Trans. 2628 (1975).

36. T.S. Roche and J.F. Endicott, Inorg. Chem. 13:1575 (1974).

37. H. Elroi and D. Meyerstein, J. Am. Chem. Soc. 100:5540 (1978).

38. G.N. Schrauzer and M. Mashimoto, J. Am. Chem. Soc. 101:4593 (1979).

39. M. Naumberg, K.N.V. Duong and A. Gaudemer, J. Organomet. Chem. 25:231 (1970).

40. K.N.V. Duong, A. Ahond, C. Merienne and A. Gaudemer, J. Organomet. Chem. 56:375 (1973).

41. S.M. Chemaly and J.M. Pratt, J. C. S. Dalton Trans. 2274 (1980).

42. J.N. Grate and G.N. Schrauzer, J. Am. Chem. Soc. 101:4601 (1979).

43. S.M. Chemaly and J.M. Pratt, J. C. S. Dalton Trans. 2259 (1980).

44. S.M. Chemaly and J.M. Pratt, J. C. S. Dalton Trans. 2267 (1980).

45. H. Eckert, D. Lenoir and I. Ugi, J. Organomet. Chem. 141:C23 (1977).

46. P. Bougeard, M.D. Johnson and M. Lewin, unpublished work.

47. J.F. Endicott and G.J. Ferraudi, J. Am. Chem. Soc. 99:243 (1977).

48. J.F. Endicott and C.Y. Mok, J. Am. Chem. Soc. 100:123 (1978).

49. J.F. Endicott and T.L. Netzel, J. Am. Chem. Soc. 101:4000 (1979).

50. K.N. Joblin, A.W. Johnson, M.F. Lappert and B.K. Nicholson, J. C. S. Chem. Commun. 441 (1975).

51. G. Roewer and D. Rehorek, J. Prakt. Chem. 320:566 (1978).

52. Ph. Maillard, J.C. Massot and C. Giannotti, J. Organomet. Chem. 159:219 (1979).

53. C. Giannotti, G. Merle and J.R. Bolton, J. Organomet. Chem. 99:145 (1975).

54. G.N. Schrauzer and L.P. Lee, J. Am. Chem. Soc. 92:1551 (1970).

55. J.H. Bayston, F.D. Looney. J.R. Pilbrow and M.E. Winfield, Biochemistry 9:2164 (1970).

56. V.D. Ghanekar and R.E. Coffman, J. Organomet. Chem. 198:C15 (1980).

57. B.T. Golding, T.J. Kemp, P.J. Sellars and E. Nocchi, J. C. S. Dalton Trans. 1266 (1977).

58. C.W. Fong and M.D. Johnson, unpublished work.

59. B.T. Golding, "Vitamin B$_{12}$", in "Comprehensive Organic Chemistry", Pergamon Press, 1979, Chapter 24.4.

60. B.M. Babior, Acc. Chem. Res. 8:376 (1975).

61. A.J. Hartshorn, A.W. Johnson, S.M. Kennedy, M.F. Lappert and J.J. MacQuitty, J. C. S. Chem. Commun. 643 (1978).

62. I.P. Rudakova, T.E. Ershova, A.B. Belikov and A.M. Yurkevich, J. C. S. Chem. Commun. 592 (1978).

63. B.T. Golding, C.S. Sell and P.J. Sellars, J. C. S. Chem. Commun. 693 (1977).

64. B.T. Golding, T.J. Kemp, C.S. Sell, P.J. Sellars and W.P. Watson, J. C. S. Perkin Trans. II 839 (1978).

65. D.A. Clarke, D. Dolphin, R. Grigg, A.W. Johnson and H.A. Pinnock, J. C. S. Perkin Trans. II 881 (1968).

66. K.L. Brown and L.L. Ingraham, J. Am. Chem. Soc. 96:7681 (1974).

67. H. Flohr, W. Pannhorst and J. Retey, Angew. Chem. Int. Ed. Engl. 15:561 (1976).

68. H. Flohr, U.M. Kempe, T. Krebs, J. Retey and G. Bridlingmaier, Angew. Chem. Int. Ed. Engl. 14:822 (1975).

69. I.A. Scott, K. Kang, D. Datton and S.K. Chung, J. Am. Chem. Soc. 100:3603 (1978).

70. P. Dowd, B.K. Trivedi, M. Shapiro and L.K. Marwaha, J. Am. Chem. Soc. 98:7875 (1976).

71. P. Dowd, M. Shapiro and K. Kang, J. Am. Chem. Soc. 97:4754 (1975).

72. A.I. Scott and K. Kang, J. Am. Chem. Soc. 99:1997 (1977).

73. P. Bougeard and M.D. Johnson, J. Organomet. Chem. 206:221 (1981).

74. A. Bury, C.J. Cooksey, T. Funabiki, B.D. Gupta and M.D. Johnson, J. C. S. Perkin Trans. II 1050 (1979).

75. A.E. Crease, B.D. Gupta, M.D. Johnson, E. Bialkowska, K.N.V. Duong and A. Gaudemer, J. C. S. Perkin Trans. I 2611 (1979).

76. E. Bialkowska, K.N.V. Duong, A. Gaudemer and M.D. Johnson, unpublished work.

77. M. Veber, K.N.V. Duong, F. Gaudemer and A. Gaudemer, J. Organomet. Chem. 177:231 (1980).

78. M.R. Ashcroft, A. Bury, C.J. Cooksey, A.G. Davies, B.D. Gupta, M.D. Johnson and H. Morris, J. Organomet. Chem. 195:89 (1980).

79. M. Veber, K.N.V. Duong, A. Gaudemer and M.D. Johnson, J. Organomet. Chem. 209:393 (1981).

80. M.R. Ashcroft and M.D. Johnson, unpublished work.

81. P. Bougeard, B.D. Gupta and M.D. Johnson, J. Organomet. Chem. 206:211 (1981).

82. M. Perree-Fauvet and A. Gaudemer, unpublished work.

83. P. Bougeard, A. Bury, C.J. Cooksey, M.D. Johnson, J.M. Hungerford and G.M. Lampman, J. Am. Chem. Soc. 104:5230 (1982).

84. J. Deniau, K.N.V. Duong, A. Gaudemer, P. Bougeard and M.D. Johnson, J. C. S. Perkin Trans. II 393 (1981).

85. D. Dodd, M.D. Johnson and B.L. Lockman, J. Am. Chem. Soc. 99:3664 (1977).

86. J.H. Espenson and T.D. Sellars, J. Am. Chem. Soc. 96:94 (1974).

87. J.H. Espenson and J. Shveima, J. Am. Chem. Soc. 95:4468 (1973).

88. A. van den Bergen and B.O. West, J. Organomet. Chem. 64:125 (1974).

89. C. Cooksey, D. Dodd, M.D. Johnson and B.L. Lockman, J. C. S. Dalton Trans. 1815 (1978).

90. G. Mestroni, C. Cocevar and G. Costa, Gazz. Chim. Ital. 103:273 (1973).

91. C. Giannotti, C. Fontaine and B. Septe, J. Organomet. Chem. 71:107 (1974); C. Merienne, C. Giannotti and A. Gaudemer, J. Organomet. Chem. 54:281 (1973); C. Fontaine, K.N.V. Duong, C. Merienne, A. Gaudemer and C. Giannotti, J. Organomet. Chem. 38:167 (1972).

92. F.R. Jensen and R.C. Kiskis, J. Am. Chem. Soc. 97:5825 (1975).

93. J. Deniau and A. Gaudemer, J. Organomet. Chem. 191:C1 (1980).

94. C. Bied-Charreton and A. Gaudemer, J. Organomet. Chem. 124:299 (1977).

95. M. Perree-Fauvet and A. Gaudemer, J. Organomet. Chem. 120:439 (1976).

96. C. Cloutour, D. La Fasgne and J.C. Pommier, J. Organomet. Chem. 190:35 (1980).

97. C. Giannotti, C. Fontaine, A. Chiaroni and C. Riche, J. Organomet. Chem. 113:57 (1976).

98. A. Chiaroni and C. Pascard-Billy, Bull. Soc. Chim. France 781 (1973).

99. R.J. Cozens, G.B. Deacon, P.W. Felden, K.S. Murray and B.O. West, Aust. J. Chem. 23:481 (1970).

100. M.D. Johnson and G.J. Lewis, J. Chem. Soc. (A) 2153 (1970).

101. C.J. Cooksey, D. Dodd, C. Gatford, M.D. Johnson and G.J. Lewis, J. C. S. Perkin Trans. II 655 (1972).

102. A.E. Crease and M.D. Johnson, J. Am. Chem. Soc. 100:8013 (1978).

103. C. Giannotti and G. Merle, J. Organomet. Chem. 113:45 (1976).

104. N.W. Alcock, M.P. Atkins, E.H. Cyrzon, B.T. Golding and P.J. Sellars, J. C. S. Chem. Commun. 1238 (1980).

105. A.L. Crumbliss, S.T. Bowman, P.L. Gaus and A.T. McPhail, J. C. S. Chem. Commun. 415 (1973).

106. A.L. Crumbliss and P.L. Gaus, Inorg. Chem. 14:486 (1975).

107. B.T. Golding, H.L. Holland, U. Horn and S. Sakrikar, Agnew. Chem. Int. Ed. Engl. 9:959 (1970).

108. R.B. Silverman and D. Dolphin, J. Am. Chem. Soc. 98:4626 (1976).

109. J.H. Espenson and D.M. Wang, Inorg. Chem. 18:2853 (1979).

110. H.P.C. Hogenkamp, J.E. Rush and C.A. Svenson, J. Biol. Chem. 240:3641 (1965).

111. H. Eckert, I. Lagerlund and I. Ugi, Tetrahedron 33:2243 (1977).

112. R. Scheffold and E. Amble, Angew. Chem. Int. Ed. Engl. 19:629 (1979).

113. T.G. Chervyakova, E.A. Parfenov, M.G. Edelev and A.M. Yurkevich, J. Gen. Chem. USSR 44:449 (1974).

114. E.A. Parfenov, T.G. Chervyakova and M.G. Edelev, J. Gen. Chem. USSR 43:2752 (1973).

115. B.T. Golding and S. Sakrikar, J.C.S. Chem. Commun. 1183 (1972).

116. R.B. Silverman and D. Dolphin, J. Am. Chem. Soc. 98:4633 (1976).

117. T. Vickerey, R. Katz and G.N. Schrauzer, J. Am. Chem. Soc. 97:7248 (1975).

118. J. Halpern, M.S. Chan, J. Hanson and T.S. Roche, J. Am. Chem. Soc. 97:1606 (1975).

119. J. Topich and J. Halpern, Inorg. Chem. 18:1339 (1979).

120. J. Halpern, J.A. Topich and K.I. Zamaraev, Inorg. Chim. Acta 20:L21 (1976).

121. R.H. Magnusson and J. Halpern, J. C. S. Chem. Commun. 44 (1978).

122. J. Halpern, M.S. Chan, T.S. Roche and G.M. Tom, Acta Chem. Scand. A33:141 (1979).

123. I. Levitin, A.L. Sigan and M.E. Volpin, J. C. S. Chem. Commun. 469 (1975).

124. G. Costa, A. Puxxedu and E. Reisenhofer, J. C. S. Dalton Trans. 1519 (1972).

125. I.Y. Levitin, A.L. Sigan and M.E. Volpin, Izv. Akad. Nauk SSSR 1205 (1974).

126. D. Mansuy, M. Lange and J.C. Chottard, J. Am. Chem. Soc. 101:6437 (1979).

127. J.M. Wood, Environment 14:33 (1972); Science 183:1049 (1974).

128. J.M. Wood, F.S. Kennedy and C.G. Rosen, Nature 220:173 (1968).

129. J.H. Espenson, W.R. Bushey and M.E. Chmielewski, Inorg. Chem. 14:1302 (1975).

130. J.H. Espenson and T.H. Chao, Inorg. Chem. 16:2553 (1977).

131. R.J. Allen and C.J. Bunton, Bioinorg. Chem. 5:311 (1976).

132. P. Abley, E.R. Dockal and J. Halpern. J. Am. Chem. Soc. 95:3166 (1973).

133. G.N. Schrauzer, J.H. Weber, T.M. Beckman and R.K.Y. Ho, Tetrahedron Letters 275 (1971).

134. G. Tauzher, R. Dreos, G. Costa and M. Green, J. Organomet. Chem. 81:107 (1974).

135. V.E. Magnusson and J.H. Weber, J. Organomet. Chem. 74:135 (1974).

136. J.H. Espenson, H.L. Fritz, R.A. Heckman and C. Nicolini, Inorg. Chem. 15:906 (1976).

137. G.J. Samuels and J.H. Espenson, Inorg. Chem. 19:233 (1980).

138. H. Fritz, J.H. Espenson, D. Williams and G.A. Molander, J. Am. Chem. Soc. 96:2378 (1974).

139. H. Shinozaki, H. Ogawa and M. Tada, Bull. Chem. Soc. Jpn. 49:775 (1976).

140. G. Costa, A. Camus, G. Zassinovich and G. Mestroni, Transition Metal Chem. 1:32 (1975).

141. M.D. Johnson, D. Dodd, B.S. Meeks, D.M. Titchmarsh, K.N.V. Duong and A. Gaudemer, J. C. S. Perkin Trans. II 1261 (1976).

142. H. Shinozaki, M. Kubota, O. Yagi and M. Tada, Bull. Chem. Soc. Jpn. 49:2280 (1976).

143. M. Tada, M. Kubota and H. Shinozaki, Bull. Chem. Soc. Jnp. 49:1097 (1976).

144. J.S. Thayer, Inorg. Chem. 18:1171 (1979).

145. R.E. De Simone, M.W. Penley, L. Charbonneau, S.G. Smith, J.M. Wood, H.A.O. Hill, J.M. Pratt, S. Ridsdale and R.J.P. Williams, Biochem. Biophys. Acta 304:851 (1973).

146. V.C.W. Chu and D.W. Gruenwedel, Bioinorg. Chem. 7:169 (1977).

147. P.J. Craig and S.F. Morton, J. Organomet. Chem. 145:79 (1978).

148. M. Yakamoto, T. Yokoyama, J-L. Chen and T. Kwan, Bull. Chem. Soc. Jpn. 48:844 (1975).

149. N. Imura, E. Sukegawa, S-K. Pan, K. Nagao, J-Y. Kim, T. Kwan and T. Ukita, Science 172:1248 (1971).

150. A. Adin and J.H. Espenson, J. C. S. Chem. Commun. 653 (1971).

151. G.C. Robinson, F. Nome and J.H. Fendler, J. Am. Chem. Soc. 99:4969 (1977).

152. W.M. Scovell, J. Am. Chem. Soc. 96:3451 (1976).

153. E.G. Chauser, I.P. Rudakova and A.M. Yurkevich, J. Gen. Chem. USSR 46:356 (1976).

154. Y.T. Fanchiang, W.P. Ridley and J.M. Wood, J. Am. Chem. Soc. 101:1442 (1979).

155. G. Agnes, S. Bendle, H.A.O. Hill, F.R. Williams and R.J.P. Williams, J. C. S. Chem. Commun. 850 (1971).

156. R.T. Taylor and M.L. Hanna, Bioinorg. Chem. 6:281 (1976).

157. J.Y. Kim, H. Yamamoto and T. Kwan, Chem. Pharm. Bull. 23:1091 (1975).

158. M.E. Volpin, A.M. Yurkevich, L.G. Volkova, E.G. Chauser, I.P. Rudakova, I.Y. Levitin, E.M. Tachkova and T.M. Ushaleova, J. Gen. Chem. USSR 45:150, 164 (1975).

159. J.M. Wood, W.P. Ridley and L.J. Dizikes, J. Am. Chem. Soc. 100:1010 (1978).

160. M.W. Witman and J.H. Weber, Inorg. Chem. 16:2512 (1977).

161. R.T. Taylor and L.M. Hanna, J. Environ. Sci. Health A11:201 (1976).

162. B.C. McBride and R.S. Wolfe, Biochemistry 10:4312 (1971).

163. A. Bakac and J.H. Espenson, Inorg. Chem. 19:242 (1980).

164. H. Yamamoto, T. Yokoyama and T. Kwan, Chem. Pharm. Bull. 23:2186 (1975).

165. A. Gaudemer and F. Gaudemer, Tetrahedron. Letters 1445 (1980).

166. D. Dodd, M.D. Johnson and E.D. McKenzie, J. Am. Chem. Soc. 98:6399 (1976).

167. G.L. Blackmer, T.M. Vickerey and J.N. Marx, J. Organomet. Chem. 72:261 (1974).

168. S.N. Anderson, D.H. Ballard, J.Z. Chrzastowski, D. Dodd and M.D. Johnson, J. C. S. Chem. Commun. 685 (1972).

169. R.D. Garlatti, G. Tauzher and G. Costa, J. Organomet. Chem. 182:409 (1979).

170. R.D. Garlatti, G. Tauzher and G. Costa, J. Organomet. Chem. 139:179 (1977).

171. G. Tauzher, N. Marisch and G. Costa, J. Organomet. Chem. 108:235 (1976).

172. J.P. Kitchin and D.A. Widdowson, J. C. S. Perkin Trans. II 1384 (1979).

173. R.D. Garlatti, G. Tauzher, N. Marisch and G. Costa, J. Organomet. Chem. 92:227 (1975).

174. R. Breslow and P. Khanna, J. Am. Chem. Soc. 98:1297 (1976).

175. S.N. Anderson, D.H. Ballard and M.D. Johnson, J. C. S. Perkin Trans. II 311 (1972).

176. M.R. Ashcroft, B.D. Gupta and M.D. Johnson, J. C. S. Perkin Trans. II 1050 (1980).

177. D. Dodd and M.D. Johnson, J. Organomet. Chem. 52:1 (1973).

178. K.L. Brown and R.J. Bacquet, J. Organomet. Chem. 172:C23 (1979).

179. K.L. Brown, D. Lyles, M. Pencovia and R.G. Kallen, J. Am. Chem. Soc. 97:7338 (1975).

180. K.L. Brown, A. Awtrey and R. LeGates, J. Am. Chem. Soc. 100:823 (1978).

181. H.P.C. Hogenkamp, J.E. Rush and C.A. Swenson, J. Biol. Chem. 240:3641 (1965).

182. R. Barnett, H.P.C. Hogenkamp and R.H. Abeles, J. Biol. Chem. 241:1483 (1966).

183. G.N. Schrauzer, J.H. Weber and T.M. Beckham, J. Am. Chem. Soc. 92:7078 (1970).

184. J.D. Brodie, Proc. Nat. Acad. Sci. USA 62:461 (1969).

185. H.P.C. Hogenkamp, Fed. Proc. 25:1623 (1966).

186. G.N. Schrauzer and R.J. Windgassen, J. Am. Chem. Soc. 89:143, 4250 (1967).

187. K.L. Brown, J. Am. Chem. Soc. 101:6600 (1978).

188. A.V. Cartarno and L.L. Ingraham, Bioinorg. Chem. 7:351 (1977).

189. R.G. Finke and W. McKenna, J. C. S. Chem. Commun. 460 (1980).

190. G.N. Schrauzer, A. Ribero, L.P. Lee and R.K.Y. Ho, Angew. Chem. Int. Ed. Engl. 10:807 (1971).

191. E.A. Parfenov, T.G. Chervyakova, M.G. Edelev and I.K. Shmyrev, J. Gen. Chem. USSR 44:1778 (1974).

192. D. Dodd and M.D. Johnson, J. C. S. Dalton Trans. 1218 (1973).

193. A.L. Crumbliss and P.L. Gaus, Inorg. Nuc. Chem. Letters 10:485 (1974).

194. E. Stadlbauer, R.J. Holland, F.P. Lamm and G.N. Schrauzer, Bioinorg. Chem. 4:67 (1974).

195. I.P. Rudakova, T.A. Pospelova, V.I. Borofulina-Shvets, B.I. Kurganov and A.M. Yurkevich, J. Organomet. Chem. 61:389 (1973).

196. A.W. Johnson and N. Shaw, J. Chem. Soc. 4608 (1962).

197. S.S. Kerwar, T.A. Smith and R.H. Abeles, J. Biol. Chem. 245:1169 (1970).

198. M.N. Richroch and A. Gaudemer, J. Organomet. Chem. 67:119 (1974).

199. M.W. Penley, D.G. Brown and J.M. Wood, Biochemistry 9:4302 (1970).

200. G.N. Schrauzer and E.A. Stadlbauer, Bioinorg. Chem. 3:353 (1974).

201. K.L. Brown and R.G. Kallen, J. Am. Chem. Soc. 94:1894 (1972).

202. G.N. Schrauzer, J.A. Seck, R.J. Holland, T.M. Beckham, E.M. Rubin and J.W. Sibert, Bioinorg. Chem. 2:93 (1972).

203. G.N. Schrauzer, J.H. Grate and R.N. Katz, Bioinorg. Chem. 8:1 (1978).

204. P. Law and J.M. Wood, J. Am. Chem. Soc. 95:914 (1973).

205. G. Agnes, H.A.O. Hill, J.M. Pratt, S.C. Ridsdale, F.S. Kennedy and R.J.P. Williams, Biochem. Biophys. Acta 262:207 (1971).

206. G.N. Schrauzer, J.A. Seck and T.M. Beckham, Bioinorg. Chem. 2:211 (1973).

207. G.L. Blackmer and C-W. Tsai, J. Organomet. Chem. 155:C17 (1978).

208. J.W. Sibert and G.N. Schrauzer, J. Am. Chem. Soc. 92:1421 (1970).

209. D.J. Brockway, B.O. West and A.M. Bond, J. C. S. Dalton Trans. 1891 (1979).

210. P. Boucly, J. Devynck, M. Perree-Fauvet and A. Gaudemer, J. Organomet. Chem. 149:65 (1978).

211. H.P.C. Hogenkamp and S. Holmes, Biochemistry 9:1816 (1970).

212. R. Scheffold, M. Dike, S. Dike, T. Herold and L. Walder, J. Am. Chem Soc. 102:3602 (1980).

213. D. Dodd and M.D. Johnson, J. Am. Chem. Soc. 96:2279 (1974).

214. S. Chemaly and J.M. Pratt, J. C. S. Chem. Commun. 988 (1976).

215. M.P. Atkins, B.T. Golding and P.J. Sellars, J. C. S. Chem Commun. 952 (1976).

216. A. Bury, M.R. Ashcroft and M.D. Johnson, J. Am. Chem Soc. 100:3217 (1978).

217. M.P. Atkins, B.T. Golding, A. Bury, M.D. Johnson and P.J. Sellars, J. Am. Chem. Soc. 102:3630 (1980).

ALKYLIDENE COMPLEXES OF THE EARLIER TRANSITION METALS

Richard R. Schrock

Department of Chemistry

Massachusetts Institute of Technology

1. INTRODUCTION

The first carbene complex, $W(CO)_5(C(OMe)Ph)$, was prepared by Fischer in 1964.[1] Hundreds are now known.[2] Usually, the carbene carbon atom has at least one heteroatom (O or N) bound to it and no α-hydrogen atom. [Exceptions are $W(CO)_5(CPh_2)$,[3] $MnCp(CO)_2(CMe_2)$,[4,5] $ReCp(CO)_2(CMePh)$,[6] $[ReCp(PPh_3)(NO)(CHR)]^+$ (R = H, Me, Ph),[7] $[FeCp(CO)(PPh_3)(CHPh)]^+$,[8] $[FeCp(dppe)(CH_2)]^+$,[9] and $Os(CHPh)(CO)Cl_2(PPh_3)_2$.[10]] All contain metals from Group 6 to 8, the vast majority with d-electron counts (counting the carbene as a neutral ligand) from d^6 to d^{10}. The chemistry of these "electrophilic" carbene complexes has been reviewed periodically,[2] and is reviewed again in Chapter 4 of this book.

$Ta(CHCMe_3)(CH_2CMe_3)_3$ was prepared in 1974.[11] At the time, it was unusual for several reasons. First, no "stabilizing" atom was bound to the α-carbon atom. Second, only metal-carbon bonds were present. Third, it reacted like an alkylidene phosphorane with ketones to give an olefin and a tantalum oxo complex, i.e., it appeared to be a "nucleophilic" carbene complex. Since the carbene-type ligand is derived from a neopentyl ligand, and since it is analogous to an alkylidene phosphorane, it was called a neopentylidene complex. One might propose that the neopentylidene ligand is a "dianion", and therefore, that the metal is in its highest oxidation state. If one wants to compare such an alkylidene complex with a Fischer-type carbene complex the alkylidene ligand could be called a neutral, two electron donor. By this criterion $Ta(CHCMe_3)Np_3$ is a d^2 complex (cf d^6 to d^{10} configurations for the Fischer-type complexes).

Since 1974, many neopentylidene, and, to a lesser extent, benzylidene, methylene, and other alkylidene complexes of Ta, Nb, W, and Ti have been prepared, structurally characterized, and studied. The chemistry of such species is the subject of this chapter. I will take the liberty of including unpublished work which may not appear elsewhere, as well as new work which is not yet in print, in order to make this account as complete and up to date as possible.

2.   TYPES AND METHODS OF PREPARATION

In this section, I discuss the major types of alkylidene complexes and how they are prepared. They are all listed in Table 1.

2.1   Trialkylalkylidene Complexes

The reaction between $TaNp_3Cl_2$ and two equivalents of LiNp in pentane or ether gives sublimable, orange $Ta(CHCMe_3)Np_3$, quantitatively.[11,12] The rate determining step is believed to be formation of thermally unstable $TaNp_4Cl$. $TaNp_4Cl$ can be prepared from $Ta(CHCMe_3)Np_3$ and HCl at $-78°C$. In the absence of LiNp, it decomposes to thermally unstable $Ta(CHCMe_3)Np_2Cl$. Although it is reasonable to propose that $Ta(CHCMe_3)Np_3$ forms by the relatively rapid reaction of $Ta(CHCMe_3)Np_2Cl$ with LiNp, the results of deuterium labeling studies suggest that it also forms either by dehydrohalogenation of $TaNp_4Cl$ or by loss of neopentane from $TaNp_5$ (eq 1). Loss of neopentane

$$TaNp_3Cl_2 \xrightarrow[\text{slow}]{\text{LiNp}} TaNp_4Cl \xrightarrow{-CMe_4} Ta(CHCMe_3)Np_2Cl$$

$$LiNp \searrow \xrightarrow{-CMe_4} \qquad \downarrow LiNp$$

$$Ta(CHCMe_3)Np_3 \qquad (1)$$

Table 1. Known "High Oxidation State" Alkylidene Complexes

		Ref.
M(CHCMe₃)Np₃₋ₓ(OCMe₃)ₓ	x = 0, 1, or 3	11, 12, 13
M(η^5-C₅R₅)(CHCMe₃)X₂	R = H or Meᵃ	14, 15
Ta(η^5-C₅R₅)(CHCMe₃)Cl₂₋ₓR'ₓ	x = 1 or 2; R = H or Me, R' = Np or Me	14
TaCp(CHCMe₃)Cl₂(PMe₃)		14
TaCp*(CHPh)(CH₂Ph)ₓCl₂₋ₓ	x = 0ᵇ, 1ᵇ, or 2	16
TaCp*(CHPh)(CH=PPh₃)Cl		16
TaCp(CH₂)Me₂(dmpe)		17
TaCp₂(CHR)X	R = CMe₃, X = Clᵃ or CHPh₂; R = Ph, X = CH₂Ph or Me; R=SiMe₃, X=Me or CH₂R; R=H, X=Meᵃ ; R=Me, X=Me	18, 19, 20, 21
NbCp₂(CHCMe₃)Cl		20
Ta(η^5-C₅H₅)(η^5-C₅H₄Me)(CH₂)Me		19
M(CHCMe₃)L₂X₃	M = Ta, X = Cl, L = PMe₃, PPhMe₂, PPh₂Me, py, THF L₂ = dmpe, bipy, tmeda, diars; M = Ta, X = Br, L = PMe₃, THF, L₂ = dmpe, diphos,; M = Nb, X = Cl, L = PMe₃ PPhMe₂, THF, L₂ = dmpe	15
[M(CHCMe₃)LX₃]₂	M = Ta, X = Cl, L = PMe₃, PPhMe₂, PPh₂Me; M = Ta X = Br, L = PMe₃; M = Nb, X = Cl, L = PMe₃, PPhMe₂	15

Table 1.

Compound	Conditions	Ref.
Ta(CHR)(PMe$_3$)$_2$X$_3$	R = SiMe$_3$, X = Cl; R = Ph or C$_6$H$_3$Me$_2$	15
M(CHCMe$_3$)(OCMe$_3$)$_{3-x}$Cl$_x$(PMe$_3$)$_y$ x = 1, y = 1; x = 2, y = 2		13
M(CHPh)(OCMe$_3$)$_2$Cl(PMe$_3$)$_2$		13
M(CHCMe$_3$)$_2$L$_2$X	L = PMe$_3$ or PPhMe$_2$, X = Np or Cl; M = Ta, L = PMe$_3$, X = Me, Et, mesityl	22, 23
Ta(η^5-C$_5$R$_5$)(CHCMe$_3$)$_2$(PMe$_3$)	R = H or Me	22
W(O)(CHR)L$_x$X$_2$	R = H, CMe$_3$, Et, Ph, L = PEt$_3$ or PMe$_3$, x = 1 or 2	24, 25, 26^c
W(O)(CHCMe$_3$)(OCMe$_3$)$_2$(PEt$_3$)		25^c
[W(O)(CHCMe$_3$)(PEt$_3$)$_2$Cl]$^+$AlCl$_4^-$		25^c
[W(O)(CHCMe$_3$)(PEt$_3$)$_2$]$^+$(AlCl$_4^-$)$_2$		25^c
Ta(NSiMe$_3$)(CHCMe$_3$)(PMe$_3$)$_2$Cl		27
Ta(CHCMe$_3$)(PMe$_3$)$_4$X		28
TaCp*(CHCMe$_3$)(H)(PMe$_3$)Cl		28
Ta(CHCMe$_3$)(H)(PMe$_3$)$_3$Cl$_2$		28
Ta(CHCMe$_3$)R(C$_2$H$_4$)(PMe$_3$)$_2$	R = Et or Np	23
TaCp*(CHCMe$_3$)(C$_2$H$_4$)(PMe$_3$)		29

$[Ta(CHCMe_3)(PMe_3)_2X]_2N_2$	$X = Cl, Br, CH_2CMe_3, Me$	30
$\overline{Cp_2TiCH_2AlR_2Cl}$	$R = Cl, Me, Np$	31
$Cp_2Ta(CH_2AlMe_3)Me$		19
$Ta(CH_2CMe_3)_3[C(LiL_2)(CMe_3)]$	$L_2 = tmeda, N,N'-dimethylpiperazine$	32
$W(CCMe_3)(CHCMe_3)(CH_2CMe_3)(dmpe)$		33
$Ta(CHSiMe_3)[N(SiMe_3)_2]_2(CH_2SiMe_3)$		34

[a] Also $\eta^5\text{-}C_5H_4Me$ analogs

[b] Also $\eta^5\text{-}C_5Me_4Et$ analogs

[c] Also W=NPh analogs

from $TaNp_4Cl$ or hypothetical $TaNp_5$ was called "α-hydrogen atom abstraction" rather than "α-hydride elimination" (cf "β-hydride elimination"[35]) since it was felt that direct removal of the α-hydrogen atom by a neopentyl leaving group was more reasonable for a d^0 metal than α-elimination to give a neopentylidene hydride intermediate followed by reductive elimination of neopentane.

A better example of α-abstraction is the first-order decomposition of $TaNp_4(OCMe_3)$ to $Ta(CHCMe_3)Np_2(OCMe_3)$.[36] Like $TaNp_4Cl$, $TaNp_4(OCMe_3)$ contains three neopentyl groups of one type, probably equatorial neopentyl groups in a trigonal bipyramid. It is not clear whether $TaNp_4(OCMe_3)$ is a better model for $TaNp_4Cl$ or $TaNp_5$, but an important observation is that $TaNp_4(OCMe_3)$ is much more stable (decomp. $50°$) than $TaNp_4Cl$ or hypothetical $TaNp_5$.

$Ta(CHSiMe_3)(CH_2SiMe_3)_3$ can be prepared smoothly from $Ta(CH_2SiMe_3)_3Cl_2$ and $Mg(CH_2SiMe_3)_2(diox)$ in ether at $-30°$ ($\delta H_\alpha = 5.70$, $\delta C_\alpha = 241$).[37] It cannot be isolated in pure form because it is very soluble in pentane, and because it decomposes to $[Ta(\mu\text{-}CSiMe_3)(CH_2SiMe_3)_2]_2$[38] smoothly and quantitatively above ca. $0°C$. This decomposition reaction is almost certainly intermolecular since it can be blocked by adding PMe_3 (cf the reaction between $Ta(CHCMe_3)Np_3$ and PMe_3 later). In the presence of a two-fold excess of PMe_3, $Ta(CH_2SiMe_3)_3Cl_2$ reacts with $Mg(CH_2SiMe_3)_2(diox)$ to give $Ta(CHSiMe_3)(CH_2SiMe_3)_3(PMe_3)$ quantitatively ($\delta H_\alpha = 6.14$, $\delta C_\alpha = 252$, $J_{CH} = 96$ Hz).

$TaBz_5$ is an isolable, red crystalline species.[39] It decomposes at $40°$ in a first-order reaction ($k \approx 4 \times 10^{-5}$ sec^{-1}) which is unaffected by the addition of radical initiators or traps.[40] The proposed initial product, $Ta(CHPh)Bz_3$, can be trapped with acetonitrile[36] (a characteristic reaction; see 6.4) to give $Ta[N(Me)C=CHPh]Bz_3(MeCN)$, but in the absence of acetonitrile $Ta(CHPh)Bz_3$ decomposes to an intractable oil. This rapid second phase of the decomposition of $TaBz_5$ probably is intermolecular. Only toluene is produced when $TaBz_5$ decomposes, a total of 2.6 equivalents. $Ta(CD_2Ph)_5$ decomposes considerably more slowly ($k_H/k_D \approx 3$).

TaMe$_5$ can be prepared from TaMe$_3$Cl$_2$ and two equivalents of methyl lithium in ether.[41,39]   It is a volatile, pale yellow, crystalline complex which melts at about room temperature to an oil which rapidly turns dark as 3.4 ± 0.1 equivalents of methane evolve.   The non-free radical decomposition is autocatalytic.   It, too, shows a deuterium isotope effect.   The non-hydrolyzable metal-containing residue has the approximate composition TaC$_{1.5}$H.   Since TaMe$_5$ is more stable in dilute solutions, the first and slowest step of the reaction is probably bimolecular.   We might speculate that it is an intermolecular version of the α-abstraction reaction, the first step of which, formation of a CH$_2$ fragment, is the slowest.   The decomposition can be blocked by adding a chelating phosphine (dmpe) to give stable, white, crystalline TaMe$_5$(dmpe).   As we will see later, intramolecular α-abstraction reactions appear to be accelerated when donor ligands are added, in contrast to the behaviour observed here.   At this juncture, a Ta(CH$_2$)Me$_3$ is purely hypothetical.

Bright yellow NbMe$_5$ decomposes at low temperatures (ca. -50°) in ether to give only methane.   Only NbMe$_5$(dmpe) could be isolated.[41,39]

Donor ligand-free trialkylalkylidene complexes almost certainly will continue to be rare.   Intermolecular decomposition reactions appear to be fast (see 5.2) and only when this decomposition is blocked by substitutents on the alkyl and alkylidene ligands (or, in theory, by adding donor ligands) can such a complex, e.g. Ta(CHCMe$_3$)(CH$_2$CMe$_3$)$_3$, be isolated.

2.2   Monocyclopentadienyl Complexes

The best studied example so far of α-abstraction is the decomposition of TaCpNp$_2$Cl$_2$ to TaCp(CHCMe$_3$)Cl$_2$ (eq 2).[14]   It was shown that

$$TaNp_2X_3 \xrightarrow{\ LiC_5R_5\ } Ta(\eta^5\text{-}C_5R_5)Np_2X_2 \xrightarrow{\ -CMe_4\ } Ta(\eta^5\text{-}C_5R_5)(CHCMe_3)X_2$$
$$\text{(only cis)} \hspace{4cm} (2)$$

(i) the α-abstraction reaction is intramolecular;   (ii) it is much faster in dichloromethane than pentane and when the halide is Br instead of Cl;   (iii) the rate of decomposition of the η^5-C$_5$Me$_5$ complex is 10^{-3} that of the η^5-C_5H_5 complex; and (iv) the deuterium isotope effect is about six. These results were explained by postulating that only cis-Ta(η^5-C_5R_5)Np$_2$X$_2$ (X = Cl or Br) is prone to α-abstraction. This postulate was supported by spectroscopic observation of the cis/trans equilibrium in the case of TaCpNpBzX$_2$. Yields of the corresponding Nb complexes are disastrously low in what are comparatively complex reactions; the dineopentyl intermediates could not be isolated. The TaCpNp$_2$X$_2$ system was the first in which we noticed that light markedly accelerated the -abstraction reaction.

Analogous benzylidene complexes are more difficult to prepare.[16] TaBz$_3$Cl$_2$ reacts with NaCp to give TaCpBz$_3$Cl. This complex decomposes readily, but hypothetical, presumably unstable TaCp(CHPh)BzCl (cf TaCp(CHCMe$_3$)NpCl[14]) could not be observed. However, TaBz$_3$Cl$_2$ reacted with LiC$_5$Me$_5$ to give isolable TaCp*(CHPh)(CH$_2$Ph)Cl. TaCp*Bz$_2$Cl$_2$ can be photolyzed to give TaCp*(CHPh)Cl$_2$ in poor yields or dehydrohalogenated with an alkylidene phosporane to give TaCp*(CHPh)BzCl.

TaMe$_3$Cl$_2$ reacts with TlCp to give TaCpMe$_3$Cl and with LiC$_5$Me$_5$ to give TaCp*Me$_3$Cl. There is no indication that these molecules decompose smoothly to give methane and Ta(η^5-C_5R_5)(CH$_2$)Me$_2$, at least under conditions where such a species might be stable.

2.3 Biscyclopentadienyl Complexes

This class was one of the earliest to be explored since its members have 18 valence electrons and are, therefore, comparatively stable towards bimolecular decomposition.

TaCp$_2$(CHCMe$_3$)Cl can be prepared straightforwardly from TaCp(CHCMe$_3$)Cl$_2$ and TlCp,[20,18] and NbCp$_2$(CHCMe$_3$)Cl, in moderate yield, from NbNp$_2$Cl$_3$ and two equivalents of TlCp. While the reaction between TaBz$_3$Cl$_2$ and two equivalents of TlCp gives TaCp$_2$(CHPh)Bz[20,18] in moderate yield, it is uncertain whether this product forms from TaCp$_2$Bz$_3$ or intermediate (and unstable) TaCp(CHPh)BzCl and TlCp. TaCp$_2$Me$_3$ slowly decomposes in the solid state or solution to give methane (1-3 equivalents) but no readily characterizeable product. Evidently, CH bonds in the Cp ring are involved since decomposition of $TaCp_2(CD_3)_3$ does not give solely CD_4. Interestingly, photolysis of $TaCp_2Me_3$ (medium pressure Hg lamp) yields $TaCp_2(CH_2)Me$ in fair estimated yield (observed by ^{1}H NMR[42]) but it is too unstable to be prepared in pure form in this manner.

$TaCp_2(CH_2)Me$ can be prepared from $TaCp_2Me_3$ by an indirect method (eq 3).[18,19] Trityl removes the central methyl group in $TaCp_2Me_3$

$$TaCp_2Me_3 \xrightarrow[-Ph_3CMe]{Ph_3C^+BF_4^-} [TaCp_2Me_2]^+BF_4^- \xrightarrow{base} TaCp_2(CH_2)Me \qquad (3)$$

selectively as shown by deuterium labeling. Deprotonation of $[TaCp_2Me_2]^+$ is formally analogous to deprotonation of a methyl phosphonium salt to give a methylenephosphorane. This preparative method is potentially general. In practice, however, it has been of limited utility since cations without two Cp groups abstract F^- from BF_4^- to give neutral fluoride complexes.[17] Other examples are deprotonations of $[MCp_2(CH_2SiMe_3)(R)]^+$ to give $MCp_2(CHSiMe_3)(R)$ (M = Ta, R = Me;[20] M = Nb or Ta, R = CH_2SiMe_3[21]). The success of dehydrohalogenation reactions (e.g. of $TaCp*Bz_2Cl_2$ by $Ph_3P=CH_2$ to give $TaCp*(CHPh)BzCl$[16]) may rely on formation of cationic intermediates.

An interesting variation of the deprotonation reaction is the reaction between $TaCp_2Me_3$ and $AlMe_3$ (eq 4).[19] The postulated intermediate,

$$TaCp_2Me_3 \xrightarrow{AlMe_3} [TaCp_2Me_2]^+AlMe_4^- \xrightarrow{-CH_4} Cp_2Ta(CH_2AlMe_3)Me \qquad (4)$$

$[TaCp_2Me_2]^+AlMe_4^-$, forms an oil in toluene which redissolves slowly to give the final product. This reaction is similar to the Tebbe reaction of $AlMe_3$ with $TiCp_2Cl_2$ to give $TiCp_2(CH_2AlMe_2Cl)$ (see 2.7).

An alternative route to several members of this class of complexes is via transfer of an alkylidene from a phosphorane to Ta(III) as shown in eq 5.[43a] The methylene complex can be observed but cannot be

$$TaCp_2(PR_3')Me + R_3''P=CHR \xrightarrow[-PR_3']{-PR_3''} TaCp_2(CHR)Me$$
$$R = H, Ph, Me \qquad (5)$$

isolated since it reacts with methylene phosphoranes (see later). The most important feature of this reaction is that an ethylidene complex can be prepared. It is the only one of Ta known so far. Contrary to what one might expect, it does not rearrange to give the known tautomer, $TaCp_2(CH_2CH_2)Me$ (see 5.3).

The only example of a heteroatom-substituted carbene complex of an early transition metal in a high oxidation state is $Cp_2^*Zr(H)(OCH=Nb(H)Cp_2)$, prepared by adding $Cp_2^*ZrH_2$ to $Cp_2Nb(H)(CO)$.[43b] The x-ray structure of a related species, $Cp_2^*Zr(H)(OCH=WCp_2)$, shows that the plane of the zirconoxy carbene ligand is perpendicular to the two Cp ligand planes, and that the W=C bond length is 2.005(13)Å.

2.4 Complexes of the Type $M(CHR)X_3L_x$ (X = halide or alkoxide)

The simplest type of neopentylidene complex discovered so far is $M(CHCMe_3)X_3(THF)_2$ (M = Nb or Ta, X = Cl or Br).[15] They can be prepared in good (M = Nb) to quantitative (M = Ta) yield simply by adding THF to MNp_2X_3. From them one can prepare a variety of phosphine adducts, many of which can be prepared directly from $TaNp_2X_3$ and phosphine.[15] As in the case of the formation of $TaCp(CHCMe_3)Cl_2$, the α-abstraction reaction is markedly accelerated in a chlorinated solvent (chloroform or dichloromethane), or when X = Br (vs when X = Cl). The reaction is believed to involve an intermediate seven-coordinate species, $MNp_2X_3L_2$, in which the two neopentyl groups are <u>cis</u> to one another. If L = PR_3 (e.g. PMe_3 or PEt_3) dimeric complexes of the type $[M(CHCMe_3)X_3L]_2$ form when $M(CHCMe_3)X_3L_2$ loses L, or when MNp_2X_3 reacts with only one equivalent of PR_3. The THF adducts also offer a simple route to complexes such as $TaCp^*(CHCMe_3)Cl_2$ or $NbCp^*(CHCMe_3)Cl_2$ which cannot be prepared directly from MNp_2Cl_3 and LiC_5Me_5.[14]

The benzylidene complex, $Ta(CHPh)X_3(PMe_3)_2$, prepared analogously from $TaBz_2X_3$ and PMe_3 in dichloromethane,[15] is considerably less stable than the analogous neopentylidene complex. When it loses one PMe_3 ligand, it decomposes to give stilbenes rather than forming dimeric

$[Ta(CHPh)X_3(PMe_3)]_2$. Two seven-coordinate species could be observed by NMR, $TaBz_2Cl_3(PMe_3)_2$ and $Ta(CHPh)Cl_3(PMe_3)_3$. The first is the type of species (but possibly not the species) which is believed to be the direct precursor to $Ta(CHPh)X_3(PMe_3)_2$. Formation of $Ta(CHPh)Cl_3(PMe_3)_3$ could explain why the yield of $Ta(CHPh)X_3(PMe_3)_2$ is good only when excess PMe_3 is present.

Attempts to prepare analogous methylene complexes so far have failed. Complexes such as $TaMe_3Cl_2(PMe_3)_2$ (like $TaMe_5(dmpe)$; see above) are quite stable and only decompose at temperatures where any methylene complex almost certainly would be unstable. A good candidate for "forcing" methylene formation, we felt, was $TaMe(mesityl)X_3(PMe_3)_2$. These complexes do yield an alkylidene complex smoothly when they decompose (especially when X = Br) but it is not a methylene complex. Methane is given off and, ultimately, a substituted benzylidene complex is formed, probably via γ-hydrogen abstraction (eq 6)[44], although an alternative possibility is shown in eq 7. Preferential abstraction of a benzyl α-hydrogen atom would give the observed product.

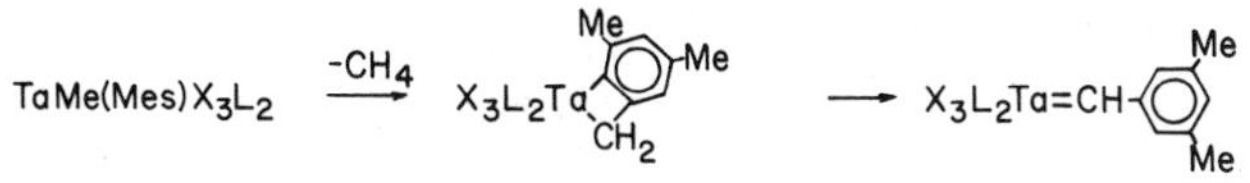

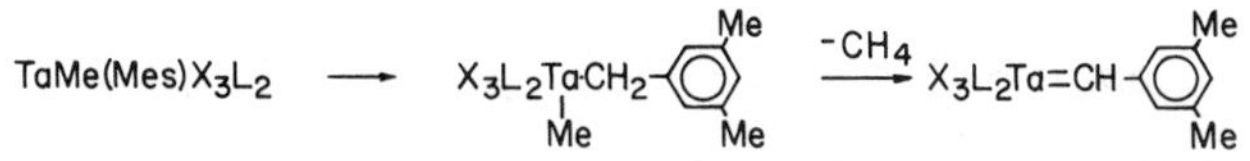

The only other known halo/alkylidene complex of this type is $Ta(CHSiMe_3)Cl_3(PMe_3)_2$.[15] It forms from $Ta(CH_2SiMe_3)_2Cl_3$ and PMe_3 in dichloromethane more slowly than does $Ta(CHCMe_3)Cl_3(PMe_3)_2$. This was taken as evidence that abstraction of an α-hydrogen from a trimethylsilylmethyl group by a trimethylsilylmethyl group is not as easy as from a neopentyl group by a neopentyl group. This assumes that the same type of immediate precursor to the alkylidene complex is present in about the same concentration in each case.

Di-t-butoxy/neopentylidene complexes can be prepared as shown in eq 8.[13] One PMe_3 ligand is lost in the process, but the monophosphine

$$M(CHCMe_3)Cl_3(PMe_3)_2 + 2LiOCMe_3 \xrightarrow{-PMe_3} M(CHCMe_3)(OCMe_3)_2Cl(PMe_3)$$

$$M = Nb \text{ or } Ta \qquad (8)$$

complex will add a second equivalent of PMe_3 at low temperature to give $M(CHCMe_3)(OCMe_3)_2Cl(PMe_3)_2$. Monotertiarybutoxide complexes are best prepared by decomposition of $M(CH_2CMe_3)_2(OCMe_3)Cl_2$ in the presence of PMe_3. $Ta(CHCMe_3)(OCMe_3)_3$ can   be   prepared   from $Ta(CHCMe_3)Cl_3(THF)_2$.   It is a thermally unstable oil which was only characterized by ^1H and ^{13}C NMR.

Di-t-butoxy/benzylidene complexes can be prepared by a metathesis-type reaction[13] (eq 9; see 6.3). One PMe_3 ligand is lost readily to give

$$M(CHCMe_3)(OCMe_3)_2Cl(PMe_3) \xrightarrow[PhCH=CH_2]{PMe_3} M(CHPh)(OCMe_3)_2Cl(PMe_3)_2 \quad (9)$$

monophosphine complexes which can be observed by ^{13}C and ^{31}P NMR but which are unstable with respect to formation of stilbenes. We do not know whether these complexes are better formulated as dimers or monomers. Although both ethylene and cis-2-pentene react readily in a metathesis fashion (see 6.3), the expected alkylidene complexes could not be observed.

2.5 Bisalkylidene Complexes

The first complex of this type was prepared by the reaction of $Ta(CHCMe_3)Np_3$ with PMe_3[22] (eq 10). We proposed that the first

$$M(CHCMe_3)Np_3 \xrightarrow[-CMe_4]{2L} Ta(CHCMe_3)_2NpL_2 \qquad (10)$$

intermediate was a five-coordinate adduct from which neopentane was lost in an α-hydrogen abstraction reaction. However, it is uncertain whether the new neopentylidene ligand forms directly, or whether the neopentylidene ligand is converted into a neopentylidyne ligand which then abstracts an α-hydrogen atom from a neopentyl group.   (Alkylidyne complexes are discussed in Section 5.1.) An analogous chloro derivative is the result of the reaction between MNp_4Cl and a phosphine (eq 11). MNp_4Cl decomposes to unstable $M(CHCMe_3)Np_2Cl$, and it is probably

$$MNp_4Cl \xrightarrow[]{-CMe_4} M(CHCMe_3)Np_2Cl \xrightarrow[-CMe_4]{2L} M(CHCMe_3)_2ClL_2 \qquad (11)$$

this species which is attacked by L to yield the final product. Several alkyl derivatives can be prepared from $Ta(CHCMe_3)_2Cl(PMe_3)_2$, among them $Ta(CHCMe_3)_2(CD_2CMe_3)(PMe_3)_2$ and $Ta(CHCMe_3)_2Et(PMe_3)_2$. The deuterium labels do not scramble in the former and the latter will not interconvert with its tautomer, $Ta(CHCMe_3)(C_2H_4)Np(PMe_3)_2$ (see 2.7).

It is interesting to note that $Ta(CHSiMe_3)(CH_2SiMe_3)_3(PMe_3)$ is stable at room temperature.[37] It is not known whether it decomposes to a bisalkylidene complex in the presence of PMe_3 at elevated temperatures, or whether it simply decomposes the way $Ta(CHSiMe_3)(CH_2SiMe_3)_3$ does to give $[Ta(\mu\text{-}CSiMe_3)(CH_2SiMe_3)_2]_2$ and TMS.

$TaCp(CHCMe_3)_2(PMe_3)$ and $TaCp^*(CHCMe_3)_2(PMe_3)$ were first prepared by the sequence shown in eq 12.[22] They also can be prepared from

$$Ta(\eta^5\text{-}C_5R_5)(CHCMe_3)NpCl + 2PMe_3 \rightarrow Ta(\eta^5\text{-}C_5R_5)(CCMe_3)Cl(PMe_3)_2 \xrightarrow[LiNp]{-PMe_3}$$

$$Ta(\eta^5\text{-}C_5R_5)(CHCMe_3)_2(PMe_3) \qquad (12)$$

$Ta(CHCMe_3)_2Cl(PMe_3)_2$ and LiC_5R_5.

So far no bisbenzylidene complexes have been prepared nor any bisneopentylidene complex of any metal other than Nb or Ta.

2.6 Oxo and Imido Complexes

Tungsten oxo neopentylidene complexes can be prepared by the alkylidene transfer reaction shown in eq 13.[24] $Ta(OCMe_3)_4X$ is left behind

$$Ta(CHCMe_3)X_3L_2 + W(O)(OCMe_3)_4 \xrightarrow{pentane} Ta(OCMe_3)_4X + W(O)(CHCMe_3)X_2L_2$$

$$L = PMe_3 \text{ or } PEt_3; \ X = Cl \text{ or } Br \quad (13)$$

when the tungsten complex crystallizes out in high yield.   Several methylene, propylidene, and benzylidene complexes have been prepared by the $AlCl_3$ catalyzed metathesis reaction shown in eq 14 (see 6.3). All are thermally stable, yellow, crystalline complexes.

$$W(O)(CHCMe_3)X_2L_2+CH_2=CHR \xrightarrow[PhCl]{AlCl_3} Me_3CCH=CH_2+W(O)(CHR)X_2L_2$$

$$R = H, Et, Ph \qquad (14)$$

There are two plausible intermediates in the $AlCl_3$ catalyzed reactions: cationic complexes, and five-coordinate, netral species.   In dichloromethane, not only monocationic but dicationic complexes, in fact, can be prepared smoothly (e.g. eq 15).   On adding some solvent that

$$W(O)(CHCMe_3)Cl_2L_2 \xrightarrow[CH_2Cl_2]{AlCl_3} [W(O)(CHCMe_3)ClL_2]^+AlCl_4^- \xrightarrow[CH_2Cl_2]{AlCl_3}$$

$$[W(O)(CHCMe_3)L_2]^{2+}(AlCl_4^-)_2 \qquad (15)$$

coordinates to $AlCl_3$ (e.g. TMEDA or THF) these reactions are reversed. A monomeric, five-coordinate, neutral complex also can be prepared by removing one labile phosphine ligand with $Pd(PhCN)_2Cl_2$[26] or, more economically, with $CuCl$[45] (eq 16).

$$W(O)(CHCMe_3)Cl_2(PEt_3)_2 \xrightarrow[-M(PEt_3)]{M} W(O)(CHCMe_3)Cl_2(PEt_3) \qquad (16)$$

A tertiarybutoxide complex can be prepared as shown in eq 17.[25] This species must lose PEt_3 readily since its ^1H NMR spectrum shows

$$W(O)(CHCMe_3)Cl_2(PEt_3)_2+2LiOCMe_3 \rightarrow W(O)(CHCMe_3)(OCMe_3)_2(PEt_3)$$
$$+ PEt_3 \qquad (17)$$

loss of coupling of the alkylidene's α-hydrogen to phosphorus at $25°$.

All of the above oxo complexes have phenyl imido analogs.[45] Since an imido group is isoelectronic with an oxo group and usually at least as good a π-electron donor, the chemistry of the imido complexes is largely similar to that of the oxo complexes.

One example of a tantalum imido neopentylidene complex is known. It was prepared by reacting $Ta(CHCMe_3)(PMe_3)_4Cl$ (see below) with Me_3SiN_3 (eq 18).[27] It is a relative of the bisneopentylidene complexes

$$Ta(CHCMe_3)L_4Cl + Me_3SiN_3 \xrightarrow[-2L]{-N_2} Ta(NSiMe_3)(CHCMe_3)L_2Cl \qquad (18)$$

mentioned above. It, too, is thermally stable and sublimes unchanged.

2.7 Other Alkylidene Complexes

When $Ta(CHCMe_3)(PMe_3)_2X_3$ is reduced by two electrons it gives in good yield the only reduced alkylidene complex which does not also contain an olefin ligand (eq 19).[28] The neopentylidene ligand is the

$$Ta(CHCMe_3)L_2X_3 \xrightarrow[\text{excess L, Ar}]{2Na/Hg} Ta(CHCMe_3)L_4X \qquad (19)$$

most distorted yet encountered (see Section 4.), as if the α-hydrogen atom is in the process of transferring to the metal. In another instance this formally does occur to give an alkylidyne hydride complex (eq 20). This "α-elimination" principle can be applied to prepare neopentylidene

$$TaCp^*(CHCMe_3)Cl_2 \xrightarrow[\text{excess PMe}_3]{2Na/Hg} \text{trans-}TaCp^*(CCMe_3)(PMe_3)_2(H) \quad (20)$$

hydride complexes from incipient Ta(III) neopentyl complexes (eqs 21 and 22).

$$TaCp^*(CH_2CMe_3)Cl_3 \xrightarrow[\text{excess PMe}_3]{2Na/Hg} TaCp^*(CHCMe_3)(PMe_3)(H)Cl \qquad (21)$$

$$Ta(CH_2CMe_3)Cl_4 \xrightarrow[\text{excess PMe}_3]{2Na/Hg} Ta(CHCMe_3)(H)(PMe_3)_3Cl_2 \qquad (22)$$

"Reduced" neopentylidene complexes that contain an ethylene ligand are shown in eqs 23 and 24. It is interesting to note that the

$$Ta(C_2H_4)(PMe_3)_2Cl_3 \xrightarrow{1.5\ MgNp_2(diox)} Ta(CHCMe_3)(C_2H_4)Np(PMe_3)_2 \quad^{36}(23)$$

$$\text{TaCp*(CCMe}_3\text{)L}_2\text{Cl} \xrightarrow{\text{0.5 MgEt}_2\text{(diox)}} \text{TaCp*(CHCMe}_3\text{)(C}_2\text{H}_4\text{)L}^{29} \qquad (24)$$

tautomer of the first, $\text{Ta(CHCMe}_3\text{)}_2\text{(C}_2\text{H}_5\text{)(PMe}_3\text{)}_2$, can be prepared straightforwardly from $\text{Ta(CHCMe}_3\text{)}_2\text{Cl(PMe}_3\text{)}_2$. They do not interconvert under conditions where both are stable.[23]

The reduction of $\text{Ta(CHCMe}_3\text{)L}_2\text{X}_3$ under N_2 in the presence of L yields $[\text{Ta(CHCMe}_3\text{)L}_2\text{X]}_2\text{N}_2$ (L = PMe_3, X = Cl or Br).[30] X can be replaced by alkyl groups. These molecules are related to the imido/neopentylidene complex mentioned earlier since an x-ray study shows the N-N bond to be unusually long (1.298Å; see 4.).

The reaction of PMe_3 or dmpe with $\text{W(CCMe}_3\text{)Np}_3$ gives the complex shown in eq 25.[33] This reaction is formally an α-abstraction process of the type which has yielded Nb and Ta alkylidene complexes.

$$\text{W(CCMe}_3\text{)Np}_3 + 2\text{L} \longrightarrow \text{W(CCMe}_3\text{)(CHCMe}_3\text{)NpL}_2 \qquad (25)$$

$$\text{L = PMe}_3 \text{ or 1/2 dmpe}$$

A related reaction is the formation of $\text{Ta(CHCMe}_3\text{)}_2\text{Np(PMe}_3\text{)}_2$ from $\text{Ta(CHCMe}_3\text{)Np}_3$. $\text{W(CCMe}_3\text{)(CHCMe}_3\text{)Np(dmpe)}$ reacts with more dmpe at elevated temperatures to give a complex which is empirically $\text{W(dmpe)}_2\text{(CHCMe}_3\text{)}$ but which actually is $\underline{\text{trans}}\text{-W(dmpe)}_2\text{(CCMe}_3\text{)(H)}$.[46] It is unknown whether the neopentylidyne hydride complex is preferred for kinetic or thermodynamic reasons.

The reaction between $\text{Ta(CHCMe}_3\text{)Np}_3$ and $\text{LiBu}^\cdot\text{L}_2$ yields the "lithio carbene" complex, $\text{TaNp}_3\text{[C(LiL}_2\text{)(CMe}_3\text{)]}$.[32] This was originally called an anionic carbyne complex, $[\text{TaNp}_3\text{(CCMe}_3\text{)]}^-\text{LiL}_2^+$, according to structural studies (see 4.), but from what is now known about distortion of alkylidene complexes the α-lithio-substituted carbene description may be more accurate.

The reaction between TiCp_2Cl_2 and AlMe_3 gave the complex shown in eq 26.[31] Several analogous species can be prepared by exchanging the

$$\text{TiCp}_2\text{Cl}_2 + \text{AlMe}_3 \longrightarrow \text{Cp}_2\text{Ti}\overset{\text{CH}_2}{\underset{\text{Cl}}{\diagdown\diagup}}\text{AlMe}_2 \qquad (26)$$

$AlMe_2Cl$ group for others (e.g. $AlCl_3$ or $AlNp_2Cl$). This species can be called a Lewis acid protected methylene complex. It decomposes when base is added, presumably by forming unstable $TiCp_2(CH_2)$. Tebbe proposes that coordination of $AlMe_3$ polarizes $TiCp_2MeCl$ so that the reaction is related to a deprotonation reaction (eq 27).

$$TiCp_2Cl_2 + 2\,AlMe_3 \xrightarrow[-\,AlMe_2Cl]{} Cp_2Ti\overset{\delta+}{\underset{Cl\rightarrow AlMe_3}{\diagdown}}\overset{Me}{\diagup}\underset{\delta-}{} \xrightarrow{-CH_4} Cp_2Ti\overset{CH_2}{\underset{Cl}{\diagdown}}\diagup AlMe_2 \qquad (27)$$

A related Zr complex has been observed recently by ^{13}C NMR.[47] It was prepared by alkylidene transfer from an alkylidene phosphorane to Zr(II) (eq 28; cf preparation of $TaCp_2(CHMe)Me$, eq 5).

$$Cp_2ZrL_2 + CH_2PPh_3 \xrightarrow[-PPh_3]{-L} Cp_2Zr(CH_2)(L) \quad (L = PPh_2Me) \qquad (28)$$

$[WCp_2(CH_2)(H)]^+$ is the proposed intermediate (eq 29) in the reaction between $[WCp_2(C_2H_4)(CH_3)]^+$ and $PPhMe_2$ to give first $[WCp_2(CH_2CH_2PPhMe_2)(CH_3)]^+$, then $[WCp_2(CH_2PPhMe_2)(H)]^+$, and finally, $[WCp_2(PPhMe_2)(CH_3)]^+$.[48]

$$[WCp_2(C_2H_4)(CH_3)]^+ \xrightarrow{-C_2H_4} \text{"}[WCp_2(CH_3)]^{+}\text{"} \rightarrow \text{"}[Cp_2W(CH_2)(H)]^{+}\text{"} \qquad (29)$$

The reaction between $Ta[N(SiMe_3)_2]_2Cl_3$ and three equivalents of $LiCH_2SiMe_3$ gives an air-stable complex with the formula $Ta(CHSiMe_3)[N(SiMe_3)_2]_2(CH_2SiMe_3)$.[34] It is not clear when and how the alkylidene ligand forms.

3.   NMR AND IR STUDIES

The ^{13}C NMR spectrum of any alkylidene complex shows a peak for the α-carbon atom between 210 and 320 ppm downfield from TMS. (The data for complexes whose structures are known can be found in Table 2.) The chemical shift does not correlate especially well with any of the structural features we will discuss in the next section. It is true, however, that the signal for the α-carbon atom in "electrophilic" alkylidene complexes

Table 2. NMR, IR, and Structural Data for Alkylidene Complexes

Complex (Ref., method)	δH_α	δC_α	J_{CH_α}	νCH_α	M-C	M=C	$\angle M{=}C{-}R$
TaCp₂(CH₂)(CH₃)(49,b) (50,d)	10.22	228	132	n.o.[a]	2.246(12) 2.269(3)	2.026(10) 2.041(3)	126.4 123.8(123.5)(2)[c]
TaCp₂(CHPh)(CH₂Ph)(20,b)	10.86	246	127	n.o.	2.30(1)	2.07(1)	135.2(7)
TaCp₂(CHCMe₃)Cl (51,52,b)	10.10	274	121	2900[e]		2.030(6)	150.4(7)
TaCp(CHCMe₃)Cl₂(53,b)	6.38	246	84	2510		1.75(6)	165(3)
TaCp*(CHPh)(CH₂Ph)(16,b)	~7	220	82	n.o.	2.211(15)[f]	1.883(14)	166.0(10)
TaCp*(CHCMe₃)(C₂H₄)(PMe₃)(29,d)	-2.86	223	74	2480		1.946(3)	170.0(2)[g]
TaCp*(CPh)(PMe₃)₂Cl (54,b)		347				[1.849(8)][h]	[171.8(6)][h]
[Ta(CHCMe₃)(PMe₃)Cl₃]₂(55,29,d)	5.30	276	101	2605		1.898(2)	161.2(1)[i]
Ta(CHCMe₃)₂(mesityl)(PMe₃)₂ (56,b)	2.00 6.72	243 275	91 104		2.303(6)	1.932(7) 1.955(7)	168.9(6) 154.0(6)
[Ta(CHCMe₃)(CH₂CMe₃)(PMe₃)₂]₂N₂ (30,b)	6.22	271	88	2600	2.299(10) 2.285(10)	1.932(9) 1.937(9)	158.5(7) 160.3(7)
W(O)(CHCMe₃)(PMe₃)₂Cl₂(57,b)	11.89	316	121			2.006(15)	141.1(16)
W(O)(CHCMe₃)(PEt₃)Cl₂(26,b)	9.80	295	115			1.882(14)	140.6(11)

W(CCMe$_3$)(CHCMe$_3$)(CH$_2$CMe$_3$)(dmpe) (58,b)	8.15	256	84	2580	2.258(8)	1.942(9)[j] 150.4(8)[j]
Ta(CH$_2$CMe$_3$)$_3$[C(Lipip)(CMe$_3$)]^k (32,b)					2.23(2)[f]	1.76(2) 165(1)

[a] Not observable

[b] X-ray diffraction

[c] $\angle$H-C-H = 112.3(2), C-H = 1.079(1.081)(3)

[d] Neutron diffraction

[e] Based on ν_{CD_α} = 2150 cm^{-1} and ν_H/ν_D = 1.35[14]

[f] Average

[g] $\angle$M-C-H = 78.1(3)°, $\angle$H-C-R = 111.5(3)°, C-H = 1.135(5)Å

[h] These values are for the Ta$\equiv$C bond length and Ta$\equiv$C$_\alpha$-C$_\beta$ bond angle

[i] $\angle$M-C-H = 84.8(2)°, $\angle$H-C-R = 113.7(2)°, CH = 1.131(3)Å

[j] W$\equiv$C = 1.785(8)Å, W$\equiv$C$_\alpha$-C$_\beta$ = 175.34(69)°, W-C$_\alpha$-C$_\beta$ = 124.53(69)°

[k] pip = N,N'-dimethylpiperazine

(e.g. $[ReCp(PPh_3)(NO)(CHPh)]^{+}$[7] or $[FeCp(dppe)(CH_2)]^{+}$,[9] or Fischer-type carbene complex) often is found further downfield, consistent, at least, with the α-carbon atom being most positively charged.

One feature of the ^{13}C NMR spectra which will turn out to be important is the alkylidene CH_α coupling constant. In $Ta(CHCMe_3)Np_3$, the first neopentylidene complex, J_{CH_α} = 90 Hz. This value is low for an "olefinic" CH_α bond. We now know that J_{CH_α} ranges from ca. 60 Hz to ca. 130 Hz and tells us a good deal about the structure of the alkylidene ligand and the nature of the CH bond. When J_{CH_α} is low, the CH_α bond is long and weak since H_α is pulled into a position where it is nearly bridging between C_α and M (see 4.). When J_{CH_α} is high, the alkylidene ligand is comparatively undistorted and the CH_α bond more "normal," i.e. olefin-like.

The chemical shift, δ(ppm), of H_α in the ^1H NMR spectra also correlates fairly well with the structure of the alkylidene ligand (Table 2); the chemical shift for H_α in a "normal" or undistorted alkylidene ligand is 11-12 ppm. For a severely distorted one the signal for H_α may be found as high as -8 ppm. When H_α actually transfers to the metal the hydride resonance is usually found at relatively low field again (2-8 ppm).

The third spectroscopic parameter which correlates with the structure of the alkylidene ligand is the CH_α stretching frequency. A low value for v_{CH_α} was first observed (at 2510 cm^{-1}) in the IR spectrum of $TaCp(CHCMe_3)Cl_2$.[14] It is a weak to medium strength peak which shifts to 1855 cm^{-1} in the spectrum of $TaCp(CDCMe_3)Cl_2$. It can be quite a strong peak (sharper and stronger at lower frequencies) but also curiously weak or absent, especially in benzylidene complexes.[16] Such low v_{CH_α} values are believed due to long, weak CH_α bonds (see next section), although it is not clear at present why the intensities are often so low. A few examples are listed in Table 2.

Examples of the two extreme types of complexes (as measured by J_{CH}, δH, and v_{CH}; Table 2) are $Ta(CHCMe_3)(PMe_3)_4Cl$ and $W(O)(CHCMe_3)(PMe_3)_2Cl_2$. In $Ta(CHCMe_3)(PMe_3)_4Cl$, J_{CH} = 69 Hz, δH = -7.4 ppm and v_{CH} = 2200 cm^{-1}. In $W(O)(CHCMe_3)(PMe_3)_2Cl_2$, J_{CH} = 121 Hz, δH = 11.89, and v_{CH} was not observable in Nujol mull.

4.   STRUCTURAL STUDIES

Table 2 lists the compounds whose structures have been determined, and the most relevant structural data for each.   In many cases, we can compare directly a metal-carbon single bond length with a metal-carbon "double" bond length in the same environment.   A benzylidyne complex is also listed as an example of a complex with a true metal-carbon triple bond.

The structure of $TaCp_2(CH_2)(CH_3)$ has been determined by both x-ray[49] and neutron diffraction.[50]   It and the two other complexes of this family are typical pseudo-tetrahedral "bent sandwich" MCp_2L_2 complexes[59]

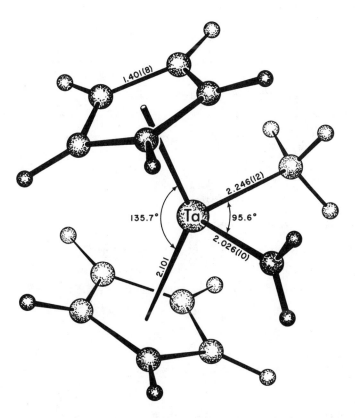

Fig. 1   The Structure of $TaCp_2(CH_2)(CH_3)$

(Fig. 1).   The methylene ligand and the metal lie in a plane that is perpendicular to the plane containing the metal and the methyl and methylene carbon atoms.   This is what one would expect since olefin ligands in such complexes lie in the plane which passes through the metal and between the two Cp ligands, and each ligand uses the same π-type metal orbital to form a π-type bond.   Two pieces of data suggest that the methylene ligand is doubly bound to Ta fairly strongly.   First, the difference between the M-C(methyl) and M=C(methylene) bond lengths is 0.22Å. Second, by labeling one of the Cp ligands with a methyl group (thereby making the methylene ligand protons inequivalent) it was shown that the barrier to rotation of this methylene ligand about the M=C bond was greater than ca. 20 kcal mol^{-1}.[19]

Although the structure of TaCp$_2$(CHPh)(CH$_2$Ph) is similar,[20] there are two important differences.   First, the benzylidene ligand plane is not strictly perpendicular to the pseudo-plane that passes through the metal and between the two Cp rings, but it is tipped ~4° so that the phenyl substituent points slightly toward the benzyl ligand.   Second, the M=C$_\alpha$-C$_\beta$ angle is about 10° larger than the Ta=C-H angle in TaCp$_2$(CH$_2$)(CH$_3$).   Both results could be explained by steric crowding.   It was suggested that a barrier to rotation of the ligand into the TaC$_2$ pseudo-plane of only 19.2 kcal mol^{-1} was an important consequence of the benzylidene ligand already being "tipped" in that direction.

The structure of TaCp$_2$(CHCMe$_3$)Cl is consistent with yet further steric crowding within the molecule.[52]   The Ta=C$_\alpha$-C$_\beta$ angle is now ~25° larger than the Ta=C-H angle in TaCp$_2$(CH$_2$)(CH$_3$) and the neopentylidene ligand plane is now tipped ~10° from being perpendicular to the C$_\alpha$-Ta-Cl plane with the t-butyl group pointing away from the Cl ligand.   The barrier to rotating the neopentylidene ligand into the C$_\alpha$-Ta-Cl plane consequently is only 16.8 kcal mol^{-1}.

An interesting fact is that the CH coupling constant correlates inversely with the Ta=C$_\alpha$-R angle in the three biscyclopentadienyl compounds: 132 Hz and 126° for R = H; 127 Hz and 135° for R = Ph; 121 Hz and 150° for R = CMe$_3$.   One might think this fortuitous, were it not for

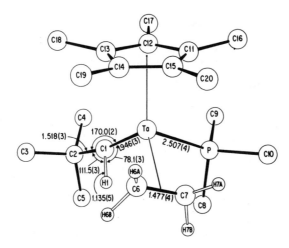

<u>Fig. 2</u>   The Structure of TaCp*(CHCMe$_3$)(C$_2$H$_4$)(PMe$_3$)

much additional data for electron deficient (less than 18 valence electrons) compounds which support this correlation.

Three structures of monocyclopentadienyl alkylidene complexes have been done.   Although the results for TaCp(CHCMe$_3$)Cl$_2$ were inaccurate due to librational motion of the Cp ring (at 25°C), the basic features of this pseudo-tetrahedral molecule are clear.[53]   The t-butyl group of the neopentylidene ligand points toward the Cp ring, but the M=C$_\alpha$-C$_\beta$ plane is

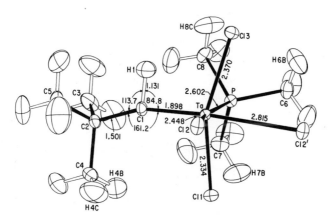

<u>Fig. 3</u>   Half the Structure of [Ta(CHCMe$_3$)(PMe$_3$)Cl$_3$]$_2$

not strictly perpendicular to the plane of the cyclopentadienyl ring (estimated dihedral angle = 80°). The $M=C_\alpha-C_\beta$ angle $(165(3)^\circ)$ is extraordinarily large, but almost certainly not for steric reasons. If the $H-C_\alpha-C_\beta$ angle is on the order of 112° then the $Ta-C_\alpha-H_\alpha$ angle must be less than 90°. (Note that J_{CH_α} = 84 Hz and ν_{CH_α} = 2510 cm^{-1}.) The short $Ta=C$ bond (1.75 Å) is far too short to account for by a change in the coordination number about the metal alone.

In TaCp*(CHPh)Bz$_2$ the phenyl substituent on the benzylidene ligand is pointed toward the Cp* ring, but the dihedral angle between the plane of the benzylidene ligand and the Cp* ring is only 78°. The $Ta=C-Ph$ angle is, again, large (166°), a fact which suggests that steric bulk of an alkylidene ligand's substituent is not the sole cause of a large $M=C_\alpha-C_\beta$ angle. The $Ta=C$ bond is 0.33 Å shorter than the average $Ta-C$ bond length, although it is not as short as in TaCp(CHCMe$_3$)Cl$_2$. In this molecule, too, J_{CH_α} is low (84 Hz). However, ν_{CH_α} was too weak to be located unambiguously.

TaCp*(CHCMe$_3$)(C$_2$H$_4$)(PMe$_3$) is unusual in the sense that the metal is formally in a lower oxidation state (by two electrons) than in the complexes we have discussed so far. Since J_{CH_α} = 74 Hz, ν_{CH_α} = 2480 cm^{-1}, and δH_α = -2.86, we suspected the neopentylidene ligand to be even more distorted than in the above two complexes. A neutron diffraction study[29] showed the $Ta=C_\alpha-C_\beta$ angle to be 170.0°, the $C_\beta-C_\alpha-H$ angle to be 111.5°, and the $Ta=C_\alpha-H_\alpha$ angle to be only 78.1° (Fig. 2). The C-H bond is long, a fact which (it was proposed) accounts for the low values for J_{CH_α} and ν_{CH_α}. H_α is only 2.042Å from Ta, a figure which is well within the sum of the van der Waals radii for Ta and H. The $Ta=C$ bond is markedly longer than it was in the first two complexes, a fact which seems inconsistent with the neopentylidene ligand being the most distorted. The t-butyl group is pointing roughly toward the Cp* ligand, but the neopentylidene ligand is rotated ca. 38° from the "perpendicular" position, i.e., one where the dihedral angle between the neopentylidene ligand plane and the Cp* ring would be 90°.

A hypothesis which is based on the above (and later) studies is that in electron deficient complexes the metal attracts electron density from the

C-H$_\alpha$ bond, much as it would attract (no doubt more easily) the electron pair(s) in isoelectronic imido[60] or oxo[61] ligands, thereby making the Ta=C bond shorter and the C-H$_\alpha$ bond longer than each would otherwise be. The extreme version is for H$_\alpha$ actually to transfer to the metal to give an alkylidyne/hydride complex, a reaction that has been observed when the metal is reduced (see 2.7).

The reaction between TaCp*(CHPh)(CH$_2$Ph)Cl and excess PMe$_3$ yields TaCp*(CPh)(PMe$_3$)$_2$Cl. We include a discussion of it here in order to compare the benzylidyne ligand with distorted alkylidene ligands. TaCp*(CPh)(PMe$_3$)$_2$Cl is a fairly regular tetragonal pyramid containing trans PMe$_3$ ligands. The Ta≡C bond (1.849(8)Å) is shorter than most Ta=C bonds. Since it is becoming apparent that Ta=C bond lengths vary considerably and do not correlate well with J$_{CH_\alpha}$ or ν_{CH_α} we might expect Ta≡C bond lengths to vary significantly also. Therefore 1.85Å must be regarded as an example at this time. Deviation of the Ta≡C$_\alpha$-C$_\beta$ angle (171.8(6)°) from 180° is not uncommon in other types of alkylidyne complexes.[62a]

[Ta(CHCMe$_3$)(PMe$_3$)Cl$_3$]$_2$ is a member of a large class of octahedral complexes,[15] all of which show low values for J$_{CH_\alpha}$ and ν_{CH_α}. It is a dimeric complex which contains two Ta(CHCMe$_3$)(PMe$_3$)Cl$_3$ units linked by two bridging chloride ligands. A view of half the molecule through one of the bridging chlorides is shown in Figure 3. The plane of the neopentylidene ligand contains the Cl-Ta-Cl axis. The neopentylidene ligand is, again, highly distorted. In fact, H$_\alpha$ appears to be more demanding sterically than the t-butyl group. Note that the Ta-Cl(2)' bond is long. This is probably the reason why these dimers break into monomers (which can be captured by L to give M(CHCMe$_3$)L$_2$Cl$_3$ species) so readily.

No x-ray structural study of a molecule of the type M(CHR)L$_2$X$_3$ has been done. However, the general features have been elucidated by NMR studies. It is assumed that the plane of the alkylidene ligand contains one of the two axes which is perpendicular to the M=C bond axis. The two types of complexes which are found are cis, mer and trans, mer. Unless L$_2$ is a chelating ligand, these two types usually interconvert rapidly on the NMR

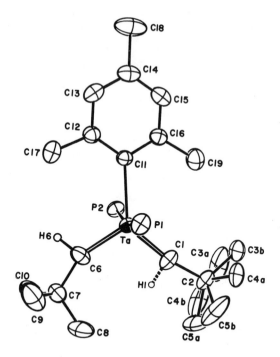

Fig. 4   The Structure of Ta(CHCMe$_3$)$_2$(C$_6$H$_2$Me$_3$)(PMe$_3$)$_2$.

(One methyl group is disordered.   H(1) was not located.)

time scale near room temperature by loss of L.   It was postulated that these
molecules, like [Ta(CHCMe$_3$)(PMe$_3$)Cl$_3$]$_2$, are distorted significantly from
an octahedral geometry due to the interaction of the C-H$_\alpha$ bond with the
metal, and that such a distortion makes L more labile.   In what should be a
relatively undistorted octahedral molecule, <u>trans</u>, <u>mer</u>-Ta(olefin)L$_2$X$_3$
(see 6.3), L is relatively firmly bound.

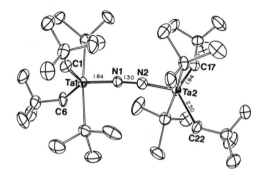

Fig. 5   The Structure of [Ta(CHCMe$_3$)(CH$_2$CMe$_3$)(PMe$_3$)$_2$]$_2$N$_2$

The structure of $Ta(CHCMe_3)_2(mesityl)(PMe_3)_2$[56] is shown in Figure 4. The two $CHCMe_3$ ligands lie in the trigonal plane and point in the same direction. Therefore they are inequivalent. One is also more distorted ($Ta=C_\alpha-C_\beta = 168.9°$, $Ta=C = 1.932Å$, $J_{CH_\alpha} = 91$ Hz, $\delta H_\alpha = 2.00$ ppm) than the other ($Ta=C_\alpha-C_\beta = 154.0°$, $Ta=C = 1.955$ A, $J_{CH_\alpha} = 104$, $\delta H_\alpha = 6.72$ ppm). They interconvert rapidly on the NMR time scale at $100°$ by a process that involves loss of PMe_3. Therefore the two most likely equilibrate in some four-coordinate intermediate in this case. (In at least one other case, PMe_3 is not lost during equilibration.) Note that the $Ta=C$ bond length in each neopentylidene ligand is $\sim0.35Å$ shorter than the $Ta-C(mesityl)$ bond length. The fact that the neopentylidene ligands lie in the trigonal plane was initially surprising. Later we found that olefin ligands in TBP analogs where the olefin has formally replaced the alkylidene prefer to orient perpendicular to the trigonal plane. Apparently, the π-type orbitals that are best suited to π-bond to alkylidene or olefin ligands in these complexes are those whose lobes project above and below the trigonal plane.

The structure of $[Ta(CHCMe_3)Np(PMe_3)_2]_2N_2$ (Fig. 5) is actually closely related to that of $Ta(CHCMe_3)_2(mesityl)(PMe_3)_2$, since the bridging dinitrogen ligand is "imido-like," $Ta(CHCMe_3)(NSiMe_3)Np(PMe_3)_2$ is known,[27] and an imido ligand is isoelectronic with an alkylidene ligand. Each neopentylidene ligand in $[Ta(CHCMe_3)Np(PMe_3)_2]_2N_2$ is distorted to a degree about halfway between the two in $Ta(CHCMe_3)_2(mesityl)(PMe_3)_2$. This suggests that the imido-like $\mu-N_2$ is not as good a π-electron donor as a neopentylidene ligand, and probably not as good a π-electron donor as an imido ligand.

The structure of $W(O)(CHCMe_3)Cl_2(PMe_3)_2$, a distorted octahedron, is shown in Figure 6.[57] The neopentylidene lgand is the least distorted one we have seen so far ($W=C_\alpha-C_\beta = 141.(6)°$), consistent with the relatively high J_{CH_α} (121 Hz) and δH_α (11.89 ppm), and long $W=C$ bond ($2.006(15)Å$). It is interesting to note that the oxo and neopentylidene ligands are cis and that the neopentylidene ligand must lie in the $C_\alpha-W-O$ plane in order to form a π-bond; the other two d orbitals are used to form the "triple" bond[61] between W and O.

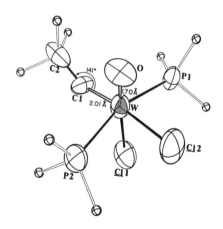

<u>Fig. 6</u>   The Structure of W(O)(CHCMe₃)(PMe₃)₂Cl₂

W(O)(CHCMe₃)(PEt₃)Cl₂ is a distorted trigonal bipyramid (Fig. 7) in which the neopentylidene and oxo ligands lie in the pseudo trigonal plane. The neopentylidene ligand is, again, relatively undistorted but the W=C bond is more than 0.12Å shorter than in the six-coordinate species. The W=O bond length (1.66Å) is also shorter than that in the six-coordinate species (1.70Å). The shorter bonds could be ascribed solely to the presence of only one phosphine ligand and a different coordination geometry.

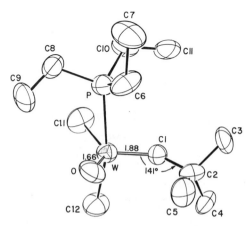

<u>Fig. 7</u>   The Structure of W(O)(CHCMe₃)(PEt₃)Cl₂

W(CCMe$_3$)(CHCMe$_3$)(CH$_2$CMe$_3$)(dmpe) offers what is so far a unique opportunity to compare metal-carbon bond distances and M-C$_\alpha$-C$_\beta$ angles in a neopentyl, a neopentylidene, and a neopentylidyne ligand, all bound to the same metal. The complex is a distorted tetragonal pyramid (Fig. 8). The metal-carbon bond lengths are 2.258(8) (yl), 1.942(9) (ene), and 1.785(8)Å (yne), and the M-C$_\alpha$-C$_\beta$ angles 124.53(69), 150.4(8), and 175.34(69)$^\circ$, respectively.

The last complex listed in Table 2 was actually the first to have its structure determined.[32] It was called a carbyne complex because of the short Ta=C$_\alpha$ bond (1.76(2)Å) and large Ta=C$_\alpha$-C$_\beta$ angle (165(1)$^\circ$). Now it is perhaps better viewed as an α-lithio alkylidene complex in which lithium behaves as a pseudo proton. Additional evidence is the fact that a TMEDA analog by ^{13}C NMR studies is a mixture of 80% TaNp$_3$[C(Li.TMEDA)CMe$_3$] and 20% TaNp$_2$(CHCMe$_3$)[CH(Li.TMEDA)CMe$_3$] (δC$_\alpha$ = 223 ppm. J$_{CH_\alpha}$ = 98 Hz).[46] We believe this is an equilibrium mixture since it could not be enriched significantly in one tautomeric form by crystallization techniques

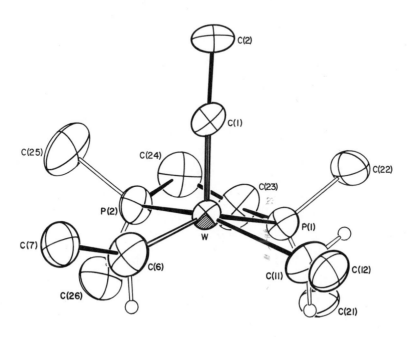

Fig. 8   The Structure of
W(CCMe$_3$)(CHCMe$_3$)(CH$_2$CMe$_3$)(Me$_2$PCH$_2$CH$_2$PMe$_2$)

or changes in temperature. Analogous complexes with dimethylpiperazine, dioxane, or dmpe bound to lithium consist of only the α-lithioneopentylidene tautomer.

5.   DECOMPOSITION STUDIES

5.1   Formation of Alkylidyne Complexes

Alkylidyne complexes are formed in reactions that are related to those which give alkylidene complexes. Two documented types are the reactions shown in eqs 30[54a] and 31.[30] There also is evidence that

$$RCH_2M=CHR \longrightarrow RCH_3 + M\equiv CR \tag{30}$$

$$M=CHR \xrightarrow{\text{2e reduction}} H-M\equiv CR \tag{31}$$

cationic neopentylidene complexes can be deprotonated[54] (eq 32).

$$[M=CHR]^+ \xrightarrow{\text{base}} M\equiv CR \tag{32}$$

$$Cl-M=CHR \xrightarrow{\text{base}} M\equiv CR \tag{33}$$

Finally, dehydrohalogenation (eq 33), although it has not yet been demonstrated, is a plausible variation of deprotonation; it is related to the demonstrable dehydrohalogenation of alkyls to give alkylidenes. There is increasing evidence, however, that the metal complex itself may act as a base or a reducing agent in complex intermolecular "decomposition" reactions related to those shown in eqs 30 to 33. Of course, unless the initial decomposition product is stable or can be trapped, it cannot be identified unambiguously. For example, one possible mechanism for decomposition of $Ta(CHSiMe_3)(CH_2SiMe_3)_3$[37] is shown in Figure 9. Formation of $W(CCMe_3)(CH_2CMe_3)_3$[33] in 25% yield from "$W(CH_2CMe_3)_3$"[46] must also involve several complex intermolecular α-hydrogen abstraction reactions. Main group alkyls such as aluminum alkyls should also be able to take part in such reactions as shown by the recent isolation of $W(CH)Cl(PMe_3)_4$ in 60% yield from the reaction shown in eq 34.[62b] Almost certainly decomposition of early transition metal alkyl complexes[35] often involves complex intermolecular reactions.

$$WCl_2(PMe_3)_4 \xrightarrow[\text{2. 2tmeda}]{\text{1. 2AlMe}_3} W(CH)Cl(PMe_3)_4 \qquad (34)$$

5.2 Bimolecular Olefin Formation

Bimolecular decomposition of alkylidene complexes to give free olefins or olefin complexes appears to be a general reaction type which is most facile for methylene complexes. The best documented example is the decomposition of $TaCp_2(CH_2)Me$[19]. The rate of decomposition is second order in tantalum and zero order in added tertiary phosphine. A plausible transition state is the double bridging methylene complex shown in eq 35. (One multiple bridging methylene complex is known.[63])

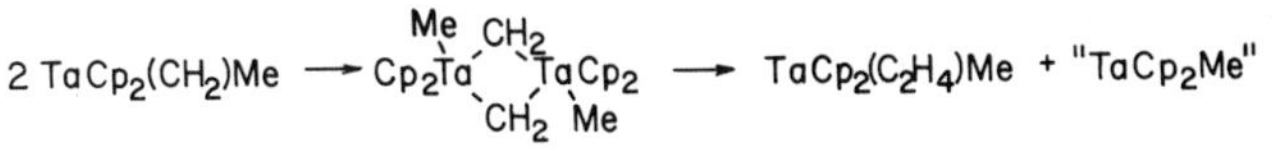

$$2\,TaCp_2(CH_2)Me \longrightarrow Cp_2Ta \overset{\overset{\displaystyle Me}{\underset{\displaystyle CH_2}{\underset{\displaystyle \,}{CH_2}}}{\underset{\displaystyle Me}{\,}} TaCp_2 \longrightarrow TaCp_2(C_2H_4)Me + "TaCp_2Me" \qquad (35)$$

"$TaCp_2Me$" can be trapped by L (C_2D_4, CO, PMe_3) to give $TaCp_2(L)Me$. The fact that $TaCp_2(CH_2)Me$ also reacts with $Me_3P{=}CH_2$ as shown in eq 36 is good evidence that the $Ta{=}CH_2$ group is "ylide-like" (see 6.). There is also

$$TaCp_2(CH_2)Me + Me_3P{=}CH_2 \longrightarrow TaCp_2(C_2H_4)Me + PMe_3 \qquad (36)$$

evidence that $Nb(CH_2)(OCMe_3)_2Cl(PMe_3)$ rapidly decomposes bimolecularly to give ethylene.[24,13]

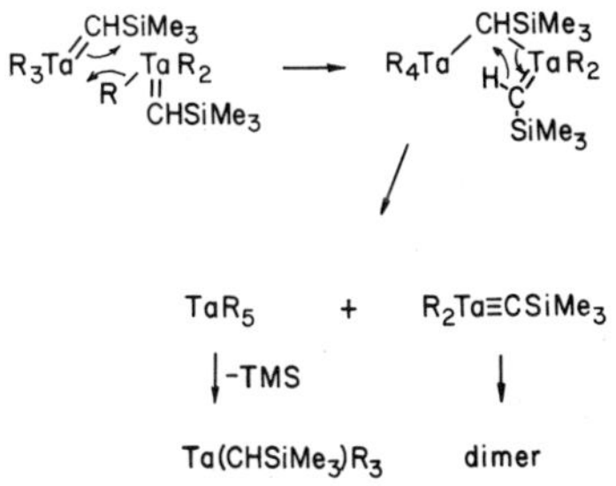

Fig. 9   A Plausible Mechanism for Decomposition of $Ta(CHSiMe_3)(CH_2SiMe_3)_3$

Benzylidene complexes are less susceptible to bimolecular olefin formation.   For example.   TaCp(CHPh)(CH$_2$Ph) decomposes only at high temperatures to a complex mixture of organic products, little of which consists of stilbenes.[43]   Ta(CHPh)(PMe$_3$)$_2$Cl$_3$, however, decomposes readily at 50° in the absence of PMe$_3$ to give stilbenes,[15] as do Ta(CHPh)(OCMe$_3$)$_2$Cl(PMe$_3$) and Nb(CHPh)(OCMe$_3$)$_2$Cl(PMe$_3$).[13]

Neopentylidene complexes are considerably more stable than benzylidene complexes.   Ta(CHCMe$_3$)(PMe$_3$)$_2$Cl$_3$ decomposes slowly only at elevated temperatures to give mostly neopentane.[15]   TaCp(CHPh)Cl$_2$ decomposes to give stilbenes[16] while TaCp(CHCMe$_3$)Cl$_2$ can be sublimed.[14] Decreasing susceptibility to olefin formation in the order R = H > Ph > CMe$_3$ probably can be ascribed   largely to steric hindrance in a double bridging alkylidene intermediate.   Other modes of decomposition such as intermolecular deprotonation by the alkylidene ligand could be slowed for similar reasons.   Steric hindrance must be the primary reason why many terminal neopentylidene complexes can be isolated while isolable terminal methylene complexes are rare.

5.3   Intramolecular Olefin Formation

Complexes in which β-hydrogens are present in the alkylidene ligand have not been prepared by α-abstraction techniques since β-hydrogen atoms appear to be abstracted more easily.   [However, note that recent work on the reaction of ReCp(NO)(PPh$_3$)(CH$_2$R) with CPh$_3$$^+$ suggests that "α-hydride abstraction" is sometimes preferred over "β-hydrogen abstraction".[7]]   They have been made indirectly by metathesis-type reactions (see below).   In one system they were found to rearrange readily to olefins[24,13] (eq 37). Complexes of the type W(O)(CHCH$_2$R)L$_2$Cl$_2$ (L = e.g. PMe$_3$ or PEt$_3$),

$$M(CHCH_2R)(OCMe_3)_2Cl(PMe_3) \longrightarrow CH_2{=}CHR \text{ (R=H or Me, M=Nb or Ta)} \quad (37)$$

however, are fairly stable toward alkylidene ligand rearrangement, possibly because they are "18 electron" complexes (if the oxo ligand is called a four electron donor[61]).   Five-coordinate W(O)(CHCH$_2$R)LCl$_2$ is less stable.[25] It is not yet known how much more stable such species are simply because a good π-electron donor ligand is present.

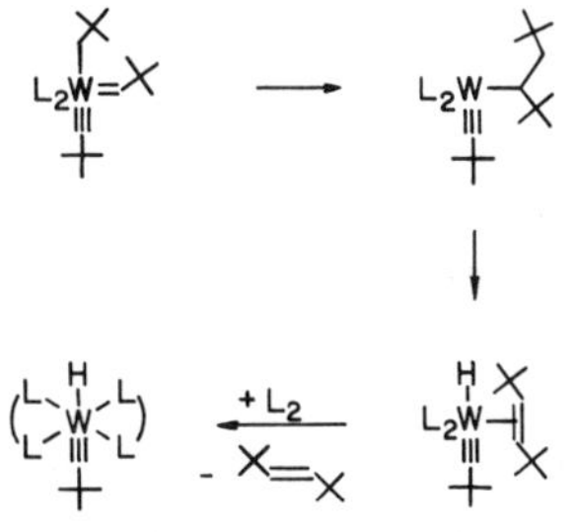

L_2 = dmpe

Fig. 10   A Plausible Mechanism for Forming $W(CCMe_3)(H)(dmpe)_2$

$TaCp_2(CHMe)Me$ is stable toward rearrangement to the known $TaCp_2(C_2H_4)Me$, possibly also because $TaCp_2(CHMe)Me$ is an 18-electron complex. Interestingly, when it does decompose it gives exo and endo $TaCp_2(propylene)(H)$,[43] possibly via intermediate $TaCp_2(CHMe_2)$. This is an example of what one might think should be a more common reaction of alkyl/alkylidene complexes, migration of an alkyl to an alkylidene ligand. Steric hindrance by t-butyl or phenyl substituents may block this reaction in most complexes we have discussed here, or the rates of other decomposition reactions may simply be faster in most circumstances. Recently, however, we have found that the complex with the formula $W(dmpe)_2(C_5H_{10})$[33] is a neopentylidyne hydride complex[46]; one explanation of its formation is by migration of a neopentyl ligand to a neopentylidene ligand followed by β-elimination (Fig. 10).

6.   REACTIONS

6.1   Acids

Several alkylidene complexes react with HCl to give alkyl/chloride complexes (e.g. eqs 38 and 39). This should be expected since they are

$$M(CHCMe_3)Np_3 \xrightarrow[-78^\circ]{HCl} TaNp_4Cl \quad (M = Nb \text{ or } Ta)[12] \qquad (38)$$

$$TaCp(CHCMe_3)Cl_2 \xrightarrow[-78^\circ]{HCl} TaCp(CH_2CMe_3)Cl_3[64] \qquad (39)$$

formed by reactions which are, or are related to, deprotonations.   It is almost certainly a general reaction.   Therefore, an alkylidene complex should be able to function as a strong base in intermolecular "deprotonation" or "α-abstraction" reactions, though there is yet no unambiguous example.

"Stablization" of methylene complexes by Lewis acids, especially aluminum reagents, has already been mentioned.   It may be an important feature of the chemistry of alkylidene complexes in reactions such as olefin metathesis (see 6.3).   There is no apparent reason why a transition metal complex could not similarly stabilize an alkylidene ligand (eq 40).   However, if M' can also form an alkylidene complex, then by exchanging ligands

$$L_yM=CHR + M'L_x \rightleftharpoons L_{y+1}M-CHR-M'L_{x-1} \rightleftharpoons L_{y+2}M + RHC=M'L_{x-2} \quad (40)$$

(e.g. alkyls, halides, etc.) from M' to M, a bridging alkylidene complex and, finally, a new alkylidene complex can form (cf eq 13).

6.2   Carbonyls and Imines

Several alkylidene complexes are known to react readily with the carbonyl function.   The first reported reaction of this type involved $M(CHCMe_3)Np_3$ (M = Nb or Ta)[65] (eq 41).   A plausible intermediate is an

$$M(CHCMe_3)Np_3 + RC(O)R' \rightarrow RR'C=CHCMe_3 + [M(O)Np_3]_x \quad (41)$$

oxymetallacyclobutane complex which loses olefin to give a metal oxide. Interestingly, this Wittig-type reaction is successful not only with ketones and aldehydes but esters, and even amides.   The strength of the Ta=O bond (or Ta-O-Ta bond) must be an important driving force.

The reaction between $TiCp_2(CH_2AlMe_2Cl)$ and ketones, aldehydes, and esters also gives Wittig-type products.[31]   Since a base such as pyridine is beneficial, the reaction probably proceeds via intermediate $TiCp_2(CH_2)$. Recently, this reaction has been developed as a synthetic technique.[66] Since $TiCp_2(CH_2AlMe_2Cl)$ is easily prepared from $TiCp_2Cl_2$ and $AlMe_3$, is thermally stable, and is inexpensive, it should be of considerable importance as a reagent for converting esters into vinyl ethers.

In general, this type of reaction has been used as a test and as a means of quantifying alkylidene complexes in solution.   Although we know that alkylidene complexes such as $M(CHR)L_2X_3$ (M = Nb or Ta, R = CMe_3 or Ph, L = P, O, or N donor, X = Cl or Br)[15] or $M(CHCMe_3)(OCMe_3)_2Cl(PMe_3)$[13] react readily with acetone, little is known about the reactions of tungsten oxo or imido alkylidene complexes with the carbonyl function. We do know that 18e complexes such as $TaCp_2(CH_2)Me$ do not react readily with acetone. Either the molecule is too crowded for the methylene ligand to attack the carbonyl carbon atom, or acetone must first coordinate to the metal so as to become more polarized, and therefore more susceptible to attack at the carbonyl carbon atom by the relatively nucleophilic methylene ligand.

The reaction between imines and neopentylidene complexes proceeds analogously (eq 42).[27] In this case, the primary goal was the preparation

$$M(CHCMe_3)(THF)_2Cl_3 + RN=CHPh \rightarrow M(NR)(THF)_2Cl_3 + Me_3CHC=CHPh \quad (42)$$

$$(M = Nb \text{ or } Ta, R = Ph, Me, CMe_3)$$

of the imido complex.   The reaction is slower when R = CMe_3 or when M = Nb.   As in the case of reactions with the carbonyl function, the strength of the M=N bond must be an important driving force, in spite of the fact that the imido complexes also react with ketones to give oxo complexes and the expected imine.[27]   An analogous reaction employing PhCH=N-N=CHPh gave "imido-like" $\mu\text{-}N_2$ complexes.[30]

6.3   Olefins

It has been known for some time that ethylene reacts rapidly with $Ta(CHCMe_3)Np_3$, but the reaction is complex.[23]   The first neopentylidene complex that reacted cleanly with olefins was $TaCp(CHCMe_3)Cl_2$.[67] With ethylene, the products were almost exclusively 4,4-dimethyl-1-pentene and a tantalacyclopentane complex (eq 43). Propylene reacted with

$$TaCp(CHCMe_3)Cl_2 + \text{excess } C_2H_4 \rightarrow Me_3CCH_2CH=CH_2 +$$

$$CpCl_2\overline{TaCH_2CH_2CH_2CH_2} \quad (43)$$

$$TaCp(CHCMe_3)Cl_2 + \text{excess } CH_3CH=CH_2 \rightarrow Me_3CCH_2(Me)C=CH_2 +$$

$$CpCl_2\overline{TaCH_2CHMeCHMeCH_2} \quad (44)$$

$TaCp(CHCMe_3)Cl_2$ + excess $PhCH=CH_2 \rightarrow t\text{-}PhCH=CHCH_2CMe_3$ + ?   (45)

$TaCp(CHCMe_3)Cl_2$ as shown in eq 44 and styrene as in eq 45. These results were consistent with attack by a nucleophilic neopentylidene α-carbon atom on an olefin so as to produce the most stable carbanion before the ring closed to give a neutral intermediate tantalacyclobutane complex (e.g.

$$Ta=CHCMe_3 + PhCH=CH_2 \longrightarrow Ta \overset{-}{\underset{CMe_3}{\overset{+}{\diagup}}}^{Ph} \longrightarrow Ta \overset{Ph}{\underset{CMe_3}{\diamond}} \longrightarrow Ta \overset{Ph}{\underset{CH_2CMe_3}{\parallel}} \qquad (46)$$

eq 46).   Rearrangement by a β-elimination process could then give, in theory, two possible olefin products. $NbCp(CHCMe_3)Cl_2$ reacts with these olefins to give analogous organic products but no stable metallacyclopentane complexes.   No metathesis products (e.g. $Me_3CCH=CH_2$ with ethylene) or cyclopropanes were found in either the Nb or Ta reaction.

The interpretation in eq 46 changed after studying the reactions between complexes of the type $M(CHR)L_2X_3$ (M = Nb or Ta, R = CMe_3 or Ph, L = a phosphine, X = Cl or Br) and ethylene, propylene, or styrene, since all four possible rearrangement products of the two possible intermediate metallacyclobutane complexes were observed.[13]   Here olefin complexes were isolated in high yield when M = Ta.   Again, no olefin metathesis products nor cyclopropanes were found.   Significantly, complexes that contain a chelating L_2 ligand do not react with olefins readily, presumably because the olefin must first coordinate to the metal, and no octahedral coordination site is available.   (Monodentate ligands in $M(CHR)L_2X_3$ complexes are known to be labile.[15])

By changing two halide ligands in complexes of the type $M(CHR)L_2X_3$ to alkoxides, rearrangement of the intermediate metallacyclobutane ring could be slowed down relative to metathesis to give a new olefin and a new alkylidene complex (eq 47; L = PMe_3).[13]   When $RCH=CH_2$ is

$M(CHCMe_3)(OBu^t)_2(Cl)L + RCH=CH_2 \rightarrow Me_3CCH=CH_2 + Me_3CCH=CHR +$

$$M(CHR')(OBu^t)_2(Cl)L \qquad (47)$$

$$R' = H \text{ or } R$$

ethylene, metallacyclobutane rearrangement is still the major reaction pathway. On the other hand, the metallacyclobutane complex made from a substituted olefin only metathesizes. When R = Ph the new olefin is almost exclusively $Me_3CCH=CH_2$, and the benzylidene complex can be trapped by added L to give green $M(CHPh)(OBu^t)_2(Cl)L_2$. Orange $[M(CHPh)(OBu^t)_2(Cl)L]_x$ can be observed spectroscopically, but it decomposes readily to stilbenes. When R = Et, both $Me_3CCH=CH_2$ (~75%) and $Me_3CCH=CHR$ are found. However, neither the methylene nor the propylidene complex could be observed spectroscopically; the methylene complex decomposes bimolecularly, and the propylidene complex is captured by 1-butene to give a diethylmetallacyclobutane complex, which, unlike the initial t-butylethylmetallacyclobutane complex, rearranges to an olefin. By adding L to the Nb sytem the decomposition of the methylene complex could be slowed so that ~1 turnover of 1-butene to ethylene and 3-hexene could be observed before activity ceased. The data suggest that in each catalytic cycle the intermediate niobium methylene complex decomposes 20% of the time, and that the ratio of the rate of forming and metathesizing the intermediate α-ethylniobacyclobutane complex to that of forming and rearranging the intermediate β-ethylniobacyclobutane complex is 65:35. (It was assumed that the propylidene complex did not decompose bimolecularly before it reacted with more 1-butene to give metathesis products.)

M(CHCMe$_3$)(OBu^t)$_2$(Cl)L also reacts readily with cis-2-pentene to give the two expected initial metathesis products in roughly equal amounts.[13] The resulting ethylidene and propylidene intermediates cannot be observed spectroscopically. They react with more cis-2-pentene to give metallacyclobutane complexes, which, apparently because they are trisubstituted, rearrange slowly relative to the rate at which they metathesize. Productive metathesis of cis-2-pentene to ~35 equivalents of 2-butenes and 3-hexenes can be realized before catalytic activity ceases. The primary chain terminating step in this case is rearrangement of the intermediate ethylidene and propylidene complexes (eq 48). Ethylene and propylene are found as soon as the reaction commences, and well before

$$M(CHCH_2R)(OBu^t)_2(Cl)L \longrightarrow M(CH_2CHR)(OBu^t)_2(Cl)L \qquad (48)$$

formation of the initial (t-butyl-containing) metathesis products is complete.

When $Ta(CHCMe_3)(OBu^t)Cl_2L_2$ reacts with 1-butene, a mixture of metathesis (32%) and rearrangement products (57%) is found.[13]   Evidently, one t-butoxide ligand is not enough to slow down metallacyclobutane rearrangement relative to metathesis.

$Ta(CHCMe_3)(OBu^t)_3$ reacts with 1-butene to give mostly rearrangment products.[13]   It was postulated that an empty coordination site allows ready β-elimination, in spite of the fact that three t-butoxide ligands are present. $Ta(CHCMe_3)(OBu^t)_3$ is also not a good catalyst for metathesizing cis-2-pentene.   The initial metallacyclobutane complexes metathesize, but the resulting propylidene and ethylidene complexes apparently decompose readily bimolecularly.   They must not have time to rearrange since only traces of ethylene and propylene are found.

Tungsten complexes of the type $W(O)(CHCMe_3)(PR_3)_2Cl_2$ do not react readily with olefins in the presence of added PR_3 (R = Me, Et).[24] They do react in the presence of a trace of $AlCl_3$ to give metathesis products (eqs 49 and 50).   Although only one initial metathesis product is found

$$W(O)(CHCMe_3)L_2Cl_2 + RCH{=}CH_2 \xrightarrow[C_6H_6]{AlCl_3 \ cat.} Me_3CCH{=}CH_2 +$$

$$W(O)(CHR)L_2Cl_2 \qquad (49)$$

$$R = H, \ Et, \ Ph$$

$$W(O)(CHCMe_3)L_2Cl_2 + c\text{-}MeCH{=}CHEt \rightarrow Me_3CCH{=}CHR + W(O)(CHR')L_2Cl_2$$

$$R = Me, \ Et; \ R' = Et, \ Me, \ resp. \qquad (50)$$

when $RCH{=}CH_2$ is used, both possible $W(O)(CHR)L_2Cl_2$ products actually are present.   The methylene complex must form by reaction of $W(O)(CHR)L_2Cl_2$ with $RCH{=}CH_2$.[26] In fact, these species are convincing metathesis catalysts for terminal olefins (~25 turnovers in two days at 25°C) and cis-2-pentene (~50 turnovers in two days at 25°).

The five-coordinate complex, $W(O)(CHCMe_3)(PEt_3)Cl_2$, reacts with cis-2-pentene in the absence of aluminum reagents to give 2-butenes and 3-hexenes at a rate at least as good as the rate of the system described

immediately above, but the activity is shorter-lived.[26]   The mono- and
dicationic   complexes, $[W(O)(CHCMe_3)(PEt_3)_2Cl]^+AlCl_4^-$ and
$[W(O)(CHCMe_3)(PEt_3)_2]^{2+}(AlCl_4^-)_2$, also react with cis-2-pentene to give
productive metathesis products.[25]   In each system, however, no new
alkylidene products have been identified.

The chain-terminating step(s) in the metathesis reaction using tungsten
oxo alkylidene complexes is(are) not yet known.   Only traces of the
alkylidene →olefin rearrangement products are observed.   One interesting
recent result is shown in eq 51.[25]   For this reason it would be

$$W(O)(CH_2)(PEt_3)_2Cl_2 + C_2H_4 \rightarrow W(O)(C_2H_4)(CH_2PEt_3)(PEt_3)Cl_2 \qquad (51)$$

desirable to avoid phosphine ligands in these systems.

We now feel that the relatively slow rearrangement of
metallacyclobutane rings and alkylidene ligands to olefins is the major
reason why metathesis reactions with tungsten oxo alkylidene complexes are
successful.   Apparently, however, bimolecular decomposition of
intermediate alkylidene complexes, or as yet unknown alternative
decomposition pathways, is(are) still reasonably fast.   Therefore, an
alkylidene ligand in a "classical" metathesis system perhaps must be
"protected" as, for example, a Lewis acid adduct (cf $TiCp_2(CH_2AlMe_2Cl)$
above).   High dilution or anchoring a catalyst to a solid support also should
discourage bimolecular decomposition reactions.

Some recent results in the $Ti=CH_2$ system firmly establish that a
metallacyclobutane complex is, indeed, the product of a reaction between an
alkylidene complex and an olefin (e.g. eq 52).[68]   We can speculate that this

$$TiCp_2(CH_2AlMe_2Cl) + Me_3CCH=CH_2 \underset{AlMe_2Cl}{\overset{py}{\rightleftarrows}} Cp_2Ti \langle \rangle CMe_3 \qquad (52)$$

metallacyclobutane complex does not rearrange to an olefin complex
because the MC_3 ring cannot bend sufficiently to put the β-hydrogen atom
close to the metal.   The $AlMe_2Cl$ adduct is regenerated on adding
$AlMe_2Cl$.

6.4   Carbon Monoxide, Nitriles, Acetylenes, and Hydrogen

Carbon monoxide reacts with several complexes of the type $M(CHR)X_3(PR_3)_2$ to give ketene complexes.[69,37]   An example is shown in eq 53.   The ketene is believed to bond to Ta through the C=O bond since tantalum should

$$Ta(CHPh)(PMe_3)_2Cl_3 + CO \xrightarrow[\text{l atm}]{25^\circ} Ta(OCCHPh)(PMe_3)_2Cl_3 \qquad (53)$$

have a higher affinity for O than C, and since two isomers are present. The basic structure is believed to be _trans_, _mer_, and the isomers to be those containing a transoid and a cisoid arrangement about the C=C bond.   Carbon monoxide must first bind to the metal since no reaction is observed in similar complexes which contain strongly bound chelating phosphine or amine ligands.   Another example of a reaction which yields a ketene complex is shown in eq 54.[69]

$$TaCp_2(CHCMe_3)Cl + CO \longrightarrow [TaCp_2(OCCHCMe_3)]^+Cl^- \qquad (54)$$

Nitriles react immediately with $M(CHCMe_3)Np_3$ (M = Nb or Ta),[12] $TaCp(CHCMe_3)Cl_2$,[14] or complexes of the type $M(CHR)X_3L_2$[69,37] to give alkenyl imido complexes (both E and Z isomers, eq 55).   Relatively few

$$M{=}CHR + R'CN \longrightarrow \overset{\displaystyle N{=}C^{R'}}{\underset{\displaystyle M{-}CHR}{|\diagup|}} \longrightarrow M{\equiv}N{\underset{R'}{\diagdown}}C{=}CHR \qquad (55)$$

examples are known since this reaction, like many others, has not yet been explored systematically.

$TiCp_2(CH_2AlMe_2Cl)$ reacts with an acetylene to give a metallacyclobutene complex (eq 56).[70]   The x-ray structure of the

$$TiCp_2(CH_2AlMe_2Cl) + RC{\equiv}CR \longrightarrow Cp_2Ti\overset{CH_2}{\underset{C-R}{\diagdown\diagup}} \qquad (56)$$

diphenylacetylene product shows the expected Ti-C, C-C, and C=C bond lengths.   In contrast, the reaction between $TaCp(CHCMe_3)Cl_2$ and diphenylacetylene produces a new alkylidene complex (one isomer; eq 57), presumably via a metallacyclobutene complex analogous to that shown   in

$$\text{TaCp(CHCMe}_3)\text{Cl}_2 + \text{PhC}\equiv\text{CPh} \longrightarrow \text{CpCl}_2\text{Ta}\overset{\text{CMe}_3}{\underset{\text{Ph}}{\langle\rangle}}\text{Ph} \longrightarrow \text{CpCl}_2\text{Ta}=\text{C}\overset{\text{Ph}}{\underset{\text{C=CHCMe}_3}{\big\langle}} \qquad (57)$$

eq 56.   More than one equivalent of several other acetylenes (dimethyl, methyl phenyl, etc.) is consumed rapidly, even at low temperatures, to give as yet uncharacterized products.[14,64] Rapid successive addition of the acetylene to the new alkylidene is a postulated mechanism of polymerizing acetylenes with carbene complex catalysts.[71]   The titanacyclobutene complex may not rearrange to a new alkylidene ligand due to the fact that the rather bulky substituents on the new alkylidene's α-carbon atom would be pointing toward the two cyclopentadienyl rings.

Several alkylidene complexes are known to react readily with H_2 to give alkane (neopentane or toluene) and metal hydrides, most of which are too unstable to characterize.   Unlike similar reactions involving Fischer-type carbene complexes where high temperatures and pressures are required,[72] 2 atm and $25°$ are sufficient, possibly because no ligand need be lost.   Examples are shown in eqs 58[73] and 59.[37]

$$\text{TaCp(CHCMe}_3)\text{Cl}_2 + H_2 \longrightarrow \text{CMe}_4 + 0.5[\text{TaCpCl}_2H]_2 \qquad (58)$$

$$\text{Ta(CHCMe}_3)(\text{PMe}_3)_2\text{Cl}_3 + H_2 \longrightarrow \text{CMe}_4 + "\text{TaH}_2(\text{PMe}_3)_2\text{Cl}_3" \text{ (unstable)} \quad (59)$$

7.   CALCULATIONS

Hoffmann has used the extended-Hückel method to explore unusual M-C-H angles, C-H$_\alpha$ bond activation, and α-hydrogen abstraction in alkylidene complexes.[74]   In hypothetical $[\text{TaH}_4(\text{CH}_2)]^{3-}$ he finds that a symmetrically bound methylene lying in the trigonal plane of a trigonal bipyramidal molecule is disfavored relative to a methylene which is distorted by pivoting in the trigonal plane about the α-carbon atom.   As the methylene ligand pivots, the metal interacts with H_α in a bonding manner.   The methylene distorts more when more electron density is withdrawn onto the four hydride ligands, and it distorts less when π-electron density is added to the metal via two NH_2 groups in place of two equatorial hydrides. In five-coordinate, 14-electron $[\text{TaH}_4(\text{CH}_2)]^{3-}$, H_α may not actually transfer to Ta to give $[\text{TaH}_5(\text{CH})]^{3-}$ due to a crossing of two levels, one of which is occupied, but it will transfer to a methyl group in hypothetical $[TaH_3Me(CH_2)]^{3-}$. When the methylene ligand is substituted, he believes the substituent's bulk helps determine the $Ta=C_\alpha-C_\beta$ angle.

Hoffmann also tries to answer questions concerning charge distribution in the M=CHR bond and nucleophilicity of the alkylidene ligand. He finds that in $[TaH_4(CH_2)]^{3-}$ the methylene $2p_z$ orbital is occupied by 0.82 electrons compared with ~0.5 electrons in the methylene $2p_z$ orbital in hypothetical $Cr(CO)_5(CH_2)$ or $FeCp(CO)_2(CH_2)$. (Nucleophilic behaviour could be ascribed to high $2p_z$ orbital electron density.) The tantalum's π-bonding orbital and the carbon's $2p_z$ orbital overlap especially well compared to similar overlap in the Cr and Fe compounds. Alternatively, one could attempt to explain differences in nucleophilicity by focusing on the frontier molecular orbitals. In the Fe and Cr complexes, the LUMO is made up largely of the carbon $2p_z$ orbital; therefore, it is to this carbon that PR_3 adds. In the Ta complex $d_\pi-p_\pi$ bonding is so strong that the antibonding orbital is pushed to high energy. It is still open to attack by a nucleophile but this should be considerably more difficult than in the Fe or Cr complexes.

Ab initio calculations by Goddard address the question of reactivity patterns of oxo/alkylidene complexes with olefins.[75] He calculates that the methylene ligand in an $M(O)(CH_2)Cl_2$ complex (M = Cr or Mo) reacts preferentially with ethylene to give a metallacyclobutane complex. When M = Cr, reductive elimination to give $Cr(O)Cl_2$ and cyclopropene is preferred over the back reaction to reform the methylene complex and ethylene. When M = Mo, the back reaction is slightly favored over reductive elimination. He concludes that oxo complexes could be important in olefin metathesis reactions.

8.   RECENT DEVELOPMENTS

The facile metathesis of acetylenes by tungsten(VI) alkylidyne complexes has recently been reported[76]. This chemistry is relevant to the chemistry of alkylidene complexes in two ways. First, $W(CCMe_3)Cl_3(Et_3PO)$ was initially prepared by reacting $W(O)(CHCMe_3)(PEt_3)_2Cl_2$ with C_2Cl_6. It was postulated that

$W(O)(CHCMe_3)(PEt_3)Cl_2$ and $Et_3PCl^+Cl^-$ were the first (unobserved) products, and that these subsequently reacted with each other to form the unstable intermediate shown in eq 60.   Loss of a proton from the neopentylidene ligand in this intermediate would lead to the final product.

$$(60)$$

Secondly, $[W(CCMe_3)Cl_4]^-$ was   prepared   by   reacting $W(CCMe_3)(CH_2CMe_3)_3$ with three equivalents of HCl in the presence of $Et_4N^+Cl^-$, and it is stable in the presence of HCl in dichloromethane[77]. These results all suggest that neopentylidene ligands can be quite acidic if only weak π-bonding or π-donor ligands are present; $W(CHCMe_3)Cl_4$, for example, probably would not be stable toward loss of a proton to give $[W(CCMe_3)Cl_4]^-$.

One of the important questions in alkylidene chemistry has been whether a methylene ligand would also distort toward an alkylidyne hydride complex through interaction of one of its α-hydrogen atoms (or the C-H$_\alpha$ electron pair) with the metal in a manner analgous to that which is now well-documented for certain neopentylidene and benzylidene complexes. NMR studies of trans-$[W(CH_2)L_4Cl]^+$ (L = PMe_3) and related species have now shown that this is the case[78a,b].   At low temperatures the methylene protons are inequivalent.   It was proposed that one of them was sitting on a $C_\alpha L_2$ face of the octahedron.   When L_4 = 2dmpe the α-hydrogen atom actually transferred to the metal to give isolable $[W(CH)(H)(dmpe)_2Cl]^+$, probably because two dmpe ligands can spread apart more readily than PMe_3 ligands (for steric reasons) in order to receive the α-hydrogen atom and form a pentagonal bipyramidal molecule (see below).   In this particular example the metal is d^2 in "$[W(CH_2)L_4Cl]^+$" and d^0 in the product.   There probably will be some methylene complexes in which the metal is formally in the d^0 state (counting the methylene as a dianionic ligand) and in which the methylene ligand will distort in a similar manner.

In another recent paper it has been possible to observe equilibria between tantalum neopentylidene complexes and neopentylidyne hydride complexes, and an x-ray structure of one of the products, $Ta(CCMe_3)(H)(dmpe)_2(ClAlMe_3)$, has shown it to be a pentagonal bipyramidal molecule[79]. The neopentylidyne ligand is trans to the weakly bound $ClAlMe_3$ ligand ($Ta\cdots Cl = 2.758(2)$Å) and the hydride was located in the TaL_4H plane. $Ta(CCMe_3)(H)(dmpe)_2(ClAlMe_3)$ was prepared by adding $AlMe_3$ to $Ta(CHCMe_3)(dmpe)_2Cl$, a molecule in which the neopentylidene ligand is extremely distorted towards a neopentylidyne hydride system ($J_{CH_\alpha} = 57$ Hz), but which at low temperatures shows no evidence of freezing out in one configuration with H_α on an $MC_\alpha L_2$ face. "$Ta(CHCMe_3)(dmpe)_2I$" is actually approximately a 50:50 mixture of rapidly interconverting $Ta(CHCMe_3)(dmpe)_2I$ and $Ta(CCMe_3)(H)(dmpe)_2I$ at $25°C$; at $-75°C$ the mixture largely consists of $Ta(CCMe_3)(H)(dmpe)_2I$. These studies further confirm that α-hydride elimination from an alkylidene ligand can be extremely facile in some circumstances, especially if as a result the metal is "oxidized" from "d^2" to "d^0".

$W(CHCMe_3)L_2(CO)Cl_2$ (L = PMe_3) was prepared by reacting $W(CCMe_3)(H)L_3Cl_2$ or $W(CHCMe_3)(H)L_2Cl_3$ with carbon monoxide[80]. It is roughly an octahedral molecule with trans PMe_3 ligands and cis chloride ligands. Its structure is interesting in three respects. First, the neopentylidene ligand is distorted so that the $WC_\alpha H_\alpha$ angle is $<90°$. Second, the α-hydrogen atom is capping a $C_\alpha PCl$ face. Third, a carbon monoxide ligand, which one might expect to combine with the alkylidene ligand to form a ketene (see earlier section), is present in the coordination sphere.

An example of competition between an α-hydride elimination process and a β-hydride elimination process is the equilibrium between $Ta(CHCMe_3)(C_2H_5)L_2Cl_2$ (L = PMe_3) and $Ta(C_2H_4)(CH_2CMe_3)L_2Cl_2$[81]. In $Ta(C_2H_4)(CH_2CMe_3)L_2Cl_2$ the neopentyl ligand appears to be close to being neopentylidene hydride. The two tautomers were shown to interconvert by ^{13}C NMR magnetization transfer experiments.

A full paper has recently appeared which describes the preparation and properties of the oxo neopentylidene complexes of tungsten(VI)[82]. In addition to complexes of the type $W(O)(CHCMe_3)L_2Cl_2$ and $W(O)(CHCMe_3)LCl_2$ (L = a phosphine), cationic complexes such as $[W(O)(CHCMe_3)(PEt_3)_2]^{2+}(AlCl_4^-)_3$ have been prepared. In none of these complexes is the $W=C_\alpha-C_\beta$ angle large, as judged by the relatively high values for J_{CH_α} (115-130 Hz).

A recent paper discusses the phenomenon of migratory insertion reactions in zirconoxy carbene complexes, $Cp_2^*Zr(H)(OCH=Nb(R)Cp_2)$ (R = H or alkyl)[83]. Alkyl migration to give intermediate $Cp_2^*Zr(H)(OCHRNbCp_2)$ was found to be much slower than hydride migration to give intermediate $Cp_2^*Zr(H)(OCH_2NbCp_2)$. This was attributed to the greater ease of forming a three-center, two-electron Nb(H)(C) bond than a Nb(R)(C) bond. Interaction of an α-hydrogen atom with a metal to give a similar three-center, two-electron bond has been proposed as the means of "activating" an α-hydrogen atom toward abstraction by an alkyl ligand in metal alkyl complexes.[84]

The reaction between $[WCp_2Me_2]^+$ and the dimer of the trityl radical gives $[WCp_2(C_2H_4)H]^+$.[85] In the presence of PMe_3Ph the first product is $[WCp_2(CH_2PMe_2Ph)Me]^+$, which, on heating, is converted into a mixture of $[WCp_2(PMe_3Ph)Me]^+$ and $[WCp_2(C_2H_4)H]^+$. These results can be explained by proposing that the reaction between $[WCp_2Me_2]^+$ and $Ph_3C^·$ yields $[WCp_2(CH_2)Me]^+$ and Ph_3CH, and that in the absence of phosphine $[WCp_2(CH_2)Me]^+$ rearranges to $[WCp_2Et]^+$ (cf the zirconium system above).

The photolysis of $WMe_6(PMe_3)$ in the presence of PMe_3 gives $W(CMe)(PMe_3)_4Me$ via the proposed sequence shown in eq 61.[86] It also

$$L_xWMe_6 \xrightarrow{-2\ CH_4} L_yMe_3W\equiv \longrightarrow L_zMe_2W=CHMe \xrightarrow{-CH_4} L_4MeW\equiv CMe \quad (61)$$

seems possible that the C-C bond is formed by migration of a methyl group to an intermediate methylene ligand followed by a double α-hydrogen abstraction as shown in eq 62:

$$L_xWMe_6 \xrightarrow{-CH_4} L_yMe_4W{=}CH_2 \longrightarrow L_zMe_3WCH_2Me \xrightarrow{-2\,CH_4} L_4MeW{\equiv}CMe \quad (62)$$

Recent reports explore further the hypothesis that W(VI) alkylidene complexes are the generic catalysts in the "classical" black box olefin metathesis systems. One[87] is a full-paper version of an earlier communication in which the energies of various types of alkylidene and metallacyclobutane complexes are compared. In another[88a] Osborn proposes that formation of an adduct at an oxo ligand by AlX_3 (e.g. X = Cl or Br) induces formation of a neopentylidene complex by α-hydrogen abstraction and also greatly enhances the reactivity of that neopentylidene and subsequent alkylidene ligands in the olefin metathesis reaction. Some of the most intriguing observations are found in a publication the following year[88b] where neopentylidene complexes of the type $W(CHCMe_3)$ $(OCH_2CMe_3)_2X_2$ were isolated from reactions between $W(O)(OCH_2CMe_3)_2$ $(CH_2CMe_3)_2$ and aluminum halides. Although these neopentylidene complexes themselves do not react with cis-2-pentene, they will react with cis-2-pentene in the presence of aluminum halides. Two types of intermediate ethylidene and propylidene complexes can be observed in low temperature NMR spectra. Both are thought to be AlX_3 adducts, i.e. $W(CHR)(OCH_2CMe_3)_2X_2(AlX_3)$ and $W(CHR)(OCH_2CMe_3)X_3(AlX_3)$, in which AlX_3 is bound to X. The metal is rendered relatively electrophilic when an OCH_2CMe_3 ligand is replaced by X, and especially when AlX_3 coordinates to X to form what can be viewed as an incipient cationic complex with AlX_4^- as the anion.

Several neopentylidene hydride complexes of tantalum have been reported.[28] It has now been found that $Ta(CHCMe_3)(H)(PMe_3)_3I_2$ will polymerize ethylene slowly.[89] A new species can be observed by NMR which is believed to be $Ta[CH(CH_2CH_2)_nCMe_3](H)(PMe_3)_3I_2$, the result of a reaction of n equivalents of ethylene with $Ta(CHCMe_3)(H)(PMe_3)_3I_2$. Since all Ta(V) alkylidene hydride complexes appear to be in rapid equilibrium with their Ta(III) alkyl tautomers, it is not possible to say for certain whether the Ta(V) alkylidene hydride complex or the Ta(III) alkyl complex reacts more rapidly with ethylene. However, the alkylidene mechanism was preferred on the basis of rapid reactions between ethylene and other d^0 alkylidene complexes.

Further studies of the chemistry of the $AlMe_2Cl$ "adduct" of $Ti(\eta^5-C_5H_5)_2(CH_2)$, $TiCp_2(CH_2)(AlMe_2Cl)$, have shown that it reacts with $CH_2=C(^iPr)Me$ in the presence of paradimethylaminopyridine to give an unstable titanacyclobutane complex which loses 2,3-dimethyl-1-butene above $0°$ to give the purple dimer $[TiCp_2(CH_2)]_2$, a 1,3-dititanacyclobutane complex.[90a] It is not clear why $[TiCp_2(CH_2)]_2$ does not disproportionate to $TiCp_2(C_2H_4)$ and the $TiCp_2$ fragment, except that each product of the disproportionation is likely to be a relatively high energy species (cf. the disproportionation of the postulated intermediate $[TaCp_2(Me)(CH_2)]_2$; see main part of review). A later study of the kinetics and stereochemistry of the titanacyclobutane-titanamethylene interconversion through the reaction of titanacyclobutane complexes with diphenylacetylene revealed that the diphenylcyclobutene complex, $Cp_2\overline{Ti CH_2C(Ph)}=C(Ph)$, is formed in a reaction that is first order in Ti and zeroth order in diphenylacetylene, consistent with the reaction proceeding by a rate-limiting titanacyclobutene ring opening to give $TiCp_2(CH_2)$ and free or complexed olefin.[90b] Certain titanacyclobutane complexes are effective catalysts for the degenerate metathesis of terminal olefins, a reaction that presumably proceeds via an analogous ring opening to give incipient $TiCp_2(CH_2)$. Finally, further synthetic applications of $TiCp_2(CH_2)$ consist of reactions between $Cp_2\overline{TiCH_2CMe_2CH_2}$, a source of $TiCp_2(CH_2)$ above $0°$, and acyl chlorides to give enolates of methyl ketones, i.e. $TiCp_2Cl[OC(R)=CH_2]$.[90c] These enolates are likely to be of utility in organic synthesis since they can be formed regiospecifically.

Several new reports of α-hydrogen abstraction to give isolable alkylidyne complexes have appeared. Chisholm[91] reports that addition of PMe_3 to $[MoBr(CH_2SiMe_3)_2]_2$ produces two equivalents of TMS and amber, crystalline $[MoBr(CHSiMe_3)(PMe_3)_2]_2$ in 60% yield. An xray structure shows the Mo≡C distance to be 1.949(5)Å and the Mo≡C_α-Si angle to be

$129.8(3)^\circ$, both indicative of a "normal" $=CHSiMe_3$ ligand in which there is little attraction of the C-H electron pair to the metal. Rothwell[92a] has shown that addition of three equivalents of $LiCH_2SiMe_3$ to $Ta(OR)_2Cl_3$ (OR = $O-2,6-C_6H_3{}^tBu_2$) yields $Ta(CHSiMe_3)(OR)_2(CH_2SiMe_3)$ as yellow crystals (δC_α = 237.1; δH_α = 8.60 ppm). An xray structural study has shown $Ta(CHSiMe_3)(OR)_2(CH_2SiMe_3)$ to be a pseudo-tetrahedral molecule with the O-Ta-O angle opened up to 127° and Ta=C = $1.888(29)$Å. Upon thermolysis $Ta(CHSiMe_3)(OR)_2(CH_2SiMe_3)$ loses TMS to give the bis-cyclometalated complex $(Me_3SiCH_2)\overline{Ta(O-2,6-C_6H_3{}^tBuCMe_2)_2}$.[92b] It is proposed that $(Me_3SiCH_2)_2(OR)\overline{Ta(O-2,6-C_6H_3{}^tBuCMe_2)}$ and $(Me_3SiCH)(OR)\overline{Ta(O-2,6-C_6H_3{}^tBuCMe_2)}$ are intermediates. In another publication Rothwell reports an efficient photochemical generation of a methylene ligand by photolysis of $Ta(2,6-di-t-butylphenoxide)_2(CD_3)_3$ (eq 63).[93] Although the photochemical generation of a tantalum alkylidene complex has been noted before, this is the first unequivocal demonstration of a photochemically induced α-abstraction to give a tantalum alkylidene complex that is inaccessible by a thermal reaction, and the first efficient formation of a methylene ligand by an α-abstraction reaction. The methylene ligand subsequently removes a proton from a t-butyl group in the phenoxide ligand. In the thermal reaction the products are CD_3H and $(OR)(CD_3)_2\overline{Ta(O-2,6-C_6H_3{}^tBuCMe_2)}$.

$$Ti(OR)_2(CD_3)_3 \xrightarrow[-CD_4]{h\nu} Ta(OR)_2(CD_3)(CD_2) \longrightarrow$$

$$(OR)(CD_3)(CD_2H)\overline{Ta(O-2,6-C_6H_3{}^tBuCMe_2)}$$

$$(OR = O-2,6-C_6H_3{}^tBu_2) \tag{63}$$

Many tungsten neopentylidene complexes containing a phenylimido ligand have been prepared.[94] In general, the imido complexes are more easily prepared and are more stable than their oxo analogues.[82] One class was first prepared via neopentylidene ligand transfer as shown in equation 64

$$W(NPh)(OCMe_3)_4 + Ta(CHCMe_3)L_2Cl_3 \longrightarrow Ta(OCMe_3)_4Cl +$$

$$W(NPh)(CHCMe_3)L_2Cl_2 \tag{64}$$

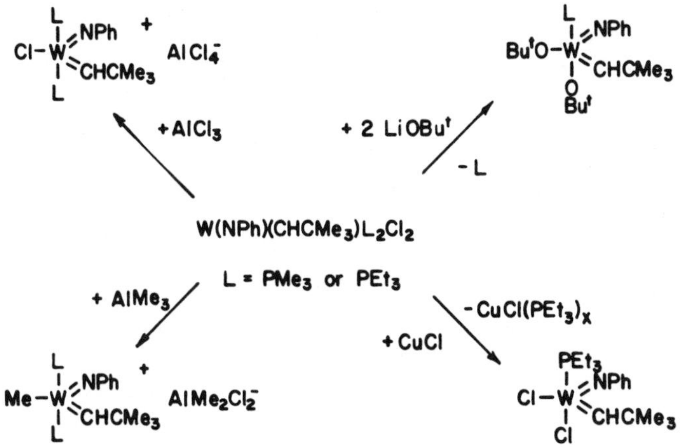

$$\begin{array}{ccc}
\underset{\underset{L}{|}}{\overset{\overset{L}{|}}{Cl-W}}{\overset{NPh}{\underset{CHCMe_3}{<}}} \; AlCl_4^- & & Bu^tO-\underset{\underset{Bu^t}{\overset{|}{O}}}{\overset{\overset{L}{|}}{W}}{\overset{NPh}{\underset{CHCMe_3}{<}}}
\end{array}$$

$$\nwarrow \; +AlCl_3 \qquad +2\ LiOBu^t \nearrow \\ -L$$

$$W(NPh)(CHCMe_3)L_2Cl_2$$

$$L = PMe_3 \text{ or } PEt_3$$

$$+AlMe_3 \swarrow \qquad \qquad \searrow -CuCl(PEt_3)_x \\ +CuCl$$

$$\begin{array}{ccc}
Me-\underset{\underset{L}{|}}{\overset{\overset{L}{|}}{W}}{\overset{NPh}{\underset{CHCMe_3}{<}}} \; AlMe_2Cl_2^- & & Cl-\underset{\underset{Cl}{|}}{\overset{\overset{PEt_3}{|}}{W}}{\overset{NPh}{\underset{CHCMe_3}{<}}}
\end{array}$$

Figure 11. Preparation of some phenylimido alkylidene complexes from W(NPh)(CHCMe$_3$)L$_2$Cl$_2$.

(L = PMe$_3$ or PEt$_3$). The NPh and CHCMe$_3$ ligands are mutually <u>cis</u> and the phosphine ligands mutually <u>trans</u>. From W(NPh)(CHCMe$_3$)L$_2$Cl$_2$ several related complexes were prepared as shown in Figure 11. Alternative methods of preparing neopentylidene complexes consist of α-hydrogen abstraction reactions. For example, the reaction between W(NPh)(CH$_2$CMe$_3$)$_3$Cl and LiCH$_2$CMe$_3$ yields W(NPh)(CHCMe$_3$) (CH$_2$CMe$_3$)$_2$ as a red oil. When NaC$_5$H$_5$ is added to W(NPh)(CH$_2$CMe$_3$)$_3$Cl the product is WCp(NPh)(CHCMe$_3$)(CH$_2$CMe$_3$), and when W(NPh)(CH$_2$CMe$_3$)$_3$Cl is treated with PMe$_3$ and Me$_3$PHCl in chloroform at 60-80° a high yield of W(NPh)(CHCMe$_3$)(PMe$_3$)$_2$Cl$_2$ results, possibly by the sequence of reactions shown in equation 65. Yellow, crystalline

$$W(NPh)(CH_2CMe_3)_3Cl \xrightarrow{PMe_3} W(NPh)(CHCMe_3)(CH_2CMe_3)(PMe_3)Cl \quad (65)$$

$$\xrightarrow{Me_3PHCl} W(NPh)(CH_2CMe_3)_2(PMe_3)Cl_2 \xrightarrow{PMe_3} W(NPh)(CHCMe_3)(PMe_3)_2Cl_2$$

$W(NPh)(CH_2SiMe_3)_4$ can be prepared from $W(NPh)(CH_2SiMe_3)_3Cl$ and $LiCH_2SiMe_3$. When it is heated to $60°$ in toluene 1 eq of TMS is evolved and $W(NPh)(CHSiMe_3)(CH_2SiMe_3)_2$ can be isolated in high yield as a dark red oil. The reaction is first order in tungsten with $\Delta H^{\ddagger}$ = 92+8 kJ mol^{-1} and $\Delta S^{\ddagger}$ = -30+15 JK^{-1} mol^{-1}. Finally, $W(NPh)(CH_2SiMe_3)_2Cl_2$ reacts with 2 eq of PMe_3 in dichloromethane to give $W(NPh)(CHSiMe_3)(PMe_3)_2Cl_2$ in 88% isolated yield. The NMR data listed in Table 3 suggest that the neopentylidene ligand in many of these complexes is distorted (large $W=C_\alpha-C_\beta$ angle) as judged by the relatively low value for J_{CH_α}. These studies serve to illustrate the similarity between Ta(V) and W(VI) alkyl and alkylidene chemistry. In particular, tungsten neopentyl complexes seem most susceptible to intramolecular α-hydrogen abstraction reactions, alkoxide ligands in place of chloride ligands slow down α-hydrogen abstraction reactions, and α-hydrogen abstraction can be induced by adding donor ligands.

The chemistry of some rhenium(VII) imido alkyl complexes has also been explored in some detail.[95] $Re(NCMe_3)_2(CH_2SiMe_3)_3$ can be prepared in good yield from $Re(NCMe_3)_2Cl_3$ and 1.5 eq of $Zn(CH_2SiMe_3)_2$. Photolysis of $Re(NCMe_3)_2(CH_2SiMe_3)_3$ in pentane through Pyrex with a high pressure mercury lamp yields yellow, crystalline $Re(NCMe_3)_2(CHSiMe_3)$ (CH_2SiMe_3) virtually quantitatively (δC_α = 237.5 ppm, J_{CH} = 128 Hz). All attempts to prepare $Re(NCMe_3)_2(CH_2CMe_3)_3$ from $Re(NCMe_3)_2Cl_3$ failed; the product was $Re(NCMe_3)_2(CHCMe_3)(CH_2CMe_3)$ (yellow oil; δC_α = 262.2 ppm, J_{CH} = 134 Hz). Attempts to prepare analogous benzylidene or methylidene complexes by photolyzing $Re(NCMe_3)_2(CH_2Ph)_3$ or $Re(NCMe_3)_2Me_3$, respectively, failed. Protonation of $Re(NCMe_3)_2$ $(CHCMe_3)(CH_2CMe_3)$ with gaseous HCl (3 eq) in pentane at $-78°$ produced two isomers of a molecule with the formula $[Re(CCMe_3)(CHCMe_3)$ $(NH_2CMe_3)Cl_2]_2$. Surprisingly, the neopentylidene C_α signal for the major isomer (298.4 ppm, J_{CH} = 128 Hz) occurs <u>downfield</u> of the neopentylidyne C_α signal (294.3 ppm). The same is true for the relative position of the two signals for the minor isomer. In $Re(CCMe_3)(CHCMe_3)(OCMe_3)_2$ and $Re(CCMe_3)(CHCMe_3)(OSiMe_3)_2$, prepared straightforwardly from the dimer

Table 3. Pertinent ^{1}H and ^{13}C NMR Data for Phenylimidoalkylidene Complexes.

Compound	δ^1H_α (ppm)	$\delta^{13}C_\alpha$ (ppm)	J_{CH_α} (Hz)
W(NPh)(CHCMe$_3$)(PMe$_3$)$_2$Cl$_2$	10.92	307	123
W(NPh)(CHCMe$_3$)(PEt$_3$)$_2$Cl$_2$	11.92	304	119
[W(NPh)(CHCMe$_3$)(PMe$_3$)$_2$Cl][AlCl$_4$]	10.39	303	106
[W(NPh)(CHCMe$_3$)(PEt$_3$)$_2$Cl][AlCl$_4$]	9.6	301	106
W(NPh)(CHCMe$_3$)(PMe$_3$)(OCMe$_3$)$_2$	10.17	265	114
W(NPh)(CHCMe$_3$)(PEt$_3$)(OCMe$_3$)$_2$	10.27	266	111
[W(NPh)(CHCMe$_3$)(PMe$_3$)$_2$Me][AlMe$_2$Cl$_2$]	8.40	303	106
[W(NPh)(CHCMe$_3$)(PEt$_3$)$_2$Me][AlMe$_2$Cl$_2$]	7.84	301	105
W(NPh)(CHCMe$_3$)(PEt$_3$)Cl$_2$	10.8	301	106
W(NPh)(CHCMe$_3$)(CH$_2$CMe$_3$)$_2$	6.61	246	106
W(NPh)(CHSiMe$_3$)(CH$_2$SiMe$_3$)$_2$	7.79	230	108
W(η^5-C$_5$H$_5$)(NPh)(CHCMe$_3$)(CH$_2$CMe$_3$)	9.81	269	117
W(NPh)(CHSiMe$_3$)(PMe$_3$)$_2$Cl$_2$	12.75	293	119
W(NPh)(CHCMe$_3$)(py)$_2$Cl$_2$	11.3	303	121

above, the neopentyli dyne C$_\alpha$ signals (287.4, 290.5) are downfield of the neopentylidene C$_\alpha$ signals (229.9, J_{CH} = 129 Hz; 231.3, J_{CH} = 121 Hz), as usual. An analogous iodide complex can be prepared by treating [Re(CCMe$_3$)(CHCMe$_3$)(NH$_2$CMe$_3$)Cl$_2$]$_2$ with Me$_3$SiI. A single isomer of a bispyridine adduct of the iodide complex can be prepared straightforwardly. The neopentylidene C$_\alpha$ signal (δC_α = 307.1 ppm, J_{CH} = 123 Hz) is again found downfield of the neopentylidyne C$_\alpha$ signal (δC_α = 299.5 ppm). An xray structural study revealed the structure shown in Figure 12. As expected, the neopentylidene and neopentylidyne ligands are mutually cis,

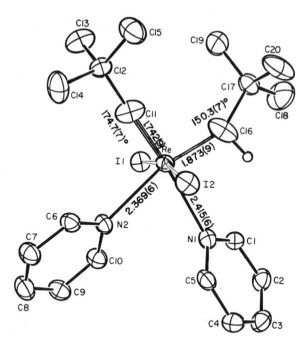

Figure 12. Molecular geometry and labeling of non-hydrogen atoms in the
Re(CCMe$_3$)(CHCMe$_3$)(C$_5$H$_5$N)$_2$I$_2$ molecule.
(ORTEP-II diagram; 30% probability ellipsoids).

the neopentylidene ligand lies in the C$_\alpha$≡Re=C$_\alpha$ plane, and the t-butyl group
of the neopentylidene ligand is pointed toward the neopentylidyne ligand (cf.
W(O)(CHCMe$_3$)(PMe$_3$)$_2$Cl$_2$ earlier).   The distances and angles shown in
Figure 12 are not unusual. The Re-carbon double bond is quite short for an
"undistorted" neopentylidene ligand, as is the Re-carbon triple bond, most
likely as a result of the comparatively high oxidation state of the metal.

The proposed mechanism of formation of [Re(CCMe$_3$)(CHCMe$_3$)
(NH$_2$CMe$_3$)Cl$_2$]$_2$ shown in Figure 13 is based on the deuterium labeling
results. The unusual feature of this reaction is the overall replacement of
Re-nitrogen multiple bonds by Re-carbon multiple bonds.   One might

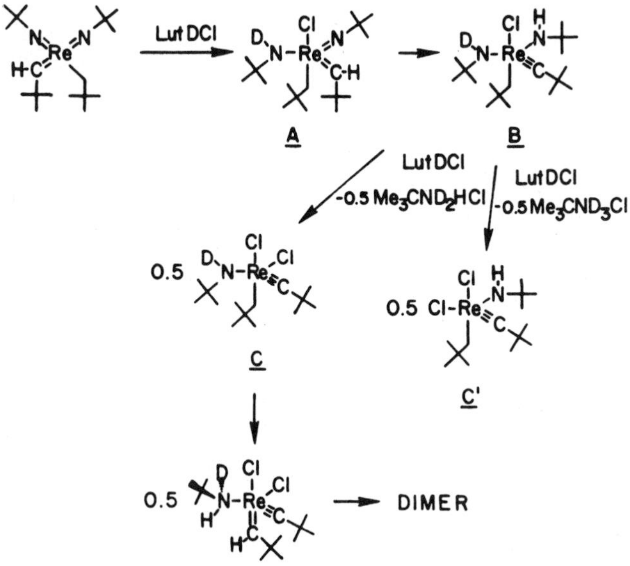

Figure 13.   Proposed mechanism of protonating
[Re(CCMe$_3$)(CHCMe$_3$)(NH$_2$CMe$_3$)Cl$_2$]$_2$.

speculate that a complex containing Re bonded to the more electronegative element (N) would be thermodynamically preferred, but since the product is not an isomer of the starting material we do not know whether the product results from thermodynamic or kinetic control.

The formation of neopentylidyne complexes[96] by what can be viewed as a double deprotonation of a neopentyl complex suggests that under the right circumstances a neopentylidyne ligand should be protonated to give a neopentylidene ligand.   Although [W(CCMe$_3$)Cl$_4$]$^-$ does not readily react with HCl in dichloromethane[96] it will react with water in the presence of base and a phosphine ligand, as shown in eq 66.[97] An analogous reaction with ammonia or aniline yields intermediate W(CCMe$_3$)(NHR)(PR$_3$)$_2$Cl$_2$.   When W(CCMe$_3$)(NHPh)(PEt$_3$)$_2$Cl$_2$ is heated to 70° in toluene it is converted into

$$[NEt_4][W(CCMe_3)Cl_4] + H_2O \xrightarrow[+ 2PR_3 - NEt_4Cl]{+ NEt_3 - NEt_3HCl} W(O)(CHCMe_3)(PR_3)_2Cl_2 \qquad (66)$$

$W(CHCMe_3)(NPh)(PEt_3)_2Cl_2$ in a first order reaction with an activation energy of 142 ± 12 kJ mol^{-1}. The rate of conversion is the same in chloroform or chlorobenzene as in toluene, and the same in the presence of added PEt_3 as in its absence, but the rate increases ten-fold in the presence of NEt_3. Reversible dehydrohalogenation by NEt_3 may be the explanation, as $W(CCMe_3)(NPh)L_2Cl$ is formed upon addition of $Ph_3P=CH_2$ to $W(CCMe_3)$ $(NHPh)L_2Cl_2$ (L = PEt_3). The reaction between $[NEt_4][W(CCMe_3)Cl_4]$, $PhPH_2$, PEt_3, and NEt_3 yields $W(CCMe_3)(PHPh)(PEt_3)_2Cl_2$. In contrast to the above results $W(CCMe_3)(PHPh)(PEt_3)_2Cl_2$ is unchanged upon heating in benzene at $75°$ for 24 h.

The addition of excess water to $W(CCMe_3)(CH_2CMe_3)_3$ and $W(CCMe_3)(OCMe_3)_3$ (in the presence of NEt_4OH) yields the neopentyl complexes, $[W(O)(CH_2CMe_3)_3]_2(\mu\text{-}O)$ and $[NEt_4][WO_3(CH_2CMe_3)]$, respectively.[98] In each case an early step, if not the first, is thought to be addition of water across the W≡C bond to give a hydroxo neopentylidene complex. The neopentyl ligand is formed by a subsequent transfer of a proton from the hydroxide oxygen to carbon, or by addition of water across the W=C bond at a later stage. In these, as well as the above systems, addition of an O or N donor to the metal is believed to enhance the nucleophilicity of the alkylidyne α-carbon atom. Alternatively, one could speculate that if $W(CHCMe_3)Cl_4$ were somehow prepared in a circuitous fashion the neopentylidene ligand would be so distorted by the electrophilic metal attracting the CH_α electron pair that an α proton would be lost to a relatively weak base.

A further study of reactions between $W(CH)(PMe_3)_4X$ (X = I or triflate) and protic or Lewis acids has appeared.[99] As in the case of $W(CH)(PMe_3)_4Cl$, addition of triflic acid to $W(CH)(PMe_3)_4X$ yields a face-capped methylidyne hydride or severely distorted methylene complex. At room temperature the "methylene" protons equilibrate rapidly and intramolecularly on the ^{1}H NMR time scale. Addition of AlMe$_3$ yields what could be called an aluminomethylene complex, W(CHAlMe$_2$)(Me)L$_3$Cl. In the presence of ethylene the α proton is lost to give W(CAlMe$_2$)(AlMe$_2$Cl)(C$_2$H$_4$)MeL$_2$, an AlMe$_2$Cl adduct of the hypothetical "aluminomethylidyne" complex, W(CAlMe$_2$)(C$_2$H$_4$)MeL$_2$. An attempt to prepare the methylene analogue of W(CHCMe$_3$)(PMe$_3$)$_2$(CO)Cl$_2$ by treating W(CH)(H)(PMe$_3$)$_3$Cl$_2$ with CO led to W(CH$_2$PMe$_3$)(PMe$_3$)$_2$(CO)$_2$Cl$_2$ instead; presumably the analogous CH(CMe$_3$)(PMe$_3$) complex cannot form in the former case for steric reasons. A related Me$_3$P=CH$_2$ complex can be prepared as shown in eq 67; an xray study[100] elucidated the structure.

$$[W(CH_2)(PMe_3)_4Cl][OTf] + CO \longrightarrow [W(CH_2PMe_3)(PMe_3)_3(CO)_2Cl][OTf]$$

$$(67)$$

An xray study of [W(CH$_2$)(PMe$_3$)$_4$I][OTf] confirmed the composition shown and the short, methylidyne-like character of the W–carbon bond (1.83(2)Å).[101] Unfortunately, an undetermined disorder problem prevented a more accurate determination of the W–carbon bond length. A large crystal also proved unsuitable for neutron diffraction studies.

Grubbs has investigated the reaction between TiCp$_2$Cl$_2$ and AlMe$_3$ to give TiCp$_2$(CH$_2$AlMe$_2$Cl) in some detail.[102] The rate-limiting step is the bimolecular reaction between TiCp$_2$Me(ClAlMe$_2$Cl) and AlMe$_3$. The data led to $\Delta H^{\ddagger}$ = 67±1 kJ mol^{-1} and $\Delta S^{\ddagger}$ = -109±4 JK^{-1} mol^{-1} with an isotope effect (k$_H$/k$_D$) of 2.9. The reaction rate in dichloromethane was ~1.5 times that in benzene. It is believed that formation of an incipient [TiCp$_2$Me]$^+$ ion in the highly polarized TiCp$_2$Me(ClAlMe$_2$Cl) complex is sufficient to render the methyl group acidic enough to be deprotonated by AlMe$_3$. Alternatively, the author favors deprotonation of TiCH$_3$ by a methyl group bound to Al in TiCp$_2$Me(ClAlMe$_3$), the bimolecular stage of the reaction being replacement of AlMe$_2$Cl in TiCp$_2$Me(ClAlMe$_2$Cl) by AlMe$_3$. It is interesting to compare these results with those obtained for the reaction between TaCp$_2$Me$_3$ and AlMe$_3$ to give TaCp$_2$Me(CH$_2$AlMe$_3$)[19] where the rate limiting step was believed to be the bimolecular reaction between

[TaCp$_2$Me$_2$][AlMe$_4$] and AlMe$_3$. The isotope effect for the reaction between [TaCl$_2$Me$_2$][BF$_4$] and Me$_3$P=CH$_2$ was shown to be 3.4±0.3.

A paper has appeared[103] in which the presence of Mo(CHSiMe$_3$)(CH$_2$SiMe$_3$)$_3$ mixed with yellow Mo(CSiMe$_3$)(CH$_2$SiMe$_3$)$_3$ is postulated in a volatile purple oil formed in the reaction between MoCl$_5$ and five equivalents of LiCH$_2$SiMe$_3$. Addition of LiCH$_2$SiMe$_3$ to the purple oil converts Mo(CHSiMe$_3$)(CH$_2$SiMe$_3$)$_3$ into [Mo(CHSiMe$_3$)(CH$_2$SiMe$_3$)$_4$]$^-$. Both alkylidene complexes are proposed on the basis of ESR spectra characteristic of a d^1 Mo complex. The ESR results do not rule out alternative possibilities such as Mo(O)(CH$_2$SiMe$_3$)$_3$ and [Mo(O)(CH$_2$SiMe$_3$)$_4$]$^-$, respectively.

Schwartz recently reported[104a] the preparation of AlR$_2$X adducts of "Cp$_2$M(CHR')" where R' $\neq$ H and M = Ti, Zr, analogous to Cp$_2$Ti(CH$_2$AlMe$_2$Cl), by adding HAlR$_2$ to an alkenyl complex (e.g., eq 68) or an alkenyl complex of Al to Cp$_2$ZrHCl (e.g., eq 69). For zirconium a sometimes significant reaction pathway is addition of the hydride to the

$$Cp_2Ti(CH=CHMe)(Cl) + HAl^iBu_2 \longrightarrow Cp_2Ti(CHEtAl^iBu_2Cl) \qquad (68)$$
$$Cp_2ZrH(Cl) + CH_3CH=CHAlMe_2 \longrightarrow Cp_2Zr(CHEtAlMe_2Cl) \qquad (69)$$

double bond the other way to give a non-alkylidene-bridged species. Cp$_2$Zr(CHCH$_2$CMe$_3$Al^{i}Bu$_2$Cl) reacts with HMPA to give what is proposed to be a mixture of cis and trans isomers of [Cp$_2$Zr(CHCH$_2$CMe$_3$)]$_2$.[104b] If PPh$_3$ is present a compound is obtained whose ^{1}H and ^{13}C NMR spectra are consistent with the formulation Cp$_2$Zr(CHCH$_2$CMe$_3$)(PPh$_3$) (δ H$_\alpha$ = 11.71, δ C$_\alpha$ = 270 ppm). An analogous reaction produced Cp$_2$Ti(CH$_2$)(PEt$_3$) (δ H$_\alpha$ = 12.04, δ C$_\alpha$ = 287 ppm). In the absence of phosphine Cp$_2$Zr(CHCH$_2$CMe$_3$ Al^{i}Bu$_2$Cl) reacts with HMPA to give Cp$_2$Zr(CHCH$_2$CMe$_3$)(HMPA) (δ H$_\alpha$ = 9.88, δ C$_\alpha$ = 230 ppm). Both Zr and Ti complexes of the type Cp$_2$Zr(CHR)(L) react with the carbonyl function in esters, ketones, etc. to give the Wittig-like olefin products in high yield. Reactions of the Zr species require 6 h at ~70° while the Ti species react with ketones in 2-3 h at 25°C.

Cooper has published two further studies in the cationic tungstenocene system. He previously reported that $[WCp_2Me_2][PF_6]$ was converted into $[WCp_2(C_2H_4)(H)][PF_6]$ upon treatment with trityl radical. He now finds that the same product is obtained by treating WCp_2Me_2 with Ph_3CBF_4 in dichloromethane, $[WCp_2Me_2]^+$ being formed by electron transfer as the first product.[105a] An analogous reaction between WCp_2MeEt and Ph_3CBF_4 yielded $[WCp_2(CH_2CHMe)(H)][BF_4]$ via formation of $[WCp_2(Me)(Et)]^+$ followed by abstraction of a hydrogen atom from an α-carbon atom by trityl radical. Labeling studies showed that $[WCp_2(CH_2)(Et)]^+$ was the intermediate. Cooper argues that loss of a hydrogen atom from the methyl group instead of from the β-carbon of the ethyl group can be ascribed to partial formation of a metal-carbon π bond in the transition state ("hyperconjugation"). Presumably loss of H· from the ethyl group is sterically blocked. More recently[105b] Cooper has shown that $[WCp_2MePh]^+$ reacts with trityl radical in the presence of acetonitrile to give $[WCp_2(CH_2Ph)(CH_3CN)]^+$, presumably via $[WCp_2(CH_2)Ph]^+$ followed by migration of the phenyl group and coordination of acetonitrile. Many of the results in the tungstenocene system are summarized in a short review.[105c]

A mechanistic study of the decomposition of $Ti(\eta^5-C_5Me_5)_2(CH_3)_2$ has shown that $Ti(\eta^5-C_5Me_5)(C_5Me_4CH_2)(CH_3)$ is formed cleanly in an intramolecular reaction with $\Delta H^{\ddagger} = 115\pm12$ kJ mol^{-1} and $\Delta S^{\ddagger} = -12\pm3$ J K^{-1} mol^{-1}.[106] $Ti(\eta^5-C_5Me_5)_2(CD_3)_2$ decomposes at about one third the rate of $Ti(\eta^5-C_5Me_5)_2Me_2$, while $Ti(\eta^5-C_5(CD_3)_5]_2(CH_3)_2$ decomposes at the same rate as $Ti(\eta^5-C_5Me_5)_2Me_2$. All data are consistent with $Ti(\eta^5-C_5Me_5)_2(CH_2)$ being formed in the first step followed by addition of a C-H bond from a ring methyl group across the Ti=C bond.

A study of olefin metathesis by catalysts prepared by adding $Mo(\eta^3-C_3H_5)_4$ to alumina gave some unusual results relating to the formation of alkylidene ligands from olefins.[107] Good evidence was obtained for addition of a vinylic C-H bond in propene to Mo(IV) to give a propenyl hydride intermediate. Migration of the hydride to the β-carbon atom would then yield a propylidene complex.

Acknowledgements I thank the National Science Foundation for its financial support and each student whose name has appeared in the references for his or her dedication, enthusiasm, and individual contributions. I also want to thank Professor Melvyn Churchill, Dr. Galen Stucky, and Dr. Art Schultz for permission to use figures that have appeared elsewhere.

REFERENCES

1. E.O. Fischer and A. Maasböl, Angew. Chem. Int. Ed. Engl. 3:580 (1964).

2. (a) D.J. Cardin, B. Cetinkaya, M.J. Doyle and M.F. Lappert, Chem. Soc. Rev. 2:99 (1973); (b) D.J. Cardin, B. Cetinkaya and M.F. Lappert, Chem. Rev. 72:545 (1972); (c) F.A. Cotton and C.M. Lukehart, Prog. Inorg. Chem. 16:487 (1972).

3. C.P. Casey, T.J. Burkhardt, C.A. Bunnell and J.C. Calabrese, J. Am. Chem. Soc. 99:2127 (1977).

4. P. Friedrich, G. Besl, E.O. Fischer and G. Huttner, J. Organomet. Chem. 139:C68 (1977).

5. E.O. Fischer, R.L. Clough, G. Besl and F.R. Kreissl, Angew. Chem. Int. Ed. Engl. 15:543 (1976).

6. E.O. Fischer, R.L. Clough and P. Stückler, J. Organomet. Chem. 120:C6 (1976).

7. W.A.Kiel, G.-Y. Lin and J.A. Gladysz, J. Am. Chem Soc. 102:3299 (1980).

8. M. Brookhart and G.O. Nelson, J. Am. Chem. Soc. 99:6099 (1977).

9. M. Brookhart, J.R. Tuchin, T.C. Flood and J. Jensen, J. Am. Chem. Soc. 102:1203 (1980).

10. G.C. Clark, K. Marsden, W.R. Roper and L.J. Wright, J. Am. Chem. Soc. 102:6570 (1980).

11. R.R. Schrock, J. Am. Chem. Soc. 96:6796 (1974).

12. R.R. Schrock and J.D. Fellmann, J. Am. Chem. Soc. 100:3359 (1978).

13. S.M. Rocklage, J.D. Fellmann, G.A. Rupprecht, L.W. Messerle and R.R. Schrock, J. Am. Chem. Soc. 103:1440 (1981).

14. C.D. Wood, S.J. McLain and R.R. Schrock, J. Am. Chem. Soc. 101:3210 (1979).

15. G.A. Rupprecht, L.W. Messerle, J.D. Fellmann and R.R. Schrock, J. Am. Chem. Soc. 102:6236 (1980).

16. L.W. Messerle, P. Jennische, R.R. Schrock and G.D. Stucky, J. Am. Chem. Soc. 102:6744 (1980).

17. R.R. Schrock, unpublished results.

18. R.R. Schrock, J. Am. Chem. Soc. 97:6577 (1975).

19. R.R. Schrock and P.R. Sharp, J. Am. Chem. Soc. 100:2389 (1978).

20. R.R. Schrock, L.J. Guggenberger, L.W. Messerle and C.D. Wood, J. Am. Chem. Soc. 100:3793 (1978).

21. M.F. Lappert and C.R.C. Milne, Chem. Commun. 925 (1978).

22. J.D. Fellmann, G.A. Rupprecht, C.D. Wood and R.R. Schrock, J. Am. Chem. Soc. 100:5964 (1978).

23. J.D. Fellmann, R.R. Schrock and G.W. Rupprecht, J. Am. Chem. Soc. 103:5752 (1981).

24. R.R. Schrock, S. Rocklage, J. Wengrovius, G. Rupprecht and J. Fellmann, J. Mol. Catal. 8:73 (1980).

25. J. Wengrovius, unpublished results.

26. J. Wengrovius, R.R. Schrock, M.R. Churchill, J.R. Missert and W.J. Youngs, J. Am. Chem. Soc. 102:4515 (1980).

27. (a) S. Rocklage and R.R. Schrock, J. Am. Chem. Soc. 102:7808 (1980); (b) idem, ibid, 104:3077 (1982).

28. J.D. Fellmann, H.W. Turner and R.R. Schrock, J. Am. Chem. Soc. 102:6608 (1980).

29. A.J. Schultz, R.K. Brown, J.M. Williams and R.R. Schrock, J. Am. Chem. Soc. 103:169 (1981).

30. (a) H.W. Turner, J.D. Fellmann, S.M. Rocklage, R.R. Schrock, M.R. Churchill and H. Wasserman, J. Am. Chem. Soc. 102:7809 (1980); (b) S.M. Rocklage, H.W. Turner, J.D. Fellman and R.R. Schrock, Organometallics 1:703 (1982).

31. F.N. Tebbe, G.W. Parshall and G.S. Reddy, J. Am. Chem. Soc. 100:3611 (1978).

32. R.R. Schrock and LJ. Guggenberger, J. Am. Chem. Soc. 97:2935 (1975).

33. D.N. Clark and R.R. Schrock, J. Am. Chem. Soc. 100:6774 (1978).

34. R.A. Andersen, Inorg. Chem. 18:3622 (1979).

35. (a) R.R. Schrock and G.W. Parshall, Chem. Rev. 76:243 (1976); (b) P.J. Davidson, M.F. Lappert and R. Pearce, Chem. Rev.76:219 (1979).

36. J.D. Fellmann, Ph.D. Thesis, MIT, 1980.

37. G.A. Rupprecht, Ph.D. Thesis, MIT, 1979.

38. W. Mowat and G. Wilkinson, J. C. S. Dalton 1120 (1973).

39. R.R. Schrock, J. Organomet. Chem. 122:209 (1976).

40. V. Malatesta, K.U. Ingold and R.R. Schrock, J. Organomet. Chem. 152:C53 (1978).

41. R.R. Schrock and P. Meakin, J. Am. Chem. Soc. 96:5288 (1974).

42. P.R. Sharp, Ph.D. Thesis, MIT, 1980.

43. (a) P.R. Sharp and R.R. Schrock, J. Organomet. Chem. 171:43 (1979); (b) P.T. Wolczanski, R.S. Threlkel and J.E. Bercaw, J. Am. Chem. Soc. 101:218 (1979).

44. P.R. Sharp, D. Astruc and R.R. Schrock, J. Organomet. Chem. 182:477(1979).

45. S. Pedersen, unpublished results.

46. D.N. Clark, Ph.D. Thesis, MIT, 1979.

47. J. Schwartz and K.I. Gell, J. Organomet. Chem. 184:C1 (1980).

48. N.J. Cooper and M.L.H. Green, J. C. S. Dalton 1121 (1979).

49. R.R. Schrock and L.J. Guggenberger, J. Am. Chem. Soc. 97:6578 (1975).

50. F. Takusagawa and T. Koetzle, unpublished results.

51. M.R. Churchill, F.J. Hollander and R.R. Schrock, J. Am. Chem. Soc. 100:647 (1978).

52. M.R. Churchill and F.J. Hollander, Inorg. Chem. 17:1957 (1978).

53. G.D. Stucky, unpublished results.

54. (a) S.J. McLain, C.D. Wood, L.W. Messerle, R.R. Schrock, F.J. Hollander, W.J. Youngs and M.R. Churchill, J. Am. Chem. Soc. 100:5962 (1978); (b) M.R. Churchill and W.J. Youngs, Inorg. Chem. 18:171 (1979).

55. A.J. Schultz, J.M. Williams, R.R. Schrock, G.A. Rupprecht and J.D. Fellmann, J. Am. Chem. Soc. 101:1593 (1979).

56. M.R. Churchill and W.J. Youngs, Inorg. Chem. 18:1930 (1979).

57. M.R. Churchill, A.L. Rheingold, W.J. Youngs, R.R. Schrock and J.H. Wengrovius, J. Organomet. Chem. 204:C17 (1981).

58. M.R. Churchill and W.J. Youngs, Inorg. Chem. 18:2454 (1979).

59. J.W. Lauher and R. Hoffmann, J. Am. Chem. Soc. 98:1729 (1976).

60. W.A. Nugent and B.L. Haymore, Coord. Chem. Rev. 31:123 (1980).

61. W.P. Griffith, Coord. Chem. Rev. 5:459 (1970).

62. (a) E.O. Fischer and U. Schubert, J. Organomet. Chem. 100:59 (1975); (b) P.R. Sharp, S.J. Holmes, R.R. Schrock, M.R. Churchill and H. Wasserman, J. Am. Chem. Soc. 103:965 (1981).

63. (a) M.B. Hursthouse, R.A. Jones, K.M.A. Malik and G. Wilkinson, J. Am. Chem. Soc. 101:4128 (1979); (b) R.A. Andersen, R.A. Jones, G. Wilkinson, M.B. Hursthouse and K.M.A. Malik, Chem. Commun. 865 (1977).

64. C.D. Wood, Ph.D. Thesis, MIT, 1979.

65. R.R. Schrock, J. Am. Chem. Soc. 98:5399 (1976).

66. S.H. Pine, R. Zahler, D.A. Evans and R.H. Grubbs, J. Am. Chem. Soc. 102:3270 (1980).

67. S.J. McLain, C.D. Wood and R.R. Schrock, J. Am. Chem. Soc. 99:3519 (1977).

68. T.R. Howard, J.B. Lee and R.H. Grubbs, J. Am. Chem. Soc. 102:6876 (1980).

69. L.W. Messerle, Ph.D. Thesis, MIT, 1979.

70. F.N. Tebbe and R.L. Harlow, J. Am. Chem. Soc. 102:6149 (1980).

71. T.J. Katz and S.J. Lee, J. Am. Chem. Soc. 102:422 (1980).

72. C.P. Casey and S.M. Neuman, J. Am. Chem. Soc. 99:1651 (1977).

73. P. Belmonte, unpublished observations.

74. R.J. Goddard, R. Hoffmann and E.D. Jemmis, J. Am. Chem. Soc. 102:7676 (1980).

75. A.K. Rappe and W.A. Goddard III, J. Am. Chem. Soc. 102:5114 (1980).

76. J.H. Wengrovius, J. Sancho and R.R. Schrock, J. Am. Chem. Soc. 103:3932 (1981).

77. Jose Sancho, unpublished observations.

78. (a) S.J. Holmes and R.R. Schrock, J. Am. Chem. Soc. 103:4599 (1981); (b) S.J. Holmes, D.N. Clark, H.W. Turner and R.R. Schrock, J. Am. Chem. Soc. 104:6332 (1982).

79. M.R. Churchill, H.J. Wasserman, H.W. Turner and R.R. Schrock, J. Am. Chem. Soc. 104:1710 (1982).

80. J.H. Wengrovius, R.R. Schrock, M.R. Churchill and H.J. Wasserman, J. Am. Chem. Soc. 104:1739 (1982).

81. J.D. Fellmann, R.R. Schrock and D.D. Traficante, Organometallics 1:481 (1982).

82. J.H. Wengrovius and R.R. Schrock, Organometallics 1:148 (1982).

83. R.S. Threlkel and J.E. Bercaw, J. Am. Chem. Soc. 103:2650 (1981).

84. R.R. Schrock, Acc. Chem. Res. 12:98 (1979).

85. J.C. Hayes, G.D.N. Pearson and N.J. Cooper, J. Am. Chem. Soc. 103:4648 (1981).

86. K.W. Chiu, R.A. Jones, G. Wilkinson, A.M.R. Galas, M.B. Hursthouse and K.M.A. Malik, J. C. S. Dalton 1204 (1981).

87. A.K. Rappe and W.A. Goddard, III, J. Am. Chem. Soc. 104:448 (1982).

88. (a) J. Kress, M. Wesolek, J.-P. LeNy and J.A. Osborn, J.C.S. Chem. Commun. 1039 (1981); (b) J. Kress, M. Wesolek and J.A. Osborn, J.C.S. Chem. Commun. 514 (1982).

89. (a) H.W. Turner and R.R. Schrock, J. Am. Chem. Soc. 104:2331 (1982); (b) H.W. Turner, R.R. Schrock, J.D. Fellmann and S.J. Holmes, J. Am. Chem. Soc. 105:4942 (1983).

90. (a) K.C. Ott and R.H. Grubbs, J. Am. Chem. Soc. 103:5922 (1981); (b) J.B. Lee, K.C. Ott and R.H. Grubbs, J. Am. Chem. Soc. 104:7491 (1982); (c) J.R. Stille and R.H. Grubbs, J. Am. Chem. Soc. 105:1664 (1983).

91. K.J. Ahmed, M.H. Chisholm, I.P. Rothwell and J.C. Huffman, J. Am. Chem. Soc. 104:6453 (1982).

92. (a) L. Chamberlain, J. Keddington and I.P. Rothwell, Organometallics 1:1098 (1982); (b) L. Chamberlain, I.P. Rothwell and J.C. Huffman, J. Am. Chem. Soc. 104:7338 (1982).

93. L.R. Chamberlain, A.P. Rothwell and I.P. Rothwell, J. Am. Chem. Soc. 106:1847 (1984).

94. S.F. Pedersen and R.R. Schrock, J. Am. Chem. Soc. 104:7483 (1982).

95. (a) D.S. Edwards and R.R. Schrock, J. Am. Chem. Soc. 104:6806 (1982); (b) D.S. Edwards, L.V. Bianchi, J.W. Ziller, M.R. Churchill and R.R. Schrock, Organometallics 2:1505 (1983).

96. R.R. Schrock, D.N. Clark, J. Sancho, J.H. Wengrovius, S.M. Rocklage and S.F. Pedersen, Organometallics 1:1645 (1982).

97. S.M. Rocklage, R.R. Schrock, M.R. Churchill and H.J. Wasserman, Organometallics 1:1332 (1982).

98. I. Feinstein-Jaffe, S.F. Pedersen and R.R. Schrock, J. Am. Chem. Soc. 105:7176 (1983).

99. S.J. Holmes, R.R. Schrock, M.R. Churchill and H.J. Wasserman, Organometallics 3:476 (1984).

100. M.R. Churchill and H.J. Wasserman, Inorg. Chem. 21:3913 (1982).

101. A.J. Schultz, J.M. Williams, R.R. Schrock and S.J. Holmes, Acta Cryst. C40:590 (1984).

102. K.C. Ott, E.J.M. DeBoer and R.H. Grubbs, Organometallics 3:223 (1984).

103. R.A. Andersen, M.H. Chisholm, J.F. Gibson, W.W. Reichert, I.P. Rothwell and G. Wilkinson, Inorg. Chem. 20:3934 (1981).

104. (a) F.W. Hartner, Jr. and J. Schwartz, J. Am. Chem. Soc. 103:4979 (1981); (b) F.W. Hartner, Jr., J. Schwartz and S.M. Clift, J. Am. Chem. Soc. 105:640 (1983).

105. (a) J.C. Hayes and N.J. Cooper, J. Am. Chem. Soc. 104:5570 (1982); (b) P. Jernakoff and N.J. Cooper, J. Am. Chem. Soc. 106:3026 (1984); (c) J.C. Hayes, P. Jernakoff, G.A. Miller and N.J. Cooper, Pure Appl. Chem. 56:25 (1984).

106. C. McDade, J.C. Green and J.E. Bercaw, Organometallics 1:1629 (1982).

107. Y. Iwasawa and H. Hamamura, J.C.S. Chem. Commun. 130 (1983).

CARBENE COMPLEXES OF GROUPS VIA, VIIA AND VIII

K.H. Dötz

Technical University of Munich

1. INTRODUCTION

Metal coordinated carbenes were isolated as early as 1915,[1-2] but these compounds were recognized as such only 55 years later.[3] Increased interest in the coordination chemistry of short-lived intermediates led to the first planned synthesis and well-documented characterization of a stable transition metal carbene complex $(OC)_5W:C(C_6H_5)OMe$ **(I)** by E.O. Fischer and A. Massböl in 1964.[4]

Since then, this field of organometallic chemistry has expanded at a rapid rate, and a great variety of preparative routes have led to the synthesis of many hundreds of compounds. Several reviews have covered the whole area[5-8] or parts of it.[9-19]

Because isolable complexes provide mechanistic models for important metal-catalyzed reactions, such as olefin metathesis (cf Chapter 5), much attention has shifted from synthesis to reactivity. So this chapter will give only a brief summary of synthetic methods (which have been reviewed very recently[7-8]) but will discuss reactions and mechanisms in some detail. Since Chapter 3 deals with the early transition elements, this chapter will be confined to the VIA, VIIA and VIII elements in complexes bearing a metal coordinated, sp^2-hybridized carbene carbon atom. Thus, compounds with bridging carbene ligands or with a coordinated sp^2-hybridized carbon atom formally doubly bonded to one of its two nonmetal neighbours (e.g. acyl-, alkenyl- or imino metal complexes) are excluded. The majority of carbene complexes are mononuclear and neutral and contain a single terminal unidentate carbene ligand coordinated to a transition metal in a low oxidation state. Among them, most of the reactivity studies have been done with the Fischer-type complexes related to **I**.

$$(OC)_5W = C(C_6H_5)OMe$$
$$\mathbf{I}$$

2.   SYNTHESIS

The synthesis of carbene complexes is mainly based on three strategies, (1) the modification of an existing metal to carbon bond, (2) the coordination of a carbene group originating from organic precursors, and (3) the modification of carbenes already coordinated. The latter will be dealt with in parts 4., 6., 7. below. The first two principles are covered in a short summary, since they have been reviewed very recently.[7-8] Special emphasis is on the Fischer-type complexes $(CO)_5M=C(R)R'$, the chemistry of which has been studied most thoroughly.

2.1   Nucleophilic Addition to Carbonyl Ligands

The first stable transition metal carbene complex recognised as such was prepared by E.O. Fischer and A. Maasböl by reacting tungsten hexacarbonyl with phenyllithium. Subsequent acidification and alkylation with diazomethane led to the tungsten carbene complex **I** (eq 1).[4]

$$W(CO)_6 + C_6H_5Li \longrightarrow [(CO)_5W=C(C_6H_5)O]^- \ Li^+ \quad \xrightarrow[\text{2. } CH_2N_2]{\text{1. } H^+} \quad \mathbf{I} \qquad (1)$$

In spite of a great variety of alternative methods now available the original concept (eq 2) has remained the most useful and the most general, and high yield one-pot syntheses have been developed starting from metal carbonyl derivatives of VIA, VIIA and VIII elements and alkyl,[20-4] aryl,[20,25] alkenyl,[26-7] alkynyl,[28] furyl,[26,29-30] thienyl,[26,29] pyrrolyl,[26,29] metallocenyl,[31-3] cymantrenyl $[-C_5H_4Mn(CO)_3]$[34] and benchrotrenyl $[-C_6H_5Cr(CO)_3]$[35] lithium reagents. Binary metal carbonyls

$M(CO)_6$ (M = Cr, Mo, W),[20] $M_2(CO)_{10}$ (M = Mn, Tc, Re),[36] $Fe(CO)_5$[37] and $Ni(CO)_4$[38] may be used as well as their substitution products $ArCr(CO)_3$ (Ar = π-$C_6H_{6-n}(CH_3)_n$, n = 0-3),[39] π-$C_5H_5M(CO)_2(NO)$ (M = Cr, Mo, W),[40] $Ph_3XM(CO)_5$ (M = Cr, Mo, W; X = P, As, Sb),[41] π-$C_5H_5M(CO)_3$ M = Mn, Re),[20,42] $R_3GeMn(CO)_5$,[43] $(Ph_3P)_2Cl(N_2)Re(CO)_2$,[44] $Ph_3PFe(CO)_4$,[45] $(NO)_2Fe(CO)_2$,[38] $R_3MCo(CO)_4$ (M = Ge, Sn)[45-6] and $(NO)Co(CO)_3$.[38]

$$L_mM(CO)_n + \text{Li-R} \longrightarrow [L_m(CO)_{n-1}M=C(R)O]^- \text{Li}^+ \qquad (2)$$

The range of nucleophiles may be expanded to include heteroatom compounds like triorganyl silyl,[47-9] dialkyl amido,[37-8,50] dithianyl lithium derivatives[51] and potassium ethoxide,[52] the latter leading to a carbene complex in extremely low yield. In comparison with lithium analogues, Grignard reagents have proved to react more slowly.[53]

The acylmetallate intermediates are alkylated most effectively by trialkyl oxonium salts[21,54] or fluorosulfonates,[55] but diazonium salts,[56] acyl halides,[57-9] halogenosilanes[60-2] and chlorometal complexes[63-4] have also been used (eq 3).

$$[L_m(CO)_{n-1}M=C(R)O]^- \text{Li}^+ + R'\text{-}X \xrightarrow[-\text{LiX}]{} L_m(CO)_{n-1}M=C(R)OR' \quad (3)$$

In accord with the greater electrophilicity of the carbene carbon compared with the coordinated carbonyl carbon atom, this route is generally confined to monocarbene metal complexes even though an excess of the nucleophile in question is used. Till now only very few exceptions to this rule are known. Due to electron transfer from the amino substituents, the carbene carbon in the cyclic diaminocarbene complex **II** becomes less electrophilic than the coordinated carbonyl carbon atoms. So subsequent carbanion-carbonium ion addition leads to the biscarbene complex **III** (eq 4).[65]

$$(4)$$

A rather unusual bis-addition of a nucleophile occurs, although in very low yield, by the reaction of lithium dimethyl-phosphide with chromium and tungsten hexacarbonyl followed by alkylation (eq 5).[66] Quite recently a chelating biscarbene ligand was synthesized in a one-pot reaction starting from ortho-bislithiated benzene and VIA hexacarbonyl complexes (eq 6).[67]

$$M(CO)_6 \xrightarrow[\text{2. } (C_2H_5)_3OBF_4]{\text{1. } LiP(CH_3)_2} \underline{cis} - (CO)_4M\{=C[P(CH_3)_2]OC_2H_5\}_2 \qquad (5)$$

$$M(CO)_6 \xrightarrow[\text{2. } (Et)_3O^+BF_4^-]{\text{1. } o\text{-}LiC_6H_4Li} \qquad \qquad (6)$$

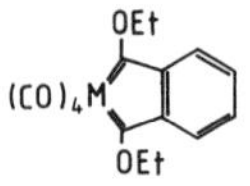

IV

Tetracarbonyl complexes accessible from monosubstituted hexacarbonyl compounds prefer to assume the cis-configuration. Warming in solution, however, leads to a cis-trans-equilibrium the isomers of which can be isolated (eq 7). The equilibrium ratio depends on the steric requirements of the ligands as well as on the size of the central metal and on the solvent used.[22,68-9]

$$\underline{cis}\text{-}(CO)_4(XR_3)Cr=C(CH_3)OCH_3 \rightleftharpoons \underline{trans} - (CO)_4(XR_3)Cr=C(CH_3)OCH_3$$
$$X = P, As \qquad (7)$$

The carbanion-carbonium ion addition can be realized in a one-step reaction using metal amides or hydrides. By this route mono- and binuclear metalloxycarbene complexes have been obtained from the addition of a metal amide or hydride across the carbonyl C-O bond (eqs 8-10).[70-4]

$$M(CO)_n + Ti[N(CH_3)_2]_4 \longrightarrow (CO)_{n-1}M=C[N(CH_3)_2]OTi[N(CH_3)_2]_3$$
$$M = Cr, W \ (n = 6); \ M = Fe \ (n = 5) \qquad (8)$$

$$Cp_2M(CO) + Cp_2ZrH_2 \longrightarrow Cp_2M=C(H)OZr(H)Cp_2 \qquad (9)$$
$$M = Cr, Mo, W$$

$$2 \ M(CO)_n + 2 \ Al[N(CH_3)_2]_3 \longrightarrow \{(CO)_{n-1}M=C[N(CH_3)_2]OAl[N(CH_3)_2]\}_2$$

M = Fe (n = 5); M = Ni (n = 4)   (10)

Chiral carbene complexes are accessible either by starting from optically active nucleophiles, e.g. (+)-menthyllithium, or from substituted carbonyl compounds containing chiral ligands such as suitable tertiary phosphines.[75-6]   Modification of carbene ligands leading to optical antipodes is discussed in (4.2) below.

2.2   Nucleophilic Addition to Carbyne Ligands

The   nucleophilic   addition to   carbonyl   ligands   fails   if weak nucleophiles are used.   In such cases the by-pass via carbyne complexes, now available by standard   procedures   starting   from carbene complexes,[77] is necessary. Carbyne complexes contain a metal to carbon triple bond the carbon atom   of which   has   proved   to be   a strongly electrophilic center.   As precursors to carbene complexes two types of cationic   carbyne   compounds   have   been   used   so   far: (1) pentacarbonyl(dialkylaminocarbyne)chromium tetrafluoroborate, (V) and (2) arene (carbonyl) carbyne complexes of chromium, manganese and rhenium $[Ar(CO)_n M \equiv C\text{-}R]^+ BX_4^-$, M = Cr, n = 3, Ar = $-C_6H_{6-m}R_m$, m = 0-3; X = F, Cl; M = Mn, Re, n = 2; Ar = Cp, MeCp, X = F, Cl.

$$[(CO)_5 Cr \equiv C\text{-}NR_2]^+ BF_4^-$$

V

The   aminocarbyne   cation **V** is   synthesized   by   reaction   of pentacarbonyl[(diethylamino)ethoxycarbene]chromium   with   boron trifluoride   at   very   low   temperature[78]   or   starting   from pentacarbonyl[chloro(dimethylamino)carbene]chromium using a silver salt as halogen acceptor.[79]   The carbyne cation adds weak nucleophiles like halogen and pseudohalogen anions, and thus offers a route to halogeno- and pseudohalogenocarbene   complexes   (eq   11).[79-81] The   cyanate   and thiocyanate groups are bonded to the carbene carbon through nitrogen.

$$[(CO)_5Cr \equiv C\text{-}NR_2]^+ + X^- \longrightarrow (CO)_5Cr = C(NR_2)X \qquad (11)$$

X = F, Cl, Br, I, CN, OCN, SCN

The addition of these nucleophiles to the trans-tetracarbonyl(phosphine)aminocarbyne chromium cation, however, could not be realized. Instead, the carbonyl substitution product mer-$X(PR_3)(CO)_3Cr \equiv C\text{-}NR_2$ was formed.[82]

The pentacarbonylcarbyne cation may be regarded as a key compound in the synthesis of carbene complexes containing heavier heteroatom substituents of the IVB and VB groups. Thus, by reaction with triphenylstannyl potassium and potassium diphenylarsenide, stannyl and arsenido carbene compounds are obtained (eq 12).[83-4] By analogy, addition of lithium dimethylamide leads to the acyclic diaminocarbene complex (eq 13).[79]

$$[(CO)_5Cr \equiv C\text{-}N(C_2H_5)_2]^+ BF_4^- + K^+E(C_6H_5)_n^- \xrightarrow[- KBF_4]{}$$

$$(CO)_5Cr = C[N(C_2H_5)_2]E(C_6H_5)_n \qquad (12)$$

E = Sn, n = 3; E = As, n = 2

$$[(CO)_5Cr \equiv C\text{-}N(CH_3)_2]^+ BF_4^- + LiN(CH_3)_2 \xrightarrow[- LiBF_4]{}$$

$$(CO)_5Cr = C[N(CH_3)_2]_2 \qquad (13)$$

Pseudohalogenocarbene complexes are obtained by addition of cyanide, cyanate and isothiocyanate to dicarbonyl(cyclopentadienyl)phenylcarbyne compounds of manganese and rhenium (eq 14).[85-6] Sodium alcoholates, phenolates and naphtholate may be added in a similar way to form alkoxy- or aryloxycarbene ligands.[87]

$$[Cp(CO)_2M \equiv C\text{-}C_6H_5]^+ + X^- \longrightarrow Cp(CO)_2M = C(C_6H_5)X \qquad (14)$$

M = Mn, Re; X = CN, OCN, SCN, OR

The introduction of the cyanocarbene substituent has also been achieved by reaction with t-butyl isocyanide followed by decomposition at room temperature (eq 15).[88]

$$[Cp(CO)_2Mn\equiv C\text{-}C_6H_5]^+ + t\text{-BuNC} \longrightarrow$$

$$[Cp(CO)_2Mn=C(C_6H_5)CNBu\text{-}t]^+ \longrightarrow Cp(CO)_2Mn=C(C_6H_5)CN+\dots \quad (15)$$

The addition of lithium reagents such as methyllithium[89-90] or cyclopentadienyllithium[86] to the cationic carbyne compounds provides a route to carbene complexes which are not stablized by heteroatoms bonded to the carbene carbon (eq 16).

$$[Cp(CO)_2M\equiv C\text{-}R]^+BCl_4^- + LiCH_3 \longrightarrow$$

$$- LiBCl_4$$

$$Cp(CO)_2M=C(R)CH_3 \quad (16)$$

M = Mn, Re; R = CH_3, C_6H_5

The addition of amines to arene(dicarbonyl) (phenylcarbyne)chromium cations leads to neutral amino(phenyl)carbene complexes[39] which may also, however, be synthesized more conviently by direct aminolysis of alkoxycarbene compounds (4.2).

Several attempts have been made to reduce the metal carbyne triple bond using boron and aluminium hydrides.   By this method secondary carbene complexes containing a hydrogen atom bonded to the carbene carbon - which in general are difficult to obtain - become accessible (eq 17).[49,91]

$$[Cp(CO)_2Re\equiv C\text{-}R]^+ \xrightarrow[\;(C_4H_9)_4NBH_4\;]{(C_2H_5)_2AlH \text{ or}} Cp(CO)_2Re=C(R)H \quad (17)$$

2.3   Nucleophilic Addition to Isocyanide Ligands

The concept of nucleophilic addition to a metal-bonded carbon atom may also be extended to coordinated isocyanides.   This method has been preferentially employed in the synthesis of aminocarbene complexes containing a group VIIIA metal in the oxidation state +2.

Alcohols[92-5] and thiols[95] as well as primary and secondary amines[95-7] have been found to add across the carbon-nitrogen bond of coordinated isocyanides.   In this way alkoxy(amino)-, amino(thio)- and diaminocarbene complexes can be obtained (eqs 18-20).

$$\underline{cis} - Cl_2[P(C_2H_5)_3]Pt\text{-}CNR + HOR' \longrightarrow$$

$$\underline{cis} - Cl_2[P(C_2H_5)_3]Pt\text{=}C(NHR)OR' \qquad (18)$$

$$\underline{trans} - (PR_3)_2Pt(CNC_2H_5)_2 + C_6H_5CH_2SH \longrightarrow$$

$$\underline{trans} - (PR_3)_2(H_5C_2NC)Pt\text{=}C(NHC_2H_5)SCH_2C_6H_5 \qquad (19)$$

$$\underline{cis} - Cl_2(PR_3)Pd\text{-}CNC_6H_4\text{-}X + Y\text{-}C_6H_4NH_2 \longrightarrow$$

$$\underline{cis} - Cl_2(PR_3)Pd\text{=}C(NHC_6H_4\text{-}X)NHC_6H_4\text{-}Y \qquad (20)$$

A nucleophilic attack of the amine at the coordinated carbon atom is verified by the finding that the reaction rate is increased by electron-withdrawing groups X in the isocyanide ligand and by electron-releasing subsituituents Y of the amine.[96]   The susceptibility of coordinated isocyanides to nucleophilic attack seems to be governed by a balance of steric and electronic effects.   Nevertheless, even more than one isocyanide ligand can be transformed into a carbene ligand.   The tetraisocyanide complexes of nickel(II), palladium(II) and platinum(II) react with methylamine to give the percarbene compounds (eq 21).[98]

$$[M(CNCH_3)_4]^{2+} + 4\,CH_3NH_2 \longrightarrow M[\text{=}C(NHCH_3)_2]_4{}^{2+} \qquad (21)$$

M = Ni, Pd, Pt

Addition of hydrazines yields cyclic biscarbene complexes (eq 22).[99-100] This type of reaction led in 1915 to the first isolated carbene complexes,[1-2] the structures of which, however, have only recently been elucidated.[3]

$$[Fe(CNCH_3)_6]^{2+} + RNH\text{-}NH_2 \longrightarrow (MeNC)_4Fe\overset{NHMe}{\underset{NHMe}{\overset{\displaystyle\diagdown\!NR}{\diagdown\!NH}}} \quad \textbf{VI} \qquad (22)$$

Chelating N,O-carbene ligands have been obtained quantitatively by the reaction of α,ω-hydroxyisocyanides with palladium(II) iodide (eq 23).[101]

$$Pd^{2+} + 4\,HO\text{-}(CH_2)_n\text{-}NC \longrightarrow \{Pd[\text{=}\overline{CNH(CH_2)_n O}]_4\}^{2+} \qquad (23)$$

Isocyanide derivatives of chromium and molybdenum carbonyls failed to undergo nucleophilic addition of alcohols and amines.[102]   The pentaisocyanide(nitrosyl)molybdenum(I) cation, however, is found to give monocarbene compounds (eq 24).[103]

$$[(NO)Mo(CNR)_5]^+ + R'NH_2 \longrightarrow [(NO)(RNC)_4Mo=C(NHR)NHR']^+ \quad (24)$$

2.4   Electrophilic Addition to Acyl and Related Ligands

The electrophilic addition to a metal coordinated acyl group, originally demonstrated in the classical Fischer synthesis, may be extended to neutral acyl and related complexes as well.   Some of the first examples are given by the alkylation of dicarbamoyl mercury[104] and the protonation of acetyl(dicarbonyl)cyclopentadienyl iron to a cationic hydroxycarbene complex (eqs 25-6).[105]   Similar protonation reactions have been performed to give closely related compounds of iron, ruthenium and molybdenum.[106-7] The protonation is easily reversed, even by water.   Alkylation of the hydroxycarbene compounds using diazomethane leads to methoxycarbene complexes, which are also accessible by direct methylation with oxonium salts.

$$Hg(CONR_2)_2 + 2(CH_3)_3OBF_4 \longrightarrow Hg[=C(NR_2)OCH_3]_2^{2+}(BF_4^-)_2 = 2(CH_3)_2O \quad (25)$$

$$Cp(CO)_2Fe-COCH_3 + HCl \longrightarrow [Cp(CO)_2Fe=C(CH_3)OH]^+ Cl^- \quad (26)$$

Bishydroxycarbene complexes have been obtained by protonation of bisacetyl compounds of rhenium (eq 27).[108]

$$\underline{fac}[X(CO)_3Re(COCH_3)_2]^{2-} + 2H^+ \longrightarrow \underline{fac}\text{-}X(CO)_3Re[=C(CH_3)OH]_2 \quad (27)$$

Similar O-alkylations of acyl complexes using methylfluorosulfonate have been performed to give a variety of methoxycarbene compounds containing iron and nickel metal centers (e.g. eq 28).[109-111]

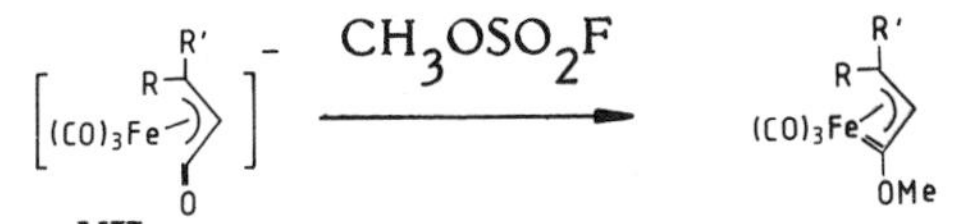

$$(28)$$

VII   **VIII**

Similarly, dialkoxycarbene complexes, which are difficult to prepare by nucleophilic attack at carbonyl ligands,[52] become accessible from coordinated alkyl carboxylate precursors (eq 29).[110]

$$trans - (Cl_5C_6)[P(CH_3)_2C_6H_5]_2Ni-COOCH_3 \xrightarrow{\quad CH_3OSO_2F \quad} \qquad (29)$$

$$trans - (Cl_5C_6)[P(CH_3)_2C_6H_5]_2Ni=C(OCH_3)_2$$

The unusual reaction between a rhenium carbonyl compound and methylfluorosulfonate leads to a rare example of a metal coordinated nonbridging methylene, which although unstable at room temperature could be trapped by addition of phosphines (eq 30).[112]

$$Cp(NO)PR_3Re(CO) \xrightarrow{\quad CH_3OSO_2F \quad} [Cp(NO)PR_3Re=CH_2]^+ \qquad (30)$$

$$\xrightarrow{\quad PR'_3 \quad} [Cp(NO)PR_3ReCH_2PR'_3]^+$$

An interesting variation of this type of reaction has been observed in acyl ligands containing γ-halogenated longer chain substituents. Initiation by silver salts or nucleophiles like iodide and pentacarbonylmanganate leads to intramolecular cyclization yielding cyclic carbene ligands (eqs 31-2).[113-4]

$$(CO)_5Mn-C(O)O(CH_2)_2Cl \xrightarrow[-AgCl]{\quad Ag^+ \quad} [(CO)_5Mn=\overset{\frown}{C-O(CH_2)_2-O}]^+ \qquad (31)$$

$$(CO)_5Mn-C(O)(CH_2)_3Cl \xrightarrow[-CO,-Cl^-]{\quad X^- \quad} cis - X(CO)_4Mn=\overset{\frown}{C(CH_2)_3O} \qquad (32)$$

Intramolecular alkylation depends on the nucleophilicity of the acyl oxygen. For instance, silver tetrafluoroborate fails to initiate the cyclization of the pentacarbonyl-δ-chlorobutyryl manganese mentioned in eq 32, but does transform the cis-substituted derivative, containing a weaker acceptor ligand, into the cyclic carbene compound (eq 33).

$$\underline{cis} - L(CO)_4Mn-C(O)(CH_2)_3Cl \xrightarrow{\text{Ag}^+} [\underline{cis} - L(CO)_4Mn=\overline{C(CH_2)_3O}]^+ \quad (33)$$

$$L = P(OCH_3)_3$$

The acyl manganese compounds can be obtained as alkyl migration intermediates in the reaction of the pentacarbonyl manganese anion with α,ω-dibromoalkanes. By this procedure, a binuclear complex was prepared,[115] which later had to be reformulated as a cyclic carbene complex (eq 34).[116]

$$[(CO)_5Mn]^- + Br(CH_2)_3Br \xrightarrow[-Br^-]{} (CO)_5Mn-(CH_2)_3Br \xrightarrow{[(CO)_5Mn]^-}$$

$$(34)$$

$$[(CO)_5Mn-(CO)_4Mn-C(O)(CH_2)_3Br]^- \xrightarrow[-Br^-]{} (CO)_5Mn-(CO)_4Mn=\overline{C(CH_2)_3O}$$

Intramolecular alkylation has been used mostly in the synthesis of cyclic carbene ligands coordinated to manganese, but related reactions have also been performed with molybdenum, iron and ruthenium compounds.[113,117-8]

The O-alkylation of acyl compounds may be extended to thioacyl complexes as well. This route, starting from neutral thiocarboxylate, dithiocarboxylate and thiocarbamoyl complexes of platinum, leads to a variety of cationic compounds containing two different heteroatoms bonded to the carbene carbon (eq 35).[119-120]

$$\underline{trans}-Cl(PR_3)_2Pt-C(S)X \xrightarrow{CH_3OSO_2F} [\underline{trans}-Cl(PR_3)_2Pt=C(SCH_3)X]^+$$

$$X = OCH_3, SC_2H_5, N(CH_3)_2 \quad (35)$$

Like a coordinated acyl group, the nitrogen atom of an isoelectronic imidoyl ligand is susceptible to electrophilic attack.   Thus reversible protonation[121-2] as well as alkylation using dimethyl sulfate or methyl iodide[109,123-4] leads to aminocarbene complexes.   This procedure has also been employed to synthesize secondary   carbene complexes, of which only very few examples are known (eqs 36-7).

$$\text{Cp(RNC)Ni-C(=NR)R'} \xrightarrow{\text{HBF}_4} \text{[Cp(RNC)Ni=C(NHR)R']}^{+} \qquad (36)$$

$$\text{(CO)(PR}_3)_2\text{(RNC)(CH}_3\text{CO}_2)\text{Ru-C(=NR)H} \xrightarrow{\text{CH}_3\text{I}}$$
$$\{\text{(CO)(PR}_3)_2\text{(RNC)(CH}_3\text{CO}_2)\text{Ru=C[N(CH}_3)\,\text{R]H}\}^{+}\,\text{I}^{-} \qquad (37)$$

Electrophilic addition is not confined to heteroatoms doubly bonded to a metal coordinated carbon atom.   Thus, very recently, vinyl conpounds of the nickel triad, which can be prepared from metal halides and vinyl lithium reagents, have been protonated to give cationic methylcarbene complexes (eq 38).[125]

$$\underline{\text{trans-}}\text{Cl}_5\text{C}_6[\text{P(CH}_3)_2\text{C}_6\text{H}_5]_2\text{M-C(=CH}_2)\text{OCH}_3 \xrightarrow{\text{H}^+}$$
$$\{\underline{\text{trans-}}\text{Cl}_5\text{C}_6[\text{P(CH}_3)_2\text{C}_6\text{H}_5]_2\text{M=C(CH}_3)\text{OCH}_3\}^{+} \qquad (38)$$
M = Ni, Pd, Pt

2.5   From Alkynes or Acetylide Ligands

A route   to   alkoxycarbene   complexes,   especially those of platinum and nickel, has been described starting from metal(II) compounds by reacting with terminal alkynes and alcohols in the presence of silver salts.   If ω-hydroxyalkynes are used, cyclic carbene ligands are obtained (eqs 39-40).[126-9]

$$\text{trans-(CH}_3\text{)[P(CH}_3\text{)}_2\text{C}_6\text{H}_5\text{]}_2\text{PtCl} + \text{H-C}\equiv\text{C-R} + \text{CH}_3\text{OH} \xrightarrow[\text{-AgCl}]{\text{AgPF}_6} \quad (39)$$

$$\text{\{ trans-(CH}_3\text{)[P(CH}_3\text{)}_2\text{C}_6\text{H}_5\text{]}_2\text{Pt=C(CH}_2\text{R)OCH}_3\}^+\text{PF}_6^-$$

$$\text{trans-(CH}_3\text{)}_2\text{(CF}_3\text{)[P(CH}_3\text{)}_2\text{C}_6\text{H}_5\text{]}_2\text{PtI} + \text{H-C=C-(CH}_2\text{)}_2\text{OH} \xrightarrow[\text{-AgI}]{\text{AgPF}_6} \quad (40)$$

$$\text{\{trans-(CH}_3\text{)}_2\text{(CF}_3\text{)[P(CH}_3\text{)}_2\text{C}_6\text{H}_5\text{]}_2\text{Pt=C(CH}_2\text{)}_3\text{O\}}^+ \text{PF}_6^-$$

A proposed reaction mechanism involving vinyl ether complex intermediates could not be verified and had to be reformulated later in terms of vinylidene complex intermediates.

Recently this route has been employed to synthesize neutral alkoxy(alkyl)carbene compounds from halogen-bridged binuclear platinum complexes (eq 41).[130]

$$\text{L(X)Pt(}\mu\text{-X)}_2\text{(X)L} + 2 \text{ H-C}\equiv\text{C-R} + 2 \text{ R'OH} \longrightarrow 2 \text{ LX}_2\text{Pt=C(CH}_2\text{R)OR'} \quad (41)$$

Similar carbene complexes can be obtained from acetylide complexes when treated with alcohol in the presence of acid (eq 42).[131-3]

$$\text{trans-(Cl}_5\text{C}_6\text{)[P(CH}_3\text{)}_2\text{C}_6\text{H}_5\text{]}_2\text{Ni-C}\equiv\text{CH} + \text{ROH} \xrightarrow{\text{HClO}_4} \quad (42)$$

$$\text{[trans-(Cl}_5\text{C}_6\text{)[P(CH}_3\text{)}_2\text{C}_6\text{H}_5\text{]}_2\text{Ni=C(CH}_3\text{)OR]}^+\text{ClO}_4^-$$

If good nucleophiles are present, no carbene complex is isolated; instead conversion to a neutral acyl complex takes place.

2.6 Scission of Electron-Rich Alkenes

Lappert and coworkers discovered that cyclic tetraaminoalkenes are suitable carbene precursors when reacted with transition metal complexes susceptible to nucleophilic attack. Thus, cleavage of bridged binuclear complexes or ligand substitution reactions of mononuclear compounds have been used to synthesize carbene complexes containing nearly all of the VIA, VIIA and VIII metals. The first reported example is shown in eq 43.[134]

$$[Cl_2(PR_3)Pt]_2 + R'\overline{N(CH_2)_2(R')NC=CN(R')CH_2)_2}NR' \longrightarrow \quad (43)$$

$$2 \ \underline{trans}-Cl_2(PR_3)Pt=\overline{CN(R')(CH_2)_2}NR'$$

In substitution reactions, ligands like carbon monoxide, phosphines, isocyanides, halides or cyclic alkenes may be used as leaving groups. By this method mono-, bis-, tris-, and tetrakis-carbene complexes are accessible (eq 44).[65,134-9]

$$Cl_2(PR_3)_3Ru + CH_3\overline{N(CH_2)_2(CH_3)NC=CN(CH_3)(CH_2)_2}NCH_3 \longrightarrow (44)$$

$$\underline{trans}-Cl_2Ru[=\overline{CN(CH_3)(CH_2)_2}NCH_3]_4 + \ldots$$

With N-phenylated olefins, orthometallation is also observed (eq 45).[140]

$$Cl_2L_3Ru + C_6H_5\overline{N(CH_2)_2(C_6H_5)NC=CN(C_6H_5)(CH_2)_2}NC_6H_5 \longrightarrow (45)$$

$$IX + \ldots \quad (L = PR_3)$$

N,S-olefins affording benzothiazolylidene fragments have also been employed. The reactivity of the alkene has been shown to depend both on the heteroatom and its substituents and on the ring size. It decreases in the order **X > XI > XII.**

The participation of free carbenes could not be corroborated. A plausible mechanistic scheme involves a ligand substitution by the alkene which, originally coordinated via the heteroatom, rearranges under fragmentation to give the metal carbene complex.[65]

Two heteroatom carbene substituents stabilize the metal carbene unit. As a consequence reactions of these complexes are generally confined to the metal ligand framework.

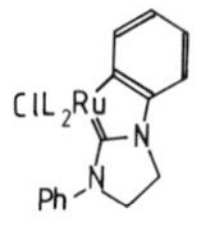

IX 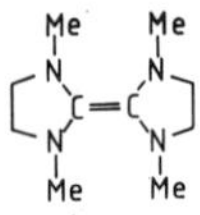   X 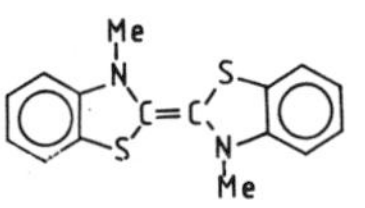   XI 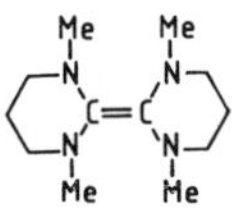   XII

2.7 From Other Carbene Precursors

The well-known fact that diazo compounds are able to act as carbene sources has been used for the synthesis of metal coordinated carbenes from metal complexes susceptible to ligand dissociation. By this route several manganese complexes have been prepared (eq 46).[141-4]

$$Mecp(CO)_2Mn(THF) + N_2\text{-}CRR' \xrightarrow[-N_2 \quad -THF]{} Mecp(CO)_2Mn\text{=}CRR' \qquad (46)$$

$R = C_6H_5, p\text{-}NO_2\text{-}C_6H_4, R' = C_6H_5, C(O)C_6H_5, CO_2C_2H_5$

In a similar way a dithiocarbene derivative accessible from the base-induced decomposition of $(C_6H_5S)_3CH$ has been trapped by pentacarbonyl chromium and tungsten complexes (eq 47).[145]

$$(CO)_5M(THF) + :C(SC_6H_5)_2 \longrightarrow (CO)_5M\text{=}C(SC_6H_5)_2 \qquad (47)$$

$M = Cr, W \qquad\qquad -THF$

Iron carbonyl complexes may be used in the desulfuration reactions of thiocarbonyl compounds to carbene complexes (eq 48).[146-7] By this route the parent diaminocarbene ligand could be coordinated to a metal center.

$$Fe_2(CO)_9 + S\text{=}C(NH_2)_2 \longrightarrow (CO)_4Fe\text{=}C(NH_2)_2 + \ldots \qquad (48)$$

Another approach to the metal carbene structure is the abstraction of substituents bonded to the α-carbon atom of a metal alkyl group, especially in the dicarbonyl(cyclopentadienyl)iron system. Both alkoxide and hydride abstraction have been performed by triphenylmethyl salts (eqs 49-50).[148-9]

$$Cp(CO)_2Fe\text{-}CH(C_6H_5)OCH_3 \xrightarrow{Ph_3C^+} [Cp(CO)_2Fe\text{=}C(H)C_6H_5]^+ +\ldots(49)$$

$$Cp(CO)_2Fe\text{-}CH_2OCH_3 \xrightarrow{Ph_3C^+} [Cp(CO)_2Fe\text{=}C(H)OCH_3]^+ + \ldots \qquad (50)$$

Another type of hydrogen abstraction is observed in a trisneopentyl carbyne complex of tungsten. When reacted with trimethylphosphine at very low temperature a unique compound containing alkyl, carbene and carbyne ligands is obtained (eq 51).[150]

$$[(CH_3)_3CH_2]_3W\text{=}C\text{-}C(CH_3)_3 \xrightarrow{(CH_3)_3P}$$

$$[(CH_3)_3P]_2[(CH_3)_3CH_2]W[\text{=}C(H)C(CH_3)_3][\equiv C\text{-}C(CH_3)_3] + \qquad (51)$$

Dichlorocarbene, which is readily accessible by base-induced β-dehydrochlorination of chloroform, was successfuly stablized in a transition metal complex from 1977, when tetraphenylporphyrin iron(II) compounds were allowed to react with polychlorinated reagents in a reducing medium. This reaction could be generalized to provide a route to various iron carbene complexes (eq 52).[151-3]

$$\text{TPPFe(II)} + \text{Cl}_2\text{CRR'} \xrightarrow[-2\ \text{Cl}^-]{+2e^-} \text{TPPFe=C(R)R'} \qquad (52)$$

TPP = tetraphenylporphyrin

R = Cl, CN, $CO_2C_2H_5$, $SCH_2C_6H_5$, alkyl

R' = Cl, alkyl

Quite recently vinylidene complexes have been used as precursors for carbene complexes. Addition of nucleophiles, eg. OCH_3^-, NH_2^-, to a vinylidene manganese compound followed by acidification led to the isolation of methoxy- and aminocarbene complexes (eq 53).[154]

$$\text{Cp(CO)}_2\text{Mn=C=C(H)CO}_2\text{R} \xrightarrow[2)\ H^+]{1)\ X^-} \text{Cp(CO)}_2\text{Mn=C(CH}_2\text{CO}_2\text{R)X} \quad (53)$$

X = NH_2, OCH_3

Oxidative addition has proved to be a valuable principle in the coordination of carbene ligands. The addition of immonium halides to a wide variety of both anionic and neutral metal compounds provides an interesting route to halogenocarbene and secondary carbene complexes (eqs 54-5).[14,155-9]

$$[(CO)_5Cr]^{2-} + [(CH_3)_2NCCl_2]^+ \xrightarrow[-Cl^-]{} (CO)_5Cr=C(Cl)N(CH_3)_2 \qquad (54)$$

$$[(CO)_4Fe]^{2-} + [(CH_3)_2NCHCl]^+ \xrightarrow[-Cl^-]{} (CO)_4Fe=C(H)N(CH_3)_2 \qquad (55)$$

A similar reaction type has been used to prepare cyclic diaminocarbene complexes from imidazolium salts and carbonyl hydrido metal anions (eq 56). As early as 1968 the first carbene complex not stabilized by heteroatom carbene substituents was synthesized in a similar way (eq 57).[160-4] The surprising stability of cyclopropenylidene complexes has been referred to as the aromatic character of the carbene ligand.

The synthesis of cycloheptatrienylidene complexes has been established by reaction of cycloheptatrienyllithium with halogenometal compounds followed by hydride abstraction (eq 58).[165]

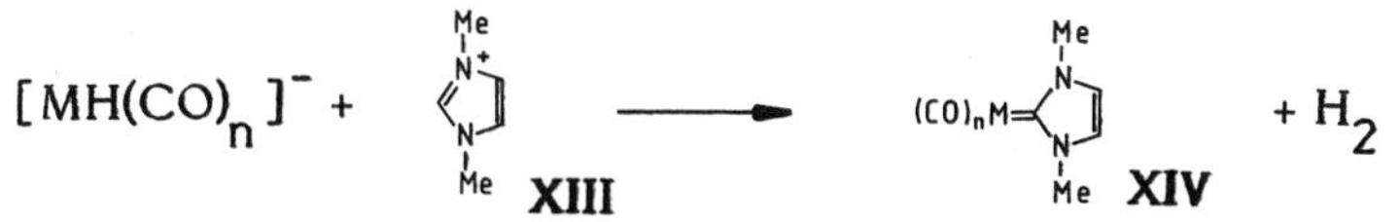

$$[MH(CO)_n]^- + \text{XIII} \longrightarrow \text{XIV} + H_2 \tag{56}$$

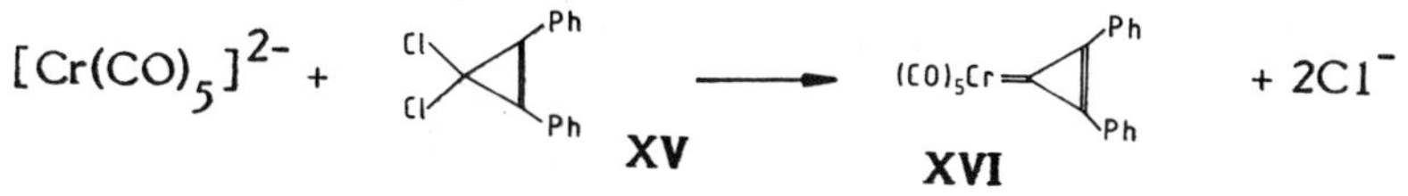

$$[Cr(CO)_5]^{2-} + \text{XV} \longrightarrow \text{XVI} + 2Cl^- \tag{57}$$

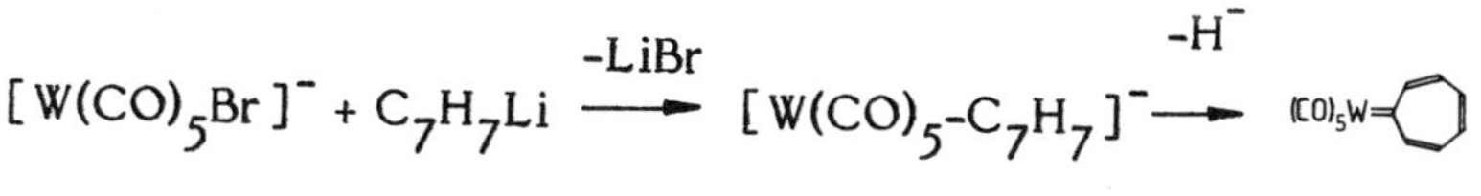

$$[W(CO)_5Br]^- + C_7H_7Li \xrightarrow{-LiBr} [W(CO)_5-C_7H_7]^- \xrightarrow{-H^-} (CO)_5W=\text{XVII} \tag{58}$$

2.8   Modification of Carbene Ligands

The variety of carbene complex syntheses includes also the modification of carbene ligands. This route has been used in the Fischer-type series to some extent, since the carbene carbon atom has proved to be an electrophilic center ready for nucleophilic attack (Section 4.). Experimental procedures have been elaborated to modify methoxycarbene ligands to yield amino-, thio-, seleno- or phenoxycarbene ligands. Analogously, using phenyllithium as a nucleophile gave the first diphenylcarbene complex.[166] A similar first step is found to occur in the insertion of aminoacetylenes into the metal-alkoxycarbene bond to yield β-alkoxyalkenyl(amino)carbene compounds.[167]

Another interesting feature is the remarkable acidity of hydrogen atoms attached to the α-carbon in carbene ligands.[168] This property may be used for the modification, prolongation or functionalization of the carbene side chain (Section 6.).

3.   STRUCTURE AND BONDING

A metal coordinated carbene carbon atom is best regarded as an sp^2-hybridized carbon atom with a vacant p_z orbital, which is able to accept electron density from both the carbene substituents and the metal by competing π-overlap.[5,12,15,169]   These features are supported by a lot of x-ray structural data by which a trigonal planar environment of the carbene carbon is confirmed.

Some   insight   may   be   provided   by   pentacarbonyl[methoxy-(phenyl)carbene]chromium (**XVIII**), the first carbene complex for which an x-ray structure was established.[170]   The bonds involving the carbene carbon atom are coplanar and bond angles were found to be 122, 104 and 134°.   The phenyl substituent is approximately perpendicular to the carbene plane, and its distance to the carbene carbon corresponds to a bond between two sp^2-hybridized carbon atoms.   The C(carbene)-O bond, however, is considerably shortened (133 pm) and is even shorter than corresponding bond lengths in esters.   On the other hand, the metal carbene distance (204 pm) represents a bond order somewhat greater than 1, and thus is significantly longer than the average chromium-carbonyl distances.   This has led to the conclusion that carbene ligands are better donors and weaker acceptors than carbon monoxide.   The same results could be obtained from infrared spectroscopic studies.

The competition of the groups bonded to the carbene carbon for its positive charge can be verified for the related π-arene- and aminocarbene complexes **XIX** and **XX**. Because the electron donor ability of the metal is enhanced by a partial replacement of the good acceptor ligands CO, the metal carbene bond is considerably shortened (to 194 pm).[169]   On the other hand replacement of the methoxy group by amino substituents leads to a lengthening of the chromium carbene distance (215 pm).[171]

$$(CO)_5Cr=C(C_6H_5)OCH_3$$

XVIII

$$\pi\text{-}C_6H_6(CO)_2Cr=C(C_6H_5)OCH_3 \qquad (CO)_5Cr=C(CH_3)N(C_2H_5)_2$$

XIX   **XX**

Similar characteristics have been found for platinum carbene complexes. If two heteroatom substituents are able to build up considerable π-bonding to the carbene carbon, the platinum carbene distance is very close to the expected single bond length.[134] In square planar complexes,[92] though less in octahedral pentacarbonyl carbene complexes,[172] the carbene ligands shows a <u>trans</u>-influence comparable to that of a tertiary phosphine. In a valence bond description the bonding may be represented by canonical forms the uncharged of which has been used for clarity in this chapter although, in hetero-substituted carbene ligands, **XXIc** and **XXId** will become the most important contributions. In this formalism alkoxy-, amino-and thiocarbene complexes may be regarded as related to carboxylic esters, amides and thioesters.

XXIa   **XXIb**   **XXIc**   **XXId**

The participation of π-bonding in the stabilization of the carbene carbon atom is reflected in temperature dependent nmr spectroscopy. In the pentacarbonylchromium series the barrier to rotation about the C(carbene)-heterosubstituent bond decreases from amino to thio and alkoxy compounds. For the dimethylamino(methyl)carbene compounds **XXII** even at +200°C no rotation could be observed, indicating an activation energy of more than 100 kJ mole^{-1}.[10,173] The rotational barriers of the thiomethyl(methyl)carbene and methoxy(methyl)carbene complexes **XXIII** and **XXIV** have been determined to be 63-71 kJ mole^{-1} and 57 kJ mole^{-1} respectively.[10,174]

$$(CO)_5Cr=C(CH_3)N(CH_3)_2 \quad (CO)_5Cr=C(CH_3)SCH_3 \quad (CO)_5Cr=C(CH_3)OCH_3$$

XXII   **XXIII**   **XXIV**

The rotation about the metal-carbene bond is very rapid because this process does not give a variation in net p-d overlap. Thus rotational isomers can only be detected by nmr techniques if rotation slows down as a consequence of steric hindrance. An example is the <u>cis</u>-biscarbene complex **XXV**, the rotational barrier of which was found to be 39 kJ mole^{-1}.[10,175] π-Arene(carbonyl)carbene complexes like **XXVI** and **XXVII**, however, were shown to exist as pairs of rotational isomers on the infrared time scale.[176]

$$\pi\text{-}C_6H_{6-m}(CH_3)_m(CO)_2Cr=C(C_6H_5)OCH_3 \qquad \pi\text{-}C_5H_5(CO)_2Mn=C(C_6H_5)OCH_3$$
$$m = 0\text{-}3$$

XXVI   **XXVII**

In spite of the longer metal-carbene distance established by x-ray studies, the carbene ligand is more tightly attached to the metal than the CO groups. This is illustrated in substitution reactions with VB donors which give mainly <u>cis</u>-carbene complexes (eq 59).[22,69] Similarly, the mass spectra of carbonyl carbene compounds are dominated by the primary elimination of the CO ligands.[177]

$$(CO)_5M=C(CH_3)OCH_3 + XR_3 \xrightarrow[-CO]{} \underline{cis}\text{-}(CO)_4(XR_3)M=C(CH_3)OCH_3 \quad (59)$$

M = Cr, W;

X = P, As, Sb

The participation of resonance form **XXIb**, which exhibits a positive charge on the carbene carbon atom, is reflected by its marked ^{13}C-nmr deshielding.[178-80] For example, the chemical shifts of the carbene carbon in pentacarbonyl[methoxy(phenyl)carbene]chromium **XVIII** and pentacarbonyl(diphenylcarbene)chromium occur at 351.4 and 399.4 ppm downfield from tetramethylsilane[29,178] and exceed in part the shifts known for related carbonium ions. Although at present the enormous downfield shifts are not fully understood, they are useful in the identification of carbene complexes.

XXV   cis-(CO)₄Cr(=C(N-Me)(N-Me))₂

4.   REACTIONS WITH NUCLEOPHILES

ESCA measurements[181] and Mulliken population analyses[182-3] have led to the conclusion that in pentacarbonylcarbene complexes the carbonyl carbon atoms carry a greater positive charge than the carbene carbon. Molecular orbital calculations have explained nucleophilic attack in terms of frontier orbital control.[182]   Thus the lowest unoccupied MO in pentacarbonyl[methoxy(methyl)carbene]chromium was found to be predominantly localized on the carbene carbon atom. It is moreover a well established experimental fact that the carbene carbon atom is highly susceptible to nucleophilic attack.

4.1   Formation of Ylid Complexes

Pentacarbonyl[methoxy(phenyl)carbene]chromium was found to react with dimethylphosphine at -60°C to give a 1:1 ylid type addition product.[184] Later, this type of reaction was extended both to related complexes of chromium and tungsten and to a series of secondary and tertiary phosphines (eq 60).[59,61,185-7] It was shown by a thermodynamic study that the adduct formation is a reversible equilibrium reaction the association-dissociation constants of which depend on the phosphine and the metal as well as on the temperature and the solvent used.[186] In the cases of triaryl-, mixed arylalkylphosphines and tertiary phosphines containing C-branched substituents dissociation predominates and the addition products could be identified only by spectroscopic methods. If however, trialkyl phosphines are used and steric crowding is avoided, the ylid complexes can be isolated. For example, at +10°C the equilibrium constant for tri-n-butylphosphine exceeds that for diethylphenylphosphine by more than two orders of magnitude.[186]   The ylid complex formation may be followed by a marked decrease in the carbonyl ligand C-O stretching frequencies and an extraordinary shielding of approximately 250 ppm for the metal coordinated carbon atom in ^{13}C-nmr spectroscopy.[185]

$$(CO)_5M{=}C(R^1)R^2 + PR_3^3 \rightleftharpoons (CO)_5M{-}C(R_1)(R_2)PR_3 \qquad (60)$$

$$M = Cr, W \qquad\qquad R^1 = CH_3, C_6H_5, p{-}CH_3C_6H_4, furyl, thienyl$$

R^2 = C_6H_5, furyl, thienyl, OCH_3, OC_6H_5, $OSi(CH_3)_3$, $OCOCH_3$, SCH_3,

$SeCH_3$

R_3^3 = $(CH_3)_3$, $(C_2H_5)_3$, $(i-C_3H_7)_3$, $(C_4H_9)_3$, $(CH_3)_2H$, $(C_2H_5)_2C_6H_5$

Whereas ylid complexes could be prepared from oxy-, thio- and selenocarbene precursors,[185,188] aminocarbene complexes failed to react due to the reduced electrophilicity of the aminocarbene carbon atom. Similarly, no ylid complex formation could be observed on reacting pentacarbonyl[methoxy(methyl)carbene]chromium **XXIV** with the less basic tertiary arsines, stibines and bismuthines.[69]

Closely related nitrogen-ylid complexes have been obtained using rigid tertiary amines. Pentacarbonyl[methoxy(phenyl)carbene] complexes of chromium and tungsten react with 1-azabicyclo[2.2.2]octane and 1,4-diazabicyclo[2.2.2]octane at ambient temperature to give 1:1 adducts (eq 61).[189-90] The C-O stretching frequencies of the pentacarbonyl metal fragment fall even more than in the corresponding phosphorous ylid complexes.

$$(CO)_5M=C(C_6H_5)OCH_3 + NR_3 \longrightarrow (CO)_5MC(Ph)(OCH_3)-NR_3 \qquad (61)$$

$$M = Cr, W; NR_3 = N(CH_2CH_2)_3N, N(CH_2CH_2)_3CH$$

On treatment with aqueous $KHSO_4$ the amine addition is reversed. In contrast to phosphorous ylid complexes, however, instead of the original carbene compounds pentacarbonyl[hydroxy(phenyl)carbene] complexes are formed (eq 62).

$$\overset{KHSO_4}{(CO)_5MC(Ph)(OCH_3)-NR_3 \longrightarrow (CO)_5M=C(Ph)OH + NR_3} \qquad (62)$$

Ylid formation has been used to trap carbene complexes which are unstable under the reaction conditions. Thus by addition of triphenyl- or tri-n-butylphosphine the existence of a cationic rhenium methylene complex and a tungsten phenylcarbene compound could by verified (eq 30).[112,191-2] A similar adduct formation distinguished between a benzocyclobutylidene and an alternative π-benzocyclobutene complex structure (eq 63).[193]

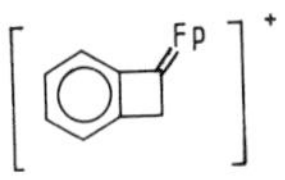

XXVIII

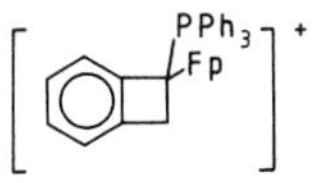

XXIX

$$\text{XXVIII} + P(C_6H_5)_3 \longrightarrow \text{XXIX} + \dots \qquad (63)$$

$Fp = CpFe(CO)_2$

4.2   Aminolysis and Related Reactions

Alkoxy substituents in carbenes are excellent leaving groups. Thus alkoxycarbene complexes, which are readily accessible both from carbonyl ligands and from 1-alkynes and alcohols, may serve as suitable precursors for the introduction of nucleophilic heteroatoms at the electrophilic carbene carbon atom.   Most of this work has been done using nitrogen nucleophiles. Ammonia, primary and less crowded secondary amines were soon found to give aminocarbene complexes in excellent yields (eq 64).[194-201]

$$L(CO)_4 M{=}C(R)OCH_3 + R^1R^2NH \xrightarrow[-CH_3OH]{} L(CO)_4 M{=}C(R)NR^1R^2 \qquad (64)$$

XXX   XXXI

$M = Cr, W; R = CH_3, C_6H_5, C_6H_4X \qquad L = CO, PR_3$

In general, primary amines lead to a mixture of E/Z-isomers (with respect to the restricted rotation about the C(carbene)-N bond). ^{1}H-nmr studies have indicated that the E-isomer predominates.   Due to its negligible steric requirements, however, methylamine yields exclusively the E-compound, the structure of which has been established by x-ray crystallography.[200] Strong bases effect isomerization which is assumed to proceed via abstraction of the amino proton.[173]

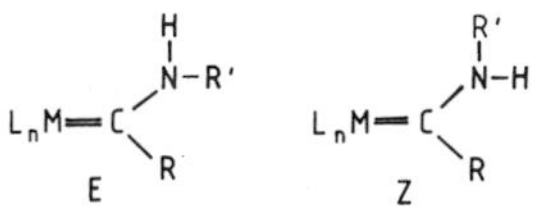

$L_nM{=}C$ with N-R' and H, R (E) $L_nM{=}C$ with N-H and R', R (Z)

With secondary amines, steric factors become more important. Cyclic amines e.g. pyrrolidine or piperidine give the expected complexes containing cyclic carbene substituents.[198,201]   Crowded compounds such as diisopropylamine undergo alkene elimination to give the mono-alkylaminocarbene complexes (eq 65).[202]

$$(CO)_5Cr=C(CH_3)OCH_3 + (i-C_3H_7)_2NH \longrightarrow (CO)_5Cr=C(CH_3)NH(i-C_3H_7)$$

$$+ CH_3OH + i-C_3H_6 \qquad (65)$$

Bridged dinuclear aminocarbene complexes are obtained by reaction with aliphatic diamines,[203] whereas aromatic diamines e.g. benzidine lead only to the mononuclear carbene complex. This may be explained by the reduced basicity of aromatic amines, especially if one amino group is already bonded to the carbene carbon atom. The same reason explains the failure to observe aminocarbene formation from secondary aromatic amines and carboxylic amides.[195]

The kinetics of the aminolysis reaction indicate a mechanistic relationship to the aminolysis of carboxylic esters.[204-5] In inert solvents such as hexane the reaction is first order in carbene complex and third order in amine. In proton-active solvents such as dioxan, the amine order is reduced to two. Comparing eq 64, the rate is given by the expression

$$\frac{d[\textbf{XXXI}]}{dt} = k \cdot [amine][HX][Y][\textbf{XXX}]$$

where HX represents a proton-donating and Y a proton accepting reagent. A mechanism has been proposed which involves initial addition of the proton donor to the methoxy substituent thus increasing the electrophilicity of the carbene carbon atom. This is then attacked by the amine, the nucleophilicity of which has been enhanced by hydrogen bonding to the proton acceptor Y (amine or dioxan). The existence of hydrogen bonding has been verified by nmr spectroscopy. The mechanism accounts for the observed fall in reaction rate with increasing temperature.

An interesting competition between aminolysis and conjugate addition reactions has been observed with the alkynylcarbene complex **XXXII** (eqs 66-7).[28,206] On addition of dimethylamine at very low temperature (-115°C) only aminolysis takes place, while at -20°C the conjugate addition product is isolated. At ambient temperature both processes occur to yield the amino(aminoalkenyl)carbene complex, which is also accessible from the aminolysis product.   In contrast, the aminoalkenylcarbene compound remains unchanged on addition of excess amine.   The homologous tungsten complex shows similar behaviour.

$$(CO)_5Cr=C(C\equiv CPh)OEt + Me_2NH \xrightarrow[-EtOH]{-115^\circ C} (CO)_5Cr=C(C\equiv CPh)NMe_2 \quad (66)$$

XXXII   **XXXIII**

$$\xrightarrow{+ Me_2NH} (CO)_5Cr=C[CH=CPh(NMe_2)]NMe$$

XXXIV

$$(CO)_5Cr=C(C\equiv CPh)OEt + Me_2NH \xrightarrow{-20^\circ C} (CO)_5Cr=C[CH=CPh(NMe_2)]OEt$$

XXXV   (67)

The negative activation energy found for aminolysis may be held responsible for these unusual results.   With decreasing temperature the aminolysis reaction rate increases and aminocarbene complex formation predominates.   For the activation enthalpy parameter of the conjugate addition reaction, however, a positive sign is to be expected.   In these terms, increasing temperature favours the addition of amine across the $C\equiv C$ triple bond.   This conclusion is consistent with a mixture of **XXXIII** and **XXXV** obtained at $-78^\circ C$.   Due to the aminoalk·   electron deficiency in **XXXV** seems to be drastically   minolysis becomes impossible.

The aminolysis reaction has been used to synthesize optically active carbene complexes by means of chiral amine components (eq 68).[207-9] Thus pentacarbonyl[phenyl-(R)-α-phenylethylaminocarbene]chromium (with an E:Z ratio of 3:1 due to the C(carbene)-N bond) is obtained in 93% yield from the methoxycarbene complex **XVIII** and R-(+)-α-phenylethylamine. Aminolysis proceeds under mild reaction conditions.   Thus metal carbene moieties   can   be   useful   as   amino   protecting   groups   in peptide synthesis [210-1] (Section 12).

$$(CO)_5Cr=C(R)OCH_3 + R\text{-}(+)\text{-}\alpha\text{-}C_6H_5(CH_3)C^*HNH_2 \xrightarrow[-CH_3OH]{} \quad (68)$$

$$(CO)_5Cr=C(R)NHC^*H(CH_3)C_6H_5$$

If rotation is restricted about the aryl-carbene bond as well as the carbon-nitrogen axis, additional atropisomerism is to be expected. This phenomenon is well-known in biphenyl derivatives and could be demonstrated in the 1-naphthylcarbene complex as well. No interconversion of the atropisomers was observed up to 80°C. Separation of the atropisomers was effected by repeated crystallization. The analogous 2-naphthylcarbene compound, however, shows free rotation about the naphthylcarbene carbon bond, even down to -90°C, as indicated by its ^{1}H-nmr spectrum. There are separate signals for the E/Z-isomers, but no further splitting of signals occurs. Similarily atropisomerism was ascertained in pentacarbonyl[diethylamino(styryl)carbene] complexes **XXXVI** of chromium and tungsten.[212]

$$(CO)_5M=C\underset{Me}{\overset{NEt_2}{\diagup}}\,\underset{}{C}=C\underset{R}{\overset{Ph}{\diagup}}$$

XXXVI

Aminolysis is not restricted to Fischer-type compounds and has also been observed with other types of carbene complexes, although competing pathways occur, as in the platinum complex series. Whereas the cationic methyl compound **XXXVII** yields on addition of methylamine the expected aminocarbene complex (in which, in contrast to the chromium complexes, the Z-configuration prevails), with the closely related chloro compound **XXXVIII** formation of the neutral acetyl complex takes place (eqs 69-70).[213] The latter reaction seems to be favoured by bulky amines.

$$[\underline{trans} - L_2(CH_3)Pt=C(CH_3)OCH_3]^+ + CH_3NH_2 \xrightarrow[-CH_3OH]{} \quad (69)$$

XXXVII

$$[\underline{trans} - L_2(CH_3)Pt=C(CH_3)NHCH_3]^+$$

$$[\underline{trans} - L_2(Cl)Pt=C(CH_3)OCH_3]^+ \xrightarrow{RNH_2} \underline{trans} - L_2(Cl)Pt\text{-}COCH_3 + \ldots$$

XXXVIII   (70)

Cyclic oxycarbene ligands undergo ring-opening in aminolysis reactions (eq 71).[213]

$$[\underline{trans} - L_2(CH_3)Pt=\overline{C(CH_2)_3O}]^+ + (CH_3)_2NH \longrightarrow$$

$$\underline{trans} - L_2(CH_3)Pt=C[(CH_2)_3OH]N(CH_3)_2 \qquad (71)$$

An interesting reaction, which might become important in the synthesis of isotope-labeled isocyanides from carbon tetrachloride and amines, has been reported for a porphyrin iron carbene complex. Addition of a primary amine modifies a dichlorocarbene ligand to give a coordinated isocyanide, which can be released from the metal on heating in excellent yields (eqs 52,72).[214-5]

$$TPPFe=CCl_2 + RNH_2 \xrightarrow[-2 HCl]{} TPP(RNH_2)Fe-CNR \qquad (72)$$

The formation of cyanide ligands from coordinated carbenes may be achieved by addition of hydrazines to alkoxycarbene complexes (eq 73).[216] Reacting an acetoxycarbene complex with hydrazoic acid leads to a similar result.[57] Hydrazinocarbene complexes are supposed to be intermediates which rearrange under cleavage of the nitrogen-nitrogen bond. In another reaction a hydrazone derivative $C_6H_5CH=NNHC_6H_5$ has been obtained from pentacarbonyl[ethoxy(phenyl)carbene]chromium and phenylhydrazine.[217] Later stable palladium hydrazinocarbene complexes were isolated (eq 74).[218]

$$(CO)_5Cr=C(CH_3)OCH_3 + H_2NN(CH_3)_2 \rightarrow (CO)_5CrNCCH_3 + (CH_3)_2NH + CH_3OH$$

XXIV   (73)

$$Cl_2(RNC)Pd=C(NHR)OR' + H_2NNHC_6H_5 \xrightarrow{-HOR'}$$

XXXIX $Cl_2(RNC)Pd=C(NHR)NHNHC_6H_5$ (74)

Other nitrogen nucleophiles such as benzaldoxime, benzophenonimine or 1-aminoethanol react with alkoxycarbene complexes to give alkylideneaminocarbene complexes (eq 75).[219-20] The expected product, similar to aminolysis, was also obtained from the palladium complex **XXXIX** and benzylidenehydrazone,[218] whereas ketoximes and hydroxylamine were found to give carbene substitution or addition-rearrangement reactions (Section 7.[216,221])

$$(CO)_5Cr=C(R)OCH_3 + HN=C(C_6H_5)_2 \rightleftharpoons (CO)_5Cr=C(R)N=C(C_6H_5)_2 + CH_3OH$$

$$(75)$$

Group VIB nucleophiles too can add to the carbene carbon atom. For instance, the alkoxycarbene substituent may be replaced by alkoxide[132] or enolate anions.[222] If we include more reactive acyloxycarbene complexes[58,187] carbene ligands can be modified in a great variety (eqs 76-8); for the deuteration of the methyl group in eq 76, see 6.2 below. Silyloxycarbene ligands were found to provide two competing electrophilic centers for nucleophilic attack. Whereas trimethylphosphine adds to the carbene carbon leading to an ylid complex, the silicium center is attacked by nucleophiles such as alkoxides, alcohols, dimethylamine and phenyllithium (eq 79).[61]

$$(CO)_5Cr=C(CH_3)OCH_3 \xrightarrow[\text{,OD}]{\text{NaOCD}_3} (CO)_5Cr=C(CD_3)OCD_3 \qquad (76)$$

$$(CO)_5W=C \qquad \xrightarrow[\text{2)} \quad \text{H}^+]{\text{-}_2C=C(C_6H_5)O^-} (CO)_5W=C(C_6H_5)OC(C_6H_5)=CR_2 \qquad (77)$$

$$(CO)_5Cr=C(C_6H_5)OCOCH_3 \xrightarrow{\text{NaOC}_6H_5} (CO)_5Cr=C(C_6H_5)OC_6H_5 \qquad (78)$$

$$(CO)_5W=C(R)OSi(CH_3)_3 + (CH_3)_2NH \rightarrow [(CH_3)_2NH_2]^+ \ [(CO)_5W=C(R)O]^- + ...$$

$$(79)$$

Thiols or thiolates with subsequent acidification and methylselenol react in a similar way to give thio- and selenocarbene complexes (eqs 80-1).[188,223-4] Phenylselenol, however, which is known to be more acidic, yields a rearranged complex containing a selenium-metal bond (cf Section 7.1).[225]

$$(CO)_5M=C(R)OCH_3 + R'SH \longrightarrow (CO)_5M=C(R)SR' + CH_3OH \qquad (80)$$

$$M = Cr, W; \quad R = CH_3, C_6H_5$$

$$R' = alkyl, aryl$$

$$(CO)_5Cr=C(C_6H_5)OCH_3 + CH_3SeH \longrightarrow (CO)_5Cr=C(C_6H_5)SeCH_3 + CH_3OH$$
$$(81)$$

4.3   Addition of Carbon Nucleophiles

The addition of carbon nucleophiles to the coordinated carbene carbon atom leads to the formation of a new carbon-carbon bond. Thus this type of reaction is very useful in organic synthesis via carbene complexes.

Vinyl ethers, diazoalkanes and phosphorous ylids are believed to attack the coordinated carbene carbon atom. Due to the thermal instability of the addition products fragmentation occurs leading to alkenes (eqs 82-3).[226-9] The reaction involving vinyl ethers has attracted much attention as a model for olefin metathesis. Phosphorous ylids have been found to convert alkoxy(aryl)carbene ligands into enol ethers in high yields. However as a consequence of steric hindrance and stabilization by the carbonyl function no reaction was observed with dimethylmethylenephosphorane and carbonyl ylids. Alkoxy(alkyl)carbene complexes undergo a competing α-hydrogen abstraction by Wittig reagents to produce the conjugate base (eq 84); in this case the less basic diazoalkanes may be used in the synthesis of alkyl enol ethers.

$$(CO)_5Cr=C(C_6H_5)OCH_3 + CH_2=CHOR \longrightarrow CH_2=C(C_6H_5)OCH_3 + \ldots \qquad (82)$$

$$(CO)_5W=C(C_6H_5)OCH_3 + CH_2=PR_3 \longrightarrow CH_2=C(C_6H_5)OCH_3 + (CO)_5WPR_3$$
$$(83)$$

$$(CO)_5W=C(CH_3)OCH_3 + CH_2=PR_3 \rightarrow [(CO)_5W-C(=CH_2)OCH_3]^- + \ldots \quad (84)$$

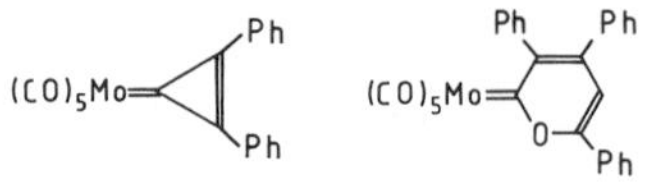

XL   **XLI**

In contrast to phosphorous ylids a carbonyl nitrogen ylid is reported to yield an isolable unique metal complex by nucleophilic attack at the carbene carbon of a cyclopropenylidene compound and ring expansion (eq 85).[230]

$$\textbf{XL} + C_6H_5C(O)CH.NC_5H_5 \longrightarrow \textbf{XLI} + \ldots \quad (85)$$

Adducts which are obtained from carbene complexes and organolithium reagents decompose at ambient temperature to give products characteristic of free radical involvement. Thus a persubstituted ethane derivative is isolated from the reaction of pentacarbonyl[methoxy(phenyl)carbene]chromium **XVIII** and phenyllithium (eq 86).[231] At low temperature, however, decomposition is avoided and the adducts can be isolated preferentially as bis(triphenylphosphine)iminium salts (eq 87).[166,232] Consistent with the transformation of the carbene carbon atom into a sp^3-hydridized metal coordinated carbon atom, a dramatic upfield shift of 268 ppm is observed for the phenyllithium adduct of **XVIII.** Similar metalates have been obtained from pentacarbonyl(chloro)tungstate and lithium organyls.[233]

$$(CO)_5Cr=C(C_6H_5)OCH_3 \xrightarrow{\ C_6H_5Li\ } CH_3O(C_6H_5)_2C-C(C_6H_5)_2OCH_3 + \ldots (86)$$

$$(CO)_5M=C(C_6H_5)OCH_3 \xrightarrow[\text{2) PPNCl}]{\text{1) RLi -78}^\circ\text{C}} [PPN]^+[(CO)_5M-C(C_6H_5)(R)OCH_3]^- + \ldots$$
$$(87)$$

$$M = Cr, W; \quad R = CH_3, C_6H_5; \quad [PPN]^+ = [(Ph_3P)_2N]^+$$

The metalate(-i) compounds can be used in the modification of carbene ligands. Treatment with acid at low temperature or simple chromatography on silica gel induces the elimination of methoxide to yield carbene complexes which are no longer stabilized by heteroatom substituents (eq 88).[29,166,229] By this route diarylcarbene complexes containing phenyl and five-membered heterocyclic carbene substituents have been isolated. Attempts to prepare alkyl(aryl)carbene compounds by addition of alkyllithium reagents failed. Instead, alkene products orginating from rearrangement or decomposition were isolated (eqs 89-90).[234-5]

$$(CO)_5M=C(R)OCH_3 \xrightarrow[\substack{2) SiO_2 \\ -LiOCH_3}]{1) R'Li} (CO)_5M=C(R)R' \qquad (88)$$

M = Cr, W; R = phenyl, furyl, thienyl; R' = aryl, furyl, thienyl

$$(CO)_5W=C(C_6H_5)OCH_3 \xrightarrow[\substack{2) SiO_2 \\ -LiOCH_3}]{1) RCH_2Li} (CO)_5W(\pi\text{-}RCH=CHC_6H_5) \qquad (89)$$

$$(CO)_5Cr=C(C_6H_5)OCH_3 \xrightarrow[\substack{2) HCl}]{1) CH_2=CHLi} C_6H_5(OCH_3)C=CHCH_3 +... \qquad (90)$$

Alkoxy substituents are the preferred leaving groups in carbene ligands. Nevertheless, substitution of chloride by cyanide has been achieved in a chloro(dimethyl)aminocarbene complex (eq 91).[79]

$$(CO)_5Cr=C(Cl)N(CH_3)_2 + KCN \rightarrow (CO)_5Cr=C(CN)N(CH_3)_2 + KCl \qquad (91)$$

Conjugate addition of phenyllithium or lithium diphenylcuprate to a styrylcarbene ligand was found to occur in low to moderate yields giving a 2,2-diphenylethylcarbene complex after acidification along with an alkene component arising from decomposition (eq 92).[235] Enolates derived from cyclopentanone or isobutyrophenone and dimethyl malonate have been added to alkenylcarbene ligands in a similar way to achieve a prolongation and functionalization of the carbene side chain.[222] With less bulky enolates such as the conjugate base of acetone decomposition products are found due to a primary attack of the nucleophile at the carbene carbon atom.

$$\text{(CO)}_5\text{Cr=C(CH=CHC}_6\text{H}_5\text{)OCH}_3 \xrightarrow[\text{2) HCl}]{\text{1) C}_6\text{H}_5\text{Li}} \text{(CO)}_5\text{Cr=C[CH}_2\text{CH(C}_6\text{H}_5\text{)}_2\text{]OCH}_3$$

$$+ \text{ C}_6\text{H}_5\text{(OCH}_3\text{)C=CHCH}_2\text{C}_6\text{H}_5 + \ldots \quad (92)$$

In aminocarbene complexes the electron deficiency of the carbene carbon atom is reduced and the resonance forms **XXIc** and **XXId** predominate. Thus strong nucleophiles are expected to add to the nitrogen atom rather than to the carbene carbon atom. This supposition has been verified by the reaction of 1-amino-1-alkynes (ynamines) with aminocarbene complexes of manganese, chromium, molybdenum and tungsten to give alkylideneaminocarbene complexes (eq 93).[236-7] The reaction may be understood in terms of a 1,2-addition of the aminocarbene substituent to the triple bond of the alkyne, followed by a 1,3-hydrogen shift. It should be noted, however, that alkylaminocarbene complexes undergo addition-rearrangement reactions on addition of ynamines (cf Section 7.2).[238]

$$\text{(CO)}_5\text{M=C(R)NH}_2 + \text{CH}_3\text{C≡C-N(C}_2\text{H}_5\text{)} \quad \longrightarrow$$

$$\text{(CO)}_5\text{M=C(R)N=C(C}_2\text{H}_5\text{)N(C}_2\text{H}_5\text{)}_2 \quad (93)$$

$$\text{M = Cr, Mo, W}$$

4.4   Addition of Transition Metal Nucleophiles

The well-known coordination ability of nucleophilic metal d^8 or d^{10} complexes, notably those derived from platinum(0), towards alkenes or alkynes may be also extended to metal-metal multiple bond systems affording cluster compounds. Keeping this in mind, reactions between such nucleophiles and the metal-carbon multiple bonds, exemplified in carbene and carbyne complexes, have been studied by Stone and coworkers. Bis(cycloocta-1,5-diene)platinum or mixed ethylene-trimethylphosphine complexes of platinum were found to add to the metal carbene bond of Fischer-type compounds yielding dimetal compounds with bridging carbene ligands. In this manner a series of complexes containing nickel, palladium and platinum as well as chromium, molybdenum, tungsten and manganese have been prepared (eq 94).[239-44] X-ray studies indicate that the carbene bridge originating from the VIA or VIIA metal complex is shifted towards the VIII metal. The bonding in such a platinum-tungsten compound may be described by the electron-pair donation of the electron-rich d^{10}-platinum atom to the 16-electron d^6-tungsten atom which relieves itself of excess of charge by electron transfer to the carbene π^*- orbital.

$$(CO)_5M=C(C_6H_5)OCH_3 + M'L_n \longrightarrow (CO)_5\overline{M-C(C_6H_5)(OCH_3)}M'L_n \qquad (94)$$

$$M = Cr, Mo, W; \quad M' = Ni, Pd, Pt; \quad L_n = (cod)_2, (C_2H_4)[P(CH_3)_3]_2$$

A mononuclear dicarbonyl (π-cyclopentadienyl)carbene complex of manganese reacts in a similar way. Enneacarbonyl[(2-oxacyclopentylidene)]dimanganese, however, affords on reaction with the platinum nucleophile a binuclear compound containing only one manganese atom. A modified 2-oxacyclopentenylidene carbene ligand bridges the manganese-platinum bond so that the manganese atom is η^2-coordinated by the cycloalkene group which in turn is σ-bonded to the platinum. Effectively, the carbene ligand has been transferred to the platinum. In addition, a trinuclear platinum complex without carbene ligands is isolated (eq 95).[239,241] The formation of trimetal compounds containing μ-CO or μ-C(C_6H_5)OCH_3 groups is favored if bulky phosphine ligands are coordinated to platinum.[243] A complete carbene transfer has been observed starting from $I(CO)_4Mn=\overline{C(CH_2)_3O}$ and a bisethylene(phosphine)platinum complex (eq 96).[244]

$$(CO)_5Mn-(CO)_4Mn=\langle\!\!\!\!\!\!\!\!\!\!\!_O$$

XLII

$$(CO)_4Mn-Pt(PMe_3)_2$$

XLIII

$$\textbf{XLII} + Pt(C_2H_4)(PMe_3)_2 \longrightarrow \textbf{XLIII} + (\mu\text{-}CO)_3Pt_3(PMe_3)_3 +... \qquad (95)$$

$$I(CO)_4Mn=\overline{C(CH_2)_3O} + Pt(C_2H_4)_2L \longrightarrow (CO)_4Mn \overset{I}{-} PtL[=\overline{C(CH_2)_3O}]$$

$$L = P(t\text{-}Bu)_2Me \qquad\qquad\qquad + 2C_2H_4 \qquad (96)$$

5.   REACTIONS WITH ELECTROPHILES

Carbene complexes bearing heteroatom carbene substituents react with Lewis acids to give a novel class of organometallic compounds known as carbyne complexes.

5.1   Lewis Acids

When pentacarbonyl alkoxycarbene complexes of the VIA metals are treated with boron trihalides elimination both of the alkoxy substituent and of the trans-CO ligand occurs to give trans-halogeno(tetracarbonyl)carbyne complexes (eq 97).   This type of reaction is quite general[18,77,245] and has been extended from alkyl- and arylcarbene ligands[246] to complexes containing silyl-,[247] cycloalkenyl-,[248] alkynyl-,[249] metallocenyl-[33-5,250] and dialkylaminocarbene substituents.[78]   The reactivity of the boron halides depends on their Lewis acidity.   As a consequence the use of the weak Lewis acid boron trifluoride is restricted to complexes bearing a high electron density on the alkoxy atom, such as amino(ethoxy)carbene complexes.[78]   Even though in most cases boron compounds have been employed for convenience, the halides of aluminum or gallium work as well and sometimes even offer a preparative advantage due to their graduated reactivity.[33,250-2]

$$(CO)_5M=C(R)OR' + M'X_3 \longrightarrow \underline{trans} - X(CO)_4M \equiv C-R + CO + MX_2OR' \quad (97)$$

As well as alkoxy groups, siloxy-,[61] acetoxy-,[253] amino-[211] and thiosubstituents[254] have been used as leaving groups.   Cis-substituted carbene complexes are transformed into mer-carbyne compounds.[255]   Trans-ligands such as phosphines or π-bonded aromatic ligands, however, being less effective than CO in withdrawing charge from the metal, remain attached and give rise to cationic carbyne complexes (eq 98).[39,49,87,256]

$$\pi-C_nH_n(CO)_mM=C(R)OCH_3 + BX_3 \rightarrow [\pi-C_nH_n(CO)_mM \equiv C-R]^+ BX_4^- + ...(98)$$

M = Cr, n = 6, m = 3; M = Mn, Re, n = 5, m = 2

If there are two heteroatom substituents present in the carbene ligand, as in pentacarbonyl[dialkylamino(ethoxy)carbene]chromium, the alkoxy group is removed without loss of carbon monoxide, if boron trifluoride is used as the Lewis acid (eq 99).[78]

$$(CO)_5Cr=C(NR_2)OC_2H_5 + BF_3 \longrightarrow [(CO)_5Cr\equiv C-NR_2]^+BX_4^- + \ldots \qquad (99)$$

The mechanism of the carbyne formation is thought to involve addition of the Lewis acid to the heteroatom followed by a cleavage of the heteroatom to carbene carbon bond. A related adduct has been isolated from a hydroxycarbene complex of manganese (eq 100).[257]

$$\underline{cis} - Br(CO)_4Mn=C(CH_3)OH + BBr_3 \xrightarrow[- HBr]{} \underline{cis} - Br(CO)_4Mn=C(CH_3)OBBr_2 \qquad (100)$$

The possible intermediacy of halogenocarbene compounds must be considered. Support for this idea comes from the rearrangement of amino(halogeno)carbene complexes to give carbyne complexes (eq 101).[258] A kinetic study established a first order rate law for this reaction and excluded an influence of free CO on the reaction rate.[258-9]

$$(CO)_5Cr=C(NR_2)X \xrightarrow[- CO]{} trans - X(CO)_4Cr\equiv C-NR_2 \qquad (101)$$

The heterolytic elimination of the original carbene substituent leads to a cationic carbyne intermediate. If the positive charge can be delocalized into the ligand framework, as expected for phosphine, π-bonded aromatic, or even dialkylaminocarbyne ligands, such cationic compounds become isolable. If, however, no such ligands are present the decrease in backbonding especially to the trans-CO ligand causes substitution by the halide to give a neutral carbyne complex.

5.2 Miscellaneous

The rearrangement of amino(halogeno)carbene complexes to aminocarbyne compounds can also be achieved by silver(I) salts (eq 102).[79]

$$(CO)_5Cr=C[N(CH_3)_2]Cl + AgY \longrightarrow [(CO)_5Cr\equiv C-N(CH_3)_2]^+Y^- + ... \quad (102)$$
$$Y = BF_4, PF_6, ClO_4$$

Hydroxycarbene tungsten carbonyl complexes react with dicyclohexylcarbodiimide (DCCD) to give binuclear carbene-carbyne compounds (eq 103).[260-1]

$$2(CO)_5W=C(C_6H_5)OH \xrightarrow[\text{- CO, - H}_2\text{O}]{H_{11}C_6N=C=NC_6H_{11}} (CO)_5W=C(C_6H_5)O-W(CO)_4(\equiv C-C_6H_5)$$

$$(103)$$

The reactions of carbene complexes with electrophilic agents such as group VIB elements, leading to cleavage of the metal carbene bond, are mentioned in Section 9.5.

6.   REACTIONS OF TRANSITION METAL CARBENE ANIONS

6.1   Acidity of Carbene Complexes

Alkoxy(alkyl)carbene complexes show a marked acidity.   Base-catalyzed hydrogen-deuterium exchange within the methyl carbene substituent of **XXIV** occurs immediately.[168]   Even if no base is added, the half-life for the exchange of the α-hydrogen atoms in a binuclear 2-oxacyclopentylidene manganese complex is 23 min at 40°C.[116]   Similar results have been obtained from both neutral and cationic alkyl(methoxy)carbene complexes containing metals of the nickel triad.[125,133,262]   In the series of homologous compounds of type **XLIV**, the relative Bronsted acidity was found to be in the order Ni Pd Pt.[125] Employing deuterated alcohols, any exchange of the methoxy group is so slow as to be virtually non-existent.[262]

$$\underline{trans} -\{Cl_6C_5[P(CH_3)_2C_6H_5]_2M=C(CH_3)OCH_3\}^+$$
XLIV

These results imply the existence of an intermediate carbene anion. This suggestion is corroborated by the reaction of **XXIV** with a stoichiometric amount of n-butyl lithium at low temperature to give the conjugate base **XLV** (eq 104).[263-4]   As has been concluded from spectroscopic studies, the anion, which can be isolated as bis(triphenylphosphine)iminium salt in 95% yield, is to be regarded as a vinyl chromium anion **XLVb** rather than as a carbanion **XLVa.**

$$(CO)_5Cr=C(CH_3)OCH_3 \xrightarrow[-78°C]{n-C_4H_9Li} [(CO)_5CrC(CH_2)OCH_3]^- \qquad (104)$$

XXIV   **XLV**

$$\left[(CO)_5Cr=C \begin{smallmatrix} OMe \\ CH_2^- \end{smallmatrix} \right]^- \qquad\qquad \left[(CO)_5Cr-C \begin{smallmatrix} OMe \\ CH_2 \end{smallmatrix} \right]^-$$

XLV a   **XLV b**

Compound **XXIV** is among the most acidic neutral carbon acids known. **XLV** is not measurably protonated by excess methanol, although the addition of one equivalent of p-cyanophenol (pK_a = 8 in water) leads to a 55% protonation indicating a similar acidity for **XXIV.** The high acidity can be attributed in part to the delocalization of negative charge in **XLV.**[264] The effect of α-alkyl substitution on the kinetic and the thermodynamic acidity has been studied using pentacarbonyl(2-oxacyclopentylidene)chromium complexes.[265] The parent compound is found to be only about 5% more acidic than the 5-methyl substituted analogue.

6.2   Modification of the Carbene Side Chain

The acidity of carbene complexes may be used in the synthesis of deuterated analogues. Thus, by addition of excess DCl to **XLV,** the d_1-methylcarbene complex is obtained with 90% monodeuteration.[264] On the other hand, **XXIV** is converted to the methoxy(trideuteromethyl)carbene complex on reaction with d_1-methanol containing amounts of sodium methylate.[168]

The facile preparation and the great thermodynamic stability of carbene anions has been broadly used to modify the carbene ligands. This route is especially useful for the synthesis of carbene ligands containing functional groups, such as esters or ketones, which are incompatible with the

lithium reagents used in the original Fischer synthesis.   Strong alkylating
agents such as oxonium salts, fluorosulfonates, allyl and benzyl halides, as
well as α-bromoesters and aldehydes, react with carbene anions at room
temperature.   Less active reagents like alkyl halides require moderate
heating and esters or ketones do not react at all (eq 105-6).[17,263,265-7]

$$(CO)_5W=C(CH_3)OCH_3 \xrightarrow[\text{2) } (CH_3)_3OBF_4]{\text{1) } n\text{-}C_4H_9Li} (CO)_5W=C(C_2H_5)OCH_3 \qquad (105)$$

$$(CO)_5Cr=\overline{CO(CH_2)_2}CHCH_3 \xrightarrow[\text{2) } (CH_3)_2C=CHCH_2Br]{\text{1) } n\text{-}C_4H_9Li}$$

$$(CO)_5Cr=\overline{CO(CH_2)_2}C(CH_3)CH_2CH = C(CH_3)_2 \qquad (106)$$

Acylation is achieved by reaction with acyl chlorides (eq 107).[268] If
the initial acylated product, however, contains an enolizable hydrogen atom,
the corresponding enol ester is isolated (eq 108).[263]   Aldehydes lead to the
formation of alkenylcarbene complexes (eq 109)[235,266] or, in the case of
formaldehyde, to a methylene-bridged binuclear product, presumably formed
via an intermediate vinylcarbene complex.[266]

$$(CO)_5Cr=\overline{CO(CH_2)_2}CHCH_3 \xrightarrow[\text{2) } CH_3COCl]{\text{1) } n\text{-}C_4H_9Li} (CO)_5Cr=\overline{CO(CH_2)_2}C(CH_3)COCH_3 \qquad (107)$$

$$(CO)_5W=C(CH_3)OCH_3 \xrightarrow[\text{2) } CH_3COCl]{\text{1) } n\text{-}C_4H_9Li} (CO)_5W=C[CH=C(CH_3)OCOCH_3]OCH_3 \qquad (108)$$

$$(CO)_5Cr=C(CH_3)OCH_3 \xrightarrow[\text{2) } C_6H_5CHO]{\text{1) } n\text{-}C_4H_9Li} (CO)_5Cr=C(CH=CHC_6H_5)OCH_3 \qquad (109)$$

Methoxy(methyl)carbene complexes can be transformed into 2-oxacyclopentylidene compounds by reaction with epoxides. Substituted epoxides are attacked by the carbene anion at the least crowded carbon atom (eq 110).[267] Ethylene sulfide reacts in a similar way to give a cyclic thiocarbene complex.[268]

$$(CO)_5Cr=C(CH_3)OCH_3 \xrightarrow[\text{2) } \overline{CH_2CH(R)O}]{\text{1) } n\text{-}C_4H_9Li} (CO)_5Cr=\overline{CO(CH_2)_2CHR} \qquad (110)$$

In the alkylation of carbene anions dialkylation often becomes an important side reaction[265,267] which is governed by the dependence of the acidity and reactivity of carbene anions on the degree of substitution of the α-carbon atom. In the pentacarbonyl(2-oxacyclopentylidene)chromium series the secondary anion **XLVI** and the methylated tertiary anion **XLVII** were found to be of equal basicity. In spite of the increased hindrance the more substituted species reacts three times faster with benzyl halides. Even though this fact is not understood it must be held responsible for the tendency of carbene anions to undergo dialkylation.

XLVI R = H

XLVII R = Me

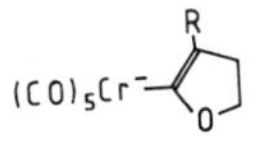

Carbene anions may also be employed in conjugate addition to α,β-unsaturated esters and ketones. Michael-type products are obtained if secondary carbene anions such as **XLVI** are used (eq 111).[268] In the absence of β-hydrogen atoms, e.g. in **XLVII**, the Michael addition is blocked.

$$(CO)_5Cr=\overline{CO(CH_2)_2CH_2} \xrightarrow[\text{2) } CH_3CH=CHCO_2CH_3]{\text{1) } n\text{-}C_4H_9Li} (CO)_5Cr=\overline{CO(CH_2)_2CHCH(CH_3)CH_2CO_2CH_3} \qquad (111)$$

Halogenation of the carbene anion **XLVI** yielded a cyclic β-bromocarbene complex.[268]

7.   ADDITION-REARRANGEMENT REACTIONS

A few reactions of carbene complexes with compounds containing heteroatoms or multiple bonds are known which result either in the formation of a metal-heteroatom bond or of a new metal-carbene bond.

7.1   Formation of Metal-Heteroatom Bonds

Unlike methyl selenol, which undergoes a selenolysis reaction with the methoxy(methyl)carbene complex **XXIV**, the more acidic phenylselenol leads to a selenol ether complex (eq 112).[225]   The reaction may be formally referred to as an insertion of the selenol into the metal-carbene bond and is supposed to involve nucleophilic addition of the selenol to the carbene carbon atom, followed by rearrangement to yield a more stable metal-selenium bond.

$$(CO)_5Cr=C(CH_3)OCH_3 + C_6H_5SeH \longrightarrow (CO)_5Cr-Se(C_6H_5)CH(CH_3)OCH_3$$

XXIV   (112)

Similar reactions are observed when calcium cyanide[269] or dimethylphosphine[185] are reacted with the methoxy(phenyl)carbene complex **XVIII**.   The isolation of an intermediate phosphorous ylid complex in the latter reaction supports the mechanism outlined above (eq 113).[185]

$$(CO)_5Cr^--C(C_6H_5)(OCH_3)PH^+(CH_3)_2 \xrightarrow{25°C} (CO)_5Cr-P(CH_3)_2CH(C_6H_5)OCH_3$$

(113)

A formal insertion of an NH group occurs in the reaction with hydroxylamine giving an imido complex (eq 114).[216]

$$(CO)_5Cr=C(CH_3)OCH_3 + NH_2OH \longrightarrow (CO)_5Cr-NH=C(CH_3)OCH_3 + ... \quad (114)$$

Closely related are reactions of heteroatom-substituted carbene complexes with hydrogen halides.   Thus on reaction with hydrogen bromide at low temperature a thiocarbene complex rearranges to give a thioether complex (eq 115).[270]

$$(CO)_5Cr=C(CH_3)SCH_3 + HBr \xrightarrow[-30°C]{} (CO)_5Cr-S(CH_3)CH(CH_3)Br \qquad (115)$$

A similar reaction seems to occur with dialkylaminocarbene complexes.   The addition of hydrogen bromide or hydrogen chloride may involve α-halogenoamine complexes as possible intermediates which, however, could not be isolated.   Instead immonium halogenopentacarbonyl chromates have been obtained (eq 116).[271]   Alkoxycarbene complexes may react in the same fashion.   At -78°C pentacarbonyl (hydrogen halide) complexes are obtained.[272]   These compounds are strong acids which account for the formation of the immonium ion from the aminocarbene ligand in eq 116.   Under the reaction conditions the alkoxycarbene ligand is released from the metal and is converted into an aldehyde (eq 117).   If the reaction and workup are performed at higher temperatures mononuclear and halogen-bridged dinuclear carbonyl metallates are obtained.[273]

$$(CO)_5Cr=C(C_6H_5)NR_2 + HX \rightarrow [R_2N=CHC_6H_5]^+[(CO)_5CrX]^- \qquad (116)$$

$$(CO)_5W=C(C_6H_5)OCH_3 + HX \xrightarrow[-78°C]{} (CO)_5WXH + C_6H_5CHO + ... \qquad (117)$$

7.2   Reactions with Multiple Bond Systems

Alkoxycarbene and diarylcarbene complexes of manganese and the VIA metals react with 1-aminoalkynes (ynamines) to give alkenyl(amino)carbene compounds (eq 118).[212,274-6]   The reaction, which may formally be described as insertion of the alkyne into the metal-carbene bond, is highly stereoselective.   The alkenyl substituent is formed predominantly in the E-configuration.[167,276]

$$L_n(CO)_2M=C(R^1)R^2 + CH_3-C\equiv C-NR_2 \rightarrow L_n(CO)_2M=C[C(CH_3)=CR^1R^2]NR_2$$

$$(118)$$

$M = Cr, Mo, W, L_n = (CO)_3$; $M = Mn, L_n = Mecp$;

$R^1 = CH_3, C_6H_5$; $R^2 = C_6H_5, OCH_3$; $R = CH_3, C_2H_5$

A kinetic study indicated a second-order rate law, low activation enthalpies (25-40 kJ mol^{-1}) and marked negative activation entropies (- 129 to - 145 J mol^{-1}K^{-1}). These results support a nucleophilic attack of the alkyne at the carbene carbon atom in the first step. The site of the attack at the carbene carbon, with subsequent formation of a four-membered metallacycle followed by ring opening leading to the aminocarbene complex, have been held responsible for the stereoselectivity of the reaction. The E-configuration of the styrylcarbene chromium complex was established by heating in decane. Cyclization of the carbene ligand occurs to give indanone and indenone complexes. The Z-isomer fails to react.[276]

An ynamine insertion into the metal-carbene bond is also found in the reaction with monoalkylamino carbene compounds giving a chelating aminocarbene ligand (eq 119).[238] Dialkylamino carbene complexes, however, fail to react due to the reduced electrophilicity of the carbene carbon atom, and amino carbene ligands bearing a NH_2-substituent add to the alkyne to give alkylideneaminocarbene complexes (cf. Section 4.3, eq 93).

$$(CO)_5Cr=C(C_6H_5)NHMe + Me-C\equiv C-NR_2 \xrightarrow[-CO]{}$$

(CO)$_4$Cr structure labeled NR$_2$, H, Me, Me, N, Ph   **XLVIII** (119)

Ethoxyacetylene was found to insert into the metal-carbene bond in a similar manner (eq 120).[192]

$$(CO)_5W=C(C_6H_5)R + H-C\equiv C-OC_2H_5 \longrightarrow$$

$$(CO)_5W=C[CH=C(C_6H_5)R]OC_2H_5 \qquad (120)$$

Closely related products have been obtained starting from cyclic enol ethers and enamines (eqs 121-2).[277-9] The effect of ring size has been studied for monocyclic enol ethers. Ethoxycyclopentene gave the expected insertion product, while the six- and eight-membered homologues failed to react.[277]   Ring opening of the cycloalkenes and insertion into the metal-carbene bond are thought to occur via a metallacyclobutane intermediate. When acyclic enol ethers and enamines are used, a similar intermediate can be held responsible for the formation both of metathetic products and of cyclopropanes.[279-80]

$$(CO)_5W=CPh_2 + \underset{XLIX}{\overset{OEt}{\triangle}} \rightarrow (CO)_5W=C[(CH_2)_3CH=CPh_2]OEt \qquad (121)$$

$$(CO)_5Cr=C(Ph)OMe + \underset{L}{\triangle} \longrightarrow \underset{(CO)_5Cr=C}{\overset{N}{\triangle}} \overset{LI}{\underset{(CH_2)_3CH=C(Ph)OMe}{}} \qquad (122)$$

The insertion of cyclopentenes into the metal-carbene bond has been referred to as the chain-carrying process in the transition metal catalyzed formation of polypentenamers.[277]   A similar step might be involved in the polymerization of alkynes initiated by tungsten carbene complexes.[281]

Bis(diethylamino)acetylene can be incorporated into the indene skeleton on reaction with pentacarbonyl[methoxy(phenyl)]carbene chromium XVIII.[282-3]   The first step of a temperature-controlled reaction sequence carried out at +14°C involves insertion of the yndiamine into the metal-carbene bond (eq 123). Warming to 70°C induces an intramolecular substitution reaction affording a chelating carbene ligand (eq 124). By the coordination of the enamine nitrogen atom to the metal the carbene ligand is held in a conformation which favours cyclization to the indene system on further heating (eq 125). The tungsten compound I, homologous to XVIII, is observed to give the corresponding aminocarbene complexes, but fails to undergo the final cyclization step. The reaction sequence demonstrates clearly the role of the metal as template.

$$(CO)_5Cr=C(C_6H_5)OCH_3 + R_2N-C\equiv C-NR_2 \xrightarrow{14°C} LII \qquad \underset{R_2N}{\overset{(CO)_5Cr=C}{\underset{}{\overset{NR_2}{\diagup}}}} \overset{Ph}{\underset{OMe}{C=C}} \qquad (123)$$

XVIII

$$\text{LII} \xrightarrow[\substack{-CO}]{70^\circ C} \quad (CO)_4Cr=C\begin{smallmatrix} \nearrow NR_2 \\ R_2N-C \\ C-Ph \\ OMe \end{smallmatrix} \quad \text{LIII} \qquad (124)$$

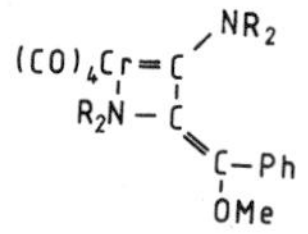

$$\text{LIII} \xrightarrow[\substack{-CO}]{125^\circ C} \quad \text{LIV} \qquad (125)$$

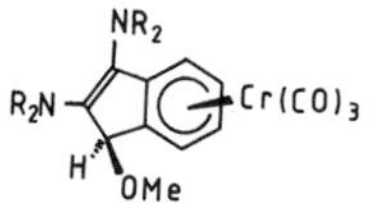

The insertion of alkynes into the metal-carbene bond can be extended to triple bond systems containing heteroatoms. Dimethyl cyanamide is found to yield alkylideneamino(dimethylamino)carbene compounds (eq 126).[284] Based on a kinetic study a mechanism similar to that for the insertion of ynamines was proposed.[285]

$$(CO)_5M=C(R^1)R^2 + (CH_3)_2N-C\equiv N \longrightarrow (CO)_5M=C[N=C(R^1)R^2]N(CH_3)_2$$

$$M = Cr, W; \qquad (126)$$

$$R^1 = CH_3, C_6H_5; \quad R^2 = p\text{-}XC_6H_4, OCH_3$$

On reaction with the carbene complex **XXIV**, cyclohexyl isocyanide yields a 1:1 product which is considered to be a N-coordinated ketenimine complex on the basis of spectroscopic and chemical evidence (eq 127).[286-7] The product is transformed into various cyclohexylaminocarbene complexes when treated with acid, methanol or water.

$$(CO)_5Cr=C(CH_3)OCH_3 + C_6H_{11}N\equiv C \longrightarrow$$
$$(CO)_5Cr-N(C_6H_{11})=C=C(CH_3)OCH_3 \quad (127)$$

The addition of some heterocumulenes to carbene complexes has led to unexpected results. Alkyl isothiocyanates give thioketone complexes along with coordinated ketenimes (eq 128).[288] Dicyclohexylcarbodiimide (DCCD), known as a strong condensation reagent, reacts with pentacarbonyl[hydroxy(methyl)carbene]chromium to give an acyclic and a four-membered ring aminocarbene complex. The latter might arise from the

[2+2] cycloaddition of the heterocumulene to an intermediate vinylidene complex (eq 129).[289]

The reaction of DCCD with hydroxycarbene complexes is strongly dependent both on the metal and on the carbene substituents. Thus, the hydroxy(phenyl)carbene complex of chromium undergoes an intermolecular condensation reaction yielding the binuclear anhydride (eq 130).[260,290] The homologous tungsten complex, however, is transformed into a binuclear carbene-carbyne compound under the same conditions (cf Section 5.2, eq 103).[260-1]

$$2(CO)_5W=C(C_6H_5)_2 \xrightarrow{\text{RN=C=S}} (CO)_5W\text{-}S=C(C_6H_5)_2 +$$

$$(CO)_5W\text{-}N(R)=C=C(C_6H_5)_2 \qquad (128)$$

$$2(CO)_5Cr=C(CH_3)OH + 2\,H_{11}C_6N=C=NC_6H_{11} \longrightarrow$$
$$-\,H_2O$$

$$(CO)_5Cr=C(CH_3)N(C_6H_{11})CONHC_6H_{11} + (CO)_5Cr=C\begin{smallmatrix}NC_6H_{11}\\ \diagup \\ \diagdown \\ CH_2\end{smallmatrix}C=NC_6H_{11} \qquad (129)$$

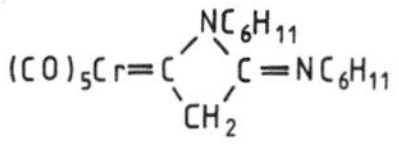

LV

$$2(CO)_5Cr=C(C_6H_5)OH \xrightarrow[-\,H_2O]{H_{11}C_6N=C=NC_6H_{11}} (CO)_5Cr=C(C_6H_5)\text{-}O(C_6H_5)C=Cr(CO)_5 \qquad (130)$$

7.3 Miscellaneous

On addition of sodium alkoxides to methoxy(phenyl)carbene complexes of chromium and tungsten a rearrangement reaction is observed leading to a formal CO insertion into the metal carbene bond. Subsequent alkylation yields alkoxybenzylcarbene derivatives (eq 131).[291-2] Monoalkoxybenzylcarbene complexes are also formed. The first step of the reaction is assumed to be nucleophilic attack of the alkoxide at the carbene carbon atom followed by an alkyl-acyl rearrangement.

$$(CO)_5M=C(C_6H_5)OCH_3 \xrightarrow[\text{2) R}_3'OBF_4]{\text{1) NaOR}} (CO)_5M=C[C(OR)_2C_6H_5]OR' \qquad (131)$$

8.   REDOX REACTIONS WITHOUT METAL CARBENE CLEAVAGE

There are a fair number of reactions of carbene complexes in which redox processes are involved.   Under suitable conditions a metal carbene cleavage can be avoided.   Besides oxidation and reduction accomplished either by chemical reagents or by electrochemical procedures some examples of oxidative additions will be briefly mentioned.

8.1   Reduction Reactions

Radical anions have been prepared by reduction of pentacarbonyl carbene complexes of chromium, molybdenum and tungsten using Na/K alloy at low temperature.[293]   Esr data indicate that the delocalization of the unpaired electron is dependent upon the nature of the carbene ligand. Whereas   the   odd   electron   in   the anionic methoxy(phenyl)carbene complex of chromium resides predominantly on the carbene ligand, considerable spin delocalization is indicated in the diphenylcarbene tungsten anion by the large g value and [183]W splitting.   Reduction of the carbene complexes   was   also   accomplished   by   sodium   naphthalide, dicarbonyl(cyclopentadienyl)ferrate(-1),   pentacarbonylmanganate(-1)   and organolithium reagents.   The latter are supposed to effect homolysis of the metal to carbon bond yielding the pentacarbonylchromate(-1) species in a side reaction.

Cyclic voltammetry techniques have been employed to study the reduction of a series of arylmethoxy carbene complexes.[294]   The half-wave potential has been found to depend on the conformation of the aryl ring.

Reduction of the metal carbene bond can be achieved by complex hydrides or diborane.[91,106-7,209]   This type of reaction has been used to convert a prochiral carbene carbon atom such as in alkoxy(alkyl)carbene complexes into an asymmetric center.   Whereas the reaction outlined in eq 132 results in the formation of both enantiomers in approximately equal amounts,[209] a remarkable 95% enantioselectivity has been observed in the reduction of an iron carbene chelate (eq 133).[295]

$$[Cp(CO)PR_3Fe=C(CH_3)OC_2H_5]^+ \xrightarrow{\text{B}_2\text{H}_6}$$

$$Cp(CO)PR_3Fe-C(H)(CH_3)OC_2H_5 \qquad (132)$$

$$[Cp(CO)\overline{Fe=C(C_6H_5)N(CH_3)C(C_6H_5)=NCH_3}]^+ \xrightarrow{\text{NaBH}_4}$$

$$Cp(CO)\overline{Fe-C(H)(C_6H_5)N(CH_3)C(C_6H_5)=NCH_3} \quad (133)$$

A reversible addition of dihydrogen across the metal carbene bond was recently achieved using an iridium complex which may be formally regarded as an iridium(I) carbene or an iridium(III) ylid compound (eq 143).[296]

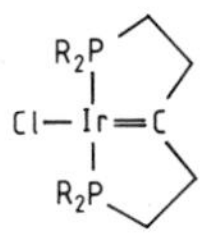

$$(134)$$

8.2   Oxidation Reactions

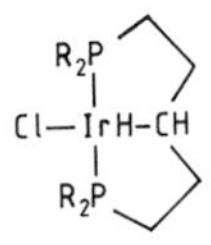

One-electron oxidation reactions of carbene complexes have been performed using silver(I) compounds as oxidizing agents.   By this route, starting from neutral compounds which contain cyclic diaminocarbene or β-silylated alkoxycarbene ligands, cationic paramagnetic carbene complexes of chromium and iron have been obtained (eq 135).[24,297] The thermal and aerial stability of the cationic compounds reflect the electron density at the metal center depending on the nature of ligands. Esr data indicate that in the iron(I) complex of eq 135 the odd electron is mainly localized on the metal center.   On the other hand, where chromium(I) complexes containing phosphorus ligands are concerned, substantial delocalization of the unpaired electron seems probable.

$$(CO)_3Fe[=\overline{CN(CH_3)(CH_2)_2NCH_3}]_2 \xrightarrow{\text{AgBF}_4}$$

$$\{(CO)_3Fe[=\overline{CN(CH_3)(CH_2)_2NCH_3}]_2\}^+ \text{ BF}_4^- \qquad (135)$$

The electrochemical oxidation of molybdenum and tungsten complexes containing cyclic diaminocarbene ligands has been studied by cyclic

voltammetry. Starting with <u>trans</u> - $(CO)_4M[=\overline{CN(R)\rightsquigarrow NR}]_2$ or <u>mer</u> -
$L(CO)_3M[=\overline{CN(R)\rightsquigarrow NR}]_2$, conversion to the corresponding <u>cis</u>- and <u>fac</u>-
compounds has been observed. These results have been explained in terms of
a rapid isomerization of the cationic intermediates.[298-9]

8.3   Oxidative Addition Reactions

Oxidative addition has been used to obtain carbene complexes
containing a metal, preferably of group VIII, in a higher oxidation state.
For example, addition of methyl iodide to a platinum(II) carbene complex
leads to the corresponding platinum(IV) compound (eq 136).[213] The reaction
fails if the methyl ligand is replaced by a chlorine group, indicating that a
high electron density is essential for the reaction.

$$\{\underline{trans} - CH_3[P(CH_3)_2C_6H_5]_2Pt=\overline{C(CH_2)_3O}\}^+ + CH_3I \longrightarrow$$
$$\{I(CH_3)_2[P(CH_3)_2C_6H_5]_2Pt=\overline{C(CH_2)_3O}\}^+ \qquad (136)$$

Oxidative addition may result in unusual coordination numbers. Thus a
seven coordinate complex has been synthesized by the reaction of eq 137.[137]

$$(CO)_5W=\overline{CN(CH_3)(CH_2)_2NCH_3} + I_2 \xrightarrow[-CO]{} I_2(CO)_4W=\overline{CN(CH_3)(CH_2)_2NCH_3} \qquad (137)$$

9.   CLEAVAGE OF THE METAL CARBENE BOND

For some time it was an open question whether carbene complexes
might be regarded as sources of free carbenes. Thus several methods have
been developed to achieve the cleavage of the metal carbene bond; none of
them, however, has been shown to involve free carbenes.

9.1   Ligand Substitution

Pentacarbonyl(alkoxycarbene)chromium complexes react with suitable
ligands such as phosphines (including phosphorus trihalides),[300-2] or
dialkylsulfides,[303] with substitution of the carbene ligand. Products arising
from the carbene ligands, however, have been identified only in one
reaction. A pentafluorophenyl(ethoxy)carbene ligand coordinated to iron is

found to undergo an alkyl migration in the presence of triphenylphosphine to yield an alkylaryl ketone (eq 138).[304] This type of rearrangement is in contrast to complexes of chromium and tungsten which are known to give dimers of the carbene ligand on thermal decomposition (cf Section 9.2).

$$[(H_5C_6)_3P](CO)_3Fe=C(C_6F_5)OC_2H_5 + P(C_6H_5)_3 \xrightarrow[35^\circ C]{CH_3COOH}$$

$$C_6F_5C(O)C_2H_5 + [(H_5C_6)_3P]_2Fe(CO)_3 \qquad\qquad (138)$$

Release of the diaminocarbene ligand from a group VIII metal is effected by treatment with triphenylphosphine.[218]   Ligands containing suitably functionalized aminosubstituents are cyclized to give heterocyclic compounds along with carbodiimides or cyanamides (cf Section 10.4).

Carbon monoxide and pyridine have been used to cleave the metal aminocarbene bond.   Imines formed by a 1,2-hydrogen shift have been obtained in high yield (eq 139).[198,217]   The formation of enol ethers from the reaction of alkoxy(alkyl)carbene complexes with pyridine may be explained by a similar base-induced hydrogen shift (cf Section 9.3).

$$(CO)_5Cr=C(C_6H_5)NHCH_3 \xrightarrow[160^\circ C]{100\ atm\ CO} C_6H_5CH=NCH_3 + Cr(CO)_6 \quad (139)$$

9.2   Thermal Decomposition

On heating, the Fischer-type carbene complexes decompose yielding the parent carbonyls and the dimers of the carbene ligand as main products (eq 140).[23,305]   As expected, the decomposition occurs more readily in the absence of stabilizing heteroatom carbene substituents (eq 141).[148,166,229]

$$(CO)_5Cr=C(R)OCH_3 \xrightarrow{>130^\circ C} \text{cis- or trans- } R(CH_3O)C=C(OCH_3)R + Cr(CO)_6 +$$

$$R = CH_3, C_6H_5 \qquad\qquad\qquad\qquad\qquad ... (140)$$

$$[Cp(CO)_2Fe=C(H)C_6H_5]^+ \xrightarrow{25^\circ C} \underline{trans}\text{-} C_6H_5CH=CHC_6H_5 + \ldots \qquad (141)$$

In some cases, instead of the olefinic carbene dimers the corresponding alkanes have been obtained.[306-7]

The thermolysis of pentacarbonyl(2-oxacyclopentylidene)chromium has been studied to test for the intermediacy of free carbenes.[308]   As main product the carbene dimer was obtained along with a small amount of dihydrofuran.   The absence of cyclobutanone, which is known to be formed in significant amounts on pyrolysis of the sodium salt of butyrolactone tosylhydrazone, rules out the intervention of an uncoordinated carbene. A kinetic investigation of the decomposition (eq 142) established that the reaction is second order in the carbene complex and inhibited by carbon monoxide. A mechanism was proposed involving a primary dissociation of a CO ligand, followed by the formation of a biscarbene complex to account for the formation of the carbene dimer.

$$(CO)_5Cr=\overline{C(CH_2)_3O} \xrightarrow{180^\circ C} \overline{O(CH_2)_3C}=\overline{C(CH_2)_3O} + \overline{O(CH_2)_2CH}=CH + \ldots$$
$$\qquad\qquad\qquad\qquad\qquad (50\%) \qquad\qquad\qquad (5\%) \qquad (142)$$

It is surprising, however, that no dimerization of the carbene ligand occurs in the decomposition of an ethoxy(pentafluorophenyl)carbene complex of iron.   Instead a novel alkyl migration to the carbene carbon leads to the formation of an alkylaryl ketone (eq 138).[304]   Similarly the decomposition of a hydroxy(methyl)carbene complex of rhenium leads to the formation of acetaldehyde along with the parent metal carbonyl (eq 143).[42]

$$Cp(CO)_2Re=C(CH_3)OH \xrightarrow[50^\circ C]{} CH_3CHO + CpRe(CO)_3 + \ldots \qquad (143)$$

9.3   Base-Induced Decomposition

Pyridine and related basic nitrogen compounds have been found to induce the decomposition of alkoxy(alkyl)carbene and 2-oxacyclopentylidene complexes to give enol ethers (eq 144).[23,54,308]   Since arylcarbene complexes are considerably more stable in the presence of pyridine and decompose finally to give the carbene dimers,[305] a mechanistic scheme involving carbene anions appears attractive.[16]

$$(CO)_5Cr=C(CHR^1R^2)OCH_3+NC_5H_5 \underset{80^\circ C}{\longrightarrow} CH_3OCH=CR^1R^2+(CO)_5CrNC_5H_5 \qquad (144)$$

A similar hydrogen shift is observed in the reaction of pyridine with aminocarbene complexes derived from primary amines.   Along with carbonyl(pyridine) complexes, imines have been obtained as main products (eq 145).   In a closely related reaction a tetrahydroharman derivative has been isolated starting from a tryptaminocarbene compound (eq 146).[217] It should be noted that even if no base is present the thermolysis of aminocarbene complexes may also lead to imines.[198]

$$(CO)_5Cr=C(C_6H_5)NHC_6H_5 + 3 NC_5H_5 \xrightarrow[- 2 CO]{68^\circ C}$$

$$C_6H_5CH=NC_6H_5 + (C_5H_5N)_3Cr(CO)_3 \qquad (145)$$

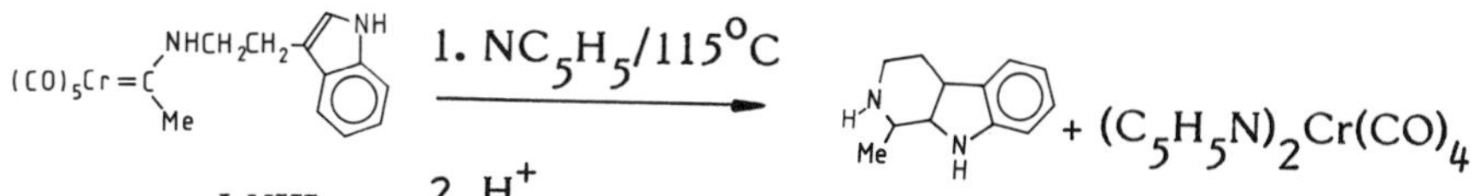

$$+ (C_5H_5N)_2Cr(CO)_4 \qquad (146)$$

LVIII   2. H$^+$ LIX

9.4   Reductive Cleavage

Whereas complex hydrides are preferred for reducing the metal carbene bond to give alkyl compounds, the cleavage of the metal-carbene bond is better achieved by hydrogenolysis.   In this manner the saturated hydrocarbon derived from the carbene ligand can be obtained in high yield (eq 147).[309]   The hydrogenolysis conditions depend on the carbene substituents.   For instance, rather drastic conditions are necessary to cleave off the 2-oxacyclopentylidene ligand.   The formation of carbene dimers may be suppressed if a low concentration of the carbene complex and a high hydrogen pressure are employed.

$$(CO)_5Cr=\overline{C(CH_2)_3O} \xrightarrow[170°C]{69 \text{ atm } H_2} thf + Cr(CO)_6 + \ldots \qquad (147)$$

9.5   Oxidative Cleavage

The metal-carbene bond can be replaced by an oxygen-carbon double bond by oxidative cleavage.   This procedure has proved useful in the characterization of carbene complexes.   Early work indicated that oxygen could be employed directly as an oxidizing agent.[310]   Later this type of reaction has been extended to sulfur and selenium providing an interesting route to thiocarbonyl and selenocarbonyl compounds (eq 148).[311]

$$(CO)_5Cr=C(C_6H_5)OCH_3 + X_n \longrightarrow C_6H_5C(X)OCH_3 + \ldots \qquad (148)$$
$$X_n = O_2, S_8, Se_n$$

The metal-carbene cleavage occurs under mild conditions if dipolar reagents such as sulfoxides, amine-N-oxides or iodosobenzene are used (eq 149).[117,229,312-3]   The latter is known to transform aminocarbene ligands into amides even though in low yield.

$$(CO)_5W=C(C_6H_5)_2 \xrightarrow{(CH_3)_2S=O} (C_6H_5)_2C=O + \ldots \qquad (149)$$

A versatile method is the use of ceric compounds.   For example, the oxidation of an enol ester substituted carbene complex of tungsten affords the enediol diester in high yield, the olefinic double bond being unaffected (eq 150).[166,229,263]

$$(CO)_5W=C[CH=C(CH_3)O_2CCH_3]OCH_3 \xrightarrow{(NH_4)_2Ce(NO_3)_6}$$

$$CH_3(CH_3CO_2)C=CHCO_2CH_3 + \ldots \quad (150)$$

Diaminocarbene complexes of palladium are oxidised by silver(I) oxide. In this manner a variety of carbodiimide derivatives have been obtained (eq 151).[314]

$$Cl_2(RNC)Pd=C(NHR)NHR' \xrightarrow{Ag_2O} RN=C=NR' + \ldots \quad (151)$$

9.6   Miscellaneous

Hydrogen halides have been found to cleave the metal-carbene bond even at low temperatures.   Depending on the nature of the carbene ligand different products have been obtained (eqs 115-7).[270-1,315] Phosphonium salts could be isolated by reacting alkoxycarbene complexes with hydrogen chloride and an excess of triphenylphosphine (eq 152).[315-6]

$$\underline{cis} - (C_6H_5)_3P(CO)_4Cr=C(R)OCH_3 + HCl \xrightarrow{(C_6H_5)_3P}$$

$$\underline{trans} - [(C_6H_5)_3P]_2Cr(CO)_4 + [C_6H_5)_3P-CH(R)OCH_3]^+Cl^- \quad (152)$$

A similar phosphonium salt formation has been observed in the reaction of a methyl(phenyl)carbene complex intermediate with trifluoroacetic acid in the presence of phosphine.[317]   Non-halogenated carboxylic acids, however, react with alkoxycarbene complexes to yield esters derived from the insertion of the carbene ligand into the O-H bond.[315] Similar insertion reactions are known to occur into IVB element-hydrogen bonds as well (cf. Section 10.3).

The reaction of alkoxycarbene complexes with trihalomethyl mercury compounds results in the formation of dihalo-enol ethers (eq 153).[318]

$$(CO)_5Cr=C(C_6H_5)OCH_3+C_6H_5HgCX_2Y \rightarrow X_2C=C(C_6H_5)OCH_3+C_6H_5HgY +$$

$$\ldots \quad (153)$$

Further carbene transfer reactions are treated in Sections 4. and 10.

10.   CARBENE TRANSFER REACTIONS

An interesting reaction of carbene complexes from a synthetic point of view is the transfer of the carbene ligand to other substrates, especially unsaturated organic compounds.   Intermetallic transfer has also been established.

10.1 Intermetallic Transfer

The transfer of a carbene ligand from one metal to another can occur under both photochemical and thermal reaction conditions.   This type of reaction is thought to proceed via a binuclear complex containing a bridging carbene ligand.[16,244,308]   The first use of the photosensitivity of a molybdenum carbene complex to synthesize an iron carbene compound is outlined in eq 154.[40,319]

$$Cp(NO)(CO)Mo=C(C_6H_5)OCH_3 + Fe(CO)_5 \xrightarrow{h\nu} (CO)_4Fe=C(C_6H_5)OCH_3 +$$
$$Cp(NO)Mo(CO)_2 \qquad\qquad (154)$$

A reaction rate similar to that for CO exchange and thermal decomposition of the carbene complex has been found for the transfer of the 2-oxacyclopentylidene ligand from chromium to tungsten (eq 155).[16,308] Whereas considerable heating is needed for this reaction to proceed, remarkably mild conditions can be employed to effect a carbene transfer from tungsten to gold. The reaction occurs with retention of configuration within the carbene ligand (eq 156).[320]

$$(CO)_5Cr=\overline{C(CH_2)_3O} + W(CO)_6 \underset{K_{eq}=3}{\overset{140°C}{\rightleftharpoons}} (CO)_5W=\overline{C(CH_2)_3O} + Cr(CO)_6 \quad (155)$$

$$(CO)_5W=C(R)X + HAuCl_4 \xrightarrow[-HCl]{-CO} ClAu=C(R)X + (CO)_4WCl_2 \qquad (156)$$

$R = CH_3, C_6H_5$

$X = OCH_3, NH_2, E-NHCH_3, Z-NHCH_3, N(CH_3)_2$

Closely related is the disproportionation of complexes containing cyclic diaminocarbene ligands. This type of reaction (eq 157) was first achieved by photoassistance, and later by heating, or by reaction with basic ligands such as tricyclohexylphosphine or pyridine, giving the products in higher yields.[175,321-2]

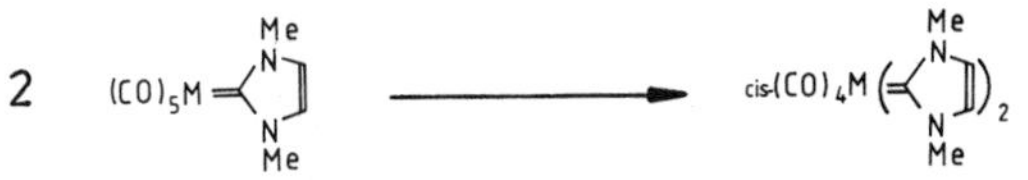

$$2 \quad (CO)_5M=\overset{\overset{Me}{N}}{\underset{\underset{Me}{N}}{\langle}} \longrightarrow cis-(CO)_4M(=\overset{\overset{Me}{N}}{\underset{\underset{Me}{N}}{\langle}})_2 + M(CO)_6 \qquad (157)$$

LX $\qquad\qquad$ LXI $\qquad$ M = Cr, Mo, W

The cis - biscarbene complexes undergo a photo-induced isomerization to the trans - compounds which in turn may be converted irreversibly into the thermodynamically more stable cis - compounds under thermal reaction conditions.[322] Similar isomerization reactions have been observed with palladium and platinum complexes.[136,323]

Carbene transfer to a transition metal nucleophile, leading to binuclear compounds, is discussed in 4.4 above.

10.2 Cyclopropanation

The cyclopropanation of olefins is the most useful reaction of carbenes. Early in the history of carbene complex chemistry carbene ligands were shown to add to two different classes of alkenes - the electron-deficient α,β-unsaturated esters and the electron-rich vinyl ethers (eqs 158-9).[226,324-5] Different reaction conditions are required for these two groups of alkenes, and hence different mechanisms are supposed to be operative, neither, however, involving free carbenes. The ratios of syn:anti-cyclopropanes obtained from methoxy(phenyl)carbene complexes of chromium, molybdenum and tungsten, and trans - methyl crotonate or ethyl vinyl ether, were found to depend on the nature of the metal. Further evidence for the active role of the metal was provided by the formation of an optically active cyclopropane when diethyl fumarate was reacted with a chiral metal carbene complex (eq 160).[76]

$$(CO)_5M=C(C_6H_5)OCH_3 + \underline{trans} - CH_3CH=CHCO_2CH_3 \longrightarrow$$

M = Cr, Mo, W

$$90-140°C$$

$$\textbf{LXII} + \textbf{LXIII} + \ldots \qquad (158)$$

$$2 (CO)_5M=C(C_6H_5)OCH_3 + 2 CH_2=CHOC_2H_5 + 2 CO \xrightarrow[50°C]{100 \text{ atm CO}}$$

M = Cr, Mo, W

$$\textbf{LXIV} + \textbf{LXV} + 2M(CO)_6 \qquad (159)$$

$$[(H_3C)(H_5C_6)(H_7C_3)P](CO)_4Cr=C(C_6H_5)OCH_3 +$$

$$\underline{trans} - H_5C_2O_2CCH=CHCO_2C_2H_5$$

$$\longrightarrow \textbf{LXVI} \qquad + \ldots \qquad (160)$$

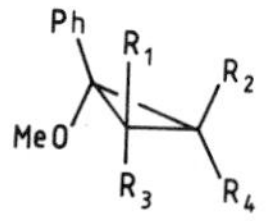

LXII $R_1=R_4=H$, $R_2=CO_2Me$, $R_3=Me$; **LXIII** $R_1=R_4=H$, $R_2=Me$, $R_3=CO_2Me$;

LXIV $R_1=R_2=R_3=H$, $R_4=OEt$; **LXV** $R_1=OEt$, $R_2=R_3=R_4=H$;

LXVI $R_1=R_4=CO_2Et$, $R_2=R_3=H$

In the cyclopropanation of α, β-unsaturated esters reaction conditions are employed under which CO ligand exchange is known to occur in carbonyl carbene complexes. A reasonable mechanism has been suggested involving CO elimination as the first step followed by coordination of the alkene. The alkene carbene complex may rearrange to a metallacyclobutane which is supposed to undergo reductive elimination to give the cyclopropane (eq 161). Complexes containing both a carbene and an alkene ligand, as well as metallacyclobutanes, have been proposed as key intermediates in the olefin metathesis reaction (Chapter 5).[280,326-7]

$$(CO)_5M=C(R)OR' \; \underset{+CO}{\overset{-CO}{\rightleftharpoons}} \; (CO)_4M=C(R)OR' \xrightarrow{\overset{C=C}{}} (CO)_4M=C(R)OR' \longrightarrow$$

$$\underset{C=C}{|} \qquad (161)$$

LXVII $(CO)_4M - C(-OR') \longrightarrow$ **LXVIII**

Carbene transfer to vinyl ethers occurs under conditions milder than those required for ligand substitution of carbene complexes. The reaction products depend strongly on the reaction conditions. Under CO pressure cyclopropanes are formed in good yields; in the absence of external carbon monoxide, however, alkene scission products predominate.[226] One fragment of the alkene is bonded to the former carbene ligand to give an olefinic metathesis product, while the other fragment is incorporated into a new metal carbene complex (eq 162).[280] The first step of these reactions is presumed to be a nucleophilic addition of the alkene to the coordinated carbene carbon atom. Alkene carbene complex intermediates, as well as the intermediacy of metallacyclobutanes containing six- and seven-coordinated metal centers, have been invoked to explain the formation of cyclopropanes and olefinic products.[16]

$$(CO)_5W=C(C_6H_5)_2 + CH_2=C(C_6H_5)OCH_3 \longrightarrow$$

$$CH_2=C(C_6H_5)_2 + (CO)_5W=C(C_6H_5)OCH_3 \quad (162)$$

A similar cyclopropanation and alkene scission has been observed in the reaction of carbene complexes with enamines.[328-30]

Coordinated carbene ligands which are not stablized by heteroatoms attached to the carbene carbon can be transferred even to non-activated alkenes. The diphenylcarbene complex of tungsten has been found to react with isobutene yielding diphenylethene as the major product along with the cyclopropane derivative (eq 163).[280]

$$(CO)_5W=C(C_6H_5)_2 + CH_2=C(CH_3)_2 \rightarrow CH_2=C(C_6H_5)_2 + \underset{\textbf{LXIX}}{\triangle} + W(CO)_6 + ..(163)$$

In contrast, no metathesis-like products have been observed in the reaction of pentacarbonyl(phenylcarbene)tungsten with alkenes. This complex, which decomposes rapidly at -50°C, was found in competition experiments at -78°C to be 318 times more reactive towards isobutene than towards propene. Hence it was concluded that the reaction is initiated by an electrophilic attack of the carbene carbon atom upon the alkene without prior loss of carbon monoxide.[192]

The cyclopropanation of alkenes is stereospecific, the configuration of the alkene being maintained in the cyclopropane.[192,325] In the reaction of the phenylcarbene complex of tungsten with propene, styrene and cis - 2-butene, the cyclopropanes are preferentially formed in the thermodynamically less stable syn - configuration (eq 164).[192] This result was initially interpreted in terms of the preferred formation of a puckered metallacyclobutane intermediate in which repulsive 1,3-diaxial interactions are minimized. An anti - cyclopropane would then be expected as the major product from 2-methyl-2-butene. However, a 94:1 excess of the syn-isomer was found, showing that the stereochemistry of the cyclopropanes is determined at a transition state substantially earlier than the metallacyclobutane.[17]

$$(CO)_5W{=}C(C_6H_5)H \; + \quad \underset{H}{\overset{R^1}{\diagdown}}C{=}C\underset{H}{\overset{R^2}{\diagup}} \quad \longrightarrow \quad \text{(structure LXX)} \qquad (164)$$

$R^1 = CH_3, C_6H_5; \; R^2 = H$

$R^1 = R^2 = CH_3$

LXX + ...

10.3 Synthesis of Heterocycles

Cycloaddition reactions of carbene ligands have been extended to the synthesis of heterocycles. Fischer-type alkoxy- and amino-carbene complexes react with N-acyl imines, which are known to add to diazomethane, isocyanides and carbenes to give [4+1] cycloaddition products (eq 165).[331] Thiocarbene complexes, however, fail to give 2-oxazolines. Instead, addition of the thioalkyl group to the imine carbon atom occurs, with cleavage of the sulfur-carbene carbon bond.

$$(CO)_5Cr=C(R)X + (F_3C)_2C=NC(O)R' \longrightarrow \quad\quad\quad\quad (165)$$

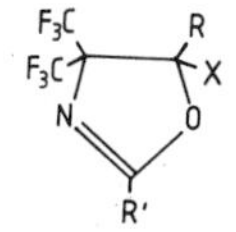

$$R = CH_3, C_6H_5$$

$$X = OCH_3, N(CH_3)_2 \qquad\qquad\qquad\qquad \textbf{LXXI} + \ldots$$

The principle of α-addition in diaminocarbene complexes of palladium has led to a variety of nitrogen heterocycles. The carbene ligands bearing γ- or δ-functionalized substituents are released from the metal either by ligand substitution or by oxidation. Imidazolones, along with carbodiimides, iminooxazolidines, dihydrooxadiazinones, cyanamides and 2-oxazolines have been obtained by this route (eqs 166-8).[218,332-3] The heterocycles can also be synthesized with catalytic amounts of palladium(II) starting from isocyanides and β-aminoalcohols or α-aminoesters.

$$\overset{\displaystyle PR_3}{Cl_2(RNC)Pd=C(NHR)NHCHRCO_2R' \longrightarrow} \qquad\qquad (166)$$

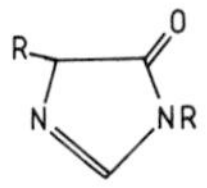

$$\textbf{LXXII} + \ldots$$

$$\overset{\displaystyle R_3P}{Cl_2(RNC)Pd=C(NHR)NHNHC(O)CH(CH_3)OH \longrightarrow}$$

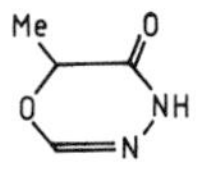

$$\textbf{LXXIII} + RNHCN + \ldots$$

$$(167)$$

$$\overset{\displaystyle Ag_2O}{Cl_2(RNC)Pd=C(NHR)NH(CH_2)_2OH \longrightarrow}$$

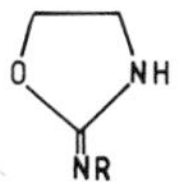

$$\textbf{LXXIV} + \ldots \qquad\qquad (168)$$

10.4 Insertion Reactions

Besides cycloaddition to unsaturated substrates insertion reactions are another characteristic feature of carbenes. The insertion of free electrophilic carbenes into the carbon-hydrogen bond has been studied in particular detail.[334] Metal complex derived carbenes, however, have been found to attack only the reactive silicon, germanium and tin to hydrogen bonds. Insertions into oxygen-hydrogen bonds are also known.

Carboxylic acids react with alkoxycarbene ligands to give the corresponding esters (eq 169.)[315] Triphenylphosphine is added to avoid decomposition of the metal-carbonyl fragment. Insertion of the chromium coordinated methoxy(phenyl)carbene into the O-H bond of alcohols has been achieved in a CO atmosphere.[13]

$$\underline{cis} - [(H_5C_6)_3P](CO)_4Cr=C(R)OCH_3 + R'CO_2H \xrightarrow{(C_6H_5)_3P}$$
$$R'CO_2C(H)(R)OCH_3 + \underline{trans}-[(H_5C_6)_3P]_2Cr(CO)_4 \qquad (169)$$

Alkoxy- and aminocarbene complexes of chromium give element-hydrogen insertion products on reaction with organyl hydrides of silicon, germanium and tin (eq 170).[335-9] Increased yields have been obtained in the presence of pyridine. It is suggested from a kinetic study of the reaction that the significant contributions to the insertion process comprise a first-order dissociative pathway involving an internal rearrangement of the carbene complex, and a second-order associative path. For the latter a nucleophilic attack of the hydride at the carbene carbon atom has been postulated. The reaction rate increases in the order Si<Ge«Sn. This result is believed to reflect the increasing ability of the IVB element to achieve five-coordination.[338-9]

$$(CO)_5Cr=C(X)p-C_6H_4Y + R_3EH \xrightarrow{NC_5H_5} R_3E-C(H)(X)p-C_6H_4Y$$
$$+ (CO)_5CrNC_5H_5 \qquad (170)$$
$$X = OCH_3, \overline{N(CH_2)_3CH_2}; \ Y = OCH_3, Cl, H; \ R = alkyl, C_6H_5; \ E = Si, Ge, Sn$$

Insertion reactions of carbene ligands are so far restricted to Fischer-type compounds. Dichlorophosphine complexes of platinum containing imidazolidinylidene ligands undergo halogen-hydrogen ligand exchange when reacted with triethylsilane.[340]

11.   CARBONYL CARBENE COMPLEXES AS BIFUNCTIONAL REAGENTS

Metal carbonyls are the reagents of choice for the carbonylation of organic substrates.[341]   Thus there have been attempts to make simultaneous use of both the carbene and the carbonyl ligands of carbonyl carbene complexes.

11.1 Synthesis of Naphthols and Related Systems

Pentacarbonyl[methoxy(phenyl)carbene]chromium (**XVIII**) reacts with a variety of alkynes under mild conditions in ethers to give 4-methoxy-1-naphthols coordinated to a tricarbonyl chromium fragment.[30,342-5]   The unsubstituted naphthol ring and the C-4 methoxy part originate from the carbene ligand, while the naphthol group is formed from a CO ligand.   The incorporation of the alkyne into the naphthol skeleton is shown to give the 2-alkyl compounds regiospecifically if unbranched 1-alkynes are used.   p-Substituted tolans, however, lead to the formation of both stereoisomers arising from a different orientation of the alkyne component.[345]   If the reaction proceeds under kinetic control the metal is coordinated exclusively to the substituted naphthol ring (eq 171). On heating the chromium carbonyl fragment migrates to the unsubstituted naphthol ring to give the thermodynamically more stable $5\text{-}10\text{-}\eta^6$-isomers (eq 172).   Naphthol formation requires the presence of effective acceptor ligands in the carbene complex.   If one carbonyl group is replaced with tri(n-butyl)phosphine the yields fall drastically even under more vigorous reaction conditions.[344]

$$(CO)_5Cr=C(p\text{-}XC_6H_4)OCH_3 + R^1\text{-}C\equiv C\text{-}R^2 \xrightarrow[\text{- CO}]{45^\circ C} \text{LXXV} \qquad (171)$$

$X = H, CH_3, CF_3; R^1 = \text{alkyl, aryl}; R^2 = H, \text{alkyl, aryl}$ **LXXV**

$$\textbf{LXXV} \xrightarrow{70\text{-}90^\circ C} \textbf{LXXVI} \qquad (172)$$

The annelation of the carbene substituent is not restricted to the phenyl group.   Carbene ligands containing naphthyl groups, heterocyclic systems such as furyl or thienyl, and cycloalkenyl groups also react to form phenanthrene, benzofuran, benzothiophene and indan complexes (eqs 173-4).[30] The annelation is regioselective. Thus 2-naphthyl- or 3-furylcarbene ligands which offer two possible modes of cyclization lead exclusively to phenanthrene or benzofuran derivatives.

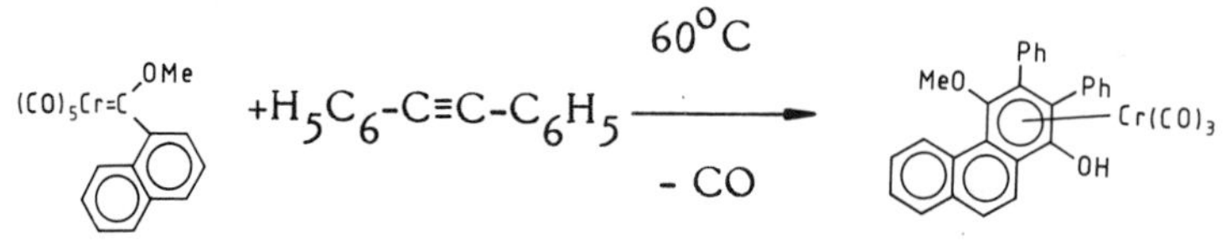

$$(CO)_5Cr=C \quad +H_5C_6-C\equiv C-C_6H_5 \xrightarrow[-CO]{60^{\circ}C} \quad \text{LXXVIII} \quad (173)$$

LXXVII

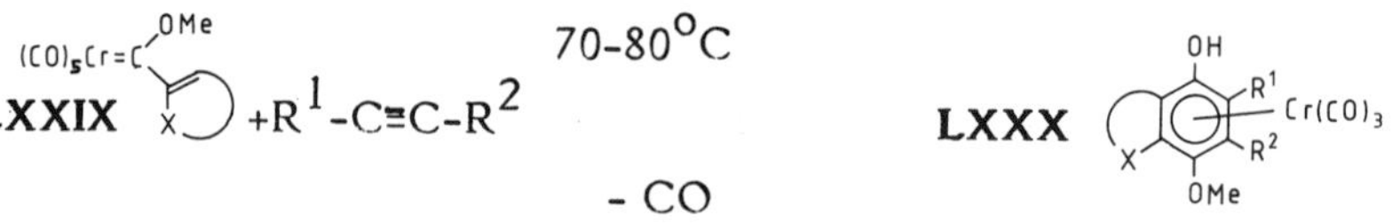

$$(CO)_5Cr=C \quad +R^1-C\equiv C-R^2 \xrightarrow[-CO]{70-80^{\circ}C} \quad \text{LXXX} \quad (174)$$

LXXIX

X = OCH=CH, CH=CHO, SCH=CH, $(CH_2)_3$

The cyclization reaction is favored, if the electron deficiency at the carbene carbon atom is increased. Thus diarylcarbene complexes are found to react at room temperature, even with acetylene and phenyl propynoic acid ester which fail to react with the methoxycarbene compound **XVIII**. Carbene complexes containing two different aromatic substituents have been used for competition studies.[30] The annelation is favored by electron-accepting substitution of the aromatic group as shown for the p-tolyl(p-trifluoromethylphenyl)carbene complex (eq 175). Furthermore, phenyl rings are annelated in preference to furyl and naphthyl groups.

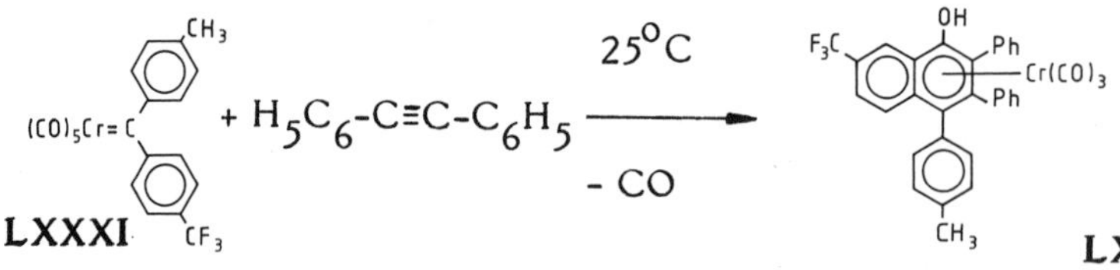

$$(CO)_5Cr=C \quad + H_5C_6-C\equiv C-C_6H_5 \xrightarrow[-CO]{25^{\circ}C} \quad \text{LXXXII} \quad (175)$$

LXXXI

An unusual 1:2-cycloaddition product has been obtained from the methoxy(methyl)carbene complex **XXIV** when reacted with an excess of phenylacetylene (eq 176).[346]

$$(CO)_5Cr=C(CH_3)OCH_3 + 2\ H_5C_6\text{-}C\equiv C\text{-}H \longrightarrow \quad (176)$$

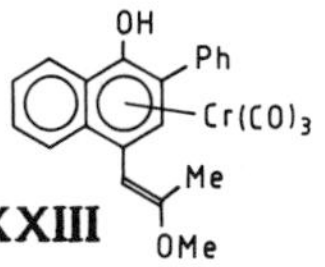

XXIV $-\ CO$

LXXXIII

A kinetic study of naphthol formation from the methoxy(phenyl)carbene complex **XVIII** and tolan derivatives **LXXXIV** in di-n-butyl ether solution established the rate law:[347]

$$-\frac{d[\text{XVIII}]}{dt} = \frac{k_1 \cdot k_2 \cdot [\text{LXXXIV}]}{k_1[\text{CO}] + k_2[\text{LXXXIV}]} \cdot [\text{XVIII}]$$

Based on this result, and on activation parameters similar to those for dissociative ligand substitution reactions of carbene complexes, the suggested mechanism involves initial loss of a CO ligand followed by the coordination of the alkyne. The subsequent steps are supposed to occur within the coordination sphere of the metal, which is believed to act as a template holding the reactants in a position favorable for cyclization. This idea is supported by the formation of the thermodynamically less stable product under kinetically controlled conditions and by the fact that chromium cannot be replaced by molybdenum or tungsten in these reactions.[348]

The existence of coordinatively unsaturated intermediates is probably responsible for the dramatically reduced product selectivity if non-coordinating solvents such as n-heptane are used. In addition to naphthol complexes obtained in moderate yield, indene and furan derivatives have been isolated from the reaction of the methoxy(phenyl)carbene complex **XVIII** and tolan (eq 177).[349]

The naphthol ligands can be released from the metal either by ligand substitutition using carbon monoxide or by oxidation (eqs 178-9).[349] Under

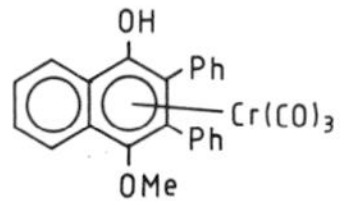

LXXXV 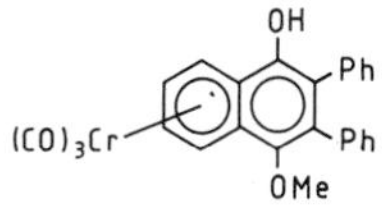   **LXXXVI** 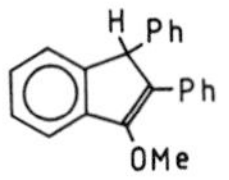   **LXXXVII** 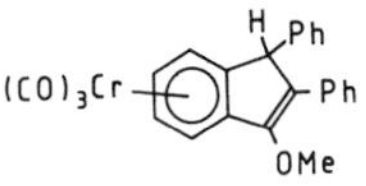   **LXXXVIII**

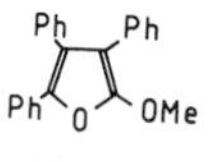

LXXXIX 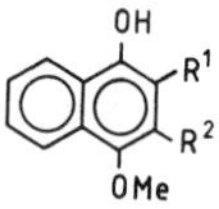   **XC** 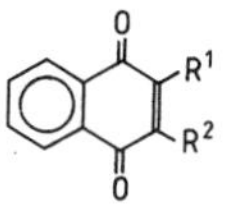   **XCI** 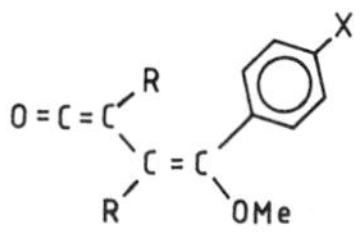   **XCII**

$$\text{heptane}$$

$$(CO)_5Cr=C(C_6H_5)OCH_3 + H_5C_6-C\equiv C-C_6H_5 \xrightarrow[80°C]{} \text{LXXXV} + \text{LXXXVI}$$

$$+ \text{ LXXXVII } + \text{LXXXVIII} + \text{LXXXIX} + \text{LXXXIX.Cr(CO)}_3 + ... \qquad (177)$$

$$\text{XC.Cr(CO)}_3 + 3\text{ CO} \xrightarrow[80°C]{50 \text{ atm CO}} \text{XC} + \text{Cr(CO)}_6 \qquad (178)$$

$$\text{XC.Cr(CO)}_3 \xrightarrow{HNO_3/CH_3COOH} \text{XCI} \qquad +... \qquad (179)$$

mild oxidizing conditions the metal-ring bond is maintained in the 5-10-η^6-complexes and naphthoquinone complexes are obtained (eq 180).[350]

$$\text{XC.5-10-}\eta^6\text{-Cr(CO)}_3 \xrightarrow{Ag_2O} \text{XCI.}\eta^6\text{-Cr(CO)}_3 \quad +... \qquad (180)$$

11.2 Synthesis of Vinyl Ketenes

In contrast to alkyl alkynes, bis(trimethylsilyl)acetylene reacts with aryl(methoxy)carbene complexes of chromium to give silyl substituted vinyl ketenes in which the aryl group is coordinated to a tricarbonyl chromium fragment. The free ligands are also obtained (eq 181).[351-2] By this route vinyl ketenes known as intermediates in cyclization reactions become isolable. An x-ray analysis established a transoid conformation of the vinyl ketene, possibly due to the large silyl substituents.[353] Cisoid vinyl ketenes coordinated to a carbonyl chromium fragment in a diene-like manner have been suggested as intermediates in the synthesis of naphthols and cyclobutenones (cf. the next Section).[352]

$$(CO)_5Cr=C(p-XC_6H_4)OCH_3 + R-C\equiv C-R \longrightarrow \text{XCII} + \text{XCII.Cr(CO)}_3 + ... \qquad (181)$$

$$R = Si(CH_3)_3; \quad X = H, CH_3, OCH_3, CF_3$$

The formation of the products may formally be regarded as a head-to-tail addition of a carbene and a carbonyl ligand to the alkyne. Related carbonylations of a carbene ligand yielding a coordinated ketene had previously been observed for complexes of manganese and iron.[354-5]

11.3 Miscellaneous

Instead of naphthols or vinyl ketenes cyclobutenones and furans arising from the transfer of the carbene and one carbonyl ligand to the alkyne are sometimes obtained. The methoxy(methyl)carbene complex of chromium **XXIV**, lacking an aromatic or olefinic carbene substituent, reacts with tolan to give a cyclobutenone ligand bonded to the metal (eq 182).[356] Similarly the four-membered ring system is formed if the ortho-positions of the phenyl carbene substituent are blocked for annelation as in pentacarbonyl[methoxy(2,6-difluorophenyl)carbene]chromium.[357] Blocking of the phenyl ring by 2,6-dimethyl substitution, however, is not effective. On reaction with tolan migration of a methyl group occurs leading to an indene derivative without incorporation of a carbonyl ligand (eq 183).[358]

$$(CO)_5Cr=C(CH_3)OCH_3 + H_5C_6-C\equiv C-C_6H_5 \xrightarrow{-CO} \quad \textbf{XCIII} \qquad (182)$$

$$\textbf{XCIV} + H_5C_6-C\equiv C-C_6H_5 \longrightarrow \textbf{XCV} + \ldots \qquad (183)$$

An unusual formation of a furan ring has been observed in the reaction of a ferrocenylcarbene complex with tolan (eq 184).[359] Related furan derivatives have been obtained along with other products from the methoxy(phenyl)carbene complex **XVIII** using a non-coordinating solvent.[349]

$$(CO)_5Cr=C(Fc)OCH_3 + H_5C_6-C\equiv C-C_6H_5 \xrightarrow{-CO} \textbf{XCVI} \qquad (184)$$

Fc = Ferrocenyl

12. CARBONYL CARBENE COMPLEXES IN ORGANIC SYNTHESIS

The role of carbene complex intermediates in the olefin metathesis reaction is treated elsewhere in this volume. Other stoichiometric reactions of carbonyl carbene complexes leading to interesting organic products will be discussed below.   These reactions are based on a) the electrophilicity of alkoxy-substituted carbene carbon atoms, b) the acidity of hydrogen atoms bonded to an α-carbon atom, and c) the addition of a carbene and a carbonyl ligand to unsaturated hydrocarbons.

12.1 Synthesis of Peptides

The facile and high yield aminolysis reaction has been used in peptide synthesis.   Like primary and secondary amines, amino acid esters react with pentacarbonyl(alkoxycarbene) complexes of chromium and tungsten to give aminocarbene complexes (eq 185).[210-1]   Due to the nucleophilicity of carboxylate anions the aminolysis reaction fails if amino acid carboxylates are employed instead of esters.   As a consequence using the glutamic acid component both carboxylate groups have to be protected by esterification. However, no further protection is required either for heterocyclic nitrogen functions or for hydroxy groups such as those in histidine and serine esters. The aminolysis reaction is not confined to the α-amino group, both the α- and the ε-amino group being reactive in lysine esters.   In the aminocarbene complex the nitrogen atom is no longer basic or nucleophilic and thus the pentacarbonyl(organylcarbenyl)metal fragment may be used as an amino-protecting group.   Alkaline hydrolysis of the ester leads to the free carboxyl group at which coupling of a second amino ester is achieved using the dicyclohexylcarbodiimide/N-hydroxysuccinimide   (DCCD/HOSU)   method (eqs 186-7).

$$(CO)_5M=C(R)OCH_3 + H_2NCH(R^1)CO_2CH_3 \xrightarrow[- CH_3OH]{(C_2H_5)_2O/20^\circ C}$$

$$(CO)_5M=C(R)NHCH(R^1)CO_2CH_3 \qquad\qquad (185)$$

$$M = Cr, W$$

$$(CO)_5M=C(R)NHCH(R^1)CO_2CH_3 \xrightarrow[\text{2) HCl}]{\text{1) NaOH/dioxane}}$$

$$(CO)_5M=C(R)NHCH(R^1)CO_2H \qquad\qquad (186)$$

$$(CO)_5M=C(R)NHCH(R^1)CO_2H + H_2NCH(R^2)CO_2CH_3 \xrightarrow[-H_2O]{\text{DCCD/HOSU}}$$

$$(CO)_5M=C(R)NHCH(R^1)C(O)NHCH(R^2)CO_2CH_3 \qquad (187)$$

The carbenyl N-protected peptide gly-gly-pro-gly-OCH$_3$, occuring as sequence 14-17 of human proinsuline, has been synthesized by this route from the methoxy(phenyl)carbene complex **XVIII** and the methyl ester of glycine in an overall yield of 14%. The peptide can be released from the carbene complex by mild acid hydrolysis using trifluoroacetic acid/20°C or 80% acetic acid/80°C or by treatment with boron trifluoride at -25°C (eq 188). While the above-mentioned tetrapeptide is obtained from the chromium complex in a clean reaction, cleavage of the pentacarbonyl-(methylcarbenyl)tungsten protected ile-phe-OCH$_3$ occurs with side reactions in the N-terminal amino acid isoleucine.

$$(CO)_5M=C(R)NHCH(R^1)C(O)NHCH(R^2)CO_2CH_3 \xrightarrow{CF_3CO_2H/20°C}$$
$$\overset{+}{H_3}NCH(R^1)C(O)NHCH(R^2)CO_2CH_3 + RCHO + M(CO)_6 + ..(188)$$

Carbenyl protection provides a method by which a) peptide derivatives acquire a yellow color which can be recognized easily when using techniques such as chromatography, and b) heavy metal atoms may be introduced into peptides for labeling of amino groups.

12.2 Synthesis of Vitamin K

Pentacarbonyl[methoxy(phenyl)carbene]chromium **XVIII** has been found to react with alkenes and alkynes in different ways leading to cyclopropanes and metathesis like products on the one hand and naphthol complexes on the other. A competitive study using 1,4-enynes established that this class of compounds behaves regiospecifically as alkynes when reacted with carbonyl carbene complexes. By this route metal coordinated allyl-substituted 4-methoxy-1-naphthols become accessible, which may be oxidized to 1,4-naphthoquinones by silver(I) oxide (eqs 189-90).[350]

$$(CO)_5Cr=C(C_6H_5)OCH_3 + R-C\equiv C-CH_2CH=CH_2 \longrightarrow \textbf{XCVII} \qquad (189)$$
$$\textbf{XVIII} \qquad\qquad R = CH_3, C_2H_5, n-C_4H_9 \quad - CO$$

$$\textbf{XCVII} \xrightarrow{Ag_2O} \textbf{XCVIII} \qquad + \textbf{XCVIII.}Cr(CO)_3 \qquad (190)$$

This route has been used for the synthesis of vitamin $K_{(20)}$ containing a methyl and a phytyl side chain in the quinoid ring.[348,360] Similarly 1,4-naphthoquinones with oligo-enyl substituents have been obtained.[348,361] The enyne required for the incorporation into the naphthol ring is readily accessible from the isoprenoid alcohol, which is converted into the bromide and reacted with the propynyl Grignard reagent (eq 191).

$$R-CH_2C(CH_3)=CHCH_2OH \xrightarrow{HBr \text{ or } PBr_3} R-CH_2C(CH)_3=CHCH_2Br$$

$$\xrightarrow[-MgBr_2]{BrMg-C\equiv C-CH_3} \quad R-CH_2C(CH)_3=CHCH_2-C\equiv C-CH_3 \qquad (191)$$

$$R = [(CH_2)_2CH(CH_3)CH_2]_3H, \ [CH_2CH=C(CH_3)CH_2]_nH \ n = 0-2$$

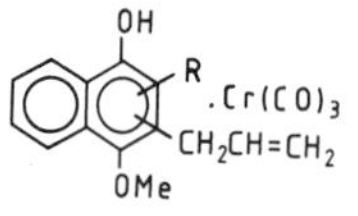

XCVII

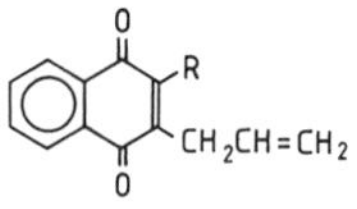

XCVIII

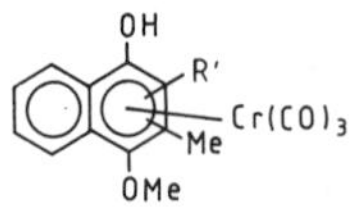

XCIX

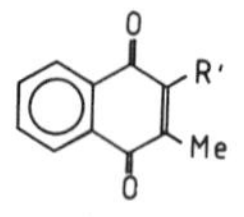

C

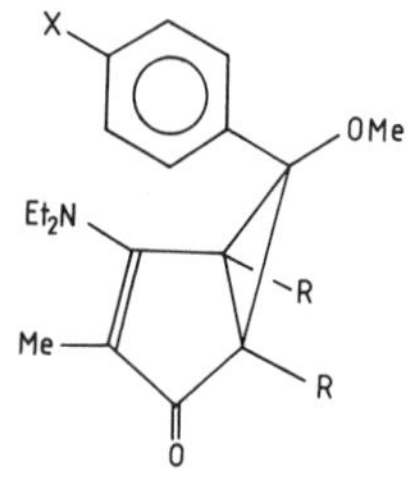

CI

If the reaction of the carbene complex **XVIII** and the enyne is carried out under mild conditions using t-butyl methyl ether as solvent, isomerization of the $1\text{-}4\text{:}9,10\text{-}\eta^6$-naphthol complex to give the $5\text{-}10\text{-}\eta^6$ compound can be avoided (eq 192). Subsequent oxidation with silver(I) oxide affords the vitamin K compounds in overall yields of 50% (eq 193).

$$\textbf{XVIII} + R\text{-}CH_2C(CH_3)=CHCH_2\text{-}C\equiv C\text{-}CH_3$$

$$\xrightarrow[\text{-CO}]{\text{t-BuOMe},45^\circ C} \qquad \textbf{XCIX}; \ R' = CH_2CH=C(CH_3)CH_2R \quad (192)$$

$$\textbf{XCIX} \xrightarrow{\ Ag_2O\ } \textbf{C} \qquad\qquad\qquad (193)$$

Unlike the usual syntheses, based on the condensation of the isoprenoid side chain to the ring system under acidic conditions, this route proceeds with complete retention of configuration at the allylic double bond.

12.3 Miscellaneous

The facile synthesis of stable vinyl ketenes via carbonyl carbene complexes (cf. Section 11.2) has prompted an investigation of the reactivity of this class of compounds. On addition of ynamines an unusual cycloaddition-rearrangement reaction leads to bicyclo[3.1.0]-3-hexene-2-ones (eq 194).[362-3] The structure of the products has been confirmed by an x-ray study of the endo-aryl isomer.[362]

$$\textbf{XCII} \quad + CH_3C\equiv C\text{-}N(C_2H_5)_2 \longrightarrow \textbf{CI} \qquad\qquad (194)$$

$$R = Si(CH_3)_3; \ X = H, CH_3, CF_3$$

Other interesting applications of carbene complexes in organic synthesis are exemplified by the preparation of an indole alkaloid (cf. Section 9.3) and of α-methylene-γ-butyrolactone using carbene complex anions. The 2-oxacyclopentylidene complex of chromium, accessible (Section 6.2) from the methoxy(methyl)carbene complex and epoxide, is alkylated at the α-carbon atom using chloromethyl methyl ether in the presence of butyllithium. Elimination of methanol on alumina followed by oxidation gave the lactone in an overall yield of 20% (eq 195).[266]

$$(CO)_5Cr=C(CH_3)OCH_3 \xrightarrow[\text{2) } \overline{CH_2\text{-}CH_2\text{-}O}]{\text{1) } n\text{-}C_4H_9Li} (CO)_5Cr=\langle O \rangle \quad \textbf{CII} \xrightarrow[\text{2) } ClCH_2OCH_3]{\text{1) } n\text{-}C_4H_9Li}$$

$$\textbf{CIII} \xrightarrow{Al_2O_3} \textbf{CIV} \xrightarrow{(NH_4)_2Ce(NO_3)_6} \textbf{CV} \quad + \ldots \quad (195)$$

13. MORE RECENT DEVELOPMENTS

A review by C.P. Casey[364] deals with reactions of carbene complexes. Selective exchange of ^{13}CO into the cis position of pentacarbonyl[methoxy(phenyl)carbene] complexes of chromium, molybdenum and tungsten is observed, whereas the diphenylcarbene tungsten compound shows no stereoselectivity.[365] The trans-cis isomerization of bis(1,3-imidazolidin-2-ylidene-) molybdenum complexes has been studied.[366] Temperature dependent nmr spectra of the cationic species $\{[P(CH_3)_3]_4ClW=CH_2\}^+$ have been explained by invoking a T-shaped methylene ligand.[367]

Novel complexes bearing bridging or chelating biscarbene ligands have been prepared according to standard procedures (e.g. eq 6) and studied by x-rays.[67,368-9] Addition of amines and phosphines to dithiocarbene ligands coordinated to iron and tungsten leads to substitution and addition-rearrangement products.[370-2] The isomerization of 1,3-dithiol-2-ylidene iron compounds forming heterometallacycles has been shown to be dependent on the nature of additional phosphine ligands.[373] Diaminocarbene ligands are convertible to isocyanide ligands by elimination of amine.[374] Addition of methoxide to a benzylidene rhenium complex is reported to occur with high stereoselectivity.[375]

The C, Cr migration reaction, which leads to aminocarbyne complexes, has been extended from halogen to tin, lead and selenium substituted aminocarbene ligands.[259,376-7]

Thioketone complexes are formed by the reaction of diarylcarbene tungsten compounds with sulfur or isothiocyanates.[288,378-9] The insertion of cyclic enol ethers into the tungsten carbene bond provides a route to vinylcarbene complexes.[380] In the reaction with pentacarbonyl-[methoxy(phenyl)carbene] tungsten, phenylacetylene is reported to insert faster into the metal-carbene bond than cycloalkenes; non-heteroatom-stabilized metal carbenes show an opposite behaviour.[381]

The photo-induced degradation of diphenylcarbene complexes in hexane leads to tetraphenylethylene and hydrogen abstraction products.[382] A nickel coordinated methylene, generated from a nickelacyclobutane precursor, has been converted to methane and ketene on reaction with hydrogen and carbon monoxide respectively.[383] Migration of a methyl ligand to a coordinated methylene is reported to occur in an intermediate iridium methyl methylene complex.[384] A cationic iron ethylidene complex generated by methoxide abstraction has been used as a carbene transfer reagent towards alkenes.[385] The regiochemistry of naphthol formation from pentacarbonyl[methoxy(phenyl)carbene]chromium and alkynes has been established. 1-Alkynes are found to form 2-alkyl naphthol compounds, whereas 2-alkynes give approximately 2:1 mixtures of stereoisomers, the 3-methyl derivatives predominating.[345,386] Oxidative workup leads to naphthoquinones or their monoketals, depending on the solvent used.[349,350,387]

REFERENCES

1. L. Chugaev and M. Shanavy-Grigorizeva, J. Russ.Chem. Soc. 47:776 (1915).

2. L. Chugaev, M. Shanavy-Grigorizeva and A. Posnjak, Z. Anorg. Chem. 148:37 (1925).

3. G. Rouschias and B.L. Shaw, J.C.S. Chem. Commun. 183 (1970).

4. E.O. Fischer and A. Maasböl, Angew. Chem. Int. Ed. Engl. 3:580 (1964).

5. D.L. Cardin, B. Cetinkaya and M.F. Lappert, Chem. Rev. 72:545 (1972).

6. F.A. Cotton and C.M. Lukehart, Prog. Inorg. Chem. 16:487 (1972).

7. F.J. Brown, Prog. Inorg. Chem. 27:1 (1980).

8. H. Fischer, in "The Chemistry of the Metal Carbon Bond" (S. Patai, ed.), Volume 1, (1982), Wiley-Interscience, New York, p.181.

9. E.O. Fischer, Pure Appl. Chem. 24:407 (1970).

10. C.G. Kreiter and E.O. Fischer, XXIII Int. Congr. Pure Appl. Chem. Vol. 6, p.189, Butterworth, London (1971).

11. E.O. Fischer, Pure Appl. Chem. 30:353 (1972)

12. E.O. Fischer, Angew. Chem. 86:651 (1974).

13. K.H. Dötz, Naturissenschaften 62:365 (1975).

14. M.F. Lappert, J. Organomet. Chem. 100:139 (1975).

15. E.O. Fischer, Adv. Organomet. Chem. 14:1 (1976).

16. C.P. Casey in "Transition Metal Organometallics in Organic Synthesis" (H. Alper, ed.), Vol 1, p.189, Academic Press, New York (1976).

17. C.P. Casey, Transition Metal Carbene Complexes in Organic Synthesis, Adv. Pestic. Sci., Plenary Lect. Symp. Pap., 4th Int. Congr. Prestic. Chem., Zurich, July 24-28, 1978, 2, p.148, Pergamon, Oxford (1979).

18. E.O. Fischer, U. Schubert and H. Fischer, Pure Appl. Chem. 50:857 (1978).

19. R.R. Schrock, Acc. Chem. Res. 12:98 (1979).

20. E.O. Fischer and A. Maasböl, Chem. Ber. 100:2445 (1967).

21. R. Aumann and E.O. Fischer, Angew. Chem. 79:900 (1967); Angew. Chem. Int. Ed. Engl. 6:879 (1967).

22. E.O. Fischer and H. Fischer, Chem. Ber. 107:657 (1974).

23. E.O. Fischer and D. Plabst, Chem. Ber. 107:3326 (1974).

24. M.F. Lappert, R.W. McCabe, J.J. McQuitty, P.L. Pye and P.I. Riley, J. C. S. Dalton 90 (1980).

25. E.O. Fischer, C.G. Kreiter, H-J. Kollmeier, J. Müller and R.D. Fischer, J. Organomet. Chem. 28:237 (1971).

26. J.A. Connor and E.M. Jones, J. Chem. Soc. A:1974 (1971).

27. J.W. Wilson and E.O. Fischer, J. Organomet. Chem. 57:C63 (1973).

28. E.O. Fischer and F.R. Kreissl, J. Organomet. Chem. 35:C47 (1972).

29. E.O. Fischer, W. Held, F.R. Kreissl, A. Frank and G. Huttner, Chem. Ber. 110:656 (1977).

30. K.H. Dötz and R. Dietz, Chem. Ber. 111:2517 (1978).

31. J.A. Connor and J.P. Lloyd, J.C.S. Dalton 1470 (1972).

32. G.A. Moser, E.O. Fischer and M.D. Rausch, J. Organomet. Chem. 27:379 (1971).

33. E.O. Fischer, F.J. Gammel, J.O. Besenhard, A. Frank and D. Neugebauer, J. Organomet. Chem. 191:261 (1980).

34. E.O. Fischer, V.N. Postnov and P.R. Kreissl , J. Organomet. Chem. 127:C19 (1977).

35. E.O. Fischer, F.J. Gammel and D. Neugebauer, Chem. Ber. 113:1010 (1980).

36. E.O. Fischer and E. Offhaus, Chem. Ber. 102:2449 (1969).

37. E.O. Fischer, H-J. Beck C.G. Kreiter, J. Lynch, J. Müller and E. Winkler, Chem. Ber. 105:162 (1972).

38. E.O. Fischer, F.R. Kreissl, E. Winkler and C.G. Kreiter, Chem. Ber. 105:588 (1972).

39. E.O. Fischer, P. Stückler, H-J. Beck and F.R. Kreissl, Chem. Ber. 109:3089 (1976).

40. E.O. Fischer and H-J. Beck, Chem. Ber. 104:3101 (1971).

41. E.O. Fischer and R. Aumann, Chem. Ber. 102:1495 (1969).

42. E.O. Fischer and A. Riedel, Chem. Ber. 101:156 (1968).

43. M.J. Webb, R.P. Stewart, jr. and W.A.G. Graham, J. Organomet. Chem. 59:C21 (1973).

44. J. Chatt, G.J. Leigh, C.J. Pickett and D.R. Stanley, J. Organomet. Chem. 184:C64 (1980).

45. D.J. Darensbourg and M.Y. Darensbourg, Inorg. Chem. 9:1691 (1970).

46. F. Carre, G. Cerveaux, E. Colomer, R.J.P. Corriu, J.C. Young, L. Ricard and R. Weiss, J. Organomet. Chem. 179:215 (1979).

47. E.O. Fischer, H. Hollfelder, F.R. Kreissl and W. Üdelhoven, J. Organomet. Chem. 113:C31 (1976).

48. E.O. Fischer, H. Hollfelder, P. Friedrich, F.R. Kreissl and G. Huttner, Chem. Ber. 110:3467 (1977).

49. E.O. Fischer, P. Rustemeyer and D. Neugebauer, Z. Naturforsch. B35:1083 (1980).

50. E.O. Fischer and H-J. Kollmeier, Angew. Chem. Int. Ed. Engl. 9:309 (1970).

51. H.G. Raubenheimer, S. Lotz, H.W. Viljoen and A.A. Chalmers, J. Organomet. Chem. 152:73 (1978).

52. E.O. Fischer, K. Scherzer and F.R. Kreissl, J. Organomet. Chem. 118:C33 (1978).

53. D.J. Darensbourg and M.Y. Darensbourg, Inorg. Chim. Acta 5:247 (1971).

54. E.O. Fischer and A. Maasböl, J. Organomet. Chem. 12:P15 (1968).

55. C.P. Casey, C.R. Cyr and R.A. Boggs, Synth. Inorg. Met.-Org. Chem. 3:249 (1973).

56. E.O. Fischer and W. Kalbfus, J. Organomet. Chem. 46:C15 (1972).

57. J.A. Connor and E.M. Jones, J.C.S. Chem. Commun. 570 (1971).

58. J.A. Connor and E.M. Jones, J. Chem. Soc. A:3368 (1971).

59. E.O. Fischer, T. Selmayr and F.R. Kreissl, Chem. Ber. 110:2947 (1977).

60. E. Moser and E.O. Fischer, J. Organomet. Chem. 12:P1 (1968).

61. E.O. Fischer, T. Selmayr, F.R. Kreissl and U. Schubert, Chem. Ber. 110:2574 (1977).

62. U. Schubert, M. Wiemer and F.H. Köhler, Chem. Ber. 112:708 (1978).

63. E.O. Fischer and S. Fontana, J. Organomet. Chem. 40:159 (1972).

64. H.G. Raubenheimer and E.O. Fischer, J. Organomet. Chem. 91:C23 (1975).

65. P.B. Hitchcock, M.F. Lappert and P.L. Pye, J. C. S. Dalton Trans. 2160 (1977).

66. E.O. Fischer, F.R. Kreissl, C.G. Kreiter and E.W. Meineke, Chem. Ber. 105:2558 (1972).

67. E.O. Fischer, W. Röll, U. Schubert and K. Ackermann, Angew. Chem. Int. Ed. Engl. 20:611 (1981).

68. H. Fischer and E.O. Fischer, Chem. Ber. 107:673 (1974).

69. E.O. Fischer and K. Richter, Chem. Ber. 109:1140 (1976).

70. W. Petz, J. Organomet. Chem. 72:369 (1974).

71. P.T. Wolczanski, R.S. Threkel and J.E. Bercaw, J. Am. Chem. Soc. 101:218 (1979).

72. W. Petz and G. Schmid, Angew. Chem. Int. Ed. Engl. 11:934 (1972).

73. W. Petz, J. Organomet. Chem. 55:C42 (1973).

74. W. Petz and A. Jonas, J. Organomet. Chem. 120:423 (1976).

75. S. Fontana, U. Schubert and E.O. Fischer, J. Organomet. Chem. 146:39 (1978).

76. M.D. Cooke and E.O. Fischer, J. Organomet. Chem. 56:279 (1973).

77. U. Schubert, in "The Chemistry of the Metal Carbon Bond" (S. Patai, ed.), Volume 1, (1982) Wiley-Interscience, New York, p.233.

78. E.O. Fischer, W. Kleine, G. Kreis and F.R. Kreissl, Chem. Ber. 111:3542 (1978).

79. A.J. Hartshorn and M.F. Lappert, J. C. S. Chem. Commun. 761 (1976).

80. E.O. Fischer, W. Kleine and F.R. Kreissl, Angew. Chem. Int. Ed. Engl 15:616 (1976).

81. E.O. Fischer, W. Kleine, F.R. Kreissl, H. Fischer, P. Friedrich and G. Huttner, J. Organomet. Chem. 128:C49 (1977).

82. H. Fischer, A. Motsch, U. Schubert and D. Neugebauer, Angew. Chem. Int. Ed. Engl. 20:463 (1981).

83. U. Schubert, E.O. Fischer and D. Wittmann, Angew. Chem. Int. Ed. Engl. 19:643 (1980).

84. E.O. Fischer, R.B.A. Pardy and U. Schubert, J. Organomet. Chem. 181:37 (1979).

85. E.O. Fischer, P. Stückler and F.R. Kreissl, J. Organomet. Chem. 129:197 (1977).

86. E.O. Fischer and G. Besl, Z. Naturforsch. B34:1186 (1979).

87. E.O. Fischer, E.W. Meineke and F.R. Kreissl, Chem. Ber. 110:1140 (1977).

88. E.O. Fischer, W. Schambeck and F.R. Kreissl, J. Organomet. Chem. 169:C27 (1979).

89. E.O. Fischer, R.L. Clough, G. Besl and F.R. Kreissl, Angew. Chem. Int. Ed. Engl. 15:543 (1976).

90. E.O. Fischer, R.L. Clough and P. Stückler, J. Organomet. Chem. 120:C6 (1976).

91. E.O. Fischer and A. Frank, Chem. Ber. 111:3740 (1978).

92. E.M. Badley, J. Chatt, R.L. Richards and G.A. Sim, Chem. Commun. 1322 (1969).

93. E.M. Badley, J. Chatt and R.L. Richards, J. Chem. Soc. A:21 (1971).

94. F. Bonati and G. Minghetti, J. Organomet. Chem. 24:251 (1970).

95. H.C. Clark and L.E. Manzer, Inorg. Chem. 11:503 (1972).

96. B. Crociani, T. Boschi, M. Nicolini and U. Belluco, Inorg. Chem. 11:1292 (1972).

97. C.H. Davies, C.H. Game, M. Green and F.G.A. Stone, J. C. S. Dalton 357 (1974).

98. J. Miller amd A.L. Balch, Inorg. Chem. 11:2069 (1972).

99. A.L. Balch and J. Miller, J. Am. Chem. Soc. 94:417 (1972).

100. D.J. Doonan and A.L. Balch, Inorg. Chem. 13:921 (1974).

101. K. Bartel and W.P. Fehlhammer, Angew. Chem. Int. Ed. Engl. 13:600 (1974).

102. J.A. Connor, E.M. Jones, G.K. McEven, M.K. Lloyd and J.A. McCleverty, J. C. S. Dalton 1246 (1972).

103. J.A. McCleverty and J. Williams, Transition Met. Chem. 1:288 (1976).

104. U. Schöllkopf and F. Gerhart, Angew. Chem. Int. Ed. Engl. 6:560 (1967).

105. M.L.H. Green and C.R. Hurley, J. Organomet. Chem. 10:188 (1967).

106. M.L.H. Green, L.C. Mitchard and M.G. Swanwick, J. Chem. Soc. A:794 (1971).

107. H. Felkin, B. Meunier, C. Pascard and T. Prange, J. Organomet. Chem. 135:361 (1977).

108. K.P. Darst, P.G. Lenhert, C.M. Lukehart and L.T. Warfield, J. Organomet. Chem. 195:317 (1980).

109. M. Wada, N. Asada and K. Oguro, Inorg. Chem. 17:2353 (1978).

110. M. Wada, S. Kanai, R. Maeda, M. Kinoshita and K. Oguro, Inorg. Chem. 18:417 (1979).

111. T-A. Mitsudo, H. Watanabe, Y. Watanabe, N. Nitani and Y. Takegami, J. C. S. Dalton 395 (1979).

112. W-K. Wong, W. Tam and J.A. Gladysz, J. Am. Chem. Soc. 101:5440 (1979).

113. C.H. Game, M. Green, J.R. Moss and F.G.A. Stone, J. C. S. Dalton 351 (1974).

114. D.H. Bowen, M. Green, D.M. Grove, J.R. Moss and F.G.A. Stone, J. C. S. Dalton 1189 (1974).

115. R.B. King, J. Am. Chem. Soc. 85:1922 (1963).

116. C.P. Casey, J. C. S. Chem. Commun. 1220 (1970).

117. F.A. Cotton and C.M. Lukehart, J. Am. Chem. Soc. 93:2672 (1971).

118. T. Kruck and L. Liebig, Chem. Ber. 106:1055 (1973).

119. E.D. Dobrzynski and R.J. Angelici, Inorg. Chem. 14:1513 (1975).

120. F.B. McCormick and R.J. Angelici, Inorg. Chem. 18:1231 (1979).

121. D.F. Christian, G.R. Clark, W.R. Roper, J.M. Waters and K.R. Whittle, J. C. S. Chem. Commun. 458 (1972).

122. Y. Yamamoto and H. Yamazaki, Bull. Chem. Soc. Jpn. 48:3691 (1975).

123. D.F. Christian, G.R. Clark and W.R. Roper, J. Organomet. Chem. 81:C7 (1974).

124. D.F. Christian, H.C. Clark and R.F. Stepaniah, J. Organomet. Chem. 112:227 (1976).

125. M. Wada and Y. Koyama, J. Organomet. Chem. 201:477 (1980).

126. M.H. Chisholm and H.C. Clark, J. C. S. Chem. Commun. 763 (1970).

127. M.H. Chisholm and H.C. Clark, J. C. S. Chem. Commun. 1484 (1971).

128. M.H. Chishom and H.C. Clark, Inorg. Chem. 10:1711 (1971).

129. M.H. Chisholm and H.C. Clark, J. Am. Chem. Soc. 94:1532 (1972).

130. G.K. Anderson, R.J. Cross, L. Manojlović-Muir, K.W. Muir and R.A. Wales, J. C. S. Dalton 684 (1979).

131. R.A. Bell and M.H. Chisholm, J. C. S. Chem. Commun. 818 (1974).

132. R.A. Bell, M.H. Chisholm, D.A. Couch and L.A. Rankel, Inorg. Chem. 16:677 (1977).

133. K. Oguro, M. Wada and R. Okawara, J. C. S. Chem. Commun. 899 (1975).

134. D.J. Cardin, B. Cetinkaya, M.F. Lappert, L. Manojlovic-Muir and K.W. Muir, J. C. S. Chem. Commun. 400 (1971).

135. B. Cetinkaya, P. Dixneuf and M.F. Lappert, J. C. S. Dalton 1827 (1974).

136. M.F. Lappert, P.L. Pye and G.M. McLaughlin, J. C. S. Dalton 1272 (1977).

137. M.F. Lappert and P.L. Pye, J. C. S. Dalton 1283 (1977).

138. M.F. Lappert and P.L. Pye, J. C. S. Dalton 2172 (1977).

139. M.F. Lappert and P.L. Pye, J. C. S. Dalton 837 (1978).

140. P.B. Hitchcock, M.F. Lappert, P.L. Pye and S. Thomas, J. C. S. Dalton 1929 (1979).

141. W.A. Herrmann, Angew. Chem. Int. Ed. Engl. 13:599 (1974).

142. W.A. Herrmann, Chem Ber. 108:486 (1975).

143. W.A. Herrmann, Chem. Ber. 108:3412 (1975).

144. W.A. Herrmann, J. Plank, M.L. Ziegler and K. Weidenhammer, Angew. Chem. Int. Ed. Engl. 17:777 (1978).

145. H.G. Raubenheimer and H.E. Swanepoel, J. Organomet. Chem. 141:C21 (1977).

146. J. Daub and J. Kappler, J. Organomet. Chem. 80:C5 (1974).

147. W. Petz, J. Organomet. Chem. 172:415 (1979).

148. M. Brookhart and G.O. Nelson, J. Am. Chem. Soc. 99:6099 (1977).

149. A.R. Cutler, J. Am. Chem. Soc. 101:604 (1979).

150. D.N. Clark and R.R. Schrock, J. Am. Chem. Soc. 100:6774 (1978).

151. D. Mansuy, M. Lange, J-C. Chottard, P. Guerin, P. Morliere, D. Brault and M. Rougee, J. C. S. Chem. Commun. 648 (1977).

152. D. Mansuy, P. Guerin and J-C. Chottard, J. Organomet. Chem. 171:195 (1979).

153. J-P. Battioni, D. Mansuy and J-C. Chottard, Inorg. Chem. 19:791 (1980).

154. H. Berke, Z. Naturforsch B35:86 (1980).

155. B. Cetinkaya, M.F. Lappert and K. Turner, J. C. S. Chem. Commun. 851 (1972).

156. B. Cetinkaya, M.F. Lappert, G.M. McLaughlin and K. Turner, J. C. S. Dalton 1591 (1974).

157. A.J. Hartshorn, M.F. Lappert and K. Turner, J. C. S. Chem. Commun. 929 (1975).

158. C.W. Fong and G. Wilkinson, J. C. S. Dalton 1100 (1975).

159. A.J. Hartshorn, M.F. Lappert and K. Turner, J. C. S. Dalton 348 (1978).

160. K. Öfele, J. Organomet. Chem. 12:P42 (1968).

161. K. Öfele, Angew. Chem. Int. Ed. Engl. 8:916 (1969).

162. K. Öfele, Angew. Chem. Int. Ed. Engl. 7:950 (1968).

163. K. Öfele, J. Organomet. Chem. 22:C9 (1970).

164. Y. Kawada and W.M. Jones, J. Organomet. Chem. 192:87 (1980).

165. N.T. Allison, Y. Kawada and W.M. Jones, J. Am. Chem. Soc. 100:5224 (1978).

166. T.J. Burkhardt and C.P. Casey, J. Am. Chem. Soc. 95:5833 (1973).

167. H. Fischer and K.H. Dötz, Chem. Ber. 113:193 (1980).

168. C.G. Kreiter, Angew. Chem. Int. Ed. Engl. 7:390 (1968).

169. U. Schubert, J. Organomet. Chem. 185:373 (1980).

170. O.S. Mills and A.D. Redhouse, J. Chem. Soc. A:642 (1968).

171. J.A. Connor and O.S. Mills, J. Chem. Soc. A:334 (1969).

172. E.O. Fischer, E. Winkler, C.G. Kreiter, G. Huttner and B. Krieg, Angew. Chem. Int. Ed. Engl. 10:922 (1971).

173. E. Moser and E.O. Fischer, J. Organomet. Chem. 15:147 (1968).

174. C.G. Kreiter and E.O. Fischer, Angew. Chem. Int. Ed. Engl. 8:761 (1969).

175. C.G. Kreiter, K. Öfele and G.W. Wieser, Chem. Ber. 109:1749 (1976).

176. H-J. Beck, E.O. Fischer and C.G. Kreiter, J. Organomet. Chem. 26:C41 (1971).

177. J. Müller and J.A. Connor, Chem. Ber. 102:1148 (1969).

178. C.G. Kreiter and V. Formacek, Angew. Chem. Int. Ed. Engl. 11:141 (1972).

179. L.F. Farnell, E.W. Randall and E. Rosenberg, J. C. S. Chem. Commun. 1078 (1971).

180. J.A. Connor, E.M. Jones, E.W. Randall and E. Rosenberg, J. C. S. Dalton 2419 (1972).

181. W.B. Perry, T.F. Schaaf, W.L. Jolly, L.J. Todd and D.L. Cronin, Inorg. Chem. 13:2038 (1974).

182. T.F. Block, R.F. Fenske and C.P. Casey, J. Am. Chem. Soc. 98:441 (1976).

183. T.F. Block, and R.F. Fenske J. Organomet Chem. 139:235 (1977).

184. F.R. Kreissl, C.G. Kreiter and E.O. Fischer, Angew. Chem. Int. Ed. Engl. 11:643 (1972).

185. F.R. Kreissl, E.O. Fischer, C.G. Kreiter and H. Fischer, Chem. Ber. 106:1262 (1973).

186. H. Fischer, E.O. Fischer, C.G. Kreiter and H. Werner, Chem. Ber. 107:2459 (1974).

187. E.O. Fischer, T. Selmayr and F.R. Kreissl, Monatsh. Chem. 108:759 (1977).

188. E.O. Fischer, G. Kreis, F.R. Kreissl, C.G. Kreiter and J. Müller, Chem. Ber. 106:3910 (1973).

189. F.R. Kreissl, E.O. Fischer, C.G. Kreiter and K. Weiss, Angew. Chem. Int. Ed. Engl. 12:563 (1973).

190. F.R. Kreissl and E.O. Fischer, Chem. Ber. 107:183 (1974).

191. C.P. Casey and S.W. Polichnowski, J. Am. Chem. Soc. 99:6097 (1977).

192. C.P. Casey, S.W. Polichnowski, A.J. Shusterman and C.R. Jones, J. Am. Chem. Soc. 101:7282 (1979).

193. A. Sanders, L. Cohen, W.P. Giering, D. Kenedy and C.V. Magatti, J. Am. Chem. Soc. 95:5430 (1973).

194. U. Klabunde and E.O. Fischer, J. Am. Chem. Soc. 89:7141 (1967).

195. J.A. Connor and E.O. Fischer, J. Chem. Soc A:578 (1969).

196. E. Moser and E.O. Fischer, J. Organomet. Chem. 16:275 (1969).

197. E.O. Fischer and H-J. Kollmeier, Chem. Ber. 104:1339 (1971).

198. E.O. Fischer and M. Leupold, Chem. Ber. 105:599 (1972).

199. E.O. Fischer, B. Heckl and H. Werner, J. Organomet. Chem. 28:359 (1971).

200. P.E. Baikie, E.O. Fischer and O.S. Mills, Chem. Commun. 1199 (1967).

201. J.A. Connor and P.D. Rose, Tetrahedron Lett. 3623 (1970).

202. J.A. Connor and E.O. Fischer, Chem. Commun. 1024 (1967).

203. E.O. Fischer and S. Fontana, J. Organomet. Chem. 40:367 (1972).

204. B. Heckl, H. Werner and E.O. Fischer, Angew. Chem. Int. Ed. Engl. 7:817 (1968).

205. H. Werner, E.O. Fischer, B. Heckl and C.G. Kreiter, J. Organomet. Chem. 28:367 (1971).

206. E.O. Fischer and H-J. Kalder, J. Organomet. Chem. 131:57 (1977).

207. H. Brunner, E.O. Fischer and M. Lappus, Angew. Chem. Int. Ed. Engl 10:924 (1971).

208. H. Brunner, J. Doppelberger, E.O. Fischer and M. Lappus, J. Organomet. Chem. 112:65 (1976).

209. A. Davison and D.L. Reger, J. Am. Chem. Soc. 94:9237 (1972).

210. K. Weiss and E.O. Fischer, Chem. Ber. 106:1277 (1973).

211. K. Weiss and E.O. Fischer, Chem. Ber. 109:1868 (1976).

212. K.H. Dötz, Chem. Ber. 110:78 (1977).

213. M.H. Chisholm, H.C. Clark, W.S. Johns, J.E.H. Ward and K. Yasufuku, Inorg. Chem. 14:900 (1975).

214. D. Mansuy, M. Lange, J-C. Chottard and J.F. Bartoli, Tetrahedron Lett. 3027 (1978).

215. D. Mansuy, Pure Appl. Chem. 52:681 (1980).

216. E.O. Fischer and R. Aumann, Chem. Ber. 101:963 (1968).

217. J.A. Connor and P.D. Rose, J. Organomet. Chem. 46:329 (1972).

218. Y. Ito, T. Hirao and T. Saegusa, J. Organomet. Chem. 131:121 (1977).

219. L. Knauss and E.O. Fischer, Chem. Ber. 103:3744 (1970).

220. L. Knauss and E.O. Fischer, J. Organomet. Chem. 31:C68 (1971).

221. E.O. Fischer and L. Knauss, Chem. Ber. 103:1262 (1970).

222. C.P. Casey and W.R. Brunsvold, Inorg. Chem. 16:391 (1977).

223. E.O. Fischer, M. Leupold, C.G. Kreiter and J. Muller, Chem. Ber. 105:150 (1972).

224. C.T. Lam, C.V. Senoff and J.E.H. Ward, J. Organomet. Chem. 70:273 (1974).

225. E.O. Fischer and V. Kiener, Angew. Chem. Int. Ed. Engl. 6:961 (1967).

226. E.O. Fischer and K.H. Dötz, Chem. Ber. 105:3966 (1972).

227. C.P. Casey and T.J. Burkhardt, J. Am. Chem. Soc. 94:6543 (1972).

228. C.P. Casey, S.H. Bertz and T.J. Burkhardt, Tetrahedron Lett. 1421 (1973).

229. C.P. Casey, T.J. Burkhardt, C.A. Bunnell and J.C. Calabrese, J. Am. Chem. Soc. 99:2127 (1977).

230. C.W. Rees and E.v. Angerer, J. C. S. Chem. Commun. 420 (1972).

231. E.O. Fischer and S. Riedmüller, Chem. Ber. 109:3358 (1976).

232. E.O. Fischer, W. Held and F.R. Kreissl, Chem. Ber. 110:3842 (1977).

233. C.P. Casey, S.W. Polichnowski and R.L. Anderson, J. Am. Chem. Soc. 97:7375 (1975).

234. E.O. Fischer and W. Held, J. Organomet. Chem. 112:C59 (1976).

235. C.P. Casey and W.R. Brunsvold, J. Organomet. Chem. 77:345 (1974).

236. K.H. Dötz, J. Organomet. Chem. 118:C13 (1976).

237. K.H. Dötz, Chem. Ber. 113:3597 (1980).

238. K.H. Dötz, B. Fügen-Köster and D. Neugebauer, J. Organomet. Chem. 182: 489 (1979).

239. T.V. Ashworth, M. Berry, J.A.K. Howard, M. Laguna and F.G.A. Stone, J. C. S. Chem. Commun. 43 (1979).

240. T.V. Ashworth, J.A.K. Howard, M. Laguna and F.G.A. Stone, J. C. S. Dalton 1593 (1980).

241. M. Berry, J.A.K. Howard and F.G.A. Stone, J. C. S. Dalton 1601 (1980).

242. T.V. Ashworth, M. Berry, J.A.K. Howard, M. Laguna and F.G.A. Stone, J. C. S. Chem. Commun. 45 (1979).

243. T.V. Ashworth, M. Berry, J.A.K. Howard. M. Laguna and F.G.A. Stone, J. C. S. Dalton 1615 (1980).

244. M. Berry, J. Martin-Gil, J.A.K. Howard and F.G.A. Stone, J. C. S. Dalton 1625 (1980).

245. E.O. Fischer and U. Schubert, J. Organomet. Chem. 100:59 (1975).

246. E.O. Fischer, G. Kreis, C.G. Kreiter, J. Müller, G. Huttner and H. Lorenz, Angew. Chem. Int. Ed. Engl. 12:564 (1973).

247. E.O. Fischer, H. Hollfelder and F.R. Kreissl, Chem. Ber. 112:2177 (1979).

248. E.O. Fischer, W.R. Wagner, F.R. Kreissl and D. Neugebauer, Chem. Ber. 112:1320 (1979).

249. E.O. Fischer, H-J. Kalder and F.H. Köhler, J. Organomet. Chem. 81:C23 (1974).

250. E.O. Fischer, M. Schluge, J.O. Besenhard, P. Friedrich, G. Huttner and F.R. Kreissl, Chem. Ber. 111:3530 (1978).

251. E.O. Fischer, A. Schwanzer, H. Fischer, D. Neugebauer and G. Huttner, Chem. Ber. 110:53 (1977).

252. E.O. Fischer, S. Walz and W.R. Wagner, J. Organomet. Chem. 134:C37 (1977).

253. E.O. Fischer and T. Selmayr, Z. Naturforsch B32:105 (1977).

254. A. Schwanzer, Dissertation Techn. Univ. Munich, 1976.

255. E.O. Fischer and K. Richter, Chem. Ber. 109:2547 (1976).

256. E.O. Fischer and K. Richter, Chem. Ber. 109:3079 (1976).

257. E.W. Meineke, Dissertation Techn. Univ. Munich, 1976.

258. H. Fischer, A. Motsch and W. Kleine, Angew. Chem. Int. Ed. Engl. 17:842 (1978).

259. H. Fischer, J. Organomet. Chem. 195:55 (1980).

260. E.O. Fischer, K. Weiss and C.G. Kreiter, Chem. Ber. 107:3554 (1974).

261. E.O. Fischer and K. Weiss, Chem. Ber. 109:1128 (1976).

262. G.K. Anderson and R.J. Cross, J. C. S. Dalton 690 (1979).

263. C.P. Casey, R.A. Boggs and R.L. Anderson, J. Am. Chem. Soc. 94:8947 (1972).

264. C.P. Casey and R.L. Anderson, J. Am. Chem. Soc. 96:1230 (1974).

265. C.P. Casey and W.R. Brunsvold, J. Organomet. Chem. 118:309 (1976).

266. C.P. Casey and W.R. Brunsvold, J. Organomet. Chem. 102:175 (1975).

267. C.P. Casey and R.L. Anderson, J. Organomet. Chem. 73:C28 (1974).

268. C.P. Casey, W.R. Brunsvold and D.M. Scheck, Inorg. Chem. 16:3059 (1977).

269. E.O. Fischer, S. Fontana and U. Schubert, J. Organomet. Chem. 91:C7 (1975).

270. E.O. Fischer and G. Kreis, Chem. Ber. 106:2310 (1973).

271. E.O. Fischer, K.R. Schmid, W. Kalbfus and C.G. Kreiter, Chem. Ber. 106:3893 (1973).

272. E.O. Fischer, S. Walz, G. Kreis and F.R. Kreissl, Chem. Ber. 110:1651 (1977).

273. S. Fontana, Dissertation Techn. Univ. Munich, 1971.

274. K.H. Dötz and C.G. Kreiter, J. Organomet. Chem. 99:309 (1975).

275. K.H. Dötz and I. Pruskil, J. Organomet. Chem. 132:115 (1977).

276. K.H. Dötz and I. Pruskil, Chem. Ber. 111:2059 (1978).

277. J. Levisalles, H. Rudler and D. Villemin, J. Organomet. Chem. 146:259 (1978).

278. J. Levisalles, H. Rudler, D. Villemin, J. Daran, Y. Jeannin and L. Martin, J. Organomet. Chem. 155:C1 (1978).

279. K.H. Dötz and I. Pruskil, Chem. Ber. 114:1980 (1981).

280. C.P. Casey and T.J. Burkhardt, J. Am. Chem. Soc. 96:7808 (1974).

281. T.J. Katz and S.J. Lee, J. Am. Chem. Soc. 102:422 (1980).

282. K.H. Dötz and C.G. Kreiter, Chem. Ber. 109:2026 (1976).

283. K.H. Dötz and D. Neugebauer, Angew. Chem. Int. Ed. Engl. 17:851 (1978).

284. H. Fischer and U. Schubert, Angew. Chem. Int. Ed. Engl. 20:461 (1981).

285. H. Fischer, J. Organomet. Chem. 197:303 (1980).

286. R. Aumann and E.O. Fischer, Chem. Ber. 101:954 (1968).

287. C.G. Kreiter and R. Aumann, Chem. Ber. 111:1223 (1978).

288. H. Fischer and R. Märkl, Chem. Ber. 115:1349 (1982).

289. K. Weiss, E.O. Fischer and J. Müller, Chem. Ber. 107:3548 (1974).

290. K. Weiss and E.O. Fischer, Chem. Ber. 109:1120 (1976).

291. U. Schubert and E.O. Fischer, Liebigs Ann. Chem. 393 (1975).

292. E.O. Fischer, U. Schubert, W. Kalbfus and C.G. Kreiter, Z. Anorg. Allg. Chem. 416:135 (1975).

293. P.J. Krusic, U. Klabunde, C.P. Casey and T.F. Block, J. Am. Chem. Soc. 98:2015 (1976).

294. C.P. Casey, L.D. Albin, M.C. Saeman and D.H. Evans, J. Organomet. Chem. 155:C37 (1978).

295. H. Brunner and J. Wachter, J. Organomet. Chem. 201:453 (1980).

296. H.D. Empsall, E.M. Hyde, R. Markham, W.S. McDonald, M.C. Norton, B.L. Shaw and B. Weeks, J. C. S. Chem. Commun. 589 (1977).

297. M.F. Lappert, J.J. MacQuitty and P.L. Pye, J. C. S. Chem. Commun. 411 (1977).

298. R.D. Rieke, H. Kojima and K. Öfele, J. Am. Chem. Soc. 98:6735 (1976).

299. R.D. Rieke, H. Kojima and K. Öfele, Angew. Chem. Int. Ed. Engl. 19:538 (1980).

300. H. Werner and H. Rascher, Inorg. Chim. Acta 2:181 (1968).

301. E.O. Fischer, E. Louis and W. Bathelt, J. Organomet. Chem. 20:147 (1969).

302. E.O. Fischer and L. Knauss, Chem. Ber. 102:223 (1969).

303. H.G. Raubenheimer, J.C.A. Boeyens and S. Lotz, J. Organomet. Chem. 112:145 (1976).

304. U. Schubert and E.O. Fischer, J. Organomet. Chem. 170:C13 (1979).

305. E.O. Fischer, B. Heckl, K.H. Dötz, J. Müller and H. Werner, J. Organomet. Chem. 16:P29 (1969).

306. E.O. Fischer and T. Selmayr, Z. Naturforsch B32:238 (1977).

307. J.A. Connor and J.P. Lloyd, J. Chem. Soc. A:3237 (1970).

308. C.P. Casey and R.L. Anderson, J. C. S. Chem. Commun. 895 (1975).

309. C.P. Casey and S.M. Neumann, J. Am. Chem. Soc. 99:1651 (1977).

310. R.B. Silverman and R.A. Olofson, Chem Commun. 1313 (1968).

311. E.O. Fischer and S. Riedmüller, Chem. Ber. 107:915 (1974).

312. C.M. Lukehart and J.V. Zeile, J. Organomet. Chem. 97:421 (1975).

313. C.M. Lukehart and J.V. Zeile, Inorg. Chim. Acta 17:L7 (1976).

314. Y. Ito, T. Hirao and T. Saegusa, J. Org. Chem. 40:2981 (1975).

315. U. Schubert and E.O. Fischer, Chem. Ber. 106:3882 (1973).

316. U. Schubert and E.O. Fischer, J. Organomet. Chem. 51:C11 (1973).

317. C.P. Casey, L.D. Albin and T.J. Burkhardt, J. Am. Chem. Soc. 99:2533 (1977).

318. A. de Renzi and E.O. Fischer, Inorg. Chim. Acta 8:185 (1974).

319. E.O. Fischer and H-J. Beck, Angew. Chem. Int. Ed. Engl. 9:72 (1970).

320. R. Aumann and E.O. Fischer, Chem. Ber. 114:1853 (1981).

321. K. Öfele and M. Herberhold, Angew. Chem. Int. Ed. Engl. 9:739 (1970).

322. K. Öfele, E. Roos and M. Herberhold, Z. Naturforsch B31:1070 (1976).

323. D.J. Cardin, B. Cetinkaya, E. Cetinkaya, M.F. Lappert, L.J. Manojlovic-Muir and K.W. Muir, J. Organomet. Chem. 44:C59 (1972).

324. E.O. Fischer and K.H. Dötz, Chem. Ber. 103:1273 (1970).

325. K.H. Dötz and E.O. Fischer, Chem. Ber. 105:1356 (1972).

326. J.L. Herrison and Y. Chauvin, Makromol. Chem. 141:161 (1970).

327. J.P. Soufflet, D. Commereuc and Y. Chauvin, C.R. Acad. Sci. Ser. C 276:169 (1973).

328. E.O. Fischer and B. Dorrer, Chem. Ber. 107:1156 (1974).

329. B. Dorrer and E.O. Fischer, Chem. Ber. 107:2683 (1974).

330. B. Dorrer, E.O. Fischer and W. Kalbfus, J. Organomet. Chem. 81:C20 (1974).

331. E.O. Fischer, K. Weiss and K. Burger, Chem. Ber. 106:1581 (1973).

332. Y. Ito, I. Ito, T. Hirao and T. Saegusa, Synth. Commun. 4:97 (1974).

333. Y. Ito, T. Hirao and T. Saegusa, J. Organomet, Chem. 82:C47 (1974).

334. W. Kirmse, "Carbene Chemistry", 2nd ed., Academic Press, New York 1971.

335. J.A. Connor and P.D. Rose, J. Organomet. Chem. 24:C45 (1970).

336. E.O. Fischer and K.H. Dötz, J. Organomet. Chem. 36:C4 (1972).

337. J.A. Connor, P.D. Rose and R.M. Turner, J. Organomet. Chem. 55:111 (1973).

338. J.A. Connor, J.P. Day and R.M. Turner, J. C. S. Dalton 108 (1976).

339. J.A. Connor, J.P. Day and R.M. Turner, J. C. S. Dalton 283 (1976).

340. B. Cetinkaya, E. Cetinkaya and M.F. Lappert, J. C. S. Dalton 906 (1973).

341. Organic Sytheses via Metal Carbonyls (I. Wender and P. Pino, ed.) Wiley-Interscience, New York 1968, 1977.

342. K.H. Dötz, Angew. Chem. Int. Ed. Engl. 14:644 (1975).

343. K.H. Dötz, R. Dietz, A.v. Imhof, H. Lorenz and G. Huttner, Chem. Ber. 109:2033 (1976).

344. K.H. Dötz and R. Dietz, Chem. Ber. 110:1555 (1977).

345. K.H. Dötz, J. Mühlemeier, M. Schubert and O. Orama, J. Organomet. Chem. 247:187 (1983).

346. R. Dietz, K.H. Dötz and D. Neugebauer, Nouveau. J. Chim. 2:59 (1978).

347. H. Fischer and K.H. Dötz, Chem. Ber. 115:1355 (1982).

348. K.H. Dötz, Habilitationsschrift, Techn. Univ. Munich, 1979.

349. K.H. Dötz, J. Organomet. Chem. 140:177 (1977).

350. K.H. Dötz and I. Pruskil, Chem. Ber. 113:2876 (1980).

351. K.H. Dötz, Angew. Chem. Int. Ed. Engl. 18:954 (1979).

352. K.H. Dötz and B. Fügen-Köster, Chem. Ber. 113:1449 (1980).

353. U. Schubert and K.H. Dötz, Cryst. Struct. Commun. 8:989 (1979).

354. W.A. Herrmann and J. Plank, Angew. Chem. Int. Ed. Engl. 17:525 (1978).

355. T-A. Mitsudo, T. Sasaki, Y. Watanabe, Y. Takegami, S. Nishigaki and K. Nakatsu, J. C. S. Chem. Commun. 252 (1978).

356. K.H. Dötz and R. Dietz, J. Organomet. Chem. 157:C55 (1978).

357. K.H. Dötz and G. Raudaschl, unpublished results.

358. K.H. Dötz, R. Dietz, C. Kappenstein, D. Neugebauer and U. Schubert, Chem. Ber. 112:3682 (1979).

359. K.H. Dötz, R. Dietz and D. Neugebauer, Chem. Ber. 112:1486 (1979).

360. K.H. Dötz and I. Pruskil, J. Organomet. Chem. 209:C4 (1981).

361. K.H. Dötz, I. Pruskil and J. Mühlemeier, Chem. Ber. 115:1278 (1982).

362. K.H. Dötz, B. Trenkle and U. Schubert, Angew. Chem. Int. Ed. Engl. 20:287 (1981).

363. K.H. Dötz and B. Trenkle, unpublished results.

364. C.P. Casey in "Reactive Intermediates" (M. Jones and R.A. Moss, ed.), Vol. II, Wiley-Interscience, New York (1981).

365. C.P. Casey and M.C. Cesa, Organometallics 1:87 (1982).

366. M.F. Lappert, P.L. Pye, A.J. Rogers and G.M. McLaughlin, J. C. S. Dalton 701 (1981).

367. S.J. Holmes and R.R. Schrock, J. Am. Chem. Soc. 103:4599 (1981).

368. E.O. Fischer, W. Röll, N. Hoa Tran Huy and K. Ackermann, Chem. Ber. 115:2951 (1982).

369. U. Schubert, K. Ackermann, N. Hoa Tran Huy and W. Röll, J. Organomet. Chem. 232:151 (1982).

370. R.A. Pickering, R.A. Jacobson and R.J. Angelici, J. Am. Chem. Soc. 103:817 (1981).

371. F.B. McCormick and R.J. Angelici, Inorg. Chem. 20:1118 (1981).

372. F.B. McCormick and R.J. Angelici, J. Organomet. Chem. 205:79 (1981).

373. H. Le Bozec, A. Gorgues and P.H. Dixneuf, Inorg. Chem. 20:2486 (1981).

374. W.P. Fehlhammer, A. Mayr and G. Christian, J. Organomet. Chem. 209:57 (1981).

375. A.G. Constable and J.A. Gladysz, J. Organomet. Chem. 202:C21 (1980).

376. H. Fischer, E.O. Fischer, D. Himmelreich, R. Cai, U. Schubert and K. Ackermann, Chem. Ber. 114:3220 (1981).

377. H. Fischer, E.O. Fischer and R. Cai, Chem. Ber. 115:2707 (1982).

378. H. Fischer, J. Organomet. Chem. 219:C34 (1981).

379. H. Fischer, J. Organomet. Chem. 222:241 (1981).

380. H. Rudler, J. Organomet, Chem. 212:203 (1981).

381. T.J. Katz, E.B. Savage, S.J. Lee and M. Naír, J. Am. Chem. Soc. 102:7942 (1980).

382. B.H. Edwards and M.D. Rausch, J. Organomet. Chem. 210:91 (1981).

383. A. Miyashita and R.H. Grubbs, Tetrahedron Lett. 22:1255 (1981).

384. D.L. Thorn and T.H. Tulip, J. Am. Chem. Soc. 103:5984 (1981).

385. M. Brookhart, J.R. Tucker and G.R. Husk, J. Am. Chem. Soc. 103:979 (1981).

386. K.H. Dötz, I. Pruskil and J. Mühlemeier, Chemiedozententagung Tübingen, 25.3.1981.

387. W.D. Wulff, P-C. Tang and J.S. McCallum, J. Am. Chem. Soc. 103:7677 (1981).

MECHANISTIC ASPECTS OF THE OLEFIN METATHESIS REACTION

M. Leconte, J.M. Basset, F. Quignard and C. Larroche

Institut de Recherches sur la Catalyse

Villeurbanne, France

1. INTRODUCTION

Since the discovery of the olefin metathesis reaction[1] and the ring opening polymerisation of cyclo-olefins,[2-3] the interest of industrial and academic researchers in such catalytic reactions has never declined. The metathesis reaction has constituted and still constitutes both industrial and academic challenges.

The industrial challenges were multiple: to try to make new olefins or known olefins and new polymers or known polymers, carry out metathesis of olefins or cyclo-olefins having functional substituents, apply the reaction to organic synthesis, try to control the molecular weight, the tacticity and the stereochemistry of the polymers, carry out metathesis of acetylenes, and metathesis of other functions such as aldehydes, ketones etc.

The mechanistic challenges were also multiple: it was a typical case where the same reaction could be carried out with the same transition metal by homogeneous[4] or heterogeneous catalysis.[1] The cleavage of a double bond of an olefin could not fit with the elementary steps known at the time of the discovery (metal carbenes were not well known at that period and there was only one reported case of a metallocyclobutane[5]).

Both challenges are still not fully met which explains the high number of industrial and academic laboratories working in the field.

The application of metathesis in industry is still limited.   Three processes are using the metathesis reaction: CdF- Chimie makes polynorbornene,[6] Chemische Werke Hüls makes poly-octenamer[7] and Shell makes α-olefins.[8] Two processes use homogeneous catalysts and one process uses heterogeneous catalysis.

The mechanistic features have been partially elucidated in the last ten years, mainly by organometallic chemists.   One should notice the parallel development in organometallic chemistry of metal carbenes[9] and metallocycles (cf. Chapters 3,4,10).   However many points still remain unclear.   Among them is the determination of the coordination sphere of a real catalyst, and the choice of a metal or of a ligand environment which could tolerate functional groups or would induce a given stereochemistry.

The success of this reaction is such that an International Symposium on Metathesis, initiated by H. Höcker[10] in Germany, is held every two years.[11]

The aim of this review is to give a simple account of recent work devoted to the mechanism of the reaction.   Many excellent reviews have already described various aspects of the reaction,[12-23] and so we will limit this review to some recent mechanistic and stereochemical features.

The occurence of a "non-pairwise" mechanism being now commonly admitted, we will consider in this review the various aspects of the metal carbene mechanism[24]: (i) what are the various facts in favor of a carbene mechanism; (ii) how the initial metal carbene can be formed; (iii) what evidence is there for a metallacyclobutane intermediate; (iv) what mechanistic information can be derived from the stereochemistry of the reaction; and (v) what is the electronic nature of the chain carrying carbene.

Wherever possible we will separate the fields of homogeneous and heterogeneous catalysis.   We will try also, when possible, to separate the case of acyclic and cyclic olefins, which show different reactivity and stereoselectivity.   It should be mentioned here that most of the progress regarding the mechanism of the reaction has been achieved in the field of organometallic chemistry and homogeneous catalysis rather than in the field of heterogeneous catalysis and surface science.

2.   THE METALLO-CARBENE MECHANISM

This mechanism has been progressively accepted for three main reasons.[22] First it could account for most of the experiment kinetic results on product distribution, secondly it could be verified stoichiometrically by model reactions between metal carbenes and olefins, and finally it was found that a metallo-carbene of tungsten could initiate metathesis.

2.1   The ring opening polymerization of cyclic olefins

It has been observed by many authors that cyclic olefins, when in contact with metathesis catalysts, can be polymerized by opening of the ring.[2,3,25-8] Any mechanistic path considered for this reaction should account for the following observations: at the beginning of the polymerization there is formation of a high molecular weight polymer, this polymer is linear, and the only cyclic olefins are oligomers.

These observations cannot fit with a diolefin mechanism. According to Chauvin[29] and others [30] such a mechanism should lead to the initial formation of a cyclic product of low molecular weight which should lead later on to a high molecular weight polymer (eq 1).

$$(CH_2)_n \begin{array}{c} CH \\ \| \\ CH \end{array} + \begin{array}{c} CH \\ \| \\ CH \end{array} (CH_2)_m \xrightarrow{M} (CH_2)_n \begin{array}{c} CH=CH \\ \\ CH=CH \end{array} (CH_2)_m \longrightarrow etc \qquad (1)$$

In contrast, the carbene mechanism gives the products which are in fact observed at the early stages of the reaction (eq 2).

$$M=C \begin{array}{c} H \\ \\ R \end{array} + \begin{array}{c} CH \\ \| \\ CH \end{array} (CH_2)_n \longrightarrow M=C \begin{array}{c} H \\ \\ (CH_2)_n \end{array} CH=CHR \longrightarrow etc \qquad (2)$$

It should be pointed out, however, that a diolefin mechanism may also show chain reaction character if the metathesis step is more rapid than the exchange step between a coordinated olefin and a free olefin. On such an hypothesis, the metal olefin complex becomes the growing center for the chain reaction and a polymer of high molecular weight can be formed in the early stage of the reaction.[15,17,31]

The metallo-carbene leads, therefore, by ring opening polymerization, to the initial formation of a linear polymer. The cyclic oligomers which are

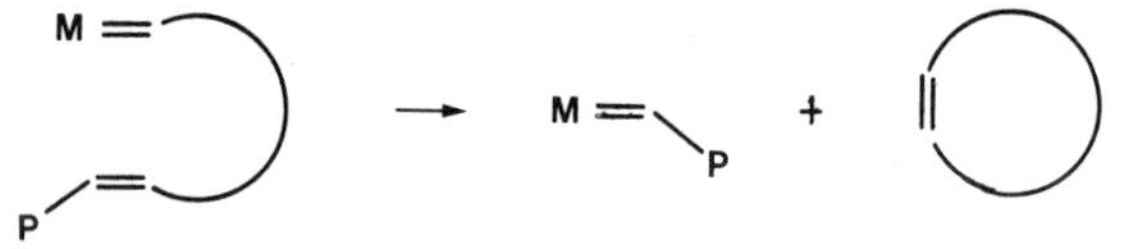

$$(3)$$

in fact observed would be due to intramolecular secondary reactions (eq 3).[26,28-9,31]

A diolefin mechanism implies a cyclic polymer as a primary product, in apparent contradiction to the experimental facts. Let us mention, however, that if traces of acyclic olefins are present in the reactants or are produced in the various steps of formation of the catalyst, a high percentage of linear products may be formed.[14,27,31] Simultaneously the molecular weight of the polymer may be considerably decreased.

In conclusion, most work on the ring opening polymerization of cyclic olefins and in particular that of Calderon,[31] Höcker[28] and Chauvin[29] favor a metallo-carbene mechanism. However such studies could not completely reject a concerted diolefin mechanism.

2.2   The cross metathesis between cyclic and acyclic olefins

Herisson and Chauvin[24] were the first to propose the carbene mechanism, to account for the formation of the primary products of the reaction between cyclic olefin and a non symmetrical acyclic olefin (cyclopentene and cis-2-pentene). More recently Katz and McGinnis[32] carried out the study of a similar reaction (cyclooctene and 2-hexene). Those reactions may be written as in eq 4.

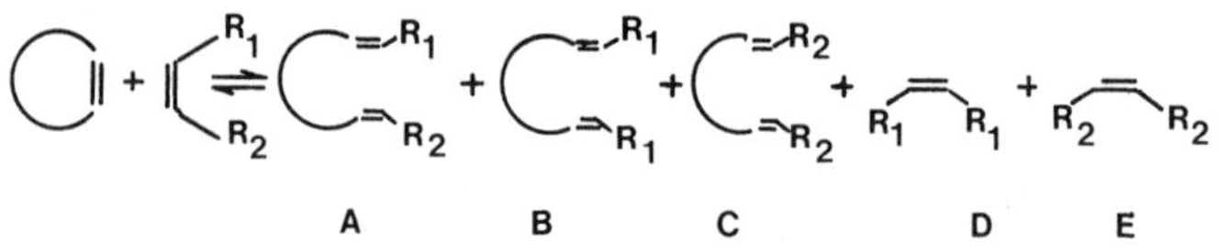

$$(4)$$

Both groups were able to observe at the early stages of the reaction significant amounts of symmetrical products such as B and C. This result cannot be explained in the frame of a concerted diolefin mechanism if exchange between a coordinated olefin and a free olefin is more rapid than metathesis. With such an hypothesis products B and C would be produced only by a secondary metathesis reaction of D and E with the cyclic olefin.

In fact the experiments show clearly that significant amounts of B and C are formed before the concentration of D and E is important. These results are easily explained by a metallo-carbene mechanism in which symmetrical products can be formed in a single step.

However the results are also in agreement with a concerted diolefin mechanism if one considers that metathesis is more rapid than exchange. For example it is possible to explain the formation of B and C at the early stages of the reaction as in eq 5.[21]

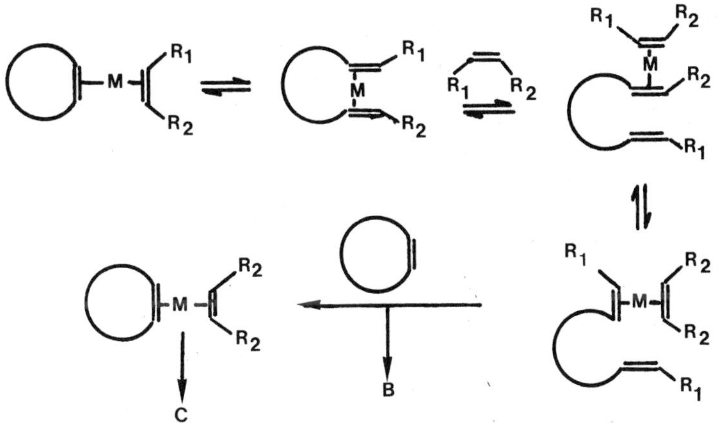

$$(5)$$

In conclusion, cross metathesis reactions between cyclic and acyclic olefins, as well as polymerization of cyclic olefins, do not allow us to decide on the validity of the metallo-carbene mechanism. The main difficulty in this type of approach arises from the fact that in this type of cross metathesis one reaches, via secondary reactions, an equilibrium between the products which may be explained in a satisfactory way by either mechanism. This difficulty could be overcome by choosing a reaction in which the products could not equilibrate via secondary reactions with the starting olefin. This was achieved with α,ω-dienes such as α,ω-1,7-dienes.

2.3 Metathesis of (α,ω)-1,7-dienes

1,7-octadiene gives ethylene and cyclohexene almost quantitatively by metathesis (eq 6).

$$(6)$$

The main advantages of this reaction are the following: (i) high yield; (ii) no metathesis reaction of cyclohexene; (iii) high volatility of ethylene. Since one of the products does not show metathesis and since it is theoretically possible to determine by labeling experiments the fate of the four carbons of the double bonds, the mechanistic information deduced from such a reaction is of primary importance.

The first study to be done on this type of olefin was carried out by Grubbs[33] who considered the metathesis of a mixture of 1,7-octadiene and 1,1,8,8,-d_4-(1,7)-octadiene (eq 7).

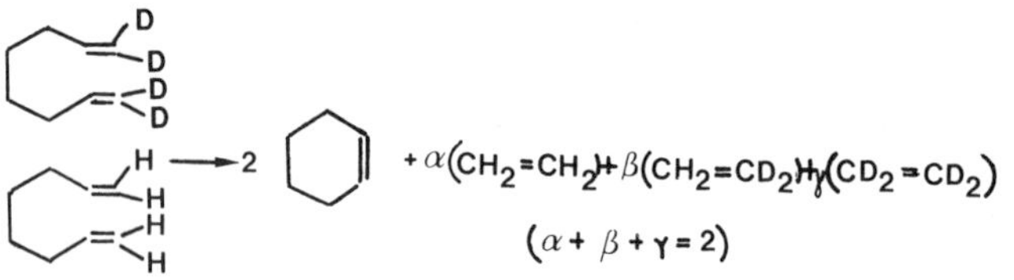

$$+\alpha(CH_2=CH_2)+\beta(CH_2=CD_2)+\gamma(CD_2=CD_2) \qquad (7)$$
$$(\alpha + \beta + \gamma = 2)$$

In a similar way, Katz[34] has studied the metathesis of 2,2'-divinyl-biphenyl (eq 8).

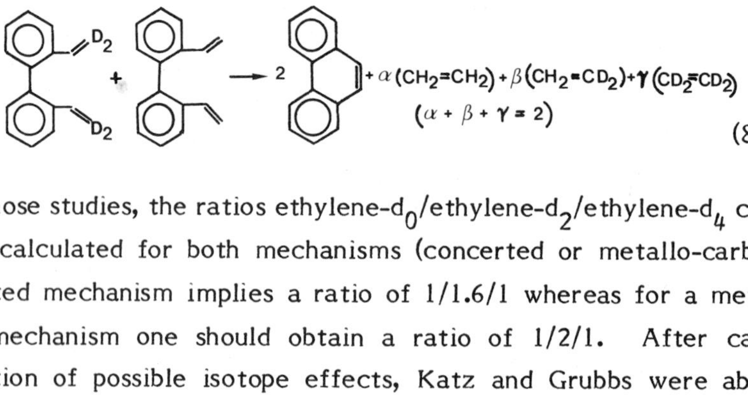

$$+\alpha(CH_2=CH_2)+\beta(CH_2=CD_2)+\gamma(CD_2=CD_2)$$
$$(\alpha + \beta + \gamma = 2) \qquad (8)$$

In those studies, the ratios ethylene-d_0/ethylene-d_2/ethylene-d_4 can be precisely calculated for both mechanisms (concerted or metallo-carbene). A concerted mechanism implies a ratio of 1/1.6/1 whereas for a metallo-carbene mechanism one should obtain a ratio of 1/2/1. After careful consideration of possible isotope effects, Katz and Grubbs were able to demonstrate that the product distribution at low conversion was in agreement with the metallo-carbene mechanism. More recently Grubbs has studied metathesis of a mixture of 2,8-cis,cis-decadiene and 1,1,1,10,10,10-d_6-2,8-cis,cis-decadiene.[35-6] He has studied the ratios butene-2-d_0/butene-2-d_3/butene-2-d_6 as a function of the trans/cis ratio of the 2-butene. By extrapolation to zero percent conversion, he was able to determine the isotopic ratios of 2-butenes. The values are again those expected from a carbene mechanism.

2.4   Model reactions and catalytic reactions with metallo–carbenes

Important work has been carried out by Casey[37-8] concerning the synthesis and reactivity of metallo–carbenes.   Casey was able to carry out the synthesis of $W(CO)_5CPh_2$.   By reaction of this metallo-carbene complex with an olefin, there is cyclopropanation of the olefin and formation of a new olefin, which can be explained by the exchange of an alkylidene fragment of the olefin with the carbene moiety of the metallo-carbene (eq 9).

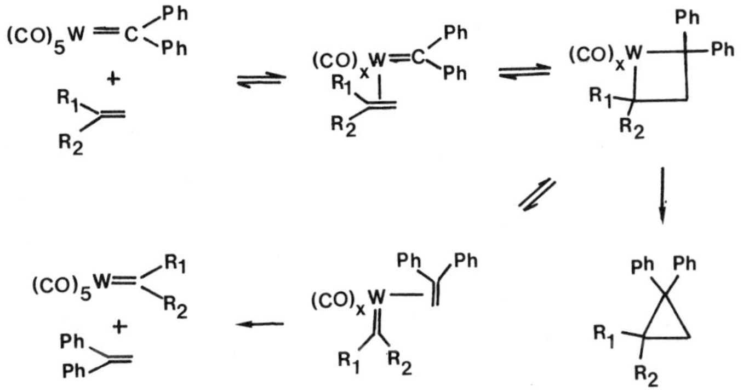

$$\tag{9}$$

The first methylene carbene was isolated by Schrock[39] as a tantalum complex: $Ta(\eta^5\text{-}C_5H_5)_2(CH_3)(CH_2)$.   The isolation of this carbene was a real argument in favor of the possibility of such an intermediate in metathesis.   Unfortunately this type of compound does not lead to metathesis.   Later on many alkylidenes of niobium and tantalum were isolated by Schrock.   Most of them react with olefins to give homologation of the olefin (eq 10).[40-1]

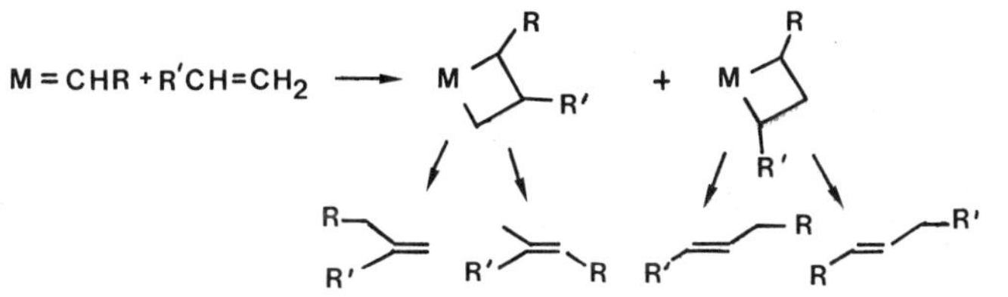

$$M=CHR + R'CH=CH_2 \longrightarrow \tag{10}$$

The organometallic products are olefin complexes (olefin = ethylene, propylene, styrene) of the type: trans, mer, $Ta(\text{olefin})L_2X_3$ (if M = Ta) and olefin free complexes of the type: $[NbL_2X_3]_2$ (if M = Nb).   Internal olefins do not react readily with the alkylidene complexes.

However when the alkylidene complex has t-butoxy ligands rather than halogen (e.g. $Ta(CHCMe_3)(OCMe_3)_2(PMe_3)Cl$) it reacts with olefins to give metathesis-like products (eqs 11,12).

$$Ta(CHCMe_3)(OCMe_3)_2(PMe_3)Cl + C_2H_4 \rightarrow CMe_3CH=CH_2 + CH_3CH=CH_2 \quad (11)$$

$$Ta(CHCMe_3)(OCMe_3)_2(PMe_3)Cl + PhCH=CH_2 \rightarrow CMe_3CH=CH_2 \qquad (12)$$

Schrock[41] was also able to prepare tungsten (VI) analogues of the tantalum complexes by a remarkable ligand exchange reaction (eq 13).

$$W(O)(OCMe_3)_4 + Ta(CHCMe_3)(PEt_3)_2Cl_3 \rightarrow Ta(OCMe_3)_4Cl +$$
$$+ W(O)(CHCMe_3)(PEt_3)_2Cl_2 \quad (13)$$

Terminal olefins react slowly with this compound and give rise to metathesis. $AlCl_3$ increases the rate of the reaction, probably by favoring phosphine dissociation, as in eq 14.

$$W(O)\,(CHCMe_3)\,(PEt_3)_2\,Cl_2 + CH_2 = CHR \longrightarrow \qquad (14)$$

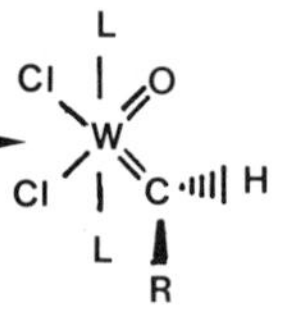

R = Ph, Et or H $+ Me_3CCH=CH_2$

A five-coordinate active metathesis catalyst was also prepared (eq 15) by Schrock[42] by removal of one PEt_3 ligand with a phosphine scavenger such as $Pd(PhCN)_2Cl_2$:

$$W(O)(CHCMe_3)(PEt_3)_2Cl_2 + 0.5\ Pd(PhCN)_2Cl_2 \longrightarrow$$

$$0.5Pd(PEt_3)Cl_2 + W(O)(CHCMe_3)(PEt_3)Cl_2 \qquad (15)$$

The new complex is slightly active for metathesis in the absence of a Lewis acid but it has a short life time. Interestingly the x-ray structure has been obtained. The molecule is a distorted trigonal bipyramid in which the oxo ligand, C_α and C_β of the neopentilydene ligand and the chloride ligand

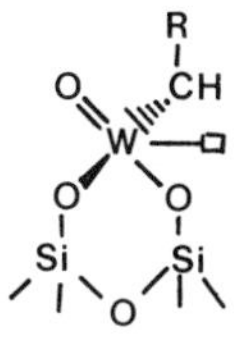

I 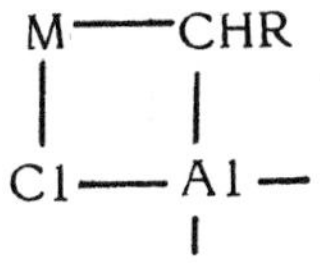   **II**

all lie in the equatorial plane.   It is interesting to compare this molecule with what could be the active site in heterogeneous metathesis of olefins. One could speculate on the occurence of a surface compound such as **I**.   But, on the surface, the tungsten oxo species would be stable as coordinatively unsaturated.   In solution, coordination of the phosphine may prevent olefin coordination and justifies therefore the existence of the Lewis acid as a necessary component.

However the role of the Lewis acid in homogeneous catalysis may be more complex:

- it may favor ligand dissociation by acid-base equilibrium;[43]
- it may stabilize a nucleophilic carbene via bridges, as in **II**,[44] (see later);
- it may coordinate to the olefin[45] by creation of a cationic adduct.

Recently metallo-carbenes were used to initiate ring opening polymerization of cyclo-olefins.   One effective initiator is (diphenylcarbene) - pentacarbonyl-tungsten[46,47d] which induces a variety of cyclo-alkene metatheses to give polyalkenamers (eq 16).

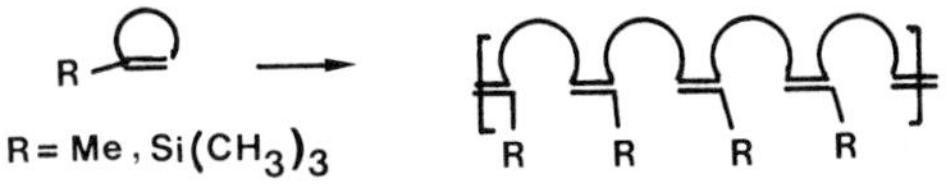

$$R = Me, Si(CH_3)_3 \tag{16}$$

Interestingly the same catalyst, $W(CO)_5CPh_2$, also initiates the polymerization of acetylenes (eq 17).[47d]   Using the same catalyst Levisalles, Rudler et al[48] were able to demonstrate that the first step of the ring opening polymerization of cyclo-olefins occurs via the carbene mechanism; for example, reaction of cyclopentanone-enol-ether with $W(CO)_5CPh_2$ leads inter alia to the stoichiometric opening of the ring by metathesis (eq 18).

$$(17)$$

$$(18)$$

This new carbene-olefin complex is stabilized by the presence of the hetero-atom on the new carbene.   Norcamphor enol ether reacts with Casey's complex in the same way (eq 19).

$$(19)$$

The isolation of these two compounds illustrates very clearly how a metallo-carbene can initiate ring opening.   It also indicates, in this particular case, that the first double bond in the polymer chain need not interact with tungsten, at least with these 18 e$^-$ complexes, although many authors have suggested that with cyclopentene or functionalized olefins, intramolecular coordination of a double bond, an aromatic ring or nitrogen lone pairs could stabilize the vacant site during propagation (Fig. 1).

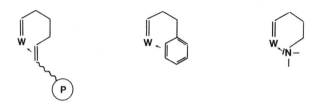

Figure 1:   Suggested models of stabilization of vacant site during metathesis of cyclopentene,[49,50] ω-aryl-olefins[51] and olefinic amines[52]

Ring opening polymerization of cyclopentene can be carried out with an extremely high activity by metallo-carbenes of the type $(CO)_5W=C\,{}^{OEt}_{\,R}$ activated by an excess of $TiCl_4$ and uv light. For example, Chauvin[45,53-4] reported in 1975 that equilibrium can be reached in cyclopentene polymerization with an olefin/W ratio as high as 10^6.

It should be pointed out also here that Tebbe and Parshall[44] were able to carry out the synthesis of **III**, a carbene complex of titanium in which the carbene ligand is coordinated to an alkyl-aluminum ligand.

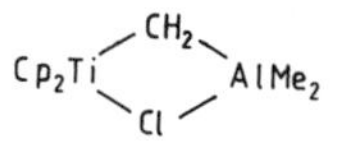

$$Cp_2Ti\underset{Cl}{\overset{CH_2}{<\,\!>}}AlMe_2 \qquad\qquad \textbf{III}$$

This compound is able to catalyse metathetical exchange between methylene fragments of isobutylene and methylene-cyclohexane (eq 20).

The authors were able to show that the carbene fragment on titanium could exchange with that of isobutylene via the mechanism of eq 21.

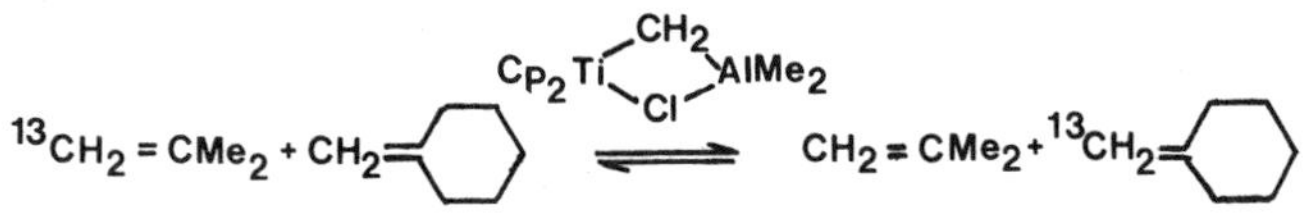

$$^{13}CH_2=CMe_2 + CH_2{=}\!\!\bigcirc \;\;\overset{Cp_2Ti\underset{Cl}{\overset{CH_2}{<\,\!>}}AlMe_2}{\rightleftharpoons}\;\; CH_2=CMe_2 + {}^{13}CH_2{=}\!\!\bigcirc \tag{20}$$

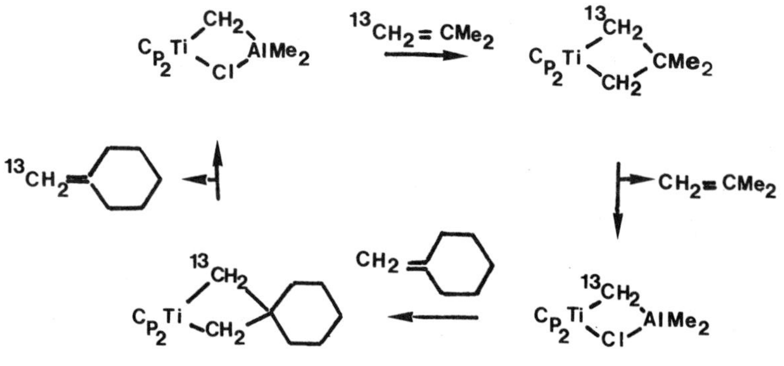

$$\tag{21}$$

3.   **MODE OF FORMATION OF THE FIRST METALLO-CARBENE**

If one accepts the hypothesis of a metallo-carbene mechanism in metathesis, then one has to solve the problem of the formation of the first metallo-carbene.   The solution is obvious when the catalyst is already a metallo-carbene (vide infra).   However those metallo-carbene complexes are far inferior to the two component or three component catalysts, which can be extremely active for cyclic or acyclic olefins.   We will consider separately the cases of homogeneous and heterogeneous catalysis.

3.1   Homogeneous catalysts with organometallic co-catalysts

The alkylation of transition metals by organometallic compounds so as to form stable or unstable metal-alkyl bonds is well known (see for example the review by Schrock and Parshall;[55] also Chapter 1).

Alkylation may obey the general equation:

$$L_nM\text{-}X + R\text{-}M' \rightarrow L_nM\text{-}R + M'\text{-}X \tag{22}$$

This alkylation may be simple, or multiple:

$$WCl_6 + 6\ CH_3\text{-}Li \rightarrow W(CH_3)_6 + 6\ LiCl \tag{23}$$

There are many examples of organometallic co-catalysts in metathesis;[20] some typical examples are: $WCl_6 + R_mAlCl_n$ (m+n = 3); $WCl_6 + RLi$; $WCl_6 + R_4Sn$; $WCl_6 + R_2Zn$; $MoCl_5 + R_3Al$; $Mo(NO)_2Cl_2(PPh_3)_2 + R_3Al_2Cl_3$.

The formation of a metallo-carbene from metal-alkyl bonds may proceed via α-elimination of hydrogen on the α-carbon.   For example, Muetterties[56-7] observed the formation of methane in the reaction between $(CH_3)_2Zn$ and WCl_6; the schemes of eqs 24-5 were suggested.

$$WCl_6 + (CH_3)_2Zn \xrightarrow{\ -ZnCl_2\ } Cl_4W(CH_3)_2 \rightleftharpoons Cl_4W(CH_3)(H)=CH_2 \tag{24}$$

$$\downarrow$$

$$Cl_4W=CH_2 + CH_4$$

$$WCl_6 + (CH_3)_2Zn \xrightarrow{\ -\ CH_3ZnCl\ } Cl_5WCH_3 \rightleftharpoons Cl_5W(H)=CH_2 \tag{25}$$

$$CH_3Zn^+ \nearrow \quad -ZnCl^+$$

Such a mechanism was also suggested by Chauvin[58] in 1973 to account for the initiation process of metallo-carbenes starting from WCl_6 + $M(CH_3)_n$. The equilibrium of eq 26 has been demonstrated by M.L.H. Green.[59-60] The α-elimination of hydrogen as a way to produce alkylidene complexes of tantalum was also studied in detail by Schrock[39,61-4] (cf. Chapter 3).

$$M-CH_3 \rightleftharpoons H-M=CH_2 \qquad\qquad (26)$$

Recent work by Grubbs[65] gives additional evidence for the formation of metallo-carbenes via the alkylating co-catalyst. With the particular catalytic systems $Mo(NO)_2Cl_2(PPh_3)_2$ + $(CH_3)_3Al_2Cl_3$ and WCl_6 + $Sn(CH_3)_4$, there is formation of methane and ethylene at the early stage of the catalyst co-catalyst interaction. When the co-catalyst is $Sn(CD_3)_4$, the methane and ethylene are per-deuterated:

$$WCl_6 + (CD_3)_4Sn \rightarrow CD_4 + CD_2=CD_2 + [W] \qquad\qquad (27)$$

Grubbs explains ethylene formation by a dimerisation of the metal-methylene complex. In addition, the hypothesis of the formation of an initial methylene complex accounts for the formation of propylene at the early stage of metathesis of 2,8-decadiene with the catalytic system WCl_6 + $Sn(CH_3)_4$ (eq 28). Experiments carried out with $Sn(CD_3)_4$ and 1,1,1,10,10,10-d_6-decadiene-2,8 confirm the fact that propylene is the first formed olefin via the $=CD_2$ and $=CD-CD_3$ alkylidene fragments.

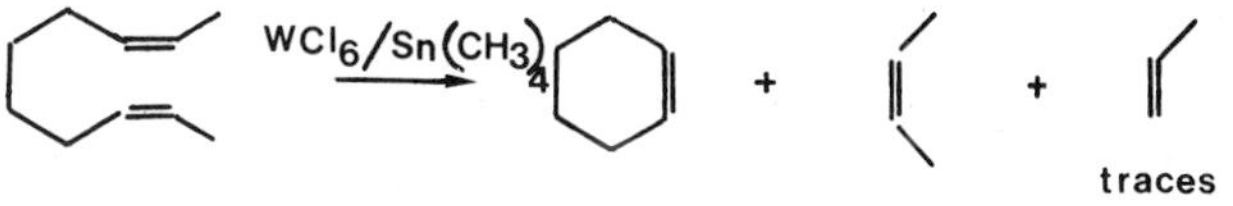

traces

The prospect of metallo-carbene formation by α-elimination suggested the synthesis of new group VI metal-alkyl complexes such as **IV** or **V**. These compounds are not themselves very good catalysts for olefin metathesis but they become very active in the presence of a Lewis acid.[66]

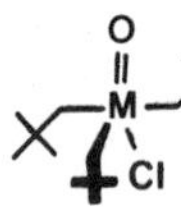

IV 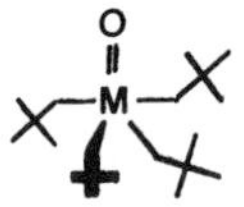   V   M = Mo,W

It is interesting to mention here that titanium also forms methylene complexes. An early indication of the potential of titanium for generating methylene complexes was obtained by de Vries[67] who found that CD_3TiCl_3 in normal hydrocarbon solvents decomposes to CD_4. Formation of $[TiCD_2]$ as a coproduct was suggested. Later on, Feay[68-9] reported that the reaction of titanocene dichloride with trimethylaluminum yielded methane and decomposition products which were thought to contain $[TiCH_2Al]$ units. Sinn and coworkers[70] reinforced this suggestion by labeling studies which showed that the methane produced contained hydrogen only from the methyl groups:

$$Cp_2TiCl_2 + (CD_3)_3Al \longrightarrow CD_4 + [TiCD_2Al] \qquad\qquad (29)$$

Parshall and Tebbe[44] isolated the Ti-Al adduct and proposed a stoichiometry for its formation:

$$Cp_2TiCl_2 + 2\ Me_3Al \rightarrow Cp_2Ti(CH_2)(Cl)AlMe_2\ \textbf{(III)} + CH_4 + ClAlMe_2\ (30)$$

It is possible that there is a double alkylation of titanium followed by α-H elimination and reductive elimination of CH_4.

Finally it should be mentioned that α-elimination from metal-alkyl is probably not the only way to generate metallo-carbenes in metathesis. For example $WCl_6 + SnPh_4$ is a catalyst for metathesis and one does not see easily how the first metallo-carbene can be formed. Abnormal behavior for initiation has been observed by Levisalles[71] and Farona.[72] In one case[71] the authors suggest that α-cleavage of a C-C bond of a metal alkyl may be an important path for producing a metallo-carbene. Farona[72] suggests, for the particular system $RAlCl_2 + Re(CO)_5Cl$, that the key step is the formation of an acyl intermediate.

3.2   Homogeneous catalysts with no organometallic co-catalysts

The formation of a metallo-carbene can be achieved by direct interaction of the transition metal and a diazo-alkane. Dolgoplosk[73] and Herman[74] have shown that phenyl-diazomethane is decomposed by tungsten hexachloride into nitrogen and stilbene. If this reaction is carried out in the presence of a cyclic olefin there is formation of a high molecular weight polymer by ring opening (eq 31).

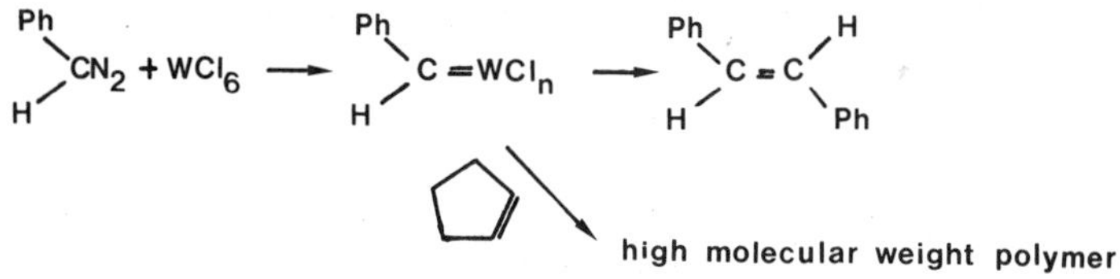

$$\tag{31}$$

high molecular weight polymer

With catalytic systems which contain a transition metal complex and a Lewis acid, it seems necessary to consider that the olefin itself is responsible for the formation of the first metallo-carbene. Some recent results have shown that hydrogen transfer can occur between a hydride and an olefin coordinated to Mo or W. Osborn[75] has shown that it was possible to protonate the olefinic complex $Mo(C_2H_4)(diphos)_2$ by reaction with CF_3COOH (Fig. 2):

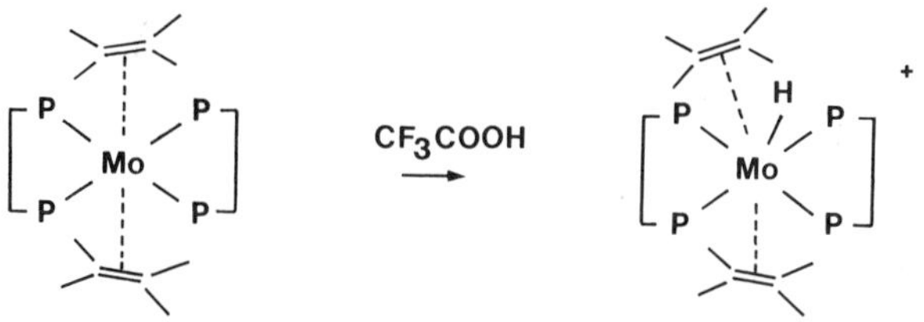

Figure 2

It was possible to demonstrate by NMR that a rapid insertion-de-insertion of the olefin into the metal-hydride could occur (eq 32), while with propylene there is formation of a stable -allyl-hydride complex (eq 33).

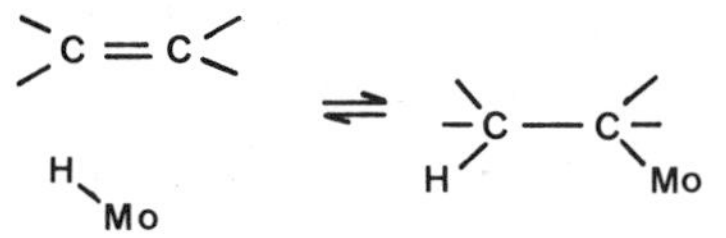

$$\tag{32}$$

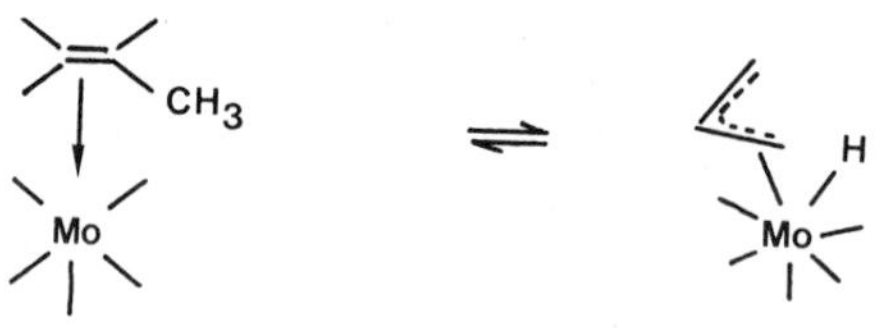

$$\text{(33)}$$

M.L.H. Green[76-9] has also shown that a π-allyl complex of tungsten can undergo nucleophilic attack by H^- on the central carbon atom of the π-allyl; this reaction leads to a stable metalla-cyclobutane complex which upon irradiation can give metathesis-like products (Fig. 3).

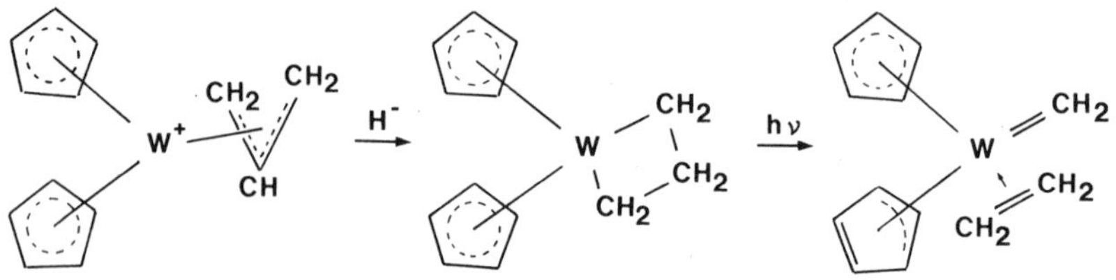

Figure 3

One may therefore consider the scheme of eq 34 for generating a metallo-carbene from an olefin.

Another way to make metallo-carbenes from two olefins comes from the work of Schrock[134] with tantalum complexes (eq 35).

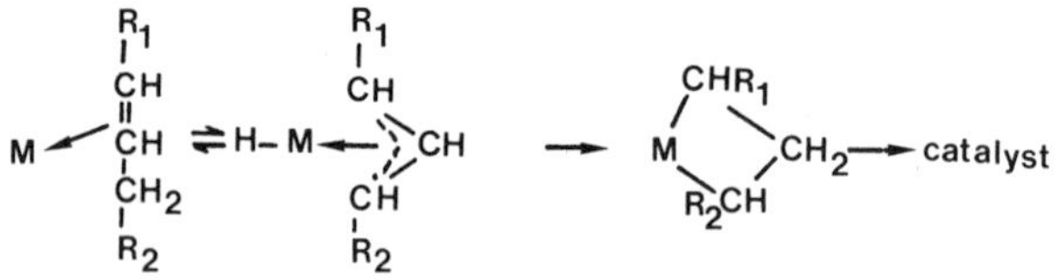

$$\text{(34)}$$

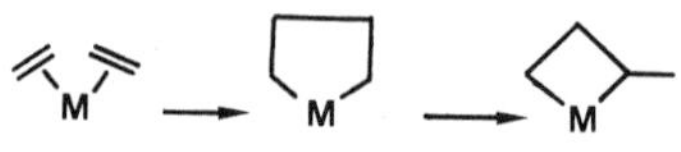

$$\text{(35)}$$

Here the formation of the metallo-carbene occurs by ring contraction of the metallacyclopentane. Generally speaking, from a metallacyclopentane it is possible to consider three different paths; α-H elimination,[21] ring contraction,[21,80] and β-H elimination followed by cis-migration of hydrogen[41,134] (Fig. 4).

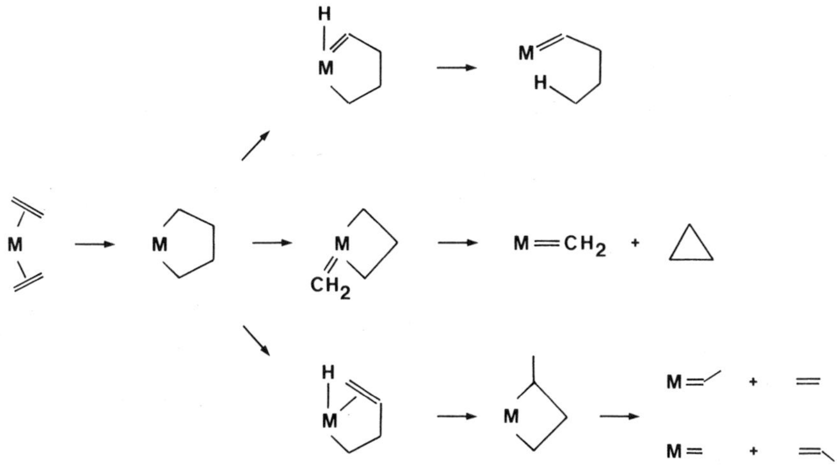

Figure 4:   Possible modes of formation of metallo-carbenes from olefins only, according to references[21] and[41]

3.3   Heterogeneous catalysts activated by an organometallic compound

Various heterogeneous catalysts prepared by interaction of alumina or silica with organometallic compounds are excellent catalysts for metathesis. In particular, the interaction of Mo^{IV} (π-allyl)$_4$ with the surface of silica was first studied by the Ballard group at ICI,[81] and by Yermakov et al.[85] It is likely that an organometallic fragment left on the surface is the origin of the metallo-carbene (eq 36).

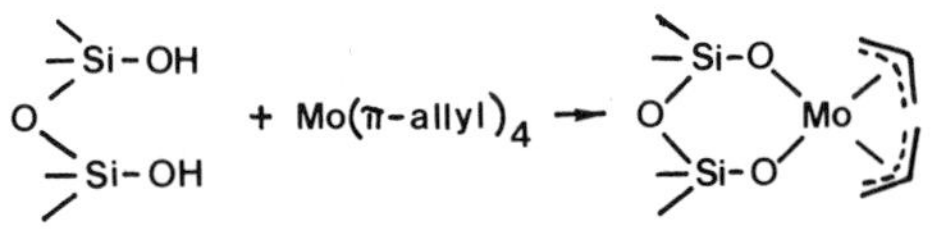

$$ (36) $$

The surface complex which is formally a Mo^{IV} complex is very active at room temperature. After reduction under H_2 at 600°C followed by partial oxygen treatment at room temperature, one may speculate[82] that the new active catalyst is **VI**, which is not very different from the oxo catalysts described by Schrock,[41] but for which immobilization results in a coordinatively unsaturated stabilized Mo^{IV} species. We will speculate later on how the metallo-carbene can be formed on such systems.

$$\begin{array}{c} \text{≥Si-O} \\ \text{Mo} \overset{IV}{\diagdown} \overset{O}{\diagup} \\ \text{≥Si-O} \end{array} \qquad \textbf{VI}$$

Boelhouwer et al,[83-4] who have studied in detail the metathesis of unsaturated fatty acid esters, have recently used the catalytic systems $Re_2O_7/Al_2O_3/Sn(CH_3)_4$ and $WOCl_4/SiO_2/Sn(CH_3)_4$. Here also the role of the alkylating agent is to favor the formation of the metallo-carbene on the surface by alkylation of Re or W. It should be pointed out here that Yermakov et al[85-6] do not consider that the π-allyl fragment could be responsible for the formation of the metallo-carbene. They rather focus their research on the determination of the oxidation state of the active catalyst. However, considering the work of Green,[77-9] the possibility that the π-allyl fragment on the surface is at the origin of the metallo-carbene deserves further study; preliminary results on this point have recently been published.[82]

The role of the co-catalyst $Sn(CD_3)_4$ was also studied by Grubbs[36] with the catalysts $MoO_3/CoO/Al_2O_3$. In the absence of the alkylating agent there is an induction period which is explained by the fact that metallo-carbenes are formed from the olefin (vide supra). In the presence of the alkylation agent, the induction period is suppressed and labeling experiments indicate that the olefin formed in the early stages of the reaction comes from the alkylating agent.

3.4   Heterogeneous catalysts without alkylating agent

These catalysts are the most commonly used in heterogeneous catalysis. They usually exhibit a break-in period that we may attribute to the formation of the organometallic bond necessary to the chain reaction. In fact there are many possibilities which are not really completely satisfactory.

First Rooney[87] has studied the reaction of cycloheptatriene with ethylene with the catalysts Re_2O_7, MoO_3 and WO_3 supported on Al_2O_3. Among the products, Rooney could observe the formation of propylene and methane, which was explained by the initiation scheme of Fig. 5.

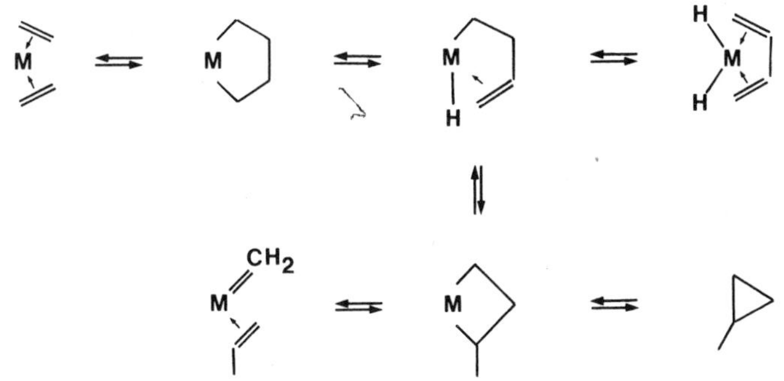

Figure 5:   Mechanism for initiation in heterogeneous catalysis according
            to Rooney[87]

The same authors, using a mixture of ethylene-d_0 and ethylene-d_4, were able
to show that propylene could be formed before the exchange could occur.[88]

        More recently, Rooney[89] has proposed the formation of metal hydride
which could arise from a surface OH group (eq 37).

According to Rooney the metallo-carbene would then be formed according
to eq 38.   However once it is considered that hydride is present on the
metal, the mechanism of M.L.H. Green[77-8] from a π-allyl intermediate is
also very possible (eq 39).

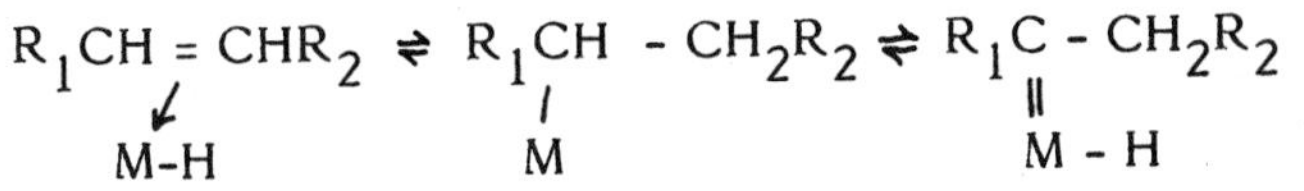

$$R_1CH = CHR_2 \rightleftharpoons R_1CH - CH_2R_2 \rightleftharpoons R_1C - CH_2R_2 \qquad (38)$$
$$\quad\;\; M\text{-}H \qquad\qquad\quad M \qquad\qquad\quad M\text{-}H$$

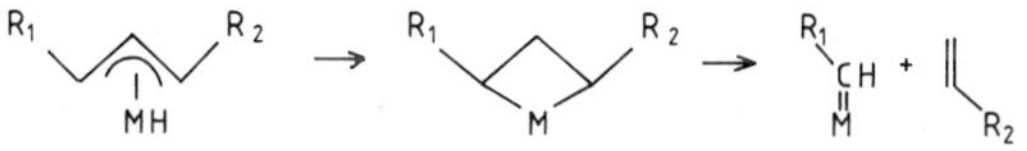

$$(39)$$

In fact the intermediacy of π-allyl intermediates has become more and more
likely since the recent work of Boelhouwer,[90] Farona[91] and Grubbs.[36] Using
the catalyst Re_2O_7/Al_2O_3, Boelhouwer[90] studied the formation of
$CH_2=CD_2$ from a mixture of $CH_2=CH_2$ and $CD_2=CD_2$. He found that it was

necessary that the catalyst was contacted with propylene or 1-butene before ethylene could undergo metathesis; this agrees with the requirement of an allylic hydrogen to make the π-allyl. At the same time, Grubbs[36] observed, with the catalyst $MoO_3/CoO/Al_2O_3$, that there was a strong isotopic effect in the metathesis of a mixture of decadiene-2,8 and 1,1,1,10,10,10-d_6-decadiene-2,8. This isotopic effect is suppressed if a co-catalyst such as $Sn(CD_3)_4$ is used. The isotopic effect could be due to an initiation step which occurs via a π-allyl intermediate (eq 40).

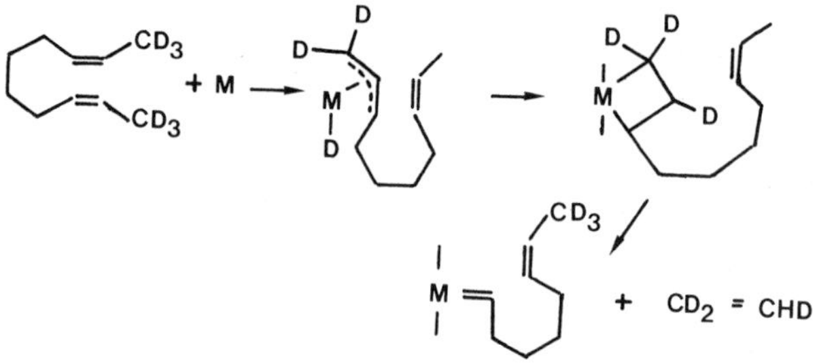

$$(40)$$

Theoretical calculations by Goddard[92] have recently considered the thermochemistry and mechanism for epoxidation and metathesis of ethylene by Cr and Mo complexes having oxo ligands. One possible path to metallo-carbenes from metal-oxo bonds is shown in Fig 6.

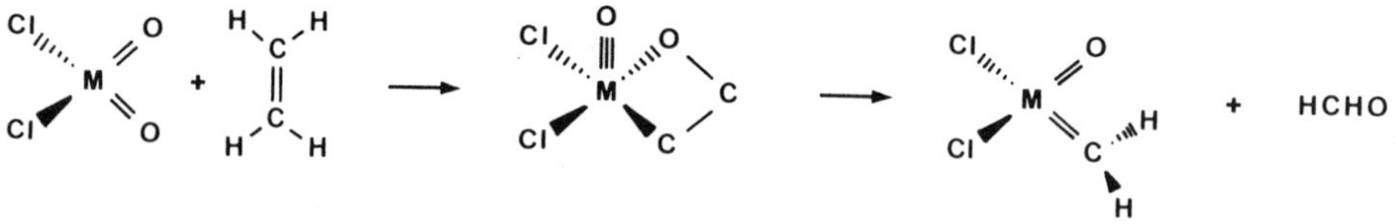

Figure 6

It is quite possible that this process occurs on surfaces which contain metal-oxo species. In this connection Kim and coworkers[93] at Shell reported that ferric molybdate catalyzed the conversion of ethylene and oxygen to formaldehyde. The mechanism proposed would involve a 4-member transition state similar to the dioxetane intermediate proposed in singlet oxygen reactions (eq 41).

However it is not unreasonable to assume a mechanism similar to that proposed by Goddard.

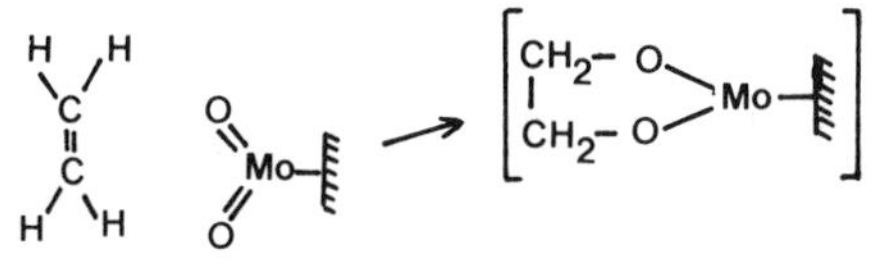

$$
\begin{array}{c}
\underset{H}{\overset{H}{\diagdown}}C\!=\!\underset{H}{\overset{H}{C}}\diagup \quad \overset{O}{\underset{O}{\diagdown}}Mo\!-\!\xi \quad \longrightarrow \quad \left[\begin{array}{c} CH_2\!-O \\ | \qquad \diagdown Mo\!-\!\xi \\ CH_2\!-O \diagup \end{array} \right]
\end{array}
\qquad (41)
$$

4. THE METALLA–CYCLOBUTANE AND METALLA–CYCLOBUTENE INTERMEDIATES

Many recent studies indicate that metallo-carbenes can react with olefins to give three types of reaction which occur via metalla-cyclobutane intermediates: cyclopropanation, homologation and metathesis (Fig. 7).

Work in this field has examined the reactivity of various olefins or acetylenic compounds with metallo-carbenes, the polarity of the metal-carbene bond, and the prior coordination of the olefin to the metallo-carbene or the direct interaction of the olefin with the carbon atom of the carbene, as well as steric factors.

Schrock[40] has studied the reaction of various olefins with the metallo-carbene $(\eta^5\text{-}C_5H_5)\,TaCl_2\,(=CH\text{-}C\text{-}(CH_3)_3)$. He was able to show that the olefin was coordinated to the metal before undergoing an intramolecular nucleophilic attack by the carbene. A metalla-cyclobutane intermediate led to homologation. Casey[94-5] has studied the reaction of various olefins with $W(CO)_5CPh_2$. The reactivity of the various olefins does not follow the

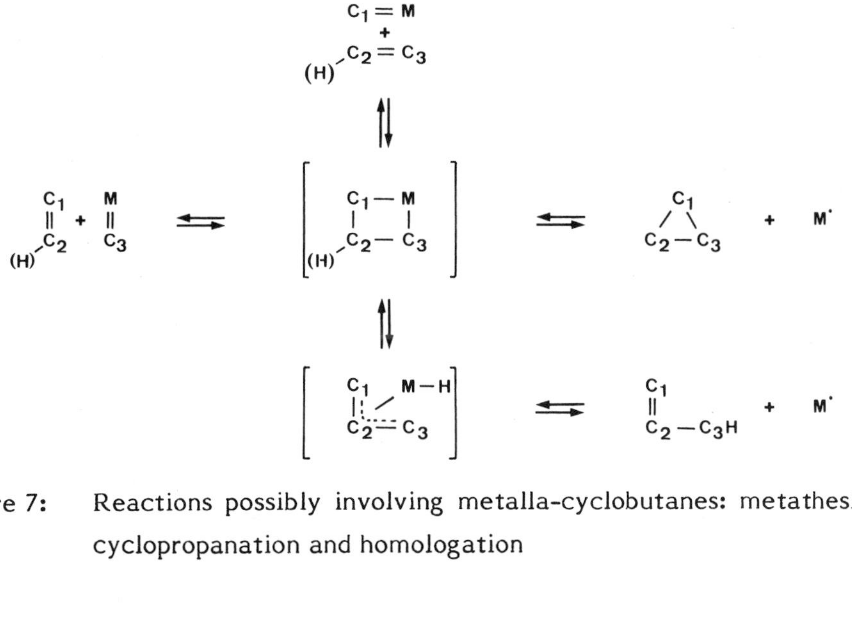

Figure 7:   Reactions possibly involving metalla-cyclobutanes: metathesis, cyclopropanation and homologation

same trend as that reported by Schrock with carbenes of tantalum. Casey was also able to compare the reactivity of $(CO)_5W=CPh_2$ with that of $(CO)_5W=CHPh$. Significant differences could be observed. $(CO)_5W=CHPh$ reacts very quickly with olefins even at $-78°C$ to give almost exclusively cyclopropane derivatives; apparently there is a concerted insertion of the carbene into the olefinic double bond, either by direct interaction or _via_ a metalla-cyclobutane. In contrast, $(CO)_5W=CPh_2$ reacts only at relatively high temperature $(50-100°C)$, and yields mainly metathesis olefins. Besides, $(CO)_5W=CHPh$ reacts 350 times faster with isobutylene than with propylene which suggests an electrophilic carbene; but with $(CO)_5WCTol_2$ the order of reactivity is 1-pentene > isobutene > 2-butene >> 2-methyl-2-hexene.[96] This order is not understood on the basis of a nucleophilic or electrophilic carbene.

It should be mentioned here that the intermediacy of a metalla-cyclobutane in metathesis was first strongly supported by the work of Ephritikhine and Green who were able to cleave a tungstacyclobutane photochemically into metathesis-like olefins.[77-8] The decomposition of metalla-cyclobutanes has been studied by many workers[40,97] and in most cases decomposition results in the preservation of the three carbons of the ring: formation of cyclopropane or formation of an η^3-allyl intermediate. A few examples are given in Figs 8 and 9.

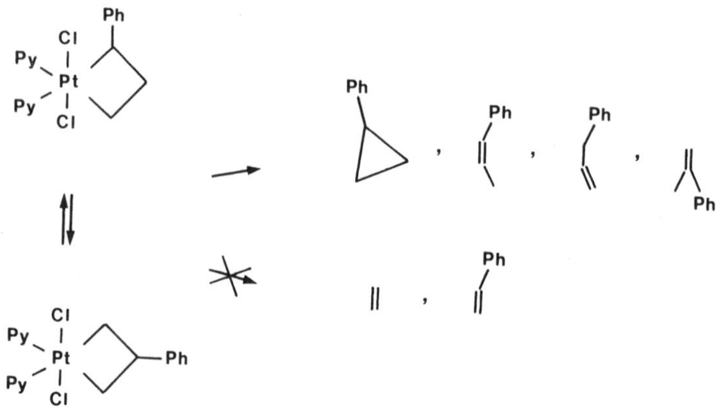

Figure 8:   Decomposition of a metalla-cyclobutane[97]

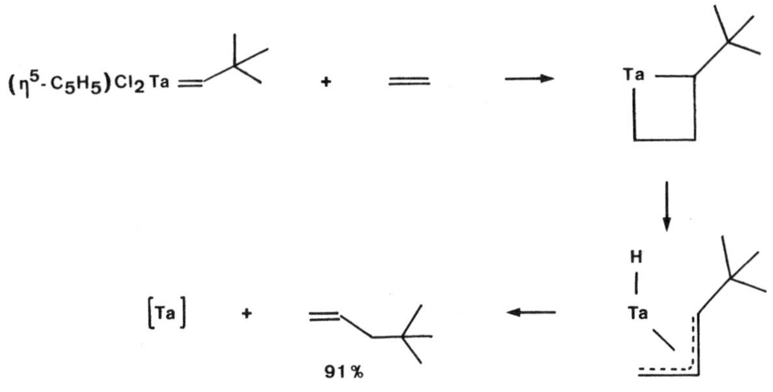

$$(\eta^5\text{-}C_5H_5)Cl_2\,Ta = \underset{}{\overset{}{\diagdown}} \quad + \quad = \quad \longrightarrow \quad Ta\underset{}{\overset{}{\square}}$$

$$[\text{Ta}] \quad + \quad =\diagup\diagdown \quad \longleftarrow \quad \overset{H}{\underset{\text{Ta}}{|}}\diagup\diagdown$$

91%

Figure 9:   Alternative decomposition of a metalla-cyclobutane ^{40}

Recently a metalla-cyclobutane of titanium able to carry out metathesis of olefins has been isolated by Grubbs[98] from the reaction of the Tebbe reagent with neohexene in THF in the presence of pyridine (eq 42).

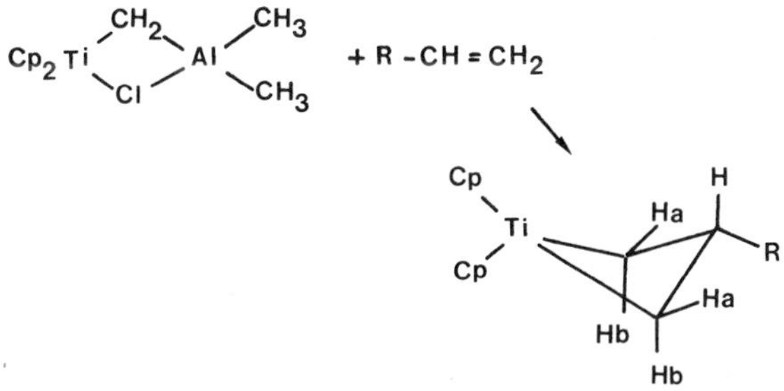

$$Cp_2\,Ti\underset{Cl}{\overset{CH_2}{\diagup}}\,Al\underset{CH_3}{\overset{CH_3}{\diagup}} \quad + R\text{-}CH = CH_2 \tag{42}$$

The isolation of this compound constitutes a major discovery in metathesis, not only because it is a titanium analog of the Green complex but also because it has a reactivity which explains many unanswered questions related to metathesis (Fig. 10). Firstly, this metalla-cyclobutane decomposes in toluene at 85°C to give mainly metathesis-like products. Secondly, the neohexene fragment of the metalla-cycle can be exchanged at 25°C for a trans-neohexene-1-d$_1$. Thirdly, the neohexene can be exchanged for a Lewis acid to give back the Tebbe reagent. This last experiment explains in a very satisfactory manner the role of the Lewis acid co-catalyst in metathesis: this Lewis acid facilitates the carbene-metallacycle mechanism. Finally, the metalla-cyclobutane complex reacts with diphenyl acetylene to give a metalla-cyclobutene which was postulated in metathesis

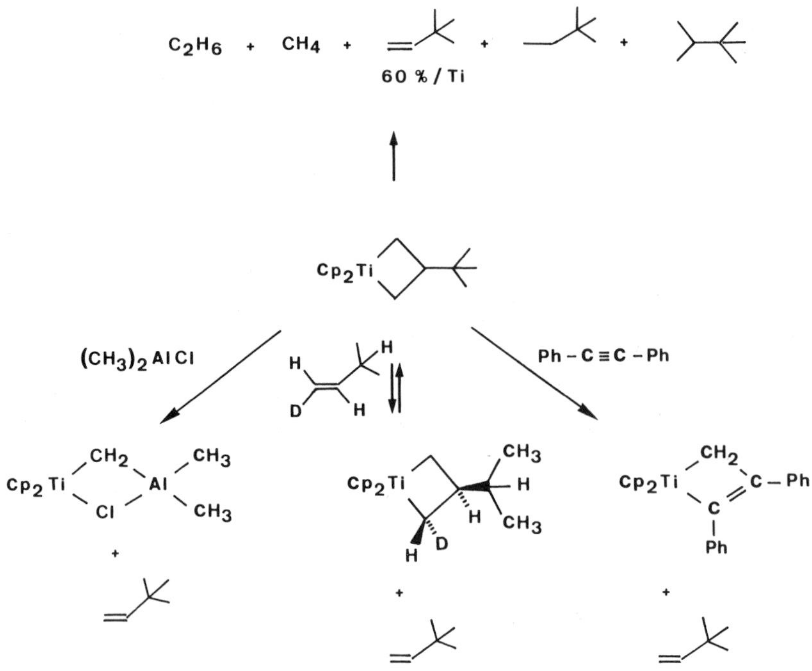

Figure 10:   Some reactions of titana-cyclobutanes[98]

of acetylenic compounds,[47d] first carried out, with homogeneous catalysts, by Blanchard and Mortreux.[47a-c]

Interestingly, a metalla-cyclobutene of titanium was isolated by Tebbe[44] from the reaction of the Tebbe reagent with diphenylacetylene (or bis (trimethyl-silyl)acetylene) (eq 43).   An x-ray crystal structure of this compound showed that the titana-cyclobutene was planar.   The intermediacy of a metalla-cyclobutene was postulated by Katz[47d] in the polymerization of acetylenes induced by $(OC)_5W=CPh_2$.

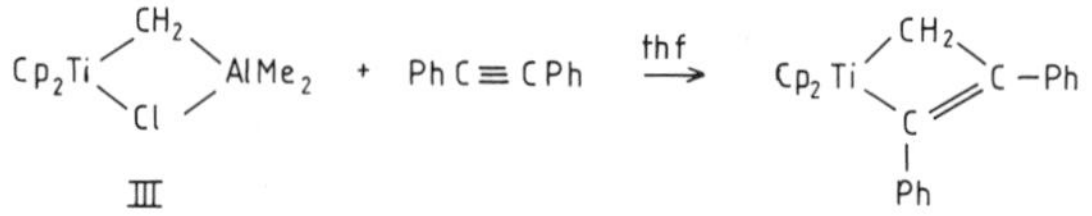

$$(43)$$

Ab inito calculations were recently performed by Goddard[92] in order to understand the relative behavior of Cr, Mo and W with respect to oxidation or metathesis and also to understand the role of oxo ligands in metathesis and in oxidation.   With molybdenum, the reaction paths depicted in Fig. 11 involving metalla-cyclobutanes were calculated. According to these calculations the metallo-carbene-olefin complex has the same energy as the metalla-cyclobutane.

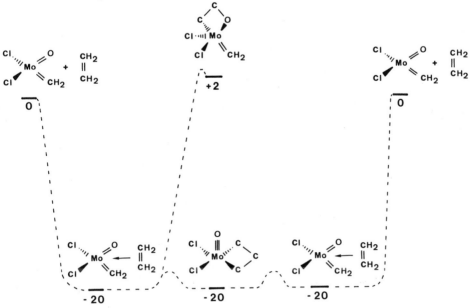

Figure 11: Schematic potential curves for the reaction of $Cl_2Mo(O)(CH_2)$ with C_2H_4 (energy in kcal/mol)[92]

5. STEREOCHEMISTRY OF THE REACTION

In the early studies of acyclic olefin metathesis, three experimental facts gave information regarding the stereochemistry of the reaction:

1 - In most cases and with most catalysts, thermodynamic cis- trans equilibrium of a given olefin was reached at high conversion;

2 - At low conversion, the trans/cis ratio of the products could differ significantly from the thermodynamic equilibrium;

3 - At low conversion, the reaction exhibited a degree of stereoselectivity: a cis olefin gave preferentially a cis olefin and a trans olefin gave a trans olefin.

It appeared progressively that the stereochemical behavior of metathesis of acyclic and cyclic olefins was able to supply useful information on the reaction mechanism.

In order to account for the preliminary studies, a concerted di-olefin process was first considered.[99] However it is only recently that a tentative rationalization of the stereochemistry of metathesis has been proposed on the basis of the metallo-carbene mechanism of Chauvin.[24,58]

Metathesis of acyclic olefins may be regarded as a combination of three competitive or consecutive reactions: formal metathesis, degenerate metathesis, and cis-trans isomerization. These three reactions proceed at comparable rates in the case of internal olefins such as cis-2-pentene with many metathesis catalysts (Fig. 12).

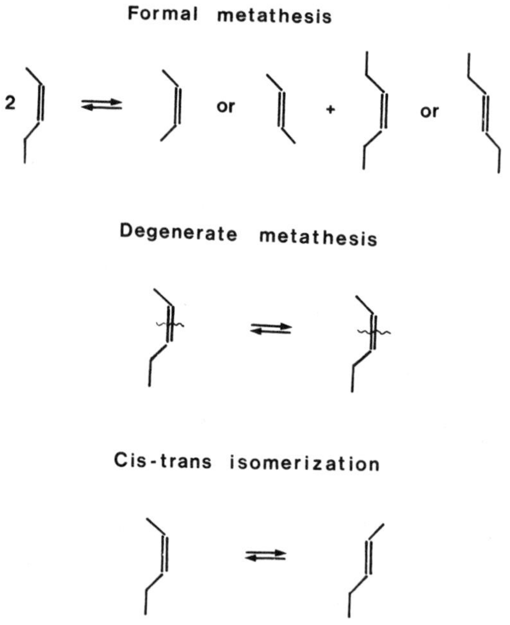

Formal metathesis

Degenerate metathesis

Figure 12

Cis-trans isomerization

On the basis of the metallo-carbene mechanism proposed by Chauvin, Bilhou et al.[100-101] have studied the metathesis of cis-2-pentene with the catalyst $W(CO)_5PPh_3$ + $EtAlCl_2$ + O_2. A stereochemical model has been proposed which accounts for most of the kinetic data observed with acyclic internal olefins. According to this model, the mode of olefin coordination towards the metallo-carbene will determine the product distribution. For a given metallo-carbene there are four modes of olefin coordination leading to cis-$(R_1)(R_2)C = C(R_1)(R_2)$, trans-$(R_1)(R_2)C = C(R_1)(R_2)$ and to cis-$(R_3)(R_4)C = C(R_1)(R_2)$ and trans-$(R_3)(R_4)C = C(R_1)(R_2)$ (Fig. 13).

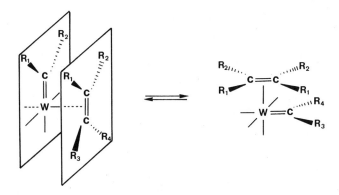

Figure 13

In the model depicted in Fig. 13,[101] it is assumed that the carbon atom of the coordinated carbene has sp^2 character with a metal-carbon bond order equal to two, as has been shown in many alkylidene complexes.[102] In some cases the double bond between the transition metal and the carbene may result in a rigid conformation with a high barrier to rotation as determined by NMR spectroscopy. The model is probably oversimplified since it is possible that: (i) the coordinated carbene rotates around the metal-carbon bond;[62] (ii) the alkyl-aluminum is still coordinated to the carbene and prevents its rotation, as in **III** (iii) the olefin coordinates to the transition metal prior to the transition state which would result from the rotation of the coordinated carbene and the olefin before reaching the overlap of appropriate orbitals.

Nevertheless, the model is kinetically useful. It accounts for the respective rates of formal metathesis, degenerate metathesis and cis-trans isomerization for internal olefins of the type R-CH = CH-R' where R and R' are sterically and electronically not very different. (The case of degenerate metathesis will be considered later on).

If the type of olefin coordination determines the product distribution between formal metathesis, degenerate metathesis, and cis- trans isomerization, there must be a simple kinetic relation between the relative rates of these reactions and the various probabilities of coordination of the olefin leading to these three types of reaction.

Such a relation occurs with many catalytic systems derived from tungsten which exhibit a very weak stereoselectivity. In the case of cis-2-pentene metathesis,[101] random olefin coordination to the metallo-carbene would give the calculated data of Table 1 at zero conversion. The experimental values are not far from the theoretical ones.

Table 1

	observed	expected from random coordination
trans C_4/cis C_4	0.73	1
trans C_6/cis C_6	0.88	1
trans C_5/trans C_4	2.30	2

From the model of Fig. 13, it is possible to derive a mathematical equation which could apply to metathesis of any internal olefin, taking into account the real stereoselectivities. When applied to cis-2-pentene metathesis it is observed that there is a linear relationship between trans C_4/cis C_4 (C_4 = 2-butene) and trans C_5/cis C_5 (C_5 = 2-pentene). The linear relationship observed indicates in fact that cis-trans isomerization of the starting olefin and cis- trans isomerization of the products are both competing with formal metathesis during propagation. The extrapolation of the experimental data to zero conversion is the only possible way of measuring the stereoselectivity of the propagation of the reaction.

The same type of model has been used by Grubbs[35] in the case of metathesis of cis,cis-2,8-decadiene (Fig. 14).

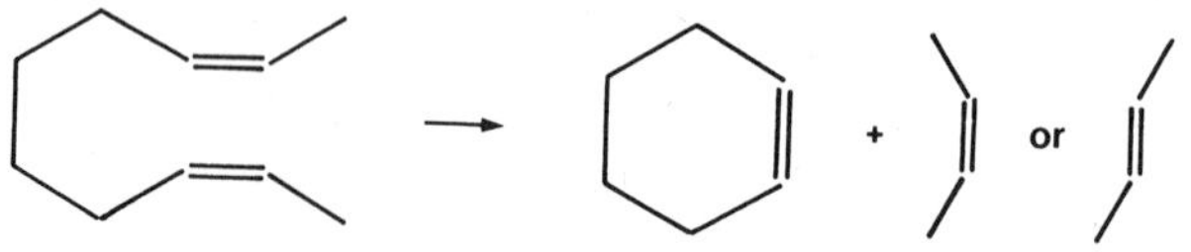

Figure 14

The mathematical equations derived from the model were

$$\text{trans } C_4/\text{cis } C_4 = (0.25 + 0.5\alpha)/(1 - 0.34\alpha)$$

$$\text{trans } C_{10}/\text{cis } C_{10} = (0.25\alpha - 0.21\alpha^2)/(1 - 0.34\alpha)$$

where $\alpha = \underline{cis}\, C_4 / \underline{cis}\, C_{10}$

α is a measure of the amount of productive metathesis. On varying α, there is a linear relationship between $\underline{trans}\, C_4 / \underline{cis}\, C_4$ and $\underline{trans}\, C_{10} / \underline{cis}\, C_{10}$. Such a procedure was used to definitively confirm the non-pairwise mechanism even at high conversion;[35] it is also very useful in the determination of the stereoselectivity of the propagation step with acyclic olefins.

5.1   Stereoselectivity with acyclic olefins

Katz et al.[32,103] were the first to consider that the stereoselectivity of the metathesis reaction was related to the stability of the metalla-cyclobutane intermediate. Using $W(CO)_5=CPh_2$, Katz[103] studied metathesis of cis and trans-2-pentenes. With the cis olefin he found trans/cis ratios for C_4 and C_6 of 0.060 and 0.081 respectively. Even at 90% conversion, the amount of cis isomer was still very high. With trans olefins, thermodynamic values were observed. Katz suggested that the weak stereoselectivities observed with the catalytic systems having a Lewis acid co-catalyst were due to the fact that the Lewis acid facilitates the cleavage of the metal-carbon bond of the metalla-cyclobutane, forming a 3-metalla-propyl cation in which the remaining bonds rotate.

There were a priori two possible explanations for the early results on the stereochemistry of metathesis of acyclic olefins. The first hypothesis, given by Casey[94,104] and Basset,[101] is that the stereoselectivity may be explained by the favored formation of the metalla-cyclobutane in which the 1-3 diaxial interactions are minimized. According to this hypothesis a substituent in a β-axial position would be less destabilizing than a substituent in an α-axial position (Fig. 15).

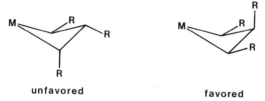

unfavored   favored

Figure 15:   Role of substituents R in the stability of metalla-cyclobutanes according to Casey[104]

The second hypothesis, given by Bilhou et al.[101] and by Casey,[105] would assume an asymmetric ligand arrangement around the metallo-carbene: the bulky R group of the carbene and the bulky R group of the incoming olefin would be far from the bulky ligand (or ligands) on the metal (Fig. 16).

Figure 16:   Possible   explanation   for   stereochemistry   involving   an asymmetric ligand arrangement according to [101, 105]

Two types of approach were considered to try to distinguish between these possibilities. Casey[104,106] studied in detail model reactions between metallo-carbenes and olefins, whereas Basset and co-workers[49,107-110] studied the real stereoselectivities of metathesis with various olefins and various catalysts.

A study of the model reactions of cis and trans disubstituted alkenes with an isolable metal-carbene complex in which the metal-carbene ligand possesses one sterically large and one sterically small substituent was essential to test the stereochemical implication of a puckered metalla-cyclobutane intermediate. The first "large-small" carbene complex, $(CO)_5W=C(C_6H_5)(CH_3)$, was obtained by Casey[104] as an intermediate. However, its decomposition was much more rapid than its reaction with alkenes. This high reactivity was probably due to the easy β-H elimination from the methyl substituent. To avoid such β-H elimination Casey et al.[106] prepared $(CO)_5W=CH(C_6H_5)$. which is not able to decompose by a similar pathway. The phenyl-carbene complex was observed by ^{1}H NMR at -78°C but decomposed readily at -50°C. Unfortunately no metathesis products have been observed in the reaction of $(CO)_5WCH(C_6H_5)$ with alkenes. Cyclopropane derivatives are the major organic products observed in all cases.

The reactions of $(CO)_5WCH(C_6H_5)$ with propylene, styrene and cis-2-butene lead to the predominant formation of the thermodynamically less stable syn cyclopropanes. This result was initially interpreted in terms of formation of the more stable puckered metalla-cyclobutane intermediate. However, the reaction of 2-methyl-2-butene with $(CO)_5WCH(C_6H_5)$ gave a 94: 1 excess of the syn isomer. This result, according to Casey,[106] is inconsistent with the stereochemistry being controlled by preferential formation of the sterically prefered metalla-cyclobutane since the metalla-cyclobutane which leads to the syn product has an axial methyl group on the central carbon atom. It should be pointed out here that Casey did not consider the possibility of a 1-2 (a-e or e-a) interaction in the metalla-cyclobutane, as we shall later on.

According to Casey, the stereochemistry of cyclopropane formation is determined at an earlier stage than the metalla-cyclobutane. The preference for formation of syn cyclopropane from 2-methyl-2-butene is determined by the approach of the alkene to the carbene carbon atom, so that steric interactions between $W(CO)_5$ and the methyl group are minimized (Fig 17).

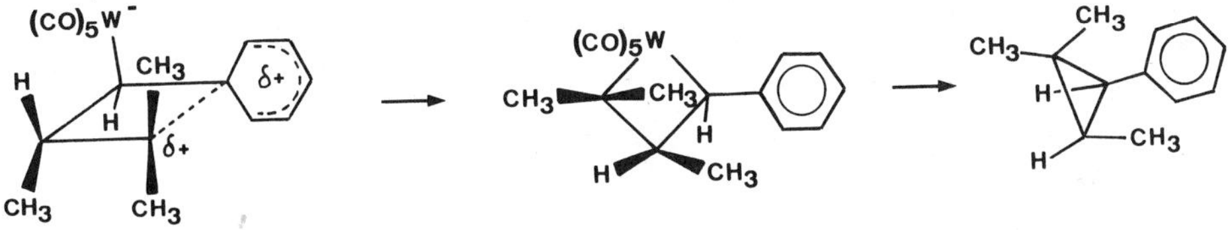

Figure 17:   Explanation for preferential formation of syn cyclopropane on reaction of 2-methyl-2-butene with $(CO)_5W=CH(C_6H_5)$ according to Casey[106]

The second type of approach was to study in detail the various parameters which may affect the stereoselectivity of real metathesis reactions with acyclic olefins. Such studies were carried out by Basset et al..[107-111] The difficulty in such an approach arises from the cis-trans isomerization which is competing with formal metathesis: any determination of the stereoselectivity should be the result of careful kinetic measurements where the stereoselectivity should be extrapolated after considering a wide range of conversion. In this respect not all the published results are reliable. Among the parameters which play a role on the stereoselectivity of the reaction, the following were studied in detail: the cis or trans nature of the starting olefin, the bulkiness of the substituents α to the double bond, the nature of the transition metal, and the ligands on the transition metal. We will consider these parameters separately.

5.1.1 The cis or trans nature of the olefin[49,99,103,107,109]

In most cases a cis olefin gives a cis olefin and a trans olefin gives a trans olefin. The retention of configuration is higher for trans than for cis olefins. The explanation for this retention of configuration may be deduced from the possible conformation of the metalla-cyclobutane transition state: among the four transition states leading to C_4 observed in cis-2-pentene metathesis, the transition state (●) leading to cis C_4 is the most favored. In the case of trans-2-pentene metathesis, the transition state (●) leading to trans C_4 is the most favored (Fig. 18).

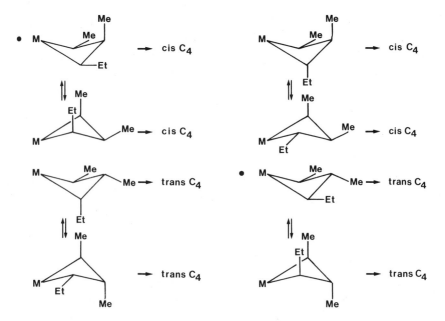

Figure 18:   Metalla-cyclobutane intermediates involved in metathesis of cis-2-pentene (left) and trans-2-pentene (right) (leading to C_4)

5.1.2 Substituents α to the double bond

The steric influence of R in R-CH=CH-CH$_3$ has been studied in detail with various molybdenum and tungsten based catalysts.[107-9] For R=H, no stereoselectivity occurs (trans C$_4$/cis C$_4$ ~ 1). For R=Et, Propyl, n-Butyl..., stereoselectivity appears with a retention of configuration of the starting olefin. When the R size increases above a certain value, the stereoselectivity decreases as well as the overall rate of productive metathesis (Fig. 19).

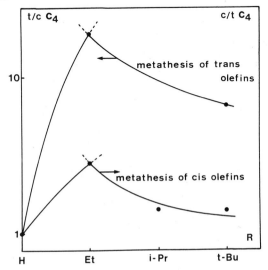

Figure 19: Stereoselectivities in metathesis of R-CH=CH-CH$_3$ with the catalytic system: Mo(NO)$_2$Cl$_2$(PPh$_3$)$_2$/EtAlCl$_2$

The results are explained on the basis of a change in the rate-determining step. When R is large, there is a non-stereoselective repulsive interaction between R and the transition metal and/or its ligands; such an effect accounts for the very drastic effect of bulky substituents of the olefins on the rate of productive metathesis (Fig. 20).

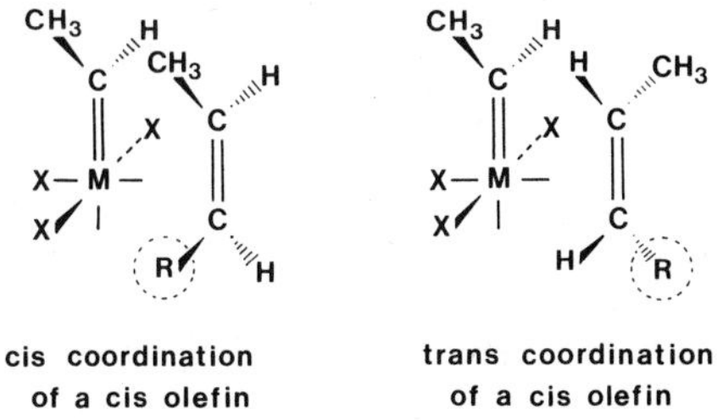

Figure 20

With the α olefins $CH_2=CH-R$, the <u>trans/cis</u> ratio of R-CH=CH-R increases when the R size increases.[109,111] This is related to the occurence of a 1-2 (equatorial-axial, axial-equatorial) repulsive effect in the metalla-cyclobutane transition state. Fig. 21 depicts the various transition states to be considered in the case of α olefins. The 1-3 diaxial interaction is negligible. Let us consider the 1-2 effect: upon increasing the R size, transition states C and D leading to <u>trans</u> olefins will be progressively favored.

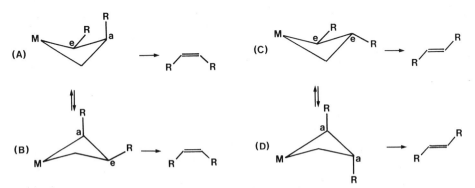

Figure 21:   Metalla-cyclobutanes involved in metathesis of α-olefins.

To what extent are the various effects competing? In most cases there is retention of the configuration of the starting olefin: with the <u>cis</u> olefins, the 1-3 diaxial interaction will be more effective than the 1-2 (e-a or a-e) interaction; with the <u>trans</u> olefins, the two simultaneous interactions will favor the formation of <u>trans</u> products. This accounts for the fact that <u>trans</u> olefins exhibit a higher degree of stereoselectivity than <u>cis</u> olefins. For very bulky olefins (R=isopropyl, <u>t</u>-butyl), the 1-3 diaxial interaction can be erased by the steric repulsion between the R group and the transition metal.

Finally, there are some cases for which the 1-2 interaction is the most important effect[22,50] (Fig. 22).

This is an extreme case where a <u>cis</u> olefin gives preferentially a <u>trans</u> olefin in the products due to a predominant 1-2 effect: the most favored transition states are those represented as (●) in Fig. 23.

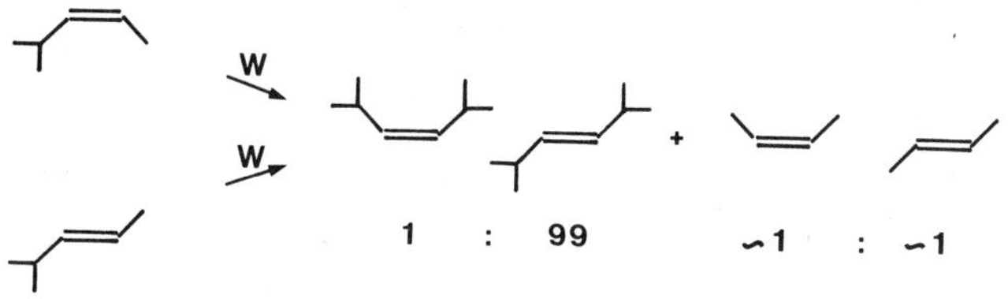

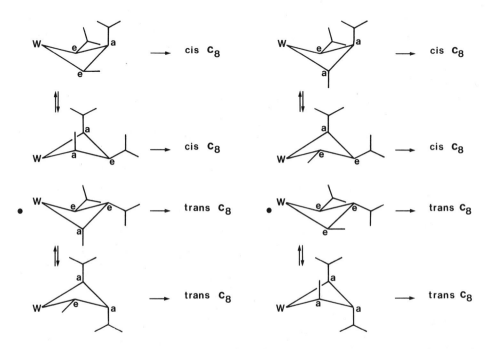

Figure 22:   Metathesis of <u>cis</u> and <u>trans</u> 4-methyl-2-pentene according to
              Calderon[22,50]

Figure 23:   Metalla-cyclobutanes involved in metathesis of <u>cis</u> (left) and
              <u>trans</u> (right) 4-methyl-2-pentene (leading to C_8)

5.1.3 The nature of the transition metal

Chromium, molybdenum, tungsten and rhenium complexes have been so
far the most studied,[20] although recent studies cover titanium,[112]
nickel,[113] niobium and tantalum[41] complexes as catalysts for metathesis.
In group VI the order of activity for formal metathesis may be considered
as Cr ≪ Mo < W. Most chromium complexes were found to be almost
inactive if one excepts the catalytic systems (Arene)Cr(CO)$_3$ + EtAlCl$_2$.
Most tungsten complexes lead to degenerate metathesis with α-olefins. The

stereoselectivities follow the reverse order compared with activity:[109] Cr >
Mo > W. The rather weak stereoselectivities observed with tungsten based
catalysts may be explained as follows: in the transition state $\overline{M-C_1-C_2-C_3}$,
the $M-C_1$, and $M-C_3$ bond lengths should increase from Cr to W and result in
a decrease of the 1-3 diaxial interaction due to a larger distance between C_1
and C_3.

5.1.4 Ligands on the transition metal

In many cases it was found that the ligands on the precursor complex
had no effect on the stereoselectivity of the propagation.[49,110] There are a
few exceptions with some catalysts of very weak activity such as
$W(CO)_5CPh_2$[103] or with nitrosyl complexes such as $M(NO)_2X_2(PPh_3)_2$
associated with $EtAlCl_2$ (X=Cl, Br, I; M = W,Mo).[99,109,110]

This lack of effects of ligands on the stereoselectivity is a frustrating
situation in homogeneous catalysis where the ligands usually induce a given
stereochemistry. The absence of steric effects of the catalytic systems
studied so far has many different possible origins, which are not mutually
exclusive:

- as previously observed, the stereoselectivity reflects mainly the steric
requirements of the metalla-cyclobutane (the C-C distances in the metallacycle are shorter than the metal-C distances);

- many catalytic systems may have the same coordination sphere which
would include only halogen or oxo ligands; many groups assume that the
active catalysts include the following coordination spheres[41,114-116] (Fig.
24):

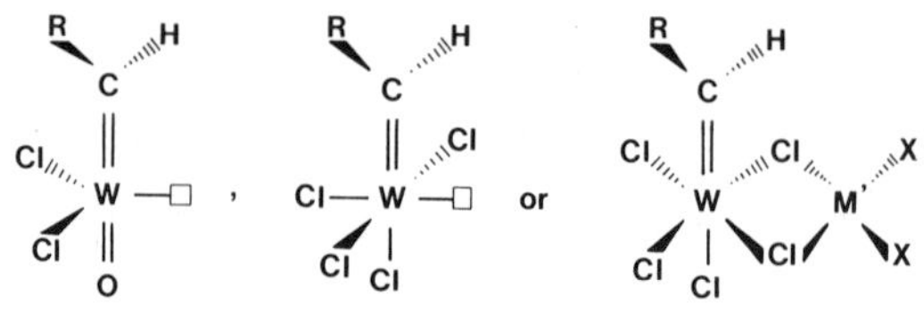

Figure 24:   Possible coordination spheres of active metathesis catalysts

5.2   Stereoselectivity with cyclic olefins

The problem of the stereochemistry with cyclic olefins seems to be, a priori, more simple than with acyclic olefins since the structure of the polymer is a fingerprint and is directly related to the steric path of the elementary steps occuring on the metallo-carbene. In reality the situation is more complicated. The number of studies related to the mechanistic aspect of the stereochemistry of the ring opening polymerization is therefore limited.[22,50,117-121]

Most cyclic olefins are strained, except cyclohexene,[133] and have a cis double bond. If we apply the results obtained with acyclic olefins, the polymer should be a cis polymer. In fact there are two types of catalytic systems: some catalysts lead to high cis content and some catalysts lead to a high trans content of the polymer chain. For example, high cis polypentenamer can be prepared with the following catalysts: $MoCl_5/R_3Al$, WF_6/R_2AlCl or $WCl_6/R_3Al/(BzO)_2$ associated with a low temperature.[2,3,122-126] High trans polypentenamer, which is thermodynamically favored, can be obtained with a great variety of catalytic systems.[117,118,127,128]

Recent studies by Calderon[22,50] and by Ivin and Rooney[121] have tried to rationalize the behaviour of such systems. Calderon has observed that the cis polymer remains cis during the overall propagation and has a very high molecular weight compared with the trans polymer. This phenomenon is observed although the double bond in the polymer chain is acyclic. This acyclic double bond could therefore coordinate to the metallo-carbene to give cyclic oligomers and could decrease the molecular weight of the polymer. In fact Calderon found that cis-stereoselective catalysts are unable to carry out metathesis of acyclic olefins, in agreement with the resulting high molecular weight of this particular polymer. Besides, whereas it is quite easy to regulate the molecular weight with most ring opening polymerization catalysts by adding a small amount of α-olefin to the monomer, this procedure does not work with the cis stereoselective catalysts. In contrast to α-olefins, an α,ω-diene such as 1,6-heptadiene can be an excellent chain scission agent; other dienes, such as 1,5-hexadiene or

1,7-heptadiene, which unlike **VII** do not exhibit the well known chelate effect of 3 x CH_2 units (vide supra), are much less efficient molecular weight regulators.

VII

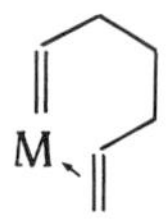

All these findings indicate that two types of catalyst or mechanism may be operative; the <u>cis</u> stereoselective catalyst (or mechanism) would be (or would correspond to) a cage-like complex in which a double bond of the polymer would be always coordinated to the metallo-carbene (Fig. 25). This cage effect would not occur with non-stereoselective catalysts (Fig. 26).

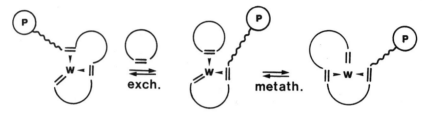

Figure 25:   Selective cyclo-olefin metathesis according to Calderon[22,50]

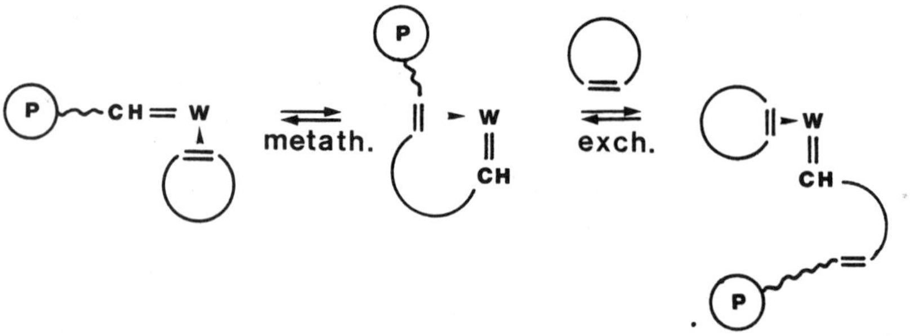

Figure 26:   Non-selective cyclo-olefin metathesis according to Calderon[22,50]

Such observations had already been made by Pampus[125] who called the two catalysts CIK and TRAK, the CIK catalyst being operative at -30°C and the TRAK above 25°C. It is logical to assume that low temperature will favor a well-ordered cage-like complex.

Recently, Rooney and Ivin,[121] studying the polymerization of norbornene substituted with an alkyl group, were able to carry out very elegant work on the stereochemistry of the polymer.

For polymers of norbornene and its derivatives, there is a trend from blocky to random cis/ trans double bond distribution and from fully syndiotatic to atactic ring dyads as the cis content of the polymer decreases from 100% to 35%. With a cis stereoselective catalyst, even the trans double bonds tend to occur in pairs; in other words, a trans double bond next to the metallo-carbene will possibly induce trans stereoselectivity in the following propagation step: this can be interpreted by assuming that the cis stereoselective catalysts retain one double bond during the coordination of the monomer.

Recent work by Larroche et al.[135] leads to a common interpretation for the stereochemistry of both cyclic and acyclic olefins. These authors have studied the stereochemistry of norbornene metathesis with a great variety of group VI metal based catalysts. With all the catalysts studied the percentage of cis double bonds in the polymer was found to be equal to $50 \pm 5\%$, this value corresponding to a random coordination of the exo double bond of the norbornene to the metallo-carbene. The same catalysts exhibited a varying degree of stereoselectivity with acyclic olefins, in sharp contrast with the results obtained with norbornene. A general explanation is given for the origin of the stereoselectivity in metathesis of acyclic and cyclic olefins: if the coordinated olefin has an energy which is smaller than that of the two possible metallacyclobutanes leading to the cis or the trans isomers, the resulting stereoselectivity will be governed by the energy levels of these two metallacycles; if the coordinated olefin has an energy which is higher than that of the cis and trans directing metallacyclobutane, the system will lose its stereoselectivity with a trans/cis ratio of unity. Most acyclic olefins belong to the first category (Fig. 27). Highly strained olefins belong to the second category due to the exothermic strain release which is already partially achieved in the metallacyclobutane intermediate (Fig. 28). The ligands do not govern the stereoselectivity by their own steric requirements but rather by their electronic effect. Lewis acids decrease the

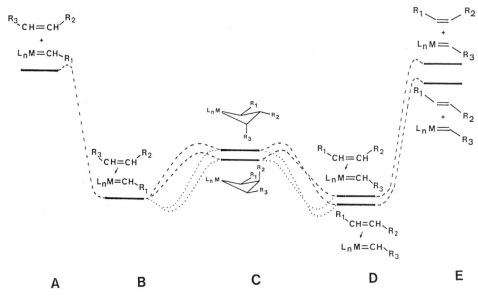

| A | B | C | D | E |

Figure 27:   Schematic energy levels for the various intermediates involved
in metathesis of a <u>cis</u> acyclic olefin

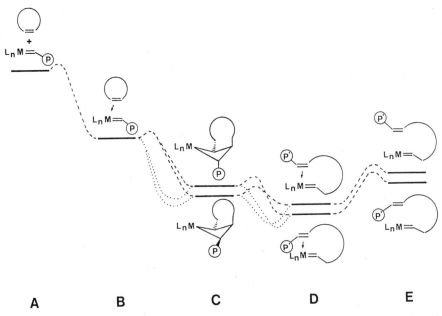

| A | B | C | D | E |

Figure 28:   Schematic energy levels for the various intermediates involved
in metathesis of a <u>cis</u> strained cyclic olefin

energy levels of the metallacyclobutane by their ability to coordinate to the metal and/or its ligands. The increase of stereoselectivity observed between W and Cr in the metathesis of acyclic olefins, which parallels the decrease of activity, would be due to the high energy level of the chromiacyclobutane.

6. ELECTRONIC NATURE OF THE CHAIN CARRYING CARBENE IN OLEFIN METATHESIS

For α-olefins, a degenerate metathesis occurs which regenerates the starting α-olefin. In some cases, degenerate metathesis of linear α-olefins can be at least one hundred times as fast as productive metathesis.[105,129-131] It is usually considered that a detailed understanding of the degenerate metathesis of terminal alkenes will provide information on the polarity of the chain carrying metal-carbene complex. There are two possible explanations for the degenerate metathesis of terminal alkenes: (i) The reaction may proceed via a $M=CH_2$ complex which reacts selectively with a terminal alkene to transfer the most substituted alkylidene and to regenerate a $M=CH_2$ carbene complex (eq 44). Such a selectivity could arise from steric effects and/or from electronic stabilization of a negatively polarized carbene ligand $\overset{+}{M} - \overset{-}{CH_2}$. (ii) Alternatively, the reaction may proceed via a $M=CH-R$ complex which reacts selectively with a terminal alkene to transfer the least substituted alkylidene unit and to regenerate a $M=CH-R$ carbene complex (eq 45). Such a selectivity could be attributed to electronic stabilization of a positively polarized carbene ligand $\overset{-}{M} - \overset{+}{CHR}$.

$$\overset{\delta^-}{\underset{M}{\overset{\|}{CH_2}}} \quad \overset{R}{\overset{\|}{\diagup}} \quad \longrightarrow \quad \overset{R}{\diagdown}\!= \; + \; M=CH_2 \tag{44}$$

$$\overset{\delta^+}{\underset{M}{\overset{R}{\overset{\|}{\diagup}}}} \quad \overset{R}{\underset{R}{\overset{\|}{\diagdown}}} \quad \longrightarrow \quad \overset{R}{\diagdown}\!= \; + \; M\!\!=\!\!\overset{\diagup}{\underset{R}{\diagdown}} \tag{45}$$

Model reactions with known metallo-carbenes complexes do not give the expected results. For example, the reaction of the electrophilic carbene complex $(CO)_5W=C(C_6H_4\text{-}\underline{p}\text{-}CH_3)_2$ with alkenes[105] led to the selective transfer of the least substituted alkylidene unit of a terminal alkene to the diaryl carbene providing a good model for the reaction of eq 46.

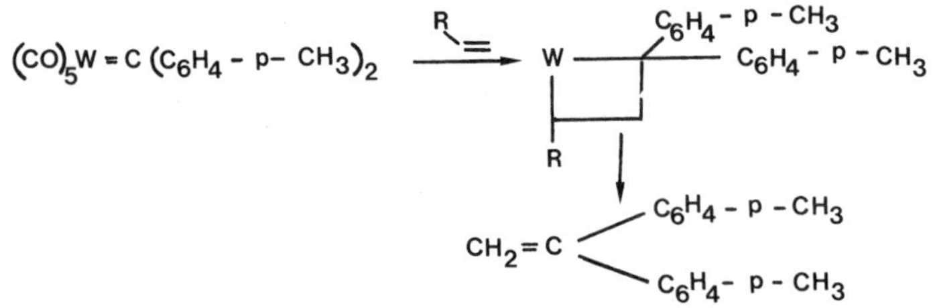

$$(CO)_5W=C(C_6H_4-p-CH_3)_2 \longrightarrow$$

$$(46)$$

Conversely the carbene complex $(C_5H_5)Cl_2Ta=CHC(CH_3)_3$[40] gives with propylene 95% $(CH_3)_3CCH_2C(CH_3)=CH_2$ which can be explained in terms of a nucleophilic carbene (eq 47).

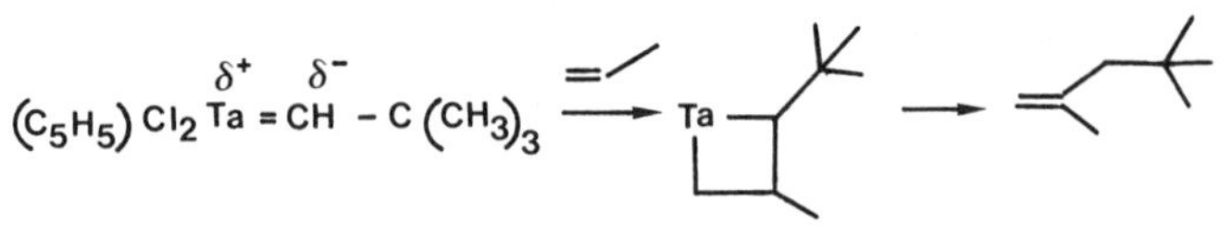

$$(C_5H_5)\,Cl_2\,\overset{\delta^+}{Ta}=CH-\overset{\delta^-}{C}(CH_3)_3 \longrightarrow$$

$$(47)$$

To distinguish between these two possible polarizations of the chain-carrying metal-carbene complex, Casey[105] has studied the metathesis of a mixture of $CD_2=CH-C_6H_{13}$ and (Z)-$CDH=CH-C_8H_{17}$. If the reaction proceeds via sequence (i), the ratio of (Z) - to (E) - $CHD=CH-C_6H_{13}$ should be 1:1 since neither the metal-carbene-alkene complex nor the metalla-cyclobutane are capable of retaining stereochemical information. However, if degenerate metathesis proceeds via reaction sequence (ii), both the metal-carbene-alkene complex and the metalla-cyclobutane retain stereochemical information; a preference for retention of stereochemistry would be expected since the reaction should proceed through the more stable puckered metalla-cyclobutane.   The experimental results indicate a retention of stereochemistry. Such observation of retention of configuration is consistent with M=CH-R being the chain-carrying carbene complex in degenerate metathesis.

Nevertheless, Casey pointed out that the observed stereochemical results may also be consistent with a $M=CH_2$ chain-carrying carbene, but only if the metathesis catalyst is asymmetric and if there is a substantial barrier to rotation about the M=CHD bond.

More recently Rooney et al.[132] reinvestigated the problem of chain-carrying species in degenerate metathesis by studying cross metathesis between norbornadiene and 1-hexene. They found that the major products were **VIII** and **IX**, which are unsymmetrical.

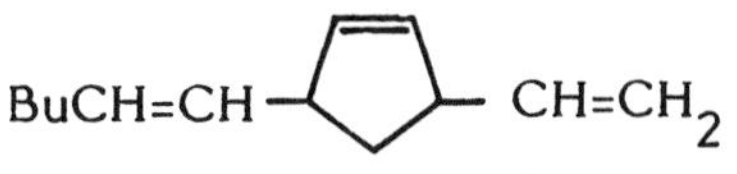

$$BuCH=CH \text{—⬠—} CH=CH_2 \qquad\qquad BuCH{=\!\!\!\!|}\, CH \text{—⬠—} CH{=\!\!\!\!|}CH_2$$

VIII 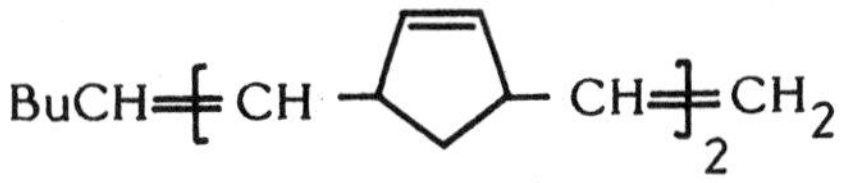   **IX**

After careful consideration of the kinetic parameters and the product distribution, they concluded that the substituted metallo-carbene $Bu\,^+CH = M$ is the major chain carrier. This conclusion is confirmed by experiments carried out with cyclopentene and 1,7-octadiene.

7. CONCLUSIONS

In the course of this review we have considered various mechanistic features of the olefin metathesis reaction. The metallo-carbene mechanism seems to be now well accepted. Some parameters which govern the stereoselectivity of the reaction have also been defined. Nevertheless a number of questions remain unanswered and deserve further study.

It is our belief that a full understanding of the olefin metathesis mechanism cannot be achieved unless a well defined catalyst with a known coordination sphere is found. It is probably via this kind of approach that it will be possible to design a metal and/or ligand environment which will tolerate functional substituents, which remains a challenge for synthetic organic chemists. Such approach could also, hopefully, lead to catalysts able to induce a given stereochemistry. The problem of stereochemistry with cyclic olefins is probably even more complex than with acyclic olefins.

In our opinion, the discovery of the metallo-carbene mechanism and of the metalla-cyclobutane intermediate in metathesis is at the origin of new ideas regarding the mechanism of C-C bond formation. Metallo-carbenes have been recently invoked in olefin polymerization, olefin homologation, Fischer-Tropsch synthesis, cyclopropanation, and acetylene polymerization.

Theoretical studies have also suggested that metal-oxo complexes or metal-oxo surface complexes can induce epoxidation, metathesis, or C-C bond breakage depending on the group VI transition metal used. Interestingly the transition state is a metalla-cycle which may contain one hetero atom. These studies will probably stimulate new research regarding metathesis-like reactions between C=X double bonds in which X is O, N. etc.

REFERENCES

1. R.L. Banks and G.C. Bailey, Ind. Eng. Chem., Prod. Res. Dev., 3:170 (1964).

2. G. Natta, G. Dall'Asta, G. Mazzanti and G. Motroni, Makromol. Chem. 69:163 (1963).

3. G. Natta, G. Dall'Asta and G. Mazzanti, Angew. Chem., 76:765 (1964).

4. N. Calderon, E.A. Ofstead, J.P. Ward, W.A. Judy and K.W. Scott, J. Am. Chem. Soc. 90:4133 (1968).

5. D.M. Adams, J. Chatt, R.G. Guy and N. Sheppard, J. Chem. Soc. 738 (1961).

6. Informations Chimie 179:83 (1978).

7. (a) R. Streck, Private communication; (b) Der Lichtbogen 198:1 (1980).

8. (a) S.R.I. (Linear Alpha Olefins), Dossier n° 12, B.I. (1974); (b) Chem. Eng. News. 48(10):60 (1970).

9. (a) E.O. Fischer, Adv. Organomet. Chem. 14:1 (1976); (b) C.P. Casey, Transition metal organometallics in organic synthesis 1:189 (1976); (c) F.J. Brown, Progress in inorganic chemistry 27:1 (1980).

10. First International Symposium on Metathesis, Mainz, January, 22-23, (1976).

11. Fourth International Symposium on Metathesis, Belfast, September (1981).

12. G.C. Bailey, Catalysis Rev. 3:37 (1969).

13. R.L. Banks, Top. Curr. Chem. 25:39 (1972).

14. N. Calderon, Acc. Chem. Res. 5:127 (1972).

15. W.B. Hughes, Organometal. Chem. Synth. 1:341 (1972).

16. W.B. Hughes, Adv. Chem. Ser. 132:192 (1974).

17. J.C. Mol and J.A. Mouljin, Adv. Catalysis 24:131 (1975).

18. R.J. Haines and G.J. Leigh, Chem. Soc. Rev. 4:155 (1975).

19. L. Hocks, Bull. Soc. Chim. Fr. 1893 (1975).

20. J.J. Rooney and A. Stewart, Catalysis, Chem. Soc., Spec. Period Rep. 1:277 (1977).

21. R.H. Grubbs, Prog. Inorg. Chem. 24:1 (1978).

22. N. Calderon, J.P. Lawrence and E.A. Ofstead, Adv. Organometal. Chem. 17:449 (1979).

23. J. Otton, Informations Chimie 201:161 (1980).

24. J.L. Herisson and Y. Chauvin, Makromol. Chem. 141:161 (1970).

25. K.W. Scott, N. Calderon, E.A. Ofstead, W.A. Judy and J.P. Ward, Rubber Chem. Technol. 44:1341 (1971).

26. N. Calderon, E.A. Ofstead and W.A. Judy, J. Polymer. Sci., Part A-1, Polymer Chem. 5:2209 (1967).

27. W.B. Hughes, Chem. Tech. 486 (1975).

28. H. Höcker, W. Reimann, K. Riebel and Z. Szentivanyi, Makromol. Chem. 177:1707 (1976).

29. (a) J.P. Arlie, Y. Chauvin, D. Commereuc and J.P. Soufflet, Makromol. Chem. 175:861 (1974); (b) Y. Chauvin, D. Commereuc and G. Zaborowski, Rec. Trav. Chim. Pays-Bas 96:M131 (1977).

30. B.A. Dolgoplosk, R.L. Makovetsty, T.G. Golenko, Y.U. Korshak and E.I. Timyakova, Eur. Pol. J. 10:901 (1974).

31. N. Calderon, E.A. Ofstead and W.A. Judy, Angew. Chem., Int. Ed. Engl. 15:401 (1976).

32. T.J. Katz and J. Mc Ginnis, J. Am. Chem. Soc. 97:1592 (1975).

33. R.H. Grubbs, P.L. Burk and D.D. Carr, J. Am. Chem. Soc. 97:3265 (1975).

34. T.J. Katz and R. Rothchild, J. Am. Chem. Soc. 98:2519 (1976).

35. R.H. Grubbs and C.R. Hoppin, J. Am. Chem. Soc. 101:1499 (1979).

36. R.H. Grubbs and S.J. Swetnick, J. Mol. Catal. 8:25 (1980).

37. C.P. Casey and T.J. Burkhardt, J. Am. Chem. Soc. 95:5833 (1973).

38. C.P. Casey and T.J. Burkhardt, J. Am. Chem. Soc. 96:7808 (1974).

39. R.R. Schrock, J. Am. Chem. Soc. 97:6577 (1975).

40. (a) S.J. McLain, C.D. Wood and R.R. Schrock, J. Am. Chem. Soc. 99:3519 (1977); (b) S.J. McLain, C.D. Wood and R.R. Schrock, J. Am. Chem. Soc. 101:4558 (1979).

41. R. Schrock, S. Rocklage, J. Wengrovius, G. Rupprecht and J. Fellmann, J. Mol. Catal. 8:73 (1980).

42. J.H. Wengrovius, R.R. Schrock, M.R. Churchill, J.R. Missert and W.J. Youngs, J. Am. Chem. Soc. 102:4515 (1980).

43. Y. Ben Taarit, J.L. Bilhou, M. Leconte and J.M. Basset, J.C.S. Chem. Commun. 38 (1978).

44. (a) F.N. Tebbe, G.W. Parshall and D.W. Ovenall, J. Am. Chem. Soc. 101:5074 (1979); (b) U. Klabunde, F.N. Tebbe, G.W. Parshall and R.L. Harlow, J. Mol. Catal. 8:37 (1980).

45. Y. Chauvin, D. Commereuc and D. Cruypelinck, Makromol. Chem. 177:2637 (1976).

46. T.J. Katz, S.J. Lee and N. Acton, Tetrahedron Lett. 4247 (1976).

47. (a) A. Mortreux and M. Blanchard, J.C.S. Chem. Commun. 787 (1974); (b) A. Mortreux, N. Dy and M. Blanchard, J. Mol. Catal. 1:101 (1975/1976); (c) A. Mortreux, J.C. Delgrange, M. Blanchard and B. Lubochinsky, J. Mol. Catal. 2:73 (1977); (d) T.J. Katz, S.J. Lee and M.A. Shippey, J. Mol. Catal. 8:219 (1980).

48. H. Rudler, J. Mol. Catal. 8:53 (1980).

49. J.M. Basset, J.L. Bilhou, R. Mutin and A. Theolier, J. Am. Chem. Soc. 97:7376 (1975).

50. E.A. Ofstead, J.P. Lawrence, M.L. Senyek and M. Calderon, J. Mol. Catal. 8:227 (1980).

51. P. Chevalier, D. Sinou, G. Descotes, R. Mutin and J.M. Basset, J. Organomet. Chem. 113:1 (1976).

52. C. Edwige, A. Lattes, J.P. Laval, R. Mutin, J.M. Basset and R. Nouguier, J. Mol. Catal. 8:297 (1980).

53. Y. Chauvin, D. Commereuc, D. Cruypelinck and J.P. Soufflet, German Patent, 2 503 943 (1975).

54. Y. Chauvin and D. Commereuc, First International Symposium on Metathesis, Mainz, January, 22-23, 1976.

55. R.R. Schrock and G.W. Parshall, Chem. Rev. 76:243 (1976).

56. E.L. Muetterties and M.A. Busch, J.C.S. Chem. Commun. 754 (1974).

57. E.L. Muetterties, Inorg. Chem. 14:951 (1975).

58. J.P. Soufflet, D. Commereuc and Y. Chauvin, C.R. Acad. Sci. Paris, Ser. C 276:169 (1973).

59. N.J. Cooper and M.L.H. Green, J.C.S. Chem. Commun. 761 (1974).

60. N.J. Cooper and M.L.H. Green, J.C.S. Dalton 1121 (1979).

61. R.R. Schrock, J. Am. Chem. Soc. 96:6796 (1974).

62. J.D. Fellman, G.A. Rupprecht, C.D. Wood and R.R. Schrock, J. Am. Chem. Soc. 100:5964 (1978).

63. D.N. Clark and R.R. Schrock, J. Am. Chem. Soc. 100:6774 (1978).

64. C.D. Wood, S.J. McLain and R.R. Schrock, J. Am. Chem. Soc. 101:3210 (1979).

65. R.H. Grubbs and C.R. Hoppin, J.C.S. Chem. Commun. 634 (1977).

66. J.R.M. Kress, M.J.M. Russell, M.G. Wesolek and J.A. Osborn, J.C.S. Chem. Commun. 431 (1980).

67. H. de Vries, Rec. Trav. Chim. Pays-Bas 80:866 (1961).

68. D.C. Feay and S. Fujioka, Proc. 10th Int. Conf. Coord. Chem., September 12-16, 1967, Butterworths, London, p.1.

69. U.S. Patent, 3 274 223 (1966) to D.C. Feay.

70. E. Heins, H. Hinck, W. Kaminsky, G. Oppermann, P. Raulinat and E. Sinn, Makromol. Chem. 134:1 (1970).

71. J. Levisalles, H. Rudler and D. Villemin, J. Organomet. Chem. 192:375 (1980).

72. (a) M.F. Farona and V.W. Motz, J.C.S. Chem. Commun. 930 (1976); (b) W.S. Greenlee and M.F. Farona, Inorg. Chem. 15:2129 (1976); (c) V.W. Motz and M.F. Farona, Inorg. Chem. 16:2545 (1977).

73. B.A. Dolgoplosk, Dokl. Chem. 216:380 (1974).

74. W.A. Herman, Angew, Chem., Int. Ed. Engl. 13:599 (1974).

75. J.W. Byrne, H.V. Blaser and J.A. Osborn, J. Am. Chem. Soc. 97:3871 (1975).

76. M. Ephritikhine, M.L.H. Green and R.E. McKenzie, J.C.S. Chem. Commun. 619 (1976).

77. M. Ephritikhine and M.L.H. Green, J.C.S. Chem. Commun. 926 (1976).

78. M.L.H. Green, Pure Appl. Chem. 50:27 (1978).

79. G.J.A. Adam, S.G. Davies, K.A. Ford, M. Ephritikhine, P.F. Todd and M.L.H. Green, J. Mol. Catal. 8:15 (1980).

80. C.G. Biefield, H.A. Eick and R.H. Grubbs, Inorg. Chem. 12:2166 (1973).

81. (a) A. Morris, H. Thomas and C.J. Attridge, German Patent 2 213 948 (1972); (b) J.P. Candlin, A.H. Mawby and H. Thomas, German Patent 2 213 947 (1972); (c) I.C.I. Ltd., French Patent 2 120 509 (1972); (d) J.P. Candlin and H. Thomas, Adv. Chem. Ser. 132:212 (1974).

82. Y. Iwasawa, M. Yamagishi and S. Ogasawara, J.C.S. Chem. Commun. 871 (1980).

83. E. Verkuijlen, F. Kapteijn, J.C. Mol and C. Boelhouwer, J.C.S. Chem. Commun. 198 (1977).

84. A.J. Van Roosmalen, K. Polder and J.C. Mol, J. Mol. Catal. 8:185 (1980).

85. (a) Yu. I. Yermakov, B.N. Kuznetsov and A.N. Startsev, React. Kinetics Catalysis Letters 2:151 (1975); (b) Yu. I. Yermakov, B.N. Kuznetsov and A.N. Startsev, Kinetica i Kataliz 15:556 (1974); (c) A.N. Startsev, B.N. Kuznetsov and Yu. I. Yermakov, React. Kinetics Catalysis Letters 3:321 (1976).

86. B.N. Kuznetsov, A.N. Startsev and Yu. I. Yermakov, J. Mol. Catal. 8:135 (1980).

87. J.I.C. Archibald, J.J. Rooney and A. Stewart, J.C.S. Chem. Commun. 547 (1975).

88. P.P. O'Neill and J.J. Rooney, J. Am. Chem. Soc. 94:4383 (1972).

89. D.T. Laverty, J.J. Rooney and A. Stewart, J. Catal. 45:110 (1976).

90. A.A. Olsthoorn and C. Boelhouwer, J. Catal. 44:207 (1976).

91. M.F. Farona and R.L. Tucker J. Mol. Catal. 8:85 (1980).

92. A.K. Rappe and W.A. Goddard III, J. Am. Chem. Soc. 102:5114 (1980).

93. L. Kim, J.H. Raley and C.S. Bell, Rec. Trav. Chim. Pays-Bas 96:M136 (1977).

94. C.P. Casey and S.W. Polichnowski, J. Am. Chem. Soc. 99:6097 (1977).

95. C.P. Casey, Chem. Tech. 378 (1979).

96. C.P. Casey, H.E. Tuinstra and M.C. Saeman, J. Am. Chem. Soc. 98:608 (1976).

97. R.J. Puddephatt, M.A. Quyser and C.F.H. Tipper, J.C.S. Chem. Commun. 626 (1976).

98. T.R. Howard, J.B. Lee and R.H. Grubbs, J. Am. Chem. Soc. 102:6876 (1980).

99. W.B. Hughes, J.C.S. Chem. Commun. 431 (1969).

100. J.L. Bilhou, J.M. Basset, R. Mutin and W.F. Graydon, J.C.S. Chem. Commun. 970 (1976).

101. J.L. Bilhou, J.M. Basset, R. Mutin and W.F. Graydon, J. Am. Chem. Soc. 99:4083 (1977).

102. R.R. Schrock, L.W. Messerle, C.D. Wood and L.J. Guggenberger, J. Am. Chem. Soc. 100:3793 (1978).

103. T.J. Katz and W.H. Hersch, Tetrahedron Letters 6:585 (1977).

104. C.P. Casey, L.D. Albin and T.J. Burkhardt, J. Am. Chem. Soc. 99:2533 (1977).

105. C.P. Casey and H.E. Tuinstra, J. Am. Chem. Soc. 100:2270 (1978).

106. C.P. Casey, S.W. Polichnowski, A.J. Shusterman and C.R. Jones, J. Am. Chem. Soc. 101:7282 (1979).

107. M. Leconte, J.L. Bilhou, W. Reimann and J.M. Basset, J.C.S. Chem. Commun. 341 (1978).

108. M. Leconte and J.M. Basset, Nouveau J. Chim. 3:429 (1979).

109. M. Leconte and J.M. Basset, J. Am. Chem. Soc. 101:7296 (1979).

110. M. Leconte, Y. Ben Taarit, J.L. Bilhou and J.M. Basset, J. Mol. Catal. 8:263 (1980).

111. M. Leconte and J.M. Basset, Annals New York Acad. Sci. 333:165 (1980).

112. F. Tebbe, G.W. Parshall and G.S. Reddy, J. Am. Chem. Soc. 100:3611 (1978).

113. R.H. Grubbs and A. Miyashita, J. Am. Chem. Soc. 100:7418 (1978).

114. E.L. Muetterties and E. Bund, J. Am. Chem. Soc. 102:6572 (1980).

115. E. Verkuijlen, J. Mol. Catal. 8:107 (1980).

116. F. Garnier and P. Krausz, J. Mol. Catal. 8:91 (1980).

117. E.A. Ofstead and M.L. Senyek, 8th Cent. Reg. Meet. Am. Chem. Soc., 1976.

118. N. Calderon, E.A. Ofstead and M.L. Senyek, Jpn - US. Joint Seminar on Elastomers, 1977.

119. K.J. Ivin, D.T. Laverty and J.J. Rooney, Makromol. Chem. 178:1545 (1977).

120. K.J. Ivin, D.T. Laverty, J.J. Rooney and P. Watt, Rec. Trav. Chim. Pays-Bas 96:M54 (1977).

121. K.J. Ivin, G. Lapienis, J.J. Rooney and C.D. Stewart, J. Mol. Catal. 8:203 (1980).

122. F. Haas, K. Nützel, G. Pampus and D. Theisen, Rubber Chem. Tech. 43:1116 (1970).

123. G. Dall'Asta and G. Motroni, Angew. Makromol. Chem. 16/17:51 (1971).

124. P. Günther, F. Haas, G. Marwede, K. Nützel, W. Oberkirch, G. Pampus, N. Schön and J. Witte, Angew. Makromol. Chem. 14:87 (1970).

125. G. Pampus and G. Lehnert, Makromol. Chem. 175:2605 (1974).

126. R.J. Minchak and H. Tucker, Polym. Prepr., Amer. Chem. Soc., Div. Polym. Chem. 13:885 (1972).

127. G. Dall'Asta and P. Scaglione, Rubber Chem. Tech. 42:1235 (1969).

128. G. Dall'Asta, Rubber Chem. Tech. 47:511 (1974).

420

129. W.J. Kelly and N. Calderon, J. Macromol. Sci., Chem. 9:911 (1975).

130. J. McGinnis, T.J. Katz and S. Hurwitz, J. Am. Chem Soc. 98:605 (1976).

131. M.T. Mocella, M.A. Busch and E.L. Muetterties, J. Am. Chem. Soc. 98:1283 (1976).

132. L. Bencze, K.J. Ivin and J.J. Rooney, J.C.S. Chem. Commun. 834 (1980).

133. L. Hocks, D. Berck, A.J. Hubert, Ph. Teyssie, J. Polym. Sci., Polym. Letters, 13:391 (1975).

134. S.J. McLain, J. Sancho and R.R. Schrock, J. Am. Chem. Soc. 101:5451 (1979).

135. C. Larroche, J.P. Laval, A. Lattes, M. Leconte, F. Quignard and J.M. Basset J. Org. Chem. 47:2019 (1982).

CARBONYLATION AND RELATED CHEMISTRY: SOME GENERAL ASPECTS

P.S. Braterman

Department of Chemistry

University of Glasgow

1. GENERAL

Carbon monoxide is among the most studied of all ligands. Its insertion into metal-carbon bonds may lead to esters, aldehydes, alcohols, ketones and other products, and is the basis of major industrial processes, many of them using relatively cheap mixtures of carbon monoxide and hydrogen (synthesis gas). In addition, reactions involving CO hydrogenation, and the use of CO as a reducing agent, have attracted growing attention since the 1973 oil price rise. The ambidentate character of the CO grouping has led in the past decade to much fascinating chemistry among the early transition elements. Finally, the spectroscopic properties of coordinated CO provide a particularly useful probe of its environment.[1] The following two Chapters describe aspects of carbonylation, CO reduction, and related reactions. This Chapter is limited to a general discussion of mechanisms, and to aspects of the chemistry of CO and related ligands not discussed elsewhere in this volume.

In addition to the references cited in the following Chapters, books and symposium collections have appeared on the catalytic activation of CO,[2] new syntheses with CO,[3] catalysis in C(1) chemistry[4] (with coverage[5] of chemical production from synthesis gas), the methanation of synthesis gas,[6] and hydrocarbon synthesis from CO and H_2.[7] Review topics include CO insertion into metal-carbon bonds,[8] the carbonylation of olefins,[9] hydroformylation,[10] large metal carbonyl clusters,[11] metal-bound CO as a Lewis acid,[12] transition metal formyls,[13] $HCo(CO)_4$ as catalyst,[14] transition metal catalyzed reactions of organic halides with CO,[15] organometallic aspects of the Fischer-Tropsch reaction,[16] homogeneous catalysis of the water gas shift reaction by ruthenium and other carbonyls,[17] asymmetric hydroformylation,[18] and immobilized metal carbonyls as catalysts.[19]

2.   THE CARBON MONOXIDE INSERTION REACTION

In general, the reaction step

$$RM(CO) \rightleftharpoons M.COR \qquad \textbf{(I)} \tag{1}$$

is reversible, and leads to isolable acyls only if the primary product is stabilized by addition of a further carbon monoxide or other ligand:

$$M.COR + L \longrightarrow LM.COR \tag{2}$$

Similarly, loss of a supporting ligand from metal promotes decarbonylation acyls [reverse of step (2)]. Thus the overall reaction

$$RM(CO) + L \rightleftharpoons LM.COR \qquad \textbf{(II)} \tag{3}$$

is important both in carbonylation and in decarbonylation processes.

Reaction (1) has long been regarded as internal nucleophilic attack by coordinated electron-rich R on the electron-attracting carbon of bound CO, and labeling experiments confirm that in eq (4) the entering CO does not form part of the acyl group:[20]

$$RM(CO) + {}^*CO \rightleftharpoons ({}^*CO)M.COR \tag{4}$$

The rates of the reactions

$$RCH_2M + CO \longrightarrow RCH_2.CO.M \tag{5}$$

$[M = Mn(CO)_5, Fe(CO)_4^-]$ display linear free energy relationships, correlating with $\sigma*$ so that more electron-donating R gives a faster reaction.[21] Hard Lewis acids promote the reaction by coordinating to ligand CO, according to Scheme 1,[22] and rate studies show[23] that aluminum tribromide does indeed accelerate the conversion of **III** to **IV**, as well as

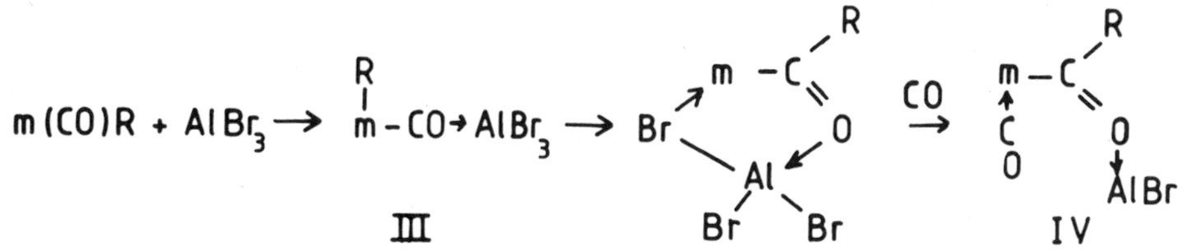

$$m(CO)R + AlBr_3 \rightarrow m\text{-}CO\text{→}AlBr_3 \rightarrow$$

Scheme 1

trapping the CO-poor intermediate. Oxidation of the central metal enhances the electronegativity of the CO ligand, leading to acceleration of Reaction (1).[24] This can lead to redox catalysis of Reaction (6):[25-6]

$$CpFe(CO)_2CH_3 + PPh_3 \longrightarrow CpFe(CO)(PPh_3).COMe \qquad (6)$$

or in alcoholic solvents to the formation of formyl[27] or other[24] esters from $CpFe(CO)_2H$ or $CpFe(CO)_2R$ and Cu(II). Bulky groups R can enhance reactivity,[28] but <u>two</u> large ortho substituents impede the reactivity of species $CpMo(CO)_3CH_2Ar$ towards phosphines.[29] Donor solvents catalyze the reaction of $CpMo(CO)_3CH_3$ with PPh_2Me; the catalysis can itself be sterically blocked and thus represents a specific interaction.[30] It is suggested that metal-acyl π-donation stabilizes intermediate I and thus promotes reaction. Cp_2VCH_3 reacts with CO to give $Cp_2V(CO).COMe$,[31] but $Cp_2Nb(CO)CH_3$ does not react further with CO, a fact ascribed to higher electronegativity of the central atom.[32]

Reaction (3) occurs with retention at R in the carbonylation of species $CpFe(CO)_2R$ (whether or not assisted by oxidation)[24,33] and the carbonylation of palladium alkyls,[34] but retention merely predominates in the decarbonylation of aldehydes[35] and acyl halides by $Rh(PPh_3)_3Cl$, and the corresponding reaction of active $PhCH(CF_3)COCl$ gives racemic product.[36] The loss of steric information is ascribed to incomplete stereoselectivity in a reversible acyl - carbonyl alkyl step.[37]

The formation of intermediate **I** may formally be assigned to ligand CO or to ligand R migration. The latter description was shown to be appropriate for the interconversion of $Mn(CO)_5CH_3$ and cis-$Mn(CO)_4(L)(CH_3)$, both by ^{13}C ir monitoring of forward and reverse reactions $(L = {}^{13}CO)$[20] and by ^{13}C nmr of the products formed from cis-$Mn(CO)_4({}^{13}CO)CH_3$ $[L = {}^{13}CO, P(OCH_2)_3CMe]$.[38] In HMPA as solvent, however, randomization of the label occurred.[38] The stereochemistry at metal for related reactions of species $CpM(CO)(L)R$ is less clearcut. Photodecarbonylation of $CpFe(CO)(PPh_3)COEt$ occurs with formal inversion, as expected for simple CO loss followed by alkyl migration:[39]

$$CpFe(CO)(L)CoEt \xrightarrow[-CO]{h\nu} CpFe(vacancy)(L)COEt$$

$$\longrightarrow CpFe(Et)(L)(CO) \tag{7}$$

[If the photodecarbonylation step were associated with pseudorotation, as in the reactions of Group VI hexacarbonyls and their derivatives,[40] the interpretation of this work would need to be precisely reversed. Photolysis of species $CpM(CO)(L)R$ (L or R optically active) in a donor matrix might help resolve this point]. On the other hand, Brunner and Vogt have reported evidence for formal retention (i.e. CO migration) in directly related thermal reactions.[41] For a while, the situation appeared very confused, following a report[42] of specific methyl migration in the reaction

$$CpRh(I)(L).COCH_3 + Ag^+ \longrightarrow [CpRh(CH_3)(L)(CO)]^+ + AgI \tag{8}$$

It is now, however, agreed[43-4] that the degree and direction of specificity of Reaction (7) depends on the solvent used. Once again (cf Chapter 1) the notion of a single reaction mechanism proves too simple. Solvent donor power and/or the freedom of rotation of an acyl intermediate may be crucial in the crowded[45] environment of these complexes, and (especially in donor solvents) the simple unsaturated fragment assumed for photodecarbonylation may not lie on the thermal reaction path at all. A final complication comes from the existence of kinetic evidence which is consistent either with competition between forward reaction (2) and back-reaction (1),[46] or with an <u>associative</u> pathway,[47] and for a photochemically generated species (η^3-$C_5H_5)Fe(CO)_3.CH_3$) which could be an intermediate in such a route.[47]

3. EXTERNAL NUCLEOPHILIC ATTACK ON CO

The conversion of mCO to $[mCOOR]^-$ is a possible key step in the carboxylation of alkenes (Chapters 7, 11), while for R = H the reaction is the so-called oxidative hydrolysis step of the catalyzed water gas shift reaction (Chapter 7). Such reactions are sometimes reversible, as in the solvent-dependent equilibria (9) (R = H, Me), in which the left hand side is favored by high dielectric constant solvents:[48]

$$[CpFeL_1L_2CO]^+ + OR^- \rightleftharpoons CpFeL_1L_2.COOR \qquad (9)$$

More surprising is the conversion of CO to carbonyl halides:[49,50]

$$[V(CO)_4 diars]^- + 3HX \longrightarrow VH_3(CO)_3 diars + COX_2 + X^- \qquad (10)$$
$$Fe(CO)_5 + 2HgCl_2 \longrightarrow \underline{cis}\text{-}Fe(CO)_4(HgCl)_2 + COCl_2 \qquad (11)$$

Attack at carbonyl may be the first step in other processes, as in the sequence

$$Re(CO)_4(PPh_3)Br + CH_3Li \longrightarrow Re(CO)_3(PPh_3)(Br).C(OLi)CH_3$$
$$\longrightarrow Re(CO)_4(PPh_3)CH_3 + LiBr \qquad (12)$$

where the acyl-like intermediate has been identified spectroscopically.[51]

4. METAL ACYLS AS BASES

Metal-acyl bonding lowers ν(acyl C = O) below that of ketones; in valence bond terms there is a contribution from the structure $M^+ = C(O^-)R$ as well as from M - C(=O)R. Metal acyl carbonyl seems to be more basic than metal-metal bridging CO in a comparable environment. Thus $CpFe(CO)_2.COR$ is more basic than $Cp_2Fe_2(CO)_4$, though less basic than acetone, towards BF_3,[52] while the more electron-rich $CpFe(dppe)COPh$ [ν(C=O) = 1510 cm^{-1}] is methylated by Me_3OBF_4, and reversibly protonated by HBF_4, to give oxycarbenes $[CpFe(dppe):C(OR)Ph]^+$ (R = Me, H respectively).[53] The acyl group of CpFe(PPh$_3$)(CO)COMe also acts as a ligand towards [CpFe(CO)$_2$]$^+$ and [CpMo(CO)$_3$]$^+$.[54]

5. η^2-BOUND ACYL LIGANDS AND RELATED SPECIES

In many (though by no means all)[55] formally unsaturated acyl complexes, the acyl group is bound to the metal through both carbon and oxygen. This situation is most common early in the transition series, where metal-oxygen bonds are strong. The bonding may formally be regarded as a special case of acyl oxygen basicity; but early in the transition series, where this kind of bonding is not common, there may be a considerable contribution from an alkoxycarbene structure:

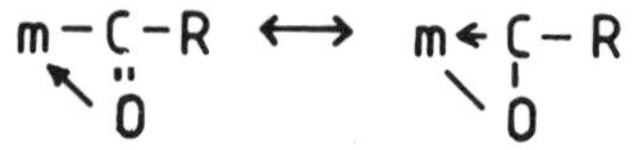

$$m-\underset{\overset{\|}{O}}{C}-R \longleftrightarrow m\leftarrow\underset{\overset{|}{O}}{C}-R$$

leading to further additions and insertions, as in Schemes 2[56] and 3.[57]

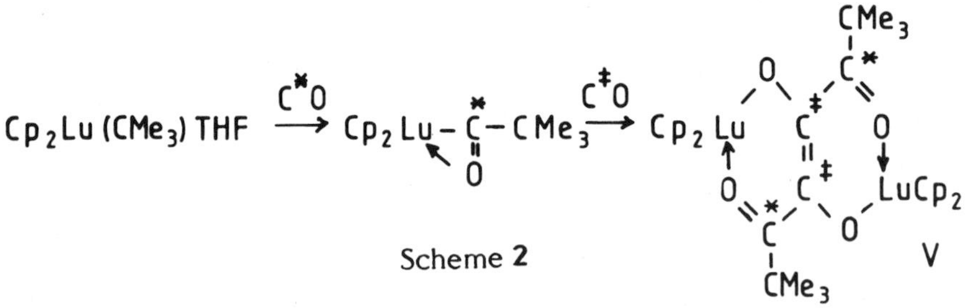

Scheme 2

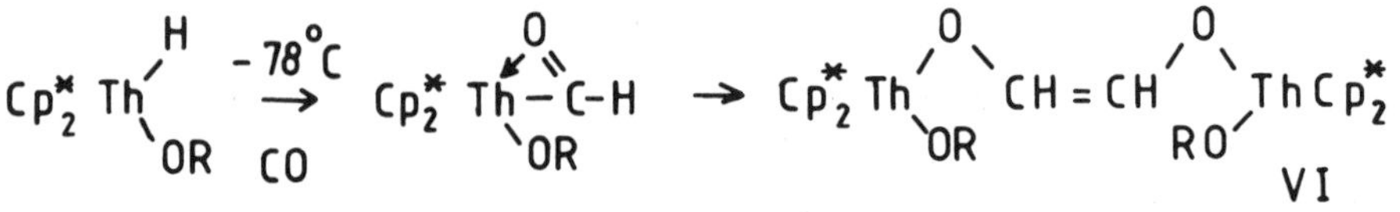

Scheme 3

Cp$_2$ZrMe$_2$ reacts with CO at -130°C to give Cp$_2$Zr(CO)Me$_2$, a d^0 carbonyl, which on warming gives first **VII** and then the isomer **VIII**.[58] Cp*$_2$ZrMe$_2$ gives **IX**, which reacts with a further mole of CO to give **X**. The corresponding reactions of the hydride are complex, Cp*$_2$Zr(H)$_2$CO being converted to **XI** or, in the presence of excess Cp*$_2$ZrH$_2$, to **XII**. The latter reacts further with excess H_2 to give $Cp^*_2Zr(H)OMe$.[59-60] Species **XIII**, closely related to **XII**, are obtained from $Cp_2Zr(COR)Cl$ and $Cp_2Zr(H)Cl$.[61] These show degenerate isomerism, with inversion at C, via the aldehyde-bridged structure **XIV**.[62] Related transformations involving migration of various groups to the η^2-acyl (or oxycarbene) ligand, and the formation of binuclear intermediates involving similar or dissimilar metals, give rise to a chemistry of great complexity.[59-64] The species $MoCl(PMe_3)R_3$ (R = CH_2SiMe_3) is carbonylated to the chloride-bridged dimer $[Mo(CO)_2(PMe_3)$ $(\eta^2\text{-}RCO)Cl]$.[65] The possible role of η^2-acyl intermediates in carbonylation reactions was discussed in Section 2, and indeed the species $Ru(CO)_2(PPh_3)_2(CH_3)I$ exists in the solid (and to a small extent in solution) as $Ru(\eta^2\text{-}COCH_3)I(CO)(PPh_3)_2$.[66]

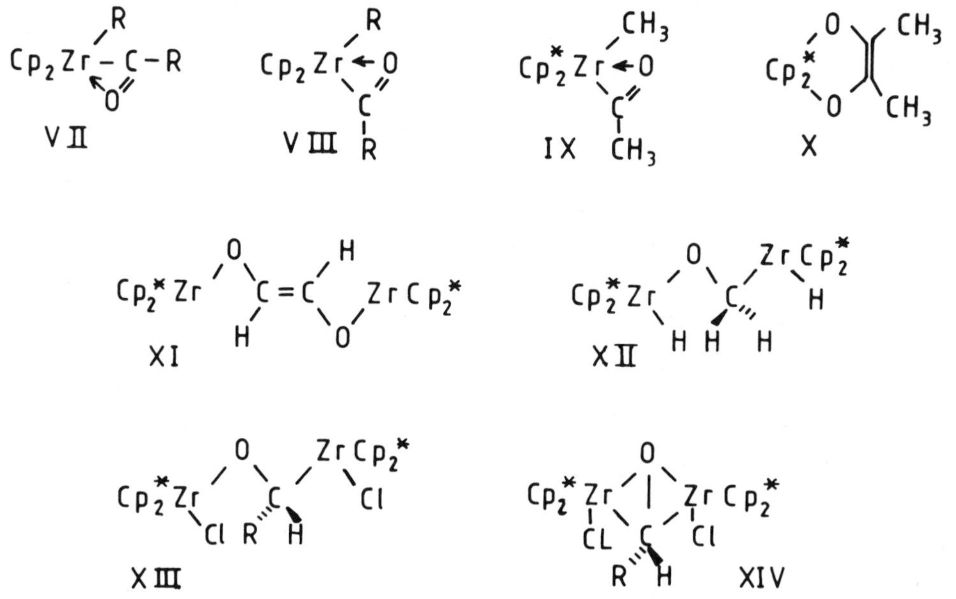

Somewhat akin to the η^2-acyls are η^2-ketone complexes such as $Cp_2\overline{Zr.C(Ph)_2.O}$ (from Cp_2ZrPh_2)[67] and $Cp^*\overline{Ta(Me)_2.C(Me)_2.O}$ (from Cp^*TaMe_4),[68] which undergo further interesting reactions with unsaturated species.

Vanadocene with paraformaldehyde gives $\overline{Cp_2V.CH_2.O}$, which reacts with CO to form $Cp_2V(CO)$ and half a mole of methyl formate, and with benzoyl chloride to give $Cp_2V(Cl).CH_2O.CO.Ph.$[69] $Os(CO)_2(PPh_3)_3$ reacts with formaldehyde to give $\overline{Os(CO)_2(PPh_3)_2.CH_2O}$, which can be methylated to $[Os(CO)_2(PPh_3)_2(H_2O).CH_2OMe]^+.$[70] Other transformations of η^2-formaldehyde and η^2-formyl ligands are discussed in Chapter 8.

6.   ALDEHYDE-HYDRIDOACYL INTERCONVERSIONS

Aldehydic precursors of chelating ligands can give isolable acyl hydrides, which serve as model intermediates for the metal-catalyzed decarbonylation of aldehydes by the sequence

$$RCHO + m \rightleftharpoons RCO.m.H \rightleftharpoons R.m(CO)H \longrightarrow RH^+ + \ldots \qquad (13)$$

Examples are the formation of **XV** from quinoline-8-carboxaldehyde,[71] and of **XVI** from o-(diphenylphosphino)benzaldehyde. **XV** gives quinoline on heating,[71] while **XVI** gives trans-$Ir(PPh_3)_2(CO)Cl.$[72] Elimination from metal can however involve other ligands, as in the formation of **XVII** from salicylaldehyde, K_2PtCl_4 and the phosphine using potassium carbonate as an acid sink.[73]

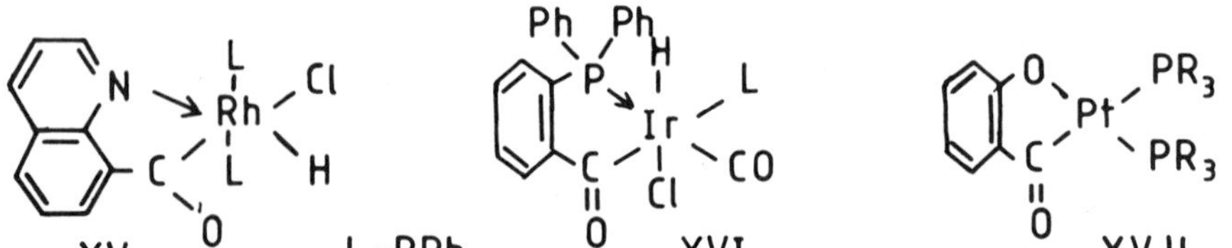

XV   L =PPh₃   XVI   XVII

7. KETONE ELIMINATION REACTIONS

As pointed out in Chapter 1, ketone elimination reactions

$$RmCOR' \longrightarrow m + RCOR' \qquad (14)$$

are a particularly facile subset of carbon-carbon bond-forming reductive eliminations. Even so, the reaction may fail when it could leave a highly unsaturated fragment, as in the case of the equilibrating anions $[Re(CO)_4(COMe)Ph]^-$ and $[Re(CO)_4(Me)COPh]^-$.[74] The reaction has been shown to occur with retention in the sequence

$$Co(Si^*)(CO)_4 \xrightarrow{\text{PhLi}} [Co(Si^*)(COPh)(CO)_3]^- \longrightarrow Si^*COPh +... \qquad (15)$$

$[Si^* = (Me)(Ph)(\alpha\text{-naphthyl})Si]$.[75]
In the reaction

$$(Me_3P)_3CoMe_3 + 4CO \longrightarrow (Me_3P)_2(CO)_2Co.COMe + MeCOMe + PMe_3 \qquad (16)$$

the elimination is presumably unimolecular.[76] However, the elimination of ketones from species $RNi(acac)(PPh_3)_n$ (R=Me, n=2; R=Et, n=1) is thought to involve elimination between Ni-R and Ni-COR species, since it leaves $Ni(acac)_2$ and $Ni(CO)_2(PPh_3)_2$ as the other products.[77]

Ketone formation can usefully be combined with other reactions. The formation of diethyl ether from ethane, carbon monoxide and water using a $Co_2(CO)_8$-dppe catalyst, combines features of the water gas shift reaction, hydroformylation, and alkene insertion.[78] A similar reaction converts methyl acrylate to dimethyl 4-oxopimelate in up to 42% yield.[79] Combined reductive coupling (to give a metallacyclopentene) and carbonylation (leading to reductive elimination at the metal) can be used to form fused ring cyclopentanone systems from cyclic alkenes with high specificity. The selectivity depends on electronic and steric effects, and is affected by bonding between the metal and substituents.[80] Aldehydes react in the presence of Rh(I) with alkenes to give ketones by a sequence such as

$$R.CO.H + m + C_2H_4 \longrightarrow R.CO.m(H)(C_2H_4)$$
$$\rightleftharpoons R.CO.m.Et \longrightarrow RCOEt + m \qquad\qquad (17)$$

The intermediate equilibrium has been demonstrated by scrambling of the label in R.CO.D over both positions in the ethyl group of the product,[81] while aldehydes unsaturated at the 4 position can be cyclized to cyclopentanones.[82]

8.   MISCELLANEOUS

The carbene complex $MecpMn(CO)_2CPh_2$ reacts with carbon monoxide to give the ketone complex $MecpMn(CO)_2(\eta^2\text{-}Ph_2C=CO)$. With hydrogen, this forms diphenylacetaldehyde and 2,2-diphenylethanol.[83] The carbyne species $CpW(CO)(PPh_3)\equiv CAr$ (Ar = p-tolyl) reacts with carbon monoxide to give $CpW(CO)_2(PPh_3)\text{-}C(Ar)=C=O$. This at 60°C and 60 atm CO gives $CpW(CO)_2(PPh_3)\text{-}C\equiv CAr$ and CO_2.[84]

The β-hydrogen of $CpFe(CO)(PPh_3).COCH_3$ is sufficiently acidic to react with strong bases (such as $LiNPr\text{-}i_2$). The resultant carbon nucleophile attacks species RX [giving $CpFe(CO)(PPh_3).COCH_2R$] or acetone [giving $CpFe(CO)(PPh_3).COCH_2.CMe_2OH$].[85] These reactions are also shown by derivatives $CpFe(CO)(PPh_3).COCH_2R$, and because of steric crowding are highly stereoselective. The starting material contains an asymmetric center at iron, and is readily obtained in resolved form; thus these reactions show great promise in the synthesis of optically active compounds.[45]

9.   MORE RECENT DEVELOPMENTS

Reviews have appeared on the role of metal carbonyl clusters in catalysis[86] and on CO insertion in square planar complexes.[87]

The 18-electron complex $Cp_2MoCH_2CH_2CH_2CH_2$ reacts with CO to give cyclopentanone; it is not known whether the reaction proceeds through an $\eta^3\text{-}C_5H_5$ complex, or by alkyl carbon to ring migration.[88] Labeling experiments at -30°C in hexane show that the conversion of species cis-$(CO)_2$,trans-(phosphine)$_2$M(Me)I to cis-$(CO)_2$,trans-(phosphine)$_2$M(COMe)I (M = Fe,Ru) occurs by formal CO rather than formal methyl migration.

Above $-10^{\circ}C$, rearrangement occurs, probably via species [mer-$(CO)_3$,trans-(phosphine)$_2$.COMe]$^+$, to give trans-$(CO)_2$, trans-(phosphine)$_2$M(COMe)I, the product of apparent methyl migration. Thus arguments about detailed mechanism from formal stereochemistry can be seriously misleading.[89]

The palladium(II)-catalyzed co-polymerization of ethylene and carbon monoxide to [-$CH_2CH_2C(O)$-] proceeds by a simple chain growth mechanism, with (presumably) alternate CO and C_2H_4 insertion.[90] The apparent insertion of CO into iridium(I) alkoxide to give RO.CO.Ir complexes proceeds, however, by displacement of alkoxide from metal followed by nucleophilic attack by alkoxide on CO in complexes of type [Ir$(CO)_3L_2$]$^+$.[91] CO can undergo formal nucleophilic attack by ylids, as in the reaction[92]

$$[Cp*Fe(CO)_3]^+ + 2\ Me_3P{:}CH_2 \longrightarrow Cp*Fe(CO)_2\text{-}C(O)CHPMe_3 + etc. \quad (18)$$

Carbon monoxide inserts into species Cp_3ThR to give η^2-acyls $Cp_3Th(COR)$, at a rate dependent on R.[93] η^2-Acyls also arise by the reaction[94]

$$Mo(Cl)_2(CO)_2(PMe_3)_3 + MeLi \longrightarrow Mo(\eta^2\text{-}COMe)(Cl)(CO)(PMe_3)_3 + LiCl \quad (19)$$

A range of metal carbonyls react with species $Cp*_2ZrHX$ to give products of type M-CHO $\longrightarrow$ ZrCp$*_2$X.[95] Zirconocene dihydride itself reacts with CO to give a range of products, including $(Cp_2ZrCH_2O)_3$, in which the oxygens of the η^2-formaldehyde ligands bridge the metals.[96] Low-temperature borohydride reduction[97] of $CpMo(CO)_3Me$ gives initially the formyl complex [CpMo$(CO)_2$(CHO)Me]$^-$, which on warming rearranges successively to the acetyl [CpMo$(CO)_2$(COMe)H]$^-$, and the acetaldehyde complex [CpMo$(CO)_2(\eta^2$-MeCHO)]$^-$.

Interest continues in the incorporation of CO into C_2 or larger products. Reaction of [Cp*RhCl$_2$]$_2$ with CO, base and ethanol gives Cp*Rh(CO)(COOEt)$_2$, which is cleaved by iodine to give ethyl oxalate.[98] Species trans-PdR(X)(PMePh$_2$)$_2$ react with CO and dialkylamines to give products RCO.CO.NR'$_2$. This reaction may well proceed via intermediates of type $RCO.Pd(L)_2.CONR'_2$; species containing the $Pd.CO.COR$ grouping, which may be prepared via $RCO.COCl$, readily lose CO and give amides (containing only one CO) with amines.[99]

The oxidative conversion of $CpFe(CO)_2CH_2C(Me)NHCOPh$ to $O = \overline{C.CH_2.C(Me).N}(COPh)$ (**XVIII**) proceeds via an iron(IV) species; oxidation to iron(III) gives the expected metallacycle $[\overline{CpFe(CO).C(O).CH_2.CMe.N}(COPh)]^+$, but this decomposes only slowly, and does not give **XVIII**.[100]

Ni(0) inserts into the CO-O bond of phthalic anhydride.[101] Co(I) inserts into the CO-CO bond of cyclobutenediones. Treatment with dimethylglyoxime and pyridine, followed by acetylenes, gives a route to benzoquinones.[102] The reactions of resolved species of type $CpFe(CO)(L)(COR)$[45,85] have been further extended and improved,[103] and the observed stereoselectivities systematized.[104]

REFERENCES

1. P.S. Braterman, "Metal Carbonyl Spectra", Academic Press, London and New York, 1975.

2. "Catalytic Activation of Carbon Monoxide, P.C. Ford, ed., ACS Symp. Ser. 152 (1981).

3. "New Syntheses with Carbon Monoxide", J. Falbe, ed., Springer-Verlag, Berlin, 1980.

4. "Catalysis in C(1) Chemistry", W. Keim, ed., D. Reidel, Dortrecht, 1983.

5. R.A. Sheldon, "Chemicals from Synthesis Gas", D. Reidel, Dortrecht, 1983.

6. "Methanation of Synthesis Gas", R.F. Gould, ed., ACS Advan. Chem. Ser. 146 (1975).

7. "Hydrocarbon Synthesis from Carbon Monoxide and Hydrogen", E.L. Kigler and F.W. Steffgen, eds., ACS Advan. Chem. Ser. 178 (1979).

8. E.J. Kuhlmann and J.J. Alexander, Coord. Chem. Rev. 33:195 (1980).

9. J.K. Stille and D.E. James in "The Chemistry of Functional Groups", S. Patai, ed., Suppl. A part 2, John Wiley and Sons, Chichester, 1977, p. 725.

10. P.J. Davidson, R.R. Hignett and D.T. Thompson, Catalysis, Chem. Soc. Spec. Per. Rep. 1:369 (1977); H. Siegel and W. Himmele, Angew. Chem. Int. Ed. Engl. 19:178 (1980); P. Pino, J. Organomet. Chem. 200:223 (1980).

11. P. Chini, J. Organomet. Chem. 200:37 (1980).

12. D.F. Shriver and A. Alich, Coord. Chem. Rev. 8:15 (1972).

13. J.A. Gladysz, Adv. Organomet. Chem. 20:1 (1982).

14. M. Orchin, Acc. Chem. Res. 14:259 (1981).

15. R.F. Heck, Adv. Catal. 26:323 (1977).

16. W.A. Herrmann, Angew. Chem. Int. Ed. Engl. 21:117 (1982).

17. P.C. Ford, Acc. Chem. Res. 14:31 (1981).

18. G. Consiglio and P. Pino, Top. Curr. Chem. 105:77 (1982).

19. D.C. Bailey and S.H. Langer, Chem. Rev. 81:109 (1981).

20. K. Noack and F. Calderazzo, J. Organomet. Chem. 10:101 (1967).

21. J.N. Cawss, R.A. Fiato and R.L. Pruett, J. Organomet. Chem. 172:405 (1979).

22. S.B. Butts, S.H. Strauss, E.M. Holt, R.E. Stimson, N.W. Alcock and D.F. Shriver, J. Am. Chem. Soc. 102:5093 (1980).

23. T.G. Richmond, F. Basolo and D.F. Shriver, Inorg. Chem. 21:1272 (1982).

24. K.M. Nicholas and M. Rosenblum, J. Am. Chem. Soc. 95:4449 (1973); P.L. Bock, D.J. Boschetto, J.R. Rosmussen, J.P. Demers and G.M. Whitesides, J. Am. Chem. Soc. 96:2814 (1974).

25. D. Miholova and A.A. Vlcek, J. Organomet. Chem. 240:413 (1982).

26. R.H. Magnuson, R. Meirowitz, S.J. Zulu and W.P. Giering, Organometallics 2:460 (1983).

27. A. Cameron, V.H. Smith and M.C. Baird, Organometallics 2:465 (1983).

28. J.D. Cotton, G.T. Crisp and L. Latif, Inorg. Chim. Acta 47:171 (1981).

29. J.D. Cotton, H.A. Kimlin and R.D. Markwell, J. Organomet. Chem. 232:C75 (1982).

30. M.J. Wax and R.G. Bergman, J. Am. Chem. Soc. 103:7028 (1981).

31. G. Fachinetti, S. Del Nero and C. Floriani, J.C.S. Dalton Trans. 203 (1976).

32. E.E.H. Otto and H.H. Brintzinger, J. Organomet. Chem. 170:209 (1979).

33. G.M. Whitesides and D.J. Boschetto, J. Am. Chem. Soc. 91:4313 (1969).

34. L.F. Hines and J.K. Stille, J. Am. Chem. Soc. 94:485 (1972).

35. H.M. Walborsky and L.E. Allen, J. Am. Chem. Soc. 93:5465 (1971).

36. K.S.Y. Lau, Y. Becker, F. Huang, N. Baenziger and J.K. Stille, J. Am. Chem. Soc. 99:5664 (1977).

37. J.K. Stille and R.W. Fries, J. Am. Chem. Soc. 96:1514 (1974).

38. T.C. Flood, J.E. Jensen and J.A. Statler, J. Am. Chem. Soc. 103:4410 (1981).

39. A. Davison and N. Martinez, J. Organomet. Chem. 74:C17 (1974).

40. J.D. Black and P.S. Braterman, J. Organomet. Chem. 63:C19 (1973); M. Poliakoff, Inorg. Chem. 15:2022,2892 (1976).

41. H. Brunner and H. Vogt, Angew. Chem. Int. Ed. Engl. 20:405 (1981).

42. S. Quinn, A. Shaver and V.W. Day, J. Am. Chem. Soc. 104:1096 (1982).

43. T.C. Flood, K.D. Campbell, H.H. Downs and S. Nakanishi, Organometallics 2:1590 (1983).

44. H. Brunner, B. Hammer, I. Bernal and M. Draux, Organometallics 2:1599 (1983).

45. G.J. Baird and S.G. Davies, J. Organomet. Chem. 248:C1 (1983).

46. I.S. Butler, F. Basolo and R.G. Pearson, Inorg. Chem. 6:2074 (1967); M. Green and D.J. Westlake, J. Chem. Soc. (A) 367 (1971).

47. D.J. Fettes, R. Narayanaswami and A.J. Rest, J.C.S. Dalton Trans. 2311 (1981).

48. N. Grice, S.C. Rao and R. Pettit, J. Am. Chem. Soc. 101:1627 (1979).

49. J.E. Ellis, R.A. Faltynek and S.G. Hentges, J. Am. Chem. Soc. 99:626 (1977).

50. J.E. Pardue and G.R. Dobson, J. Organomet. Chem. 132:121 (1977).

51. D.W. Parker, M. Marsi and J.A. Gladysz, J. Organomet. Chem. 194:C1 (1980).

52. R.E. Stimson and D.F. Shriver, Inorg. Chem. 19:1141 (1980).

53. H. Felkin, B. Meunier, C. Pascard and T. Prange, J. Organomet. Chem. 135:361 (1977).

54. S.J. La Croce and A.R. Cutler, J. Am. Chem. Soc. 104:2312 (1982).

55. R.B. Hitam, R. Narayanaswamy and A.J. Rest, J.C.S. Dalton Trans. 615 (1983).

56. W.J. Evans, A.L. Wayda, W.E. Hunter and J.L. Atwood, J.C.S. Chem. Commun. 706 (1981).

57. P.J. Fagan, K.G. Moloy and T.J. Marks, J. Am. Chem. Soc. 103:6959 (1981).

58. G. Erker and F. Rosenfeldt, Angew. Chem. Int. Ed. Engl. 17:605 (1978); idem, J. Organometal. Chem. 188:C1 (1980).

59. J.M. Manriquez, D.R. McAlister, R.D. Sanner and J.E. Bercaw, J. Am. Chem. Soc. 100:2716 (1978).

60. J.A. Marsella, C.J. Curtis, J.E. Bercaw and K.G. Caulton, J. Am. Chem. Soc. 102:7244 (1980).

61. K.I. Gell and J.Schwartz, J. Organomet. Chem. 162:C11 (1978).

62. K.I. Gell, G.M. Williams and J. Schwartz, J.C.S. Chem. Commun. 550 (1980).

63. B. Longato, J.R. Norton, J.C. Huffman, J.A. Marsella and K.G. Caulton, J. Am. Chem. Soc. 103:209 (1981).

64. J.A. Marsella, K. Folting, J.C. Huffman and K.G. Caulton, J. Am. Chem. Soc. 103:5596 (1981).

65. E. Carmona Guzman, G. Wilkinson, J.L. Atwood, R.D. Rogers, W.E. Hunter and M.J. Zaworotko, J.C.S. Chem. Commun. 465 (1978).

66. W.R. Roper and L.J. Wright, J. Organomet. Chem. 142:C1 (1977); W.R. Roper, G.E. Taylor, J.M. Waters and L.J. Wright, J. Organomet. Chem. 182:C46 (1979).

67. G. Erker and F. Rosenfeldt, J. Organomet. Chem. 224:29 (1982).

68. C.D. Wood and R.R. Schrock, J. Am. Chem. Soc. 101:5421 (1979).

69. S. Gambarotta, C. Floriani, A. Chiesi-Villa and C. Guastini, J. Am. Chem. Soc. 104:2019 (1982).

70. G.R. Clark, C.E.L. Headford, K. Marsden and W.R. Roper, J. Organomet. Chem. 231:335 (1982).

71. J.W. Suggs, J. Am. Chem. Soc. 100:640 (1978).

72. T.B. Rauchfuss, J. Am. Chem. Soc. 101:1045 (1979).

73. H. Motschi, P.S. Pregosin and H. Ruegger, J. Organomet. Chem. 193:397 (1980).

74. C.P. Casey and D.M. Scheck, J. Organomet. Chem. 142:C12 (1977); idem, J. Am. Chem. Soc. 102:2723,2728 (1980).

75. E. Colomer, R.J.P. Corriu and J.C. Young, J.C.S. Chem. Commun. 73 (1977).

76. H-F. Klein and H.H. Karsch, Chem. Ber. 108:956 (1975).

77. T. Saruyama, T. Yamamoto and A. Yamamoto, Bull. Chem. Soc. Jpn 49:546 (1976).

78. K. Murata and A. Matsuda, Bull. Chem. Soc. Jpn 54:249,2089 (1981).

79. K. Murata and A. Matsuda, Bull. Chem. Soc. Jpn 55:2195 (1982).

80. B.E. Foulger, F-W. Grevels, D. Hess, E.A. Koerner von Gustorf and J. Leitich, J.C.S. Dalton Trans. 1451 (1979), and references therein.

81. K.P. Vora, C.F. Lochow and R.G. Miller, J. Organomet. Chem. 192:257 (1980).

82. R.C. Larock, K. Oertle and G.F. Potter, J. Am. Chem. Soc. 102:190 (1982); C.F. Lochow and R.G. Miller, J. Am. Chem. Soc. 98:1281 (1976); R.E. Campbell, jr., and R.G. Miller, J. Organomet. Chem. 186:C27 (1980).

83. W.A. Herrmann and J. Plank, Angew. Chem. Int. Ed. Engl. 17:525 (1978).

84. F.R. Kreissl, W. Uedelhoven and K. Eberl, Angew. Chem. Int. Ed. Engl. 17:859 (1978); F.R. Kreissl, K. Eberl and W. Uedelhoven, ibid. 17:860 (1978).

85. L.S. Liebeskind and M.E. Welker, Organometallics 2:194 (1983).

86. R. Ugo and R. Psaro, J. Mol. Catal. 20:53 (1983).

87. G.K. Anderson and R.J. Cross, Acc. Chem. Res. 17:67 (1984).

88. P. Diversi, G. Ingrosso, A. Lucherini, W. Porzio and M. Zocchi, J.C.S. Dalton Trans. 967 (1983).

89. M. Pankowski and M. Bigorne, J. Organomet. Chem. 251:333 (1983).

90. T-W. Lai and A. Sen, Organometallics 3:866 (1984).

91. W.M. Rees and J.D. Atwood, Organometallics 4:402 (1985).

92. A. Stasunik and W. Malisch, J. Organomet Chem. 247:C47 (1983).

93. D.C. Sonnenberger, E.A. Mintz and T.J. Marks, J. Am. Chem. Soc. 106:3484 (1984).

94. E. Carmona, L. Sanchez, J.M. Marin, M.L. Poveda, J.L. Atwood, R.D. Priester and R.D. Rogers, J. Am. Chem. Soc. 106:3214 (1984).

95. P.T. Barger and J.E. Bercaw, Organometallics 3:278 (1984).

96. K. Kropp, V. Skibbe, G. Erker and C. Krüger, J. Am. Chem. Soc. 105:3353 (1983).

97. J.T. Gauntlett, B.F. Taylor and M.J. Winter, J.C.S. Chem. Commun. 420 (1984).

98. P.L. Burk, D. Van Engen and K.S. Campo, Organometallics 3:493 (1984).

99. F. Ozawa and A. Yamamoto, Chem. Lett. 865 (1982); F. Ozawa, T. Sugimoto, Y. Yuasa, M. Santra, T. Yamamoto and A. Yamamoto, Organometallics 3:683 (1984); F. Ozawa, T. Sugimoto, T. Yamamoto and A. Yamamoto, ibid. 3:692 (1984).

100. S. Javaheri and W.P. Giering, Organometallics 3:1927 (1984).

101. E. Uhlig and O. Reitman, Z. Chem. 24:103 (1984).

102. L.S. Liebeskind, J.P. Leeds, S.J. Blayson and S. Iyer, J. Am. Chem. Soc. 106:6451 (1984).

103. G.J. Baird, J.A. Bandy, S.G. Davies and K. Prout, J.C.S. Chem. Commun. 1202 (1983); N. Aktogu, H. Felkin, G.J. Baird, S.G. Davies and O. Watts, J. Organomet. Chem. 262:49 (1984); L.S. Liebeskind, M.E. Welker and V. Goedken, J. Am. Chem. Soc. 106:441 (1984); S.G. Davies, I.M. Dordor and P. Warner, J.C.S. Chem. Commun. 956 (1984).

104. S.G. Davies and J.I. Seeman, Tetrahedron Lett. 25:1845 (1984).

PROMOTION EFFECTS IN TRANSITION METAL-CATALYZED

CARBONYLATION

G.P. Chiusoli and G. Salerno
Istituto di Chimica Organica dell'Universita, Via d'Azeglio 85
43100, Parma, Italy

and M. Foa
Istituto Donegani SpA, Novara, Italy

1. INTRODUCTION

Catalysis of carbonylation by transition metal complexes has received the attention of many research groups in the last decades, owing to its scientific as well as industrial importance.

As for all catalytic processes occurring with transition metals, forming a species R able to react with a coordinated ligand (in the present case carbon monoxide), causing R to undergo external attack or migratory insertion on the ligand:

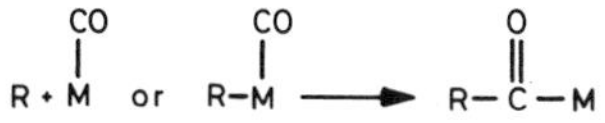

$$\underset{R \cdot M}{\overset{\overset{\displaystyle CO}{|}}{}} \quad \text{or} \quad \underset{R-M}{\overset{\overset{\displaystyle CO}{|}}{}} \longrightarrow \underset{R-C-M}{\overset{\overset{\displaystyle O}{\|}}{}}$$

and detaching the metal from the organic chain thus formed, are crucial problems. Furthermore, the metal-bonded carbonylation intermediate can be caused to react with other groups and molecules before final elimination of the metal and the problem of stabilizing this intermediate to allow further reaction also appears very important.

The aim of this Chapter is to offer a survey of the most important tools that have been evolved to promote carbonylation catalysis both in terms of rate acceleration and of product selectivity.

Transition metal complexes used for catalytic carbonylation are derived from a restricted number of metals, because different requirements have to be met by the same metal, such as ability to coordinate substrates

437

and to form metal carbonyls, to favor attack at coordinated carbon monoxide, and to allow elimination. The metal must therefore be sufficiently 'noble' to be able to revert to its initial oxidation state, which has to be low to allow metal carbonyl formation, and sufficiently oxidizable to pass to the oxidation state required by the catalytic process.

For most carbonylations the choice of metal is restricted to Ni, Pd, Co and Rh, followed by Pt. Fe and Ru, which are less noble than the other group VIII metals, can be used in a limited number of cases. The use of Ir, Mo and Mn for catalytic carbonylation reactions in solution is even more restricted.

There are many ways by which metal complexes can be modified to enhance their reactivity and specificity. For example, the nucleophilic character of a metal complex can often be increased through formation of an anionic complex, and the electrophilic character through formation of a cationic complex. Many other transformations can be caused on the complex, however, depending on coordinative unsaturation, ligand stereochemistry, polarizability and electronegativity, nature of the counter-ion, etc. We shall see how modification of the environment can generate more efficient catalytic species. The effect of changing the metal itself will not be considered in this review, which is devoted to ligand, substrate and medium effects.

In examining promotion effects in carbonylation reactions, we shall not distinguish among the many types of carbonylation (e.g. carbon monoxide hydrogenation, reductive or oxidative carbonylation, substitutive carbonylation, cyclocarbonylation). We shall only report a series of promotion effects due to substrates, ligands, acids and bases, solvents and ion-pairs, oxidants and reductants. The effect of temperature, pressure and concentration will be reported in each Section in the appropriate context. We shall choose suitable examples to show the various promotion effects, the catalytic species supposed to be active, and how we think it works. Generalizations are quite difficult and we believe that to use the results the reader must take into account the occurrence of many steps and of opposing effects, the combination of which often causes unpredictable consequences in catalytic processes.

Owing to the huge amount of material available, this relatively short Chapter will not have the character of an exhaustive review, but rather that of a collection of selected examples, mainly referring to recent literature. Also for reasons of space, it has been deemed necessary to limit the subject to reactions in solution.

The first Section, which is devoted to substrate effects, is divided into five subsections corresponding to substrate effects on single steps of the catalytic process. The following Sections (ligands, solvents, etc.), will also refer to these stages of the catalytic process, but owing to the complexity of the effects examined, they will be articulated in different ways.

The authors fully realize that the distinctions between substrates, ligands or solvents are at times arbitrary, yet the common practice of making such distinctions appears to be useful and justifies the classification chosen.

2.   SUBSTRATE STRUCTURE AND SUBSTITUENT EFFECTS

As with the other factors considered, substrates can play a very important role in each stage of the carbonylation process. We shall discuss the specific action exerted by substrates in forming the reactive intermediate and in influencing rate and/or selectivity in the various steps of the catalytic process.

2.1   Effect on formation of the reactive intermediate

Formation of a metal coordinated group can occur in several ways, for example by oxidative addition of organic compounds to metals in their low oxidation states.[1,1a-d] (Hereinafter R indicates an aliphatic or aromatic residue, M a metal, and X a leaving group. Non-essential ligands are omitted):

$$RX + M \longrightarrow R-M-X$$

Other ways to form the reactive intermediate are replacement of group X by metal anions:[2,2a]

$$RX + M^- \longrightarrow RM + X^-$$

coordination of unsaturated substrates S followed by protonation:[3]

$$\underset{M}{\overset{S}{|}} + H^+ \longrightarrow \underset{M^+}{\overset{SH}{|}}$$

insertion into a M-H bond or into a M-R bond:[4,4a-d]

$$\overset{S}{\underset{M-H}{|}} \longrightarrow \overset{SH}{\underset{M}{|}} \qquad \overset{S}{\underset{R-M}{|}} \longrightarrow \overset{SR}{\underset{M}{|}}$$

and other types of nucleophilic or electrophilic attack on coordinated substrates.[5]

Some metal-bonded intermediates able to react with CO are listed in Table 1, which gives examples of representative carbonylation processes.

Table 1. Some substrates undergoing catalytic carbonylation

	Reactions (5a-g)	Metal-bonded group undergoing first CO attack
1	$CO + 3H_2 \longrightarrow CH_4 + H_2O$	H—M
2	$CO + 2H_2 \longrightarrow MeOH$	H—M
3	$2CO + 3H_2 \longrightarrow HOCH_2CH_2OH$	H—M
4	$MeOH + CO + 2H_2 \longrightarrow MeCH_2OH + H_2O$	Me—M
5	$MeOH + CO \longrightarrow MeCOOH$	Me—M
6	$2MeOH + 2CO \longrightarrow MeOOC-COOMe + 2H$	MeO—M or MeO⁻ + M—CO
7	$H_2C{=}CH_2 + CO + H_2 \longrightarrow MeCH_2CHO$	MeCH$_2$—M
8	$HC{\equiv}CH + CO + MeOH \longrightarrow CH_2{=}CHCOOMe$	CH$_2$=CH—M
9	$HC{\equiv}CH + 2CO + 2MeOH \longrightarrow MeOOCCH{=}CHCOOMe + 2H$	MeO—M or MeO⁻ + M—CO
10	$CH_2{=}CHCH{=}CH_2 + CO + MeOH \longrightarrow CH_3CH{=}CHCH_2COOMe$	⌡—M
11	$2CH_2{=}CHCH{=}CH_2 + CO \longrightarrow O{=}C\underset{CH_2CH{=}CHCH_2CH_2}{\overset{\textstyle CHCH{=}CH_2}{\rule{3cm}{0.4pt}}}$	⟨M⟩
12	$PhCH_2Cl + CO + MeOH \longrightarrow PhCH_2COOMe + HCl$	PhCH$_2$—M
13	$PhCl + CO + MeOH \longrightarrow PhCOOMe + HCl$	Ph—M
14	$CH_2{=}CHCH_2Cl + HC{\equiv}CH + CO + MeOH \longrightarrow$ $\longrightarrow CH_2{=}CHCH_2CH{=}CHCOOMe + HCl$	CH$_2$=CHCH$_2$CH=CH—M
15	$CH_2{=}CHCH_2Cl + CH_2{=}CHCH{=}CH_2 + CO + MeOH \longrightarrow$ $\longrightarrow CH_2{=}CHCH_2CH_2CH{=}CHCH_2COOMe + HCl$	⟨M⟩
16	$CH_2{=}CHCH_2Cl + CH_2{=}CHCH_2CH_2CH{=}CH_2 + CO + H_2O \longrightarrow$ $\longrightarrow CH_2{=}CHCH_2CH_2{-}\square{-}CH_2COOH + HCl$	$CH_2{=}CH_2{-}CH_2CH_2\overset{CH_2-CH_2}{\underset{\textstyle M}{\diagdown}}\overset{CH}{\underset{CH_2}{\diagup}}$
17	$2HC{\equiv}CH + 2CO + H_2 \longrightarrow OH{-}\hexagon{-}OH$	‖-M-‖ or ⎢◁M◁⎢
18	$(Ph)_2CN_2 + CO \longrightarrow (Ph)_2C{=}C{=}O + N_2$	$(Ph)_2C{=}M$
19	$2PhNH_2 + CO \longrightarrow PhNHCONHPh + H_2$	PhNH—M or PhNH⁻ + M—CO
20	$PhNO_2 + 3CO \longrightarrow PhNCO + 2CO_2$	PhN=M?

Formation of the metal-bonded species has a remarkable effect on the following carbonylation step and is often the rate determining step. Thus, although the subject has been separately treated in reviews of transition metal catalyzed reactions (see for example[4,4a-d,5a-g]), we shall briefly summarize some important findings referring to substrate effects on formation of the species able to react with CO.

A wide range of organic compounds, particularly halides (Cl, Br, I) (Table 1; Nos. 12-16) give rise to oxidative addition to transition metal complexes.[1] Although quantitative data are lacking in most cases, it can be stated that allylic halides are generally more reactive than vinylic and aromatic halides. Qualitative data are available for $Ni(CO)_4$,[4c,6-7] $Ni(PR_3)_4$[8-9] and $Pd(CO)(PR_3)_3$.[10-11] Iodides react easier than bromides and bromides easier than chlorides.[1b] With aromatic halides oxidative addition[1,1a-d] is favored by electron-withdrawing substituents on the aromatic ring.[12-13] Among saturated halides the benzylic ones certainly are the most reactive,[14] whereas non-activated aliphatic halides are rather inert in oxidative addition to d^8 and d^{10} metal complexes.

The situation is different for those oxidative addition or replacement reactions which involve anionic carbonylic species. The mechanism approaches S_N and substitution of aliphatic halides is favored. Thus work on reaction of benzyl halides with cobalt tetracarbonyl anion has shown that electron-releasing substituents favor the reaction because the departure of halide ions is facilitated.[15] With iron tetracarbonyl anion, however, the reaction seems to be a true S_N2 reaction and electronic substrate effects do not appear as important as steric effects.[2] This type of oxidative addition has also been shown to give rise to a preferred stereochemistry, derived from inversion of configuration at the metal-bonded carbon.[2] Oxidative addition to neutral complexes offers examples of both inversion (benzylic halides) and retention (vinylic halides) of configuration.[1a,9,16]

Radical ($S_N R$) processes can activate otherwise unreactive substrates towards carbonylation.[16a] For example, aromatic halides react with cobalt carbonyl anion under the influence of light.[16b] Data on substituent effects are not available.

Diolefins such as butadiene can easily give rise to oxidative addition to low-valent metal complexes, because the addition product of two or more molecules can be stabilized by formation of two allylic bonds, for example (L = ligand)[5g,17,17a] (Table 1, No. 11):

$$2 \diagup\!\!\!\diagdown \cdot Ni + L \longrightarrow \boxed{Ni-L}$$

When the reactive intermediate is formed from a metal-coordinated unsaturated substrate, as in hydroformylation or hydrocarboxylation reactions (Table 1, Nos. 7-10), the substrate can be attacked either by electrophiles or by nucleophiles depending on metal-substrate charge distribution. In fact both direct ligand protonation and insertion into a metal-hydride bond can occur,[1,1a-d,4a-c] for example:

$$H^+ + M \longrightarrow M^+ \longleftarrow M^+ -H$$

Other factors being equal, terminal olefins react faster than internal or cyclic olefins. Branching at the double bonded carbon atom has a negative effect, which increases with the groups' bulkiness. The presence of carbonylic functions on the double bond favors hydrogenolysis rather than CO insertion unless double bond isomerization occurs.[4a,17b] Cyclic olefins insert into the H-metal bond in the order: norbornene > cyclopentene ~ cycloheptene > cyclooctene > cyclohexene, probably because their steric strain increases the advantage in steric relief by -alkyl group formation.[4,18-19]

Alternative sources of M-C bonds are (Table 1, Nos. 4-5) alcohols, ethers, esters, acetals, orthoformates, and epoxides, which easily give rise to reactive intermediates by C-O bond cleavage. For example, $HCo(CO)_4$ is thought to attack the oxygenated substrate to form an alkylcobalt carbonyl:[4,4a]

$$ROH + HCo(CO)_4 \longrightarrow RCo(CO)_4 + H_2O$$

Homologation of benzyl alcohols to phenylethyl alcohols has been shown to be promoted by electron-releasing substituents. For example p-methoxybenzyl alcohol homologates 10,000 times faster than benzyl alcohol, whereas the relative rate for trifluoromethylbenzyl alcohol is only 0.01.[20] Although these are overall rates it is probable that the main substrate action is exerted on formation of the intermediate undergoing insertion. Many substrate effects of oxygenated compounds are discussed in the literature.[5a] Other ways to form this intermediate consist of protonation of oxygenated compounds with formation of organic halides, which easily give rise to oxidative addition[21-2] as mentioned above:

$$MeOH + HI \rightleftharpoons MeI + H_2O$$

The alcoholic substrate for carbonylation can also be prepared by interaction of aldehydic substrates with amides,[23-4] as exemplified for acetaldehyde and acetamide:

$$MeCHO + MeCONH_2 \longrightarrow MeCH(OH)NHCOMe \xrightarrow[CO_2(CO)_8 /dioxane]{CO/H_2 , 120°} MeCH(COOH)NHCOMe$$

Other recent methods to form coordinated groups from oxygenated compounds are based on the use of hydrosilanes and are mentioned in Section 5.2.

Alkynes or alkenes, as well as other unsaturated substrates, can insert into metal-coordinated organic groups[4c] or can be attacked by nucleophiles[24a] to form a new σ-bond able to react with CO. It has been shown that these reactions proceed particularly well in the case of terminal alkynes or alkenes not containing too bulky groups and able to coordinate to metal easily. In particular, strained olefins and chelated olefins are suitable for insertion reactions, for example:[4c,25]

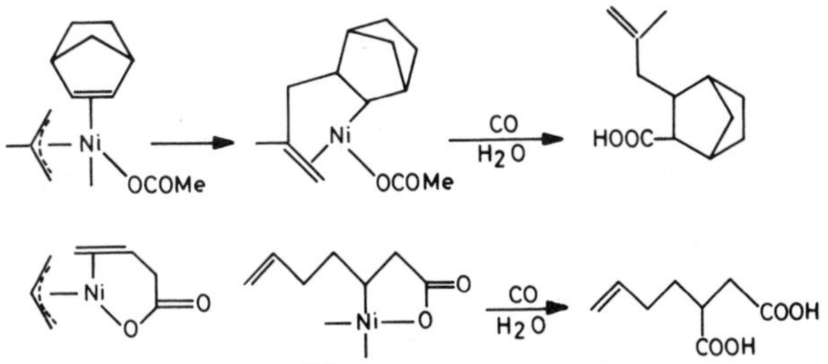

Strained rings such as cubane,[25a] vinylcyclopropane and pinene[25b,c] can be opened by Rh or Fe carbonyls to give C-metal bonds.

Other substrates easily forming metal-bonded groups able to react with carbon monoxide are those which give rise to fragments such as carbenes or nitrenes. These fragments are stabilized by the metal complex (see for example Refs. 26-7)(Table 1, Nos. 18, 20). Metal alkoxides and amides, formed from aldehydes or alcohols[3] and amines[28,28a] respectively, can also react with CO (Table 1, Nos. 6, 9, 19).

Coordinated nitrogen-containing compounds also form N-metal bonds via hydride attack on multiple C-N bonds (aldimines, nitriles, semicarbazones, azobenzenes, etc.) or OH elimination (oximes):

$$RC{\equiv}N \xrightarrow{M-H} RCH{=}N{-}M \qquad RCH{=}NOH \xrightarrow[-H_2O]{M,H^+} RCH{=}N{-}M$$

Cobalt carbonyl and its derivatives and to a lesser extent iron and nickel carbonyl have been used.[5c,28-9] High temperatures (200-300°) and pressures (100-300 atm) are generally required for the overall carbonylation process. Evidence for formation of the N-bonded metal-substrate complex is available in a few cases.

Substrates for carbonylation can also be formed easily by reaction of CCl_4 with anilines and Mo, Co, Cr carbonyls at 150° and 50 atm of CO, probably via formation of radicals.[30]

2.2 Effect on insertion or nucleophilic attack at CO

Insertion or nucleophilic attack at coordinated CO are in general reversible reactions and are followed by an irreversible step involving the cleavage of acyl-metal bonds. Alternatively, acylation of other molecules and groups, followed by various types of termination processes, can be at work.

Substrate effects on insertion are generally difficult to separate from other effects. Acyl-metal formation being reversible, the subsequent steps (elimination or further reaction) can influence the feasibility of the insertion step. For example, the catalytic carbonylation reaction of aromatic halides has been shown to be favored by electron-withdrawing substituents on the ring both for Ni[7] and Pd[31] complexes. This, however, is an overall effect which results from a combination of effects in different stages of the catalytic process. Thus, to understand substrate effects on the single insertion step, it is useful to refer to studies of stoichiometric reactions on metal complexes:

$$RM(CO) \rightleftharpoons RCOM \xrightleftharpoons{L} RCOML$$

Groups able to attack carbon in coordinated CO are nucleophilic in character. Electron-releasing substituents have been shown to exert a positive action on aryl group migration to CO in arylpalladium or nickel bistriarylphosphine complexes;[31] thus the overall positive effect of electron-withdrawing substituents in carbonylation of aryl halides appears to be due to their predominant effect on oxidative addition. Analogous behaviour of electron-releasing and electron-withdrawing groups on CO insertion has been observed for $RMn(CO)_5$ (R = CH_3 CH_2F CF_3) and $CpFe(CO)_2R$ (R = $CHMe_2$ Et Me).[4b] Pruett[32] has recently plotted log k vs σ^* for reactions of complexes $RFe(CO)_4^-$ and $RMn(CO)_5$ with different substituents in R and has found the same large negative ρ^* (-8.8 and -8.7, respectively) in spite of the differences in oxidation state, charge, and d-electron configuration. It has been observed[33] that electron-withdrawing substituents in R stabilize both R and RCO, but R is stabilized to a greater extent. On the base of the ρ^* obtained, Pruett argues that a slight change in σ^*, such as for the difference between methyl and ethyl, would stabilize the alkyl significantly. According to the results of extended CNDO calculations,[34] however, it is likely that alkyl "stabilization" should be considered as a kinetic phenomenon. See also Ref. 34a for further discussion.

As to the relative ability of alkyl, benzyl, and phenyl groups to promote insertion, the observed sequence is not always the same. Thus for $R-Pd(X)(PR'_3)_2$ complexes (PR'_3 = tertiary phosphine) the observed reactivities of R towards CO in tetrachloroethane are in the order: methyl > benzyl > phenyl[31] whereas for $RMn(CO)_5$ in 2,2'-diethoxydiethyl ether the order turns out to be n-propyl > ethyl > phenyl > methyl ≫ benzyl, CF_3. For $RMo(CO)_3Cp$ complexes in MeCN one finds ethyl > methyl > benzyl > allyl.[4b]

Another recent study of $R-PtCl(CO)(MePPh_2)$[35] with R trans to $MePPh_2$ shows that insertion decreases from ethyl to phenyl to methyl to benzyl (~0) (the ligand effect is further discussed in Section 3).

Vinylic substrates appear to have a marked tendency to react with CO to form metal-bonded acryloyl groups. Accordingly, the well-known acrylic ester synthesis from acetylene and CO,[3] involving formation of a nickel-coordinated vinyl group, occurs under very mild conditions:

$$HC\equiv CH + CO + MeOH \xrightarrow[\text{Ni cat. } 40°]{H^+} CH_2=CHCOOMe$$

Allylic compounds in the presence of acetylene prefer to form a nickel-bonded allylacryloyl group $CH_2=CHCH_2CH=CHCO-Ni$ first, rather than a butenoyl group $CH_2=CHCH_2CO-Ni$:[36]

$$CH_2=CHCH_2Cl + HC\equiv CH + CO + MeOH \xrightarrow[\text{r.t.}]{Ni\ cat.} CH_2=CHCH_2CH=CHCOOMe + HCl$$

Acetylenic substrates lead to <u>cis</u> double bonds stereoselectively.

As mentioned above (Sec. 2.1) the vinyl group can also be formed by oxidative addition of vinyl halides to metals. CO has been shown to insert invariably with retention of configuration.[1a] Since the vinyl group oxidatively adds with retention of configuration, the overall catalytic carbonylation process of vinyl halides occurs with high degree of stereospecificity.[9,11]

Other substrates can be induced to react more easily if formation of chelated rings is favored. Again the stereochemistry is determined by CO insertion with retention of configuration as in the case of hydroxycyclooctenylpalladium intermediates[37] formed in hydroxycarbonylation of cyclooctadiene:

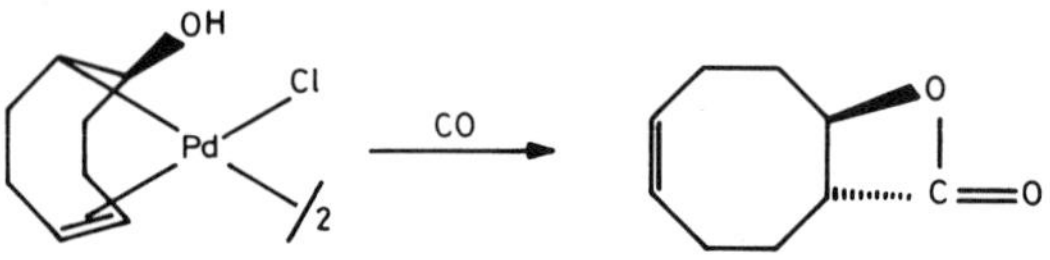

Another example is offered by 1,5-hexadiene, which reacts with CO in cationic complexes of Ni in methanol at room temperature to give dienoic esters[38] much more easily than 1-hexene, because of the assistance of the second double bond to insertion:[39]

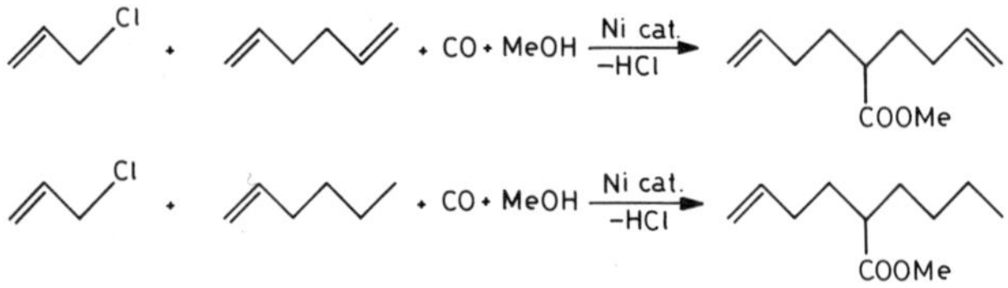

Other substrates able to attack coordinated CO (possibly via "external" nucleophilic attack) are alcohols and amines. In general an oxidizing agent (Sec. 5.) is needed for catalysis with Pd as in the examples:[40-1]

$$2ROH + CO \longrightarrow ROCOOR + (2H); \quad 2RNH_2 + CO \longrightarrow RNHCONHR + (2H)$$

It is worth noting, however, that ureas can also be prepared directly from primary amines and CO with H_2 evolution using manganese carbonyl as catalyst.[42] Carbalkoxy or carbamoyl complexes appear to be the intermediates.[43]

Isocyanates can also be obtained:[44]

$$RNH_2 + CO \xrightarrow[65°]{Pd} RNCO + H_2$$

Formation of nitrogen-bonded metallacycles can favor insertion as in the case of benzaldimines, and of other nitrogen derivatives containing aromatic rings in an appropriate position:[45]

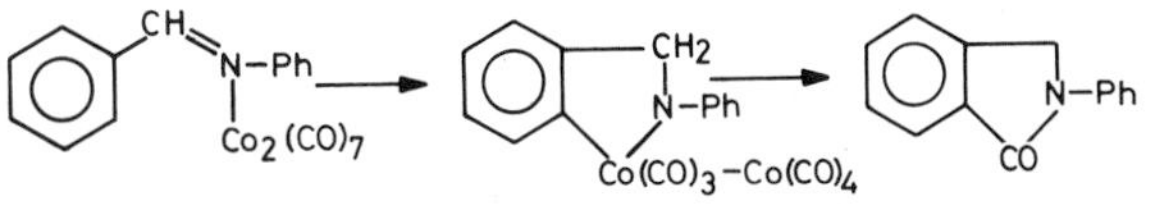

2.3 Effect on metal elimination from acylmetal complexes

The elimination step has a profound influence on carbonylation reactions because it drives the preceding reversible steps towards the final product.

Three main pathways are generally followed in catalytic reactions: uptake of hydrogen to form aldehydes and alcohols, nucleophilic attack on the carbonyl group by a variety of compounds (water, alcohols, amines, etc.), and reductive coupling. Simple acyl groups, RCO, not containing other groups able to stabilize the acyl by steric or chelating effects, generally split from metal easily:

$$RCOM + R'OH \longrightarrow RCOOR' + MH$$
$$RCOM + H_2 \longrightarrow RCHO + MH$$
$$RCOMR' \longrightarrow RCOR' + M$$

Substrates possessing a hydroxyl group, able to give rise to a lactone ring with the acyl group formed by CO insertion, prefer internal attack to give the lactone rather than attack by external nucleophiles:[4c,5c]

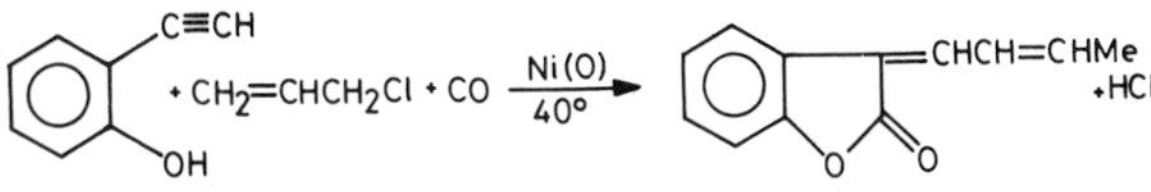

Cycloalkanones are easily formed from strained olefins and CO.[45a,5d]

Another instance of facile reductive elimination by internal attack at an acyl group is offered by the synthesis of butenylsuccinic anhydride in a two-phase system:[46]

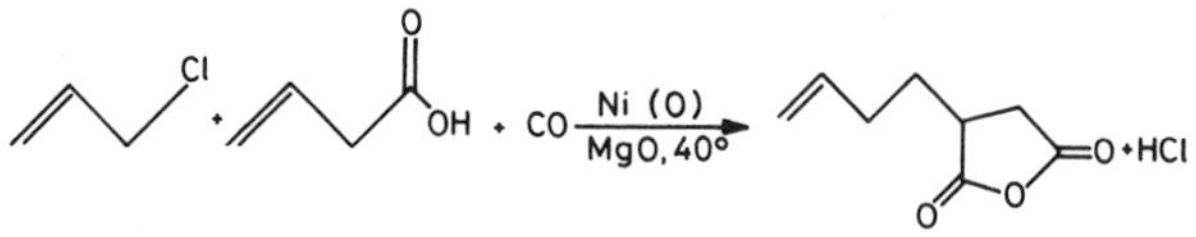

The reaction must involve intermediates of type:

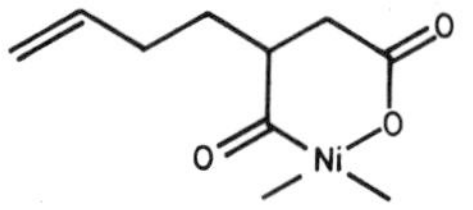

The presence of amino groups in the substrates can promote elimination to lactams.[29,47] The synthesis of β-lactams:[48]

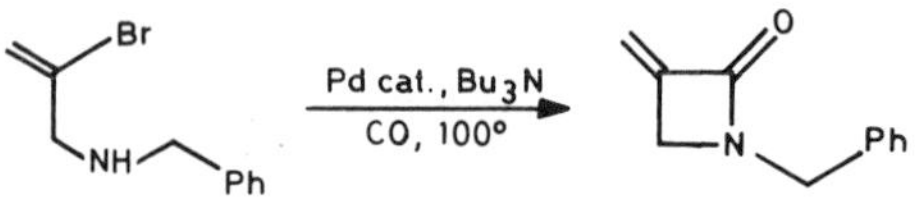

and of other benzolactams[48a] and of cyclic ureas[48b] has also been reported.

Reductive elimination can also be induced by ring formation without splitting of any fragment. Acetylenic substrates favor this type of elimination:[49-53]

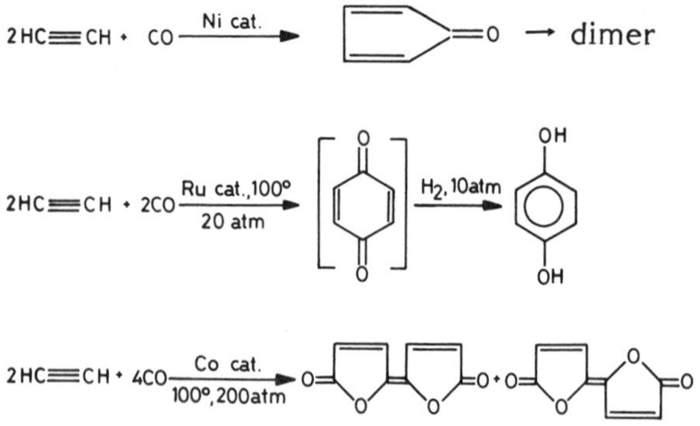

In the same category should be included reactions of butadiene and CO where metal-bonded allyl groups can easily give rise to CO insertion and reductive elimination:[5g,54]

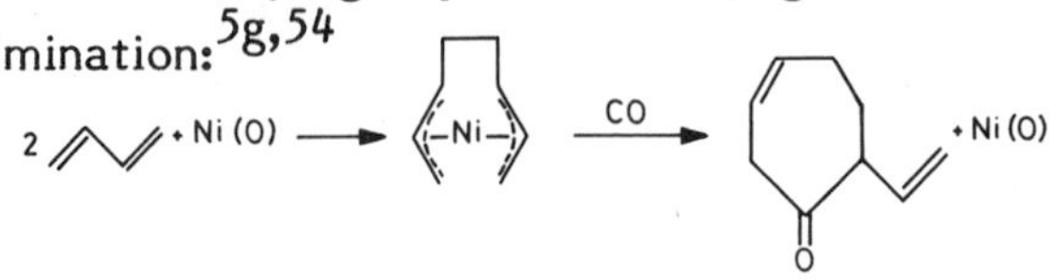

Formation of a double bond from a carbenic[26] or nitrenic[55] intermediate and CO also promotes reductive elimination:

$$Ph_2CN_2 \xrightarrow[\text{Ni cat.}]{CO} Ph_2C{=}C{=}O \; {+}\, N_2$$

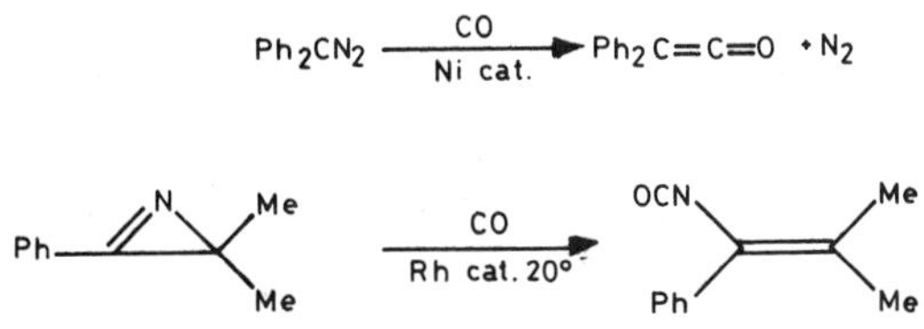

The latter case is also noteworthy as an example of oxidative addition favored by steric relief of the substrate.

Aromatization seems to be the driving force for reductive elimination following CO insertion in quinone methides:[56]

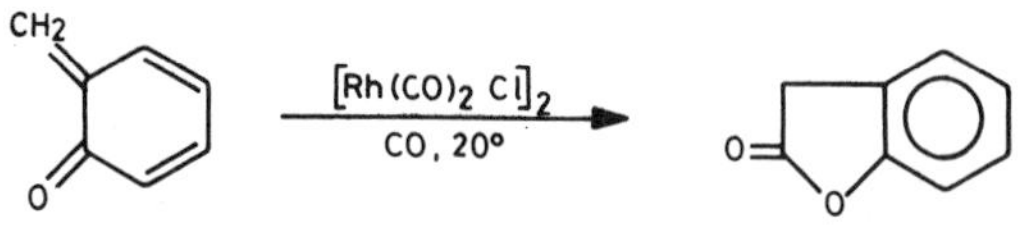

Heterocycle ring formation is another instance of reductive elimination promoted by the substrate. We report here only the cyclization of azobenzene to 2-phenyl-indazolone:

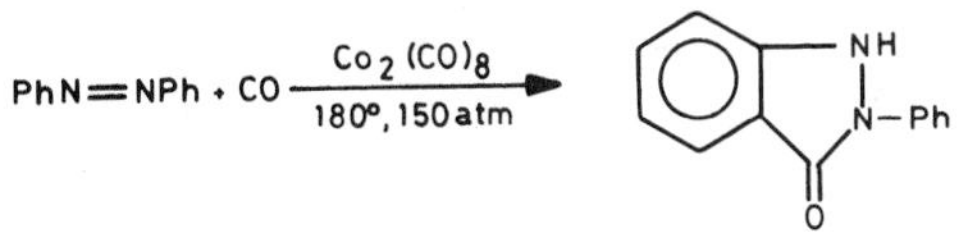

$$PhN{=}NPh + CO \xrightarrow[\text{180°, 150 atm}]{Co_2(CO)_8}$$

For other examples see 5d, 24, 29, 45, 57.

A substrate-promoted ring forming elimination has been postulated[57] to explain the absence of aldehydes in carbonylation of α-acetamido alcohols with CO/H_2 (Sec. 2.1):

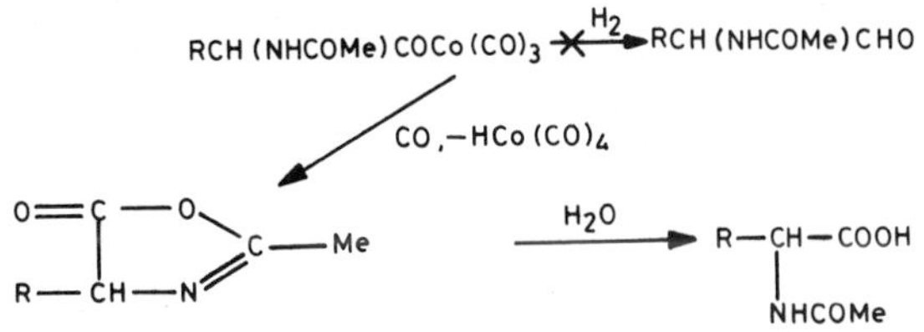

If the hypothesis is confirmed, this could be an interesting example of substrate-promoted elimination leading to hydrocarboxylation instead of hydroformylation.

2.4   Effect on insertion into acyl metal bonds

Instead of eliminating the metal, the acyl metal bond formed by CO insertion can undergo further reaction with suitable substrates. The same substrates exemplified as promoters of CO insertion often favor further reaction; for example, the second double bond of 1,5-hexadiene lends itself to further insertion after reaction of the first with methallyl chloride and CO:[58-9]

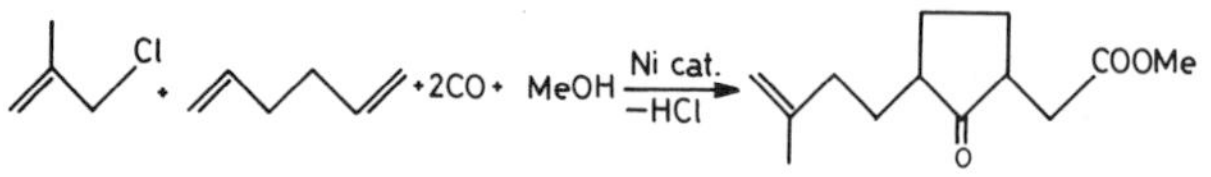

The effect seems to be due to acyl group stabilization by chelation:

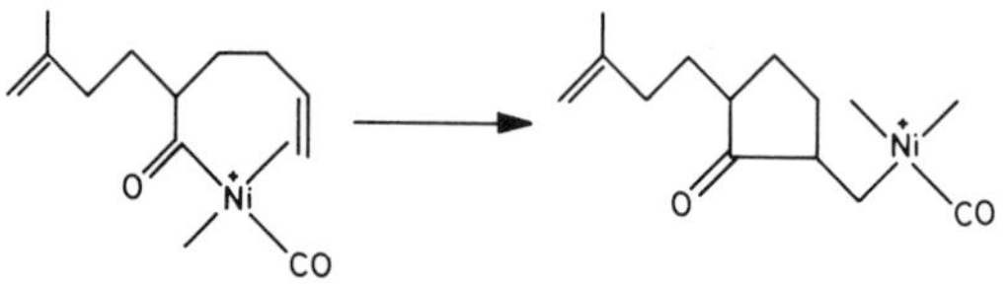

It is not observed with 1,4-pentadiene or 1,6-heptadiene. When the hexadiene system bears a bulky substituent, attack of the methallyl group preferentially occurs at the site which is furthest from the substituent. The effect has been exploited for the synthesis of steroid analogue precursors.[60]

Acylnickel bonds suitable for further cyclization can also be formed from alkynes, allyl halides and CO.[4c,61] The presence of alkyl or aryl substituents R in 1-alkynes favors the intramolecular reaction of hexadienoylnickel complexes with double bonds. Acylnickel derivatives are generally solvolyzed very easily by alcohols, but the presence of a substituent α to the carbonyl group probably stabilizes the intermediate against attack by alcohols. A cyclopentenonic or cyclohexenonic ring is closed and further addition of CO or HNiX elimination takes place:[61]

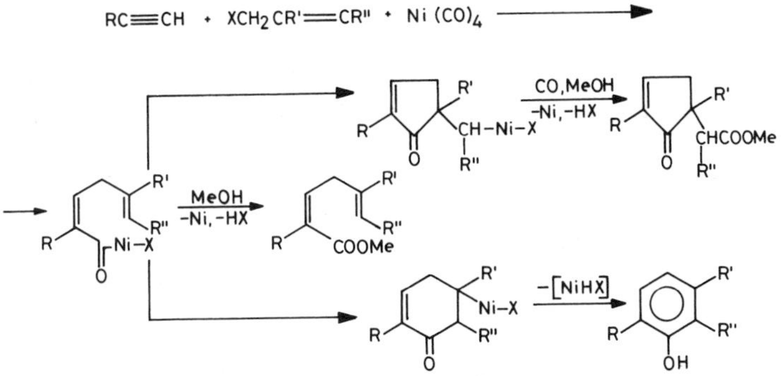

$$RC\!\equiv\!CH \;+\; XCH_2CR'\!\equiv\!CR'' \;+\; Ni\,(CO)_4 \;\longrightarrow$$

If R' is H and R" is an alkyl the attack of the acyl group occurs on the most internal carbon atom of the double bond. The reverse is true in the case that R' = alkyl and R" = H. A ring closure leading to an aromatic cycle occurs in this case. Apparently the energy gain from the aromatization process is quite important.

Insertion of two CO molecules in series into metal-alkyl groups cannot be achieved easily.[32,62] Double carbonylation is facile with benzylic substrates:[63-6]

$$PhCH_2X + 2CO + OH^- \xrightarrow{\text{Co cat.}} PhCH_2COCOOH + X^-$$

possibly because[66] they form intermediates of the vinylic type shown below, which easily insert CO:

$$PhCH\!=\!C\,(OH)\!-\!\underset{|}{\overset{}{M}} \longrightarrow PhCH\!=\!C\,(OH)\!-\!\underset{\parallel}{\overset{}{C}}\!-\!M$$
$$\quad\qquad CO \qquad\qquad\qquad\qquad\quad O$$

The use of $Ca(OH)_2$ and high CO pressure[63] and of electron-releasing substituents in the phenyl ring under phase transfer conditions (Sec. 4.4)[65-6] have been reported to favor double carbonylation.

Another way to cause further reaction of the group resulting from a first attack on CO by a nucleophile is to trap it with reactive unsaturated substrates.

In oxidative carbonylation of methanol with olefin insertion, in the presence of Pd catalysts and of an excess of $CuCl_2$ as Pd reoxidant (see Sec. 5. for other examples) and of carboxylate buffers, the carbomethoxy group attacks olefins regiospecifically in an anti-Markovnikov direction.[67] Interesting substrate effects have been detected. Strained cyclic olefins favor formation of dicarboxylated products. The reactivity order: norbornene > cyclopentene > cycloheptene > cyclooctene > cyclohexene is the same as that reported for hydroformylation. Probably steric strain relief on passing from η^2-coordination of the olefin bond to the σ-alkylpalladium bond favors carbomethoxylation, for example in the case of cyclopentene:

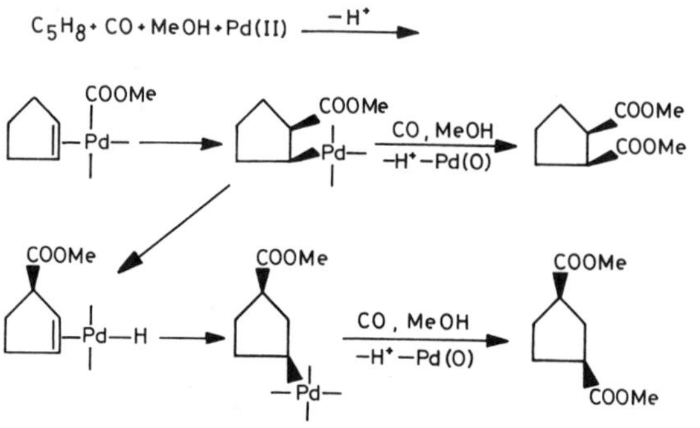

$$C_5H_8 + CO + MeOH + Pd(II) \xrightarrow{-H^+}$$

The relative position of the carbomethoxy groups in dicarboxylation products of alkyl-substituted olefins depends on the ease of hydride abstraction from C atoms adjacent to the Pd-C bond (tertiary > secondary > primary). Cis dicarbomethoxylation is observed, in contrast to the trans methoxylation-carbomethoxylation found for acyclic olefins under neutral conditions:[67]

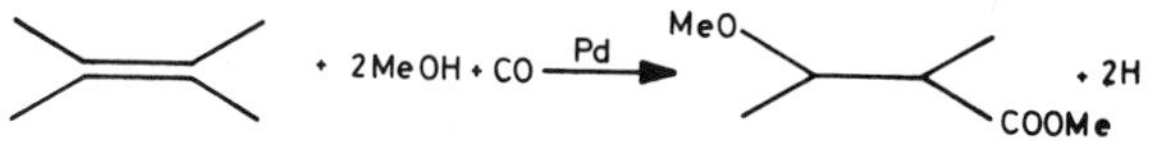

Diene and acetylenic substrates have also been successfully employed in oxidative dicarbomethoxylation.[5e,17a,68-9]

2.5 Effect on metal elimination from organic groups formed by further insertion

The substrate effect to be expected in this case is common to all elimination reactions. Since we are dealing with catalytic processes generally taking place on low valent metals, the elimination has to be of the reductive type to allow formation of a metal species able to start a new catalytic cycle.

If further addition of molecules and groups to the acyl group initially formed does not lead to a new acyl group, the elimination process generally consists of H elimination or transfer (both from metal-bonded hydrides and organic compounds having active hydrogen atoms), coupling, disproportionation, or X⁻ attack:

$$R-M \xrightarrow{-H} R(-H) + M$$
$$\xrightarrow{+H} RH + M$$
$$\xrightarrow{RM} R-R + 2M$$
$$\xrightarrow{RM} R(-H) + RH + 2M$$
$$\xrightarrow{X^-} RX + M^-$$

The type of substrate can influence the nature of the metal bond with the organic groups to the point that one of these termination types can predominate. An example of elimination promoted by a substrate having active hydrogen is offered by the following phenylacetylene and CO insertion into a carbomethoxy substituted η^3-allylnickel bond. The acyl group first formed attacks a double bond to give a cyclopentenonyl-alkylnickel intermediate, which takes up hydrogen from the solvent stoichiometrically:[70]

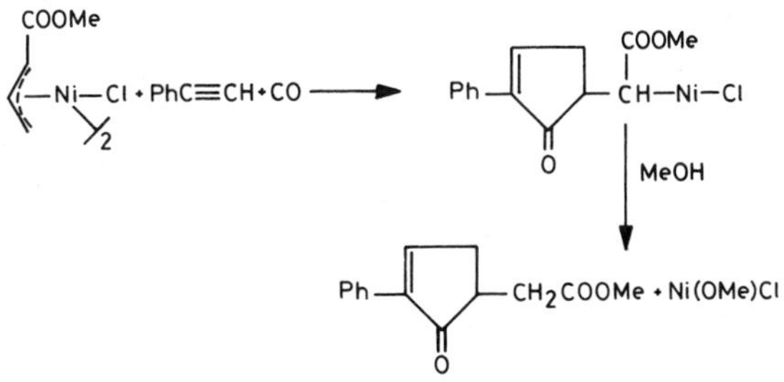

In the absence of the carbomethoxy group a new CO insertion into the C-Ni bond takes place, followed by hydrolytic cleavage of the acylnickel bond (R = H or alkyl):

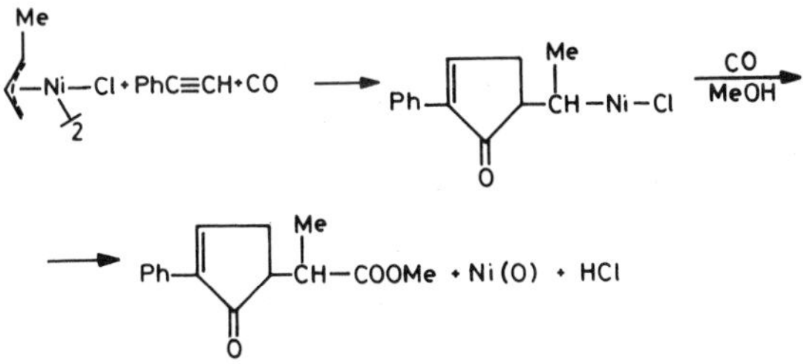

3.   LIGANDS

3.1   General behavior

Promotion of carbonylation by ligands can be effected in many ways. We shall first distinguish between reactive and unreactive ligands. CO, H_2, C_2H_2, and many other molecules, including substrates for carbonylation, not only take part in the reaction, but also play a role as ligands.

Unreactive ligands can be part of catalyst precursors which may retain them during the entire catalytic process or lose one or more of them to give rise to the true catalyst. Furthermore, ligands can be lost or acquired at any stage (associative and dissociative processes) or can be exchanged with other ligands or with substrates. Thus a distinction has to be made between easily dissociable ligands and ligands which do not dissociate or dissociate with difficulty, and between fast and slow exchanging ligands. A classification of this type has been made by Tolman[71] for phosphorus-containing ligands and the reader is referred to his useful review.

Ligands can influence the various steps of the carbonylation process in several ways:

a)   by activating metals, for example towards oxidative addition[1,1a]

b)   by stabilizing complexes deriving for example from oxidative addition or reductive elimination[1,72]

c)   by favoring coordinative unsaturation by dissociation or steric hindrance[1,1a,d,72a]

d)   by favoring changes of stereochemistry also by dissociation or association[31,35,73]

e)   by favoring migration of _trans_ ligands[1,1a,4a,b,35,74-6]

f)   by occupying the coordination site left free by a migrating group[4a,76a]

g)   by protecting coordination sites from inactivation by other ligands[77]

h)   by providing hydride ligands by internal metalation[78,78a]

i)   by favoring formation of cationic or anionic complexes[79]

j)   by acting as a "bridge" in electron transfer processes and ligand transfer processes[80,80a]

k)   by favoring cluster formation[81]

l)   by inducing regio- and stereoselectivity[82]

m)   by inducing asymmetry.[83,83a,b]

3.2 Reactive ligands

Most of the reactive ligands have been considered among substrates. We therefore shall consider here only CO and H_2 (including both their interaction with substrates and their mutual interaction) in view of their relevance both as reactive and non-reactive ligands.

3.2.1 Carbon monoxide and hydrogen

CO plays a number of roles as ligand and as reagent. It stabilizes low oxidation states of metals,[79] thus favoring reductive elimination processes; it occupies coordination sites left free by migrating groups, thus stabilizing insertion products;[4a,76a] it dissociates causing coordinative unsaturation and facilitating substrate coordination;[2,2a] it associates favoring group migration[4a,d] it exerts a high field favoring the square planar stereochemistry, which is the most appropriate for oxidative addition and insertion;[1,2a] it also promotes migratory insertion by virtue of its <u>trans</u> effect;[74,76] being both a terminal and a bridging ligand, it also helps cluster formation[81] and controls metal-metal bond distances;[84] CO pressure can also induce regioselectivity in hydride attack at olefins and in other reactions;[4] in some cases the insertion product can be stabilized as an η^2 carbonylic ligand $M\begin{smallmatrix} C-R \\ O \end{smallmatrix}$;[85] CO insertion being a stereospecific reaction,[4a] products of well-defined stereochemistry can be obtained.

Hydrogen also behaves as ligand and as reagent. Among the many effects of the hydride ligand it is worth recalling that, due to its high <u>trans</u> effect, it often helps substitution reactions of other ligands even if it is present in catalytic amounts.[86]

Examples of promoting effects of CO, generally involving more than one of the effects listed above, are quite numerous. Simple variation of CO partial pressure can have important consequences both on rate and selectivity. Some of the first examples of the importance of coordinative unsaturation were noted many years ago in the case of hydroformylation of olefinic substrates with cobalt carbonyl,[87] in the case of carboxylation of allylic halides with tetracarbonylnickel,[36] and in many other instances. In all cases a high CO pressure inhibited the reaction because in a rate-determining step CO had to be evolved from complexes to allow coordination of substrates or ligands.

In hydroformylation with cobalt the real catalyst (or catalysts) in the presence of CO only, without added ligands, probably changes with CO pressure.[4] The mononuclear mechanism proposed by Heck in 1963, however, explains CO and H_2 effects satisfactorily:[2a]

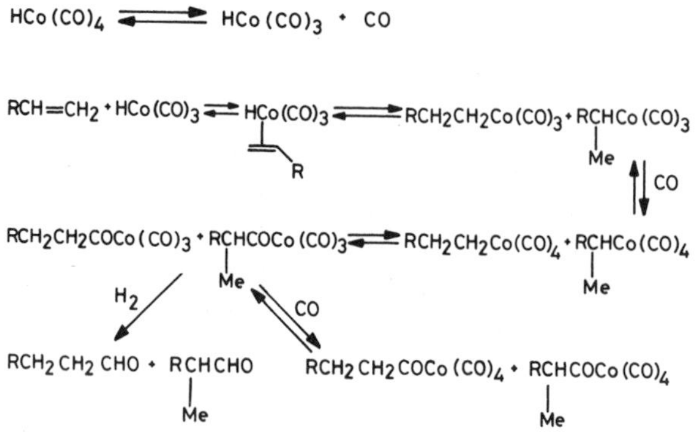

As shown in the Scheme, hydrogen enters the process in two different stages in the form of hydride (the final splitting is possibly preceded by oxidative addition of hydrogen to the acylcobalt carbonyl complex). Competition with CO clearly results from the Scheme and therefore relative partial pressures of the two gases have an important effect on rate. A similar mechanism seems to be at work with Rh complexes[4,88,88a] although multinuclear clusters seem to be present.[19] As to stereochemistry, cis addition of H and CO to double bonds has been shown to occur.[4]

Cluster complexes can catalyze hydroformylation under relatively mild conditions ($110°$-$130°$ and 230-1160 psi initial pressure for the cluster $PhCCo_3(CO)_9$[89]) to form linear aldehydes preferentially. The cluster $PhCCo_3(CO)_9$ is thought to activate hydrogen by opening its structure to give[90]

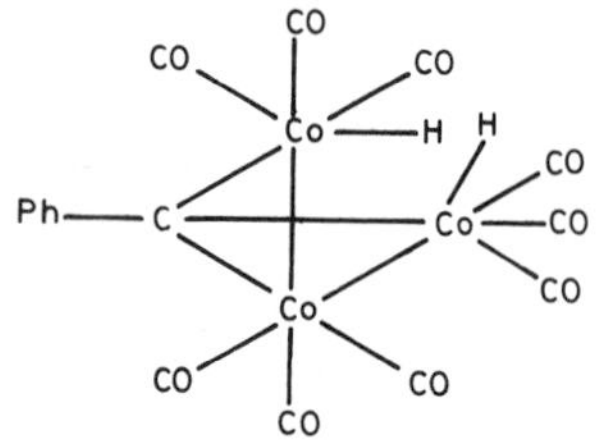

The effect of CO partial pressure was particularly studied in relation to the problem of regioselectivity in hydroformylation. It was soon realized that linear products in hydroformylation with Co could be made to predominate by using high CO pressure. A mechanism implying restriction of olefin coordination and preferential hydride attack at the most hindered site of the olefin was shown to explain this result.[4] A low partial pressure was beneficial with rhodium carbonyl catalysts.[19] Further studies in the presence of tertiary phosphines as ligands have thrown light on the mechanism and will be discussed in Sec. 3.3.1.

Another example of CO competition with substrates is offered by the palladium-catalyzed carbomethoxylation of olefins in the presence of $CuCl_2$ as reoxidant (Sec. 5.) and carboxylate buffers. Dicarboxylated products are formed from monoolefins. Non-conjugated chelating diolefins give different products depending on CO pressure. Thus 1,5-hexadiene gives the tetraester at 1-2 atm of CO, but at higher pressure (6 atm) the other olefinic double bond is displaced by CO and only the butenylsuccinic diester is formed:[68]

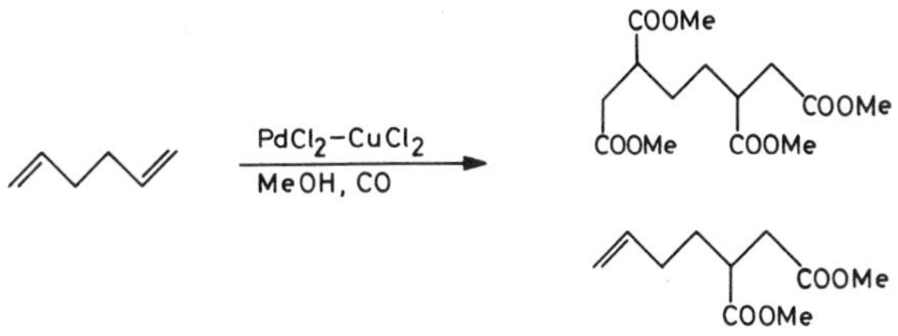

Alcohol (ether etc.) homologation presents features similar to hydroformylation once the metal-bonded alkyl group is formed:[91-3]

$$MeOH + CO + H_2 \xrightarrow{Co_2(CO)_8} MeCHO + H_2O$$

$$MeOH + CO + 2H_2 \xrightarrow{Co_2(CO)_8} MeCH_2OH + H_2O$$

At high H_2 partial pressure ethanol can be obtained in a 50% yield.[93]

Under sufficiently high pressure CO can also stabilize the reactive intermediate in the CO hydrogenation reaction which leads to methanol, glycol and hydrocarbons. Hydrogen is also present as a reactive ligand, directly attacking CO. In the case of the synthesis of methanol and ethylene glycol with rhodium catalysts:[94]

$$CO + 2H_2 \longrightarrow MeOH \qquad 2CO + 3H_2 \longrightarrow HOCH_2CH_2OH$$

the active catalytic intermediates at $220°$ and 550 atm of CO/H_2 seem to be rhodium carbonyl clusters, containing terminal and bridging carbonyl groups, which appear to be activated towards reduction by hydrides. I.r. evidence indicates the pressence of the species $[Rh_6(CO)_{15}]^-$ (formerly considered $[Rh_{12}(CO)_{34}]^{2-})$[95] at 500 atm and $160°$.[96-9] Other species, however, such as rhodium carbonyl hydrides $[Rh_{13}(CO)_{24}H_3]^{2-}$ and $[Rh_{13}(CO)_{24}H_2]^{3-}$, might be the active ones, although present in very low concentration.[96-9] When methanol is the main product the species $[Rh(CO)_4]^-$ predominates in solution.[100] Rhodium clusters containing encapsulated atoms, like $[Rh_6(CO)_{15}C]^{2-}$, $[Rh_9P(CO)_{21}]^{2-}$, and $[Rh_{17}S_2(CO)_{32}]^{3-}$, show lower catalytic activity, behaving as relatively poisoned catalysts as in heterogeneous catalysis.

Another polymetallic complex

catalyzes formation of methanol from CO and H_2 at $80-100°$ and 20 atm.[101]

The subject of CO hydrogenation on metal clusters will be discussed further in Sec 4.1.

Examples of methanol and methyl formate synthesis,[102-3] of carbonylation of formaldehyde to glycolaldehyde and glycol,[104,104a] and of carbonylation of acetals to diols[104b] with CO/H_2 in the presence of mononuclear species, such as $HCo(CO)_4$ and $Ru(CO)_5$, as such or phosphine-modified, have been reported, however; thus mononuclear as well as polynuclear species have been suggested as catalysts.

Metallic clusters have been invoked[105] to explain formation of hydrocarbons. Methane has been obtained from CO/H_2 at $140°$ and 2 atm[105] with $Os_3(CO)_{12}$ and $Ir_4(CO)_{12}$. Better results have been reached using $Ir_4(CO)_{12}$ in a melt of $NaCl.2AlCl_3$ at $180°$ and $1-2$ atm of CO/H_2.[106] Under these conditions ethane is the main product along with methane, propane and isobutane. Complexes $Os_3(CO)_{12}$, $Ir_4(CO)_{12}$, and above all $Ru_3(CO)_{12}$ (as such or modified by $(Bu_2P)_3SiMe$) have been claimed to form alkanes C_1-C_{30} under more drastic conditions ($300°$ and 1000 atm),[107] but they do not appear to be truly homogeneous.[103-8]

In the attempt to clarify catalytic mechanisms some model reactions have been studied. Reduction of coordinated CO by molecular H_2 has been reported for a Zr complex.[109] Formation of a formyl group by CO insertion into a Zr-H bond has been postulated[110] and has been suggested indirectly by other model reactions,[111,111a] in particular by the reduction of a cobalt-bonded CO with $HCo(CO)_4$.[112] The Fischer-Tropsch synthesis possibly involves such species.[113,113a] Activation of the oxygen atom of the formyl group and more generally of the oxygen atom of the CO group itself by coordination with metals has been shown by other model reactions[114-118b] and by isolation of certain carbonylic complexes containing the acyl groups.[114,119-20] Metal-oxygen bond formation appears to be an essential step in both homogeneous and heterogeneous hydrocarbon formation.[121] Recent work[122] has shown that it is possible to form a C-C bond between CO and the C atom of a carbidocarbonyl cluster in a homogeneous system, analogous to that originally proposed by Fischer and Tropsch.[123] Thus tools for homogeneous CO hydrogenation are likely to be developed from observation of the behavior of CO-activating clusters or surfaces. For mechanistic aspects the reader is referred to a recent review on CO hydrogenation,[124] and to the following Chapter.

Carbonylic clusters such as $Rh_4(CO)_{12}$ and $Rh_6(CO)_{16}$ are also able to activate protic solvents, thus generating the hydrogen necessary for hydrocarbonylation reactions.[125] Hydrogen can even be provided by aromatic C-H activation with concomitant C-C coupling, for example:[126]

$$C_6H_6 + 3CH_2{=}CH_2 + CO \xrightarrow[\text{220°, 25atm CO, 30atm } C_2H_4]{Rh_4(CO)_{12}} PhCH{=}CH_2 + Et_2CO$$

3.2.3 Halides, carboxylates and other reactive ligands

Among numerous reactive ligands that can be used with CO and H_2, halides (Cl,Br,I) and carboxylates are widely employed and will be briefly treated here, most of the others having been considered in Sec. 2. as substrates.

Halides can react both by forming the reactive intermediate by nucleophilic attack on suitable substrates, and by favoring reductive elimination, as in the following stoichiometric reaction,[41] which includes both effects:

$$Cl-Pd-Cl + CO + C_2H_4 \longrightarrow ClOCCH_2CH_2Cl + Pd$$

Halide ions catalyze the disproportionation of $Co_2(CO)_8$ and $Co_4(CO)_{12}$ to $[Co(CO)_4]^-$.[127]

Halides can stabilize to varying extents the products of oxidative addition or insertion,[128] sometimes via dimeric structures, and recombine either with acyl groups or with protons or cations in the reductive elimination step; for example:

$$PhX + M \longrightarrow Ph-M-X \xrightarrow{CO} PhCO-M-X \longrightarrow PhCOX + M$$

Halides can sometimes take part in ligand transfer reactions.[80,80a]

An interesting example of the different effects caused by complexes, with or without Cl, is offered by carbonylation of butadiene with Pd-tertiary phosphine complexes at 110° and 50 atm of CO.[129] Chloride-containing complexes are thought to react via η^3-crotyl intermediates (selectivities are of the order of 90% for conversions of the order of 30% to 3-pentenoic esters):

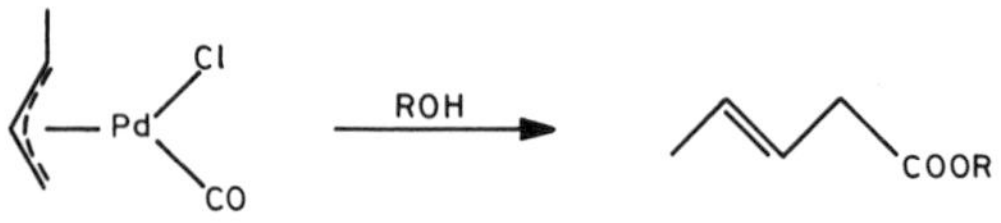

whereas other ligands such as acetate can be replaced by carbon (allylic) ligands formed by butadiene itself (yields of nonadienoic esters are of the order of 90%):

In other instances carboxylate ligands appear to work analogously to halides, forming anhydrides by combination with acyl groups or weak acids by proton capture. In addition, carboxylate ligands can favor insertion,[130] probably because their ambidentate system allows them to occupy the coordination site left free by migrating groups:

$$RCOO-\overset{\overset{\text{CO}}{|}}{M}-R' \longrightarrow R'CO-M\overset{O}{\underset{O}{\diagup}}C-R$$

Direct activation of the aromatic C-H bond to form benzoic acids or esters by carbonylation with CO at 15 atm and H_2O or alcohols at $100°$ can be obtained using Pd acetate in a stoichiometric reaction.[130a]

Many other ligands are eliminated by protonation during the catalytic process to give rise to carbonylic species by reaction with CO; for example Co or Rh acetylacetonates give rise to carbonyls under the reaction conditions required for various types of carbonylation.

3.3 Unreactive ligands

The most used in carbonylation are tertiary phosphines, $SnCl_3^-$, and I^-.

3.3.1 Phosphorus ligands

Tertiary phosphines are versatile ligands[131] forming kinetically stable complexes with various oxidation states of metals. As monodentate ligands phosphines have been mainly used in carbonylation of organic halides and in hydroformylation and hydrocarboxylation. In the latter cases attention was mainly focused on obtaining high regioselectivity. In general tertiary phosphines, particularly the aromatic ones, offer the best compromise of properties to promote carbonylation, because they favor oxidative addition, insertion, and reductive elimination.[72,131-2]

With organic halides oxidative addition is favored by electron-releasing groups on aromatic phosphines in most cases,[1,1a] although a recent report on cationic iridium chemistry shows the opposite behavior.[133] Chelating phosphines, such as tris(o-methoxyphenyl)phosphine, have been shown to favor oxidative addition to Rh complexes.[134]

Phosphines stabilize pentacoordinated intermediates, present in carbonylation of allylic halides with Ni complexes[135] and in carbonylation of aromatic halides with Ni, Pd, and Pt complexes.[31] The best results have been obtained with phosphines containing electron-releasing groups on the aromatic ring:[31]

$$M(PR_3)_2(R)X + CO \longrightarrow M(PR_3)_2(CO)(R)X$$

Added phosphine retards insertion, which is therefore supposed to pass mainly through a tetracoordinated intermediate formed by phosphine dissociation:

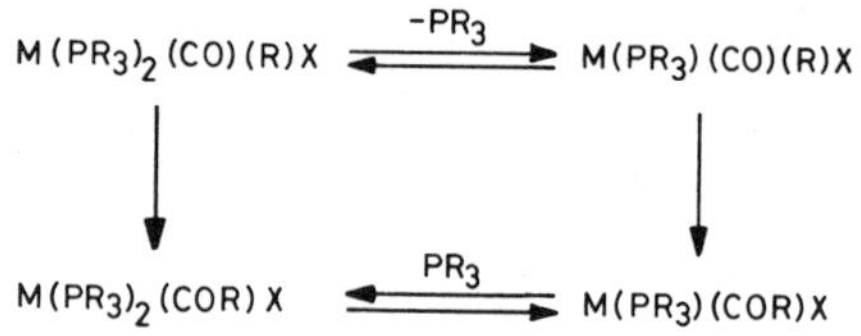

$$M(PR_3)_2(CO)(R)X \underset{-PR_3}{\rightleftharpoons} M(PR_3)(CO)(R)X$$

$$M(PR_3)_2(COR)X \underset{PR_3}{\rightleftharpoons} M(PR_3)(COR)X$$

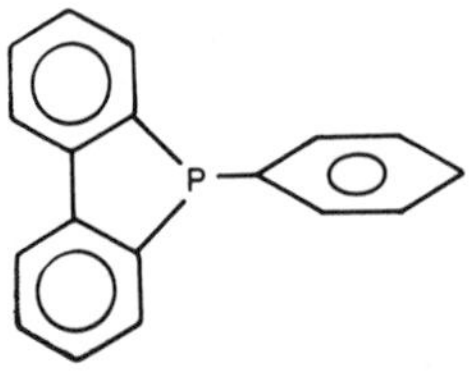

Using PR_3 = PMe_2Ph Anderson and Cross[35] have shown that in the case of Pt the tetracoordinated intermediate undergoing insertion is that containing R and X in mutually cis positions with PR_3 trans to the migrating group R. Less basic phosphines (PPh_3) may act on other square planar isomers favoring isomerization to the above mentioned one. Basic phosphines may induce CO elimination rather than insertion. The use of phosphines as promoters of the insertion step thus appears to require an accurate control of their molar ratio to the complex, basicity and ability to dissociate. If the overall effect of phosphines on catalytic carbonylation of organic halides is considered, one has to take into account also their already mentioned effect on oxidative addition and their effect on reductive elimination. The latter is favored by phosphines able to stabilize the lower oxidation states of metals; thus opposite properties in respect to those required for oxidative addition are needed. This explains why, all in all, aromatic phosphines are the most used ligands. Recently dibenzophosphole ligands have been shown to enhance hydroformylation rates better than the corresponding triarylphosphines, probably because they are more efficient "buffers" of the charge fluctuations occurring on the catalyst in each step of the catalytic process.[136]

The effect of phosphines on hydroformylation rate has been studied with both Co and Rh catalysts. Retardation has been reported for Co and acceleration, passing through a maximum corresponding to an excess of 5-10 mol per mol of complex, has been noted for $RhH(CO)(PPh_3)_2$.[137-8]

Particular attention has been devoted to the regioselectivity problem. Phosphine-modified Co catalysts lead to an increased proportion of linear aldehydes.[138] For example, hexanal with higher than 90% linearity is formed from 1-pentene at about $150°$ and 35 atm $1/1$ CO/H_2. Lower partial pressures of CO are needed with these catalysts because the stability of the Co carbonyl is higher. Phosphine ligands, however, render the metal-hydrogen bond more hydridic and substantial amounts of alcohols and of saturated hydrocarbons are formed. N-hexanol is mainly formed under the conditions reported above using a CO/H_2 ratio of $1/2$. High linear vs branched aldehyde ratios are obtained with phosphine-modified Rh catalysts. Conditions are much milder than those required by Co catalysts (for example $90°$ and 6-8 atm for 1-octene).[88a,138-140a] The origin of regioselectivity has been shown to be essentially steric, although basic phosphine ligands, which render metal-bonded hydrogen more hydridic, favor formation of linear products. Olefinic substrates prefer the type of coordination that places the most hindered group CXY at the side nearest to the less crowded part of the complex (M-H):[4]

Chelating ditertiary phosphines have recently been reported to be effective as ligands to form complexes with $Co_2(CO)_8$ able to catalyze hydroformylation from olefins, CO and H_2O in ethereal solvents,[141] for example:

$$CH_2=CH_2 + 2CO + H_2O \longrightarrow EtCHO + CO_2$$

Optically active chelating ditertiary phosphines are efficient ligands for asymmetric carbonylation. A 27% optical yield has been obtained in carbonylation of styrene to α-methylphenylacetaldehyde in the presence of Rh(DIOP)Cl:[4,90,142]

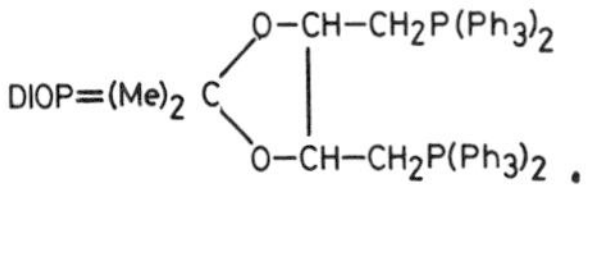

DIOP=(Me)$_2$ C $\begin{array}{l} O-CH-CH_2P(Ph_3)_2 \\ | \\ O-CH-CH_2P(Ph_3)_2 \end{array}$.   With   (structure with P P)   (P) = P (structure)

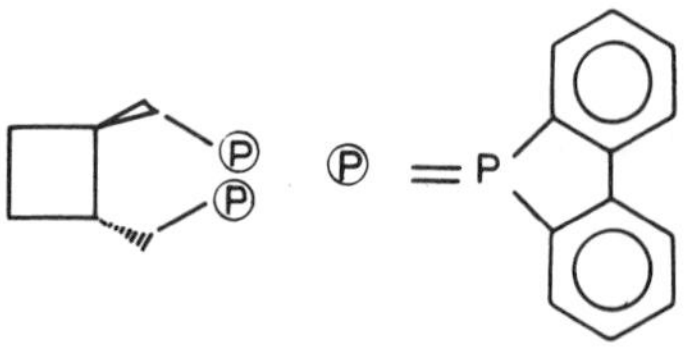

a 40% optical yield has been claimed in hydroformylation of styrene[143] at 80° and 100 atm of 1/1 CO/H_2. A 44% optical yield has been reached in hydroesterification[143a] of α-methylstyrene with a DIO (P) type ligand with (P) as above.

Asymmetric hydroformylation of vinyl acetate to α-acetoxypropionaldehyde has been obtained with Rh-phosphine catalysts.[144,144a] This is a way to optically active lactic acid.

Isoprene has been hydroformylated (95° and 90 atm 1/1 $CO/H)_2$ with $RhH(CO)(PPh_3)_2$ + (-DIOP) to S-3-methylpentanal (32% o.p.).[145] Interestingly, the same aldehyde can be obtained from 2-methyl-1-butene with inverted (R) configuration. This fact has been attributed to different mechanisms, the first arising from an enantioface discriminating hydrogenation of a mixture of α,β-unsaturated stereomeric aldehydes deriving from a double bond shift, the second resulting from an enantioface discriminating hydroformylation:

$$CH_2=C(Me)CH=CH_2 \xrightarrow{CO/H_2} OHCCH=C(Me)CH_2Me \xrightarrow{H_2} (S)- OHCCH_2CH(Me)CH_2Me$$

$$CH_2=C(Me)CH_2Me \xrightarrow{CO/H_2} (R)- OHCCH_2CH(Me)CH_2Me$$

As shown above, double bonds conjugated to carbonyl groups are hydrogenated rather than hydroformylated; thus bis-hydroformylation of conjugated dienes does not take place if, after the introduction of the first formyl group, the remaining double bond shifts to a position conjugated with that group. A large excess of tertiary phosphine prevents isomerization and allows formation of bis-hydroformylation products, for example adipaldehyde diacetal from butadiene in the presence of $RhCl(CO)(PPh_3)_2$ at $120°$ and 280 atm CO/H_2 in MeOH.[146-7]

Acrylic and methacrylic esters are preferentially α-hydroformylated (98% selectivity) in the presence of chelating ditertiary phosphines.[148]

Tertiary phosphines have also been used to direct hydrocarboxylation of olefins in the presence of Pd complexes[149-50] at $120°$ and 200 atm of CO in benzene ethanol:

$$PhCH{=}CH_2 \longrightarrow \begin{cases} PhCH_2CH_2\,COOEt \\ \\ PhCH(Me)\,COOEt \end{cases}$$

PPh_3 and PPh_2Bu give mainly 2-phenylpropionate, whereas chelating diphosphines give predominantly 3-phenylpropionate if the number of CH_2 in $Ph_2P(CH_2)_nPPh_2$ is 3-5 and 2-phenylpropionate if n is 1, 6 or 10. The result has been rationalized in terms of steric effects leading to the branched isomer with monodentate phosphines and to the linear isomer with stable chelating rings (n = 3-5).[150] As mentioned in Sec. 3.2.2., Pd acetate or halide complexes with tertiary phosphines have been used to carbonylate dienes to 3-pentenoic esters in the presence of halide ligands and to 3,8-nonadienoic esters in their absence.[129] In quinoline as solvent the following order of ligand activity, corresponding to increased basicity, has been observed for the formation of nonadienoic esters[151]: $P(OPh)_3 < P(p\text{-}$ $ClPh)_3 < PPh_3 < P(p\text{-tol})_3 < P(n\text{-Bu})_3$. Certain types of phosphine-based "constraining" ligands, i.e. ligands able to hold together metal atoms in a cluster or in a bridge complex, might be useful in CO/H_2 chemistry (Fischer-Tropsch process in homogeneous phase), where the presence of more than one atom in a complex of Ru, Os, Ir seems to be useful to activate both CO and H_2. One such ligand is $(PBu_2)_3SiMe$.[99]

Replacing phosphines with weaker ligands such as arsines may have important consequences. A catalytic system consisting of phosphine-modified Rh and Co complexes was found to be highly selective for ethanol formation from methanol (homologation reaction), while the corresponding catalysts with arsine ligands oriented the reaction towards acetaldehyde:[152]

$$MeOH + CO + H_2 \quad \underset{AsPh_3cat.}{\overset{PPh_3cat.}{<}} \quad \begin{array}{l} MeCH_2OH \\ MeCHO \end{array}$$

3.3.2 $SnCl_3^-$

The association of tertiary phosphines with $SnCl_3^-$ as ligand has proved to be extremely effective in hydroformylation[153,133a] and hydrocarboxylation[154] because of the high linear content of the aldehydic or carboxylic product. From 1-heptene at $66°$ and ca. 100 atm of CO/H_2 in methylisobutylketone as solvent, aldehydes were obtained in 85% yield and 90% selectivity in 1-octanal. The best catalyst turned out to be formed from a 5:2:1 mixture of $SnCl_2.2H_2O$, PPh_3, and $PtCl_2$.[153] Similarly, $SnCl_2$, PPh_3, and $PdCl_2$ in the same ratio at $70°$ and 136 atm in methanol-methyl isobutyl ketone formed methyl octanoate with 88% selectivity. The complex trans-$HPd(PPh_3)_2(SnCl_3)$ was isolated and found to be an effective catalyst. The $SnCl_2$ excess is thought to cause steric crowding, so that preferential olefin coordination to form linear isomers prevails.[154] Both electron-withdrawing para substituents on the phosphine aromatic ring, and trialkylphosphines, lower the ester yield.[154] The situation is different with Pt in place of Pd; electron-withdrawing substituents and, in general, decreasing ligand basicity favor ester formation. Triphenylarsine gives the best result with Pt, but forms an inactive complex with Pd;[154] thus the size of the metal-bonded atom of the ligand plays an important role.

If $SnCl_2$ is added to $PdCl_2(PPh_3)_2$ in anhydrous MeCN, the cation $[Pd(SnCl_3)(MeCN)(PPh_3)_2]^+$ is formed. This is an effective catalyst for cyclocarbonylation of $H\equiv CCH_2CH_2OH$ to

$$\begin{array}{c} CH_2\text{——}CH_2 \\ | \quad\quad | \\ CH_2=C \quad\quad C=O \\ \diagdown \quad \diagup \\ O \end{array}$$

at $75°$ and ca. 8 atm of CO.[155] Due to its high trans effect[156] the $SnCl_3^-$ ligand (a strong π-acceptor and a weak σ-donor)[157-8] labilizes Cl^- and favors coordination of the substrate and CO uptake.

The use of $PdCl_2 \cdot 2PPh_3 + SnCl_2$ catalyst has recently been found advantageous in many other carbonylation reactions of allylic and benzylic substrates.[47]

3.3.3 I⁻, Cl⁻, and Br⁻

We consider these charged ligands separately from the same ligands acting as reactive groups, although the former become scrambled with the latter in some anionic complexes.

An efficient ligand is I⁻. It has been used in many catalytic processes with Ni, for example in carboxylation of benzyl chloride, a reaction which does not occur with $Ni(CO)_4$ alone under mild conditions, but does occur in high yield in the presence of the more nucleophilic $[Ni(CO)_3I]^-$.[159] The synthesis of vinylacrylic esters is also catalyzed by the same anion:[160]

$$2HC\equiv CH + CO + MeOH \xrightarrow{Cat.} CH_2=CHCH=CHCOOMe$$

Another promotion effect by I⁻ has been noted in homologation reactions of alcohols and ethers.[161-3] The first step involves formation of an alkylmetal carbonyl which probably is facilitated by I⁻. In the case of Ru the following sequence has been suggested for homologation of ethers to alcohols or esters, taking also into account the need for the presence of a proton donor such as HI:[162]

$$\left[Ru I_x(CO)_y\right]^- \left[RO(Me)H\right]^+ \xrightarrow[-ROH]{-CO} \left[Ru(Me)I_x(CO)_{y-1}\right] \xrightarrow{CO} \left[Ru(MeCO)I_x(CO)_{y-1}\right] \begin{array}{c} \nearrow \overset{MeCOOR'}{\underset{R'OH}{}} \\ \searrow \underset{MeCH_2OH}{\overset{H_2}{}} \end{array}$$

The most important industrial application of I⁻ as ligand is in the synthesis of acetic acid from methanol and CO. The rate determining step of this reaction has been shown to be oxidative addition of MeI, formed from MeOH and HI, to a $[Rh(CO)_2I_2]^-$ complex. The presence of I⁻ causes Rh to undergo oxidative addition more easily:[164]

$$MeOH \xrightarrow{HI} MeI \xrightarrow{\left[Rh(CO)_2 I_2\right]^-} \left[Me-Rh(CO)_2I_3\right]^- \xrightarrow{+CO} \left[MeCO-Rh(CO)_2I_3\right]^-$$
$$\underset{-MeCOI}{\longleftarrow}$$

It is worth noting that iodide plays different roles in the process as neutral group in oxidative addition and as charged ligand in complex anionization. Another observation is in order: the effect of iodide ion varies on passing to complexes of different metals, depending on the ease with which single steps occur. For example, methanol carbonylation to acetic acid or ester in the presence of Ir complexes involves different species, which insert CO with different ease. CO insertion can even be inhibited by excess iodide when the predominant species in solution is $[MeIr(CO)_2I_3]^-$. On the other hand, neutral $Ir(CO)_3I$ is much more nucleophilic than analogous Rh(I) complexes and oxidatively adds MeI more easily. The insertion step being slower with Ir(III) than with Rh(III), the overall rates are similar for the two metal complexes.[165]

It is interesting to note that Cl^- and Br^- can become more effective than I^- in forming the catalytic species when the latter originates from precursors that are more reactive if containing Cl or Br in place of I. For example, nickelates $[Ni_x(CO)_y]^{2-}$, which have been shown to attack aromatic halides in carbonylation reactions, form more easily from the anion $[Ni(CO)_3Cl]^-$ than from $[Ni(CO)_3I]^-$. Accordingly the rate is increased by addition of Cl^- ions to $Ni(CO)_4$.[7]

A positive effect of I^- has been reported in the Rh-catalyzed carbonylation of allyl halides and amines to pyrrolidones at $120°$ and 153 atm.[47]

3.3.4 Other ligands

While in certain cases formation of anionic carbonyl species is profitable, in other cases cationic species are more efficient. Cationic complexes can be generated by replacement of coordinated anions with less strongly coordinated ones. This is the case for cyclopentanone synthesis by carbonylation of 1,5-hexadiene and allyl compounds, mentioned in Sec. 2. The need for coordination of the two hexadiene double bonds is met by the use of "cationizing" reagents, such as large and poorly coordinated anions like PF_6^-, able to exchange with chloride.[39]

Another ligand which has been successfully employed in carbonylation processes is thiourea. It forms complexes with Ni and Pd that allow carbonylation under mild conditions. Thus allylic halides are easily carbonylated in methanol in the presence of η^3-allyl(chloro)-(thiourea)nickel,[166,166a] and acetylenes are carboxylated to unsaturated dicarboxylic esters[69] or to lactones.[167] Aceto- and benzonitrile have also been largely used in Pd chemistry (see for references [5f]).

Co carbonyl complexes with ethylenebis[3(2-pyridyl)propionate] have recently been proved to be particularly active in catalytic hydroformylation of olefins at 50° and less than 1 atm of CO partial pressure. Linear aldehydes are preferentially formed when the H_2 partial pressure is high.[168]

Phthalocyanine complexes of Ru and of other noble metals have been used to activate CO for the synthesis of isocyanates from nitro derivatives[168a] (Sec. 5.2.).

Although widely used for other reactions, η^5-cyclopentadienyl ligands so far have not offered special advantages in carbonylation reactions (see for example [168b]).

4.   SOLVENT, ION PAIR, ACID AND BASE EFFECTS

Like ligands, solvents can affect each step of the catalytic process in many ways, depending on their polarity, coordination power, acidity and basicity, stereochemistry etc.[169] The difficulty of separating solvent from other effects accounts for the limited number of studies reporting reliable correlations. We shall mention examples of overall solvent effects as well as of effects on single stages of the catalytic process.

Solvents may intervene as reactive substrates, as previously mentioned in the formation of acetic acid from methanol and CO, ethyl alcohol from methanol, CO and H_2 etc., and as reagents, either initially as sources of hydrogen or of species able to attack CO or in the last stage, when attack on acyl groups or hydrogen transfer is involved. Of particular importance is the action of water, which often is dominant even in small traces, for example in carbonylation of olefins with $PdCl_2$- $SnCl_2$,[154] where it acts as a source of H.

Solvents can also react as acids and bases and we treat these classes together with solvents, because their use as solvents is not easy to separate from that as reagents.

Even when they do not act as reagents, solvents can play a role as ligands that, although weak, are present in large excess.[170] This fact can offer some advantages, because dissociation and consequent coordinative unsaturation is easily controlled. In some cases alteration of the metal complex to obtain the catalytically active species can be caused.

As is well known, polar solvents stabilize transition states when these are more polar than ground states; thus all catalytic steps occurring with charge separation should be favored by polar solvents. In particular, reactions occurring via ion pairs can be affected by solvents in respect to both rate and selectivity.

Ion pair extraction can be effected by two-solvent systems. This phase-transfer technique has also been applied to carbonylation.

Homogeneous phase catalysts can also be heterogenized by supporting them on insoluble substrates and this technique too has proved useful for carbonylation.

The present Section will be divided in the following sub-sections, the first three regarding solvents, acid and base, and ion pair effects on different stages of the catalytic process: 4.1 Effect on formation of the reactive intermediates; 4.2 Effect on CO insertion; 4.3 Effect on metal elimination and acyl trapping; 4.4 Phase transfer techniques; 4.5 Heterogenization techniques.

4.1   Effect on formation of the reactive intermediate generated:

4.1.1 By oxidative addition or replacement reactions

The medium can exert a marked influence on formation of the catalytic species which gives rise to metal-bonded reactive intermediates by oxidative addition or replacement reactions. The type of catalytic species varies according to the wide range of possible mechanisms, from the radical to the S_N2 type.[1,1a-d,2,2a] For example, in some cases the active catalytic species derives from a neutral complex and ionic species are inactive. Thus polar solvents retard hydroformylation with $[Co_2(CO)_6(PR_3)_2]$ because they stabilize the inactive species $[Co(CO)_3(PR_3)_2]^+[Co(CO)_4]^-$.[170a]

Polar solvents appear suitable for oxidative addition of organic halides to neutral complexes, in accord with a more polar transition state.

Neutral Pd complexes able to undergo oxidative addition of aromatic halides can be generated in the presence of alcohols or amines:[31]

$$PdX_2 (PPh_3)_2 + 2CO + 2n\text{-}BuOH \longrightarrow Pd(CO)(PPh_3)_2 + (n\text{-}BuO)_2CO + 2HX$$

Oxidative addition of aromatic halides to Ni[7,171] or Co[172] carbonyl complexes is favored by alkali, probably via formation of anionic species, for example[79,173] (pyr = pyridine):

$$3 Co_2(CO)_8 + 12\ pyr \longrightarrow 2\left\{(Co\ pyr_6)^{++}\left[Co(CO)_4\right]_2^{-}\right\}$$

$$Ni(CO)_4 \xrightarrow{base} \left[Ni_5(CO)_{12}\right]^{--} + \left[Ni_6(CO)_{12}\right]^{--}$$

Interesting solvent effects have been described: alkali alkoxides in alcohols generate nickel species which are reactive with vinylic bromides and iodides and with aromatic iodides but not with aromatic bromides.[171] In aprotic dipolar solvents with $Ca(OH)_2$ as base bromides are more reactive than chlorides.[7]

A quantitative evaluation of solvent and ion pair effects on the rate of carbonylation reactions requires kinetic studies which are available only in a limited number of cases, particularly referring to carbonylation of organic halides with metal carbonyl anions.

Opposing effects of solvent and ion pairs have been described with tetracarbonylferrate and tetracarbonylcobaltate. In the first case the reaction:

$$RX + Na_2 Fe(CO)_4 \longrightarrow \left[R\text{-}Fe(CO)_4\right] Na + NaX$$

is favored by all the factors that help to generate more dissociated species in the reaction medium.[174] Thus benzyl chloride reacts 2×10^4 times faster in N-methylpyrrolidone than in tetrahydrofuran. In the first solvent the following equilibrium is assumed (S = solvent):[175]

$$(Na^+ S)_2 \left[Fe(CO)_4\right]^{--} \underset{S}{\rightleftharpoons} Na^+ + Na^+ S\left[Fe(CO)_4\right]^{--}$$

where the most dissociated species is the solvent-separated ion pair $Na^+SFe(CO)_4^{2-} + Na^+$. In tetrahydrofuran the predominant equilibrium should be:

$$Na_2 Fe(CO)_4 + Na^+ S\left[NaFe(CO)_4\right]^{-} \underset{S}{\rightleftharpoons} 2Na^+ + 2NaFe(CO)_4^{-}$$

and the most dissociated species is the contact ion pair. Addition of 10% of N-methylpyrrolidone to tetrahydrofuran is sufficient to increase the rate by a factor of about 20. Analogously, one equivalent of dicyclohexano-18-crown-6 increases the rate in tetrahydrofuran by 60 times. This behavior is that expected for a S_N2 reaction. Further support comes from inversion of configuration at C and sensitivity to steric hindrance on the substrate (Me RCH_2 R_2CH).

With tetracarbonylcobaltate, benzyl chloride gives the reaction:

$$PhCH_2Cl + NaCo(CO)_4 \longrightarrow PhCH_2Co(CO)_4 + NaCl$$

for which the following order of aprotic solvent reactivity has been established:[176] anisole > diisopropyl ether > di-n-propyl ether > tetrahydrofuran > acetonitrile > dimethylformamide, with k(anisole)/k(dimethylformamide)~1200. Alcohols, particularly those having low dielectric constant, are among the best solvents. Contrary to what is observed with tetracarbonylferrate, crown ethers and coordinating solvents in general decrease the reaction rate. This behavior has been attributed[176] to the dominant influence of the assistance to the leaving group (Cl) by cations in contact ion pairs (Na > K > Cs > Li > NR_4 in tetrahydrofuran) requiring non-coordinating aprotic solvents[176a] and by hydrogen bridges in protic solvents. A similar effect can be inferred from literature data on the reaction:[177]

$$PhCH_2Cl + NaMn(CO)_5 \longrightarrow PhCH_2Mn(CO)_5 + NaCl$$

The different behavior of carbonylferrates and cobaltates or manganates is likely to be due to the much higher nucleophilicity of the former, which has the consequence that formation of the C-M bond is more crucial than assistance in breaking the C-Cl bond. In some cases opposite counterion effects have been observed depending on the type of RX reagent. Thus $[CpMo(CO)_3]^-$ behaves with $PhCH_2Cl$ as $[Mn(CO)_5]^-$ whereas the opposite is true with BuI.[177a]

4.1.2 By H-addition to unsaturated substrates

When the reactive intermediate is formed by proton addition to unsaturated substrates, protic solvents and acid addition can help the process, as in the well-known acrylic synthesis, which is believed to involve formation of a cationic Ni complex:[178]

$$Ni(CO)_4 + HC\equiv CH + HCl \longrightarrow CH_2=CHNi(CO)_3^- Cl^-$$

Other examples are offered by the role played by Lewis acids in olefin carbonylation with Pd complexes.[179]

Hydridic complexes are easily formed from transition metal halides in the presence of bases[180] and are intermediates in hydroformylation or hydrocarboxylation:

$$PdCl_2(PPh_3)_2 + EtOH + NaOH \longrightarrow Pd(H)(Cl)(PPh_3)_2 + MeCHO + NaCl + H_2O$$

Enhanced hydroformylation rates have been observed on addition of weak bases to Co carbonyls, and inhibition with stronger bases such as triethylamine.[181] By contrast a one-step formation of alcohols has been noted with Rh carbonyls in the presence of amines, which are believed to enhance the hydridic character of an intermediate Rh hydride.[181a]

It has been observed that when $Fe(CO)_5$ dissolves in aqueous KOH solution at 100° and at 15 atm of ethylene and 22 atm of CO to give hydroformylation to propionaldehyde and propyl alcohol, the anion $HFe(CO)_4^-$ so formed:[50]

$$Fe(CO)_5 + OH^- \xrightarrow{-CO_2} HFe(CO)_4^-$$

does not interact with olefins (although it is able to reduce aldehydes) unless the pH is reduced from 12.0 to a lower value (10.7). Under these conditions the reactive species is $H_2Fe(CO)_4$,[182] which reacts with olefins and CO according to patterns generally accepted for hydroformylation. $Ir_4(CO)_{12}$ and $Rh_6(CO)_{16}$ are also attacked by mild bases with formation of metal carbonyl hydride anions,[183] which are effective in hydroformylation with CO and H_2O.[182]

With pyridine at 150° and ca. 60 atm CO, $Rh_6(CO)_{16}$ gives rise to partial hydrogenation of the aromatic nucleus, followed by hydroformylation (and/or denitrogenation as a secondary path), condensation and hydrogenation, leading eventually to the following main product:[184]

This is an instance of a rather complex action of a base.

4.1.3 By CO-H cluster formation

We have seen that in certain instances the separation of ion pairs is critical to the formation of the reactive species. We mention other examples of solvent and ion pair influence on formation of Rh clusters, believed to be the active species in the synthesis of ethylene glycol and methanol from CO and H_2, in view of the special interest in this industrially important reaction. Solvents having high dielectric constant (e.g. sulfolane) and good coordinating power for cations (e.g. tetraglyme, crown ethers) allow the highest reaction rates to be attained.[185] The reaction is generally carried out in the presence of promoters such as salts or alkali or alkaline earth metals and amines. Cesium benzoate and pyridine are particularly effective. The action of these promoters is quite complex, because it depends not only on the type of cation but also on its ratio to Rh, on the basicity of the counter-ion, and on the complexing power of the solvent. For example, selectivity for ethylene glycol, using tetraglyme as solvent, increases with Cs^+ present until a maximum is reached (70-80%) for a Cs/Rh ratio of 1/6. Beyond this ratio selectivity decreases until for a 2/6 ratio methanol is formed selectively. The inhibiting action of Cs^+ towards glycol formation, however, decreases with increasing basicity of the associated anion and with increasing coordinating power of the solvent. If crown ethers are employed as solvents, high selectivity and rate of formation of ethylene glycol are attained and the inhibiting effect of Cs^+ is less pronounced.[185] A tentative model to explain the main feature of this complex reaction assumes that the Rh complex taken is transformed by the Lewis base and/or by alkali salts in the presence of CO/H_2 into anionic species, which give rise to glycol in the best way when they are held separate from the counter-ion as $[Rh]^- M^+$. The Lewis base, if used beyond a certain limit, has an inhibiting effect, probably because it deactivates the separate ion pairs by some association mechanism not yet clearly understood.[185] Phosphine oxides, which are solvents having both high complexing power for cations and good polarity, have proved best for ion pair separation. Binary solvents (one component of high dielectric constant and the other of high coordinating power), such as tetraglyme and sulfolane are also effective to this end. Cations like Cs^+ also give the least interaction with the Rh carbonyl anion. Interestingly, it has been noted that formation of strong complexes with crown ethers is not sufficient to obtain the desired separation of the ion pair, because interaction with $[Rh]^-$ anion may still be possible. Thus 15-crown-5 fits Na^+ but leaves the possibility for axial complexation by the anion, whereas the same crown does not fit K^+ and Cs^+, which are too large, but form sandwiches wherein these metals are inaccessible.[185] Recent patents disclose the use of other promoters to obtain increased amounts of other alcohols. Thus triethanolamine vanadate, associated with ammonium benzoate, gives propyl alcohol as by-product, together with ethylene glycol, methanol being the major and butyl alcohol and glycerol being minor products.[98] Glycerol can be formed in substantial amounts at high CO/H_2 pressure (1000-4000 atm) in the presence of added methanol.[186] Propylene glycol can also be formed in the presence of aluminum compounds as promoters.[187]

4.2   Effect on CO insertion

The CO insertion step is subject to different and often opposite requirements for the formation of the intermediate undergoing insertion. This is true also for solvent and ion pair effects and is clearly shown by contrast with the process of oxidative addition of organic halides mentioned in Sec 4.1.1 above. The migrating group should be able to move intramolecularly to coordinated CO and solvents must allow this process. CO can be activated by Lewis acids and the reaction products can also be stabilized by Lewis acids or by tight ion pairs between the carbonylic oxygen and an alkaline cation. Studies on stoichiometric carbonyl insertion reactions have shown the dual role of Lewis acids in activating the carbonyl group towards insertion and in stabilizing the resulting acyl group, e.g.[188]

$$
RMn(CO)_5 \xrightarrow{\ AlBr_3\ } R-\overset{\overset{\textstyle O}{\|}}{C}-Mn(CO)_4 \cdots Al\begin{matrix} Br \\ -Br \\ Br \end{matrix}
$$

A solvent-assisted migration was postulated[189] to explain promotion of carbonyl insertion in methylmanganese pentacarbonyl by polar solvents and among the latter by methanol more than by nitromethane.

The following stoichiometric reaction takes place under the influence of a tight ion pair:[2]

$$Na^+ RFe(CO)_{\overline{4}} \; \underset{}{\overset{CO}{\rightleftharpoons}} \; R-\overset{\overset{\textstyle ONa^+}{\|}}{C}-Fe^-(CO)_4$$

and the solvent effect is the opposite to that observed in formation of $[RFe(CO)_4]^-$ (Sec. 4.1.1). Thus the reaction is faster in tetrahydrofuran than in dipolar aprotic solvents such as N-methylpyrrolidone or hexamethylphosphoramide and it is sufficient to add 10% of N-methypyrrolidone to tetrahydrofuran to decrease the rate 100 times. A 150-fold decrease is also caused by addition of a crown ether.[2,174] The cation associated with the alkyl complex also favors tight ion pair formation $[Li > Na \gg (Ph_3P)_2N^+]$. These examples of stoichiometric CO insertion show how the overall effect of solvents, acids and bases, and ion pairs can be modified by opposite effects in different stages of the catalytic process. In practice it is not easy to effect the insertion step by Lewis acids without adversely affecting other stages of a catalytic reaction. The fact that alcohols are good solvents for hydroformylation can possibly be explained by assistance to insertion by hydrogen bonding to the carbonylic oxygen.

Base effects can also be at work in certain types of insertion. In olefin carbomethoxylation with the Pd-CuCl$_2$ system it has been shown[67] that alkaline salts of carboxylic acids shift the reaction from methoxyesters to diesters (from 70% of methoxyester to 94% of diester with sodium isobutyrate in the case of 1-hexene):

$$RCH=CH_2 \; + \; \overset{CO,\,MeOH}{\underset{}{\diagup\diagdown}} \; \begin{array}{l} RCH(OMe)CH_2COOMe \\[1em] RCH(COOMe)CH_2COOMe \end{array}$$

The result has been attributed to base catalysis of formation of the palladium bonded carbomethoxy group:

$$\|-\underset{\underset{CO}{|}}{\overset{|}{Pd}}-X \; \underset{}{\overset{MeOH}{\rightleftharpoons}} \; \|-\underset{\underset{COOMe}{|}}{\overset{|}{Pd}} \; + HX$$

Alcohols, amines or other hydrogen-active compounds can be used in hydrocarboxylation reactions with PdX$_2$-tertiary phosphine catalysts. A possible pattern implies reaction of O-H, N-H or of other bonds with the metal, followed by CO insertion. Alternatively, hydrogen-active compounds might attack CO directly, without previous interaction with the metal [routes (a)]. The coordinated carbalkoxy or carbamido or similar group should then attack olefins to give the same product that would be formed by the classical hydrocarboxylation route (b):

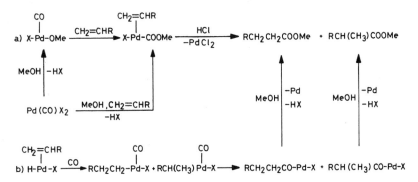

It has been shown[190] that on simply passing from methanol to butanol in the $PdCl_2$-triphenylphosphine-catalyzed reaction different complexes are obtained from the reaction mixture, $Pd(COOMe)(Cl)(PPh_3)_2$ in the first case and $Pd(COPr-n)(Cl)(PPh_3)_2$ in the second. This probably corresponds to a change of mechanism from route (a) to route (b). Thus alcohol properties play an important role in this type of reaction.

4.3   Effect on metal elimination and acyl trapping

As already mentioned, solvents can also act as reagents in the final step of a catalytic process, detaching the acyl group from the metal by nucleophilic attack,[3,4c,5g] or by accepting[191] or donating[93] hydrogen, as in the following examples:

$$RCO-Ni-X \xrightarrow{R'OH} RCOOR' + HX + Ni$$

$$RCO-Co(CO)_4 + \text{/\\/} \xrightarrow{NR'_3} RCO\text{/\\/\\} + R'_3NH^+Co(CO)_4^-$$

$$MeOH + 3CO + H_2O \xrightarrow{cat.} MeCH_2OH + 2CO_2$$

Solvents can also stabilize the complex resulting from the elimination step. When the $Pd(OAc)_2$-2PBu_3$-catalyzed carbonylation of butadiene (Sec. 3.2.2) to linear esters at $110°$ and ca. 50 atm of CO is carried out in heterocyclic basis (pyridine, quinoline, etc.) as solvents, a better yield (60%), a higher than 95/5 ratio of 3,8-nonadienoic ester to 3-pentenoic ester, and good ester/oligomer ratio are obtained, with greater complex stability. Yields improve if Pd salts are solubilized in tertiary amines having pKa$<$7. Apparently these solvents do not block coordination sites, but seem to stabilize Pd after elimination.[151]

The use of a melt of tetraalkylammonium trichlorostannate to disperse $PdCl_2(PR_3)_2$ catalysts has been shown to offer further advantages in addition to those connected with catalyst modification by the $SnCl_3^-$ ligand, cited in Sec. 3.2.2. Improved catalyst stability, higher concentration of reagents and ease of separation of catalyst from product are obtained.[151]

Another less-known solvent effect is the blocking of reactive groups, thus allowing selective reactions. Acyl groups can be trapped by aldehydic or ketonic solvents, for example:[192-3]

$$4PhI + 2Ni(CO)_4 \longrightarrow PhC(OCOPh)=C(OCOPh)Ph + 2NiI_2 + 4CO$$

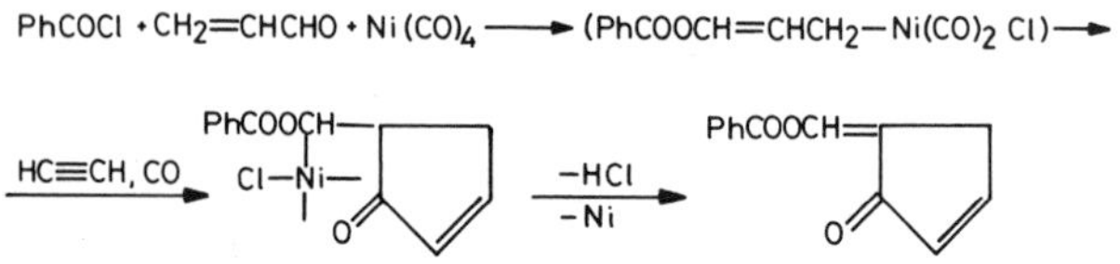

$$PhCOCl + CH_2=CHCHO + Ni(CO)_4 \longrightarrow (PhCOOCH=CHCH_2-Ni(CO)_2 Cl) \longrightarrow$$

Other reactions could be performed successfully in ketonic solvents, without incorporation of the latter in the product. The ketone action was attributed to a reversible trapping of acyl groups,[193] for example:

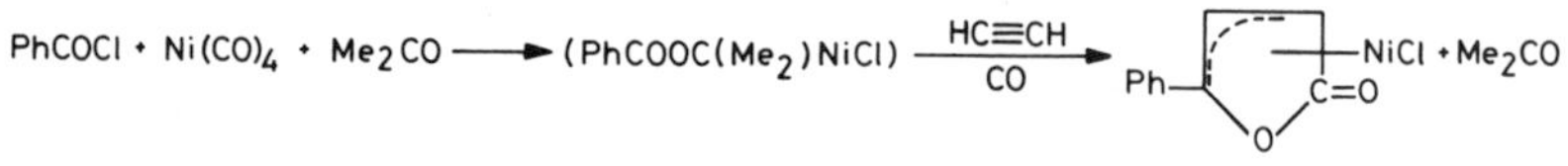

A recent report shows that, if hydroformylation is carried out in benzil as solvent, a high yield of the hydrogenolysis product of the complex resulting from trapping acyl groups on benzil is obtained[194] (HRh = Rh hydride complex):

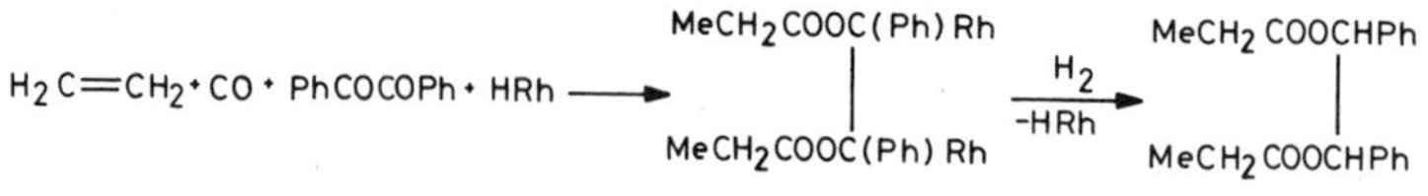

4.4   Phase transfer techniques

The new phase transfer techniques [195-8] have been applied successfully to carbonylation reactions. Organic halides are carbonylated under mild conditions in a two-phase system, containing a water soluble quaternary ammonium salt and sodium hydroxide in the presence of Co, Ni, or Pd catalysts:[199-202]

$$RX + CO \xrightarrow[\text{NR}_4'\,X\,,\,\text{cat.}]{\text{Aq. NaOH / solvent}} RCOONa$$

An advantage of the phase transfer carbonylation process lies in the improved catalytic activity. This is due to the fact that the catalyst works in the organic phase whilst the acid formed is continuously extracted by aqueous sodium hydroxide. Phenylacetic acid is obtained from benzyl chloride with $PdCl_2(PPh_3)_2$ as catalyst in the ratio of 4000 mol per mol of complex.[199] Another advantage offered by phase transfer conditions is the possibility of generating more easily the catalytically active species which is extracted by the organic phase. Thus the $[Co(CO)_4]^-$ anion used as catalyst in carbonylation of organic halides can easily be generated by phase transfer.[201] Analogously, anionic clusters are formed from $Ni(CO)_4$. These clusters, of formula $[Ni_x(CO)_y]^{2-}$,[180] are involved in carbonylation of allylic halides.[202]

The use of phase transfer techniques also makes it possible to influence the selectivity of carbonylation reactions. Thus p-bromobenzoic acid and 3,5-dichlorobenzoic acid are selectively obtained from p-dibromobenzene and 1,3,5,-trichlorobenzene, respectively, in the presence of $PdCl_2(PPh_3)_2$ as catalyst.[199]

Another interesting fact is that, while in homogeneous systems allyl halides are carbonylated with insertion of acetylene to form hexadienoic acids, under phase transfer conditions only butenoic acids are formed, probably owing to formation of anionic carbonylic complexes with little tendency to replace CO with acetylene.[202]

Stabilization of acylmetal complexes relative to homogeneous phase conditions is another important feature of the phase transfer technique. Insertion of a new molecule of CO,[202] acetylene or dienes[203-4] is easily obtained:

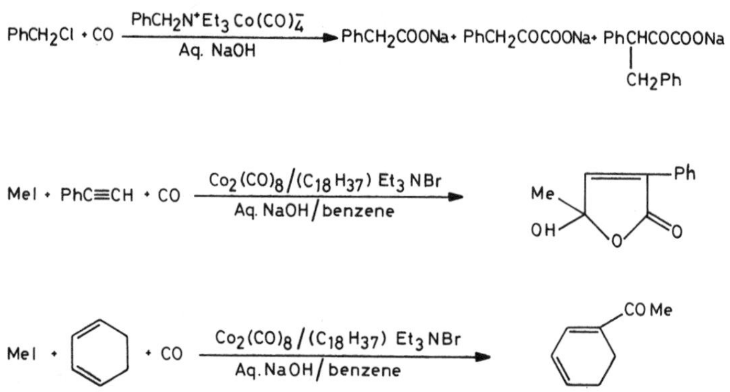

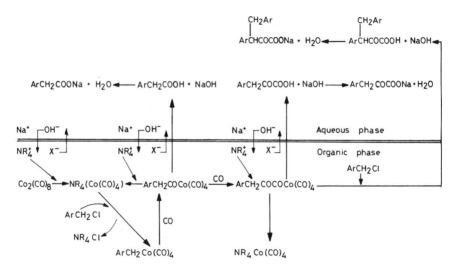

The mechanism assumed for phase transfer carbonylation is shown in the Scheme (for $[Co(CO)_4]^-$ and benzyl halides[200]):

A semicatalytic carbonylation (2 mol per mol of complex) of alkyl halides to ketones under phase transfer conditions has been reported:[205]

$$RBr \xrightarrow[\text{Bu}_4\text{NBr}]{\text{Fe(CO)}_5, \text{NaOH}} RFe(CO)_4^- \xrightarrow{R'I} RCOR'$$

4.5 Heterogenization techniques

Although the present review is not concerned with heterogeneous catalysis, the heterogenization technique on polymers (heterogenization on metal oxides, zeolites and inorganic supports will not be considered here) is worth mentioning, because it generally preserves the main features of homogeneous catalysis by transferring homogeneous complexes to polymers possessing suitable groups to "anchor" them.   General advantages lie in easier catalyst recovery and in higher selectivity towards substrates.[206] In some cases protection against adverse influence of water has been observed.[207]

Phosphorus-containing compounds have been supported on different materials.   Triphenyl   phosphite   on   carbon   has   been   used   for hydroformylation with Rh.[208] Supporting phosphine on polystyrene (P) to form (P) PPh_2 gives higher selectivity than that observed with the corresponding homogeneous complexes.   Thus a supported catalyst (P) $PPh_2)_3Rh(CO)$ with a high ratio of supported phosphines to Rh (19:1) at $140°$ and 16 atm gives a 16:1 ratio of linear to branched aldehyde in hydroformylation of 1-pentene.   The corresponding value for the homogeneous catalyst (same PPh_3:Rh ratio) was found to lie between 2 and 3.[209] The result has been attributed to high ligand concentration in the polymer swollen beads. In this way the homogeneous phase equilibrium:

$$(PPh_3)_2 RhH(CO)_2 \rightleftharpoons PPh_3 RhH(CO)_2 \cdot PPh_3$$

should be displaced to left, thus favoring the associative pathway, which according to Wilkinson[140] gives a higher selectivity for linear product.

The first indication that a polymer supported homogeneous catalyst can change its catalytic activity comes from recent work[210] showing that the cyclopentadienylcobalt unit, supported on a styrene-divinylbenzene polymer, is active as a Fischer Tropsch catalyst. By contrast, the same unit is not active in homogeneous systems.

Asymmetric hydroformylation on supported Rh(I) carbonyl complexes, containing the optically active DIOP ligand, has been achieved. An optical yield of 28%[211] has been reached with 2-butene (27% in homogeneous phase).

In addition to the just mentioned heterogenization, other techniques have   allowed   improvement   in   catalyst   recovery.   The   use   of aminophosphines,[212] which are extractable with acids, and of sulfonated phosphines, which are water soluble,[213] has been described.

Reverse osmosis through polyimide membranes has also been shown to be a useful way to separate catalysts from carbonylation products.[214]

5. OXIDANTS AND REDUCTANTS

Catalytic processes often benefit from the presence of reagents able to push to the right unfavorable steps and to restore metals to their original oxidation state. Oxidants and reductants have been used to this end.

5.1 Oxidants

Oxidative carbonylation is based on the reaction of active hydrogen compounds such as alcohols or amines with CO to form metal-coordinated carbalkoxy or carbamido groups,[43] which can further react in many ways. Oxidative carbonylation generally can proceed if hydrogen is transferred to suitable acceptors. In the simplest case two molecules of methanol and two of CO combine to give dimethyl oxalate.[215] The two hydrogen atoms from methanol are transferred to oxygen or to a reducible compound:[215-6]

$$\text{2MeOH + 2CO} \xrightarrow[\text{oxidant}]{\text{Pd cat.}} \text{MeOOCCOOMe}$$

Quinones[217] or Cu(II) salts can be used as oxidants. Dialkylcarbonates can also be formed.[218]

Ethylenethiocarbonate has been obtained from $HSCH_2CH_2OH$ with $Ni(CO)_4$, CO and O_2.[218a] Unsaturated substrates such as acetylene, ethylene, or butadiene can be inserted between the two carboxylic groups to give dimethylmaleate, dimethylsuccinate, and dimethyl β-dihydromuconate, respectively.[67-9,219,219a]

The use of oxygen to reoxidize Cu salts generally causes problems because it gives rise to water, which to its turn can attack CO to give CO_2; thus dehydrating agents such as acetals or orthoformates have been added.[220-1] Cu(II) salts in the presence of ammonia, amines or pyridines in methanol appear to be suitable, probably because they give rise to Cu(OMe)Cl.[222,222a] The latter has been shown to be very effective.[223] The type of complex on which oxidation takes place is not known. Mixed Cu-Pd complexes with halide bridges are a possibility.[221]

There is also uncertainty about the way by which hydrogen is transferred to such organic substrates as quinones.[217] These coupled oxidation-reduction processes have a high significance in relation to analogous enzymatic processes as well. An interesting observation has been made on the effect of the oxidation potential of different quinones on yield of dimethyl oxalate from methanol and CO.[217] Only quinones with high oxidation potentials are able to cause formation of dimethyl oxalate of 45° and 100 atm of CO with palladium acetylacetonate as catalyst. Thus benzoquinone (0.72V) gave a 61% yield, tetrachloroquinone (0.68V) 41%, and naphthoquinone (0.48V), tetramethylquinone (0.47V) and anthraquinone (0.40V) gave no dimethyl oxalate. Secondary reactions were favored by low oxidation potentials, however, and quinones disappeared as well. In the presence of water the oxidation potential required for secondary reaction, in particular for CO_2 formation, is further lowered. This points to the importance of maintaining a delicate balance between oxidizing agent and solvent to achieve an efficient carbonylation process.

Oxidative carbonylation of amines to ureas at room temperature, a reaction which is catalyzed by both Cu and Pd, has been found to proceed much better in the presence of the Pd(II)-Cu(II) system in pyridine.[224] A cooperative action between Pd(II) (able to coordinate CO whereas Cu(II) is not) and Cu(II), which oxidizes a Pd(II) carbamoyl complex, has been suggested.[224]

More recently new oxidative agents have been proved effective, e.g. nitric and nitrosoderivatives.[225-6]

Oxidative agents have also been successfully employed to promote carbonylation of olefins and acetylenes, followed by hydrogen elimination with formation of unsaturated acids or esters:[69,227-9]

$$CH_2=CH_2 \ + \ CO \ + \ H_2O \ \xrightarrow{\text{Pd cat.}} \ CH_2=CHCOOH$$

$$PhCH=CH_2 \ + \ CO \ + \ MeOH \ \xrightarrow{\text{Pd cat.}} \ \begin{cases} PhCH=CHCOOMe \\ PhCH(OMe)CH_2COOMe \\ PhCH(COOMe)CH_2COOMe \end{cases}$$

$$PhC{\equiv}CH \ + \ CO \ + \ MeOH \ \xrightarrow{\text{Pd cat.}} \ PhC{\equiv}CCOOMe$$

Combination of the oxidative action with the appropriate ligand, solvent, and base optimizes the effect. Thus dimethyl phenylsuccinate is the main product of the reaction of styrene with CO and methanol under slight CO pressure, but methyl cinnamate predominates using a large excess of $CuCl_2$ + MeCOONa in respect of Pd, with added $MgCl_2$ and under lower than atmospheric CO pressure.[228]

An interesting effect of oxidizing agents on CO insertion is worth noting here, although it refers to a stoichiometric process, because it indicates that the insertion process itself can be affected by oxidizing agents. Some alkylmetal carbonyl complexes undergo easier CO insertion on oxidation with Ce(IV) ion in methanol at room temperature.[230] Oxidatively induced nucleophilic cleavage of acylmetal complexes was also observed with Ce(IV).   Nitroderivatives have also been reported to induce insertion.[231]

5.2   Reductants (other than H_2)

We do not consider hydrogen here, having treated CO reduction, hydroformylation and homologation in preceding Sections.   Other reducing agents can affect the carbonylation process at different stages, for example at the stage of formation of the intermediate undergoing carbonylation, as in the case of CO, which acts as a reducing agent of nitro- or nitroso-derivatives, possibly forming a metal-coordinated nitrenic species.   The latter then reacts with CO to give isocyanates[232-3] or ureas.[27,234]

The reactive intermediate can also be trapped with CO and diphenylacetylene in the presence of $Rh_6(CO)_{16}$ at $150°$ and 150 atm of CO. From nitrobenzene the compound

is obtained.[234a]   Powdered iron or manganese-iron alloys can be used as reducing agents in many catalytic carbonylation processes of allylic halides with Ni or Co catalysts.   The powdered metal reduces nickel or cobalt halides to a low-valent state, thus enabling these metals to form the active carbonylic species.   Iron powder has also been used to push forward the catalytic carbonylation process leading to m-cresol from methallyl chloride, acetylene and CO, probably by helping cyclization of the intermediate 5-methyl-2,5-hexadienoylnickel halide.[4c,235]

The replacement of hydrogen with hydrosilane in hydroformylation with Co, Ru, and Rh complexes has been described.[236-8] Although the overall effect of hydrosilanes $HSiR_3$ is that of a reducing agent like hydrogen, it should be attributed to a combination of different reactions, based on the ability of hydrosilanes to attack metal-metal bonds, for example:

$$HSiEt_2Me + Co_2(CO)_8 \longrightarrow HCo(CO)_4 + MeEt_2SiCo(CO)_4$$

or acylmetal complexes:

$$RCOCo(CO)_3 + HSiEt_2Me \longrightarrow RCO-Co(H)(SiEt_2Me)(CO)_3 \longrightarrow RCHO + MeEt_2SiCo(CO)_3$$

and to bind preferentially to oxygen.   The following representative reactions:[236]

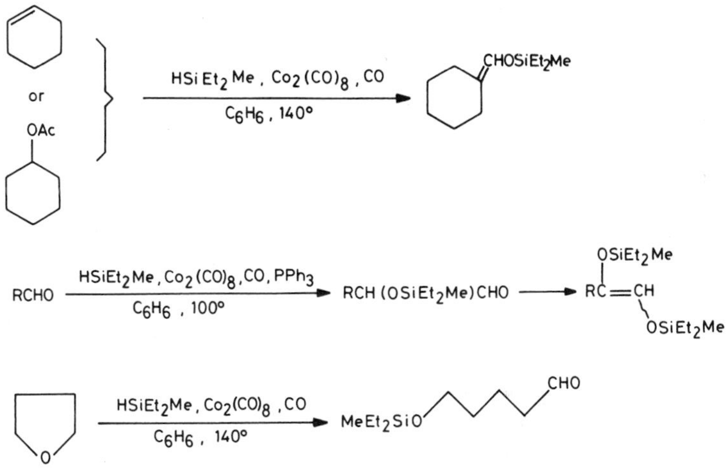

appear to involve formation of $HCo(CO)_4$ and $MeEt_2SiCo(CO)_4$ from $Co_2(CO)_8$ as shown above and, in the first case, $MeEt_2SiCo(CO)_4$ addition to the aldehyde initially formed:

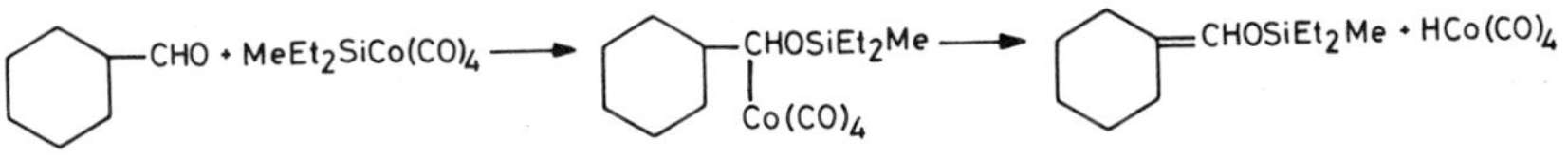

CO insertion instead of elimination can convert the addition intermediate into an aldehyde with one carbon atom more:

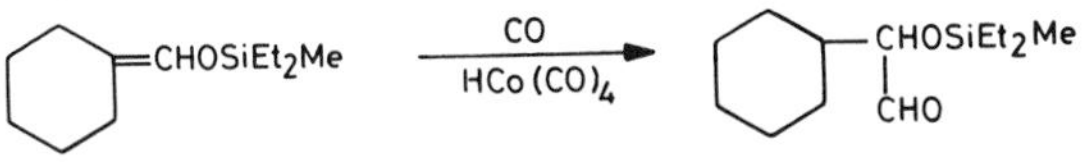

It is interesting to note that the use of hydrosilanes also modifies the regioselectivity of CO attack at the aldehydic group;[236] when $HCo(CO)_4$ adds to RCHO in the absence of hydrosilane the oxygen atom is linked to Co and CO insertion gives formates:

$$RCHO + HCo(CO)_4 \longrightarrow RCH_2OCo(CO)_4 \longrightarrow RCH_2OCOCo(CO)_3 \xrightarrow{H_2} RCH_2OCHO + HCo(CO)_3$$

Carbonylation of cyclic ethers is also likely to involve insertion into the Si-Co bond:

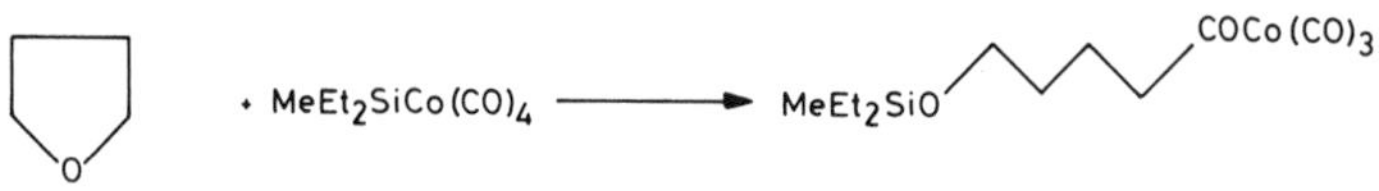

The results obtained with hydrosilanes point to the wide possibilities offered by changing reducing agents in reductive carbonylation.

Other ways to couple an organometallic compound as stoichiometric reducing agent with a transition metal complex acting as catalyst have been reported. Organomagnesium compounds give ketones and alcohols on reflux in diethyl ether with organic halides under CO in the presence of NiX_2R_2 complexes,[239] for example (bpy = 2,2'-bipyridine):

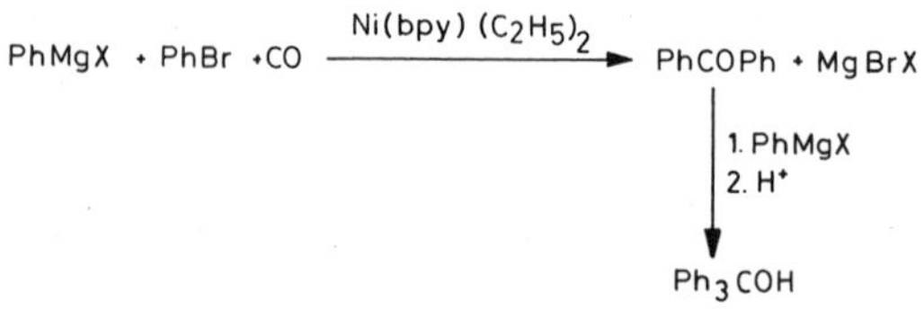

$$PhMgX + PhBr + CO \xrightarrow{Ni(bpy)(C_2H_5)_2} PhCOPh + MgBrX$$

$$\downarrow \begin{array}{l} 1.\ PhMgX \\ 2.\ H^+ \end{array}$$

$$Ph_3COH$$

Organotin compounds react with alkyl halides and CO in the presence of $PhPdI(PPh_3)_3$ and hexamethylphosphoramide to give reductive elimination of ketones:[240]

$$PhI + CO + SnBu_4 \xrightarrow[30\ atm,\ 120°]{Pd\ cat.} PhCOBu + SnIBu_3$$

6.   MORE RECENT DEVELOPMENTS

A large amount of work on carbonylation appeared after completion of the main text. Owing to the limited space available only some papers are reported and the reader is referred to the literature cited therein.

6.1   Substrates

Hydroformylation has been reviewed,[241] as has the use of several olefinic substrates for the synthesis of important natural products by hydroformylation with Rh complexes.[242]

Substituents with increasing σ-values raise selectivity in hydroformylation of styrene to 2-phenylpropanaldehyde when a Rh complex with tertiary phosphines is used. The opposite effect is observed in the absence of phosphines.[243]

Like aromatic and vinylic halides,[243a] allylic halides can be hydroformylated with Pd complexes.[243b] Allylic carbonates are an efficient source of allyl groups, which are generated by Pd-catalyzed decarboxylation.[243c]

Annulation and carbonylation in sequence, to form condensed α-methylene-γ-lactones stereospecifically, can be accomplished via 2-bromoallylnickel complexes.[244] Annulation can also be accomplished by intramolecular alkoxypalladation of alkenes of appropriate length, followed by carbonylation.[245] Lactones have been prepared from haloalcohols with Pd catalysts.[246]

Cyclopropanes can be carbonylated to ketones with Rh catalysts.[247] See also Ref. 247 for cyclobutanes.

Carbonylation of acetylenic substrates[49-53] continues to afford a variety of different products[248-9] depending on catalysts and conditions used. Formation of a phenylcyclobutenedione from two molecules of CO and one of phenylacetylene on Ni throws light on the mechanism.[250]

Sequential insertion of diphenylacetylene and CO into the Rh-COOEt bond, terminated by protonation at 125°, gives α,β-diphenyl-γ-ethoxy-butenolactone.[251] At higher temperature (220°) and in the presence of ethylene, sequential insertion of ethylene, CO, diphenylacetylene and CO into the Rh-H bond, terminated by H-transfer from ethanol, gives a γ-ethyl-γ-butenolactone.[252]

Sequential nucleophilic attack of secondary amines on Pd-coordinated olefins and CO insertion have been reported.[253] Pd-catalyzed sequential olefin and CO addition to the CCl_3 radical (from CCl_4) has also been obtained.[254] Another reaction sequence, leading to carboxylic acid anhydrides, has been performed on Pd catalysts using organic halides, norbornene or strained olefins, and carbon monoxide.[255] An inverted order of insertion (CO before norbornene) takes place in a manganese complex.[255a]

Cyclocarbonylation of strained olefins to condensed cyclopentenone and cyclohexenone systems by means of Ni catalysts has been described.[256] Similar reactions can be carried out with $Co_2(CO)_8$ on norbornene, alkynes and CO, to give cyclopentenones condensed with norbornene and norbornadiene.[257] Alkylidenecyclopentanones, condensed with strained olefins, have been obtained in a Pd-catalyzed reaction. Some products can undergo retro-Diels-Alder reaction to prostaglandins or their synthons.[257a] Ring formation has been achieved with special chelating diynes and a $PdCl_2$-thiourea catalyst at room temperature. Carbon monoxide induces cyclization and forms an exocyclic bond without entering the ring.[257b]

Aromatic ring formation from carbene-chromium complexes, alkynes and CO[258-9] offers an interesting model for catalytic reactions.

A one-step synthesis of bicyclic β-lactams from aziridines has been achieved on a Pd(0) catalyst.[260] For α-methylenelactams see 260a.

Pd-catalyzed cyanocarbonylation of aryl iodides[261] and formation of chlorobutyl esters from organic halides, CO and THF have been reported.[262]

6.2 Ligands

Ligand effects in metal carbonyl substitution processes have been discussed.[262a]

The interaction of coordinated CO and H_2 has been the object of several studies pertaining to the water gas shift reaction.[263] Support for the original Fischer-Tropsch hypothesis of methylene polymerization[126] comes from isolation of a carbene intermediate from reduction and deoxygenation of an acetylzirconium complex,[264] from analogy between product distribution in diazomethane pyrolysis on transition metal salts and in the Fischer-Tropsch reaction,[265] and from isolation of propene from pyrolysis of bimetallic Rh complexes, containing both Me and bridging CH_2 groups.[266] Metal-coordinated carbynes have been shown to be reactive with CO.[266a-c] For interpretations of the Fischer-Tropsch reaction see References 267-8.

The role played by metal-oxygen bonds in the synthesis of methyl and ethyl alcohol from CO and H_2 on Co catalysts has been stressed and interesting mechanistic interpretations have been put forward.[269] The possible intermediacy of the hydroxy derivative of methylidynetricobalt-nonacarbonyl in the cobalt carbonyl catalyzed hydrogenation of CO to methanol and ethylene glycol has been pointed out.[270-1] Reduction of CO to coordinated CHO has been achieved, for example on Ir and Mn model compounds[271a,b] and on rhodium octaethylporphyrin hydride.[271c] The mechanism of reductive elimination has been discussed.[272]

Formation of C_2 from C_1 via reduction of coordinated CO to oxygenated species finds support in several studies.[273-5] Ruthenium catalysts have been shown to be effective, especially in the presence of I^- as promoter.[274] A recently proposed mechanism via coordinated CH_2O[276] also explains homologation reactions. Homologation of methanol to acetaldehyde (and ethanol) has been achieved with $[FeCo_3(CO)_{12}]^-$ activated by MeI[277] or by $RhCl_3$ activated by I^-.[277a] Ethanol is obtained from methanol with another cluster, $[RuCo_3(CO)_{12}]^-$.[278] Acids can be homologated with Ru complexes.[279]

Ligand and substrate effects on the insertion reaction[35,128a] have been clarified by further studies.[280-1] The complex $PtX(R)L_2$ [X = halide, R = migrating group, L = ligand (tertiary phosphine)] inserts CO to give the halide-bridged dimer $[PtX(COR)L]_2$ by two different paths, the former involving a pentacoordinated intermediate $PtX(CO)(R)L_2$, the latter via a square planar complex $PtX(CO)(R)L$. Only the isomer having R and L trans to each other undergoes insertion. Electron-releasing groups in R favor insertion. Octahedral methyliron carbonyl complexes insert CO via a pentacoordinated intermediate.[282] A model for olefin and CO insertion into a Ru hydride has been reported.[282a] 3,3-Dimethyl-2-butene hydroformylates stoichiometrically with $HCo_3(CO)_9$ and $HCo(CO)_4$, the alkyl H coming from the former and the aldehydic H from the latter hydride.[282b] A review on insertion reactions with transition metal complexes has appeared.[283]

The function of $HCo(CO)_4$ in hydroformylation[284] and the chemistry of alkylcobalt tetracarbonyls[285] have been reviewed. A recent study[240a] on hydroformylation of 1-dodecene shows that regiospecificity, as measured by the linear-branched product ratio, varies inversely to increased ligand basicity.

The use of the $SnCl_3$ ligand in Pt-catalyzed hydroformylation has been described.[286] The system $Pt(PPh_3)(CO)Cl_2/SnCl_2.2H_2O$ (1:2) gives a high normal to isoaldehyde ratio and appears to behave differently from the $Pt(PR_3)_2Cl_2/SnCl_2.2H_2O$ (1:5) system previously mentioned.[154]

Catalysis of cyclocarbonylation of acetylenes by means of [_trans_ - $Pd(PPh_3)_2(SnCl_3)(L)]^+$ (L = ligand) has been extended to many hydroxyacetylenic substrates which appear to form a carbalkoxo complex with Pd, followed by alkyne insertion and protonation. With the same catalyst terminal alkynes mainly give branched carboxylic esters, whereas $HPd(PPh_3)_2SnCl_3$ gives linear esters predominantly.[287] On the other hand, catalysis of olefin hydroformylation by a hydride mechanism involving the $HPd(PPh_3)_2SnCl_3$ complex has been confirmed by the isolation and crystal structure determination of $n-PrCOPt(PPh_3)_2SnCl_3$.[288] Other interesting effects caused by the $SnCl_3$ ligand can be found in the hydroformylation of internal olefins to form linear aldehydes with $[PtCl(CO)(PPh_3)_2]^+/SnCl_2$[289] and in the enhancement of hydroformylation rate with $PtCl_2(DIOP)/SnCl_2$ complexes.[290] A reinterpretation of the role of the $SnCl_3$ ligand in homogeneous catalysis has appeared.[291] An investigation of the species formed in solution from _cis_-$PtCl_2(CO)(PEt_3)$ and $SnCl_2.2H_2O$ has established that the first step is $SnCl_2$ insertion into the PtCl bond. This step can be blocked in $CHCl_3$ as solvent. In more polar solvents several rearrangements occur, leading to both cationic and anionic complexes such as $[Pt(SnCl_3)_5]^{3-}$ and $[PtCl(CO)(PEt_3)_2]^+$.[291a]

Other ligands such as nitrogen heterocycles have found some applications.[241] (The action of nitrogen bases was considered in Sec. 4.)

A fairly high asymmetric induction has been reported for hydroformylation of vinyl acetate with Rh catalysts complexed with DIOP-type ligands (51% ee).[292] For a recent review on asymmetric hydroformylation see 292a. The highest optical yield obtained so far in hydroformylation of styrene with $PtCl_2$-$SnCl_2$ and (-)-2,2-dimethyl-4,5-bis(5-H-dibenzophosphol-5-ylmethyl)-3,3-dioxolane reaches ca 73%ee.[292b] Asymmetric hydroesterification of vinylaromatics to α-arylpropionates (anti-inflammatory agents) has been achieved with 42-52% ee at room temperature and atmospheric pressure, using Pd complexes with (+)-neomenthyldiphenylphosphine in the presence of trifluoroacetic acid.[292c] New anionic cobalt carbonyl complexes such as $[NCCH_2Co(CO)_3COOCH_3]^-$ have been obtained[292d] and shown to be exceedingly active in carbonylation of aliphatic and aromatic halides.[292f]

Some organic reactions, among them carbonylations, with rhodium carbonyl clusters have been reviewed.[292f]

6.3 Solvents

Evidence has been given for solvent (α-substituted THF) coordination to the metal center in migratory insertion of the methyl group of $CpMo(CO)_3Me$.[293] Protonic acids accelerate alkyl migratory insertion of CO into Mn-CO bonds.[294] The use of molten phosphonium salts to dissolve such ruthenium carbonyl catalysts as $[HRu_3(CO)_{11}]^-$ at 220° and 430 atm of CO/H_2 (1:1) allows formation of ethylene glycol, methanol and ethanol.[295]

The association of alkoxides with hydrides and cobalt generates a very active catalyst for carbonylation of aromatic halides at atmospheric pressure.[295a]

Phase transfer has been applied to the synthesis of ketoacids from methyl iodide, alkynes and CO in presence of cobalt and ruthenium carbonyl complexes, the former promoting carbonylation and the latter hydrogenation.[295b] Hydroacylation of allenes[296] and carbonylation of primary benzyl halides with $Co(CO)_3NO$ or $Fe(CO)_5$/bases as catalysts can be performed under phase transfer conditions.[297]

Secondary benzyl halides are also efficiently carbonylated either under phase transfer or homogeneous conditions to give α-arylpropionic acids and, for the first time, arylalkylpyruvic acids, resulting from double carbonylation.[298] Double carbonylation of aromatic compounds has been described.[298a,b] Formation of α-ketoamides occurs by coupling between aroyl and carbamido groups, the latter being formed by amine attack on Pd-coordinated CO. Co catalysts are also active for α-ketoacid formation.[298c]

Phase transfer catalysis of carbonylation of aryl halides has been obtained under photostimulation with $NaCo(CO)_4$ as catalyst. A $S_{RN}1$ mechanism was proposed.[16b]

Heterogenization of the tetracarbonylcobaltate anion on anion exchange resins allows carbonylation of organic halides without significant catalyst losses.[299] See also references contained in this paper for other carbonylations in heterogeneous systems.

6.4   Oxidation and reduction

Oxidative carbonylation has been applied to aromatic substrates. The industrially important p-toluic acid can be obtained from toluene with rhodium trifluoromethanesulfonate as catalyst.[299b] Aromatic acids also form using a $Pd(OAc)_2$/t-BuOOH/$CH_2=CHCH_2Cl$ system.[299c] Aromatic hydrocarbons can be carbonylated to anhydrides[299d] or anthraquinones[299e] in low yield with Pd complexes. Dibromoethane is added in excess in the former case, whilst $PdCl_2$/$CuCl_2$ is used in the latter.

Oxidative carbonylation of allene with $PdCl_2$-$CuCl_2$ in methanol gives methyl 2-methoxymethylacrylate in good yield.[299f]

Hydrocarboxylation or hydroesterification of olefins can also occur with $PdCl_2$-$CuCl_2$ in the presence of added HCl.[299g]

Carbamates can be prepared at 160-170° from amines and CO (80 atm) with Pd/C as catalyst and KI as promoter. Other systems consisting of noble metals and promoters are also active.[300]

Oxidative carbonylation of phenols to diarylcarbonates has been reported with Pd as catalyst and a promoter such as $Mn(OAc)_2$ or under phase transfer conditions.[301-2]

Reduction of nitroarenes RNO_2 with CO/H_2 and $M_3(CO)_{12}$ (M=Fe, Ru) gives carbamates with Fe and formamides with Ru. A species [RN-$M_3(CO)_{11}$COOMe]$^-$ has been proposed as the intermediate.[303] Further improvements have been obtained in the presence of quaternary ammonium halides.[304]

Nitroderivatives can be transformed into carbamates with CO in the presence of palladium complexes with chelating nitrogen ligands and added Brönsted acids.[305]

The use of $HSiR_3$ (R=alkyl), CO and cobalt carbonyl allows the transformation of epoxides into 1,3-propandiols.[306]

6.5   Other reagents

Carbonylation of benzylic and aromatic halides has been shown to occur with high yields with rhodium and palladium catalysts in the presence of aluminum alkoxides or borate esters.[307-8]

REFERENCES

1. J. Halpern, Acc. Chem. Res. 3:386 (1970).

1a. J.K. Stille and K.S.Y. Lau, Acc. Chem. Res. 10:434 (1977).

1b. S.D. Robinson, Organomet. Chem. 7:348 (1978).

1c. J.K. Kochi, "Organometallic Mechanisms and Catalysis", Academic Press, New York (1978).

1d. J.P. Collman, Acc. Chem. Res. 1:136 (1968).

1e. T.T. Tsou and J.K. Kochi, J. Am. Chem. Soc. 101:6319 (1979).

2. J.P. Collman, Acc. Chem. Res. 8:342 (1975).

2a. R.F. Heck and D.S. Breslow, J. Am. Chem. Soc. 85:2779 (1963).

3. W. Reppe, Ann. Chem. 582:1 (1953); W. Reppe and H. Kröper, Ann. Chem. 582:38 (1953); W. Reppe, H. Kröper, N.v. Kutepow and H.J. Pistor, Ann. Chem. 582:72 (1953); W. Reppe, H. Kröper, H.J. Pistor and O. Weissbarth, Ann. Chem. 582:87 (1953).

4. P. Pino, F. Piacenti and M. Bianchi, in "Organic Syntheses via Metal Carbonyls", (I. Wender and P. Pino, eds.), Vol. 2, p. 43, Wiley-Interscience, New York (1977).

4a. F. Calderazzo, Angew. Chem. Int. Ed. Engl. 16:299 (1977).

4b. A. Wojcicki, Adv. Organomet. Chem. 11:87 (1973).

4c. G.P. Chiusoli and L. Cassar, in "Organic Syntheses via Metal Carbonyls", (I. Wender and P. Pino, eds.), Vol. 2, p. 297, Wiley-Interscience, New York (1977).

4d. R.F. Heck, Acc. Chem. Res. 2:10 (1969); R.F. Heck, "Organotransition Metal Chemistry", Academic Press, New York (1974).

5. J.K. Stille, L.F. Hines, R.W. Fries, P.K. Wong, D.E. James and K. Lau, Adv. Chem. Ser. 132:90 (1974).

5a. "Organic Syntheses via Metal Carbonyls", (I. Wender and P. Pino, eds.), Vols. 1 and 2, Wiley, New York (1977).

5b. Advan. Organometal. Chem. (F.G.A. Stone and R. West, eds.), Vol. 17, Academic Press, New York (1979).

5c. J. Falbe, "Synthesen mit Kohlenmonoxyd", Springer, Berlin (1967); J. Falbe, "Neue Synthesen mit Kohlenmonoxyd", Springer, Berlin (1980).

5d. C.W. Bird, "Transition Metal Intermediates in Organic Synthesis", Academic Press, New York (1967).

5e. J. Tsuji, "Organic Syntheses by Means of Transition Metal Complexes", Springer, Berlin (1975).

5f. P.M. Maitlis, "The Organic Chemistry of Palladium", Vol. 2, Academic Press, New York (1971).

5g. P.W. Jolly and G. Wilke, "The Organic Chemistry of Nickel", Vol. 2, Academic Press, New York (1975).

6. L. Cassar and M. Foà, Chim. Ind. (Milan) 51:673 (1969).

7. L. Cassar and M. Foà, J. Organomet. Chem. 51:381 (1973).

8. M. Hidai, T. Izuchi and Y. Uchida, J. Organomet. Chem. 30:279 (1971).

9. L. Cassar and A. Giarrusso, Gazz. Chim. Ital. 103:793 (1973).

10. K. Kudo, M. Sato, M. Hidai and Y. Uchida, Bull. Chem. Soc. Jpn. 46:2820 (1973).

11. A. Shoenberg, I. Bartoletti and R.F. Heck, J. Org. Chem. 39:3118 (1974).

12. P. Fitton and E.A. Ric, J. Organomet. Chem. 28:287 (1971).

13. M. Foà and L. Cassar, J.C.S. Dalton 2572 (1975).

14. J.K. Stille and K.S.Y. Lau, J. Am. Chem. Soc. 98:5841 (1976).

15. A. Moro, M. Foa and L. Cassar, J. Organomet. Chem. 185:79 (1980).

16. Y. Becker and J.K. Stille, J. Am. Chem. Soc. 100:838 (1978).

16a. P.J. Krusic, P.J. Fagan and J. San Filippo, Jr., J. Am. Chem. Soc. 99:250 (1977).

16b. J.J. Brunet, C. Sidot and P. Caubere, J. Organomet. Chem. 204:229 (1980).

17. P. Heimbach, P. Jolly and G. Wilke, Adv. Organomet. Chem. 8:29 (1970).

17a. J. Tsuji, Adv. Organomet. Chem. 17:141 (1979) and references therein.

17b. R. Lai and E. Ucciani, Adv. Chem. Ser. 132:1 (1974).

18. I. Wender, S. Metlin, E. Ergun, H.W. Sternberg and H. Greenfield, J. Am. Chem. Soc. 78:5401 (1956).

19. R. Pruett, Adv. Organomet. Chem. 17:1 (1979).

20. I. Wender, H. Greenfield, S. Metlin and M. Orchin, J. Am. Chem. Soc. 74:4079 (1952).

21. D. Forster, Adv. Organomet. Chem. 17:255 (1979).

22. H. Hohenschutz, N.v. Kutepow and W. Himmele, Hydrocarbon Proc. 45:141 (1966).

23. H. Wakamatsu, J. Uda and N. Yamagami, J.C.S. Chem. Commun. 1540 (1971).

24. J. Parraud, G. Campari and P. Pino, J. Mol. Cat. 6:341 (1979).

24a. L.S. Hegedus and W.H. Darlington, J. Am. Chem. Soc. 102:4980 (1980).

25.   M.C. Gallazzi, L. Porri and G. Vitulli, J. Organomet. Chem. 97:131 (1975); N.C. Gallazzi, P.L. Hanlon, G. Vitulli and L. Porri, J. Organomet. Chem. 33:C45 (1971).

25a.   L. Cassar and J. Halpern, J. Am. Chem. Soc. 92:3515 (1970); J. Halpern in "Organic Syntheses via Metal Carbonyls", Vol. 2, p. 705, (I. Wender and P. Pino, eds.), Wiley, New York (1977).

25b.   R. Victor, R. Ben-Shoshan and S. Sarel, Tetrahedron Lett. 4253 (1970).

25c.   A. Stockis and E. Weissberger, J. Am. Chem. Soc. 97:4288 (1975).

26.   C. Rüchardt and G. Schrauzer, Chem. Ber. 93:1840 (1960).

27.   F. L'Eplattenier, P. Matthys and F. Calderazzo, Inorg. Chem. 9:342 (1970).

28.   A. Rosenthal and I. Wender in "Organic Syntheses via Metal Carbonyls", Vol. 1 p. 405 (I. Wender and P. Pino, eds.), Wiley, New York (1968).

28a.   G.L. Rempel, W.K. Ted, B.R. James and D.V. Plackett, Adv. Chem. Ser. 132:166 (1974).

29.   J.M. Thompson and R.F. Heck, J. Org. Chem. 40:2667 (1975).

30.   Y. Mori and J. Tsuji, Tetrahedron 27:3811 (1971).

31.   P.E. Garrou and R.F. Heck, J. Am. Chem. Soc. 98:4115 (1976).

32.   J.N. Cause, R.A. Fiato and R.L. Pruett, J. Organomet. Chem. 172:405 (1979).

33.   A.J. Chalk and J.F. Harrod, Adv. Organomet. Chem. 6:119 (1968).

34.   D. Saddei, H.J. Freund and G. Hohlneicher, J. Organomet. Chem. 186:63 (1980).

34a.   C.P. Casey and D.M. Scheck, J. Am. Chem. Soc. 102:2723 (1980).

35.   G.K. Anderson and R.J. Cross, J.C.S. Dalton 1246 (1979).

36.   G.P. Chiusoli, Gazz. Chim. Ital. 89:1332 (1959).

37.   J.K. Stille and D.E. James, J. Organomet. Chem. 108:401 (1976).

38.   G.P. Chiusoli and G. Cometti, J. Chem. Soc. 1051 (1972).

39.   G.P. Chiusoli, G. Cometti and V. Bellotti, Gazz. Chim. Ital. 103:569 (1973); G.P. Chiusoli, G. Cometti and V. Bellotti, Gazz. Chim. Ital. 104:259 (1974).

40.   J.L. Mador and U.A. Blackam, U.S. Pat. 3,114,762; CA 60:9149a (1964).

41.   J. Tsuji and N. Iwamoto, Chem. Commun. 828 (1966).

42.   F. Calderazzo, Inorg. Chem. 4:293 (1965).

43. B.D. Dombeck and R.J. Angelici, J. Organomet. Chem. 134:203 (1977); R.J. Angelici, Acc. Chem. Res. 5:335 (1979).

44. E.W. Stern and M.L. Spector, J. Org. Chem. 31:596 (1966).

45. S. Horie and S. Murahashi, Bull. Chem. Soc. Jpn. 33:247 (1960).

45a. J. Mantzaris and E. Weissberger, J. Am. Chem. Soc. 96:1880 (1974).

46. G.P. Chiusoli, G. Cometti and S. Merzoni, Organomet. Chem. Syn. 1:439 (1972).

47. J.F. Knifton, J. Organomet. Chem. 188:223 (1980).

48. M. Mori, K. Chiba, M. Okita and Y. Ban, J.C.S. Chem. Commun. 698 (1979).

48a. M. Mori, K. Chiba and Y. Ban, J. Org. Chem. 43:1684 (1978).

48b. T. Yamamoto, K. Igarashi, J. Ishizu and A. Yamamoto, J. Chem. Soc. Chem. Commun. 554 (1979).

49. C.W. Bird and J. Hudec, Chem. Ind. (London) 570 (1959).

50. W. Reppe and H. Vetter, Ann. Chem. 582:133 (1953).

51. P. Pino, G. Braca, G. Sbrana and A. Cuccurru, Chem. Ind. (London) 1732 (1968).

52. G. Albanesi and M. Tovaglieri, Chim. Ind. (Milan) 41:189 (1959).

53. G.N. Schrauzer, Adv. Organomet. Chem. 2:1 (1964).

54. G. Wilke, F.M. Müller, M. Kröner and B. Bogdanovic, Angew. Chem. Int. Ed. Engl. 2:105 (1963).

55. H. Alper, Pure Appl. Chem. 52:607 (1980).

56. R.F. Heldeweg and H. Hogeveen, J. Am. Chem. Soc. 98:6040 (1976).

57. S. Murahashi and S. Horiie, J. Am. Chem. Soc. 78:4816 (1956).

58. G.P. Chiusoli and G. Cometti, J.C.S. Chem. Commun. 1015 (1972).

59. G.P. Chiusoli, Acc. Chem. Res. 6:422 (1973).

60. G.P. Chiusoli, G. Salerno, E. Bergamaschi, G.D. Andreetti, G. Bocelli and P. Sgarabotto, J. Organomet. Chem. 177:245 (1979).

61. G.P. Chiusoli, Pure Appl. Chem. Suppl. XXIII Int. Congress, Boston 1971, Special Lectures No. 6, 169 (1971).

62. C.P. Casey, C.A. Bunnel and J.C. Calabrese, J. Am. Chem. Soc. 98:1166 (1976).

63. R. Perron, Ger. Offen. 2,600,541 (1976); CA 86:55171n (1977).

64. H. Alper and H. Des Abbayes, J. Organomet. Chem. 134:C11 (1977).

65. H. Des Abbayes and A. Buloup, J.C.S. Chem. Commun. 1090 (1978).

66. L. Cassar, Ann. N.Y. Acad. Sci. 333:208 (1980).

67. D.E. James and J.K. Stille, J. Am. Chem. Soc. 98:1806, 1810 (1976).

68. J.K. Stille and R. Divakaruni, J. Org. Chem. 44:3474 (1979).

69. G.P. Chiusoli, C. Venturello and S. Merzoni, Chem. Ind. (London) 977 (1968).

70. G.P. Chiusoli, G. Bottaccio and C. Venturello, unpublished results.

71. C.A. Tolman, Chem. Rev. 77:313 (1977).

72. J. Halpern, Pure Appl. Chem. 20:59 (1969).

72a. C.A. Tolman, Chem. Soc. Rev. 1:337 (1972).

73. H. Berke and R. Hoffman, J. Am. Chem. Soc. 100:7224 (1978); D.L. Thorn and R. Hoffman, J. Am. Chem. Soc. 100:2079 (1978).

74. F.R. Hartley, Chem. Soc. Rev. 2:163 (1973).

75. H.B. Gray and C.H. Langford, Chem. Eng. News 46:68 (1968).

76. H.C. Clark, C.R. Jablonski and C.S. Wong, Inorg. Chem. 14:1332 (1975).

76a. M.H. Chisholm and H.C. Clark, Acc. Chem. Res. 6:202 (1973).

77. See for example W.C. Drinkard Jr. and R.J. Kassal, Belg. Pat. 723382 (1969); CA 71:30092q (1969).

78. G. Parshall, Acc. Chem. Res. 8:113 (1975).

78a. R.D. Gillard, Communication at the EUCHEM Conference on Oxidative Addition and Reductive Elimination, Bressanone, Italy, September 1977.

79. G. Natta, R. Ercoli and F. Calderazzo, in "Organic Synthesis via Metal Carbonyls", (I. Wender and P. Pino, eds.), Vol. 1 p. 1, Wiley, New York (1968).

80. H. Taube, Adv. Inorg. Chem. Radiochem. 1:1 (1959); N. Sutin, Acc. Chem. Res. 1:225 (1968); A. Haim, Acc. Chem. Res. 8:264 (1975).

80a. J.K. Kochi, Pure Appl. Chem. 52:571 (1980).

81. P. Chini, G. Longoni and V.G. Albano, Adv. Organomet. Chem. 14:285 (1976); P. Chini, Inorg. Chim. Acta Rev. 2:31 (1968).

82. See for numerous examples 5a.

83. B. Bogdanovic, B. Henc, A. Lösler, B. Meister, H. Pauling and G. Wilke, Angew. Chem. Int. Ed. Engl. 12:954 (1973).

83a. G. Consiglio, C. Botteghi, C. Solomon and P. Pino, Angew. Chem. Int. Ed. Engl. 12:669 (1973).

83b. J.W. ApSimon and R.P. Seguin, Tetrahedron 35:2797 (1979).

84. R.G. Woolley, Inorg. Chem. 18:2945 (1979).

85. G. Fachinetti, C. Floriani, F. Marchetti and S. Merlino, J.C.S. Chem. Commun. 522 (1976).

86. F. Basolo and R.G. Pearson, Prog. Inorg. Chem. 4:381 (1962).

87. G. Natta, R. Ercoli, S. Castellano and F. Barbieri, J. Am. Chem. Soc. 76:4049 (1954).

88. R.B. King, A.D. King, Jr. and M.Z. Iqbal, J. Am. Chem. Soc. 101:4893 (1979).

88a. B. Heil and L. Marko, Chem. Ber. 101:2209 (1968); L. Marko, in "Aspects of Homogeneous Catalysis", (R. Ugo, ed.), Vol. 2, p. 3, Reidel Publ. Co., Dordrecht (Holland) (1974).

89. K. Nicholas, M.O. Nestle and D. Seyferth, in "Transition Metal Organometallics in Organic Synthesis", (H. Alper, ed.), Vol. 2, p. 1, Academic Press, New York (1978).

90. R.C. Ryan, C.U. Pittman, Jr. and J.P. O'Connor, J. Am. Chem. Soc. 99:1986 (1977).

91. I. Wender, R. Levine and M. Orchin, J. Am. Chem. Soc. 71:4160 (1949).

91a. G. Albanesi, Chim. Ind. (Milan) 55:319 (1973).

92. G. Braca, G. Sbrana, G. Valentini, G. Andrich and G. Gregorio, in "Fundamental Research in Homogeneous Catalysis", (M. Tsutsui, ed.), Vol. 3, p. 221, Plenum Press, New York (1979), and references therein.

93. H. Dumas, J. Levisalles, and H. Rudler, J. Organomet. Chem. 177:239 (1979).

94. R.L. Pruett, Ann. N.Y. Acad. Sci. 295:239 (1977).

95. A. Fumagalli, T.F. Koetzle, F. Takusagawa, P. Chini, S. Martinengo and B.T. Heaton, J. Am. Chem. Soc. 102:1740 (1980); P. Chini and S. Martinengo, Inorg. Chim. Acta 3:299 (1969).

96. R.L. Pruett and W.E. Walker, Ger. Offen. 2,262,318 (1973); CA 79:78088k (1973).

96a. E.S. Brown, U.S. Pat. 3,929,969 (1975); CA 84:127210h (1976).

97. J.N. Cawse, Ger. Offen. 2,531,070 (1976); CA 85:20594e (1976).

98. L. Kaplan, U.S. Pat. 4,151,192 (1979); CA 91:125984j (1979).

99. C. Masters, Adv. Organomet. Chem. 17:61 (1979).

100. J.L. Vidal, W.E. Walker, R.L. Pruett, R.C. Shoening and R.A. Fiato, in "Fundamental Research in Homogeneous Catalysis", (M. Tsutsui, ed.), Vol. 3, p. 499, Plenum Press, New York (1979).

101. J. Tkatchenko, in "Fundamental Research in Homogeneous Catalysis", (M. Tsutsui, ed.), Vol. 3, p. 119, Plenum Press, New York (1979).

102. J.W. Rathke and H.M. Feder, J. Am. Chem. Soc. 100:3623 (1978).

103. J.S. Bradley, in "Fundamental Research in Homogeneous Catalysis", (M. Tsutsui, ed.), Vol. 3, p. 165, Plenum Press, New York (1979); J.S. Bradley, J. Am. Chem. Soc. 101:7419 (1979).

104. J.A. Roth and M. Orchin, J. Organomet. Chem. 172:C27 (1979).

104a. R.W. Goetz, Ger. Offen. 2,741,589 (1978); CA 88:190089y (1978).

104b. T. Onoda and S. Tomita, Ger. Offen. 2,655,406 (1978); CA 87:117575d (1977).

105. M.G. Thomas, B.F. Beier and E.L. Muetterties, J. Am. Chem. Soc. 98:1296 (1976).

106. G.C. Demitras and E.L. Muetterties, J. Am. Chem. Soc. 99:2796 (1977).

107. C. Masters and J.A. van Doorn, Ger. Offen. 2,644,185 (1977); CA 87:41656h (1977); C. Masters and J.A. van Doorn, U.K. Pat. 1,573,422 (1980); CA 95:9715f (1981).

108. M. Doyle, A. Kouwenhoven, C.A. Schaap and B. van Oort, J. Organomet. Chem. 174:C55 (1979).

109. J.M. Manriquez, D.R. McAlister, R.D. Sanner and J.E. Bercaw, J. Am. Chem. Soc. 98:6733 (1976).

110. J.A. Gladysz, Adv. Organomet. Chem. 20:1 (1982) and references therein.

111. P.T. Wolczanski and J.E. Bercaw, J. Am. Chem. Soc. 101:6450 (1979).

111a. G. Fachinetti, C. Floriani, A. Rosselli and S. Pucci, J.C.S. Chem. Commun. 269 (1978).

112. G. Fachinetti, J.C.S. Chem. Commun. 396 (1979); G. Fachinetti, S. Pucci, P.F. Zanassi and U. Methong, Angew. Chem. Int. Ed. Engl. 18:619 (1979); G. Fachinetti, communication at the Int. Symp. Solute Solute Solvent Interactions, Florence (Italy), June 1980.

113. H. Pichler and H. Schulz, Chem. Ing. Tech. 42:1162 (1970).

113a. H. Olive and S. Olive, Angew. Chem. Int. Ed. Engl. 15:136 (1976).

114. C.P. Casey and S.M. Neumann, J. Am. Chem. Soc. 98:5395 (1976); C.P. Casey and S.M. Neumann, J. Am. Chem. Soc. 100:2544 (1978).

114a. J.A. Gladysz and W. Tam, J. Am. Chem. Soc. 100:2545 (1978).

114b. C.P. Casey, S.M. Neumann, M.A. Andrews and D.R. McAlister, Pure Appl. Chem. 52:625 (1980).

115. J.C. Huffman, J.G. Stone, W.C. Krusell and K.C. Caulton, J. Am. Chem. Soc. 99:5829 (1977).

116. C. Masters, C. van der Woude and J.A. van Doorn, J. Am. Chem. Soc. 101:1633 (1979).

117. L.I. Shoer and J. Schwartz, J. Am. Chem. Soc. 99:5831 (1977).

118. C.D. Wood and R.R. Schrock, J. Am. Chem. Soc. 101:5421 (1979).

118a. J.R. Sweet and W.A.G. Graham, J. Organomet. Chem. 173:C9 (1979).

118b. K. Whitmire and D.F. Shriver, J. Am. Chem. Soc. 102:1456 (1980).

119. K.A. Azam and A.J. Deeming, J.C.S. Chem. Commun. 472 (1977).

120. K.L. Brown, G.R. Clark, C.E.L. Headford, K. Marsden and W.R. Roper, J. Am. Chem. Soc. 101:503 (1979).

121. R. Sapienza, M.J. Sansone, L.D. Spaulding and J.F. Linch, in "Fundamental Research in Homogeneous Catalysis", (M. Tsutsui, ed.), Vol. 3, p. 179, Plenum Press, New York (1979).

122. J.S. Bradley, G.B. Ansell and E.W. Hill, J. Am. Chem. Soc. 101:7417 (1979).

123. F. Fischer and H. Tropsch, Chem. Ber. 59:830 (1926).

124. E.L. Muetterties and J. Stein, Chem. Rev. 79:479 (1979).

125. I. Moritani and Y. Fujiwara, Synthesis 524 (1973).

126. P. Hong and H. Yamazaki, Chem. Lett. 1335 (1979).

127. P.S. Braterman, B.S. Walker and T.H. Robertson, J.C.S. Chem. Commun. 651 (1977).

128. G.K. Anderson and R.J. Cross, J.C.S. Dalton 712 (1980).

129. S. Hosaka and J. Tsuji, Tetrahedron 27:3821 (1971); J. Tsuji, Y. Mori and M. Hara, Tetrahedron 28:3721 (1972); J. Tsuji, Adv. Organomet. Chem. 17:141 (1979).

130. U. Bersellini, M. Catellani, G.P. Chiusoli, W. Giroldini and G. Salerno, in "Fundamental Research in Homogeneous Catalysis", (M. Tsutsui, ed.), Vol. 3, p. 893, Plenum Press, New York (1979).

130a. Y. Fujiwara, T. Kawauchi and H. Taniguchi, J.C.S. Chem. Commun. 220 (1980).

131. J. Chatt, Adv. Organomet. Chem. 12:1 (1974).

132. R. Mason and D.W. Meek, Angew. Chem. Int. Ed. Engl. 17:183 (1978).

133. R. Crabtree, Acc. Chem. Res. 12:331 (1979).

134. H.D. Empsall, E.M. Hyde, C.E. Jones and B.L. Shaw, J.C.S. Dalton 1980 (1974).

135. F. Guerrieri and G.P. Chiusoli, J. Organomet. Chem. 15:209 (1968).

136. T. Hayashi, M. Tanak and I. Ogota, J. Mol. Catal. 6:1 (1979).

137. K.L. Oliver and L. Booth, Hydrocarbon Process. 49:112 (1970).

138. B. Cornils, R. Payer and K.C. Traenkner, Hydrocarbon Process. 54:83 (1975).

139. L.H. Slaugh and R.D. Mullineaux, J. Organomet. Chem. 13:469 (1968).

140.   D. Evans, J.A. Osborn and G. Wilkinson, J. Chem. Soc.(A) 3133 (1968).

140a.   C.K. Brown and G. Wilkinson, J. Chem. Soc.(A) 2753 (1970).

141.   K. Murata, A. Matsuda, K. Bando and Y. Sugi, J.C.S. Chem. Commun. 785 (1979).

142.   G. Consiglio, C. Botteghi, C. Salomon and P. Pino, Adv. Chem. Ser. 132:295 (1974).

143.   T. Hayashi, M. Tanaka, Y. Ikeda and I. Ogata, Bull. Chem. Soc. Jpn. 52:2605 (1979).

143a.   T. Hayashi, M. Tanaka and I. Ogata, Tetrahedron Lett. 3925 (1978).

144.   H.B. Tinker and A.J. Solodar, Can. Pat. 1,027,141; CA 89:42440m (1978).

144a.   Y. Watanabe, T. Mitsudo, Y. Yasunori, J. Kikuchi and Y. Takegami, Bull. Chem. Soc. Jpn. 52:2735 (1979).

145.   C. Botteghi, M. Branca and A. Saba, J. Organomet. Chem. 184:C17 (1980).

146.   R. Kummer and F.J. Weiss, Proc. Int. Symp. Rhodium Homog. Catal., Veszprem (Hungary), p. 87 (1978).

147.   B. Fell and H. Bahrmann, J. Mol. Catal. 2:211 (1977).

148.   M. Tanaka, T. Hayashi and I. Ogata, Bull. Chem. Soc. Jpn. 50:2351 (1977).

149.   K. Bittler, N. v. Kutepow, D. Neubauer and H. Reis, Angew. Chem. Int. Ed. Engl. 7:329 (1968).

150.   Y. Sugi and K. Bando, Chem. Lett. 727 (1976).

151.   J.F. Knifton, in "Fundamental Research in Homogeneous Catalysis", (M. Tsutsui, ed.), Vol. 3, p. 199, Plenum Press, New York (1979).

152.   T.P. Kobylinski and W.R. Pretzer, Abs. IX Int. Conf. Organomet. Chem., Dijon (France), C12 (1979).

153.   C. Hsu and M. Orchin, J. Am. Chem. Soc. 97:3553 (1975).

153a.   I. Schwager and J.F. Knifton, J. Catal. 45:256 (1976).

154.   J.F. Knifton, J. Org. Chem. 41:793,2885 (1976).

155.   T.F. Murray and J.R. Norton, J. Am. Chem. Soc. 101:4107 (1979).

156.   H.A. Tayim and J.C. Bailar, Jr., J. Am. Chem. Soc. 89:3420,4330 (1967).

157.   R.D. Cramer, E.L. Jenner, R.V. Lindsey, Jr. and U.G. Stolberg, J. Am. Chem. Soc. 85:1691 (1963).

158.   A.G. Davies, G. Wilkinson and J.F. Young, J. Am. Chem. Soc. 85:1692 (1963).

159.   L. Cassar and M. Foà, Inorg. Nucl. Chem. Lett. 6:291 (1970); Ital. Pat. 845993 (1969), CA 72:43198z (1970).

160. M. Foà and L. Cassar, Gazz. Chim. Ital. 102:85 (1972).

161. I. Wender, H. Greenfield and M. Orchin, J. Am. Chem. Soc. 73:2656 (1951).

161a. J. Berty, L. Marko and D. Kallo, Chem. Tech. (Berlin) 8:260 (1956).

162. G. Braca, G. Sbrana, G. Valentini, G. Andrich and G. Gregorio, J. Am. Chem. Soc. 100:6238 (1978).

163. F. Piacenti and M. Bianchi, in "Organic Syntheses via Metal Carbonyls", (I. Wender and P. Pino, eds.), Vol. 2, p. 1, Wiley, New York (1977), and references therein.

164. D. Forster, Adv. Organomet. Chem. 17:255 (1979); D. Forster, J. Am. Chem. Soc. 98:846 (1976).

165. D. Forster, J.C.S. Dalton 1639 (1979).

166. G.P. Chiusoli, S. Merzoni and G. Mondelli, Chim. Ind. (Milan) 46:743 (1964); F.Guerrieri, Chem. Commun. 983 (1968).

166a. L. Cassar, G.P. Chiusoli and F. Guerrieri, Synthesis 509 (1973).

167. J.R. Norton, K.E. Shenton and J. Schwartz, Tetrahedron Lett. 51 (1975).

168. A. Matsuda, S. Shin, J. Nakayama, K. Bando and K. Murata, Bull. Chem. Soc. Jpn. 51:3016 (1978).

168a. M. Hiraoka and M. Ito, Jap. Pat. 7,414,731 (1974); CA 81:120195m (1974).

168b. E. Cesarotti, A. Fusi, R. Ugo and G.M. Zanderighi, J. Mol. Catal. 4:205 (1978).

169. V. Gutman, Coord. Chem. Rev. 18:225 (1976); E. Uhlig, Coord Chem. Rev. 10:227 (1973).

170. F.R. Hartley, Chem. Rev. 81:79 (1981).

170a. G.F. Pregaglia, A. Andreetta, G. Gregorio, G. Ferrari, G. Montrasi and R. Ugo, Chim. Ind. (Milan) 55:203 (1973); R. Ugo, Communication at the V Int. Symp. on Solute Solute Solvent Interactions, Florence (Italy) June 1980.

171. E.J. Corey and L.S. Hegedus, J. Am. Chem. Soc. 91:1233 (1969).

172. J.J. Brunet, C. Sidot, B. Loubinoux and P. Caubere, J. Org. Chem. 44:2199 (1979).

173. G. Longoni, P. Chini and A. Cavalieri, Inorg. Chem. 15:3025 (1976).

174. J.P. Collman, J.N. Cawse and J.I. Brauman, J. Am. Chem. Soc. 94:5905 (1972).

175. J.P. Collman, R.G. Finke, J.N. Cawse and J.I. Brauman, J. Am. Chem. Soc. 99:2515 (1977).

176. A. Moro, M. Foa and L. Cassar, J. Organomet. Chem. 185:79 (1980).

176a. W.F. Edgell, in "Ions and Ion-Pairs in Organic Reactions", (M. Szwarc, ed.), Vol. 1, p. 153, Interscience, New York (1972).

177. M.Y. Darensbourg, D.J. Darensbourg, D. Burns and D.A. Drew, J. Am. Chem. Soc. 98:3127 (1976).

177a. M.Y. Darensbourg, P. Jimenez and J.R. Sackett, J. Organomet. Chem. 202:C68 (1980).

178. H. Krüper, in "Houben-Weil Methoden der Organischen Chemie", (E. Müller, ed.), Vol. IV.2, p. 414, Thieme, Stuttgart (1955).

179. M. Takeda, M. Uchide and H. Iwane, French Pat. Appl. 32143 (1976).

180. J. Chatt, Science 160:723 (1968).

181. R.H. Hasek and C.W. Wayman, U.S. Pat. 2,820,059 (1958); CA 53:13040c (1959); R.Iwanaga, Bull. Chem. Soc. Jpn. 35:865 (1962).

181a. A.T. Jurewicz, L.D. Rollmann and D. Duayne Whitehurst, Adv. Chem. Ser. 132:240 (1974).

182. H. Kang, C.H. Mauldin, T. Cole, W. Slegeir, K. Cann and R. Pettit, J. Am. Chem. Soc. 99:8323 (1977).

183. L. Malatesta and G. Caglio, Chem. Commun. 420 (1967).

184. R.M. Laine, D.W. Thomas and L.W. Cary, J. Org. Chem. 44:4964 (1979).

185. L. Kaplan, Proc. IX Int. Conf. Organomet. Chem., Dijon (France), Sept. 1979, C11; L. Kaplan, Communication at the V Int. Symp. on Solute Solute Solvent Interactions, Florence (Italy), June 1980.

186. A. Deluzarche, F. Fonseca, G. Jenner and A. Kiennemann, Erdoel Kohle Erdgas Petrolchem. 32:313 (1979).

187. L. Kaplan, W.E. Walker and G.L. O'Connor, U.S. Pat. 4,153,623 (1979); CA 91:107657a (1979).

188. S.B. Butts, E.M. Holt, S.H. Strauss, N.W. Alcock, R.E. Stimson and D.F. Shriver, J. Am. Chem. Soc. 101:5864 (1979).

189. R.J. Mawby, F. Basolo and R.G. Pearson, J. Am. Chem. Soc. 86:3994 (1964).

190. G. Cavinato and L. Toniolo, J. Mol. Catal. 6:111 (1979); R. Bardi, A. Del Pra, A.M. Piazzesi and L. Toniolo, Inorg. Chim. Acta 35:L345 (1979).

191. R.F. Heck, in "Organic Syntheses via Metal Carbonyls", (I. Wender and P. Pino, eds.), Vol. 1, p. 373, Wiley, New York (1968).

192. N.L. Bauld, Tetrehedron Lett. 1841 (1963).

193. L. Cassar and G.P. Chiusoli, Tetrahedron Lett. 2805 (1966).

194. K. Bott, Proc. Symp. Rhodium Homogeneous Catal., Veszprem (Hungary), Sept. 1980, p. 106.

195. W.P. Weber and G.W. Gokel, "Phase Transfer Catalysis in Organic Synthesis", Springer, New York (1977).

196. C.M. Starks and C.L. Liotta, "Phase Transfer Catalysis; Principles and Methods", Academic Press, New York (1978); C.M. Starks, Chem. Tech. 110 (1980).

197. D. Landini, A. Maia and F. Montanari, J. Am. Chem. Soc. 100:2796 (1978).

198. M. Makosza and M. Ludvikow, Angew. Chem. Int. Ed. Engl. 13:665 (1974).

199. L. Cassar, M. Foa and A. Gardano, J. Organomet. Chem. 121:C55 (1976).

200. L. Cassar and M. Foa, J. Organomet. Chem. 134:C15 (1977).

201. H. Alper and H. Des Abbayes, J. Organomet. Chem. 134:C11 (1977).

202. M. Foa and L. Cassar, Gazz. Chim. Ital. 109:619 (1979).

203. H. Alper, J.K. Currie and H. Des Abbayes, J.C.S. Chem. Commun. 311 (1978).

204. H. Alper and J.K. Currie, Tetrahedron Lett. 2665 (1979).

205. Y. Kimura, Y. Tomita, S. Nakanishi and Y. Otsuji, Chem. Lett. 321 (1979).

206. R.H. Grubbs, C. Gibbons, L.C. Kroll, W.D. Bonds, Jr. and C.H. Brubaker, Jr., J. Am. Chem. Soc. 95:2373 (1973).

207. D.C. Neckers, D.A. Kooistra and G.W. Green, J. Am. Chem. Soc. 94:9284 (1972).

208. R.L. Pruett and J.A. Smith, J. Org. Chem. 34:327 (1969).

209. C.U. Pittman, Jr. and R.M. Hanes, J. Am. Chem. Soc. 98:5402 (1976).

210. P. Perkins and K.P.C. Vollhardt, J. Am. Chem. Soc. 101:3985 (1979).

211. S.J. Fritschel, J.J.H. Ackerman, T. Keyser and J.K. Stille, J. Org. Chem. 44:3152 (1979).

212. A. Andreetta, G. Barberis and G. Gregorio, Chim. Ind. (Milan) 60:887 (1978).

213. A. Borovski, D.J. Cole-Hamilton and G. Wilkinson, Nouveau J. Chim. 2:137 (1978).

214. L.W. Gosser, W.H. Knoth and G.W. Parshall, J. Mol. Catal. 2:253 (1977).

215. D.M. Fenton and P.J. Steinwand, U.S. Pat. 3,393,136 (1968); CA 69:95982h (1968).

216. W. Gaenzler, K. Kabs and G. Schroeder, Ger. Offen. 2,213,435 (1973); CA 80:3114c (1974).

217. G. Agnes, F. Guerrieri and G. Rucci, private communication.

218. M. Graziani, P. Uguagliati and G. Carturan, J. Organomet. Chem. 27:275 (1971).

218a. P. Koch and E. Perrotti, J. Organomet. Chem. 81:111 (1974).

219. J. Tsuji, M. Morikawa and N. Iwamoto, J. Am. Chem. Soc. 86:2095 (1964).

219a. J.K. Stille, Belg. Pat. 877,770 (1979); CA 92:214907z (1980).

220. H.S. Kesling, Jr. and L.R. Zehner, Ger. Offen. 2,827,453 (1979); CA 90:121035g (1979).

221. D.M. Fenton and P.J. Steinwand, J. Org. Chem. 39:701 (1974).

222. T. Yamazaki, M. Eguchi, S. Uchiumi, A. Iwayama, M. Takahashi and M. Kurahashi, Ger. Offen. 2,514,685 (1975); CA 84:16780z (1976).

222a. L. Cassar and A. Gardano, Ger. Offen. 2,601,139 (1976); CA 85:123378s (1976).

223. G. Agnes, G. Bimbi, G. Guerrieri and G. Rucci, Ger. Offen. 2,716,170 (1977); CA 88:50307t (1978).

224. V.A. Golodov, Yu. L. Sheludyakov and D.V. Sokolski, in "Fundamental Research in Homogeneous Catalysis", (M. Tsutsui, ed.), Vol. 3, p. 239, Plenum Press, New York (1979).

225. M. Suizu, K. Fuji, K. Nishihira, K. Mizutare and M. Yamashita, Jpn. Kokai Tokkyo Koho 84514-84517 (1979); CA 91:157294a-157298e (1979).

226. T. Yamazaki, M. Eguchi, S. Uchiumi, K. Nishihira, M. Yamashita and H. Itatani, Ger. Offen. 2,733,730 (1978); CA 88:152047 (1978).

227. R.F. Heck, J. Am. Chem. Soc. 91:6707 (1969).

228. G. Cometti and G.P. Chiusoli, J. Organomet. Chem. 181:C14 (1979).

229. J. Tsuji, M. Takahashi and T. Takahashi, Tetrahedron Lett. 849 (1980).

230. S.N. Anderson, C.W. Fong and M.D. Johnson, J.C.S. Chem. Commun. 163 (1973).

231. Y. Watanabe, K. Taniguchi, M. Suga, T. Mitsudo and Y. Takegami, Bull. Chem. Soc. Jpn. 52:1869 (1979).

232. E.H. Kober, W.J. Schnabel and P.D. Hammond, French Pat. 1,558,898 (1969); CA 72:43191s (1970).

233. B.K. Nefedov, V.I. Manov-Yuvenskii, A.L. Chimishkyan and V.M. Englin, Kinet. Katal. 19:1065 (1978).

234. H.A. Dieck, R.M. Laine and R.F. Heck, J. Org. Chem. 40:2819 (1975).

234a. A.F.M. Iqbal, Angew. Chem. Int. Ed. Engl. 11:634 (1972).

235. L. Cassar, M. Foa and G.P. Chiusoli, Organomet. Chem. Syn. 1:302 (1971).

236. S. Murai and N. Sonoda, Angew. Chem. Int. Ed. Engl. 18:837 (1979) and references therein.

237. J.F. Harrod and A.J. Chalk, in "Organic Syntheses via Metal Carbonyls", (I. Wender and P. Pino, eds.), Vol. 2, p. 673, Wiley, New York (1977).

238. Y. Colleuille, Holl. Pat. Appl. 6,513,584 (1966); CA 65:8959d (1966).

239. T. Yamamoto, T. Kohara and A. Yamamoto, Chem. Lett. 1217 (1976).

240. M. Tanaka, Tetrahedron Lett. 2601 (1979).

240a. J.T. Carlock, Tetrahedron 40:185 (1984).

241. C. Botteghi, S. Gladiali, V. Bellagamba, R. Ercoli and A. Gamba, Chim. Ind. (Milan) 62:604,757 (1980); 63:29 (1981).

242. H. Siegel and W. Himmele, Angew. Chem. Int. Ed. Engl. 19:178 (1980).

243. T. Hayashi, M. Tanaka and I. Ogata, J. Mol. Catal. 13:323 (1981).

243a. A. Schoenberg and R.F. Heck, J. Am. Chem. Soc. 96:7761 (1974).

243b. A. Kasahara, T. Izumi and H. Yanai, Chem. and Ind. 898 (1984).

243c. J. Tsuji, K. Sato and H. Okumoto, J. Org. Chem. 49:1341 (1984).

244. M.F. Semmelhack and S.J. Brickner, J. Am. Chem. Soc. 103:3945 (1981).

245. M.F. Semmelhack and C. Bodurow, J. Am. Chem. Soc. 106:1496 (1984).

246. A. Cowell and J.K. Stille, J. Am. Chem. Soc. 102:4193 (1980).

247. M. Hidai, M. Orisaka and Y. Uchida, Chem. Lett. 753 (1980).

248. S. McVey and P.M. Maitlis, J. Organomet. Chem. 19:169 (1969).

249. J. Kiji, S. Yoshikawa and J. Furukawa, Bull. Chem. Soc. Jpn. 47:490 (1976).

250. A. Herrera, H. Hoberg and R. Mynott, J. Organomet. Chem. 222:331 (1981).

251. T. Mise, P. Hong and H. Yamazaki, Chem. Lett. 993 (1981).

252. P. Hong, T. Mise and H. Yamazaki, Chem. Lett. 989 (1981).

253. L.S. Hegedus and J.M. McKearin, J. Am. Chem. Soc. 104:2444 (1982).

254. J. Tsuji, K. Sato and H. Nagastime, Tetrahedron Lett. 893 (1982).

255. M. Catellani, G.P. Chiusoli and C. Peloso, Tetrahedron Lett. 24:813 (1983).

255a. P. De Shong and G.A. Slough, Organometallics 3:636 (1984).

256. E. Amari, M. Catellani, G.P. Chiusoli and S. Salerno, Chim. Ind. (Milan) 64:240 (1982).

257.   P.L. Pauson, I.U. Khand, G.R. Knox, W.E. Watts and M.I. Foreman,
        J.C.S. Perkin I 977 (1973); Abs. IX Int. Conf. Organomet. Chem.,
        Dijion 1979, C31.

257a.   E. Amari, M. Catellani and G.P. Chiusoli, J. Organomet. Chem. in
        press.

257b.   G.P. Chiusoli, M. Costa, E. Masarati and G. Salerno, J. Organomet.
        Chem. 255:C35 (1983).

258.   K.H. Dötz, Angew. Chem. Int. Ed. Engl. 14:644 (1975); K.H. Dötz and
        W. Kuhn, Angew. Chem. Int. Ed. Engl. 22:732 (1983).

259.   M.F. Semmelhack, J.J. Bozell, T. Sato, W. Wulff, E. Spiess and A.
        Zask, J. Am. Chem. Soc. 104:5850 (1982).

260.   H. Alper, C. Perera and F.R. Ahmed, J. Am. Chem. Soc. 103:1289
        (1981).

260a.   M. Mori, Y. Washioka, T. Urayama, K. Yoshiura, K. Chiba and Y. Ban,
        J. Org. Chem. 48:4058 (1983).

261.   M. Tanaka, Bull. Chem. Soc. Jpn. 54:637 (1981).

262.   M. Tanaka, M. Koyanagi and T. Kobayashi, Tetrahedron Lett. 3875
        (1981).

262a.   D.J. Darensbourg, Adv. Organomet. Chem. 21:113 (1982).

263.   P.C. Ford, Acc. Chem. Res. 14:37 (1981).

264.   J.A. Marsella, K. Folting, J.C. Huffman and K.G. Caulton, J. Am.
        Chem. Soc. 103:5596 (1981).

265.   R.C. Brady and R. Pettit, J. Am. Chem. Soc. 102:6181 (1980).

266.   K. Isobe, B. Andrews, B.E. Mann and P.M. Maitlis, J.C.S. Chem.
        Commun. 809 (1981).

266a.   F.R. Kreissl, K. Eberl and W. Uedelhoven, Chem. Ber. 110:3782
        (1977); F.R. Kreissl, W.J. Sieber, M. Wolfgruber and J. Riede, Angew.
        Chem. Int. Ed. Engl. 23:640 (1984).

266b.   M.R. Churchill, H.J. Wasserman, S.J. Holmes and R.R. Schrock,
        Organometallics 1:766 (1982).

266c.   C.P. Casey and P.J. Fagan, J. Am. Chem. Soc. 104:7360 (1982).

267.   C.K. Roper De Poorter, Chem. Rev. 81:447 (1981).

268.   C. Masters, Adv. Organomet. Chem. 17:61 (1979); G.H. Olive' and S.
        Olive', Angew. Chem. Int. Ed. Eng. 15:316 (1976); W.A. Herrmann,
        Angew. Chem. Int. Ed. Eng. 21:17 (1982); C.P. Casey, P.J. Fagan,
        W.H. Miles and S.R. Marder, J. Mol. Catal. 21:173 (1980).

269.   R.J. Daroda, J.R. Blackborrow and G. Wilkinson, J.C.S. Chem.
        Commun. 1098,1101 (1980).

270.   G. Fachinetti, L. Balocchi, F. Secco and M. Venturini, Angew. Chem.
        Int. Ed. Engl. 20:204 (1981).

271. H.N. Adams, G. Fachinetti and J. Strähle, Angew. Chem. Int. Ed. Eng. 20:125 (1981).

271a. D.L. Thorn, Organometallics 1:197 (1982).

271b. H. Berke, G. Huttner, O. Scheidsteger and G. Weiler, Angew. Chem. Int. Ed. Eng. 23:735 (1984).

272. J. Halpern, Acc. Chem. Res. 15:332 (1982).

272a. B.B. Wayland, B.A. Woods and R. Pierce, J. Am. Chem. Soc. 104:302 (1982).

273. T. Bodnar, C. Coman, S. La Croce, C. Lambert, K. Menard and A. Cutler, J. Am. Chem. Soc. 103:2471 (1981).

274. B.D. Dombek, J. Am. Chem. Soc. 102:6855 (1980); 103:6508 (1981); J. Organomet. Chem. 250:467 (1983).

275. W. Keim, M. Berger, A. Eisenbeis, J. Kadelka and J. Schlupp, J. Mol. Catal. 13:95 (1981).

276. D.R. Fahey, J. Am. Chem. Soc. 103:136 (1981).

277. G. Doyle, J. Mol. Catal. 13:237 (1981).

277a. G.R. Steinmetz and T.H. Larkins, Organometallics 2:1879 (1983).

278. M. Hidai, M. Orisaku, M. Ue, Y. Uchida, K. Yasukufu and H. Yamazaki, Chem. Lett. 143 (1981).

279. J.F. Knifton, J. Mol. Catal. 11:91 (1981).

280. R.J. Cross and J. Gemmill, J.C.S. Dalton 2317 (1981).

281. G.K. Anderson and R.J. Cross, J.C.S. Dalton 1434 (1981).

282. G. Reichenbach, G. Cardaci and G. Bellachioma, J.C.S. Dalton 847 (1982).

282a. T.W. Dekleva and B.R. James, J. Chem. Soc. Chem. Commun. 1350 (1983).

282b. P. Bradamante, A. Stefani and G. Fachinetti, J. Organomet. Chem. 266:303 (1984).

283. E.J. Kuhlmann and J.J. Alexander, Coord. Chem. Rev. 33:195 (1980).

284. M. Orchin, Acc. Chem. Res. 14:259 (1981).

284a. H. Des Abbayes, A. Buloup and G. Tanguy, Organometallics 2:1730 (1983).

285. V. Galamb and G. Pàlyi, Coord. Chem. Rev. 59:203 (1984).

286. H.C. Clark and J.A. Davies, J. Organomet. Chem. 213:503 (1981).

287. T.F. Murray, E.G. Samsel, V. Varna and J.R. Norton, J. Am. Chem. Soc. 103:7520 (1981).

288. R. Bardi, A.M. Piazzesi, G. Cavinato, P. Cavoli and L. Toniolo, J. Organomet. Chem. 224:407 (1982).

289. S.C. Tang and L. Kim, J. Mol. Catal. 14:231 (1982).

290. T. Hayashi, S. Kawabata, T. Isoyama and J. Ogata, Bull. Chem. Soc. Jpn. 54:3438 (1981).

291. G.K. Anderson, H.C. Clark and J.A. Davies, Organometallics 1:64 (1982).

291a. G.K. Anderson, H.C. Clark and J.A. Davies, Inorg. Chem. 22:427,434 (1983).

292. C.F. Hobbs and W.S. Knowles, J. Org. Chem. 46:4422 (1981).

292a. G. Consiglio and P. Pino, Adv. Chem. Ser. 196:371 (1982).

292b. G. Consiglio, P. Pino, L.I. Flowers and C.U. Pittman, Jr., J.C.S. Chem. Commun. 612 (1983).

292c. G. Cometti and G.P. Chiusoli, J. Organomet. Chem. 236:C31 (1982).

292d. F. Francalanci, A. Gardano, L. Abis and M. Foà, J. Organomet. Chem. 251:C5 (1983).

292e. M. Foà, F. Francalanci, E. Bencini and A. Gardano, J. Organomet. Chem. in press.

292f. H. Yamazaki and P. Hong, J. Mol. Catal. 21:133 (1983).

293. M.J. Wax and R.G. Bergman, J. Am. Chem. Soc. 103:7028 (1981).

294. S.B. Butts, T.G. Richmond and D.F. Shriver, Inorg. Chem. 20:278 (1981).

294a. S.J. La Croce and D.R. Cutler, J. Am. Chem. Soc. 104:2312 (1982).

295. J.F. Knifton, J. Am. Chem. Soc. 103:3959 (1981).

295a. J.J. Brunet, C. Sidot, B. Loubinoux and P. Caubere, J. Org. Chem. 44:2199 (1979).

295b. H. Alper and J. Petrignani, J.C.S. Chem. Commun. 1154 (1983).

296. S. Gambarotta and H. Alper, J. Org. Chem. 46:2142 (1981).

297. S. Gambarotta and H. Alper, J. Organomet. Chem. 212:C23 (1981); G. Tustin and R.T. Hembre, J. Org. Chem. 49:1761 (1984); G. Tanguy, B. Weinberger and H. Des Abbayes, Tetrahedron Lett. 4005 (1983).

298. F. Francalanci, A. Gardano, L. Abis, T. Fiorani and M. Foa, J. Organomet. Chem. 243:87 (1983); F. Francalanci and M. Foa, J. Organomet. Chem. 232:59 (1982); F. Francalanci, A. Gardano and M. Foa, J. Organomet. Chem. in press and unpublished results.

298a. T. Kobayashi and M. Tanaka, J. Organomet. Chem. 233:C64 (1982); F. Ozawa, T. Sugimoto, T. Yamamoto and A. Yamamoto, Organometallics 3:692 (1984); F. Ozawa, T. Sugimoto, Y. Yuesa, M. Sentra, T. Yamamoto and A. Yamamoto, Organometallics 3:683 (1984) and references therein.

298b. J. Chen and A. Sen, J. Am. Chem. Soc. 106:1506 (1984).

298c. M. Foà, F. Francalanci and G. Gardano, unpublished results.

299.   M. Foa, F. Francalanci, A. Gardano, G. Cainelli and A. Umani Ronchi, J. Organomet. Chem. 248:255 (1983).

299b. F.I. Waller (Du Pont) Eur. Pat. Appl. 82,633 (1983); CA 99:194623 (1983).

299c. Y. Fujiwara, I. Kawata, H. Sujimoto and T. Taniguchi, J. Organomet. Chem. 256:C35 (1983).

299d. Y. Fujiwara, I. Kawata, T. Kawauchi and H. Taniguchi, J.C.S. Chem. Commun. 132 (1982).

299e. G.C. Arzoumanides and F.C. Rauch, J. Mol. Catal. 9:336 (1980).

299f. H. Alper, F.W. Hartstock and B. Despeyroux, J.C.S. Chem. Commun. 905 (1984).

299g. H. Alper, J.B. Woell, B. Despeyroux and D.J.H. Smith, J.C.S. Chem. Commun. 1270 (1983); S.B. Ferguson and H. Alper, J.C.S. Chem. Commun. 1349 (1984).

300.   S. Fukuoka, M. Chono and M. Kohno, J. Org. Chem. 49:1458 (1984).

301.   J.E. Hallgreen, G.M. Lucas and R.O. Matthews, J. Organomet. Chem. 204:135 (1981).

302.   J.E. Hallgreen and G.M. Lucas, J. Organomet. Chem. 212:135 (1981).

303.   H. Alper and K.E. Hashem, J. Am. Chem. Soc. 103:6514 (1981).

304.   S. Cenini, M. Pizzotti, C. Crotti, F. Porta and G. LaMonica, J.C.S. Chem. Commun. 1286 (1984) and references therein.

305.   E. Alessio and G. Mestroni, J. Mol. Catal. 26:337 (1984).

306.   T. Murai, S. Kato, S. Murai, T. Toki, S. Suzuki and N. Sonoda, J. Am. Chem. Soc. 106:6093 (1984).

307.   H. Alper, S. Antebi and J.B. Woell, Angew. Chem. Int. Ed. Engl. 23:732 (1984).

308.   J.B. Woell and H. Alper, Tetrahedron Lett. 25:4879 (1984).

HYDRIDE TRANSFER TO COORDINATED CARBON MONOXIDE

AND RELATED LIGANDS

Christine E.L. Headford and Warren R. Roper

University of Auckland

New Zealand

1. INTRODUCTION

An interesting special case of the organometallic migratory-insertion reaction is that in which the migrating group is H and the inserting ligand is one of the carbon-donor ligands CO, CS, CSe, CTe or CNR. The new ligand formed in the reaction is formyl, thioformyl etc. Complexes with the same CHX-ligands are often more accessible through direct nucleophilic addition of H^- to coordinated CX, particularly in cationic species.

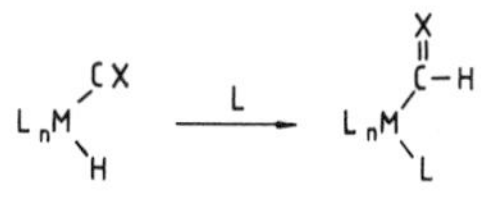

$$L_nM \overset{CX}{\underset{H}{\diagdown}} \xrightarrow{\quad L \quad} L_nM \overset{\overset{\displaystyle X}{\overset{\|}{C-H}}}{\underset{L}{\diagdown}} \qquad (1)$$

$$[L_nM\text{-}CX]^+ \xrightarrow{\quad H^- \quad} L_nM\text{-}C(=X)H \qquad (2)$$

$$X = O, S, Se, Te, NR$$

This chapter will be concerned with the formation of CHX-ligands either by the intramolecular migratory-insertion process of eq 1 or by the intermolecular addition process of eq 2. The chemistry of these ligands, especially further reduction to CH_2X and CH_2XH (or CH_3X-), will also be considered.

The reduction products of CO, formyl-, formaldehyde-, and hydroxymethyl-ligands have been widely studied in recent years both as models to help understand heterogeneously catalysed CO reduction and because such species will undoubtedly be involved in any homogeneously catalysed process which may be discovered for CO hydrogenation to methanol, higher alcohols or hydrocarbons.[1-2]

An advantage of examining all the related CX ligands together is that clear-cut examples of certain bonding modes for CHX- and CH_2X-ligands are easily observed, e.g. both monohapto- and dihapto-thioformyl, whereas the corresponding ligands from reduced CO are not so easily studied.

In order to provide a background for discussion the established features of the general insertion reaction will be briefly reviewed for each ligand where appropriate and since there are examples of the migratory-insertion process being catalysed by electrophiles there will also be some discussion of the interaction of CX-ligands with various electrophiles. With very few exceptions the discussion will be confined to mononuclear complexes. The material has been organised in the following way: each of the ligands CO, CS, CSe , CTe and CNR will be examined in turn and for each ligand there will be further subdivision under the headings, nucleophilic addition, migratory-insertion, interaction with electrophiles.

2.   THE CARBON MONOXIDE LIGAND

Both nucleophilic attack at the carbon atom of coordinated carbon monoxide and migratory-insertion reactions affordingly acyl complexes are well-documented reactions of metal carbonyls. Electrophilic attack at the oxygen atom of a terminal carbonyl ligand, however, is only known to occur in a few instances.

2.1 Nucleophilic Addition

For heteronuclear diatomic π-acceptor ligands such as CO, CS and CSe the π-bonding orbital is localised on the more electronegative atom, i.e. O, S and Se while the π*-antibonding orbital is concentrated at the less electronegative atom, namely carbon.[3] Therefore, a decrease in metal to ligand π-backbonding lowers the electron density in the π*-orbital, making the carbon atom more susceptible to nucleophilic attack. This condition is enhanced in cationic complexes with a maximum number of π-acceptor ligands.

The reaction is postulated to occur via attack of the incoming nucleophile at the carbon atom to form an intermediate species of the form M—C(=X)Nu (X = O, S or Se; Nu = nucleophile). This intermediate is either itself stable or else undergoes rearrangement or elimination of a small molecule to give the observed product.

Nucleophiles such as carbanions,[4] azide ion,[5] hydrazine, primary and secondary amines, ammonia, alcohols[6] and water[7] are known to attack coordinated carbon monoxide. The addition of carbanions to carbonyl ligands represents a convenient route into metal carbene chemistry and has been exploited extensively.[8]

In the light of the above observations, nucleophilic attack by the hydride ion, H$^-$, on coordinated carbon monoxide was expected to yield the formyl ligand according to eq 2, but early attempts to obtain evidence of such a species uniformly met with failure.[9] The reactions of borohydride reagents with transition metal carbonyl complexes were used to facilitate the substitution of one or more ligands for CO,[10-11] and to synthesize certain Group VI dinuclear metal carbonyl hydrides.[12] The intermediacy of formyl complexes is proposed for these reactions.

In 1976, Casey and Neumann observed that reaction of trialkoxyborohydride reagents with various neutral metal carbonyl compounds led to the formation of transition metal formyl anions (eq 3).[13]

$$L_nM(CO) + (OR)_3BH^- \rightarrow [L_nM(CHO)]^- + B(OR)_3 \qquad (3)$$

$$L = PPh_3, P(OPh)_3, CO$$
$$R = -CH_3, -CH(CH_3)_2$$

In particular, reaction of $Na(OCH_3)_3BH$ with $Fe(CO)_5$ gave $[Fe(CHO)(CO)_4]^-$ (I), previously prepared by formylation of $[Fe(CO)_4]^{2-}$ with acetic formic anhydride to be the first reported formyl complex.[14] Winter et al subsequently noted that several boron and aluminum hydride reagents reacted with $Fe(CO)_5$ to afford the formyl complex **I**, but in widely varying yields.[15] N.m.r. evidence was obtained for the formation of formyl complexes from the reactions of $Na(OCH_3)_3BH$ and $K(O\text{-}\underline{i}\text{-}Pr)_3BH$ with the carbonyl complexes $Cr(CO)_6$, $W(CO)_6$, $Cr(CO)_5(PPh_3)$, $W(CO)_5(PPh_3)$ and $Fe(CO)_4(PPh_3)$.[13] Reaction of the phosphite-containing complex $Fe(CO)_4(P[OPh]_3)$ with the more reactive hydride $K(O\text{-}\underline{i}\text{-}Pr)_3BH$ afforded the isolable formyl species $[Et_4N]$ $\underline{trans}\text{-}[Fe(CHO)(CO)_3P(OPh)_3]$. Results of studies on several transition metal complexes containing both carbonyl and acyl ligands showed a kinetic preference for nucleophilic attack by H^- at coordinated CO rather than at metal-bound acyl groups.[16,17] For example,

$$Mn(C[O]R)(CO)_5 \xrightarrow{\text{LiEt}_3\text{BH}} [Mn(CHO)(C[O]R)(CO)_4]^-$$

$$R = C_6H_5, CH_2OMe, C[O]OMe, CF_3$$

The formation of a bimetallic manganese-formyl intermediate $[Mn_2(CHO)(CO)_9]^-$ **(II)** was observed during cleavage of a metal carbonyl dimer by trialkylborohydride reagents[18] (eq 4). That an anionic formyl

$$(CO)_5Mn - Mn(CO)_5 + R_3BH^- \rightarrow [(CO)_5Mn - Mn(CHO)(CO)_4]^- \quad (4)$$

$$\textbf{II}$$

$$R = Et, s\text{-}Bu$$

complex could act as an hydride donor (nucleophile) towards a metal carbonyl compound, and thus produce a new metal formyl complex, was demonstrated by the reaction of **II** with $Fe(CO)_5$ to yield the known iron formyl $[Fe(CHO)(CO)_4]^-$.[13-15] Such "transformylation" reactions between various metal formyl and metal carbonyl compounds have been used to determine the order of stability of a series of metal formyl complexes relative to the corresponding metal carbonyl precursors:[19] $[(Ph_3P)_2N][Fe(CHO)(CO)_4]$ > $[Et_4N]\underline{cis}\text{-}[Re_2(CHO)(CO)_9]$ > $[Et_4N]\underline{trans}\text{-}[Fe(CHO)(CO)_3(P[OPh]_3)]$. Thus, trans-$[Fe(CHO)(CO)_3(P[OPh]_3)]^- + Re_2(CO)_{10}$

$\Downarrow\Uparrow$

$Fe(CO)_4(P[OPh]_3)$ + cis-$[Re_2(CHO)(CO)_9]^-$ + $Fe(CO)_5$

III $\Downarrow\Uparrow$

$Re_2(CO)_{10}$ + $[Fe(CHO)(CO)_4]^-$

The rhenium formyl complex **III** was independently synthesized through the reaction of $K(O-i-Pr)_3BH$,[19] $LiEt_3BH$ or $K(s-Bu)_3BH$[20] with $Re_2(CO)_{10}$.

Although anionic formyl complexes may act as hydride donors (nucleophiles),[18-19] further nucleophilic attack at the formyl ligand by H^- is suggested by the formation of formaldehyde and methanol respectively upon the addition of two and three equivalents of $K(s-Bu)_3BH$ to $Re_2(CO)_{10}$.[20] Methanol was produced under similar conditions from $[Fe(CHO)(CO)_4]^-$.[21]

Anionic rhenium formyl complexes have been prepared[17] (eq 5). These are, in effect, substituted forms of the dimeric anion **III** (the $Re(CO)_5$ group being replaced by other one-electron donor ligands such as halogen). It was thought that a cis-halo ligand might stabilize, through hydrogen bonding,[22]

$$ReX(CO)_5 \xrightarrow[\text{[LiX]}]{LiEt_3BH} \text{cis-}[Re(CHO)X(CO)_4]^- \tag{5}$$

X = Cl,Br,I

$$\left[(CO)_4Re\overset{Cl}{\underset{C=O}{<}}\atop{H} \right]^- \xrightarrow{H^+} (CO)_4Re\overset{Cl\cdots H}{\underset{C-O}{<}}\atop{H} \tag{6}$$

the hydroxymethylene complexes which would result from protonation of the formyl anions[17] (eq 6). All attempts to protonate the formyl ligands led to decomposition.

Reaction of the mixed dimer $ReMn(CO)_{10}$ with $LiEt_3BH$ afforded the unstable formyl complex $[ReMn(CHO)(CO)_9]^-$[20] (**IV**) and completed a series of three homologous binuclear formyl complexes (**II, III** and **IV**). The kinetic stability of these complexes increases with "rhenium content"; in general, rhenium forms stronger bonds to carbon than does manganese.[23]

With the synthesis of anionic metal formyl complexes from the reaction of hydride donors with neutral carbonyl complexes firmly established, this potentially general methodology was extended to the preparation of neutral metal formyl complexes from cationic metal carbonyl precursors.[24-5]   Reaction of $LiEt_3BH$ with metal carbonyl cations successfully effected the reduction of coordinated CO and provided a convenient entry into a number of neutral formyl systems (eq 7):

$$[L_nM(CO)]^+ + Et_3BH^- \longrightarrow L_nM(CHO) + BEt_3 \qquad (7)$$

Cationic carbonyl salts which were reduced in this manner include $[CpRe(CO)_2(NO)]BF_4$,[24-5] $[CpRe(CO)(PPh_3)(NO)]BF_4$, $[CpMn(CO)_2(NO)]PF_6$, $[Re(CO)_5(PPh_3)]BF_4$, $[Mn(CO)_4(PPh_3)]PF_6$, $[Ir(CO)_3(PPh_3)_2]PF_6$ and $[CpMo(CO)_3(PPh_3)]PF_6$.[25] Similarly, $NaBH_4$ was found to be effective for converting a series of osmium cations $[OsX(CO)_3(PPh_3)_2]^+$ to the corrresponding formyl complexes, $Os(CHO)X(CO)_2(PPh_3)_2$ ($X = Cl,Br,H,CH_3,CH_2OMe$).[26] Reductive transformation of the neutral formyl complex $CpRe(CHO)(PPh_3)(NO)$ (**V**) to $CpRe(CH_3)(PPh_3)(NO)$ was accomplished[27] with $BH_3.THF$ or Cp_2ZrHCl.[28-9] The molecular structure of $CpRe(CHO)(PPh_3)(NO)$ was determined by x-ray diffraction methods,[27] and the Re-C formyl bond length found to be unusually short [2.055 $\overset{o}{A}$]; the rhenium-acyl bond length in $Re(\eta^1-C[O]-p-C_6H_4Cl)(CO)_5$ is 2.22 $\overset{o}{A}$,[30] and the average rhenium-acyl/carbene) bond length in the "cyclic" complex $\overline{Re[C(Me)O---H---OCMe]}(CO)_4$ is 2.16 $\overset{o}{A}$.[31] A substantial contribution to the structure of **V** by the resonance form **Vb** is implied.   The electron rich nature of the formyl ligand in **V** was evinced by

$$\textbf{Va} \qquad M-C\overset{\displaystyle O}{\underset{\displaystyle H}{\diagup}} \longleftrightarrow \overset{+}{M}=C\overset{\displaystyle O^-}{\underset{\displaystyle H}{\diagup}} \qquad \textbf{Vb}$$

the formation of a species formulated as $[CpRe(CHOH)(PPh_3)(NO)]^+$ (i.e. hydroxymethylene complex) in a reversible protonation reaction with CF_3CO_2H[32] (eq 8):

$$Cp(PPh_3)(NO)\ Re-CHO \overset{CF_3CO_2H}{\underset{\longleftarrow}{\rightleftharpoons}} Cp(PPh_3)(NO)\ \overset{+}{Re} = C(H)OH \qquad (8)$$

The stoichiometric reduction of a coordinated carbonyl to methyl via nucleophilic attack by H^- was first demonstrated by Treichel and Shubkin.[33] Treatment of $[CpM(CO)_3(PPh_3)]^+$ (M = Mo, W) with $NaBH_4$ in THF gave

$CpM(CH_3)(CO)_2(PPh_3)$.   A possible mechanism proposed for this reduction is shown in eq 9. $(L=PPh_3)$

$$[CpML(CO)_3]^+ \rightarrow CpML(CO)_2CHO \rightarrow CpML(CO)_2CH_2OH \rightarrow CpML(CO)_2CH_3$$
$$(9)$$

The overall reaction was viewed as a stepwise reduction of the CO ligand, with the stability of the formyl derivative to decarbonylation being considered significant.   Reaction of the analogous compounds $[CpM(CO)_4]^+$ (M = Mo, W)[33] and $[CpFe(CO)_3]^+$ [34] with $NaBH_4$ yielded the hydrido-containing complexes $CpMH(CO)_3$ and $CpFeH(CO)_2$ respectively. For these reactions initial hydride attack at the CO ligand is presumed, but the formyl ligands thus formed are considered unstable to loss of CO and formation of the hydrides.

Graham et al[35] initially found that the reaction of the cationic carbonyl complex $[CpRe(CO)_2(NO)]^+$ (VI) with $NaBH_4$ in THF resulted in the formation of the methyl derivative $CpRe(CH_3)(CO)(NO)$.   Later Casey[24] and Gladysz[25] independently showed that $[CpRe(CO)_2(NO)]^+$ reacted with $K(O\text{-}\underline{i}\text{-}Pr)_3BH$ or $LiEt_3BH$ to yield the formyl complex $CpRe(CHO)(CO)(NO)$, VII, suggesting that $NaBH_4$ effects the initial formation of a formyl complex, but that the concomitant generation of BH_3 from $NaBH_4$ provides a second reductant which may further reduce the formyl ligand.   Accordingly, $CpRe(CHO)(CO)(NO)$ reacted with $BH_3 \cdot THF$ to yield $CpRe(CH_3)(CO)(NO)$, but no intermediates such as a hydroxymethyl-containing complex were detected.[24-5]

Nesmeyanov et al claimed the reduction of a carbonyl ligand in $[CpRe(CO)_2(NO)]^+$ to hydroxymethyl using $NaBH_4$ in benzene-water.[36] Independent attempts[24-5,37] to reproduce this reaction under the reported conditions led only to the expected conversion of the cation VI into the formyl complex VII, since the BH_3 generated is rapidly hydrolysed in the water layer, thus allowing little chance for the formyl ligand to be further reduced.

Conversion of a coordinated formyl into a hydroxymethyl ligand was first achieved in a rather unusual reaction sequence.[38]   Neat

CpRe(CHO)(CO)(NO) was found to undergo a Cannizzaro-like disproportionation[39] to give the dimeric metallo-ester Cp(CO)(NO)ReCH$_2$OC(O)Re(NO)(CO)Cp (**VIII**) as a mixture of the two possible diastereomers (Scheme 1). Methanolysis of **VIII** proceeded to afford the hydroxymethyl complex **IX** and the metal ester **X**.

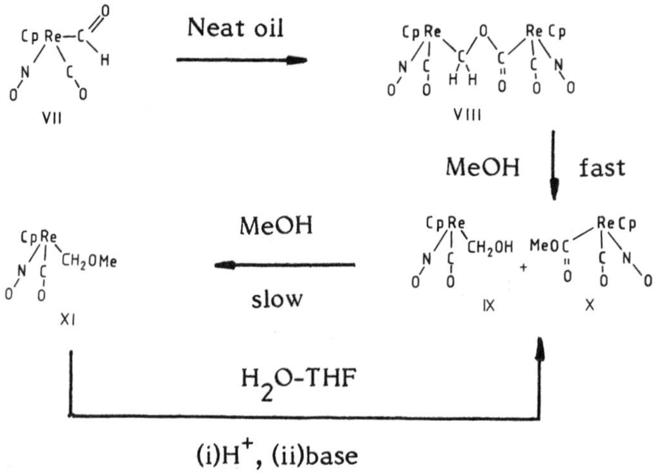

Scheme 1   Synthesis of CpRe(CH$_2$OH)(CO)(NO)

Continued methanolysis led to the formation of the methoxymethyl complex CpRe(CH$_2$OMe)(CO)(NO) (**XI**). The hydroxymethyl complex was obtained in a pure form by base quenching the acid-catalysed hydrolysis of **XI**. The reduction of [CpRe(CO)$_2$(NO)]$^+$ with NaAlH$_2$(CH$_2$CH$_3$)$_2$ in THF provided a more direct synthesis of **IX**.[38]

A study by Graham et al of the reduction of [CpRe(CO)$_2$(NO)]$^+$, using a NaBH$_4$-THF-H$_2$O system, demonstrated the sequential reduction of coordinated carbon monoxide[37] (eqs 10-16).

$$[CpRe(CO)_2(NO)]^+ + NaBH_4 \xrightleftharpoons[\substack{THF/H_2O \\ 0^\circ C,\ 15\ min}]{\substack{THF \\ 25^\circ C,\ 1hr}} \begin{array}{l} CpRe(CH_3)(CO)(NO) \qquad (10) \\[2ex] CpRe(CHO)(CO)(NO) \qquad (11) \end{array}$$

VI   **VII**

$$[CpRe(CO)_2(NO)]^+ + 2NaBH_4 \xrightarrow[\text{0°C, 30 min}]{\text{THF/H}_2\text{O}} CpRe(CH_2OH)(CO)(NO) \quad (12)$$

IX

$$CpRe(CHO)(CO)(NO) + NaBH_4 \xrightarrow[\text{25°C, 1hr}]{\text{THF}} CpRe(CH_3)(CO)(NO) \quad (13)$$

VII

$$\xrightarrow[\text{0°C, 15 min}]{\text{THF/H}_2\text{O}} CpRe(CH_2OH)(CO)(NO) \quad (14)$$

$$CpRe(CH_2OH)(CO)(NO) + NaBH_4$$

IX

$$\xrightarrow[]{\text{THF}\atop\text{25°C, 5hrs}} CpRe(CH_3)(CO)(NO) \quad (15)$$

$$\xrightarrow[\text{25°C, 5hrs}]{\text{THF/H}_2\text{O}} CpRe(CH_3)(CO)(NO) \quad (16)$$

The first direct reduction of coordinated formyl to hydroxymethyl was observed (eq 14); the reduction of the hydroxymethyl ligand to the methyl group (eqs 15-6) provides a striking example of activation of a coordinated ligand, since normal alcohols are not reduced by $NaBH_4$. Graham et al regard suggestions[24,33,35] that formyl or hydroxymethyl derivatives are intermediates in the three-step $NaBH_4$ reduction of $[M-CO]^+$ to $M-CH_3$ under anhydrous conditions as imprecise, in that such intermediates are most probably present as borane adducts or esters, differing from the "free" formyl or hydroxymethyl compounds obtained on hydrolysis. The rate of reaction of pure $CpRe(CH_2OH)(CO)(NO)$ in THF (eq 15) is qualitatively much slower than the overall reduction of the carbonyl cation VI to the methyl derivative $CpRe(CH_3)(CO(NO)$ (eq 10); hence the hydroxymethyl complex IX as such cannot be an intermediate in the overall reduction shown in eq 10.

Other stable hydroxymethyl complexes result from protonation of $Os(\eta^2\text{-}CH_2O)(CO)_2(PPh_3)_2$, e.g. $Os(CH_2OH)Cl(CO)_2(PPh_3)_2$ and $Os(CH_2OH)H(CO)_2(PPh_3)_2$.[40]

Reduction of a second coordinated carbonyl ligand in a complex was realised by treatment of CpRe(CHO)(CO)(NO) with one equivalent of LiEt$_3$BH[24-5] (or [CpRe(CO)$_2$(NO)]$^+$ with two equivalents of hydride[24]) to yield the first, but unstable, bis(formyl) complex [CpRe(CHO)$_2$(NO)]$^-$ (eq 17).   Other anionic bis(formyl) complexes were prepared similarly; for example, reaction of [Re(CO)$_5$(PPh$_3$)]$^+$ with two equivalents of LiEt$_3$BH afforded the corresponding bis(formyl) species [Re(CHO)$_2$(CO)$_3$(PPh$_3$)]$^-$.[25]

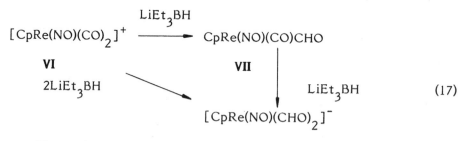

$$[CpRe(NO)(CO)_2]^+ \xrightarrow{\text{LiEt}_3\text{BH}} CpRe(NO)(CO)CHO$$

VI   **VII**

Thermal   decomposition   of   the   bis(formyl)   complex [CpRe(CHO)$_2$(NO)]$^-$led   to   the   formation   of   the   methyl   complex CpRe(CH$_3$)(CO)(NO) (16%).[25] Analogously, in low temperature reactions of the   formyl   complex   CpRe(CHO)(PPh$_3$)(NO)   with   electrophiles   (e.g. CH$_3$SO$_3$F, (CH$_3$)$_3$SiCl and CF$_3$CO$_2$H), a stoichiometric fraction of the formyl ligands were transformed, without the addition of a reductant, into methyl ligands[32] (eq 18).

$$CpRe(CHO)(PPh_3)(NO) \xrightarrow{\text{CH}_3\text{SO}_3\text{F}} [CpRe(CO)(PPh_3)(NO)]SO_3F \ (59\%)$$

$$+ \ CpRe(CH_3)(PPh_3)(NO) \ (29\%) \qquad (18)$$

The detailed mechanism proposed[32] for this process is shown in Scheme I of reference 32.

The hydrogen ligands of Cp$_2^*$ZrH$_2$ (Cp* = pentamethylcyclopentadienyl) are distinctly hydridic; comparable, in fact, with trialkylborohydrides in reactivity.   Reaction of Cp$_2^*$Zr(CO)$_2$ with Cp$_2^*$ZrH$_2$ afforded the reduced species   cis-[Cp$_2^*$ZrH]$_2$-(μ-CHO=CHO), presumably by direct hydride transfer.[41]   The generality of this type of reduction reaction was demonstrated by successful hydride transfer to carbonyl ligands of other transition metal complexes, to form "zirconoxy" carbene-containing complexes (eq 19).

$$L_nM\text{----}CO + Cp_2^*ZrH_2 \longrightarrow L_nM=C \begin{smallmatrix} O-ZrCp_2^* \\ H \\ H \end{smallmatrix} \qquad (19)$$

XII

Reaction of $Cp_2M(CO)$ (M = Cr, Mo or W) and $Cp_2NbH(CO)$ with $Cp_2^*ZrH_2$ yielded complexes of this type. The structure of the tungsten containing member of the series, $Cp_2W{=}{=}CHOZr(H)Cp_2^*$, was confirmed by x-ray diffraction methods.

2.2   Migratory-Insertion

A large number of reactions with certain common features have been recognised for organometallic compounds and have been called migratory-insertion reactions. The general reaction (eq 20) is a 1,2-shift of a ligand L to an unsaturated fragment XY, both coordinated to a metal M. The "insertion" of CO into a metal-carbon σ-bond to form a acyl complex has been the subject of numerous synthetic and mechanistic investigations.[42-3] It is generally accepted that the mechanism involves a 1,2-migration of the R group between a coordinated carbonyl group and a cis-coordination site on the metal (eq 21). This mechanism is consistent with a large number of kinetic studies,[42] with the observation that the reaction appears always to proceed with retention of configuration at the α-carbon of the R group,[44] and with the observed stereochemical changes at the metal in octahedral manganese complexes[45] and pseudo-tetrahedral iron complexes.[46]

$$\begin{matrix} L \\ | \\ M-X-Y \end{matrix} \xrightarrow{L'} \begin{matrix} L' \\ | \\ M-X \end{matrix}\begin{smallmatrix} \diagup L \\ \diagdown Y \end{smallmatrix} \qquad (20)$$

$$\begin{matrix} R \\ | \\ M-C\equiv O \end{matrix} \rightleftharpoons \begin{matrix} R \\ \diagup\diagdown \\ M\text{----}C{\equiv}{\equiv}O \end{matrix} \rightleftharpoons M-C\begin{smallmatrix} \diagup R \\ \diagdown O \end{smallmatrix} \qquad (21)$$

A special example of migratory insertion reactions is intramolecular transfer involving the hydride ligand. Hydride transfer reactions from metal to ligand have been achieved for coordinated isocyanide and thiocarbonyl ligands and will be described later. It would therefore appear not unreasonable to assume that the carbonyl ligand, which is similar to these ligands, might also undergo "migratory-insertion" into a metal-hydride bond. Such a process is often invoked as a fundamental step in the Fisher-Tropsch synthesis.[47-8] However, hydride migration to coordinated carbon

$$H-C\equiv O \longrightarrow M-C\begin{smallmatrix}H\\\diagdown O\end{smallmatrix} \tag{22}$$

monoxide to yield a formyl ligand (eq 22) in most cases appears thermodynamically unfavorable although two recent reports, of spontaneous insertion of CO into a metal-hydride bond, one involving an organo-actinide hydride[123] and the other a rhodium porphyrin hydride[124] may be examples of this reaction. In this section other possible examples of this process are examined.

Transition metal carbonyl hydrides are generally considered quite labile toward substitution, and Basolo and Pearson[49] have suggested that the apparently anomalous lability of some carbonyl hydrides might be accounted for in terms of a hydride migration mechanism. The results of both substitution and ^{13}CO exchange studies for $MnH(CO)_5$ are consistent with a hydride migration mechanism involving intermediate formation of a formyl species which undergoes rapid decarbonylation[50] (eq 23).

$$MnH(CO)_5 + {}^{13}CO \longrightarrow Mn(CHO)(CO)_4({}^{13}CO)$$
$$Mn(CHO)(CO)_4({}^{13}CO) \longrightarrow MnH(CO)_4({}^{13}CO) + CO \tag{23}$$

Dombek[51] performed several experiments with $MnH(CO)_5$ under H_2 pressure, suggesting that if, as had been proposed,[50] this complex was in equilibrium with a coordinatively unsaturated formyl complex, $Mn(CHO)(CO)_4$, the latter might be expected to yield formaldehyde or methanol upon hydrogenation. Experiments at temperatures up to $250°C$ and H_2 pressures up to 400atm. gave traces of methanol, but only at the higher temperature and pressure levels. (Similar conditions were required for catalytic methanol formation for CO/H_2 with $MnH(CO)_5$[52]).

Studies on the mechanism of substitution of other transition metal carbonyl hydrides, for example $ReH(CO)_5$[52] and $CpWH(CO)_3$,[54] have provided evidence for a radical chain pathway for substitution. Indeed, even in the case of $MnH(CO)_5$, difficulties in obtaining reproducible kinetic data were encountered owing to contributions from a radical chain process caused by trace radical impurities.[50]   A radical chain pathway may thus have considerable generality for substitution reactions of transition metal carbonyl hydrides compared with a hydride migration mechanism.

Metal formyl complexes have approximately the same kinetic stability as the corresponding metal acetyl complexes[55] and Hoffmann and Berke[56] found that the computed activation energy for formyl complex formation by hydride migration to a carbonyl ligand in $MnH(CO)_5$ was actually smaller than for acetyl formation in $Mn(CH_3)(CO)_5$.   However, whereas acetyl complexes are greatly favored thermodynamically relative to methyl complexes, formyl complexes are much less thermodynamically stable than the corresponding metal hydrides.[55]   Thus, the failure to observe metal formyl species in the reactions of metal hydrides with CO is ascribed to the thermodynamic instability of the formyl complexes and not to greater kinetic lability in such species.   The notably different thermodynamic stabilities of metal acyl and metal formyl compounds is probably attributable to the greater strength of the M-H bond (estimated 200-240 kJ mol^{-1}) compared with the M-C bond (estimated 120-160 kJ mol^{-1}).[57]

The reverse of intramolecular hydride transfer to a coordinated carbonyl ligand, CO "deinsertion" of a formyl group to a hydride ligand, represents an established decomposition pathway for formyl complexes. Solutions of the formyl complex $[Fe(CHO)(CO)_4]^-$ slowly decompose to $[FeH(CO)_4]^-$, presumably by initial loss of CO to give the unsaturated complex **XIII**, followed by hydride migration to give $[FeH(CO)_4]^-$ [14-5] (eq 24).

$$[Fe(CHO)(CO)_4]^- \underset{}{\overset{-CO}{\rightleftharpoons}} [Fe(CHO)(CO)_3]^- \rightarrow [FeH(CO)_4]^- \qquad (24)$$

$$\text{XIII} \quad (t_{1/2} \ 25°C > 12 \text{ days [Na}^+ \text{ salt, THF]})$$

Decomposition of $[Et_4N]$ trans-$[Fe(CHO)(CO)_3(P[OPh]_3)]$ to $[Et_4N]$-$[FeH(CO)_4]$ and $P[OPh]_3$ ($t_{1/2}$ 67°C ~ 1 hr) was observed, and photolysis of solutions of $[Et_4N]$cis-$[Re_2(CHO)(CO)_9]$ gave the dinuclear rhenium hydride $[Et_4N]$cis-$[Re_2H(CO)_9]$.[19]   The iridium formyl complex

$Ir(CHO)(CO)_2(PPh_3)_2$ decomposes through loss of CO to the corresponding metal hydride species,[25] as does the neutral osmium formyl $OsCl(CHO)(CO)_2(PPh_3)_2$.[58] In solution $CpRe(CHO)(CO)(NO)$[24-5] and cis-$[Re(CHO)(\eta^1-C[O]CH_3)(CO)_4]^-$[17] decompose to afford $CpReH(CO)(NO)$ and cis-$[ReH(\eta^1-C[O]CH_3)(CO)_4]^-$ respectively.

Recently, however, studies concerned with the stoichiometric and catalytic reduction of ligated carbon monoxide by hydrogen have furnished evidence for intramolecular hydride migration to the carbonyl ligand. Investigations by Bercaw et al,[59-60] revealing that mononuclear carbonyl and hydrido complexes of bis(pentamenthylcyclopentadienyl) zirconium were capable of promoting the stoichiometric hydrogen reduction of carbon monoxide to methoxide under very mild conditions, constituted the first report of the stoichiometric reduction of coodinated CO by molecular H_2. Treatment of the zirconium dicarbonyl complexes $Cp_2^*Zr(CO)_2$ and $[Cp_2^*Zr(CO)]_2N_2$ with H_2 afforded the methoxy-hydrido species $Cp_2^*ZrH(OCH_3)$ **XIV** (eq 25). Hydrolysis of **XIV** with aqueous HCl led to the formation of $Cp_2^*ZrCl_2$, H_2 and methanol.

$$Cp_2^*Zr(CO)_2 \; + \; H_2 \longrightarrow Cp_2^*ZrH(OCH_3) \qquad\qquad (25)$$
$$\textbf{XIV}$$

The mechanism proposed by Bercaw et al for the reduction of a ligated CO in $Cp_2^*Zr(CO)_2$ by molecular H_2 is depicted in Scheme 2.

The reduction to the methoxy-hydrido species **XIV** is thought to proceed via initial loss of CO from the dicarbonyl complex followed by addition of hydrogen to give the dihydrido carbonyl complex $Cp_2^*ZrH_2(CO)$. The next step proposed is an intramolecular hydride transfer to the carbonyl carbon to afford the dihapto formyl species $Cp_2^*Zr(\eta^2-CHO)H$, **XV**. The η^2-coordination mode has been established by x-ray crystallographic methods for the related acyl complexes $Cp_2Zr(\eta^2-C[O]CH_3)CH_3$[61] and $Cp_2Ti(\eta^2-C[O]CH_3)Cl$.[62] This type of "side-on" bonding imparts an oxycarbenoid character to the carbonyl carbon which allows rationalisation of the reaction patterns observed such as dimerization of **XV** to give the product trans-$[Cp_2^*ZrH]_2(\mu-OCH=CHO)$, **XVI**. In the presence of $Cp_2^*ZrH_2$, insertion of

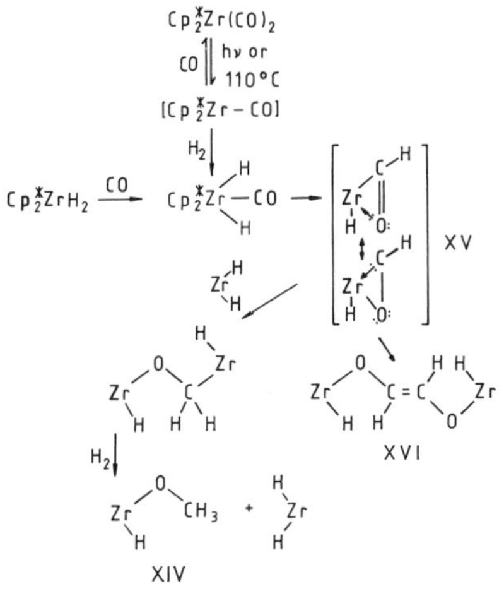

Scheme 2 Proposed Mechanism of the Reaction of $Cp_2^*Zr(CO)_2$ with H_2. (pentamethylcyclopentadienyl ligands are omitted from part of scheme for clarity and compounds in brackets were not observed).

the oxymethylene into a Zr-H bond would be expected to compete with dimerization, and further reduction of the ligated carbonyl-containing moiety to coordinated methoxide is observed. With no direct evidence of the presence of $Cp_2^*ZrH_2$ in this system, the proposed mechanism is questionable; however, under the reaction conditions (~ 1.5 atm. of H_2, $110°C$ or hν) sufficient $Cp_2^*ZrH_2$ may be formed to catalyse the reduction of **XV**.[59]

An examination of the reactivity of hydrido-alkyl derivatives of the type Cp_2^*ZrHR towards CO proved to be mechanistically informative.[60] In particular, the rapid reaction of the iso-butylhydrido complex $Cp_2^*ZrH(CH_2CHMe_2)$ **(XVII)** with CO at $25°C$ afforded $Cp_2^*ZrH(OCH=CHCHMe_2)$ **(XVIII)**. However, at $-50°C$ under 1 atm. of CO, **XVII** was slowly converted to the acyl complex $Cp_2^*ZrH(\eta^2-C[O]CH_2CHMe_2)$, which then rearranged upon warming to the final product **XVIII** (eq 26).

$$Cp_2^x\overset{\overset{H}{|}}{Zr}(CH_2CHMe_2) \xrightarrow[25°C]{CO} Cp_2^xZr\overset{O}{\diagup}\diagdown \begin{smallmatrix}H\\H\end{smallmatrix}\diagdown\diagup\diagup^{\diagdown}\begin{smallmatrix}H\\CHMe_2\end{smallmatrix} \quad XVIII$$

XVII

$$CO \quad\big\downarrow\qquad\diagup\; -20°C$$

$$-50°C\diagdown\qquad\diagup\; CH_2CHMe_2$$

$$Cp_2^xZr\diagdown\diagup^{\diagup C}\overset{\diagdown}{\underset{O}{|}}$$

$$H$$

(26)

Results of labeling studies using deuterium and ^{13}CO suggest the mechanism outlined in Scheme **3**. In the initial step alkyl migration is favored over the also possible hydride migration. However, the final location of the deuterium (which was originally coordinated to the Zr centre) indicates that the oxycarbenoid species thus formed prefers to insert into the Zr-H bond (rather than an adjacent carbon-hydrogen bond as is observed for free oxycarbenes), reminiscent of the proposed reaction of the hydrido formyl species $Cp_2^*Zr(\eta^2\text{-CHO})H$ with $Cp_2^*ZrH_2$ in Scheme **2**. Such an insertion provides the aldehyde adduct, which can undergo subsequent β-hydride abstraction to yield the enolate hydride **XVIII**. The relative ease of alkyl as compared to hydride migration in this system is evinced by the initial quantitative conversion of $Cp_2^*ZrH_2$ to $Cp_2^*ZrH_2(CO)$, in contrast to formation of the acyl complex from CO and $Cp_2^*ZrH(CH_2CHMe_2)$ without the formation of a detectable concentration of the carbonyl adduct $Cp_2^*ZrH(CH_2CHMe_2)(CO)$.

In a related system, Caulton et al[63] have reported that when a solution of the titanium complex $Cp_2Ti(CO)_2$ was treated with a mixture of H_2 and CO at 150°C methane was produced. The methanation was not catalytic in titanium; only the carbonyls originally present in the complex were reduced. As in the zirconium system[59-60] metal-oxygen bonding (i.e. dihapto coordination of the carbonyl moiety) was considered a mechanistic feature critical to the activation of CO toward reduction by hydrogen. Thus, possible intermediates which were proposed for this reaction included an η^2-formyl species ($\overline{Ti-CH-O}$) formed by intramolecular hydride migration.

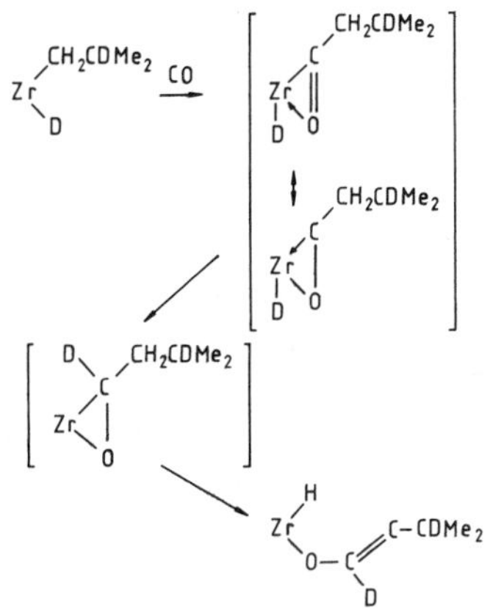

XVIII-d²

Scheme 3   Proposed Mechanism of the Reaction of $Cp_2^*ZrH(CH_2CHMe_2)$ with CO.   (pentamethylcyclopentadienyl ligands are omitted for clarity and complexes in brackets were not observed).

Intramolecular hydride migration was also proposed as a step in the mechanism postulated by Schwartz and Shoer[64] for the catalytic reductive polymerization of CO by aluminum hydrides and zirconium complex catalysts.   In the presence of Cp_2ZrCl_2 as a catalyst DIBAH ($i\text{-}Bu_2AlH$) will reduce CO at room temperature and 1-4 atm. to give, on hydrolysis, a mixture of linear aliphatic alcohols. ^{13}C labeling of CO established that these linear alcohols were derived by incorporation of CO into each carbon unit of the alkyl chain.

$$Cp_2ZrCl_2 + 3 \text{ DIBAH} \longrightarrow \textbf{XIX}, R = i\text{-}C_4H_9$$

$$\begin{array}{l} \text{1. CO} \\ \hphantom{xx} \longrightarrow CH_3OH + CH_3(CH_2)_nOH + i\text{-}C_4H_8 \\ \text{2. } H_3O^+ \\ \hphantom{xxxxxx}(n = 1\text{-}3) \end{array}$$

The mechanistic details of this system are not yet clear, but a suggested mechanism for the initial reduction of CO involves dissociation of DIBAH from **XIX** to provide a vacant coordination site on Zr.   Reduction of coordinated CO could then occur either by intramolecular hydride migration from Zr or by (intramolecular) attack by "external" Al-H.

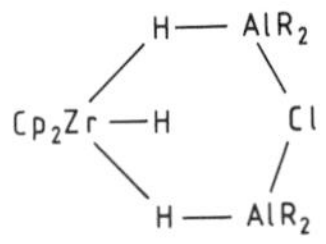

$$\textbf{XIX} \qquad\qquad \textbf{XX} \qquad\qquad \textbf{XXII}$$

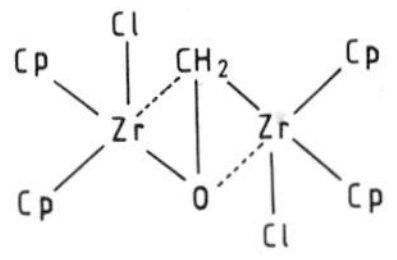

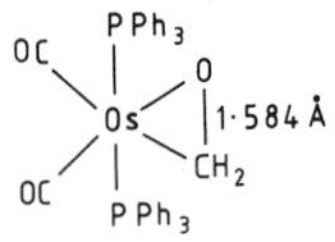

Floriani and coworkers[29] also executed the stoichiometric reduction of carbon monoxide to methanol and cited evidence for CO insertion into the Zr-H bond. The reaction of CO with Cp_2ZrHCl yielded the dinuclear Zr(IV) derivative $(Cp_2ZrCl)_2CH_2O$ (**XX**). The mechanism of this reaction was postulated to involve initial insertion of CO into the Zr-H bond, followed by further reduction of the formyl intermediate thus formed by another molecular of Cp_2ZrHCl[28] (eq 27).

$$Cp_2ZrHCl + CO \longrightarrow [Cp_2ZrCl(CHO)] \xrightarrow{\;Cp_2ZrHCl\;} \textbf{XX} \qquad (27)$$

The reactions of **XX** are consistent with a structure in which the CH_2O group is C- and O-bonded to two different Zr atoms. For example,

$$(Cp_2ZrCl)_2CH_2O \xrightarrow[\text{THF}]{HCl} 2Cp_2ZrCl_2 + MeOH$$

$$(Cp_2ZrCl)_2CH_2O \xrightarrow[\text{1 mol/Zr}]{EtOH} Cp_2Zr(OEt)Cl + Cp_2Zr(OMe)Cl$$

The arrangement of the CH_2O unit between the two Zr atoms is, therefore, assumed to be the same as that found for $CH_2=CH_2$ in $(Cp_2ZrClAlEt_3)_2(CH_2CH_2)$,[65] giving the structure shown.

Reaction of the niobium zirconoxycarbene containing species $Cp_2(H)Nb=CHOZr(H)Cp_2^*$ (**XXI**) with CO afforded the tautomerised species $Cp_2(CO)Nb-CH_2OZr(H)Cp_2^*$ by intramolecular hydride migration.[41] Smooth reaction of **XXI** with H_2 gave the further reduced species $Cp_2^*ZrH(OCH_3)$ and Cp_2NbH_3; in contrast, the tungsten complex $Cp_2W=CHOZr(H)Cp_2^*$ was observed to react only slowly with H_2, to yield $Cp_2^*ZrH(OCH_3)$ and Cp_2WH_2.

The facility of the reaction of the niobium complex with H_2 may be attributed to the availability of a tautomerization pathway for **XXI** which generates a formal 16-electron Nb(III) species capable of oxidatively adding H_2.

A well-defined example of a mononuclear formaldehyde complex is provided by $Os(\eta^2-CH_2O)(CO)_2(PPh_3)_2$ (**XXII**). This compound was formed not by successive hydride transfers but by direct reaction of formaldehyde with the zerovalent complex $Os(CO)_2(PPh_3)_3$;[66] x-ray diffraction study confirms the structure.[26] This compound rearranges to the formyl $Os(CHO)H(CO)_2(PPh_3)_2$. In a related study direct reaction of formaldehyde with the iridium cation $[Ir(PMe_3)_4]^+$ afforded the first cationic formyl complex $[Ir(CHO)H(PMe_3)_4]^+$.[67] Formaldehyde complexes of iron have also been reported.[125]

2.3 Interaction with Electrophiles

In the realm of organic chemistry the effect of electrophiles on carbonyl reactions is well known.[68] In the organometallic area, despite the positive character of the carbon atom in most, if not all, carbonyls, only in certain anionic or donor substituted carbonyl complexes is the carbonyl oxygen atom sufficiently basic to complex with Lewis acids. A high electron density on the carbonyl ligand, indicated by a low carbonyl stretching frequency (<1900 cm^{-1}), is a prerequisite for adduct formation. Activation of transition metal carbonyls by Lewis acids has been suggested as being important for the catalytic reduction of carbon monoxide[69] and this section, therefore, examines the interaction between the carbonyl ligand and various electrophiles.

Organoaluminum compounds have been found to coordinate to the oxygen atom of ligated carbonyl in transition metal complexes, presumably as a result of the inherent strength of aluminum-oxygen bonds. Trimethylalane (Al_2Me_6) reacted with the molybdenum complex $(Ph_3PC_5H_4)Mo(CO)_3$ to form a 1:1 complex $(Ph_3PC_5H_4)Mo(CO)_3 \cdot AlMe_3$ of structure **XXIII**.[70] Similarly the 1:1 complex $[CpW(CO)_3 \cdot AlPh_3]^-$ formed through reaction of the anionic complex $[CpW(CO)_3]^-$ with triphenylaluminum.[71] Reaction of the complexes $Hg[CpM(CO)_3]_2$ (M = Cr,

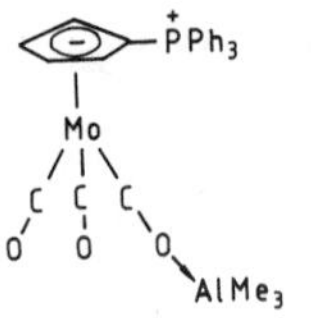

XXIII 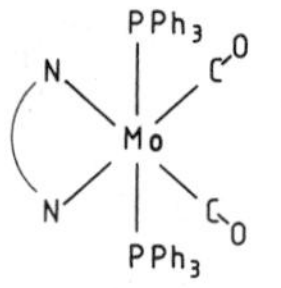   XXIV 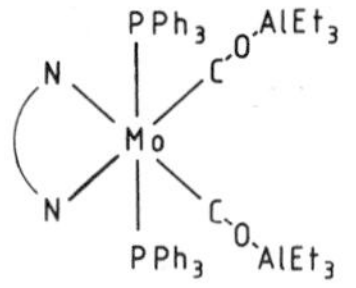   XXV

Mo, W) in THF with excess powdered aluminum metal afforded
$[CpM(CO)_3]_3 \cdot Al(THF)_3$.[72]   The structure of the tungsten derivative was
determined by x-ray diffraction methods and shown to contain an octahedral
aluminum atom bonded to three $CpW(CO)_3$ ligands (a CO ligand on each
$CpW(CO)_3$ unit directly links the two metal atoms through an Al-O-C-W
bonded sequence) with three normal THF ligands in a mer configuration
completing the coordination sphere.   Two analogous series of compounds of
general formula $[CpM(CO)_3]_3 \cdot AlL_3$ (M = Cr, Mo, W; L = py, isoquinoline)
have also been prepared.[72]   Electron donor-substituted metal carbonyls
$Mo(phen)(CO)_2(PPh_3)_2$ and $Mo(phen)_2(CO)_2$ were found to interact with
$AlEt_3$ and $Al(i-Bu)_3$ to afford 1:2 adducts[73] (eq 28).

$$XXIV \ + 2AlEt_3 \xrightarrow{\hspace{3cm}} XXV \hspace{2cm} (28)$$

NN = 1.10-phenanthroline

Other Main Group metals have been observed to undergo O-coordination to
ligated carbon monoxide.   A metal exchange reaction of magnesium and
$Hg[CpMo(CO)_3]_2$ resulted in the formation of $[CpMo(CO)_3]_2 \cdot Mgpy_4$.[14]
Related compounds with the general formula m_2Mgpy_4 (m = $CpFe(CO)_2$,
$Co(CO)_4$ or $Mn(CO)_5$) are also known. $Ga(CH_3)_3$ interacts with the electron
donor-substituted metal carbonyl $Mo(phen)(CO)_2(PPh_3)_2$ to yield the 1:1
adduct XXVI[73] (eq 29).

$$XXIV \ + \ GaMe_3 \xrightarrow{\hspace{3cm}} XXVI \hspace{2cm} (29)$$

XXVI   XXVII

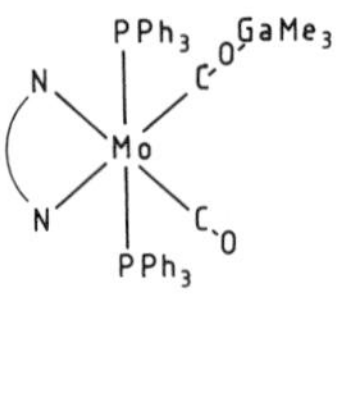

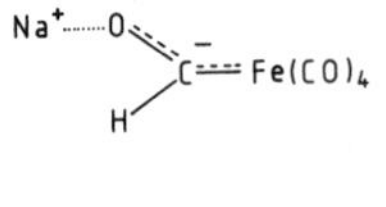

Evidence for the interaction of a transition metal centre with the oxygen atom of a terminal metal carbonyl group was initially afforded by the reaction of $[Fe(CN)_5(CO)]^{3-}$ with aqueous $CoCl_2$ which yielded solid $Co_3[Fe(CN)_5(CO)]_2$.[75]   O-coordination of a metal carbonyl to another transition metal was also demonstrated by the preparation of the complexes $[CpM(CO)_3]_2 \cdot Mnpy_4$ (M = Cr, Mo, W) from $Hg[CpM(CO)_3]_2$ and manganese metal.[76]

Thus far, only strong and specific interactions with the carbonyl oxygen have been discussed.   In addition, there is some evidence for weaker interactions, such as ion pairing with metal carbonyl anions.   Edgell and coworkers have shown by infrared and Raman spectroscopy that $NaCo(CO)_4$ exists in part as an associated ion pair in solvents such as THF and pyridine; the $Co(CO)_4^-$ anion has C_{3v} symmetry with the sodium ion adjacent to a single CO group.[77]   Ion pairing has also been demonstrated for certain alkali metal and Mg^{2+} salts of $Mn(CO)_5^-$ [78] and of trans-$[Fe(\eta^1-C[O]Ph)(CO)_3(PPh_3)]^-$ [79] in diethyl ether solution.   Solvent- and cation-induced shifts of the carbonyl stretching frequency of the formyl ligand in the complexes $M[Fe(CHO)(CO)_4]$ [M = Li^+, Na^+, K^+, $N(PPh_3)_2^+$][14-5] strongly suggest ion pairing involving association primarily through the formyl oxygen, as in **XXVII**.

Ion association of this type appears to be highly influential on some reactions of some metal carbonyl anions.   The rates of migratory insertion reactions are dramatically influenced by the nature of the countercation and added solvents.[80]   Migration of an alkyl group to an adjacent terminal carbonyl is apparently facilitated by small cations such as Li^+ and Na^+ (eq 30).

$$M[Fe(R)(CO)_4] \; + \; L \longrightarrow M[Fe(C[O]R)(CO)_3L] \qquad\qquad (30)$$

$$\text{Rate: } M = Li^+ \; > Na^+ \; \gg N(PPh_3)_2^+$$

Larger, more diffuse cations such as $N(PPh_3)_2^+$, or solvent separated cations, are less effective.   As summarized by Calderazzo, the mechanism for alkyl migration in $Mn(CH_3)(CO)_5$ involves the initial formation of a transient coordinatively unsaturated intermediate (eq 31) which rapidly adds ligand.[43]

$$L_nM(R)(CO) \rightleftharpoons L_nM-COR \xrightarrow{L} L_{n+1}M-COR \tag{31}$$

The steady-state concentration of this intermediate is very low, and the evidence for its existence rests on kinetic analysis. A mechanism proposed for "assisted" alkyl migrations on the anion $[Fe(R)(CO)_4]$[81] is shown in eq 32.

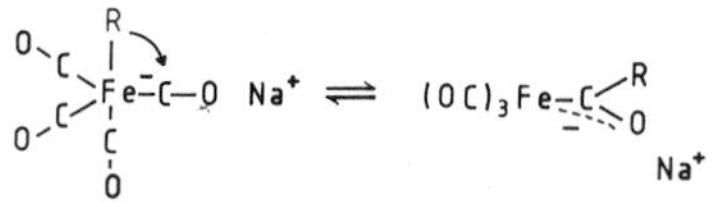

$$\underset{Na^+}{\text{(OC)}_3Fe} \rightleftharpoons \text{(OC)}_3Fe \xrightarrow{L} Na[Fe(C[O]R)(CO)_3L] \tag{32}$$

This mechanism involves $Na[Fe(R)(CO)_4]$ ion pairs in a rapid initial equilibrium with the tight ion-paired coordinatively unsaturated acyl iron tricarbonyl intermediates. The electrostatic interaction of small polarizing cations with the electron-rich carbonyl oxygen atom presumably helps to stabilize the extra negative charge which accompanies the alkyl migrating group. Addition of L follows as the rate-determining step.

In a theoretical discussion of the methyl migration in $Mn(CH_3)(CO)_5$ Hoffmann and Berke[56] considered the influence of attached H^+ or Li^+ to model the observed counterion effect.[80-81] Even though the calculated effects were small, a stabilization of one orbital believed to be crucial to the migration step was noted.

Molecular Lewis acid assistance of alkyl migration has recently been demonstrated in the absence of added CO.[82] The addition of $AlBr_3$ to $Mn(CH_3)(CO)_5$ or $CpMo(CH_3)(CO)_3$ yielded products with spectral features (i r and ^1H nmr) characteristic of Lewis acid coordinated metal acetyls[83] (eqs 33-4).

$$CH_3Mn(CO)_5 + AlBr_3 \longrightarrow \qquad\qquad \textbf{XXVIII} \tag{33}$$

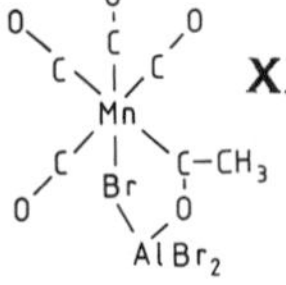

$$CpMo(CO)_3CH_3 + AlBr_3 \longrightarrow \qquad\qquad \textbf{XXIX} \tag{34}$$

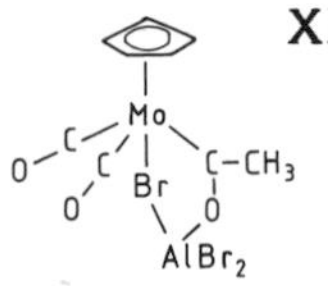

The structure of **XXVIII** was confirmed by x-ray crystal structure analysis. Kinetic data suggest that $AlBr_3$ increases the rate of the primary methyl migration step rather than simply stabilizing an otherwise unstable acetyl intermediate. Thus, the initial formation of an $AlBr_3$ complex with a terminal carbonyl ($L_nM-CO-AlBr_3$), which greatly accelerates the methyl migration, is proposed. Although not detected, the presumed interaction with the terminal CO has precedent in more basic metal carbonyls.[84]

Facile reaction of **XXVIII** and **XXIX** with CO afforded $Mn(\eta^1-C[OAlBr_3]CH_3)(CO)_5$ and $CpMo(\eta^1-C[OAlBr_3]CH_3)(CO)_3$ (**XXX**) respectively, and hydrolysis effected the removal of $AlBr_3$ from these complexes to yield the corresponding simple metal acyls $Mn(\eta^1-C[O]CH_3)(CO)_5$ and $CpMo(\eta^1-C[O]CH_3)(CO)_3$. The isolable adduct **XXX** is obviously stabilized by the Lewis acid, since the parent acetyl $CpMo(\eta^1-C[O]CH_3)(CO)_3$ is unstable with respect to CO loss[85] and is difficult or impossible to produce by high pressure CO insertion on $CpMo(CH_3)(CO)_3$.[86]

The rate of decomposition of the formyl anion $[Fe(CHO)(CO)_4]^-$ to the hydrido-containing anion $[Fe(H)(CO)_4]^-$ has also been found to be dependent upon the countercation[15] (eq 35).

$$[Fe(CHO)(CO)_4]^- \xrightleftharpoons{\quad -CO \quad} [Fe(CHO)(CO)_3]^- \longrightarrow [Fe(H)(CO)_4]^- \qquad (35)$$

$$\text{Rate: } Li^+ > Na^+ > K^+ \gg N(PPh_3)_2^+$$

Acceleration of the decomposition process by the smaller cations is thought to be due to the increased ability of the smaller cations to ion pair through the formyl oxygen atom. The resultant lowering of the electron density on the metal and decreased π-backbonding to the terminal carbonyls then increase the lability of the carbonyl ligands and hence the rate of decomposition.

3.   THE CARBON MONOSULFIDE LIGAND

Complexes containing coordinated carbon monosulfide have been comprehensively reviewed relatively recently.[87] An examination of the different types of CS ligand reactivity reveals an increased propensity for the CS, compared to the CO ligand, to undergo reactions which decrease the multiplicity of the C-X [X = O, S] bond. This reflects the decrease in the tendency to form $p(\pi)$-$p(\pi)$ multiple bonds down any group of the periodic table.

3.1   Nucleophilic Addition

The carbon atom of a coordinated thiocarbonyl is considered to be significantly more electrophilic than that of a ligated carbonyl and hence the thiocarbonyl ligand appears to be more reactive than the carbonyl ligand towards nucleophiles.[88]   There is, however, a significant variation in reactivity from system to system.   For example, the thiocarbonyl ligand in $[CpFe(CO)_2(CS)]^+$ was attacked by the methoxide ion to afford the thioester complex $CpFe(\eta^1\text{-}C[S]OMe)(CO)_2$[89] while the same nucleophile preferentially attacked the carbonyl ligand in $[Ir(CO)_2(CS)(PPh_3)_2]^+$ to give the ester-containing complex $Ir(\eta^1\text{-}C[O]OMe)(CO)(CS)(PPh_3)_2$.[90] $[OsCl(CO)_2(CNR)(PPh_3)_2]^+$ (R = p-tolyl) reacted with OMe^- to yield $OsCl(\eta^1\text{-}C[O]OMe)(CO)(CNR)(PPh_3)_2$, but with SH^- to give the dihapto-p-tolylisothiocyanato complex $Os(\eta^2\text{-}SCNR)(CO)_2(PPh_3)_2$.[91]   Treatment of $[CpFe(CNPh)_2(CS)]^+$ with Na_2CO_3 resulted in attack at the CS carbon atom.[88]

These few examples illustrate the complexity of the relative susceptibility to nucleophilic attack of the three ligands CO, CS and CNR. Thus the results of studies on the reactions of the nucleophiles SH^- and SeH^- with the mixed carbonyl-thiocarbonyl-isocyanide containing cations $[MCl(CO)(CS)(CNR)(PPh_3)_2]^+$ (M = Ru, Os) are particularly interesting.

The reaction of $[OsCl(CO)(CS)(CNR)(PPh_3)_2]^+$ (R = p-tolyl) with SH^- was complicated and the products showed a solvent dependence.[92] In acetone attack occurred exclusively at the isocyanide ligand to afford $Os(\eta^2\text{-}SCNR)(CO)(CS)(PPh_3)_2$, but in dichloromethane-ethanol the products were $Os(\eta^2\text{-}SCNR)(CO)(CS)(PPh_3)_2$ (ca. 85%) and $Os(\eta^2\text{-}CS_2)(CO)(CNR)(PPh_3)_2$ (ca. 15%) indicating attack by the nucleophile at both CNR and CS ligands.

Treatment of $[RuCl(CO)(CS)(CNR)(PPh_3)_2]^+$ (R = p-tolyl) (**XXXI**) with SH^- yielded $Ru(\eta^2\text{-}CS_2)(CO)(CNR)(PPh_3)_2$ as the sole product.[93] Similarly the nucleophile SeH^- has been shown to attack the carbon atom of the thiocarbonyl ligand in **XXXI** to afford novel η^2-CSeS-containing complexes. The initial reaction product is a mixture containing two isomeric forms, namely $\overline{Ru(\eta^2\text{-}CSeS)}(CO)(CNR)(PPh_3)_2$ and $\overline{Ru(\eta^2\text{-}CSeS)}(CO)(CNR)(PPh_3)_2$. The distribution of these isomers in the product was observed to be solvent dependent. Reaction of the bis(isocyanide) cations $[RuCl(CS)(CNR)_2(PPh_3)_2]^+$ (R = p-tolyl, p-chlorophenyl) with SeH^- gave only one isomeric form, namely $\overline{Ru(\eta^2\text{-}CSeS)}(CNR)_2(PPh_3)_2$.

The bis(thiocarbonyl) cation $[Os(\eta^2\text{-}S_2CNEt_2)(CS)_2(PPh_3)_2]^+$ contains only the thiocarbonyl ligand as a site for nucleophilic attack and, as an illustration of this point, reacted with sodium borohydride to afford the stable thioformyl-containing complex $Os(\eta^2\text{-}S_2CNEt_2)(\eta^1\text{-}CHS)(CS)(PPh_3)_2$[94] (eq 36).

$$\text{(36)}$$

The iridium CS cations $[IrClX(CO)(CS)(PPh_3)_2]^+$ (X = Cl,[95] CN[96]) reacted with borohydride to give the iridium(III) thioformyl complexes $IrClX(\eta^1\text{-}CHS)(CO)(PPh_3)_2$. Thus, coordinated carbon monosulfide is reduce by borohydride reagents in a manner analogous to that observed for the carbonyl ligand.[24-5]

3.2 Migratory Insertion

While migratory insertion reactions involving the carbonyl ligand have elicited much study, only recently have reactions of the ligand-transfer type been reported for the thiocarbonyl ligand, primarily because complexes containing suitable ligand combinations such as hydrido, alkyl or aryl ligands cis to CS were rare or unknown. The special case of hydride migration to the carbonyl is supported by indirect evidence only, as mentioned earlier (Section 2.2). For carbon monosulfide, migration reactions to the ligand were considered more likely because the reduction in C-S bond multiplicity which accompanies such processes is more favorable. The lower stability of

C-S multiple bonds when compared to C-O multiple bonds could also serve to stabilize η^2-bonded intermediates in the sulfur case, and because sulfur is not "hard" like oxygen, transition metal-sulfur bonds were expected to be more stable than analogous transition metal-oxygen bonds thus adding to the stability of an η^2-bonded C-S compound.

Hydride transfer to the CS ligand to form a thioformyl ligand represents a logical first step in the complete reduction of carbon monosulfide to methyl thiol. The complexes MHX(CS)(PPh$_3$)$_3$ (M = Os; X = Cl, Br[58] and M = Ru; X = Cl),[97] and OsHCl(CS)(CNR)(PPh$_3$)$_2$ (R = p-tolyl),[58] containing mutually cis hydrido and thiocarbonyl ligands, reacted with CO to yield the monohapto-thioformyl-containing complexes MX(η^1-CHS)(CO)$_2$(PPh$_3$)$_2$ and OsCl(η^1-CHS)(CO)(CNR)(PPh$_3$)$_2$ formed by hydride transfer from the metal centre to the thiocarbonyl ligand (eqs 37-8). The addition of only one equivalent of carbon monoxide to OsHX(CS)(PPh$_3$)$_3$ produced the intermediate complexes OsHX(CO)(CS)(PPh$_3$)$_2$, but in the presence of excess CO the thioformyl complexes were formed exclusively.

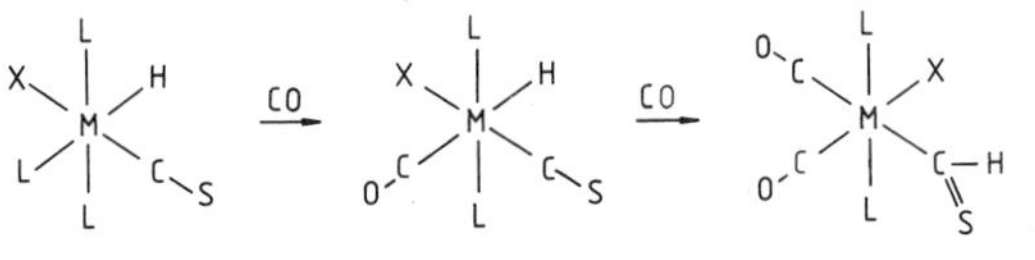

$$(37)$$

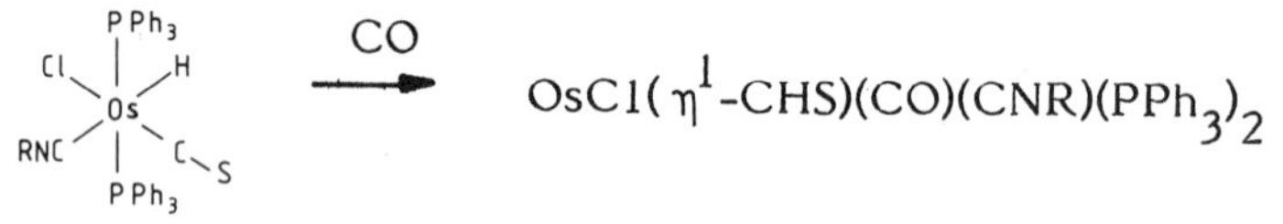

$$OsCl(\eta^1\text{-CHS})(CO)(CNR)(PPh_3)_2 \qquad (38)$$

Regeneration of OsHCl(CO)(CS)(PPh$_3$)$_2$ by prolonged heating of OsCl(η^1-CHS)(CO)$_2$(PPh$_3$)$_2$ in dichloromethane demonstrated that the hydride transfer was reversible.[58]   Reaction of RuHCl(CS)(PPh$_3$)$_3$ with two equivalents of p-tolylisocyanide afforded the thioformyl complex RuCl(η^1-CHS)(CNR)$_2$(PPh$_3$)$_2$.[97]

Dissolution of the monohapto-thioformyl complexes in polar solvents was found to result in a rapid color change accompanied by the partial disappearance of the thioformyl proton resonance in the ^{1}H nmr spectra, and, in the case of $OsCl(\eta^1\text{-CHS})(CO)(CNR)(PPh_3)_2$, the appearance of a broad weak signal at τ -3.23.[98]   This behaviour was attributed to a reversible displacement of a halo ligand in solution concomitant with the formation of an η^2-bonded thioformyl ligand (eq 39).

$$L_nOs\overset{\overset{\overset{H}{|}}{C=S}}{\underset{X}{\diagup}} \rightleftharpoons \left[L_nOs\diagup\overset{\overset{C\diagdown H}{|}}{\underset{S}{}} \right]^+ + X^- \qquad (39)$$

Heating the η^1-thioformyl complex $RuCl(\eta^1\text{-CHS})(CO)_2(PPh_3)_2$ at $\sim 60°C$ for several hours resulted in a novel rearrangement to form the neutral ruthenium dihapto-thioformyl complex $RuCl(\eta^2\text{-CHS})(CO)(PPh_3)_2$.[97] A multiplet at τ -4.0 in the ^1H nmr spectrum has been assigned to the thioformyl proton.   Dihapto coordination of the formyl ligand has been considered to be an important mechanistic feature in the reduction of carbon monoxide[47,69] and the formation of dihapto-formyl species in the stoichiometric and catalytic reduction of ligated carbon monoxide by hydrogen has been proposed.[59,60,63]

The product of hydrogen addition to $IrH(CS)(PPh_3)_3$ was, surprisingly, not $IrH_3(CS)(PPh_3)_2$ but instead, the compound $IrH_2(SMe)(PPh_3)_3$.[95] (H_2 and $IrH(CO)(PPh_3)_3$ are in equilibrium with $IrH_3(CO)(PPh_3)_2$ and PPh_3.[99]) It was postulated that this product resulted from a series of hydride transfers from the metal to the CS moiety (Scheme 4).   Although under the reaction conditions none of the intermediates could be isolated, that a thioformyl complex is not an unreasonable intermediate was demonstrated by the independent preparation $IrCl_2(\eta^1\text{-CHS})(CO)(PPh_3)_2$.   The same sequence of reduction steps as depicted in Scheme 4 for carbon monosulfide has been proposed by some authors to occur in the Fischer-Tropsch reduction of carbon monoxide to methanol.[48,100]

The proposal that a thioformaldehyde complex was not an unreasonable intermediate for the observed reduction of CS by hydrogen is supported by the independent syntheses of two analogous complexes containing the unstable thioformaldehyde monomer viz $M(\eta^2\text{-CH}_2S)(CO)_2(PPh_3)_2$ (M = Ru,[97] Os.[58,101])   Three separate syntheses of the osmium complex were developed;[101,26] (i) inducing the intramolecular transfer of both hydrides of

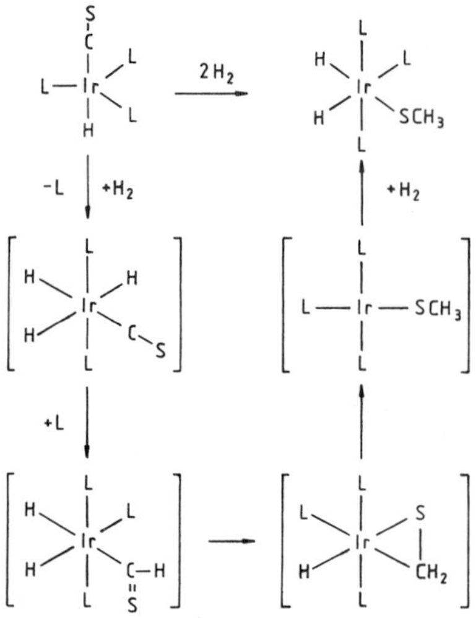

Scheme 4 Proposed Mechanism for the Preparation of $IrH_2(SMe)L_3$ from $IrH(CS)L_3$ and H_2.   (L = PPh_3; bracketed compounds not isolated)

$OsH_2(CS)(PPh_3)_3$ through reaction with CO (eq 40),   (ii) reaction of the thioformyl complex $OsCl(\eta^1\text{-}CHS)(CO)_2(PPh_3)_2$ with $NaBH_4$ (eq 41), and (iii) reaction of $Os(CH_2I)I(CO)_2(PPh_3)_2$[40] with SH^- (eq 42).

$$OsH_2(CS)(PPh_3)_3 \xrightarrow[\text{140}^\circ\text{C}]{\text{4atm.CO}} Os(\eta^2\text{-}CH_2S)(CO)_2(PPh_3)_2 \qquad (40)$$

$$OsCl(\eta^1\text{-}CHS)(CO)_2(PPh_3)_2 \xrightarrow{\text{NaBH}_4} Os(\eta^2\text{-}CH_2S)(CO)_2(PPh_3)_2 \qquad (41)$$

$$Os(CH_2I)I(CO)_2(PPh_3)_2 \xrightarrow{\text{SH}^-} Os(\eta^2\text{-}CH_2S)(CO)_2(PPh_3)_2 \qquad (42)$$

The ruthenium thioformaldehyde complex was prepared by the $NaBH_4$ route only.[97]

Treatment of $Os(\eta^2-CH_2S)(CO)_2(PPh_3)_2$ with HCl rapidly yielded the methylthiolato-containing complex $OsCl(SMe)(CO)_2(PPh_3)_2$ (probably via initial protonation at the metal). Subsequent acid cleavage of the SMe ligand afforded $OsCl_2(CO)_2(PPh_3)_2$.[58,101] An analogue of each intermediate in the stepwise reduction of CS to the CH_3SH adduct $IrH_2(SMe)(PPh_3)_3$[95] has thus been isolated and characterised.

The feasibility of intramolecular transfer of an alkyl ligand to coordinated carbon monsulfide was initially examined by Efraty et al.[102] Reaction of the alkyl complexes $(RC_5H_4)Mn(\eta^1-C_4F_7)(CS)(NO)$ (R = H, CH_3) with PPh_3 was proposed to yield the monohapto-thioacyl complexes $(RC_5H_4)Mn(\eta^1-C[S]C_4F_7)(PPh_3)(NO)$; the tentative structural assignments were based primarily on infrared data.

Octahedral osmium complexes of the type $OsX(R)(L)(CS)(PPh_3)_2$ which have adjacent R and CS ligands, undergo rearrangement to dihapto-thioacyl complexes $OsX(\eta^2-C[S]R)(L)(PPh_3)_2$[103] (eq 43).

$$OsX(R)(L)(CS)(PPh_3)_2 \xrightarrow{\Delta} OsX(\eta^2-C[S]R)(L)(PPh_3)_2 \qquad (43)$$

R = p-tolyl

L = CO; X = Cl, Br, O_2CCF_3

L = CNR; X = Cl

L = CN-p-C_6H_4Cl; X = O_2CCF_3

This formulation was confirmed by an x-ray structure determination of $Os(\eta^1-O_2CCF_3)(\eta^2-C[S]R)(CO)(PPh_3)_2$. Slow alkylation of the sulfur atom of the thioacyl group by methyl triflate provided a route to various thiocarbene complexes, a typical example of which is $OsCl_2(C[SMe]R)(CNR)(PPh_3)_2$. The ruthenium thiocarbonyl complex $RuCl(R)(CS)(PPh_3)_2$ (R = p-tolyl) undergoes a facile rearrangement in the presence of CO to yield the dihapto-thioacyl complex $RuCl(\eta^2-C[S]R)(CO)(PPh_3)_2$.[97]

Efforts to convert the $\eta^2-C[S]R$ ligand to an $\eta^1-C[S]R$ ligand in the osmium complex $OsCl(\eta^2-C[S]R)(CO)(PPh_3)_2$ (the analogous $\eta^1-C[O]R$ coordination mode is common for acyl complexes[104]) through reaction with the powerful chelating ligand diethyldithiocarbamate resulted in the retention of the dihapto-thioacyl and displacement of a phosphine ligand to give $Os(\eta^2-S_2CNEt_2)(\eta^2-C[S]R)(CO)(PPh_3)$.[103] Reaction of the iridium(III) complex $IrCl(CH_3)(CF_3SO_3)(CS)(PPh_3)_2$ with diethyldithiocarbamate, however, yielded the monohapto-thioacyl complex $Ir(\eta^1-S_2CNEt_2)(\eta^1-C[S]CH_3)(PPh_3)_2$.[96]

$CpIr(CS)(PPh_3)$ was found to react readily with CH_3I to give the thiocarbene complex $[CpIr(C[SCH_3]CH_3)I(PPh_3)]I$, the structure of which has been confirmed by x-ray diffraction methods.[105] A reasonable mechanism proposed for this reaction invokes initial oxidative addition of methyl iodide followed by a migratory-insertion reaction involving the thiocarbonyl and methyl ligands for the second step in the sequence. A similar mechanism has been proposed for the reaction between $Ir(C_6F_5)$-$(CS)(PPh_3)_2$ and CH_3I which yielded $Ir(C_6F_5)(C[SCH_3]CH_3)I_2(PPh_3)_2$.[106]

3.3 Interaction with Electrophiles

The thiocarbonyl ligand has been found to be capable of developing a more substantial negative charge on the sulfur atom than the CO ligand can develop on the oxygen atom.[107] In these circumstances the sulfur atom of the thiocarbonyl ligand is susceptible to electrophilic attack. In order to activate the CS ligand in this way an electron-rich metal center with the minimum number of competing π-acceptor ligands is required, and thiocarbonyl stretching frequency must be below 1200 cm^{-1}.

Methyl fluorosulfonate (eq 44) and triethyloxoniumtetrafluoroborate, $[Et_3O]BF_4$, reacted rapidly with cis-$W(CO)(CS)(diphos)_2$ (**XXXII**) to afford the corresponding thiocarbyne complexes.[108]

$$cis\text{-}W(CO)(CS)(diphos)_2 + CH_3SO_3F \rightarrow [W(CSCH_3)(CO)(diphos)_2]SO_3F \quad (44)$$

In contrast, the carbonyl analogue of **XXXII**, cis-$W(CO)_2(diphos)_2$, reacted with $[Et_3O]BF_4$ to yield the alkyl complex $[W(C_2H_5)(CO)_2(diphos)_2]BF_4$. The anionic thiocarbonyl complex trans-$[WI(CO)_4(CS)]^-$ is alkylated at sulfur with CH_3SO_3F or $[Et_3O]BF_4$ to give trans-$WI(CSR)(CO)_4$ products and is acylated by carboxylic acid anhydrides to yield trans-$WI(CSC[O]R)(CO)_4$ (R = CH_3, CF_3).

Analogous $\equiv$C-OR (R = alkyl or acyl) containing complexes derived from metal carbonyl complexes are unknown.   Recently Shriver and coworkers have demonstrated O-protonation and O-alkylation of bridging carbonyl ligands for the iron complexes $[Fe_3(CO)_{10}]^{2-}$ and $[HFe_3(CO)_{11}]^-$ to afford "carbyne" clusters.[109] $[HM_3(\eta_2\text{-COMe})(CO)_{10}]$ (M = Ru, Os) were similarly formed by methylation of the bridging carbonyl ligand in the corresponding $[HM_3(CO)_{11})^-$ anion.[110]   Adducts of terminal CO ligands which are similar to the $\equiv$ CSR carbyne species have been suggested as being reasonable intermediates in catalytic reactions[111] of carbon monoxide which lead to the cleavage of the carbon-oxygen bond.

4.   THE CARBON MONOSELENIDE AND CARBON MONOTELLURIDE LIGANDS

The first complexes incorporating coordinated carbon monoselenide were prepared by Butler et al[112] in 1975, but the study of complexes containing the CSe ligand has been hindered by the lack of uncomplicated synthetic routes of general applicability.   The first ligand reaction of the CSe entity was reported for the selenocarbonyl-containing cation $[RuCl(CO)(CSe)(CNR)(PPh_3)_2]^+$ (R = p-tolyl).[93]   Nucleophilic attack at the carbon atom of the CSe ligand in this substrate by SeH$^-$ led to the novel synthesis of the carbon diselenide complex $Ru(\eta^2\text{-CSe}_2)(CO)(CNR)(PPh_3)_2$ (eq 45).

$$[RuCl(CO)(CSe)(CNR)(PPh_3)_2]^+ \xrightarrow{\text{SeH}^-} \underset{\substack{\text{PPh}_3}}{\overset{\text{PPh}_3}{\underset{\text{RNC}}{\overset{O_{\diagdown}C}{\text{Ru}}}}} \overset{Se}{\underset{Se}{\diagup}} \qquad (45)$$

There are only a few examples of η^2-bonded carbon diselenide, and all other such complexes were prepared by coordination of molecular carbon diselenide to a suitable substrate.[113]

Treatment of $IrCl(CSe)(PPh_3)_2$ (the selenium analogue of Vaska's compound) with $NaBH_4$ in the presence of excess triphenylphosphine led to the formation of $IrH_2(SeMe)(PPh_3)_3$[96] (eq 46).

$$IrCl(CSe)(PPh_3)_2 \xrightarrow[\text{PPh}_3]{\text{NaBH}_4} \underset{\text{not observed}}{[IrH(CSe)(PPh_3)_3]} \rightarrow IrH_2(SeMe)(PPh_3)_3 \qquad (46)$$

It is postulated that the first step involved formation of the hydrido complex $IrH(CSe)(PPh_3)_3$ analogous to the thiocarbonyl case[114] (eq 47). The observed product then resulted from a series of hydride transfers from the metal to the CSe moiety as proposed for the addition of hydrogen to $IrH(CS)(PPh_3)_3$.[95]

$$IrCl(CS)(PPh_3)_2 \xrightarrow{\text{NaBH}_4/\text{PPh}_3} IrH(CS)(PPh_3)_3 \qquad (47)$$

Other reduced forms of the selenocarbonyl ligand which have been reported include an η^2-selenoformaldehyde complex and an η^2-selenoacyl but in neither case was the ligand the result of a migratory-insertion reaction. $Os(\eta^2\text{-}CH_2Se)(CO)_2(PPh_3)_2$ results from reaction of $Os(CH_2I)I(CO)_2(PPh_3)_2$[40] with SeH^- in a manner exactly similar to the sulfur analogue.[126] Two different but related routes led to the synthesis of the η^2-selenoacyl complex; (i) reaction of the chlorotolyl carbene complex $OsCl_2(C[Cl]R)(CO)(PPh_3)_2$ with the nucleophile SeH^- and (ii) reaction of the novel osmium carbyne complex $OsCl(CR)(CO)(PPh_3)_2$ (R = p-tolyl) with elemental selenium[115] (eq 48).

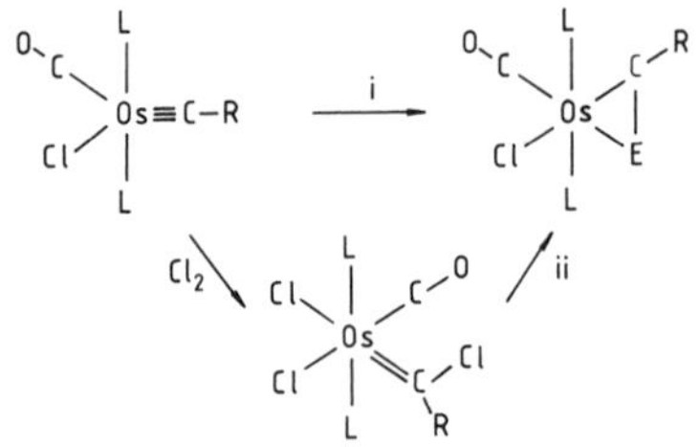

(48)

(R= p-tolyl; E = Se, Te; L = PPh_3; i = Se or Te; ii = SeH^- or TeH^-)

The first tellurocarbonyl complex, $OsCl_2(CO)(CTe)(PPh_3)_2$ was reported in 1980.[116] No ligand reactions of the CTe group have been reported although again a dihapto-telluroacyl complex of osmium has been synthesised, analogous to the selenium case[115] (eq 48) as has a telluroformaldehyde complex $Os(\eta^2\text{-}CH_2Te)(CO)_2(PPh_3)_2$.[126]

5.   THE ISOCYANIDE LIGAND

There is a close similarity between the ligand behavior of coordinated isocyanide and the behavior of the chalcocarbonyl ligands which have been discussed in the previous sections.   This similarity extends to hydride migration reactions and the first examples of such a process were provided by ruthenium hydride-isocyanide complexes.   Excess acetate with the appropriate isomer of $RuHI(CNR)(CO)(PPh_3)_2$ (R = p-tolyl) replaced iodide and the acetate adopted the dihapto-arrangement accompanied by hydride migration to the isocyanide forming an iminoformyl ligand in **XXXIV** (eq 49).

XXXIII $\underset{I^-}{\overset{OAc^-}{\rightleftharpoons}}$ **XXXIV**   (49)

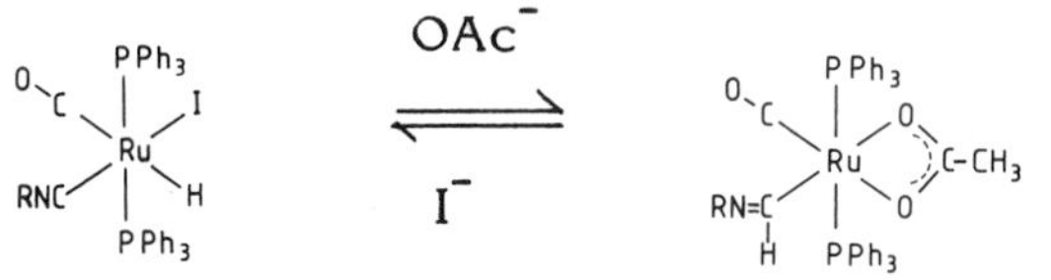

This reaction is easily reversed and excess iodide with **XXXIV** regenerates **XXXIII.**   The structure of **XXXIV** has been determined by an x-ray diffraction study.[117]   **XXXIV** is also formed by hydride migration in $RuH(OAc)(CNR)(PPh_3)_2$ induced by CO coordination (eq 50).[118]

$$RuH(\eta^2\text{-}OAc)(CNR)\underline{trans}\text{-}(PPh_3)_2 \xrightarrow{\ CO\ } \textbf{XXXIV} \qquad (50)$$

Similar migratory-insertions of isocyanide into Pt-H[119] and Zr-H[120] bonds are known.   An attempt to model some of the steps proposed in Scheme 2 led to direct observation of the iminoformyl species **XXXV** (eq 51) formed by facile migratory-insertion of isocyanide into a Zr-H bond of the unstable adduct $Cp_2^*ZrH_2(CNR)$.[120]

$$Cp_2^*ZrH_2 \ + \ RNC \ \longrightarrow \ \textbf{XXXV} \qquad (51)$$

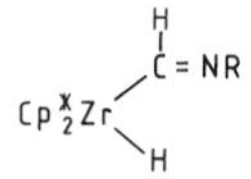

R = CH_3 or 2,6-dimethylphenyl

6.   CONCLUSION

The neutral carbon donor ligands CX (X=O,S,Se,NR) are converted to anionic CHX ligands through external attack of H$^-$ or by migratory-insertion into an existing M-H bond.   The migration reaction appears thermodynamically favorable for all the above ligands except CO.   The situation for H transfer to CO will become more favorable when the following conditions are met (i) the hydride ligand is particularly electron rich as indicated for hydrides of zirconium on the basis of ^{1}H nmr[59-60] and reactivity criteria,[121] (ii) the CO ligand has minimal back-bonding from the metal thus maximising the positive charge on carbon, a situation which could be brought about by competing π-acceptor ligands, a high oxidation state or a low d^n configuration, (iii) association of the carbonyl O atom with a suitable electrophile.   An example of this last requirement is provided by the organo-actinide system.[124]   Both monohapto- and dihapto-arrangements have been recognised for the CHX-ligands.

Further reduced ligands CH$_2$X have all been recognised in complexes with a dihapto-arrangement of the ligand.   The identification of η^2-formaldehyde in a stable complex indicates the possibility of such an intermediate in CO reduction and indeed a recent report describing ruthenium compounds as homogeneous catalysts for the hydrogenation of CO to methanol postulates such an intermediate.[122]

Several complexes containing the hydroxy-methyl ligand (the final reduction product CH$_2$XH), have now been thoroughly characterised.

7.   MORE RECENT DEVELOPMENTS

Since this manuscript was written there have been few developments relating to the ligands CS, CSe and CTe.   However, steady interest has been maintained and some progress has occurred in work directed towards the reduction of the CO ligand.   The reader is referred in particular to three reviews[127-9] relevant to this topic.

REFERENCES

1. R. Eisenberg and D.E. Hendriksen, Advan. Catal. 28:79 (1979).

2. E.L. Muetterties and J. Stein, Chem. Rev. 79:479 (1979).

3. R. Hoffmann, M.M.-L. Chen and D.L. Thorn, Inorg. Chem. 16:503 (1977).

4. (a) D.J. Cardin, B. Cetinkaya and M.F. Lappert, Chem. Rev. 72:545 (1972).
   (b) E.O. Fischer, Advan. Organomet. Chem. 14:1 (1976).

5. R.J. Angelici and L. Busetto, J. Am. Chem. Soc. 91:3197 (1969).

6. R.J. Angelici, Acc. Chem. Res. 5:335 (1972).

7. A.J. Deeming and B.L. Shaw, J. Chem. Soc. (A) 443 (1969).

8. (a) E.O. Fischer and A. Maasböl, Chem. Ber. 100:2445 (1967).
   (b) D.J. Darensbourg and M.Y. Darensbourg, Inorg. Chem. 9:1691 (1970).
   (c) E.O. Fischer, U. Schubert and H. Fischer, Inorg. Synth. 19:172 (1979).

9. A. Berry and T.L. Brown, J. Organomet. Chem. 33:C67 (1971).

10. J. Chatt, G.J. Leigh and N. Thankarajan, J. Organomet. Chem. 29:105 (1971).

11. W.O. Siegl, J. Organomet. Chem. 92:321 (1975).

12. (a) R.G. Hayter, J. Am. Chem. Soc. 88:4376 (1966).
    (b) M.R. Churchill, S.W.-Y. Ni Chang, M.L. Berch and A. Davison, J. C. S. Chem. Commun. 691 (1973).

13. C.P. Casey and S.M. Neumann, J. Am. Chem. Soc. 98:5395 (1976).

14. J.P. Collman and S.R. Winter, J. Am. Chem. Soc. 95:4089 (1973).

15. S.R. Winter, G.W. Cornett and E.A. Thompson, J. Organomet. Chem. 133:339 (1977).

16. J.A. Gladysz and J.C. Selover, Tetrahedron Letters 319 (1978).

17. K.P. Darst and C.M. Lukehart, J. Organomet. Chem. 171:65 (1979).

18. J.A. Gladysz, G.M. Williams, W. Tam and D.L. Johnson, J. Organomet. Chem. 140:C1 (1977).

19. C.P. Casey and S.M. Neumann, J. Am. Chem. Soc. 100:2544 (1978).

20. J.A. Gladysz and W. Tam, J. Am. Chem. Soc. 100:2545 (1978).

21. J.A. Gladysz and W. Tam, J. Organomet. Chem. 43:2279 (1978).

22. J.R. Moss, M. Green and F.G.A. Stone, J. C. S. Dalton Trans 975 (1973).

23. (a) H.J. Svec and G.A. Junk, J. Am. Chem. Soc. 89:2836 (1967).
    (b) D.L.S. Brown, J.A. Connor and H.A. Skinner, J. Organomet. Chem. 81:403 (1974).

24. C.P. Casey, M.A. Andrews and J.E. Rinz, J. Am. Chem. Soc. 101:741 (1979).

25. W. Tam, W.-K. Wong and J.A. Gladysz, J. Am. Chem. Soc. 101:1589 (1979).

26. G.R. Clark, C.E.L. Headford, K. Marsden and W.R. Roper, J. Organomet. Chem, 231:335 (1982).

27. W.-K. Wong, W. Tam, C.E. Strouse and J.A. Gladysz, J. C. S. Chem. Commun. 530 (1979).

28. J. Schwartz and J.A. Labinger, Angew. Chem. Int. Ed. Engl. 15:333 (1976).

29. G. Fachinetti, C. Floriani, A. Roselli and S. Pucci, J. C. .S. Chem. Commun. 269 (1978).

30. I.S. Astakhova, A.A. Johannson, V.A. Semion, Yu. T. Struchkov, K.N. Anisimov and N.E. Kolobova, J. C. S. Chem. Commun. 488 (1969).

31. C.M. Lukehart and J.V. Zeile, J. Am. Chem. Soc. 98:2365 (1976).

32. W.-K. Wong, W. Tam and J.A. Gladysz, J. Am. Chem. Soc. 101:5440 (1979).

33. P.M. Treichel and R.L. Shubkin, Inorg. Chem. 6:1328 (1967).

34. A. Davison, M.L.H. Green and G. Wilkinson, J. Chem. Soc. (A) 3172 (1961).

35. R.P. Stewart, N. Okamoto and W.A.G. Graham, J. Organomet. Chem. 42:C32 (1972).

36. A.N. Nesmeyanov, K.N. Anisimov, N.E. Kolobova and L.L. Krasnoslobodskaya, Bull. Acad. Sci. U.S.S.R. 807 (1970).

37. J.R. Sweet and W.A.G. Graham, J. Organomet. Chem. 173:C9 (1979); idem, J. Am. Chem. Soc. 104:2811 (1982).

38. C.P. Casey, M.A. Andrews and D.R. McAlister, J. Am. Chem. Soc. 101:3371 (1979).

39. J. March, "Advanced Organic Chemistry: Reactions, Mechanisms and Structure", McGraw-Hill, New York, (1968).

40. C.E.L. Headford and W.R. Roper, J. Organomet. Chem. 198:C7 (1980).

41. P.T. Wolczanski, R.S. Threlkel and J.E. Bercaw, J. Am. Chem. Soc. 101:218 (1979).

42. A. Wojcicki, Advan. Organomet. Chem. 11:87 (1973).

43. (a) G. Henrici - Olive and S. Olive, Transition Met. Chem. 1:77 (1976).
    (b) F. Calderazzo, Angew. Chem. Int. Ed. Engl. 16:299 (1977).

44. (a) L.F. Hines and J.K. Stille, J. Am. Chem. Soc. 94:485 (1972)
    (b) N.A. Dunham and M.C. Baird, J. C. S. Dalton Trans 774 (1975)
    (c) J.A. Labinger, D.W. Hart, W.E. Seibert III and J. Schwartz, J. Am. Chem. Soc. 97:3851 (1975).

45. K. Noack and F. Calderazzo, J. Organomet. Chem. 10:101 (1967).

46. H. Brunner and J. Strutz, Z. Naturforsch. B29:446 (1974).

47. H. Pichler and H. Schulz, Chem.-Ing.-Tech. 42:1162 (1970).

48. G. Henrici-Olive and S. Olive, Angew. Chem. Int. Ed. Engl. 15:136 (1976).

49. F. Basolo and R.G. Pearson, "Mechanisms of Inorganic Reactions" 2nd ed., Wiley, New York, 1967, p.555.

50. (a) A. Berry and T.L. Brown, J. Organomet. Chem. 33:C67 (1971).
    (b) B.H. Byers and T.L. Brown, J. Organomet. Chem. 127:181 (1977).

51. B.D. Dombek, J. Am. Chem. Soc. 101:6466 (1979).

52. J.W. Rathke and H.M. Feder, J. Am. Chem. Soc. 100:3623 (1978).

53. B.H. Byers and T.L. Brown, J. Am. Chem. Soc. 97:947 (1975).

54. N.W. Hoffman and T.L. Brown, Inorg. Chem. 17:613 (1978).

55. C.P. Casey and S.M. Neumann, Adv. Chem. Ser. 173:131 (1978).

56. H. Berke and R. Hoffmann, J. Am. Chem. Soc. 100:7224 (1978).

57. (a) F.A. Adedeji, J.A. Connor, H.A. Skinner, L. Galyer and G. Wilkinson, J. C. S. Chem. Commun. 159 (1976).
    (b) M.F. Lappert, D.S. Patil and J.B. Pedley, J. C. S. Chem. Commun. 830 (1975).
    (c) J.A. Connor, Topics Current Chem. 71:71 (1977).

58. (a) T.J. Collins and W.R. Roper, J. C. S. Chem. Commun. 1044 (1976)
    (b) T.J. Collins and W.R. Roper, J. Organomet. Chem. 159:73 (1978).

59. J.M. Manriquez, D.R. McAlister, R.D. Sanner and J.E. Bercaw, J. Am. Chem. Soc. 98:6733 (1976).

60. J.M. Manriquez, D.R. McAlister, R.D. Sanner and J.E. Bercaw, J. Am. Chem. Soc. 100:2716 (1978).

61. G. Fachinetti, C. Floriani, F. Marchetti and S. Merlino, J. C. S. Chem. Commun. 522 (1976).

62. G. Fachinetti, C. Floriani and H. Stoeckli-Evans, J. C. S. Dalton Trans 2297 (1977).

63. J.C. Huffman, J.G. Stone, W.C. Krusell and K.G. Caulton, J. Am. Chem. Soc. 99:5829 (1977).

64. L.I. Shoer and J. Schwartz, J. Am. Chem. Soc. 99:5831 (1977).

65. W. Kaminsky, J. Kopf, H. Sinn and H.-J. Vollmer, Angew. Chem. Int. Ed. Engl. 15:629 (1976).

66. K.L. Brown, G.R. Clark, C.E.L. Headford, K. Marsden and W.R. Roper, J. Am. Chem. Soc. 101:503 (1979).

67. D.L. Thorn, J. Am. Chem. Soc. 102:7109 (1980).

68. J.L. Pierre and H. Handel, Tetrahedron Letters 2317 (1974).

69. G.C. Demitras and E.L. Muetterties, J. Am. Chem. Soc. 99:2796 (1977).

70. J.C. Kotz and C.D. Turnipseed, J. C. S. Chem. Commun. 41 (1970).

71. J.M. Burlitch and R.B. Petersen, J. Organomet. Chem. 24:C65 (1970).

72. R.B. Petersen, J.J. Stezowski, C. Wan, J.M. Burlitch and R.E. Hughes, J. Am. Chem. Soc. 93:3532 (1971).

73. D.F. Shriver and A. Alich, Inorg. Chem. 11:2984 (1972).

74. S.W. Ulmer, P.M. Skarstad, J.M. Burlitch and R.E. Hughes, J. Am. Chem. Soc. 95:4469 (1973).

75. E.L. Brown and D.B. Brown, J. C. S. Chem. Commun. 67 (1971).

76. T. Blackmore and J.M. Burlitch, J. C. S. Chem. Commun. 405 (1973).

77. (a) W.F. Edgell, J. Lyford IV, A. Barbetta and C.I. Jose, J. Am. Chem. Soc. 93:6403 (1971).
    (b) W.F. Edgell and J. Lyford IV, J. Am. Chem. Soc. 93:6407 (1971).

78. C.D. Pribula and T.L. Brown, J. Organomet. Chem. 71:415 (1974).

79. M.Y. Darensbourg and D. Burns, Inorg. Chem. 13:2970 (1974).

80. J.P. Collman, J.N. Cawse and J.I. Brauman, J. Am. Chem. Soc. 94:5905 (1972).

81. J.P. Collman, R.G. Finke, J.N. Cawse and J.I. Brauman, J. Am. Chem. Soc. 100:4766 (1978).

82. S.B. Butts, E.M. Holt, S.H. Strauss, N.W. Alcock, R.E. Stimson and D.F. Shriver, J. Am. Chem. Soc. 101:5864 (1979).

83. R.E. Stimson and D.F. Shriver, Inorg. Chem. 19:1141 (1980).

84. D.F. Shriver, J. Organomet. Chem. 94:259 (1975).

85. K.W. Barnett and D.W. Slocum, J. Organomet. Chem. 44:1 (1972).

86. R.B. King, A.D. King, jr., M.Z. Iqbal and C.C. Frazier, J. Am. Chem. Soc. 100:1687 (1978).

87. P.V. Yaneff, Coord. Chem. Rev. 23:183 (1977).

88. (a) I.S. Butler and A.E. Fenster, J. Organomet. Chem. 66:161 (1974)
    (b) L. Busetto and A. Palazzi, Inorg. Chim. Acta 19:233 (1976).

89. L. Busetto, M. Graziani and U. Belluco, Inorg. Chem. 10:78 (1971).

90. M.J. Mays and F.P. Stefanini, J. Chem. Soc. (A) 2747 (1971).

91. K.R. Grundy and W.R. Roper, J. Organomet. Chem. 113:C45 (1976).

92. G.R. Clark, T.J. Collins, D. Hall, S.M. James and W.R. Roper, J. Organomet. Chem. 141:C5 (1977).

93. P.J. Brothers, C.E.L. Headford and W.R. Roper, J. Organomet. Chem. 195:C29 (1980).

94. T.J. Collins and W.R. Roper, J. Organomet. Chem. 139:C9 (1977).

95. W.R. Roper and K.G. Town, J. C. S. Chem. Commun. 781 (1977).

96. W.R. Roper and K.G. Town, J. Organomet. Chem. 252:C97 (1983).

97. S.V. Hoskins and W.R. Roper, unpublished work.

98. T.J. Collins, Ph.D. Thesis, University of Auckland (1977).

99. M.G. Burnett and R.J. Morrison, J. Chem. Soc. (A) 2325 (1971).

100. M.A. Vannice, Catal. Rev.-Sci. Eng. 14:153 (1976).

101. T.J. Collins and W.R. Roper, J. C. S. Chem. Commun. 901 (1977).

102. A. Efraty, R. Arneri and W.A. Ruda, Inorg. Chem. 16:3124 (1977).

103. G.R. Clark, T.J. Collins, K. Marsden and W.R. Roper, J. Organomet. Chem. 157:C23 (1978); idem, ibid. 259:215 (1983).

104. (a) D.J. Egglestone, M.C. Baird, C.J.L. Lock and G. Turner, J. C. S. Dalton Trans 1576 (1977).
(b) C.M. Lukehart and J.V. Zeile, J. Organomet. Chem. 140:309 (1977).
(c) C.-H. Cheng, D.E. Hendriksen and R. Eisenberg, J. Organomet. Chem. 142:C65 (1977).

105. F. Faraone, G. Tresoldi and G.A. Loprete, J. C. S. Dalton Trans 933 (1979).

106. G. Tresoldi, F. Faraone and P. Piraino, J. C. S. Dalton Trans 1053 (1979).

107. (a) B.D. Dombek and R.J. Angelici, Inorg. Chem. 15:2397 (1976)
(b) B.D. Dombek and R.J. Angelici, J. Am. Chem. Soc. 97:7568 (1974).

108. B.D. Dombek and R.J. Angelici, J. Am. Chem. Soc. 97:1261 (1975).

109. (a) D.F. Shriver, D. Lehman and D. Strope, J. Am. Chem Soc. 97:1594 (1975)
(b) H.A. Hodali, D.F. Shriver and C.A. Ammlung, J. Am. Chem. Soc. 100:5239 (1978).

110. J.B. Keister, J. C. S. Chem. Commun. 214 (1979).

111. C. Masters, Advan. Organomet. Chem. 17:61 (1979).

112. I.S. Butler, D. Cozak and S.R. Stobart, J. C. S. Chem. Commun. 103 (1975).

113. (a) K.A. Jensen and E. Huge-Jensen, Acta Chem. Scand. 27:3605 (1973)
(b) K. Kawakami, Y. Ozaki and T. Tanaka, J. Organomet. Chem. 69:151 (1974)
(c) G.R. Clark, K.R. Grundy, R.O. Harris, S.M. James and W.R. Roper, J. Organomet. Chem. 90:C37 (1975).

114. M.P. Yagupsky and G. Wilkinson, J. Chem. Soc. (A) 2813 (1968).

115. G.R. Clark, K. Marsden, W.R. Roper and L.J. Wright, J. Am. Chem. Soc. 102:6570 (1980).

116. G.R. Clark, K. Marsden, W.R. Roper and L.J. Wright, J. Am. Chem. Soc. 102:1206 (1980).

117. D.F. Christian, G.R. Clark, W.R. Roper, J.M. Waters and K.R. Whittle, J. C. S. Chem. Commun. 458 (1972).

118. D.F. Christian and W.R. Roper, J. Organomet. Chem. 80:C35 (1974).

119. D.F. Christian, H.C. Clark and R.F. Stepaniak, J. Organomet. Chem. 112:209 (1976).

120. P.T. Wolczanski and J.E. Bercaw, J. Am. Chem. Soc. 101:6450 (1979).

121. J.A. Labinger and K. Komadina, J. Organomet. Chem. 155:C25 (1978).

122. B.D. Dombek, J. Am. Chem. Soc. 102:6855 (1980).

123. P.J. Fagan, K.G. Moloy and T.J. Marks, J. Am. Chem. Soc. 103:6959 (1981).

124. B.B. Wayland, B.A. Woods and R. Pierce, J. Am. Chem. Soc. 104:302 (1982).

125. H. Berke, W. Bankhardt, G. Huttner, J.V. Seyerl and L. Zsolnai, Chem. Ber. 114:2754 (1981).

126. C.E.L. Headford and W.R. Roper, J. Organomet. Chem. 244:C53 (1983).

127. J.A. Gladysz, Advan. Organomet. Chem. 20:1 (1982).

128. W.A. Herrmann, Angew. Chem. Int. Ed. Engl. 21:117 (1982).

129. C. Floriani, Pure and Appl. Chem. 55:1 (1983).

REACTIONS OF COORDINATED ISOCYANIDES

B. Crociani

Istituto di Chimica Generale

University of Palermo

1. INTRODUCTION

Isocyanides or isonitriles, CNR, represent a class of unsaturated carbon ligands, isoelectronic with carbon monoxide, which in most of their transition metal derivatives act as σ-donors through the lone pair of electrons on the terminal carbon atom and as π-acceptors through the π^* antibonding orbitals of the CN multiple bond system. Compared to carbon monoxide, isocyanides exhibit a much greater versatility since their electronic and steric properties may be widely varied depending on the nature of the substituent R. Despite this feature, the chemistry of transition metal-isocyanide complexes began to be extensively investigated rather recently, in the late sixties, when a unique concomitance of favorable circumstances occurred: (i) availability of easy and general synthetic methods by dehydration of N-monosubstituted formamides;[1] (ii) discovery of important reactions, such as insertion into a metal-carbon σ bond,[2] the formation of metal-stabilized carbene ligands by attack of protic nucleophiles[3] and the copper-catalyzed 1,1-addition of active hydrogen compounds;[4] (iii) publication of an excellent monograph on isocyanide complexes by Malatesta and Bonati.[5] As a consequence, the rate of publication in this field increased to such extent that a number of reviews appeared in the early seventies,[6-10] the last of which dates back to 1974. A

553

review of isocyanide complexes with zerovalent transition metals has been published more recently.[11]

The aim of this chapter is to survey most of the literature on reactions of coordinated isocyanides from 1974 onwards with a special emphasis on mechanistic aspects and the use of coordinated isocyanides in organic synthesis. References to papers which have been already reported in the earlier reviews are kept to a minimum, but are given each time they are required. On the other hand, synthetic aspects concerning the preparation of new types of functionalized CNR ligands or novel routes to isocyanide complexes, although an active and interesting field of research,[11a] are in general beyond the scope of this chapter. Some additional, more recent, references do not affect the main arguments and are listed without discussion.

2. BONDING AND REACTIVITY

Recent xray structural investigations have shown that, in addition to the well-known η^1-terminal and μ_2, η^1-bridging coordination modes, the isocyanide ligands may also act as μ_2,η^2-bridging and μ_3,η^2-bridging four-electron donors, in remarkable analogy to the carbonyl systems in transition metal complexes.[12]

2.1 Terminal isocyanides

Terminal CNR ligands are bound to transition metals through the carbon atom of the isocyano group (η^1- or "head on" coordination). No η^2- or "side on" coordination has been so far reported in mononuclear compounds. This type of bonding has been described by many authors in terms of σ and π contributions, represented by mesomerism between Ia and Ib.

$$\bar{M} - C \equiv \overset{+}{N} - R \qquad M = C = N \diagdown_R$$
$$\text{Ia} \qquad\qquad\qquad \text{Ib}$$

While oversimplified, this scheme is generally used to interpret most of the spectral and structural data, and is particularly convenient in explaining the different reactivity of coordinated isocyanides. The bent limiting formula Ib would result only if one of the two orthogonal π^*orbitals of the isocyano group is preferentially involved in the back-donation.

An approximate molecular orbital study has shown that σ-donation by the CNR carbon lone pair will bring about an increased CN bond order, since these electrons occupy an orbital which is somewhat antibonding in the CN region.[13] This will result in a shorter CN bond and in a higher (CN) frequency relative to the free ligand. A rather short CN bond (1.11 Å) was actually found for the linearly bound isocyanide in the complex, [PrCp$_3$(CNCy)] (Cy=cyclohexyl), in which an almost pure Pr-CNCy bond is presumably present (Pr-C bond distance 2.65 Å).[14] The microwave spectrum of the adduct MeNC-BH$_3$ gives a carbon-nitrogen distance of 1.155 Å, which is 0.011 Å shorter than that in the free ligand.[15]

In transition metal complexes, π back-donation of metal d electrons of appropriate symmetry into the antibonding orbitals of the CN multiple bond system (formula Ib) will increase the bond order in the metal-carbon region, with a concomitant bond order decrease in the CN region. Furthermore, when extensive and uneven population of the antibonding orbitals occurs, as appears to be the case for some "electron rich" metal sites (Table 1), a marked deviation from linearity at the nitrogen atom of the M-CNR unit is observed.

A considerable number of xray structural analyses on terminally bound isocyanide complexes have been published.[16] In general, the shortening of the M-C bond relative to the sum of covalent radii was ascribed to π bonding effects, although no appreciable concomitant increase in the CN bond distance was noticed. Except for the complex [FeCl(CNC$_6$H$_4$Me-4)$_2$[PPh (OEt)$_2$]$_3$]ClO$_4$, where the rather short Fe-C bond (1.73 Å) is accompanied by a longer CN bond (1.23 Å),[17] the CN distances in the complexes are not significantly different from those in the free ligands. If such a difference is likely to be small for bonds of order 2-3, one must also take into account that σ and π contributions have opposite effects on the CN bond length.

In most cases, the M-C-N-R arrangement was found to be nearly linear with angles at nitrogen higher than 170°, the small distortions being attributed to crystal packing interactions. In Table 1 some complexes are listed for which the C-N-R angles of terminal isocyanides are significantly lower than 170°.

Table 1. Structural Data for Complexes with Terminal Bent Isocyanide Ligands.

Compound	M-C(Å)	C≡N(Å)	∠C-N-R(°)	Position of CNR Ligands	Reference
Fe(TPP)(CNBu-t)$_2$·2PhMe[a]	1.901(3) 1.900(3)	1.152(3) 1.162(3)	159.1(3) 165.2(2)	trans to each other	16
Fe(η^4-C$_6$H$_8$)(CO)$_2$(CNEt)[b]	1.83(2)	1.15(3)	162(2)	basal	18
Co$_2$(μ-CNBu-t)$_2$(CNBu-t)$_6$	1.813(15) 1.842(14) 1.856(14)	1.153(20) 1.169(20) 1.153(19)	160.5(16) 154.6(17) 164.1(17)	terminal	19
Cr(η^6-PhCO$_2$Me)(CO)$_2$(CNCOPh)	1.85(1)	c	168(1)		20
Fe(η^4-Diiminodiene)(CNBu-t)$_3$[d]	1.864(11) 1.862(8)	1.16(1) 1.17(1)	175.4(8) 177(1)	basal apical	21
	1.813(8)	1.18(1)	165(1)		
[FeCl(CNC$_6$H$_4$Me-4)$_3$(PPh$_3$)$_2$][FeCl$_4$]	1.874(4) 1.860(4)	1.147(5) 1.143(5)	175.0(8) 178.1(8)	trans to each other	22
	1.813(4)	1.158(5)	167.1(8)	trans to Cl	
Mo(CNMe)$_2$(dppe)$_2$[e]	2.101(7)	1.10(1)	156(1)	trans to each other	23
Ru(CNBu-t)$_4$(PPh$_3$)	2.00(3) 1.92(2)	1.17(4) 1.19(4)	167(3) 172(3)	axial	24
	1.86(2) 1.91(2)	1.28(3) 1.22(3)	131(2) 129(2)	equatorial	
Fe(CNBu-t)$_5$	1.813(7) 1.803(10) 1.835(8)	1.21(1) 1.22(1) 1.19(1)	133.1(8) 134.6(8) 169.2(7)	radial	24
	1.811(7) 1.837(7)	1.20(1) 1.16(1)	151.6(9) 177.1(8)	axial	

(a) TPP = Meso-tetraphenylporphyrinato (b) C_6H_8 = 1,3-Cyclohexadiene (c) Not reported (d) Diiminodiene = 2,3-diphenyl-N,N'-di-t-butyl-buta-1,3-diene-1,4-diimine (e) dppe = 1,2-bis(diphenylphosphino)ethane.

Table 2. Structural Data for Complexes with μ_2,η^1- bridging isocyanides.[a]

Compound	M-C(Å)	C=N(Å)	M-M(Å)	∠C-N-R(°)	∠M-C-M(°)	Reference
<u>cis</u>-Fe₂Cp₂(CO)₂(μ-CO)(μ-CNPh)	1.93	1.28	2.532	131	83.6	78
<u>cis-anti</u>-Fe₂Cp₂(CO)₂(μ-CNMe)₂	1.94av.	1.22	2.538	125.5	81.9	79
<u>trans-anti</u>-Fe₂Cp₂(CO)₂(μ-CNPh)₂	1.93	1.23	2.525	129.0	81.7	80
<u>trans-anti</u>-Fe₂Cp₂(CNPh)₂(μ-CNPh)₂	1.92 (1.775)	1.24 (1.16)	2.523	126.4 (171.1)	82.2	81
Fe₂(CNEt)₆(μ-CNEt)₃	1.96 (1.84)	1.26 (1.17)	2.462	123-124 (171)	b	82
Co₂(CNBu-t)₆(μ-CNBu-t)₂[c]	1.95	1.21	2.457	131.2	78.0	19
Pt₃(CNBu-t)₃(μ-CNBu-t)₃	2.08 (1.90)	1.21 (1.15)	2.632av.	132.7 (175)	78.6	83
[Pd₂(CNMe)₂(μ-CNMe)(dppm)₂] [PF₆]₂[d]	2.01 (2.075)	1.23 (1.15)	3.215	131 (179)	106	84

(a) Values in parentheses refer to terminal isocyanides (b) Not reported (c) Structural data of the terminally bound isocyanides are reported in Table 1 (d) dppm = bis(diphenylphosphino)methane.

Whereas the more or less pronounced deviations from linearity in Fe(TPP)(CNBu-t)$_2$·2PhMe,[16] Fe(η^4-C$_6$H$_8$)(CO)$_2$(CNEt)[18] and Co$_2$(μ-CNBu-t)$_2$(CNBu-t)$_6$[19] have also been explained in terms of steric effects (demonstrated by calculations of interligand contacts in the latter compound) electronic factors have been claimed to play an important role for the marked distortions in the other derivatives of Table 1. Strong π back-donation (i.e., large contribution of the "carbene-like" formula **Ib**) occurs particularly for <u>trans</u>-Mo(CNMe)$_2$(dppe)$_2$[23] and for two equatorial isocyanide groups of the distorted trigonal-bipyramidal complexes, Ru(CNBu-t)$_4$(PPh$_3$)[24] and Fe(CNBu-t)$_5$.[24] The essentially electronic nature of these unusual bendings is confirmed by the low CN stretching frequencies, which range from 1886 cm^{-1} for <u>trans</u>-Mo(CNMe)$_2$(dppe)$_2$[23,25] to 1830 cm^{-1} for Ru(CNBu-t)$_4$(PPh$_3$) and Fe(CNBu-t)$_5$,[24] both in the solid and in solution.

Infrared spectroscopy has often been used to assess the σ/π ratio in the M-CNR bond. The ν(CN) vibrations of the coordinated isocyanides are markedly influenced by the electronic density on the central metal. Any increase in the formal oxidation state will enhance the σ contribution with a consequent shift of these bands to higher frequencies. An example is given by the six-coordinate complexes [Cr(CNPh)$_6$]$^{n+}$ [n=0, ν(CN)=2040,1975 cm^{-1}; n=1, ν(CN)=2065,1985 cm^{-1}; n=2, ν(CN)=2161 cm^{-1}][26-7] and <u>trans</u>-[Mo(CNC$_6$H$_4$Me-4)$_2$(dppe)$_2$]$^{n+}$[n=0, ν(CN)=1860 cm^{-1}; n=1, ν(CN)=1953 cm^{-1}; n=2, ν(CN)=2052 cm^{-1}].[25,28] It has been shown, however, that even in the higher oxidation states some π back-bonding contribution is still present.[13] In some organometallic anions, containing both CO and CNR ligands, the extent of π bonding is so large as to produce very low ν(CO) and ν(CN) values; for [MoCp(CO)$_2$(CNMe)]$^-$, 1875,1765,1710 cm^{-1}; for [MoCp(CO)(CNMe)$_2$]$^-$, 1855,1745,1690 cm^{-1}.[29] The electron density on the metal is also influenced by the bonding properties of the other ancillary ligands. In mixed complexes, good π-accepting ligands, such as CO and electronegatively substituted olefins, will reduce the metal electron density and correspondingly increase the σ/π ratio in the M-CNR bond, whereas good σ-donors, such as tertiary phosphines or phosphites, have the opposite effect. In the complexes $M(CO)_{6-n}(CNBu-t)_n$ (n=1-3, M=Cr,Mo,W; n=4, M=Mo,W), both the CO and CN stretching force constants decrease on substitution of carbonyl groups by CNBu-t,[30] while in the series $MnBr(CO)_{5-n}(CNMe)_n$ (n=0-4) and $[Mn(CO)_{6-n}(CNMe)_n]^+$ (n=0-6), linear correlations are found between k_{CO} or k_{CN} and the electronic occupations of the σ and π^* orbitals of the ligands.[13] In Ni(CNBu-t)$_2(\eta^2$-un) (un = substituted olefins, acetylenes, nitrosobenzenes, azobenzenes and carboxo organic molecules) the ν(CN) values increase with electron-withdrawing ability and number of substituents on the η^2 ligand.[31-3]

Furthermore, when a substituted phenyl group is present, linear correlations exist between ν(CN) values and the appropriate Hammett σ parameters, implying a delocalization of electron density throughout the entire system. Progressive substitution of CNMe by the better net donor PPh$_3$ brings about a lowering of both ν(CN) and k_{CN} values in the complexes $[FeCp(CNMe)_{3-n}(PPh_3)_n]^+$ (n=0-2).[34]

In all these correlations, approximate stretching force constant values have been used, as obtained by the Cotton-Kraihanzel method. A further approximation is also introduced since the contribution of the steric interaction term to the total bonding energies of the M-CO and M-CNR bonds is generally neglected.[35] However, in the absence of more detailed calculations, the observed regular trends seem to justify these approximations, at least within a series of closely related compounds.

The donor/acceptor properties of CNR ligands are also affected by the nature of R. Whereas the isocyanides CNCH$_2$MMe$_3$ (M=Si,Ge,Sn,Pb) have a higher donor character than CNMe,[36] ligands CNCOR' (R'=Ph,C$_6$H$_4$Cl-4,NMe$_2$,OEt,SEt) have been found to have acceptor abilities similar to those of carbon monoxide.[37-8] Aryl isocyanides are in general better π-acceptors than their alkyl analogues, due to the presence of lower energy π-accepting orbitals resulting from conjugation of the out-of-plane π system of the isocyano group with phenyl ring π orbitals.[39,40]

The different electronic properties of the CNR ligands are reflected in the different frequencies of the intense metal-to-ligand, $d(\pi) \rightarrow \pi^*$, charge transfer (MLCT) bands in the electronic spectra of the corresponding metal complexes. A decrease in frequency is generally observed when alkyl isocyanides are replaced by their aryl analogues.[28,41-4] The MLCT bands depend also on the nature and oxidation state of the central metal[42,45-6] and, for some mono- and binuclear Rh(I)-isocyanide complexes, on the extent of association in solution.[47-50] In trans-$M(CNC_6H_4X-4)_2(dppe)_2$ (M=Mo,W; X=Cl,H,Me,OMe)[28] the frequency of the lowest energy MLCT band decreases with increasing electron-withdrawing properties of the para substituent X, due to a greater stabilization of the ligand π^* orbitals.[40] However, in $[Ru(bipy)_2(CNR)_2]^{2+}$ (R=$CH_2Ph,C_6H_4OMe-4,C_6H_4Cl-4$)[51] the lowest energy band arises from d (Ru(II)) $\longrightarrow \pi^*$(bipy) transitions and shows an increase in frequency as the electron-accepting character of CNR ligand increases: $CNC_6H_4Cl-4 > CNC_6H_4OMe-4 > CNCH_2Ph$, in accord with a greater stabilization of the metal d(π) levels through π-bonding interaction.

The Mössbauer center shift parameters for closely related series of neutral or cationic Fe(II)-isocyanide complexes have permitted comparison of relative ($\sigma+\pi$) bonding abilities of a variety of ligands.[34,52-5] In particular, isocyanides have bonding properties similar to trialkyl phosphites,[52] while carbenes are better σ-donors and poorer π-acceptors.[53] In cationic cyclopentadienyliron complexes, the ligand ($\sigma+\pi$) abilities increase in the order $PPh_3 < CNMe \sim C(NHMe)_2 < CO$.[34] The [129]I Mössbauer spectra of trans-$[Pt^{129}I(L)_2L']^+$ (L=PMe_2Ph) give a trans-influence series for L' in the order:[56]

$$P(OMe)_3 > EtNC > P(OMe)_2Ph \sim 4\text{-}MeOC_6H_4NC > AsPh_3 > PPh_3$$

The results of several electrochemical investigations may be summarized as follows:[25-8,41,51,57-69] (i) alkyl isocyanide derivatives are more easily oxidized than their aryl isocyanide counterparts; furthermore, for substituted phenyl isocyanide complexes, the $E_{\frac{1}{2}}$ values for one-electron oxidation processes increase with increasing electron-withdrawing properties of the substituents (in general, linear correlations are observed between $E_{\frac{1}{2}}$ values and Hammett σ parameters). (ii) in mixed CO/CNR compounds the ease of oxidation decreases as the number of carbonyl ligands increases. (iii) in isomeric complexes the $E_{\frac{1}{2}}$ parameters may differ substantially. All these data may be rationalized by assuming that, unless there are unusual changes due to free energies of solvation, the highest occupied molecular orbital (HOMO) in the complex is also the redox orbital. This is strongly suggested by an approximate molecular orbital calculation on the series $[Mn(CO)_{6-n}(CNMe)_n]^+$ (n=0-6), for which the electrochemical $E_{\frac{1}{2}}$ potentials for the +1 $\longrightarrow$ +2 process were found to correlate very well with the HOMO energies,[13] and also by the linear relationship between $E_{\frac{1}{2}}$ and the frequency of the $d(\pi) \longrightarrow \pi^*$ charge transfer transitions in the series $[Co(CNR)_3(PPh_3)_2]^+$ (R=Me,Pr-i,Bu-t,C_6H_4Me-4,C_6H_4Cl-4)[41] and trans-$M(CNC_6H_4X-4)_2(dppe)_2$ (M=Mo,W; X=Cl,H,Me,MeO).[28] Since the HOMO consists essentially of metal d-orbitals stabilized by interaction with π^* orbitals of coordinated ligands, it follows that either replacement of an alkyl isocyanide by a better π-acceptor aryl analogue or an increasing number of better acceptor ligands, such as carbon monoxide, would result in a lower HOMO energy and hence in a higher $E_{\frac{1}{2}}$ oxidation potential. Similar arguments account for the different $E_{\frac{1}{2}}$ values in isomeric complexes.[61,66]

   The different acceptor ability of CO and CNR ligands is clearly reflected by the different tendency to stabilize anionic complexes with metals in formal negative oxidation states. Whereas homoleptic carbonyl anions can easily be prepared and are well documented in the literature,[70] no isocyanide analogue has so far been reported. A markedly negative reduction potential (-2.1 V) has been reported for the complex $Co_2(CNR)_8$ (R=C_6H_3Me$_2$-2,6),[71] compared to the corresponding value (-0.4 V) for $Co_2(CO)_8$.[72] By contrast, isocyanide complexes are much more easily oxidized (e.g., by stoichiometric amounts of Ag^+[27-8,41,60,64] or by photochemical processes).[73] The reactions of electrophiles, such as protic acids or alkylating agents, on zero-valent metal complexes generally occur at the metal center (oxidative addition), but in some cases, where the formula Ib predominates, attack at the isocyanide nitrogen atom takes place (Section 5). Conversely, in complexes with enhanced σ donation (higher contribution of formula Ia) the isocyanide ligand becomes susceptible to nucleophilic attack (Section 4). The site of this attack is the metal bound carbon atom, even though a formal positive charge is located on the nitrogen atom. It must be remembered that the mesomeric formula $\bar{M}-C\equiv\overset{+}{N}-R$ depicts only the charge distribution of the π electrons of the CN multiple bond, and tells nothing about the distribution of the σ electrons (another limitation of the valence bond scheme). Calculations carried out on free isocyanides have shown that the σ charge distribution overcompensates the π dipole moment of the mesomeric formula $\bar{C}\equiv\overset{+}{N}-R$, resulting in a gross dipole moment with a reverse polarity.[74] The σ donation of the terminal lone pair will bring about a higher electron deficiency at the carbon atom rather than at the nitrogen atom of the isocyano group.

The substitution of coordinated isocyanides appears to be mainly influenced by the oxidation state of the central metal and by the nature of the entering ligand. This reaction is generally favored by a reduced π contribution to the M-CNR bond (Section 3).

Isocyanide insertion into the metal-carbon σ bond was found to occur more readily when the contact carbon of the coordinated isocyano group is more electrophilic,[75] suggesting that activation of the isocyanide ligand by predominant σ donation is required in this type of reaction as well. This is indirectly confirmed by the fact that no insertion takes place, even under forcing conditions, in the complexes $MRCp_2(CNR')$ (M=Nb,Ta; R=alkyl; R'=Me,Bu-t,Cy,$C_6H_3Me_2$-2.6), in which an extensive π contribution to the metal-isocyanide bond occurs.[76-7]

2.2 Bridging isocyanides

In polynuclear complexes the CNR ligands may adopt different bridging coordination modes which can involve two or three metal centers, through the terminal carbon or through both carbon and nitrogen atoms of the isocyano group.

2.2.1 μ_2,η^1-Bridging isocyanides

In this coordination mode, two metal centers are linked to the terminal carbon atom of the CNR group with two essentially single bonds (**Ic**).
The xray structural data of Table 2 indicate a considerable bending of the CNR unit and lengthening of the CN bond, suggesting sp^2 hybridization of

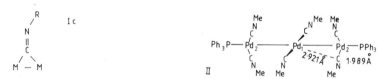

the nitrogen atom with a CN bond order close to two. A metal-metal bond is generally present in these systems, the three-membered cycle being characterized by an M-C-M angle in the range 78-83.6°. When both terminal and bridging isocyanides are present in the same molecule, the M-μC distances are ca. 0.15 Å longer than the terminal M-C bonds, indicating reduced π back-donation in the M-μCNR bond. A noteworthy exception is the binuclear complex, $[Pd_2(\mu\text{-}CNMe)(CNMe)_2(dppm)_2](PF_6)_2$ in which the Pd-μC distance is shorter than the terminal Pd-C bond and the Pd-Pd separation (3.215 Å) appears too large for a direct metal-metal bond.[84] This distance should be compared to the Pd-Pd bond lengths in the compounds $[Pd_2(CNMe)_6]^{2+}$ (2.531 Å)[85-6] and $[Pd_3(CNMe)_6(PPh_3)_2]^{2+}$ (2.592 Å),[87] which contain essentially linear MeNC-Pd-Pd-CNMe and Ph_3P-Pd-Pd-Pd-PPh_3 arrangements, respectively, and no bridging isocyanide. However, in the latter compound (structure II) the four equatorial CNMe ligands have been considered "semibridging" because of their bending towards the central Pd_1 atom. The inward bend (average Pd_1-Pd_2-C angle 80.0(2)°) was in part attributed to electronic interaction between filled d orbitals of Pd_1 and empty π^* orbitals of the isocyanide ligands of the adjacent Pd_2 atoms.

The ν(CN) bands of η^1-bridging isocyanides fall in the range 1845-1580 cm^{-1}, the highest values being observed for $[NiCp(\mu\text{-}CNR)]_2$ (R=Me,CD_3,Pr-i; ν(CN)=1795-1845 cm^{-1})[88] and the lowest ones for $Pd_2X_2(\mu\text{-}CNR)(dppm)_2$ (X=Cl,Br; R=Me,Cy,Ph,C_6H_4Me-4; ν(CN)=1624-1677 cm^{-1})[84] and for $Ru_2(CNPr\text{-}i)_9$ (ν(CN)=1580 and 1648 cm^{-1}).[89] The marked low-frequency shift relative to free ligand is a further indication of the reduced CN bond order in this type of coordination. However, the occurrence of ν(CN) frequencies in this range cannot be regarded as proof of the presence of bridging isocyanides since ν(CN) values below 1800 cm^{-1} have also been found in terminal CNR ligands bound to electron rich metal sites.[24-5,76-7,90] Much more diagnostic are the ^{13}C resonances, which occur in the low-field range 220-260 ppm (from $SiMe_4$) in the ^{13}C nmr spectra of a variety of bridged complexes.[19,82-3,91-2]

In solution, the mixed complexes $M_2(\eta^5\text{-dienyl})_2(CO)_3(CNR)$ (M=Fe, dienyl=C_5H_5,C_5H_4Me,C_9H_7; M=Ru,dienyl=C_5H_5) exist as mixtures of cis and trans isomers containing bridging and terminal isocyanides, which interconvert rapidly (on the nmr time scale) at room temperature.[92-6] The bridging position of the CNR ligand depends on the electronic and steric properties of the substituent R, as well as on the central metal, solvent and temperature.   The position of the equilibria shifts away from μ-CNR isomers, as the steric requirements of R increase, in the order:

$$R=Me \;>\; Et \;>\; Bu\text{-}n \;>\; Bu\text{-}i \;>\; Bu\text{-}s \;>\; C_6H_{11} \;>\; Bu\text{-}t$$

For the bulky Bu-t substituent, no isocyanide-bridged species is present in detectable amount in solution. However, according to the Adams-Cotton mechanism,[97] a μ-CNBu-t transient (or transition state) is traversed in the rapid passage of the CNBu-t ligand from one metal atom to the other at room temperature.[91,94] The influence of steric effects is also evident in the cationic complexes $[Pt_2H(CNR)(dppe)_2]^+$ where the CNR group is bridging for R=Me,C_6H_4Me-4 and terminal for R=Bu-t.[98] The μ-CNR isomers are favored by more electron-withdrawing R. For R=$CH_2C_6H_4X$-4 and M=Fe, the proportion of CNR-bridged species follows the order: X=Cl>H>Me>OMe.[95] The bridging position is so stabilized when R=Ph that the complex $Fe_2Cp_2(\mu\text{-CO})(\mu\text{-CNPh})(CO)_2$ does not show any fluxional behavior.[91] Interestingly, the iron-isocyanide derivatives crystallize as CO-bridged forms when R is an alkyl group,[94-5,99] whereas isocyanide-bridged compounds are generally obtained when R=$CH_2C_6H_4X$-4[95] or Ph.[79,80] The ruthenium complexes have a greater tendency towards occupation of the bridge position by CNR, and lower activation energies for isomer interconversion.[92,96] The cis-trans isomerization and the terminal-bridging ligand exchange are generally interpreted on the basis of the Adams-Cotton mechanism, which involves initial pairwise opening of the bridge to form a transient intermediate, which may undergo rotation about the M-M bond, followed by pairwise reclosure to yield the doubly bridged derivative.[97] This mechanism accounts for the fluxional behavior of other binuclear doubly bridged complexes, such as $Fe_2Cp_2(CO)_2(CNR)_2$ (R=Me,[100] Bu-t[101]) and $Co_2(CNBu-t)_8$,[19] and also of polynuclear compounds with no bridging ligand, such as $Mo_2Cp_2(CO)_5(CNMe)$,[102] $Mn_2(CO)_7(CNMe)_3$,[103] $Os_3(CO)_{12-n}(CNR)_n$ (n=1,2; R=Me,C_6H_4OMe-4; n=1-4, R=Bu-n,Bu-t)[104] and possibly, $Ru_3(CO)_{12-n}(CNBu-t)_n$ (n=1,2).[105] Fast terminal-bridge ligand exchange at room temperature is likely to occur in the bridged complexes $Co_2(CNC_6H_3Me_2$-2,6$)_8$,[71] $Fe_2(CNEt)_9$[82] and $Ru_2(CNPr-i)_9$,[89] although no detailed discussion of the variable temperature ^1H and ^{13}C nmr spectra has been reported.   By contrast, the dynamic behavior of the binuclear compounds $[Pd_2(CNMe)_5(PPh_3)]^{2+}$ and $[Pd_2(CNMe)_6]^{2+}$ is better interpreted in terms of tetrahedral distortions at the metal centers and rotation about the Pd-Pd bond rather than by an isocyanide bridged transition state.[86,106] The corresponding $[Pt_2(CNMe)_6]^{2+}$ derivative is a stereochemically rigid molecule in the temperature range 20-140°C.[106] The trinuclear isocyanide-bridged complex $Pt_3(\mu$-CNBu-t$)_3(CNBu-t)_3$ is also rigid at room temperature, but in the presence of trace amounts of free CNBu-t at higher temperatures it undergoes a dynamic intermolecular process involving terminal and bridge isocyanide site exchange catalyzed by the free ligand.[83]

In accord with their non-linear structures, the bridging isocyanide ligands may undergo another type of dynamic rearrangement in solution, involving inversion at the nitrogen atom, analogous to that reported for organic imino derivatives.[107] This process has been observed in the variable temperature ^1H nmr spectra of the systems $Fe_2Cp_2(CO)(CNMe)_3$ and $Fe_2Cp_2(CO)_2(CNMe)_2$, for which calculations have given activation barriers of ca. 40 kJ mol^{-1} (much lower than the bridge-terminal rearrangement barriers of 60-80 kJ mol^{-1}).[108] Due to the low activation energies, in many bridged systems the inversion at nitrogen is still rapid at the lowest explored temperatures.[19,82,92]

The analogy in structural and dynamic properties between bridging isocyanides and compounds containing imino groups is also reflected in their chemical reactivity.   Metal imino complexes are known to undergo electrophilic attack (protonation and/or alkylation) at nitrogen to yield metal-carbene complexes[9,109-112] (eq 1).

$$M - C \overset{Y}{\underset{\underset{R}{|}}{\overset{}{N}}} \qquad \xrightarrow{H^+} \qquad M \overset{+}{=} C \overset{Y}{\underset{\underset{R}{|}}{N - H}}$$

(Y=H, alkyl, aryl, OR', NR'R", etc.)   (1)

A similar electrophilic attack involving the nitrogen lone pair of the bent μ-CNR ligands has been reported for $M_2(\eta^5\text{-dienyl})_2(CO)_{4-n}(CNR)_n$ (M=Fe,Ru; n=1,2) (Section 5.2).

2.2.2 μ_2,η^2-Bridging isocyanides

In the binuclear comples $Mn_2(CO)_4(CNC_6H_4Me\text{-}4)(dppm)_2$,[113-4] the two Mn centers are linked by a metal-metal bond (Mn-Mn distance of 2.936 Å), by two bridging dppm ligands (with the P donor atoms mutually trans to each other) and by a bridging isocyanide, the bonding mode of which is shown in structure III (R=C_6H_4Me-4).

The structural unit of III is planar with Mn_1-C-N and C-N-R angles of $168.1°$ and $137.2°$ respectively, and with a $\nu(CN)$ band at 1661 cm^{-1}. The isocyanide behaves as a four-electron ligand by donating the carbon lone pair to Mn_1 and an additional pair of electrons to Mn_2 (either the nitrogen lone pair of the bent valence bond structure Ib or the in-plane π-bond of the linear structure Ia). The short Mn_1-C bond and the bending of the C-N-R group suggest a significant contribution of Ib (Section 2.1).

μ_2,η^2-Bridging isocyanides are probably present in the cluster compound $Ni_4(CNBu\text{-}t)_7$, which contains a trigonally compressed tetrahedral arrangement of nickel atoms[115] (Fig. 1). Each nickel atom is bound to a terminal two-electron donating isocyanide. The three CNBu-t ligands that bridge the basal nickel atoms act as four-electron donors. Because of disorder in the crystal, the positions of the bent bridging isocyanides could not be exactly determined. This is a stereochemically non-rigid molecule which undergoes two intramolecular dynamic processes at different temperatures.[115-7] In the low-temperature process, equilibration of the terminal, apical and basal isocyanides occurs, while, in the high-temperature

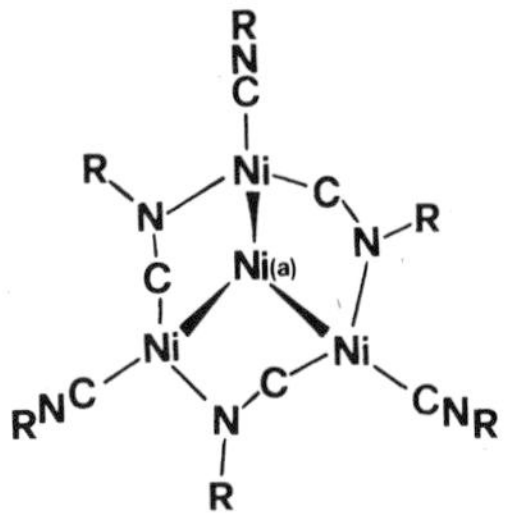

Figure 1

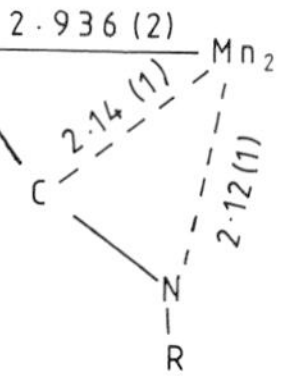

one, all ligand environments become equivalent (on the nmr time scale) possibly through terminal-bridge site exchange. A third, intermolecular, process involving ligand exchange between $Ni_4(CNBu-t)_7$ and traces of $Ni(CNBu-t)_4$ is also apparent. The bridging isocyanides can be replaced by aryl acetylenes to form the cluster $Ni_4(CNBu-t)_4$ (μ_3,η^2-ArC≡CAr).[118] Both types of clusters are catalyst precursors in oligomerization and hydrogenation reactions (Section 7).

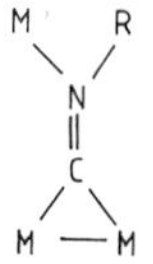

2.2.3 μ_3,η^2-Bridging isocyanides

This bonding mode has been observed in the clusters $Os_6(CO)_{18}(CNC_6H_4Me-4)_2$[119-20] and $Pt_7(CNC_6H_3Me_2-2,6)_{12}$[121] where a unique isocyanide behaves as a bent four-electron ligand, bridging three metal centers in a quasi-planar arrangement of type **IV**. The bond distances and angles are consistent with metal-carbon and metal-nitrogen single bonds and with approximate sp^2 hybridization at the N and C atoms of the double C=N bond. The three bridging isocyanides in the disordered $Ni_4(CNBu-t)_7$ cluster have also been alternatively interpreted as face-bridging μ_3,η^2-ligands.[115]

3. SUBSTITUTION REACTIONS

Several substitution reactions of coordinated isocyanides have been reported, but no systematic study has hitherto been carried out. From the available data it appears that substitution is mainly influenced by the oxidation state of the central metal, and by the electronic and steric properties of the entering ligand, and, to a lesser extent, of the leaving CNR group. In low-valent metal complexes, isocyanides are usually replaced by good π-acceptors, such as activated olefins, acetylenes and carbon monoxide. Ligands with good σ-donating and π-accepting properties, such as phosphines, phosphites and isocyanides themselves, can displace the CNR groups from metal complexes in both lower and higher oxidation states. Substitutions by N-donor ligands generally require photo-activation.

Whereas kinetic studies on the exchange of carbon monoxide with metal carbonyls have been reported,[122] there are only a few investigations on isocyanide complexes. An early report showed that the rate of exchange of CNPh with $Cr(CNPh)_6$ is much higher than that of CO with $Cr(CO)_6$.[123] A fast exchange also occurs in the system $Ni(CNBu-t)_4$/CNBu-t, probably through a dissociative mechanism of the type:[116]

$$Ni(CNBu-t)_4 \rightleftharpoons Ni(CNBu-t)_3 + CNBu-t \qquad (2)$$

Dissociation of the $Ni(CNBu-t)_4$ complex has been confirmed by molecular weight measurements[124] and electronic spectra.[117]

On the other hand, ligand exchange between the 16-electron species $[Rh(CNBu-t)_4]^+$ and CNBu-t takes place via an associative step in which an intermediate of low concentration is formed.[125]

$$[Rh(CNBu-t)_4]^+ + CNBu-t \rightleftharpoons [Rh(CNBu-t)_5]^+ \qquad (3)$$

For a 1:1 mixture, the ^1H nmr spectra indicate that the five-coordinate complex is present in substantial concentration at temperatures below $-30^\circ C$. This compound exhibits ligand lability even at $-160^\circ C$. By extrapolation of the shift data, an upper limit for the heat of dissociation of ca. 25 kJ mol^{-1} was derived. A rapid intermolecular isocyanide exchange is also reported for the systems $Ni(\eta^2-ONC_6H_4Cl-4)(CNBu-t)_2$/CNBu-t[31] and $[M_2(CNMe)_6]^{2+}$/CNMe (M=Pd,Pt).[106] In the latter case, when M=Pd, the exchange remains fast (on the nmr time scale) even at temperatures as low as $-50^\circ C$. Treatment of this compound with an excess of either CNBu-t or CNPh results in complete displacement of CNMe and formation of

$[Pd_2(CNBu-t)_6]^{2+}$ and $[Pd_2(CNPh)_6]^{2+}$. The cobalt complex [Co(CNBu-t)$_5$]$^+$ is a stereochemically non-rigid species that undergoes fast scrambling of ligands through intra- and intermolecular processes, the latter involving a ligand dissociation corresponding to the reverse of reaction (3).[126] Isocyanide exchange in the bridged system $Co_2(CNR)_8$ [R=Bu-t,2,6-Me$_2$C$_6$H$_2$X-4 (X=Me,H,Br)] is slow at room temperature,[19,71,127] probably due to steric bulkiness of the ligands, whereas the rate of ^{14}CO exchange with $Co_2(CO)_8$ is extremely fast.[128] In the ^1H nmr spectra of the <u>para</u>-substituted 2,6-xylyl isocyanide derivatives, coalescence temperatures >115°C were observed with $\Delta G^{\neq} > 87$ kJ mol^{-1}.[71] In the mixture Pt$_3$(μ_2, η^1-CNR)$_3$(CNR)$_3$/CNR (R=Bu-t), intermolecular exchange is faster for the terminal than for the bridging groups.[83] Two CNBu-t ligands are easily replaced in the reaction between Fe(CNBu-t)$_5$ and CNC$_6$H$_3$Me$_2$-2,6 (1/2 molar ratio).[24]

In low-valent metal complexes, terminal and/or bridging isocyanides can be displaced by π-accepting ligands:

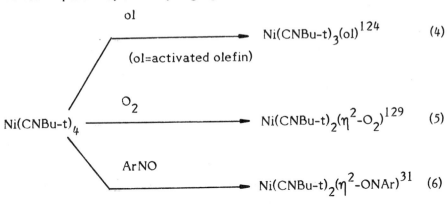

$$Ni(CNBu-t)_4 \xrightarrow{\ ol\ } Ni(CNBu-t)_3(ol)^{[124]} \qquad (4)$$
(ol=activated olefin)

$$\xrightarrow{\ O_2\ } Ni(CNBu-t)_2(\eta^2-O_2)^{[129]} \qquad (5)$$

$$\xrightarrow{\ ArNO\ } Ni(CNBu-t)_2(\eta^2-ONAr)^{[31]} \qquad (6)$$

Reaction (4) occurs with activated olefins such as maleic anhydride, fumaronitrile and TCNE, but not with ethylene. The mechanism is presumed to be of S_N1 type involving the preliminary dissociation equilibrium (2). Similarly, Ni(CNAr)$_4$ reacts with TCNE yielding Ni(CNAr)$_3$(TCNE).[130] The O_2 adduct, Ni(CNBu-t)$_2$(η^2-O$_2$), has a square-planar configuration with η^2-bonded dioxygen.[131] One isocyanide ligand is also displaced in the reactions of $[Co(CNR)_5]^+$ with an equimolar amount of TCNE.[132] Photo-activation is required for the substitution reactions:[133]

$$Cr(CNAr)_6 + ol \xrightarrow{h\nu} Cr(CNAr)_5(ol) + CNAr \tag{7}$$

(ol = dimethyl fumarate, fumaronitrile and maleic anhydride). No reaction takes place for acrylonitrile and dimethyl maleate, whereas the more electron-withdrawing TCNE reacts thermally without irradiation.

Diarylacetylenes react with the cluster compound $Ni_4(CNBu-t)_7$ according to eq 8:[117]

$$4Ni_4(CNBu-t)_7 + 15\ ArC\equiv CAr \longrightarrow 12\ Ni(CNBu-t)_2(ArC\equiv CAr)$$
$$+\ Ni_4(CNBu-t)_4(ArC\equiv CAr)_3 \tag{8}$$

An xray structure analysis of the tetranuclear product with diphenylacetylene has shown that the three acetylenes are each μ_3, η^2-bonded, having replaced the three bridging CNBu-t groups of the starting complex.[118] The bridging CNBu-t ligands of $Co_2(CNBu-t)_8$ are also displaced by a molecule of PhC≡CPh to yield $Co_2(CNBu-t)_6(\mu_2, \eta^2$-PhC≡CPh).[19]

Monosubstitution occurs in the reactions of $Pt(CNBu-t)_2(PPh_3)_2$,[134] $RhCl(CNC_6H_3Me_2-2,6)_3$[135] and $Co(\eta^5-C_5H_5)(CNC_6H_3Me_2-2,6)_2$[136] with carbon monoxide. However, exchange reactions between CO and $Ni(CO)_3(CNMe)$, where the rate-determining step is dissociation of one ligand from the complex, involve only the CO molecules rather than CNMe, indicating that the Ni-CNMe bond is stronger than the Ni-CO bonds in this compound.[35]

Photolysis of the mixed derivatives $[FeCp(CO)_{3-n}(CNMe)_n]PF_6$ (n=1,2) in the presence of excess nucleophile L (L=PPh_3, $AsPh_3$, $SbPh_3$, py, MeCN, $CH_2=CH_2$) leads only to replacement of the carbonyl groups,[137] whereas irradiation of $Cr(\eta^6$-benzonorbornadiene)$(CO)_2(CNR)$ (R=Me,Bu-t, Cy) caused the preferential displacement of

CNR by the olefinic double bond of benzonorbornadiene.[138] This different behavior was ascribed to changes in the σ/π ratio of M-CO and M-CNR bonds resulting from changes in the electron density on the central metal.[137]

Good σ-donor and π-acceptor ligands, such as phosphine and phosphites, can displace one or more isocyanide ligands from complexes of metals in various oxidation states.[24,84,87,89,106,135,139-44] An example is given by the reactions of PPh$_3$ with $[Pd(CNR)_2]_x$,[145-6] $[Pd_3(CNMe)_8]^{2+}$,[87] $[Pd_2(CNMe)_6]^{2+}$[106,139] and $[Pd(CNMe)_4]^{2+}$.[106] In the linear bi- and trinuclear complexes $[MM'(CNMe)_6]^{2+}$ (M=M'=Pd=Pt; M=Pd,M'=Pt) and $[Pd_3(CNMe)_8]^{2+}$ the PPh$_3$ substitution occurs preferentially on the axial position.[106,139,141] In the cationic compounds $[Co(CNAr)_5]^+$, disubstitution is generally observed with trialkylphosphites and triphenylphosphine, except for sterically hindered aryl isocyanides (e.g. CNC$_6$H$_3$Et$_2$-2,6 and CNC$_6$H$_2$Me$_3$-2,4,6) which favor monosubstitution.[147-9] Monosubstitution also occurs with triarylphosphites and P(C$_6$H$_4$Cl-4)$_3$.[149-50] Aryl isocyanides are not displaced by AsPh$_3$.[150-1] However, $[Co(CNBu-t)_4(AsPh_3)]^+$ can be obtained by substitution under prolonged refluxing in ethanol.[152] Photochemical substitution by P(OPh)$_3$ on Cr(η^6-arene)(CO)$_2$(CNR) yields Cr(η^6-arene)(CO)$_2$[P(OPh)$_3$] when R=Me, and Cr(η^6-arene)(CO)(CNR)[P(OPh)$_3$] when R=CNCOPh.[37-8] These results are in agreement with spectroscopic studies and confirm that the order of increasing strength of the chromium-ligand bond is CNMe < CO < CNCOPh.

In cationic palladium and platinum complexes (16-electron species), the terminal isocyanide ligands may be replaced even by weak nucleophiles such as halides and pseudo-halides.[84,106,139,153-4]

Substitution by nitrogen ligands usually occurs under photolytic conditions:

$$M(CNAr)_6 + py \xrightarrow{h\nu} M(CNAr)_5(py) + CNAr \qquad (9)$$

(M=Cr,Mo,W)[73]

$$Fe(Pc)(CNCH_2Ph)(L) + L \underset{dark}{\overset{light}{\rightleftharpoons}} Fe(Pc)(L)_2 + CNCH_2Ph \qquad (10)$$

(Pc=phthalocyanine; L=piperidine)[155]

$$Fe(N_4)(CNCH_2Ph)(L) + L \overset{h\nu}{\longrightarrow} Fe(N_4)(L)_2 + CNCH_2Ph \qquad (11)$$

(L=nitrogen base; N_4=tetradentate macrocycle)[156]

In reaction (9), the quantum yields decrease from Cr to W and with increasing steric hindrance of the isocyanide ligand. This suggests that for the molybdenum and particularly for the tungsten complexes the mechanism has associative character, involving a direct nucleophilic attack on the positively charged metal center in a MLCT state, $[M^+(CNAr)^-]$.[73] In reaction (11), the dissociation of $CNCH_2Ph$ could come from a MLCT state or from a triplet ligand field state reached by a rapid non-radiative decay from the MLCT state.[156] Reaction (11) also occurs thermally, but at much lower rates.[157-9] The substitution of the axial isocyanide ligand (trans to L) in these tetragonal low-spin iron(II) complexes proceeds by a dissociative mechanism in nonaqueous solvents:

$$Fe\,(N_4)(CNCH_2Ph)(L) \underset{k_{-1}}{\overset{k_1}{\rightleftharpoons}} Fe(N_4)(L) + CNCH_2Ph$$

$$-L \Big\Updownarrow +L$$
$$k_{-2} \quad k_2 \qquad\qquad (12)$$

$$Fe(N_4)(L)_2$$

For such a mechanism the observed pseudo-first-order rate constant (neglecting the reverse reaction k_{-2}) is given by eq (13):

$$k(obs) = k_1 k_2 [L]/(k_{-1}[CNCH_2Ph] + k_2[L])$$ (13)

At high concentration of L, the observed rate becomes independent of [L] and $k(obs) = k_1$. The dissociative character was confirmed by the positive values of the activation entropies ($\Delta S^{\ddagger} = 40\text{-}100$ J K^{-1} mol^{-1}) and also by the fact that the reaction of the diisocyanide complex $Fe(Pc)(CNCH_2Ph)_2$ with various amines was found to be independent of the nature or concentration of the entering ligand under the experimental conditions.[157] Although the k_1 values cannot be directly compared, as they were obtained at different temperatures and/or different solvents, the lability of the axial isocyanide in these "heme model molecules" depends on both the equatorial N_4 macrocycle and the trans ligand L in the order:

$$N_4 = TPP > (dmgH)_2 > Pc \sim TAAB$$

(TPP = Tetraphenylporphyrin; $dmgH_2$ = dimethylglyoxime; Pc = phthalocyanine; TAAB = tetrabenzo [b,f,j,n][1,5,9,13] tetraazacyclohexadecane).

$$L = CNCH_2Ph \gg piperidine > pyridine > methylimidazole, \text{ for } N_4 = Pc$$

$$L = pyridine > methylimidazole \gg CNCH_2Ph, \text{ for } N_4 = (dmgH)_2$$

Equilibrium constants show that the stability of $Fe(N_4)(CNCH_2Ph)(L)$ complexes towards isocyanide replacement follows the order:

$$N_4 = TPP > Pc > TAAB$$

The same order was observed for the corresponding CO derivatives. The isocyanide and carbonyl complexes have comparable stability in the TTP system, whereas in Pc and TAAB systems the $CNCH_2Ph$ ligand is much more strongly bound than CO.

4.   REACTIONS WITH NUCLEOPHILES

The reactions of isocyanide-metal complexes with nucleophiles may yield different products via three general routes: (i) ligand substitution[9,10,160] (see also Section 3); (ii) reduction of the metal complex;[9,145,161-2] (iii) attack at the isocyanide carbon atom.[9,10,109-11,163] In some cases, these reactions may occur simultaneously; e.g. in the reaction of $[NiCp(CNCy)(PPh_3)]^+$ with LiPh both the substitution product $Ni(Ph)Cp(PPh_3)$ and the imino derivative $Ni[C(=NCy)Ph]Cp(CNCy)$ were obtained.[2] Nucleophilic attack on the coordinated isocyanides depends on a delicate balance of electronic and steric factors, as will be discussed in detail further on. A necessary, but not sufficient, condition is the activation of the CNR ligand by coordination to a metal center with reduced electron density (generally, metal ions in the higher oxidation states), so as to enhance the σ/π ratio in the M-CNR bond (Section 2.1.). Since a high contribution of the limiting formula **Ia**, $\bar{M}-C\equiv\overset{+}{N}-R$, will shift the $\nu(CN)$ band of the coordinated isocyanide to higher frequencies relative to the free ligand, a positive value of the term $\Delta\nu = \nu(CN)_{coord} - \nu(CN)_{free} \gtrsim 40$ cm^{-1} was assumed indicative of CNR group susceptibility to nucleophilic attack.[164]

4.1   Reactions with anionic nucleophiles

These reactions have been discussed in previous reviews,[9,10] and only the general pattern will be described here:

$$\overset{+}{M} \leftarrow CNR \quad + \quad LiR' \quad \longrightarrow \quad M-C\overset{R'}{\underset{NR}{\diagdown}} \qquad (14)$$

$R' = Ph, C_6F_5$

$$\overset{+}{M}\overset{CNR}{\underset{CNR}{\diagup}} \quad + \quad BH_4^- \quad \longrightarrow \quad M \cdots \overset{\overset{H}{C}-\overset{R}{N}}{\underset{\underset{H}{C}-\underset{R}{N}}{\diagup}} BH_2 \qquad (15)$$

$$\overset{+}{M} \leftarrow CNR \quad + \quad OH^- \quad \longrightarrow \quad M-C\overset{\diagup O}{\underset{NHR}{\diagdown}} \qquad (16)$$

$$\overset{+}{M}{\leftarrow}CNR \quad + \quad HS^{-} \quad \longrightarrow \quad M-C\overset{\displaystyle S}{\underset{\displaystyle NHR}{\diagup}} \qquad (17)$$

$$\overset{+}{M}{\leftarrow}CNR \quad + \quad R'O^{-} \quad \longrightarrow \quad M-C\overset{\displaystyle OR'}{\underset{\displaystyle NR}{\diagup}} \qquad (18)$$

$$\overset{+}{M}{\leftarrow}CNR \quad + \quad R'NH^{-} \quad \longrightarrow \quad M-C\overset{\displaystyle NHR'}{\underset{\displaystyle NR}{\diagup}} \qquad (19)$$

$$\overset{+}{M}{\leftarrow}CNR \quad + \quad N_3^{-} \quad \longrightarrow \quad M-C\overset{\displaystyle \underset{R}{N}-N}{\underset{\displaystyle N-N}{\diagup}} \qquad (20)$$

The imino nitrogen atom in the products of reactions (14), (18) and (19) is easily protonated to yield the corresponding carbene complex.   When reactions (14), (16-20) are carried out on isocyanide complexes containing other unsaturated ligands, such as CO and/or CS, the site of nucleophilic attack cannot be predicted a priori unless the charge distribution and the LUMO energy and localization are known from theoretical calculations on the substrate.[165]   On the other hand, if the attack on a particular site is kinetically controlled, thermodynamic factors may bring about isomerization to more stable products by shift of the nucleophile onto a different ligand.[166]   Experimental evidence suggests that attack on isocyanide is favored by increasing the number of CNR ligands in the molecule.[9] For the mono-isocyanide derivatives, $Cr(CO)_5(CNBu\text{-}t)$[167] and cis-$Mn(COMe)(CO)_4$ (CNR) (R=Me, Bu-t, Cy),[168] the reactions with LiR' (R' = Bu-n and Me, respectively) occur at the carbonyl group.

The cation $[OsCl(CO)_2(CNC_6H_4Me\text{-}4)(PPh_3)_2]^+$ reacts with MeO^- to give $OsCl(CO_2Me)(CO)(CNC_6H_4Me\text{-}4)(PPh_3)_2$, whereas the reaction with HS^- yields the isothiocyanate complex, $Os(\eta^2\text{-}SCNC_6H_4Me\text{-}4)(CO)_2(PPh_3)_2$.[169] The isocyanide ligand is also preferentially attacked by HS^- in the complex

$[OsCl(CS)(CO)(CNR)(PPh_3)_2]^+$ $(R=C_6H_4Me-4)$.[170] In acetone the product is exclusively $Os(\eta^2\text{-SCNR})(CS)(CO)(PPh_3)_2$, but in dichloromethane/ethanol (1/1) a second product, $Os(\eta^2\text{-CS}_2)(CO)(CNR)(PPh_3)_2$, is formed, indicating attack by HS^- also at the CS ligand.

In the presence of a base, organic molecules with active methylene groups, such as malodinitrile, methyl cyanoacetate and ethyl acetylacetate, give 1,2-additions on coordinated isocyanides, yielding α-metalated enamines:[171]

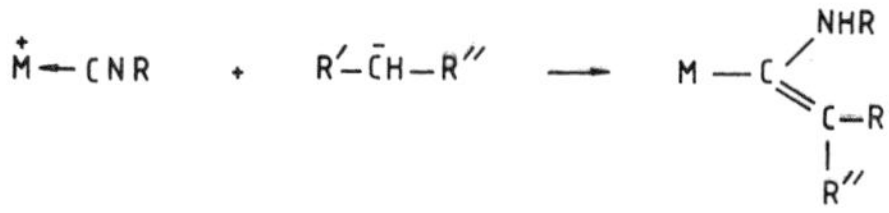

$$\overset{+}{M}-CNR \quad + \quad R'-\overset{-}{C}H-R'' \quad \longrightarrow \quad M-C\overset{\displaystyle NHR}{\underset{\underset{R''}{\overset{|}{C-R'}}}{\diagdown}}$$

$(M=Pd,Pt; \ R'=R''=CN, R''=CO_2Me; \ R'=CO_2Et, R''=COMe)$ (21)

Reaction (18) has been extensively studied by Bonati and Minghetti,[10,172-4] who found that the (alkoxy)imino group can be also prepared by reaction (22):

$$L_nM\text{-}X + CNR + R'OH + KOH \longrightarrow L_nM\text{-}C(=NR)OR' + KX + H_2O \qquad (22)$$

$[M=Ag(SMe_2), X=NO_3; M=Au(PPh_3),Au(SMe_2),X=Cl]$. The trimeric nature of the products $[M(C(=NR)OR')]_3$ (M=Ag,Au) was established by molecular weight measurements and by an xray structural analysis of $[Au(C(=NC_6H_4Me-4)OEt)]_3$.[175] Complexes with two (alkoxy)imino ligands, such as $Hg[C(=NR)OR']_2$ and $Pt[C(=NR)OR']_2(CNR)(PPh_3)$, can be obtained by analogous reactions on $HgCl_2(PPh_3)_2$ and $PtCl_2(PPh_3)_2$, respectively. The (alkoxy)imino moiety can act as N-ligand towards various metal ions, yielding di- and trinuclear compounds.[176-9]

4.2   Reactions with protic nucleophiles

The reactions of coordination-activated terminal isocyanides with protic nucleophiles, such as primary and secondary amines, alcohols and thiols, lead to 1,2-addition according to the general equation (23):[9,10,109-11,163]

$$L_nM\!\leftarrow\!CNR \quad + \quad HY \quad \longrightarrow \quad L_nM\!-\!C\!\!\begin{array}{c}^{NHR}\\[-2pt]_{Y}\end{array}$$

(Y=NHR', NR'R", OR', SR')   (23)

The products have been alternatively described as metal carbenes[109-11] **Va**, stabilized carbonium ions **Vb**[180-1] or metalated amidinium ions **Vc-d** (Y=NHR'):[182-3]

$$M\!\leftarrow\!C\!\!\begin{array}{c}^{NHR}\\[-2pt]_{NHR'}\end{array} \;\longleftrightarrow\; \bar{M}\!-\!\overset{+}{C}\!\!\begin{array}{c}^{NHR}\\[-2pt]_{NHR'}\end{array} \;\longleftrightarrow\; \bar{M}\!-\!C\!\!\begin{array}{c}^{\overset{+}{N}HR}\\[-2pt]_{NHR'}\end{array} \;\longleftrightarrow\; \bar{M}\!-\!C\!\!\begin{array}{c}^{NHR}\\[-2pt]_{\overset{+}{N}HR'}\end{array}$$

Va   Vb   Vc   Vd

Although the limiting structure **Va** does not represent the actual electronic configuration of the system, we will adopt the more generally used term of "carbene" (or resonance stabilized carbene) and the symbolism of eq (23) for this type of ligand. Xray photoelectron spectra for a series of palladium and platinum complexes indicate that the carbene groups are better donors of electron density to the metal than methyl isocyanide, and the carbene carbon atoms are less positively charged than the isocyanide carbon of coordinated CNMe.[184] The good σ donor character also affects the [57]Fe Mössbauer spectra[34,53] and accounts for the markedly high <u>trans</u>-influence exerted by these ligands.[185-6] Reaction (23) was first studied for Hg(II),[187] Pt(II),[3] Pd(II),[188] Fe(II)[182] and Au(I)[189] isocyanide complexes, and later extended to Ni(II),[190-1] Rh(I) and Rh(III),[192-3] Co(III),[194-5] Ru(II),[59,196] Os(II),[164] Mn(I)[166] and Mo(II)[197] substrates. Some Rh(III),[192] Pt(IV)[198-200] and Au(III)[201-3] derivatives could be prepared by oxidative addition to the corresponding Rh(I), Pt(II) and Au(I) complexes. Xray structures of carbene complexes have shown that the carbene carbon is essentially sp^2 hybridized with C-N and C-Y bond distances shorter than those expected for single bonds.[9,109-11,163,183,186,199,204-7] Due to the partial double bond character, restricted rotation about these bonds may give rise to rotational isomers in solution. For example, when Y = NHR, the carbene moiety may assume three different configurations:

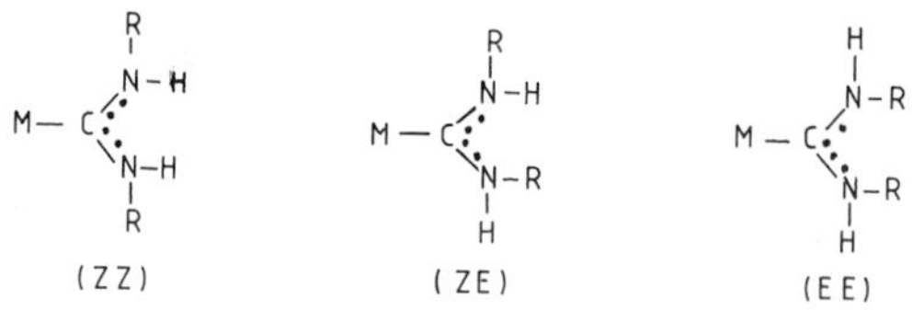

(ZZ)   (ZE)   (EE)

On the basis of ^{1}H and ^{13}C nmr spectra, the isomer distribution appears to depend on the solvent and, to greater extent, on steric requirements, the amphi configuration (ZE) being largely preferred.[59,191,194,208-9] Rotational free energy barriers of 70-75 and 55-60 kJ mol^{-1} have been reported for disubstituted and trisubstituted diaminocarbene-iron(II) derivatives, respectively.[208,210] If reaction (23) is carried out with bifunctional amines, such as 2-aminopyridines, ethylenediamine and o-phenylenediamine, or amidines, chelating carbene ligands of type **VI** can be prepared:[200,211-2]

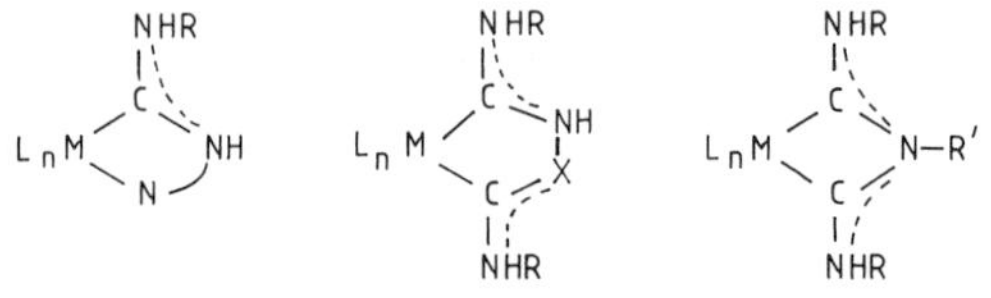

VI   VII   VIII

On the other hand, bifunctional nucleophiles such as hydrazines and hydroxylamine will add to a pair of cis isocyanide ligands to produce five-membered bis-carbene compounds of type **VII** [X=NH or O; M=Pd(II),Pt(II),[183,213-6] Fe(II),Ru(II),[59,217] Rh(III)[193]]. The four-membered bis-carbene complexes **VIII** have been obtained from the reaction of six-coordinate compounds, such as $[Fe(CNMe)_6]^{2+}$,[218] cis-$[Mn(CO)_4(CNMe)_2]^+$ and mer-$[Mn(CO)_3(CNMe)_3]^+$,[166] trans-[RhX(R)(CNBu-t)$_4$]$^+$ (X=I,R=Me,Pr-n),[193] cis-$[Fe(L-L)_2(CNR)_2]^{2+}$ (L-L = 2,2'-bipyridine, 1,10-phenanthroline; R=Me,Pr-i),[193] with an excess of primary amine R'NH$_2$ (isomers with interchanged position of R and R' substituents may be present).[196] Formation of the four-membered cycle probably occurs through the normal attack of R'NH$_2$ on a coordinated isocyanide to yield a mono-carbene group, which, owing to the close proximity of ligands in this six-coordinate intermediate, can act as a nucleophile towards a second <u>cis</u> isocyanide to yield **VIII**.[59,166,196] Such intramolecular ring closure is catalyzed by free amine, $R'NH_2$, and requires rather stringent steric properties of both the central metal and the carbene substituents, although, in some cases, electronic and solvent effects may also be important.[193] In the series $[M(CNMe)_6]^{2+}$ (M= Fe,Ru,Os),[59,164,196,218] only the complex $[Fe(CNMe)_6]^{2+}$ reacts with methyl- and ethylamine to yield products of type **VIII**. (However, with ammonia a mono-carbene derivative is obtained).[196] On the other hand, with the square-planar complexes $[M(CNMe)_4]^{2+}$ (M=Pd,Pt) it is possible to convert all of the isocyanide ligands into carbene ligands, but no four-membered bis-carbene group is formed.[215] Upon heating the complexes **IX** (R=Me,R'=Et; R=Et,R'=Me), Doonan and Balch observed a random scrambling of the ethyl group throughout the molecule with formation of coordinated CNEt.[196] Similarly, prolonged heating of the mono-carbene derivative, $\{Ru[C(NHEt)NHMe](CNMe)_5\}^{2+}$ resulted in isomerization to $\{Ru[C(NHMe)_2](CNMe)_4(CNEt)\}^{2+}$. Statistical scrambling of the amine and isocyanide substituents also took place in the reactions of $[M(CNMe)_6)]^{2+}$ (M=Fe,Ru) with CD_3NH_2. These results were interpreted on the basis of a migration mechanism involving reversible opening and closing of the four-membered cycle. An irreversible exchange of substituents has been observed in the reaction of $[Fe(L)(CO)(CNMe)_2]^+$ (L=Cp,η^5-C_9H_7) with $R'NH_2$(R'=Et,Pr-i,Bu-t), in which the structural isomers, $\{Fe(L)[C(NHR')NHMe](CO)(CNMe)\}^+$ and $\{Fe(L)[C(NHMe)_2](CO)(CNR')\}^+$, are produced.[208]

Cyclic carbene ligands can be obtained in the reactions with bifunctional amines, such as α-aminoacetals or α-aminoketals:[219]

$$PdCl_2(CNR)_2 + H_2N.CHR'.CR''(OEt)_2 \xrightarrow{} X \xrightarrow{H^+} XI + 2EtOH \qquad (24)$$

(R'=R''=H; R'=Me,R''=H; R'=H,R''=Me)

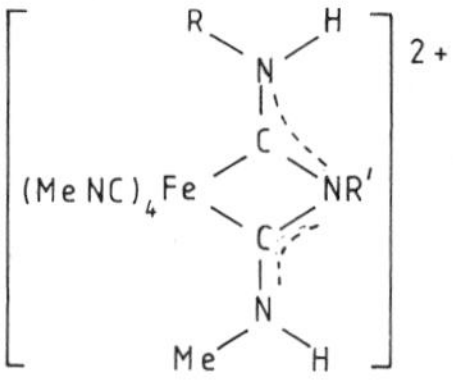

IX 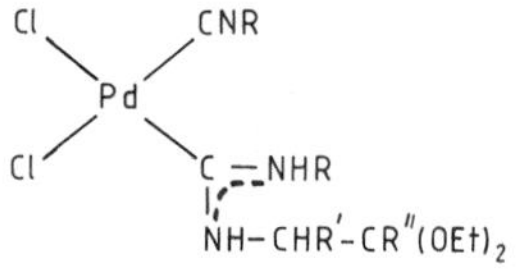   X 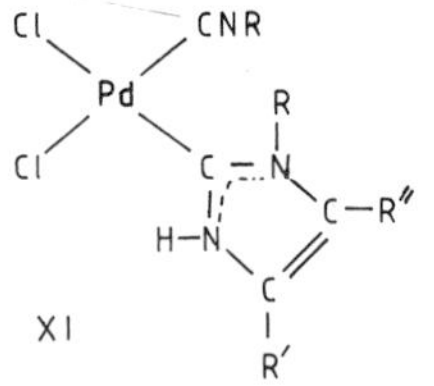   XI

On the other hand, functionalized isocyanides, such as 2-isocyanoethanol or 3-isocyanopropanol, undergo intramolecular addition upon coordination of the isocyano group onto a metal center in the higher oxidation states:

$$HO-(CH_2)_n-NC-ML_m \longrightarrow L_mM-C \overset{NH}{\underset{O}{\diagup}}(CH_2)_n \quad \textbf{XIII} \qquad (25)$$

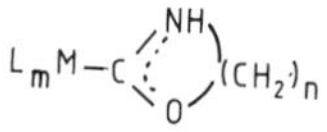

$$\textbf{XII}$$

[M=Pd(II),Pt(II),Au(III),Au(I);[220] M=Co(III);[195] n=2,3]

In the reaction of $ZnCl_2$ with two equivalents of $CN(CH_2)_2OH$, the cyclic carbene intermediate is not stable and rearranges to yield N-coordinated 2-oxazoline as in **XIV**. If the isocyano function is not sufficiently activated, open-chain isocyanide derivatives of the type **XII** are obtained (e.g., the reaction of the zero-valent metal complexes $M(CO)_5(THF)$ (M=Cr,W) with $CN(CH_2)_nOH$ (n=2,3) affords $M(CO)_5(CN(CH_2)_nOH)$).[220]

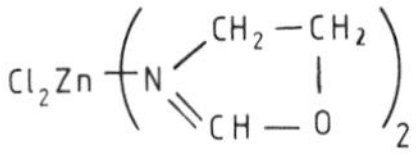

$$Cl_2Zn \left(N \overset{CH_2-CH_2}{\underset{CH-O}{\diagup}} \right)_2 \qquad XIV$$

Reaction (23) appears to be quite general, provided that electronic and steric factors are favorable. In general, for a given M-CNR substrate, amines (better nucleophiles) react faster than alcohols and, for a given nucleophile HY, aryl isocyanides (more electron-withdrawing R) react faster than the alkyl analogues. The latter case is clearly illustrated by reaction (26):

cis-$PdCl_2(CNC_6H_4Me-4)(CNCy)$ + $H_2NC_6H_4Me-4$ $\longrightarrow$
cis-$PdCl_2[C(NHC_6H_4Me-4)_2](CNCy)$ (26)

in which only the CNC_6H_4Me-4 group is attacked by p-toluidine.[212]

The reactivity of the complexes $Au(C_6X_5)(CNC_6H_4Me\text{-}4)$ (X=F,Cl,Br) towards nucleophiles was found to decrease in the order:[221]

primary amines > ammonia > secondary amines > aromatic amines > alcohols

The carbene derivatives are generally obtained from neutral or cationic complexes of metals in the higher oxidation states. However, in the anionic compound $[PdCl_3(CNPh)]^-$, the CNPh ligand is still sufficiently activated $(\Delta\nu = 80\ cm^{-1})$ to undergo a slow nucleophilic attack by methanol, yielding $[PdCl_3[C(NHPh)OMe]]^-$.[212] Steric effects are apparent in the ease of formation of bis-carbene derivatives of Au(I) and Pd(II): for the linear complex, $[Au(CNC_6H_4Me\text{-}4)_2]^+$, both isocyanide ligands easily react with primary and secondary anilines (e.g. p-toluidine and N-methylaniline),[222] whereas, for the cis-$PdCl_2(CNAr)_2$ compounds, the same reaction goes to completion only with primary anilines.[223] Furthermore, the rates of reaction (27):

$$[Fe(L)(CO)(CNR)_2]^+ + H_2NR' \longrightarrow [Fe(L)[C(NHR)NHR'](CO)(CNR)]^+$$
$$(L=Cp, \eta^5\text{-}C_9H_7;\ R=Me,Et,Pr\text{-}i,Bu\text{-}t) \hspace{3cm} (27)$$

depend primarily on the nature of the amine and decrease markedly in the order R'=Me > Et > Pr-i > Bu-t.[208] Reaction (27) also occurs with dimethylamine, but secondary amines with larger alkyl groups are unreactive under the same conditions.[210] The resulting trisubstituted diaminocarbenes are not stable and dissociate to the starting materials:

$$[FeCp[C(NHR)NMe_2](CO)(CNR')]^+ \longrightarrow [FeCp(CO)(CNR')(CNR)]^+$$
$$+ NHMe_2 \hspace{3cm} (28)$$

The course of reaction (28) was followed by ^1H nmr spectroscopy and the first-order rate constants at 300°K were found to increase with increasing bulkiness of both substituents R and R':

$k(sec^{-1})$	R	R'
4.3×10^{-4}	Me	Me
5.3×10^{-3}	Me	Pr-i
2.7×10^{-2}	Pr-i	Pr-i

A detailed kinetic study of the mechanism of carbene formation has been carried out on reaction (29):[224-9]

$$cis\text{-}PdCl_2(CNC_6H_4Y\text{-}4)(L) + RHNC_6H_4Z\text{-}4$$
$$\longrightarrow cis\text{-}PdCl_2[C(NHC_6H_4Y\text{-}4)N(R)C_6H_4Z\text{-}4](L) \qquad (29)$$

(Y=MeO,Me,H,Cl,NO$_2$; Z=MeO,Me,H,Cl; L=phosphine, phosphite, CNC$_6$H$_4$Y-4, diaminocarbene)

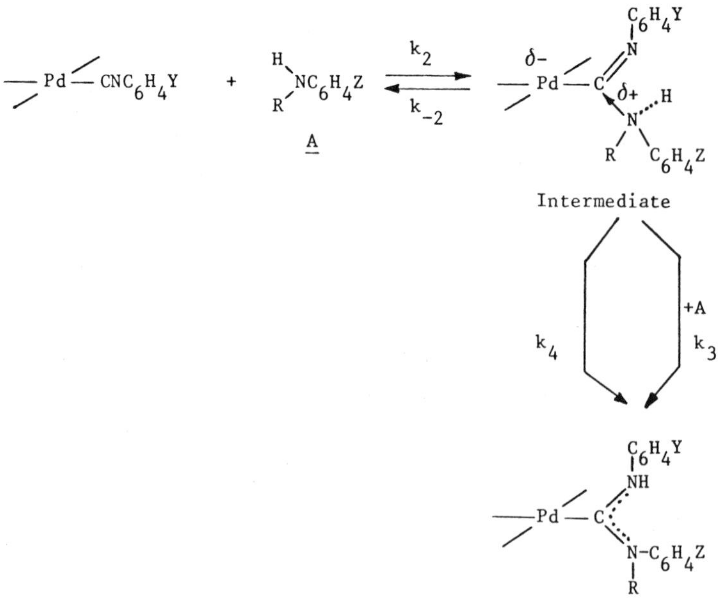

Scheme 1

All the kinetic data can be interpreted on the basis of the step-wise mechanism of Scheme 1.

On applying the steady-state approach, the following general rate law is obtained:

$$k(obs)/[A] = k_2(k_4 + k_3[A])/(k_{-2} + k_4 + k_3[A]) \tag{30}$$

Depending on the relative magnitude of k_{-2}, k_4 and $k_3[A]$ terms, three different types of $k_{obs}/[A]$ vs $[A]$ plots may occur. When all the terms are of comparable value, curvilinear plots are observed, whereas, when $k_{-2} \gg k_4 + k_3[A]$, the rate law is expressed by eq (31), giving a linear dependence of $k_{obs}/[A]$ on $[A]$:

$$k_{obs}/[A] = (k_2 k_4 + k_2 k_3[A])/k_{-2} \tag{31}$$

Finally, when $k_{-2} \ll k_4 + k_3[A]$, the rate law takes the simple form:

$$k_{obs}/[A] = k_2 \tag{32}$$

The first step (k_2) of the mechanism in Scheme 1 is nucleophilic attack of the entering amine at the carbon atom of the coordinated isocyanide, producing an intermediate which rearranges to the final carbene product by proton transfer between the nitrogen atoms. This may occur either intramolecularly (step k_4) in a four-membered cyclic transition state, or by the agency of one further amine molecule (A) (step k_3) serving as a proton acceptor-donor in a six-membered cyclic transition state:

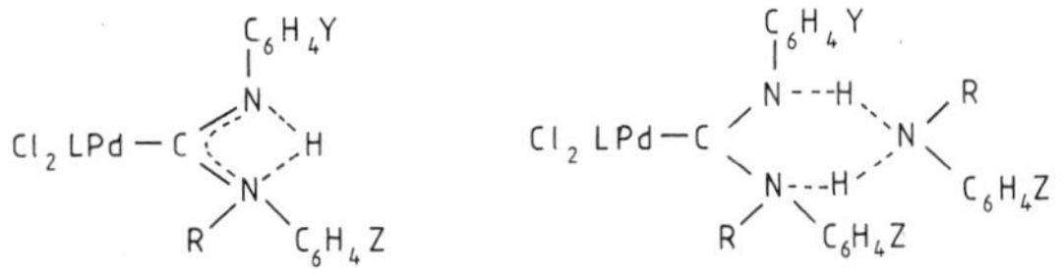

This has been confirmed by the fact that reaction (29) (Y=H, L=PPh$_3$, R=Et, Z=Me) is catalyzed by primary anilines, such as H$_2$NC$_6$H$_4$Br-4, which is itself too weak a nucleophile to compete with the secondary amine in forming a palladium-carbene species.[225] On the other hand, tertiary amines, such as N,N-dimethylaniline, have no appreciable effect on the rates of reaction (29), lacking a transferable proton for step k_3. The overall reaction rate ($k_{obs}/[A]$) is markedly influenced by solvent in the order: benzene > 1,2-dichloroethane > 1,4-dioxane ~ acetone (i.e., $k_{obs}/[A]$ decreases with increasing ability of the solvent to form hydrogen bonds with the amine.)[226] The nucleophilic attack is favored by decreasing steric hindrance and increasing π-accepting capability of the ancillary ligand L.[226] When L=PCy$_3$, the conditions are so unfavorable that no reaction occurs. On the contrary, when L=CNC$_6$H$_4$Z-4 (a ligand less sterically demanding than phosphines and phosphites and, in general, with better π-accepting properties) reaction (29) is so fast that stopped-flow techniques must be employed.[228] In the reactions of para-substituted isocyanide complexes with para-substituted primary anilines the kinetic law (32) is observed.[224] This is due mainly to the reduced steric hindrance at the amino nitrogen atom (R=H), which results in a low value of the k_{-2} term (dissociation of the intermediate to the starting reactants) so that $k_{-2} \ll k_4 + k_3[A]$. In other words, the bimolecular nucleophilic attack (k_2) becomes the rate determining step. However, an increased steric strain, brought about by ortho-substituents either on the incoming primary aniline or on the coordinated isocyanide, produces an overall decrease in rate and makes the k_{-2} term comparable with the subsequent steps ($k_4 + k_3[A]$).[227] Consequently, even for primary amines kinetic laws of type (30) and (31) are observed.

In general, k_2 increases with increasing basicity of the amine. For para-substitued anilines, RHNC$_6$H$_4$Z-4, reactivity is in the order:

Z = MeO > Me > H > Cl

The k_2 values also follow the electron-withdrawing abilities of the isocyanide para-substituent Y in the order:

Y = NO$_2$ > Cl > H > Me > MeO

This is also the order of deshielding of the isocyano carbon atoms in the ^{13}C nmr spectra of the complexes cis-PtCl$_2$(CNC$_6$H$_4$Y-4)(PEt$_3$) and trans-[PtCl(CNC$_6$H$_4$Y-4)(PEt$_3$)$_2$]$^+$.[230]

The reaction of the chloro-bridged binuclear compounds [PdCl$_2$(CNAr)]$_2$ (Ar=C$_6$H$_4$Me-4, C$_6$H$_4$OMe-4) leads initially to the rapid formation of trans-PdCl$_2$(CNAr)(RHNC$_6$H$_4$Z-4), which is then attacked at the isocyano group by a second molecule of amine to yield the carbene derivative trans-PdCl$_2$[C(NHAr)N(R)C$_6$H$_4$Z-4](RHNC$_6$H$_4$Z-4).[229]

Slightly different mechanisms have been proposed for the reactions of hydrazines with Pt(II) and Fe(II) isocyanide complexes.[231-2] In the reaction of cis-PtCl$_2$(CNAr)$_2$ with MeNHNH$_2$, a mono-carbene complex, cis-PtCl$_2$[C(NHAr)(NMeNH$_2$)](CNAr) is formed, with a kinetic law:

$$k_{obs} = k_1 + k_2[\text{MeNHNH}_2] \tag{33}$$

The rate constant k_2 corresponds to the initial bimolecular nucleophilic attack of reaction (29). The k_1 term has been related to a parallel path involving a rather unusual addition of the solvent (CH$_2$Cl$_2$) to the substrate, followed by a fast reaction with methylhydrazine to yield the final carbene product.[231] In the reaction of [Fe(CNMe)$_6$]$^{2+}$ with NH$_2$NH$_2$ a chelating bis-carbene complex of type **VII** is formed, with a kinetic law:

$$k_{obs} = (K_1k_2[\text{H}^+] + k_3)[\text{NH}_2\text{NH}_2] \tag{34}$$

where K_1 is the equilibrium constant of eq (35)

$$\text{NH}_2\text{NH}_2 + \text{H}^+ \rightleftharpoons \text{NH}_2\text{NH}_3^+ \tag{35}$$

and k_2 and k_3 refer to the bimolecular nucleophilic attack of NH$_2$NH$_3^+$ and NH$_2$NH$_2$, respectively, on one isocyanide ligand of [Fe(CNMe)$_6$]$^{2+}$, to be followed by a rapid ring closure to yield the final cyclic bis-carbene product.[232]

In Au(I) derivatives, the diamino- or (alkoxy)amino carbene groups can easily be displaced, even in very mild conditions, by other ligands, such as PPh$_3$, CN$^-$ or CNR, to yield formamidines or formimino alkyl esters:[201-2,222,233]

$$\{Au[C(NHAr)_2]_2\}^+ + PPh_3 \longrightarrow \{Au[C(NHAr)](PPh_3)\}^+ + \qquad (36)$$

$$(Ar=C_6H_4Me-4) \qquad\qquad\qquad HC(=NAr)NHAr$$

$$AuCl[C(NHMe)OMe] + 2\ CN^- \longrightarrow [Au(CN)_2]^- + Cl^- + HC(=NMe)OMe \quad (37)$$

On the other hand, the carbene complex, CoMe(dmgH)$_2$[C(NHMe)$_2$] undergoes irreversible thermal isomerization in solution to the isomeric N-bound formamidine compound.[194] All these reactions suggest that labile carbene intermediates may be involved in the 1,1-addition of protic nucleophiles to isocyanides, catalyzed by metal ions of groups IB and IIB (Section 7.1). Initial formation of a carbene ligand also occurs in the synthesis of heterocyclic compounds from isocyanides and bifunctional amines.[201]

5. REACTIONS WITH ELECTROPHILES

The reactions of electrophiles with low-valent metal-isocyanide complexes may occur either at the central metal (oxidative addition) or at the nitrogen atom of the isocyano group. In some cases, both sites of attack are involved, as shown by the course of the protonation reactions on trans-M(CNMe)$_2$(dppe)$_2$ (M=Mo,W).[234] The proton attack is kinetically controlled and occurs initially at the coordinated isocyanide to yield the compounds trans-[M(CNHMe)(CNMe)(dppe)$_2$]$^+$, which can be isolated as solids, but rearrange in solution to the thermodynamically more stable hydride species [MH(CNMe)$_2$(dppe)$_2$]$^+$.[235]

5.1 Oxidative addition reactions

These reactions are of great importance in organometallic chemistry and homogeneous catalysis, and have been extensively studied in phosphine, carbonyl and isocyanide complexes. However, since they have already been reviewed in some detail,[9-11] only a brief account of the most recent advances and mechanistic aspects will be given here.

In accord with bonding properties and electrochemical investigations (Section 2.1), the isocyanide complexes undergo oxidative additions more easily than the corresponding carbonyl derivatives. A higher reactivity is usually observed for alkyl than for aryl isocyanides. In general, the attack of electrophilic addends, such as halogens, protic acids, alkyl and aryl halides, molecular oxygen, cyano- and fluoroolefins etc., brings about an increase in oxidation state (reflected in a corresponding increase in the CN stretching frequencies of the coordinated CNR ligands) and in coordination number of the central metal. In this way, seven-coordinate Mo(II),[28,235] W(II),[28,235-6] Re(III)[237] and five-coordinate Pt(II)[134] isocyanide complexes have been prepared.

A variety of interesting reactions have been carried out on the trinuclear complex, $Pt_3(CNBu-t)_6$: with unsaturated fluorocarbons three-, five- and six-membered metallacyclic derivatives are obtained,[238] e.g., the "chair" diplatinacycle XV. The reaction with diacetylenes yields the four-membered metallacycles XVI,[239] while with hexakis(tris-fluoromethyl)benzene a C-C bond cleavage occurs with formation of the seven-membered metallacycle XVII.[240] The reaction with diphenylcyclopropenone results in carbon-carbon double bond cleavage to give the complex XVIII.[241] In the reaction with Fe(CNBu-t)$_5$, both C-C bond cleavage and isocyanide insertion occur.[242] Reaction (38) is just one of the many oxidative additions in which single and/or multiple successive insertions of coordinated CNR ligands are involved,[21,134,243-5] as will be discussed in Section 6.

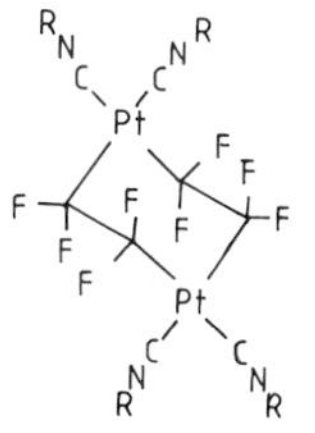

XV 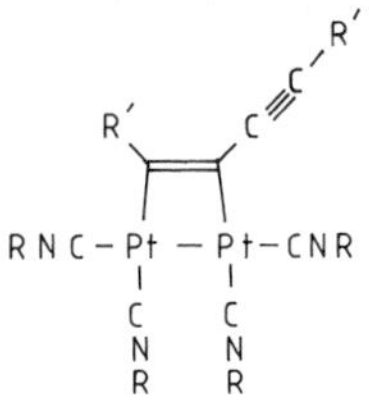   XVI 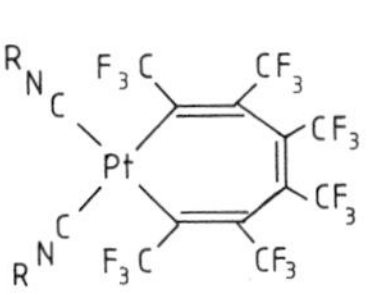   XVII 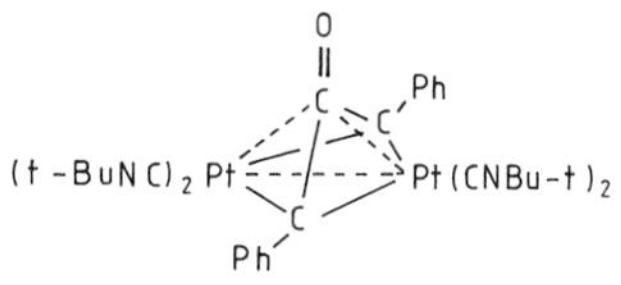   XVIII

$$Fe(CNBu\text{-}t)_5 \ + \ \begin{array}{c} Ph \\ {}^{\diagdown}C \\ \parallel \quad {>}C{=}O \\ {}^{\diagup}C \\ Ph \end{array} \ \longrightarrow \ (t\text{-}BuNC)_4 Fe \begin{array}{c} O \quad Ph \\ \parallel \quad \diagup \\ C \ - \ C \\ \parallel \\ C \ - \ C \\ \parallel \quad \diagdown \\ NBu\text{-}t \quad Ph \end{array} \tag{38}$$

Two- and three-center oxidative addition reactions have been reported for the complexes $[Rh(CNR)_4]^+$:[246-7]

$$2[Rh(CNR)_4]^+ + X_2 \quad \xrightarrow{\hspace{3cm}} \quad \textbf{XIX} \tag{39}$$
$$(R=Cy, C_6H_4Me\text{-}4; \ X=Br,I)$$

$$3[Rh(CNR)_4]^+ + I_2 \quad \xrightarrow{\hspace{3cm}} \quad \textbf{XX} \tag{40}$$
$$(R=CH_2Ph)$$

XIX   XX

The crystal structures of **XIX** $(R=C_6H_4Me\text{-}4; \ X=I)^{248}$ and $\textbf{XX}^{247}$ show the presence of essentially linear I-Rh-Rh-I and I-Rh-Rh-Rh-I arrangements with direct metal-metal bonds [Rh-Rh, 2.785(2) and 2.796(1)Å, respectively]. The coordination about each rhodium is pseudooctahedral; the $Rh(CNR)_4$ units are staggered $26°$ in **XIX** and $38°$ in **XX**. The complex **XIX** can be considered as containing two Rh(II) centers: the position of the $\gamma(CN)$ band $(R = Cy, 2214 \ cm^{-1})$ is in fact intermediate between those of $[Rh(CNCy)_4]^+$ $(2170 \ cm^{-1})$ and $[RhX_2(CNCy)_4]^+$ $(2239 \ cm^{-1})$. Similar two-center oxidative additions also occur for binuclear Rh(I) compounds, such as $[Rh_2(CNR)_4(L\text{-}L)_2]^{2+}$ $(L\text{-}L=$ bis(diphenylphosphino)methane and bis(diphenylarsino)methane$)^{140-1}$ and $[Rh_2(bridge)_4]^{2+}$ (bridge = 1,3-diisocyanopropane,[48] where the two Rh(I) centers are held together by bridging chelating ligands. The Rh-Rh distance of 3.242(1)Å in $[Rh_2(bridge)_4]^{2+}$ [44] is reduced to 2.837(1)Å in $[Rh_2Cl_2(bridge)_4]^{2+}$,[249] indicating the formation of a direct metal-metal bond in the oxidized Rh(II) dimer. The latter compound is also produced by visible irradiation (546 nm) of $[Rh_2(bridge)_4]^{2+}$ in acidic aqueous solution[250] (eq 41):

$$[Rh_2(bridge)_4]^{2+} + 2HCl \xrightarrow[\text{12M HCl}]{\text{546 nm}} [Rh_2Cl(bridge)_4]^{2+} + H_2 \quad (41)$$

This interesting reaction represents a way of converting solar (visible) light into chemical energy. The overall reaction (41) was found to proceed in two steps:[251]

$$[Rh_2(bridge)_4]^{2+} + HCl \xrightarrow[\text{12M HCl}]{\Delta} (1/n)[Rh_2(bridge)_4Cl^{2+}]_n + \tfrac{1}{2}H_2 \quad (42)$$

$$(1/n)[Rh_2(bridge)_4Cl^{2+}]_n + HCl \xrightarrow[\text{12M HCl}]{\text{546 mm}} [Rh_2Cl_2(bridge)_4]^{2+} + \tfrac{1}{2}H_2 \quad (43)$$

The thermal step (42) yields a blue photoactive species $[Rh_2(bridge)_4Cl^{2+}]_n$, which was shown to be tetranuclear (n=2) by flash photolysis and quenching studies on the analogue $[Rh_2(bridge)_4]_2^{6+}$ (prepared by oxidation of $[Rh_2(bridge)_4]^{2+}$ in H_2SO_4 solution).[251] This was later confirmed by an xray crystallographic analysis of the related compound, $H_3[Rh_4(bridge)_8Cl][CoCl_4]_4 \cdot nH_2O$, obtained on addition of $CoCl_2 \cdot 6H_2O$ to the photoactive solution of $[Rh_2(bridge)_4](BF_4)_2$ in 12M HCl.[252]

The rates of the reaction (44)

$$\underline{cis}\text{-}[PtMe_2(CNR)_2] + X\text{-}Y \xrightarrow{k} \underline{cis}\text{-}Me_2, \underline{cis}\text{-}(CNR)_2Pt \text{ } \underline{trans} \text{ } XY \quad (44)$$
$$(R = C_6H_4Me\text{-}4; \text{ } X\text{-}Y = MeI, MeCOCl)$$

were followed by nmr spectroscopy in chloroform.[253] The kinetic data obey a second-order rate law:

$$-d[PtMe_2(CNR)_2]/dt = k[X-Y][PtMe_2(CNR)_2] \qquad (45)$$

The low activation energies (36 and 35.5 kJ mol^{-1} for the addition of MeI and MeCOCl respectively) suggest that the metal center is rather electron rich. Similar reactions with cis-[PtMe$_2$L$_2$] (L=PMe$_2$Ph and AsMe$_3$) were found to proceed even faster than for the isocyanide substrate. A marked solvent dependence was observed: in benzene no oxidative addition occurred under identical conditions, suggesting that a polar transition state is probably involved. The oxidative addition of chiral α-bromo esters, RBr, (R=PhCHCO$_2$Et, MeCHCO$_2$Et) to [Rh(CNR')$_4$]$^+$ (R'=Bu-t, C$_6$H$_4$Me-4) yields the optically inactive adducts, trans-[RhBr(R)(CNR')$_4$]$^+$.[254] The reaction of PhCHBrCO$_2$Et with [Rh(CNBu-t)$_4$]$^+$ starts immediately in the absence of light, with a rate law of the type v = k[Rh(I)]^2[RBr]. The addition of PhCHBrCO$_2$Et to [Rh(CNC$_6$H$_4$Me-4)$_4$]$^+$ proceeds slowly in the dark, but rapidly under irradiation (440 nm) with a high quantum yield (ϕ=4.8). Photo-initiation is required for the addition of MeCHBrCO$_2$Et. The optical activity of (S)-(-)-MeCHBrCO$_2$Et was completely lost prior to the oxidative addition, and esr experiments using t-BuNO as a spin trap indicated the presence of a radical, t-Bu(R)NO$^{\cdot}$. However, a radical scavenger, such as duroquinone, does not inhibit the reaction. All these results were rationalized on the basis of the chain mechanism (46):

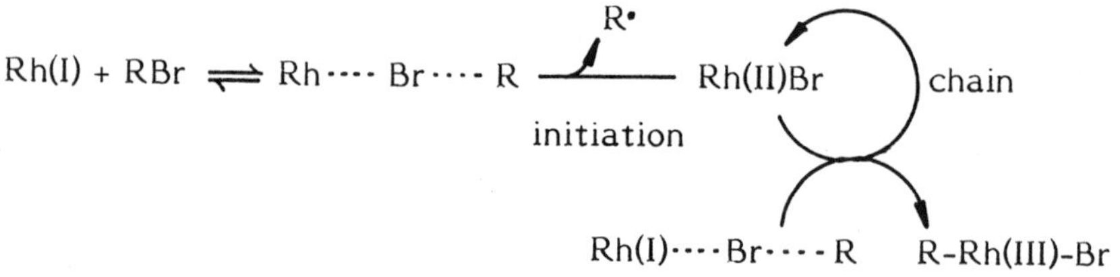

The initial reversible step, in which the intermediate Rh$\cdots$Br$\cdots$R is formed, accounts for the racemization preceding the addition. No radical mechanism was observed for the reversible addition of XSnPh$_3$ (X=Cl,Br) to [Rh(CNBu-t)$_4$]$^+$:[255]

$$[Rh(CNBu\text{-}t)_4]^+ + XSnPh_3 \underset{k_b}{\overset{k_f}{\rightleftharpoons}} \underline{cis}\text{-}[RhX(SnPh_3)(CNBu\text{-}t)_4]^+ \qquad (47)$$

Under pseudo-first-order conditions, the k_{obs} is expressed as

$$k_{obs} = k_f[XSnPh_3] + k_b \qquad (48)$$

The rate constant of the forward reaction (k_f) is strongly dependent on the solvent and increases in the order: 1,2-dichloroethane acetone acetonitrile. On the contrary, the rate constant of the backward reaction depends little on solvent polarity. The activation entropy $\Delta S^{\ddagger}$ for the forward reaction (X=Cl) is -100 J K^{-1} mol^{-1}, while that of the backward reaction is 25 J K^{-1} mol^{-1}. On this basis, a polar three-center transition state **XXI** was proposed.

L = CNBu-t

Similar transition states have been proposed for the reactions of $IrCl(CO)(PR_3)_2$ with alkyl halides[256] and also for the addition of TCNE to $\underline{trans}\text{-}[Rh(CNR)_2(PR'_3)_2]^+$ (R = Cy,C_6H_4Cl-4,C_6H_4OMe-4; R' = Ph,OPh).[257] In the latter case, the reaction is first-order with respect to both substrate and TCNE, and the rates increase with increasing solvent polarity (THF < acetone < acetonitrile) and with increasing electron density on the metal (R = Cy > C_6H_4OMe-4 > C_6H_4Cl-4, for R' = Ph). The addition of I_2 to square-planar Rh(I)-isocyanide complexes has been reported to proceed via a $\underline{cis}$-adduct, which then rearranges to the final $\underline{trans}$ product.[258-9] The $\underline{cis}$-$\underline{trans}$ isomerization is zero-order with respect to the I_2 concentration and has small $\Delta S^{\ddagger}$ values in the range $8\overset{+}{-}10$ to $-3\overset{+}{-}5$ JK^{-1} mol^{-1}, suggesting an intramolecular process, probably via a twist mechanism.

5.2 Electrophilic attack on coordinate isocyanides

Electrophilic attack at the nitrogen atom of the isocyano group may occur either on terminal isocyanides coordinated to electron rich metal sites, such as M(dppe)$_2$ (M=Mo,W), or on bridging isocyanides. The electronic factors which favor these reactions have been already discussed (Section 2.). This interesting type of ligand reactivity has so far been limited to protonation and alkylation reactions.

The addition of strong mineral acids to trans-M(CNR)$_2$(dppe)$_2$ (M=Mo,W) affords a variety of products depending on the reaction conditions (solvent, temperature, rate of acid addition, nature of M and R).[23,234-5,260] For R=Me, the initial protonation always occurs at the isocyanide (Scheme 2).

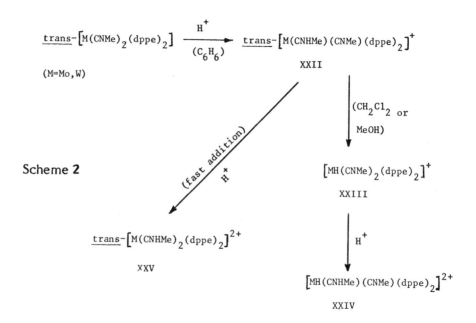

Scheme 2

The complexes **XXII** precipitate from reactions in benzene, whereas in polar solvents they convert into hydride species **XXIII**, which are thermodynamically more stable. Fast acid addition to **XXII** affords the dicarbene complexes **XXV**. Complexes of type **XXII** cannot be obtained when the starting compounds contain aryl isocyanide ligands. In this case, only the hydrides **XXIII** and dicarbene species **XXV** have been isolated. On the basis of ir and nmr (^{1}H and ^{13}C) spectral data, the structure of the CNHR ligands can be described in valence-bond terms by the limiting formulae:

$$[\overset{+}{M}\equiv C\text{-}NHR]$$

(carbyne)

$$\left[\overset{+}{M}\equiv C-N\overset{H}{\underset{R}{\diagdown}}\right]$$

XXVIa

$$\left[M=C=\overset{+}{N}\overset{H}{\underset{R}{\diagdown}}\right]$$

XXVIb

$$[M=C=\overset{+}{N}HR]$$

(carbene)

XXVIc

with a higher contribution of the carbene structure **XXVIb**. The weight of the canonical formula **XXVIb** is even greater in the diprotonated complexes **XXV**, which are best regarded as bis(iminiocarbene) compounds. The CNHR groups are strong electron-withdrawing ligands, as can be inferred from the high values of $\nu(CN)$ of coordinated isocyanides in **XXII**. Protonation at the nitrogen atom is also reported for the complex <u>trans</u>-ReCl(CNMe)(dppe)$_2$, characterized by a very low $\nu(CN)$ value of 1840 cm^{-1}.[261] Due to the relatively high electron density on the isocyanide nitrogen atom, the complexes <u>trans</u>-W(CNMe)$_2$(dppe)$_2$ and Ta(Pr-n)Cp$_2$(CNMe) react with Lewis acids, such as Al$_2$Et$_6$, yielding N-adducts of the type W[CN(AlEt$_3$)Me]$_2$(dppe)$_2$[260] and Ta(Pr-n)Cp$_2$[CN(AlEt$_3$)Me],[76] which have been studied only in solution.

The reaction of <u>trans</u>-M(CNMe)$_2$(dppe)$_2$ (M=Mo,W) with alkylating agents, such as MeFSO$_3$, Me$_2$SO$_4$ and [Et$_3$O]BF$_4$, proceeds according to eq (49):[262]

(i)

<u>trans</u>-M(CNMe)$_2$(dppe)$_2$ + RX $\longrightarrow$ <u>trans</u>-{M[CN(R)Me](CNMe)(dppe)$_2$}X

(R=Me,Et; X=FSO$_3$,MeSO$_4$,BF$_4$)

(i) benzene

(ii) chlorinated solvents or THF

$$\Big\updownarrow \text{(ii)}$$

(49)

<u>cis</u>-{M[CN(R)Me](CNMe)(dppe)$_2$}X

Only one CNMe ligand is alkylated at nitrogen, even in the presence of a large excess of RX. The structure and binding properties of the CN(R)Me ligands are very close to those of the protonated CNHMe group, (strong electron-withdrawing ability and significant weight of canonical form **XXVIb**).

In the complexes M(CNBu-t)$_5$ (M=Fe,Ru), which also exhibit very low CN stretching frequencies and structures (M=Fe) with terminal bent isocyanides,[24] protonation and alkylation occur only at the central metal.[21,263-4] Further work is required towards understanding the balance of kinetic and thermodynamic factors which control the site of electrophilic attack in these low-valent metal-isocyanide complexes.

The protonation of a bridging isocyanide ligand was first reported for the complex, $Fe_2Cp_2(CO)_3(CNMe)$, which gave $[Fe_2Cp_2(CO)_2(\mu\text{-}CO)(\mu\text{-}CNHMe)]^+$ on treatment with dilute mineral acids.[265] This reaction was later extended to a variety of compounds of the type $Fe_2(\eta^5\text{-}dienyl)_2(CO)_{4-n}(CNR)_n$ (dienyl=C_5H_5 or C_9H_7; R=alkyl or benzyl; n=1,2)[266] and $M_2Cp_2(CO)_3(CNR)$ (M=Fe, R=Me,Et,Pr-i,Bu-t,Ph; M=Ru,R=Et).[92] The protonated CNHR groups were found to assume a bridging position, as in **XXVIc.**

When two isocyanide ligands are present, they are both protonated yielding $[Fe_2(\eta^5\text{-}dienyl)_2(CO)_2(\mu\text{-}CNHR)_2]^{2+}$.[266] The basic nitrogen atoms in the μ-CNR isomers of $[Fe_2(\eta^5\text{-}dienyl)_2(CO)_{4-n}(CNR)_n]$ interact with protic solvents, such as ethanol or phenols, to give N-hydrogen-bonded species and also μ-CNHR species in the more acidic phenols.[267] The reaction with alkyl halides, R'X, proceeds as follows:[268-9]

$$[Fe_2(\eta^5\text{-}dienyl)_2(CO)_2(L)(CNR)] \xrightarrow{R'X} [Fe_2(\eta^5\text{-}dienyl)_2(CO)(L)(\mu\text{-}CO)[\mu\text{-}CN(R')R]]^+X^- \qquad (50)$$

(L=CO,CNR)

For L = CNR, the reaction occurs more readily than for L = CO, but in contrast to the protonation reaction, only one isocyanide ligand is alkylated. The reactivity of the halides R'X depends on both the alkyl group R' and the halogen X in the orders: R' = allyl~benzyl~Me>Et>Pr-n>Bu-n≫Pr-i≫Ph or Bu-t and X=I>Br≫Cl, as found for most nucleophilic substitution reactions (the anomalous position of Bu-t is probably due to steric factors). Some alkylation reactions by $[Et_3O]BF_4$ are also reported.[92] Both μ-CNHR and μ-CN(R')R groups are strong electron-withdrawing ligands and better π-acceptors than either μ-CNR or μ-CO. Xray structural analyses of <u>cis-</u>

$[Fe_2Cp_2(CO)_2(\mu\text{-}CO)(\mu\text{-}CNHMe)]BF_4{}^{266}$ and $\underline{cis}$-$[Fe_2Mecp_2(CO)_2(\mu\text{-}CO)(\mu\text{-}CNMe_2)]I^{269}$ indicate that the μ-CNHMe and μ-CNMe$_2$ groups are better described as bridging iminiomethylene ligands than as bridging carbynes (sp^2 hybridization at nitrogen and C-N bond lengths of 1.28 and 1.30 $\overset{o}{A}$, respectively). Methyl iodide was found to react with $[Co_2(CNR)_6(\mu\text{-}CNR)_2]$ (R=$C_6H_3Me_2$-2,6) to give the N-methylated product **XXVII**:[127]

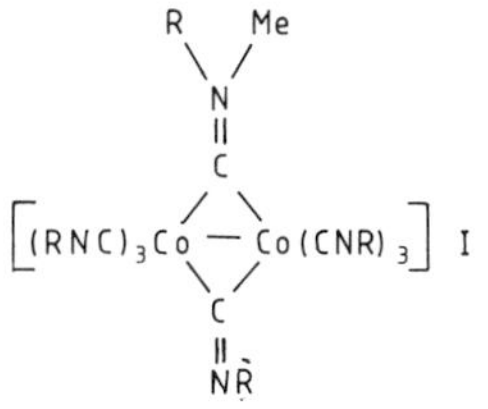

XXVII

6.   INSERTION REACTIONS

The insertion reaction is of great importance in organometallic chemistry and represents a fundamental step in many catalytic processes. The insertion of carbon monoxide into metal-carbon σ bonds has been extensively studied and shown to involve an intramolecular rearrangement, which may be assisted by solvent or by an incoming ligand, whereby the σ-bonded organic moiety migrates onto the carbon atom of a $\underline{cis}$-coordinated CO group.[270-2] (c.f. also Chapters 6-8) The isoelectronic isocyanide molecules are also known to give similar insertion reactions and appear to be even more reactive than carbon monoxide.[8-10] In addition to readily showing multiple successive insertion with various transition metal substrates, CNR ligands also insert into M-H, M-N, M-O and metal-halogen bonds.

6.1   Insertion into metal-carbon bonds

Several mechanistic and kinetic studies on CO insertion have shown that this reaction may proceed by either of two pathways:[270-2]

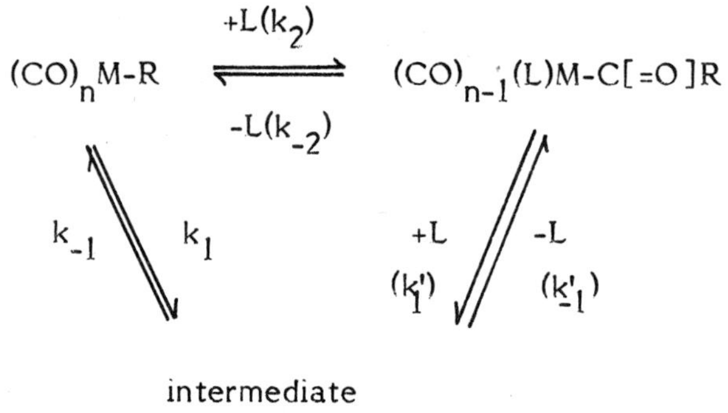

$$(CO)_n M-R \underset{-L(k_{-2})}{\overset{+L(k_2)}{\rightleftharpoons}} (CO)_{n-1}(L)M-C[=O]R \qquad (51)$$

$$k_{-1} \diagdown k_1 \qquad \qquad \overset{+L}{\underset{(k_1')}{\diagup}} \overset{-L}{\underset{(\underline{k}_1')}{\diagup}}$$

intermediate

In step k_2, the insertion is promoted by the incoming ligand L and second-order kinetics are observed. In the k_1 step, the starting compound rearranges slowly to give an intermediate, which then reacts rapidly with L affording the final acyl product. This pathway may be influenced by the coordinating properties of the solvent. Only a few mechanistic investigations have been carried out on isocyanide insertion. However, the available data indicate that the reaction follows pathways essentially analogous to steps k_1 or k_2 of mechanism (51), suggesting that a similar intramolecular cis-migratory rearrangement is involved:

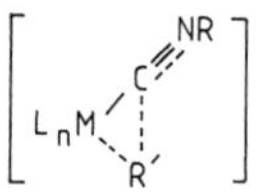

$$\left[L_n M \underset{R'}{\overset{C\overset{NR}{\nearrow}}{\diagup}} \right]$$

A kinetic study of the reaction between trans-PdI(Me)(CNBu-t)$_2$ and a nucleophile L, such as CNBu-t, PPh$_3$ or P(OPh)$_3$, showed that the rate was independent of the nature and concentration of L:[75]

$$Rate = k(obs)[PdI(Me)(CNBu-t)_2] \qquad (52)$$

This was interpreted on the basis of mechanism (53):

$$\text{trans-PdI(Me)(CNBu-t)}_2 \xrightarrow[k(obs)]{slow} \text{PdI[C(=NBu-t)Me](CNBu-t)}$$

$$\text{(intermediate)}$$

$$\Big\downarrow \begin{array}{l} \text{fast} \\ \text{+ L} \end{array} \qquad\qquad (53)$$

$$\text{PdI[C(=NBu-t)Me](CNBu-t)(L)}$$

where the rate-determining methyl migration step $k(obs)$ leads to a coordinatively unsaturated intermediate, which reacts rapidly with L to give the final product. The slow step corresponds to the k_1 step of the CO insertion mechanism (51). The solvent effect [the $k(obs)$ term increases with increasing solvent polarity] and the negative entropy value of -92 ± 4 J K^{-1} mol^{-1} suggest a solvent assisted migratory rearrangement in a polar transition state. Alkyl migration through a five-coordinate transition state was proposed for reaction (54) (M=Pd,Pt; R'=cis-CH=CHCO$_2$Me; R=C$_6$H$_4$Me-4, Bu-t):[75]

$$\text{trans-MBr(R')(PPh}_3)_2 \xrightarrow{\text{CNR}} \text{trans-[M(R')(CNR)(PPh}_3)_2\text{]Br}$$

$$\textbf{XXVIII} \quad \text{(M=Pt; R=C}_6\text{H}_4\text{Me-4,Bu-t)} \quad (54)$$

$$\Big\downarrow$$

$$\text{trans-MBr[C(=NR)R'](PPh}_3)_2$$

The ionic intermediates **XXVIII** were obtained only for M=Pt. This result parallels those previously reported for the reaction of CNR (R=Me,C$_6$H$_4$Cl-4) with PtX(R')(L)$_2$ (X=Br,I; R'=Me,Ph; L=tertiary phosphine), for which the following mechanism was proposed:[273]

$$\text{PtX(R')(L)}_2 \xrightarrow{\text{CNR}} [\text{Pt(R')(CNR)(L)}_2]\text{X} \xrightarrow{\hspace{1cm}}$$

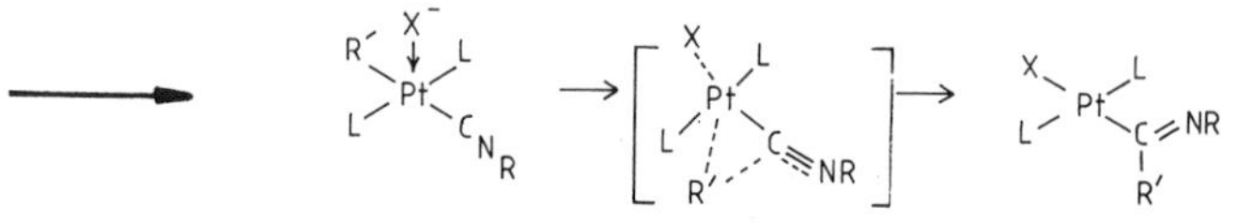

$$(55)$$

The rates of insertion were found to depend on isocyanide ($\text{CNC}_6\text{H}_4\text{Cl-}$4>CNMe) and on halide (I>Br). This insertion mechanism, promoted by halide attack at the metal (as indicated by the rate dependence on X^-), can be related to the second-order k_2 step of the CO insertion mechanism (51). Subsequently, for the reactions of $[\text{Pt(R')}_2\text{(L)}_2]$ complexes with isocyanides, the intermediacy of a three-coordinate species, corresponding to the coordinatively unsaturated intermediate of reaction (53), was also proposed:[274]

$$\text{Pt(R')}_2\text{(L)}_2 + \text{CNR}$$

$$\text{Pt(R')}_2\text{(CNR)(L)} + \text{L} \longrightarrow \quad \xrightarrow{(+ \text{L})} \text{Pt(R')}[\text{C(=NR)R'}]\text{(L)}_2 \quad (56)$$

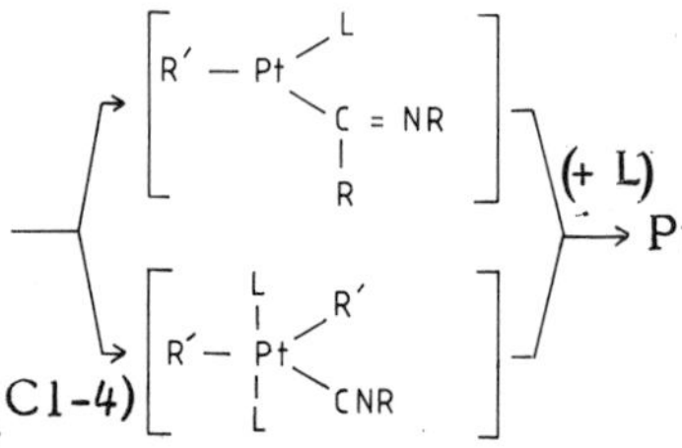

$$(\text{L}=\text{PEt}_3; \text{R}=\text{Me},\text{C}_6\text{H}_4\text{Cl-4})$$
$$(\text{L}=\text{PPhMe}_2; \text{R}=\text{C}_6\text{H}_4\text{Cl-4})$$

In general, the insertion rates increase with increasing nucleophilicity of the migrating organic moiety and with increasing electrophilicity of the isocyanide contact carbon atom. These electronic effects suggest that the insertion reaction may also be viewed as an <u>intramolecular</u> nucleophilic attack on the coordination-activated <u>cis</u>-isocyanide. Most insertion reactions of unsaturated ligands are in fact promoted by lower electron density on the metal.[270,275-6]

Those σ-bonded organic groups with electron-withdrawing substituents, such as cyanides or halides, are generally resistant[8,9,21,29,273,277-8] to insertion. A notable exception is represented by reaction (57a):[279]

CNCy

Ti(Me)[C(=NCy)C$_6$F$_5$]Cp$_2$ (57a)

Ti(Me)(C$_6$F$_5$)Cp$_2$

CO

Ti(C$_6$F$_5$)(COMe)Cp$_2$ (57b)

Whereas isocyanides are known to insert into M-alkenyl σ bonds,[75,280-1] no such reaction has yet been reported for complexes with σ-bonded alkynyl groups.[280,282] The influence of the CNR substituent is clearly reflected by the higher reactivity of aryl isocyanides as compared to the alkyl analogues.[75,273,283-4]

On the other hand, increasing steric requirements of both the migrating σ-bonded group and the recipient isocyanide ligand reduce the ease of insertion to such extent that, in some cases, no reaction occurs, even though it would be favored on purely electronic grounds.[8,9,21,75,278,285-7]

A marked influence is also exerted by the central metal, as shown by the different behaviour of the Ni,Pd,Pt triad. The Ni(II)-isocyanide complexes readily give successive multiple insertion reactions [8-10] and are among the most active catalysts of isocyanide polymerization.[288] Double[283,285,289-90] and even triple insertion[75,285] are reported also for Pd(II) derivatives. In the case of Pt(II), however, the insertion reaction proceeds at much slower rate if at all and affords only mono-inserted products.[8,9,75,134,273-4,280,291] This seems to be an essentially kinetic effect, since the 1,2-bis(imino)propyl complex XXIX, which may formally be derived from successive insertion of CNR and CNR' into the Pt-Me bond, can be prepared by a different route and is stable at ambient conditions.[292] Some W(II)-isocyanide complexes are also resistent to insertion, unlike their Mo(II) analogues.[293-4] Only single insertions are reported for metal complexes of the triad Ti,Zr,Hf.[278-9,281,286,295-8] In reaction (58), the insertion of CNBu-t involves both Hf-C bonds but not the Hf-N bonds:[286]

$$\underline{cis}[(Me_3Si)_2N]_2HfR_2 + 2CNBu\text{-}t \longrightarrow \underline{cis}[(Me_3Si)_2N]_2Hf[C(R)=NBu\text{-}t]_2 \quad (58)$$

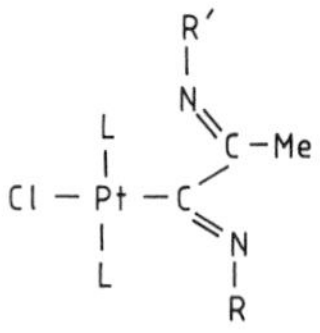

XXIX

All the reactions (54-58) have been carried out by adding free isocyanide to complexes with one or two M-C σ bonds. Another general route to insertion products is to form an M-C σ bond in an existing metal-isocyanide compound. Up to now, this has been achieved either by oxidative addition[21,75,134,242-4,293-4,299] (Scheme 3)[21] or by exchange[283,289] (Scheme 4).

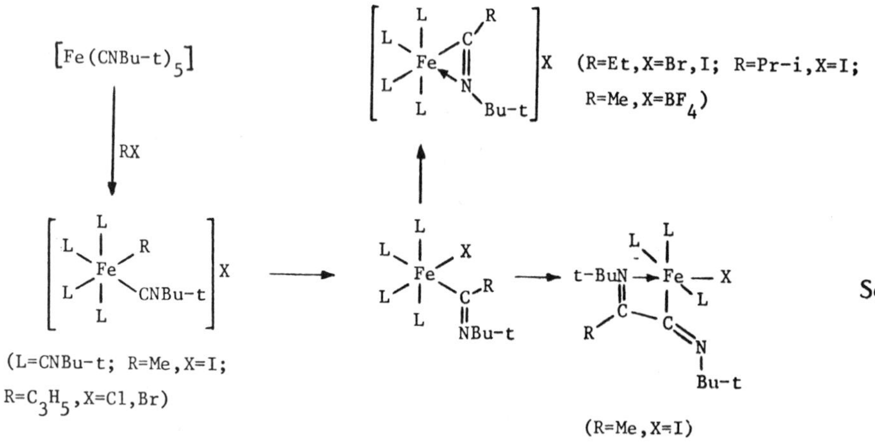

Scheme 3

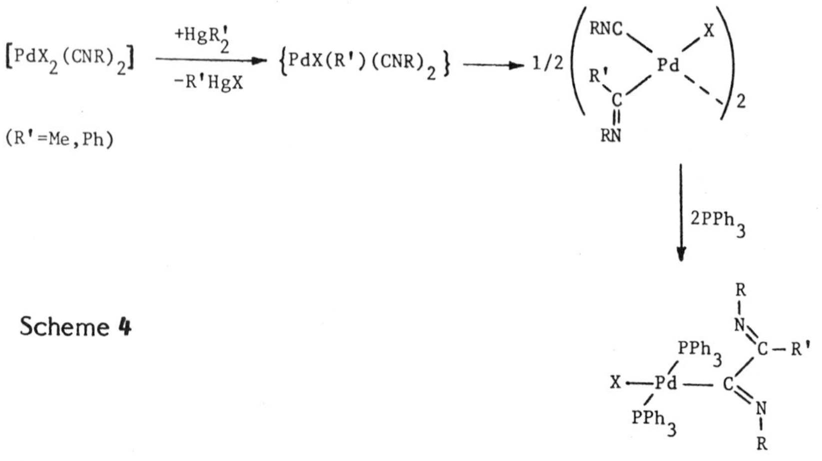

Scheme 4

In general, single insertion reactions afford either the η^1-[300] or the η^2-iminoalkyl[299,301-3] ligands, **XXX** and **XXXI** respectively:

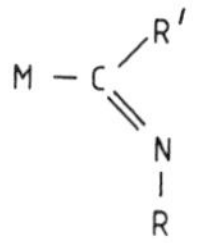
XXX

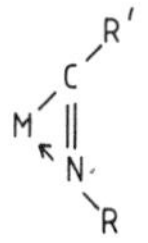
XXXI

It was suggested that the three-electron donor η^2-iminoalkyl groups may be involved as intermediates in the insertion reaction mechanism[294] [for instance, an η^2-1-(t-butylimino)ethyl ligand might be present in the unsaturated intermediate of reaction (53)]. Support for this hypothesis comes from the ready conversion of the complexes MoCp [η^2-C(=NR)Me](CO)$_2$ into MoCp [η^1-C(=NR)Me](CO)$_2$(L) (R=Me,Ph; L=P(OMe)$_3$) upon addition of a donor L molecule.[301]

Successive double or triple insertion gives 1,2-bis(imino)alkyl or 1,2,3-tris(imino)alkyl moieties, respectively, which can act either as η^1-C monodentate (structures **XXXII** or **XXXIV**)[75,285,287,304] or as C,N bidentate ligands (structures **XXXIII** or **XXXV**:[21,243,285,305]

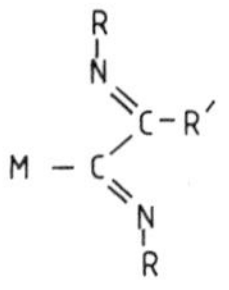
XXXII

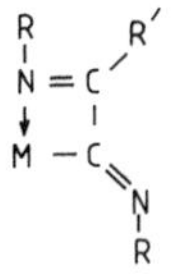
XXXIII

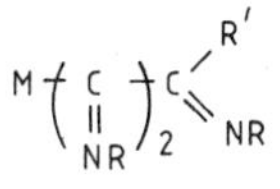
XXXIV

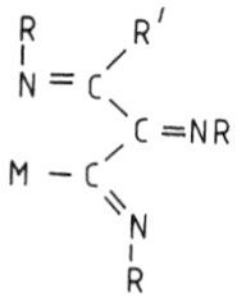
XXXV

An interesting reaction occurs in some 1,2-bis(imino)alkyl-Fe(II) derivatives (eq 59; R=C_6H_{11},R'=Bu-t).[287,306-7]

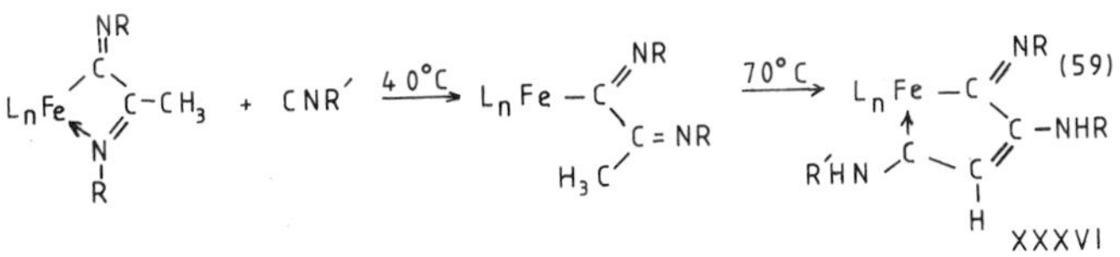
XXXVI

The formation of the chelating carbene ligand in **XXXVI** was interpreted as resulting from 1,2-addition of a methyl C-H bond to the coordinated CNR' group. This is probably related to slight acidic character in the methyl protons of the 1,2-bis(imino)propyl moiety, as suggested by the occurrence of a ketimine-enamine tautomeric equilibrium in <u>trans</u>-PdCl[C(=NC$_6$H$_4$Me-4)Me](PEt$_3$)$_2$:[308]

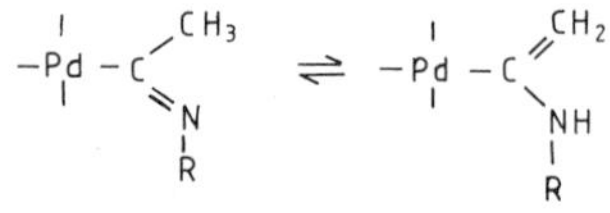

$$-\overset{|}{\underset{|}{Pd}}-C\begin{array}{l}CH_3\\\diagdown\\\underset{\underset{R}{|}}{N}\end{array} \rightleftharpoons -\overset{|}{\underset{|}{Pd}}-C\begin{array}{l}CH_2\\\diagup\\\underset{\underset{R}{|}}{NH}\end{array} \qquad (60)$$

Isocyanides insert also into the metal-carbon bond of unsaturated ligands, such as carbenes and acetylenes. Reaction of Cr[C(OMe)Me](CO)$_5$ with an equivalent of CNCy gives a 1:1 adduct **XXXVII**, which upon addition of methanol rearranges to the carbene species **XXXVIII**:[309]

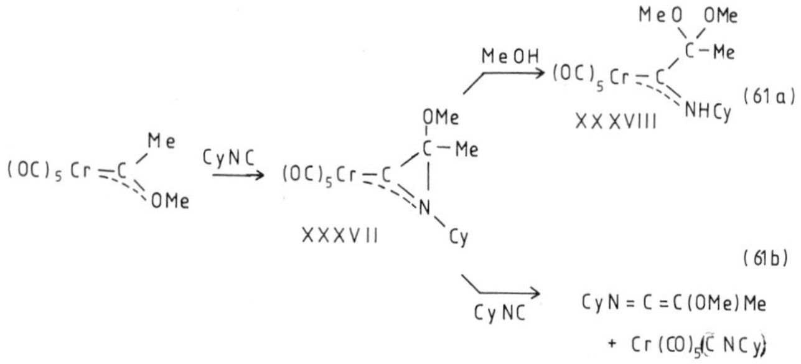

If **XXXVII** is further treated with CNCy (eq 61b), ketenimines are obtained in quantitative yield.[310] Carbene complexes of type **XXXVIII** are also formed in the reaction of {Mn[C(OMe)Me](CO)$_3$(P(OMe)$_3$)$_2$}$^+$ with CNR (R=Me,C$_6$H$_4$Cl-4) in methanol.[311] The reaction of the η^3-(vinyl)(methoxy)carbene tricarbonyliron complex **XXXIXa** with isocyanides affords a new type of insertion product, **XXXIXb**, containing an η^4-vinylketenimine ligand (R=Bu-t,Cy,Ph).[312]

Insertion into the metal-acetylene bond yields interesting types of either iminometalacycle compounds or cyclic imino derivatives. The reactions of [CoCp(η^2-PhC≡CPh)(PPh$_3$)] are summarised in Scheme 5.[313] The bis(t-butylimino)diphenylcyclobutenes and the tris(imino)cyclopentenes can be produced only stoichiometrically. A four-membered platinacycle **XL** has been obtained from the reaction of [Pt(η^2-cyclohexyne)(PPh$_3$)$_2$] with CNBu-t.[314] No further insertion of CNBu-t into the Pt-C σ bond of **XL** occurs. Similarly, no isocyanide insertion was observed in the reactions of

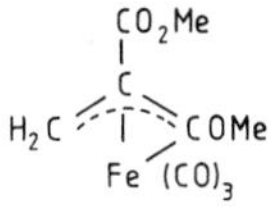

XXXIXa

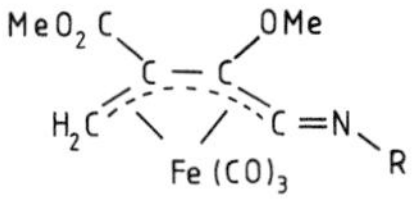

XXXIXb

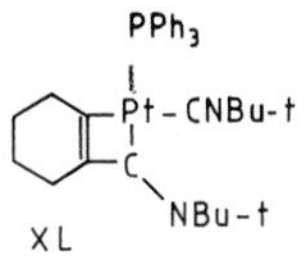

XL

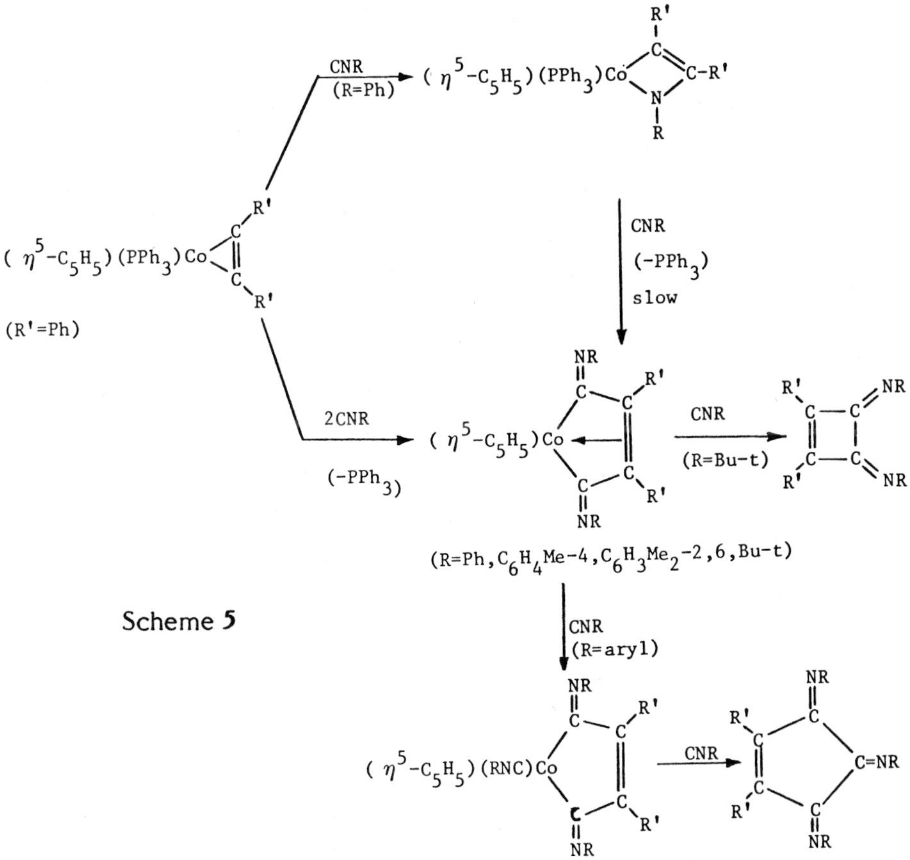

$(\eta^5\text{-}C_5H_5)(PPh_3)Co$

(R'=Ph)

CNR
(R=Ph)

$(\eta^5\text{-}C_5H_5)(PPh_3)Co$

CNR
(-PPh$_3$)
slow

2CNR

(-PPh$_3$)

$(\eta^5\text{-}C_5H_5)Co$

CNR
(R=Bu-t)

$(R=Ph, C_6H_4Me\text{-}4, C_6H_3Me_2\text{-}2,6, Bu\text{-}t)$

Scheme 5

CNR
(R=aryl)

$(\eta^5\text{-}C_5H_5)(RNC)Co$

CNR

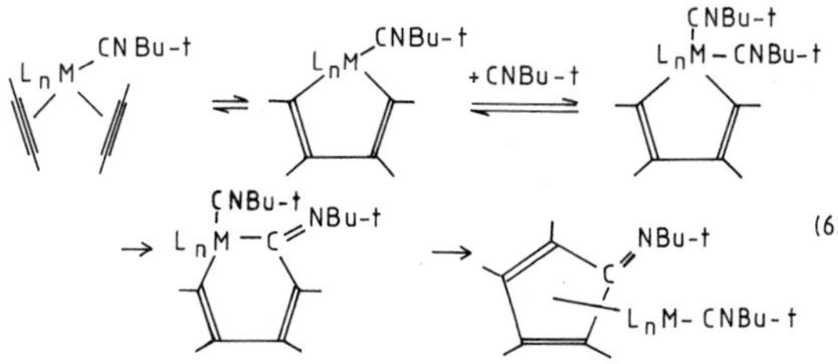

(62)

$Pt_3(CNBu-t)_6$ with some diacetylenes.[239] The reaction of $MXCp(\eta^2-$ $CF_3C\equiv CCF_3)_2(CNBu-t)$ (M=Mo, X=CF_3; M=W, X=Cl) with an excess of CNBu-t involves an initial reversible conversion of the bis(acetylene) complex into a metallacyclopentadiene species, which undergoes insertion, followed by reductive cyclization to form the cyclopentadienimine ligand (eq 62).[315] Further support for mechanism (62) comes from the formation of $(\eta^5$-cyclopentadienyl)(η^4-cyclopentadieneimine)cobalt in the reaction of $(\eta^5$-cyclopentadienyl)(triphenylphosphine)cobaltacyclopentadienes with iso-cyanides.[316] These reactions provide an important insight into the mechanism of the co-cyclization of acetylenes with isocyanides promoted by transition metal complexes.[317] The formation of the η^4-(1,3-diene-1,4-diimine)-complex **XLI** in the reaction of $Fe(CNBu-t)_5$ with $PhC\equiv CPh$ is also explained in terms of successive insertions of CNBu-t into the iron-acetylene bond.[21] In the final step, however, the formation of an open-chain 1,3-diene-1,4-diimine is preferred to the cyclization coupling observed in the previous reactions.

XLI   (L = t-BuNC)   **XLII**   (L = PPh_3, R = C_6H_4Me-4)

6.2   Insertion into metal-hydrogen bonds

The insertion of unsaturated molecules into metal-hydrogen bonds is of great importance in catalytic hydrogenation. For this reason, interest has grown over recent years in such reactions of mono- and polynuclear complexes, the latter being regarded as possible models for catalytic metal surfaces.

Isocyanide insertion into an M-H bond was first reported for Ru(II) complexes.[318-9] An xray structural analysis of compound **XLII** showed that the (N-p-tolylimino)methyl group is η^1-bonded to the metal.[320] The presence of a chelating ligand in **XLII** seems to be important in stabilizing the η^1-iminomethyl ligand, since its replacement by iodide brings about hydride transfer to the metal and formation of the isocyanide complex

RuHI(CO)(CNC$_6$H$_4$Me-4)(PPh$_3$)$_2$, which in turn reacts with acetate to reform **XLII**.[318] Mechanistic studies on reaction (63) indicate that the insertion is favored by low polarity solvents and coordinating anions, suggesting that a neutral intermediate is traversed:[321-2]

$$L_2Pt(H)(X)(CNR)$$

trans-[PtH(CNR) L$_2$]X $\longrightarrow$ or $\longrightarrow$ trans-PtX(CH=NR)(L)$_2$

$$HPt(X)(L)(CNR) + L \qquad (63)$$

The five-coordinate intermediate is preferred since it may assume a cis-hydride-isocyanide configuration without further rearrangement. This is supported by the course of reaction (64), where replacement of PPh$_3$ by an isocyanide ligand yields a product with trans-hydride-isocyanide geometry, which does not undergo insertion even in the presence of an excess of CNR:[323]

trans-PtH(CH$_2$CN)(PPh$_3$)$_2$ + CNR(excess) $\longrightarrow$ trans[PtH(CH$_2$CN)(PPh$_3$)

(R=Bu-t,C$_6$H$_3$Me$_2$-2,6)   (CNR)] + PPh$_3$ (64)

The formation of complexes **XLIII** and **XLIV** has been interpreted on the basis of insertion reactions of the coordinated isocyanide into M-H bonds formed in a preliminary oxidative addition:[245,324]

$$\overset{HSiR_3}{}$$

[Pt$_3$(CNBu-t)$_6$] $\longrightarrow$ **XLIII** (65)

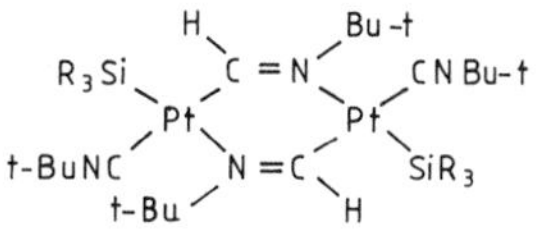

[SiR$_3$=SiMe$_3$,SiEt$_3$,SiMe$_2$Ph,

SiMePh$_2$,SiPh$_3$,Si(OEt)$_3$]

$$\overset{2\ CNR,\ HCl}{}$$

[Pd(PPh$_3$)$_4$] $\longrightarrow$ **XLIV** (66)

$$-2PPh_3$$

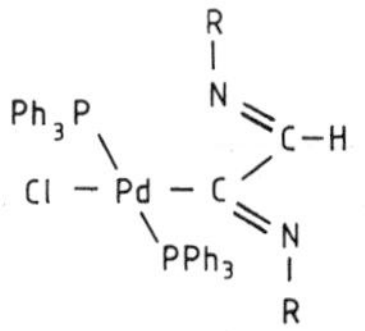

(R=C$_6$H$_4$OMe-4)

The xray structure of **XLIII** (SiR$_3$=SiMePh$_2$) shows the presence of a six-membered ring in the boat conformation.

Addition of H$_2$ to Ru$_3$(CO)$_{11}$(CNBu-t) gives HRu$_3$(CO)$_9$ (CH=NBu-t) as the major product.[325] This interesting result, representing a partial hydrogenation of coordinated isocyanide, can be related to the catalytic properties of cluster compounds in the reduction of heteronuclear multiple bonds.[326] In fact, complete hydrogenation of isocyanides has been achieved with the tetranuclear complex $Ni_4(CNBu-t)_7$.[327]

In this connection, the reactions of $H_2Os_3(CO)_{10}$ with CNR molecules (R=Ph,Me,Bu-t) have been studied in detail (Scheme 6).[328-9]

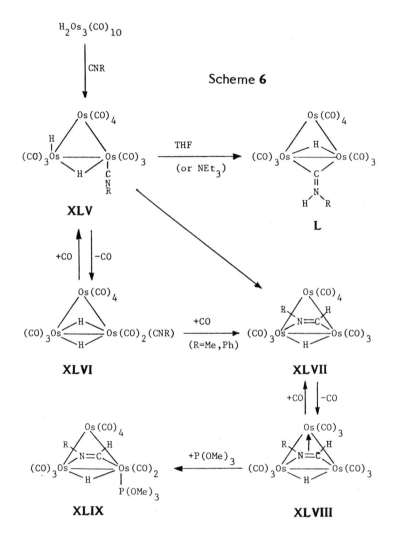

Scheme **6**

The adduct **XLV** is formed first at room temperature. Upon heating in poor donor solvents, this compound rearranges to **XLVI**. On further heating, insertion products **XLVII** and **XLVIII** are both obtained (R=Me,Ph). The presence of a face-bridged μ_3,η^2-CH=NR ligand in **XLVIII** (R=Ph) and of an edge-bridged μ-η^2-CH=NR group in **XLIX** (R=Ph) was confirmed by xray crystallographic analyses. In THF (or with trace amounts of NEt$_3$) a quite different product, **L**, is obtained. The formation of the bridging iminiomethylene in **L** was explained as a proton abstraction from **XLV** and subsequent protonation of a bridging isocyanide in the resulting transient anion $[(H)Os_3(CO)_{10}(CNR)]^-$.

The inserted products, Re$_2$H(CH=NR)(CO)$_6$(dppm) (R=Bu-n,Bu-t,C$_6$H$_4$OMe-4), have been reported to result from the reaction of the binuclear complex, Re$_2$(μ-H)$_2$(CO)$_6$(dppm), with isocyanides.[330] Reduction of methyl isocyanide to dimethylamido ligand occurs in the reaction with the mononuclear compound ZrH$_2$Cp$_2^*$ under H$_2$:[297]

$$Cp_2^*ZrH_2 + CNMe \xrightarrow{-65^\circ} Cp_2^*Zr(H)CH{:}NMe \xrightarrow[25^\circ C]{H_2} Cp_2^*Zr(H)NMe_2 \quad (67)$$

$$\xrightarrow{\text{MeI}} Cp_2^*Zr(I)NMe_2 + CH_4$$

Similar reactions take place with CNC$_6$H$_3$Me$_2$-2,6.

6.3 Other insertion reactions

Aliphatic isocyanides, CNR (R = Me, Bu-t), insert into the metal-halogen bonds of halides of early transition metals in their higher oxidation states.[331-4] Reaction (68) is typical:

$$2MCl_4 + 4\,CNBu\text{-}t \xrightarrow[20^\circ C]{CH_2Cl_2} [\text{yellow intermediate}]$$

$$(M{=}Ti,Hf) \qquad [MCl_3(CCl{=}NBu\text{-}t)(CNBu\text{-}t)]_2 \quad (68)$$

The insertion appears to proceed via initial formation of an unstable adduct, followed presumably by intramolecular halide transfer to isocyanide carbon. Whereas t-butyl isocyanide gives only single insertion into an M-X bond, with the less sterically hindered methyl isocyanide further reaction can occur with the other M-X bonds of the same molecule, yielding complexes of type $MX_{4-n}(CX=NMe)_n(CNMe)$ (n=1,2,3). The reaction seems to be confined to the transition metals of groups IV and V. The relative inertness of the later transition metal halides is in contrast with the widely observed insertion into metal-carbon bonds, and is probably related to higher metal-halogen bond energies.[162]

Insertion of isocyanides into the metal-nitrogen bond of Pd(II),Pt(II),Au(I),Au(III) and Hg(II) azide complexes readily gives C-bonded tetrazole derivatives:[335]

$$L_nM-N_3 + CNR \longrightarrow \quad\quad\quad L_nM-C\underset{\underset{N\diagdown_N}{\diagdown}}{\overset{\overset{R}{\underset{|}{N}}}{\diagup}}\overset{N}{\underset{||}{}}N \quad\quad (69)$$

This reaction has been recently extended to six-coordinate cobalt(III) complexes, $CoN_3(chelate)(L)$ (L=pyridine, PPh_3, N_3), where L and N_3 ligands are trans to each other.[195] On the basis of steric and electronic factors, an ionic intermediate is proposed:

$$L_nM-N_3 + CNR \longrightarrow [L_nM-CNR]^+N_3^- \longrightarrow \quad L_nM-C\underset{\underset{N\diagdown_N}{\diagdown}}{\overset{\overset{R}{\underset{|}{N}}}{\diagup}}\overset{N}{\underset{||}{}}N \quad (70)$$
(M=Co)

The final tetrazole product would then result from nucleophilic attack of N_3^- at the coordinated isocyanide. Isocyanide insertions into metal-nitrogen bonds have also been reported for palladium(II)-ylide,[336] copper(I)-amide[337] and nickel(II)-triazenido[338] complexes. In the latter case, fluxional five-membered metallacycles LI are obtained.

A preliminary report on insertion into platinium-oxygen bonds has recently appeared.[339] Even the palladium-palladium bond of some palladium(I) dimers is cleaved by isocyanides to give the inserted products LII (L-L=dppm, L'= CNR, halide or pseudo-halide).[84,340]

7. ORGANIC SYNTHESIS VIA METAL-ISOCYANIDE COMPLEXES

Since the pioneering work of Saegusa and coworkers on copper-catalyzed 1,1-addition of protic nucleophiles to isocyanide molecules,[4,6] metal isocyanide derivatives have become increasingly important in organic

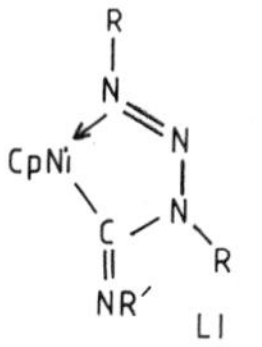

LI

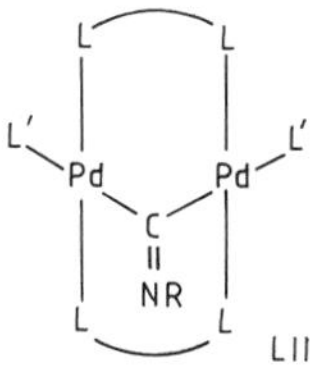

LII

chemistry, and their potential is far from exhausted. Several synthetic applications of α-metalated isocyanides, such as the lithium compounds RR'C(Li)NC, have been reviewed,[341-3] but this Section concentrates on organic reactions promoted by transition metal complexes. Most of the synthetic processes can be classified either as intermolecular nucleophilic attack, or as intramolecular migratory rearrangement (insertion) of the coordination-activated isocyanides.

7.1 Syntheses via nucleophilic attack on coordinated isocyanides

It is well known that 1,1 additions of compounds HY, containing a heteroatom-hydrogen bond (Y = NR_2', PR_2', R'O, R'S, $R_3'Si$), are effectively catalyzed by group IB and IIB metal ions:[6]

$$RNC + HY \longrightarrow Y-CH=NR \qquad (71)$$

This reaction is generally regarded as nucleophilic attack of HY on the coordination activated isocyanide group. For the $[AuCl_4]^-$ catalyzed addition of secondary amines,[344] the intermediacy of carbene species was inferred from the formation of formamidines in the ligand displacement reactions of $[Au(diaminocarbene)_2]^{+201-2,222,233}$ complexes (Section 4.2). It was also suggested that labile carbene intermediates may be formed in the other catalytic systems, the higher activity of Cu(I) derivatives being related to the lower stability of the copper-carbon bonds.[201,233] In a recent report, the AgCl-catalyzed addition of secondary amines (eq 72) was found to be stereospecific at low temperatures, giving the Z isomer, which isomerizes to the thermodynamically more stable E form upon heating (or in the presence of acid):[345]

$$R_2'NH + CNR \xrightarrow{\quad AgCl \quad} \underset{Z}{\overset{H}{\underset{R_2'N}{\diagdown}}C=N\overset{}{\diagdown}_R} \longrightarrow \underset{E}{\overset{H}{\underset{R_2'N}{\diagdown}}C=N\overset{R}{\diagup}} \tag{72}$$

Since this stereospecific addition can hardly be rationalized as a rearrangement of a displaced carbene ligand, a new mechanism was proposed involving stereospecific attack by the amine at the coordinated isocyanide and direct nitrogen-carbon proton transfer. A complex mechanism was proposed for the reaction of aryl isocyanides with ethanol under the influence of base and $[Cu(CNAr)_4]BF_4$ as catalyst.[346] In the rate-determining step, intramolecular nucleophilic attack by the ethoxide ion occurs in the ion pair $Cu(CNAr)_4^+\ldots OEt$ as confirmed by the dependence of reaction rates on the phenyl ring para-substituents and by the high positive ρ-value in the Hammett equation. Other catalyzed 1,1-additions of active hydrogen compounds to isocyanide have been reported.[347-8] A quantitive and specific addition of thiols to CNR substrates yields thioformimidates through the action of the clusters $[Fe_4S_4(SR')_4]^{n-}$ (n=2-4), the more reduced species being the more effective catalysts.[349] The proposed mechanism involves insertion of CNR into an Fe-SR' bond and subsequent release of thioformimidate by reaction with free thiol.

Carbodiimides are conveniently prepared by oxidation of Pd(II)-N,N'-disubstituted diaminocarbenes complexes with Ag_2O:[350]

$$PdCl_2[C(NHR)NHR'](CNR) \xrightarrow[\substack{(-H_2O)}]{\quad Ag_2O \quad} RN=C=NR' \tag{73}$$

Reaction (73) can be run catalytically in a heterogeneous mixture of primary amine RNH_2, isocyanide CNR, Ag_2O and a small amount of $PdCl_2$. In this case, the corresponding ureas, RNHCONHR', are also formed from the reaction of carbodiimides with the water produced. N,N'-Disubstituted ureas (or N-substituted urethanes) are formed in the stoichiometric reaction of Au(III)-bis(carbene) complexes with water:[202,351]

$\{AuI_2[C(NHR)Y]_2\}^+ + H_2O \longrightarrow RNH\text{-}CO\text{-}Y + 2H^+ + I^- + AuI[C(NHR)Y]$ (74)
(Y=OEt,NHR)

Carbodiimides are also formed in the reactions of mercuric chloride-isocyanide compounds with $H_2NBu\text{-}n$, in the presence of triethylamine:[352]

$$1/2[HgCl_2(CNR)]_2 + H_2NBu\text{-}n \xrightarrow{\quad NEt_3 \quad} RN=C=NBu\text{-}n + Hg + 2\,NEt_3HCl\,(75)$$
$(R=Bu\text{-}t,C_6H_3Me_2\text{-}2,6)$

In the absence of NEt_3, guanidines are obtained through a similar redox decomposition reaction. A mechanism is proposed in which either product, carbodiimide or guanidine, can be derived by reactions of a Hg(II)-(diaminocarbene) intermediate.

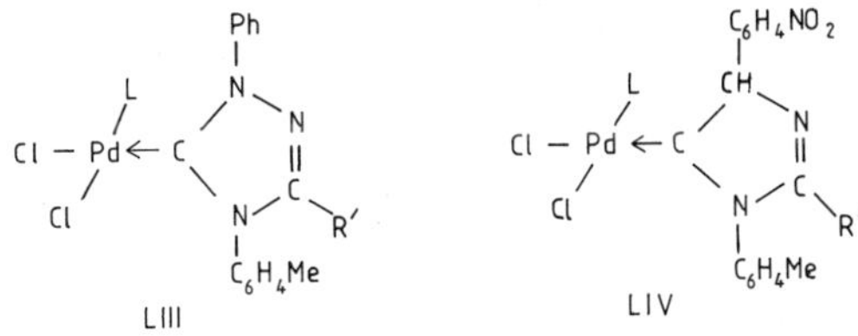

LIII   LIV

Heterocyclic carbenes of type **LIII** or **LIV** can be obtained from the reaction of nitrilimines, $R'\text{-}\overset{+}{C}{=}N\text{-}\bar{N}\text{-}Ph,$[353] or nitrilylides, $R''\text{-}\overset{+}{C}{=}N\text{-}\bar{C}HC_6H_4NO_2\text{-}4,$[354] respectively, with cis-$PdCl_2(CNC_6H_4Me\text{-}4)(L)$ (L= tertiary phosphine). Other C(2)-bonded heterocycles, such as oxazoles, oxazolines, pyrrolines, pyrrole and imidazoles, have been prepared from reactions of coordinated isocyanides containing active methylenes, $CNCH_2CO_2Et$ or $CNCH_2SO_2C_6H_4Me\text{-}4$, with 1,3-dipolarophiles (RCHO, $CH_2{=}CHCN$, $CH_2{=}CHCO_2Me$, CF_3CN, $CH{\equiv}CCO_2Et$) in the presence of base[355-6] (Scheme 7).

The isocyanides $HO(CH_2)_nNC$ (n=2,3) are catalytically converted to 2-oxazoline (n=2) or to 5,6-dihydro-4H-1,3-oxazine in the presence of small amounts of $ZnSO_4$ or $PdCl_2$ via metal-carbene intermediates.[220] On the other hand, imidazole derivatives are obtained from reaction of Pd(II)-imidazolidinylidene complexes with Ag_2O:[219]

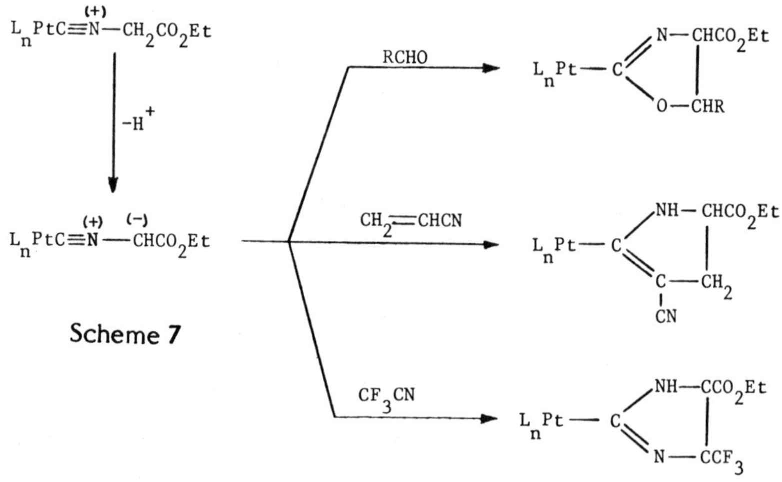

$$L_nPtC{\equiv}\overset{(+)}{N}{-}CH_2CO_2Et$$

$$\downarrow\ -H^+$$

$$L_nPtC{\equiv}\overset{(+)}{N}\ \overset{(-)}{{-}}CHCO_2Et$$

Scheme 7

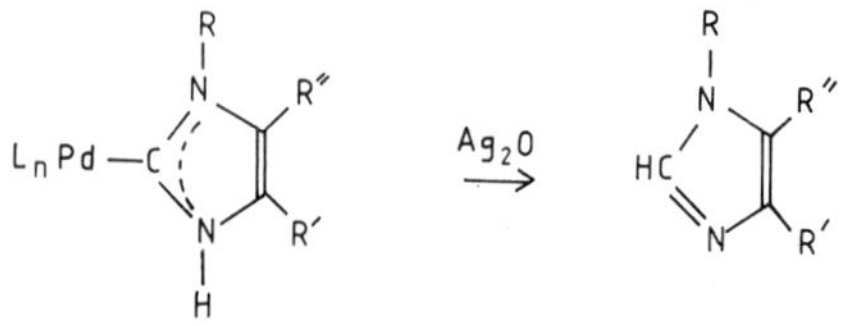

RCHO →

$$L_nPt{-}C{\big\langle}\genfrac{}{}{0pt}{}{N{-}CHCO_2Et}{O{-}CHR}$$

CH$_2$=CHCN →

$$L_nPt{-}C{\big\langle}\genfrac{}{}{0pt}{}{NH{-}CHCO_2Et}{\underset{\underset{CN}{|}}{C}{-}CH_2}$$

CF$_3$CN →

$$L_nPt{-}C{\big\langle}\genfrac{}{}{0pt}{}{NH{-}CCO_2Et}{N{-}CCF_3}$$

$$L_nPd{-}C{\langle}\ \quad\overset{Ag_2O}{\longrightarrow}\quad HC{\langle}\qquad\qquad (76)$$

Carbene compounds are also involved as key intermediates in the PdCl$_2$ catalyzed synthesis of several heterocycles from isocyanides and bifunctional amines (e.g. eq 77)[357-9]

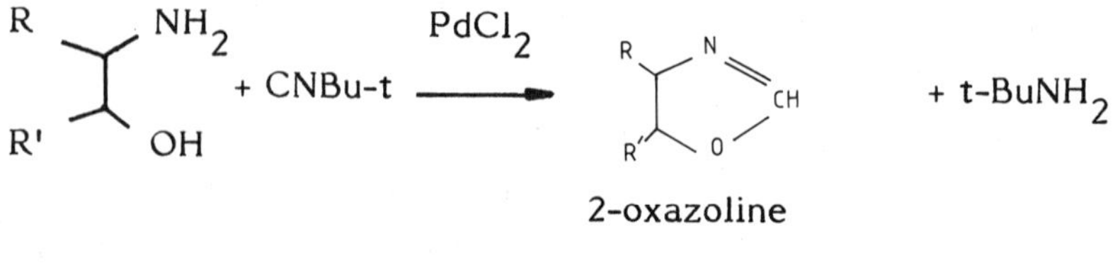

$$\underset{R'}{\overset{R}{\diagdown}}{\big\langle}\genfrac{}{}{0pt}{}{NH_2}{OH}\ +\ CNBu\text{-}t\ \xrightarrow{\ PdCl_2\ }\ \cdots\ +\ t\text{-}BuNH_2\qquad (77)$$

2-oxazoline

It was proposed that the cyclization step involves an intramolecular nucleophilic attack of the hydroxo group on the preformed carbene ligand:[201,358]

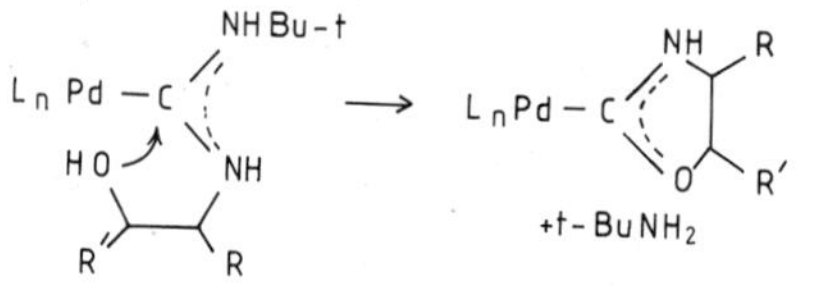

$$L_nPd{-}C\cdots\ \longrightarrow\ L_nPd{-}C\cdots\ +t\text{-}BuNH_2\qquad (78)$$

7.2   Synthesis via insertion reactions

Isocyanides are well known to undergo insertion reactions into palladium(II)-alkyl σ bonds (Section 6.1). When the resulting (imino)alkyl group has a hydrogen atom in the β position, β-elimination may occur, in the presence of base, to give keteneimines:[360]

$$L_nPd\text{-}C(\text{:}NR)CHR'R'' \xrightarrow{\text{DBU}} R\text{-}N\text{=}C\text{=}CR'R'' + [L_nPdH] \qquad (79)$$

(DBU=1,5 diazabicyclo[5.4.0]undec-5-ene)

This reaction gives further evidence of the somewhat acidic character of β-hydrogens in σ-bonded (imino)alkyl ligands, suggested also by the ketimine-enamine tautomeric equilibrium in trans-PdCl[C(=NC$_6$H$_4$Me-4)Me](PEt$_3$)$_2$[308] (eq 60, Section 6.1). Ketenimines can also be obtained in high yields from the reaction of the carbene complex Cr[C(OMe)Me](CO)$_5$ with excess of isocyanides[310] (eq 61b, Section 6.1).

Ortho-substituted N,N-dimethylbenzylamines were prepared from the insertion product LV:[284]

$$\underline{o}\text{-}C_6H_4(\overline{C\text{:}NPh})(CH_2NMe_2)Pd(II)$$
$$\mathbf{LV}$$

$$\xrightarrow{\text{LiMe}} \underline{o}\text{-}C_6H_4(CMe\text{:}NPh)(CH_2NMe_2) \quad (80a)$$
$$\xrightarrow{\text{LiAlH}_4} \underline{o}\text{-}C_6H_4(CH_2NHPh)(CH_2NMe_2) \quad (80b)$$

The synthesis of 3-imino-2-phenylindazolines **LVI** and 6H,12H-indazolo[2,1,a]-6,12-diiminoindazoles **LVII** from azobenzenes and isocyanides in the presence of Co$_2$(CO)$_8$ was interpreted as resulting from ortho-metalation of the azo compounds and insertion into the Co-C σ bonds so formed[361] (Scheme 8). The preferential formation of **LVII** with the bulkier 2-tolyl and 2,6-xylyl isocyanides suggests that double metalation occurs more readily than insertion of sterically hindered CNR ligands into the singly metalated intermediate. A product of type **LVII** (R=H, R'=C$_6$H$_3$Me$_2$-2,6) is also obtained from the reaction of Co$_2$(CNR)$_8$ (R=C$_6$H$_3$Me$_2$-2,6) with azobenzene,[127] while 3-imino-2-phenylindazolines

LVI are formed in the reactions of cyclopalladate azobenzenes with isocyanides.[362]

The Cu/CNR and Cu_2O/CNR catalytic systems have been extensively used by Saegusa and coworkers in organic syntheses. A typical feature of these systems is the ease with which an electron is released to π-substrates, such as dinitrobenzenes, benzoquinone, fluorenone and TCNE, to give the corresponding radical anions.[363] This marked reducing power (or basic character) is reflected in reaction (81) with active hydrogen compounds, (acetylacetone, acetoacetate, malonate, cyanoacetate etc.):[364]

$$RR'CH_2 \quad \xrightarrow{\quad \begin{array}{c} Cu_2O/CNBu\text{-}t \\ (-H_2O) \\ \hline Cu/CNBu\text{-}t \\ (-H_2) \end{array} \quad} \quad RR'HC\text{-}Cu(CNBu\text{-}t)_2 \qquad (81)$$

LVI

LVII

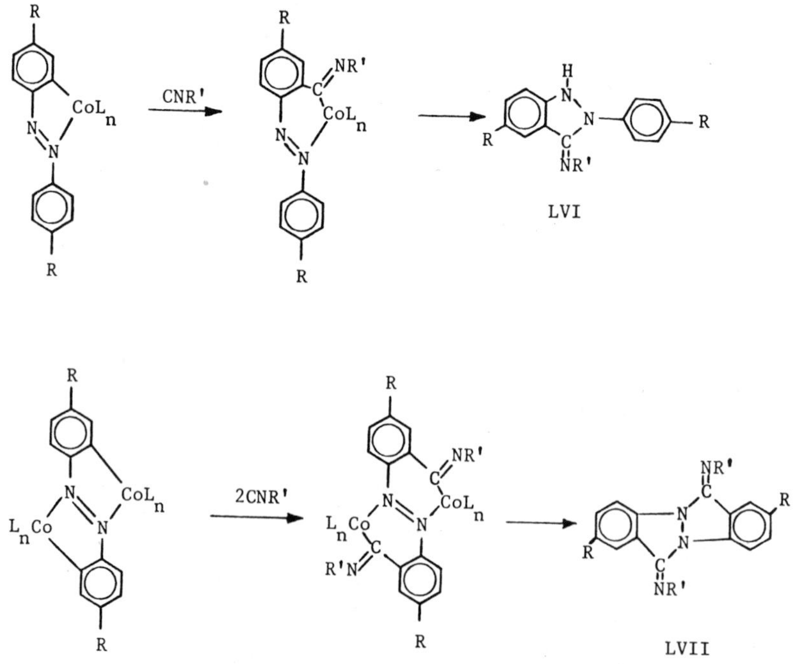

(R=H,Me,Cl; R'=Bu-t,C_6H_{11},Ph,C_6H_4Me-2,C_6H_3Me$_2$-2,6)   Scheme **8**

Further reaction with isocyanide yields insertion intermediates, which on hydrolysis give enamines:[364]

$$RR'HC\text{-}Cu(CNBu\text{-}t)_2 \xrightarrow{\text{CNBu-t}} \left[RR'CH-\underset{\underset{NBu^t}{\parallel}}{C}-Cu(CNBu\text{-}t)_n \right]$$

$$\downarrow H_2O$$

$$RR'C = CH-NHBu\text{-}t \tag{82}$$

Similar reactions have been observed with $[CuCp(CNBu\text{-}t)]$ and $[Cu(\eta^1\text{-}$ indenyl)(CNBu-t)$_3$].

Aryl isocyanides with <u>ortho</u>-substituents bearing an active α-methylene group undergo cyclization on treatment with catalytic amounts of Cu_2O.[365-7] This new synthetic route leads to 3-substituted indole derivatives **LVIII** by formation of a copper-carbon σ-bond as in reaction (81), followed by intramolecular insertion of the isocyano group (Scheme **9**). With N-substituted o-isocyanophenylacetamides (X=CONHR), the reaction with Cu_2O yields also a second product, the N-substituted 4,5-dihydro-1,3-benzodiazepin-4-one **LX**, resulting from intramolecular insertion into the copper-nitrogen bond of the intermediate **LIX**:

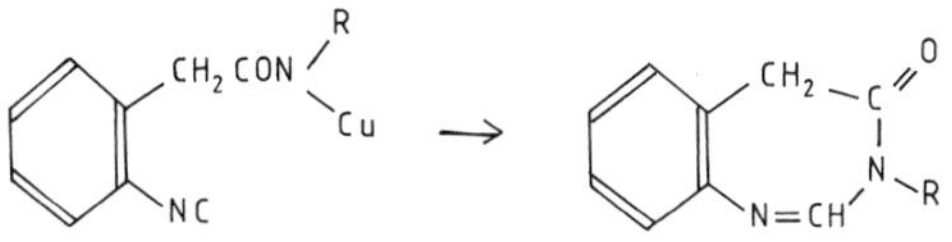

$$\text{(83)}$$

LIX   LX

The formation of **LX** is favored by N-substituents with low steric hindrance.[367] Insertion of isocyanide into a copper-nitrogen bond probably occurs in the reaction of carboxamides with $Cu_2O/CNBu\text{-}t$:[337]

$$RCONH_2 \xrightarrow[\text{(-H}_2\text{O)}]{Cu_2O/CNBu\text{-}t} [RCONHCu(CNBu\text{-}t)_n] \xrightarrow{} \quad \text{(84)}$$

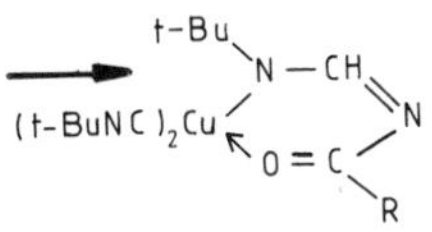

The copper-catalyzed 1,1-addition of perfluoroalkyl iodides, R_FI, to isocyanides was suggested to proceed via (i) oxidative addition, (ii) insertion

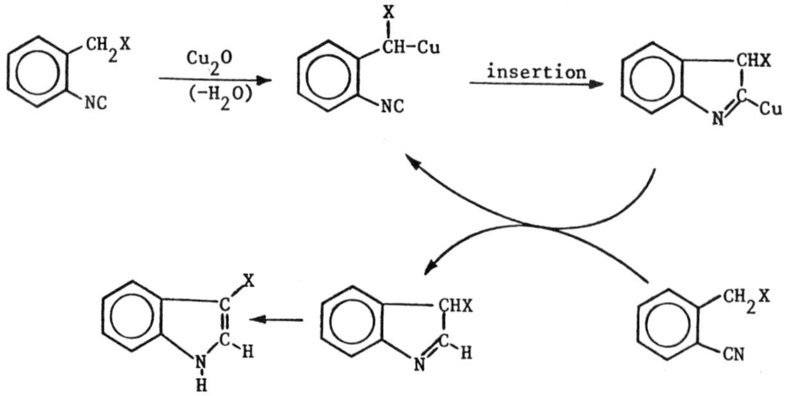

LVIII

Scheme 9

(X=CN, CO$_2$Me, COR, CONHR)

into the Cu-R$_F$ bond, (iii) reductive elimination:[368]

$$Cu(CNR)_n \xrightarrow{R_FI} [(RNC)_nCu(R_F)(I)] \longrightarrow [(RNC)_mCu(I)(CR_F\text{:}NR)]$$

$$\downarrow CNR \qquad (85)$$

$$Cu(CNR)_n + R_F\text{-}CI\text{=}NR$$

The Cu/CNR system was also used in the syntheses of substituted cyclopropanes and cyclopentanes from organic dihalides and α,β-unsaturated esters:[369-70]

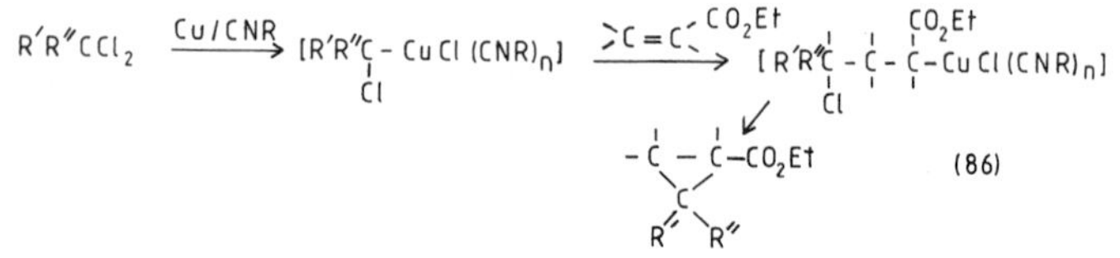

(86)

According to the proposed mechanism, the oxidative addition step is followed by insertion of the activated olefin, and then by cyclization of the resulting organic moiety through an intramolecular coupling reaction. A similar reaction with 1,3-diiodopropane yields cyclopentanes.

The reaction of alkyl isocyanides with $\alpha\omega$-bis-η^3-allylnickel complexes yields cyclic imines, together with cyclic hydrocarbons.[371-4] A typical reaction is shown in eq (87) for $\alpha\omega$-bis-η^3-pentadecatetraenediylnickel:

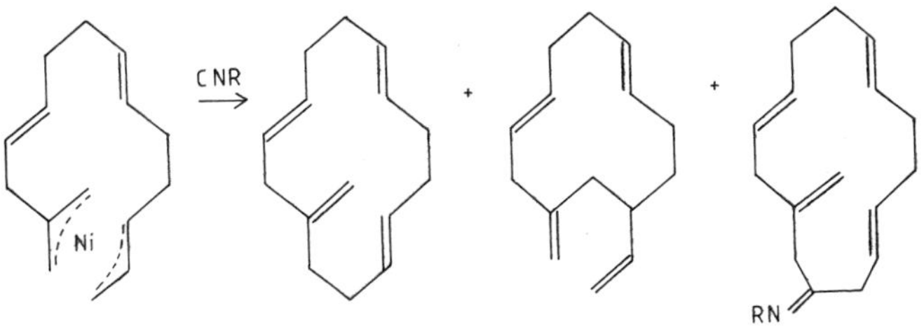

The ratio of insertion product to hydrocarbons increases with decreasing temperature and with decreasing bulkiness of the isocyanide,[371] and decreases with increasing length of the $\alpha\omega$-bis-η^3-allyl ligands.[373-4] In general, the reactions with isocyanides afford higher yields of ketone insertion products (after acid hydrolysis of the corresponding imine) than the equivalent reactions with carbon monoxide.

Isocyanides are polymerized by various transition metal ions, in particular by nickel(II) salts,[288] to poly(iminomethylenes), $(C=NR)_n$, having rigid rod helical structures with ca. four CNR units per turn.[375] (Mercuric chloride, however, has only moderate activity and gives oligomers or low molecular weight polymers ($n \sim 7,8$).)[376] The chiral properties of such helical configurations were shown by the resolution of poly(t-butyliminomethylene) into its optically active enantiomers.[377] The mechanism of polymerization, catalyzed by $[Ni(CNR)_4]^{2+}$,[378] is given in Scheme 10.

The initiation step consists of a nucleophilic attack by X^- (X=Cl,OEt or OH). The successive rate-determining insertion step (migration of the imino C(X)=NR group onto an adjacent cis CNR ligand) is promoted by the incoming C_4NR isocyanide and may involve either C_1NR (as indicated in Scheme 10) or C_3NR indifferently, when the isocyanide is achiral. Continuing the sequence of insertions $C_1 \longrightarrow C_2 \longrightarrow C_3$ (or $C_3 \longrightarrow C_2 \longrightarrow C_1$, if the first insertion step has occurred on C_3NR) a helix turn is formed, which contains four $C=NR$ units. By repeated sequences of this type, the polymer will grow in a screw movement below the metal coordination plane. This mechanism is in accord with the kinetic data for polymerization of isopropyl isocyanide: first-order in catalyst and in

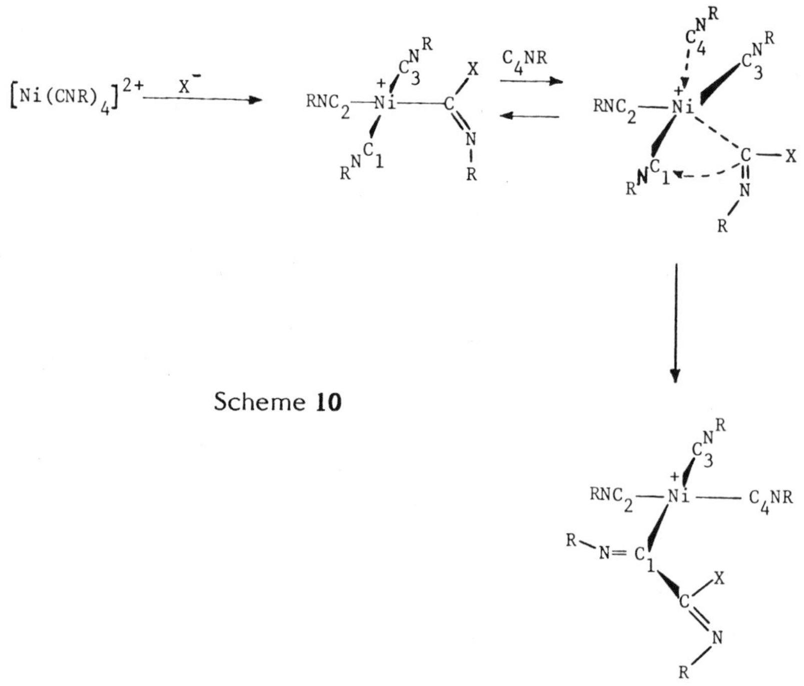

Scheme **10**

isocyanide; relatively low activation entropy of -54 ± 13 J K^{-1} mol^{-1}.[379] Enantioselective polymerization is obtained when one enantiomer of a chiral isocyanide is used.[380] According to the above mechanism, a preferential screw sense of polymer growth is induced by the steric requirements and by the coordinating properties of the different substituents on the chiral carbon atom of the R'R"R'"C*-NC molecules.

In Section 6.2 (eq 67) it was reported that methyl isocyanide is reduced to the dimethylamido ligand by ZrH_2Cp_2 and H_2, via an insertion intermediate.[297] Homogeneous and selective isocyanide hydrogenation was achieved by using the cluster $Ni_4(CNBu-t)_7$ as catalyst precursor and $Ni(CNBu-t)_4$ as source of isocyanide, according to eq (88):[117,327]

$$4Ni(CNBu-t)_4 + 18H_2 \xrightarrow{\quad Ni_4(CNBu-t)_7 \quad} 9 \text{ t-BuNHMe} + Ni_4(CNBu-t)_7 \qquad (88)$$

The products were t-butylmethylamine with about 1% t-butylamine. Similar reactions with n-butyl, isopropyl and cyclohexyl isocyanides gave selectivities of secondary to primary amines of 77, 73 and 36%, respectively. The [$Ni_4(CNBu-t)_7$] cluster is a catalyst precursor for the hydrogenation of other molecules with triple bonds, such as acetonitrile[327] and acetylenes.[115-7] In the latter case, diarylacetylenes are selectively hydrogenerated to cis-stilbenes, whereas the hydrogenation of dialkylacetylenes gives also trimerization products, such as hexaalkylbenzenes.[117]

The formation of cyclic compounds from insertion reactions of isocyanides into transition metal-acetylene bonds was discussed in Section 6.1. Both bis(imino)cyclobutene and tris(imino)cyclopentene derivatives are also obtained from the reaction of [$Co_2(CNR)_8$] ($R=C_6H_3Me_2$-2,6) and diphenylacetylene.[127]

7.3 Miscellaneous Reactions

Vinylic compounds, such as acrylonitrile and acrylates, are polymerized by the catalytic systems Cu/CNBu-t[381] or CuCp(CNBu-t)/CNBu-t.[382] At $0°C$, anionic polymerization of acrylonitrile was initiated by Cu(OBu-t) in the presence of CNBu-t. The system Cu_2O/CNR ($R=Cy,CH_2Ph$) is also a catalyst for the hydration of aromatic nitriles to amides.[384] Hydrosilylation of acrylonitrile,[385] cyclopentadiene[386] and 1,3-butadiene[387] has been reported to be catalyzed by copper, nickel and palladium isocyanide complexes. The versatile cluster compound $Ni_4(CNBu-t)_7$ catalyzes the selective trimerization of acetylene (or dialkylacetylenes) to benzene (or hexalkylbenzenes) and the stereoselective oligomerization of butadiene to 1,5-cyclooctadiene.[116] In the latter reaction, a small amount of the 1,3-isomer (ca. 5%) is also formed.

Isocyanides are oxidized to isocyanates by Hg(II), Tl(III) and Pb(IV) acetate in acetic acid, probably via an acetoxymetalation reaction.[388] Isocyanate is also formed in the reaction of $Co_2(CNR)_8$ ($R=C_6H_3Me_2$-2,6) with molecular oxygen.[127] The oxidation of CNBu-t to OCNBu-t by molecular oxygen can be carried out catalytically in the presence of Ni(CNBu-t)$_4$ or Ni(O_2)(CNBu-t)$_2$.[389] With Ni(CNBu-t)$_4$ as catalyst, nitrosobenzene and nitrobenzene can act as oxygen sources in metal-catalyzed atom-transfer reactions.[31] From the reaction of $PhNO_2$ with CNBu-t, t-butyl isocyanate is obtained in 83% yield, together with PhN=C=NBu-t, PhN=NPh and PhN=N(O)Ph. Carbon dioxide is also deoxygenated by t-butyl isocyanide in a copper(I)-t-butoxide promoted reaction:[390]

$$CO_2 + CNBu\text{-}t \xrightarrow{\text{(CuOBu-t)}} CO + ONCBu\text{-}t \qquad (89)$$

Some interesting transcarbonylation reactions of organic substrates by the complex $[Cu(O_2COH)(CNBu\text{-}t)_3]$ have been reported.[391]

REFERENCES

1. I. Ugi, U. Fetzer, E. Eholzer, H. Knupfer and K. Offermann, _Angew. Chem. Int. Ed. Engl._ 4:472 (1965).

2. Y. Yamamoto, H. Yamazaki and N. Hagihara, _Bull. Chem. Soc. Jpn._ 41:532 (1968); _J. Organomet. Chem._ 18:189 (1969).

3. E.M. Badley, J. Chatt, R.L. Richards and G.A. Sim, _Chem. Commun._ 1322 (1969).

4. T. Saegusa, Y. Ito, S. Kobayashi, K. Hirota and H. Yoshioka, _Tetrahedron Lett._ 6121 (1966).

5. L. Malatesta and F. Bonati, "Isocyanide Complexes of Metals", Wiley, New York (1969).

6. T. Saegusa and Y. Ito in "Isonitrile Chemistry", Ed. I. Ugi, Academic Press, New York (1971), p. 65.

7. A Vogler in "Isonitrile Chemistry", Ed. I. Ugi, Academic Press, New York (1971), p. 217.

8. Y. Yamamoto and H. Yamazaki, Coord. Chem. Rev. 8:225 (1972).

9. P.M. Treichel, Adv. Organomet. Chem. 11:21 (1973).

10. F. Bonati and G. Minghetti, Inorg. Chim. Acta. 9:95 (1974).

11. Y. Yamamoto, Coord. Chem. Rev. 32:193 (1980).

11a. E. Singleton and H.E. Oosthuizen, Adv. Organomet. Chem. 22:209 (1983).

12. R. Colton and M.J. McCormick, Coord. Chem. Rev. 31:1 (1980).

13. A.C. Sarapu and R.F. Fenske, Inorg. Chem. 14:247 (1975).

14. J.H. Burns and W.H. Baldwin, J. Organomet. Chem. 120:361 (1976).

15. J.F. Stevens, Jr., J.W. Bevan, R.F. Curl, Jr., R.A. Geanangel and M.G. Hu, J. Am. Chem. Soc. 99:1442 (1977).

16. G.B. Jameson and J.A. Ibers, Inorg. Chem. 18:1200 (1979), and references therein.

17. G. Albertin, A.A. Orio, S. Calogero, L. Di Sipio and G. Pelizzi, Acta Cryst. B32:3023 (1976).

18. H. Behrens, G. Thiele, A. Pürzer, P. Würstl and M. Moll, J. Organomet. Chem. 160:255 (1978).

19. W.E. Carrol, M. Green, A.M.R. Galas, M. Murray, T.W. Turney, A.J. Welch and P. Woodward, J.C.S. Dalton Trans. 80 (1980).

20. P. Le Maux, G. Simonneaux, G. Jaouen, L. Ouahab and P. Batail, J. Am. Chem. Soc. 100:4312 (1978).

21. J.M. Basset, M. Green, J.A.K. Howard and F.G.A. Stone, J.C.S. Dalton Trans. 1779 (1980).

22. G. Pelizzi, G. Albertin, E. Bordignon, A.A. Orio and S. Calogero, Acta Cryst. B33:3761 (1977).

23. J. Chatt, A.J.L. Pombeiro, R.L. Richards, G.H.D. Royston, K.W. Muir and R. Walker, J.C.S. Chem. Commun. 708 (1975).

24. J.M. Basset, D.E. Berry, G.K. Barker, M. Green, J.A.K. Howard and F.G.A. Stone, J.C.S. Dalton Trans. 1003 (1979).

25. J. Chatt, C.M. Elson, A.J.L. Pombeiro, R.L. Richards and G.H.D. Royston, J.C.S. Dalton Trans. 165 (1978).

26. G.J. Essenmacher and P.M. Treichel, Inorg. Chem. 16:800 (1977).

27. P.M. Treichel and G.J. Essenmacher, Inorg. Chem. 15:146 (1976).

28. A.J.L. Pombeiro and R.L. Richards, J. Organomet. Chem. 179:459 (1979).

29. R.D. Adams, Inorg. Chem. 15:169 (1976).

30. R.B. King and M.S. Saran, Inorg. Chem. 13:74 (1974).

31. S. Otsuka, Y. Aotani, Y. Tatsuno and T. Yoshida, Inorg. Chem. 15:656 (1976).

32. S.D. Ittel, Inorg. Chem. 16:2589 (1977).

33. S.D. Ittel, J. Organomet. Chem. 137:223 (1977).

34. B.V. Johnson, P.J. Ouseph, J.S. Hsieh, A.L. Steinmetz and J.E. Shade, Inorg. Chem. 18:1796 (1979).

35. T. Ziegler and A. Rauk, Inorg. Chem. 18:1755 (1979).

36. P.M. Treichel and D.B. Shaw, Inorg. Chim. Acta 16:199 (1976).

37. P. Le Maux, G. Simonneaux, P. Caillet and G. Jaouen, J. Organomet. Chem. 177:C1 (1979).

38. G. Simonneaux, P. Le Maux, G. Jaouen and R. Dabard, Inorg. Chem. 18:3167 (1979).

39. P. Fantucci, L. Naldini, F. Cariati, V. Valenti and C. Busetto, J. Organomet. Chem. 64:109 (1974).

40. B.E. Bursten and R.F. Fenske, Inorg. Chem. 16:963 (1977).

41. J.W. Dart, M.K. Lloyd, R. Mason and J.A. McCleverty, J.C.S. Dalton Trans. 1747 (1973).

42. K.R. Mann, M. Cimolino, G.L. Geoffroy, G.S. Hammond, A.A. Orio, G. Albertin and H.B. Gray, Inorg. Chim. Acta 16:97 (1976).

43. K.R. Mann, N.S. Lewis, R.M. Williams, H.B. Gray and J.G. Gordon II, Inorg. Chem. 17:829 (1978).

44. K.R. Mann, J.A. Thich, R.A. Bell, C.L. Coyle and H.B. Gray, Inorg. Chem. 19:2462 (1980).

45. H. Isci and W.R. Mason, Inorg. Chem. 14:913 (1975).

46. J.S. Miller and D.G. Marsh, Inorg. Chem. 15:2293 (1976).

47. K.R. Mann, J.G. Gordon II and H.B. Gray, J. Am. Chem. Soc. 97:3553 (1975).

48. N.S. Lewis, K.R. Mann, J.G. Gordon II and H.B. Gray, J. Am. Chem. Soc. 98:7461 (1976).

49. K. Kawakami, M. Okajima and T. Tanaka, Bull. Chem. Soc. Jpn. 51:2327 (1978).

50. Y. Yamamoto, K. Aoki and H. Yamazaki, Inorg. Chem. 18:1681 (1979).

51. J.A. Connor, T.J. Meyer and B.P. Sullivan, Inorg. Chem. 18:1388 (1979).

52. G.M. Bancroft and E.T. Libbey, J.C.S. Dalton Trans. 2103 (1973).

53. G.M. Bancroft and P.L. Sears, Inorg. Chem. 14:2716 (1975).

54. L. Di Sipio, S. Calogero, G. Albertin and A.A. Orio, J. Organomet. Chem. 97:257 (1975).

55. S. Calogero, U. Russo, L.L. Condorelli and I. Fragalà, Transition Met. Chem. 4:156 (1979).

56. G.M. Bancroft and K.D. Butler, J. Am. Chem. Soc. 96:7208 (1974).

57. M.K. Lloyd and J.A. McCleverty, J. Organomet. Chem. 61:261 (1973).

58. M.K. Lloyd, J.A. McCleverty, J.A. Connor, M.B. Hall, I.H. Hillier, E.M. Jones and G.K. McEwen, J.C.S. Dalton Trans. 1743 (1973).

59. D.J. Doonan and A.L. Balch, Inorg. Chem. 13:921 (1974).

60. P.M. Treichel, K.P. Wagner and H.J. Mueh, J. Organomet. Chem. 86:C13 (1975).

61. P.M. Treichel, H.J. Mueh and B.E. Bursten, J. Organomet. Chem. 110:C49 (1976).

62. P.M. Treichel and H.J. Mueh, J. Organomet. Chem. 122:229 (1976).

63. P.M. Treichel and H.J. Mueh, Inorg. Chem. 16:1167 (1977).

64. P.M. Treichel and H.J. Mueh, Inorg. Chim. Acta, 22:265 (1977).

65. P.M. Treichel and J.P. Williams, J. Organomet. Chem. 135:39 (1977).

66. P.M. Treichel, D.W. Firsich and G.P. Essenmacher, Inorg. Chem. 18:2405 (1979).

67. P.M. Treichel and D.W. Firsich, J. Organomet. Chem. 172:223 (1979).

68. P.M. Treichel and D.C. Molzahn, J. Organomet. Chem. 179:275 (1979).

69. J. Hanzlik, G. Albertin, E. Bordignon and A.A. Orio, Inorg. Chim. Acta 38:207 (1980).

70. I. Wender and P. Pino in "Organic Syntheses via Metal Carbonyls", Vol. 1, J. Wiley, New York, 1968, p. 1.

71. Y. Yamamoto and H. Yamazaki, Inorg. Chem. 17:3111 (1978).

72. W. Hieber and W. Hübel, Z. Elektrochem. 57:334 (1953).

73. K.R. Mann, H.B. Gray and G.S. Hammond, J. Am. Chem. Soc. 99:306 (1977).

74. R.W. Stephany, Thesis, Utrecht (1973).

75. S. Otsuka and K. Ataka, J.C.S. Dalton Trans. 327 (1976).

76. A.H. Klazinga and J.H. Teuben, J. Organomet. Chem. 192:75 (1980).

77. A.H. Klazinga and J.H. Teuben, J. Organomet. Chem. 194:309 (1980).

78. K.K. Joshi, O.S. Mills, P.L. Pauson, B.W. Shaw and W.H. Stubbs, Chem. Commun. 181 (1965).

79. F.A. Cotton and B.A. Frenz, Inorg. Chem. 13:253 (1974).

80. I.D. Hunt and O.S. Mills, Acta Cryst. B33:2432 (1977).

81. W.P. Fehlhammer, A. Mayr and W. Kehr, J. Organomet. Chem. 197:327 (1980).

82. J.M. Basset, M. Green, J.A.K. Howard and F.G.A. Stone, J.C.S. Chem. Commun. 1000 (1978).

83. M. Green, J.A.K. Howard, M. Murray, J.L. Spencer and F.G.A. Stone, J.C.S. Dalton Trans. 1509 (1977).

84. M.M. Olmstead, H. Hope, L.S. Benner and A.L. Balch, J. Am. Chem. Soc. 5502 (1977).

85. D.J. Doonan, A.L. Balch, S.Z. Goldberg, R. Eisenberg and J.S. Miller, J. Am. Chem. Soc. 97:1961 (1975).

86. S.Z. Goldberg and R. Eisenberg, Inorg. Chem. 15:535 (1976).

87. A.L. Balch, J.R. Boehm, H. Hope and M.M. Olmstead, J. Am. Chem. Soc. 98:7431 (1976).

88. R.D. Adams and F.A. Cotton, Synth. Inorg. Met-Org. Chem. 2:277 (1972).

89. G.K. Barker, A.M.R. Galas, M. Green, J.A.K. Howard, F.G.A. Stone, T.W. Turney, A.J. Welch and P. Woodward, J.C.S. Chem. Commun. 256 (1977).

90. A.J.L. Pombeiro, J. Chatt and R.L. Richards, J. Organomet. Chem. 190:297 (1980).

91. J.A.S. Howell, T.W. Matheson and M.J. Mays, J.C.S. Chem. Commun. 865 (1975).

92. J.A.S. Howell and A.J. Rowan, J.C.S. Dalton Trans. 503 (1980).

93. R.D. Adams and F.A. Cotton, Synth. Inorg. Met.-Org. Chem. 4:477 (1974).

94. R.D. Adams, F.A. Cotton and J.M. Troup, Inorg. Chem. 13:257 (1974).

95. J. Bellerby, M.J. Boylan, M. Ennis and A.R. Manning, J.C.S. Dalton Trans. 1185 (1978).

96. J.A.S. Howell and A.J. Rowan, J. Organomet. Chem. 165:C33 (1979).

97. R.D. Adams and F.A. Cotton in "Dynamic Nuclear Magnetic Resonance Spectroscopy", Ed. L.M. Jackman and F.A. Cotton, Academic Press, New York (1975), p. 489.

98. G. Minghetti, A.L. Bandini, G. Banditelli and F. Bonati, J. Organomet. Chem. 179:C13 (1979).

99. I.L.C. Campbell and F.S. Stephens, J.C.S. Dalton Trans. 982 (1975).

100. R.D. Adams and F.A. Cotton, J. Am. Chem. Soc. 95:6589 (1973).

101. J.A.S. Howell and P. Mathur, J. Organomet. Chem. 174:335 (1979).

102. R.D. Adams, M. Brice and F.A. Cotton, J. Am. Chem. Soc. 95:6594 (1973).

103. R.D. Adams and D.F. Chodosh, J. Organomet. Chem. 87:C48 (1975).

104. M.J. Mays and P.D. Gavens, J.C.S. Dalton Trans. 911 (1980).

105. M.I. Bruce, D. Schultz and R.C. Wallis, J. Organomet. Chem. 169:C15 (1979).

106. J.R. Boehm and A.L. Balch, Inorg. Chem. 16:778 (1977).

107. C.G. McCarty in "The Chemistry of the Carbon Nitrogen Double Bond", Ed. S. Patai, Interscience, London (1979), Chapter 9.

108. R.D. Adams and F.A. Cotton, Inorg. Chem. 13:249 (1974).

109. F.A. Cotton and C.M. Lukehart, Prog. Inorg. Chem. 16:487 (1972).

110. D.J. Cardin, B. Cetinkaya and M.F. Lappert, Chem. Rev. 72:544 (1972).

111. D.J. Cardin, B. Cetinkaya, M.J. Doyle and M.F. Lappert, Chem. Soc. Rev. 2:99 (1973).

112. H.C. Clark, Pure Appl. Chem. 50:43 (1978).

113. A.L. Balch and L.S. Benner, J. Organomet. Chem. 135:339 (1977).

114. L.S. Benner, M.M. Olmstead and A.L. Balch, J. Organomet. Chem. 159:289 (1978).

115. V.W. Day, R.O. Day, J.S. Kristoff, F.J. Hirsekorn and E.L. Muetterties, J. Am. Chem. Soc. 97:2571 (1975).

116. M.G. Thomas, W.R. Pretzer, B.F. Beier, F.J. Hirsekorn and E.L. Muetterties, J. Am. Chem. Soc. 99:743 (1977).

117. E.L. Muetterties, E. Band, A. Kokorin, W.R. Pretzer and M.G. Thomas, Inorg. Chem. 19:1552 (1980).

118. M.G.Thomas, E.L. Muetterties, R.O. Day and V.W. Day, J. Am. Chem. Soc. 98:4645 (1976).

119. C.R. Eady, P.D. Gavens, B.F.G. Johnson, J. Lewis, M.C. Malatesta, M.J. Mays, A.G. Orpen, A.V. Rivera, G.M. Sheldrick and M.B. Hursthouse, J. Organomet. Chem. 149:C43 (1978,.

120. A.V. Rivera, G.M. Sheldrick and M.B. Hursthouse, Acta Cryst. B34:1985 (1978).

121. Y. Yamamoto, K. Aoki and H. Yamazaki, Chem. Lett. 391 (1979).

122. F. Basolo and R.G. Pearson, "Mechanisms of Inorganic Reactions", Wiley, New York (1967).

123. G. Cetini and O. Gambino, Ann. Chim. (Rome) 53:236 (1963).

124. S. Otsuka, T. Yoshida and Y. Tatsuno, J. Am. Chem. Soc. 93:6462 (1971).

125. E.L. Muetterties, Inorg. Chem. 13:495 (1974).

126. E.L. Muetterties, J.C.S. Chem. Commun. 221 (1973).

127. Y. Yamamoto and H. Yamazaki, J. Organomet. Chem. 137:C31 (1977).

128. F. Basolo and A. Wojcicki, J. Am. Chem. Soc. 83:520 (1961).

129. S. Otsuka, A. Nakamura and Y. Tatsuno, J. Am. Chem. Soc. 91:6994 (1969).

130. K. Kawakami, K. Ishii and T. Tanaka, Bull. Chem. Soc. Jpn. 48:1051 (1975).

131. M. Matsumoto and K. Nakatsu, Acta Cryst. B31:2711 (1975).

132. K. Kawakami and M. Okajima, J. Inorg. Nucl. Chem. 41:1501 (1979).

133. K. Iuchi, S. Asada, T. Kinugasa, K. Kanomori and A. Sugimori, Bull. Chem. Soc. Jpn. 49:577 (1976).

134. G.A. Larkin, R. Mason and M.G.H. Wallbridge, J.C.S. Dalton Trans. 2305 (1975).

135. Y. Yamamoto and H. Yamazaki, J. Organomet. Chem. 140:C33 (1977).

136. Y. Yamamoto, T. Mise and H. Yamazaki, Bull. Chem. Soc. Jpn. 51:2743 (1978).

137. B.V. Johnson and A.L. Steinmetz, J. Organomet. Chem. 190:187 (1980).

138. B.A. Howell and W.S. Trahanovsky, J. Am. Chem. Soc. 97:2136 (1975).

139. J.R. Boehm and A.L. Balch, J. Organomet. Chem. 112:C20 (1976).

140. A.L. Balch, J. Am. Chem. Soc. 98:8049 (1976).

141. A.L. Balch, J.W. Labadie and G. Delker, Inorg. Chem. 18:1224 (1979).

142. T.E. Wood, J.C. Deaton, J. Corning, R.E. Wild and R.A. Walton, Inorg. Chem. 19:2614 (1980).

143. F. Faraone and V. Marsala, Inorg. Chim. Acta 27:L109 (1978).

144. P.V. Yaneff and J. Powell, J. Organomet. Chem. 179:101 (1979).

145. L. Malatesta, J. Chem. Soc. 3924 (1955).

146. L. Malatesta and M. Angoletta, J. Chem. Soc. 1186 (1957).

147. C.A.L. Becker, Synth. React. Inorg. Met.-Org. Chem. 4:213 (1974).

148. C.A.L. Becker, J. Organomet. Chem. 104:89 (1976).

149. C.A.L. Becker, J. Inorg. Nucl. Chem. 42:27 (1980).

150. C.A.L. Becker and B.L. Davis, J. Inorg. Nucl. Chem. 39:781 (1977).

151. C.A.L. Becker, Inorg. Chim Acta 36:L441 (1979).

152. R.B. King and M.S. Saran, Inorg. Chem. 11:2112 (1972).

153. P.M. Treichel, W.J. Knebel and R.W. Hess, J. Am. Chem. Soc. 93:5424 (1971).

154. M.F. Rettig, E.A. Kirk and P.M. Maitlis, J. Organomet. Chem. 111:113 (1976).

155. D.V. Stynes, J. Am. Chem. Soc. 96:5942 (1974).

156. C. Irwin and D.V. Stynes, Inorg. Chem. 17:2682 (1978).

157. D.V. Stynes, Inorg. Chem. 16:1170 (1977).

158. I.W. Pang and D.V. Stynes, Inorg. Chem. 16:590 (1977).

159. I.W. Pang and D.V. Stynes, Inorg. Chem. 16:2192 (1977).

160. L.E. Manzer, J.C.S. Dalton Trans. 1535 (1974).

161. E.W. Powell and M.J. Mays, J. Organomet. Chem. 66:137 (1974).

162. B. Crociani, M. Nicolini and R.L. Richards, Inorg. Chim. Acta 12:53 (1975).

163. C.P. Casey in "Transition Metal Organometallics in Organic Synthesis", Ed. H. Alper, Academic Press, New York (1976), p. 189.

164. J. Chatt, R.L. Richards and G.H.D. Royston, J.C.S. Dalton Trans. 1433 (1973).

165. T.F. Block, R.F. Fenske and C.P. Casey, J. Am. Chem. Soc. 93:441 (1976).

166. T. Sawai and R.J. Angelici, J. Organomet. Chem. 80:91 (1974).

167. J.A. Connor and E.M. Jones, J.C.S. Dalton Trans. 2119 (1973).

168. K.P. Darst and C.M. Lukehart, Inorg. Chim. Acta 27:219 (1978).

169. K.R. Grundy and W.R. Roper, J. Organomet. Chem. 113:C45 (1976).

170. G.R. Clark, T.J. Collins, D. Hall, S.M. James and W.R. Roper, J. Organomet. Chem. 141:C5 (1977).

171. W.P. Fehlhammer, K. Bartel and H. Schmidt, J. Organomet. Chem. 97:C61 (1975).

172. G. Minghetti, F. Bonati and G. Banditelli, Synth. Inorg. Met.-Org. Chem. 3:415 (1973).

173. G. Minghetti and F. Bonati, Inorg. Chem. 13:1600 (1974).

174. G. Minghetti, F. Bonati and M. Massobrio, Inorg. Chem. 14:1974 (1975).

175. A. Tiripicchio, M. Tiripicchio Camellini and G. Minghetti, J. Organomet. Chem. 171:399 (1979).

176. F. Bonati and G. Minghetti, J. Organomet. Chem. 60:C43 (1973).

177. G. Minghetti, F. Bonati and G. Banditelli, Inorg. Chem. 15:2649 (1976).

178. F. Bonati, G. Minghetti and G. Banditelli, Synth. Inorg. Met.-Org. Chem. 6:383 (1976).

179. G. Banditelli, A.L. Bandini and G. Minghetti, Synth. Inorg. Met.-Org. Chem. 9:539 (1979).

180. L.F. Farnell, E.W. Randall and E. Rosenberg, Chem. Commun. 1078 (1971).

181. G.M. Bodner, S.B. Kahl, K. Bork, B.N. Storhoff, J.E. Wuller and L.J. Todd, Inorg. Chem. 12:1071 (1973).

182. R.J. Angelici and L.M. Charley, J. Organomet. Chem. 24:205 (1970).

183. W.M. Butler, J.H. Enemark, J. Parks and A.L. Balch, Inorg. Chem. 12:451 (1973).

184. P. Brant, J.H. Enemark and A.L. Balch, J. Organomet. Chem. 114:99 (1976).

185. T.G. Appleton, H.C. Clark and L.E. Manzer, Coord. Chem. Rev. 10:335 (1973).

186. L. Manojlović-Muir, K.W. Muir and T. Solomun, J. Organomet. Chem. 142:265 (1977).

187. U. Schoellkopf and F. Gerhart, Angew. Chem. Int. Ed. Engl. 6:970 (1967).

188. B. Crociani, T. Boschi and U. Belluco, Inorg. Chem. 9:2021 (1970).

189. F. Bonati and G. Minghetti, Synth. Inorg. Met.-Org. Chem. 1:299 (1971).

190. C.H. Davies, C.H. Game, M. Green and F.G.A. Stone, J.C.S. Dalton Trans. 357 (1974).

191. M. Wada, S. Kanai, R. Maeda, M. Kinoshita and K. Oguro, Inorg. Chem. 18:417 (1979).

192. P.R. Branson and M. Green, J.C.S. Dalton Trans. 1303 (1972).

193. P.R. Branson, R.A. Cable, M. Green and M.K. Lloyd, J.C.S. Dalton Trans. 12 (1976).

194. D.J. Doonan, J.E. Parks and A.L. Balch, J. Am. Chem. Soc. 98:2129 (1976).

195. W.P. Fehlhammer, T. Kemmerich and W. Beck, Chem. Ber. 112:468 (1979).

196. D.J. Doonan and A.L. Balch, J. Am. Chem Soc. 95:4769 (1973).

197. W.G. Kita, J.A. McCleverty, B. Patel and J. Williams, J. Organomet. Chem. 74:C9 (1974).

198. A.L. Balch, J. Organomet. Chem. 37:C19 (1972).

199. K.W. Muir, R. Walker, J. Chatt, R.L. Richards and G.H.D. Royston, J. Organomet. Chem. 56:C30 (1973).

200. A.L. Balch and J.E. Parks, J. Am. Chem. Soc. 96:4114 (1974).

201. J.E. Parks and A.L. Balch, J. Organomet. Chem. 71:453 (1974).

202. G. Minghetti, F. Bonati and G. Banditelli, Inorg. Chem. 15:1718 (1976).

203. R. Usón, A. Laguna, J. Vincente, J. García, B. Bergareche and P. Brun, Inorg. Chim. Acta 28:237 (1978).

204. W.M. Butler and J.H. Enemark, Inorg. Chem. 12:540 (1973).

205. L. Manojlović-Muir and K.W. Muir, J.C.S. Dalton Trans. 2427 (1974).

206. P. Doniano, A. Muratti, M. Nardelli and G. Predieri, J.C.S. Dalton Trans. 2165 (1975).

207. D.F. Christian, D.A. Clarke, H.C. Clark, D.H. Farrar and N.C. Payne, Can. J. Chem. 56:2516 (1978).

208. B.V. Johnson, D.P. Sturtzel and J.E. Shade, J. Organomet. Chem. 154:89 (1978).

209. B. Crociani and R.L. Richards, J.C.S. Dalton Trans. 693 (1974).

210. B.V. Johnson and J.E. Shade, J. Organomet. Chem. 179:357 (1979).

211. R. Zanella, T. Boschi, B. Crociani and U. Belluco, J. Organomet. Chem. 71:135 (1974).

212. T. Boschi, B. Crociani, M. Nicolini and U. Belluco, Inorg. Chim. Acta 12:39 (1975).

213. A. Burke, A.L. Balch and J.H. Enemark, J. Am. Chem. Soc. 92:2555 (1970).

214. W.M. Butler and J.H. Enemark, Inorg. Chem. 10:2416 (1971).

215. J.S. Miller and A.L. Balch, Inorg. Chem. 11:2069 (1972).

216. G. Rouschias and B.L. Shaw, J. Chem. Soc. A 2097 (1971).

217. A.L. Balch and J.S. Miller, J. Am. Chem. Soc. 94:417 (1972).

218. J.S. Miller, A.L. Balch and J.H. Enemark, J. Am. Chem. Soc. 93:4613 (1971).

219. Y. Ito, T. Hirao, K. Tsubata and T. Saegusa, Tetrahedron Lett. 1535 (1978).

220. K. Bartel and W.P. Fehlhammer, Angew. Chem. Int. Ed. Engl. 13:599 (1974).

221. R. Usón, A. Laguna, J. Vincente, J. García and B. Bergareche, J. Organomet. Chem. 173:349 (1979).

222. G. Minghetti, L. Baratto and F. Bonati, J. Organomet. Chem. 102:397 (1975).

223. B. Crociani, T. Boschi, G.G. Troilo and U. Croatto, Inorg. Chim. Acta 6:655 (1972).

224. B. Crociani, T. Boschi, M. Nicolini and U. Belluco, Inorg. Chem. 11:1292 (1972).

225. L. Calligaro, P. Uguagliati, B. Crociani and U.Belluco, J. Organomet. Chem. 92:399 (1975).

226. P. Uguagliati, B. Crociani, L. Calligaro and U. Belluco, J. Organomet. Chem. 112:111 (1976).

227. B. Crociani, P. Uguagliati and U. Belluco, J. Organomet. Chem. 117:189 (1976).

228. E. Rotondo, M. Cusumano, B. Crociani, P. Uguagliati and U. Belluco, J. Organomet. Chem. 134:249 (1977).

229. L. Calligaro, P. Uguagliati, B. Crociani and U. Belluco, J. Organomet. Chem. 142:105 (1977).

230. B. Crociani and R.L. Richards, J. Organomet. Chem. 144:85 (1978).

231. D.H. Cuatecontzi S. and J.D. Miller, J. Chem Res. (S) 137 (1977).

232. D.H. Cuatecontzi S. and J.D. Miller, Inorg. Chim Acta 38:157 (1980).

233. J.E. Parks and A.L. Balch, J. Organomet. Chem. 57:C103 (1973).

234. A.J.L. Pombeiro and R.L. Richards, Transition Met. Chem. 5:55 (1980).

235. J. Chatt, A.J.L. Pombeiro and R.L. Richards, J.C.S. Dalton Trans. 1585 (1979).

236. E.B. Dreyer, C.T. Lam and S.J. Lippard, Inorg. Chem. 18:1904 (1979).

237. P.M. Treichel, J.P. Williams, W.A. Freeman and J.I. Gelder, J. Organomet. Chem. 170:247 (1979).

238. J. Forniés, M. Green, A. Laguna, M. Murray, J.L. Spencer and F.G.A. Stone, J.C.S. Dalton Trans. 1515 (1977).

239. J.B.B. Heyns and F.G.A. Stone, J. Organomet. Chem. 160:337 (1978).

240. J. Browning, M. Green, A. Laguna, L.E. Smart, J.L. Spencer and F.G.A. Stone, J.C.S. Chem. Commun. 723 (1975).

241. W.E. Carroll, M. Green, J.A.K. Howard, M. Pfeffer and F.G.A. Stone, J.C.S. Dalton Trans. 1472 (1978).

242. J.M. Basset, M. Green, J.A.K. Howard and F.G.A. Stone, J.C.S. Chem. Commun. 853 (1977).

243. S. Otsuka, A. Nakamura and Y. Yoshida, J. Am. Chem. Soc. 91:7196 (1969).

244. S. Otsuka, A. Nakamura, Y. Yoshida, M. Naruto and K. Ataka, J. Am. Chem. Soc. 95:3180 (1973).

245. M. Ciriano, M. Green, D. Gregson, J.A.K. Howard, J.L. Spencer, F.G.A. Stone and P. Woodward, J.C.S. Dalton Trans. 1294 (1979).

246. A.L. Balch and M.M. Olmstead, J. Am. Chem. Soc. 98:2354 (1976).

247. A.L. Balch and M.M. Olmstead, J. Am. Chem. Soc. 101:3128 (1979).

248. M.M. Olmstead and A.L. Balch, J. Organomet. Chem. 148:C15 (1978).

249. K.R. Mann, R.A. Bell and H.B. Gray, Inorg. Chem. 18:2671 (1979).

250. K.R. Mann, N.S. Lewis, V.M. Miskowski, D.K. Erwin, G.S. Hammond and H.B. Gray, J. Am. Chem. Soc. 99:5525 (1977).

251. V.M. Miskowski, I.S. Sigal, K.R. Mann and H.B. Gray, J. Am. Chem. Soc. 101:4383 (1979).

252. K.R. Mann, M.J. Di Pierro and T.P. Gill, J. Am. Chem. Soc. 102:3965 (1980).

253. T.G. Appleton, H.C. Clark and L.E. Manzer, J. Organomet. Chem. 65:275 (1974).

254. S. Otsuka and K. Ataka, Bull. Chem. Soc. Jpn. 50:1118 (1977).

255. R. Kuwae and T. Tanaka, Bull. Chem. Soc. Jpn. 53:118 (1980).

256. R. Ugo, A. Pasini, A. Fusi and S. Cenini, J. Am Chem. Soc. 94:7364 (1972).

257. M.A. Haga, K. Kawakami and T. Tanaka, Inorg. Chim. Acta 12:93 (1975).

258. R. Kuwae, T. Tanaka and K. Kawakami, Bull. Chem. Soc. Jpn. 52:437 (1979).

259. R. Kuwae and T. Tanaka, Bull. Chem. Soc. Jpn. 52:1067 (1979).

260. J. Chatt, A.J.L. Pombeiro and R.L. Richards, J.C.S. Dalton Trans. 492 (1980).

261. A.J.L. Pombeiro, R.L. Richards and J.R. Dilworth, J. Organomet. Chem. 175:C17 (1979).

262. J. Chatt, A.J.L. Pombeiro and R.L. Richards J. Organomet. Chem. 184:357 (1980).

263. G.K. Barker, M. Green, T.P. Onak, F.G.A. Stone, C.B. Ungermann and A.J. Welch, J.C.S. Chem. Commun. 169 (1978).

264. J.M. Basset, L.J. Farrugia and F.G.A. Stone, J.C.S. Dalton Trans. 1789 (1980).

265. P.M. Treichel, J.J. Benedict, R.W. Hess and J.P. Stenson, Chem. Commun. 1627 (1970).

266. S. Willis and A.R. Manning, J.C.S. Dalton Trans. 23 (1979).

267. S. Willis and A.R. Manning, J. Chem. Res. (S) 390 (1978).

268. S. Willis and A.R. Manning, J. Organomet. Chem. 97:C49 (1975).

269. S. Willis and A.R. Manning, J.C.S. Dalton Trans. 186 (1980).

270. A. Wojcicki, Adv. Organomet. Chem. 11:87 (1973).

271. F. Calderazzo, Angew. Chem. Int. Ed. Engl. 16:299 (1977).

272. E.J. Kuhlmann and J.J. Alexander, Coord. Chem. Rev. 33:195 (1980).

273. P.M. Treichel, K.P. Wagner and R.W. Hess, Inorg. Chem. 12:1471 (1973).

274. P.M. Treichel and K.P. Wagner, J. Organomet. Chem. 61:415 (1973).

275. H.C. Clark and L.E. Manzer, Inorg. Chem. 11:503 (1972).

276. I.C. Douek and G. Wilkinson, J. Chem. Soc. A. 2604 (1969).

277. R.D. Adams, Inorg. Chem. 15:174 (1976).

278. E.J.M. De Boer and J.H. Teuben, J. Organomet. Chem. 166:193 (1979).

279. A. Dormond and A. Dahchour, J. Organomet. Chem. 193:321 (1980).

280. C.J. Cardin, D.J. Cardin and M.F. Lappert, J.C.S. Dalton Trans. 767 (1977).

281. C.J. Cardin, D.J. Cardin, J.M. Kelly, R.J. Norton and A. Roy, J. Organomet. Chem. 132:C23 (1977).

282. Y. Yamamoto and H. Yamazaki, Inorg. Chem. 13:2145 (1974).

283. B. Crociani and R.L. Richards, J. Organomet. Chem. 154:65 (1978).

284. Y. Yamamoto and H. Yamazaki, Inorg. Chim. Acta 41:229 (1980).

285. Y. Yamamoto and H. Yamazaki, Inorg. Chem. 13:438 (1974).

286. R.A. Andersen, Inorg. Chem. 18:2928 (1979).

287. Y. Yamamoto and H. Yamazaki, Inorg. Chem. 16:3182 (1977).

288. W. Drenth and R.J.M. Nolte, Acc. Chem. Res. 12:30 (1979).

289. B. Crociani, M. Nicolini and R.L. Richards, J. Organomet. Chem. 104:259 (1976).

290. J.F. Van Baar, J.M. Klerks, P. Overbosh, D.J. Stufkens and K. Vrieze, J. Organomet. Chem. 112:95 (1976).

291. R. Zanella, G. Carturan, M. Graziani and U. Belluco, J. Organomet. Chem. 65:417 (1974).

292. A. Mantovani, M. Peloso, G. Bandoli and B. Crociani, J.C.S. Dalton Trans. 2223 (1984).

293. R.D. Adams and D.F. Chodosh, J. Am. Chem. Soc. 98:5391 (1976).

294. R.D. Adams and D.F. Chodosh, J. Am. Chem. Soc. 99:6544 (1977).

295. R.J.H. Clark, J.A. Stockwell and J.D. Wilkins, J.C.S. Dalton Trans. 120 (1976).

296. M.F. Lappert, N.T. Luong-Thi and C.R.C. Milne, J. Organomet. Chem. 174:C35 (1979).

297. P.T. Wolczanski and J.E. Bercaw, J. Am. Chem. Soc. 101:6450 (1979).

298. E. Klei and J.H. Teuben, J. Organomet. Chem. 188:97 (1980).

299. R.D. Adams and D.F. Chodosh, J. Organomet. Chem. 122:C11 (1976).

300. K.P. Wagner, P.M. Treichel and J.C. Calabrese, J. Organomet. Chem. 71:299 (1974).

301. R.D. Adams and D.F. Chodosh, Inorg. Chem. 17:41 (1978).

302. W.R. Roper, G.E. Taylor, J.M. Waters and L.J. Wright, J. Organomet. Chem. 157:C27 (1978).

303. F. Van Bolhuis, E.J.M. De Boer and J.H. Teuben, J. Organomet. Chem. 170:299 (1979).

304. B. Crociani, G. Bandoli and D.A. Clemente, J. Organomet. Chem. 184:269 (1980).

305. Y. Yamamoto and H. Yamazaki, J. Organomet. Chem. 90:329 (1975).

306. Y. Yamamoto, K. Aoki and H. Yamazaki, J. Am. Chem. Soc. 96:2647 (1974).

307. K. Aoki and Y. Yamamoto, Inorg. Chem. 15:48 (1976).

308. H.C. Clark, C.R.C. Milne and N.C. Payne, J. Am. Chem. Soc. 100:1164 (1978).

309. R. Aumann and E.O. Fischer, Chem. Ber. 101:954 (1968).

310. C.G. Kreiter and R. Aumann, Chem. Ber. 111:1223 (1978).

311. P.M. Treichel and K.P. Wagner, J. Organomet. Chem. 88:199 (1975).

312. T.A. Mitsudo, H. Watanabe, Y. Komiya, Y. Watanabe, Y. Takaegami, K. Nakatsu, K. Kinoshita and Y. Miyagawa, J. Organomet. Chem. 190:C39 (1980).

313. H. Yamazaki, K. Aoki, Y. Yamamoto and Y. Wakatsuki, J. Am. Chem. Soc. 97:3546 (1975).

314. M.A. Bennett and T. Yoshida, J. Am. Chem. Soc. 100:1750 (1978).

315. J.L. Davidson, M. Green, J.Z. Nyathi, F.G.A. Stone and A.J. Welch, J.C.S. Dalton Trans. 2246 (1977).

316. H. Yamazaki and Y. Wakatsuki, Bull. Chem. Soc. Jpn. 52:1239 (1979).

317. S. Otsuka and A. Nakamura, Adv. Organomet. Chem. 14:245 (1976).

318. D.F. Christian, G.R. Clark, W.R. Roper, J.M. Waters and K.R. Whittle, J.C.S. Chem. Commun. 458 (1972).

319. D.F. Christian and W.R. Roper, J. Organomet. Chem. 80:C35 (1974).

320. G.R. Clark, J.M. Waters and K.R. Whittle, J.C.S. Dalton Trans. 2556 (1975).

321. D.F. Christian and H.C. Clark, J. Organomet. Chem. 85:C9 (1975).

322. D.F. Christian, H.C. Clark and R.F. Stepaniak, J. Organomet. Chem. 112:209 (1976).

323. R. Ros, R.A. Michelin, G. Carturan and U. Belluco, J. Organomet. Chem. 133:213 (1977).

324. P.L. Sandrini, A. Mantovani and B. Crociani, J. Organomet. Chem. 185:C13 (1980).

325. M.I. Bruce and R.C. Wallis, J. Organomet. Chem. 164:C6 (1979).

326. E.L. Muetterties, Bull. Soc. Chim. Belg. 85:451 (1976).

327. E. Band, W.R. Pretzer, M.G. Thomas and E.L. Muetterties, J. Am. Chem. Soc. 99:7380 (1977).

328. R.D. Adams and N.M. Golembeski, J. Am. Chem. Soc. 100:4622 (1978).

329. R.D. Adams and N.M. Golembeski, J. Am. Chem. Soc. 101:2579 (1979).

330. M.J. Mays, D.W. Prest and P.R. Raithby, J.C.S. Chem. Commun. 171 (1980).

331. B. Crociani and R.L. Richards, J.C.S. Chem. Commun. 127 (1973).

332. B. Crociani, M. Nicolini and R.L. Richards, J. Organomet. Chem. 101:C1 (1975).

333. M. Behnam-Dahkordy, B. Crociani and R.L. Richards, J.C.S. Dalton Trans. 2015 (1977).

334. M. Behnam-Dahkordy, B. Crociani, M. Nicolini and R.L. Richards, J. Organomet. Chem. 181:69 (1979).

335. W. Beck, K. Burger and W.P. Fehlhammer, Chem. Ber. 104:1816 (1971).

336. M. Fukui, K. Itoh and Y. Ishii, Synth. React. Inorg. Met.-Org. Chem. 5:207 (1975).

337. Y. Ito, Y. Inubushi, T. Sugaya and T. Saegusa, J. Organomet. Chem. 137:1 (1977).

338. E. Pfeiffer, A. Oskam and K. Vrieze, Transition Met. Chem. 2:240 (1977).

339. R.A. Michelin and R. Ros, J. Organomet. Chem. 169:C42 (1979).

340. L.S. Benner and A.L. Balch, J. Am. Chem. Soc. 100:6099 (1978).

341. D. Hoppe, New Synth. Methods 2:17 (1975).

342. U. Schoellkopf, Angew. Chem. Int. Ed. Engl. 16:339 (1977).

343. U. Schoellkopf, Pure Appl. Chem. 51:1347 (1979).

344. T. Saegusa, Y. Ito, S. Kobayashi, K. Hirota and H. Yoshioka, Bull. Chem. Soc. Jpn. 42:3310 (1969).

345. A.E. Hegarty and A. Chandler, Tetrahedron Lett. 21, 885 (1980).

346. D. Knol, C.P.A. van Os and W. Drenth, Recl. Trav. Chim. Pays-Bas 93:314 (1974).

347. Y. Ito, Y. Inubushi and T. Saegusa, Tetrahedron Lett. 1283 (1974).

347a. P. Jakobsen, Acta Chem. Scand. B30:847 (1976).

348. S. Treppendahl, P. Jakobsen and G. Vernin, Acta Chem. Scand. B32:777 (1978).

349. A. Schwartz and E.E. van Tamelen, J. Am. Chem. Soc. 99:3189 (1977).

350. Y. Ito, T. Hirao and T. Saegusa, J. Org. Chem. 40:2981 (1975).

351. G. Minghetti and F. Bonati, J. Organomet. Chem. 73:C43 (1974).

352. H. Sawai and T. Takizawa, J. Organomet. Chem. 94:333 (1975).

353. K. Hiraki, Y. Fuchita and S. Morinaga, Chem. Lett. 1 (1978).

354. K. Hiraki and Y. Fuchita, Chem. Lett. 841 (1978).

355. W.P. Fehlhammer, K. Bartel and W. Petri, J. Organomet. Chem. 87:C34 (1975).

356. K.R. Grundy and W.R. Roper, J. Organomet. Chem. 91:C61 (1975).

357. Y. Ito, I. Ito, T. Hirao and T. Saegusa, Synth. Commun. 4:97 (1974).

358. Y. Ito, T. Hirao and T. Saegusa, J. Organomet. Chem. 82:C47 (1974).

359. Y. Ito, T. Hirao and T. Saegusa, J. Organomet. Chem. 131:121 (1977).

360. Y. Ito, T. Hirao, N. Ohta and T. Saegusa, Tetrahedron Lett. 1009 (1977).

361. Y. Yamamoto and H. Yamazaki, J. Org. Chem. 42:4136 (1977).

362. Y. Yamamoto and H. Yamazaki, Synthesis 750 (1976).

363. Y. Ito, T. Konoike and T. Saegusa, Tetrahedron Lett. 1287 (1974).

364. Y. Ito, T. Konoike and T. Saegusa, J. Organomet. Chem. 85:395 (1975).

365. Y. Ito, Y. Inubushi, T. Sugaya, K. Kobayashi and T. Saegusa, Bull. Chem. Soc. Jpn. 51:1186 (1978).

366. Y. Ito, K. Kobayashi and T. Saegusa, J. Org. Chem. 44:2030 (1979).

367. Y. Ito, K. Kobayashi and T. Saegusa, Tetrahedron Lett. 1039 (1979).

368. C. Wakselman and M. Tordeux, J. Org. Chem. 44:4219 (1979).

369. Y. Ito, K. Yonezawa and T. Saegusa, J. Org. Chem. 39:1763 (1974).

370. Y. Ito, K. Nakayama, K. Yonezawa and T. Saegusa, J. Org. Chem. 39:3273 (1974).

371. R. Baker, R.C. Cookson and J.R. Vinson, J.C.S. Chem. Commun. 515 (1974).

372. R. Baker and A.H. Copeland, Tetrahedron Lett. 4535 (1976).

373. R. Baker and A.H. Copeland, J.C.S. Perkin Trans. I 2497 (1977).

374. R. Baker and A.H. Copeland, J.C.S. Perkin Trans. I 2560 (1977).

375. F. Millich, Advan. Polym. Sci. 19:117 (1975).

376. H. Sawai and T. Takizawa, Bull. Chem. Soc. Jpn. 49:1906 (1976).

377. R.J.M. Nolte, A.J.M. van Beijnen and W. Drenth, J. Am. Chem. Soc. 96:5932 (1974).

378. R.J.M. Nolte, J.W. Zwikker, J. Reedijk and W. Drenth, J. Mol. Catal. 4:423 (1978).

379. R.J.M. Nolte and W. Drenth, Recl. Trav. Chim. Pays-Bas 92:788 (1973).

380. A.J.M. Beijnen, R.J.M. Nolte, J.W. Zwikker and W. Drenth, J. Mol. Catal. 4:427 (1978).

381. T. Saegusa and S. Horiguchi, Polym. J. 6:419 (1974).

382. T. Saegusa, S. Horiguchi and T. Tsuda, Macromolecules 8:112 (1975).

383. T. Saegusa, S. Horiguchi and T. Tsuda, Polym. J. 8:302 (1975).

384. S. Tanaka, S. Inokuma and T. Deguchi, Japan Kokai 77 65,241; Chem. Abs. 87:135102b (1977).

385. P. Svoboda and J. Hetfleis, Collect. Czech. Chem. Commun. 38:3834 (1973).

386. V. Vaisarova and J. Hetfleis, Collect. Czech. Chem. Commun. 41:1906 (1976).

387. J. Langova and J. Hetfleis, Collect. Czech. Chem. Commun. 40:432 (1975).

388. S. Tanaka, H. Kido, S. Uemura and M. Okano, Bull. Chem. Soc. Jpn. 48:3415 (1975).

389. S. Otsuka, A. Nakamura and Y. Tatsuno, Chem. Commun. 836 (1967).

390. T. Tsuda, S. Sanada and T. Saegusa, J. Organomet. Chem. 116:C10 (1976).

391. T. Tsuda, Y. Chujo and T. Saegusa, J. Am. Chem. Soc. 102:431 (1980).

ADDITIONAL REFERENCES

Section 2.1

i.   H. Umeyama and T. Nomoto, Chem. Pharm. Bull. 28:2279 (1980).

ii.   P. LeMaux, G. Simonneaux and G. Jaouen, J. Organomet. Chem. 217:61 (1981).

Section 2.2

iii. J.M. Basset, G.K. Barker, M. Green, J.A.K. Howard, F.G.A. Stone and W.C. Wolsey, J.C.S. Dalton Trans. 219 (1981).

iv. M. Ennis, R. Kumar, A.R. Manning, J.A.S. Howell, P. Mathur, A.J. Rowan and F.S. Stephens, J.C.S. Dalton Trans. 1251 (1981).

v. M.J. Bruce, J.G. Matisons, J.R. Rodgers and R.C. Wallis, J.C.S. Chem. Commun. 1070 (1981).

Section 3.

vi. L.D. Silverman, J.C. Dewan, C.M. Giandomenico and S.J. Lippard, Inorg. Chem. 19:3379 (1980).

vii. C.H. Lindsay, L.S. Benner and A.L. Balch, Inorg. Chem. 19:3503 (1980).

viii. W.S. Mialki, R.E. Wild and R.A. Walton, Inorg. Chem. 20:1380 (1981).

Section 4.1

ix F.J. Regina and A. Wojcicki, Inorg. Chem. 19:3803 (1980).

Section 4.2

x. W.P. Fehlhammer, A. Mayr and G. Christian, J. Organomet. Chem. 209:57 (1981).

xi. J.R. Briggs, C. Croker and B.L. Shaw, Inorg. Chim. Acta 51:15 (1981).

Section 5.1

xii. N.M. Boag, G.H.M. Dias, M. Green, J.L. Spencer, F.G.A. Stone and J. Vincente, J.C.S. Dalton Trans. 1981 (1981).

xiii. D.A. Bohling, T.P. Gill and K.R. Mann, Inorg. Chem. 20:194 (1981).

Section 5.2

xiv A.J.L. Pombeiro and R.L. Richards, Transition Met. Chem. 5:281 (1980).

xv. J.A.S. Howell and A.J. Rowan, J.C.S. Dalton Trans. 297 (1981).

xvi. S. Willis and A.R. Manning, J.C.S. Dalton Trans. 322 (1981).

xvii. M.H. Quick and R.J. Angelici, Inorg. Chem. 20:1123 (1981).

xviii. A.J.L. Pombeiro, M.F.N.N. Carvalho, P.B. Hitchcock and R.L. Richards, J.C.S. Dalton Trans. 1629 (1981).

xix. J.A.S. Howell and P. Mathur, J.C.S. Chem. Commun. 263 (1981).

Section 6.1

xx. R.D. Adams, J. Am. Chem. Soc. 102:7476 (1980).

xxi S.J. Simpson and R.A. Andersen, J. Am. Chem. Soc. 103:4063 (1981).

xxii. K.W. Chiu, R.A. Jones, G. Wilkinson, A.M.R. Galas and M.B. Hursthouse, J.C.S. Dalton Trans. 2088 (1981).

xxiii. D.S. Gill, P.K. Baker, M. Green, K.E. Paddick, M. Murray and A.J. Welch, J.C.S. Chem. Commun. 986 (1981).

Section 7.1

xxiv. Y. Ito, K. Kobayashi and T. Saegusa, Chem. Lett. 1563 (1980).

xxv.   Y. Fuchita, K. Hidaka, S. Morinaga and K. Hiraki, Bull. Chem. Soc. Jpn. 54:800 (1981).

Section 7.2

xxvi. A.J.M. Van Beijnen, R.J.M. Nolte, W. Drenth, A.M.F. Hezemans and P.J.F.M. Van de Coolwijk, Macromolecules 13:1386 (1980).

xxvii. Y. Wakatsuki, O. Nomura, H. Tone and H. Yamazaki, J.C.S. Perkin Trans. II 1344 (1980).

xxviii. Y. Yamamoto and H. Yamazaki, Bull. Chem. Soc. Jpn. 54:787 (1981).

Section 7.3

xxix. Y. Ito, T. Hirao, N. Ohta and T. Saegusa, Synth. Commun. 10:233 (1980).

xxx.   P.W.R. Corfield, L.M. Baltusis and S.J. Lippard, Inorg. Chem. 20:922 (1981).

xxxi. J.C. Dewan, C.M. Giandomenico and S.J. Lippard, Inorg. Chem. 20:4069 (1981).

xxxii. E.O. Fischer and W. Schambeck, J. Organomet. Chem. 201:311 (1980).

xxxiii. A.J. Carty, G.N. Mott and N.J. Taylor, J. Organomet. Chem. 212:C54 (1981).

xxxiv. K. Hargreaves, Chem. Ind. (London) 609 (1981).

xxxv. D.L. DuBois, W.K. Miller and M. Rakowski DuBois, J. Am. Chem. Soc. 103:3429 (1981).

THE FORMATION AND REACTIONS OF METALLACYCLES

G. Ingrosso

Istituto di Chimica Organica Industriale, University of Pisa

1.   INTRODUCTION

The term 'metallacycle' is scarcely specific since it denotes any cyclic arrangement of atoms at least one of which is a metal. In this chapter I will only deal with transition metal metallacarbocycles, i.e. those derivatives containing the ring system $\overline{\text{M-C......C}}$ which formally results from binding any organic diradical to a transition metal atom. I shall omit discussing metallacyclopropanes and metallacyclopropenes which are considered elsewhere in this book as olefin- and acetylene-metal complexes. Metallacyclopentadienes, the reductive coupling of conjugated dienes, and metathesis intermediates are also dealt with in other Chapters.

Transition metal metallacarbocycles were first obtained in 1936 by Buraway and Gibson[1] from the reaction of 1,4-bis(bromomagnesium)pentane with gold(III) halides. However, apart from sporadic but interesting studies,[2-9] the new area remained relatively dormant until 1970 when Cassar, Eaton and Halpern[10] published their results on the rhodium-catalyzed valence isomerization of cubane. Great interest in metallacarbocyles was aroused which caused chemists to make a variety of conjectures about their properties and, consequently, to start research in this new area.

The novel and sometimes baffling metallacarbocycle chemistry that has emerged during the past ten years has reinforced the interest in these compounds and has caused them to win a special position within the immense field of transition metal alkyls. Their versatile capacity to undergo, depending on a variety of factors, carbon-carbon cleavage, reductive elimination of carbocycles, or elimination of linear unsaturated compounds, has allowed a number of hypotheses to be put forward on their role in catalysis and in transition metal reactions, and their potential utility in metal-assisted organic synthesis to be foreseen. It is noteworthy that attention has been focused on metallacarbocycle chemistry not only in new contexts but also in the context of processes that had previously been interpreted in other terms.

Some hypotheses, indeed, fail to account even for the data generated within their chosen frame of application. This, however, should not be surprising since hypotheses are rarely discarded merely because of a poor fit with the data; they are discarded only when a better fit is provided. On the other hand, compelling evidence for the intermediacy of metallacarbocycles in several catalytic or stoichiometric reactions comes from the direct observation and even isolation of examples which are formed in analogous processes and are stabilized by virtue of their structure or their environment.

Very interestingly, theoretical studies have begun to appear in the literature,[11-13] which rationalize some of the most typical reactions of metallacarbocycles. Theories, unlike hypotheses, have strong feelings about the data and, because they make concrete predictions, they parcel out the data in the field into strong and weak facts. Strong facts are those vested with special interest and importance since they derive from experiments which test predictions from the theory and then confront it. Weak facts, although easily accomodated by the theory, are sometimes ignored, at least to a first approximation.[14]

Partial reviews of metallacycle chemistry have previously appeared.[15-18] However, no comprehensive monograph on this area has been published. In what follows I describe transition metal metallacarbocycles with emphasis on their reactions and relate their chemistry to present views of bond-making and bond-breaking processes in coordination chemistry and catalysis.

2.   FORMATION

The methods for preparing metallacarbocyclic compounds can be divided into six categories, which are individually discussed in the sections below.

2.1   Cyclodialkylation by organometallic reagents

By far the most general procedure for preparing five-, six-, and seven-membered metallacarbocycles is the reaction of a metal halide with a dilithium or a bis(halogenomagnesium) organo-compound:

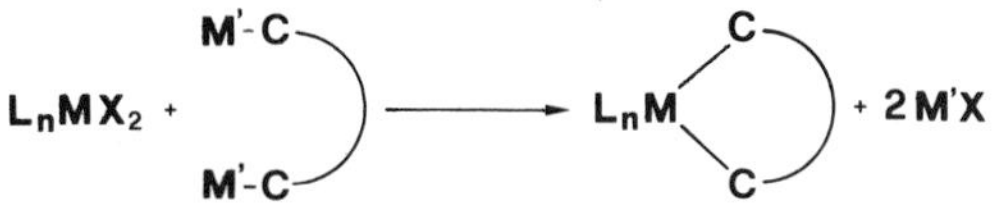

$$L_nMX_2 + \begin{array}{c} M'-C \\ \\ M'-C \end{array} \Big\rangle \longrightarrow L_nM\begin{array}{c} C \\ \\ C \end{array} \Big\rangle + 2M'X$$

This has been successfully adopted for the synthesis of a number of derivatives of titanium(IV),[19-21] zirconium(IV),[20-1,33] hafnium(IV),[21] niobium(IV),[22] molybdenum(IV),[23] cobalt(III),[24-5] rhodium(III),[25-6] iridium(III)[24-5] nickel(II),[27-8,32] palladium(II),[29-30] platinum(II),[21,31] and gold(III).[1,33] Yields were in general not very high. In some cases the molar ratio of alkylating reagent to transition metal,[25,27] as well as the reaction medium,[34] were found to be critical.

Alkylating reagents that have been used to prepare five-membered metallacycles include magnesiacyclopentane,[24,34] magnesiacyclopentene,[35] and 5,5-di-n-propyldibenzostannole.[36] No four-membered metallacycle has so far been prepared by the above procedure although one of the requisite reagents, i.e. 1,3-bis(bromomagnesium)propane, has been elegantly prepared.[37]

2.2   Cyclodialkylation by organic dihalides

This procedure has been used only for the preparation of a nickelacyclopentane complex:[38-40]

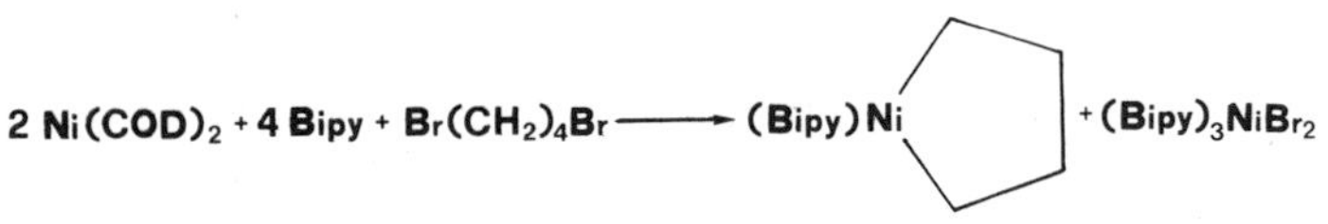

$$2\,Ni(COD)_2 + 4\,Bipy + Br(CH_2)_4Br \longrightarrow (Bipy)Ni\Big\langle\!\!\!\bigcirc + (Bipy)_3NiBr_2$$

The success of this reaction is strongly related to the ability of starting complex to undergo oxidative addition of organic halides, as is the case with nickel(0) derivatives.

2.3 Reaction of alkenyl complexes with nucleophiles

Anionic nucleophiles can attack at the unsaturated carbon atoms of alkenyl ligands thus causing a reaction of the following type to occur:

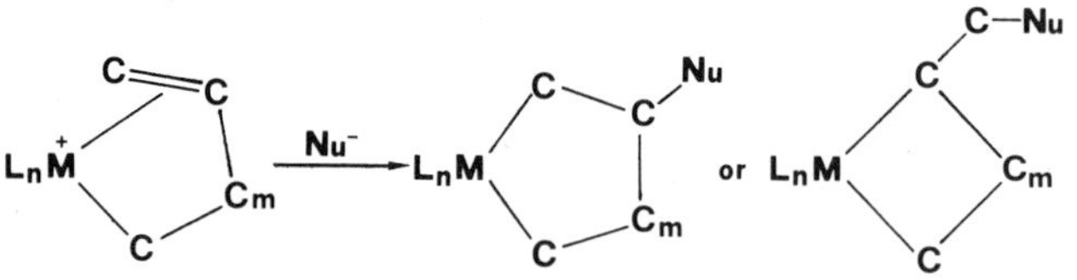

The ring size of metallacarbocycle that forms clearly depends on the value of $\underline{m}$ and on the regiospecificity of the attack. Four-, five-, and six-membered transition metal heterocycles were prepared by this way.

Green et al.[41-2] found that η^3-allyl complexes of the type **I** reacted readily with $NaBH_4$ or organometallic compounds such as alkyllithium and allylmagnesiumchloride, undergoing nucleophilic attack by H^- or $CH_2=CH-CH_2^-$ on the central carbon of the allylic group. This led to metallacyclobutanes of constitution **II**, in high yields:

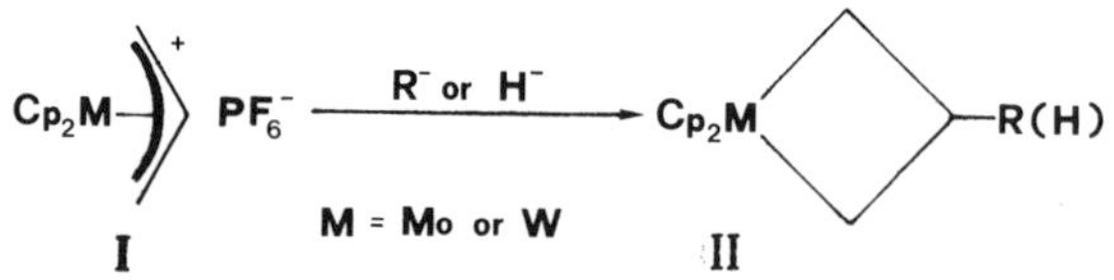

Similarly, cationic platinum(II) complexes, such as **III**, were found to react with sodium methoxide to give platinacycles such as **IV**:[43]

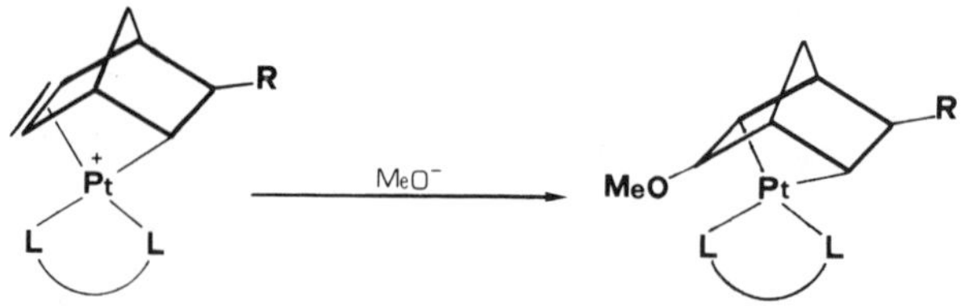

Similar products were obtained on treating complexes of the type [Pt(alkenyl)Cl(PPh$_3$)] with NaOMe in the presence of excess PPh$_3$, which presumably displaces chloride ion to form a cationic complex which then undergoes rapid nucleophilic attack.[43]   This is consistent with the observation that t-phosphines induce nucleophilic addition of supporting β-diketonato or acetato ligands to the coordinated double bond in cyclooctenyl[44] and norbornenyl[45] platinum complexes, thus giving rise to six- and five-membered metallacycles, respectively.

2.4 Ring-opening addition of carbocycles

Metallacycles can arise from reactions of carbocycles with transition metal compounds as a result of the insertion of a L_nM residue into a carbon-carbon bond:

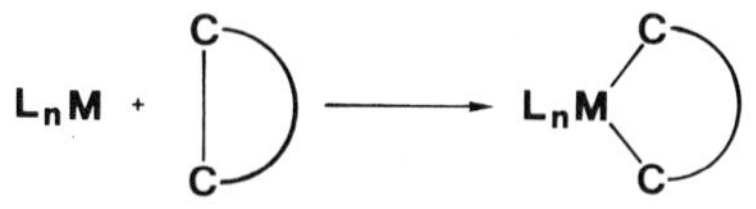

Some of these reactions have been reviewed by Rossi,[46] who covered the literature up to 1971, and more recently by Puddephatt.[18]

Carbocycles that have been used included cyclopropanes and cyclopropenes as well as cyclobutanes and cyclobutenes. The reaction of cyclopropane itself with chloroplatinic acid was first described by Tipper[2] as giving a complex of composition $PtCl_2(C_3H_6)$ which was subsequently shown[3-6,48] to be tetrameric and to contain a platinacyclobutane moiety. A number of analogous compounds, such as V,[49] VI,[50] and VII[51] have since been prepared, usually from the reaction of alkyl or aryl substituted cyclopropanes with Zeise's salt dimer.

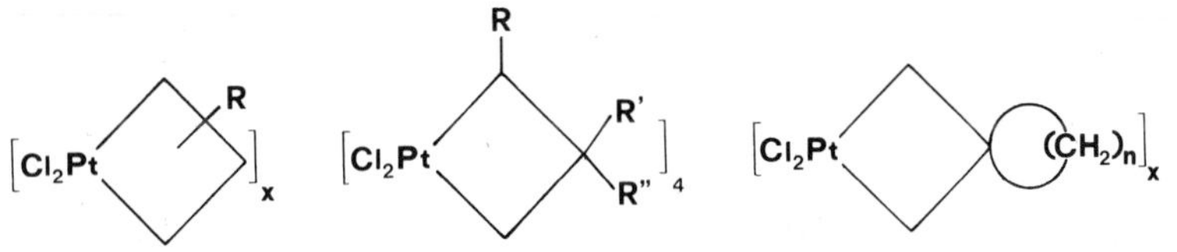

V   VI   VII   n = 0, 2, 3

In the formation of these complexes there appears to be a general preference for fission of the less substituted carbon-carbon bond; however, due to the equilibrium[52] the isolated platinacyclobutanes may represent the outcome of thermodynamic rather than kinetic reaction control.[49]

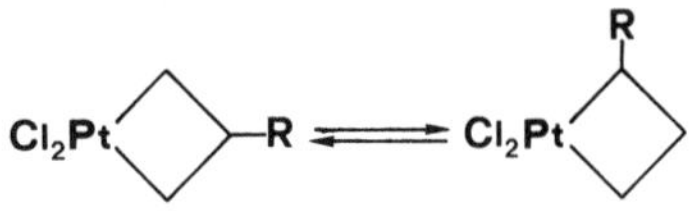

Cyclopropanes containing electron-withdrawing substituents failed to react.[49] On the other hand, the presence of a partial positive charge on carbon atoms of electronegatively substituted three- and four-membered carbocycles seems to be the driving force of ring-opening reactions of these substrates with nucleophiles such as palladium(0) and platinum(0) complexes,[53-60] or cobalt(I), rhodium(I) and iron(0) derivatives,[61] leading to four- and five-membered metallacycles.

Various metallacyclobutenyl complexes have been prepared via ring-opening of cyclopropenium ions by rhodium(I),[62-3] iridium(I),[64] palladium(0),[65] and platinum(0).[65] The bonding within these interesting units is not obvious. Frisch and Khare,[62] in discussing the structure of the rhodacyclobutenyl moiety in $[RhCl_2(PMe_2Ph)_2(C_3Ph_3)]$, suggested, in a simplified valence-bond description of the Ph_3C_3 unit, a hybrid including the following canonical forms:

The related reactions of $[Rh_2(CO)_4Cl_2]$ with cyclopropanes as well as with other compounds incorporating the cyclopropane system, led to a variety of rhodacyclopentanone complexes of type **VIII**.[66-9] Interestingly, from the reaction of some dibenzosemibullvalenes, Johnson, Lewis and

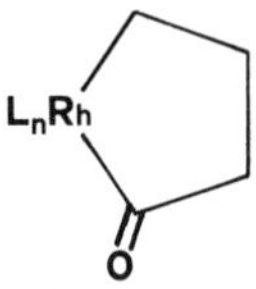

VIII

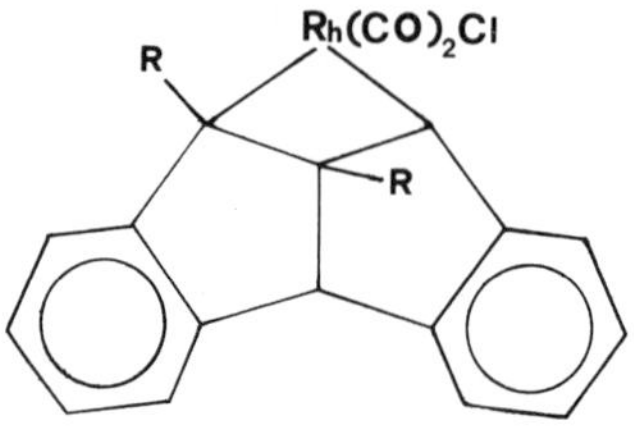

R = H or CO$_2$Me

IX

Tam[70] succeeded in obtaining stable four-membered rhodacycles **IX** which underwent CO insertion by further reaction with CO, thus providing evidence to support the proposed[67,71] intermediacy of rhodacyclobutanes in the formation of the metal-acyl complexes **VIII**.

$Fe_2(CO)_9$ reacted analogously with quadricyclene,[72] dibenzosemibullvalene,[73-4] or naphtho[b]cyclopropane,[75] yielding ferracyclopentanones. Light-induced reaction of hydrocarbons of type **X** with $Fe(CO)_5$ yielded ferracycles **XI**:[76]

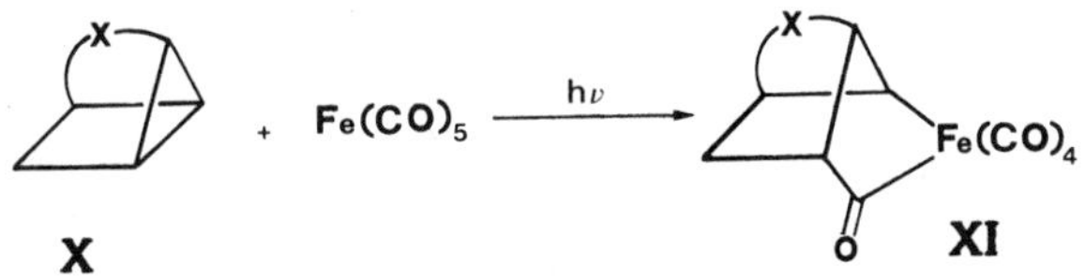

Finally, a number of rhodacyclohexanones have been obtained by reacting $[Rh_2(CO)_4Cl_2]$ with strained carbocycles containing fused cyclobutane rings.[10,77-8] $[Ir_2(CO)_6Cl_2]$ and 1,3-bishomocubane yielded the iridacyclopentane **XII**.[80]

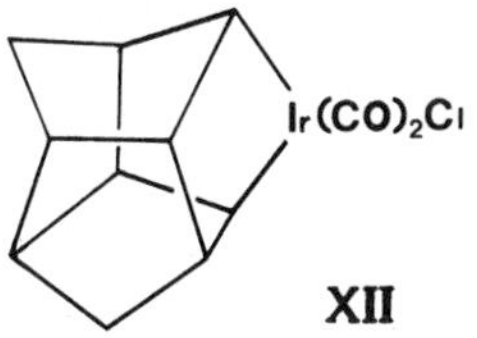

2.5   Thermal decomposition of dialkyl complexes

The thermal decomposition of transition metal dialkyl complexes via loss of alkane can lead to four- and five-membered metallacycles by intramolecular γ- or δ-hydrogen abstraction respectively.

The literature refers to neopentyl and trimethylsilylmethyl derivatives of ruthenium,[81] rhodium,[81] and platinum,[82-3] as well as some di-o-methylbenzyl platinum complexes.[84] For example:

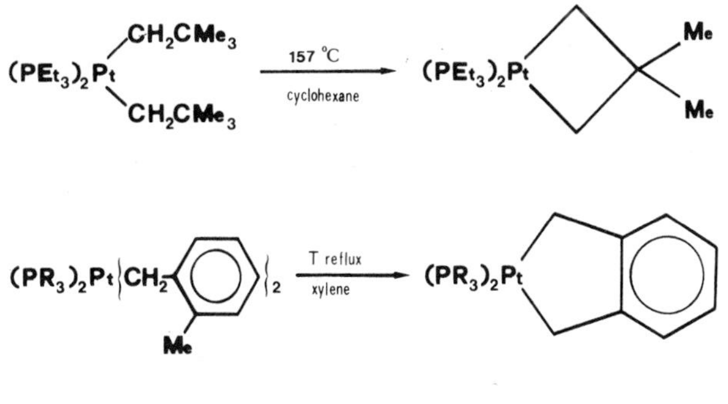

On the basis of kinetic and deuteration studies, Whitesides et al.[83] concluded that the reaction of the above dineopentyl complexes proceeded by an intramolecular oxidative addition of a C -H bond of one neopentyl group to platinum(II) followed by (1,1)-elimination of neopentane:

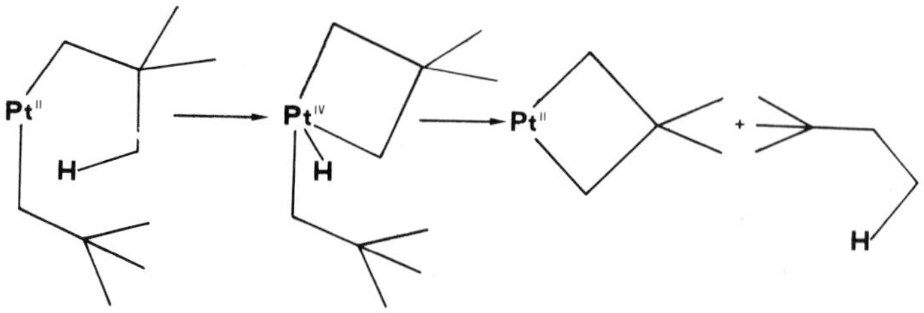

Thus, the overall process is a (1,4)-elimination, whose established pathway makes real some of the brilliant 'idealizations' put forward by Braterman and Cross.[89]

2.6   Cycloaddition reactions

Cyclization reactions of unsaturated molecules with transition metals constitute one of the most exciting paths to metallacarbocycles not only because of their synthetic importance - metallacycles are indeed obtained, usually in high yields - but also because they are the key reactions by which unsaturated molecules undergo transformations, either catalytic or stoichiometric, via metallacyclic intermediates.

These reactions have been arranged in three groups according to the skeletal constitution of the emerging metallacyclic rings.

2.6.1 Metallacyclobutanes and metallacyclobutenes

Very interesting examples have recently been given of the formation of these rings from the reaction of metallacarbene complexes with olefins[90] or acetylenes:[91]

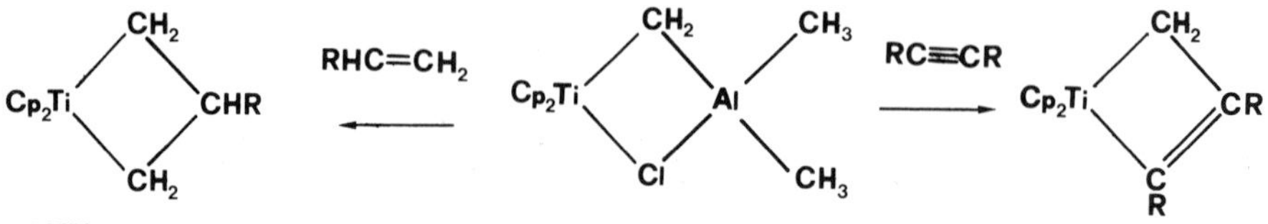

XIII and **XIV** both underwent exchange reaction with olefins or acetylenes, thus forming new four-membered metallacycles.

Compounds containing a metallacyclobutene ring have been obtained from reactions of acetylenic compounds with carbonyl[92] or isontrile[93-4] complexes.

2.6.2 Metallacyclopentanes and metallacyclopentenes

Metallacyclopentane rings result from cycloaddition reactions of the following type:

$$M + 2 = \longrightarrow M$$

Stone,[7-8] Wilkinson[9] and their co-workers had obtained such compounds by 1961, from the reaction of perfluorethylene with iron and cobalt carbonyls. Subsequently, there have been many investigations concerning reactions of low-valent transition metal complexes with olefins carrying electronegative substituents, and a number of five-membered metallacyclic derivatives of iron(II),[95-9] nickel(II),[100-1] rhodium(III),[102] platinum(II),[103-4] iridium(III),[105] and cobalt(III)[141] have been prepared. However, the most important application of such reactions to date has centered on the cyclization of CH olefins. Evidence for the formation of metallacycles from these substrates was first provided by the isolation of the irida-, rhoda-, and ruthena-cyclopentane complexes **XV-XVII**, deriving from norbornadiene (norb),[106] propadiene,[107-10] and o-styryldiphenyl-phosphine,[111] respectively. Other cyclizations of two propadiene molecules with nickel,[112] platinum,[113] or zirconium[114] have been described.

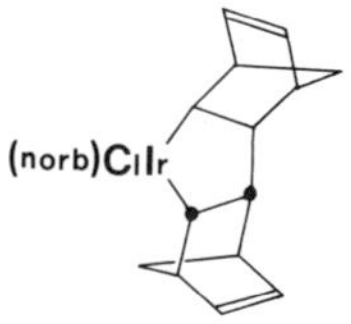

$(norb)ClIr$

$(acac)R_yLM$

M = Rh or Ir
L = Py or C_3H_4

$(CO)_2Ru$

XV 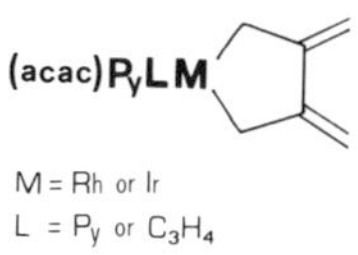   **XVI** 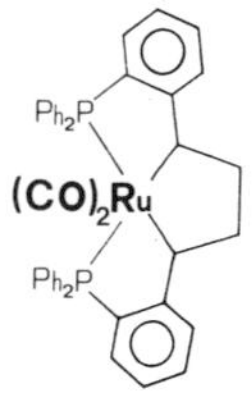   **XVII**

Most interestingly, Whitesides et al.[20] reported that several equivalents of titanocene reacted with olefins, such as ethylene, 1,7-octadiene, norbornene, and benzonorbornene, to form titanacyclopentane rings. Although pure samples were not isolated, this study provided the basis for a very versatile route to a variety of metallacarbocycles. Thus Grubbs and Miyashita[115] succeeded in isolating a mixture of (E)- and (Z)-**XVIII** (3:1) from the following reaction:

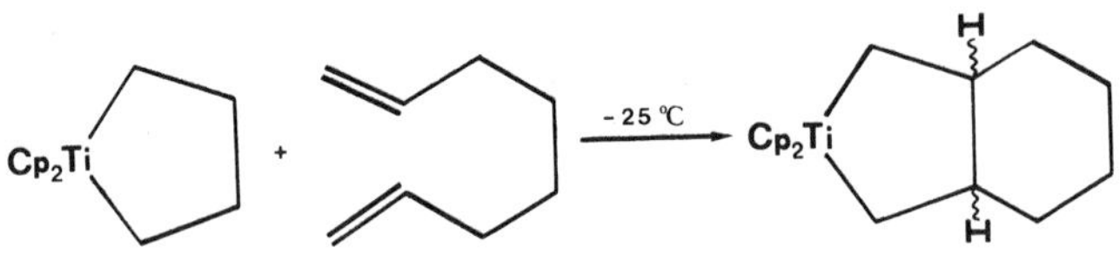

XVIII

A similar zirconium complex was obtained from $Cp_2Zr(t\text{-phosphine})_2$,[116] while the zirconacyclopentane $(\eta^5\text{-}C_5Me_5)_2\overline{ZrCH_2(CH_2)_2CH_2}$ was prepared by cyclization of two ethylene molecules with either $[(\eta^5\text{-}C_5Me_5)_2ZrN_2]_2N_2$ or $(\eta^5\text{-}C_5Me_5)_2ZrH_2$.[117]

A number of nickelacyclopentanes have been obtained by reacting nickel(0) complexes with either strained olefins,[38,118-9] 1,7-octadiene,[120] or ethylene.[121]

Several (E)-3,4-disubstituted metallacyclopentane complexes of type **XIX** were the ultimate organometallic products from reactions of α-olefins with either $(\eta^5\text{-}C_5H_5)Ta(CHCMe_3)Cl_2$ or $(\eta^5\text{-}C_5Me_5)Ta(olefin)Cl_2$.[122] No observable metallacycle was formed by styrene, (Z)-2-butene, or cyclooctene. Mixed ethylene/$RCH=CH_2$ metallacycles were obtained on adding $RCH=CH_2$ to $(\eta^5\text{-}C_5Me_5)Ta(ethylene)Cl_2$.[122] α,ω-dienes underwent intramolecular cyclization with $(\eta^5\text{-}C_5Me_5)Ta(cyclooctene)Cl_2$: 1,6-heptadiene and 1,8-nonadiene stereoselectively gave rise to (Z)-**XX** (n = 3) and (E)-**XX** (n = 5) respectively, while 1,7-octadiene gave rise to a 1:2 mixture (at 37°C) of (Z)- and (E)-**XX** (n = 4).[123]

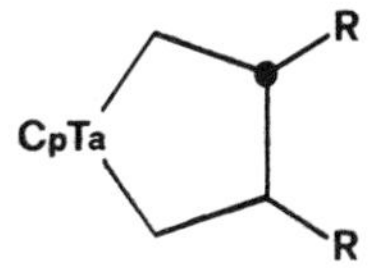

XI   **XX**

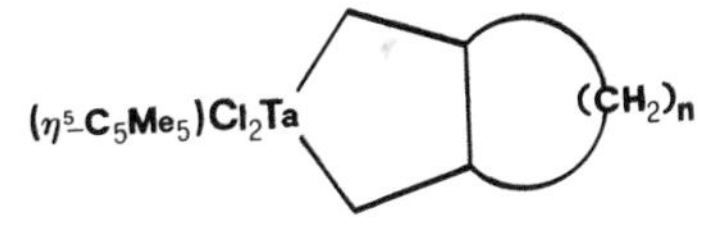

Metallacyclopentene rings have been obtained by co-cyclization of olefins and acetylenes.[124-5] Moreover, various metallaindanes were prepared by reacting dehydrobenzenezirconocene (generated from diphenylzirconocene) with olefins:[88]

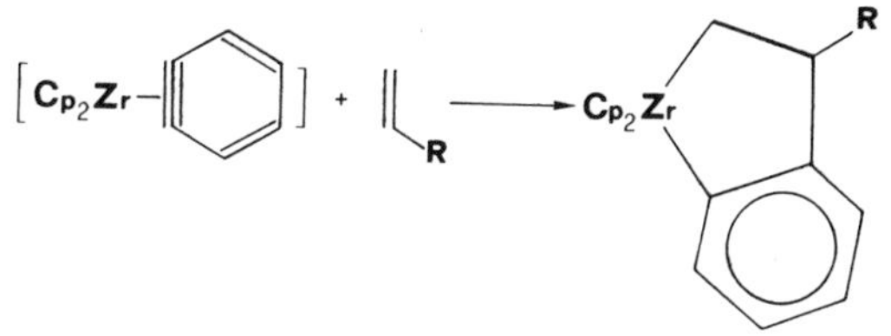

The related reactions of $Cp_2Ti(C_6H_5)_2$ with acetylenes led to titanaindenes.[85-7] Metallacyclopenten-2,5-diones[127-32] and metallacyclopent-3-enes[113,133-7] have also been obtained by cyclization reactions.

The high stereoselectivity usually exhibited by such reactions deserves comment. In this connection a recent theoretical paper by Stockis and Hoffmann[13] is relevant. They explored the reaction path for the bis(ethylene)-metallacyclopentane interconversion in the specific case of two olefins coordinated to a trigonal bipyramidal iron carbonyl. This study concluded that, in the case of unsymmetrically substituted olefins, the cyclization reaction should proceed stereoselectively placing the substituents so that the olefin antibonding orbital enters the reaction with its largest lobe in the β position in the metallacycle formed. Thus, consideration of bonding between olefin C=C orbitals and appropriate metal orbitals should allow one to predict which of the three possible five-membered rings, **XXI-XXIII**, would be mainly formed by asymmetric ethylenes, A=B.

The great majority of the above reported examples is, indeed, in accord with predictions from the theory. Apparent exceptions may easily be accommodated by invoking steric effects.

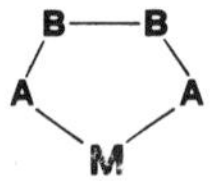

XXI 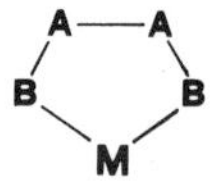 XXII 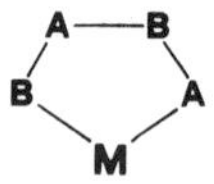 XXIII

2.6.3 Miscellaneous metallacycles

The ferracycle **XXIV** was formed from the photochemical reaction of perfluoroethylene with tricarbonyl(norbornadiene)iron.[96] Hexafluor-2-butyne reacts with $[Rh(1,5-COD)Cl]_2$[139] and $[Rh(norb)Cl]_2$[140] giving rise to interesting complexes containing the rhodacyclic moieties **XXV** and **XXVI**, respectively.

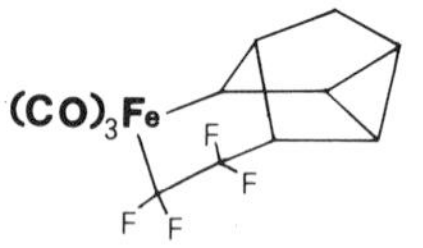

XXIV 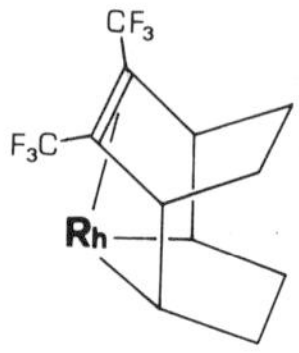 XXV 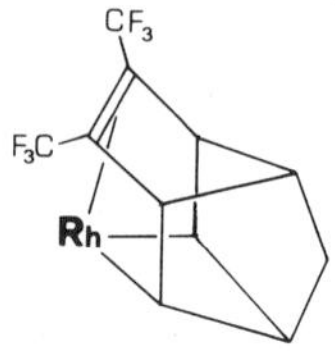 XXVI

3. REACTIONS

In the Sections that follow several reactions will be examined with emphasis on transformations peculiar to metallacycles. Although no unified view of the reactivity of these compounds has yet emerged, a rather clear rationalization of some reactions has been given, which allows a better understanding of the role of metallacycles in organometallic chemistry and catalysis.

As for metallacyclobutane reactions, these have been recently reviewed[18] and will not be examined in detail.

3.1 Thermolysis

Thermolytic processes may essentially implicate: (i) the formation of carbocycles; (ii) the formation of olefins having the same carbon number as the metallacyclic group; (iii) the formation of olefins having a carbon number lower than that of the metallacyclic group. Such events can occur in parallel, although some very selective processes are known. Thus thermal decomposition of some metallacyclopentane derivatives of titanium(IV) and zirconium(IV) was reported by Whitesides et al.[20] to yield mixtures of ethylene and 1-butene, whose composition depended markedly on reaction temperature:

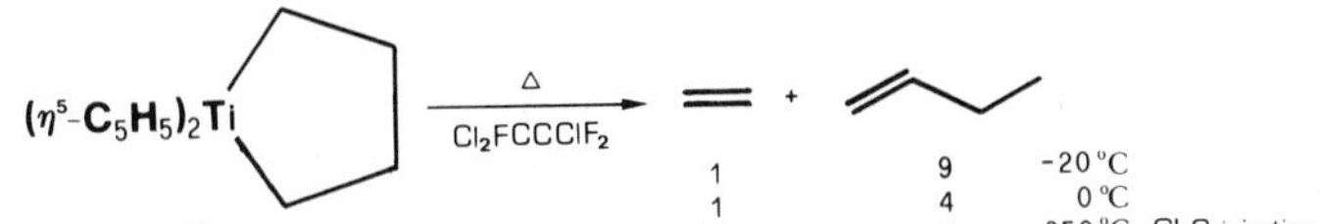

$(\eta^5\text{-}C_5H_5)_2Ti$ [cyclopentane ring] $\xrightarrow[Cl_2FCCClF_2]{\Delta}$ [ethylene] + [1-butene]

	1	9	-20 °C
	1	4	0 °C
	20	1	250 °C, GLC injection

XXVII

Interestingly, the addition of a methyl group in an α position of the metallacyclic ring, or the replacement of C_5H_5 groups by more bulky C_5R_5 ligands, both caused the new compounds to yield, if anything, less ethylene than did **XXVII**.[20]

An important point was the observation by Grubbs and Miyashita[142] that the 2,2,5,5-tetradeuterated isomer of **XXVII** underwent rapid skeletal isomerization at temperatures between -60°C and -45°C, forming the 2,2,4,4- and 3,3,4,4-d_4 isomers. This strongly suggested that **XXVII** decomposed by reversible C-C bond cleavage to produce an intermediate bis(ethylene) complex (Scheme 1).

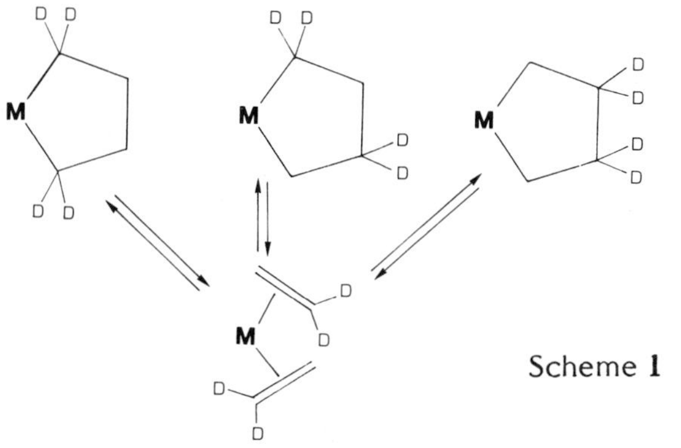

Scheme 1

On going from **XXVII** to the homologous six-membered titanacycle, the products of thermal decomposition always consisted of n-pentenes along with minor amounts of cyclopentene.[20] In contrast, the titanacyclobutane **XXVIII** gave mainly skeletal fragmentation products (t-butylethylene and methane), when heated in toluene at 85°C.[90]

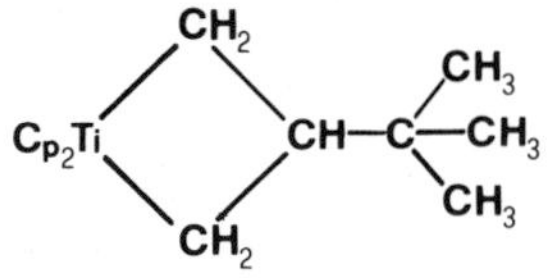

XXVIII

Metallacyclobutane derivatives of tungsten(IV) exhibited a different behavior: thermal decomposition (at ca. 80°C) yielded mixtures of alkenes and alkane having the same carbon number as the starting ring; on irradiation, skeletal fragmentation products were also formed.[41]

The influence of the supporting ligands, the metal, and the solvent on reaction pathways has emerged from studies of the thermal decomposition of a number of metallacycloalkane derivatives of the nickel triad metals. For nickel, Grubbs et al.[27] found that bis(phosphine)nickelacyclopentanes could be converted to mono(phosphine) or tris(phosphine) complexes by controlling the phosphine structure and/or concentration. Thus, it was very interestingly shown that complexes of different coordination number gave different thermal decomposition products with a very high selectivity, as illustrated by the following equations:

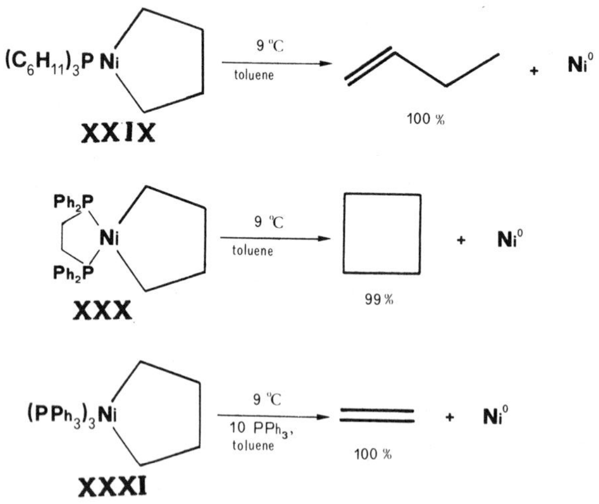

It is of interest that while **XXX** selectively underwent (1,1)-elimination, its palladium[29] and platinum[31] analogues both gave linear C-4 open chain olefins only, on thermal decomposition, at 95°C and 120°C, respectively, and (bipy)NiCH$_2$(CH$_2$)$_2$CH$_2$ yielded almost equal amounts of cyclobutane and butenes at 165°C in mesitylene.[39] On the other hand photolysis of **XXX** at -20°C resulted in high yields of ethylene (59%).[27] As far as the thermal reaction of **XXXI** was concerned, evidence for an equilibrium between a bis(ethylene) complex and a nickelacyclopentane (Scheme 1) came[142] from deuterium-labeling experiments analogous to those previously illustrated for titanacyclopentanes. Such an equilibrium also seemed to be responsible for cis-trans isomerization in 2-nickelahydrindane (Scheme 2);[120] indeed, isomerization occurred in those conditions which favored C-C cleavage, i.e. in the presence of excess ligand or light.

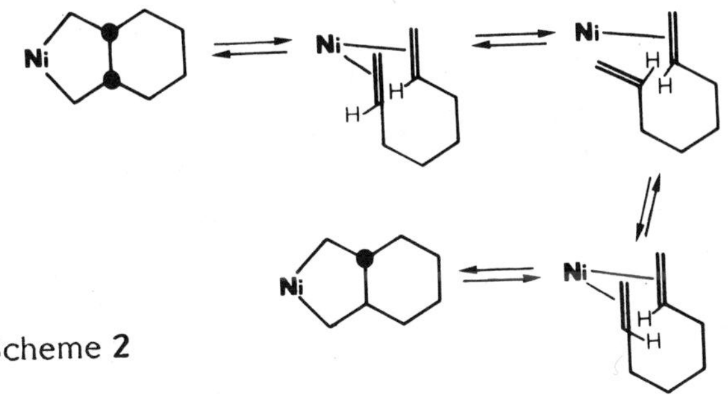

Scheme 2

The effects of supporting ligands on nickelacyclopentane thermolysis were shown by Braterman[12] to be foreseeable consequences of frontier orbital symmetry correlations, by which the (1,1)-elimination from **XXX** and the cheletropic cleavage to ethylenes for the (e, e) trigonal bipyramidal isomer of **XXXI** are both allowed processes, while the formation of ethylene from **XXX**, or of cyclobutane from **XXXI**, are forbidden.

It is noteworthy that those conditions which resulted in carbon-carbon bond cleavage in nickelacyclopentanes also gave high yields of fragmentation products in a nickelacyclohexane, probably via α-C-C bond cleavage:[28]

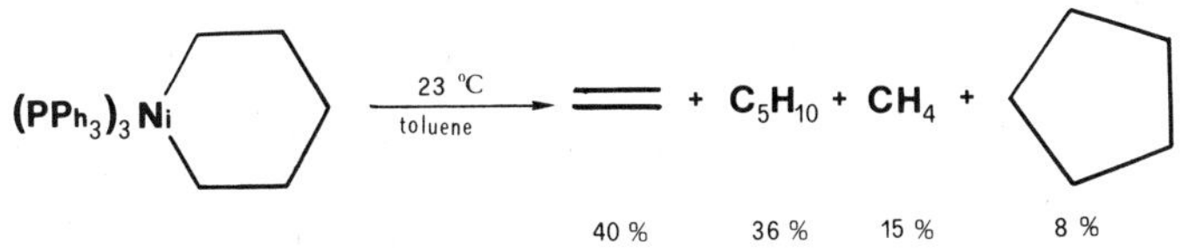

The thermal decomposition of platinum(II) metallacycloalkanes **XXXII** has been investigated in detail by Whitesides et al.[31,143] who have pointed out the important role played by the supporting ligands as well as by the reaction medium in determining the reaction pathway. In fact, while thermolysis of $L_2\overline{PtCH_2(CH_2)_nCH_2}$ (**XXXII**) (L = PPh_3, n = 2,3,4), in CH_2Cl_2 at 120°C, yielded only C-4, C-5, and C-6 open chain olefins respectively, whose composition altered neither on changing the solvent nor on adding free phosphine,[31] thermolysis of **XXXII** (L = PBu_3, n = 2,3) yielded products whose nature depended on the reaction medium. n-Alkanes having the same carbon number as the starting metallacarbocycle were formed from reactions carried out at 120°C in ethers or hydrocarbons, while cycloalkanes and alkenes having the same carbon number as the metallacyclic rings along with homologous cycloalkanes and alkenes were formed when thermolysis was carried out in CH_2Cl_2 or CH_2Br_2.[143] These results were discussed in terms of competing reaction paths (Scheme 3) according to which (1,1)-elimination could occur readily from platinum(IV) but not from platinum(II) dialkyls, the chemistry of the latter being dominated by β-hydride elimination in the absence of competing oxidative addition processes.[143] The suggested behavior of intermediate **XXXIII** (Scheme 3) was strongly supported by the thermal decomposition of model compounds.[143] Furthermore, Braterman[12] showed, from orbital symmetry correlations, that

Scheme 3

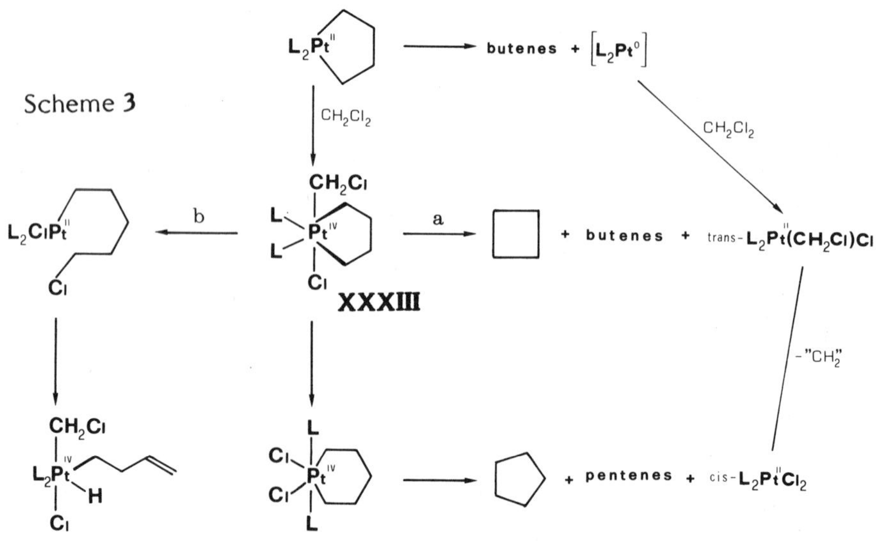

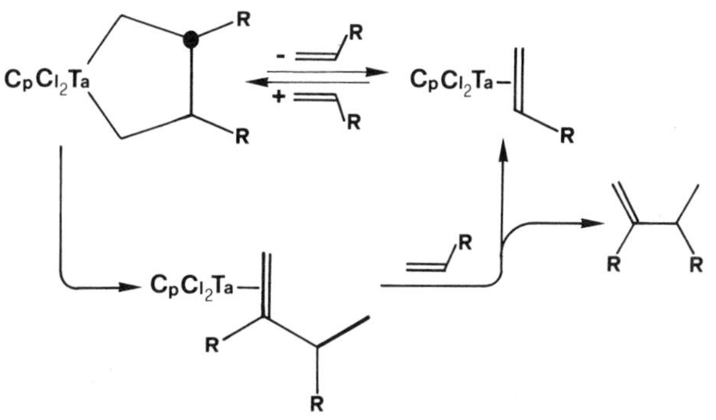

Scheme

the reductive eliminations a and b (Scheme 3) are both allowed for **XXXIII**
when L is, for example, PBu_3, but that only transformation b is energetically
feasible when L_2 is a bidentate ligand such as $Me_2PCH_2\ CH_2PMe_2$ (dmpe).
Indeed, $(dmpe)\overline{PtCH_2(CH_2)_2CH_2}$ yielded butenes and C-5 hydrocarbons, but
not cyclobutane, in CH_2Cl_2 at $120^\circ C.$[143]

Photolysis of bis(phosphine)platinacyclopentanes in CH_2Cl_2, at $25^\circ C$,
yielded ethylene as the major product along with C-4 linear hydrocarbons
and traces of C-5 hydrocarbons.[144]

Another important contribution to the understanding of mechanisms
for thermal decomposition of metallacyclopentanes came from studies by
Schrock et al.[122,145] on some tantalum(V) derivatives. 3,4-Disubstituted
metallacyclopentanes were shown to decompose in toluene, at $25^\circ C$, to give
the corresponding 2,3-disubstituted 1-butene (0.5 mol per tantalum) and
olefin complex quantitatively, consistent with the transformations outlined
in Scheme 4.[122] Accordingly, the ethylene/α-olefin metallacyclopentanes
were shown first to disproportionate to the more stable ethylene
metallacycle and the α-olefin metallacycle, and then to decompose to give
mainly the ethylene complex and the corresponding 2,3-disubstituted 1-
butene, the overall process occurring without deuterium scrambling in an α-
dideuterated metallacycle:[122]

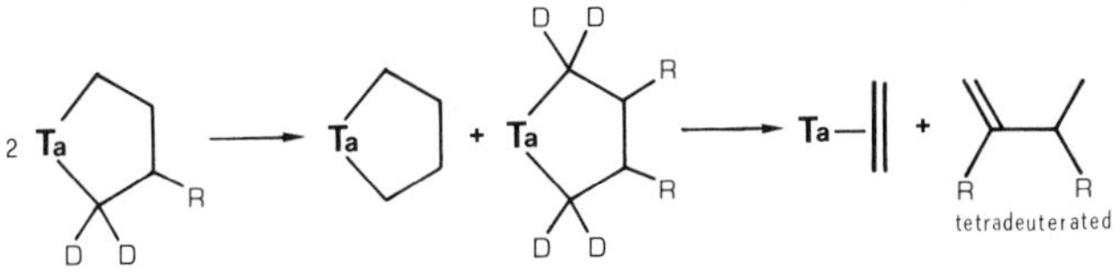

Scheme 5

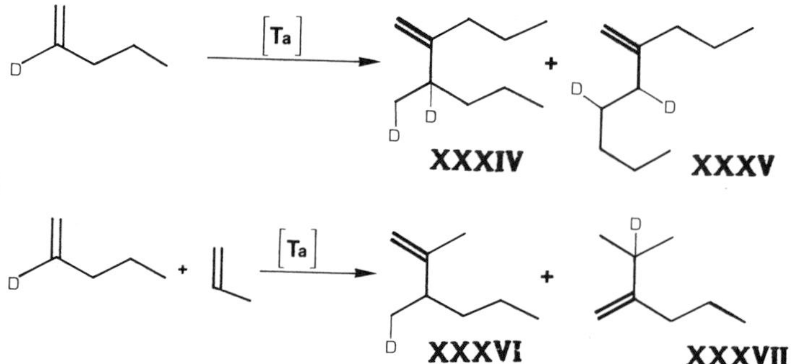

It was pointed out[146] that the observed facile β,β'-carbon bond cleavage might be accounted for by the unusual "opened-envelope" conformation of the TaC$_4$ ring.

As far as the formation of 2,3-disubstituted-1-butenes was concerned, the following mechanistic pathway implicating a metallacyclopentane-metallacyclobutane ring contraction was suggested on the basis of deuterium labeling studies:[145]

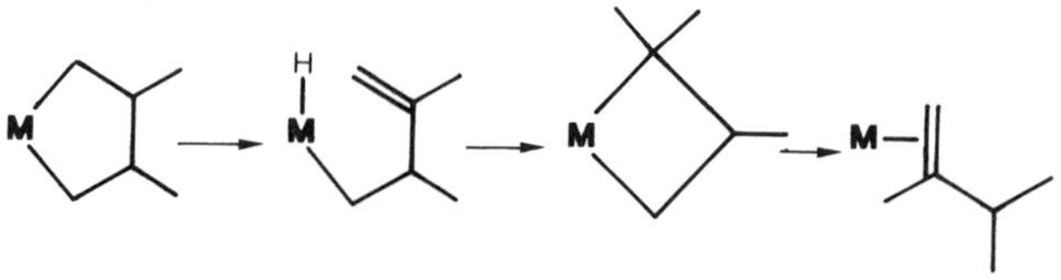

In particular, it was found that tantalum-catalyzed dimerization of 2-deuterio-1-pentene and codimerisation of propylene with 2-deuterio-1-pentene produced the C-10 olefins **XXXIV** and **XXXV** and the C-8 olefins **XXXVI** and **XXXVII** respectively (Scheme 5). While the formation of **XXXIV** could well be explained by a β-hydrogen elimination-reductive elimination reaction sequence:

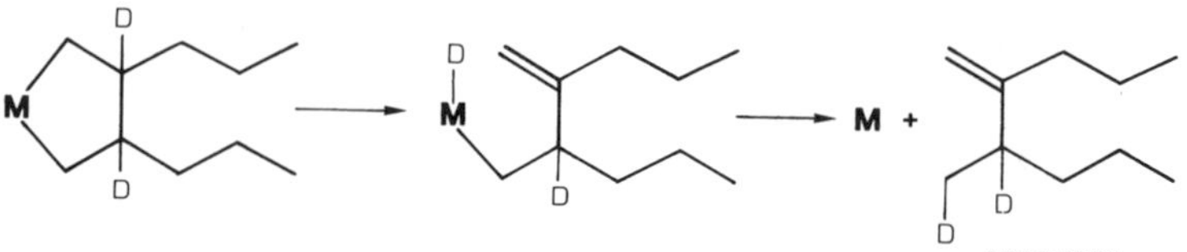

compounds **XXXV-XXXVII** were not consistent with such a reductive elimination pathway, but with the previously outlined ring-contraction pathway, head-to-tail dimer **XXXV** being formed by an initial secondary β-hydrogen elimination.[145]

I would like to point out that, although the proposed mechanism may be correct, all the above deuterated dimers and codimers are exactly those expected if the following alternative pathway, implicating a more or less concerted fast isomerization of an intermediate butenyl hydride to a σ-allyl hydride, were operating:

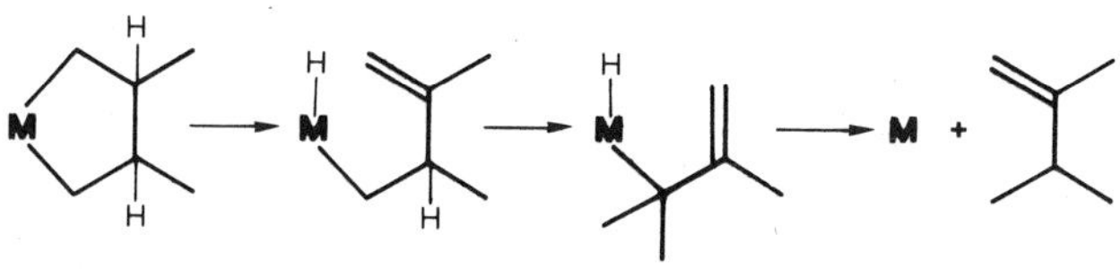

Accordingly, the head-to-tail dimer **XXXV** should be formed in the following way:

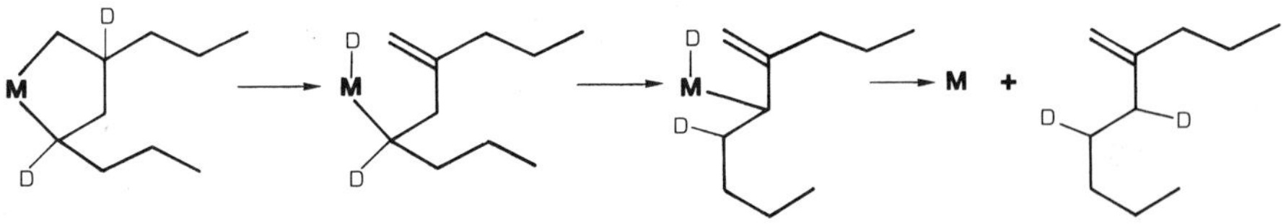

XXXV

This interpretation, though somewhat speculative, is not without foundation. In its support I should like to cite the reaction of metallacyclopentanes with the trityl cation[30] (see Section 3.3.2) as well as the following isomerisation of a butenyl-iron system:[147]

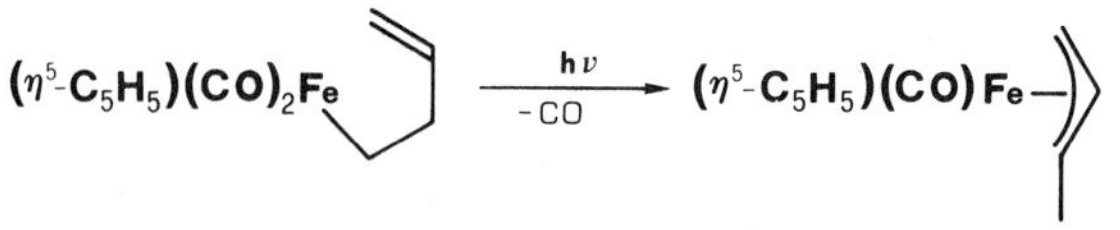

$$(\eta^5\text{-}C_5H_5)(CO)_2Fe \xrightarrow[-CO]{h\nu} (\eta^5\text{-}C_5H_5)(CO)Fe$$

Furthermore, this hypothesis lacks some of the disadvantages of the ring-contraction mechanism; thus, it does not require that head-to-tail dimer forms via an initial secondary β-hydrogen abstraction, but in all cases a tertiary β-hydrogen is initially eliminated. Furthermore, the mechanism of formation of 2,3-disubstituted 1-butenes from 2,2,3-trisubstituted platinacyclobutanes was documented by Johnson and Cheng[148] who showed, by deuterium labeling studies, that two types of β-hydrogen abstraction were operative, i.e. a β-hydrogen abstraction from the ring and a β-hydrogen abstraction from an α-methyl group, the former being favored. Indeed Schrock et al.[145] observed products only explainable on the base of the former path. However, any parallelism between platinum and tantalum may well be unfounded, at least in this context.

Finally, the solid-state thermal decomposition of five-, six-, and seven-membered metallacycles of the type $(\eta^5\text{-}C_5Me_5)\text{-}\overline{MCH_2(CH_2)_nCH_2(PPh_3)}}$ (M = Rh, Ir) has been studied by differential scanning calorimetry and thermogravimetric analysis, and shown to yield mixtures of the corresponding n-alkenes.[149] This study also allowed the first estimates of the mean Rh–C and Ir–C bond dissociation energies. An analogous study had previously been carried out for platinacyclobutanes.[150]

3.2   Reductive elimination by added molecules

(1,1)-Elimination reactions of four-membered platinacycles have been known since 1955,[2] and extensively studied.[18] Evidence for such reactions in metallacyclopentanes was first provided by Fraser et al.[106] who obtained the exo-trans-exo norbornadiene dimer from the reaction of **XV** with a five-fold excess of PPh$_3$, in refluxing CHCl$_3$:

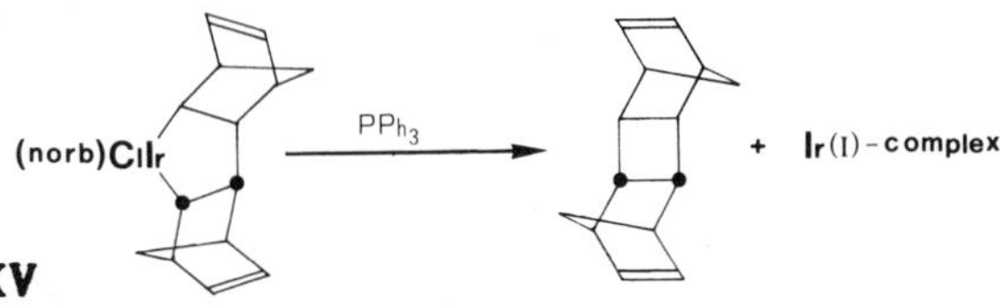

$$(norb)ClIr \quad \xrightarrow{PPh_3} \quad + \quad Ir(I)\text{-complex}$$

XV

Analogously, rhodacyclohexanones readily underwent reductive elimination of cyclopentanones on reaction with PPh$_3$.[10,77] However, **XXXVIII** reacted differently with PPh$_3$ yielding 3-cyclohexene-1-carboxaldehyde:

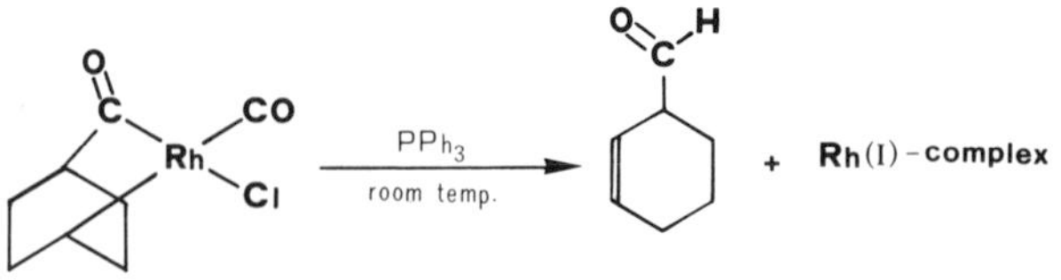

$$\xrightarrow[\text{room temp.}]{PPh_3} \quad + \quad Rh(I)\text{-complex}$$

XXXVIII

More recently, nickel(II) metallacyclopentanes have been shown to eliminate cyclobutanes on addition of a variety of agents, such as O$_2$,[16,27,151] cerium(IV) ammonium nitrate,[27] tetracyanoethylene,[27,38] maleic anhydride,[39,151] acrylonitrile,[27] methylacrylate,[39] p-benzo-quinone,[40] dimethylfumarate,[40] and hydrocarbonic olefins.[152] Both the selectivity and the yield of these reactions were generally high.

3.3 Reactions with protic acids and halogens

These reactions have in practice served as additional evidence for the presence of metallacyclic ring in a given complex. In fact, protic acids readily cleave the metal-carbon bonds of a variety of metallacycloalkanes yielding the corresponding n-alkanes along with variable amounts of n-alkene. Similarly, reaction of five-, six-, and seven-membered metallacycloalkanes with bromine resulted in the formation of the corresponding α,ω-dibromo organo-compounds.

Platinum(II) metallacycles reacted with halogens in a different manner: bis(triethylphosphine)-3,3-dimethylplatinacyclobutane mainly underwent (1,1)-elimination on reaction with Br_2 or I_2,[83] while bis(dimethyl phenylphosphine)platinacyclopentane[141] and bis(tri-n-butylphosphine) platinacyclohexane[143] both gave oxidative addition products on reaction with I_2 and Br_2, respectively.

3.4 Reactions with Lewis acids

These reactions fall into four categories that are discussed in the sections below.

3.4.1 Reactions involving displacement of the metal

Reactions of this type were reported by Binger and Doyle,[151] who found that complete replacement of nickel by aluminum occurred on reaction of **XXXIX** with $AlEt_3$ while a rearrangement-displacement reaction took place with boron trifluoride (Scheme 6). It seems likely that, in the

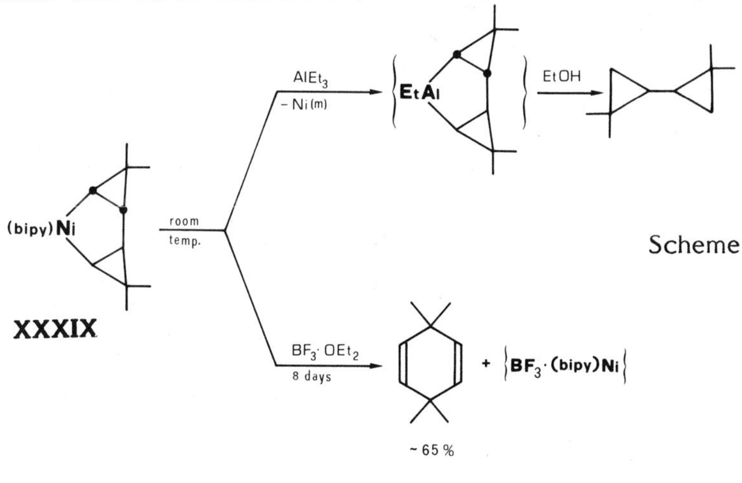

Scheme 6

latter reaction, the formation and/or the elimination of 3,3,6,6-tetramethyl-1,4-cyclohexadiene both occurred at the metal,[151] but mechanistic details are not obvious. Bipyridylnickelacyclopentane also reacted with $BF_3 \cdot OEt_2$; in this case, however, the reaction products were cyclobutane and C-4 open-chain olefins.[39]

3.4.2 Conversion to π-allyl complexes

The formation of π-allyl platinum complexes from platinacyclobutanes via an intramolecular β-hydrogen abstraction from the ring was discovered some years ago.[153-4] More recently it has been reported that palladacyclopentanes of type XL (L_2 = dppe, 2,2'-bipyridyl; L = PPh_3) readily underwent hydride abstraction by trityl cation, thus giving rise to cationic methylallyl complexes XLI, in good yields:[29-30]

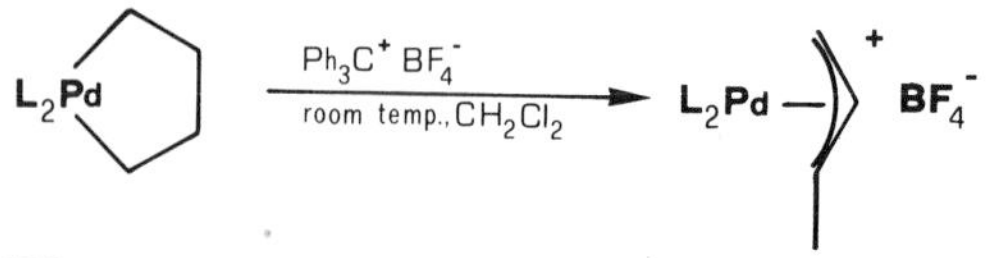

$$L_2Pd \xrightarrow[\text{room temp., }CH_2Cl_2]{Ph_3C^+ \; BF_4^-} L_2Pd \quad BF_4^-$$

XL   **XLI**

Rhodium(III) and iridium(III) metallacyclopentanes of type (η^5-C_5Me_5) $\overline{MCH_2(CH_2)_2CH_2}(PPh_3)$ reacted similarly with Ph_3C^+.[155] Both these rings exhibit two interesting structural features,[25] namely puckering of the metallacyclic ring so that one β-carbon is closer to the metal than the other, and the presence of some C-C bonds significantly shorter than the standard $C(sp^3)$-$C(sp^3)$, which resemble an early transition state for facile hydrogen release. The reaction mechanism, in fact, was assumed[30] to involve an initial β-hydride abstraction by the trityl cation (which has been supported by deuterium labeling experiments[155]) followed by a rapid isomerization of the postulated σ-but-3-enyl intermediate to the allylic complex:

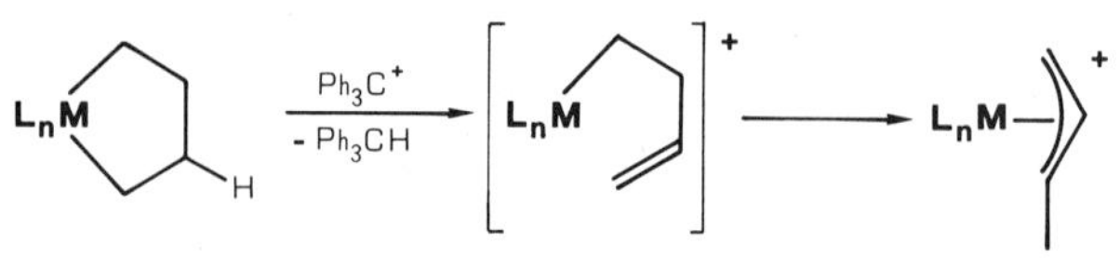

$$L_nM \xrightarrow[- Ph_3CH]{Ph_3C^+} \left[L_nM \right]^+ \longrightarrow L_nM^+$$

XLI

Such an isomerization has precedent in the related iron system (Section 3.1), and for two of the systems examined has also been documented by the following reactions from our laboratory:[155]

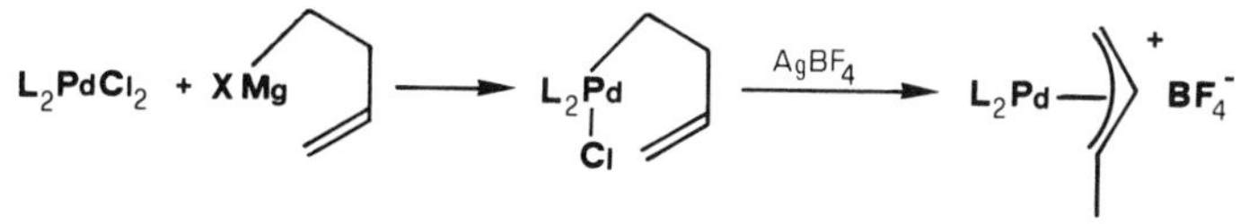

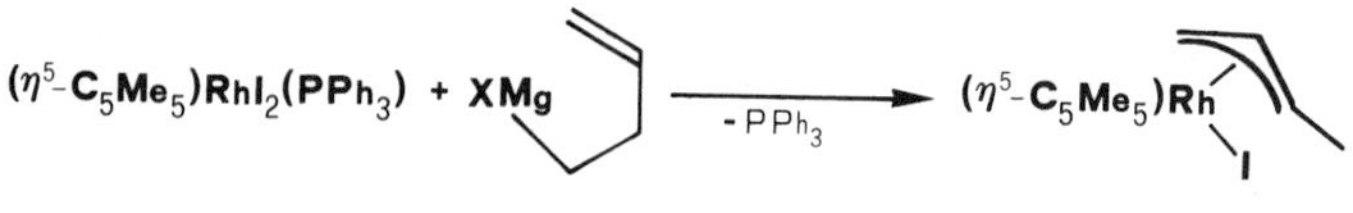

The isomerization mechanism remained uncertain.[30]

3.4.3 Insertion reactions

(Dppe)palladacyclopentane reacted rapidly with SO_2, at low temperatures, to yield the corresponding cyclic S-bonded disulfinate, thus providing the first example of insertion of two SO_2 molecules into a dialkyl palladium derivative.[30] However, it was shown that this reaction pattern was not peculiar to palladacycles: dimethyl- and dibutyl-palladium(II) complexes both reacted similarly with SO_2.[30]

3.4.4 Reactions involving C–C cleavage

Two very interesting examples of Lewis acid promoted metallacycle fragmentation have been briefly reported:[90-1]

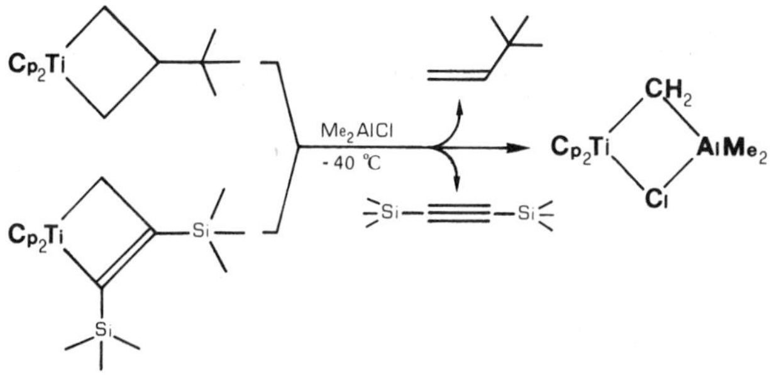

The facility of such reactions was ascribed to an easy electrophilic attack at the α-carbon by Me_2AlCl with the result that the metallacycle was activated for release of unsaturated molecules.[91]

3.5 Reactions with organic halides

Such reactions have been described for nickel(II) and platinum(II) metallacyclopentane derivatives. In the case of nickel compounds, alkyl halides[40] and gem-dihalides[16,40,51] both caused ready displacement of the metal with formation of cross-coupling products (30-70% yield), accordingly to the following general equations, along with variable amounts of cyclopentanes:

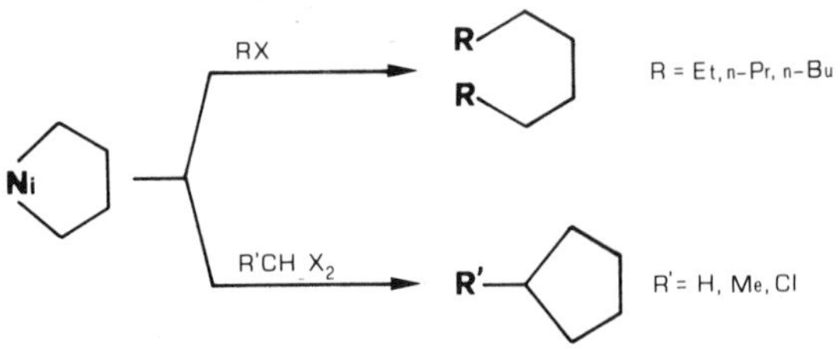

The best yields of cyclopentane derivatives were obtained by using dibromides or diodides.[40] Surprisingly, the octamethylene derivatives $RY(CH_2)_8YR$ ($Y = CO$ or SO_2) were formed from the reactions of (bipy) $\overline{NiCH_2(CH_2)_2CH_2}$ with acyl or sulfonyl halides, instead of the expected tetramethylene compounds.[40]

Platinum(II) metallacyclopentanes reacted with CH_2Cl_2 or CF_3I to give very stable oxidative addition products.[143]

3.6 Carbonylation

Five patterns have been distinguished for this reaction (Scheme 7).

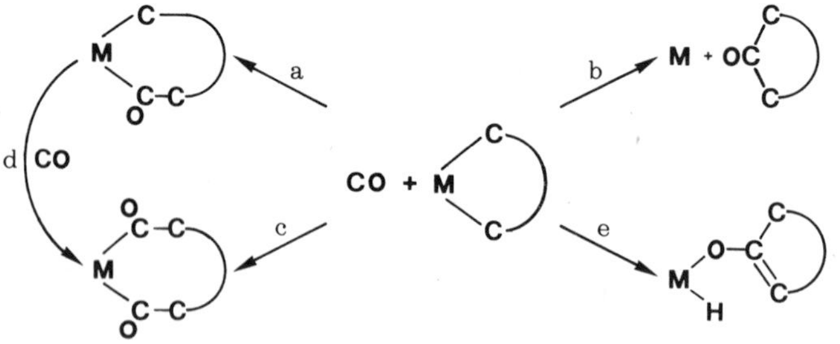

Scheme 7

Path (a) has been demonstrated for iridium(III)[106,182] and zirconium(IV)[156] five-membered metallacycles as well as for a rhodacyclobutane complex.[70] Path (b) was observed for a variety of five-membered metallacyclic derivatives of titanium(IV),[20] iron(II),[98-9] nickel(II),[27,120] and palladium(II),[30] and also for a rhodacycloheptane.[138] Diacyl derivatives were obtained either by carbonylation of a hafnium(IV)[21] metallacyclopentene (c) or, more frequently, by carbonylation of cyclic acyls (d).[68-70,76] Path (e) has been observed for two zirconium (IV) metallacycles.[117,156] Such a pattern was attributed to carbenoid character of the carbonyl carbon resulting from an unusual "side-on" co-ordination of a transient acyl compound.[117] The formation of enolate hydride was then rationalized as follows:

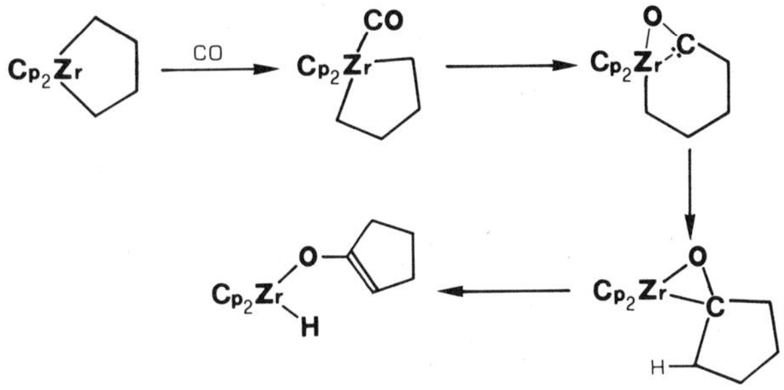

Carbonylation of bis(cyclopentadienyl)zirconaindane at 25°C, followed by acid hydrolysis, gave rise to a mixture of compounds (Scheme 8) whose formation was clearly explained on the basis of the observation and even isolation of several reaction intermediates, generated by rearrangement of an initial acyl product resulting from regiospecific insertion of CO into the Zr-C (alkyl) bond.[156]

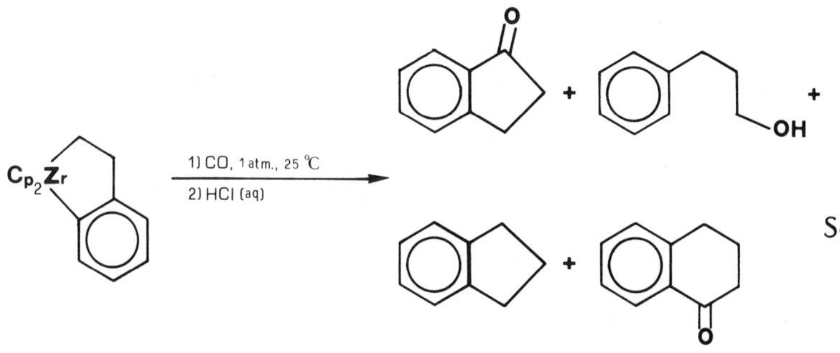

Scheme 8

3.7 Ring expansion by olefin insertion

This extremely important reaction has recently been observed for the iridacyclopentane XV.[182] This metallacycle, while giving the exo-trans-exo norbornadiene cyclodimer on reaction with PPh$_3$ (see Section 3.2), has been shown to undergo norbornadiene insertion, by cis addition, on reaction with PPhMe$_2$ in refluxing CHCl$_3$, leading to the iridacycloheptane XLIII:

XV   **XLIII**

This finding provides an unique unambiguous example of insertion of coordinated olefin into a metal-carbon σ-bond.

4.   METALLACYCLES AS INTERMEDIATES

Metallacycles have been proposed as intermediates in a variety of transition metal-promoted reactions; some selected examples are described below.

4.1   Valence isomerization reactions of strained carbocycles

The question whether these reactions are describable as concerted or stepwise overall processes has given rise to much controversy.[79,157-60] Without entering this controversy, I would like to point out that, at least in some cases[79] (e.g. the rhodium(I)-catalyzed isomerization of carbocycles containing fused cyclobutane rings) strong evidence and arguments have been put forward for a pathway involving initial breaking of a single C-C bond, through oxidative addition and the formation of a transient metallacarbocycle (see Section 2.4).

4.2   Olefin cycloaddition and related reactions

A careful survey[159] of the literature up to 1974 showed that the evidence for involvement of metallacyclopentanes in metal-catalyzed olefin [2π + 2π] cycloadditions was, at best, indirect. Considerable controversy then continued with regard to the mechanism of these reactions, the division lying once again between concerted and stepwise overall processes. I would not at present claim complete answers to this problem. However, I would submit that two primary facts have already been demonstrated: (i) simple olefins can readily undergo oxidative cyclization with transition metals (see Section 2.6); (ii) metallacyclopentanes easily undergo reductive elimination of cyclobutanes (see Section 3.2). In this connection, recent studies by Binger et al.[138] and by Grubbs and Miyashita[121] are important.

Other related reactions that have been interpreted in terms of mechanisms involving metallacyclic intermediates are: the cyclo-co-oligomerisation of olefins with acetylenes;[124-5, 161-2] the $[2\sigma + 2\pi]$ cycloaddition of strained carbocycles to electron-deficient olefins;[163-4] and the cobalt-catalyzed $[2\pi + 2\pi + 2\pi]$ cycloaddition of either acetylenes[165] or norbornene[166] to norbornadiene. The latter reactions are nicely modelled by the stoichiometric formation of the rhodacycles **XXV** and **XXVI** (Section 2.6.3).

4.3   Transition metal–allene chemistry

The stoichiometric and catalytic transition metal promoted transformations of allenes[167-70] are evidently dominated by the conjectured or proven occurrence of dimethylenemetallacyclopentanes. The role played by these species in both rhodium- and iridium-[47,107-10,171-2] and nickel-promoted[112,168-9] allene transformations has been clearly documented.

4.4   Olefin linear dimerization and related reactions

The longevity of the assumption that metal-catalyzed olefin linear dimerisation occurs via olefin insertion into metal-hydrogen and metal-carbon bonds[173-4] attests to the fact that it can account for a not insignificant proportion of experimental findings. However, it is becoming increasingly clear that alternative mechanistic pathways involving metallacycles better fit the experimental data, at least in some processes catalyzed by tantalum[145,183] (see Section 3.1), nickel(0),[121] titanium(0),[175] and zirconium(0)[175] derivatives.

Related olefin reactions that have been conjectured to occur via metallacyclic intermediates include the titanium-catalyzed conversion of ethylene into ethane and 1,3-butadiene,[176] the palladium(I)-catalyzed ethylene dimerisation,[177] and even Ziegler-Natta olefin polymerization.[178-80]

4.5   Cyclopentanone synthesis

The involvement of metallacyclopentanes in the thermally or photochemically initiated stereoselective reactions of iron carbonyls with olefins, which leads to synthetically useful yields of cyclopentanone derivatives,[181] has been strongly supported by studies of the photochemical reaction of $Fe(CO)_5$ with methylacrylate.[98] Furthermore, Whitesides et al.[20] obtained good yields of cyclopentanones by carbonylation of olefins with titanium(II).

5.   MORE RECENT DEVELOPMENTS

Since this chapter was originally completed, a number of reports on the preparation of new metallacarbocycles have appeared. In spite of this, there has been remarkably little advance in our knowledge of the mechanisms of metallacycle reactions. However, an important contribution to our understanding of the factors that control the thermal degradation of metallacyclopentanes has been made by McKinney et al.,[184] who explored the influence of geometry and coordination number on the thermolysis reaction pathways of nickelacyclopentane complexes using a semiempirical molecular orbital procedure and related their results to the experimental observations of Grubbs and co-workers (Section 3.1).

As for the preparation of new metallacarbocycles, a brief report on the synthesis and xray structure of some mono- and di-β-substituted titanacyclobutanes, obtained by reacting a well-defined titanium-alkylidene complex with terminal olefins, is of relevance.[185] Another paper,[186] which brings interesting perspectives to metallacycle chemistry, concerns the preparation of an extensive series of four- and five-membered hydridometallacyclic derivatives of iridium(III) by γ- and δ-hydrogen abstraction reactions of alkyl-iridium(I) complexes. A ruthenacyclopentene has been prepared analogously.[187]

A series of zirconacyclopentanes have been obtained by very selective oxidative coupling reactions of allene,[188] (R)(S)-1,3-dimethylallene,[188] and methylallene[189] with $[(C_5Me_5)_2ZrN_2]_2N_2$. Two nickelacyclopenten-2,5-diones have also been prepared by cyclization reactions.[190]

Depending on the experimental conditions, the reaction of $WOCl_4$ with o-$C_6H_4(CH_2MgCl)_2$ has been shown to lead to $\{[\overline{W(CH_2C_6H_4CH_2}\text{-o})_2O]_2$ $Mg(thf)_4\}$[191] or to the first metallatricycle $[\overline{W(CH_2C_6H_4CH_2}\text{-o})_3]$.[192]

Two interesting strained tantalacarbocycles, **XLV** and **XLVI**, have been obtained starting from the carbene complex **XLIV**:[193]

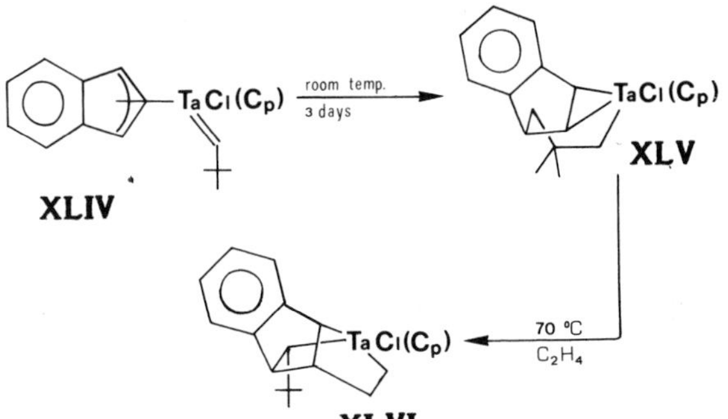

The reaction of bicyclo[1.1.0]butane with Zeise's dimer has been described as giving a complex of composition $[PtCl_2(C_4H_6)]_n$ whose pyridine derivative **XLVII** is probably in equilibrium in solution with **XLVIII**:[194]

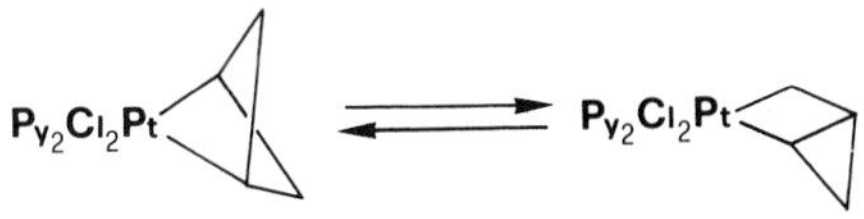

Cushman and Brown[195] have reported that platinacyclobutane complexes of general formula $L_2Cl_2\overline{PtCH_2CHR\dot{C}HR'}$ can produce, under mild conditions, either the corresponding terminal olefins or the pyridinium ylide complexes, depending on the nature of the ligand L and the substitution pattern of the ring. Both reactions were interpreted in terms of mechanisms involving a common intermediate π-allylplatinum hydride arising from a β-elimination reaction of a ring proton.[195] Furthermore, platinum(IV) metallacyclobutane complexes have been shown to undergo facile ring-cleavage reactions under certain conditions, e.g. in the presence of t-phosphines with solvents of relatively high dielectric constant (Me_2SO, MeCN)[196-7] or in the presence of tetracyanoethylene and related olefins.[197]

Metallacyclobutanes and metallacyclopentanes have been proposed as intermediate species in the reactions of atomic cobalt ions with alkenes[198] and cycloalkenes[199] in the gas phase.

6.   CONCLUSIONS

An attempt has been made to point out the most important facts about metallacycle chemistry and to emphasise that research in this area has generated new conceptual entities and provided a grounding for older ones.

Metallacarbocycles, by virtue of their structure, offer chemists the unique opportunity to study how the chemical behavior of carbon chains alters when these are constrained around a transition metal. The chemistry which results is highly versatile, and resembles only in part that of acyclic dialkyls, being rich in extremely interesting distinctive features. One of these is undoubtedly the ease with which some metallacycles undergo carbon-carbon bond cleavage. An insight into the basic preconditions for such an event might indeed bring new perspectives to research in C-C bond activation.

Furthermore, metallacarbocycle reactions often exhibit high selectivity and even stereoselectivity, and this allows one to focus on their potential synthetic utility.

Certainly, many more studies are required to understand the variables that mainly influence metallacyclic reactivity, to test rationalizations and hypotheses, and, finally, to provide an unified view of metallacycle chemistry.

ACKNOWLEDGEMENTS

I am grateful to Dr. P. Diversi and Dr. A. Lucherini for many helpful comments.

REFERENCES

1. A. Buraway and C.S. Gibson, J. Chem. Soc 324 (1936).
2. C.F.H. Tipper, J. Chem. Soc. 2045 (1955).
3. D.M. Adams, J. Chatt and R.G. Guy, Proc. Chem. Soc. 170 (1960).

4. D.M. Adams, J. Chatt, R.G. Guy and N. Sheppard, J. Chem. Soc. 738 (1961).

5. N.A. Bailey, R.D. Gillard, M. Keeton, R. Mason and D.R. Russell, J.C.S. Chem. Commun. 396 (1966).

6. S.E. Binns, R.H. Cragg, R.D. Gillard, B.T. Heaton and M.F. Pilbrow, J. Chem. Soc. (A) 1227 (1969).

7. T.A. Manuel, S.L. Stafford and F.G.A. Stone, J. Am. Chem. Soc. 83:249 (1961).

8. T.D. Coyle, R.B. King, E. Pitcher, S.L. Stafford, P. Treichel and F.G.A. Stone, J. Inorg. Nucl. Chem. 20:172 (1961).

9. H.H. Hoehn, L. Pratt, K.F. Watterson and G. Wilkinson, J. Chem. Soc. 2738 (1961).

10. L. Cassar, P.E. Eaton and J. Halpern, J. Am. Chem. Soc. 92:3515 (1970).

11. R.G. Pearson, Fortschr. Chem. Forsch. 41:75 (1973).

12. P.S. Braterman, J.C.S. Chem. Commun. 70 (1979).

13. A. Stockis and R. Hoffman, J. Am. Chem. Soc. 102:2952 (1980).

14. T.S. Kuhn, "The Structure of Scientific Revolutions", University of Chicago Press, Chicago (1970).

15. R.H. Grubbs, D.D. Carr and P.L. Burk, in "Organotransition Metal Chemistry", (Y. Ishii and M. Tsutsui, eds.), Plenum Press, New York (1975), p. 135.

16. R.H. Grubbs and A. Miyashita, in "Fundamental Research in Homogeneous Catalysis", (M. Tsutsui ed.), Vol. 3, Plenum Press, New York (1979), p. 151.

17. K. Itoh, in "Fundamental Research in Homogeneous Catalysis", (M. Tsutsui, ed.), Vol. 3, Plenum Press, New York (1979), p. 865.

18. R.J. Puddephatt, Coord. Chem. Rev. 33:149 (1980).

19. M.D. Rausch and L.P. Klemann, J.C.S. Chem. Commun. 354 (1971).

20. J.X. McDermott, M.E. Wilson and G.M. Whitesides, J. Am. Chem. Soc. 98:6529 (1976).

21. M.F. Lappert, T.R. Martin, J.L. Atwood and W.E. Hunter, J.C.S. Chem. Commun. 476 (1980).

22. M.F. Lappert, T.R. Martin, C.R. Milne, J.L. Atwood, W.E. Hunter and R.E. Pentilla, J. Organomet. Chem. 192:C35 (1980).

23. P. Diversi, G. Ingrosso, A. Lucherini, W. Porzio and M. Zocchi, J.C.S. Dalton Trans. 967 (1983).

24. P. Diversi, G. Ingrosso, A. Lucherini, W. Porzio and M. Zocchi, J.C.S. Chem. Commun. 811 (1977).

25. P. Diversi, G. Ingrosso, A. Lucherini, W. Porzio and M. Zocchi, Inorg. Chem. 19:3590 (1980).

26. P. Diversi, G. Ingrosso and A. Lucherini, J.C.S. Chem. Commun. 52 (1977).

27. R.H. Grubbs, A. Miyashita, M. Liu and P. Burk, J. Am. Chem. Soc. 100:2418 (1978).

28. R.H. Grubbs and Miyashita, J. Am. Chem. Soc. 100:7418 (1978).

29. P. Diversi, G. Ingrosso and A. Lucherini, J.C.S. Chem. Commun. 735 (1978).

30. P. Diversi, G. Ingrosso, A. Lucherini and S. Murtas, J.C.S. Dalton Trans. 1633 (1980).

31. J.X. McDermott, J.F. White and G.M. Whitesides, J. Am. Chem. Soc. 98:6521 (1976).

32. H. Hoberg and W. Richter, J. Organomet. Chem. 195:355 (1980).

33. E.H. Braye, W. Hübel and I. Caplier, J. Am. Chem. Soc. 83:4406 (1961).

34. P. Diversi, G. Ingrosso, A. Lucherini, P. Martinelli, M. Benetti and S. Pucci, J. Organomet. Chem. 165:253 (1979).

35. G. Wilke, in "Fundamental Research in Homogeneous Catalysis", (M. Tsutsui, ed.), Vol. 3, Plenum Press, New York (1979), p. 1.

36. R. Uson, J. Vincente, J.A. Cirac and M.T. Chicote, J. Organomet. Chem. 198:105 (1980).

37. L.C. Costa and G.M. Whitesides, J. Am. Chem. Soc. 99:2390 (1977).

38. M.J. Doyle, J. McMeeking and P. Binger, J.C.S. Chem. Commun. 376 (1976).

39. P. Binger, M.J. Doyle, C. Krüger and Yi-H. Tsay, Z. Naturforsch. 34B:1289 (1979).

40. S. Takahashi, Y. Suzuki, K. Sonogashira and N. Hagihara, J.C.S. Chem. Commun. 839 (1976).

41. M. Ephritikhine, M.L.H. Green and R.E. MacKenzie, J.C.S. Chem. Commun. 619 (1976).

42. M. Ephritikhine, B.R. Francis, M.L.H. Green, R.E. MacKenzie and M.J. Smith, J.C.S. Dalton Trans. 1131 (1977).

43. R.N. Haszeldine, R.V. (Dick) Parish and D.W. Robbins, J.C.S. Dalton Trans. 2355 (1976).

44. B.F. Johnson, T. Keating, J. Lewis, M.S. Subramanian and D.A. White, J. Chem. Soc. (A) 1793 (1969).

45. S.J. Betts, A. Harris, R.N. Haszeldine and R.V. Parish, J. Chem. Soc. (A) 3699 (1971).

46. R. Rossi, Atti Societa Toscana Scienze Naturali Mem. 79:101 (1972).

47. G. Ingrosso, P. Gronchi and L. Porri, J. Organomet. Chem. 86:C20 (1975).

48. R.D. Gillard, M. Keeton, R. Mason, M.F. Pilbrow and D.R. Russell, J. Organomet. Chem. 33:247 (1971).

49. F.J. McQuillin and K.G. Powell, J.C.S. Dalton Trans. 2123 (1972).

50. B.M. Cushman, S.E. Earnest and D.B. Brown, J. Organomet. Chem. 159:431 (1978).

51. D.B. Brown and V.A. Viens, J. Organomet. Chem. 142:117 (1977).

52. R.J. Al-Essa, R.J. Puddephatt, M.A. Quyser and C.F. Tipper, J. Am. Chem. Soc. 101:364 (1979).

53. M. Graziani, M. Lenarda, R. Ros and U. Belluco, Coord. Chem. Rev. 16:35 (1975).

54. M. Lenarda, R. Ros, M. Graziani and U. Belluco, J. Organomet. Chem. 65:407 (1974).

55. S. Forcolin, G. Pellizer, M. Graziani, M. Lenarda and R. Ganzerla, J. Organomet. Chem. 194:203 (1980).

56. J.P. Visser and J.E. Ramakers-Blom, J. Organomet. Chem. 44:C63 (1972).

57. W. Wong, S.J. Singer, W.D. Pitts, S.F. Watkins and W.H. Baddley, J.C.S. Chem. Commun. 672 (1972).

58. R. Ros, M. Lenarda, N. Bresciani Panor, M. Calligaris, P. Delise, L. Randaccio and M. Graziani, J.C.S. Dalton Trans. 1937 (1976).

59. J. Burgess, R.I. Haines, E.R. Hamner, R.D.W. Kemmitt and A.R. Smith, J.C.S. Dalton Trans. 2579 (1975).

60. W.E. Carroll, M. Green, J.A.K. Howard, M. Pfeffer and F.G.A. Stone, J.C.S. Dalton Trans 1472 (1978).

61. L.S. Liebeskind, S.L. Baysdon, M.S. South and J.F. Blount, J. Organomet. Chem. 202:C73 (1980).

62. P.D. Frisch and G.P. Khare, Inorg. Chem. 18:781 (1979).

63. P.D. Frisch and G.P. Khare, J. Am. Chem. Soc. 100:8267 (1978).

64. R.M. Tuggle and D.L. Weaver, Inorg. Chem. 11:2237 (1972).

65. A. Keasey and P.M. Maitlis, J.C.S. Dalton Trans. 1830 (1978).

66. D.M. Roundhill, D.N. Lawson and G. Wilkinson, J. Chem. Soc. (A) 845 (1968).

67. F.J. McQuillin and K.G. Powell, J.C.S. Dalton Trans. 2129 (1972).

68. L. Cassar and J. Halpern, J.C.S. Chem. Commun. 1082 (1970).

69. P.G. Gassman and J.A. Nikora, J. Organomet. Chem. 92:81 (1975).

70. B.F.G. Johnson, J. Lewis and S.W. Tam, J. Organomet. Chem. 105:271 (1976).

71. T.J. Katz and S.A. Cerefice, J. Am. Chem. Soc. 93:1049 (1971).

72. V. Heil, B.F.G. Johnson, J. Lewis and D.J. Thompson, J.C.S. Chem. Commun. 270 (1974).

73. R.M. Moriarty, K.N. Chen, C.L. Yeh, J.L. Flippen and J. Karle, J. Am. Chem. Soc. 94:8944 (1972).

74. S.W. Tam, Tetrahedron Lett. 2385 (1974).

75. F.A. Cotton, J.M.T. Roup, W.E. Billups, L.P. Liu and C.V. Smith, J. Organomet. Chem. 102:345 (1975).

76. R. Aumann, H. Wörmann and C. Krüger, Angew. Chem. Int. Ed. Engl. 15:609 (1976).

77. M. Sohn, J. Blum and J. Halpern, J. Am. Chem. Soc. 101:2694 (1979).

78. J. Blum, C. Zlotogorski and A. Zoran, Tetrahedron Lett. 1127 (1975).

79. J. Halpern, in "Organic Syntheses via Metal Carbonyls", (I. Wender and P. Pino, eds.), Vol. 2, Wiley, New York (1977), p. 705.

80. J. Blum and C. Zlotogorski, Tetrahedron Lett. 3501 (1978).

81. R.A. Andersen, R.A. Jones and G. Wilkinson, J.C.S. Dalton Trans. 446 (1978).

82. P. Foley and G.M. Whitesides, J. Am. Chem. Soc. 101:2732 (1979).

83. P. Foley, R. Di Cosimo and G.M. Whitesides, J. Am. Chem. Soc. 102:6173 (1980).

84. D. Chappell and D.J. Cole-Hamilton, J.C.S. Chem. Commun. 238 (1980).

85. H. Masai, K. Sonogashira and N. Hagihara, Bull. Chem. Soc. Jpn. 41:750 (1968).

86. J. Dvorak, R.J. O'Brien and W. Santo, J.C.S. Chem. Commun. 411 (1970).

87. C.P. Boekel, J.H. Teuben and H.J. De Liefde Meijer, J. Organomet. Chem. 81:371 (1974).

88. G. Erker and K. Kropp, J. Am. Chem. Soc. 101:3659 (1979).

89. P.S. Braterman and R.J. Cross, Chem. Soc. Rev. 2:271 (1973).

90. T.R. Howard, J.B. Lee and R.H. Grubbs, J. Am. Chem. Soc. 102:6876 (1980).

91. F.N. Tebbe and R.L. Harlow, J. Am. Chem. Soc. 102:6149 (1980).

92. R. Burt, M. Cooke and M. Green, J. Chem. Soc. (A) 2981 (1970).

93. H. Yamazaki, K. Aoki, Y. Yamamoto and Y. Wakatsuki, J. Am. Chem. Soc. 97:3546 (1975).

94. T.U. Harris, J.W. Rathke and E.L. Muetterties, J. Am. Chem. Soc. 100:6966 (1978).

95. R. Burt, M. Cooke and M. Green, J. Chem. Soc. (A) 2975 (1970).

96. A. Bond, B. Lewis and M. Green, J.C.S. Dalton Trans. 1109 (1975).

97. M. Green, B. Lewis, J.J. Daly and F. Sanz, J.C.S. Dalton Trans. 1118 (1975).

98. F-W. Grevels, D. Schulz and E.A. Koerner v. Gustorf, Angew. Chem. Int. Ed. Engl. 13:534 (1974).

99. B.E. Foulger, F-W. Grevels, D. Hess, E.A. Koerner v. Gustorf and J. Leitich, J.C.S. Dalton Trans. 1451 (1979).

100. J. Ashley-Smith, M. Green and F.G.A. Stone, J. Chem. Soc. (A) 3019 (1969).

101. C.S. Cundy, M. Green and F.G.A. Stone, J. Chem. Soc. (A) 1647 (1970).

102. A.J. Mukhedkar, V.A. Mukhedkar, M. Green and F.G.A. Stone, J. Chem. Soc. (A) 3166 (1970).

103. M. Green, J.A.K. Howard, A. Laguna, L.E. Smart, J.L. Spencer and F.G.A. Stone, J.C.S. Dalton Trans. 278 (1977).

104. M. Green, J.A.K. Howard, P. Mitrprachachon, M. Pfeffer, J.L. Spencer, F.G.A. Stone and P. Woodward, J.C.S. Dalton Trans. 306 (1979).

105. M. Green and S.H. Taylor, J.C.S. Dalton Trans. 1128 (1975).

106. A.R. Fraser, P.H. Bird, S.A. Bezman, J.R. Shapley, R. White and J.A. Osborn, J. Am. Chem. Soc. 95:597 (1973).

107. G. Ingrosso, A. Immirzi and L. Porri, J. Organomet. Chem. 60:C35 (1973).

108. G. Ingrosso, L. Porri, G. Pantini and P. Racanelli, J. Organomet. Chem. 84:75 (1975).

109. P. Diversi, G. Ingrosso, A. Immirzi and M. Zocchi, J. Organomet. Chem. 104:C1 (1976).

110. P. Diversi, G. Ingrosso, A. Immirzi, W. Porzio and M. Zocchi, J. Organomet. Chem. 125:253 (1977).

111. M.A. Bennett, R.N. Johnson and I.B. Tomkins, J. Am. Chem. Soc. 96:61 (1974).

112. P.W. Jolly, C. Krüger, S. Salz and J.C. Sekutowski, J. Organomet. Chem. 165:C39 (1977).

113. C.R. Barker, M. Green, J.A.K. Howard, J.L. Spencer and F.G.A. Stone, J.C.S. Dalton Trans. 1839 (1978).

114. D.M. Duggan, J. Russell Schmidt, J. Jacobson and L. Nelson, 2nd Int. Conf. Homog. Catal., Dusseldorf, Sept. 1980, Abstr. p. 92.

115. R.H. Grubbs and A. Miyashita, J.C.S. Chem. Commun. 864 (1977).

116. K.I. Gell and J. Schwartz, J.C.S. Chem. Commun. 244 (1979).

117. J.M. Manriquez, D.R. McAlister, R.D. Sanner and J.E. Bercaw, J. Am. Chem. Soc. 100:2716 (1978).

118. P. Binger, M.J. Doyle, J. McMeeking, C. Krüger and Yi-Hung Tsay, J. Organomet. Chem. 135:405 (1977).

119. J.R. Blackborow, U. Feldhoff, F-W. Grevels, R.H. Grubbs and A. Miyashita, J. Organomet. Chem. 173:253 (1979).

120. R.H. Grubbs and A. Miyashita, J. Organomet. Chem. 161:371 (1978).

121. R.H. Grubbs and A. Miyashita, J. Am. Chem. Soc. 100:7416 (1978).

122. S.J. McLain, C.D. Wood and R.R. Schrock, J. Am. Chem. Soc. 101:4558 (1979).

123. G. Smith, S.J. McLain and R.R. Schrock, J. Organomet. Chem. 202:269 (1980).

124. Y. Wakatsuki, K. Aoki and H. Yamazaki, J. Am. Chem. Soc. 101:1123 (1979).

125. C.E. Dean, R.D.W. Kemmitt, D.R. Russell and M.D. Schilling, J. Organomet. Chem. 187:C1 (1980).

126. Y. Wakatsuki and H. Yamazaki, J.C.S. Chem. Commun. 1270 (1980).

127. W. Hübel, in "Organic Syntheses Via Metal Carbonyls", (I. Wender and P. Pino, eds.), Vol. 1, Wiley, New York (1968), p. 273.

128. J.W. Kang, S. McVey and P.M. Maitlis, Can. J. Chem. 46:3189 (1968).

129. S. McVey and P.M. Maitlis, J. Organomet. Chem. 19:169 (1969).

130. J.T. Mague, M.O. Nutt and E.H. Gause, J.C.S. Dalton Trans. 2578 (1973).

131. R.S. Dickson and H.P. Kirsch, Aust. J. Chem. 27:61 (1974).

132. F. Canziani, M.C. Malatesta and G. Longoni, J.C.S. Chem. Commun. 267 (1975).

133. J. Browning, M. Green and F.G.A. Stone, J. Chem. Soc. (A) 453 (1971).

134. M. Green, S.K. Shakshooki and F.G.A. Stone, J. Chem. Soc. (A) 2828 (1981).

135. R.L. Hunt, D.M. Roundhill and G. Wilkinson, J. Chem. Soc. (A) 982 (1967).

136. P.B. Hitchcock and R. Mason, J.C.S. Chem. Commun. 242 (1967).

137. G. Erker, J. Wicher, K. Engel, F. Rosenfeldt, W. Dietrich and C. Krüger, J. Am. Chem. Soc. 102:6344 (1980).

138. P. Binger, M. Cetinkaya, M.J. Doyle, A. Germer and U. Schuchardt, in "Fundamental Research in Homogeneous Catalysis", (M. Tsutsui, ed.), Vol. 3, Plenum Press, New York (1979), p. 271.

139. D.R. Russell and P.A. Tucker, J.C.S. Dalton Trans. 841 (1976).

140. J.A. Evans, R.D.W. Kemmitt, B.Y. Kimura and D.R. Russell, J.C.S. Chem. Commun. 509 (1972).

141. A.K. Cheetham, R.J. Puddephatt, A. Zalkin, D. Templeton and L.K. Templeton, Inorg. Chem. 15:2997 (1976).

142. R.H. Grubbs and A. Miyashita, J. Am. Chem. Soc. 100:1300 (1978).

143. G.B. Young and G.M. Whitesides, J. Am. Chem. Soc. 100:5808 (1978).

144. D.C.L. Perkins, R.J. Puddephatt and C.F.H. Tipper, J. Organomet. Chem. 191:481 (1980).

145. S.J. McLain, J. Sancho and R.R. Schrock, J. Am. Chem. Soc. 102:5610 (1980).

146. M.R. Churchill and W.J. Youngs, Inorg. Chem. 19:3106 (1980).

147. M.L.H. Green and M.J. Smith, J. Chem. Soc. (A) 3220 (1971).

148. T.H. Johnson and S-S. Cheng, J. Am. Chem. Soc. 101:5277 (1979).

149. A. Cuccuru, P. Diversi, G. Ingrosso and A. Lucherini, J. Organomet. Chem. 204:123 (1981).

150. P.W. Hall, R.J. Puddephatt, K.R. Seldon and C.F.H. Tipper, J. Organomet. Chem. 81:423 (1974).

151. P. Binger and M.J. Doyle, J. Organomet. Chem. 162:195 (1978).

152. P. Binger, J. McMeeking, U. Schuchardt and M.J. Doyle, 7th Int. Conf. Organomet. Chem., Venice, 1975, Abstr. p. 129.

153. W.J. Irvin and F.J. McQuillin, Tetrahedron Lett. 1937 (1968).

154. F.J. McQuillin and K.G. Powell, J.C.S. Dalton Trans. 2123 (1972).

155. P. Barabotti, P. Diversi, G. Ingrosso, A. Lucherini and F. Nuti, J.C.S. Dalton Trans., in press.

156. G. Erker and K. Kropp, J. Organomet. Chem. 194:45 (1980).

157. L.A. Paquette, Acc. Chem. Res. 4:280 (1971).

158. D.J. Cardin, B. Cetinkaya, M.J. Doyle, and M.F. Lappert, Chem. Soc. Rev. 2:99 (1973).

159. F.D. Mango, Coord. Chem. Rev. 15:109 (1975).

160. K.C. Bishop III, Chem. Rev. 76:461 (1976).

161. T.A. Mitsudo, K. Kokuryo and Y. Takegami, J.C.S. Chem. Commun. 722 (1976).

162. P. Caddy, M. Green, E. O'Brien, L.E. Smart and P. Woodward, J.C.S. Dalton Trans. 962 (1980).

163. R. Noyori, I. Umeda, H. Kawauchi and H. Takaya, J. Am. Chem. Soc. 97:812 (1975).

164. P. Binger and U. Schuchardt, Angew. Chem. Int. Ed. Engl. 16:249 (1977).

165. J.E. Lyons, H.K. Myers and A. Schneider, J.C.S. Chem. Commun. 636 (1978).

166. J.L. Lyons, H.K. Myers and A. Schneider, J.C.S. Chem. Commun. 638 (1978).

167. B.L. Shaw and A.J. Stringer, Inorg. Chim. Acta Rev. 7:1 (1973).

168. P.W. Jolly and G. Wilke, "The Organic Chemistry of Nickel", Vol. 2, Academic Press, New York (1975).

169. S. Otsuka and A. Nakamura, Adv. Organomet. Chem. 14:245 (1976).

170. F.L. Bowden and R. Giles, Coord. Chem. Rev. 20:81 (1976).

171. A. Giarrusso, P. Gronchi, G. Ingrosso and L. Porri, Makromol. Chem. 176:1375 (1977).

172. A. Borrini and G. Ingrosso, J. Organomet. Chem. 132:275 (1977).

173. G. Lefebvre and Y. Chauvin, in "Aspects of Homogeneous Catalysis", (R. Ugo, ed.), Vol. 1, Carlo Manfredi Editore, Milano (1970), p. 107.

174. B. Bogdanovic, Adv. Organomet. Chem. 17:105 (1979).

175. S. Datta, M.B. Fischer and S.S. Wreford, J. Organomet. Chem. 188:353 (1980).

176. G.P. Pez, J.C.S. Chem. Commun. 560 (1977).

177. P. Pertici and G. Vitulli, Tetrahedron Lett. 1879 (1979).

178. K.J. Ivin, J.J. Rooney, C.D. Stewart, M.L.H. Green and R. Mahtab, J.C.S. Chem. Commun. 604 (1978).

179. M.L.H. Green, Pure Appl. Chem. 50:27 (1978).

180. R.J. McKinney, J.C.S. Chem. Commun. 490 (1980).

181. E. Weissberger and P. Laszlo, Acc. Chem. Res. 9:209 (1976).

182. S.A. Bezman, P.H. Bird, A.R. Fraser and J.A. Osborn, Inorg. Chem. 19:3755 (1980).

183. J.D. Fellmann, R.R. Schrock and G.A. Rupprecht, J. Am. Chem. Soc. 103:5752 (1981).

184. R.J. McKinney, D.L. Thorn, R. Hoffmann and A. Stockis, J. Am. Chem. Soc. 103:2595 (1981).

185. J. B. Lee, G.J. Gajda, W.P. Schaefer, T.R. Howard, T. Ikariya, D.A. Straus and R.H. Grubbs, J. Am. Chem. Soc. 103:7358 (1981).

186. T.H. Tulip and D.L. Thorn, J. Am. Chem. Soc. 103:2448 (1981).

187. S.D. Chappell and D.J. Cole-Hamilton, J.C.S. Chem. Commun. 319 (1981).

188. J.R. Schmidt and D.M. Duggan, Inorg. Chem. 20:318 (1981).

189. D.M. Duggan, Inorg. Chem. 20:1164 (1981).

190. H. Hoberg and A. Herrera, Angew. Chem. Int. Ed. Engl. 19:927 (1980).

191. M.F. Lappert, C.L. Raston, G.L. Rowbottom and A.H. White, J.C.S. Chem. Commun. 6 (1981).

192. M.F. Lappert, C.L. Raston, B.W. Skelton and A.H. White, J.C.S. Chem. Commun. 485 (1981).

193. A.W. Gal and H. v. d. Heijden, Angew. Chem. Int. Ed. Engl. 20:978 (1981).

194. A. Miyashita, M. Takahashi and H. Takaya, J. Am. Chem. Soc. 103:6257 (1981).

195. B.M. Cushman and D.B. Brown, Inorg. Chem. 20:2490 (1981).

196. R.J. Al-Essa, R.J. Puddephatt, D.C.L. Perkins, M.C. Rendle and C.F.H. Tipper, J.C.S. Dalton Trans. 1738 (1981).

197. M.C. Rendle and C.F.H. Tipper, J. Organomet. Chem. 224:321 (1982).

198. P.B. Armentrout, L.F. Halle and J.L. Beauchamp, J. Am. Chem. Soc. 103:6624 (1981).

199. P.B. Armentrout and J.L. Beauchamp, J. Am. Chem. Soc. 103:6628 (1981).

NUCLEOPHILIC ATTACK ON COORDINATED ALKENES

J.-E. Bäckvall

Department of Organic Chemistry

Royal Institute of Technology, Stockholm

1. INTRODUCTION

π-Olefin complexes of transition metals have been known for a long time. One of the first organometallic compounds known was in fact the ethene platinum complex **I**, Zeise's salt, prepared in 1827.[1] However, it was to take more than 120 years before a significant interest in the structure and reactivity of such compounds developed. In principle nucleophilic attack on an olefin coordinated to a metal has been known since 1900 when Hofmann and Sand[2] discovered the oxymercuration reaction. In this reaction e.g. water attacks a labile π-olefin complex of mercury to produce a β-hydroxyalkylmercury complex.

The early observation that palladium(II)[3] and platinum(II)[4] oxidize ethene to acetaldehyde led a German research team to developing a catalytic process for manufacture of acetaldehyde in the mid-1950s.[5-6] They recognized the importance of a nucleophilic attack by water on an intermediate π-olefin-palladium complex. At this time, a deeper understanding of the nature of π-olefin transition metal complexes emerged. The bond between the olefin and the metal was explained in terms of the Dewar-Chatt-Duncanson model[7-8] (fig. 1). According to this model there is one interaction between symmetric orbitals on the metal and the olefin and another between antisymmetrical orbitals (symmetric and antisymmetric

679

Figure 1

with respect to a plane through the metal and the midpoint of the carbon-carbon bond). In the symmetric interaction an empty orbital on the metal overlaps with the filled π-orbital. This overlap will result in an electron flow from the olefin to the metal. In the antisymmetric interaction an overlap between the empty π^*-orbital of the olefin and a filled d(xy) orbital on the metal takes place. This overlap will result in an electron flow from the metal back to the olefin (back donation). Consequently, the electron density of the olefin will depend on the relative magnitude of the two interactions. It is evident that nucleophilic attack on the olefin will be favored if the first interaction (between symmetric orbitals) predominates. If the two interactions are strong, the two new orbitals formed may very well be compared with the symmetric and antisymmetric orbitals describing the two σ-bonds in a three-membered ring (fig. 2). In some cases π-olefin complexes are indeed best explained in terms of a three-membered ring.

Since the overlap depends on the relative energy of the orbitals involved, one can distinguish three principal types of π-olefinmetal complexes in which the reactivity of the coordinated olefin will be quite different (fig. 3).[9a] In the first case (a) a strong interaction occurs between the filled d(xy) orbital of the metal and the empty π^* orbital of the olefin, since the energy differences between these two orbitals is small. On the other hand the energy difference between the empty hybrid orbital of the

Figure 2

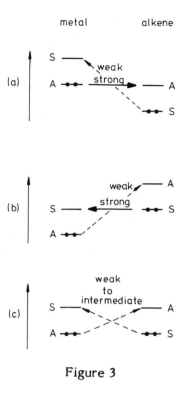

metal   alkene

Figure 3

metal and the filled π orbital of the olefin is too large to give any significant interaction. Thus, a considerable electron flow from metal to olefin occurs, resulting in a negative charge on the alkene. The olefin will therefore be activated towards electrophilic attack. In the second case (b) a strong interaction takes place between the empty orbital of the metal and the filled orbital of the olefin, resulting in considerable electron flow from the alkene to the metal. The energy difference between the filled orbital of the metal and the empty orbital of the alkene, however, is too large to give any significant "back donation" of electrons. The olefin in this case will become positively charged and hence activated towards nucleophilic attack. In the final case (c), there are weak to moderate interactions between the two symmetric orbitals as well as between the two antisymmetric orbitals. The olefin in this type of complex will be essentially neutral.

The relationship between electronic structure and reactivity of the coordinated olefin towards nucleophilic attack has recently been analyzed.[9b]

2. PALLADIUM AND PLATINUM OLEFIN COMPLEXES

Of all metal π-olefin complexes known, those of palladium and platinum have been the most studied. The first π-olefin-transition metal complex prepared, reported in 1827, was the platinum olefin complex I (Zeise's salt).[1] Generally, palladium(II) and platinum(II) olefin complexes are quite easy to prepare, and the standard procedure is by exchange of a weak ligand with the olefin.[10] Kharasch et al.[11] reported the preparation of π-olefin-palladium complexes II by reaction of the olefin with $(PhCN)_2PdCl_2$ (eq 1). The solvent used by Kharasch was benzene, but the same reaction is even more rapid in THF, where the olefin is complexed in a few minutes at $0°C$.[12]

$$2\,(PhCN)_2PdCl_2 \;+\; 2\,RCH{=}CHR' \;\longrightarrow\; \text{II} \qquad (eq\ 1)$$

$$+\,4\,PhCN$$

$$\text{(cyclooctadiene)} \;+\; MCl_4^{2-} \;\longrightarrow\; \text{III} \;+\; 2Cl^- \qquad (eq\ 2)$$

$$M = Pd, Pt$$

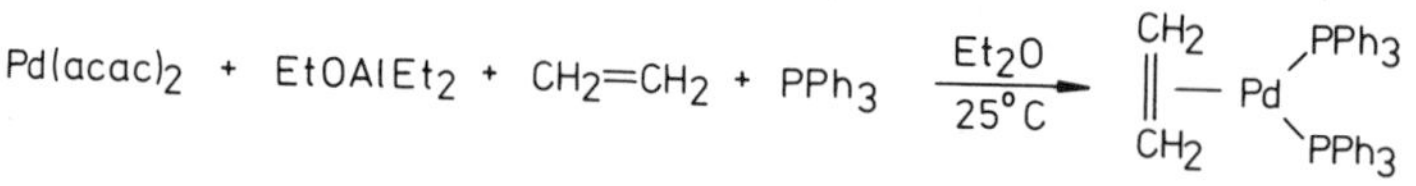

$$Pd(acac)_2 \;+\; EtOAlEt_2 \;+\; CH_2{=}CH_2 \;+\; PPh_3 \;\xrightarrow[\;25°C\;]{Et_2O}\; \text{product} \qquad (eq\ 3)$$

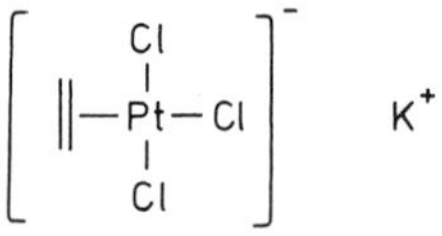

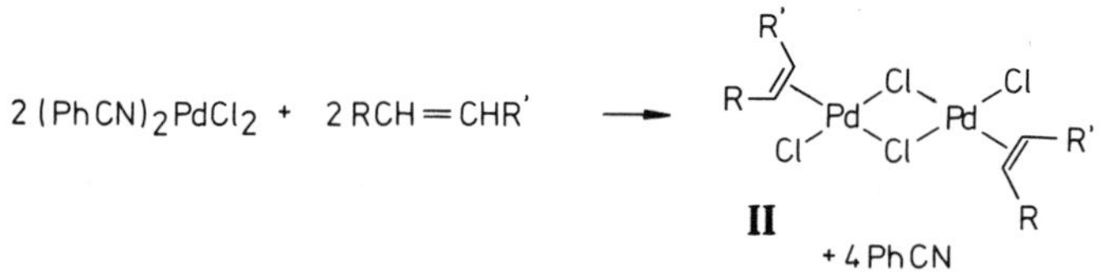

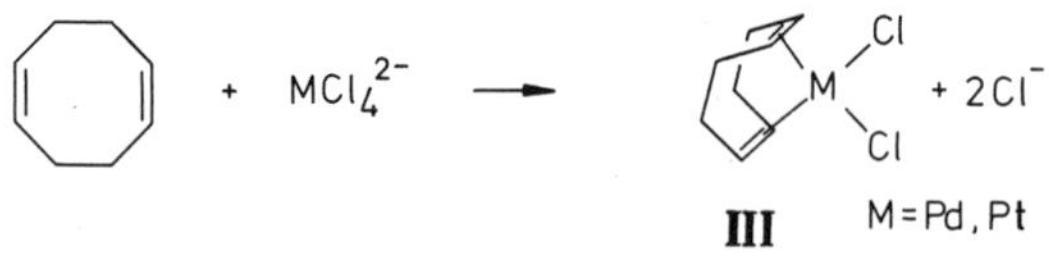

Stable diene complexes III of palladium(II) and platinum(II) were prepared by the reaction of 1,5-cyclooctadiene with sodium tetrachloropalladate and potassium tetrachloroplatinate respectively (eq 2).[13-14]

Zerovalent complexes of palladium and platinum can be prepared by reducing a metal(II) complex in the presence of an olefin (eq 3).[15-16]

There is a great difference in the reactivity of the metal(0) and the metal(II) olefin complexes. The former are d^{10} complexes with a strong back donation from the metal to the π^*-orbital of the olefin, whereas the latter are d^8-complexes with moderate to weak back donation from metal to olefin, but with an important electron flow from olefin to metal. Thus, the olefin in the metal(0) complexes will be electron rich and susceptible to electrophilic attack, and the olefin in the metal(II) complexes somewhat electron poor, depending on the remaining ligands on the metal, and consequently susceptible to nucleophilic attack. The difference between the two types of complexes is nicely illustrated by their structures. Thus the carbon-carbon double bond length in the metal(II) olefin complexes is 1.35 - 1.36Å, and in the metal(0) complexes 1.43 - 1.46Å.[16-18] The bond length in free olefins is 1.33 - 1.34Å. Furthermore, palladium(0) and platinum(0) form their strongest π-olefin complexes with electron poor olefins, whereas the reverse is true for the divalent metals.

2.1   Nucleophilic attack on palladium and platinum olefin complexes

In accordance with the electronic structure of π-olefin complexes no observation of nucleophilic attack on olefins coordinated to palladium(0) and platinum(0) has been reported. On the other hand there are numerous reports of nucleophilic additions to π-olefin-palladium(II) and -platinum(II) complexes. Phillips observed in 1894 that ethylene, when passed into an aqueous solution of palladium chloride, was oxidized to acetaldehyde with the formation of palladium metal.[3] Later, the analogous reaction was found to take place with Zeise's salt, **I**, although at a much higher temperature.[4] In the fifties, German workers developed the palladium-catalyzed oxidation of ethene into an industrial process for the manufacture of acetaldehyde (the Wacker process).[5-6] They found that the palladium(0) formed in the first cycle (eq 4) can effectively be reoxidized using $CuCl_2/O_2$ (eq 5 and 6). The first step involves nucleophilic attack by water on a coordinated ethene to give an unstable β-hydroxyethylpalladium intermediate, which decomposes to acetaldehyde (eq 7). The mechanism of this reaction is discussed further in section 2.3.

$$H_2O + CH_2=CH_2 + PdCl_2 \rightarrow Pd + CH_3CHO + 2HCl \quad \text{(eq 4)}$$

$$Pd + 2CuCl_2 \rightarrow 2CuCl + PdCl_2 \quad \text{(eq 5)}$$

$$2HCl + 2CuCl + 1/2O_2 \rightarrow CuCl_2 + H_2O \quad \text{(eq 6)}$$

$$CH_2=CH_2 + 1/2O_2 \rightarrow CH_3CHO \quad \text{(eq 4 + 5 + 6)}$$

$$HOCH_2CH_2PdCl \rightarrow CH_3CHO + Pd(0) + HCl \quad \text{(eq. 7)}$$

The first direct demonstration of nucleophilic attack on a double bond coordinated to platinum(II) and palladium(II) was in 1957 by Chatt and coworkers.[13-14] They showed that methanol or ethanol attacks one of the double bonds in a π-coordinated 1,5-cyclooctadiene or dicyclopentadiene (eq 8 and 11) to give the σ-bonded alkoxy adducts. Attack on the palladium diene complexes occurs considerably faster than on the corresponding platinum complexes. These reactions were later extended to other diene systems and other nucleophiles such as β-diketones,[19] diethyl malonate,[20] and amines.[21-2] (eq 9-11). The stereochemistry of the adducts was later established by Stille and coworkers,[23-4] and it was shown that trans-oxymetalation across the double bond had occurred.

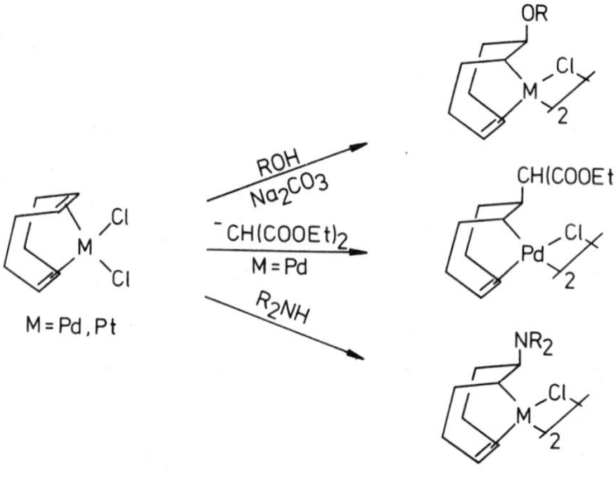

(eq 8)

(eq 9)

(eq 10)

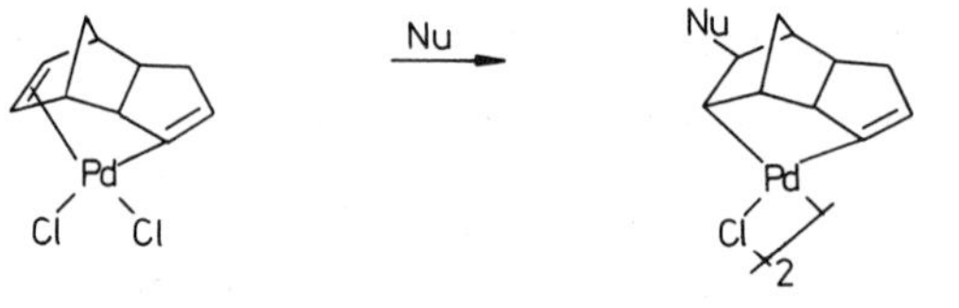

(eq 11)

$$Nu = {}^-OCH_3, {}^-CH(COOEt)_2, PhCH_2NH^-$$

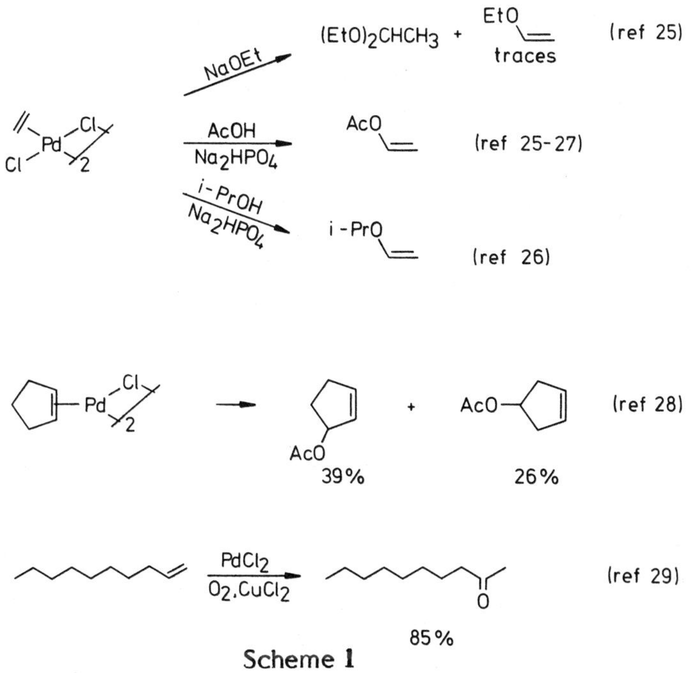

Scheme 1

$(EtO)_2CHCH_3$ + EtO— (traces)   (ref 25)

NaOEt

AcOH / Na_2HPO_4 → AcO— (ref 25-27)

i-PrOH / Na_2HPO_4 → i-PrO— (ref 26)

(ref 28)

AcO
39%   26%

$PdCl_2$ / $O_2, CuCl_2$ →   (ref 29)

O
85%

Scheme 1

In contrast to the stable adducts formed from chelated dienes on addition of a nucleophile, the corresponding adducts from monoolefins generally decompose on attempted isolation.   The most common decomposition path is β-hydride elimination, with the formation of an unsaturated organic compound and a metal hydride complex. Furthermore, addition to monoolefins coordinated to palladium or platinum is not as facile as addition to chelated dienes, since the nucleophile often attacks the metal in the monoolefin complexes to displace the olefin.   In the early studies, reactions of dienes therefore served as models for the understanding of reactions involving nucleophilic attack on monoolefin complexes.   Some reactions of monoolefinpalladium complexes related to the Wacker process are shown in Scheme 1.[25-9] All these reactions are thought to involve attack by an oxygen nucleophile on the coordinated olefin to produce a β-oxyalkylpalladium intermediate, which in none of the reactions shown has been isolated.   In later work, however, the adduct formed from methoxide attack on the cationic complex $[(ethene)Pd(Cp)PPh_3]^+$ was isolated and characterized.[30]

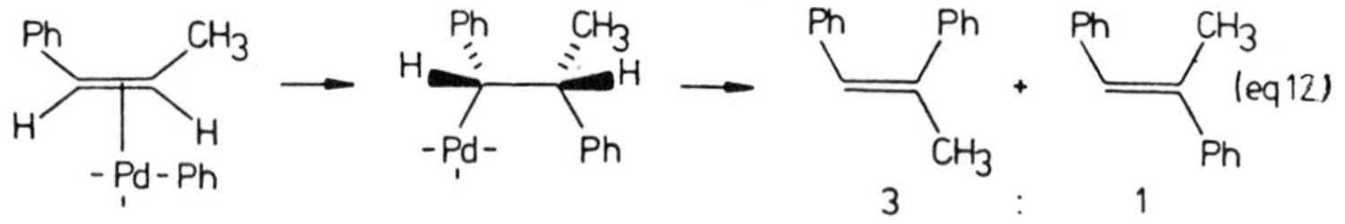

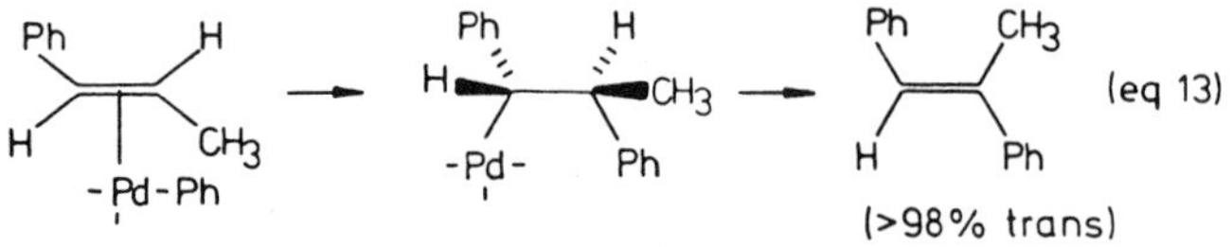

$$\text{(eq 12)}$$

$$3 \quad : \quad 1$$

$$\text{(eq 13)}$$

$$(>98\% \text{ trans})$$

Acetoxypalladation of ethene, followed by β-hydride elimination from the intermediate σ-complex, gives vinyl acetate.[25-6] This reaction is the basis for the industrially important vinyl acetate process,[27] although this process is run with a heterogeneous $PdCl_2$-$CuCl_2$-catalyst.

Arylation of olefins, involving nucleophilic attack by an aryl group on an alkene coordinated to palladium, was demonstrated independently by Heck[31] and by Fujiwara et al.[32] This reaction was developed into a useful organic synthetic method for arylation of olefins.[33] It is generally agreed that the nucleophile (the aryl group) coordinates to palladium and is intramolecularly transferred from the metal to the coordinated alkene. The products formed from β-methylstyrenes are consistent with a <u>cis</u>-addition of phenylpalladium followed by <u>cis</u>-β-elimination (eq 12 and 13).[34] Stereochemical studies on phenylpalladation of 3,3,6,6-tetradeuteriocyclohexene[35] and norbornadiene[36] also demonstrated <u>cis</u>-addition of phenyl and palladium. Methyllithium has been shown to attack styrene-palladium complexes to give β-methylstyrene in good yield.[37] Stereochemical studies using specifically deuterated styrenes indicate that the methyl nucleophile attacks the olefin from the same side as the metal (<u>cis</u>-addition), most likely via prior coordination of the methyl anion to the metal.[37]

Nucleophilic attack by amines on monoolefin complexes of platinum was shown to give stable β-aminoalkylplatinum adducts, which could be hydrolyzed to alkylamines and platinum(II).[38] Attack by amine on the optically active π-olefin complex **IV**, followed by hydrolysis of the platinum-carbon bond to give amine **V** of the S-configuration, showed that the steric course of the addition across the double bond is <u>trans</u>.[39]

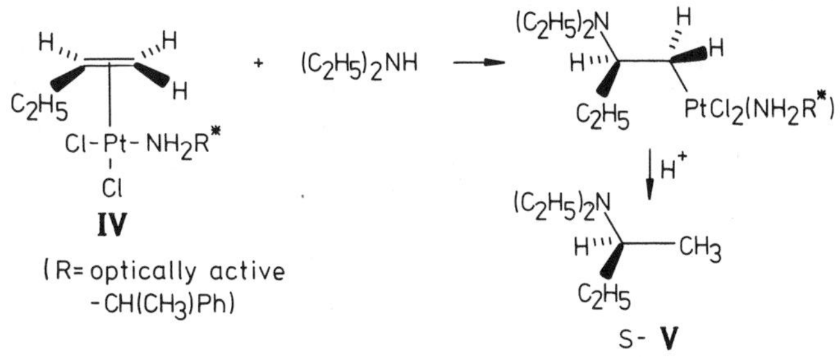

Addition of amines to monoolefin complexes of palladium under similar conditions to those employed for oxygen nucleophiles has been attempted.[26,40] Under these conditions, reaction of olefin-palladium complexes with amines gave only low yields of product amines.[40] However, more recently it was found that amines rapidly attack monoolefinpalladium complexes at low temperature (-40° to -50°) to give high yields of β-aminoalkylpalladium adducts, which on hydrogenation (-10° to 0°) give the saturated alkylamine (eq 14).[41-3] The stereochemistry of the addition was found to be <u>trans</u>-aminopalladation as shown by the reduction of the adducts from E- and Z-2-butene with LiAlD$_4$ (eq 15).[43] The configuration of the deuterated amine **VI** was established by transformation to amine oxide, followed by <u>cis</u>-Cope-elimination and analysis of the butenes formed.

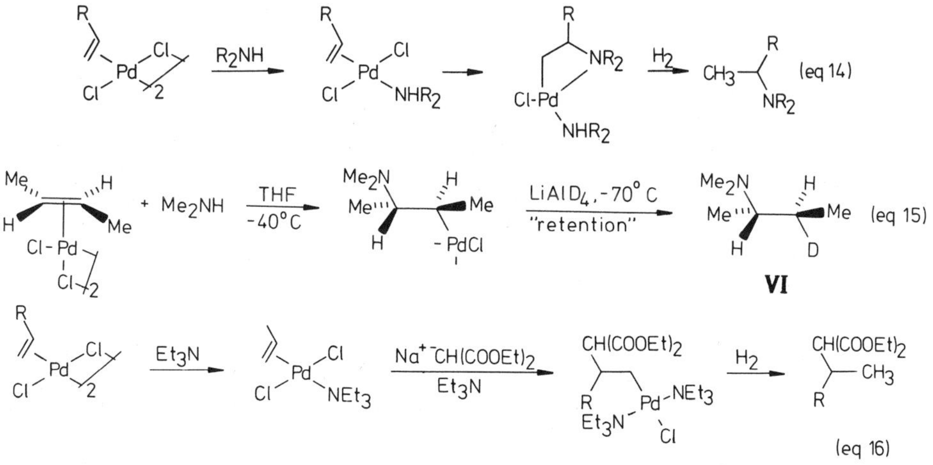

$$(eq\ 14)$$

$$(eq\ 15)$$

$$(eq\ 16)$$

By using low temperature and the presence of two equivalents of triethylamine as ligands, nucleophiles such as diethyl malonate and β-diketonates add to monoolefin complexes of palladium (eq 16).[44] Nucleophilic attack on π-olefin complex **VII** by diethyl malonate has also been reported.[45] In this case the allylic dimethylamino group serves as a ligand and in this way facilitates the nucleophilic attack.

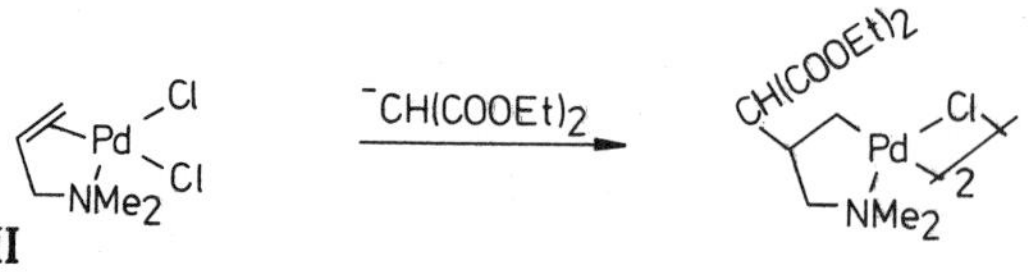

Among other nucleophiles that have been shown to attack olefins coordinated to palladium, are chloride and cyanide. Treatment of 5-vinyl-2-norbornene with $(PhCN)_2PdCl_2$ in benzene gave the complex **VIII**, whose configuration was established to be Cl- _exo_ by ^1H nmr.[46] Clearly the chloropalladation of the double bond has occurred _trans_. Similarly 7-methylene-2-norbornene underwent _trans_-chloropalladation on the exocyclic double bond.[47] Conjugated dienes react with $(PhCN)_2PdCl_2$ to give 4-chloro-π-allylic complexes **IX**.[48] The reaction proceeds via chloropalladation of one of the double bonds. Studies on this reaction with the three geometric isomers of 2,4-hexadiene show that the addition of palladium and chloride is stereospecific. Thus the E,E and Z,Z forms of 2,4 hexadiene gave one and the same diastereoisomer, whereas the E,Z form afforded another diastereoisomer, which is epimeric to the first.[49]

Reaction of ethene and propene with palladium cyanide yields products that are best explained by nucleophilic attack of cyanide on the coordinated olefin, followed by β-elimination (eq 17).[50]

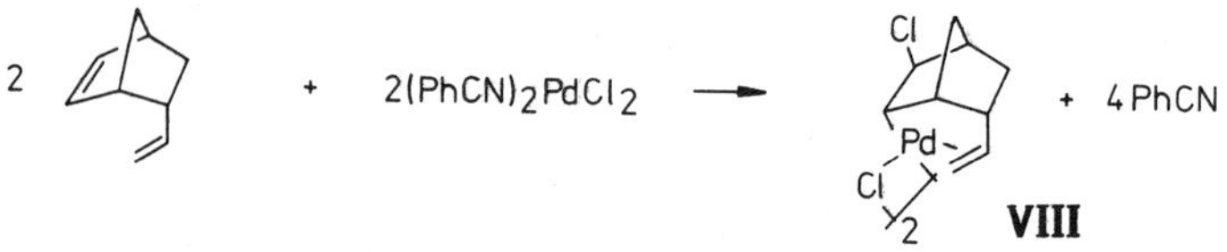

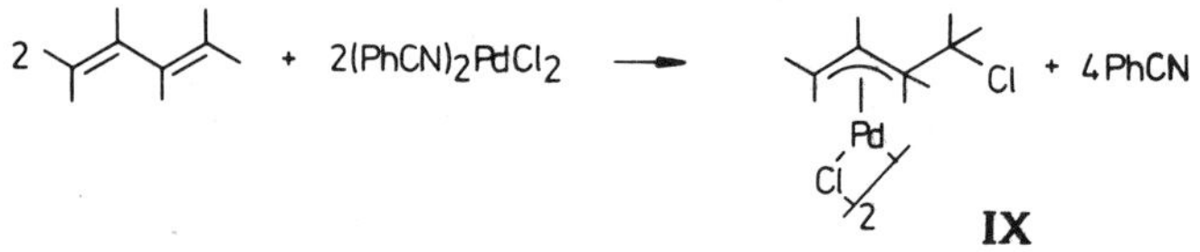

$$CH_2 = CHR \xrightarrow[\text{benzonitrile}]{Pd(CN)_2} CH_2 = C(R) - CN \quad \text{(eq 17)}$$

2.2 Stereochemistry of nucleophilic attack on olefins coordinated to palladium(II) and platinum(II)

Nucleophilic attack on an olefin coordinated to a metal may take place in two principal ways (Scheme 2). The nucleophile may attack the olefin externally on the opposite side of the metal (path A) or the nucleophile may attack the metal and then migrate to the olefin in a cis-addition process (path B). Numerous studies have been concerned with the steric course of metal-promoted nucleophilic additions, and in particular addition to palladium and platinum complexes has been thoroughly studied.[51] It has been found that additions can occur according to both pathways depending on the nature of the nucleophile.

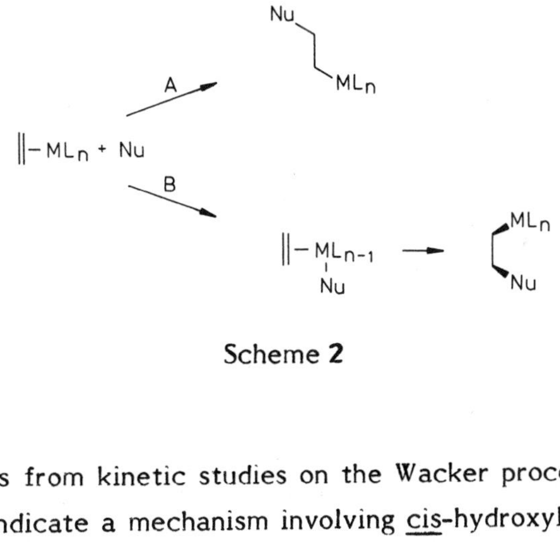

Scheme 2

The results from kinetic studies on the Wacker process were initially interpreted to indicate a mechanism involving cis-hydroxylation by transfer of a coordinated hydroxy group to a coordinated ethene (path B, M = Pd, Nu = OH).[52-3] More recent stereochemical studies have, however, shown that hydroxypalladation of E- and Z-1,2-dideuterioethene under Wacker conditions occurs trans[54] and on the basis of these results a new modified mechanism for the Wacker process was proposed (see section 2.3).

The external mode of nucleophilic attack (path A) on chelated dienepalladium complexes has been shown for several nucleophiles including

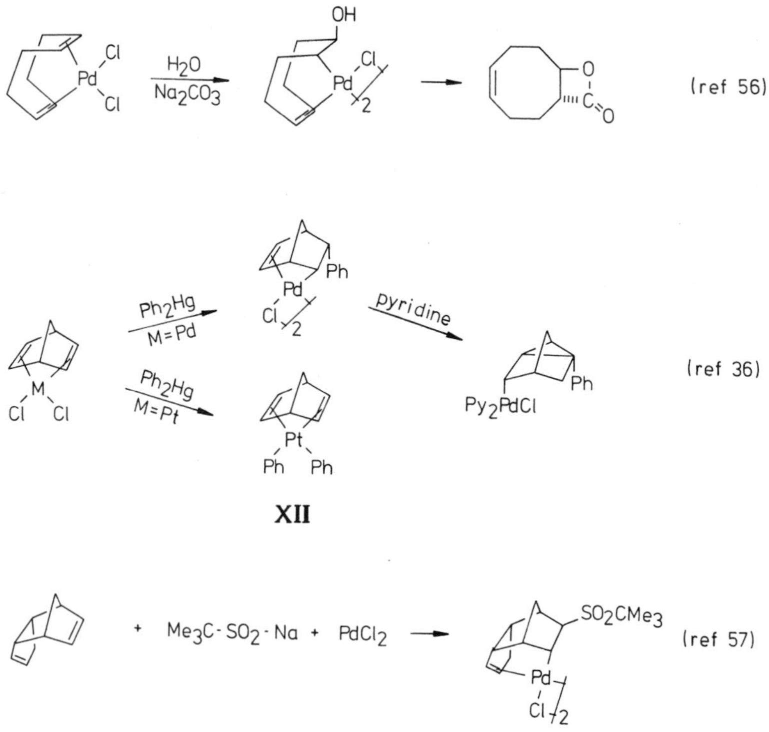

(ref 56)

(ref 36)

XII

(ref 57)

Scheme **3**

alcohols,[23] acetate,[55] hydroxide,[56] β-diketonate,[24] dialkylmalonate,[20-4] sulfinate,[57] and chloride.[46] cis-Attack (path B) on a coordinated chelated diene has been shown for phenyl.[36] Several examples were given in section 2.1 and some further examples are given in Scheme **3**.

Chloride attack on conjugated dienes coordinated to palladium has been shown to be stereospecific.[49] A cis-chloropalladation of 1,3-cyclohexadiene was suggested on the basis of nmr spectra for the adduct **X** but no conclusive evidence for such an attack was given.[49] Similar nmr arguments suggested that the analogous methoxy adduct **XI** from 1,3-cyclohexadiene, which is formed if methanol is present, has the trans-configuration. The latter assignment was recently established by stereospecific replacement (retention at carbon) of palladium by deuterium (LiAlD$_4$) in **XI**, followed by analysis of the deuterated cyclohexenyl methyl ether.[58]

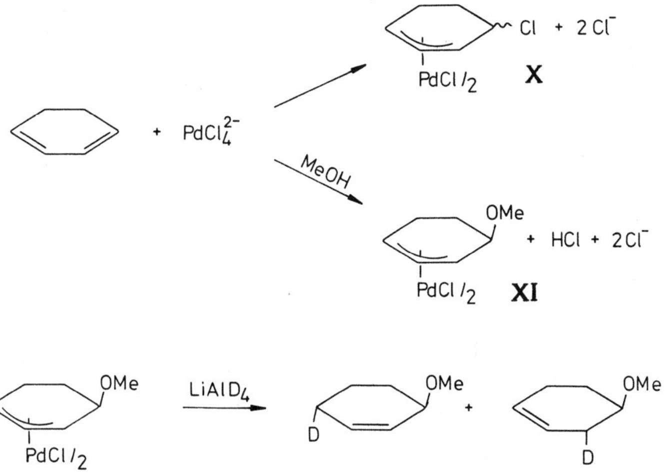

External attack has also been demonstrated for addition of oxygen nucleophiles to diene-platinum complexes (see section 2.1).[23] However, attempts to add a phenyl group to one of the double bonds in norbornadiene-platinum dichloride gave only **XII** (Scheme 3).[36]

Because adducts from monoolefins are not readily isolable, dienes for a long time served as models for monoolefins in reactions involving nucleophilic additions. However, it has been argued that the chelating dienes are atypical, since <u>cis</u>-attack by the nucleophile (path B) is expected to be disfavored for steric reasons.[59-61] The stereochemical results obtained from chelated dienes showing external <u>trans</u>-attack may therefore not be applicable to monoolefins.[60] Because of these arguments several studies of nucleophilic attack on monoolefinpalladium complexes were undertaken. Thus, nucleophiles like acetate,[62] methanol,[30,63] amine,[43] water,[54,64] and β-diketonate[65] have been shown to attack coordinated monoolefins on the side remote from the metal (path A, Scheme 2), whereas nucleophiles such as aryl,[35] methyl,[37] or methoxycarbonyl[63] have been shown to attack from the same side as the metal (path B). Some of these studies were mentioned in section 2.1. Further examples of stereochemical studies on addition to monoolefins are shown in Scheme 4.

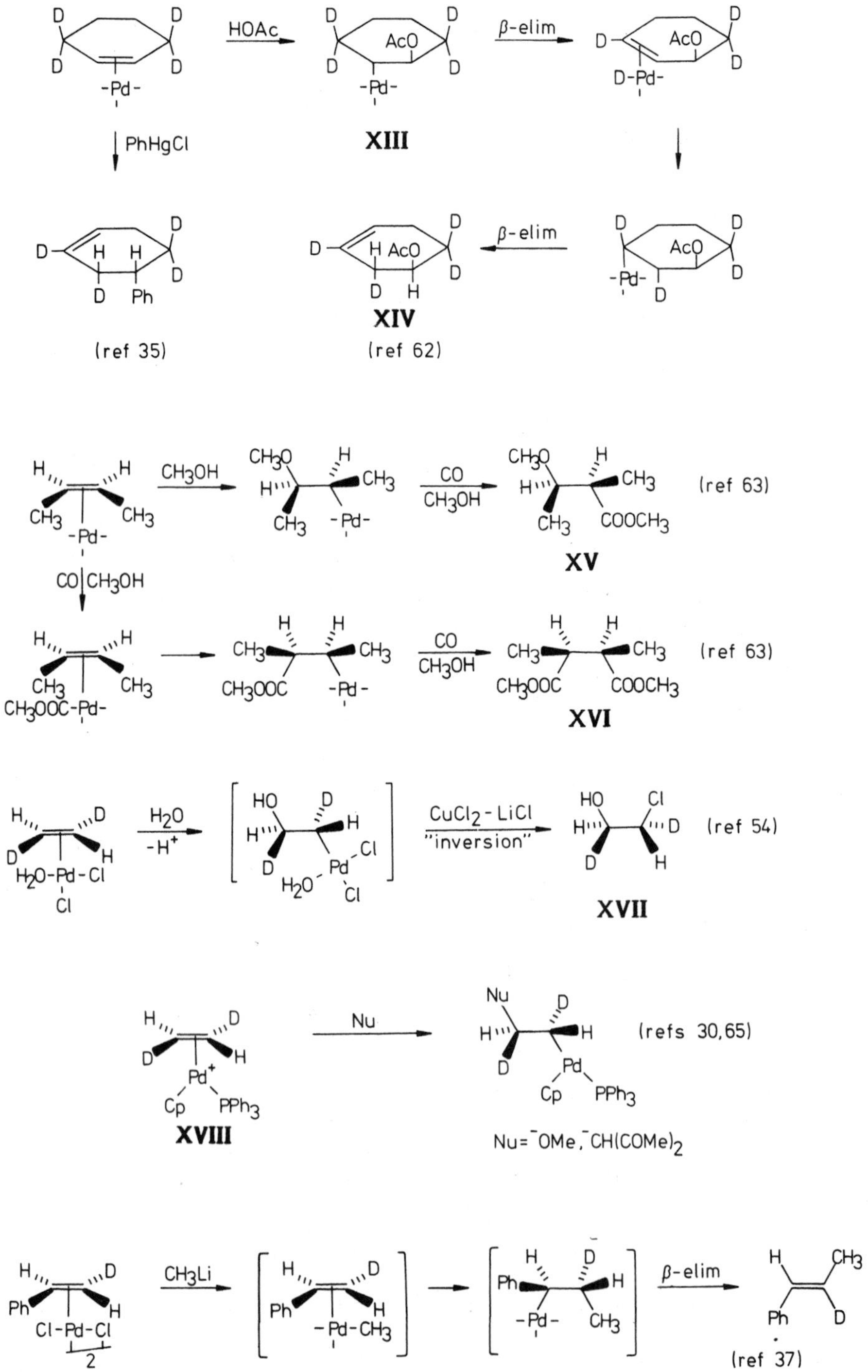
HOAC
β-elim
PhHgCl
XIII
β-elim
XIV
(ref 35)
(ref 62)
CH3OH
CO
CH3OH
(ref 63)
XV
CO|CH3OH
CO
CH3OH
(ref 63)
XVI
H2O
-H+
CuCl2-LiCl
"inversion"
(ref 54)
XVII
Nu
(refs 30,65)
XVIII
Nu= -OMe, -CH(COMe)2
CH3Li
β-elim
(ref 37)

Scheme 4

Acetoxypalladation of 3,3,6,6-tetradeuteriocyclohexene gave among other products 4-cyclohexenyl acetate **XIV**.[62] The trans-relationship between the acetoxy group and the allylic deuterium, shown by nmr, indicates that the initial adduct **XIII** also has the trans-figuration. Similarly phenylpalladation of 3,3,6,6-tetradeuteriocyclohexene affords 4-phenylcyclohexene with the phenyl group and deuterium cis to one another, indicating that the addition of palladium and phenyl has now occurred cis.[35]

Methoxypalladation of 2-butenes in the presence of carbon monoxide resulted in stereospecific formation of 3-methoxy-2-methylbutanoate (**XV**) and dimethyl-2,3-dimethyl succinate (**XVI**).[63] The formation of products **XV** and **XVI** was interpreted as a trans-methoxypalladation and a cis-methoxycarbonyl addition respectively, followed in each case by an insertion of carbon monoxide in the palladium-carbon bond.

The hydroxypalladation of Z- and E-1,2-dideuterioethene, followed by oxidative cleavage by cupric chloride of the palladium-carbon bond in the intermediate σ-complex by cupric chloride gave erythro- and threo-1,2-dideuterio-2-chloroethanol **XVII** respectively, that was analyzed by microwave spectroscopy.[54] Since the cupric chloride cleavage of the palladium-carbon bond takes place with inversion of configuration at carbon, the hydroxypalladation step must proceed trans. Nucleophilic attack on the positively charged complex **XVIII** gave stable adducts.[30,65] Nmr analysis showed that the nucleophilic attack on the coordinated ethene had occurred on the side remote from the metal.

The methyl anion from methyllithium adds cis to deuterated styrene coordinated to palladium.[37] Attack by the methyl anion on the metal followed by transfer to carbon is the most likely mechanism.

For platinum, trans nucleophilic addition to coordinated monoolefins has been demonstrated for diethylamine (see section 2.1).[39]

The stereochemical studies show that external trans-attack is strongly favored over intramolecular cis-attack for heteronucleophiles and stabilised carbon nucleophiles. Despite the fact that the nucleophile is coordinated to the metal in some cases (amine, water), external attack by another molecule of the nucleophile takes place. In contrast, another class of nucleophile, such as aryl, alkyl, and hydride, prefers the intramolecular mode of attack and does migrate from metal to carbon. It is interesting to note that a similar classification of nucleophiles and the stereochemistry obtained also appears to apply for nucleophilic attack on π-allyl- and σ-alkylpalladium complexes.[54b] There is so far no example of cis-addition of either oxygen or nitrogen nucleophiles to π-olefinpalladium complexes. However, a stereochemical study has recently shown that cis-migration of acetate from palladium to carbon takes place in π-allylpalladium complex **XIb**, when treated with carbon monoxide.[58] In this case it is possible that a σ-allyl complex is involved, and therefore a cis-addition might be favored ($S_{N'}$-type attack).

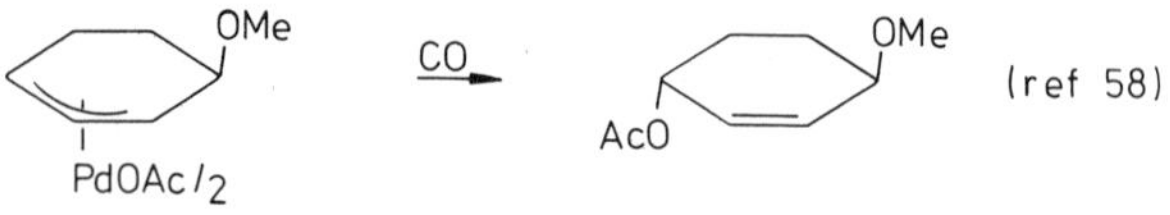

XIb

The fact that nucleophiles such as alkyl and hydride add in a cis-migration may be a result of their high energy HOMOs (highest occupied molecular orbital) which could lead to a frontier controlled process.[66a] On the other hand water, alkoxide and amine have low energy HOMOs and in these cases the nucleophilic addition is expected to be charged controlled.[66a] It is likely that charge plays a minor role in a cis-migration process and therefore this pathway could be expected to be frontier controlled.

The frontier orbitals involved in a cis-migration are shown in fig. 4. According to the method of Fukui the system has been divided into two parts.[66b,66c] The LUMO (lowest unoccupied molecular orbital) will be the π* orbital and the HOMO will be the orbital describing the metal-nucleophile bond. Now the rate of the reaction will increase with decreasing energy difference $DE = E(LUMO) - E(HOMO)$. Keeping $E(LUMO) = E(\pi^*)$ constant the rate of the reaction increases with increasing $E(HOMO) = E(M-Nu)$. We conclude that a necessary requirement for an intramolecular cis-

Fig. 4. Frontier orbitals for a cis-migration process (mixed $p_x + d_{x^2-y^2}$ on metal)

addition to occur is that the orbital describing the metal-nucleophile bond be high enough in energy, which in turn requires that the free nucleophile has a high-energy HOMO. Thus only nucleophiles with high energy HOMOs, such as alkyl anions and hydride (soft bases), would be able to add cis, whereas nucleophiles with low energy HOMOs such as hydroxide (water), alkoxide, and amine (hard bases) would react too slowly in such a process. In the latter case, external attack takes place.

2.3   The Wacker process

The industrially important process for oxidation of ethene to acetaldehyde, the so-called Wacker process, is catalyzed by palladium chloride and cupric chloride. The key step in which the carbon-oxygen bond is formed involves nucleophilic attack by water on a π-olefinpalladium complex.[5,54,67] The kinetics of the reaction have been studied by several groups,[5,52,68] and the rate expression, determined independently by these workers, is given in eq 18. The inverse dependence on acid was initially

$$d(CH_3CHO)/dt = k[ethene][PdCl_4^{2-}]/[H^+][Cl^-]^2$$

(eq. 18)

interpreted as a deprotonation of water in $PdCl_2(H_2O)(ethene)$ to give $[PdCl_2(OH)(ethene)]^-$ followed by a rate-determining cis-attack of palladium and coordinated OH^- on the coordinated ethene.[52] However, as mentioned in the previous section, hydroxypalladation of ethene-d_2 under Wacker conditions has been shown to take place with trans-stereochemistry.[54] The same stereochemistry was also found for hydroxypalladation of ethene-d_2 in aqueous acetonitrile in the presence of carbon monoxide.[64] The mechanism shown in Scheme 5[54] differs in three respects from the mechanism initially proposed on the basis of kinetic data. First, the attack by water takes place externally on a neutral π-olefinpalladium complex with trans-addition to the olefin. Second, the acid inhibition in the rate expression is a result of an equilibrium step in which reversible hydroxypalladation takes place. Finally, the rate-determining step is dissociation of a ligand, e.g. chloride, before β-elimination can take place. This mechanism for the palladium-catalyzed oxidation of ethene has been discussed[54b] and was found to be compatible

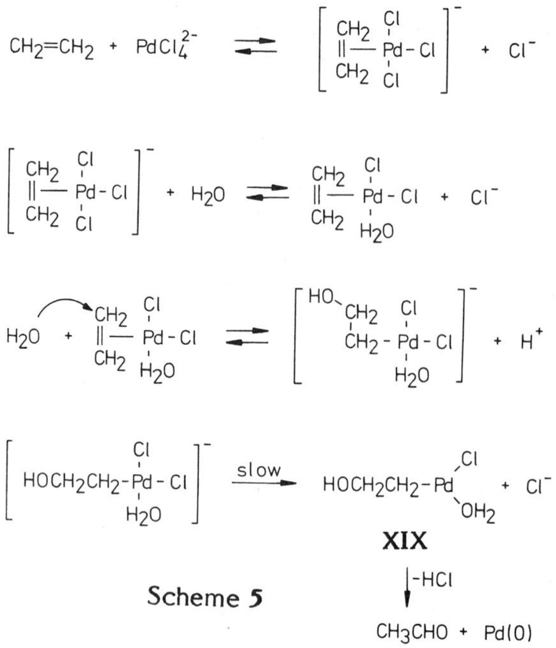

$$CH_2=CH_2 \ + \ PdCl_4^{2-} \ \rightleftharpoons \ \left[\begin{array}{c} CH_2 \\ \| \\ CH_2 \end{array}\!\!\begin{array}{c} Cl \\ Pd-Cl \\ Cl \end{array}\right]^{-} \ + \ Cl^-$$

$$\left[\begin{array}{c} CH_2 \\ \| \\ CH_2 \end{array}\!\!\begin{array}{c} Cl \\ Pd-Cl \\ Cl \end{array}\right]^{-} \ + \ H_2O \ \rightleftharpoons \ \begin{array}{c} CH_2 \\ \| \\ CH_2 \end{array}\!\!\begin{array}{c} Cl \\ Pd-Cl \\ H_2O \end{array} \ + \ Cl^-$$

$$H_2O \ + \ \begin{array}{c} CH_2 \\ \| \\ CH_2 \end{array}\!\!\begin{array}{c} Cl \\ Pd-Cl \\ H_2O \end{array} \ \rightleftharpoons \ \left[\begin{array}{c} HO{-}CH_2 \\ CH_2 \end{array}\!\!\begin{array}{c} Cl \\ Pd-Cl \\ H_2O \end{array}\right]^{-} \ + \ H^+$$

$$\left[\begin{array}{c} Cl \\ HOCH_2CH_2-Pd-Cl \\ H_2O \end{array}\right]^{-} \ \xrightarrow{\ slow\ } \ HOCH_2CH_2-Pd\!\!\begin{array}{c} Cl \\ OH_2 \end{array} \ + \ Cl^-$$

$$\mathbf{XIX}$$

$$\downarrow -HCl$$

$$CH_3CHO \ + \ Pd(0)$$

Scheme 5

with all data presently known including kinetic data,[52,67-8] isotope effects,[52a,69] and the stereochemistry[54,64] of the hydroxypalladation step.

The decomposition of the hydroxyethylpalladium complex **XIX** to acetaldehyde and palladium(0) may occur in different ways. It is, however, generally agreed that the initial step in the decomposition sequence is a β-elimination, which results in a π-vinylalcohol complex **XX**. Two pathways for the formation of acetaldehyde from **XX**, consistent with the known[67,70] formal intramolecular hydrogen shift, are possible (Scheme 6). One pathway

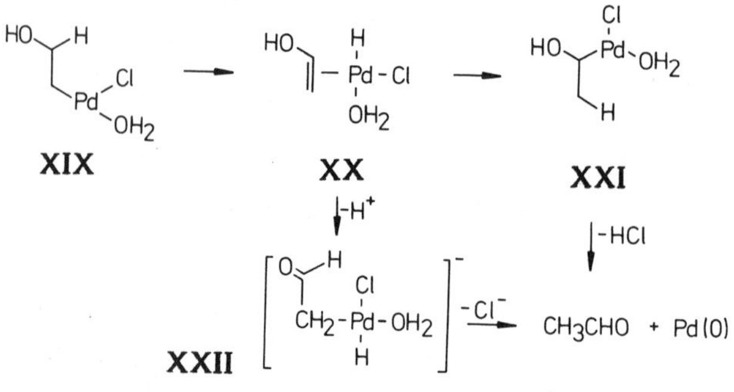

$$\begin{array}{ccc} \mathbf{XIX} & \mathbf{XX} & \mathbf{XXI} \end{array}$$

$$\mathbf{XXII}$$

$$\downarrow -H^+ \qquad\qquad \downarrow -HCl$$

$$\xrightarrow{\ -Cl^-\ } CH_3CHO \ + \ Pd(0)$$

Scheme 6

is readdition of palladium hydride to give **XXI** followed by β-elimination over oxygen and carbon, and the other is via a π-σ-rearrangement to **XXII** followed by reductive elimination.

3. NICKEL OLEFIN COMPLEXES

There are a variety of nickel(0) olefin complexes known[71] and the structure of many of these complexes has been determined by xray and neutron diffraction. On the other hand, only very few nickel(II) olefin complexes are known. The zerovalent olefin complexes are generally prepared[71] either by reduction[72] of a nickel(II) salt, usually nickel acetylacetonate, in the presence of the olefin and a stabilizing ligand or by an exchange reaction[73] of a nickel(0) complex with the appropriate olefin. Cationic diene cyclopentadienylnickel(II) complexes **XXIII** and **XXIV** have been prepared by Werner et al. from (a) $Ni_2(C_5H_5)_3$ and the diene[74] (b) nickelocene and the diene in the presence of HBF_4[75a] and (c) reaction of coordinatively unsaturated cyclopentadienyl cation and the appropriate diene.[75b] The cationic nickel ethene complex $[(PPhMe_2)CpNi(ethene)]^+$ ClO_4^- was obtained by treating $(PPhMe_2)CpNiCl$ with ethene in the presence of $AgClO_4$.[76] More recently, π-olefinnickel(II) complexes have been prepared according to eq 19.[77]

$$Cp_2Ni \; + \; CH_2{=}\overset{\underset{\displaystyle CH_3}{|}}{C}-(CH_2)_3MgCl \quad \xrightarrow[20°C]{Et_2O,\,THF} \quad CpNi\diagdown \hspace{-0.3em}\bigcirc \; + \; CpMgCl \quad (eq\ 19)$$

3.1 Nucleophilic attack on nickel(II) olefin complexes

Nucleophilic attack on olefins coordinated to nickel is known only in a few cases. The chelated diene complexes **XXIII** and **XXIV** react with excess sodium methoxide in methanol to give adducts that are formed by nucleophilic attack on one of the double bonds.[76] The adduct **XXV** from the norbornadiene complex was stable enough to be fully characterized, while the adduct **XXVI** from the 1,5-cyclooctadiene complex decomposed to **XXVII** on attempted isolation and could only be observed by nmr.

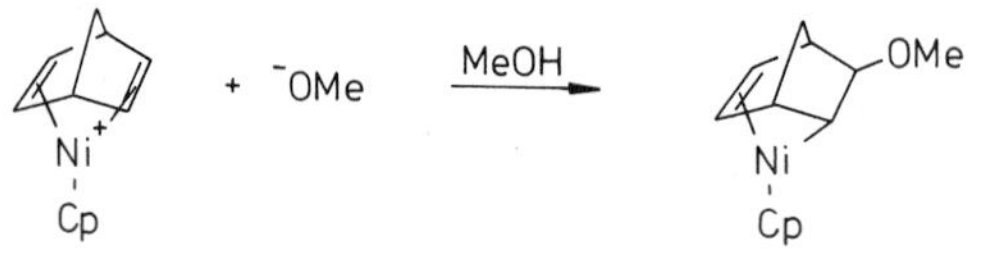

XXIII + ⁻OMe → (MeOH) → **XXV**

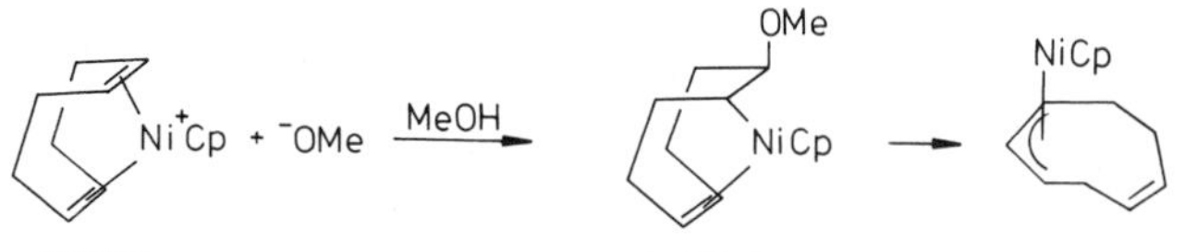

XXIV + ⁻OMe → (MeOH) → **XXVI** → **XXVII**

4.   COBALT AND RHODIUM OLEFIN COMPLEXES

Rhodium(I) olefin complexes can be prepared by the direct interaction of rhodium trichloride trihydrate and an excess of the appropriate olefin in ethanol. In this way, the 1,5-cyclooctadiene complex **XXVIII** and the bis-ethene complex **XXIX** were obtained.[78-9] In these reactions one equivalent of the olefin is oxidized; in the case of ethene, to acetaldehyde. The cobalt(I) complex **XXX** was obtained by the interaction of 1,5-cyclooctadiene with $CpCo(CO)_2$.[80] Cobalt(I) complexes **XXXI** were obtained by reaction of $Co[P(OR)_3]_4^+$ with the appropriate olefin at -78°C.[81] Other cobalt(I) olefin complexes were obtained according to eqs 20 and 21.[82] More recently, monoolefin complexes of rhodium(III) (e.g. **XXXII**) have been prepared by Werner and coworkers.[83-4]

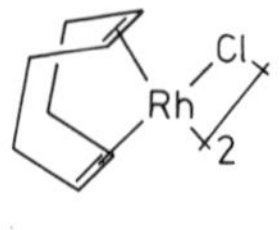

XXVIII 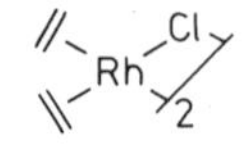   **XXIX** 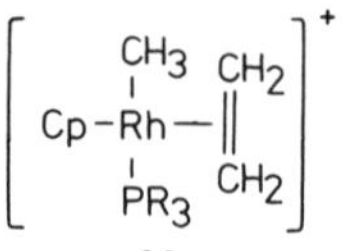   **XXXII**

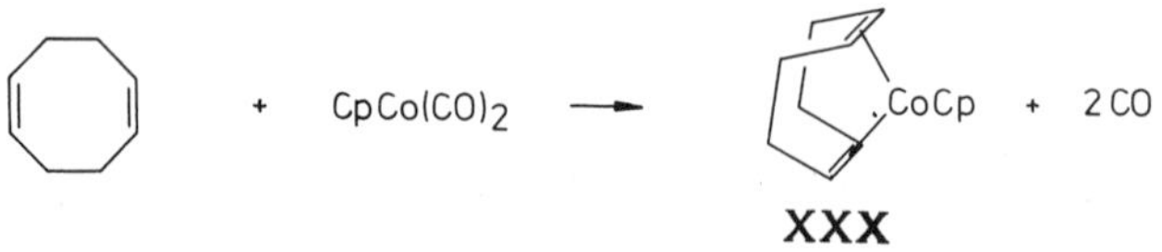

○ + $CpCo(CO)_2$ → (CoCp) + 2 CO

XXX

R—⟍= + $Co[P(OR')_3]_4^+$ → (-78°C) → R—∥—$Co^+[P(OR')_3]_4$

XXXI   R=H, CH_3

$$CoBr(PMe_3)_3$$

(eq 20)

$$\xrightarrow[\text{ethene}]{CH_3Li, THF} \quad (PMe_3)_3Co- \overset{\overset{\displaystyle CH_3}{|}}{\underset{\displaystyle CH_2}{\overset{\displaystyle CH_2}{\|}}}$$

(eq 21)

$$\xrightarrow[-70°C, CH_3CN]{Na^+BPh_4^-, ethene} \quad \left[\begin{array}{c} CH_3CN \\ | \\ \overset{}{/\!\!/-Co\overset{\cdots PMe_3}{\underset{PMe_3}{\overset{\displaystyle |}{\smash{\big|}}}}} \\ PMe_3 \end{array} \right]^+ BPh_4^-$$

4.1   Nucleophilic attack on cobalt and rhodium olefin complexes

π-Olefin complexes of cobalt and rhodium are involved in a number of homogeneous catalytic reactions, including hydrogenation and hydroformylation (the oxo process, eq 22) of olefins.[85-6] Although formal nucleophilic attack by a coordinated hydride on a coordinated alkene (cis-migration) is considered to occur in hydroformylation and hydrogenation, there are few established examples of nucleophilic additions to π-olefin complexes of cobalt and rhodium.

$$\underset{\|-Co(CO)_4}{\overset{H}{}} \longrightarrow \left[\overset{H}{\underset{Co(CO)_4}{\diagdown}} \right] \xrightarrow{CO} \left[\overset{H}{\underset{COCo(CO)_4}{\diagdown}} \right] \xrightarrow{H_2} \overset{H}{\underset{CHO}{\diagdown}} + HCo(CO)_4 \quad (eq\ 22)$$

The reaction of complex **XXIII** with ethene was studied by Evitt and Bergman.[87] The products from this reaction were propene and methane. Labeling experiments indicate that migration of the methyl group from metal to carbon takes place, rather than α-elimination followed by transfer of a methylene group.

$$\underset{\underset{33}{}}{\overset{\overset{\displaystyle CH_3}{|}}{\underset{\overset{\displaystyle |}{CH_3}}{CpCo-PR_3}}} \xrightarrow[54°C, 121\ hrs]{ethene\ (11\ atm)} \left[\underset{\overset{\displaystyle |}{CH_3}}{\overset{\overset{\displaystyle CH_3}{|}}{CpCo-\underset{CH_2}{\overset{CH_2}{\|}}}} \right] \longrightarrow \left[\underset{}{\overset{\overset{\displaystyle CH_3}{|}}{Cp-Co\diagdown_{CH_3}}} \right] \longrightarrow \diagup\!\!\diagdown^{CH_3} + CH_4$$

The reaction of ethanol with (2-acetoxyethyl)cobaloxime to give (2-ethoxyethyl)cobaloxime has been suggested to proceed via nucleophilic attack by ethanol on an intermediate π-complexed ethene coordinated to cobalt(III) in cobaloxime.[88] The intermediacy of a π-olefin complex was confirmed by the formation of a 50:50 mixture of 2-methoxyethyl-1-C_{13}- and 2-methoxyethyl-2-C_{13}(pyridine)cobaloxime from methanolysis of the corresponding 2-acetoxyethyl-2-C_{13}-cobaloxime **XXXIV**.[89] Golding et al.[90] showed independently that methanolysis of 2-acetoxyethyl-2-d_2-cobaloxime gave equal amounts of 2-methoxyethyl-1-d_2- and 2-methoxyethyl-2-d_2-cobaloxime. Furthermore, these workers showed, by using 2-acetoxy-propyl-2-(S)(pyridine)cobaloxime, that the reaction takes place with retention of configuration at carbon, consistent with a <u>trans</u>-elimination <u>trans</u>-addition sequence.[90]

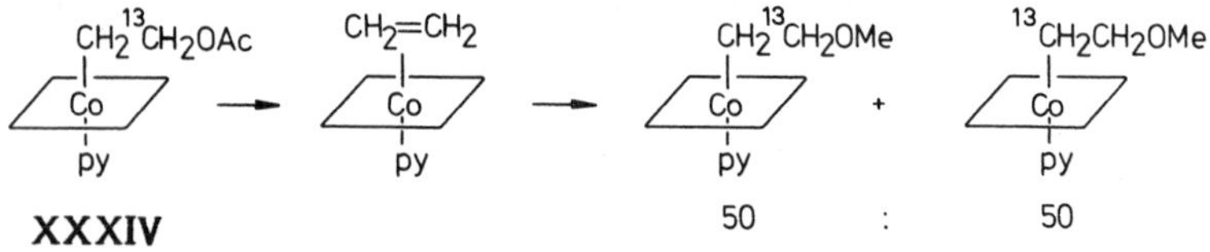

XXXIV   50   :   50

Nucleophilic attack on cycloheptatriene cobalt complex **XXXV** by nucleophiles such as H⁻, OH⁻, MeO⁻, and AcO⁻ gave stable adducts **XXXVI** in 16-65% yield.[91]

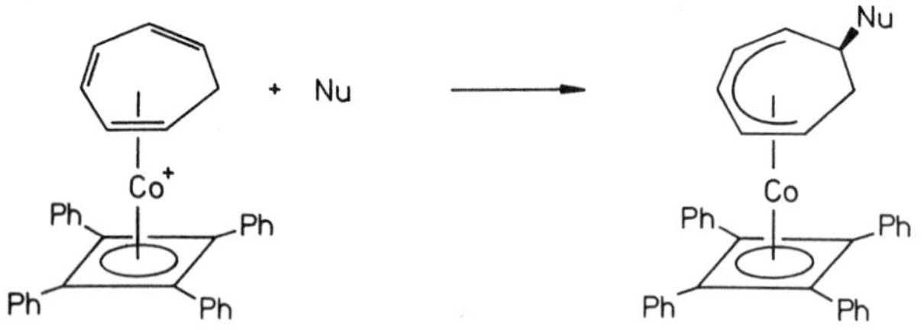

XXXV Nu= H⁻,OH⁻,MeO⁻,AcO⁻   **XXXVI**

Nucleophilic attack by water on a rhodium(III) ethene complex is probably responsible for the formation of acetaldehyde from the reaction of $RhCl_3(H_2O)$ with ethene in aqueous methanol[79] (cf. the analogous palladium-promoted reaction). In this reaction, the rhodium(I) complex **XXIX** was formed. Nucleophilic attack on a rhodium π-olefin complex probably takes place in the rhodium-catalyzed amination of ethene by morpholine.[92] A direct demonstration of nucleophilic attack on an alkene coordinated to rhodium(III) was provided by Werner and coworkers.[83-4] They showed that reaction of the cationic complex **XXXVII** with PMe₃ gives the σ-complex **XXXVIII**.[84]   The dicationic complex **XXXIX** was more reactive than complex **XXXVII** and several nucleophiles were shown to add to the coordinated alkene in **XXXIX** to give **XL**.[83]

$$
\left[
\begin{array}{c}
CH_3 \quad CH_2 \\
| \qquad \| \\
Cp-Rh- \\
| \qquad CH_2 \\
PMe_3
\end{array}
\right]^{+}
\xrightarrow{\quad PMe_3 \quad}
\begin{array}{c}
CH_3 \quad -P^{+}Me_3 \\
| \quad / \\
Cp-Rh- \\
| \\
PMe_3
\end{array}
$$

XXXVII   **XXXVIII**

$$
\left[
\begin{array}{c}
PMe_3 \; CH_2 \\
| \qquad \| \\
Cp-Rh- \\
| \qquad CH_2 \\
PMe_3
\end{array}
\right]^{2+}
\xrightarrow{\quad Nu \quad}
\begin{array}{c}
PMe_3 \quad -Nu \\
| \quad / \\
Cp-Rh^{+}- \\
| \\
PMe_3
\end{array}
$$

XXXIX   **XL**

$$Nu = Et_3N, PMe_3, P(OMe)_3$$
$$SCN^-, I^-$$

5.   IRON OLEFIN COMPLEXES

Iron olefin complexes are known for both Fe(0) and Fe(II). Their preparation, however, is not as facile as the preparation of the corresponding complexes of palladium and platinum. This rests upon the fact that ligands coordinated to iron are usually more strongly bonded to the metal than ligands coordinated to palladium and platinum. Furthermore, iron complexes in both d^6 and d^8 configurations are almost exclusively coordinatively saturated 18-electron complexes, and therefore the exchange of a ligand by an olefin must be preceded by dissociation of the ligand (dissociative exchange process).[93] The rate of the exchange will usually depend on the rate of dissociation.

[(π-Olefin)CpFe(CO)$_2$]$^+$ complexes can be prepared by the exchange of the corresponding isobutene complex with the appropriate olefin (eq 23).[94-5] Other methods include β-elimination of alkyliron complexes,[96] and treatment of an epoxide with [CpFe(CO)$_2$]$^-$ followed by protonation and elimination of water (eq 24).[97] Reaction of CpFe(CO)$_2$I with AgBF$_4$ generates the highly reactive intermediate [CpFe(CO)$_2$]$^+$, which on treatment with two or three equivalents of the appropriate olefin produces the olefin complex in good yields.[98] Similarly, {(olefin)CpFe(CO) [P(OPh)$_3$]}$^+$ complexes have been prepared by this method.[99]

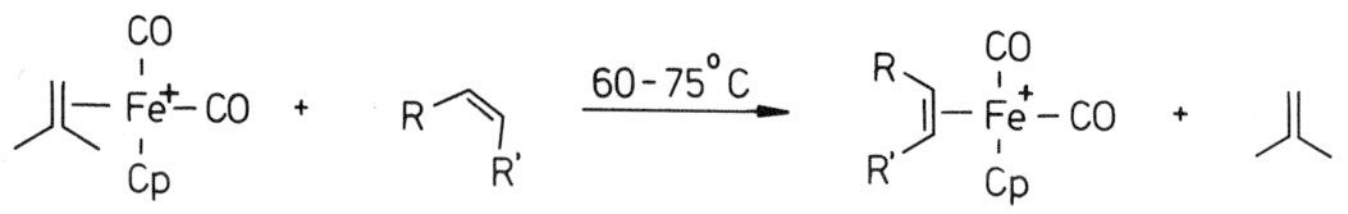

$$\text{(eq 23)}$$

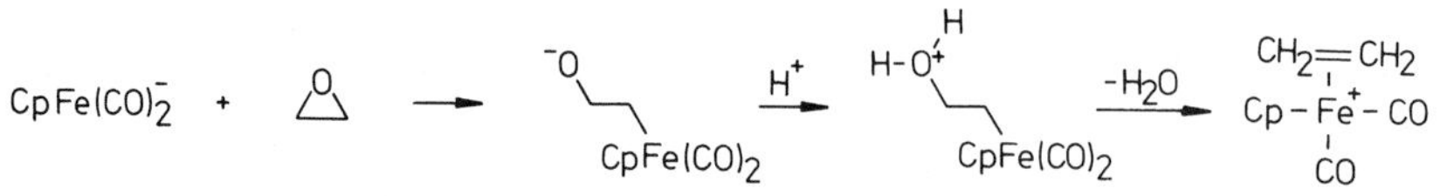

$$\text{(eq 24)}$$

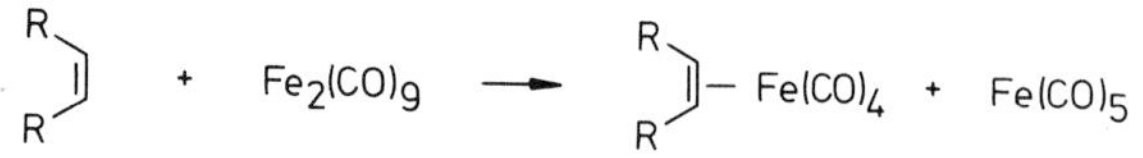

$$\text{(eq 25)}$$

Iron(0) olefin complexes can be prepared by reaction of an olefin with $Fe_2(CO)_9$ (eq 25).[100-1] The reaction with acrylic esters and maleic anhydride proceeds smoothly, while the reaction with ethene requires an ethene pressure of 50 atm for 48 hours.

5.1   Nucleophilic attack on iron olefin complexes

Nucleophilic addition to alkenes coordinated to $[CpFe(CO)_2]^+$ has been studied with a variety of different nucleophiles. One of the first examples of such a reaction was reported by Green and Nagy,[96] who found that hydride ($NaBH_4$) add to the 1-propene complex XLI (L = CO) to give a σ-isopropyliron complex XLII. Similarly, hydride attack on the ethene complex gave the corresponding ethyliron complex. More recent work has shown that $NaBH_4$ reduction of XLI (L = CO) also gives a small fraction of the σ-n-propyliron complex XLIII,[102] (XLII:XLIII=3:1), whereas $NaBH_3CN$ yielded the isopropyl derivative XLII exclusively.[103] On the other hand, in the hydride addition to the triphenylphosphite analogue XLI (L = P(OPh)$_3$) there is a predominance for formation of σ-n-propyliron complex XLIII.[99] Thus for $NaBH_4$ the ratio XLII:XLIII is 1:2 and for $NaBH_3CN$ it is 1:4. Reduction of $[(1\text{-butene})CpFe(CO)P(OPh)_3]^+$ with $NaBH_3CN$ gave the n-butyl derivative exclusively.[99]

A group of Italian workers found that reaction of the ethene complex XLIV with methoxide or amine yields the stable, though somewhat air sensitive, σ-bonded alkyliron complexes XLV.[104] Treatment of the adducts

$$\underset{\substack{\\ \textbf{XLI} \quad L = CO, P(OCH_3)_3}}{\underset{L}{\overset{CO}{Cp-\overset{+}{Fe}-\|}}{}_{CH_3}} \quad + \quad H^- \quad \longrightarrow \quad \underset{\textbf{XLII}}{\underset{L}{\overset{CO}{Cp-Fe-CH}}{}^{CH_3}_{CH_3}} \quad + \quad \underset{\textbf{XLIII}}{\underset{L}{\overset{CO}{Cp-Fe-CH_2}}{}_{\underset{CH_3}{CH_2}}}$$

XLV with HCl regenerated the cationic π-complexes **XLIV**. Other nucleophiles such as N_3^-, NCO^- and CN^- failed to give adducts from attack on olefins. N_3^- attacked on a coordinated carbon monoxide while NCO^- and CN^- displaced the olefin ligand. However, more recent work has shown that if tetraethylammonium cyanide in acetonitrile is used, an adduct $CpFe(CO)_2CH_2CH_2CN$ is formed from **XLIV** in about 21% yield.[105] Furthermore, the triphenylphosphite analogue gave an adduct $CpFe[P(OPh)_3]_2CH_2CH_2CN$ under the same reaction conditions.

Further studies on nucleophilic additions to olefin complexes such as **XLIV** and other analogues have been done by Rosenblum and coworkers.[102,106-7] A number of nucleophiles including methanol, amines, phosphines, phosphites, and thiols were found to add to a variety of different olefins coordinated to $CpFe(CO)_2^+$.[102] The addition to monosubstituted olefins is regiospecific and gives the Markovnikov adduct with a few exceptions. For example, dimethylamine and 1-butylmercaptan attack coordinated styrene at the least substituted carbon in the styrene complex **XLVI** (R = Ph). Stabilised carbon nucleophiles such as dialkyl malonate and β-diketonates add to iron olefin complexes to give σ-alkyl derivatives in which a new carbon-carbon bond has been formed (Scheme 7).[106-7] Enamines also attack the coordinated alkene to give an iminium salt, which on hydrolysis gives an organometallic ketone. The synthetic utility of these reactions will be discussed in Section 9.

$$\underset{\textbf{XLIV}}{\underset{CO}{\overset{CO}{Cp-\overset{+}{Fe}-\|}}{}^{CH_2}_{CH_2}} \quad + \quad Nu^- \quad \longrightarrow \quad \underset{\textbf{XLV} \quad Nu = OMe, NHCH_3}{\underset{CO}{\overset{CO}{Cp-Fe-CH_2CH_2Nu}}}$$

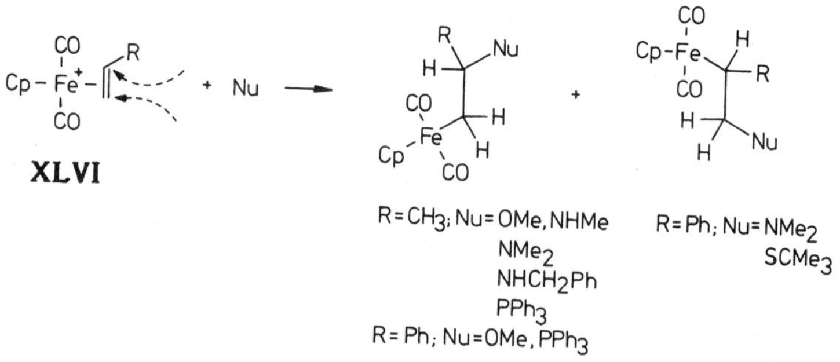

$R = CH_3$; $Nu = OMe$, $NHMe$
NMe_2
$NHCH_2Ph$
PPh_3
$R = Ph$; $Nu = OMe$, PPh_3

$R = Ph$; $Nu = NMe_2$
$SCMe_3$

Nucleophilic addition to irontetracarbonyl olefin complexes has also been reported.[108-9] Reaction of **XLVII** with dimethyl malonate gave an adduct, which on hydrolysis gave **XLVIII** in 45% yield.[108] Reaction of α,β-unsaturated compounds coordinated to irontetracarbonyl, e.g. **XLIX**, with dialkyl malonate and related stabilised carbon nucleophiles resulted in Michael adducts.[109] The α-chloromethylacrylate complex **XLIX** (R = Cl) gave products where two nucleophiles have attacked the hydrocarbon ligand. Iron carbene complex formation by α-elimination of chloride from the primary adduct **L**, followed by nucleophilic attack on the carbene carbon, was suggested to account for the product.

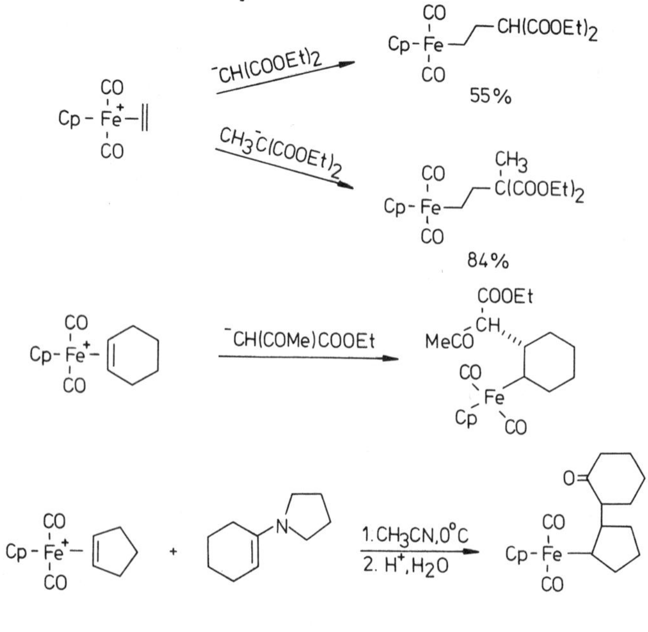

Scheme 7

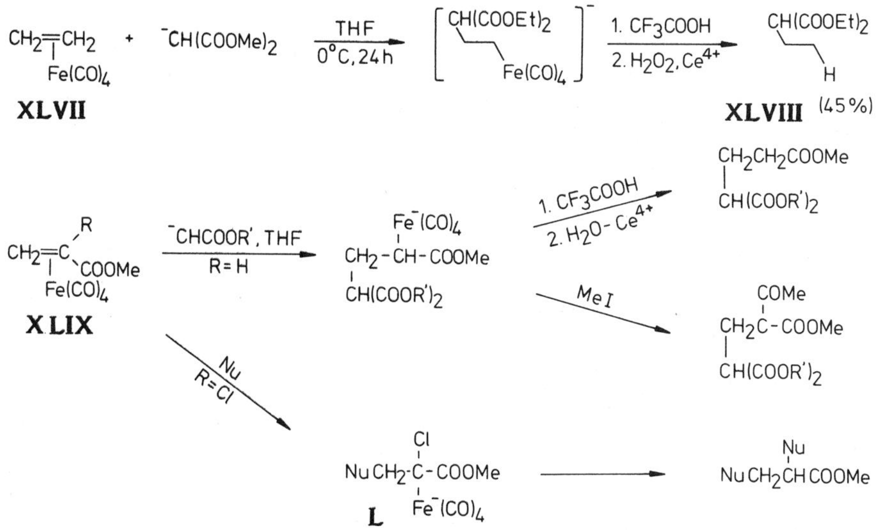

$$CH_2{=}CH_2 \ + \ ^-CH(COOMe)_2 \xrightarrow[0°C, 24h]{THF} \left[\underset{Fe(CO)_4}{\overset{CH(COOEt)_2}{\diagdown}} \right]^- \xrightarrow[2. H_2O_2, Ce^{4+}]{1. CF_3COOH} \underset{H}{\overset{CH(COOEt)_2}{\diagup}}$$

XLVII   **XLVIII** (45%)

$$\underset{Fe(CO)_4}{\overset{R}{\underset{|}{CH_2{=}C}}}{\diagdown}COOMe \xrightarrow[R=H]{^-CHCOOR', THF} \underset{\overset{|}{CH(COOR')_2}}{\overset{Fe(CO)_4}{CH_2-CH-COOMe}}$$

1. CF$_3$COOH
2. H$_2$O- Ce^{4+}

$$\underset{CH(COOR')_2}{\overset{CH_2CH_2COOMe}{\underset{|}{CH(COOR')_2}}}$$

MeI

$$\underset{\overset{|}{CH(COOR')_2}}{\overset{COMe}{\underset{|}{CH_2C-COOMe}}}$$

XLIX

Nu R=Cl

$$\underset{\overset{|}{Fe^-(CO)_4}}{NuCH_2\overset{Cl}{\underset{|}{C}}-COOMe} \longrightarrow NuCH_2\overset{Nu}{\underset{}{CH}}COOMe$$

L

Reaction of nucleophiles with unconjugated cyclic dienyl ligands coordinated to irontricarbonyl results in nucleophilic attack on the coordinated double bond.[110-1] The nucleophilic attack by cyanide takes place from the <u>exo</u>-direction.

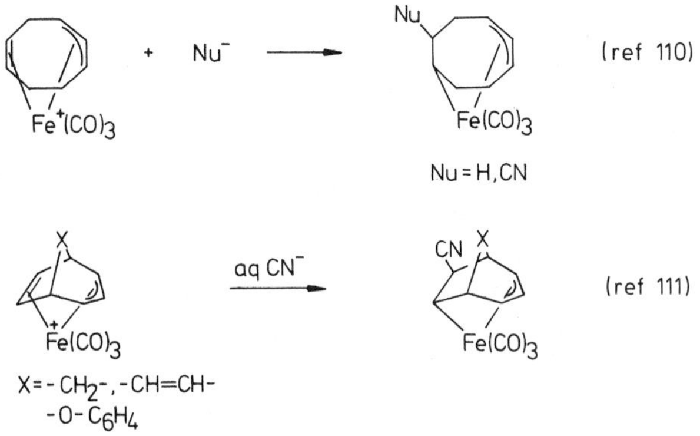

(ref 110)

Nu = H, CN

aq CN$^-$

(ref 111)

X = - CH$_2$-, -CH=CH-
-O- C$_6$H$_4$

5.2 Stereochemistry of nucleophilic attack on olefins coordinated to iron

Reaction of the acenaphthene-iron complex **LI** with a variety of different nucleophiles was studied by Nicholas and Rosan.[112] Most nucleophiles, including MeO$^-$, PPh$_3$, Et$_3$N, and I$^-$, displace the olefin. However, sterically hindered nucleophiles such as $^-$SCMe$_3$ and $^-$CMe$_2$CHO

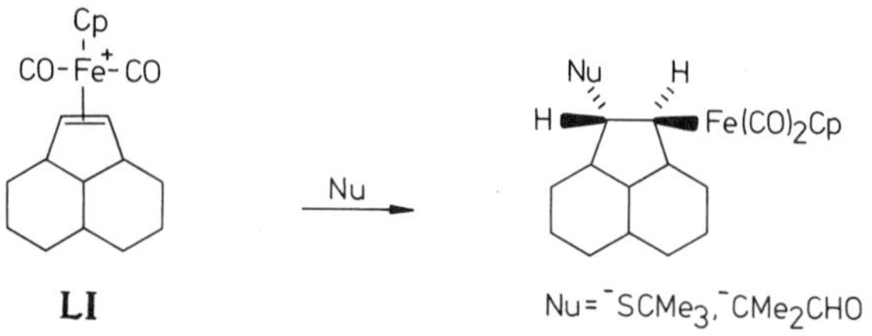

LI

Nu = $^-SCMe_3$, $^-CMe_2CHO$

gave stable adducts. The configuration of these adducts, as determined by nmr, was found to be <u>trans</u>. Addition to the corresponding iron(II) benzocyclobutadiene complex also resulted in a <u>trans</u>-adduct.[113] Only one stereochemical study of nucleophilic addition to acyclic olefins coordinated to iron has been reported. Nucleophilic attack by benzylamine on E- and Z-2-butene coordinated to $[CpFe(CO)_2]^+$ was shown to take place from the side remote from the metal.[114] Thus reaction of the Z-2-butene complex LII with the amine, followed by an oxidation induced insertion of carbon monoxide into the iron-carbon bond and ring closure, gave <u>trans</u>-β-lactam LIII. Since carbon monoxide insertion takes place with retention of configuration at carbon, the initial addition of amine must occur <u>trans</u>.

Although the stereochemistry of addition of carbon nucleophiles such as dialkyl malonate has not been demonstrated, it is generally assumed to be <u>trans</u>.[106-7] A good argument for <u>trans</u>-addition to cationic complexes $[(alkene)CpFe(CO)_2]^+$ is that the complexes are 18-electron, coordinatively saturated complexes. <u>cis</u>-Addition would require an initial coordination by the nucleophile to the metal, which can only take place if one of the ligands dissociates. Furthermore, nucleophilic additions to other related unsaturated hydrocarbon ligands coordinated to iron have been shown to take

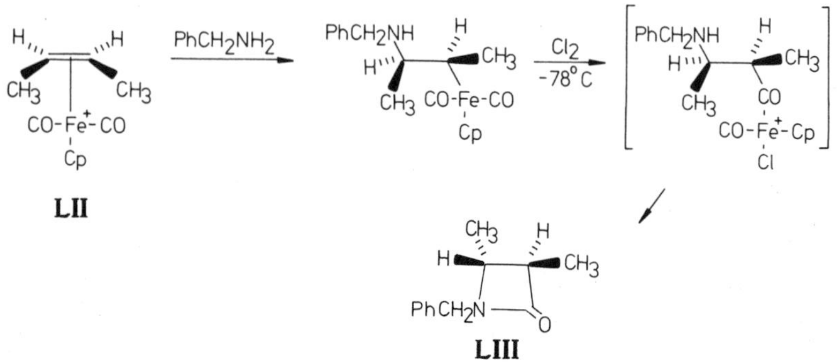

LII

LIII

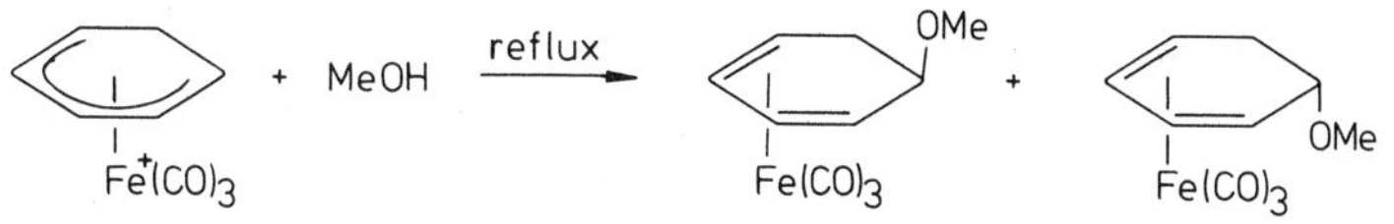

LIV   **LV**   **LVI**

place by exo-attack.[115] Even hydride and alkyl appear to attack the hydrocarbon ligand from the side remote from the metal. This is in sharp contrast to the chemistry of palladium and platinum olefin complexes, where nucleophiles such as hydride and alkyl attack the coordinatively unsaturated metal and migrate from metal to carbon.

The stereochemistry of hydride attack on olefin-iron complexes has, however, not been satisfactorily established, and one cannot completely exclude formation of a metal hydride complex followed by a cis-attack by coordinated hydride.

It has been shown that nucleophilic addition of methanol to cyclohexadienyliron tricarbonyl (**LIV**) mainly results in exo-attack to give **LV**.[116] However a small fraction (2%) of **LVI**, the product from endo attack, was observed. When the reaction was allowed to stand the product **LVI** increased and finally the thermodynamic mixture **LV:LVI** = 1:2 was obtained. It was suggested that the exo-addition is reversible and that a competing endo-addition (cis-attack) by coordinated methanol takes place.[116b] In agreement, it has been reported that methanol coordinates to iron in cycloheptadienyl iron tricarbonyl complexes prior to reaction at the unsaturated hydrocarbon ligand.[117] However, in this case, only external exo-attack by methanol was observed.

6. RUTHENIUM AND OSMIUM OLEFIN COMPLEXES

Diene ruthenium(0) olefin complexes can be prepared by the reaction of $Ru_3(CO)_{12}$ with a large excess of the diene, e.g. 1,5-cyclooctadiene in refluxing benzene (eq 26).[118] Cationic ruthenium(II) monoolefin complexes

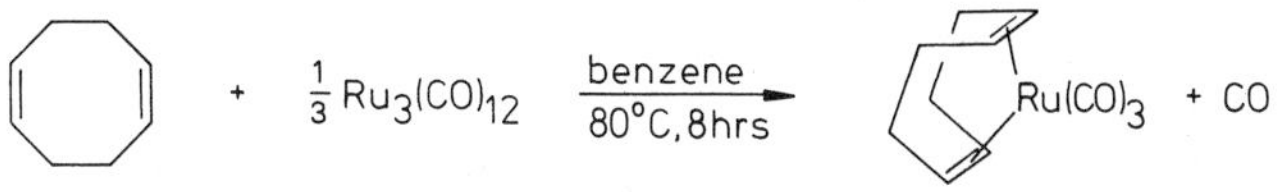

$$(C_6H_6)RuCl_2(PMe_3) \xrightarrow[\text{acetone}]{AgPF_6} \left[(C_6H_6)RuCl(acetone)PMe_3\right]^+ PF_6^- \xrightarrow{C_2H_4} \left[C_6H_6-\underset{PMe_3}{\overset{Cl}{Ru}}\underset{CH_2}{\overset{CH_2}{\parallel}}\right]^+ \quad (eq\ 27)$$

LVII

LVII have been prepared according to eq 27.[119] Reduction of **LVII** with sodium naphthalide in THF gives the ruthenium(0) olefin complex **LVIII**. Cationic ethene complexes **LIX** were obtained[120] on protonation or on alkylation (CH_3I) of **LVIII**. Cationic olefin complexes **LXI** of ruthenium were obtained by abstraction of a chloride from **LX** in the presence of the appropriate olefin.[121]

Cationic ethene complexes of osmium, analogous to the ruthenium complexes **LVIII** and **LIX** have been prepared by Werner et al.[122]

$$\textbf{LVII} \xrightarrow{NaC_{10}H_8} C_6H_6-\underset{PMe_3}{Ru}\overset{CH_2}{\underset{CH_2}{\diagdown}} \xrightarrow{RX} \left[C_6H_6-\underset{PMe_3}{\overset{R}{Ru}}\underset{CH_2}{\overset{CH_2}{\parallel}}\right]^+ \quad (eq\ 28)$$

LVIII   **LIX**

$$\underset{PPh_2}{\overset{Cp}{Ph_2P-Ru-Cl}} \xrightarrow[NH_4PF_6]{R} \left[\underset{PPh_2}{\overset{Cp}{Ph_2P-Ru}}\parallel_R\right]^+ PF_6^-$$

LX   **LXI**

6.1 Nucleophilic attack on ruthenium olefin complexes

Only a few examples of nucleophilic attack on double bonds coordinated to ruthenium and osmium have been reported. Werner et al.[83-4,120] have studied nucleophilic addition to monoolefin complexes of ruthenium. They found that trimethylphosphite adds to the cationic complex **LXII** to give **LXIII**.[120] Treatment of the analogous hydride complex **LXIV** with trimethylphosphine did not result in attack by the phosphine on the

$$\left[\underset{PMe_3}{\overset{CH_3}{Ru}}\underset{CH_2}{\overset{CH_2}{\parallel}}\right]^+ \xrightarrow{PMe_3} \underset{PMe_3}{\overset{CH_3}{Ru}}-P^+Me_3$$

LXII   **LXIII**

(eq 29)

R = Me, OMe, Ph, OPh

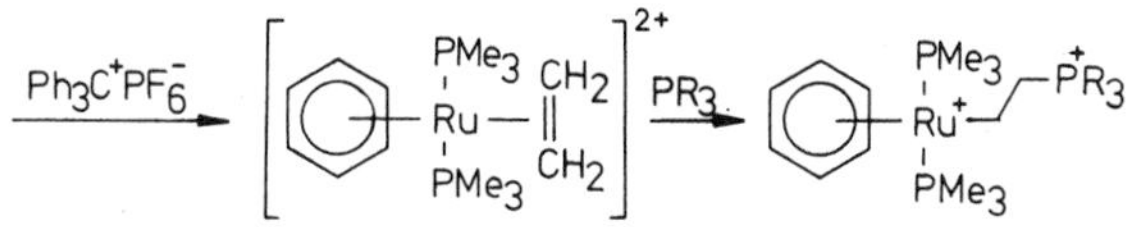

coordinated alkene, but rather in migration of the coordinated hydride to give the ethylruthenium complex **LXV**. The dicationic ethene complex **LXVI**, obtained by hydride abstraction from **LXV**, was attacked by phosphines and phosphites on the coordinated ethene (eq 29).[83]

Reaction of neutral octahedral ruthenium complexes $Ru(CO)(\eta^2\text{-}C_2H_4)Cl_2(PMe_2Ph)_2$ with different nucleophiles at $-5°C$ to $+5°C$ was recently studied.[123] Only the nucleophiles PMe_2Ph and $PMePh_2$ took part in attack on the coordinated alkene, whereas nucleophiles such as $AsMe_2Ph$, NH_2CH_2Ph and 4-methylpyridine displaced it.

Reaction of the cationic osmium ethene complex **LXVII** with trimethylphosphine resulted in nucleophilic attack on the coordinated ethene to give **LXVIII**.[122]

Nucleophilic attack on some chelated cyclic olefin complexes of ruthenium has been studied.[118,124-6] Reaction of **LXIX** with borohydride resulted in nucleophilic attack by H^- on the coordinated double bond to give **LXX**.[118,124a] This is in contrast to the isoelectronic cobalt and rhodium complexes, e.g. $[CpRh(C_8H_{11})]^+$, where the nucleophilic attack by hydride occurs on the π-allyl moiety to give the cyclooctadiene complex.[124b] Attack on the olefinic bond in complex **LXIX** by cyanide gave **LXXI**.[118,125]

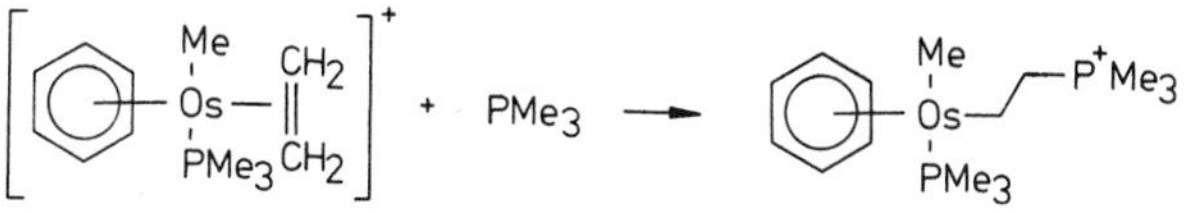

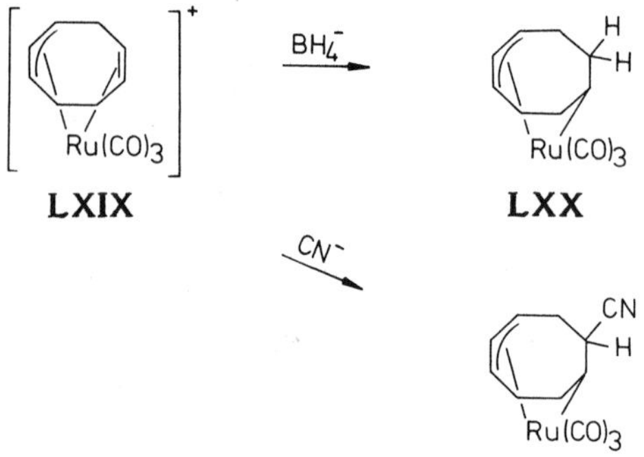

Determination of the structure of complex **LXXI** by xrays showed[125] that the cyanide attack has occurred <u>exo</u>. Reaction of **LXXII** with nucleophiles such as CN^- and $NO_2CH_2^-$ results in attack by the nucleophile on the coordinated double bond.[126]

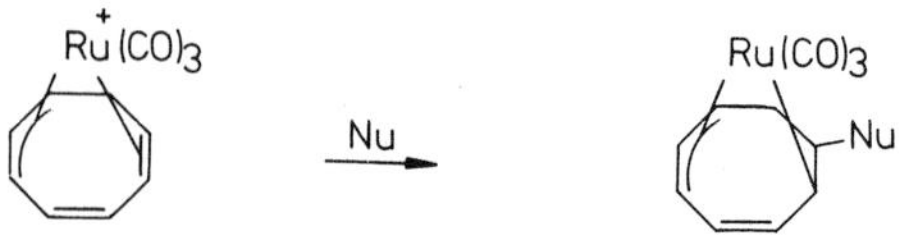

7.   MANGANESE AND RHENIUM OLEFIN COMPLEXES

Cationic ethene complexes **LXXIII** and **LXXIV** of manganese(I) and rhenium(I) are obtained from the reaction of the chloropentacarbonyl metal with $AlCl_3$ in the presence of ethene (eqs 30 and 31).[127] In a similar manner, the 1,5-cyclooctadiene complexes **LXXV** were prepared.[128] A propene analogue of **LXXIII** has been obtained by protonation of (σ-propenyl)Mn(CO)$_5$.[129]

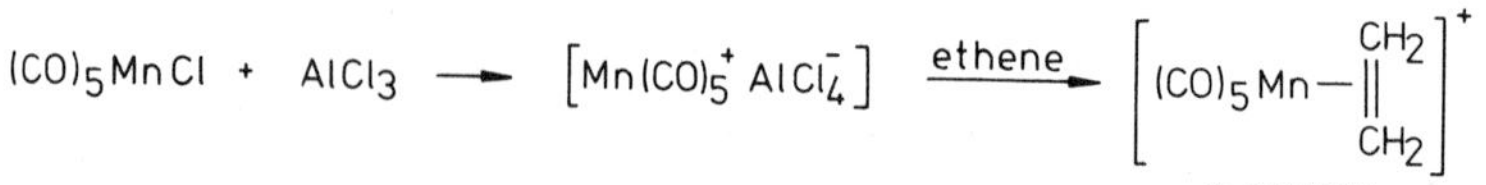

$$(CO)_5MnCl + AlCl_3 \longrightarrow \left[Mn(CO)_5^+ \, AlCl_4^-\right] \xrightarrow{\text{ethene}} \left[(CO)_5Mn-\overset{CH_2}{\underset{CH_2}{\|}}\right]^+ \quad \text{(eq 30)}$$

LXXIII

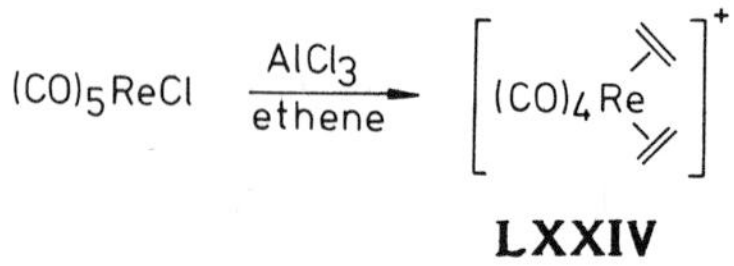

$$(CO)_5ReCl \xrightarrow[\text{ethene}]{AlCl_3} \left[(CO)_4Re\right]^+ \quad \text{(eq 31)}$$

LXXIV

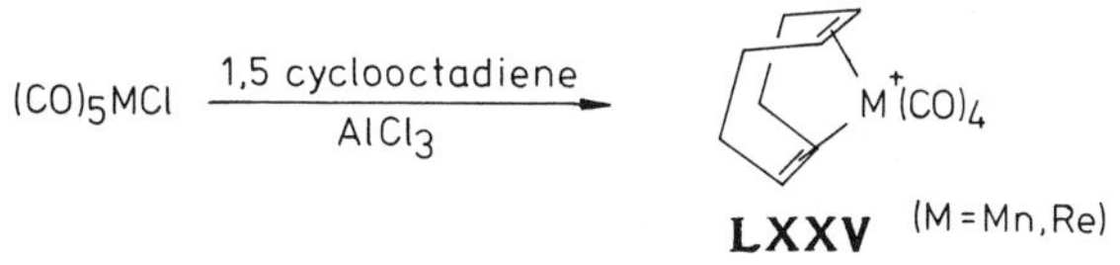

$$(CO)_5MCl \xrightarrow[AlCl_3]{1,5 \text{ cyclooctadiene}} \quad \text{LXXV} \quad (M = Mn, Re)$$

7.1 Nucleophilic attack on manganese and rhenium olefin complexes

There are no reported examples of nucleophilic attack on monoolefins coordinated to manganese or rhenium. However, nucleophilic addition to di- and triolefin complexes of these metals has been demonstrated. Thus, nucleophiles readily attack one of the coordinated double bonds in the cationic cyclooctadiene complex LXXVa.[128] The adducts LXXVI, in particular the one obtained from attack by azide, are air-sensitive and thermally unstable. Nucleophilic attack on the cycloheptatriene complex LXXVII gave aducts LXXVIII in good to high yield (67-95%).[130] Reaction of LXXIV with sodium hydroxide did not yield an adduct LXXVIII, but rather gave the dimeric ether LXXIX. All these adducts were assigned as the 6-exo-isomers on the basis of the nmr spectra of the compounds and by comparison of spectroscopic data with those of related compounds.[130-1]

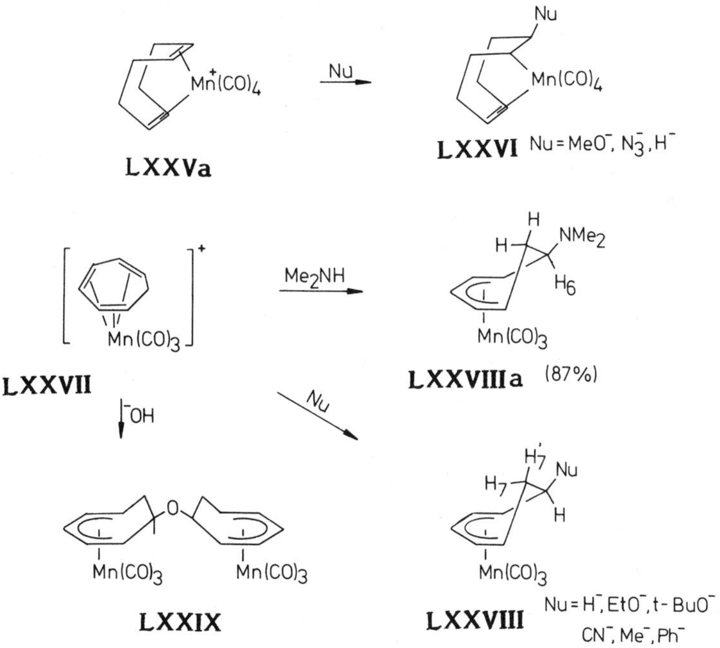

LXXVa

LXXVI Nu = MeO⁻, N₃⁻, H⁻

LXXVII

LXXVIIIa (87%)

LXXIX

LXXVIII Nu = H⁻, EtO⁻, t-BuO⁻ CN⁻, Me⁻, Ph⁻

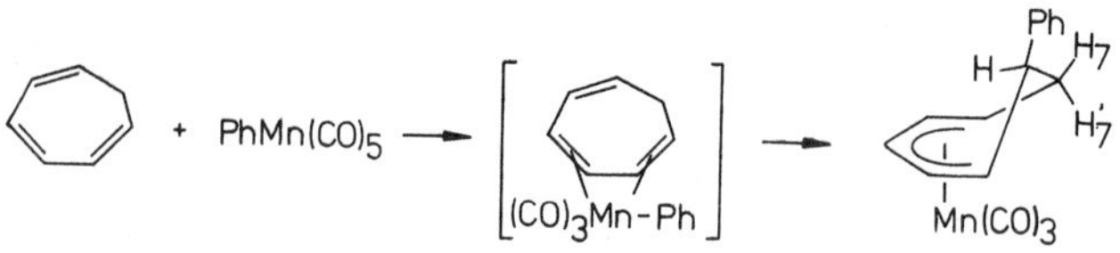

LXXX

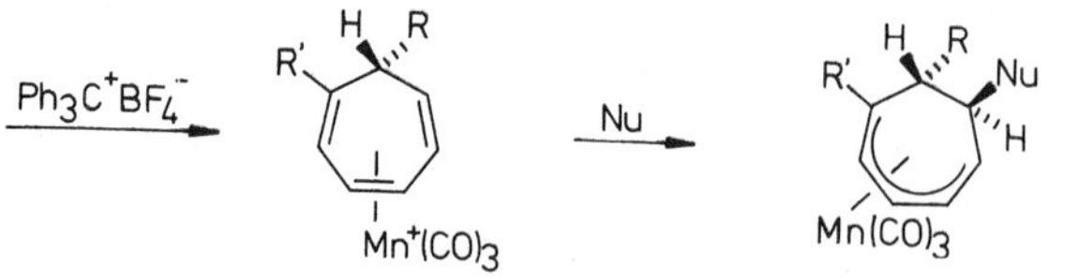

LXXXI a R=Ph, R'=H
        b R=H, R'=Ph

LXXXII Nu = ⁻H, ⁻D
           ⁻Me, ⁻OMe

Phenylmanganese pentacarbonyl reacts with cycloheptatriene in heptane at reflux temperature to give **LXXX** in 49% yield.[132] The reaction most probably involves intramolecular nucleophilic attack by the coordinated phenyl group on one of the coordinated double bonds in cycloheptatriene. A similar reaction takes place with phenylrhenium pentacarbonyl and cycloheptatriene, but the yield is much lower in this case (14%).[132] Evidence for <u>endo</u>-attack (<u>cis</u>) by the phenyl group on the cycloheptatriene was given by the proton nmr spectrum of **LXXX**, which indicates that this compound is the 6-<u>endo</u> isomer. The nmr spectrum of **LXXX** differs from that of **LXXVIII** (Nu = Ph), which is its 6-<u>exo</u> isomer. The most obvious difference is the shift of the H(7) proton. Due to a twist conformation about C_6 and C_7, the <u>exo</u>-proton H(7) in **LXXVIII** (Nu = Ph) will be within the shielding region of the cycloheptadienyl π-system and appears at $\delta = 1.1$ ppm. The proton at highest field in **LXXX** appears at $\delta = 2.1$ ppm.

Treatment of the adduct **LXXX** with $Ph_3C^+BF_4^-$ gave a mixture of π-cycloheptatrienemanganese complexes **LXXXIa** and **LXXXIb**, which react with nucleophiles such as hydride, methyl and methoxide to give complexes **LXXXII**.[132] One of these isomers, **LXXXIIa** (R =Ph, Nu = H) was identical to **LXXX**, and moreover, when deuteride was used as the nucleophile, the proton assigned as <u>exo</u>-7 had disappeared in the product **LXXXIIa** (R = Ph, Nu = D). The proton nmr spectra for all adducts **LXXXII** were consistent with an external attack by the nucleophile to give the <u>exo</u>-Nu configuration.

8.   MOLYBDENUM AND TUNGSTEN OLEFIN COMPLEXES

The preparation of molybdenum(II) and tungsten(II) monoolefin complexes resembles the preparation of the corresponding iron olefin complexes.   They generally can not be prepared by a simple exchange reaction, but rather by generation of a coordinatively unsaturated complex (usually by stripping off a halide), in the presence of the olefin. Thus the complexes **LXXXIII** are prepared according to eq 32.[133] Another route to divalent olefin complexes of molybdenum and tungsten is by hydride abstraction from the corresponding alkyl complexes (eq 33).[105,134] Zerovalent diene and triene complexes of molybdenum and tungsten have been prepared by reaction of the appropriate diene or triene with $M(CO)_6$ (M = Mo or W) at $100\text{-}120^{\circ}C$.[135-7]

$$CpM(CO)_3Cl \xrightarrow{AlCl_3} \left[CpM(CO)_3\right]^+ AlCl_4^- \xrightarrow{\text{ethene}} \left[\begin{array}{c} CO \quad CO \quad CH_2 \\ Cp-M-\;\|\; \\ CO \quad\;\; CH_2 \end{array}\right]^+ \quad \text{(eq 32)}$$

LXXXIII M = Mo,W

$$CpM(CO)_3CH_2CH_2R \xrightarrow{Ph_3C^+X^-} \left[\begin{array}{c} CO \quad CO \quad CH_2 \\ Cp-M-\;\|\; \\ CO \quad\;\; CH \\ \qquad\qquad R \end{array}\right]^+ X^- \quad \text{(eq 33)}$$

M = Mo,W $X = BF_4^-, PF_6$

8.1   Nucleophilic attack on molybdenum and tungsten olefin complexes

Addition of ammonia, amines and several other nucleophiles to coordinated ethene in the molybdenum(II) and tungsten(II) complexes **LXXXIV** was studied by Knoth.[105] Deprotonation of the adducts derived from ammonia and primary amines resulted in an insertion-cyclization reaction to give **LXXXV**. The structure of **LXXXV** (M = Mo, R = H) was confirmed by an xray study.[138] Phosphines also attack the coordinated alkene to give σ-adducts.   Reaction of **LXXXIV** (M = Mo) with tetraethylammonium cyanide gave the adduct **LXXXVI** (M = Mo, Nu = CN) in 33% yield.

Butadiene, coordinated to molybdenum in complex **LXXXVII**, reacted rapidly with the anions MeO⁻, MeS⁻, and CN⁻ to give the π-allyl derivatives **LXXXVIII**.[139] Although nucleophilic attack may in principle occur on either

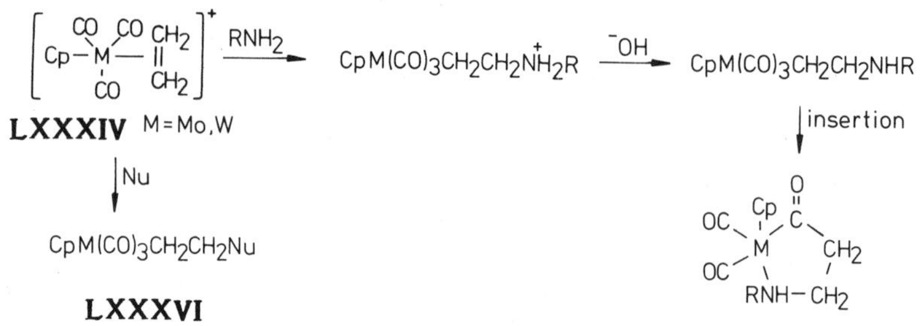

LXXXIV M = Mo,W

LXXXVI

LXXXV

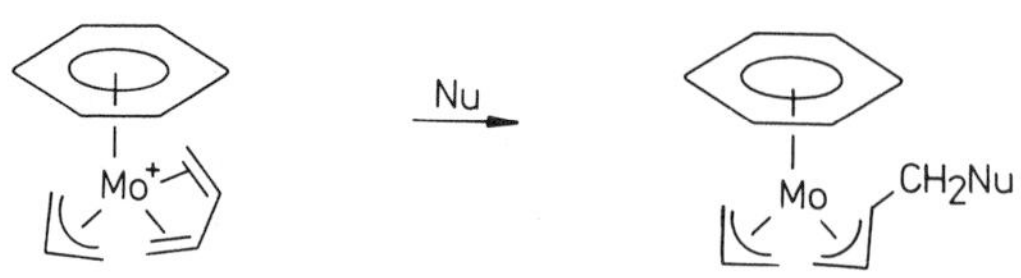

LXXXVII   **LXXXVIII**

of the three unsaturated hydrocarbon ligands, attack on the butadiene is preferred. Rules for predicting the most favorable position for nucleophilic attack in 18-electron organotransition metal cations, containing unsaturated hydrocarbon ligands, have been reviewed[140] and are discussed in Chapter 14.

9.   APPLICATIONS IN ORGANIC SYNTHESIS

Nucleophilic addition to transition metal π-olefin complexes has great potential in organic synthesis. There are numerous examples where such reactions are used to obtain carbon-oxygen, carbon-nitrogen and carbon-carbon bonds. The metals most frequently used are palladium and iron and to some extent platinum. Reviews dealing with π-olefin complexes of palladium and iron in organic synthesis have been published.[141-2]

π-Complex formation of 1,5-cyclooctadiene followed by nucleophilic addition of malonate gives a σ,π-complex **LXXXIX**, that can further react

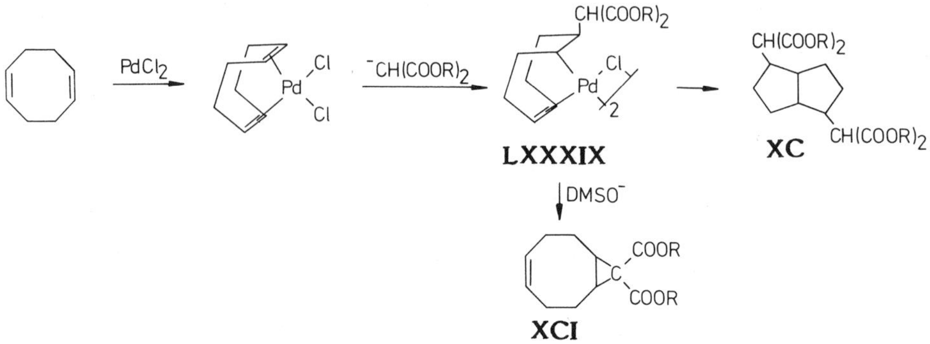

LXXXIX   **XC**

XCI

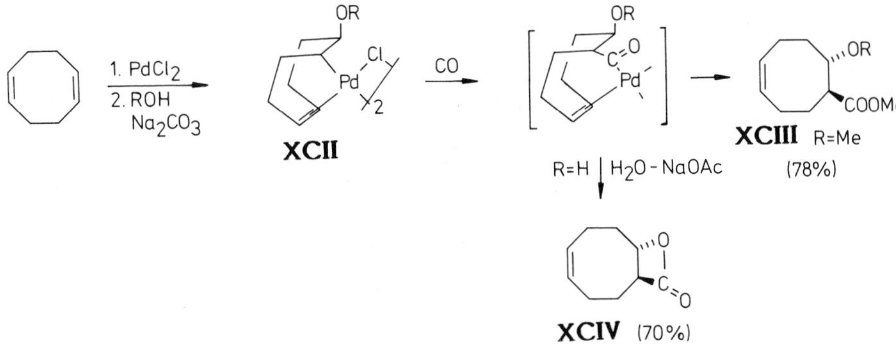

XCII

XCIII R=Me (78%)

R=H | H₂O - NaOAc

XCIV (70%)

with malonate to give **XC**.[20,143] Treatment of complex **LXXXIX** with the sodium salt of dimethyl sulfoxide gives **XCI**, via an intramolecular nucleophilic displacement of palladium by carbon.

Treatment of complexes **XCII**, related to **LXXXIX**, with carbon monoxide in methanol, using sodium carbonate as base, produces β-substituted esters **XCIII**.[144] If the nucleophile used to obtain **XCII** is hydroxide (R = H), β-lactone **XCIV** is formed in good yield when the carbonylation is performed in water containing sodium acetate. These reactions can also be extended to monoolefins, although the β-lactone reaction gives a good yield only with ethene (eq 34).[64] The oxycarbonylation reactions can also be performed with catalytic amounts of palladium using cupric chloride as reoxidant. Thus terminal olefins afford β-methoxyesters **XCV** in a yield of 50-60% in the corresponding catalytic reaction.[145] Internal and cyclic olefins gave only low yields of the oxycarbonylation adduct and the main products in these cases were dicarbonylation adducts (eq 35). Dicarbonylation to give diesters was favored by the addition of weak bases such as sodium acetate. Formation of cis-1,2-diesters is best

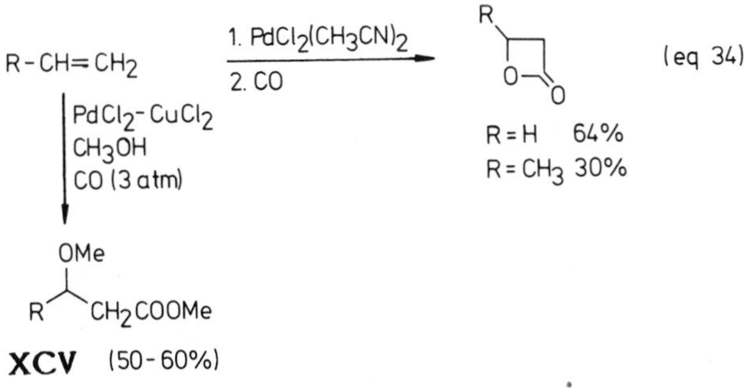

(eq 34)

R=H   64%
R = CH₃ 30%

XCV (50-60%)

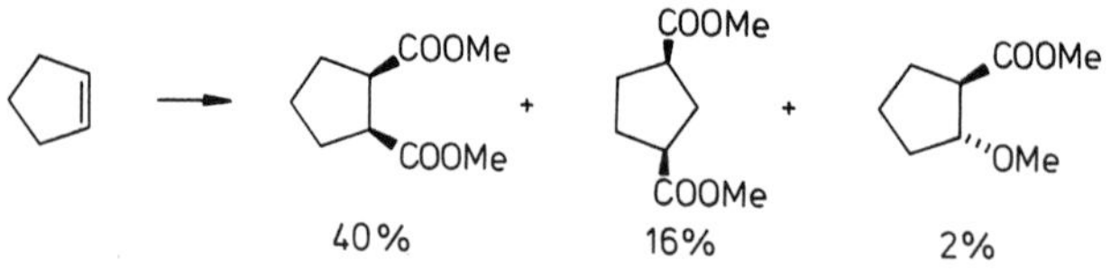

explained by <u>cis</u>-addition of a carbomethoxy-palladium species to the double bond to give a σ-complex, followed by insertion of carbon monoxide into the metal-carbon bond (cf. section 2.1).

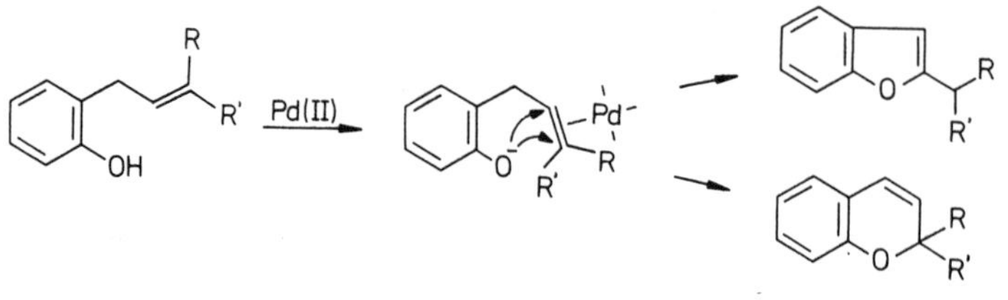

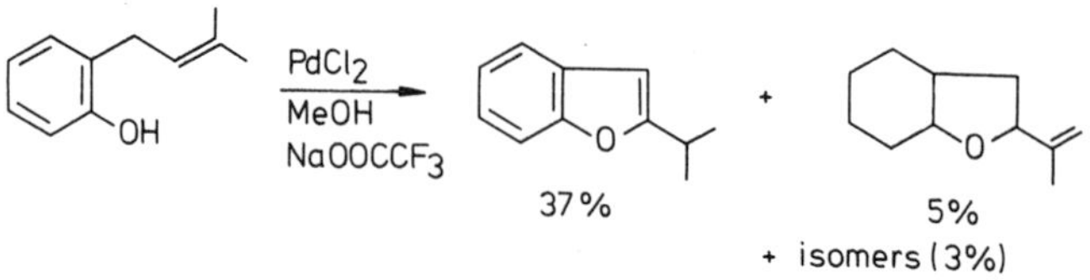

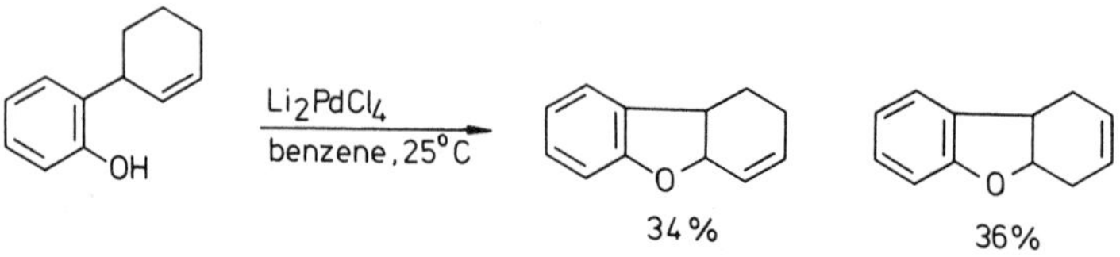

Intramolecular oxpalladation of 2-allylphenols has been used to prepare heterocyclic systems (eqs 36-9).[146-7]   Intramolecular <u>trans</u>-nucleophilic attack by phenolate anion on the coordinated double bond, followed by β-elimination, is a likely mechanism. Addition of sodium carboxylates affects the regioselectivity of the nucleophilic attack and favors the formation of the 5-membered ring (eq 38).

Nucleophilic addition of aryl- and alkylpalladium species to alkenes is a useful reaction for forming carbon-carbon bonds.[148-50] Palladium is used only in catalytic amounts and the aryl- or alkylpalladium species can be generated either from the corresponding mercury compound, the aryl or alkyl halide, or, in the aryl case, by electrophilic substitution of a hydrogen

Scheme 8

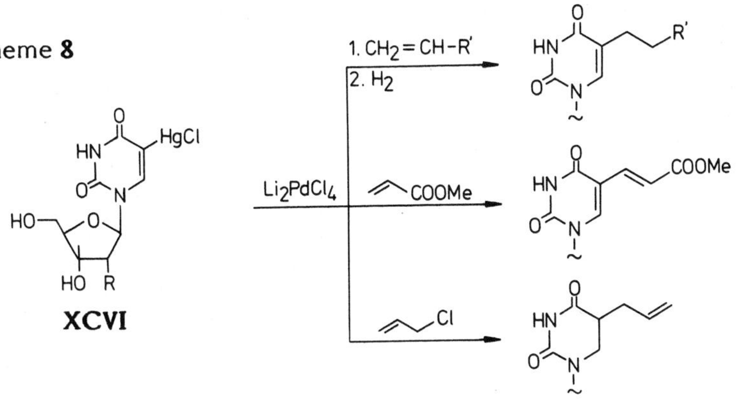

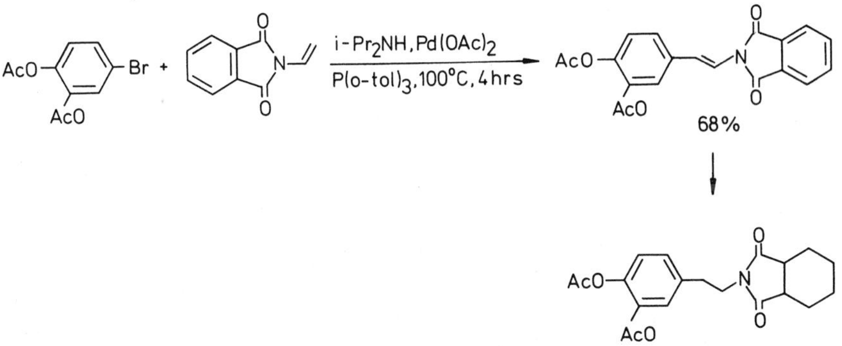

XCVII   55%

by palladium.   Some examples are given in Scheme **8.**   C-5-Substituted pyrimidine nucleosides were synthesized by reaction of the corresponding 5-chloromercuripyrimidine nucleoside **XCVI** with an olefinic compound in the presence of palladium(II).[148-9] By arylation of vinylsuccinimide followed by hydrogenation, a 2-arylethylamine was obtained, e.g. the protected dopamine **XCVII.**[150] It is interesting to note that one of the aromatic rings hydrogenated at about the same rate as the double bond.

Aminopalladation[41] of olefins followed by cleavage of the palladium-carbon bond in various ways has been used to functionalize double bonds (Scheme **9).**[41,151-4] Hydrogenation of the palladium-carbon bond gives the amine.[41] Oxidative cleavage of the palladium-carbon bond in the presence of acetate or excess amine yields aminoalcohol derivatives[151] and vicinal

Scheme 9

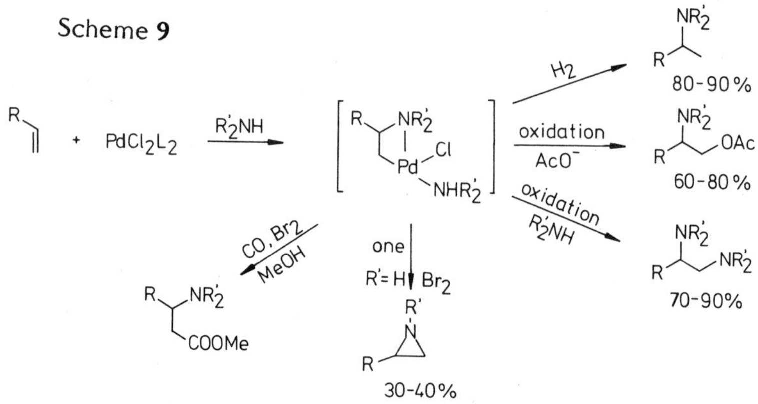

diamines[152] respectively. The reactions are stereospecific and the vicinal groups are introduced cis to one another. The cis-stereochemistry is a result of trans-aminopalladation, followed by nucleophilic displacement of palladium with inversion of configuration.[151-2] Furthermore, treatment of the aminopalladation adducts from primary amines with Br$_2$ affords aziridines.[153] Insertion of carbon monoxide into the palladium-carbon bond followed by reaction with bromine in methanol gives β-amino esters.[154]

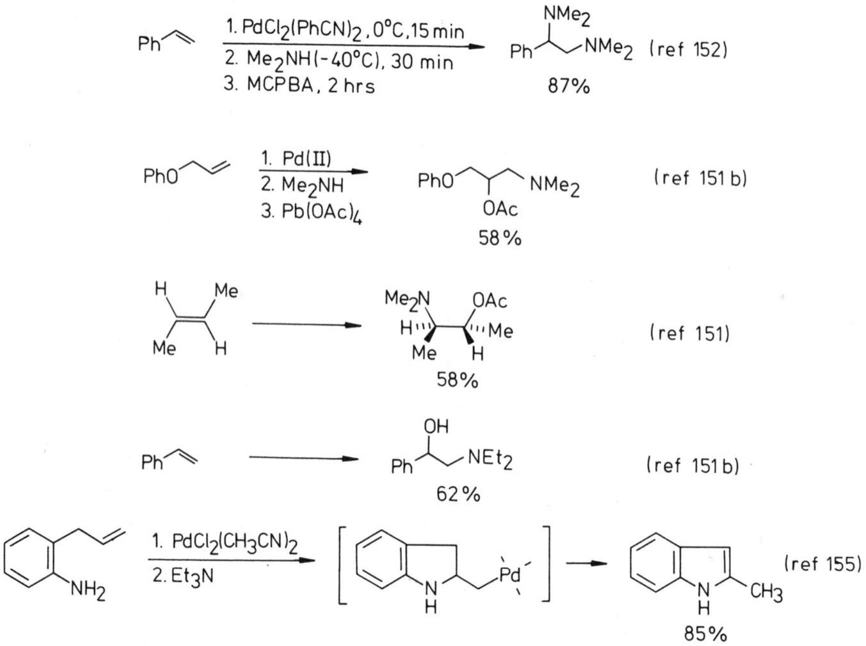

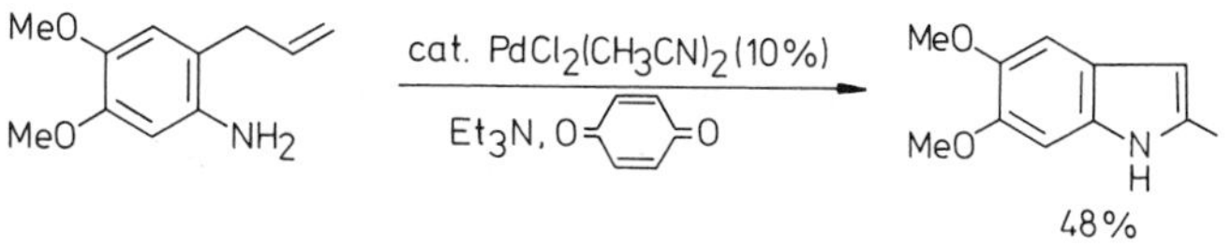

(ref 155 b)

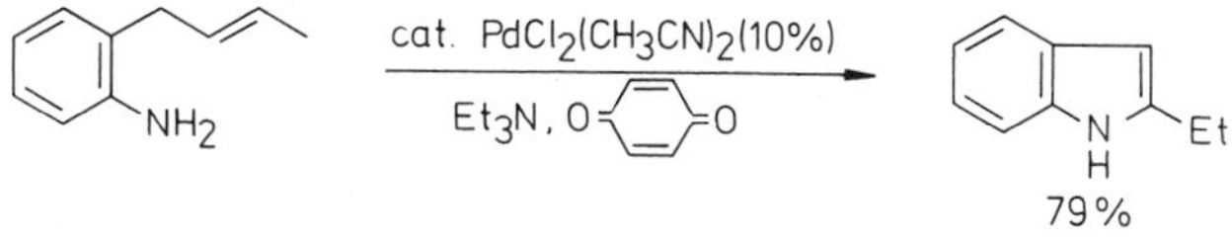

48%

(ref 155 b)

79%

Intramolecular aminopalladation of 2-allylanilines affords indoles in good yield.[155] The proposed mechanism involves <u>trans</u>-attack by the aniline nitrogen on an intermediate π-olefin triethylamine complex to give an intermediate σ-complex, which undergoes β-elimination. The reaction can also be performed with catalytic amounts of palladium complex (10%) using benzoquinone as reoxidant.

An interesting β-lactam synthesis involving nucleophilic attack by amine on an iron olefin complex has been reported.[114] Amination of the propene complex **XCVIII** followed by oxidation of the σ-complex **XCIX** by chlorine resulted in carbon monoxide insertion and formation of the β-lactam **C** in 47% yield. 6-Amino-1-hexene analogously gave **CI** by an intramolecular amination reaction. β-Lactams were also obtained via a similar sequence by amination of CpMo(CO)$_3$(ethene).[114] However, yields in the molybdenum-promoted reaction were poor.

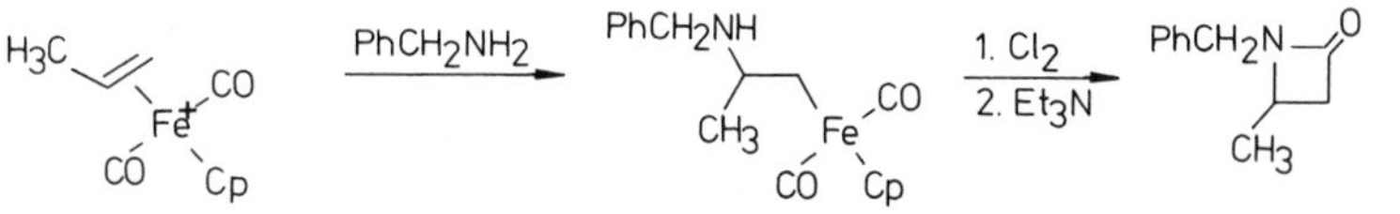

XCVIII   **XCIX**   **C**   (47%)

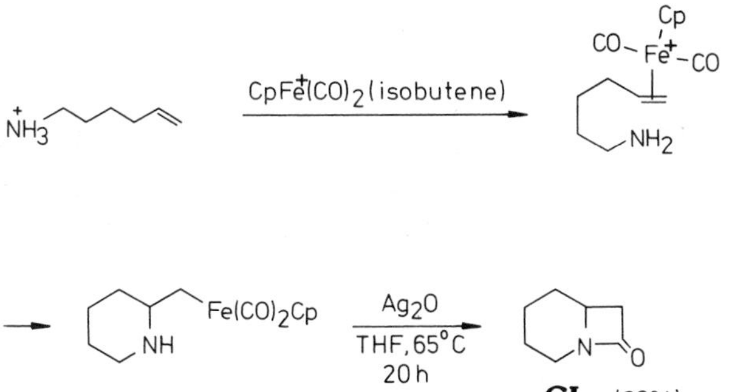

CI   (30%)

Platinum-promoted amination (see Section 2.1) has been used to cyclize 5-aminopentenes (eqs 40,41).[156] The reaction can be performed with catalytic amounts of platinum, but is limited by the rather long reaction times (several days to weeks).

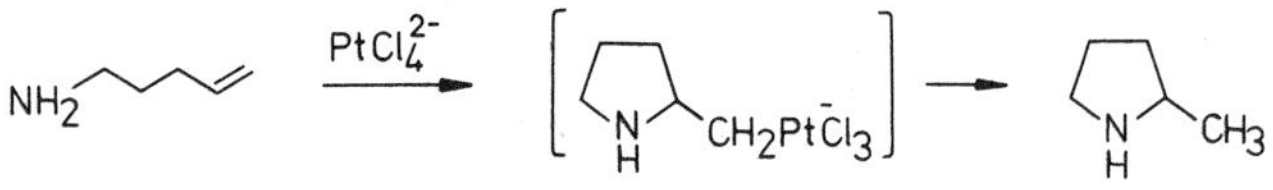

(eq 40)

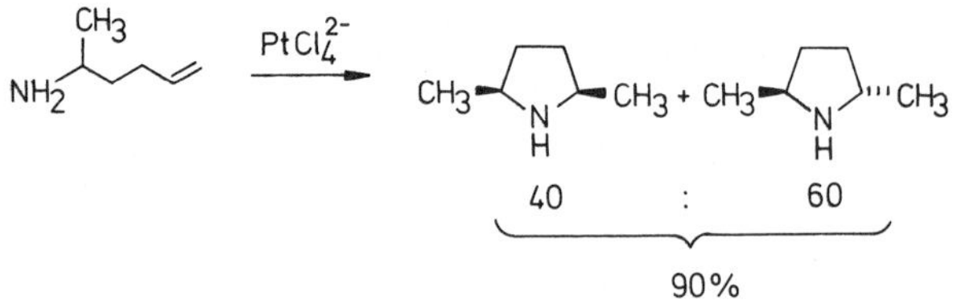

(eq 41)

Addition of dialkyl malonate, β-diketonate and other related stabilized carbanions to monoolefins coordinated to palladium[44] and iron[106] has been reported and is an important route for the formation of carbon-carbon bonds in organic synthesis (eqs 42-47). The palladium-promoted reaction has been used to prepare both β-[44] and α-aminoacid[157] precursors (eqs 43,44).

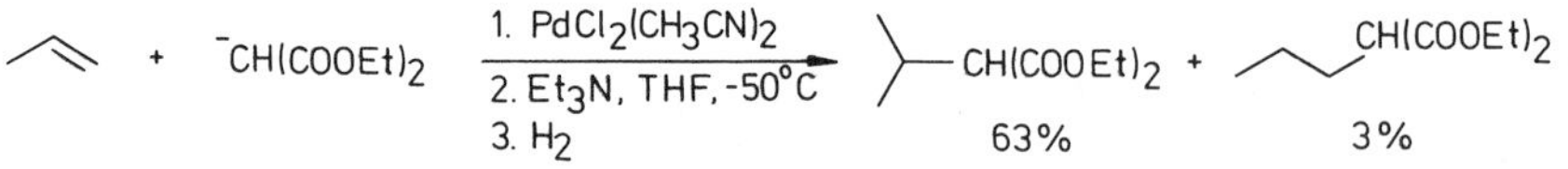

(eq 42)

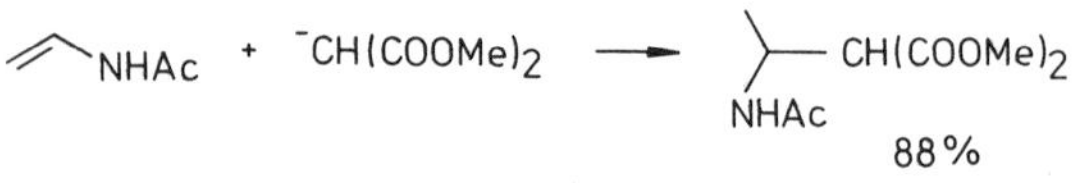

(eq 43)

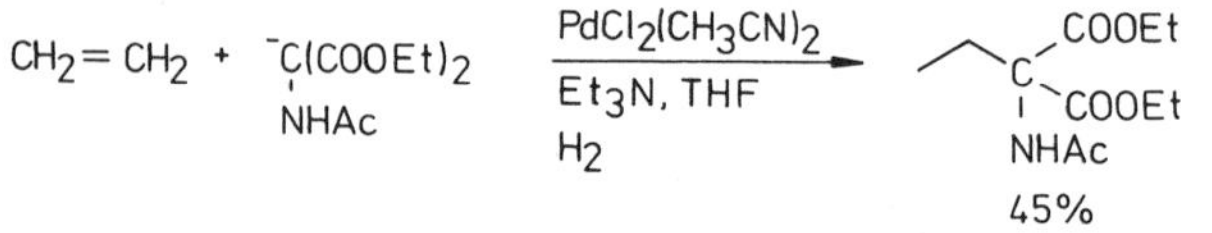

(eq 44)

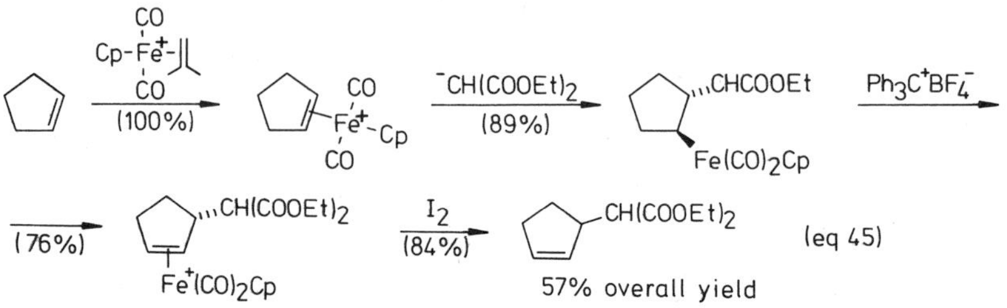

(eq 45)

57% overall yield

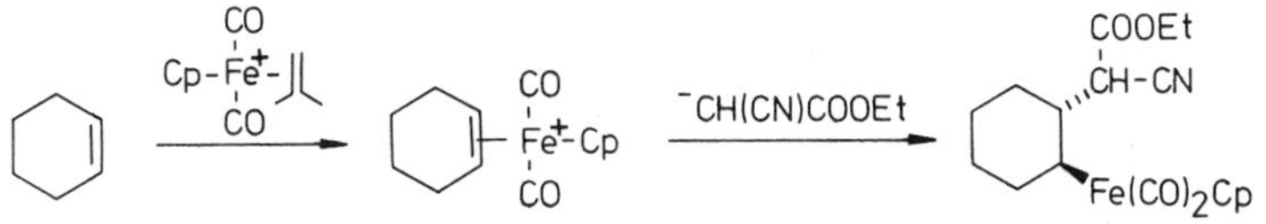

(eq 46)

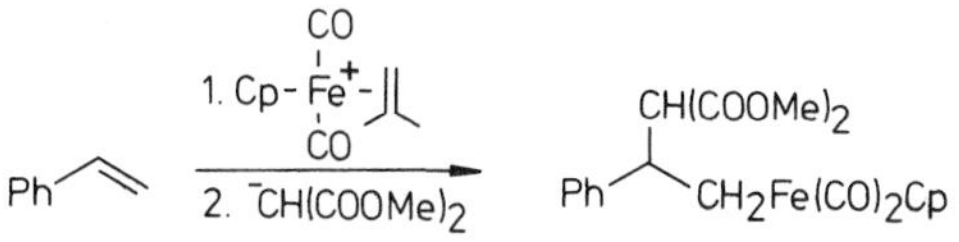

(eq 47)

Some different processes for cleaving iron-carbon bonds in adducts derived from nucleophilic addition are shown in Scheme **10**. One example of β-abstraction (1), not intramolecular by metal (as in the familiar β-elimination reaction), but intermolecular by $Ph_3C^+BF_4^-$, is given in eq 45. Thus allylic alkylation of cyclopentene in 57% overall yield can be achieved by a nucleophilic addition-elimination sequence.[106b]

Scheme **10**

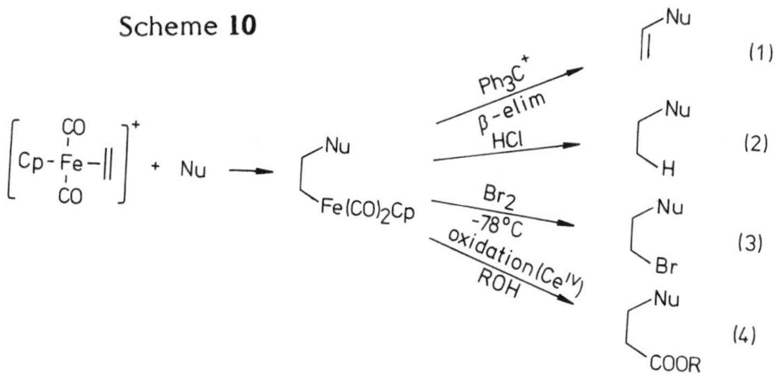

An illustrative example of the use of nucleophilic addition to coordinated double bonds in a complex synthesis directed towards prostaglandin derivatives is shown in Scheme **11**.[158] The π-complex of 3-dimethylaminocyclopentene (**CII**) with $PdCl_2$ reacts with sodium diethyl malonate to give a σ-complex **CIII,** which on treatment with diisopropylethylamine undergoes β-elimination to give **CIV** (92% yield). Internal coordination of amine to palladium results in a trans-directive effect of malonate with respect to the amine substituent. Complexation of the double bond in **CIV** to give **CV,** followed by nucleophilic attack by 2-chloroethanol, results in stereo- and regiospecific formation of palladium complex **CVI,** which was not, however, isolated. Again, chelating

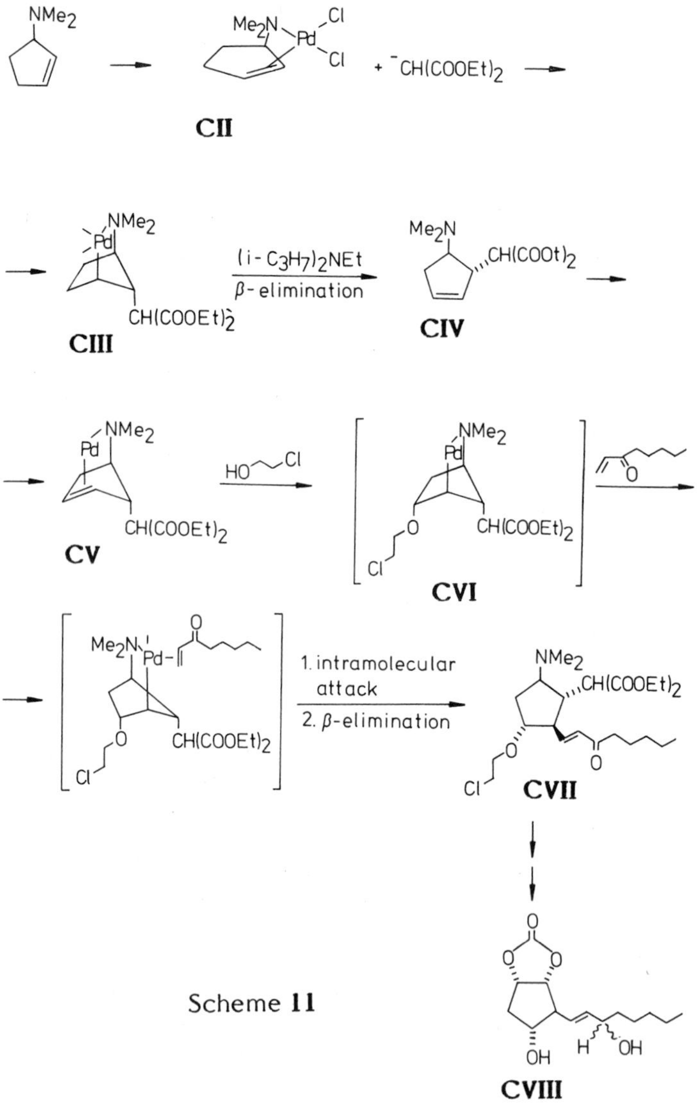

Scheme 11

coordination in π-complex **CV** directs the attack of the nucleophile. Reaction of the intermediate σ-complex with n-pentyl vinyl ketone gave **CVII** (50% yield from **CV**) via a formal <u>cis</u>-addition of the alkylpalladium complex to the vinyl group, followed by β-elimination (Heck-Fujiwara reaction). Compound **CVII** was then transformed to the important lactone **CVIII**, which can be converted to a variety of natural prostaglandins.

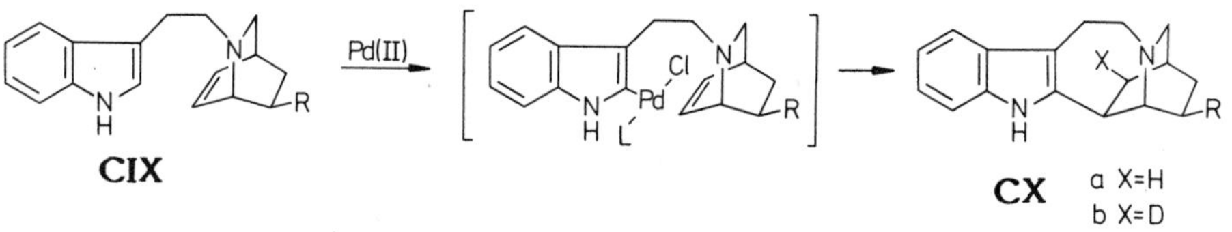

CIX   CX   a X=H   b X=D

An intramolecular Heck-Fujiwara type reaction has been applied to the cyclization of indole compound **CIX** to iboga alkaloid systems.[159-60] Formation of an indolylpalladium compound and <u>cis</u>-carbopalladation, followed by hydride cleavage, would give **CXa.** Evidence for this mechanism was provided by reductive work-up with $NaBD_4$, which gave a product **CXb**, where the deuterium and the indolyl unit are <u>cis</u>.

An example of coordinated vinyl ethers being used as vinyl cation equivalents via nucleophilic addition has been reported (Scheme 12).[161] Reaction of the lithium enolate **CXI** with the vinyl ether iron complex **CXII** resulted in nucleophilic attack by the enolate anion to give **CXIII**. Treatment of the adduct **CXIII** with HBF_4 at -78°C resulted in elimination of ethanol to give **CXIV,** which was treated with sodium iodide to release the organic product.

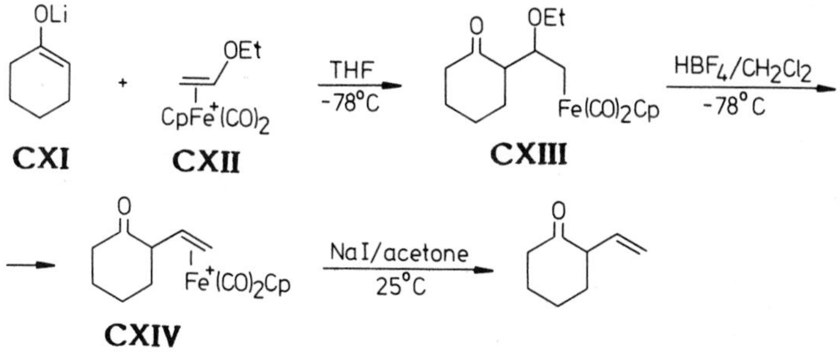

CXI   CXII   CXIII

CXIV

Scheme 12

REFERENCES

1. W.C. Zeise, Ann. Phys. Chem. 9:632 (1827).

2. K.A. Hofmann and J. Sand, Chem. Ber. 33:1340 (1900).

3. F.C. Phillips, Amer. Chem. J. 16:255 (1894).

4. J.S. Anderson, J. Chem. Soc. 971 (1934).

5. J. Smidt, W. Hafner, R. Jira, J. Sedlmeier, R. Sieber, R. Rüttinger and H. Kojer, Angew. Chem. 71:176 (1959).

6. J. Smidt, W. Hafner, R. Jira, R. Sieber, J. Sedlmeier and A. Sable, Angew. Chem. 74:93 (1962).

7. M.J.S. Dewar, Bull. Soc. Chim. Fr. 18:C71 (1951).

8. J. Chatt and L.A. Duncanson, J. Chem. Soc. 2939 (1953).

9. (a) B. Åkermark, M. Almermark, J. Almlöf, J.E. Bäckvall, B. Roos and A. Stogård, J. Am. Chem. Soc. 99:4617 (1977); (b) O. Eisenstein and R. Hoffmann, J. Am. Chem. Soc. 103:4308 (1981).

10. P.M. Maitlis, "The Organic Chemistry of Palladium", Academic Press, New York, vol. 1, 1971, chapters 3 and 4.

11. M.S. Kharasch, R.C. Seyler and F.R. Mayo, J. Am. Chem. Soc. 60:882 (1938).

12. $PdCl_2(PhCN)_2$ (1 mmol) is treated with an excess of the olefin (3 mmol) in THF (5 ml) at $0^{\circ}C$. Within one minute a color change is observed indicating the formation of the -olefin complex (cf. refs. 41-3).

13. J. Chatt, L.M. Vallerino and L.M. Venanzi, J. Chem. Soc. 2496 (1957).

14. J. Chatt, L.M. Vallerino and L.M. Venanzi, J. Chem. Soc. 3413 (1957).

15. R. Van der Linde and R.O. de Jong, J.C.S. Chem. Commun. 563 (1971).

16. F.R. Hartley, "The Chemistry of Platinum and Palladium", Appl. Sci. Publ., London (1973).

17. (a) W.C. Hamilton, K.A. Klander and R. Spratley, Acta Crystallogr. Sect. A. 25:S172 (1969); (b) N.C. Baenziger, G.F. Richards and J.R. Doyle, Acta Crystallogr. 18:924 (1965).

18. (a) P.T. Cheng, C.D. Cook, S.C. Nyburg and K.Y. Wan, Inorg. Chem. 11:2210 (1971); (b) L.J. Guggenberger, Inorg. Chem. 12:449 (1973); (c) P.T. Cheng and S.C. Nyburg, Can. J. Chem. 50:912 (1972).

19. (a) B.F.G. Johnson, J. Lewis and M.S. Subramanian, Chem. Commun. 117 (1966); (b) B.F.G. Johnson, J. Lewis and M.S. Subramanian, J. Chem Soc. (A). 1993 (1968).

20. J. Tsuji and H. Takahashi, J. Am. Chem. Soc. 90:2387 (1968).

21. G. Paiaro, A. De Renzi and R. Palumbo, Chem. Commun. 1150 (1967).

22. R. Palumbo, A. De Renzi, A. Panunzi and G. Paiaro, J. Am. Chem. Soc. 91:3874 (1969).

23. J.K. Stille and R.A. Morgan, J. Am. Chem. Soc. 88:5135 (1966).

24. J.K. Stille and D.B. Fox, J. Am. Chem. Soc. 92:1274 (1970).

25. I.I. Moiseev, M.N. Vargaftic and Y.K. Syrkin, Dokl. Akad. Nauk. SSSR (Engl. Transl.) 133:801 (1960).

26. E.W. Stern and M.L. Spector, Proc. Chem. Soc. London 370 (1961).

27. S. Nakamura and T. Yasui, J. Catal. 17:366 (1970).

28. M. Green, R.N. Haszeldine and J. Lindley, J. Organomet. Chem. 6:107 (1966).

29. W.A. Clement and C.M. Selwitz, J. Org. Chem. 29:241 (1964).

30. (a) T. Majima and H. Kurosawa, J.C.S. Chem. Commun. 610 (1977); (b) H. Kurosawa, T. Majima and N. Asada, J. Am. Chem. Soc. 102:6996 (1980).

31. (a) R.F. Heck, J. Am. Chem. Soc. 90:5518 (1968); (b) R.F. Heck, J. Am. Chem. Soc. 90:5526 (1968).

32. (a) I. Moritani and Y. Fujiwara, Tetrahedron Lett. 1119 (1967); (b) Y. Fujiwara, I. Moritani, S. Danno, R. Asamo and S. Teranishi, J. Am. Chem. Soc. 91:7166 (1969).

33. R.F. Heck, Acc. Chem. Res. 12:146 (1979).

34. R.F. Heck, J. Am. Chem. Soc. 91:6707 (1969).

35. P.M. Henry and G.A. Ward, J. Am. Chem. Soc. 94:673 (1972).

36. A. Segnitz, P.M. Bailey and P.M. Maitlis, J.C.S. Chem. Commun. 698 (1973).

37. S.I. Murahashi, M. Yamamura and N. Mita, J. Org. Chem. 42:2870 (1977).

38. A. Panunzi, A. De Renzi, R. Palumbo and G. Paiaro, J. Am. Chem. Soc. 91:3879 (1969).

39. A. Panunzi, A. De Renzi and G. Paiaro, J. Am. Chem. Soc. 92:3488 (1970).

40. H. Hirai, H. Sawai and S. Makashima, Bull. Chem. Soc. Jpn. 43:1148 (1970).

41. B. Åkermark, J.E. Bäckvall, L.S. Hegedus, K. Zetterberg, K. Siirala-Hansén and K. Sjöberg, J. Organomet. Chem. 72:127 (1974).

42. B. Åkermark and J.E. Bäckvall, Tetrahedron Lett. 819 (1975).

43. B. Åkermark, J.E. Bäckvall, K. Siirala-Hansen, K. Sjöberg and K. Zetterberg, Tetrahedron Lett. 1363 (1974).

44. (a) T. Hayashi and L.S. Hegedus, J. Am. Chem. Soc. 99:7093 (1977); (b)
    L.S. Hegedus, R.E. Williams, M.A. McGuire and T. Hayashi, J. Am.
    Chem. Soc. 102:4973 (1980).

45. R.A. Holton and R.A. Kjonaas, J. Am. Chem. Soc. 99:4179 (1977).

46. W.T. Wipke and G.L. Goeke, J. Am. Chem. Soc. 96:4244 (1974).

47. G. Wieger, G. Albelo and M.F. Rettig, J.C.S. Dalton 2242 (1974).

48. (a) B. Shaw, Chem. Ind. (London) 1190 (1962); (b) P.E. Slade and H.B.
    Jonassen, J. Am. Chem. Soc. 79:1277 (1957).

49. J. Lukas, P.W.N.M. Van Leeuwen, H.C. Volger and A.P. Kouwenhoven,
    J. Organomet. Chem. 47:153 (1973).

50. Y. Odaira, T. Oishi, T. Yukawa and S. Tsutsumi, J. Am. Chem. Soc.
    88:4105 (1966).

51. F.J. McQuillin, Tetrahedron 30:1661 (1974).

52. (a) P.M. Henry, J. Am. Chem. Soc. 86:3246 (1964); (b) P.M. Henry, J.
    Am. Chem. Soc. 88:1595 (1966).

53. P.M. Henry, Acc. Chem. Res. 6:16 (1973).

54. (a) J.E. Bäckvall, B. Åkermark and S.O. Ljunggren, J.C.S. Chem.
    Commun. 264 (1977); (b) J.E. Bäckvall, B. Åkermark and S.O.
    Ljunggren, J. Amer. Chem. Soc. 101:2411 (1979).

55. E. Vedejs and M. Salomon, J. Am. Chem. Soc. 92:6965 (1970).

56. J.K. Stille and D.E. James, J. Am. Chem. Soc. 97:674 (1975).

57. Y. Tamaru and Z. Yoshida, J. Org. Chem. 44:1189 (1979).

58. J.E. Bäckvall, R.E. Nordberg, E.E. Bjørkman and C. Moberg, J.C.S.
    Chem. Commun. 943 (1980).

59. (a) B.L. Shaw, Nature 615 (1969); (b) B.L. Shaw, Chem. Commun. 468
    (1968).

60. P.M. Henry, Adv. Chem. Ser. 70:151 (1968).

61. M. Green and R.I. Hancock, J. Chem. Soc. (A) 2054 (1967).

62. P.M. Henry and G.A. Ward, J. Am. Chem. Soc. 93:1494 (1971).

63. (a) J.K. Stille, D.E. James and L.F. Hines, J. Am. Chem. Soc. 95:5062
    (1973); (b) D.E. James, L.F. Hines and J.K. Stille, J. Am. Chem. Soc.
    98:1806 (1976).

64. (a) J.K. Stille and R. Divakaruni, J. Am. Chem. Soc. 100:1304 (1978);
    (b) J.K. Stille and R. Divakaruni, J. Organomet. Chem. 169:239 (1979).

65. H. Kurosawa and N. Asada, Tetrahedron Lett. 255 (1979).

66. (a) G. Klopman, J. Am. Chem. Soc. 90:223 (1968); (b) K. Fukui, Acc.
    Chem. Res. 4:57 (1971); (c) R.G. Pearson, "Symmetry Rules for
    Chemical Reactions", Wiley-Interscience, New York, (1976).

67. (a) P.M. Henry, Adv. Organomet. Chem. 13:363 (1975); (b) P.M. Maitlis, "The Organic Chemistry of Palladium", vol. 2, Academic Press, New York (1971).

68. (a) I.I. Moiseev, O.G. Levanda and M.N. Vargaftic, J. Am. Chem. Soc. 96:1003 (1974); (b) R. Jira and W. Freisleben, Organomet. React. 3:1 (1972).

69. (a) P.M. Henry, J. Org. Chem. 38:2415 (1973); (b) M. Kosaki, M. Isemura, Y. Kitaura, S. Shinoda and Y. Saito, J. Mol. Catal. 2:351 (1977).

70. W. Hagner, R. Jira, J. Sedlmeier and J. Smidt, Chem. Ber. 95:1575 (1962).

71. P.W. Jolly and G. Wilke, "The Organic Chemistry of Nickel", vol. 1, Academic Press, New York (1974).

72. (a) P. Heimbach and R. Traunmüller, "Chemie der Metall-Olefin-Komplexe", Verlag Chemie, Weinheim (1970); (b) E.O. Fisher and G. Burger, Chem. Ber. 94:2409 (1961).

73. S. O'Brien, M. Fishwick and B. MacDermott, Inorg. Synth. 13:73 (1972).

74. A. Salzer and H. Werner, Syn. Inorg. Metal. Org. Chem. 2:249 (1972).

75. (a) A. Salzer, T.L. Court and H. Werner, J. Organomet. Chem. 54:325 (1973); (b) G. Parker, A. Salzer and H. Werner, J. Organometal. Chem. 67:131 (1974).

76. T. Majima and H. Kurosawa, J. Organomet. Chem. 134:C45 (1977).

77. H. Lehmkuhl, A. Rufinska, R. Benn, G. Schrot and R. Mynott, J. Organomet. Chem. 188:C36 (1980).

78. J. Chatt and L.M. Venanzi, J. Chem. Soc. 4735 (1957).

79. R. Cramer, Inorg. Chem. 1:722 (1962).

80. R.B. King, P.M. Treichel and F.G.A. Stone, J. Am. Chem. Soc. 83:3593 (1961).

81. E.L. Muetterties and P.L. Watson, J. Am. Chem. Soc. 100:6978 (1978).

82. B. Capelle, M. Dartiguenave, Y. Dartiguenave, A. Beauchamp, and H.F. Klein, XXIth Int. Conf. Coord. Chem., Toulouse (1980).

83. H. Werner, R. Feser and R. Werner, J. Organomet. Chem. 181:C7 (1979).

84. H. Werner and F. Feser, Angew. Chem. Int. Edn. Engl. 18:157 (1979).

85. M.M.T. Khan and A. Martell, "Homogeneous Catalysis by Metal Complexes", vol. II, Academic Press, New York and London (1974).

86. R.F. Heck, "Organotransition Metal Chemistry", Academic Press, New York and London (1974).

87. E.R. Evitt and R.G. Bergman, J. Am. Chem. Soc. 101:3973 (1979).

88. B.T. Golding, H.L. Holland, U. Horn and S. Sakrikar, Angew. Chem. Int. Edn. Engl. 9:959 (1970).

89. (a) R.B. Silverman and D. Dolphin, J. Am. Chem. Soc. 94:4028 (1972); (b) R.B. Silverman and D. Dolphin, J. Am. Chem. Soc. 98:4626 (1976).

90. B.T. Golding and S. Sakrikar, J.C.S. Chem. Commun. 1183 (1972).

91. A. Efraty and P.M. Maitlis, Tetrahedron Lett. 4025 (1966).

92. D.R. Coulson, Tetrahedron Lett. 429 (1971).

93. (a) ref. 86, p. 16; (b) F.A. Cotton and G. Wilkinson, "Advanced Inorganic Chemistry", Interscience, New York, 4th edn. (1980), p. 1186.

94. W.P. Giering and M. Rosenblum, Chem. Commun. 441 (1971).

95. A. Cutler, D. Ehnholt, W.P. Giering, P. Lennon, S. Raghu, A. Rosan, M. Rosenblum, J. Tancrede and D. Wells, J. Am. Chem. Soc. 98:3495 (1976).

96. M.L.H. Green and P.L.I. Nagy, J. Organomet. Chem. 1:58 (1963).

97. W.P. Giering, M. Rosenblum and J. Tancrede, J. Am. Chem. Soc. 94:7170 (1972).

98. D.L. Reger, C.J. Coleman and P.J. McElligott, J. Organomet. Chem. 171:73 (1979).

99. D.L. Reger and C.J. Coleman, Inorg. Chem. 18:3155 (1979).

100. E. Weiss, K. Starck, J.E. Lancaster and H.D. Murdoch, Helv. Chim. Acta 46:288 (1963).

101. H.D. Murdoch and E. Weiss, Helv. Chim. Acta 46:1588 (1963).

102. P. Lennon, M. Madhavarao, A. Rosan and M. Rosenblum, J. Organomet. Chem. 108:93 (1976).

103. S.M. Florio and K.M. Nicholas, J. Organomet. Chem. 112:C17 (1976).

104. L. Busetto, A. Palazzi, R. Ros and U. Belluco, J. Organomet. Chem. 25:207 (1970).

105. W.H. Knoth, Inorg. Chem. 14:1566 (1975).

106. (a) A. Rosan, M. Rosenblum and J. Tancrede, J. Am. Chem. Soc. 95:3064 (1973).; (b) P. Lennon, A.M. Rosan and M. Rosenblum, J. Am. Chem. Soc. 99:8426 (1977).

107. M. Rosenblum, Acc. Chem. Res. 7:122 (1974).

108. B.W. Roberts and J. Wong, J.C.S. Chem. Commun. 20 (1977).

109. (a) M.R. Baar and B.W. Roberts, J.C.S. Chem. Commun. 1129 (1979); (b) B.W. Roberts, M. Ross and J. Wong, J.C.S. Chem. Commun. 428 (1980).

110. F.A. Cotton, A.J. Deeming, P.L. Josty, S.S. Ullah, A.J.P. Domingos, B.F.G. Johnson and J. Lewis, J. Am. Chem. Soc. 93:4624 (1971).

111. A. Eisenstadt, J. Organomet. Chem. 60:335 (1973).

112. K.M. Nicholas and A.M. Rosan, J. Organomet. Chem. 84:351 (1975).

113. A. Sanders, C.V. Magatti and W.P. Giering, J. Am. Chem. Soc. 96:1610 (1974).

114. P.K. Wong, M. Madhavarao, D.F. Marten and M. Rosenblum, J. Am. Chem. Soc. 99:2823 (1977).

115. (a) M.A. Hashmi, J.D. Munro, P.L. Pauson and J.M. Williamson, J. Chem. Soc. (A) 240 (1967); (b) A.J. Birch, K.B. Chamberlain, M.A. Haas and T.J. Thompson, J.C.S. Perkin I 1882 (1973).

116. (a) K.E. Hine, B.F.G. Johnson and J. Lewis, J.C.S. Chem. Commun. 81 (1975); (b) A.L. Burrows, K. Hine, B.F.G. Johnson, J. Lewis, D.G. Parker, A. Poë and E.J.S. Vichi, J.C.S. Dalton 1135 (1980).

117. D.A. Brown, W.K. Gloss and F.M. Hussein, J. Organomet. Chem. 186:C58 (1980).

118. A.J. Deeming, S.S. Ullah, A.J.P. Domingos, B.F.G. Johnson and J. Lewis, J.C.S. Dalton 2093 (1974).

119. H. Werner and R. Werner, Angew. Chem. Int. Edn. Engl. 17:683 (1978).

120. H. Werner and R. Werner, J. Organomet. Chem. 174:C63 (1979).

121. S.G. Davies and F. Scott, J. Organomet. Chem. 188:C41 (1980).

122. H. Werner and R. Werner, J. Organomet. Chem. 194:C7 (1980).

123. M. Stephenson and R.J. Mawby, J.C.S. Dalton 2112 (1981).

124. (a) F.A. Cotton, A.J. Deeming, P.L. Josty, S.S. Ullah, A.J.P. Domingos, B.F.G. Johnson and J. Lewis, J. Am. Chem. Soc. 93:4624 (1971); (b) J. Lewis and A.W. Parkins, J. Chem. Soc. (A) 1150 (1967); (c) J. Lewis and A.W. Parkins, J. Chem. Soc. (A) 953 (1969).

125. (a) F.A. Cotton, M.D. La Prade, B.F.G. Johnson and J. Lewis, J. Am. Chem. Soc. 93:4624 (1971); (b) F.A. Cotton and M.D. La Prade, J. Organomet. Chem. 39:345 (1972).

126. M. Cooke, P.T. Dragett, M. Green, B.F.G. Johnson, J. Lewis and D.H. Yarrow, Chem. Commun. 621 (1971).

127. E.O. Fischer and K. Ofele, Angew. Chem. Int. Edn. Engl. 1:52 (1962).

128. P.J. Harris, S.A.R. Knox and F.G.A. Stone, J. Organomet. Chem. 148:327 (1978).

129. M.L.H. Green, A.G. Massey, J.T. Moelwyn-Hughes and P.L.I. Nagy, J. Organomet. Chem. 8:511 (1967).

130. F. Haque, J. Miller, P.L. Pauson and J.B.Pd. Tripathi, J. Chem. Soc. (C) 743 (1971).

131. I.U. Khand, P.L. Pausen, and W.E. Watts, J. Chem Soc. (C) 2024 (1969).

132. J.C. Burt, S.A.R. Knox, J. McKinney and F.G.A. Stone, J.C.S. Dalton 1 (1977).

133. E.O. Fischer and K. Fichtel, Chem. Ber. 94:1200 (1961).

134. M. Cousins and M.L.H. Green, J. Chem. Soc. 889 (1963).

135. E.O. Fischer and H. Werner, "Metal π-Complexes", Elsevier, New York (1966).

136. M.A. Bennett, L. Pratt and G. Wilkinson, J. Chem. Soc. 2037 (1961).

137. E.W. Abel, M.A. Bennett, R. Burton and G. Wilkinson, J. Chem Soc. 4559 (1958).

138. G.A. Jones and L.J. Guggenberger, Acta Crystallogr. B31:900 (1975).

139. M.L.H. Green, L.C. Mitchard and W.E. Silverthorn, J.C.S. Dalton 1952 (1973).

140. S.G. Davies, M.L.H. Green and D.M.P. Mingos, Tetrahedron 34:3047 (1978).

141. (a) B.M. Trost, Tetrahedron 33:2615 (1977); (b) J. Tsuji, "Organic Synthesis with Palladium Compounds", Springer Verlag, Berlin (1980).

142. A.J. Birch and I.D. Jenkins, "Transition Metal Complexes of Olefinic Compounds", in "Transition Metal Organometallics in Organic Synthesis", Ed. H. Alper, vol. 1, Academic Press, New York (1976).

143. (a) J. Tsuji and H. Takahashi, J. Am. Chem. Soc. 87:3275 (1965); (b) J. Tsuji, Acc. Chem. Res. 2:144 (1969).

144. (a) J.K. Stille and D.E. James, J. Organomet. Chem. 108:401 (1976); (b) J.K. Stille and D.E. James, J. Am. Chem. Soc. 97:674 (1975).

145. D.E. James and J.K. Stille, J. Am. Chem. Soc. 98:1810 (1976).

146. T. Hosokawa, S. Yamashita, S.I. Murahashi, and A. Sonoda, Bull. Chem. Soc. Jpn. 49:3662 (1976).

147. T. Hosokawa, H. Ohkata and I. Moritani, Bull. Chem. Soc. Jpn. 48:1533 (1975).

148. J.L. Ruth and D.E. Bergstrom, J. Org. Chem. 43:2870 (1978).

149. C.F. Bigge, P. Kalaritis, J.R. Deck and M.P. Mertes, J. Am. Chem. Soc. 102:2033 (1980).

150. R.F. Heck, Pure Appl. Chem. 50:691 (1978).

151. (a) J.E. Bäckvall, Tetrahedron Lett. 2225 (1975); (b) J.E. Bäckvall and E.E. Björkman, J. Org. Chem. 45:2893 (1980).

152. J.E. Bäckvall, Tetrahedron Lett. 163 (1978).

153. J.E. Bäckvall, J.C.S. Chem. Commun. 413 (1977).

154. L.S. Hegedus, O.P. Anderson, K. Zetterberg, K. Siirala-Hansen, D.J. Olsen and A.B. Packard, Inorg. Chem. 16:1887 (1977).

155. (a) L.S. Hegedus, G.F. Allen, J.J. Bozell and E.L. Waterman, J. Am. Chem. Soc. 100:5800 (1978); (b) L.S. Hegedus, G.F. Allen and E.L. Waterman, J. Am. Chem. Soc. 98:2674 (1976).

156. (a) J. Ambuehl, P.S. Pregosin, L.M. Venanzi, G. Ughetto and L. Zambonelli, J. Organomet. Chem. 160:329 (1978); (b) J. Ambuehl, P.S. Pregosin, L.M. Venanzi, G. Consiglio, F. Bachechi and L. Zambonelli, J. Organomet. Chem. 181:255 (1979).

157. J.P. Haudegond, Y. Chauvin and D. Commereau, J. Org. Chem. 44:3063 (1979).

158. R.A. Holton, J. Am. Chem. Soc. 99:8083 (1977).

159. B.M. Trost and J.P. Genêt, J. Am. Chem. Soc. 98:8516 (1976).

160. B.M. Trost, S.A. Godleski and J.P. Genêt, J. Am. Chem. Soc. 100:3930 (1978).

161. T.C.T. Chang, M. Rosenblum and S.B. Samuels, J. Am. Chem. Soc. 102:5930 (1980).

ASYMMETRIC ADDITIONS TO DOUBLE BONDS

P.A. Chaloner[a] and D. Parker[b]

a) School of Molecular Science, University of Sussex, Falmer Brighton
b) Department of Chemistry, University of Durham South Road, Durham

1. INTRODUCTION

The goal of synthesizing optically active compounds has attracted chemists for almost a century. Most naturally occurring organic compounds exist as only one enantiomer, so that we see D-sugars and L-amino acids. Not surprisingly, therefore, many biologically active compounds are useful in only one enantiomeric form, the other being either inactive or actually deleterious. For example atropine (I) has about 500 times the activity as a cardiac stimulant as its enantiomer[1] and all important β-blocking agents have the same chirality as II.[2] Lysine cannot be used in its racemic form as an animal feedstock as D-lysine is toxic.[3] The synthesis of amino acids and their derivatives has so far been the most successful application of asymmetric catalysis and one large scale commercial use is for the production of L-DOPA (III) which is used for the treatment of Parkinson's disease.

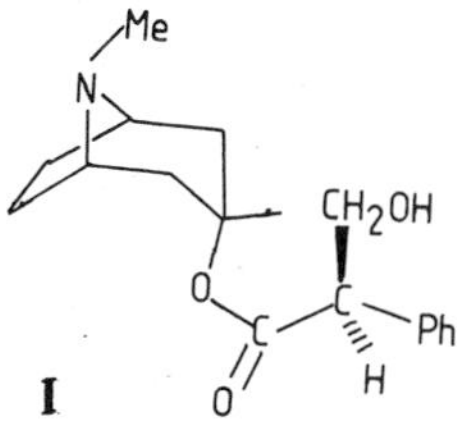

I

MeOCN—⟨benzene ring⟩—OCH$_2$—C(OH)(H)—CH$_2$NCHMe$_2$ (with H on N)

$\underline{II}$ $\underline{R}$ Practolol

HO—⟨benzene ring, HO⟩—CH$_2$—CH(NH$_2$)(CO$_2$H)

$\underline{III}$ $\underline{L}$–DOPA

Most early syntheses in this area involved the resolution either of the final product or some intermediate on the reaction pathway. This is extremely wasteful; if only one enantiomer is useful at least half the product must be discarded. Some improvement may be achieved if the unwanted material can be racemized and recycled but this is at best clumsy and laborious. With the increasing costs of petroleum chemical feedstocks for the organic chemical industry it has become commercially essential as well as intellectually appealing to introduce chirality at an early stage by means of asymmetric synthesis. Asymmetric synthesis converts a prochiral compound into a chiral unit in such a way that the stereoisomeric products are produced in differing amounts. This subject has been very extensively reviewed.[4-8]

Certain problems still remain, however. In a classic asymmetric synthesis (as in Figure 1)[9] stoichiometric amounts of a chiral auxiliary must be used and it must be possible to recover this both in high yield and unchanged optical purity. A more efficient use of a valuable chiral reagent is in a catalytic sense where usually 1% or less of the optically active material is necessary. Even if it is allowed that all the chiral starting material from a catalytic reaction is lost, a stoichiometric reaction must recover its chiral auxiliary in better than 99% yield to compete. Polymer support catalysts can give potentially even more valuable results conserving both valuable chiral material and the precious metal catalyst. Any unwanted contamination by catalyst derived products is also avoided.

Early work in catalytic chiral synthesis centred on supported catalysts. In 1956 Akabori catalyzed asymmetric hydrogenation with palladium adsorbed onto optically active silk fibroin.[10] Izumi[11] and Klabunovskii[12] later took up this idea with increasing success using metal surfaces treated

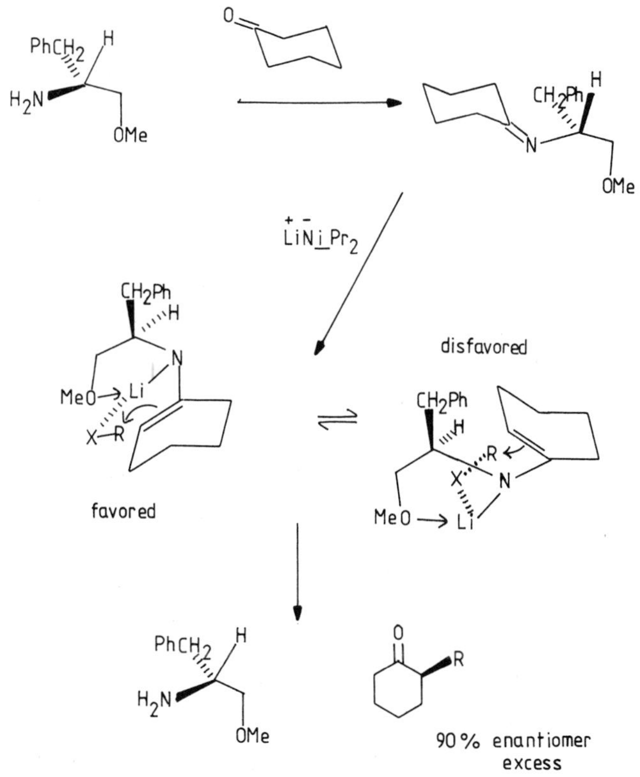

Figure 1.   Classical asymmetric synthesis using a recoverable chiral
             auxiliary.

with suitable optically active modifying agents. They have achieved very
high optical yields for the reduction of β-ketoesters but carbon-carbon
double bonds are reduced with low selectivity.

In 1966 homogeneous hydrogenation by Wilkinson's catalyst,[15]
$(Ph_3P)_3RhCl$, was first reported but it was not until ten years later that the
mechanism was fully elucidated.[16] It was quickly realised that if chiral
phosphines were used, asymmetric hydrogenation might be achieved.[17]
Early experiments with simple resolved phosphines such as Ph(iPr)MeP were
rather disappointing[18] giving low rates and poor selectivity. In the early
1970's, however, several developments occurred.

(a)   The use of more suitable substrates.

It was found that polar molecules capable of bi- or tridentate coordination
at a metal center gave better optical selectivity.

(b)   The use of more active catalytic systems.

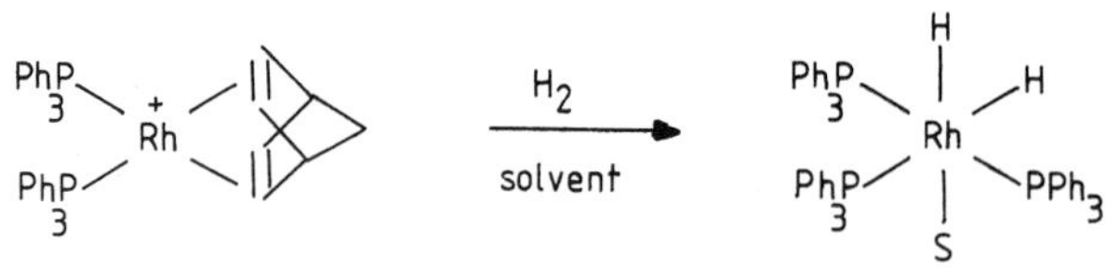

IV → V

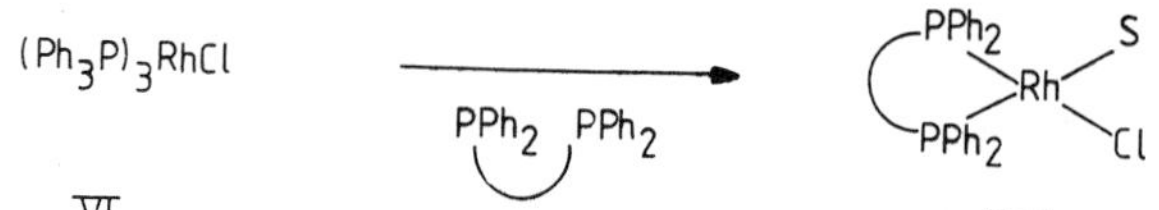

VI → VII

S = Solvent

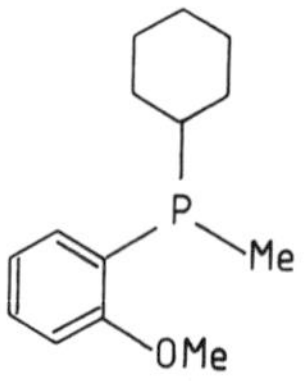

R ACMP

VIII

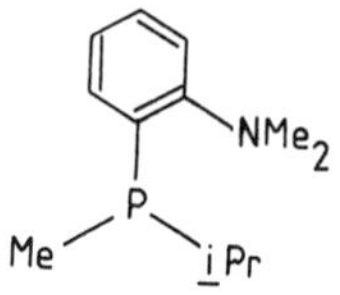

IX

Cationic complexes such as **IV** give faster rates and even chlorides such as **VII** are probably at least partly cationic in polar solvents.

(c)   Improvements in phosphine design.

An enormous variety of chiral phosphines has now been synthesized with examples of systems chiral by virtue of asymmetry either at carbon or phosphorus.   The presence of a polar ortho-substituent was found to be advantageous as in **VIII**[19] and **IX**.[20]

(d)   The use of chelating biphosphines.

There are much less flexible than their monophosphine analogues and exchange reactions are slower.

With these factors in mind it is now possible to attain close to 100% stereoselectivity for the reduction of dehydroamino acid derivatives,[21] 70-80% for unsaturated monoacids and up to 95% for carbonyl groups.

Industrially hydroformylation represents a reaction at least as important as hydrogenation but asymmetric hydroformylation has, as yet, yielded few results of practical utility.   Industrial hydroformylation is dominated by the 'oxo' process using $Co_2(CO)_8$ as catalyst and modification of this system with chiral phosphines has not yielded asymmetric products.[22] Rhodium and platinum catalysts are less widely used but operate under milder conditions and seem more promising.   Hydroesterification using palladium complexes as catalysts may give up to 50% enantiomer excess in the most favorable cases.[23]   Progress has been hampered by a lack of direct mechanistic investigation in this area; there are only empirical criteria on which to base the design of catalytic systems.

Early work on hydrosilylation reactions involved the reaction of olefins with monosilanes in the presence of rhodium monophosphine catalysts. Little success was achieved, the highest optical yield* reported being 21%.[24]   The reduction of ketones and imines offers a number of advantages

---

*) The optical yield of the reaction is defined as the ratio of the optical purity of the product to that of the optically active component in the reaction.   The terms optical yield and enantiomer excess are used interchangeably   in   the   literature   and   are   generally   determined polarimetrically. Although much careful and accurate work has been done it is necessary to treat the numbers given with a certain degree of caution. In some cases the optical rotation of the pure enantiomer is not accurately known and different authors employ different standards. Nor is there always a linear relationship between enantiomeric composition and optical rotation. Contamination with small amounts of catalyst derived material, which may have a very high rotation, is difficult to avoid and optical purity can easily be enriched by crystallization. Where it is possible the values from this determination should be checked by other methods such as the use of chiral shift reagents in nmr spectroscopy[13] or glc analysis of a volatile derivative on an optically active support.[14]   Alternatively the products may be converted into diastereomeric derivatives which may be distinguished spectroscopically.

over catalytic hydrogenation. Conditions are milder and the product alkoxysilanes may be converted to the alcohol or find some other synthetic use. The best result achieved for a simple ketone is a 60% optical yield[25] but α- and γ-keto esters may be reduced with 86% selectivity.[26] This may be compared with the 100% selectivity achieved with biological systems such as yeast.[27] The mechanism of the reaction has received little attention but work by Kagan[28] has established that a Rh-C rather than a Rh-O-C intermediate is involved; it would seem that this may be analogous to the alkyl rhodium hydride involved in hydrogenation.

Other reports of asymmetric addition to double bonds have been sparse. Epoxidation of allylic alcohols with vanadium and molybdenum complexes is successful[29] and there is one report of a reasonable optical yield in the epoxidation of a non-polar olefin.[30] Enantioselective olefin vinylation has been reported[31] and there are numerous examples of chiral polymerisation reactions[32] but there will be considered only briefly here. Reactions of π-allyl palladium complexes[33] and Grignard cross-coupling reactions[34] in the presence of chiral ligands, though they have made an important contribution to asymmetric synthesis, are outside the scope of this chapter.

2.   ASYMMETRIC HYDROGENATION

The asymmetric hydrogenation of prochiral olefins is an attractive goal for the synthetic organic chemist since in principle a wide range of tertiary asymmetric centers may be generated in a single step. Thus it is in this area that most of the effort in both synthetic and mechanistic studies has been concentrated.[21] There now exist over 100 phosphines capable of catalyzing the reduction of dehydroamino acid derivatives in good optical yield. Other substrates are normally hydrogenated with lower optical efficiency though there has been some success with prochiral αβ-unsaturated acids. Chelating biphosphines have generally proved to be more useful than monophosphines since a more rigid complex is formed; these may be chiral by virtue of asymmetry at phosphorus or in the interphosphine chain. When the asymmetric center is remote from the phosphorus atom the P-aryl groups must adopt a preferred conformation which is locally asymmetric. The types of ligand used may be conveniently divided into groups by considering the size of chelate ring formed on complexation to rhodium.

2.1 Catalysis by 5-ring chelate biphosphine rhodium complexes

While this is a small group of phosphines it has been markedly the most successful for enamide hydrogenation. Most of the examples (Figure 2.1) are chiral by virtue of asymmetry in the interphosphine chain but one, DIPAMP (and a number of analogues)[35] is chiral at phosphorus and another, (neomenthyl)PhPCH$_2$CH$_2$PPh$_2$, is chiral at one of the phosphine atoms and also by virtue of the P-neomenthyl group.

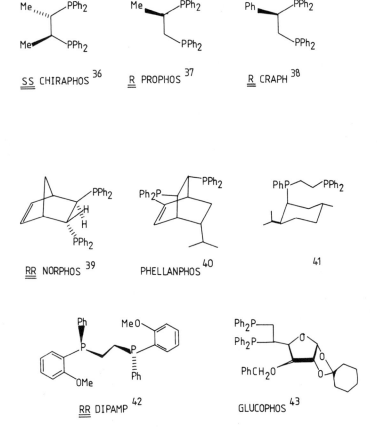

Figure 2.1   Chiral phosphines which form 5-membered chelate rings on coordination to rhodium.

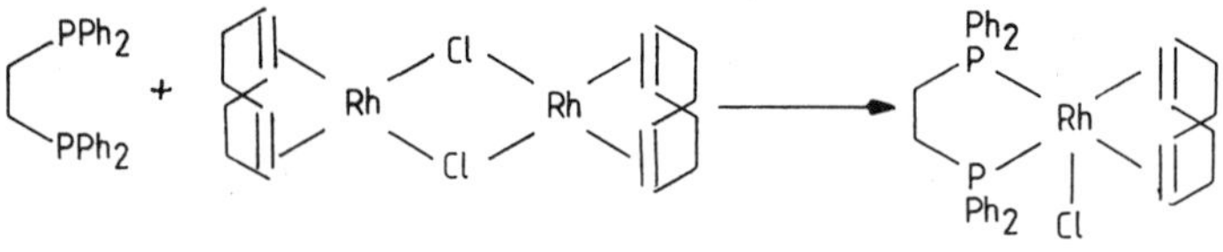

X   XI

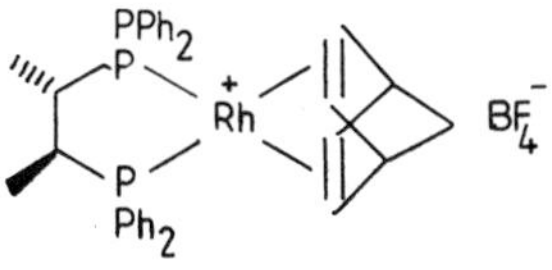

XII

Some early hydrogenations were carried out using catalysts formed in situ by the addition of the phosphine to a rhodium diene chloride complex such as **X** but in general a well defined cationic diene complex **XII** is used. These cationic complexes are reasonably air stable and give superior rates (and sometimes also optical yields) in asymmetric hydrogenation. A fairly restricted range of substrates may be hydrogenated with good optical efficiency. In addition to olefin coordination the presence of at least one polar group able to coordinate to metal is essential and many of the best results are obtained with substrates capable of tridentate binding. Some results are shown in Table 1.

Catalysis by 5-ring chelating phosphine rhodium complexes has also proved to be the most amenable to direct mechanistic investigation. Some of the intermediates established for these phosphines have proved to be common to catalysts with larger chelate rings, and only differences will be noted.

A complete description of the reaction pathway for the hydrogenation of olefins by Wilkinson's catalyst, $(Ph_3P)_3RhCl$, is comparatively recent. It had been known for some time as a result of experiments by Halpern[46] and Tolman[47] that the reaction requires hydrogen addition to the catalyst followed by loss of the phosphine <u>trans</u> to hydride and subsequent olefin

Table 1. Hydrogenations by 5-ring chelate rhodium complexes

Phosphine	Substrate	Solvent	T(°C)	P(atm)	Optical yield and configuration	Ref.
S,S-CHIRAPHOS	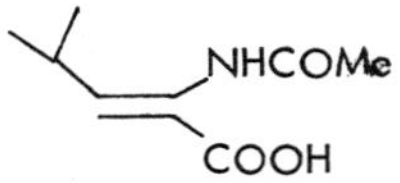 (NHCOMe, COOH)	THF	25	1	100 R	36
	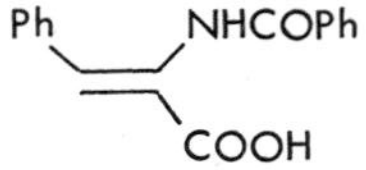 (Ph, NHCOPh, COOH)	EtOH	25	1	95 R	36
R,R-DIPAMP	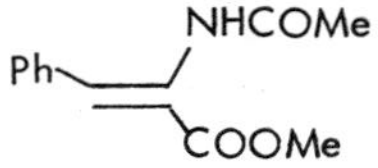 (Ph, NHCOMe, COOMe)	EtOH	50	3	96 S	42
	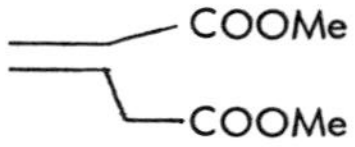 (COOMe, COOMe)	MeOH	50	3.3	88 R	44
R-PROPHOS	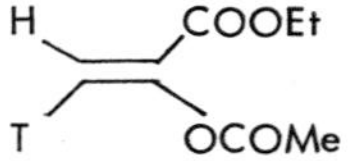 (H, COOEt, T, OCOMe)	THF	25	1(D_2)	81 S	45

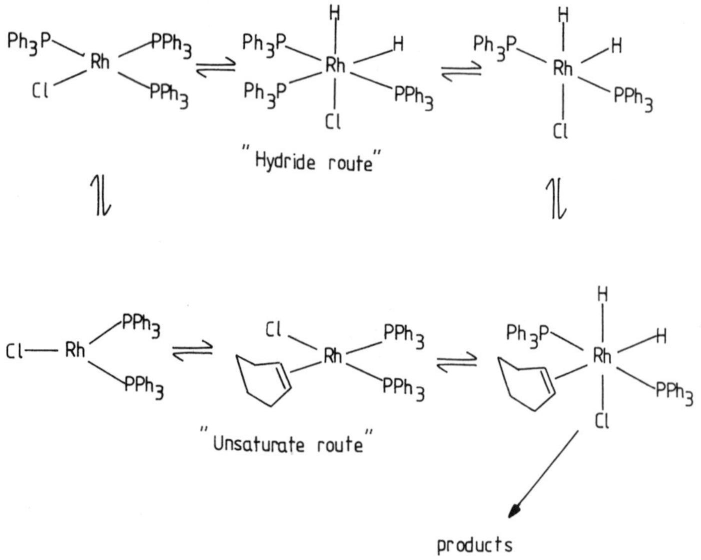

Figure 2.2   Mechanism   of   homogeneous   hydrogenation   by   Wilkinson's
              catalyst.

coordination (Figure 2.2). Recent results from Dutch and French[16] workers
permit the construction of an accurate kinetic model and rule out any role
for the "unsaturate" route involving ligand loss prior to olefin coordination.

      It was initially assumed that the mechanism followed by cationic
chelating complexes was analogous but the pathway actually observed is
significantly different. On hydrogenation of a chelating biphosphine diolefin
complex in a coordinating solvent such as methanol, thf or benzene two
moles of hydrogen per mole of rhodium are taken up and norbornane is
produced quantitatively[48] (Figure 2.3). This is in contrast to the reaction
with complexes of non-chelating phosphines such as $[(Ph_3P)_2Rh(norborna$-
diene)]$^+$ BF_4^- where three moles of hydrogen are taken up and a dihydride
disolvate is formed.[49] An appropriate enamide substrate then displaces the
solvent to form a highly air-sensitive species which can be characterized
spectroscopically. The major technique used to characterise these species
has been ^{31}P nmr spectroscopy since it has been found that each type of
complex gives rather consistent values both for phosphorus chemical shifts
and phosphorus rhodium coupling constants (both ^{31}P and ^{103}Rh have a

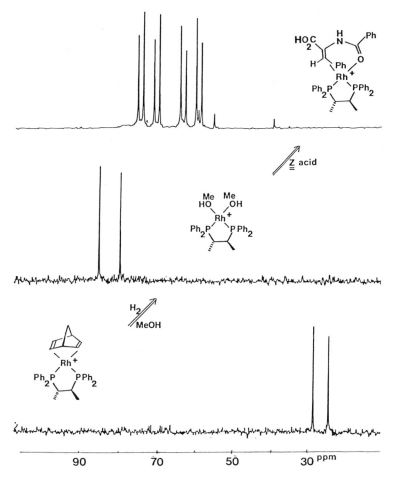

Figure 2.3   Intermediates in the course of catalysis by CHIRAPHOS complexes.

nuclear spin of $\frac{1}{2}$). Figure 2.3 shows the ^{31}P nmr spectra of each type; the norbornadiene complex has J_{Rh-P} = 155Hz while the methanol adduct has J_{Rh-P} = 200Hz. The very weak σ-bonding to methanol means that the σ-character of the phosphorus rhodium bond is increased giving rise to a larger coupling constant. Both these species have C_2 symmetry and hence the phosphorus atoms are equivalent. In the enamide complex, on the other hand, one phosphine is <u>trans</u> to a carbon-carbon double bond and the other <u>trans</u> to an amide group; they are thus inequivalent and coupled both to each other and to rhodium giving rise to an 8-line spectrum.

Ph   H
PhOCN CO₂H
H   α

XIII

Ph   H
HOOC   NCOPh
H

XIV

The mode of binding of enamide substrates to rhodium has been established by the use of ^{13}C labelled enamides.[50] For Z-enamides such as **XIII** binding is through the olefin and amide groups only. There is also a report of a crystal structure of a dppe rhodium enamide complex[51] (Figure 2.4) showing that the solution structure is preserved in the solid state. With E-enamides such as **XIV** the position is more complex. In the absence of added base isomerization occurs to give the Z-enamide complex; further enamide is then isomerized much more slowly. The isomerization of E-acids under hydrogenation conditions has also been investigated using deuterium.[52] This is found to be strongly solvent dependent and a good optical yield (85 S from R, R - DIPAMP) is obtained on hydrogenation of

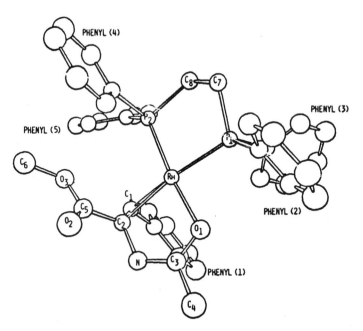

Figure 2.4   Crystal structure of complex formed between [DIPHOS-Rh]⁺ and Z-methylacetamidocinnamate.   [reproduced with permission]

XIV in benzene where little isomerization occurs. In ethanol, however, extensive isomerization is demonstrated to occur and the reduction proceeds with low optical efficiency. It seems probable that more than one mechanism may be involved both in deuterium exchange and in isomerization. It should be noted that the anion of **XIV** (as its triethylammonium salt) forms a stable carboxylate complex which does not isomerize in ethanol solution over several weeks.

Although **XIII** binds to the CHIRAPHOS rhodium solvate to form a single complex, in principle two diastereomeric species are possible, related by the binding of opposite prochiral faces of the olefin to the metal. When DIPAMP is used as the chiral ligand two diastereomeric complexes are indeed observed in the nmr spectrum (Figure 2.5). It was shown by ^{13}C labeling experiments that the substrate is coordinated in the same manner in both diasteromers.[53] The stereoselectivity under conditions of

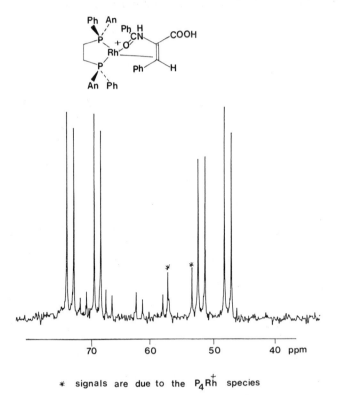

* signals are due to the P_4Rh^+ species

Figure 2.5   [31]P nmr spectrum of the complex formed between DIPAMP Rh$^+$ and methyl Z-α-benzamidocinnamate.

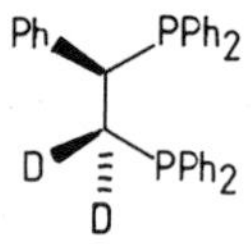

XV

thermodynamic control is high (9:1 at room temperature) but if binding is monitored by ^{31}P nmr at -40°C the kinetic pathway favors the "major enamide" by a factor of only about 2:1.

So far all the phosphines considered have been C_2 symmetric so that we have considered only two diastereomers. However, for PROPHOS and CRAPH there are four possible enamide complexes, both regioisomers and stereoisomers. Using the dideuterated species **XV** (CRAPH-d$_2$) it was possible to demonstrate that the two species observed in the spectrum with methyl-α-benzamidocinnamate were regioisomers[54] rather than steroisomers. So far the factors influencing regioselectivity in complexation are poorly understood and electronic and steric effects are difficult to distinguish.

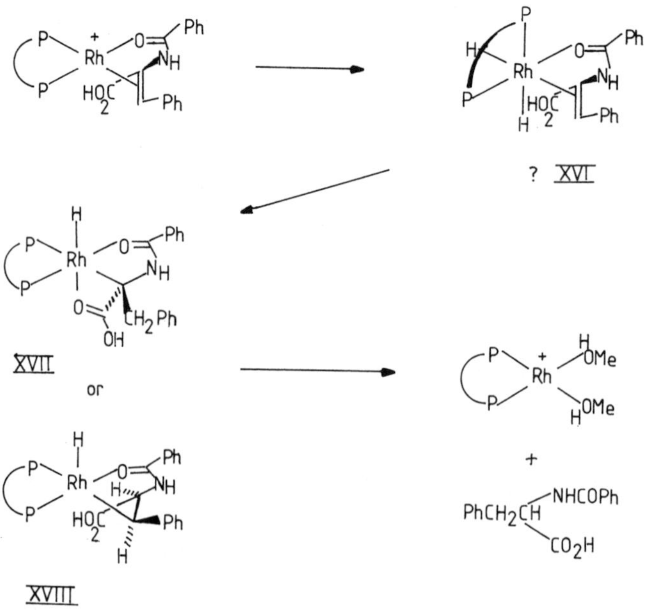

Figure 2.6   Hydrogenation of a rhodium enamide complex.

On exposure of a rhodium enamide complex to hydrogen at room temperature its characteristic deep red color fades rapidly and a pale yellow solution is obtained which contains predominantly the methanol solvate. Mechanistic considerations suggest that two intermediates, **XVI** and **XVII** or **XVIII**, should be involved in this process (Figure 2.6). The first report of the observation of a species **XVII** came from Halpern's group.[55] The ligand was dppe with methyl-Z-α-acetamidocinnamate as substrate and the alkyl hydride was characterised by ^{1}H and ^{31}P nmr spectroscopy. The presence of a mole of coordinated solvent was demonstrated by titration with ^{15}N labeled acetonitrile when an additional coupling to the hydride was observed as well as shifts in the ^{1}H and ^{31}P nmr spectra. Kinetic measurements indicated that at room temperature <u>formation</u> of the alkyl hydride is the rate-determining step in hydrogenation but that the slow step is its <u>breakdown</u> below -40°C. A similar species was obtained by Brown[56] on hydrogenation of the α-benzamidocinnamic acid rhodium DIPAMP complex. The coordination of a solvent molecule is doubtful in this case, however, since no changes are observed on addition of excess acetonitrile. Significantly the alkyl is found to be formed predominantly from the <u>minor</u> <u>enamide</u> diastereomer, contrary to all previous suppositions. It is also shown that the chelate enamide structure is maintained under hydrogenation conditions since the alternative alkyl **XVIII** is not formed. Rapid and efficient hydrogenation with DIPAMP complexes occurs only for substrates which are capable of tridentate binding[42] and it appears that the extra coordination stabilizes the rigid alkyl intermediate. Halpern[55a] has also noted that the crystal structure of the rhodium CHIRAPHOS complex of ethyl-α-acetamidocinnamate shows that the substrate is bound in the pro-S mode whereas hydrogenation gives R-amino acids. It has been noted by Ojima[57] that optical yields in hydrogenation by DIPAMP and other complexes tend to decrease as hydrogen pressure is raised. Increased pressure may be expected to raise the rate of hydrogen addition so that this may become faster than equilibration between diastereomers. The effect is much lessened in the presence of triethylamine, possibly because this may accelerate enamide formation or exchange between diastereomers. The olefin dihydride **XVI** has so far proved elusive but Crabtree's work on iridium olefin dihydrides[58] suggests that an analogous species might be observable in this series.

There have been many attempts to rationalize the optical efficiency of enamide hydrogenation by asymmetric catalysts. Knowles and co-workers have suggested that most chiral phosphines achieve their high efficiency by a common mechanism.[59] When the phosphine is chelated with rhodium the two diphenylphosphinyl groups give an array about the metal center with the phenyl rings in a pseudo axial or equatorial position. Additionally the phenyl ring can rotate to expose either its edge or its face to an approaching substrate. These two arrangements are quite different since in one case there is an unfavorable interaction of the substrate with the ortho-hydrogens and in the other only with a π-electron cloud. Since even 95% optical yield requires only a difference of about 7 kJ mol^{-1} between the diasteromeric transition states a preferred conformation would be sufficient to induce the observed selectivity. Crystallographic studies of chiral biphosphine rhodium olefin complexes abound and interesting correlations may be drawn between them. Those having a configuration such as A[42] (Figure 2.7) give rise to S-amino acids on hydrogenation while those with configuration B[60] give R-amino acids. In A the configuration is imposed by the chirality at phosphorus while in B the 5-membered ring is frozen so that the methyl groups are equatorial. The argument may successfully be extented to include phosphines such as DIOP[61] (**XIX**) and PPPM[62] (**XX**) which form 7-membered chelate rings. That this correlation exists, taken together with the results of Brown and Halpern, suggests that reversal of configuration between enamide and alkyl intermediates may indeed be a general phenomenon.

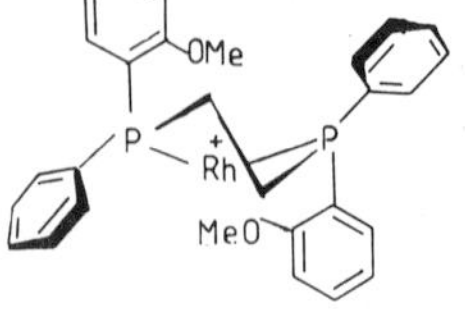

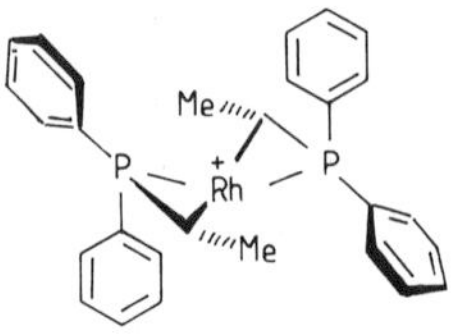

Figure 2.7   Crystal structures of phosphine rhodium diene complexes.

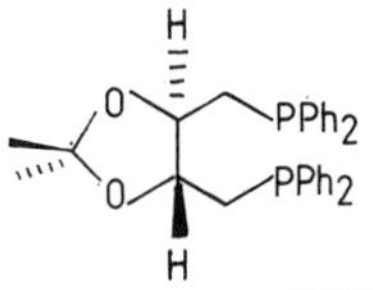

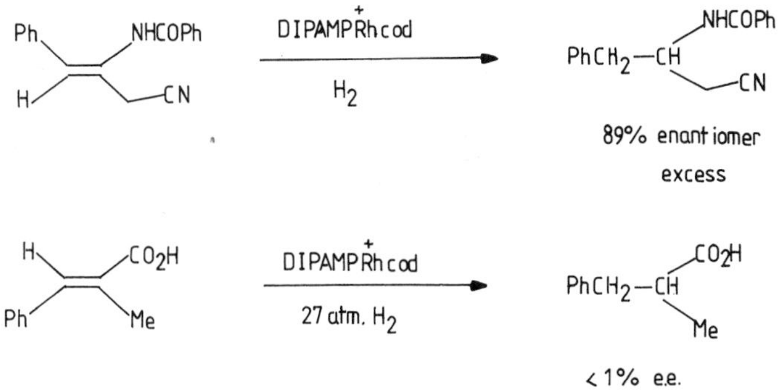

Figure 2.8   Hydrogenation of other substrates by DIPAMP complexes.

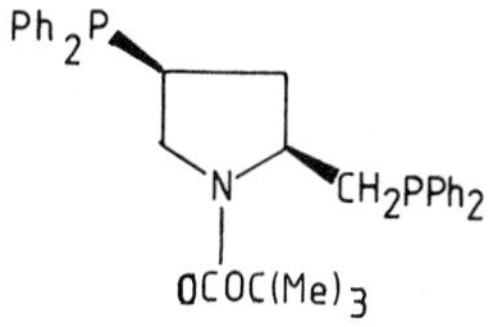

5-Membered chelate ring complexes have not so far been shown to be successful over a broad range of substrates. The replacement of the acid or ester functionality of the dehydroamino acid by a cyanide group is successful[35] but simple unsaturated acids which can give only bidentate coordination give poor optical yields (Figure 2.8). Itaconic acid and its derivatives afford the possibility of cooperative binding by both α- and β-carboxylates and some success has been achieved[44] (Figure 2.9). The enantiomer excess obtained in the hydrogenation of itaconic acid was concentration dependent and nmr studies showed that at low concentrations (or in the presence of base) bidentate complexes were formed, but in high acid concentrations tridentate species were observed.[63] However, several factors are clearly involved; dimethyl itaconate binds stereospecifically but very weakly and is hydrogenated in good optical yield.

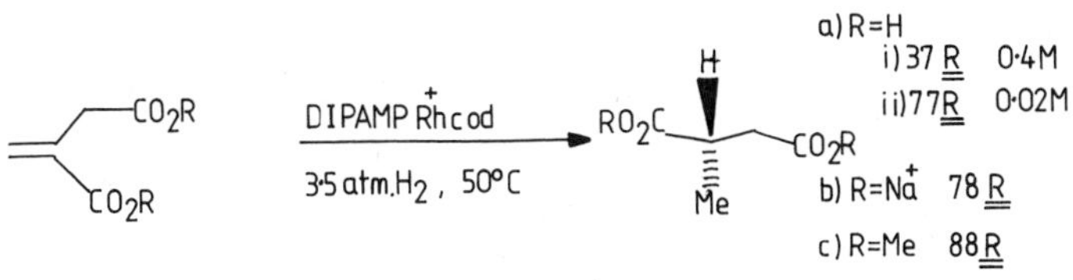

Figure 2.9   Hydrogenation of itaconic acid derivatives by [DIPAMP-Rh cod]⁺ BF₄⁻.

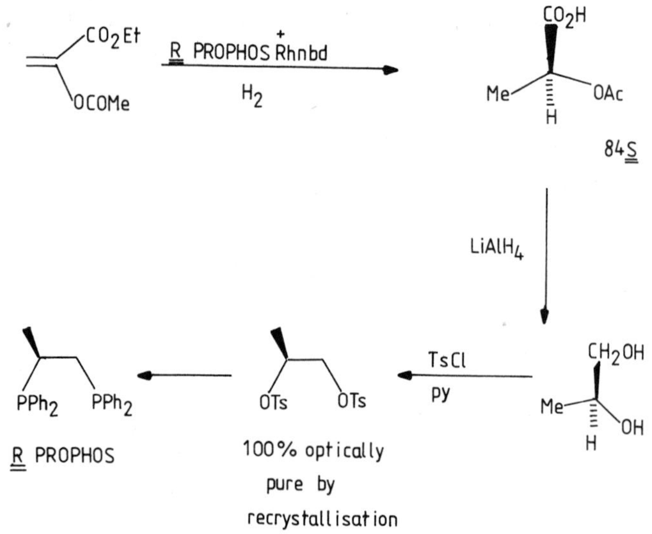

Figure 2.10 A self-generating catalyst.

An interesting report of the hydrogenation of enol acetates describes an asymmetric catalyst able to breed its own chirality (Figure 2.10).[37] This has also been used for the generation of chiral methyl chiral lactic acid. This is an important precursor for many labeled molecules which can be used to establish the stereospecificity of enzymic reactions[45] (Figure 2.11).

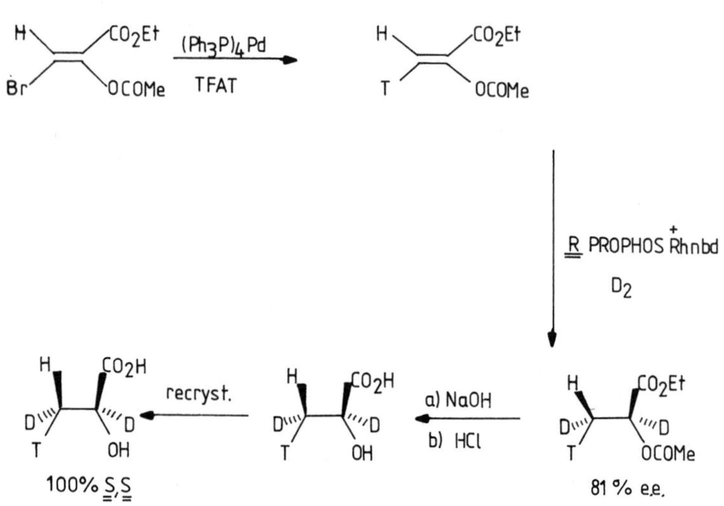

Figure 2.11 Preparation of chiral methyl chiral lactic acid (TFAT = CF_3CO_2T).

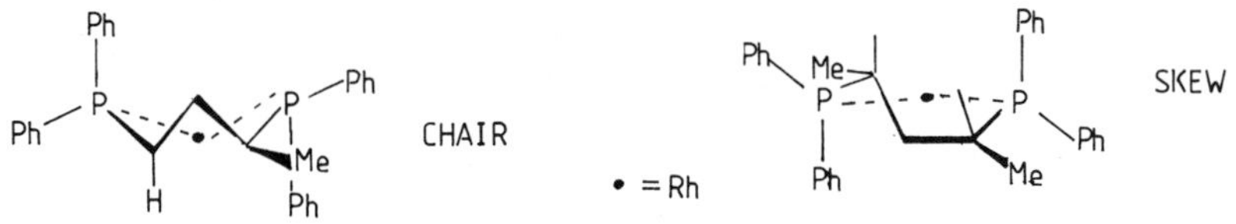

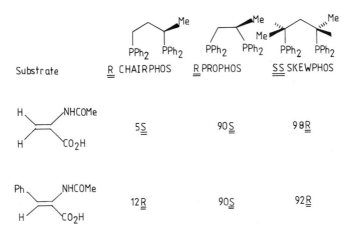

Figure 2.12 Possible   conformations   of   bis(diphenylphosphino)-propane
            complexes.

2.2   Catalysis by 6-ring chelate biphosphine rhodium complexes

Stereochemical considerations in 6-membered rings are more complex. In the absence of substituents a chair conformation is favored and with a single substituent this will also be the case since the substituent may occupy an equatorial site.   This approximates to σ-symmetry and suggests that there will be little discrimination either in binding or hydrogenation (Figure 2.12).

However, with two substituents, as in SKEWPHOS, one of these would have to be placed axial in a chair conformation so that the "skew" conformation in which both are pseudo-equatorial is favored.   The skew conformation, confirmed by xray crystallography for the norbornadiene complex, is dissymmetric and thus may be expected to give good results for hydrogenation.   These considerations are borne out in practice; CHAIRPHOS gives poor optical yields but SKEWPHOS gives good results (Figure 2.13).[64,64a]

Figure 2.13 Hydrogenation of enamides by PROPHOS, CHAIRPHOS and SKEWPHOS.

2.3   Catalysis by 7-ring chelate biphosphine rhodium complexes

By far the largest number of chiral complexes used in catalysis are those containing a seven-membered chelate ring. The class includes not only phosphines, but also numerous phosphonites, amino-phosphines and phospholes. Indeed, by 1980 there were already about one hundred of these chiral ligands which were capable of effecting a variety of asymmetric additions.

Again, the most widely studied reaction has been hydrogenation, and 7-ring chiral chelate complexes have proved remarkably versatile in hydrogenating a wide range of substrates in good optical yield. Consider the ligand DIOP (**XIX**)[65-6] (derived in high yield from tartaric acid[67]) which has been used more extensively than any other asymmetric biphosphine and is commercially available. Figure 2.14 illustrates the ability of complexes of this ligand to reduce a variety of substrates.

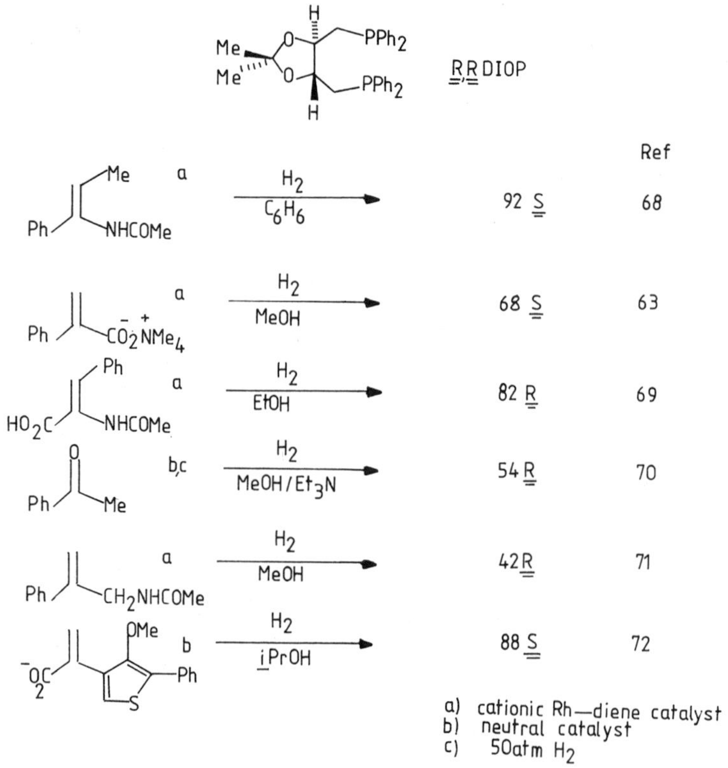

					Ref
	a	H_2 / C_6H_6		92 S	68
	a	H_2 / MeOH		68 S	63
	a	H_2 / EtOH		82 R	69
	b,c	H_2 / MeOH/Et$_3$N		54 R	70
	a	H_2 / MeOH		42 R	71
	b	H_2 / iPrOH		88 S	72

a) cationic Rh—diene catalyst
b) neutral catalyst
c) 50atm H_2

Figure 2.14 Hydrogenations with (R, R) - DIOP complexes.

While the 5-ring chelate complexes are ineffective for the reduction of bidentate substrates, this is not the case for the 7-ring systems. Indeed, simple enamides and simple αβ-unsaturated acids may be reduced in high optical yield, and asymmetric reductions of heteronuclear double bonds have also been successfully carried out.

2.3.1 Modified DIOP ligands

A number of modified DIOP ligands have been prepared with the intention of optimizing the optical yield for particular reductions. For example, the effects of substitution in each phenyl ring have been examined.[74-5] A meta-methyl group was found to increase both the rate of reduction and the optical yield in the reduction of dehydroamino acids, while para-alkyl substitution had little effect. Ortho substituents in the aromatic ring gave a more pronounced change leading to a reduction in rate and a reduction in optical yield of the product.[76] Moreover, when ortho-methoxy substituents were introduced, a reversal in chirality of the product was observed.[77] These effects are illustrated in Table 2 which shows how a series of modified DIOP ligands were examined in order to synthesize a derivative of L-tryphophan most effectively[76] (Figure 2.15).

In **XXII**, the analogue of DIOP where the phenyl rings are ortho-linked to give a phosphole system,[75] it was again found that there was a great reduction in catalytic activity and a reversal in stereospecificity. The dimethyl DIOP analogue **XXIII** has been synthesized from the sugar D-mannitol,[64] and in the reduction of dehydroamino acids gave products either with reversal of chirality (with respect to corresponding reductions with S, S- DIOP) or in very low optical yield.

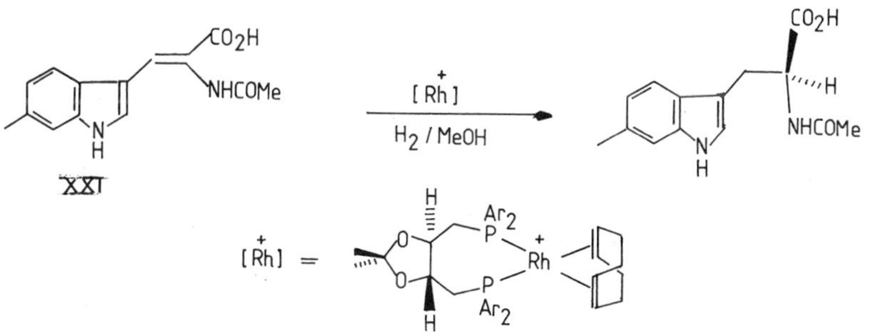

Figure 2.15 Hydrogenation of a tryptophan derivative **XXI** in the presence of modified DIOP ligands.

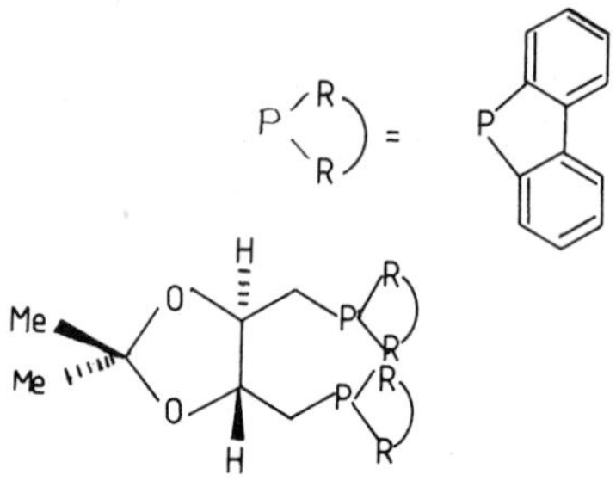

XXII

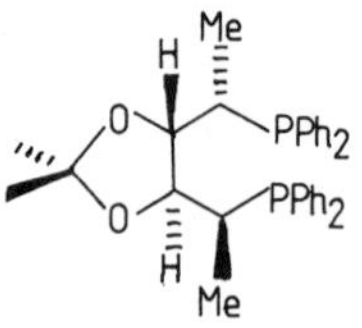

(2$\underline{S}$-5$\underline{S}$)–Dimethyl –DIOP

XXIII

Table 2. The hydrogenation of **XXII** in the presence of modified DIOP ligands $C_7H_{12}O_2(PAr_2)_2$

Phosphine (Ar)	Optical yield	Configuration	Relative rate
Ph	73	R	1
3,5-diMe-Ph	86	R	3
3-Me-Ph	83	R	2
4-Me-Ph	72	R	0.8
3-OMe-Ph	72	R	0.7
2-OMe-Ph	27	S	0.002
3-Cl-Ph	54	R	0.008

Table 3. Reduction of α-acetamidocinnamic acid by R, R carbocyclic ligands **XXIVa**(79)

Phosphine	Relative rate	Enantiomer excess	Configuration
n = 1	2	15	S
2	10	91	R
3	3	73	R
4	< 0.5	39	R

(a) Reduction with in situ catalyst from $(RhcodCl)_2$ and **XXIV**; 25° C; 1 atm H_2; MeOH: Benzene 2:1.

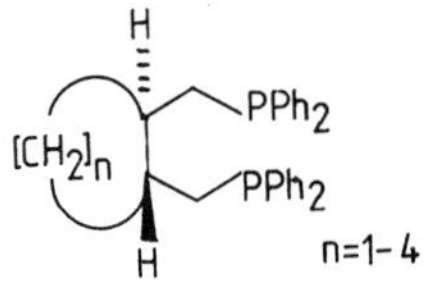

XXIV 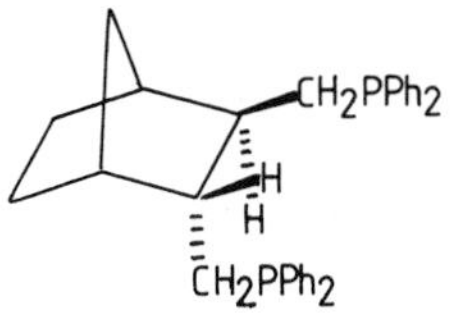   XXV 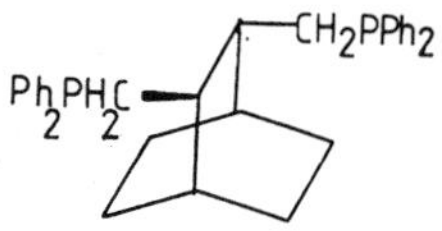   XXVI

2.3.2 Carbocyclic ligands

A number of ligands with the general structure **XXIV** have been prepared, as well as the related bis-phosphines **XXV** and **XXVI**.[78-80] In the reduction of enamides, these R, R - phosphines, like R, R - DIOP, give R - amino acids. The optical yield depends on the size of the fused ring, and decreases as the size of that ring increases (Table 3).

It is apparent then, that within the framework of modified DIOP catalysts there exists a disconcerting variation in optical yield with minor variation in ligand structure. A hypothesis has been advanced which accounts for most of the results for enamide reduction.[21]

The chelate ring of DIOP and its analogues is flexible and, to a complex with square planar geometry, several strain-free conformations are accessible. Many of these are centrosymmetric and ineffective at inducing asymmetry, but there exist two conformations of C_2 symmetry which may be related to the twist-boat and chair geometries of cycloheptane[81] (Figure 2.16). This is highlighted in an xray crystallographic study of the Group

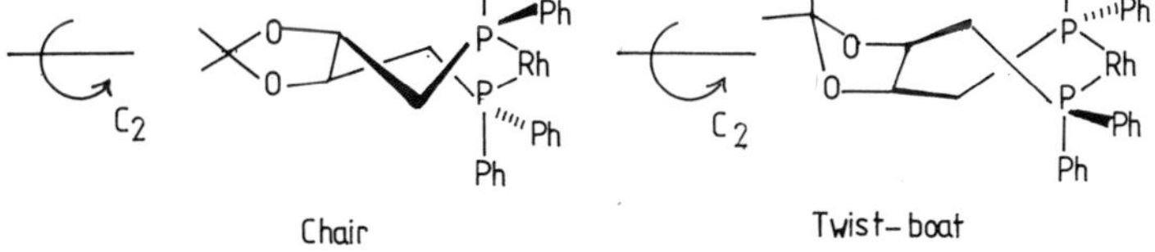

Chair   Twist-boat

Figure 2.16 Chair and boat conformations of a DIOP chelate.

VIIIC DIOP dichlorides.[82] The nickel complex is tetrahedral and has the chelate ring in a chair conformation, distorted to place a pair of phenyl rings parallel. The corresponding palladium and platinum complexes are square planar with two conformational isomers in the asymmetric unit. Again both have pairs of parallel phenyl rings, and one isomer is in a chair-conformation and the other in a boat-conformation. The crystal structure of [DIOP Rh(norbornadiene)]$^+$ BPh$_4^-$ has also been determined; on the basis of the Knowles correlation this should yield R - amino acids in hydrogenation.[59]

In the chair conformation of R, R - DIOP the pro-S phenyl rings are equatorial and the pro-R rings are axial. In the twist-boat conformer, the situation is opposite with pro-S rings axial and pro-R rings equatorial. Inspection of molecular models suggests that in the twist-boat conformation enamides prefer to bind through the si - face whereas in the chair they bind through the re - face. Hydrogen addition occurs to the face coordinated to the metal so that a si - bound enamide gives rise to R-amino acids. Thus if it is the major enamide complex which is hydrogenated to give product a twist-boat geometry will promote the formation of R-amino acids with R, R - DIOP and its analogues. If neither geometry is favored (for example, when the chelate is non-rigid) the optical yield is low.

There are a number of results supporting such a hypothesis:

(a)   In the absence of the acetonide ring, (for example XXVII), the chelate ring is non-rigid and optical yields are low.[66,75]

(b)   Dimethyl DIOP (XXIII) behaves as if it is predominantly a chair chelate, as a result of the extra stability associated with equatorial C - Me groups. In the twist-boat conformer, one of these would be axial. It was found that S, S - complexes gave R - amino acid products.[64]

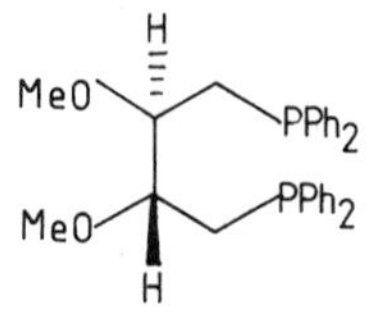

XXVII 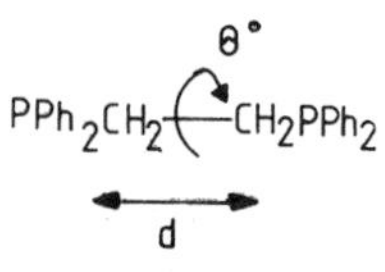   XXVIII

(c)   In the tetra-methoxy analogue of DIOP, the axial o-anisyl groups are probably weakly coordinated to rhodium, giving extra stability to the chair conformation.

(d)   When fused rings of different sizes are considered (Table 3), the optical yield decreases as the size of the fused ring increases. A small fused ring tends to stabilize the twist-boat conformer giving a higher proportion of R - product.   Alternative geometric arguments have been presented. Correlations were made between product optical yield and either the distance, d,[79] between the phosphorus bonded carbon atoms (**XXVIII**), or the angle of torsion ϑ.[83] Similar conclusions were reached using either of these interdependent parameters; for example the optical yield in a series of different reductions was found to be a maximum for d between 3.6 and 3.7Å.

2.3.3 Binding studies by ^{31}P nmr

Several studies similar to those for 5-ring chelate complexes have been carried out in this series. The binding of Z-enamide **XIII** to DIOP rhodium solvate has been shown by ^{13}C and ^{31}P nmr to involve olefin and amide - carbonyl coordination,[50] and only one diastereomer was observed at low temperatures.[84] The isomeric E - enamide (**XIV**) gave non-stereospecific olefin-carboxylate binding and isomerization occurs under hydrogenation conditions.[85]   The addition of deuterium was shown to proceed stereospecifically cis with about 5% H/D exchange at C-α.   This is consistent with an alkyl-hydride intermediate with rhodium bound to C-α, as has been demonstrated for 5-ring chelate catalysis.   Such alkyl-hydride intermediates have not yet been observed with 7-ring chelates, formation presumably being slow compared to breakdown, even at low temperatures. However, there is some evidence for the intermediacy of a rhodium alkyl with the tetra-o-methoxy analogue of DIOP.[77]

A detailed study of the binding of a range of enamides to a solvent complex of 1R,2R- trans-1,2-bis(diphenylphosphinomethyl)cyclobutane has also been made.[85] The effects of structure variation in the enamide on the ^{31}P nmr spectra of the bound species were examined. While most of the Z-enamides were bound in a bidentate manner through the olefin and amide

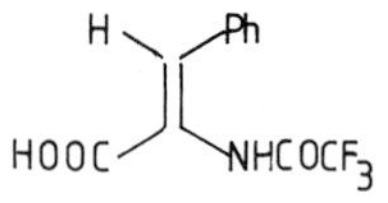

$$\underset{\text{XXIX}}{} \qquad \underset{\text{XXX}}{} \qquad \underset{\text{XXXI}}{}$$

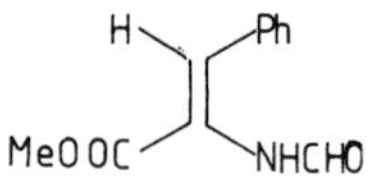

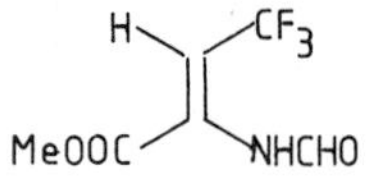

carbonyl the corresponding E-enamides were bound through olefin and carboxyl groups. Furthermore, in the case of **XXIX**, **XXX** and **XXXI**, in which the amide function is more weakly basic, the observed coupling constants (J_{Rh-P} and J_{P-P}) were substantially lower. This implies that the rhodium phosphorus π-bond is weakened, consistent with increased coordination about rhodium and it is probable that both amide and carboxyl groups are bound to rhodium simultaneously (**XXXII**). Finally the correlation between the observed enamide diastereomer excess and the measured enantiomer excess was very poor.

The binding of a range of αβ-unsaturated acids to the DIOP rhodium solvate complex has been studied by ^{31}P nmr.[63,71] It was clear that for several simple acids, such as acrylic acid, the major coordinated species was a bis-olefin complex. In the presence of base a 1:1 complex was observed involving binding of the olefin and the carboxylate anion. The bidentate 1:1 complex is generally favored under basic conditions and this correlates well with the enhanced optical yields observed generally in the asymmetric hydrogenation of unsaturated acids in basic media.

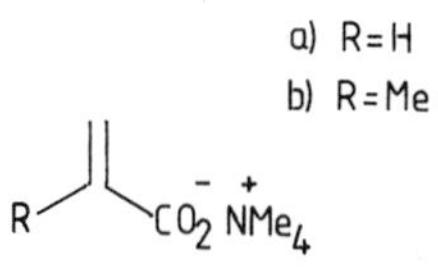

a) R=H
b) R=Me

$$\underset{\text{XXXII}}{} \qquad\qquad\qquad \underset{\text{XXXIII}}{}$$

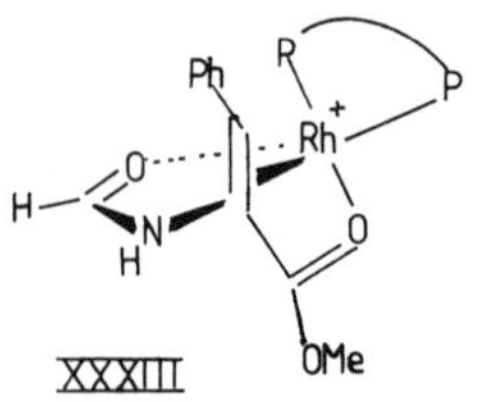

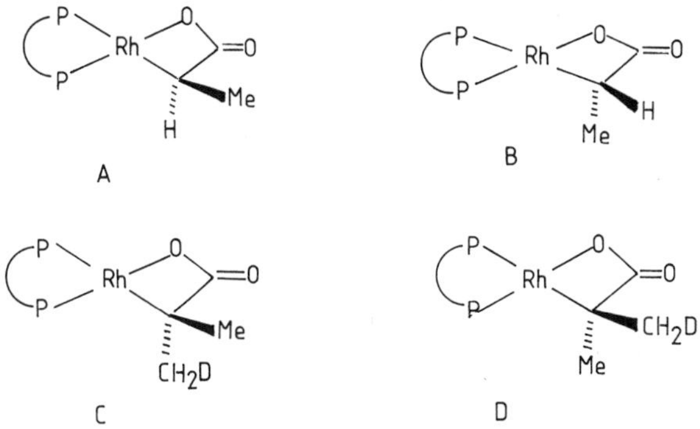

Figure 2.17 . Alkyls formed in the hydrogenation of acrylic and methacrylic acids.

A few other general points may be noted. A poor correlation exists between the diastereomer ratio of bound olefin complexes as observed by ^{31}P nmr and the ratio of enantiomers in the product. In the asymmetric deuteration of the tetramethylammonium salts of acrylic and methacrylic acids **XXXIII** the former was reduced in 51% optical yield, and the latter in 13% yield (as observed by ^{2}H nmr on derived chiral esters of methyl mandelate). Possibly in the case discrimination is at the stage of alkyl formation (Figure 2.17); A would be expected to be favored over B but discrimination between C and D is poor.

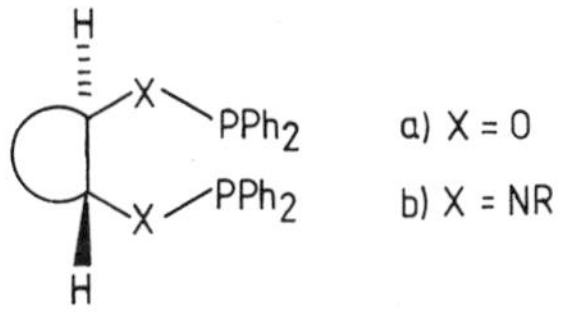

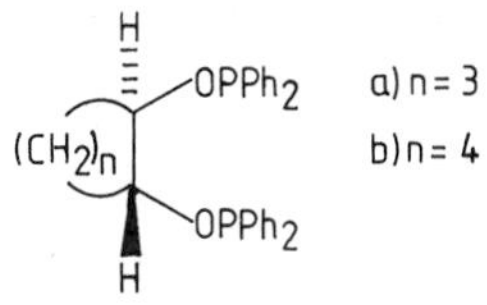

2.3.4 Phosphinites and aminophosphines

The range of seven-ring chiral ligands was expanded considerably by the introduction of chiral phosphonites **XXXIVa** and aminophosphines **XXXIVb**. After it had been shown that simple phosphinites **XXXV(a, b)**[86-7] were effective in asymmetric hydrogenations, it was clear that there were many phosphinites which might be prepared, fairly simply, from monosaccharides. A number of sugar-derived ligands were then quickly

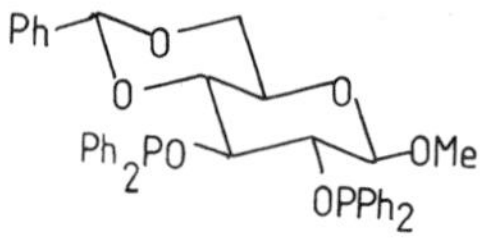

XXXVI

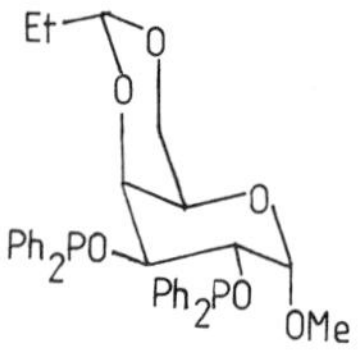

XXXVII

examined,[88-91,43] including **XXXVI**[88] derived from glucose and **XXXVII**,[89] from galactose. Complexes of these and several related ligands were found to be effective catalysts for the hydrogenation of dehydroamino acids (typically in up to 90% enantiomer excess) and also, to a lesser extent, for αβ-unsaturated acids. Surprisingly, complexes derived from **XXXV** were effective in reducing the simple olefin α-ethylstyrene in up to 60% optical yield.[87] Reaction conditions were generally much more forcing, requiring high pressures of hydrogen, high catalyst to substrate ratios and long reaction times. The mechanistic details of the reaction pathway remain unstudied although it is likely that a similar mechanism to biphosphine catalysis is operative at least in enamide reductions.

Also useful are the aminophosphines **XXXVIII**[92-3] **XXXIX**[94] and **XL**,[95] which reduce dehydro-amino acids with optical yields typically between 80 and 94%. There is a reversal of chirality, observed generally, on N-methylation of amino-phosphines such as **XXXVIII**. It is evident that the R, R - amino-phosphines (X = H) behave, like phosphinites, as if they are twist-boat chelates, while the corresponding N-alkylated phosphine behaves as if it is a chair chelate. Once again the driving force is likely to be the stability associated with placing the N-Me groups equatorial in the chair conformer, while in the twist-boat arrangement one would be axial.

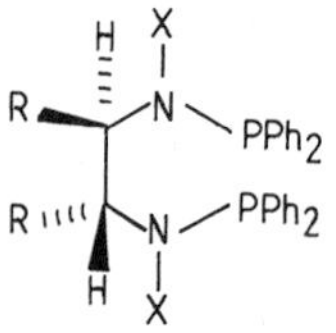

R = Me, Ph
X = H, Me

XXXVIII

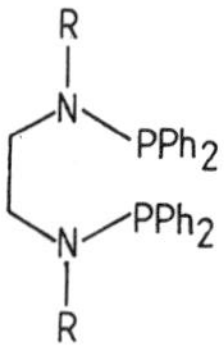

R=menthyl or
α-methylbenzyl

XXXIX

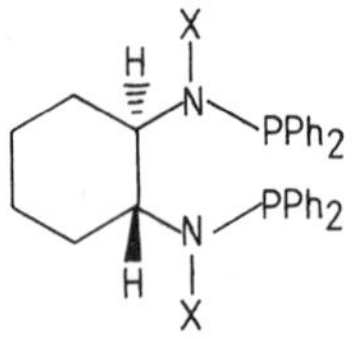

X=H, Ph, Me

XL

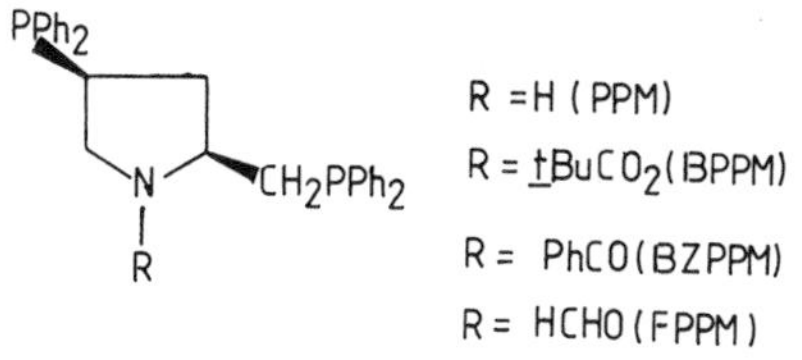
PPh2
CH2PPh2
N
R
R =H (PPM)
R = tBuCO2(BPPM)
R = PhCO(BZPPM)
R = HCHO(FPPM)

XLI

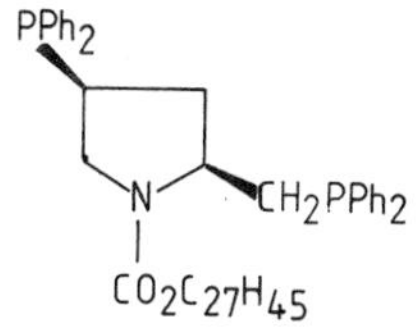
PPh2
N
CH2PPh2
CO2C27H45

XLII

2.3.5 **Chiral pyrrolidine phosphine ligands**

The synthesis and use of chiral 2S, 4S - pyrrolidine biphosphine rhodium complexes (**XLI**) has been extensively reported by Achiwa and co-workers.[96-9] As can be seen from Figure 2.18 these have proved remarkably versatile in the variety of substrates reduced. The syntheses of chiral α-amino acids (up to 98% ee),[100] α-hydroxy esters (78% ee),[101] β-amino acids (55% ee),[102] α-methylsuccinic acid (95% ee)[103] and R - (-) - pantolactone (87% ee)[104] have all been achieved.

Although a wide variety of R groups (see **XLI**) have been used to protect the N-H function in the parent biphosphine, 2S, 4S-4-diphenylphosphine-2-diphenylphosphinomethylpyrrolidine (PPM) it was found that the reduction of enamides and αβ-unsaturated acids was fairly insensitive to variation in the protecting group. Indeed this insensitivity inspired the synthesis of the lipophilized biphosphine **XLII**[105] in which the long alkyl chain confers solubility in non-polar aliphatic solvents such as n-hexane. Reductions of enamides and α-keto-esters were achieved in good optical yield. The parent phosphine PPM gave very poor optical yields in the reduction of enamides and coordination of nitrogen to the metal may be responsible for the poor result. This hypothesis is supported by an xray crystal structure of the cationic rhodium cyclooctadiene complex, in which direct coordination of nitrogen to the rhodium center was found for PPM but not for the N-acylated phosphine.[106] All that may be achieved by varying the protection at nitrogen is to "fine tune" the catalyst to optimise optical yields for a given reaction. An interesting example of this approach is provided by the reduction of the dehydrodipeptide **XLIII**. In this case it was found that the chiral substituted urea **XLIV** was most effective, catalyzing the formation of the R, S dipeptide in 96% diastereomer excess.[107]

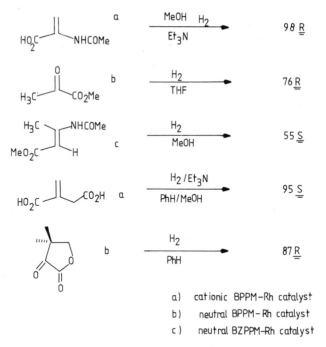

a)   cationic BPPM–Rh catalyst
b)   neutral BPPM–Rh catalyst
c)   neutral BZPPM–Rh catalyst

Figure 2.18   Hydrogenation by pyrollidine phosphine complexes.

XLIII

XLIV

There are several interesting features associated with the mechanism
of reduction catalyzed by these phosphines. For example, in complexes of
the N-acyl phosphines, 2 diastereomers may exist by virtue of restricted
rotation about the amide bond.[62,71] This has been observed in both diene
and solvent complexes for BPPM but in other cases such as BZPPM the
equilibrium is entirely biased towards one rotamer. A range of binding

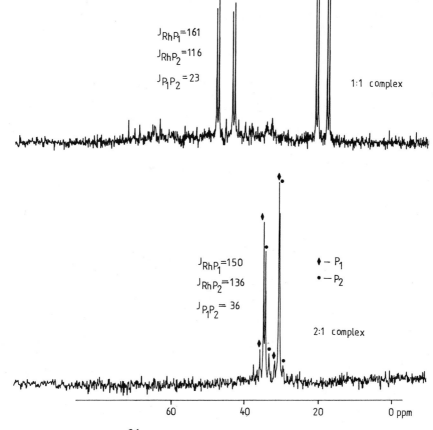

$J_{RhP_1} = 161$

$J_{RhP_2} = 116$

$J_{P_1P_2} = 23$

1:1 complex

$J_{RhP_1} = 150$

$J_{RhP_2} = 136$

$J_{P_1P_2} = 36$

♦ — P_1

● — P_2

2:1 complex

60   40   20   0 ppm

Figure 2.19 ^{31}P nmr of BPPM-itaconic acid complexes.

modes has been observed in itaconate BPPM complexes by ^{31}P nmr[71,108] (Figure 2.19). This shows that at low concentrations of added itaconic acid, the complex formed shows chemical shifts and coupling constants typical of a 1:1 complex with tridentate binding. In the presence of a large excess of itaconic acid, the spectrum observed is typical of a bis-olefin complex with reduced coupling constants.[63,71] Finally, itaconic acid in the presence of added base binds in a bidentate manner as a 1:1 olefin carboxylate complex with complexation through the α-carboxylate group.

2.4   Catalysis by large ring chelate biphosphine rhodium complexes

Fewer complexes of this type have been studied in detail although some have proved remarkably versatile in reducing a broad range of prochiral substrates. A good example is provided by chiral ferrocenylphosphines,[109] which possess a planar element of chirality. They are readily prepared by lithiation of resolved α-ferrocenylethyl-dimethylamine which proceeds stereoselectively[110] to give (R) - α -[(R)-2-lithioferrocenyl]ethyldimethylamine (Figure 2.20). This may then either be converted into the corresponding monophosphine (PPFA)[111] or further lithiated in the second cyclopentadienyl ring to give with ClPPh$_2$, (S)-α [(R)-1', 2-bis(diphenylphosphino)ferrocenyl] ethyldimethylamine (BPPFA). [112]

The adaptability of these phosphines is related to the availability of an additional functional group (amino or hydroxy) which can interact with the appropriate center of a bound substrate. Figure 2.21 illustrates a range of substrates reduced. The hydroxy-substituted phosphine, BPPFOH, is particularly successful in the reduction of carbonyl groups suggesting that hydrogen bonding between the keto-group of the bound substrate and the hydroxyl of the phosphine is important in determining selectivity. A further

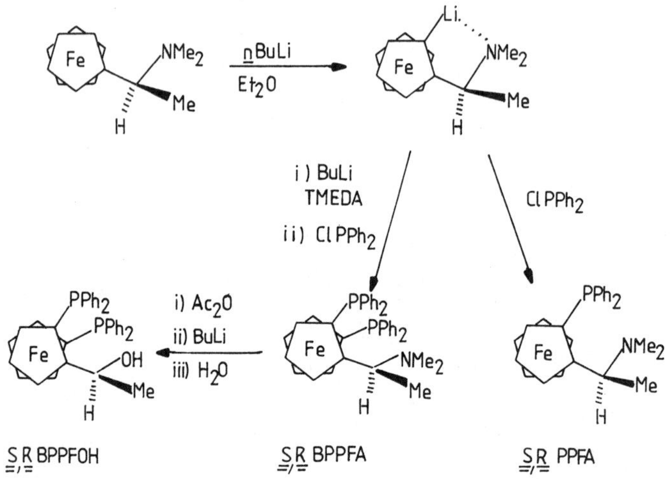

Figure 2.20 Preparation of chiral ferrocenylphosphines.

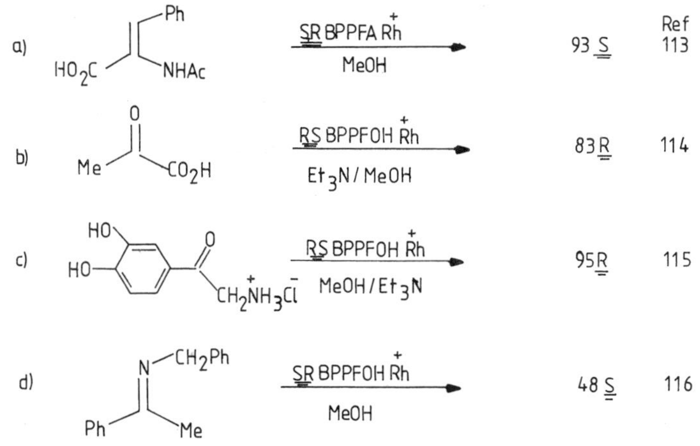

Figure 2.21 Variety of substrates reduced by chiral ferrocenylphosphines.

example is provided by the reduction of the enamide [Figure 2.21(a)]. The acid is hydrogenated in much high enantiomer excess than the corresponding methyl ester, and the optical yield is severely reduced by added triethylamine. In this case, hydrogen bonding between the carboxyl group of the bound enamide and the tertiary amine of BPPFA is probably controlling stereoselection.

Simple monosaccharides have been a prolific source of chiral products and a number of medium ring chelating biphosphines have been studied.[117-8] Particularly useful is DIOXOP, derived from D-glucose,[119] which has proved an effective catalyst for the reduction of enamides and unsaturated carboxylic acids in the presence of a tertiary amine. The course of the reaction has been shown by nmr studies[120] to differ from that with smaller ring chelates, and is summarised in Figure 2.22. This is the first example where a stable dihydride is formed by the hydrogenation of an olefin chelating biphosphine complex; this reverts slowly to a solvent adduct of the usual type. Addition of enamide causes formation of an intermediate complex whose ^{31}P and ^{13}C nmr (the latter on ^{13}C labeled enamides) indicate terdentate substrate binding. This step is catalyzed by

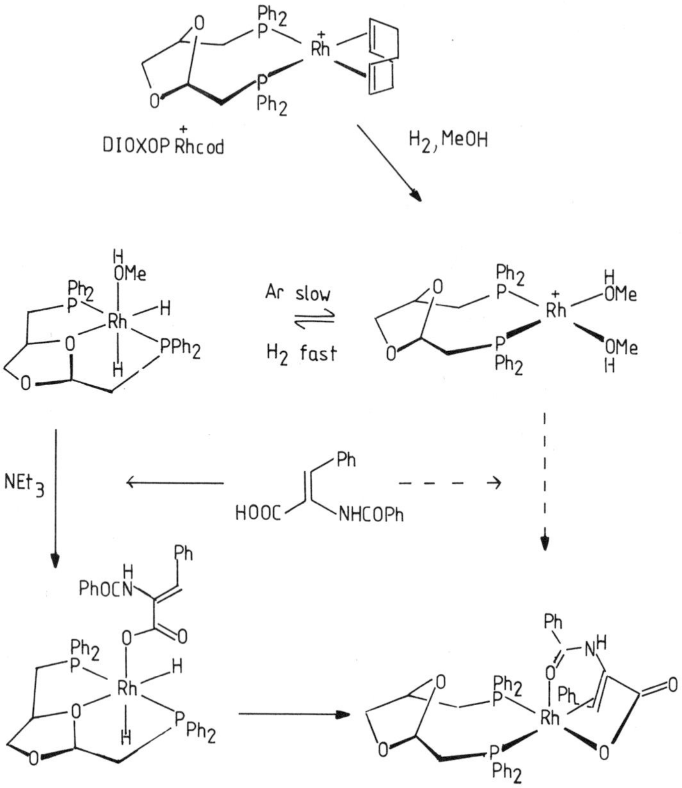

Figure 2.22 Enamide hydrogenation by [Rh-DIOXOP]⁺.

triethylamine and an intermediate hydride carboxylate complex has been observed at low temperatures. That ether oxygen coordination stabilizes the dihydride and carboxylate complexes has been confirmed by studies of model systems.[121-2]

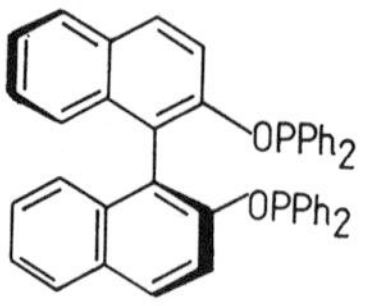

XLV

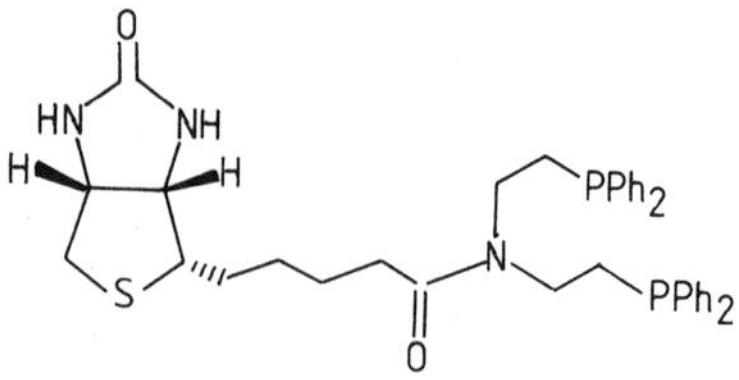

XLVI

Other large-ring chelating biphosphines have proved less interesting. Nevertheless, the atropisomeric phosphinite **XLV**[123] gives much higher optical yields in the hydrogenation of dehydroaminoacid esters than the corresponding acids. Acylaminobiphosphines also give modest optical yields in enamide reductions;[124-5] the most unusual is the biotin-derived species **XLVI** which may be irreversibly bound to the globular protein, avidin. Asymmetric homogeneous hydrogenation may be effected with the resulting "enzyme co-factor" complex in aqueous solution, thereby establishing an important concept in catalysis.

2.5 Catalysis by monophosphine rhodium complexes

Chiral monophosphines may be divided into two categories, those possessing chirality at phosphorus or carbon. The synthesis of chiral monophosphines developed by Mislow[126] and Horner[17] initiated the upsurge of activity in asymmetric hydrogenation. Although the preparation of such compounds was synthetically rather tedious, a large number have been made (**VIII, XLVII, XLVIII**). Good optical yields have been achieved in the reduction of potentially tridentate substrates (such as dehydro-amino acids) with monophosphines possessing an additional functional group which may bind to the metal. This is illustrated by the hydrogenation of α-acetamidocinnamic acid by complexes of **XLVIII a-c**.[18,20] The two ortho-substituted aryl phosphines give products with greater than 80% enantiomer excess while **XLVIIIa** and various meta- and para-substituted phosphines give optical yields below 20%. Some of the results obtained using **VIII** are shown in Table 4.[127-9,19] Cationic complexes of ACMP have also reduced saturated ketones, under more forcing conditions of temperature and pressure. The cyclopentane-trione, for example, has been reduced to an R-hydroxycyclopentanedione which is a precursor for prostaglandin E_1.[130]

Table 4. Asymmetric hydrogenation with R-ACMP as ligand (a)

Substrate	Product	$pH_2(atm)$	Optical yield
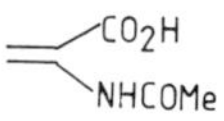	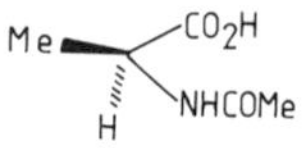	0.5	60 R
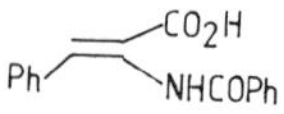	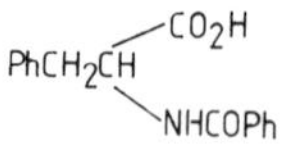	0.7	86 R
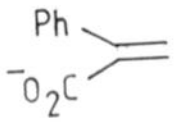	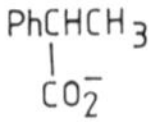	0.5	16 S
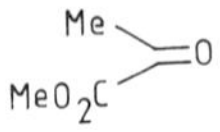	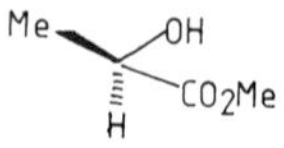	27	61 S
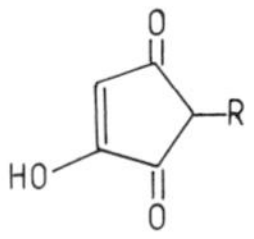	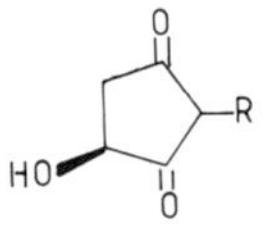	3.0	68 R

$$R = -(CH_2)_6CO_2Me$$

(a) All hydrogenations were run at 298 K, using MeOH as solvent with a cationic rhodium diene phosphine catalyst (0.05%).

The mechanism of hydrogenation using chiral monophosphines has not been fully elucidated. The first important intermediate to be identified has been a dihydride, rather than a solvent complex as is usually the case with chelating diphosphines.[53,131] The equilibrium between dihydride and solvent complex is a sensitive one, depending on the phosphine used. With phenyl methyl o-anisylphosphine (PAMP) one diastereomer of an enamide complex similar to that observed for DIPAMP may be generated but has not been studied in detail.[53]

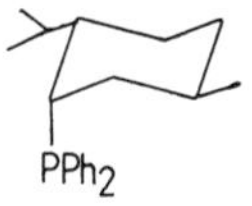

XLIX 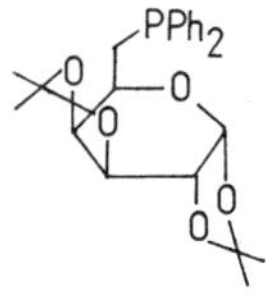   L

Chiral monophosphines which are chiral by virtue of asymmetry in one of the substituents at phosphorus are more numerous. Some have been derived from natural products, for instance, neomenthyldiphenylphosphine XLIX,[132] and L from galactose.[133-4] Neutral rhodium complexes of **XLIX** have proved particularly effective in reducing unsaturated acids in the presence of triethylamine. A wide range of alkyl and alkenyl substituted acrylic acids have been reduced and typically optical yields of 60 to 70% are found.[135-6] This reaction has been used in an effective route to chiral dihydrogeranic acid and a range of other intermediates useful in the synthesis of chiral vitamin E and citronellol. An example is shown in Figure 2.23, where the 2,3 double bond is selectively reduced.[137]

Once again, the introduction of an additional functional group into the phosphine capable of binding to the metal center, or interacting with a bound substrate, has led to improved optical yields. The ferrocenyl-derived monophosphine, PPFA, has been shown to chelate through nitrogen and phosphorus.[138] An xray structure of the cationic complex [(PPFA)Rh(nbd)]$^+$PF$_6^-$ revealed rigorous square planar geometry at rhodium with a Rh-P distance of 2.28 Å and a Rh-N bond length of 2.26Å. This complex catalyzes the hydrogenation of acylaminocinnamic and acylaminoacrylic acids with up to 84% optical yield.[139]

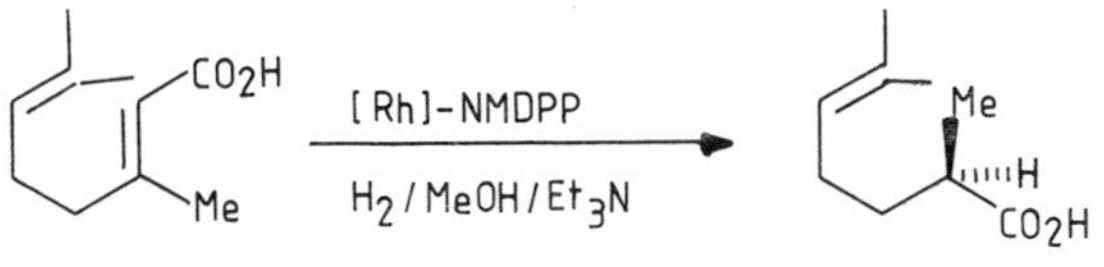

75 R

Figure 2.23 Hydrogenation with an NMDPP-Rh complex.

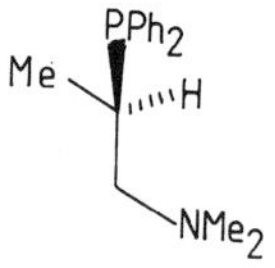

$\underline{\text{LI}}$ 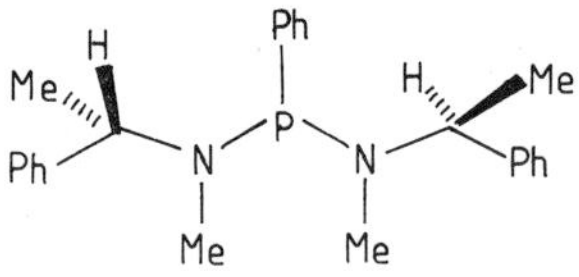 $\underline{\text{LII}}$ 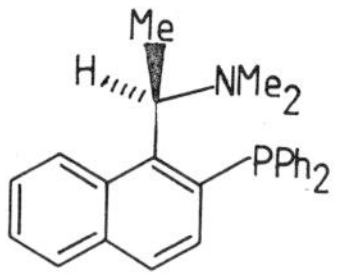 $\underline{\text{LIII}}$

The related, but less structurally rigid, rhodium complexes of the aminophosphines **LI** - **LIII** have also been studied.[64,140-1] The optical yields in the reduction of $\alpha\beta$-unsaturated acids and enamides have been disappointing; low reactivity and low stereoselectivity is associated with all of these systems. A recent report[142] on **LIV** suggests that this rather rigid system acts as a monodentate ligand but gives good optical yields.

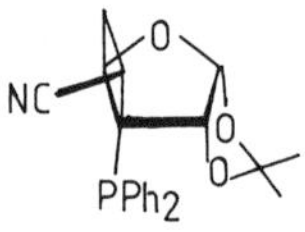

$\underline{\text{LIV}}$

2.6   Hydrogenations with chiral non-phosphine ligands

Successful asymmetric hydrogenation has also been achieved using non-phosphine ligands and metals other than rhodium. A series of papers reported homogeneous hydrogenations catalyzed by rhodium complexes of chiral amides.[143-5] Preliminary studies generated the catalyst in situ by treating trichlorotripyridylrhodium(III) with sodium borohydride in an optically active amide solvent. In later work a solution of the amide in diethylene glycol monomethyl ether was used giving products with identical optical purities. Such evidence suggested the formation of a specific rhodium-amide-substrate complex which was responsible for inducing asymmetry, rather than effects due to asymmetric solvation. Spectral and conductivity measurements supported the formulation of the complex as $[py_2(amide)RhCl(BH_4)]^+A.^-$ The results of hydrogenations effected with this catalyst are shown in Table 5.

The products from Z- and E- methyl-β-methylcinnamate had almost the same optical purity and this was also observed with other chiral amides. The hydrogenation of a carbon-nitrogen double bond in folic acid was also

Table 5. Reductions effected with Rh(III) borohydride complexes of chiral amides

Substrate	Amide	Optical purity of amide	Optical yield
E 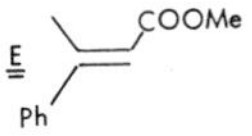 / Ph CH=CH-COOMe	R Ph–CH-NHCHO | Me	100	57 S
		96	40 S
Z 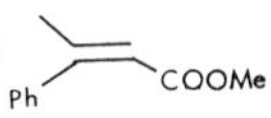 / Ph CH=C-COOMe	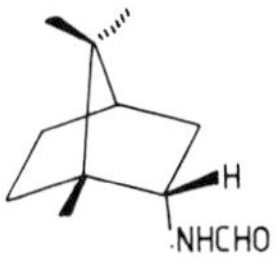 S (bornyl NHCHO structure)	96	45 S
Ph-CH=C(NHCOMe)-COOH	S PhCH-NHCHO | Me	100	74 R
(pterin structure)	R PhCH-NHCHO | Me	100	46% biological activity

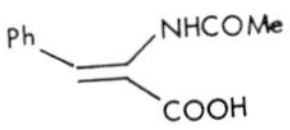

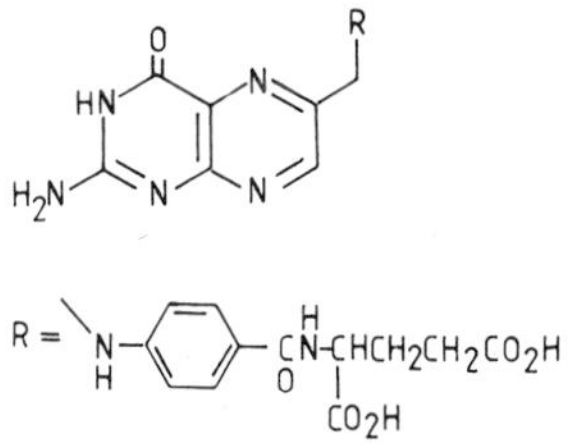

R = \N(H)-C₆H₄-CN(H)-CHCH₂CH₂CO₂H / O / CO₂H

accomplished, leading to tetrahydrofolic acid with high biological activity.[145] Although little is known about the mechanism of such hydrogenations, the insensitivity of optical yield to olefin geometry suggests that the produce determining transition state is late, resembling the structure of the product rather than the starting olefins.

A remarkable result for the hydrogenation of methyl-α-acetamidocinnamate in the presence of Wilkinson's catalyst and a Schiff's base such as benzylidene-S-alanine has been reported by Russian workers.[146] S-phenylalanine was obtained in 50% chemical yield and 95% enantiomer excess, providing another example of a catalyst able to breed its own chirality.

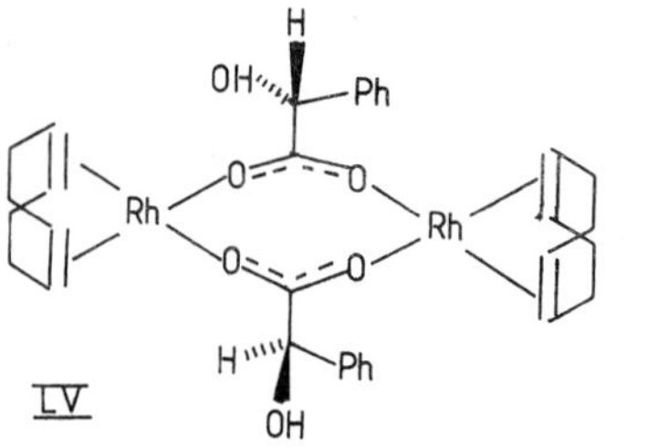

A series of very active chiral rhodium carboxylate complexes has been prepared, hydrogenations being carried out in the presence of an achiral phosphine.[147] One such complex is **LV** where the bridging carboxylate group is mandelic acid. In reduction of enamides, however, optical yields have not exceeded 10%.

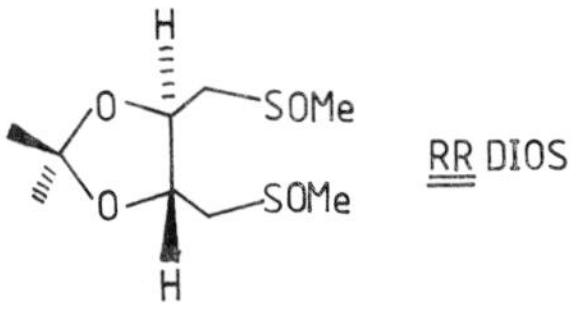

Monodentate sulfoxide ligands R_1R_2SO with chirality either at sulfur (R_1 = Me, t-Bu; R_2 = p-tolyl) or in an alkyl side chain have been coordinated to rhodium centers.[148-9] However, [(nbd)Rh(PPh$_3$) R, R - DIOS)] (DIOS = **LVI**) was found to be ineffective for asymmetric hydrogenation with unsaturated acid substrates.

Chiral cobalt catalysts

The development of chiral catalysts based on an inexpensive metal is desirable for an economic viewpoint. Some progress has been made in this area by Ohgo and co-workers with chiral cobalt catalysts.[150-1]

In early studies, a catalyst solution presumably containing a cyano-cobalt(II)-chiral amine complex was prepared (for example using R - 1,2-propane diamine), but the reduction of atropic acid was inefficient both optically and chemically.

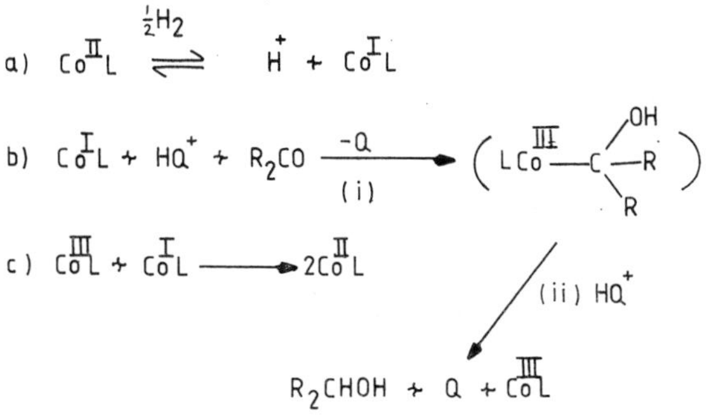

Figure 2.24 Mechanism of reduction of benzil by cobaloxime catalysts.

LVII

LVIII

RR DDIOS

More successful asymmetric reductions have been based on amine (especially alkaloid) complexes of bis(dimethylglyoximato)cobalt(II), cobaloxime(II). These complexes have been used to catalyze the hydrogenation of both olefinic and ketonic substrates. The best optical purity obtained with this catalyst is 71%, observed for the reduction of benzil in benzene solution using the hydroxyamine, quinine, as the chiral amine.[151] Using the related complex **LVII,** benzil was reduced in the presence of quinine and benzylamine in 79% optical yield, and the reduction of α-keto-esters was achieved in up to 50% enantiomer excess.[152] Some evidence was presented for the mechanism outlined in Figure 2.24. The stereochemistry of the product is determined by step (i), in which proton transfer from quinine to the coordinated carbonyl occurs. The quinine is considered to be associated with the substrate enhancing its susceptibility to nucleophilic attack by cobalt(I).

Chiral ruthenium catalysts

Some promising results have been obtained using ruthenium - chiral sulfoxide catalysts. A maximum optical yield of 15% was reported for αβ-unsaturated acid substrates using a trimeric $[RuCl_2(MBMSO)_2]_3$ complex, where MBMSO is S-bonded 2-methylbutylmethylsulfoxide, chiral at C-2 and racemic at sulfur.[150] Somewhat better yields have been obtained using ruthenium(II) complexes containing chiral chelating sulfoxides.[149] The most effective complex was $RuCl_2(DIOS)(DDIOS)$ where DDIOS is the acetal-cleaved derivative **LVIII.** This complex is thought to have a bidentate S-bonded DDIOS and a bidentate DIOS, coordinated through one sulfur and one oxygen.

Chiral titanium catalysts

Soluble Ziegler-Natta catalysts consisting of triisobutyl aluminum with the chiral alkoxide complex, titanium tetra(-)menthoxide, hydrogenated racemic terminal olefins such as 3,4 dimethylpent-1-ene but with zero enantioner excess.[153] Finally, the reduction of α-ethylstyrene by Red-Al in the presence of a chiral bis(menthyl-cyclopentadienyl) titanium dichloride catalyst gave up to 15% optical yield.[154]

2.7   Heterogeneous asymmetric hydrogenation

Heterogeneous enantioselective systems represent the very first non-enzymatic chiral catalysts. In 1940 Nakamura obtained optically active amines by platinum catalyzed hydrogenation of oximes in the presence of menthoxy acetate.[155] Most modern trends in this area, however, developed from the use of palladium on silk fibroin[10,156] which catalyzes the asymmetric hydrogenation of carbon-carbon and carbon-nitrogen double bonds. Two main groups of catalysts exist, firstly those in which a metal is adsorbed onto or coordinated with an optically active polymer and secondly those in which a metal surface is modified with a chiral substance. Both types have considerable potential for industrial applications particularly where catalyst recovery is highly desirable. Problems of contamination of the product with catalyst derived material are also largely avoided. Although the activities of polymer supported catalysts are only rarely as high as those of their homogeneous counterparts this is less of a problem on an industrial scale where reactions can easily be performed at high temperature and pressure.

Polymer supported catalysts

Akabori obtained an optical yield of 6% for the production of glutamic acid from diethyl α-acetoximinoglutarate using a catalyst obtained from the adsorption of palladium chloride onto silk fibroin.[10,156] More impressively hydrogenation of the oxazolone LIX was reported to give R-phenylalanine in 30-35% enantiomer excess.[157] However, subsequent attempts to repeat this work were less successful.[158]

Palladium on poly-S-valine and poly-S-leucine catalyzes the asymmetric hydrogenation of α-methyl and α-acetamido cinnamic acids but optical yields are low.[159] Similarly a catalyst prepared from ruthenium and

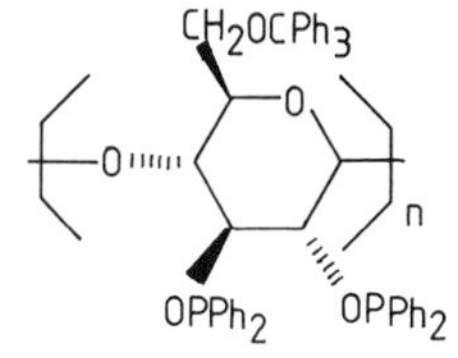

LXI

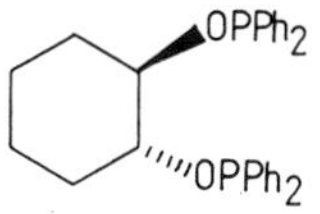

XXXV (n = 4)

poly-L-methylethylene imine catalyzes the asymmetric hydrogenation of mesityl oxide[160] and methyl acetoacetate[161] but neither the rate nor the optical yield is impressive.

There is one example of a chiral catalyst based on the cellulose backbone. **LXI** is closely analogous to the phosphinite **XXXV** (n = 4).[86-7,162] Like **XXXV** it is a rare example of a catalyst which gives a good optical yield for a non-polar substrate, α-ethylstyrene, up to 77% in the most favorable case. It is very slow to react, however; complexation with rhodium takes up to 5 days and the reaction time at 50°C and 50 atm. of hydrogen is several days.

Another group of catalysts consists of polymers modified with groups analogous to the phosphines used in homogeneous hydrogenation. There are two main synthetic approaches possible (Figure 2.25). In the first example[163] the polystyrene support is first chloromethylated, then oxidized and reacted with a diol to build the dioxolan ring of DIOP. Reaction with lithium diphenylphosphide gives the asymmetric phosphinated polymer. In the second case[164] a styrene derivative is converted to an optically active monomer and then copolymerised with methylmethacrylate. In both cases the polymer is subsequently reacted with a rhodium complex with a fairly labile ligand to give the active catalyst.

DIOP has been bonded onto several types of polymeric support, the most successful being modified polystyrenes (Table 6). It appears that the polymeric reagent may give optical yields as good as those obtained for the corresponding homogeneous system. The rate is slower, usually because diffusion of the substrate to the catalytically active site is slow. There is little loss of activity on recycling and air sensitivity is a less serious problem

Table 6. Hydrogenation with polymer bound DIOP catalysts

Polymer	Substrate	Solvent	T°C	pH$_2$	t(hrs)	Conv. %	ee	Ref.
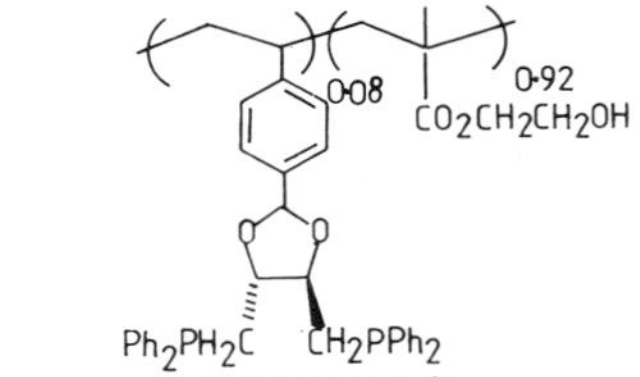 (polymer structure)	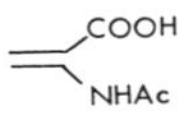 COOH / NHAc	C$_6$H$_6$/EtOH 1:1	28	1.3	5	100	52–60 R	165
	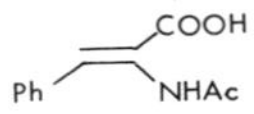 COOH / NHAc / Ph	"	25	1.3	2.4	100	86 R	165
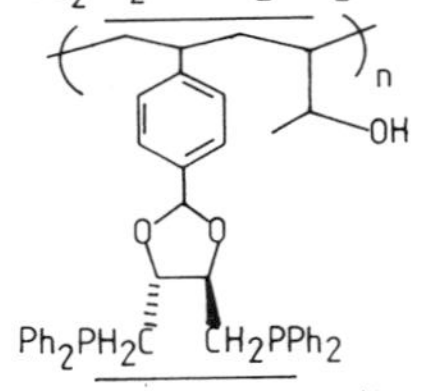 (polymer structure)	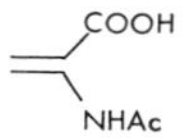 COOH / NHAc	C$_6$H$_6$/EtOH 1:2	25	1	2	100	76 R	166
	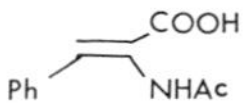 COOH / NHAc / Ph	"	25	1	24	70	83 R	166
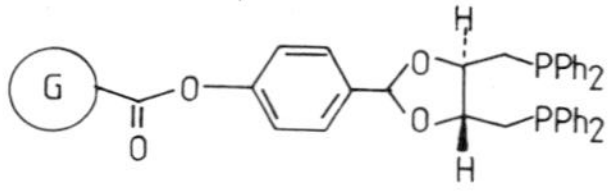 (polymer structure, G)	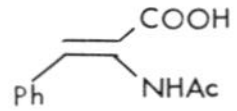 COOH / NHAc / Ph	C$_6$H$_6$/EtOH 1:2	25	1		11	7 S	167
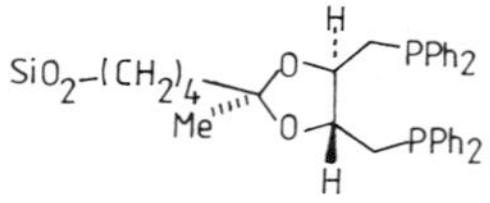 (SiO$_2$–(CH$_2$)$_4$ structure)	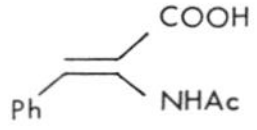 COOH / NHAc / Ph	C$_6$H$_6$/EtOH 1:1	20	1		100	63 R	168
DIOP Rh Cl S	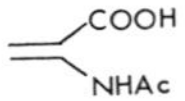 COOH / NHAc	C$_6$H$_6$/EtOH 1:1	25	1.0		100	73 R	69
	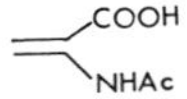 COOH / NHAc	"	25	1.0		100	81 R	69

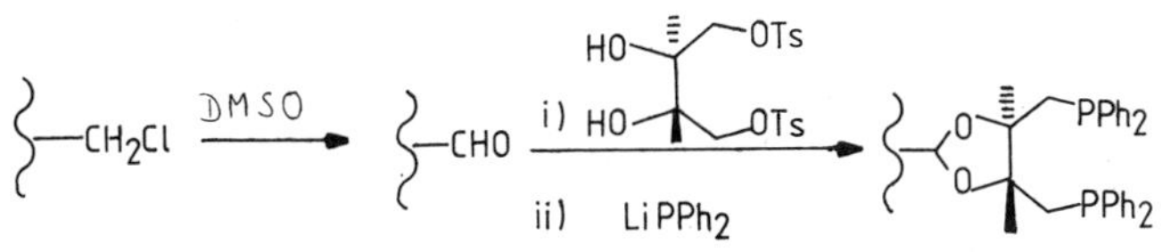

LXII

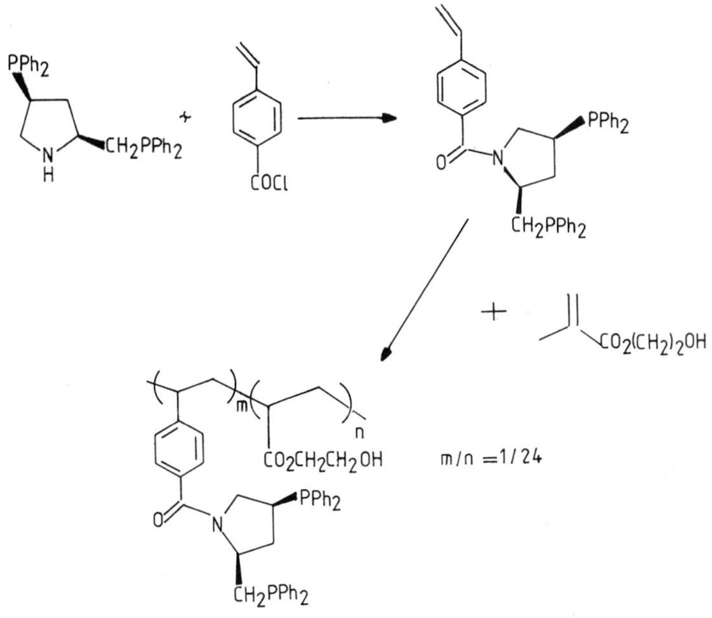

LXIII

Figure 2.25 Synthesis of optically active polymers.

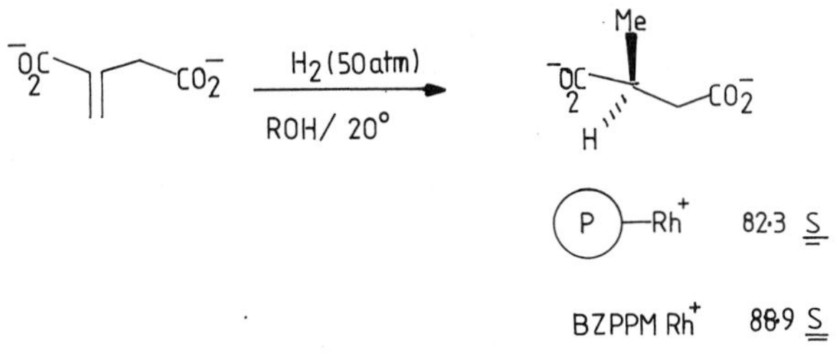

Figure 2.26 Hydrogenation with homogeneous and heterogeneous catalysts derived from pyrollidine phosphines.

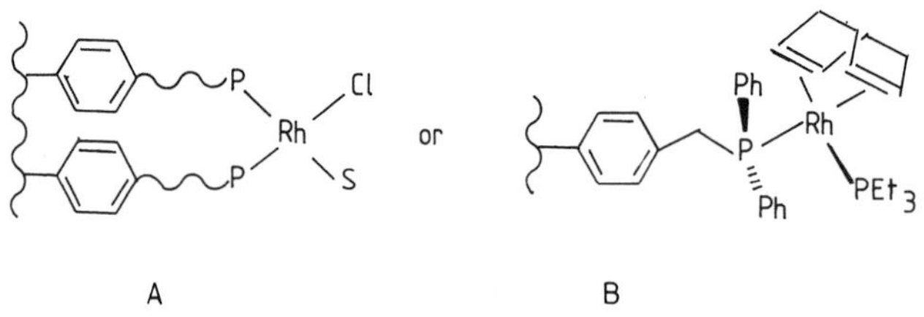

A   B

Figure 2.27 Polymer bound monophosphines.

than for the homogeneous system. The polymer **LXIII** which is analogous to BPPM[97,99] gives optical yields similar to those obtained with the homogeneous catalyst (Figure 2.26).[164]

A few catalysts utilize polymer bound analogues of non-chelating phosphines; all derive their chirality from the presence of one or more menthyl groups at phosphorus. A number of different problems arise in this case. Either the polymer must be sufficiently heavily phosphinated to ensure that two adjacent phosphorus atoms can coordinate to rhodium, or the rhodium must bear another phosphine ligand (Figure 2.27). In systems of type A the polymer backbone is sufficiently fluid to allow both cis and trans coordination of phosphorus at different sites. In B leaching of the additional phosphine ligand and disproportionation may occur and low optical purities are obtained.[169] Krause has prepared two cross-linked polymers **LXIV** and **LXV** bearing dimenthylphosphinyl groups by copolymerisation of an optically active monomer with divinyl benzene.[170] Although neither is very active, requiring a high pressure of hydrogen, optical yields of 50 - 60% can be obtained in the best cases for enamide hydrogenation which compares favorably with the analogous menthyl and neomenthyldiphenylphosphines.[132]

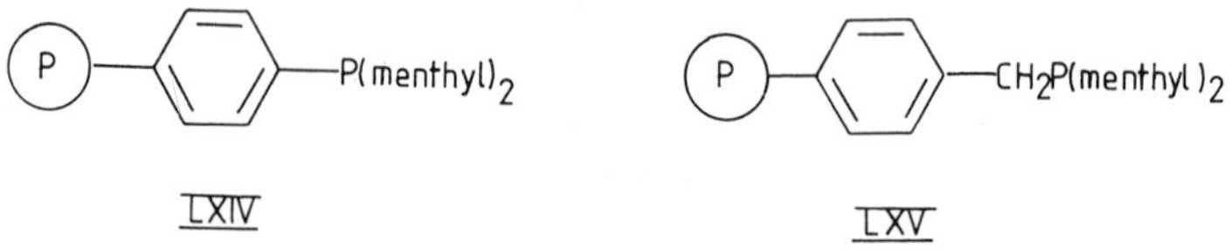

LXIV   LXV

Hydrogenation over modified metal surfaces

The modification of metal surfaces with optically active molecules to give chiral catalysts has been widely reviewed.[11-12,171] Isoda was the first to report enantioselective hydrogenation over Raney nickel modified with chiral amino acids;[172] the reduction of methyl acetoacetate **LXVI** proceeded in up to 36% optical yield. Izumi and co-workers have conducted extensive investigations into the modified Raney nickel system. Modification is achieved by soaking the activated metal in a solution of the optically active modifier and filtering and washing the catalyst before use. Modifying agents include amino acids and alcohols, mono-and dihydroxycarboxylic acids and dicarboxylic acids. Many experimental data are available but will be summarised only briefly here. Most of the results refer to the hydrogenation of **LXVI**; selectivity with simple ketones is much less good.[173]

It is observed that the best modifiers are chiral carboxylic acids, the corresponding alcohols being much less good.[174] The best second substituent at the α-position is an unsubstituted amino or hydroxyl group[175] and polar side chains also increase the optical yield.[176] For modifiers with two chiral centers bearing polar groups the best results are obtained when these have the same absolute configuration.

The optical yield increases with temperature of modification for hydroxy-acids but no generalisation can be made for amino acids.[173,175,177-8] For α-amino acids the optimal pH of modification corresponds to the isoelectric point of the acid[179] whereas for hydroxy acids a decrease was observed above pH 6.0[180] but no effect at lower pH. For hydroxydicarboxylic acids a sharp maximum occurs at pH 5.0;[181] this suggests that in more alkaline media the acid is converted to its salt and is unable to adsorb onto the metal surface. In strongly acidic media both carboxyl groups are bound to the metal. Presumably at a neutral pH one is bound and the other directs the approach of the substrate.

The direction of stereoselectivity of catalysts modified with α-amino and α-hydroxy acids of the same absolute configuration is opposite, implying that these are bound differently to the catalyst surface.[175,182] Izumi suggested therefore that the modifier was attached to the nickel as a chelate. This view was supported by demonstrating that the optical yield remained constant or increased when the nickel modifier chelate complex was used in place of the pure modifier.[183] Other studies investigated the role of the reactant and it was initially suggested that β-diketones and β-ketoesters gave better optical yields than simple ketones because of their ability to enolise and thus bind more stereoselectively.[184] However, almost the same optical yields are reported for the hydrogenation of acetylacetone and methyl acetoacetate despite the higher degree of enolisation of the former. β-Diketones are much stronger chelating agents than simple ketones and there is extensive evidence that adsorption gives a well-defined "delocalised" ring structure similar to that in the nickel acetylacetone complex.[184] Since the ultra-violet spectra of the reactant on amino- and hydroxy- acid modified surfaces differ, it was proposed that stereoselection takes place at the adsorption stage. However, it was still not clear whether the keto or enol form was hydrogenated.[185]

The effect of additives to the reaction mixture is complex and poorly understood. The presence of water increases the rate but decreases the optical yield.[186] The presence of unsaturated compounds may be beneficial[187] and the addition of acids with a pKa higher than that of the modifying tartaric acid to a nickel/palladium on Kieselguhr catalyst gives enantiomer excesses up to 89%.[188]

Russian studies, by Klabunovskii and co-workers, have concentrated on a general kinetic approach to the problem.[189] They too propose that adsorption is stereospecific and that the subsequent reaction is non-stereospecific. The stereoselective adsorption is a consequence of interactions between substrate, modifier and catalyst. Modified copper,[190] cobalt[191] and ruthenium[192] catalysts were also studied but were generally less successful.

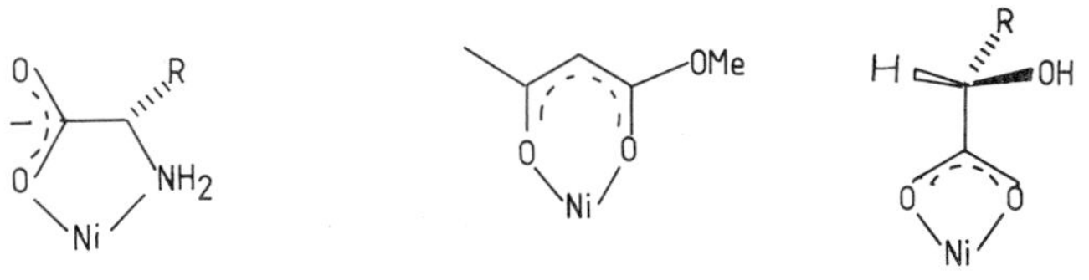

Figure 2.28 Binding of modifiers and substrate to Raney nickel.

Groenewegen and Sachtler have investigated the infrared spectrum of nickel treated with various modifying agents. This verified that bidentate chelation of both amino acids and methyl acetoacetate to a nickel atom is occurring whereas hydroxy-acids give carboxylate complexes (Figure 2.28).[193] Their model suggests that adsorption of modifier and reagent takes place on adjacent atoms, the two interacting via hydrogen bonding. The unique orientation of adsorbed modifier favors one configuration of substrate adsorption, and implies the opposite configurations of products derived from catalysts modified with α-amino and α-hydroxy acids.

A complete physical and kinetic model for this type of catalysis is lacking but it is clear that both selective and unselective adsorption sites exist on the catalyst surface. At selective sites one stereoisomer is produced exclusively; chiral differentiation takes place in the binding step rather than the rate-determining hydrogen addition.[194]

This type of catalyst has found few general applications but both β-ketoesters and β-diketones can be reduced with up to 100% enantioselectivity under appropriate conditions (Figure 2.29). In view of the importance of the stereoselective production of 1,3-diols in macrolide synthesis this area shows great potential for future applications.[195]

3.   ASYMMETRIC HYDROSILYLATION

The addition of mono- and dihydrosilanes to both homo- and heteronuclear double bonds is catalyzed by a variety of noble-metal phosphine complexes. This reaction has proved particularly effective for the asymmetric reduction of carbon-oxygen and carbon-nitrogen double bonds.[199-200]   As the resultant heteronuclear Si-X bonds are easily hydrolyzed, the hydrosilylation of ketones and imines can be considered to

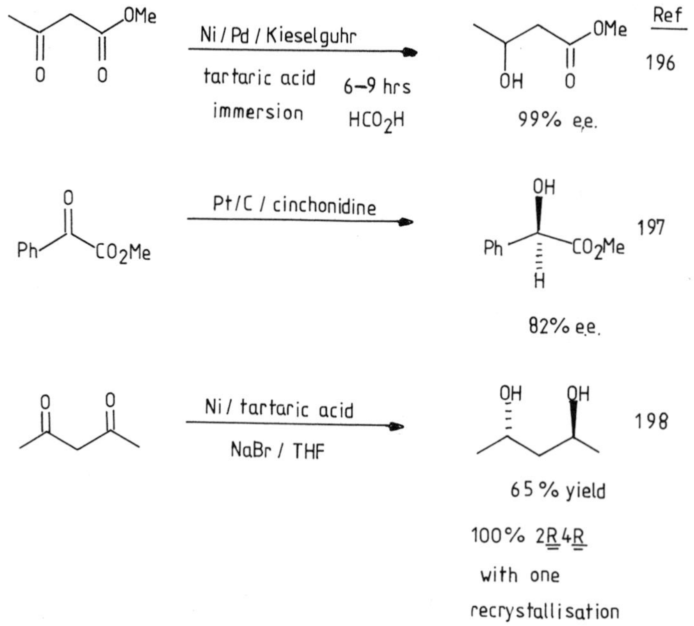

Figure 2.29 Asymmetric hydrogenation over modified metal surfaces.

be equivalent to hydrogenation (Figure 3.1). This catalytic reduction offers an attractive alternative to asymmetric hydrogenation; the reduction of the keto-ester n-propylpyruvate with a neutral rhodium-DIOP catalyst[26] illustrates this (Figure 3.2). A neutral rhodium-DIOP catalyst has also been used to generate a chiral silicon center.[201] Using the prochiral phenylnaphthylsilane, Corriu effected the catalytic reduction of diethylketone and reacted the siloxane product with methyl magnesium iodide to generate the chiral organosilane (Figure 3.3).

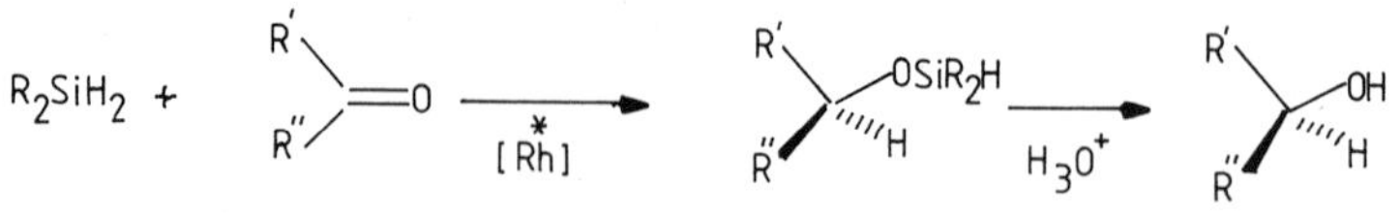

Figure 3.1 Hydrosilylation of prochiral ketones.

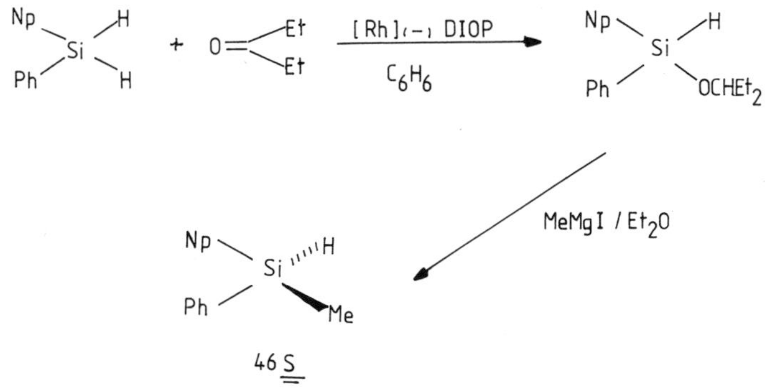

Figure 3.2 Comparison of hydrogenation and hydrosilylation of α-keto-ester.

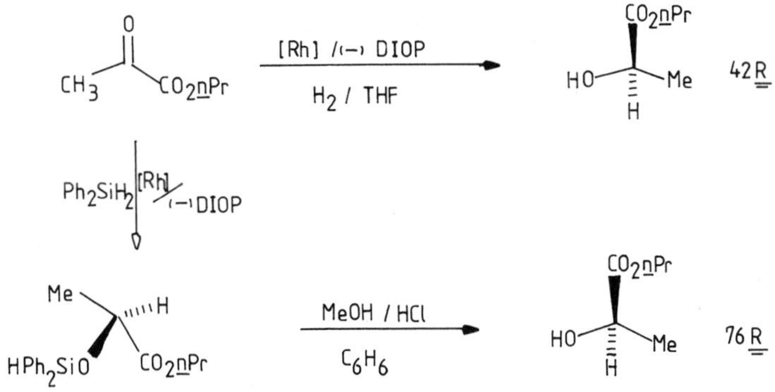

Figure 3.3 Generation of a chiral silane.

3.1 Reduction of carbon oxygen double bonds

The catalytic hydrosilylation of carbonyl compounds was first observed by Petrov[202] and Calas[203] using zinc chloride, nickel metal and chloroplatinic acid as catalyst. The reaction conditions were rather severe, leading to the formation of side products and the isomerization of both reactants and products. Moreover, the fact that the disproportionation of polyhydrosilanes was also catalyzed under these conditions,[204] restricted the use of these catalysts to addition of monohydrosilanes.

The first asymmetric hydrosilylation of prochiral ketones was effected using chiral platinum complexes.[205] Tert-butyl phenyl ketone reacts with dichloromethylsilane in the presence of an S-benzylmethylphenylphosphine-platinum complex to give, after hydrolysis, the corresponding alcohol in 19% optical yield. The disadvantages were numerous:

(a)   reaction was limited to MeSiH$_2$Cl

(b)   the chemical yield decreased markedly with more bulky ketones

(c)   dialkyl ketones were unsuitable substrates; yields were low and there were extensive side reactions.

An important development was the recognition that Wilkinson's catalyst, (Ph$_3$P)$_3$RhCl, is effective for the hydrosilylation of ketones, expecially with dihydrosilanes.[199-200] Chiral analogues were developed and some examples are given in Table 7 for the reduction of ketones using a neutral rhodium(I) complex, with a chiral monophosphine, in benzene solution.[206]

Analogous cationic rhodium complexes of the type [RhL$_2$(S)$_2$H$_2$]$^+$ (S = solvent) also catalyze asymmetric hydrosilylation of prochiral ketones.[207] A marked dependence of the optical yields on the structure of the hydrosilanes has been observed. Generally, the more bulky silanes have proved most effective; Kagan found α-naphthylphenylsilane to be the most useful of those examined, with a neutral Rh - DIOP catalyst.[208]

The best optical yields in the reduction of carbonyl compounds have been achieved using substrates with an additional polar group which may preferentially stabilize one diastereomer in the product determining transition state. In this respect the reduction of keto esters has proved particularly successful.[209-11]

3.1.1 α-keto-esters

The asummetric hydrosilylation of α-keto-esters, typically alkyl pyruvates and phenylglyoxylates has been catalyzed by rhodium complexes with BMPP, DIOP and BPPM as chiral ligands.[211] The results are shown in Table 8, which emphasizes the dependence of optical yield on the structure of the chiral ligand and hydrosilane used. Clearly then asymmetric hydrosilylation is the method of choice for the reduction of pyruvates.

3.1.2 β-keto-esters

The optical yields in the reduction of these substrates were similar to those obtained with simple ketones (Table 9). An obvious conclusion is that the ester carbonyl is not exerting any directing effect, although it has been

Table 7. Asymmetric reduction of prochiral ketones catalyzed by $(BMPP)_2RhSCl$ (S = solvent, BMPP = S benzylmethylphenylphosphine)

Ketone	Hydrosilane	Chemical yield %	Optical yield
$PhCOCH_3$	$PhMe_2SiH$	92	44 R
$PhCOiPr$	$PhMe_2SiH$	95	56 R
$PhCOtC_4H_9$	$EtMe_2SiH$	97	56 S
$PhCOcC_6H_{11}$	$PhMe_2SiH$	90	58 R
$tC_4H_9COCH_3$	Et_2SiH_2	95	39 S
$cC_6H_{11}COCH_3$	Ph_2SiH_2	92	40 S
$cC_6H_{11}COtC_4H_9$	$PhSiH_2$	90	43 S

Table 8. Asymmetric reduction of pyruvates and phenylglyoxylates via hydrosilylation

Keto-ester	Silane	Phosphine	Chemical yield %	Optical yield %
$CH_3COCO_2nPr.$	Ph_2SiH_2	S BMPP	84	60 R
	Ph_2SiH_2	S, S DIOP	82	76 S
	$NpPhSiH_2$	S, S DIOP	90	85 S
	Ph_2SiH_2	R, R BPPM	78	79 R
	$NpPhSiH_2$	R, R BPPM	85	79 R
CH_3COCO_2iBu	Ph_2SiH_2	S, S DIOP	85	67 S
	$NpPhSiH_2$	S, S DIOP	84	72 S
$PhCOCO_2Et$	Et_2SiH_2	S BMPP	80	6 S
	Ph_2SiH_2	S BMPP	82	10 R
	$NpPhSiH_2$	S, S DIOP	87	39 S

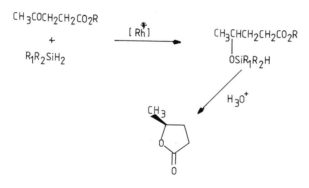

Figure 3.4   Hydrosilylation of γ-keto-ester.

suggested that the low induction may be attributed to a transfer hydrogenation process within an enol form of the substrate, coordinated through the carbon-carbon double bond, $CH_3C(OH) = CH-CO_2R$.[211]

3.1.3 γ-keto esters

The catalytic asymmetric hydrosilylation of levulinates followed by hydrolysis gives optically active 4-methyl-γ-butyrolactone through the silyl ether of 4-hydroxypentanoates (Figure 3.4).[211]   Optically active γ-valerolactone is a useful synthon for the formation of chiral terpenes or alkaloids via optically active amino-alcohols or diols. Typical results are given in Table 9.

Table 9. Asymmetric reduction of β and γ keto-esters via hydrosilylation with $NpPhSiH_2$ catalyzed by (S, S DIOP Rh S Cl)

Substrate	Chemical yield	Optical yield
$CH_3COCH_2CO_2Me$	89	21 S
$CH_3COCH_2CO_2Bu$	92	24 S
$CH_3COCH_2CH_2CO_2Me$	99	76 S
$CH_3COCH_2CH_2CO_2nPr$	94	83 S
$CH_3COCH_2CH_2CO_2iBu$	96	84 S

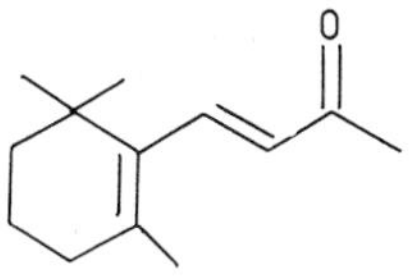

LXVIII

3.1.4 αβ-unsaturated carbonyl compounds

The hydrosilylation of αβ-unsaturated carbonyl compounds with BMPP and DIOP rhodium (I) catalysts has led to an interesting regioselectivity pattern.[212-3] The reaction with dihydrosilanes proceeds via 1,2-addition at the carbonyl group of the substrates such as 2-methylcyclohexenone and β-ionone (**LXVIII**).[214] The products are corresponding allylic alcohols with up to 43% enantiomer excess. On the other hand, monohydrosilanes undergo 1,4-addition to give products which hydrolyze to chiral saturated ketones with up to 16% enantiomer excess.

3.2   Reduction of carbon–nitrogen double bonds

Applying to imines the same procedures developed for the hydrosilylation of ketones has led to a useful synthesis of chiral secondary amines. The asymmetric hydrosilylation of prochiral imines often proceeds more efficiently than for the related ketones. Some examples are given in Figure 3.5.[215-6]

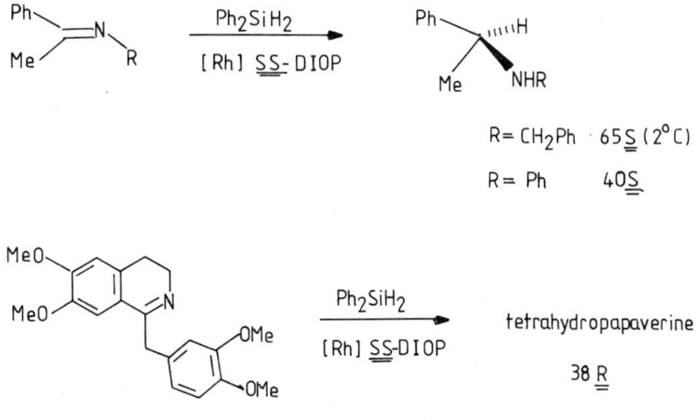

R= CH$_2$Ph · 65 S (2°C)

R= Ph   40 S

R= CH$_2$Ph

tetrahydropapaverine

38 R

Figure 3.5   Hydrosilylation of imines.

3.3   Hydrosilylation of carbon-carbon double bonds

Good asymmetric hydrosilylation of prochiral olefins has not yet been achieved.   Yamamoto has investigated the addition of dichloromethylsilane catalyzed by palladium, nickel and platinum complexes.[24,217-20]   However, catalytic activity was low and there was a lack of regioselectivity in the addition reaction and poor product optical purities.   The most promising reaction to date is the hydrosilylation of α-methylstyrene catalyzed by $Ni(BMPP)_2Cl_2$ which gave $Ph(Me)CHCH_2SiMeCl_2$ in 21% enantiomer excess and 31% chemical yield.[24]

3.4   Heterogeneous catalysts

A polystyrene supported rhodium - DIOP catalyst system was found to be very active in the hydrosilylation of ketone LXVI.[221,163]   Indeed it was almost as effective as the homogeneous analogue, although there was a diminution of reactivity and optical yield with re-use.   A graphite supported catalyst was developed to provide a rigid, solvent insensitive system where catalytic sites could be attached at given distances (LXIX) but the initial results in the hydrosilylation of ketones have been disappointing.[167]   An alternative strategy has been to suspend LXX with a macroporous silica gel in benzene.[168]   In this case the rhodium catalyzed hydrosilylation was achieved in good optical yield but the chemical yield was poor and the catalyst deteriorated rather rapidly.

3.5   Mechanism of hydrosilylation

Little evidence has been presented to establish the nature of the intermediates involved in the reaction pathway.   The first step has been demonstrated to be oxidative addition of the silane to the rhodium complex, and $Rh(PPh_3)_2H(SiEt_3)Cl$ has been isolated.[199]   A proposed mechanism is shown in Figure 3.6.

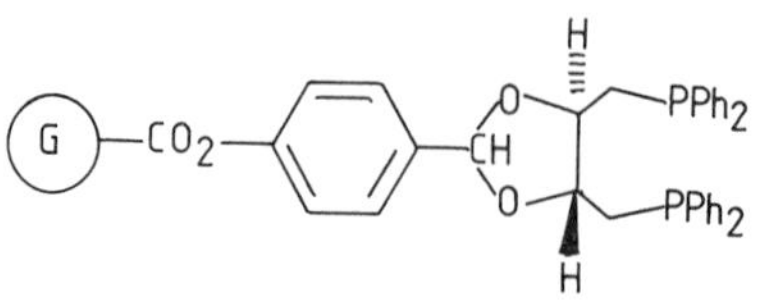

LXIX 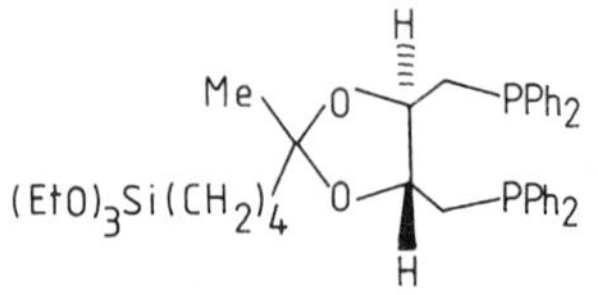   LXX

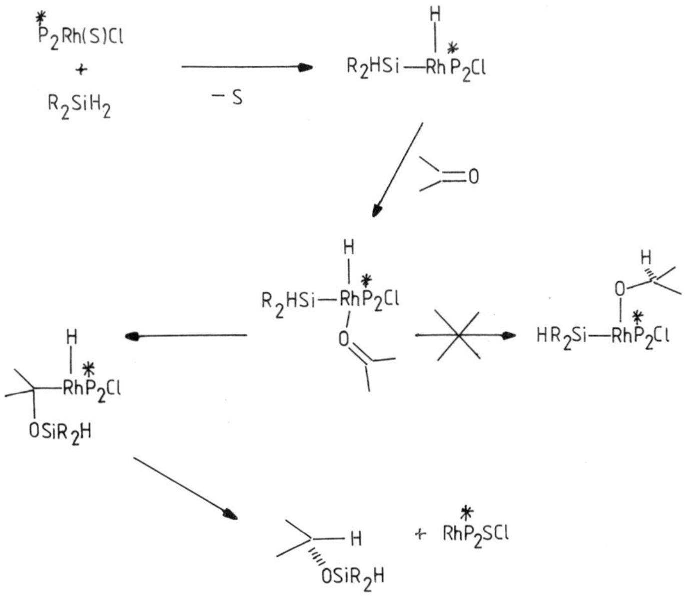

Figure 3.6   Proposed mechanism of asymmetric hydrosilylation.

There is some direct experimental evidence to support the idea of an intermediate α-siloxy alkyl-rhodium complex.   The reaction between acetophenone and α-naphthylphenylsilane with a rhodium-DIOP catalyst was studied by esr using nitrosodurene as a spin trap.   The presence of α-Npth Si(H)n-O•(Ar) was shown to derive from the induced decomposition of the hetero-atom stabilized alkylrhodium intermediate.[28]

4.   ASYMMETRIC HYDROFORMYLATION

Hydroformylation is the term applied to the addition of carbon monoxide and hydrogen to an alkene to form an aldehyde. In the presence of a suitable nucleophile such as water or an alcohol, acids and esters may be formed by an analogous reaction. Unsaturated hydrocarbons are important source materials in the petrochemical industry and oxygenated products are industrially important so that this reaction has been extensively studied and reviewed.[222-4]   The most important products of hydroformylation are straight chain aldehydes and alcohols; in 1979 nearly 3 million tonnes of butanal was produced from propene.

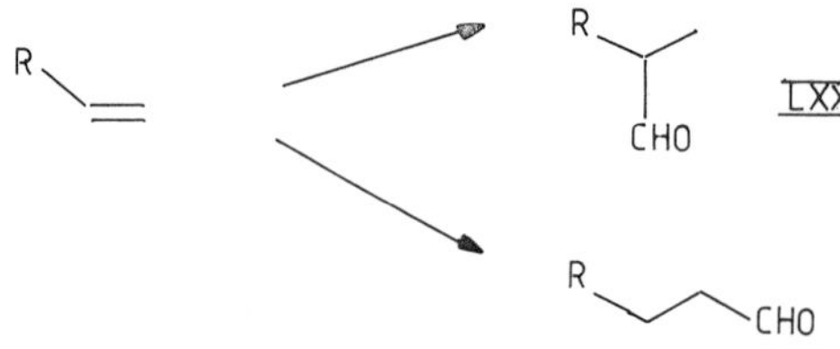

Figure 4.1   Hydroformylation of a terminal olefin.

The earliest hydroformylation catalysts were based on cobalt. The active species is thought to be $HCo(CO)_4$ and the mechanism proposed by Heck[225] is generally accepted. When the substrate is a terminal olefin two regioisomeric products are possible (Figure 4.1) resulting from the two possible insertions of the olefin into the metal hydride bond. Only the "iso-product" **LXXI** is of interest from the point of view of asymmetric hydroformylation but the linear aldehyde **LXXII** is more important industrially. Addition of phosphine ligands tends to increase the normal:iso ratio,[226] (2.5:1 to 4:1 in the unmodified system) and this has been the subject of extensive commercial research.

Catalyst systems based on rhodium, platinum and palladium have had much more success for asymmetric hydroformylation. Wilkinson first demonstrated the homogeneous hydroformylation of alkenes with hydrido carbonyl tris(triphenylphosphine)rhodium, **LXXIII**,[227] and <u>trans</u> chloro-carbonyl bis(triphenylphosphine)rhodium, **LXXIV**.[228] A series of papers[229] discussed the possible mechanisms for these reactions and both rhodium and iridium alkyls were observed by nmr spectroscopy. This led to the formulation of two possible reaction pathways involving mono- and diphosphine intermediates (Figure 4.2). It should be emphasized that little of this mechanism is supported by direct experimental observation. **LXXV** is well characterised, however, and it should be noted that the phosphine

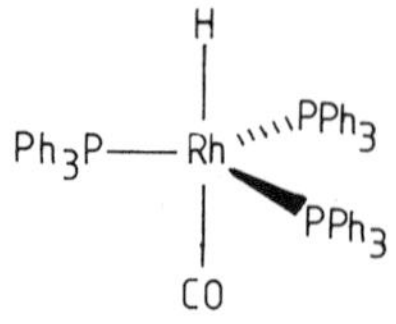

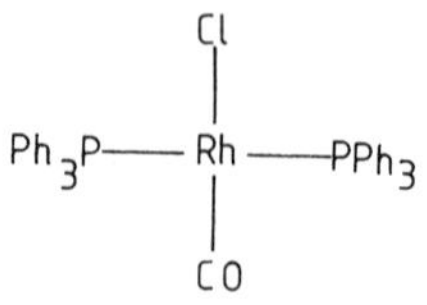

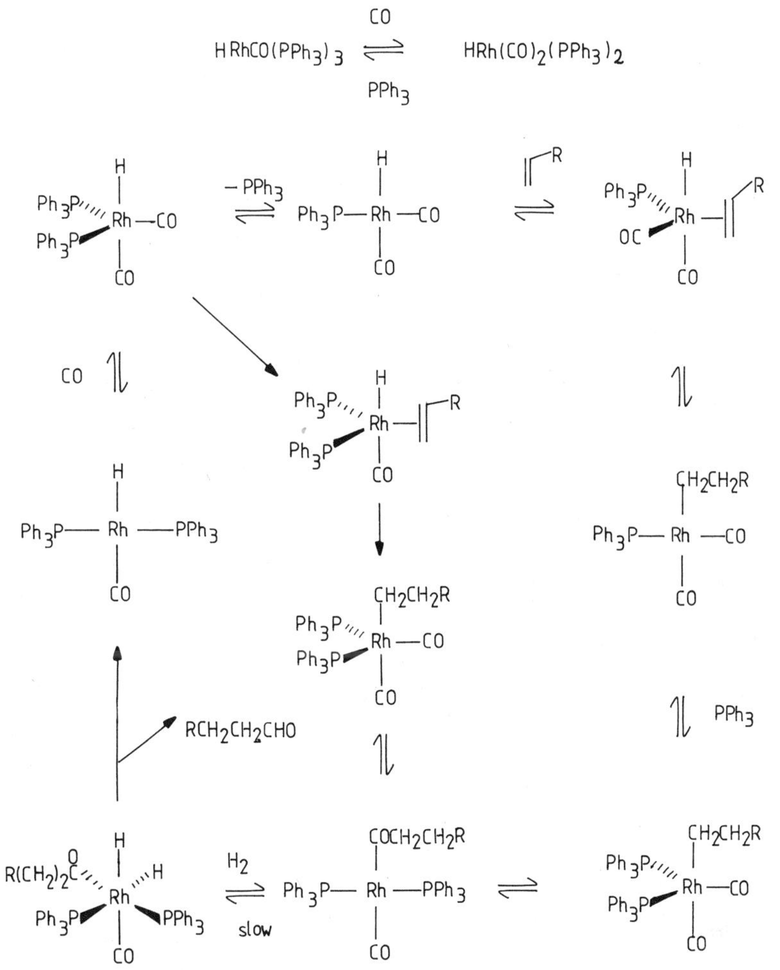

$$HRhCO(PPh_3)_3 \underset{PPh_3}{\overset{CO}{\rightleftharpoons}} HRh(CO)_2(PPh_3)_2$$

Figure 4.2 Mono- and diphosphine mechanisms for rhodium catalyzed hydroformylation.

ligands are <u>trans</u> suggesting that the chelating biphosphines which give such good results in hydrogenation might prove less successful for hydroformylation.

A number of other metal complexes catalyze hydroformylation reactions but are on the whole less active. Bis(triphenylphosphine)ruthenium tricarbonyl has been studied by Wilkinson;[230] the mechanism is thought to

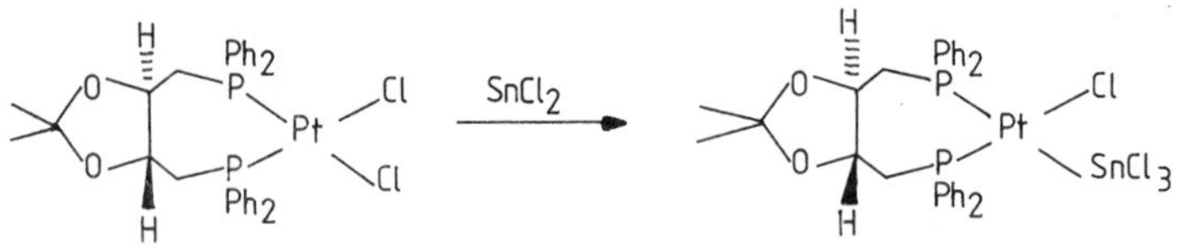

LXXVI   LXXVII

be similar to that proposed for rhodium. Iridium and osmium carbonyls are
significantly less active. A number of platinum complexes such as **LXXVI**
are good catalysts in the presence of stannous chloride.[231] Pregosin has
demonstrated by ^{31}P and [195]Pt nmr that **LXXVII** is formed on treatment of
LXXVII with stannous chloride; corresponding trialkylphosphine complexes
such as **LXXVIII** give **LXXIX** and **LXXX**.[232] The mechanism of the overall
reaction has not, however, been determined. The analogous reactions with
palladium catalysts have been used principally for the production of esters
(Figure 4.3).[233]

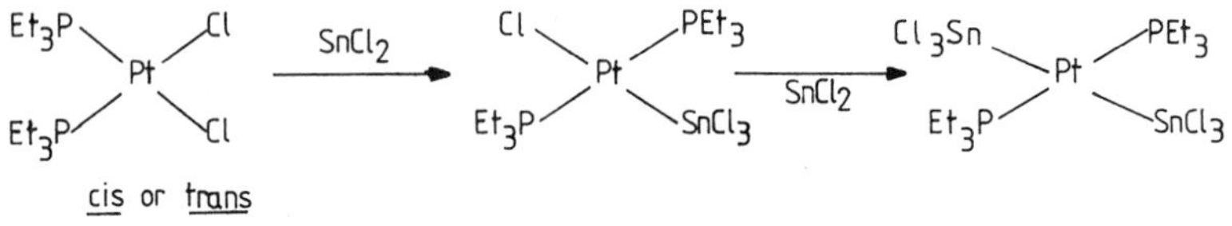

LXXVIII   LXXIX   LXXX

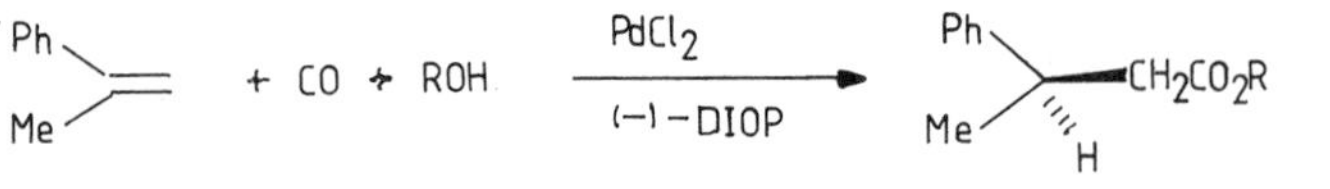

$R = $ i Pr, e.e. = 14%

Figure 4.3   Asymmetric hydroesterification of α-methyl styrene.

It is useful to consider some of the problems involved in achieving
asymmetric hydroformylation in a useful enantiomer excess. Firstly in most
cases two regioisomeric aldehydes or esters may be produced; the control of
regioselectivity is not well understood and where both isomers are optically
active the same enantioselectivity is seldom observed. Secondly the product
aldehydes (and to a lesser extent esters) are very labile with respect to

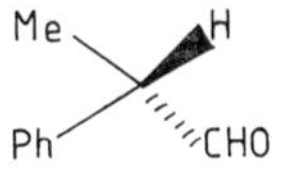

LXXXI 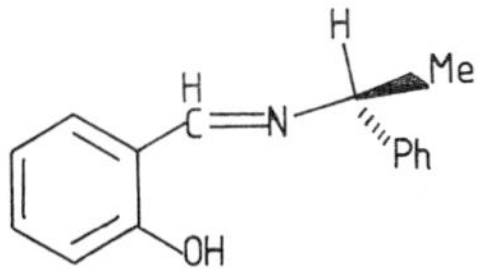   LXXXII

racemization by an enolization mechanism. Indeed the optical rotation of
hydtropaldehyde (**LXXXI**) has been the subject of a long controversy; the
value of $-238°$ for the R - enantiomer was determined in 1974,[234] but many
earlier workers in the field used a figure of $-160°$. Yet others determined
the optical purity of the corresponding acid or alcohol though lithium
aluminum hydride reduction is known to cause racemization.[235]

4.1   Cobalt catalyzed reactions

In view of the rather severe conditions associated with cobalt
catalyzed hydroformylation this is a poor candidate for modification to give
an enantioselective reaction. No asymmetric induction was achieved with
$Co_2(CO)_6(PMeEtPh)_2$ as catalyst.[236] However with a catalyst formed by
reacting $HCo(CO)_4$ with an equivalent amount of (-) - DIOP in toluene
at $-20°C$, _cis_-butene and bicyclo[2.2.2]oct-2-ene gave R-aldehydes of
measureable optical purity (1-3%).[237] With S-N-α-methylbenzyl-
salicylaldimine (**LXXXII**) as the asymmetric ligand and α-methylstyrene as
substrate, S-3-phenylbutanal (**LXXXIII**) was obtained in 2.8% optical
yield.[222,238] The low asymmetric induction may arise from racemization
during the reaction and also from the presence of more than one type of
catalytically active species, some of which may not contain the asymmetric
ligand.

LXXXIII

4.2   Rhodium catalyzed reactions

A wide variety of phosphine containing complexes have been used for rhodium catalyzed asymmetric hydroformylation but successful results achieved in only a few cases. It is convenient to divide the phosphines into two groups, chelating and non-chelating.

4.2.1 Non-chelating phosphines

(a)   Chiral at phosphorus

Most work has been done using simple phosphines bearing three different alkyl or aryl groups; the one report using ACMP where polar factors might be expected to operate is less successful. Table 10 shows a range of results for the hydroformylation of styrene.   A number of generalizations may be made. Firstly S - hydtropaldehyde is produced from catalysts containing R-phosphines.   Secondly the enantiomer excess decreases with increasing temperature and markedly with conversion; this may imply that the product is racemizing under the reaction conditions and/or that the catalyst deteriorates with time.   Results with other substrates are comparatively scarce and optical yields unimpressive. Some examples are given in Figure 4.4.

Table 10. Hydroformylation of styrene with monophosphine rhodium catalysts

Phosphine	Catalyst	$pH_2=pCO$	T^oC	n/iso	ee	conv%	Ref.
R P-Cy (C$_6$H$_4$-o-OMe) Me ACMP	Rhcod(ACMP) BF$_4$ +ACMP	60	100	1:3	15 S	-	239
R BzMePhP	[Rh(CO)$_2$Cl]$_2$ + 4 phosphine P:Rh = 4	70	60	2:98	28 S	18.5	240
R P-Ph Me nPr	[Rh(CO)$_2$Cl]$_2$+P Rh:P = 1:2.7	100	80	-	42 S	27	241

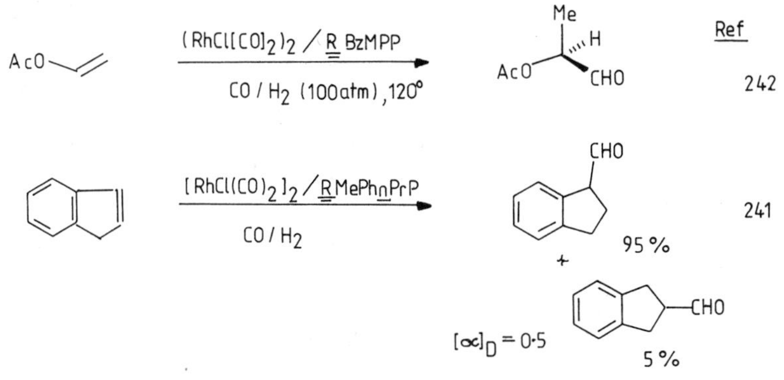

Figure 4.4 Asymmetric hydroformylation of other substrates.

(b) **Chiral at carbon**

Results with this type of ligand have been singularly unsuccessful. For the hydroformylation of styrene, neomenthyl diphenyl phosphine gives, at best, an optical yield of 1.9% R,[240] However, a product of the opposite chirality may also be obtained under other conditions.[243] α-Ethylstyrene is hydroformylated in 1% (R) and phenylvinyl ether in 0.3% (R) optical yield. Other phosphines synthesized include **LXXXIV-LXXXVII.**[244-5] Only **LXXXVII** is at all useful, giving an optical yield of 5.1% for the hydroformylation of styrene.

$(Menthyl)_3P$

LXXXIV

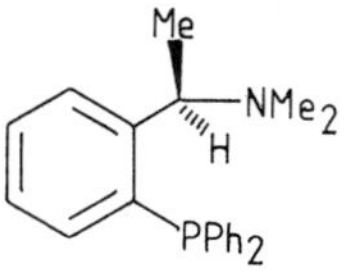

LXXXV

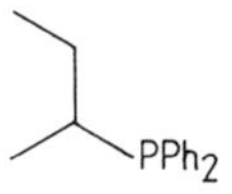

LXXXVI

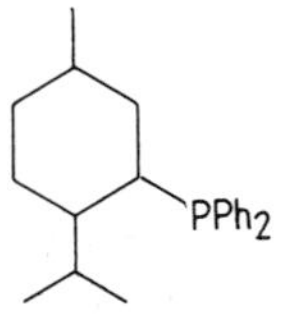

LXXXVII

4.2.2 Chelating biphosphines

When dppe is added to tris(triphenylphosphine)rhodium chlorocarbonyl the rate of hydroformylation is very drastically reduced. Thus to date no asymmetric reactions have been attempted with analogous chiral phosphines which form 5-membered chelate complexes. 7-Ring chelating biphosphines, in particular DIOP, have, however, been extensively tested. It will become evident that the distinction between chelating and non-chelating phosphines is to a certain extent an artificial one since there is evidence to suggest that the phosphine does not remain chelated throughout the catalylic cycle and the best results are obtained in the presence of a large excess of phosphine.

Examples of the hydroformylation of mono-olefins with rhodium – DIOP complexes are given in Figure 4.5. It is generally observed that optical yield decreases with increasing temperature; as well as the usual thermodynamic arguments, product racemization is accelerated. Increased pressure has little effect on the optical yield. The nature of the catalyst

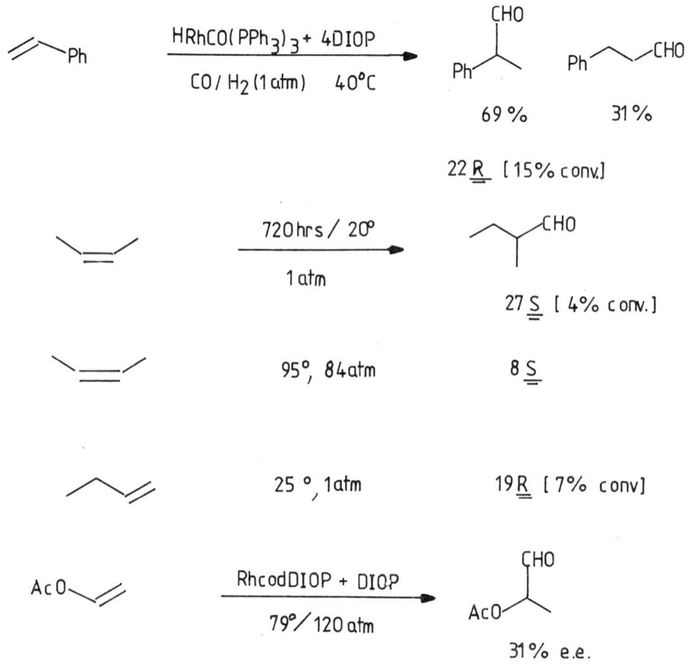

Figure 4.5   Hydroformylation with rhodium-DIOP complexes.

system remains obscure and it is unlikely that a single pathway is rigorously followed. Good optical excesses are obtained only when a substantial excess of the ligand is present, the critical ratio being 2:1 phosphine:rhodium.[242] Although it has been demonstrated by ^{31}P nmr spectroscopy that the addition of DIOP to tris(triphenylphosphine)rhodium chlorocarbonyl gives initially a chelated complex[248] it is clear that at some point in the catalytic cycle DIOP is acting as a monodentate ligand.

The fact that the C_{2v} olefins such as cis-2-butene, dihydrofuran and bicyclo[2.2.2.]octene all give optically active products on hydroformylation indicates that some chiral discrimination must occur after the olefin binding step since this necessarily produces only one complex. The fact that cis and trans - 2 -butene give different optical yields despite having the same sec-alkyl rhodium complex **LXXXVIII** as an intermediate suggests that optical induction takes place in the formation of the alkyl rather than subsequently. This does not of course preclude chiral discrimination occurring in the binding of prochiral olefins.

A few other chiral chelating biphosphines have been used in hydroformylation catalysts. The dibenzophosphole **LXXXIX** has been synthesized[249] and catalysts derived from this give the opposite configuration of the aldehyde product to that given with DIOP of the same stereochemistry and in a greater optical yield. More extensive studies showed that this was a fairly general phenomenon.[250] With **LXXXIX, XC** and **XCI** as the phosphine skeletons and styrene, 1-butene, cis-2-butene and 2-phenylpropene as substrates opposite configurations were obtained for the products of the (a) and (b) phosphines in most cases. Increased rates were also obtained with dibenzophosphole substituted ligands. The authors relate this to the planarity of the dibenzophosphole-containing ligands, considering that the possibility of a wide range of P-phenyl rotamers may be deleterious to good stereochemical control.

LXXXVIII

LXXXIX

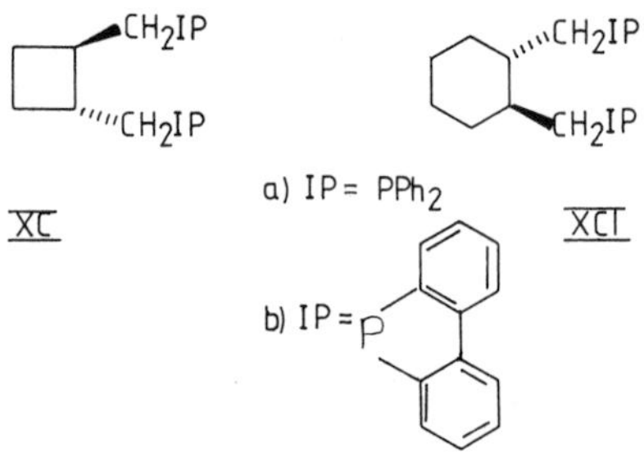

a) IP= PPh$_2$

b) IP=P

XC   XCI

Substrates other than simple mono-olefins have been little investigated. The hydroformylation of conjugated dienes can give rise to 1,6 dialdehydes, important in the plastics industry, but the only report of an asymmetric reaction is concerned with the monoaldehydes obtained (Figure 4.6).[251] The low optical induction obtained for 3-methyl-1-butene suggests

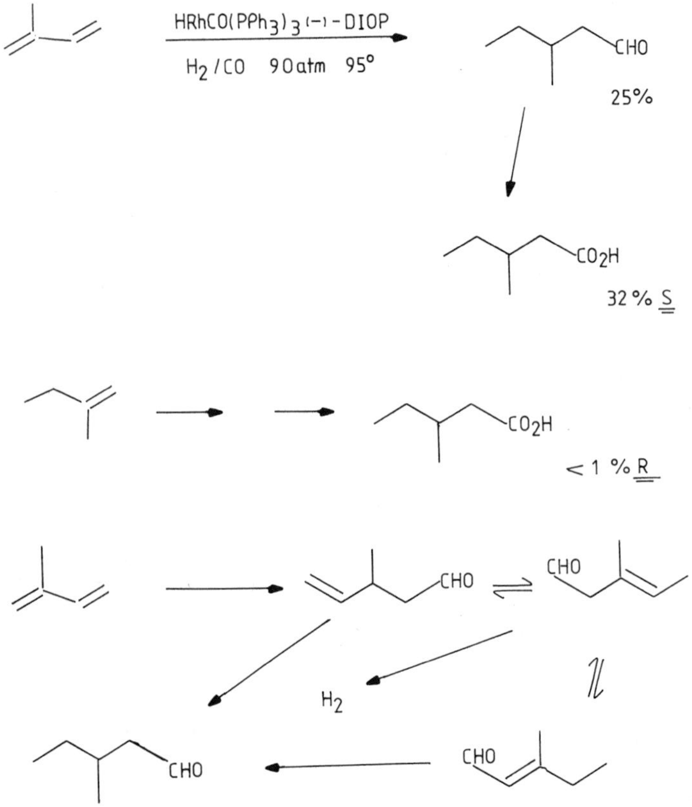

Figure 4.6   Hydroformylation of isoprene.

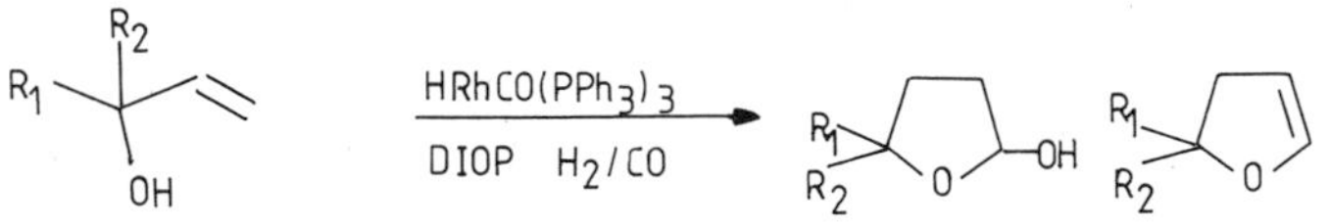

Figure 4.7   Hydroformylation of allylic alcohols.

that this is not an intermediate in the hydroformylation of isoprene and that the high optical induction arises in the hydrogenation of a mixture of prochiral aldehydes.

There is one report of a rhodium - DIOP system being used to prepare isomerically pure mono- and disubstituted dihydrofurans[252] (Figure 4.7). Optical yields did not exceed 14%.

4.3   Ruthenium catalyzed reactions

One asymmetric reaction has been noted using a catalyst derived from a ruthenium complex[237] (Figure 4.8). Addition of excess DIOP increases the proportion of hydrogenation and monophosphines give no optical induction.

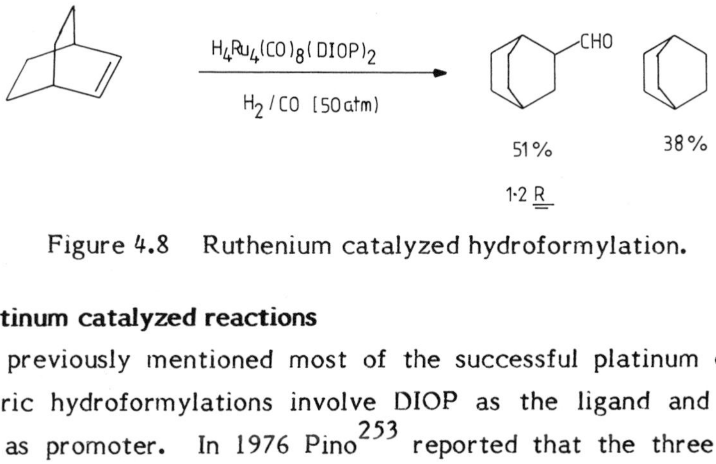

Figure 4.8   Ruthenium catalyzed hydroformylation.

4.4   Platinum catalyzed reactions

As previously mentioned most of the successful platinum catalyzed asymmetric hydroformylations involve DIOP as the ligand and stannous chloride as promoter. In 1976 Pino[253] reported that the three isomeric butanes gave the same enantiomer excess in hydroformylation (Figure 4.9). This would imply that the alkyl platinum complexes interconvert more rapidly than the subsequent step in the reaction sequence in which asymmetric induction takes place. Subsequently Japanese workers have investigated the same reaction with DIOP and its dibenzophosphole analogue

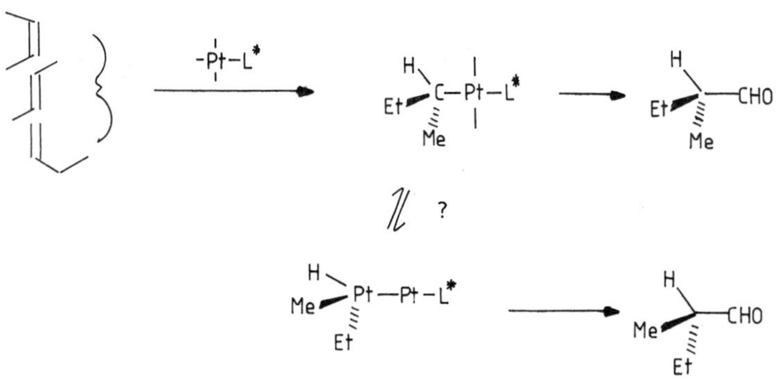

Figure 4.9   Mechanism of platinum catalyzed hydroformylation.

giving different results.[254]   They concluded that asymmetric induction took place before metal alkyl formation (as for rhodium) since the isomeric butenes in their hands gave products of different optical yields.

4.5   Palladium catalyzed hydrocarboxylation reactions

DIOP palladium dichloride and bis(triphenylphosphine)palladium dichloride are both very active catalysts for the carboxylation of olefins (Figure 4.10). Early experiments used a mixture of (-)-DIOP and palladium chloride and an optical yield of 14% was obtained for the reaction of α-methylstyrene in the presence of i-propanol.[255]   A significant optical yield was obtained for the C_{2v} cis-2-butene indicating that asymmetric induction took place after olefin coordination. A systematic investigation[23] of the effect of reaction conditions for the hydrocarboxylation of α-methylstyrene showed that the alcohol used influences the extent of the asymmetric induction, reaching a maximum for the bulkiest alcohol. The presence of a

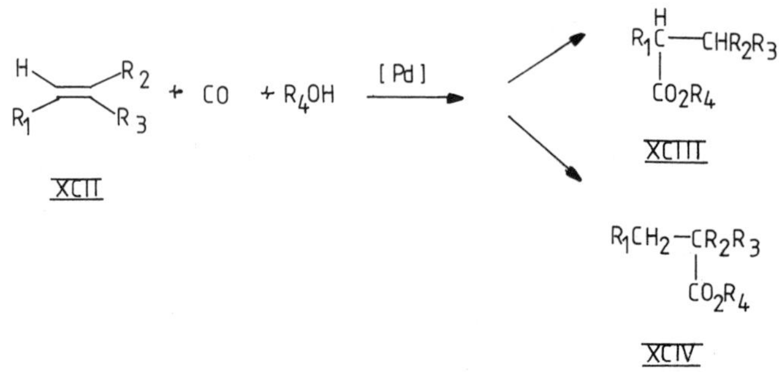

Figure 4.10 Palladium catalyzed hydrocarboxylation of olefins.

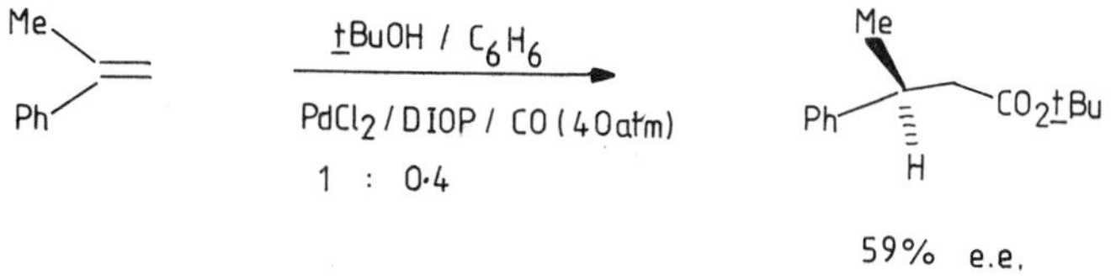

Figure 4.11 Hydroesterification of α-methyl styrene.

solvent such as benzene or thf is beneficial and the optical yield increases with carbon monoxide pressure. Decreasing the phosphine to palladium ratio also gives some improvement; despite the lack of mechanistic study, Figure 4.11 represents the most impressive result achieved for carbon monoxide addition to any double bond.

The carbocyclic analogues of DIOP, **XC** and **XCI,** have also been used for hydroesterification of α-methylstyrene.[256] **XCb,** the dibenzophosphole based on a cyclobutane skeleton, gives a good optical yield but the corresponding cyclohexyl derived ligand is not successful.   As with hydroformylation the dibenzophosphole ligands give better optical induction, but the configuration of the products is the same as for diphenylphosphinyl compounds.

4.6   Polymer supported asymmetric hydroformylation catalysts

Polymer supported catalysts have scarcely been exploited for asymmetric hydroformylation. The earliest were based on the modification of polystyrene with optically active phosphinites such as **XCV** and **XCVI.**[245] With **XCV** styrene was hydroformylated to S - hydtropaldehyde in very low enantiomer excess.

More promising results have been obtained with chiral polymers based on the DIOP skeleton. A soluble modified polystyrene catalyst gave only 2% (R) enantiomer excess[257] but a polymer **XCVIII** formed by the

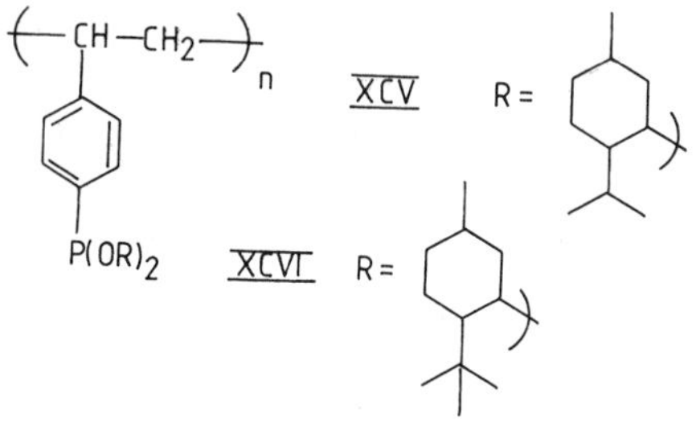

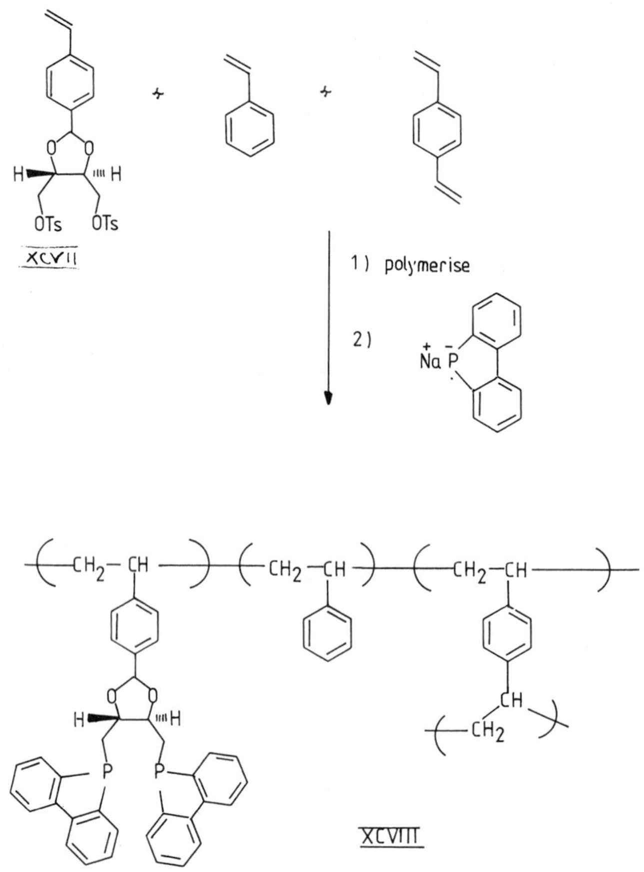

Figure 4.12 Polymer bound catalyst for asymmetric hydroformylation.

copolymerisation of the monomer **XCVII**, styrene and divinyl benzene gave up to 27.7% R - hydtropaldehyde.[258] (Figure 4.12). This is comparable with the corresponding homogeneous catalysts and the catalyst could be re-used after simple filtration without loss in rate or selectivity.

4.7   Conclusions

Asymmetric hydroformylation remains an area in which success has not been achieved. In asymmetric hydrogenation the most useful results are for olefins bearing polar groups capable of coordinating tightly to the metal center but there has been little interest in analogous substrates for hydroformylation. The rhodium catalyzed reaction appears to be the most promising since the conditions are generally milder than for platinum or cobalt. A serious problem is the design of selective catalysts; there is little direct evidence as to the mechanistic pathway followed and improvements follow only from extensive investigations as to the best reaction conditions.

5. OTHER ASYMMETRIC ADDITION REACTIONS

5.1 Asymmetric epoxidations

The importance of chiral epoxides in synthetic chemistry and biochemistry has stimulated a search for new methods of preparing this class of compounds. The abiotic asymmetric epoxidation of prochiral olefins has been carried out with chiral peracids,[259-60] or with hydrogen peroxide under phase transfer conditions using quaternized alkaloids as catalysts.[261] Promising results have recently been obtained using chiral vanadium and molybdenum complexes as catalysts for the epoxidation of olefins by alkyl hydroperoxides.

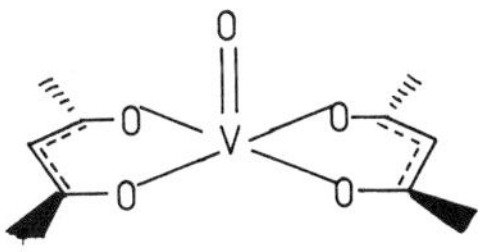

XCIX 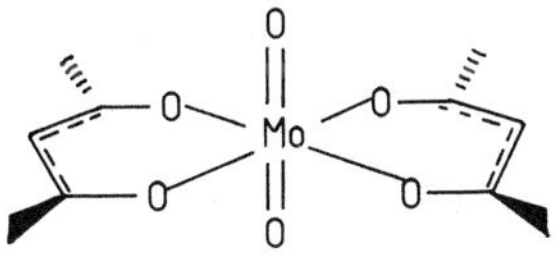   C

Sharpless has investigated the mechanism of epoxidation of allylic alcohols using **XCIX** and C.[262-4] It was found that the alcohol function was coordinated to the metal during the oxygen atom transfer step. Epoxidations carried out in the presence of $H_2{}^{18}O$ suggested that the intact alkyl hydroperoxide was present in the intermediate responsible for oxygen transfer to the olefin. This suggested that the epoxidation of prochiral olefins could be achieved asymmetrically if a suitable chiral ligand could be found. In the presence of chiral hydroxamic acids such as **CI** using a vanadium catalyst chiral epoxides were formed in up to 50% enantiomer excess and in good chemical yield (Table 11).[29]

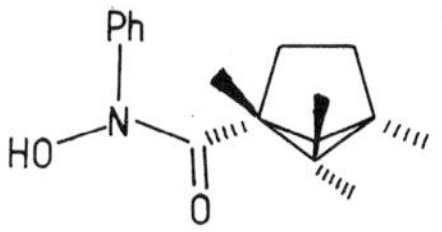

N-phenylcamphoyl-
hydroxamic acid

CI

Table 11. Asymmetric epoxidation of allylic alcohols (a)

Alcohol		T°C		Chemical yield	Optical yield (b)
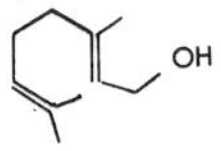 Geraniol		25		86	30
		25		80	40
		-78	25	30	50
		25		87	40

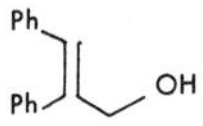

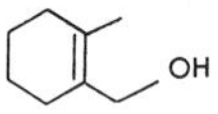

(a) Reductions were carried out with 2 molar excess of hydroxamic acid using 1% of VO(acac)$_2$ and 2 equivalents of t-butyl hydroperoxide.

(b) The enantiomeric excess was determined using the chiral shift reagent, Eu(hfac)$_3$, on the derived epoxy-acetates.

Japanese workers have used a preformed molybdenum complex with (-)-N-alkylephedrines as chiral ligands, **CII.**[265] Using cumene hydroperoxide as oxidant, geraniol was epoxidized in 33% optical yield. An interesting stoichiometric epoxidation of non-functionalized olefins has been achieved using **CIII** containing the bidentate chiral ligand (S)-N,N-dimethyl-lactamide. 1-Butene was oxidized to R - ethyloxirane in 31% enantiomer excess.[30]

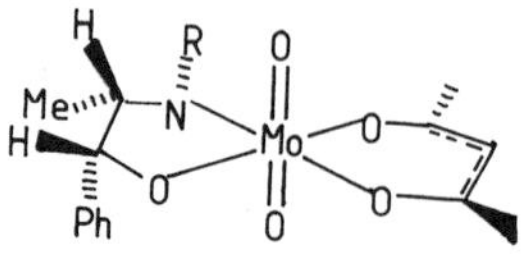

R = Me, Et

CII

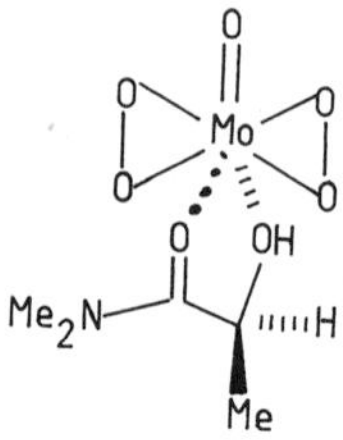

CIII

5.2   Asymmetric carbon carbon bond forming reactions

A small number of simple addition reactions have led to the formation of carbon-carbon bonds asymmetrically.   A palladium-DIOP complex has catalyzed the addition of hydrogen cyanide to norbornene giving 2- exo-cyanonorbornane exclusively and in 28% optical yield (Figure 5.1).[266] The dimerization   of   isoprene   has   been   catalyzed   by   a   palladium menthyldiphenylphosphine complex giving the chiral product shown in Figure 5.1.[267]   Wilke reported the related co-dimerization of ethylene and 1,3-cyclooctadiene, with a nickel-dimenthylphenylphosphine complex.[31]   At low temperatures   in   the   presence   of   a   greater   than   2:1   excess   of   chiral phosphine, 3-vinylcyclooctene was formed in up to 70% enantiomer excess (Figure 5.1).

Finally optically active cobalt and copper catalysts have been used in the addition of ethyldiazoacetate to olefins to form chiral cyclopropanes. Using a cobalt camphoroxime complex, a mixture of cis and trans cyclopropane products was obtained in the addition of the diazoacetate to styrene, with 70% enantiomer excess (Figure 5.2).[268]   In the presence of the chiral   copper   chelate   **CIV,**   the   catalytic   asymmetric   synthesis   of chrysanthemic acid derivatives has been successfully achieved.[269]   It is likely that both of these reactions involve the intermediacy of a metal carbene complex.

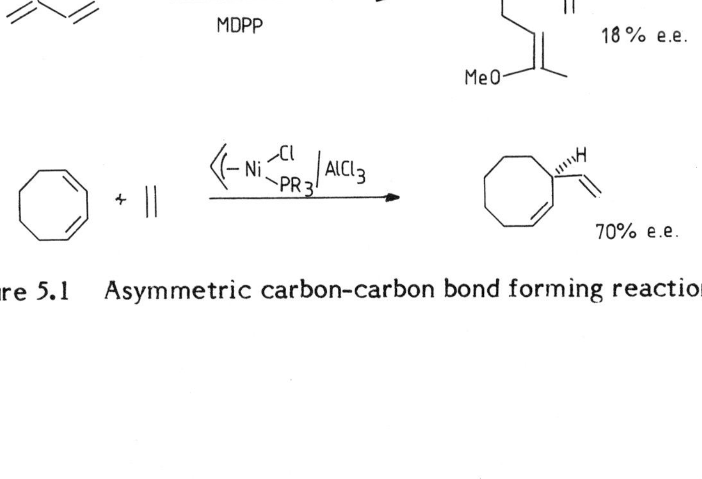

Figure 5.1   Asymmetric carbon-carbon bond forming reactions.

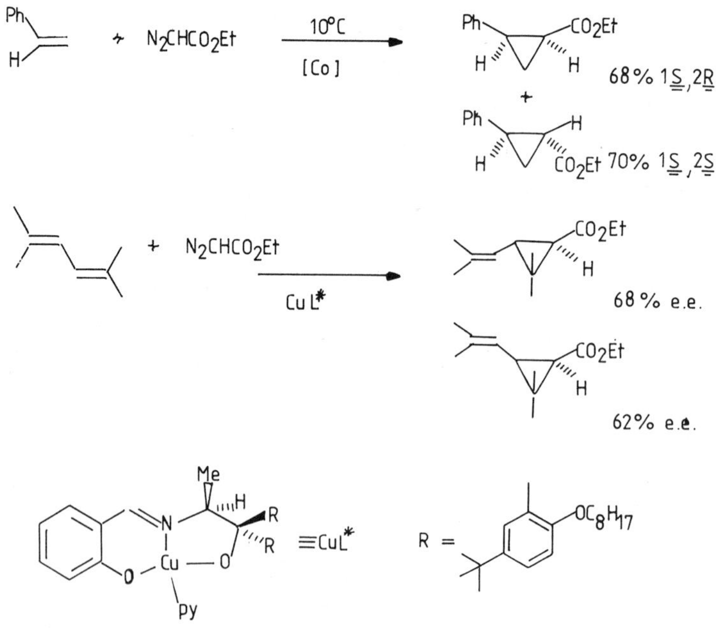
10°C
[Co]
68% 1S,2R
70% 1S,2S
CuL*
68 % e.e.
62% e.e.
≡CuL*
R =

CIV

Figure 5.2 Catalytic asymmetric cyclopropanations.

6. SOME MORE RECENT RESULTS

6.1 Hydrogenation

6.1.1 New phosphines

These continue to proliferate. **CVa-CVc** give excellent optical yields in reduction of dehydroamino acid derivatives.[270-2] **CVa** is derived from amino acids and **CVb** from mannitol. 7-ring chelating aminophosphines also continue to be prominent.[273-5] Examples include **CVI** and **CVII** and the reversal of chirality on N-methylation of this structural type is fairly general.[276] The sole exception is provided by **CVIII** which may be expected to be considerably more rigid.[227] BINAP, **CIX,** forms a 7-membered chelate

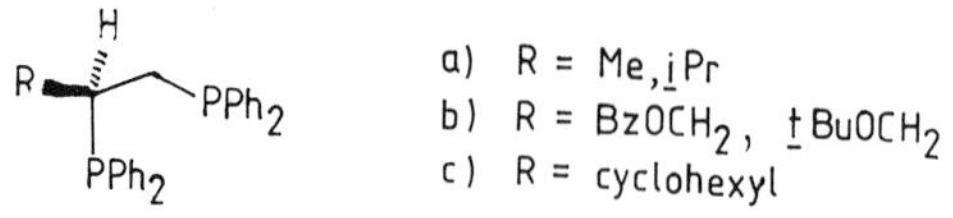
a) R = Me, iPr
b) R = BzOCH2, tBuOCH2
c) R = cyclohexyl

CV

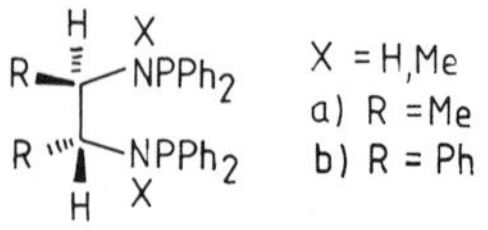
X = H,Me
a) R = Me
b) R = Ph

CVI

CVII CVIII CIX

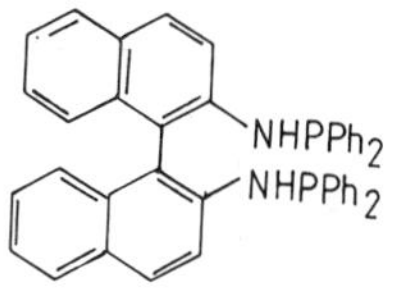

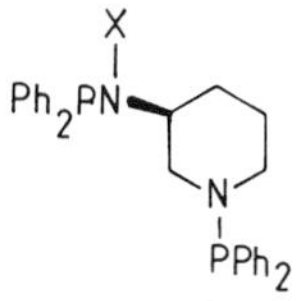

X = H,Me

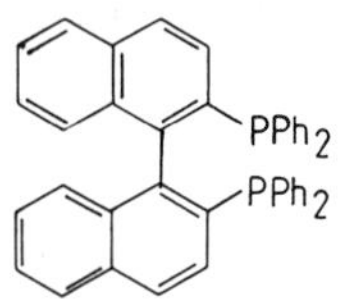

ring but its extreme rigidity ensures that optical yields in excess of 95% are obtained in enamide reduction.[278] A full catalogue of both phosphine and substrate data complete to the end of 1978[279] has been published.

6.1.2 New substrates

Studies of enamide hydrogenation continue to be numerous. Optimization of conditions and catalysts have led to efficient reduction of both Z and E-dehydroamino acid derivatives, although an explanation for the reactivity pattern observed is lacking.[280] Enol esters possess the same relative disposition of carbonyl and alkene and are reduced with fair enantioselectivity in the presence of DIPAMP complexes.[281] Enol phosphates are analogous and are also hydrogenated effectively.[282] Enamide hydrogenation has found new applications in dipeptide synthesis since dehydrodipeptides are readily available. The asymmetric center in **CX** has little effect on the configuration of the new asymmetric center when the catalyst is derived from DIPAMP,[283] DIOP[284] or the pyrrolidine biphosphines,[283] and the major diastereomer formed is that predicted by analogy with simple enamides. With DIOXOP-derived catalysts the existing chiral center plays a more important role.[285]

An interesting DIOP ruthenium catalyzed reduction of the anhydride **CXI** to give **CXII** has been reported to proceed in 20-50% optical yield[286] and ruthenium catalyzed asymmetric reactions have been comprehensively reviewed.[287]

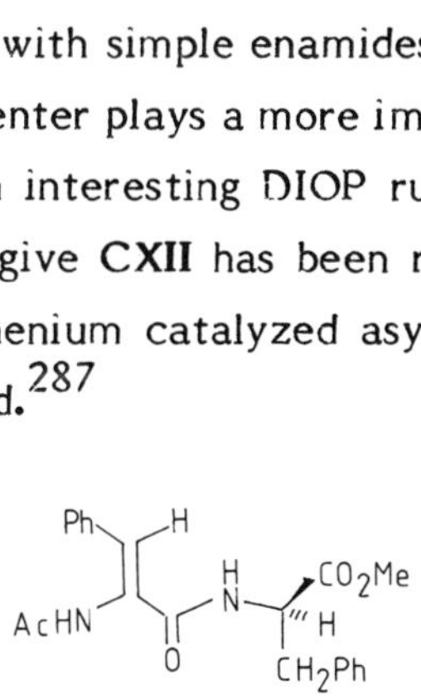

CX 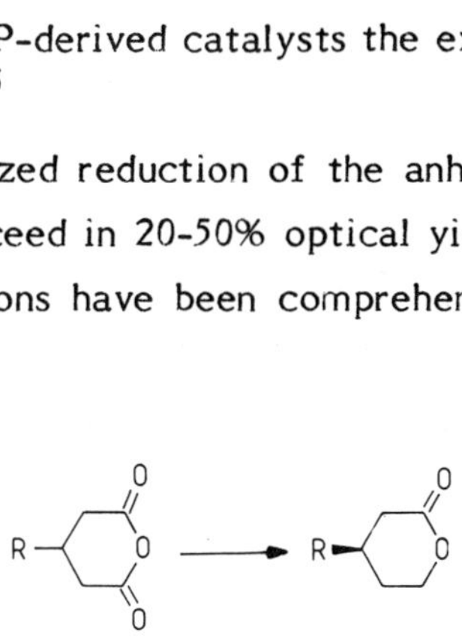 CXI CXII

6.1.3 Mechanistic studies

A range of additional crystal structures of chiral phosphine derivatives has been reported[288-91] and the solid-state structure of the CHIRAPHOS enamide complex previously published has been shown to persist in solution by analysis of its CD spectrum.[292] In rhodium DIPAMP solvates, EXAFS measurements confirm that oxygen ligands are <u>trans</u> to phosphorus in the inner coordination sphere.[293]

In DIPAMP enamide complexes it has been shown by use of the DANTE experiment[294] that <u>intramolecular</u> exchange between diastereomers is considerably more rapid than dissociative exchange of substrate and is <u>fast</u> on the time scale of catalysis.[295] Reversible addition of hydrogen to these species is ruled out by the complete lack of ortho - para hydrogen interconversion during reduction of Z - α - benzamidocinnamic acid by DIPAMP, DIPHOS or bis(diphenylphosphino)butane complexes.[296] Careful analysis of <u>intra</u> (HD) and intermolecular (H_2 versus D_2) isotope effects implies that hydrogen addition is rate-determining and is followed by rapid transfer of one H to give a metal alkyl hydride.[297]

6.2   Hydrosilylation

In the asymmetric reduction of ketones by hydrosilylation Kagan has implicated electronic as well as steric effects.[298] A particularly successful ligand AMPHOS, **CXIII**, has been found by Payne to give up to 72% enantiomer excess in reduction of **CXIV**.[299] Effective reduction of **CXV** is achieved in the presence of DIOP catalysts,[300] the chirality of the product being readily predicted from results with achiral α-keto amides. Mechanistic studies of the addition of Ph_2SiH_2 to ketones in the presence of DIOP complexes have implicated $[RhHSiPh_2H(DIOP)]^+$ as an intermediate.[301]

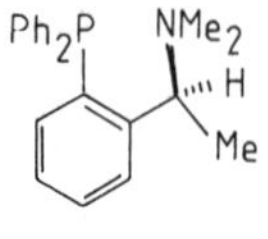

CXIII 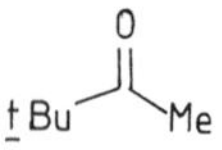   CXIV 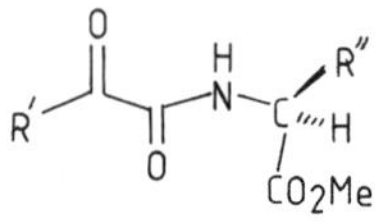   CXV

The most promising example to date of asymmetric hydrosilylation of an alkene is achieved in the presence of palladium catalyst **CXIX:**[302]

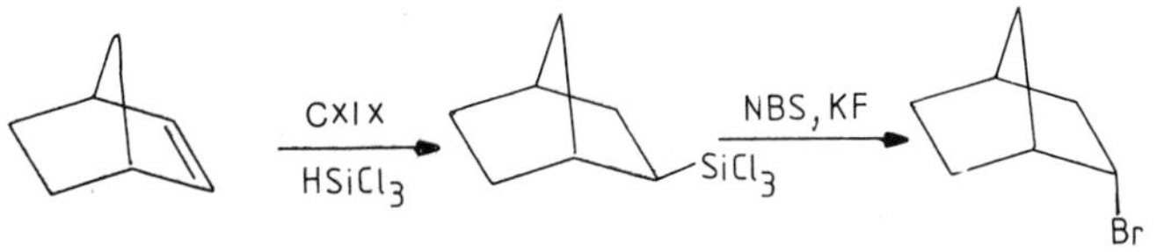

CXVI   CXVII   CXVIII

6.3 Hydroformylation

The asymmetric hydroformylation of simple alkenes depends for its effectiveness on steric factors; these have been carefully analyzed by Pino and whilst the model developed may not reflect the reaction mechanism precisely it is remarkably efficient in predicting stereochemical outcomes.[303] In an effort to introduce polar factors which might make the transition state more rigid reactions of **CXX**[304] and **CXXI**[305] have been studied. Whilst modest enantiomer excesses were achieved with **CXX, CXXI** gave the best result to date; 53% optical yield in the presence of R,R-dibenzophosphole-DIOP. The presence of the OAc group may inhibit in situ racemization of the product.

There is mounting evidence that hydroformylation in the presence of chelating biphosphine complexes proceeds via bridged intermediates such as **CXXII.**[306-7]

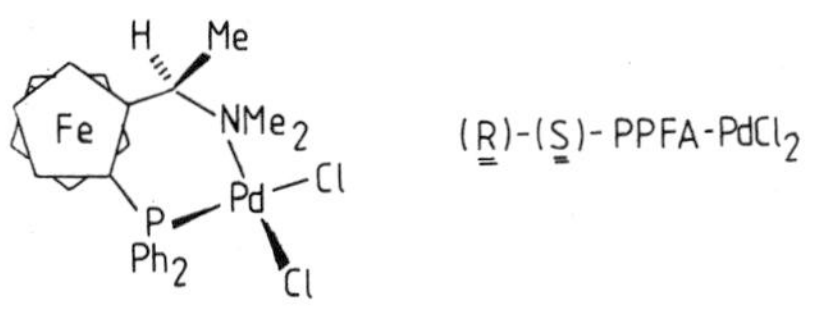

(R)-(S)-PPFA-PdCl$_2$

CXIX 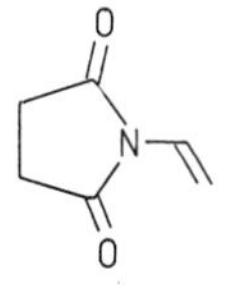   CXX

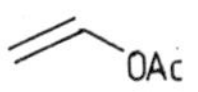

CXXI 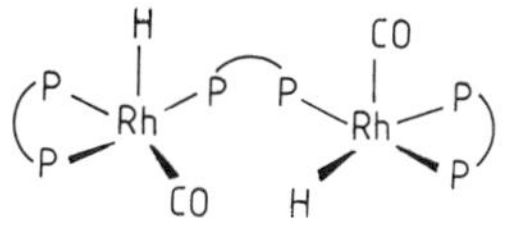   CXXII

6.4   Asymmetric epoxidation

A versatile titanium alkoxide-dialkyl tartrate epoxidation catalyst has been described for the enantioselective epoxidation of allylic alcohols.[308-310] Reaction of prochiral allylic alcohols with <u>tertiary</u> butylhydroperoxide in dichloromethane at -20°C in the presence of Ti(OiPr)$_4$/diisopropyltartrate (1:1.2) results in epoxide formation with enantiomeric purities <u>greater than</u> 95% ee.   The reaction is general and appears fairly insensitive to substitution in the substrate alkene. Key epoxy-alcohol intermediates for the synthesis of erythromycin and (+)- disparlure (the sex attractant of the gypsy moth) have been prepared using this method.[309] Furthermore racemic allylic alcohols may be kinetically resolved using this procedure to give the corresponding <u>erythro</u> epoxy alcohols in selectivities approaching 100%.[311] The xray structures of two catalytically active dimeric titanium tartrate complexes have been described[312] and the procedure has been successfully adapted for the asymmetric oxidation of aryl sulphides to give chiral sulphoxides in greater than 90% enantiomeric purity.[313]

REFERENCES

1. R.L. Cahen and K. Tvede, <u>J. Pharmacol. Exp. Ther.</u> 105:166 (1952).

2. P.N. Patil, D.D. Miller and U. Trendelenburg, <u>Pharmacol. Rev.</u> 26:323 (1975).

3. D.A. Bender, "Amino Acid Metabolism", Wiley, Chichester, 1975.

4. J.D. Morrison and H.S. Mosher, " <u>Asymmetric Organic Reactions</u>", Prentice Hall, Englewood Cliffs, N.J., 1971.

5. B. Bogdanovic, <u>Angew. Chem. Int. Ed. Engl.</u> 12:954 (1973).

6. J.W. ApSimon and R.P. Seguin, Tetrahedron 35:2797 (1979).

7. H.B. Kagan and J.C. Fiaud, Top. Stereochem. 10:175 (1978).

8. D. Valentine Jr. and J.W. Scott, Synthesis 329 (1978).

9. A.I. Meyers, G. Knaus, K. Kamata and M.E. Ford, J. Am. Chem. Soc. 98:567 (1976).

10. S. Akabori, S. Sakurai, Y. Izumi and Y. Fujii, Nature 178:323 (1956).

11. Y. Izumi, Angew. Chem. Int. Ed. Engl. 10:871 (1971).

12. E.I. Klabunovskii and E.S. Levitina, Russ. Chem. Rev. 39:1035 (1970).

13. M.D. McCreary, D.W. Lewis, D.L. Wernick and G.M. Whitesides, J. Am. Chem. Soc. 96:1038 (1974).

14. R. Charles, U. Beitler, B. Feibush and E. Gil-Av, J. Chromatogr. 112:121 (1975).

15. J.A. Osborn, F.H. Jardine, J.F. Young and G. Wilkinson, J. Chem. Soc. (A) 1711 (1966).

16. C. Rousseau, M. Evard and F. Petit, J. Mol. Catal. 3:309 (1978); M.H.J.M. de Croon, P.F.M.T. Van Nisselrooij, H.J.A.M. Kuipers and J.W.E. Coenen, J. Mol. Catal 4:325 (1978).

17. L. Horner, H. Büthe and H. Siegel, Tetrahedron Lett. 4023 (1968).

18. L. Horner, H. Siegel and H. Büthe, Angew. Chem. Int. Ed. Engl. 7:942 (1968).

19. W.S. Knowles, M.J. Sabacky and B.D. Vineyard, Ann. N.Y. Acad. Sci. 214:119 (1973).

20. L. Horner and B. Schlothauer, Phosphorus Sulfur 4:155 (1978).

21. J.M. Brown, P.A. Chaloner, B.A. Murrer and D. Parker, Advan. Chem. Ser. 119:169 (1980).

22. P. Pino, C. Botteghi, G. Consiglio and S. Pucci, Symposium of the Chemistry of Hydroformylation and Related Reactions, Veszprem Section of Hungarian Chem. Soc., 1972.

23. G. Consiglio and P. Pino, Chimia 30:193 (1976).

24. K. Yamamoto, T. Hayashi, Y. Uramoto and R. Ito, J. Organomet. Chem. 118:331 (1976).

25. T. Hayashi, K. Yamamoto, K. Kasuga, H. Omizu and M. Kumada, J. Organomet. Chem. 113:127 (1976).

26. I. Ojima and T. Kogure, Tetrahedron Lett. 1889 (1974).

27. D.D. Ridley and M. Stralow, J.C.S. Chem. Commun. 400 (1975).

28. J.F. Peyronel and H.B. Kagan, Nouv. J. Chim. 2:211 (1978).

29. R.C. Michaelson, R.E. Palermo and K.B. Sharpless, J. Am. Chem. Soc. 99:1990 (1977).

30. H.B. Kagan, H. Mimoun, C. Mark and V. Schurig, Angew. Chem. Int. Ed. Engl. 18:485 (1979).

31. B. Bogdanovic, B. Henc, B. Meister, H. Pauling and G. Wilke, Angew. Chem. Int. Ed. Engl. 11:1023 (1972).

32. A. Deffieux, M. Sepulchre, N. Spassky and P. Sigwalt, Makromol. Chem. 175:339 (1974).

33. B.M. Trost, Tetrahedron 33:2615 (1977) and references therein.

34. T. Hayashi, M. Tajika, K. Tamao and M. Kumada, J. Am. Chem. Soc. 98:3718 (1976).

35. W.S. Knowles, M.J. Sabacky and B.D. Vineyard, Ger. Offen 2,456,937 (1974), Chem. Abs. 83:164367q (1975); G.L. Bachman and B.D. Vineyard, Ger Offen 2,638,071 (1975), Chem. Abs. 87:5591z (1977).

36. M.D. Fryzuk and B. Bosnich, J. Am. Chem. Soc. 99:6262 (1977).

37. M.D. Fryzuk and B. Bosnich, J. Am. Chem. Soc. 100:5491 (1978).

38. R.B. King, J. Bakos, C.D. Hoff and L. Marko, J. Org. Chem. 44:1729 (1979).

39. H. Brunner and W. Pieronczyk, Angew. Chem. Int. Ed. Engl. 18:620 (1979).

40. M. Laver, O. Samuel and H.B. Kagan, J. Organomet. Chem. 177:309 (1979).

41. R.B. King, J. Bakos, C.D. Hoff and L. Marko, J. Org. Chem. 44:3095 (1979).

42. B.D. Vineyard, W.S. Knowles, M.J. Sabacky, G.L. Bachman and D.J. Weinkauff, J. Am. Chem. Soc. 99:5946 (1977).

43. T.H. Johnson and G. Rangarajan, J. Org. Chem. 45:62 (1980).

44. W.C. Christopfel and B.D. Vineyard, J. Am. Chem. Soc. 101:4406 (1979).

45. M.D. Fryzuk and B. Bosnich, J. Am. Chem. Soc. 101:3043 (1979).

46. J. Halpern, T. Okamoto and A. Zakhariev, J. Mol. Catal. 2:65 (1977).

47. C.A. Tolman, P.Z. Meakin, D.L. Lindner and J.P. Jesson, J. Am. Chem. Soc. 96:2762 (1974).

48. J. Halpern, D.P. Riley, A.S.C. Chan and J.J. Pluth, J. Am. Chem. Soc. 99:8055 (1977).

49. R.R. Schrock and J.A. Osborn, J. Am. Chem. Soc. 93:2397 (1971); J.M. Brown, P.A. Chaloner and P.N. Nicholson, J.C.S. Chem. Commun. 646 (1978).

50. J.M. Brown and P.A. Chaloner, J.C.S. Chem. Commun. 613 (1979).

51. A.S.C. Chan, J.J. Pluth and J. Halpern, Inorg. Chim. Acta 37:L477 (1979).

52. K.E. Koenig and W.S. Knowles, J. Am. Chem. Soc. 100:7561 (1978).

53. J.M. Brown and P.A. Chaloner, J. Am. Chem. Soc. 102:3040 (1980).

54. J.M. Brown and B.A. Murrer, Tetrahedron Lett. 4859 (1979).

55. A.S.C. Chan and J. Halpern, J. Am. Chem. Soc. 102:838 (1980).

55a. A.S.C. Chan, J.J. Pluth and J. Halpern, J. Am. Chem. Soc. 102:5952 (1980).

56. J.M. Brown and P.A. Chaloner, J.C.S. Chem. Commun. 344 (1980).

57. I. Ojima, T. Kogure and N. Yoda, Chem. Lett. 495 (1979).

58. R. Crabtree, Acc. Chem. Res. 12:331 (1979).

59. W.S. Knowles, B.D. Vineyard, M.J. Sabacky and B.R. Stults, Fundamental Research in Homogeneous Catalysis 3:537 (1979).

60. R.G. Ball and N.C. Payne, Inorg. Chem. 16:1187 (1977).

61. S. Brunie, J. Marzan, N. Langlois and H.B. Kagan, J. Organomet. Chem. 114:225 (1976).

62. K. Achiwa, Y. Ohgo and Y. Iitaka, Chem. Lett. 865 (1979).

63. J.M. Brown and D. Parker, J.C.S. Chem. Commun. 342 (1980); D. Parker and J.M. Brown, J. Org. Chem. 47:2722 (1982).

64. H.B. Kagan, J.C. Fiaud, C. Hoonaert, D. Meyer and J.C. Poulin, Bull. Soc. Chim. Belg. 88:923 (1979).

64a. P.A. MacNeil, N.K. Roberts and B. Bosnich, J. Am. Chem. Soc. 103:2273 (1981).

65. T.P. Dang and H.B. Kagan, J.C.S. Chem. Commun. 481 (1971).

66. H.B. Kagan and T.P. Dang, J. Am. Chem. Soc. 94:6429 (1972).

67. B.A. Murrer, J.M. Brown, P.A. Chaloner, P.N. Nicholson and D. Parker, Synthesis 350 (1979).

68. D. Sinou and H.B. Kagan, J. Organomet. Chem. 114:325 (1976).

69. R. Glaser, S. Geresh and J. Blumenfeld, J. Organomet. Chem. 112:355 (1976).

70. S. Torös, B. Heil and L. Marko, J. Organomet. Chem. 159:401 (1978).

71. D. Parker, D. Phil Thesis, University of Oxford, 1980.

72. A.P. Stoll and R. Süess, Helv. Chim. Acta 57:2487 (1974).

73. A. Levi, G. Modena and S. Scorrano, J.C.S. Chem. Commun. 6 (1975).

74. T.P. Dang, J.C. Poulin and H.B. Kagan, J. Organomet. Chem. 91:105 (1975).

75. H.B. Kagan, Pure. Appl. Chem. 43:401 (1975).

76. U. Hengartner, D. Valentine Jr., K.K. Johnson, M.E. Larscheid, F. Pigott, F. Schiedl, J.W. Scott, R.C. Sun, J.M. Townsend and T.H. Williams, J. Org. Chem. 44:3741 (1979).

77. J.M. Brown and B.A. Murrer, Tetrahedron Lett. 21:581 (1980).

78. P. Aviron-Violet and L. St. Genis, Ger. Offen. 2,424,543 (1974), Chem. Abs. 82:171186n (1975).

79. P. Aviron-Violet, Y. Colleuille and J. Vargynat, J. Mol. Catal. 5:41 (1979).

80. R. Glaser, M. Twaik, S. Geresh and J. Blumenfeld, Tetrahedron Lett. 4635 (1977).

81. J.M. Brown and P.A. Chaloner, J. Am. Chem. Soc. 100:4307 (1978).

82. V. Gramlich and G. Consiglio, Helv. Chim. Acta 62:1016 (1979).

83. R. Glaser, Tetrahedron Lett. 2127 (1975).

84. J.M. Brown and P.A. Chaloner, J.C.S. Chem. Commun. 321 (1978).

85. J.M. Brown, P.A. Chaloner, S. Geresh and R. Glaser, Tetrahedron 36:816 (1980).

86. M. Tanaka and I. Ogata, J.C.S. Chem. Commun 735 (1975).

87. T. Hayashi, M. Tanaka and I. Ogata, Tetrahedron Lett. 295 (1977).

88. W.R. Cullen and Y. Sugi, Tetrahedron Lett. 1635 (1978).

89. R. Jackson and D.J. Thompson, J. Organomet. Chem. 159:C29 (1978).

90. Y. Sugi and W.R. Cullen, Chem. Lett. 39 (1979).

91. R. Selke, React. Kinet. Catal. Lett. 10:135 (1979).

92. M. Fiorini and G.M. Giongo, J. Mol. Catal. 5:303 (1979).

93. K-I. Onuma, T. Ito and A. Nakamura, Chem. Lett. 905 (1979).

94. M. Fiorini, F. Marcati and G.M. Giongo, J. Mol. Catal. 4:125 (1978).

95. K. Hanaki, K. Kashiwabata and J. Fujita, Chem. Lett. 489 (1978).

96. K. Achiwa, J. Am. Chem. Soc. 98:8265 (1976).

97. K. Achiwa, J. Synth. Org. Chem. Jpn. (Yuki Gosei Kagaku Kyokai Shi) 37:119 (1979); Chem. Abs. 91:211248g (1979).

98. K. Achinami, Jpn. Kokai Tokkyo Koho, 79:66,672 (1979) Chem Abs. 91:108088c (1979).

99. K. Achiwa, Fundamental Research in Homogeneous Catalysis 3:549 (1979).

100. I. Ojima and T. Kogure, Chem. Lett. 641 (1979).

101. I. Ojima, T. Kogure and K. Achiwa, J.C.S. Chem. Commun. 428 (1978).

102. K. Achiwa and T. Soga, Tetrahedron Lett. 1119 (1978).

103. K. Achiwa, Chem. Lett. 561 (1978).

104. K. Achiwa, T. Kogure and I. Ojima, Tetrahedron Lett. 4431 (1977).

105. K. Achiwa, Tetrahedron Lett. 3735 (1977).

106. I. Ojima, Fundamental Research in Homogeneous Catalysis 2:181 (1978).

107. I. Ojima and T. Suzuki, Tetrahedron Lett. 1239 (1980).

108. K. Achiwa, P.A. Chaloner and D. Parker, J. Organomet. Chem. 218:249 (1981).

109. T. Hayashi and M. Kumada, Fundamental Research in Homogeneous Catalysis 3:159 (1979).

110. D. Marquarding, H. Klusacek, G. Gokel, P. Hoffmann and I. Ugi, J. Am. Chem. Soc. 92:5389 (1970).

111. T. Hayashi, K. Yamamoto and M. Kumada, Tetrahedron Lett. 4405 (1974).

112. T. Hayashi, T. Mise, S. Mitachi, K. Yamamoto and M. Kumada, Tetrahedron Lett. 1133 (1976).

113. M. Kumada, T. Hayashi, K. Tamao, K. Yamamoto, T. Mise, M. Tajuka and S. Mitachi, Asahi Garusu Kogyo Gijutsu Shorekai Kenkyu Hokoku 28:299 (1976); Chem. Abs. 87:151155b (1977).

114. M. Kumada and T. Hayashi, Proc. Symp. Homog. Catal. Veszprem, Hungary, 157 (1978).

115. T. Hayashi, T. Mise and M. Kumada, Tetrahedron Lett. 4351 (1976).

116. T. Hayashi, A. Katsamura, M. Konishi and M. Kumada, Tetrahedron Lett. 425 (1979).

117. H. Brunner and W. Pieronczyk, J. Chem. Res S74, M1251 (1980).

118. H. Brunner and W. Pieronczyk, J. Chem. Res S76, M1275 (1980).

119. D. Lafont, D. Sinou and G. Descotes, J. Organomet. Chem. 150:C14 (1978).

120. J.M. Brown, P.A. Chaloner, G. Descotes, R. Glaser, D. Lafont and D. Sinou, J.C.S. Chem. Commun. 611 (1979).

121. J.M. Brown, P.A. Chaloner, D. Parker and D. Sinou, Nouv. J. Chim. 5:167 (1981).

122. N.W. Alcock, J.M. Brown and J.C. Jeffery, J.C.S. Chem. Commun. 829 (1974).

123. R.H. Grubbs and R.A. de Vries, Tetrahedron Lett. 1879 (1977).

124. M.E. Wilson, R.G. Nuzzo and G.M. Whitesides, J. Am. Chem. Soc. 100:2269 (1978).

125. M.E. Wilson and G.M. Whitesides, J. Am. Chem. Soc. 100:306 (1978).

126. O. Kospium, R.A. Lewis, J. Chickos and K. Mislow, J. Am. Chem. Soc. 90:4842 (1968).

127. W.S. Knowles, M.J. Sabacky and B.D. Vineyard, J.C.S. Chem. Commun. 10 (1972).

128. W.S. Knowles, M.J. Sabacky and B.D. Vineyard, Chem. Tech. 590 (1972).

129. W.S. Knowles, M.J. Sabacky and B.D. Vineyard, Adv. Chem. Ser. 132:274 (1974).

130. C.J. Sih and J.B. Heather, U.S. Patent 3,968,141 (1976); Chem. Abs. 81:77555q (1976).

131. J.M. Brown, P.A. Chaloner and P.N. Nicholson, J.C.S. Chem. Commun. 646 (1978).

132. J.D. Morrison, R.E. Burnett, A.M. Aguiar, C.J. Morrow and C. Phillips, J. Am. Chem. Soc. 93:1301 (1971).

133. T. Iwani, T. Yoshida and M. Sato, Nippon Kagaku Kaishi 1652 (1976).

134. D. Lafont, D. Sinou and G. Descotes, J. Organomet. Chem. 169:87 (1979).

135. W.F. Masler, A.M. Aguiar, C.J. Morrow, J.D. Morrison, R.E. Burnett and N.S. Bhacca, J. Org. Chem. 41:1545 (1970).

136. W.F. Masler, Ph.D. Thesis, University of New Hampshire (1974).

137. D. Valentine, Jr., K.K. Johnson, W. Priester, R.C. Sun, K. Toth and G. Saucy, J. Org. Chem. 45:3698 (1980); D. Valentine, Jr., R.C. Sun and K. Toth, J. Org. Chem 45:3703 (1980).

138. W.R. Cullen, F.W.B. Einstein, C-H. Huang, A.C. Willis and E.S. Yeh, J. Am. Chem. Soc. 102:988 (1980).

139. W.R. Cullen and E.S. Yeh, J. Organomet. Chem. 139:C13 (1977).

140. G. Pracejus and H. Pracejus, Tetrahedron Lett. 3497 (1977).

141. K. Yamamoto, A. Tomita and J. Tsuji, Chem. Lett. 3 (1978).

142. Y. Nakamura, S. Saito and Y. Monta, Chem. Lett. 7 (1980).

143. P. Abley and F.J. McQuillin, J. Chem. Soc. (C) 844 (1971).

144. V.A. Pavlov, E.I. Klabunovskii, G.S. Barysheva, L.N. Kaigorodova and Y.S. Airapetov, Bull. Acad. Sci. USSR, Div. Chem. Sci. 24:2262 (1975).

145. P.H. Boyle and M.J. Keating, J.C.S. Chem. Commun. 375 (1974).

146. L.M. Koroleva, V.K. Latov, M.B. Saporovskaya and V.M. Belikov, Izv. Akad. Nauk SSSR, Ser. Khim. 10:2390 (1979); Chem. Abs. 92:129284z.

147. Z. Nagy-Magos, S. Vastag, B. Heil and L. Marko, J. Organomet. Chem. 171:97 (1979).

148. B.R. James. R.H. Morris and K.J. Reiner, Can. J. Chem. 55:2353 (1977).

149. B.R. James and R.S. McMillan, Can. J. Chem. 55:3927 (1977).

150. Y. Ohgo, S. Takeuchi and J. Yoshimura, Bull. Chem. Soc. Jpn. 44:583 (1971).

151. Y. Ohgo, Y. Natori, S. Takeuchi and J. Yoshimura, Chem. Lett. 709, 1327 (1974).

152. R.W. Waldron and J.H. Weber, Inorg. Chem. 16:1220 (1977).

153. C. Carlini, D. Poriti and F. Ciardelli, J.C.S. Chem. Commun. 260 (1970).

154. E. Cesarotti, R. Ugo and H.B. Kagan, Angew. Chem. Int. Ed. Engl. 18:779 (1979).

155. Y. Nakamura, J. Chem. Soc. Jpn. 61:1051 (1940); Chem. Abs. 37:377^2 (1943).

156. S. Akabori, Y. Izumi, Y. Fujii and S. Sakwai, Nippon Kakaku Zasshi 77:1374 (1956); Chem. Abs. 53:5149b (1959).

157. S. Akabori, Y. Izumi and Y. Fujii, Nippon Kakaku Zasshi 78:886 (1957); Chem. Abs. 54:9889e (1960).

158. S. Akamatsu, Bull. Chem. Soc. Jpn. 34:1067, 1032 (1961); Y. Izumi, Bull. Chem. Soc. Jpn. 32:932 (1959).

159. R.L. Beamer, C.S. Fickling and J.H. Ewing, J. Pharm. Sci. 56:1029 (1967); R.L. Beamer, R.H. Belding and C.S. Fickling, J. Pharm. Sci. 58:1419 (1969); R.L. Beamer and W.D. Brown, J. Pharm. Sci. 60:583 (1971).

160. H. Hirai and T. Funita, J. Polym. Sci. B, Polym. Lett. 9:729 (1971).

161. H. Hirai and T. Funita, J. Polym. Sci. B, Polym. Lett. 9:459 (1971).

162. Y. Kawabata, M. Tanaka and I. Ogata, Chem. Lett. 1213 (1976).

163. J-C. Poulin, W. Dumont, T.P. Dang and H.B. Kagan, C.R. Acad. Sci. Paris C 277:41 (1973).

164. K. Achiwa, Chem. Lett. 905 (1978).

165. N. Takaishi, H. Imai, C.A. Bertelo and J.K. Stille, J. Am. Chem. Soc. 100:264 (1978).

166. T. Masuda and J.K. Stille, J. Am. Chem. Soc. 100:268 (1978).

167. H.B. Kagan, T. Yamagishi, J.C. Motte and R. Setton, Isr. J. Chem. 17:274 (1978).

168. I. Kolb, M. Cerny and J. Ketflejs, React. Kinet. Catal. Lett. 7:199 (1977).

169. G. Strukul, M. Bonivento, M. Graziani, E. Cernia and N. Palladino, Inorg. Chim. Acta 12:15 (1975).

170. H.W. Krause, React. Kinet. Catal. Lett. 10:243 (1979); East German Patent 133,230 (1978); Chem. Abs. 91:158101x (1979).

171. (a) E.I. Klabunovskii and A.A. Vedenyapin, Russ. J. Phys. Chem. 51:1759 (1977); (b) M.J. Fish and D.F. Ollis, Catal. Rev. Sci. Eng. 18:259 (1978).

172. T. Isoda, A. Ichikawa and T. Shimamoto, Rikagaku Kenkyusho Hokoku 34:134 (1958); Chem. Abs. 54:287c (1960).

173. Y. Izumi, M. Imaida, T. Harada, T. Tanabe, S. Yajima and T. Ninomiya, Bull. Chem. Soc. Jpn. 42:241 (1969).

174. Y. Izumi, T. Harada, T. Tanabe and K. Okuda, Bull. Chem. Soc. Jpn. 44:1418 (1971).

175. Y. Izumi, M. Imaida, H. Fukuda and S. Akabori, Bull. Chem. Soc. Jpn. 36:155 (1963); S. Tatsumi, M. Imaida, Y. Fukuda, Y. Izumi and S. Akabori, Bull. Chem. Soc. Jpn. 37:846 (1964).

176. Y. Izumi, S. Tatsumi, M. Imaida, Y. Fukuda and S. Akabori, Bull. Chem. Soc. Jpn. 39:361 (1966); Y. Isumi, S. Tatsumi and M. Imaida, Bull. Chem. Soc. Jpn. 39:2223 (1966).

177. Y. Izumi, S. Tatsumi and M. Imaida, Bull. Chem. Soc. Jpn. 39:1087 (1966); Y. Izumi, K. Matsunaga, S. Tatsumi and M. Imaida, Bull. Chem. Soc. Jpn. 41:2515 (1968).

178. Y. Izumi and K. Okuda, Bull. Chem. Soc. Jpn. 44:1330 (1971).

179. Y. Izumi, S. Yajima, K. Okuda and K.K. Babievskii, Bull. Chem. Soc. Jpn. 44:1416 (1971).

180. T. Tanabe, K. Okuda and Y. Izumi, Bull. Chem. Soc. Jpn. 46:514 (1973).

181. Y. Orito, S. Niwa and S. Imai, Yuki Gosei Kagaku Kyokai Shi 34:672 (1976); Chem. Abs. 86:43148x (1977).

182. Y. Izumi, Proc. Int. Congr. Catal. 3rd 1364 (1964).

183. T. Ninomiya, Bull. Chem. Soc. Jpn. 45:2548 (1972); Y. Izumi and T. Ninomiya, Bull. Chem. Soc. Jpn. 43:579 (1970).

184. T. Ninomiya, Bull. Chem. Soc. Jpn. 45:2551 (1972); T. Tanabe, Bull. Chem. Soc. Jpn. 46:1482 (1973).

185. A. Tai, H. Watanabe and T. Harada, Bull. Chem. Soc. Jpn. 52:1468 (1979).

186. F. Higashi, T. Ninomiya and Y. Izumi, Bull. Chem. Soc. Jpn. 44:1333 (1971).

187. T. Ninomiya, Bull. Chem. Soc. Jpn. 45:2555 (1972).

188. Y. Orito, S. Niwa and S. Ihia, Yuki Gosei Kagaku Kyokai Shi 35:672 (1977).

189. E.I. Klabunovskii and Yu. I. Petrov, Dokl. Akad. Nauk SSSR 173:1125 (1967).

190. V.I. Neupokoev, Yu. I. Petrov and E.I. Klabunovskii, Bull. Acad. Sci. USSR, Div. Chem. Sci. 25:99 (1976); E.I. Klabunovskii, A.A. Vedenyapin, G. Kh. Areshidze and A.K. Ivanov, Bull. Acad. Sci. U.S.S.R., Div. Chem. Sci. 25:2642 (1976).

191. E.I. Klabunovskii, V.I. Neupokoev and Yu. I. Petrov, Bull. Acad. Sci. USSR, Div. Chem. Sci. 20:1954 (1971).

192. E.I. Klabunovskii, N.P. Sokolova, A.A. Vedenyapin, Yu. M. Talanov, N.D. Zubareva, V.D. Polyakova and N.V. Gorina, Bull. Acad. Sci. USSR, Div. Chem. Sci. 21:2306 (1972); E.I. Klabunovskii, N.P. Sokolova, V.I. Neupoloev, T.A. Antonova and S.E. Ternovskii, Bull. Acad. Sci. USSR, Div. Chem. Sci. 23:1280 (1974).

193. J.A. Groenewegen and W.M.H. Sachtler, J. Catal. 27:369 (1972); idem, ibid. 38:501 (1975).

194. H. Ozaki, A. Tai, S. Kobatake, H. Watanabe and Y. Izumi, Bull. Chem. Soc. Jpn. 51:3559 (1978).

195. T.G. Beck, Tetrahedron 33:3041 (1977) and references therein.

196. Y. Orito, S. Niwa and S. Imai, Yuki Gosei Kagaku Kyokai Shi 35:753 (1977); Chem. Abs. 88:37219w (1978).

197. Y. Orito, S. Imai, S. Niwa and N. Gia Hung, Yuki Gosei Kagaku Kyokai Shi 37:173 (1979); Chem. Abs. 91:140487t (1979).

198. K. Ito, T. Harada, A. Tai and Y. Izumi, Chem. Lett. 1049 (1979).

199. I. Ojima, M. Nihonyanagi and Y. Nagai, J.C.S. Chem. Commun. 938 (1972).

200. R-J.P. Corriu and J.J.E. Moreau, J.C.S. Chem. Commun. 38 (1973).

201. R-J.P. Corriu and J.J.E. Moreau, J. Organomet. Chem. 85:19 (1975).

202. S.I. Sadykh-Zade and A.D. Petrov, Zh. Obsch. Khim. 29:3194 (1959).

203. R. Calas, E. Frainnet and J. Bonastre, C.R. Hebd. Seances Acad. Sci. 251:2987 (1960).

204. M. Gilman and D.H. Miles, J. Org. Chem. 23:326 (1958).

205. K. Yamamoto, T. Hayashi and M. Kumada, J. Organomet. Chem. 46:C65 (1972).

206. I. Ojima, T. Kogure, M. Kumada, S. Horiuchi and T. Sato, J. Organomet. Chem. 122:83 (1970).

207. K. Yamamoto, T. Hayashi and M. Kumada, J. Organomet. Chem. 54:C45 (1973).

208. H.B. Kagan, N. Langlois and T.P. Dang, J. Organomet. Chem. 90:353 (1975).

209. I. Ojima, Fundamental Research in Homogeneous Catalysis 3:181 (1979).

210. I. Ojima and Y. Nagai, Chem. Lett. 191 (1975).

211. I. Ojima, T. Kogure and M. Kumagai, J. Org. Chem. 42:1671 (1977).

212. I. Ojima, T. Kogure and Y. Nagai, Chem. Lett. 985 (1975).

213. T. Hayashi, K. Yamamoto and M. Kumada, Tetrahedron Lett. 3 (1975).

214. I. Ojima, T. Kogure and Y. Nagai, Tetrahedron Lett. 5035 (1972).

215. N. Langlois, T.P. Dang and H.B. Kagan, Tetrahedron Lett. 4865 (1973).

216. H.B. Kagan, Pure Appl. Chem. 43:401 (1975).

217. K. Yamamoto, Y. Uramato and M. Kumada, J. Organomet. Chem. 31:C9 (1971).

218. Y. Kiso, K. Yamamoto, K. Tanao and M. Kumada, J. Am. Chem. Soc. 94:4373 (1972).

219. K. Yamamoto, T. Hayashi and M. Kumada, J. Am. Chem. Soc. 93:5301 (1971).

220. K. Yamamoto, T. Hayashi, M. Zambayashi and M. Kumada, J. Organomet. Chem. 118:161 (1976).

221. W. Dumont, J-C. Poulin, T-P. Dang and H.B. Kagan, J. Am. Chem. Soc. 95:8295 (1973).

222. D.T. Thompson and R. Whyman in "Transition Metals in Homogeneous Catalysis", G.N. Schrauzer ed., Marcel Dekker Inc., New York, 1971, p. 147.

223. P.J. Davidson, R.R. Hignett and D.T. Thompson, Chem. Soc. Spec. Per. Rep. Catalysis 1:369 (1977).

224. R.L. Pruett, Adv. Organomet. Chem. 17:1 (1979).

225. R.F. Heck and D.S. Breslow, J. Am. Chem. Soc. 83:4023 (1961).

226. W. Rupilius, J.J. McCoy and M. Orchin, Ind. Eng. Chem. Prod. Res. Dev. 10:142 (1971); E.R. Tucci, Ind. Eng. Chem. Prod. Res. Dev. 7:125 (1968).

227. D. Evans, G. Yagupsky and G. Wilkinson, J. Chem. Soc. (A) 2660 (1968).

228. D. Evans, J.A. Osborn and G. Wilkinson, J. Chem. Soc. (A) 3133 (1968).

229 (a) G. Yagupsky, C.K. Brown and G. Wilkinson, J. Chem. Soc. (A) 1392 (1970); (b) M. Yagupsky, C.K. Brown, G. Yagupsky and G. Wilkinson, J. Chem. Soc. (A) 937 (1970); (c) C.K. Brown and G. Wilkinson, J. Chem. Soc. (A) 2753 (1970).

230. R.A. Sanchez-Delgardo, J.S. Bradley and G. Wilkinson, J.C.S. Dalton Trans. 399 (1976).

231. G. Consiglio and P. Pino, Helv. Chim. Acta 59:642 (1976).

232. P.S. Pregosin and S.N. Sze, Helv. Chim. Acta 61:1848 (1978).

233. C. Botteghi, G. Consiglio and P. Pino, Chimia 27:477 (1973).

234. C. Botteghi, G. Consiglio and P. Pino, Annalen 864 (1974).

235. Where optical yields have been based on the rotation of the acid or alcohol they are reported here as found. Where, however, they are based on hydtropaldehyde itself they are corrected using 238° as the rotation of the pure compound.

236. P. Pino, G. Consiglio, C. Botteghi and C. Saloman, Adv. Chem. Ser. 132:295 (1974).

237. F. Piacenti, G. Menchi, P. Frediani, U. Matteoli and C. Botteghi, Chim. Ind. (Milan) 60:808 (1978).

238. C. Botteghi, G. Consiglio and P. Pino, Chimia 26:141 (1972); P. Pino, C. Saloman, C. Botteghi and G. Consiglio, Chimia 26:655 (1972).

239. H.B. Tinker and A.J. Solodar, Canadian Patent 1,027,141 (1978); Chem. Abs. 89:42440m (1978).

240. M. Tanaka, Y. Watanabe, T-A. Mitsudo and Y. Takegami, Bull. Chem. Soc. Jpn. 47:1698 (1974).

241. W. Himmele, H. Siegel. W. Aquila and F-J. Muller, Ger. Offen. 2,132,414 (1973); Chem. Abs. 78:97328j (1973).

242. Y. Watanabe, T-A. Mitsudo, Y. Yasunori, J. Kikuchi and Y. Takegami, Bull. Chem. Soc. Jpn. 52:2735 (1979).

243. M. Tanaka, Y. Watanabe, T-A. Mitsudo, K. Yamamoto and Y. Takegami, Chem. Lett. 483 (1972).

244. C.E. Clark Jr., "Hydroformylation with modified rhodium catalysts", Ph.D. Thesis, University of Cincinnati, 1974.

245. A. Bortinger, Ph. D. Thesis, University of Cincinnati, 1972.

246. C. Botteghi, M. Branca, G. Micera, F. Piacenti and G. Menchi, Chim. Ind. (Milan) 60:16 (1978).

247. G. Consiglio, C. Botteghi, C. Saloman and P. Pino, Angew. Chem. Int. Ed. Engl. 12:669 (1973).

248. J.M. Brown and P.N. Nicholson, private communication.

249. M. Tanaka, Y. Ikeda and I. Ogata, Chem. Lett. 1115 (1975).

250. T. Hayashi, M. Tanaka, Y. Ikeda and I. Ogata, Bull. Chem. Soc. Jpn. 52:2605 (1979).

251. C. Botteghi, M. Branca and A. Saba, J. Organomet. Chem. 184:C17 (1980).

252. C. Botteghi, Gazz. Chim. Ital. 105:233 (1975).

253. G. Consiglio and P. Pino, Helv. Chim. Acta 59:642 (1976).

254. Y. Kawabata, T.M. Suzuki and I. Ogata, Chem. Lett. 361 (1978).

255. C. Botteghi, G. Consiglio and P. Pino, Chimia 27:477 (1973).

256. T. Hayashi, M. Tanaka and I. Ogata, Tetrahedron Lett. 3925 (1978).

257. E. Bayer and V. Schurig, Chem. Tech. 212 (1970).

258. S.J. Fritschel, J.J.H. Ackerman, T. Keyser and J.K. Stille, J. Org. Chem. 44:3152 (1979).

259. U. Folli, D. Iarossi, F. Montanari and G. Torre, J. Chem. Soc. (C) 1317 (1968).

260. W.H. Pirkle and P.L. Rinaldi, J. Org. Chem. 42:2080 (1977).

261. R. Helder, J.C. Hummelen, R.W.P.M. Loone, J.S. Wiering and H. Wynberg, Tetrahedron Lett. 1831 (1976).

262. K.B. Sharpless and R.C. Michaelson, J. Am. Chem. Soc. 95:6136 (1973).

263. S. Tanaka, H. Yamamoto, H. Nozaki, K.B. Sharpless, R.C. Michaelson and J.D. Cutting, J. Am. Chem. Soc. 94:5254 (1974).

264. A.O. Chang and K.B. Sharpless, J. Org. Chem. 42:1587 (1977).

265. S-I. Yamada, T. Mashiko and S. Terashima, J. Am. Chem. Soc. 99:1988 (1977).

266. P.S. Elmes and W.R. Jackson, J. Am. Chem. Soc. 101:6128 (1979).

267. M. Hidai, H. Ishiwatori, H. Yogi, E. Tanaka, K. Onazawa and Y. Uchida, J.C.S. Chem. Commun. 170 (1975).

268. Y. Tatsumo, A. Komishi, A. Nakamura and S. Otsuka, J.C.S. Chem. Commun. 588 (1974).

269. T. Aratani, Y. Yoneyoshi and T. Nagase, Tetrahedron Lett. 1707 (1975).

270. W. Bergstein, A. Kleeman and J. Martens, Synthesis 76 (1981).

271. J.P. Amma and J.K. Stille, J. Org. Chem. 47:468 (1982).

272. D.P. Riley and R.E. Shumate, J. Org. Chem. 45:5187 (1980).

273. K. Kashwabara, K. Hanaki and J. Fujita, Bull. Chem. Soc. Jpn. 53:2275 (1980).

274. S. Mujario, M. Naura and H. Hashimoto, Chem. Lett. 729 (1980).

275. M. Fiorini and G.M. Giongo, J. Mol. Catal. 7:41 (1980).

276. K. Onuma, T. Ito and N. Nakamura, Bull. Chem. Soc. Jpn. 53:2012,2016 (1980).

277. K. Osakada, T. Ikariya, M. Saburi and S. Yoshikawa, Chem. Lett. 1691 (1981).

278. A. Miyashita, A. Yasuda, H. Tanaka, K. Toniimi, T. Ito, T. Souchi and R. Noyori, J. Am. Chem. Soc. 102:7932 (1980).

279. L. Marko and J. Bakos, Aspects Homogeneous Catal. 4:145 (1981).

280. J.W. Scott, D.D. Keith, G. Nix, D.R. Parrish, S. Remington, G.P. Roth, J.M. Townsend, D. Valentine Jr. and R. Yang, J. Org. Chem. 46:5086 (1981).

281. K.E. Koenig, G.L. Bachman and B.D. Vineyard, J. Org. Chem. 45:2362 (1980).

282. T. Hayashi, K. Kanehira and M. Kumada, Tetrahedron Lett. 22:4417 (1981).

283. I. Ojima and T. Suzuki, Tetrahedron Lett. 21:1239 (1980); I. Ojima, T. Kogure, N. Yoda, T. Suzuki, M. Yatabe and T. Tanaka, J. Org. Chem. 47:1329 (1982).

284. K-I. Onuma, T. Ito and A. Nakamura, Chem. Lett. 481 (1980); D. Meyer, J-C. Poulin, H.B. Kagan, H. Levine-Pinto, J-L. Morgat and P. Fromageot, J. Org. Chem. 45:4680 (1980).

285. D. Sinou, D. Lafont, G. Descotes and A.G. Kent, J. Organomet. Chem. 217:119 (1981).

286. K. Osakada, M. Obana, T. Ikariya, M. Saburi and S. Yoshikawa, Tetrahedron Lett. 22:4297 (1981).

287. U. Matteoli, P. Frediani, M. Bianchi, C. Botteghi and S. Gladiali, J. Mol. Catal. 12:265 (1981).

288. G. Balavoine, S. Brunie and H.B. Kagan, J. Organomet. Chem. 187:125 (1980).

289. R.G. Ball, B.R. James, D. Mahajan and J. Trotter, Inorg. Chem. 20:254 (1981).

290. R.G. Ball and J. Trotter, Inorg. Chem. 20:261 (1981).

291. J.M. Townsend and J.F. Blount, Inorg. Chem. 20:269 (1981).

292. P.S. Chua, N.K. Roberts, B. Bosnich, S.J. Okrasinski and J. Halpern, J.C.S. Chem. Commun. 1278 (1981).

293. B.R. Stults, R.M. Friedman, K. Koenig, W. Knowles, R.B. Gregor and F.W. Lytle, J. Am. Chem. Soc. 103:3235 (1981).

294. G.A. Morris and R.A. Freeman, J. Mag. Res. 29:433 (1978).

295. J.M. Brown, P.A. Chaloner and G.A. Morris, J.C.S. Chem. Commun. 664 (1983).

296. L. Canning, D.Phil. Thesis, University of Oxford (1983).

297. J.M. Brown and D. Parker, Organometallics 1:950 (1982).

298. J-F. Peyronel, J-C. Fiaud and H.B. Kagan, J. Chem. Res. S320 (1980).

299. N.C. Payne and D.W. Stephen, Inorg. Chem. 21:182 (1982).

300. I. Ojima, T. Tanaka and T. Kogure, Chem. Lett. 823 (1981).

301. I. Kolb and J. Hetflejs, Coll. Czech. Chem. Commun. 45:2224,2808 (1980).

302. T. Hayashi, K. Tamao, Y. Katsuro, I. Nakae and M. Kumada, Tetrahedron Lett. 21:1871 (1980).

303. G. Consiglio and P. Pino, private communication.

304. Y. Becker, A. Eisenstadt and J.K. Stille, J. Org. Chem. 45:2145 (1980).

305. C.F. Hobbs and W.S. Knowles, J. Org. Chem. 46:4422 (1981).

306. J.D. Unruh and J.R. Christenson, J. Mol. Catal. 14:19 (1982).

307. O.R. Hughes and D.A. Young, J. Am. Chem. Soc. 103:6636 (1981).

308. T. Katsuki and K.B. Sharpless, J. Am. Chem. Soc. 102:5974 (1980).

309. B.E. Rossiter, T. Katsuki and K.B. Sharpless, J. Am. Chem. Soc. 103:464 (1981).

310. K.B. Sharpless, S.S. Woodward and M.G. Finn, Pure Appl. Chem. 55:1823 (1983).

311. I.D. Williams, S.J. Pedersen, K.B. Sharpless and S.J. Lippard, J. Am. Chem. Soc. 106:6430 (1984).

312. L.D. Lu, R.A. Johnson, M.G. Finn and K.B. Sharpless, J. Org. Chem. 49:728 (1984).

313. H.B. Kagan and P. Pitchen, Tetrahedron Lett. 1049 (1984).

REACTIONS OF COORDINATED ACETYLENES

J.L. Davidson

Department of Chemistry, Heriot-Watt University

Edinburgh

1.   INTRODUCTION

Acetylenes occupy a unique place in organometallic chemistry in view of their ability, in the presence of transition metals, to undergo an unusually diverse range of reactions. On the one hand ligand substitution can occur to give simple acetylene π-complexes while on the other, complex cyclooligomerization reactions are observed which lead to cyclic acetylene dimers, trimers and tetramers some of which contain other ligands, e.g. carbon monoxide, olefins, incorporated during the cyclization process. Early studies, in particular Hubel's pioneering work in the 1960's,[1] concentrated on reactions with metal carbonyls, which unfortunately in most cases proved to be particularly complex. Thus a fascinating body of chemical knowledge was acquired, but no clear insight into the role of metals in promoting acetylene reactivity emerged. In many reactions it can reasonably be assumed that simple acetylene π-complexes function as precursors to the more complex products isolated but only in relatively few cases have they been detected or isolated. Consequently in order to begin to understand the role of the metal in manipulating the reactivity of alkynes it seems logical that studies of the reactions of coordinated alkynes be carried out. Extended systematic studies of this nature have only been undertaken infrequently and it is the purpose of this Chapter to survey the diverse reactions reported so far, in the hope that patterns of behavior may be discerned.

Useful reviews dealing with several aspects of reactions between metal complexes and alkynes have appeared in the literature.[1-7] Of particular relevance is an article by Seyferth which is concerned primarily with the synthetic aspects of reactions involving coordinated acetylenes.[7] In view of this we shall provide a complementary survey in which mechanistic features of the reactions are highlighted. Since many of the reactions promoted by coordination of an acetylene to a metal result from modifications to the electronic structure of the triple bond, we will begin with a description of the bonding and structures of metal-acetylene complexes and attempt to relate these features to chemical reactivity.

2.   STRUCTURES, BONDING AND REACTIVITY

2.1 Mononuclear complexes

Coordination of an acetylene to a transition metal results in two structural alterations to the basic carbon skeleton: (i) a deviation from linearity occurs to give a cis-bent structure I, (ii) the acetylenic $C\equiv C$

$$M - \overset{\overset{\displaystyle R}{\diagup}}{\underset{\underset{\displaystyle R}{\diagdown}}{\|}}$$

I

distance increases by an amount which approximately parallels the strength of the metal-acetylene bonding interaction.[8] The latter effect, which provides us with a semi-quantitative measure of the extent to which an acetylene is modified on coordination (the deviation from linearity is a less sensitive guide) can qualitatively be understood in terms of the orbital interactions in Fig. 1 where it is apparent that on the basis of overlap considerations alone the major contributions to the metal acetylene bond will come from interactions (a) and (b).[9] Interaction (a) involves donation from the filled π-orbital in the plane of the metal-acetylene linkage (π_y) to an empty σ-type orbital on the metal while in (b) back-donation of electron density from filled metal d-type orbitals to the antibonding counterpart of π_y occurs. As with metal-carbonyl bonding these interactions act in a

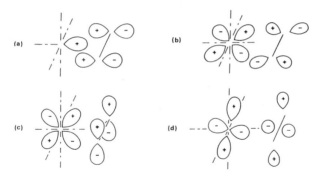

Fig. 1

synergic manner but the extent to which either contributes will obviously depend on the metal, its oxidation state, the ancillary ligands attached to the metal and the electron withdrawing or releasing ability of the acetylenic substituents. Thus with early transition metals, particularly those in high oxidation states and when electron donating substituents are attached to the alkyne, (a) will predominate, whereas π-back-donation (b) will be favored by the later transition metals, those in low oxidation states, and acetylenes bearing electron withdrawing substituents. In the former case this simple bonding scheme predicts that the acetylene will suffer a net loss of density and thus become more susceptible to nucleophilic attack whereas in the latter situation population of one of the acetylenic π^* orbitals should promote reactivity towards electrophiles.[10] In practice, however, it is rarely possible to make clear predictions about the reactivity of coordinated acetylenes towards such reagents.

The situation is also complicated in some cases by the possibility of contributions to the metal-acetylene bond by orbital interactions (c) and (d) involving the π_z and π_z^* orbitals of the C≡C bond.[9] Overlap considerations dictate that back-donation via (d) can virtually be neglected whereas a contribution from interaction (c) is not precluded on this basis. Otsuka and Nakamura[9a] have pointed out that (c) can result in a net increase in bonding where the metal has suitable empty d-type orbitals, as with the early transition metals and those in high oxidation states, whereas a destabilizing, antibonding effect is to be expected when the d orbitals are full as with low oxidation state complexes, particularly with the later transition metals.

II   III   IV

We can clearly distinguish cases where interaction (c) appears to contribute in a bonding fashion since increasing numbers of complexes containing one, two or three acetylenes coordinated to an early transition metal (Group V or VI) are being discovered which exhibit the common characteristic of at least apparent coordinative unsaturation.[9b] Complexes II,[11] III[12] and IV[13] can formally be considered as electron deficient, i.e. the metal has not attained a stable eighteen-electron configuration, unless the acetylene ligands donate more than two electrons via interaction (c). Interestingly, complexes III form 1:1 adducts V with two electron donors (L = tertiary phosphine, pyridine) in which the acetylenes must revert to simple two-electron donation,[14] thus illustrating the ability of acetylenes to adapt their bonding to the electronic requirements of the metal.

Otsuka and Nakamura[9a] have also pointed out the consequences of interaction (c) for chemical reactivity of coordinated acetylenes. The extra stability gained from a bonding interaction of this type enables the isolation of π-bonded acetylene complexes such as II, III and IV. However, with later transition metals, where the interaction becomes non-bonding or even antibonding, complexes containing more than one π-bonded acetylene are rarely isolated since acetylenes exhibit a greater propensity to undergo cyclization reactions. As a result cyclobutadiene, cyclopentadiene, benzene and related complexes are frequently obtained as a result of two or more acetylenes undergoing cyclooligomerization in the presence of the metal. Intermediates in these reactions may contain more than one acetylene coordinated to the metal and complexes II - IV can therefore act as models for such species. Significantly, addition of other ligands to complexes of type III (eq 1), can result in acetylene cyclization when the acetylenes revert to two-electron donation as in V:[14c]

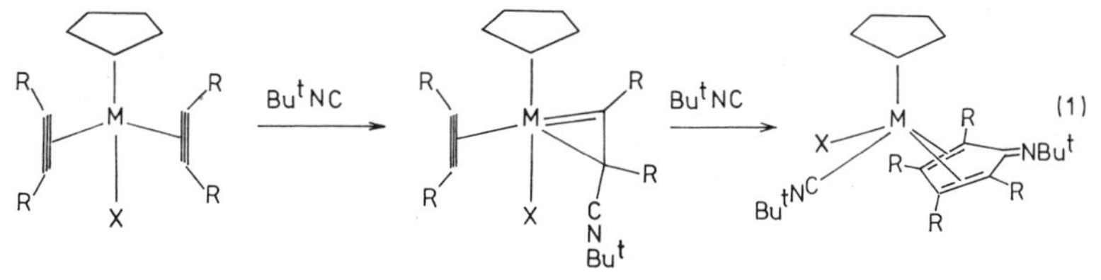

$$V \quad \begin{array}{l} M=Mo, X=CF_3 \\ M=W, X=Cl \\ R = CF_3 \end{array} \qquad\qquad VI$$

This appears to confirm Otsuka and Nakamura's concept of reactivity being related to the degree of importance of interaction (c). However it may be unwise to interpret too many observations of this type in terms of what is obviously a simple qualitative picture of metal-ligand bonding.

The activation of acetylenes by complexation to transition metals as a result of a destabilizing interaction (c) has been discussed in terms of the $\eta^2 \longrightarrow \eta^1$ transformations depicted in eqs 2 and 3.[9a] The relative amounts of radical and cationic, (eq 2) or anionic (eq 3) character of the η^1-acetylene will vary from complex to complex, and in particular, a significant contribution from the cationic form in eq 2 is to be expected with electron donating substituents R, whereas electron withdrawing groups, e.g. R = CF_3, CO_2Me, will stabilize the anionic structure in eq 3. Intermediates of the latter type are frequently invoked in reactions of hexafluorobut-2-yne and dimethylacetylene dicarboxylate with metal complexes[15] but as we shall see in Section 3.5.1.1 it is not always necessary to postulate $\eta^2 \longrightarrow \eta^1$ transformations to explain the reactivity of coordinated acetylenes.

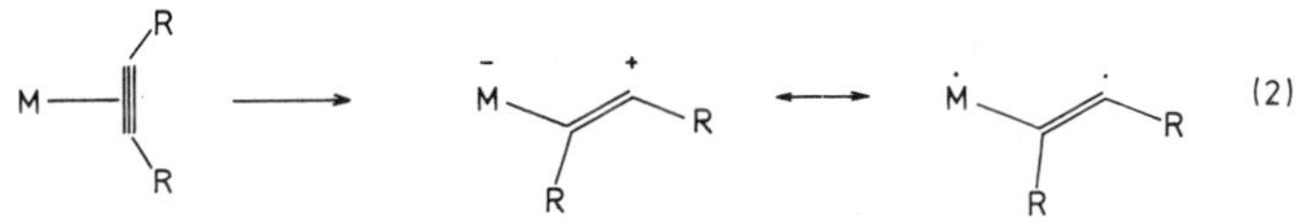

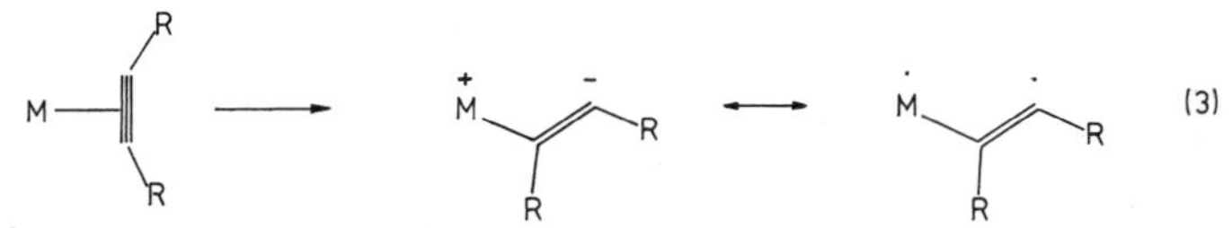

2.2 Di- and polynuclear complexes

The variety of bonding modes found in complexes formed by acetylenes in reactions with di- and polynuclear metal compounds precludes any clear relationship from being derived between structure, bonding and reactivity. An acetylene can coordinate to two, three or four different metals in several different ways as structures **VII**, **VIII**,[16] **IX**, **X**,[17] **XI**[18] and **XII**[19] illustrate. Thus in many polynuclear acetylene complexes both π-orbitals of the triple bond interact with two or more metals. Instead of thinking in terms of discrete metal-carbon linkages it may be more realistic in some cases to consider that the π-electrons of the acetylene form part of a delocalised bonding network extending over the whole cluster.

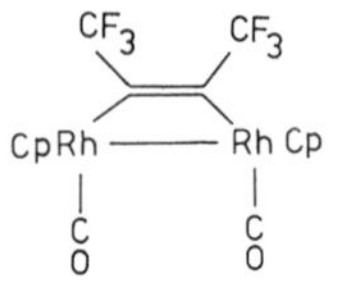

VII 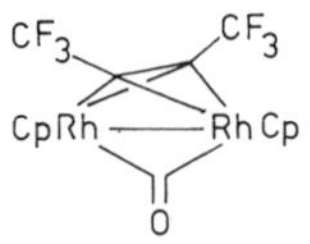   VIII 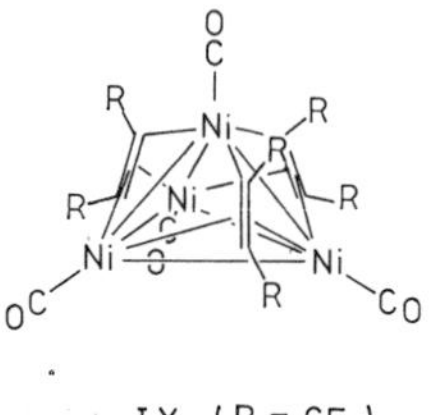   IX (R = CF₃)

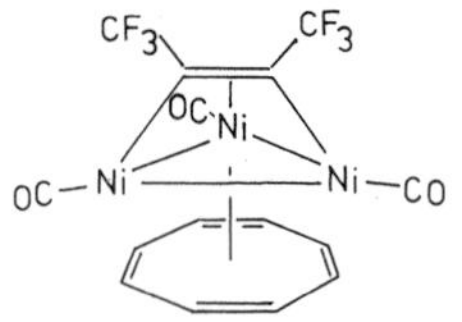

X 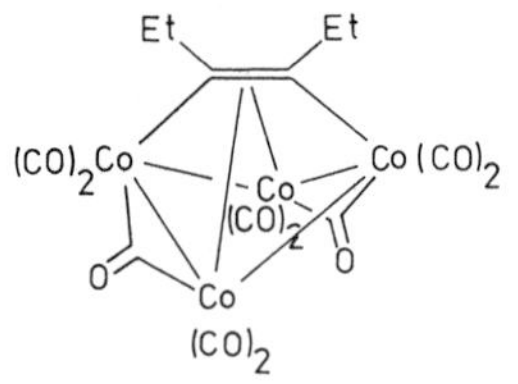   XI 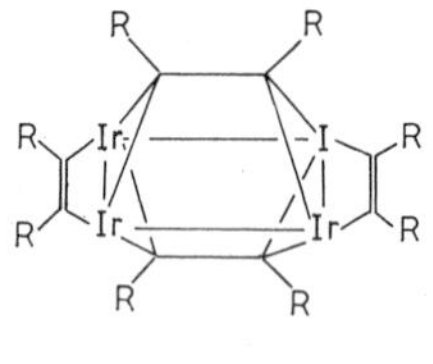   XII (R = COOMe)

As in mononuclear complexes, the coordinated acetylene retains the ability to adapt its mode of bonding to suit the electronic requirements of the metal(s), a feature exemplified by the transformations **VII** ⟶ **VIII** and **IX** ⟶ **X** promoted by Me_3NO[16] and cyclooctatetraene[17] respectively. It has been suggested that reversible interconversions of the first type might be important processes in catalysis and cluster chemistry since this would generate coordinative unsaturation at metal centres.[16,20] It may be significant that where coordinative unsaturation exists or is potentially available in bridging acetylene complexes related to **VIII**, reactivity towards molecular hydrogen is frequently found and the alkyne is readily hydrogenated to an olefin. In some cases the complexes also act as catalysts for acetylene hydrogenation.[20]

3.   REACTIONS OF COORDINATED ACETYLENES

3.1   Reactions in which the triple bond is unaffected

Although coordination of an acetylene to a transition metal can in many cases activate the carbon-carbon triple bond towards attack by a variety of substrates, there are instances in which deactivation is observed and the metal effectively acts as a protecting group. This is particularly true of cobalt complexes $Co_2(CO)_6(\eta^2$-$RC{\equiv}CR)$ (**XIII**) which have been widely used to carry out transformations on the substituents R without affecting the acetylenic function. Seyferth[7] has pointed out that the $Co_2(CO)_6$ moiety readily fulfils the essential criteria of a protecting group in organic synthesis; (a) it is easily introduced and removed, (b) it is stable throughout reactions transforming the unprotected substituents, (c) it is relatively inexpensive. The reactions described in this section illustrate these properties although we will also consider the organometallic aspects of the transformations under discussion.

XIII

$(CO)_3Co{-}Co(CO)_3$

Alkyne complexes $Co_2(CO)_6(RC{\equiv}CH)$ are not easy to obtain directly from $Co_2(CO)_8$ and monosubstituted alkynes, as a result of competing side reactions, and a useful route to such species involves treatment of the readily isolated $Co_2(CO)_6(RC{\equiv}CSiMe_3)$ with a methanolic solution of sodium hydroxide which leads to facile cleavage of the C-Si bond.[21] A related approach utilizing the well known lability of carbon-tin bonds leads to acetyl-substituted acetylene complexes of cobalt (eq 4):[22]

$$[Co_2(CO)_6 \; HC{\equiv}CSnMe_3] \xrightarrow[AlCl_3]{MeCOCl} [Co_2(CO)_6 \; HC{\equiv}C \, COMe] \qquad (4)$$

In contrast, Friedel-Crafts acylation of arylacetylenes coordinated to $Co_2(CO)_6$ results in selective acylation on the aromatic ring, as eq 5 demonstrates.[23]   In all cases the para positions are attacked almost exclusively, while release of the alkyne can be effected by treatment with Ce^{4+} giving 70-90% yield of the free acetylene.   This emphasizes the synthetic value of the process where prevention of attack at the C≡C bond is of paramount importance.

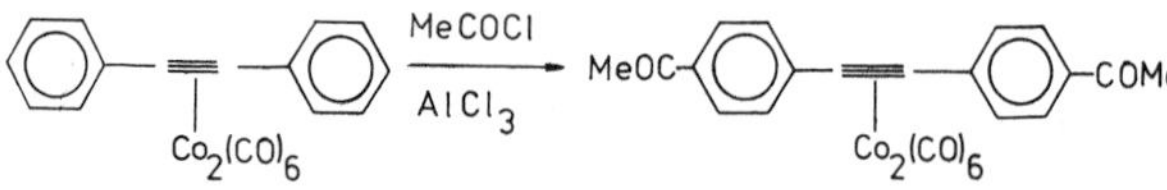

The reaction in eq 6 provides an illustration of methylene group activation in diethyl acetylene in which oxygen incorporation α to the triple bond has occurred.[24] $Co_2(CO)_6(EtC≡CEt)$ does not function as a precursor to the two other acetylenic products and the mechanism of the reaction remains obscure.   In contrast, eqs 7 and 8 illustrate cases in which hydroxyl groups α to coordinated triple bonds can be reduced to give dialkyl acetylenes:[25]

$$[Fe(CO)_5] + [Co_2(CO)_8] \xrightarrow[Acetone]{EtC≡CEt} \begin{array}{c} [Co_2(CO)_6 \, EtC≡CCOMe] \\ + \\ [Co_2(CO)_6 \, EtC≡CCH(OH)Me] \end{array} \quad (6)$$

$$[Co_2(CO)_6 \, Me_2C(OH)C≡CC(OH)Me_2] \xrightarrow[H_2SO_4]{MeOH} [Co_2(CO)_6 \, Me_2CHC≡CCHMe_2] \quad (7)$$

$$[Co_2(CO)_6 \, Me_2C(OH)C≡CH] \xrightarrow{[CoH(CO)_4]} [Co_2(CO)_6 \, Me_2CHC≡CH] \quad (8)$$

It is not clear how general these reactions may be since acid treatment of the $Ph_2C(OH)C≡CPh$ derivative gave the expected product whereas tetraphenylbutynedioldicobalt hexacarbonyl gave 1,1,4,4-tetraphenyl-1,3-butadiene.

Selective reduction of a double bond in the presence of a triple bond can be achieved by initially complexing the triple bond, as shown in eqs 9 and 10:[26]

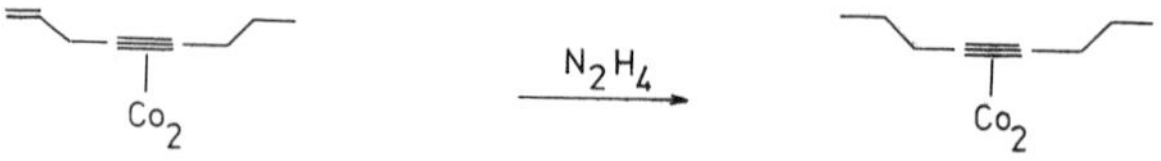

$$CO_2 = CO_2(CO)_6$$

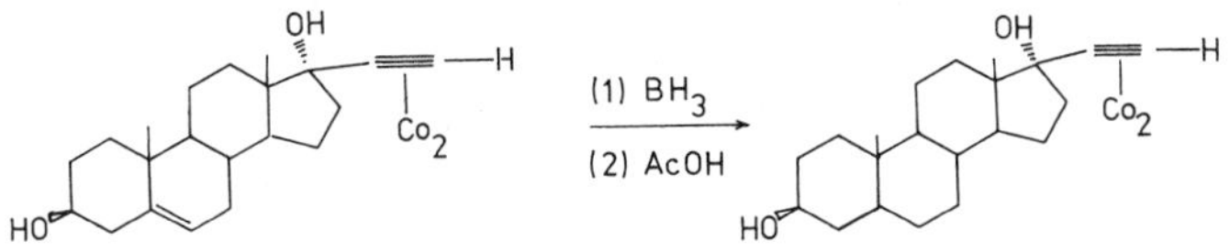

Complexation similarly enables acid catalyzed hydration of the double bond in eneynes to be effected (eq 11), a reaction which in view of its facility ($0°$, 15 min) is thought to proceed via the metal stabilized carbonium ion **XIV**.[27] Thus, in addition to deactivating the triple bond, coordination of the eneyne to the $Co_2(CO)_6$ moiety also activates the olefinic moiety towards hydration relative to the free ligand. The stability of the carbonium center adjacent to the coordinated triple bond apparently resembles that of α-carbonium ions in ferrocenyl and methinyltricobalt derivatives.

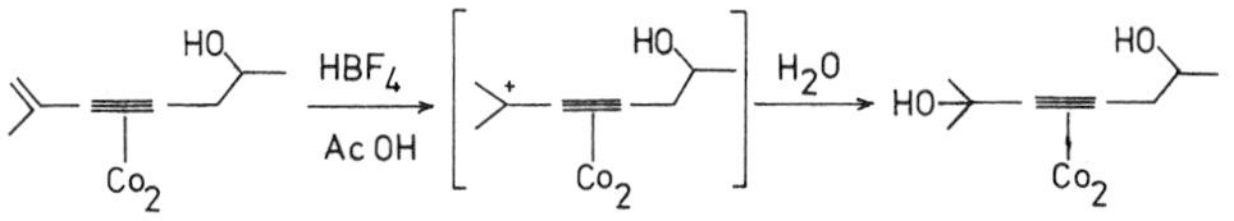

The participation of a carbonium ion related to **XIV** is probably responsible for the facility with which dehydration reactions can be effected on coordinated acetylenic alcohols, and Descoins and Semain[28] have carried out the stereoselective synthesis of E-eneynes from cyclopropyl alkynyl carbinols (eq 12) employing $HBr/ZnBr_2$ mixtures as dehydrating agents. In the absence of $Co_2(CO)_6$ the uncomplexed carbinol gives mixtures of both E and Z eneynes, thus illustrating the additional but equally valuable effect of pre-coordination in inducing stereospecificity. Nicholas has carried out several studies of the reactivity of carbonium ions **XV** including related acid-catalyzed dehydration reactions of tertiary propargyl alcohol complexes, some of which are illustrated in eqs 13 and 14.[29] The mild conditions employed contrast with the forcing conditions required with free tertiary propargyl alcohols which are, in any case, rarely successful.

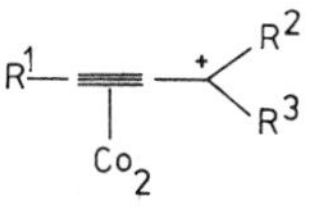

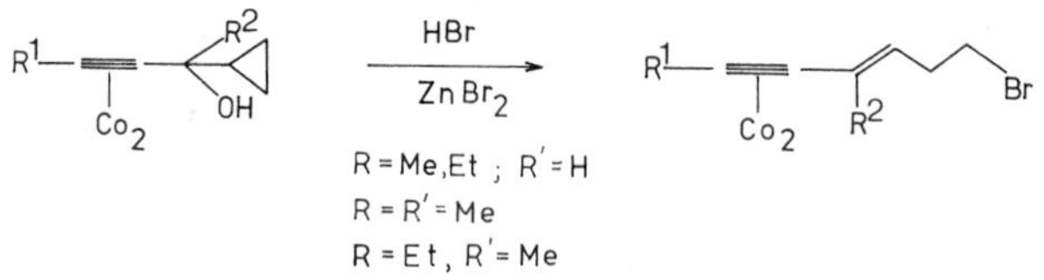

Indirect evidence was obtained for the intermediacy of carbonium ions **XV** in these reactions while their isolation as stable SbF_6^- and BF_4^- salts was reported in a later publication.[27] This involved treatment of propionic anhydride solutions of complexes $Co_2(CO)_6(HC\equiv CC(OH)R_1R_2)$ with excess $HF.SbF_5$ or $HBF_4.Et_2O$ at $-40°C$. Spectroscopic evidence [^1H nmr and ir (νCO) spectra] was obtained for extensive charge delocalization onto the $Co_2(CO)_6(HC\equiv C-)$ moiety, while pK values, which showed little dependence on R_1 and R_2, were all close to -7, i.e. approximately equal to that of the triphenylmethyl cation.

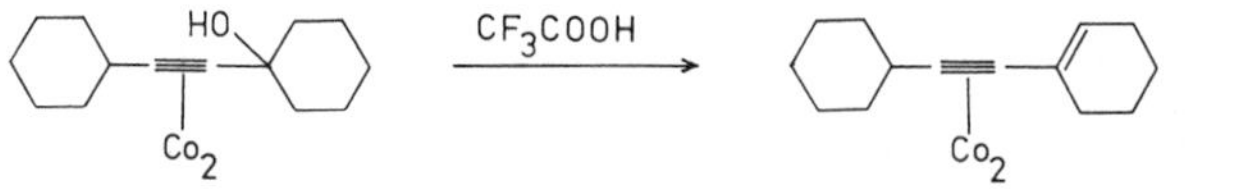

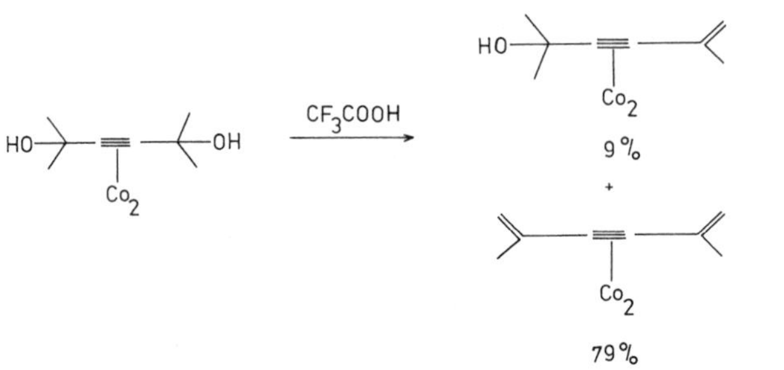

Subsequent work in this area has been directed towards developing the synthetic utility of $Co_2(CO)_6$-stabilized carbonium ions, and as reaction (15) shows it is possible for these to alkylate aromatic compounds such as anisole in the ortho and para positions.[30] It was later found that acetylacetone, 2-acetylcyclohexanone, ethylacetoacetate and benzoylacetone can be alkylated in the same way.[31]

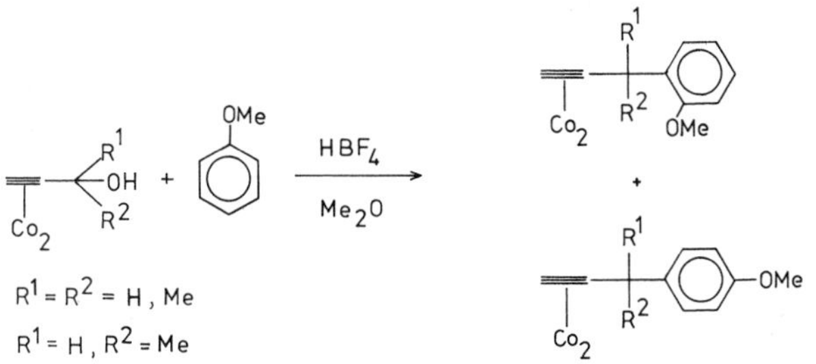

$$R^1 = R^2 = H, Me$$
$$R^1 = H, R^2 = Me$$

(15)

3.2 Isomerization and rearrangement reactions

The simplest rearrangement process induced by coordination of an acetylene to a metal is that leading to η^1-acetylide formation as a result of oxidative cleavage of a $\equiv$C-X bond. Formation of metal acetylides by addition of RC$\equiv$C-X to metal complexes is observed when the C-X bond is reactive, e.g. X = H, Cl, Br, I, SnR$_3$, and in some cases it has proved possible to isolate intermediates containing π-bonded acetylenes.[32] Kemmitt has studied the kinetics of the isomerization reaction (16) and obtained results in accord with an intramolecular mechanism in which the rate determining step is cleavage of the C-Cl bond.[32] A related reaction type is shown in eq 17 but in this case product **XVI** contains a μ_3-acetylide ligand, as frequently found in complexes obtained from the reaction of polynuclear metal carbonyls with monosubstituted alkynes.[33] The second product **XVII** contains a methinyl group bridging three metal atoms and is related to methinyltricobalt complexes described in Section 3.3.3.1.

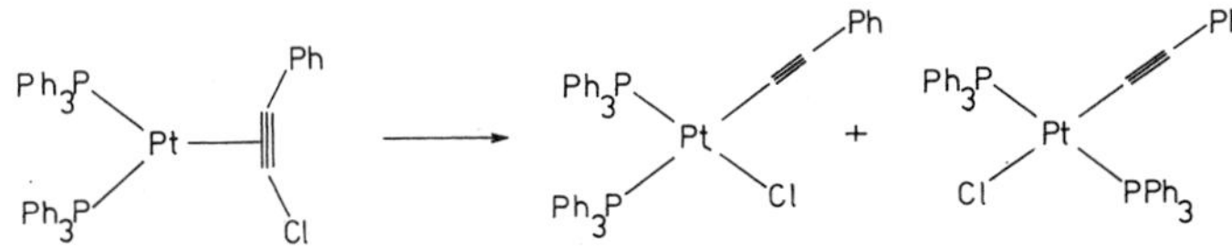

(16)

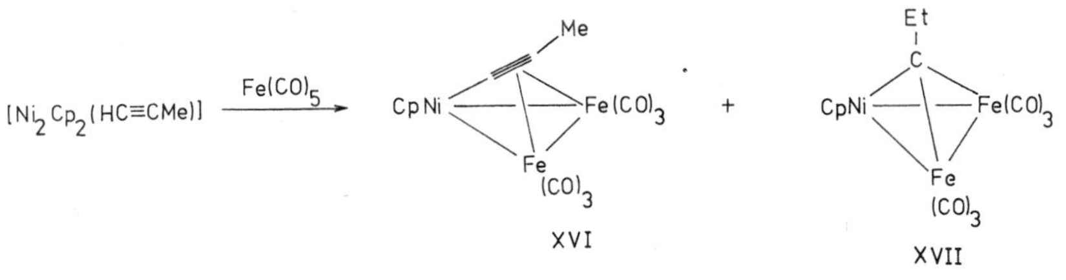

(17)

XVI   XVII

A variety of acetylenes undergo more extensive rearrangements when activated by coordination to the moiety $CpM(CO)_2$ (M = Mn,Re). Thus acetylenes RC≡CH, R = H[34] or Ph,[35] react with $CpMn(CO)_2(THF)$ to produce mono- and dinuclear vinylidene complexes via the isolable π-complex **XVIII** (Fig. 2). The proposed rearrangement pathway involves oxidative addition to produce an intermediate η^1-acetylide-hydride complex **XIX** (cf eq 16) which subsequently transfers the hydride ligand to the β-carbon of the acetylide ligand. Similar products are also obtained with $Ph_3EC≡CPh$ (E = Si, Ge, Sn), in addition to complexes assigned structures **XX** and **XXI**[36] which may result from ortho-metalation of an η^1-acetylide or a vinylidene derivative.

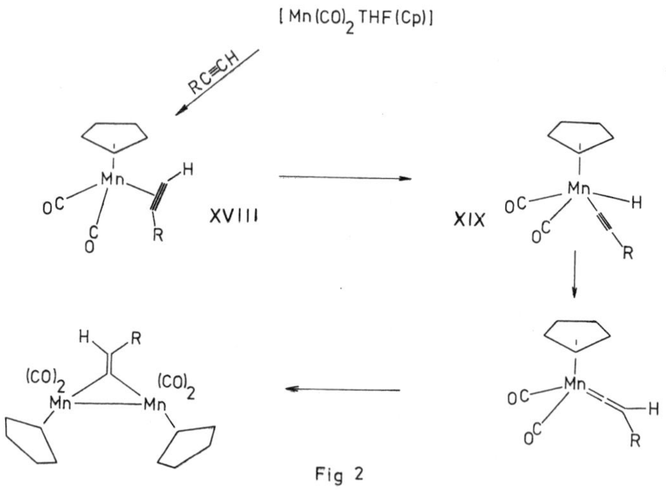

[Mn (CO)₂ THF (Cp)]

Fig 2

XX   XXI

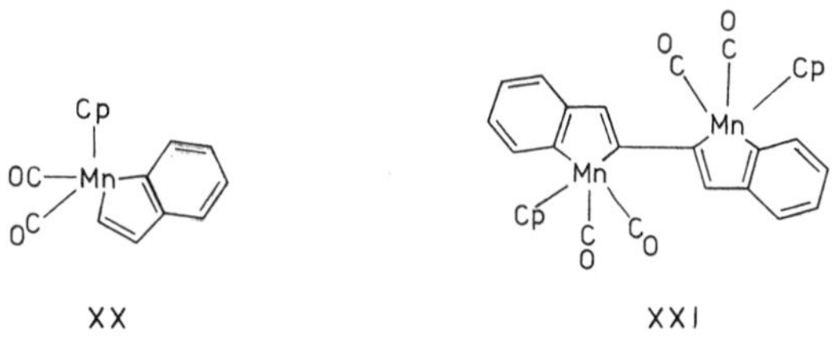

More recently it has been demonstrated[37] that protonation of electron-rich η^1-acetylide iron complexes gives η^1-vinylidene complexes, apparently as a result of attack of H^+ at the β-carbon (eq 18):

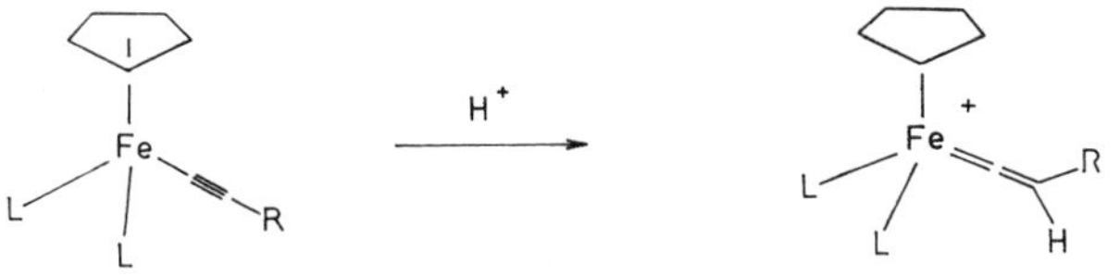

$$L_2 = (CO)_2, (CO, PPh_3), dpe$$
$$R = H, Me, Ph$$

Other iron[38] and ruthenium[39] complexes have been prepared by related methods. Interestingly, complexes of this type are unreactive towards methanol,[37-8] whereas the reaction of trans-$PtCl(Me)L_2$ (L = PMe_2Ph or $AsMe_3$) with terminal acetylenes in the presence of MeOH and $AgPF_6$ gives methoxycarbene complexes $\{PtMe[C(OMe)CH_2R]L_2\}^+$.[40] This reaction has been postulated to proceed via attack of MeO^- on vinylidene complexes produced by isomerization of platinum acetylene derivatives (see Section 3.3.1.4). Alternative isomerization reactions are observed on coordination of the triple bond of some electrophilic alkynes to the $CpMn(CO)_2$ moiety.[41] The first, a 1,5 hydrogen shift in eneyne esters, requires photolysis in acetic acid and produces the isoprene **XXII** via an isolable vinyl complex (Fig. 3).

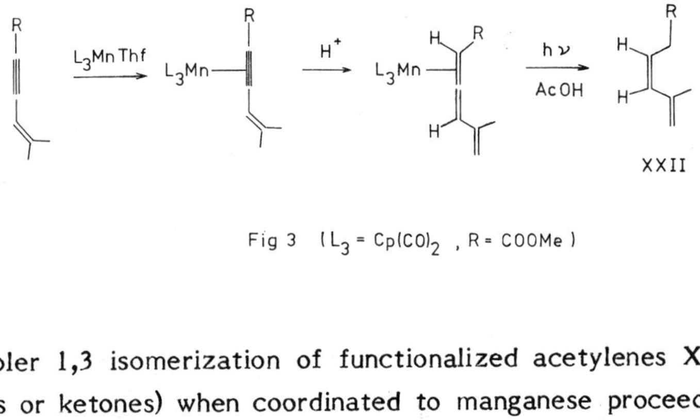

Fig 3 ($L_3 = Cp(CO)_2$, R = COOMe)

The simpler 1,3 isomerization of functionalized acetylenes **XXIII** (esters, aldehydes or ketones) when coordinated to manganese proceeds readily in the presence of basic alumina and the resulting allene can be released by treatment with ceric ion (Fig. 4).

Finally in this section we draw attention to reactions which provide the most remarkable illustrations of the ability of metals to activate acetylenic triple bonds, those in which the C≡C bond is cleaved to yield carbyne complexes. The earliest examples, reported by King, involved reactions between $Et_2NC\equiv CNEt_2$ and $Fe(CO)_5$ and $CpCo(CO)_2$[42] to give

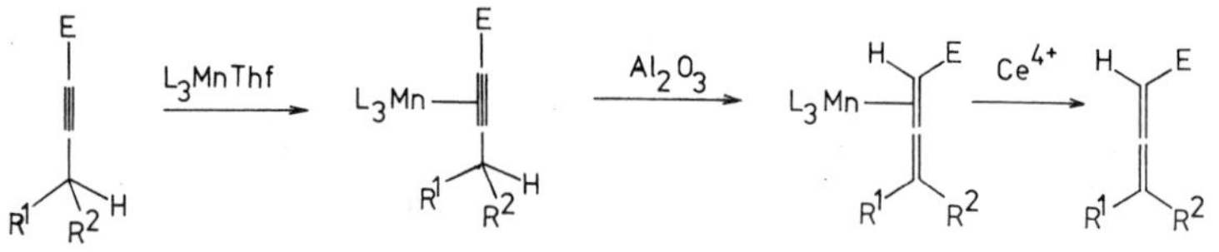

	R^1	R^2	E
	H	H	COOEt
XXIII	Bu	H	COOEt
	n-Pent.	H	CHO
	$-(CH_2)_5-$		COMe
	$-(CH_2)_5-$		CHO

Fig 4

XXIV and **XXV** (R = NEt$_2$) respectively, and in a subsequent investigation[43] Yamazaki developed a more general synthesis of the cobalt complexes in which crystals of CpCo(PPh$_3$)(R^{1}C≡CR^2) (R^1 = Ph, R^2 = Ph, CO$_2$Me, CN, R^1 = CO$_2$Me) are heated above the melting point for a few minutes under a nitrogen atmosphere. It was originally suggested[44] that carbyne complexes produced in this manner might be intermediates in acetylene metathesis reactions (eq 19) but Vollhardt has demonstrated that complexes related to

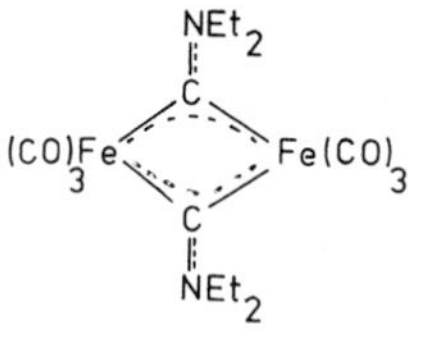

XXIV 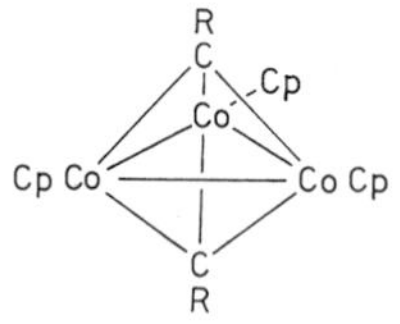   XXV

XXV are inactive in promoting metathesis and that bis-acetylene and cyclobutadiene complexes are more plausible intermediates[45] (see Section 3.5.11). However this does not exclude the possibility that mononuclear carbyne complexes may promote acetylene metathesis, since carbynes bonded to a single metal will almost certainly exhibit quite different chemical properties from those of the bridging ligand in **XXV**.

$$2\ R^1C{\equiv}CR^2 \underset{-M}{\overset{M}{\rightleftharpoons}} \text{Intermediate} \underset{M}{\overset{-M}{\rightleftharpoons}} R^1C{\equiv}CR^1 + R^2C{\equiv}CR^2 \qquad (19)$$

3.3 Reactions with nucleophiles and electrophiles

Reactions of nucleophiles and electrophiles with coordinated acetylenes frequently result in the formation of organometallic products containing a vinyl group. It has been suggested that it is possible to distinguish between two possible reaction pathways from the stereochemistry of the vinyl ligand;[7] (a) attack by the nucleophile (electrophile) X at the alkyne directly to give a _trans_ vinyl product, (eq 20a); (b) attack at the metal followed by migration of X on to the alkyne to give a _cis_ vinyl product (eq 20b). The latter process involves insertion of a

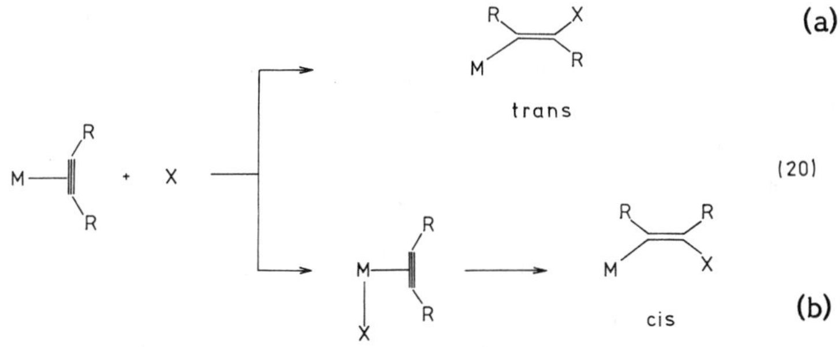

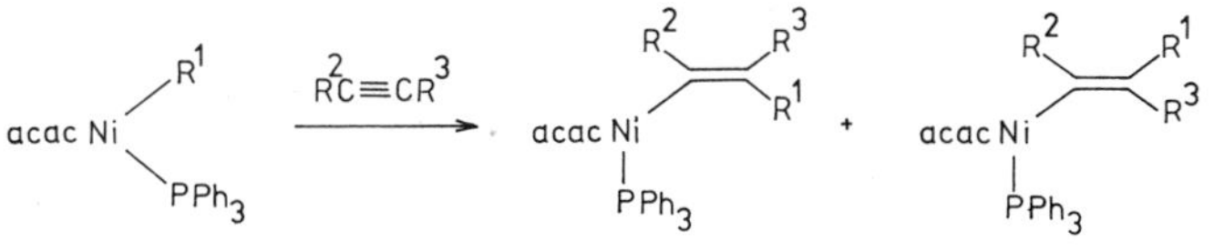

coordinated acetylene into the M-X bond and will be discussed in more detail in Section 3.4. As will become apparent the stereochemistry of insertion reactions is affected by a variety of factors. Consequently it must be stressed that it is unwise to draw firm conclusions about the mechanism of the insertion process purely on the basis of the stereochemistry of the vinyl group. Bergman's report concerning the insertion reactions of Ni(acac)Me(PPh$_3$) with acetylenes[46a] is particularly revealing in this respect and as eq 21 shows both _cis_ and _trans_ vinyl complexes can be isolated. With $R^2 = R^3 =$ Ph the _trans_ isomer is obtained according to kinetic control and undergoes isomerization to the thermodynamically favored _cis_ isomer on standing in solution. The same kinetic product, _trans_, is obtained from the reaction of Ni(acac)Ph(PPh$_3$) with PhC≡CMe, i.e. _cis_ addition occurs in this case. Thus the observation of a given stereochemical mode of addition, even when the observed complex is found to be the kinetic product of the reaction, does not necessarily mean that the crucial insertion step proceeds with that stereochemistry.

It has also become apparent that acetylenes in the presence of metal complexes are capable of forming η^2-vinyl complexes in which both carbon atoms of the vinyl ligand are bonded to the metal. These have been obtained either from insertion reactions[46b] or from addition of nucleophiles to coordinated acetylenes.[14b,49a] It has subsequently been pointed out that ring opening of η^2-vinyls could lead to either _cis_ or _trans_ η^1-vinylic products, in addition to which η^2-vinyls provide a mechanism for the isomerization of _cis_ and _trans_ η^1-vinyl complexes (eq 22).[49a] Consequently, unless the intermediacy of complexes containing the η^2-bonded form can be rigorously excluded, it is not possible to interpret the mechanism of both insertion and addition reactions of alkynes in terms of the stereochemistry of the η^1-vinyl product.

(22)

3.3.1 Reactions with nucleophiles

Uncoordinated acetylenes do not react readily with nucleophiles but reactivity can be induced by coordination to metals, particularly those in higher oxidation states or which bear a positive charge. In such circumstances acetylenes are found to react with both negatively charged and uncharged nucleophiles in reactions which we will find convenient to classify by the identity of the attacking agent.

3.3.1.1 Reactions with hydride ion:- Cationic molybdenum acetylene complexes **XXVI** undergo novel reactions with sodium borohydride, the simplest of which is hydride addition to give initially a vinyl derivative **XXVII** (eq 23).[47a] In the absence of free P(OMe)$_3$ a novel η^2-vinyl species is obtained, a result which was interpreted in terms of initial hydride attack at

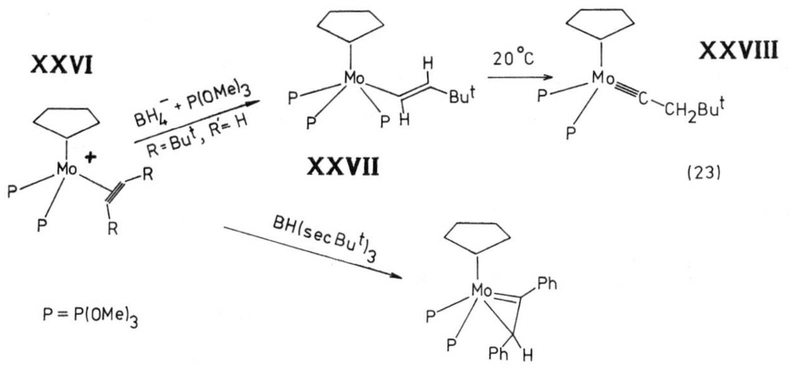

$$P = P(OMe)_3$$

the metal followed by <u>cis</u> insertion of the alkyne and subsequent $\eta^1 \longrightarrow \eta^2$ rearrangement of the η^1-vinylic product.[47b] However, related η^2-vinyl complexes have been obtained, via apparently direct nucleophilic attack at coordinated alkynes (Section 3.3.1.3, eq 28) and such a mechanism may be involved in this case also. It is interesting to note that subsequent rearrangement of the η^1-vinyl complex **XXVII** is observed to give the carbyne complex **XXVIII** as a result of a 1,2 hydrogen shift which, in view of the observed inhibition by free $P(OMe)_3$, presumably proceeds via phosphite dissociation. The but-2-yne complex **XXIX** reacts differently with $NaBH_4$[47a] giving an <u>anti</u>-crotyl derivative **XXX** which exists at room temperature in two forms, <u>exo</u> and <u>endo</u>. The proposed mechanism (Fig. 5) involves β-hydride elimination from an initially formed vinyl complex, the resulting allene-hydride intermediate transferring the hydride ligand to the central carbon of the allene to give the crotyl complex. Addition of hydride

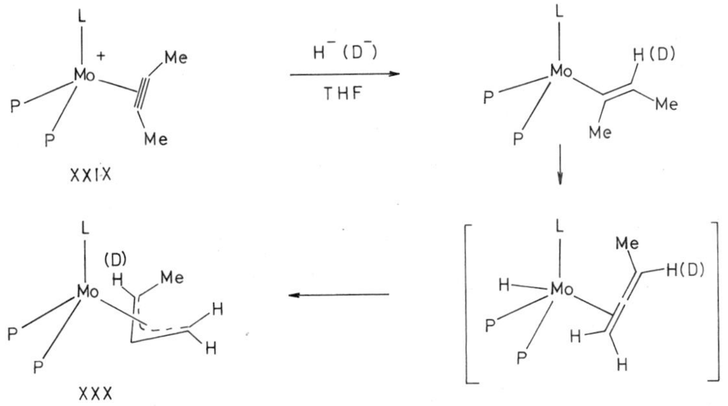

Fig 5 $P = P(OMe)_3$
          $L = \eta^5\text{-}$ indenyl

ion to η^2-acetylene-iron cations $[CpFe(CO)L(RC\dot{=}CR)]^+$ (L = CO, P(OPh)$_3$ or PPh$_3$; R = Me, Et or Ph) also gives vinyl complexes but unlike their molybdenum analogues **XXVII** they do not exhibit any tendency to undergo isomerization via β-hydride transfer.[48]

3.3.1.2 Reactions with carbon donors:- Xray diffraction studies have established structure **XXXI** for complexes which result from addition of the cyclopentadienyl anion to coordinatively unsaturated bis-hexafluorobut-2-yne derivatives **III** (M = Mo or W).[49b] This may indicate that initial attack by C$_5$H$_5^-$ occurs at the metal to give an intermediate η^1-C$_5$H$_5$ complex which subsequently undergoes acetylene insertion to give the cis-vinylcyclopentadiene ligand. However, reactions illustrated in eq 28 again suggest direct attack of the nucleophile (C$_5$H$_5^-$) at an acetylenic carbon

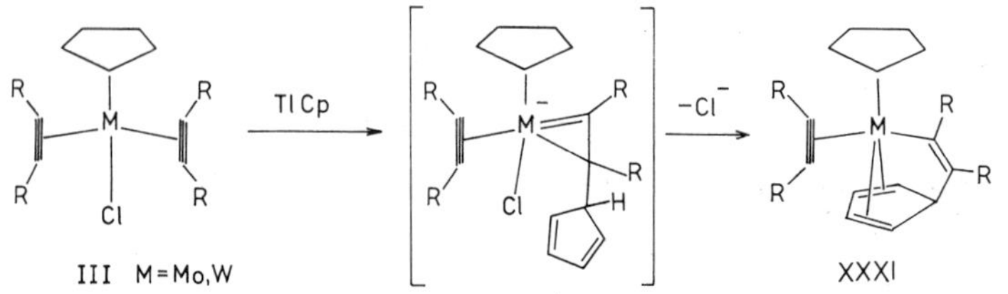

III   M=Mo,W
R = CF$_3$

(24)

XXXI

followed by rearrangement of the resulting η^2 vinyl intermediate (eq 24). Addition of tert-butyllithium to the methylpropiolate manganese derivative **XXXII** at -20 to -30°C in ether apparently produces an acetylide complex which on protolysis gives an isolable alleneylidene complex (eq 25).[50] A

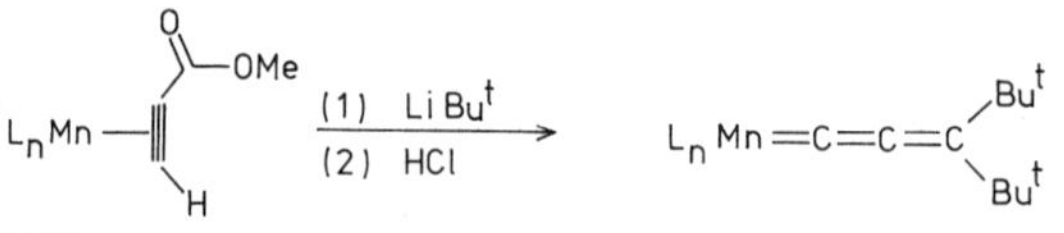

XXXII

L_n = (CO)$_2$Cp

(25)

comparison can therefore be made with the formation of related vinylidene complexes as described in Section 3.2 where acetylide intermediates are also implicated. In complete contrast the methylpropiolate ligand in the cationic iron complex $[CpFe(CO)_2(HC\dot{=}CCO_2Me)]^+$ reacts with isobutene to give lactone and hydroxyester derivatives, apparently via initial trans addition across the coordinated acetylene (Fig. 6).[51]

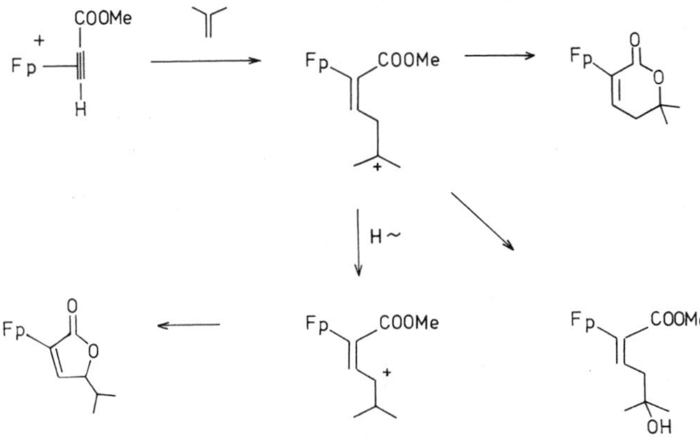

Fig 6

Acetylide complexes in which the hydrocarbon ligand is σ-bonded to one metal and π-bonded via the triple bond to another appear to be particularly reactive towards nucleophiles. Thus in reactions with electron-rich olefins, which function as a source of nucleophilic carbenes, complex **XXXIII** undergoes facile attack at the α-carbon atom of the acetylide (eq 26).[52] As will be described in the following section, reactions with nitrogen-donor nucleophiles indicate that the β-carbon also has electrophilic character.

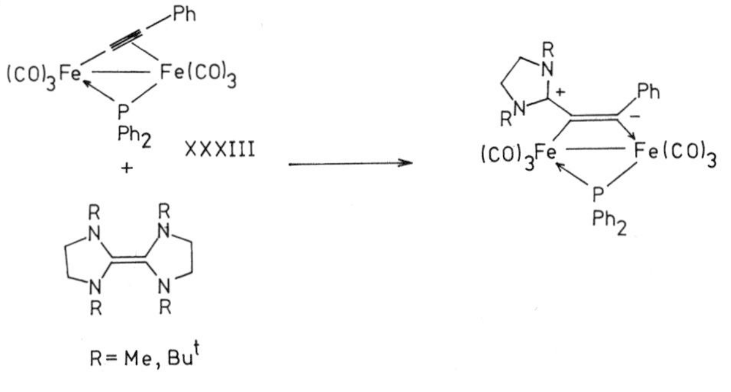

R = Me, Bu^t

3.3.1.3 Reactions with nitrogen and phosphorus donors:- Nucleophilic attack on coordinated olefins by Group V donors is generally restricted to cationic complexes, but with alkynes a number of examples of attack at neutral acetylene complexes have been observed. Thus the Pt(II) catecholato complex $Pt(C_6Cl_4O_2)(PPh_3)(PhC\equiv CPh)$ reacts with diethylamine at $0°C$ to give an enamine derivative $Pt(C_6Cl_4O_2)(PPh_3)$ $CPh=C(Ph)NEt_2$ of unknown stereochemistry,[53] although the halide

derivatives $PtX_2(PPh_3)(PhC\equiv CPh)$ (X=Cl,Br) are unreactive.   Vinylamine
complexes are also obtained from the addition of amines to Bukhovets salt
$K[PtCl_3(R^1C\equiv CR^1)]$ (R^1 = $C(OH)Me_2$) and its bromo analogue but
interestingly the _trans_ vinylamine ligand was found to result from addition
of amine to the acetylene in neutral platinum(II) intermediates (eq 27):[54]

$$R' = CMe_2OH$$
$$R'' = H, Alkyl$$
$$R''' = H, Alkyl$$

(27)

The bridging acetylene ligand in $Os_3(HC\equiv CH)(CO)_{10}$ readily adds
PMe_2Ph to give a hydrido complex assigned structure **XXXIV** or **XXXV** while
the but-2-yne analogue fails to react because of steric effects and/or the

lack of a readily transferable hydrogen atom.[55]   The hydride complex
$Os_3(C\equiv CPh)H(CO)_{10}$, containing a μ_3-η^2 bonded acetylide ligand also reacts
to give similar adducts, showing that coordination of the triple bond of an
acetylide can activate either the α- or the β-carbon atom of the ligand.
Carty has studied the reactivity of the μ_2,η^2 acetylide ligand in **XXXIII**
towards phosphites and amines and found both the α-and β-carbons to be
susceptible to nucleophilic attack (Fig. 7).[56] Thus the α-carbon is attacked
by phosphites and cyclohexylamine giving **XXXVI** and **XXXVII** respectively
while the β-position is reactive towards dicyclohexylphosphine (giving
XXXVIII), triethylamine and secondary amines (giving **XXXIX**) and
cyclohexylamine (giving **XL**).   A plausible mechanism for amine addition
involves initial nucleophilic attack at either the α- or the β-carbon of the
acetylide followed by hydrogen transfer and isomerization of the resulting
enamine.   Support for such a proposal is found in the isolation of the 1:1
adducts **XXXIX** from the reaction of secondary amines with **XXXIII** and the
observation that isomerization to **XL** occurs at higher temperatures.

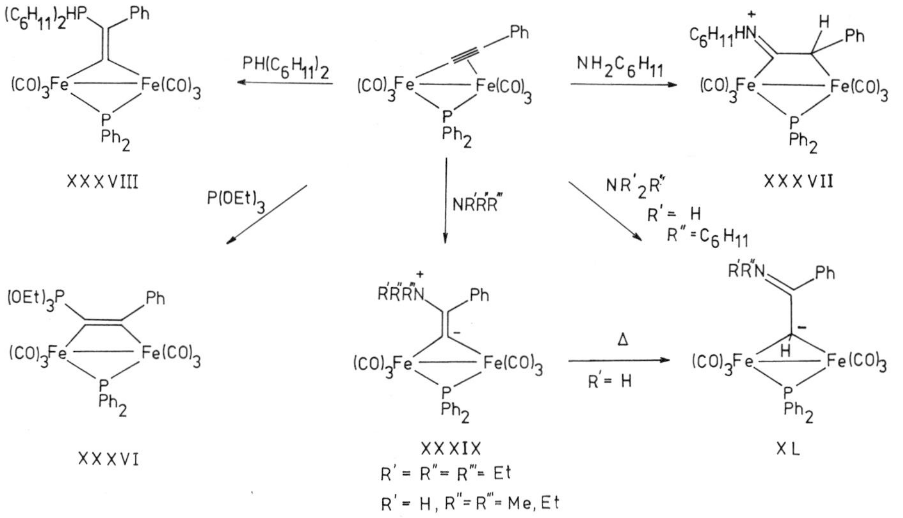

$(C_6H_{11})_2HP$ Ph

$(CO)_3Fe$————$Fe(CO)_3$

P
Ph₂

XXXVIII

$(OEt)_3P$ Ph

$(CO)_3Fe$————$Fe(CO)_3$

P
Ph₂

XXXVI

PH(C₆H₁₁)₂

P(OEt)₃

Ph

$(CO)_3Fe$————$Fe(CO)_3$

P
Ph₂

NRR'R"

$R\overset{..}{R}\overset{+}{R}N$ Ph

$(CO)_3Fe$————$Fe(CO)_3$

P
Ph₂

XXXIX
R' = R" = R"' = Et
R' = H, R"= R"'= Me, Et

NH₂C₆H₁₁

$C_6H_{11}\overset{+}{H}N$ H Ph

$(CO)_3Fe$————$Fe(CO)_3$

P
Ph₂

XXXVII

NR'₂R"
R' = H
R" = C₆H₁₁

$R\overset{..}{R'}N$ Ph

$(CO)_3Fe$———H——$Fe(CO)_3$

P
Ph₂

XL

Δ
R' = H

Fig 7

However, cyclohexylamine attacks at both α- and β-carbons to give **XXXVII** and **XL,** and the proposed mechanism may yet prove to be an over-simplification.

The reaction of coordinatively unsaturated complexes $CpMCl(CF_3C\equiv CCF_3)_2$ (M= Mo,W) with thallium(I) salts of anionic sulfur-nitrogen chelates SN⁻(SN⁻ = pyridine-2-thiolato, pyrimidine-2-thiolato or thiazoline-2-thiolato) give products shown by crystallographic studies to contain an η^2-vinyl ligand resulting from nucleophilic attack of the N-donor function of SN⁻ on an acetylenic carbon (eq 28a).[49a] This illustrates that activation of acetylenic triple bonds to attack by nucleophiles can also be promoted by coordination to a single metal atom. Since coordination normally results in activation of $CF_3C\equiv CCF_3$ towards electrophilic attack (Section 3.3.2.) it was suggested that activation of this type may result from four electron donation by the alkyne to the metal. Complexes **III** also add simple uncharged nucleophiles L (L = phosphines, phosphites, CNBu^t) to give related η^2-vinyl derivatives (eq 28b).[14c] The trispyrazolylborate anion, $[(C_3H_3N_2)_3BH]^-$, also adds to $CpMoCl(CF_3C\equiv CCF_3)_2$ but gives an unusual allyl complex **XLI** (eq 28c)[49b] as a result of addition of two pyrazole rings

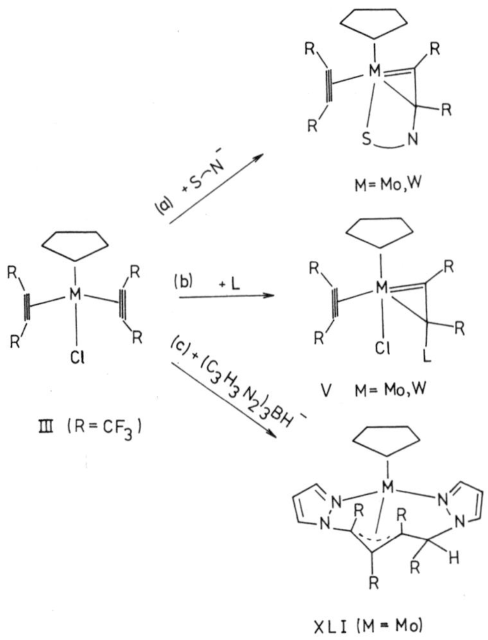

(28)

XLI (M = Mo)

and a proton to the two acetylenes which have undergone oligomerization. Conceivably this is promoted by initial nucleophilic attack at the acetylenes in view of the observed reactivity towards other nucleophiles.

3.3.1.4 Reactions with oxygen donors:- Clark and coworkers have carried out extensive studies on the interaction of acetylenes with platinum derivatives in an attempt to isolate cationic Pt(II) complexes via the following route:[57]

$$\text{trans[PtCl(Me)L}_2\text{]} + \text{R}^1\text{C}\equiv\text{CR}^2 \xrightarrow{\text{AgPF}_6} \text{trans[PtMe(R}^1\text{C}\equiv\text{CR}^2\text{)L}_2\text{]}^+\text{PF}_6^- \quad (29)$$

The products obtained, however, (Fig. 8) exhibit a distinct dependence on the nature of the alkyne and ligand L, on the nature of the solvent and on the reaction conditions. With terminal acetylenes in protic solvents carbene complexes such as XLII are obtained, while with disubstituted alkynes bearing electron withdrawing substituents vinyl ether complexes XLIII are isolated in methanol solution. These reactions have been rationalized mechanistically in terms of the intermediacy of cationic species XLIV, which are isolable in some instances. Clark and Chisholm argue that XLIV exhibits reactivity characteristic of carbonium ions and thus an alternative

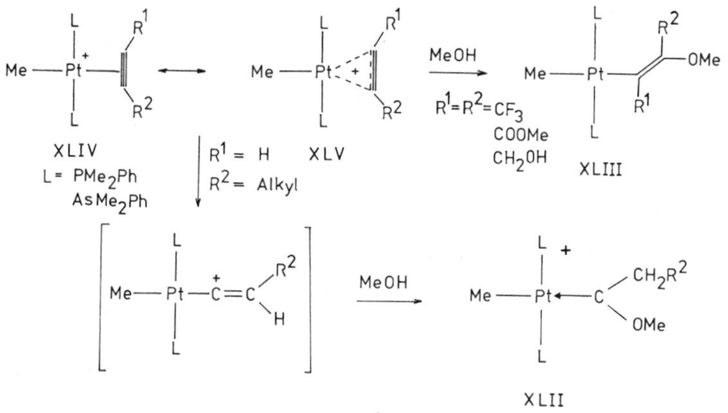

Fig 8

metal-induced carbonium structure **XLV** may be a more realistic description. Also consistent with the carbonium ion model is the formation of an alkoxycarbene platinum(IV) cation via a cationic Pt(IV) acetylene intermediate (eq 30), suggesting that cyclization proceeds via internal nucleophilic attack of the OH group on an acetylenic carbon.[58]

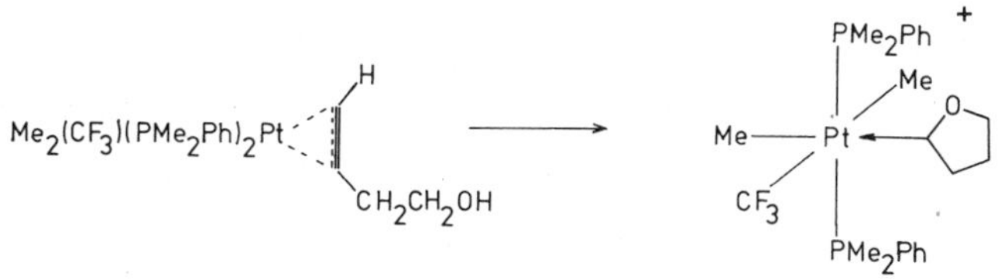

(30)

3.3.1.5 Reaction with halide ion:- Clark and Chisholm have also studied the reaction of trans-PtX(Me)L$_2$ (X=Cl, Br, I; L=PMe$_2$Ph) with RC≡CO$_2$Me, and in chloroform solution β-chlorovinyl platinum complexes trans-Pt(X)L$_2$[C(CO$_2$Me)=C(Cl)R] were isolated in the presence of a radical initiator (R = CO$_2$Me) or on addition of HCl (R = CO$_2$Me, Ph, Me, H).[59] The reaction is thought to proceed via initial formation of an acetylene complex Pt(X)MeL$_2$(RC≡CCO$_2$Me) followed by nucleophilic attack on the coordinated acetylene by Cl⁻.

3.3.2 Reactions with electrophiles

Addition of electrophiles to coordinated acetylenes frequently leads to cis and/or trans alkenes but in several cases complexes containing a vinyl ligand are isolated (eq 31). The alkenes are generally presumed to result from secondary electrophilic attack on the latter although in most instances only one product type is isolated.

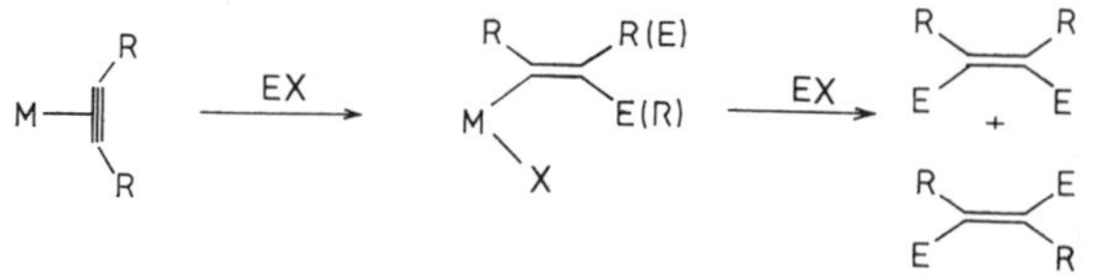

$$(31)$$

The most widely studied reactions, which involve additions of protic acids, may in certain cases proceed via protonation of the metal to give intermediate metal hydrides. Labinger and Schwartz[60] postulate such a process to account for the formation of cis-alkenes on addition of HBF_4 to the hydride complexes **XLVI** (M = Nb or Ta, R^1 and R^2 = alkyl) (Fig. 9). Protonation of the carbonyl complex **XLVII** gives the same cis-olefins, suggesting that the formation of alkenes from **XLVI** proceeds by way of vinyl intermediates. In contrast, attempted methylation of **XLVI** with methyl fluorosulfonate liberates the alkyne while simultaneously producing methane. Methylation presumably occurs at the metal and before hydrogen or methyl group migration onto the alkyne can occur reductive elimination of methane from the intermediate cationic species proceeds.

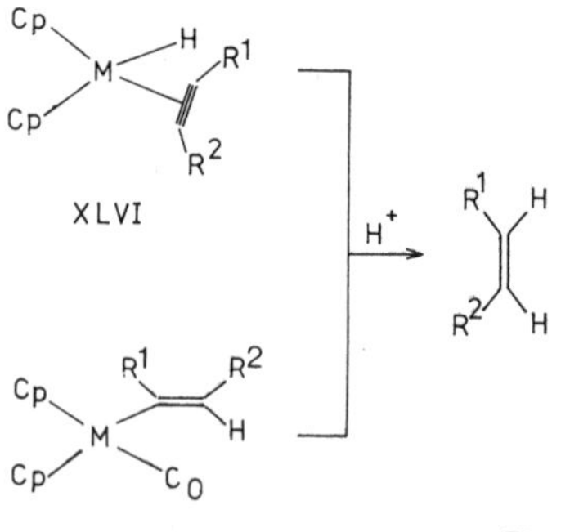

Fig. 9

Bis-cyclopentadienyl molybdenum complexes resemble their niobium and tantalum analogues in exhibiting pronounced nucleophilic character and not surprisingly $Cp_2Mo(RC\equiv CR)$ (R=H or Me) reacts readily with acids to give cis-olefins.[61] This may be of some biological significance in that a single molybdenum atom is believed to be an active site in the nitrogenase

catalyzed reduction of acetylene to ethylene.   However complexes
$Mo(HC\equiv CH)(CO)(LL)_2$ (LL = S_2CNEt_2, S_2PPr_2) give poor yields of ethylene
on protonation and cannot be considered as viable models for the nitrogenase
enzyme.[62]

Diphenylacetylene complexes $Co_2(CO)_6(\mu_2$-$PhC\equiv CPh)$ and $Co_4(CO)_{10}$
(μ_4-$PhC\equiv CPh)$ are readily pronotated in acidic methanol to give _trans_-
stilbene in 15 and 89% yield respectively.[1,73]   In contrast monosubstituted
acetylene derivatives $Co_2(CO)_6(\mu_2$-$RC\equiv CH)$, on protonation with mineral
acids in refluxing methanol (eq 32), give methinyltricobalt nonacarbonyls
XLVIII which can also be obtained from the direct reaction of $Co_2(CO)_8$ or
$Co_4(CO)_{12}$ with terminal alkynes.   In the latter reactions the acetylene
apparently acts as the acid towards the initially formed acetylene
complex.[63]

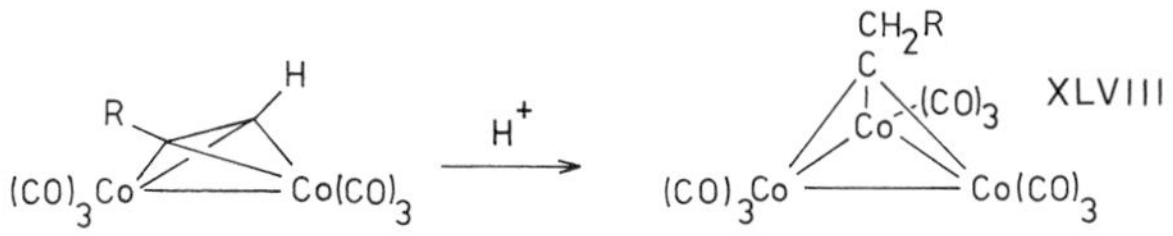

In view of the existence of a large range of simple platinum acetylene
complexes $PtL_2(RC\equiv CR)$ it is not surprising to find a variety of reports
concerning their behavior towards electrophiles.   Triphenylphosphine
derivatives for example react with one mole of acid HX (X = Cl or CO_2CF_3)
to give isolable vinyl complexes (eq 33) in which _cis_ addition of the proton
has occurred.[64]   Terminal acetylenes (R = H) undergo protonation on the
unsubstituted carbon in a Markovnikoff manner to give related products.
However, analogous but-2-yne complexes react with a variety of acids to
give ultimately both _cis_- and _trans_-butenes (eq 34).[65]   Except with the thio
acids X = MeCOS or PhCOS where ca 30% _trans_ but-2-ene was obtained, the
cis olefin generally proved to be the main product.   Moreover,

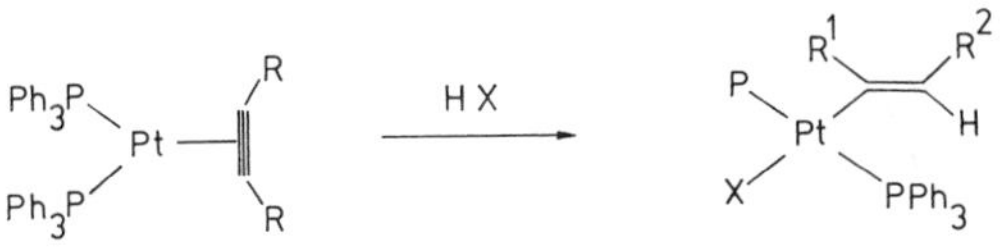

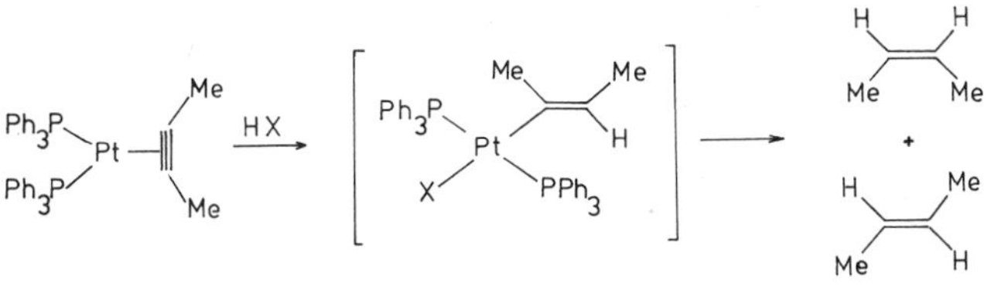

$Pt(PPh_3)_2(PhC=CPh)$ can be protonated with excess acid to give exclusively trans-stilbene,[66] while $IrCl_2(Me_2SO)_3CPh=C(H)Ph$ releases the cis form in 90% yield in methanolic HCl.[67] It would appear from these reports that retention of configuration is not necessarily observed on conversion of vinyl intermediates into free olefins.

It is particularly noticeable that although coordinated acetylenes bearing electron withdrawing substituents are readily protonated by acids, in all cases studied, e.g. $Os(CO)_2[P(OMe)_3]_2CF_3C\equiv CCF_3$,[68] $RhCl(CF_3C\equiv CCF_3)(PPh_3)_2$,[69] $IrCl(CO)(CF_3C\equiv CCF_3)(PPh_3)_2$,[69] $Pt(NCC\equiv CCN)(PPh_3)$[70] and $Pt(CF_3C\equiv CCF_3)L_2$[69] (L = tertiary phosphine), cis vinyl products are isolated which prove to be resistant to further protonolysis. This may indicate a deactivating effect of the electronegative substituents on the metal-vinyl bond towards electrophiles.

In general, relatively strong acids are required to protonate coordinated acetylenes, even when electron-withdrawing substituents are present. Consequently the ability of such weak acids as p-cresol, thiophenol, water and primarily alcohols to convert the η^2-cyclohexyne complex **XLIX** into η^1-cyclohexenyl derivatives (eq 35) requires comment.[71] The analogous

$$\textbf{XLIX} \qquad \qquad \xrightarrow{\text{HX}} \qquad \qquad \textbf{L} \qquad (35)$$

cycloheptyne complex is less reactive, suggesting that the release of ring strain is responsible for the facility with which protonation occurs. Methyl ketones also react with **XLIX** to give β-oxoalkylplatinum(II) complexes $PtCH_2C(O)R(C_6H_9)(dppe)$ but since traces of water were found to be necessary for reaction to occur the hydroxo complex **L** (X = OH) is probably the active species in such cases. Traces of water are also responsible for the transformation of $IrCl(PPh_3)_2(CF_3C\equiv CCF_3)$ into Vaska's complex $Ir(Cl)CO(PPh_3)_2$, a reaction which is thought to proceed via initial protonation of the acetylene to give a cationic vinyl intermediate.[72]

To conclude this survey of protonation reactions of coordinated acetylenes we draw attention to the one reported case in which protonation leads to an olefin complex (eq 36):[35]

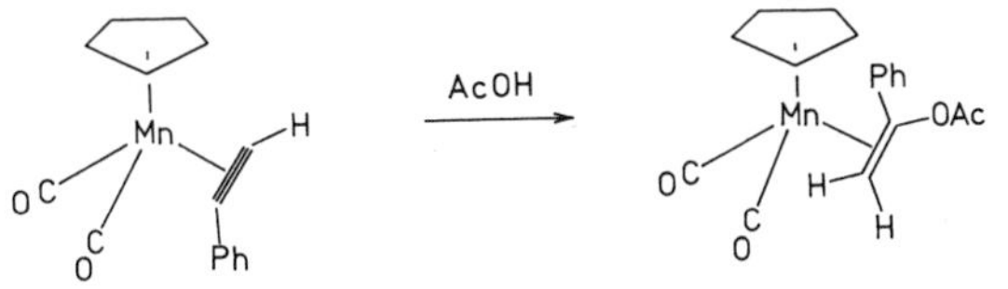

$$\text{(36)}$$

This contrasts with the mode of acetic acid addition to platinum complexes $Pt(RC\equiv CR)(PPh_3)_2$, where acetate becomes attached to the metal and Pt(II) vinyl complexes $PtOAc(CR=CHR)(PPh_3)_2$ are obtained.[69] The difference may be due to the tendency of the $CpMn(CO)_2$ moiety to bind to neutral two-electron donors whereas platinum(0) complexes exhibit a strong tendency to undergo oxidation to platinum(II).

An exception to this rule is found on addition of the mild brominating agent $C_5H_5NHBr_3$ to $Pt(NCC\equiv CCN)PPh_3)_2$ since an olefinic product $Pt[NC(Br)C=C(Br)CN](PPh_3)_2$ was apparently isolated although reaction with bromine resulted in the formation of $PtBr_2(PPh_3)_2$.[70] Bromine also displaces acetylenes from $Co_4(CO)_{10}(\mu_4\text{-}PhC\equiv CPh)$ (eq 37), and from $MoCO(\mu^2\text{-}PhC\equiv CPh)(\mu^4\text{-}C_4Ph_4)$ (eq 38), although in these cases the brominated product may result from oxidative release of diphenylacetylene followed by bromination of the unbound ligand.[73-4]

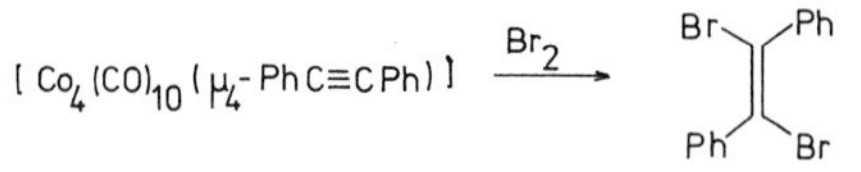

$$\text{(37)}$$

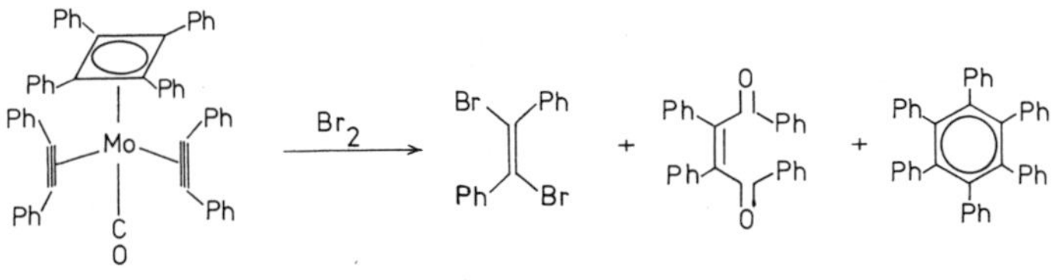

$$\text{(38)}$$

In a comparison of reactions of methyl iodide with acyclic (3-hexyne) and cyclic (cyclohexyne and cycloheptyne) platinum complexes $Pt(acetylene)L_2$ (L = PPh_3 or dppe), the acetylene-free product $PtI(Me)(PPh_3)_2$ was obtained with the hex-3-yne derivative whereas the cyclic derivatives gave mainly 2-methyl-cycloalkenylplatinum(II) complexes LI (X=Me) formed by electrophilic attack on the metal-alkyne bond.[75] With iodine analogous products were obtained, i.e. $PtI_2(PPh_3)_2$ and LI (X = I), whereas with CF_3I, $Pt(C_6H_8)(PPh_3)_2$ and $Pt(C_7H_{10})(PPh_3)_2$ gave LII and

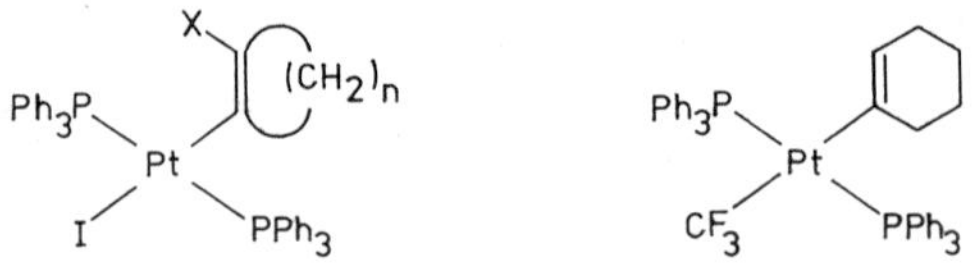

LI (n = 4,5)   LII

$Pt(CF_3)I(PPh_3)_2$ respectively. The isolation of an intermediate complex $PtI(Me)C_6H_8(PPh_3)_2$ from the reaction of MeI with $Pt(C_6H_8)(PPh_3)_2$ indicates that at least in this case oxidative addition of methyl iodide to the metal precedes methyl group migration onto the coordinated alkyne (eq 39). An analogous mechanism could also account for the formation of cis-vinyl complexes trans-$PtL_2(X)$ $[C(CF_3)=C(CF_3)HgX]$ (X=Cl, Br; L = PMe_2Ph or PPh_3) on addition of mercuric halides HgX_2 to platinum hexafluorobut-2-yne derivatives.[69] Extensive π-back-donation to the acetylene, induced by the CF_3 substituents may, however, render direct electrophilic attack at the acetylene equally plausible.

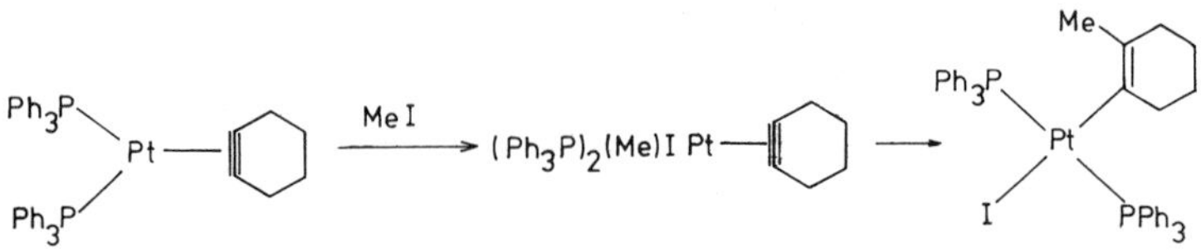

3.4   Insertion reactions

The ligand migration step in the reactions just described (more commonly referred to as an insertion reaction) is a fundamental process in organometallic chemistry and in the context of acetylene transition metal chemistry is found in several different but related guises. The most commonly observed reaction leads to linear or cyclo-oligomerization products via a sequence of consecutive acetylene insertions. However, in this section we shall deal with simple insertion reactions leading to cis and/or trans vinyl complexes (eq 40).

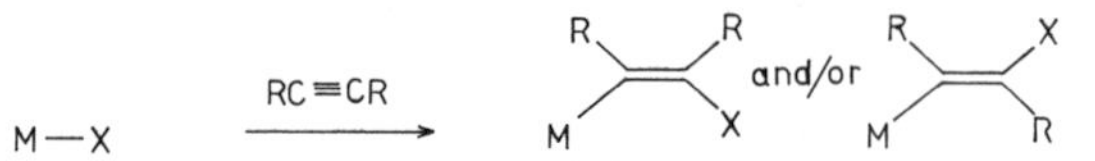

Nakamura and Otsuka[9,76] have discussed the relationship between the stereochemistry of the insertion reaction and the mechanism in some detail but as pointed out earlier[46] it is unwise to attempt to define mechanisms on the basis of product stereochemistry. Most of the simple insertion reactions reported to date do not fall within the scope of this review since the intermediacy of acetylene π-complexes has not been demonstrated. We shall survey only those reactions in which such species are either clearly involved or in which reasonable evidence for their participation has been presented.

3.4.1 Insertions into metal-hydrogen bonds

As illustrated in eq 41, cis insertion of dialkylacetylenes coordinated to niobium into Nb-H bonds is promoted by coordination of CO to the metal.[60] With $R^1 \neq R^2$ the reactions are highly regioselective since the metal preferentially remains bonded to the carbon atom bearing the smaller substituent. Insertion of acetylenes into the Pt-H bond of trans-[Pt H(PEt$_3$)$_2$L]$^+$ proceeds via a mechanism which is determined by the relative ease of displacement of the ligand L trans to hydrogen.[77] With ligands such as CO which are not readily displaced, insertion proceeds via a five coordinate intermediate (eq 42). However, weakly coordinating trans ligands

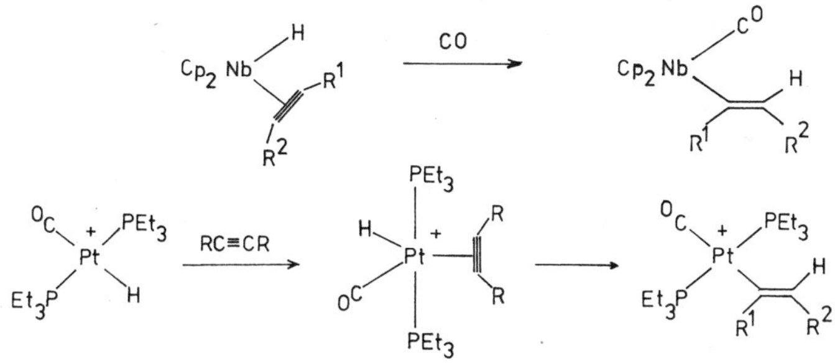

$$\text{(41)}$$

$$\text{(42)}$$

$$R^1, R^2 = \text{Alkyl, aryl}$$

such as water or acetone are readily displaced by the acetylene to give a four coordinate trans acetylene π-complex, which undergoes an insertion reaction on isomerization to the cis isomer. Subsequent recombination with a suitable ligand gives the cis or trans vinyl product (Fig. 10). With unsymmetrical acetylenes the nature of the products depends on the metal, since with [PtH(L)(PEt$_3$)$_2$]$^+$ the most electron withdrawing substituent on the acetylene (R^1) generally resides on the α carbon of the vinyl group (**LIII**) whereas with analogous Pd(II) derivatives it occupies the β-position (**LIV**).[77d] The reasons for this difference are not yet clear.

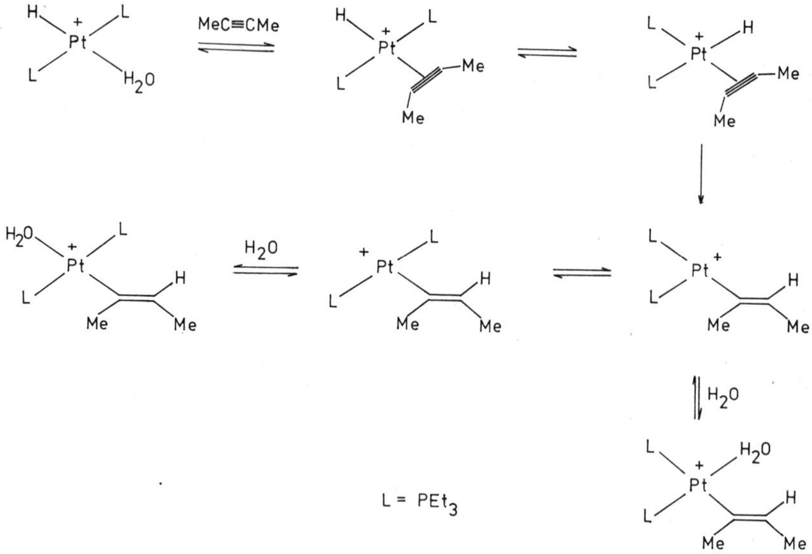

MeC≡CMe

L = PEt₃

Fig 10

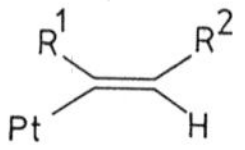

LIII

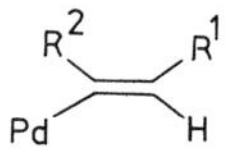

LIV

In Section 3.3.2. it was concluded that retention of configuration is not necessarily observed on acid cleavage of the metal-carbon bond in vinyl complexes, since both cis and trans olefins are obtained from cis-vinyl complexes. Similar conclusions apply to the reactions in Fig. 11, where cis

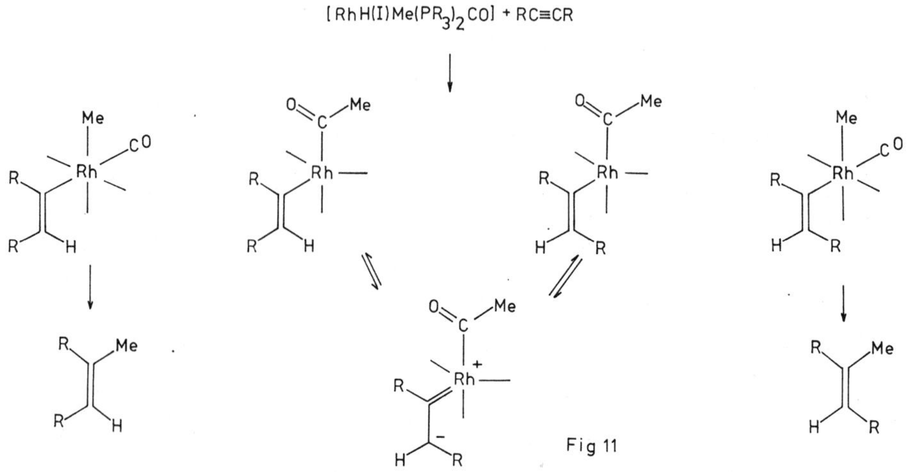

[RhH(I)Me(PR₃)₂CO] + RC≡CR

Fig 11

insertion into the rhodium hydride bond is followed by thermal elimination of cis and trans olefins.[78]   The isomerization process occurs prior to olefin elimination, possibly via dipolar intermediates in which free rotation about the C-C bond can occur.   Such a mechanism could account for the trans insertion in equation 43 which presumably involves an intermediate hydride

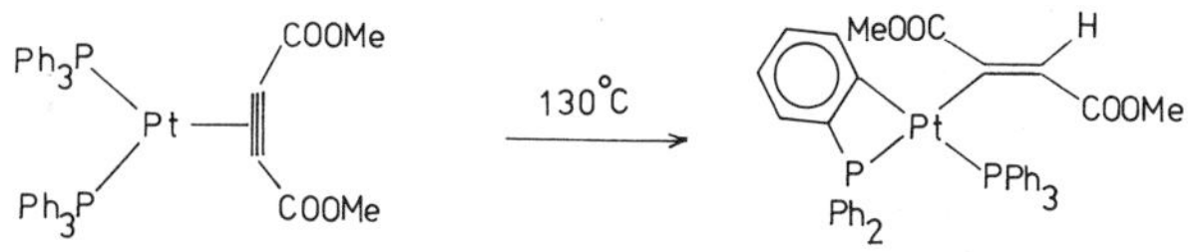

(43)

resulting from orthometalation.[79]   However, the cis vinyl complex trans-$PtCl[C(CO_2Me)=C(H)(CO_2Me)](PEt_3)_2$ does not undergo isomerization at $150°C$, which suggests that a different mechanism may be involved in this case.   Radical mechanisms are clearly implicated in the formation of trans-vinyl platinum derivatives in eq 44 where esr studies have identified the

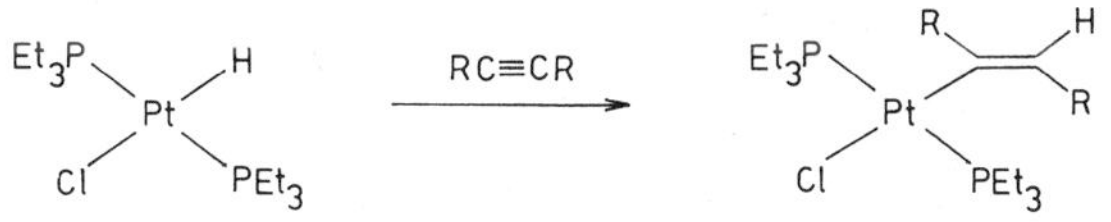

(44)

metal-based radical LV as an active species (R = CO_2Me).[80]   This abstracts a hydrogen atom from $Pt(H)Cl(PEt_3)(RC≡CR)$ to give the final product, the stereospecificity of the process arising from the fact that steric crowding in the latter renders attack by the radical LV (cis) unfavorable.

LV   trans   LV   cis

In some reactions of alkynes with metal hydrides the vinyl complex initially formed undergoes further acetylene insertion.[81]   Two independent reports[81-2] provide evidence that in certain cases the second acetylene does not simply insert into the M-C bond of the vinyl complex formed in the first step but instead apparently inserts into the C-H bond (eq 45).   To explain this unexpected result an ionic mechanism has been proposed in which the hydrogen is transferred back to the metal during the reaction (Fig. 12).

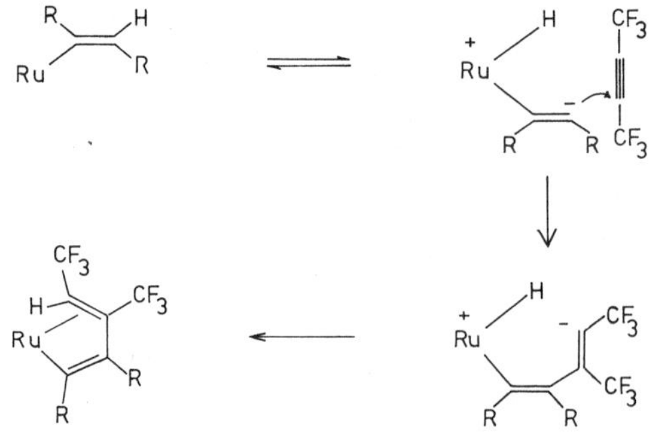

Fig 12

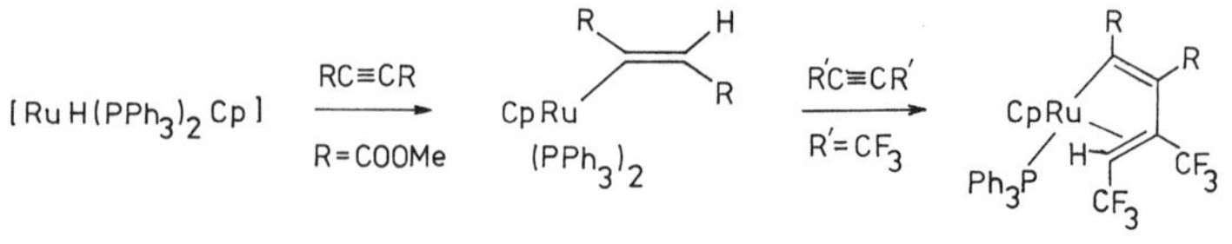

$$[RuH(PPh_3)_2Cp] \xrightarrow[R=COOMe]{RC\equiv CR} CpRu(PPh_3)_2 \xrightarrow[R'=CF_3]{R'C\equiv CR'} CpRu \quad (45)$$

3.4.2 Insertions into metal-carbon bonds

Although superficially similar, the insertion of acetylenes into metal-carbon bonds does not always exhibit close parallels with reactions involving metal-hydrogen bonds.   This is clearly illustrated by the differing stereochemistries of vinylic products **LVI** and **LVII** obtained from reactions of hexafluorobut-2-yne with MnH(CO)$_5$[83] and MnMe(CO)$_5$[84] respectively. Moreover, the reactivity of metal-carbon bonds is markedly less than that of metal-hydrogen bonds and in many cases insertion reactions are only observed with activated acetylenes such as hexafluorobut-2-yne and dimethylacetylenedicarboxylate.   Most known examples involve platinum chemistry where Clark's studies reveal that the reactivity of the Pt-C bond depends on the stability of intermediate acetylene π-complexes.

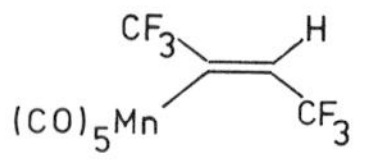

LVI 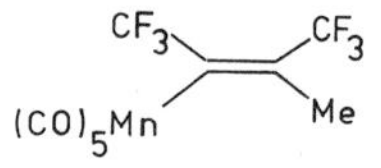   LVII

For example, the slow insertion of hexafluorobut-2-yne into the Pt-Me bond of PtX(Me)L$_2$ (X = halogen, L = tertiary phosphine, arsine or stibine) gives vinyl complexes via isolable five-coordinate acetylene π-complexes (eq 46).[85]   In contrast, the analogous five-coordinate hydrotris(pyrazolyl) borate

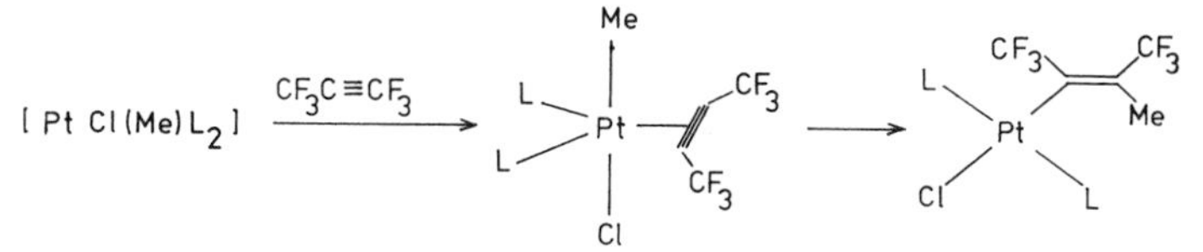

$$[\,Pt\,Cl\,(Me)\,L_2\,] \xrightarrow{\;CF_3C\equiv CF_3\;} \qquad \qquad \qquad \qquad (46)$$

complex **LVIII,** which also has the Pt-Me bond cis to the coordinated acetylene, is extremely stable and does not undergo insertion up to the decomposition point.[86] Diorganobis(pyrazolyl)borate acetylene complexes **LIX,** R^1 = Ph, R^2 = Ph or Me are also inert towards acetylene insertion whereas dimethylacetylenedicarboxylate and hexafluorobut-2-yne do not form analogous products but give methyl vinylic insertion products directly on reaction with PtMe(bpz)$_2$(cyclooctadiene).[87] The inertness of **LVIII** is not readily explained but may be related to the difficulty with which the acetylene can rotate to lie parallel to the Pt-Me bond, the orientation necessary for insertion. It is also likely that for tris(pyrazolyl)borate complexes insertion, leading with rearrangement to a square planar vinylic product, is made more difficult by the chelate nature of the ligand.

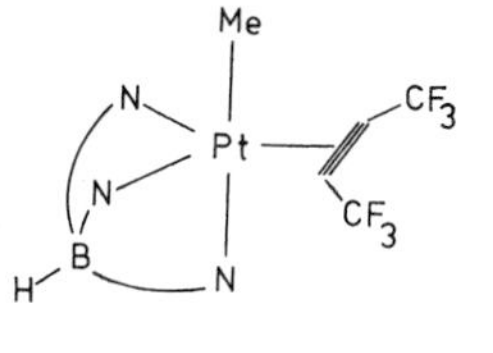

LVIII 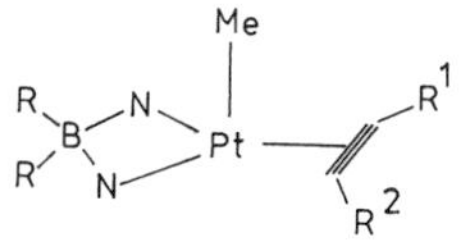   LIX

The formation of complexes containing a vinyl ketone ligand is frequently observed in insertion reactions of alkyl metal carbonyls with acetylenes. In the case of CpWMe(CO)$_3$ and HC≡CH, the vinyl ketone is formed via the reaction sequence of Fig. 13:[88]

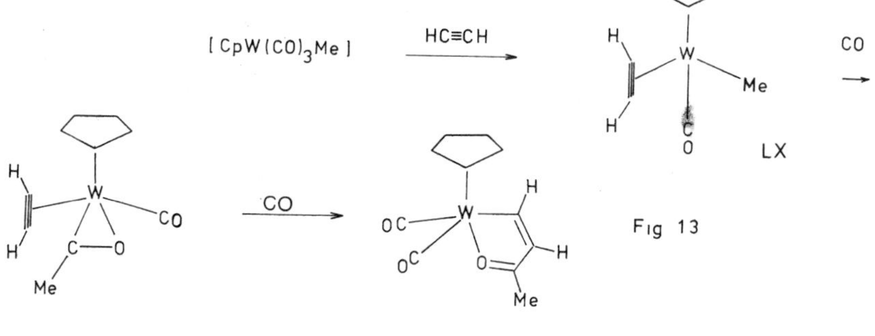

Fig 13

Coordinatively unsaturated complexes $CpM(SC_6F_5)CO(CF_3C\equiv CCF_3)$ related to **LX** undergo unusually facile CO substitution ($20°C$ in ether) in the presence of phosphines.[89]   In some cases (L = PEt_3, PBu^n) unstable intermediates were isolated, which exhibited spectral features consistent with the presence of an η^2-vinyl ligand. This was originally thought to result from insertion of the alkyne into the $M-SC_6F_5$ bond (path (a), Fig. 14) but the subsequent demonstration that η^2-vinyl complexes are formed by addition of phosphines to an acetylenic carbon in related bisalkyne derivatives[14] (Section 3.3.1.3, eq 28b), suggests that CO substitution proceeds via an η^2-vinyl derivative with the alternative structure **LXI** (path b).

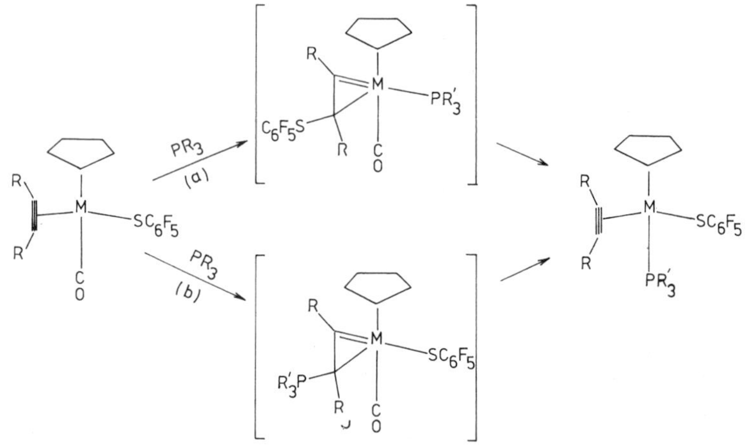

LXI

Fig 14 ($M=Mo, W; R = CF_3$)

3.5   Cyclization reactions

Although a wide range of complexes has been isolated in which one, or less commonly two or three acetylenes, are π-bonded to a metal, in many cases acetylenes undergo cyclooligomerization in the presence of metal complexes to give metallacyclopentadiene, cyclobutadiene or benzene derivatives.   Alternatively, incorporation of other ligands such as carbon monoxide can occur to give cyclopentadienone, quinone or tropone complexes. The remainder of this Chapter is concerned with reactions in which coordinated acetylenes undergo cyclization either with another acetylene or with other ligands. In many of these reactions metallacycles

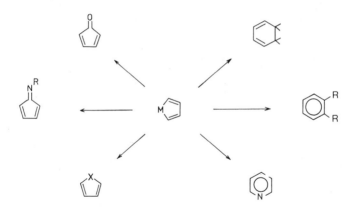

Fig 15

play an important mechanistic role, acting as precursors to a variety of
cyclic products (Fig. 15). In some instances the cyclic ligand remains
coordinated to the metal, while in others only the free ligand is isolated,
either from stoichiometric or from catalytic reactions. The following
survey includes both stoichiometric and catalytic processes where metal-
acetylene complexes are involved while the classification of reaction type is
based on the substrate undergoing cyclization with the coordinated
acetylene.

3.5.1 Reactions with acetylenes

Cyclooligomerization of acetylenes in the absence of metal complexes
is subject to the conservation of orbital symmetry if the reactions proceed
in a concerted manner, and of the three processes of interest here, (eqs 47–
49), only benzene formation is thermally allowed in the Woodward–Hoffmann

$$\| + \| \longrightarrow \square \qquad (47)$$

$$\diagdown \diagup + \diagup\diagup \longrightarrow \hexagon \qquad (48)$$

$$\| + \| \longrightarrow \octagon \qquad (49)$$

sense.[90] Despite this, dimerization reactions of acetylenes to produce cyclobutadiene complexes are well known[91] while the Reppe, nickel catalyzed, synthesis of cyclooctatetraene provides us with one of the more spectacular demonstrations of the ability of metals to catalyze forbidden pericyclic reactions.[92]

Mango proposed an ingenious explanation for metal-promoted forbidden transformations, involving direct participation of metal d orbitals in the cyclization process, but although olefin dimerization in the presence of metals can in theory become "allowed", the presence of an extra set of π-orbitals precludes this with acetylenes.[93] Consequently we expect the formally forbidden thermal cyclization reactions of acetylenes to proceed by stepwise rather than concerted mechanisms and as will become evident this appears to be the case. More surprisingly it is now becoming clear that even "allowed" $\pi_{4s} + \pi_{2s}$ and $\pi_{2s} + \pi_{2s} + \pi_{2s}$ reactions also proceed via stepwise routes in metal-promoted reactions. Moreover the mechanisms of metal-promoted acetylene cyclizations vary considerably and several different routes to cyclobutadienes and benzenes have been reported. The following Section provides a representative rather than a comprehensive survey of reactions in which coordinated acetylenes are transformed into the acetylene oligomers, cyclobutadiene, benzene and cyclooctatetraene.

3.5.1.1 Acetylene dimerization:- The most commonly observed reactions of this type lead to metallacyclopentadiene and cyclobutadiene complexes[91] which in certain cases have been shown to be capable of interconversion. Cyclobutadiene complexes are isolated from reactions of metal compounds with a variety of acetylenes but in general those bearing aromatic substituents exhibit the greatest tendency to cyclize in this manner.[91,94] With metal carbonyls a competing reaction frequently involves cyclopentadienone formation which can be explained in some cases by the fact that both products have a common precursor.

A bis-acetylene complex **III** appears to fulfill this role in the reactions of acetylenes with CpM(CO)$_4$ (M = V, Nb, Ta)[91,95] and CpM(CO)$_3$X (M = Mo, W; X = Cl, Br, I)[12,96] which leads ultimately to both types of product (Fig. 16). Although a metallacyclic intermediate may be involved in the

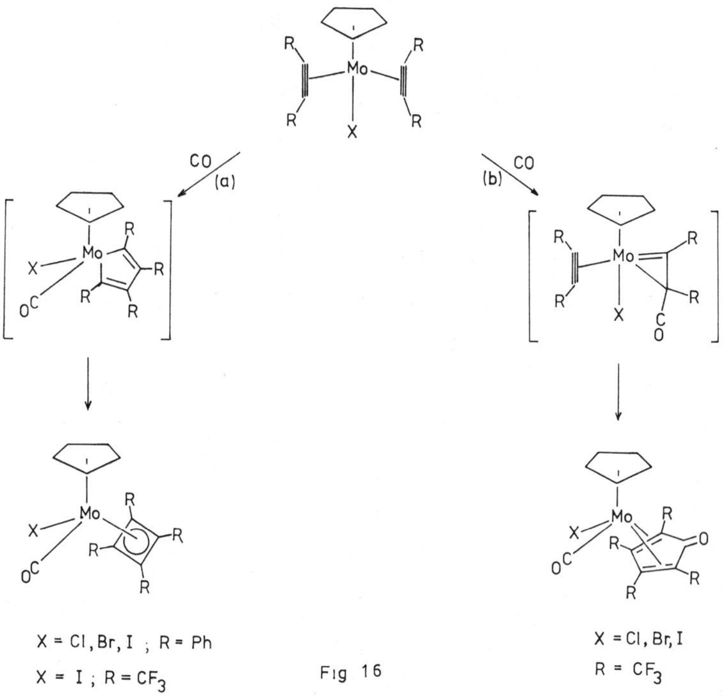

X = Cl, Br, I ; R = Ph

X = I ; R = CF$_3$ Fig 16

X = Cl, Br, I

R = CF$_3$

formation of both products from unactivated acetylenes [Fig. 16, route (a)]
reactions involving hexafluorobut-2-yne may proceed via a novel alternative
route, in view of the observed reactivity of this alkyne towards nucleophiles
when coordinated to molybdenum and tungsten.   The formation of
cyclopentadienimine derivatives structurally related to cyclopentadienones
CpMX(CO)[C$_4$(CF$_3$)$_4$CO] (M = Mo, W; X = Cl, Br, I) has been found to
involve initial attack of Bu^{t}NC on an acetylenic carbon of
CpMX(CF$_3$C≡CCF$_3$)$_2$ to give isolable η^2-vinyl complexes structurally
characterised by xray diffraction methods (eq 1, Section 2.1).   Carbon
monoxide could perhaps react similarly, to give ultimately the
cyclopentadienone products [Fig. 16, route (b)]. It is not clear whether this
route could also lead to cyclobutadiene derivatives but it may be significant
that an η^2-vinyl derivative (originally thought to be a bis-acetylene
complex) has been found to isomerize to a cyclobutadiene compound at ca
100°C, (eq 50).[14b,96]

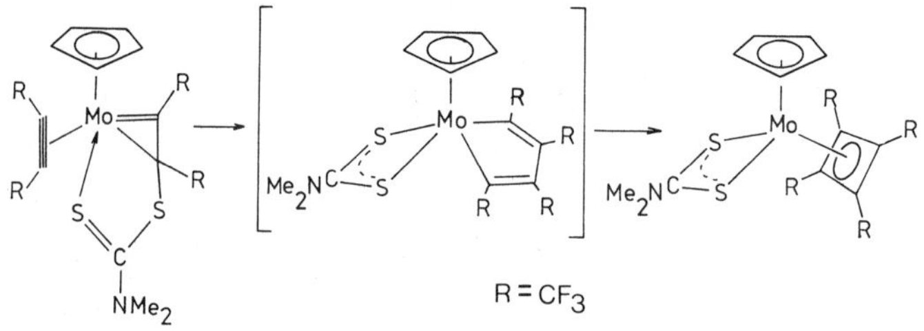

Mango has proposed an alternative theory to explain the formation of cyclobutadiene complexes via bis-acetylene intermediates such as **III**.[93] Theoretically bis-alkyne complexes can be transformed into cyclobutadiene derivatives in a concerted manner if the reaction proceeds via a dinuclear intermediate in which the acetylenes are coordinated to the two metals simultaneously. However, the reactions depicted in Fig. 17 suggest that stepwise processes involving dinuclear metallacycles may provide a lower energy pathway than Mango's concerted cyclization mechanism.[97] Alternatively, the formation of a dinuclear metallacycle may prove to be a mechanistic cul-de-sac as far as cyclobutadiene formation is concerned, since some metallacycles , e.g. $Fe_2(CO)_6C_4Ph_4$, do not react further to produce cyclobutadienes.[98]

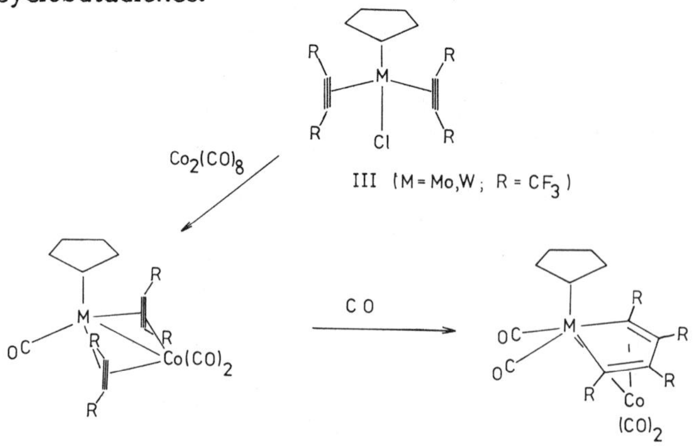

Fig 17

The reactions in Fig. 16 can be compared with the addition of acetylenes to $CpCo(CO)_2$ which also gives cyclobutadiene (**LXII**) and cyclopentadienone (**LXIII**) products. Brintzinger's studies[99] (Fig. 18) have revealed, not only the existence of both mono- and dinuclear intermediates, but also that the partial pressure of carbon monoxide in the system

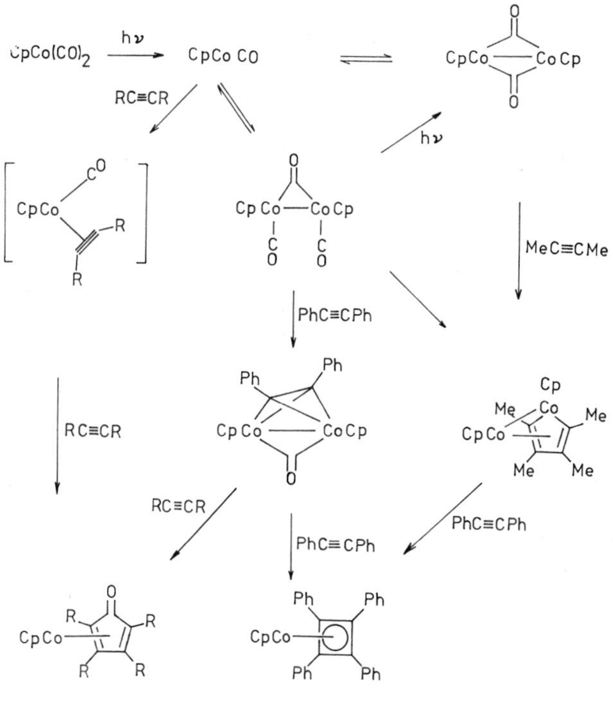

Fig 18 (R = Me, Ph)

determines the relative amounts of **LXII** and **LXIII** isolated, lower pressures resulting in preferential formation of the former. These observations are compatible with Yamazaki's report that the monoacetylene complex **LXIV** reacts with diphenylacetylene to give a metallacycle which either undergoes reductive cyclization to the cyclobutadiene or reacts with CO to give a cyclopentadienone (Fig. 19).[100]

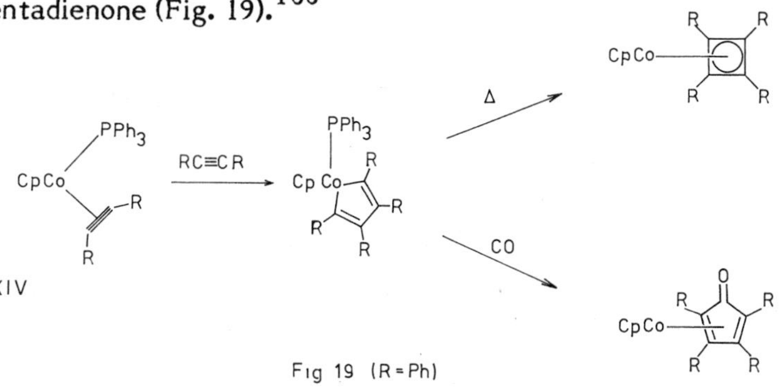

Fig 19 (R = Ph)

Yamazaki subsequently showed that complexes of type **LXIV** react with different acetylenes to produce isomeric metallacycles when $R^1 = R^2$ but of the three possible isomers only two were formed (eq 51).[101] The

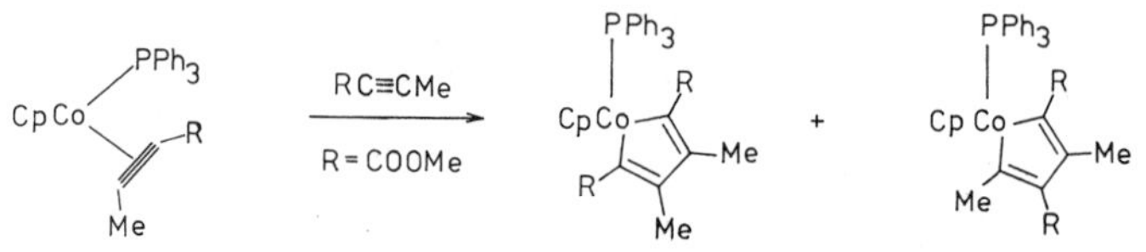

(51)

reaction of $CpCoPPh_3(PhC \equiv CCO_2Me)$ with $MeC \equiv CCO_2Me$ gave only two of the four possible isomeric products. Two mechanisms were proposed to account for these features (Fig. 20). Route (a) involves a bis-acetylene

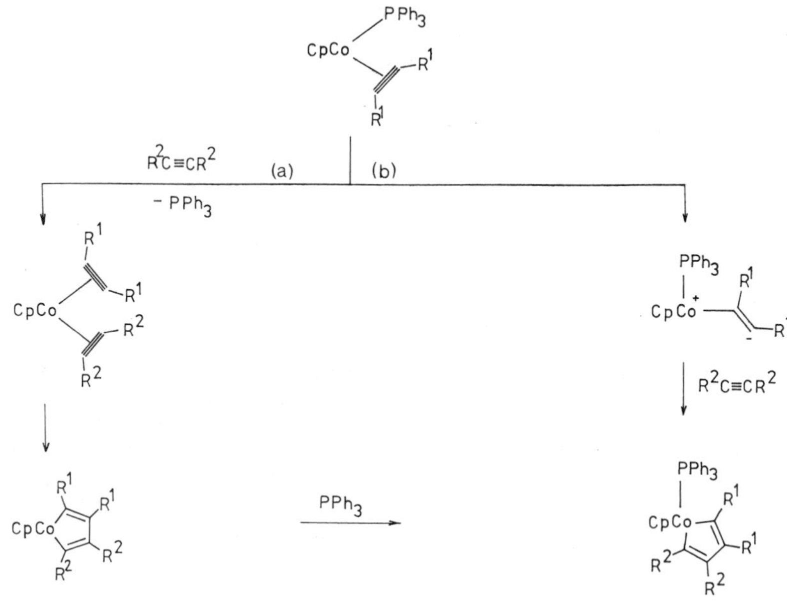

Fig 20

intermediate while route (b) proceeds via a dipolar species, (cf eq 3, Section 2). Since polar solvents produced no effect on the reaction rate, while reactions (52) and (53) gave both isomers in identical ratios, it was concluded that route (a) is the most probable. Thus the regioselectivity in such reactions is probably governed by preferential arrangements of the two acetylenes in the intermediate complex. A similar explanation accounts for the regioselectivity observed in reactions of $CpCo(CO)_2$ with unsymmetrical acetylenes, which lead to substituted cyclobutadiene and cyclopentadienone complexes of type **LXII** and **LXIII**.[102] Other interesting examples of cyclobutadiene formation are provided by the reactions of $CpCoPPh_3(R^1C\equiv CR^2)$, $(R^1, R^2 = Ph$ or $CO_2Me)$ with acetylide complexes $M-C\equiv CR^3$, $(M = Fp, CpNi(PPh_3)$; $R^3 = Ph$ or $CO_2Me)$, and ferrocenyl derivatives, $(M = Fe(\eta^5-C_5H_5)(\eta^5-C_5H_4)$, $R_3 = Ph)$ which lead to cyclobutadiene derivatives $CpCo(\eta^4-C_4R^1R^2R^3M)$, in some cases via metallacyclic intermediates.[103]

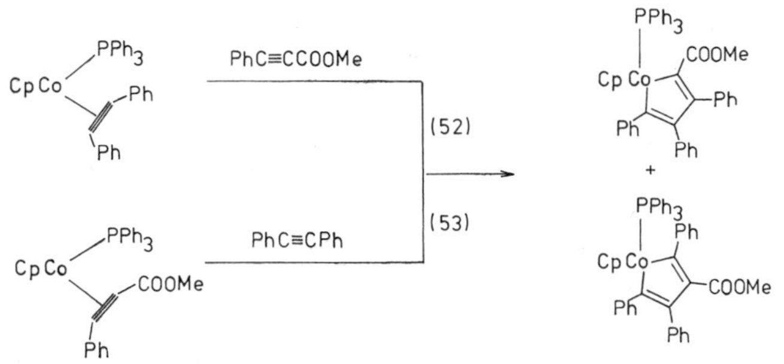

Cyclobutadiene complexes are possible intermediates in the metal-catalyzed acetylene metathesis reaction[44] (Section 3.2, eq 19) but in view of their high stability have been discounted as such by some workers. However Vollhardt has demonstrated that **LXV** undergoes thermolysis via **LXVI** and the postulated bis-alkyne intermediates **LXVII** and **LXVIII** to give acetylenes **LXIX**, **LXX** and **LXXI** (Fig. 21).[45] This clearly illustrates that some bis-acetylene to cyclobutadiene transformations are reversible and consequently that acetylene metathesis could proceed according to such a mechanism.

Reactions between acetylenes and trinuclear iron, ruthenium and osmium carbonyls $M_3(CO)_{12}$ are generally quite complex but several careful studies have resulted in elucidation of the sequence of transformations which lead to trinuclear metallacyclic derivatives.[2,104] In all cases stepwise processes are involved, the details of which differ from metal to metal, but each bears some relationship to the reaction in Fig. 22.

Fig 21 (R = Me$_3$Si)

Fig 22 (R = Ph)

A variety of other examples are known in which acetylene complexes are transformed into metallacyclopentadiene derivatives but it seems appropriate at this stage to mention only the more novel reactions. The reaction of $CF_3C \equiv CCF_3$ with $Cp_2Ni_2(\mu_2\text{-}CF_3C \equiv CCF_3)$ at ca $100^\circ C$ affords a variety of products including the novel tri- and tetranuclear metallacycles **LXXII** and **LXXIII** (Fig. 23).[105] The structure of the latter implies that two metallacyclic intermediates **LXXIV** have undergone condensation by Diels-Alder cycloaddition of one metallacycle $[NiC_4(CF_3)_4]$ to the cyclopentadienyl group of another. In contrast, the reaction of

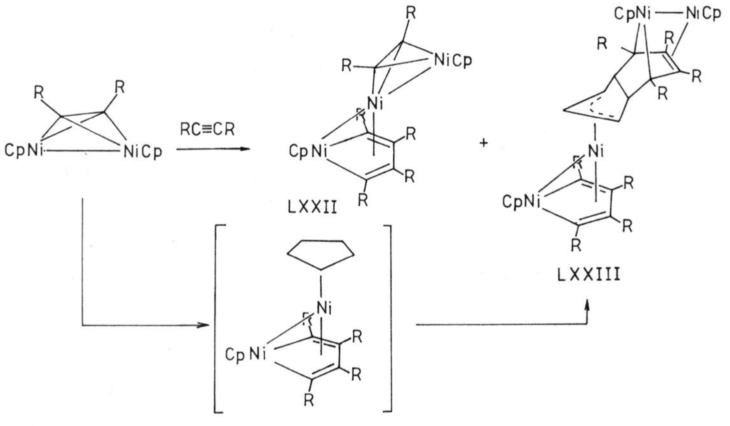

Fig 23 $(R = CF_3)$

$Pt(cod)(CF_3C{\equiv}CCF_3)$ with excess hexafluorobut-2-yne at room temperature readily affords a new type of metallacycle **LXXV** in which the ring system contains two metal atoms instead of the usual one.[106]

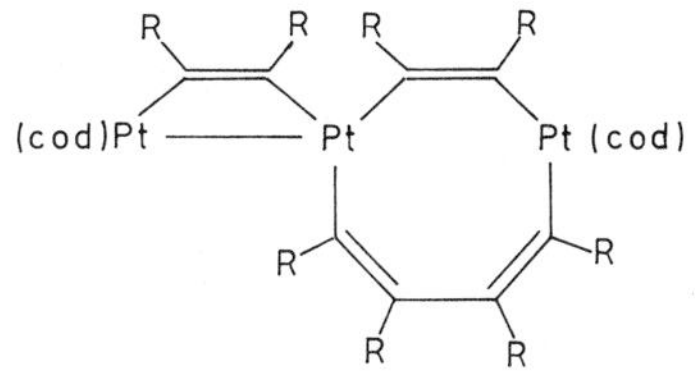

LXXV $(R = CF_3)$

Finally in this section we draw attention to the palladium promoted dimerization of acetylenes which as Fig. 24 illustrates does not involve a metallacyclic intermediate. Instead a sequence of coordination, insertion and ring closure reactions results in the formation of the cyclobutene **LXXVI** which isomerizes to the cyclobutadiene product.[6]

Fig 24

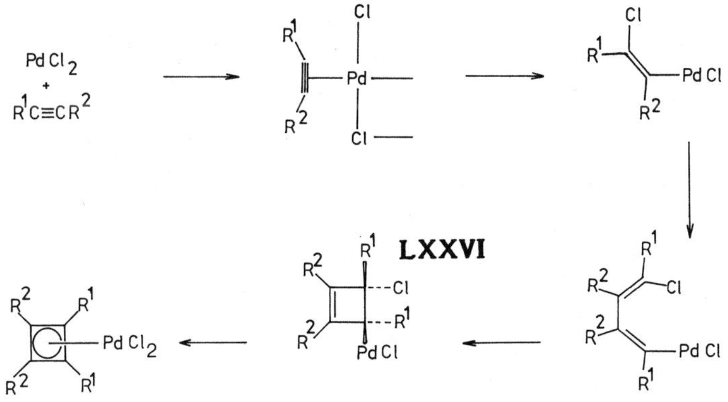

3.5.1.2 Acetylene trimerization:- Benzene π-complexes are occasionally obtained from reactions of acetylenes with metal complexes but a more frequent occurrence is the metal-catalyzed cyclotrimerization of acetylenes to give free benzenes. This is undoubtedly one of the most widely studied class of reactions in this survey and a number of reviews concerned partly or wholly with the subject have appeared.[4,5,7,107] Mono-, di- and polynuclear complexes have all been observed to promote acetylene cyclotrimerization and in many cases metallacycles again play a dominant mechanistic role.

3.5.1.2.1 Mononuclear complexes:- One of the earlier mechanistic proposals concerning metal promoted acetylene trimerization invoked cyclobutadiene complexes as reactive intermediates but this suggestion was soon refuted on a number of counts including the observed lack of reactivity of such species towards alkynes.[108] One of the few authentic cases in which a cyclobutadiene complex acts as a precursor to a benzene is illustrated in eq 54.[109] Although formally a $\pi_{4s} + \pi_{2s}$ "allowed" reaction, later work by Green[110] raises the possibility that a stepwise oxidative process proceeding via **LXXVII** is involved. The concerted cyclotrimerization of three acetylenes into a benzene is similarly "allowed" but despite this the tris-acetylene complex **LXXVIII** is unreactive under conditions which readily convert the bis-acetylene complex **LXXIX** into **LXXX** and the free benzene (eq 55).[111]

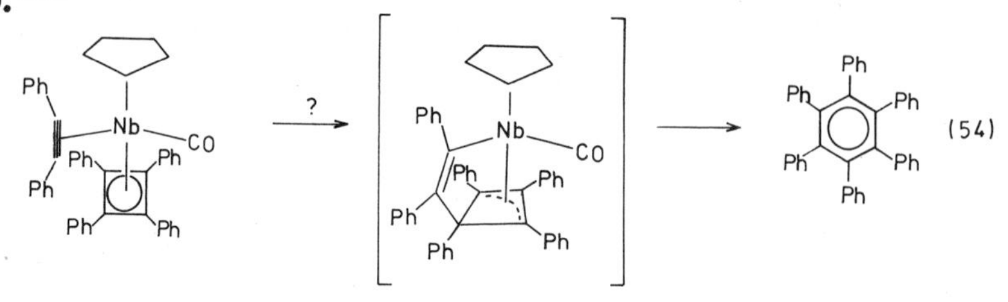

LXXVII

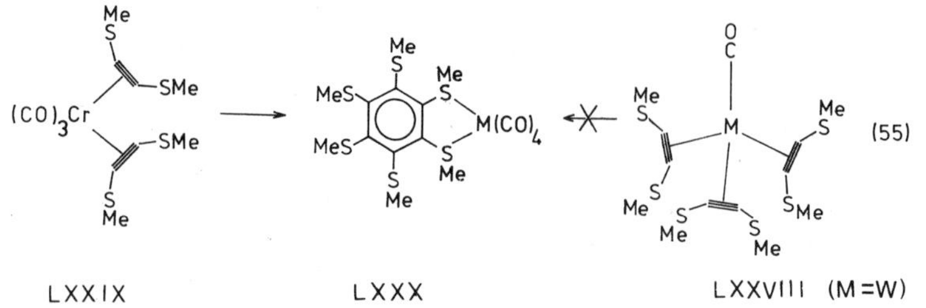

LXXIX   LXXX   LXXVIII (M=W)

Considerable effort has been expended on elucidating the role of metallacyclopentadienes in the cyclotrimerization reaction and cobalt, rhodium, iridium and palladium metallacycles have all been observed to react with acetylenes to give benzenes.[4,112] Yamazaki[113] has successfully used the ability of cobalt complexes to undergo stepwise addition of alkynes to provide routes to a range of substituted benzenes (eq 56). In this way up to three different acetylenes can be employed to produce arenes of diverse substitution patterns. Related cyclopentadienyl cobalt compounds, in particular $CpCo(CO)_2$, are useful catalysts for acetylene cyclotrimerization, and Vollhardt has developed new routes to novel cyclic organic molecules by cyclizing α,ω diynes in the presence of the latter.[114]

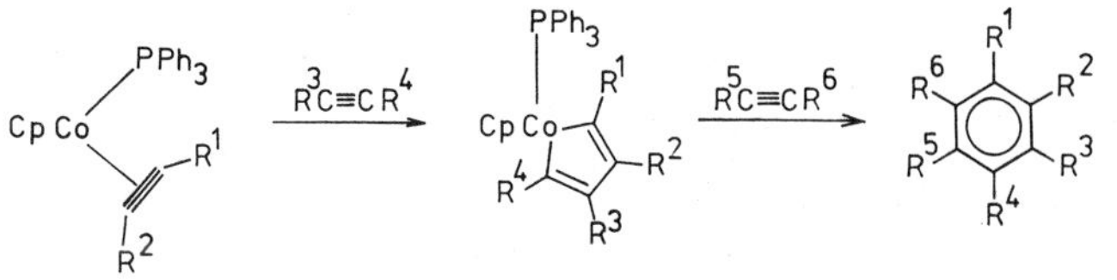

(56)

Metallacycles are again implicated in the cyclization process and of particular interest is the manner in which the metallacycle reacts with an acetylene molecule to produce the benzene. Two modes of reaction have been suggested (Fig. 25), (a) Diels-Alder addition of a coordinated acetylene to the metallacycle, and (b) insertion of a coordinated acetylene into a metal-carbon σ bond leading to a metallacycloheptatriene intermediate.[115] Reductive elimination of the benzene in either case releases the

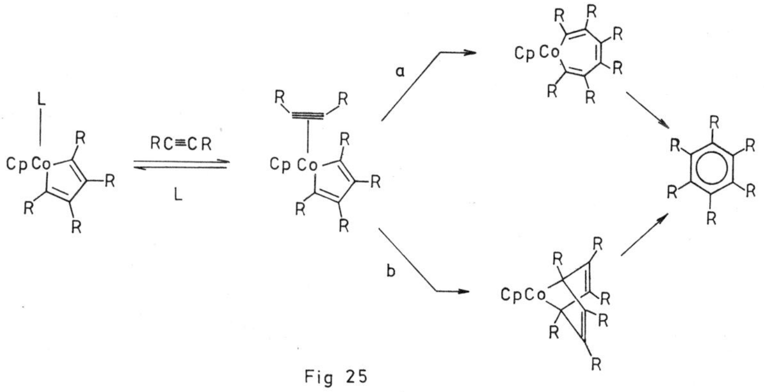

Fig 25

cyclopentadienyl cobalt moiety for further reactions. Interestingly it appears that the mechanism is critically dependent upon the nature of the alkyne.[115] Thus the metallacycle $CpCo(C_4Me_4)PPh_3$ catalyzes the cyclotrimerization of but-2-yne according to a rate law consistent with prior dissociation of the phosphine ligand, thus enabling precoordination of the acetylene to take place. In contrast, $CpCo(C_4Me_4)PMe_3$ is unreactive towards but-2-yne even at $120°C$ whereas $MeO_2CC\equiv CCO_2Me$ reacts to give 3,4,5,6-tetramethylphthalate dimethyl ester at $20°C$. The observed rate law indicates that phosphine dissociation does not occur in this case and since the reaction involves an electron rich diene and a good dienophile the evidence points to a direct Diels-Alder cycloaddition in which the acetylene does not precoordinate to the metal. Contrast $IrCl(PPh_3)_2C_4(CO_2Me)_4$, where blocking the free coordination site with CO completely inhibits cyclotrimerization of dimethyacetylenedicarboxylate by this metallacycle, showing that acetylene coordination is necessary for the reaction to proceed.[116]

Whether or not metallacycloheptatriene intermediates are formed in this and the cobalt-catalyzed but-2-yne trimerization is still a matter of conjecture and direct evidence for the involvement of such a species in any catalytic cyclotrimerization process has yet to be found. Nonetheless Stone[117] has reported the isolation of metallacycloheptatrienes (eq 57), and the subsequent demonstration that a related compound, $Ni[C_6(CF_3)_6]PCy_3$, undergoes thermolysis at $130°C$ to generate hexakis-trifluoro methylbenzene[118] indicates that this route is plausible.

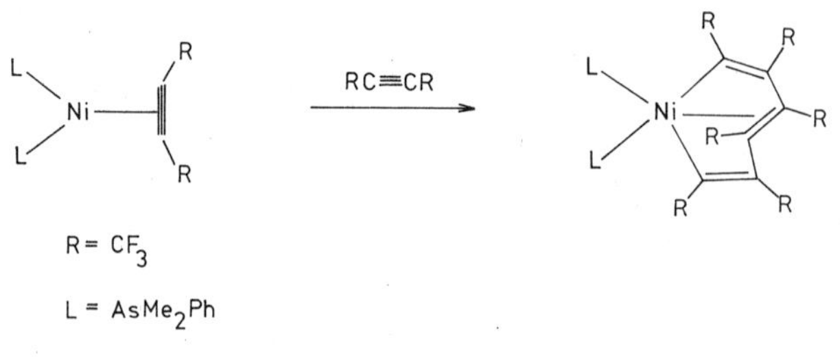

R = CF_3

L = AsMe_2Ph

As with cyclobutadiene formation, cyclotrimerization of acetylenes in the presence of palladium(II) salts does not proceed via the metallacycle route.[6] Instead a sequence of acetylene coordination and insertion reactions (cf Fig. 24) leads to **LXXXI**, which undergoes ring closure via a cyclopentadiene complex to yield free benzenes (Fig. 26).

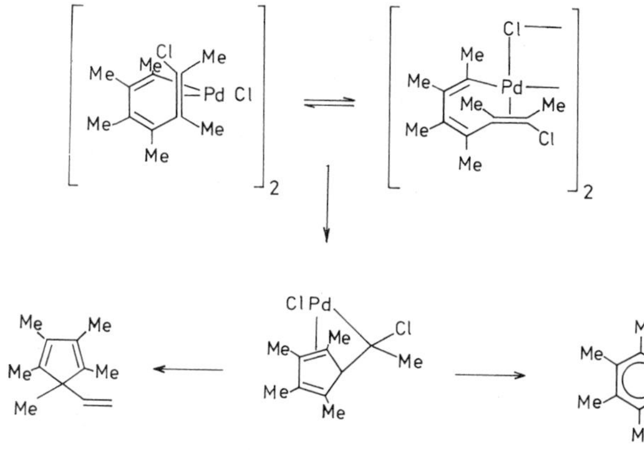

Fig 26

3.5.1.2.2 Dinuclear complexes:- The most widely studied reactions in this class involve acetylenes and dinuclear cobalt derivatives $Co_2(CO)_6$ ($RC\equiv CR$) (R=alkyl, aryl) which above ca 100°C yield free benzenes and complexes **LXXXII** exhibiting a novel flyover structure.[63] The latter undergo chemical (bromine addition) or thermal degradation to substituted benzenes and thus provide a route to arenes, such as 1,2,4-t-butylbenzene, not readily accessible by conventional synthetic methods (eq 58). Since the substituent arrangement on the benzene is related to that in the organometallic precursor it appears that benzene formation results from Co-C σ-bond cleavage and concomitant C-C bond formation.[63]

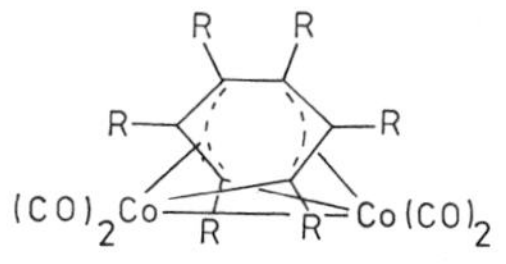

LXXXII

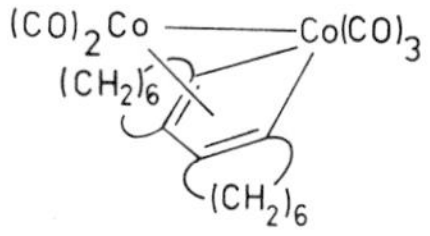

LXXXIII

$[Co_2(CO)_6(Bu^tC \equiv CH)]$
+ $\longrightarrow$ $(CO)_2Co$ —$\vert$—$\vert$—$\vert$—$\vert$—$\vert$—$\vert$— $Co(CO)_2$ $\longrightarrow$ [benzene ring with Bu^t substituents]   (58)
$Bu^tC \equiv CH$ Bu^t H Bu^t H H Bu^t

The mono-acetylene complexes $Co_2(CO)_6(RC \equiv CR)$ are also effective catalysts for acetylene cyclotrimerization but it is significant that the flyover complexes **LXXXII**[119] are not, which eliminates the possibility of their intervention in $Co_2(CO)_6(RC \equiv CR)$-catalyzed reactions. The mechanism of the catalytic process at present remains uncertain although a study of the $Co_2(CO)_8$-catalyzed cyclotrimerization of $CH_3C \equiv CCD_3$ has excluded the possibility of an intermediate cyclobutadiene complex being involved.[108]

The intervention of metallacyclopentadienes in both catalytic and stoichiometric cyclotrimerizations must again be considered in view of the isolation of **LXXXIII** from the reaction of $Co_2(CO)_8$ with cyclooctyne.[119] Both **LXXXIII** and the other reaction product, $Co_2(CO)_6\mu-C_8H_{12}$, independently catalyze the cyclotrimerization of cyclooctyne at ambient temperatures so that at least two catalytic routes to benzenes exist in cobalt carbonyl initiated cyclooctyne trimerizations. However it is not yet clear if this conclusion can be extended to reactions of other acetylenes with cobalt carbonyl.

Dinuclear complexes **LXXXIV** and **LXXXV** have been isolated from reactions of acetylenes with $CpM(CO)_2$ (M = Co, Rh, Ir),[99,120] and in some cases these react further to give a variety of products including **LXXXVI**, **LXXXVII** and free benzenes. The complexity of the reactions has in most cases prevented any detailed mechanistic relationship from being established between the acetylene complexes and the arene products. However Brintzinger has demonstrated that both **LXXXIV** and **LXXXVI** (M = Co, R = Me) are efficient cyclotrimerization catalysts, which may become deactivated by irreversible transformation into cyclobutadiene complexes $CpM(\eta^4-C_4R_4')$.

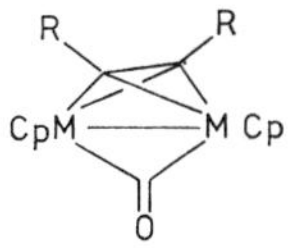

LXXXIV 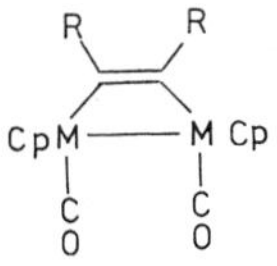   LXXXV 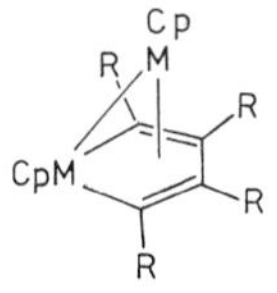   LXXXVI 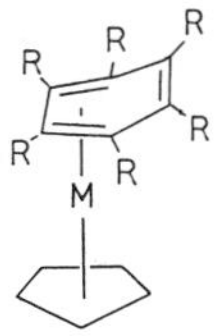   LXXXVII

3.5.1.2.3 Polynuclear complexes:- Polynuclear complexes do not normally react with alkynes to give benzene complexes and even reactions producing free arenes are not widespread. However Gambino's studies[121] indicate that $Os_3(CO)_{12}$ and diphenylacetylene react according to Fig. 27 to produce the osmiacyclopentadiene **LXXXVIII** via intermediate acetylene complexes (see ref. 104, Section 3.5.1.1.). Subsequent coordination of PhC≡CPh gives **LXXXIX** which on reaction with excess alkyne PhC≡CR (R = Ph or H) gave the free benzene via **XC** of unknown structure. The isolation of pentaphenylbenzene with R = H indicates surprisingly that it is the incoming acetylene which reacts with the metallacycle to product the arene ring rather than the alkyne already coordinated to the cluster.

3.5.1.3 Acetylene tetramerization:- The simplest products of metal promoted acetylene cyclotetramerization are cyclooctatetraenes which were first obtained catalytically by Reppe using nickel(II) salts.[92] In 1964 Schrauzer proposed a cyclization mechanism in which four acetylenes

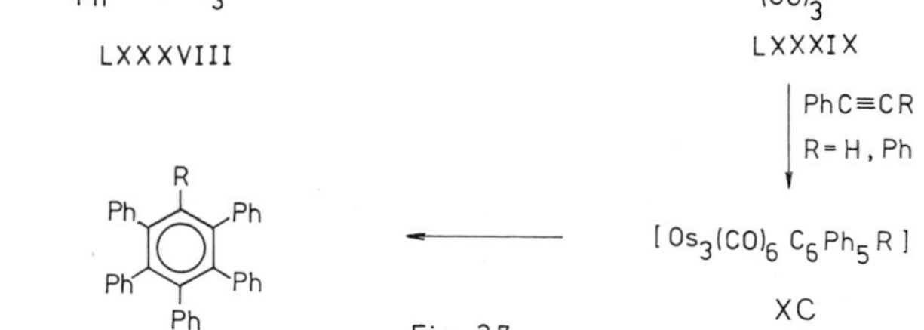

$$Os_3(CO)_{12} \quad + \quad PhC≡CPh$$

LXXXVIII   LXXXIX

PhC≡CR
R = H, Ph

$[Os_3(CO)_6 C_6 Ph_5 R]$

XC

Fig 27

simultaneously coordinated to the metal underwent concerted ring closure, but[122] this proposal can no longer be considered tenable since the reaction is formally a $\pi_{2s} + \pi_{2s} + \pi_{2s} + \pi_{2s}$ "forbidden" process. A stepwise mechanism must now be considered and the report by Hobert that nickel cyclobutadiene complexes readily release cyclooctatraenes on treatment with carbon monoxide (eq 59) suggests one possible alternative.[123]

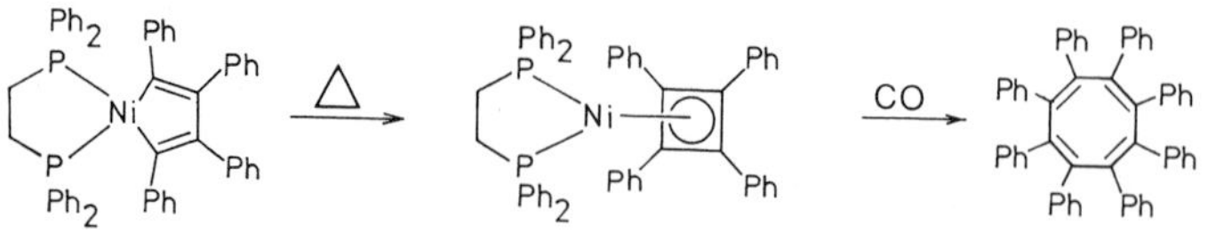

A different tetramerization process has been demonstrated by Stone and Knox who found that stepwise addition of acetylenes to $[CpM(CO)_2]_2$ (M = Cr, Mo, W) leads to XCI containing an unbranched carbon chain constructed from four acetylene units (Fig. 28).[124a] Although release of cyclic hydrocarbons from these species has not yet been achieved the similarity of XCI to cobalt flyover complexes LXXXII suggests that this may be feasible. Interestingly a complex related to XCI has also been obtained from the reduction of $[CpMo(MeCN)(MeC\equiv CMe)_2]BF_4$ with $Na[CpFe(CO)_2]$.[124b] Finally, although the involvement of acetylene complexes has yet to be demonstrated, it is worth noting that acetylene tetramers, dihydropentalenes, are obtained from the reactions of Pd(II)

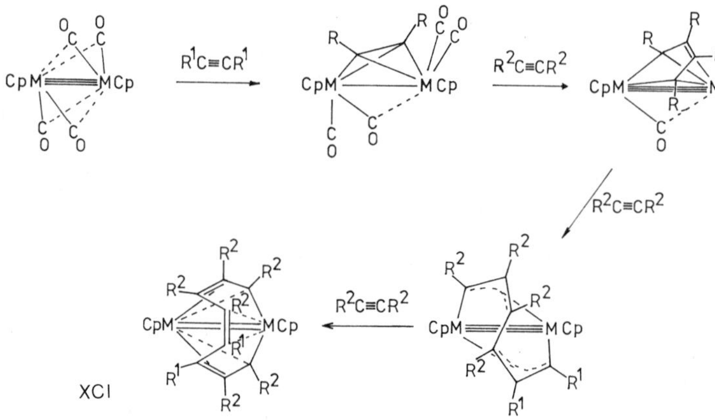

Fig 28 (R^1, R^2 = H, Ph, COOMe – not all combinations)

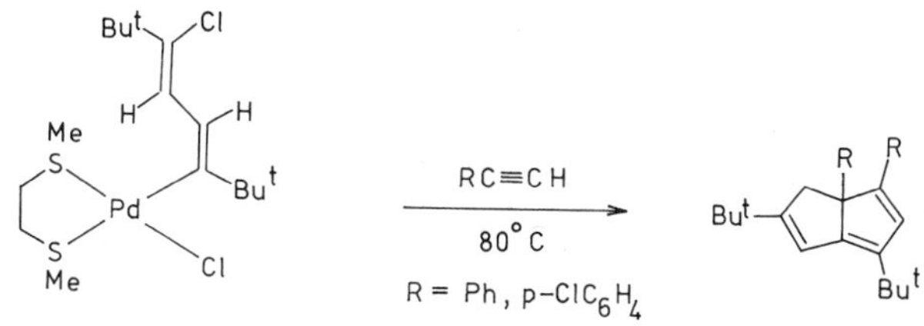

complexes with alkynes (eq 60).[125] Mechanistically their formation can be rationalized in terms of an extension of Figs. 24 and 26 which account for the di- and trimerization of alkynes in the presence of Pd(II) complexes.

3.5.2 Reactions with alkenes

Alkenes undergo condensation with coordinated acetylenes to give dienes and diene complexes. Both stoichiometric and catalytic systems have been studied from which 1:2 and 2:1 addition products have been isolated. Reactions of this kind are typified by the addition of olefins (dimethyl maleate, dimethyl fumarate, fumaronitrile, crotonitrile) to $CpCo(PPh_3)(RC\equiv CR)$ (R = Ph or CO_2Me), the features of which are summarized in Fig. 29.[126] The key process involves conversion of the acetylene complex XCII into a metallacyclopentene XCIII, which can be isolated as a triphenylphosphine adduct. Kinetic studies provide evidence for an intermediate XCIV in which both acetylene and olefin are coordinated to the metal prior to cyclization, a process which finds a precedent in reaction (61).[127] Subsequent reaction of the metallacycle XCIII with an olefin [Fig. 29, path (a)] gives linear dienes resulting from β-hydride transfer and alkyl elimination in a postulated metallacycloheptene intermediate. Alternatively reaction with an acetylene [path (b)] produces cyclic diene complexes which can also be obtained via path (c) involving condensation of an olefin with a metallacyclopentadiene (XCV).[128] The reaction of XCV with acetylenes to give benzenes obviously presents a competing pathway (cf eq 55, Section 3.5.1.2) and highly selective catalytically efficient reactions are not widespread. Ibers and Itoh have overcome this problem to some extent by employing electrophilic acetylenes

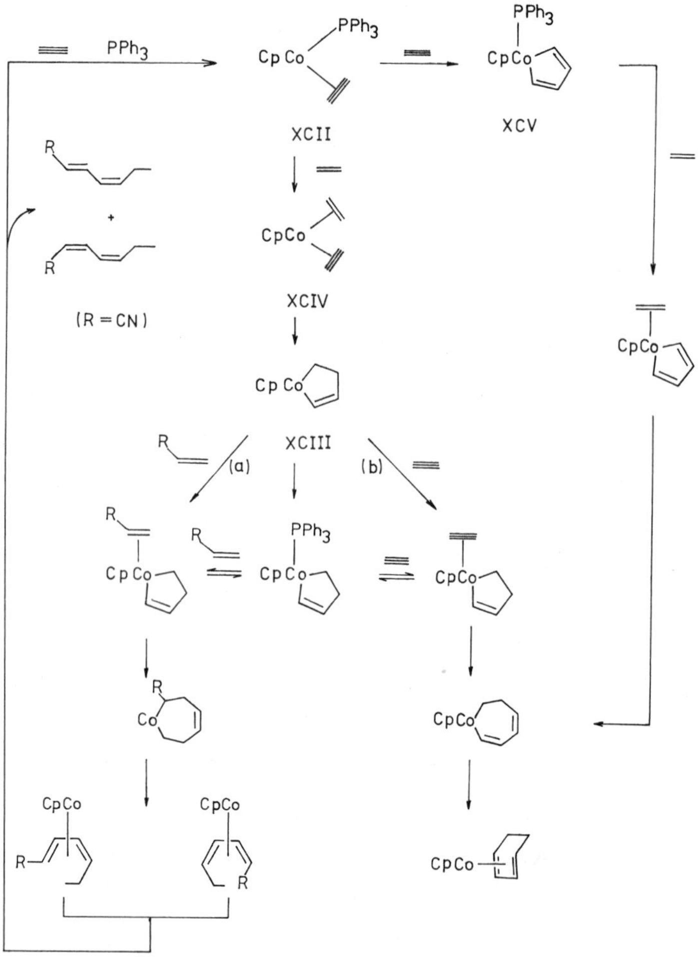

Fig 29

and electron donating olefins so that the metallacyclopentadiene shows a preference for the olefin over the electron withdrawing acetylene.[129] Thus the oligomeric metallacyclopentadiene complex $[PdC_4(CO_2Me)_4]_n$ efficiently catalyzes the 2:1 cyclooligomerization of $MeO_2CC{\equiv}CCO_2Me$ with norbornadiene to give 1,2,3,4-tetra(methoxycarbonyl)benzene as a

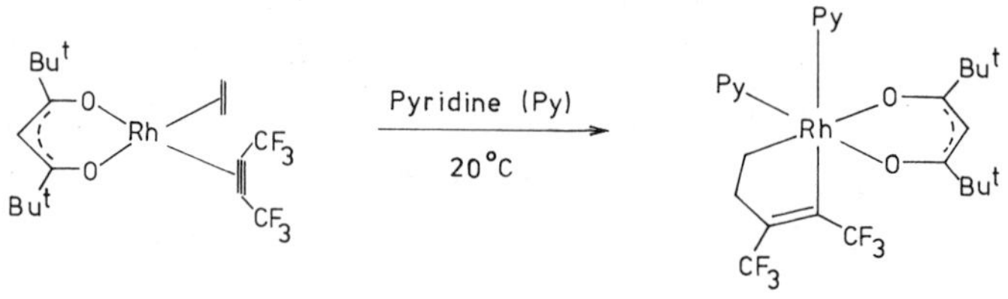

(61)

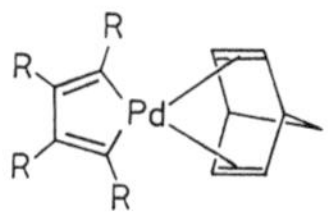

XCVI 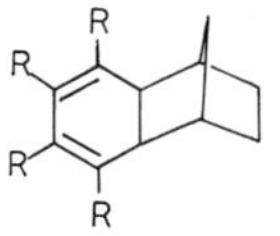   XCVII 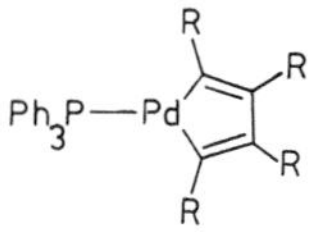   XCVIII

(R = COOMe)

result of cyclopentadiene elimination from intermediate **XCVI** whereas norbornene gives **XCVII** (R = CO_2Me) stereoselectively. Addition of one equivalent of triphenylphosphine enhances the catalytic rate of the latter reaction markedly and also extends the scope of cyclizations to include cyclopentene and cyclohexene. This enhanced reactivity is thought to result from formation of the catalytically active three coordinate intermediate **XCVIII**.

However, Green's subsequent investigation of acetylene-olefin cotrimerization promoted by indenyl rhodium complexes indicates that metallacyclopentadienes do not actively participate in the process.[130] The sole metallacyclic precursor to the cyclohexadiene products appears to be a metallacyclopentene complex (cf **XCIII**) and it was proposed that Ibers' palladiacyclopentadiene is converted into another species (not detected) and that the immediate precursor of the tetra(methoxycarbonyl)cyclohexa-1,3-dienes is a 2,3,6,7-tetra(methoxycarbonyl) palladiacyclohepta-2,6-diene and not the suggested 2,3,4,5-tetra(methoxycarbonyl)cyclohepta-2,4-diene.

Bridging acetylenes also exhibit reactivity towards alkenes, and acetylene, which is normally a poor dienophile, becomes activated towards dienes when coordinated as in **XCIX** and reacts to produce complexes C (eq 62) (X = $(-CH_2)_2$, $(-CH_2)_4$, $(-CH=CH)_2$, $-CH=CH-CH-$), structurally characterised by xray diffraction studies when X = $(-CH=CH)_2$.[131] Pauson has studied the reactivity of the acetylene ligand in structurally related cobalt complexes $Co_2(CO)_6(RC≡CR)$ towards a variety of olefins. Strained

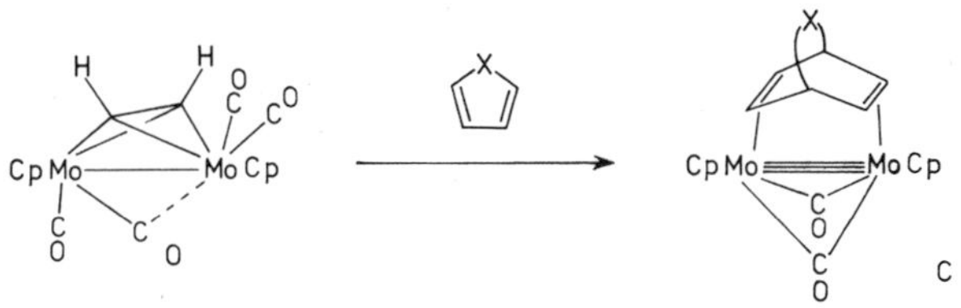

XCIX   (62)

CI $R^1 = H$, $R^2 = H, Me, Ph$ CII

$R^1 = R^2 = H, Et, Ph$

olefins norbornadiene and norbornene react at 60-80°C to give cyclopentenones CI and CII respectively, each with the exo configuration, and when unsymmetrical acetylenes ($R^1 \neq R^2$) are employed the bulky substituent R^2 was invariably found to be adjacent to the carbonyl group.[132] Cyclobutenes react similarly[133] although more drastic conditions (150-160°C) were required with unstrained olefins such as ethylene, terminal alkenes and cycloalkenes C_5H_8-C_8H_{14}.[134] The cyclopentenones CI and CII could also be obtained catalytically on combining the acetylene, norbornadiene or norbornene, carbon monoxide and catalytic amounts of dicobalt octacarbonyl.[132a]   Contrasting behavior is observed with the activated olefins ethyl crotonate and crotonaldehyde, which react with $Co_2(CO)_6(PhC\equiv CH)$ to give conjugated dienes.[135] The former gives both trans,trans and 2-cis,4-trans isomers of $PhCH=CHCMe=CHCO_2Et$ while the latter gives the trans,trans aldehyde almost exclusively.

3.5.3 Reactions with carbon monoxide

The carbonylation of acetylenes in the presence of metal compounds leads to a variety of products, some (e.g. acrylic acid and esters) of great industrial importance, but in many cases metal complexes are isolated in which the organic ligands, (quinones, cyclopentadienones, lactones and tropones) are firmly bound to the metal.   Thus reactions of CO with acetylene complexes are of interest in elucidating the manner in which carbonylation of acetylenes is promoted by metals.

The simplest reaction type observed on carbonylation of $Co(CO)_6(HC\equiv CH)$ with carbon monoxide in ethanol leads to free esters rather than cyclic products (eq 63),[136] while the ester $PhC\equiv CCO_2Et$ is

$CH_2=CHCO_2Et$ + $CH_3CH_2CO_2Et$

CH_2CO_2Et

| (63)

CH_2CO_2Et

$(CO)_3Co$———$Co(CO)_3$ $\xrightarrow{CO/EtOH}$

similarly obtained on oxidation of $Co_2(CO)_6(PhC\equiv CH)$ in ethanol.[135] α-Hydroxyacetylene complexes of platinum also carbonylate smoothly but in this case platinalactone derivatives are obtained[137] (eq 64):

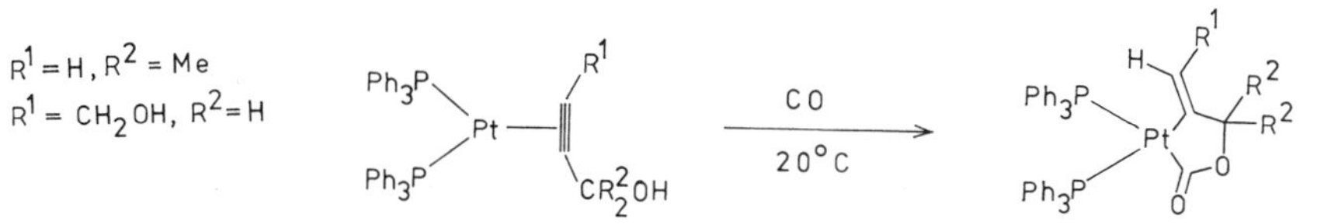

$$R^1 = H, R^2 = Me$$
$$R^1 = CH_2OH, R^2 = H$$

(64)

Lactone complexes are also generated when monosubstituted acetylene derivatives $Co_2(CO)_6(RC\equiv CH)$ are treated with carbon monoxide under pressure, the butenolide products having structure **CIII**.[138] It was subsequently demonstrated that complexes **CIII** are intermediates in the independently discovered formation of bifurandiones **CIV, CV** and **CVI** in the cobalt carbonyl catalyzed reaction of acetylenes with carbon monoxide.[139] Internal acetylenes react more slowly than terminal alkynes but in all cases the formation of the butenolide **CIII** proceeds with high regioselectivity which apparently reflects electronic rather than steric control. Thus ^{13}C nmr studies reveal that the carbon atom of the acetylene with the lower δ value invariably occupies the 3-position of the lactone ring.[138]

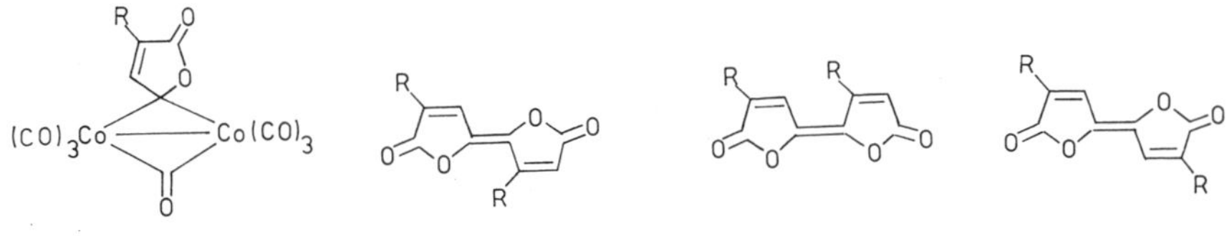

CIII   CIV   CV   CVI

A third route to lactones is provided by the carbonylation of complexes $CpWR^1(CO)(R^2C\equiv CR^2)$ (R^1 = alkyl, R^2 = H) which initially leads to vinyl ketone complexes **CVII** (Fig. 13, Section 3.4.2).[88,140-1] Subsequent reaction of this type of complex with a two electron ligand L(CO, Bu^tNC, PPh_3) promotes CO insertion to yield lactone complexes[142] but with L = $RC\equiv CR$ an alternative acetylene insertion occurs to give pyranyl complexes (eq 65):[143]

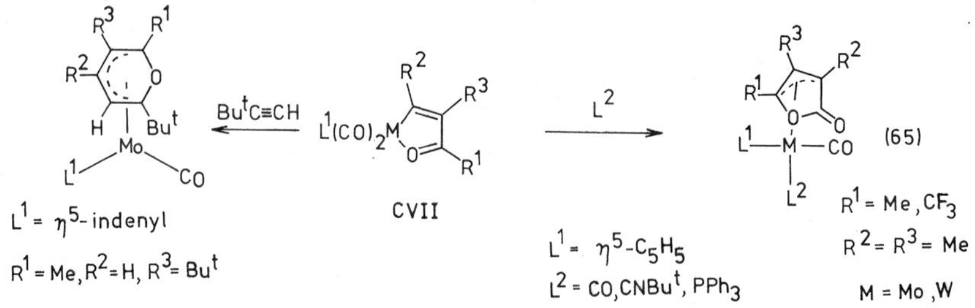
Bu^{t}C≡CH
L^2
CVII
(65)
L^1 = η^5-indenyl
R^1 = Me, R^2 = H, R^3 = Bu^t
L^1 = η^5-C_5H_5
L^2 = CO, CNBu^t, PPh_3
R^1 = Me, CF_3
R^2 = R^3 = Me
M = Mo, W

In the absence of labile alkyl or acyl ligands carbonylation reactions involving two acetylenes generally lead to cyclopentadienones or quinones, but whereas the former are obtained from a large range of acetylenes, the latter are obtained almost exclusively from dialkylacetylenes or acetylene itself. Cyclopentadienone formation is generally thought to proceed by one of two routes (Fig. 30) although a third route has been observed with palladium(II) derivatives (Fig. 31).[144] The routes in Fig. 30 depend on the

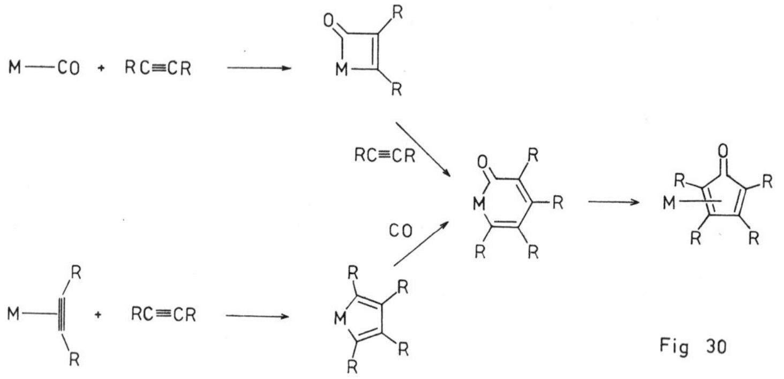
M—CO + RC≡CR
RC≡CR
CO
M + RC≡CR

Fig 30

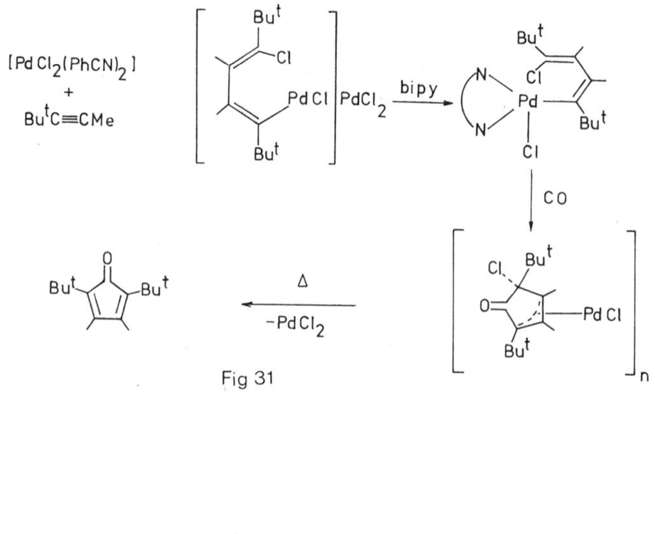
[$PdCl_2(PhCN)_2$]
+
Bu^{t}C≡CMe
PdCl | $PdCl_2$
bipy
CO
Δ
-$PdCl_2$

Fig 31

ability of the metal to undergo facile two-electron oxidative-addition, reductive-elimination reactions and not surprisingly cyclopentadienone formation is frequently observed with iron, cobalt and rhodium complexes.[1-2,4] Thus Yamazaki has reported that the cobalt complex **LXIV** is carbonylated by the metallacyclopentadiene route of Fig. 30 (eq 66):[100]

A carbonyl analogue of **LXIV** also acts as a precursor to cyclopentadienone derivatives $CpCo(C_4R_4CO)$ (R = Me, Ph) in reactions of acetylenes with $CpCo(CO)_2$, but a competing route via dinuclear intermediates was also found (Fig. 18, Section 3.5.1.1.).[99]

Yamazaki[101] has considered the possibility that monoacetylene complexes **LXIV** react with acetylenes to give metallacycles via bis-acetylene derivatives (Fig. 20, Section 3.5.1.1.). It is perhaps no surprise therefore to find that bis-acetylene complexes $CpMX(R^1C \equiv CR^2)_2$ (M = Mo, W; X = Cl, Br, I; $R^1 = R^2 = CF_3$; R^1 = Ph, R^2 = Me) are intermediates in the reactions of $CpMX(CO)_3$ with acetylenes which lead to cyclopentadienone derivatives $CpMX(CO)(C_4R^1_2R^2_2CO)$, (Fig. 16, Section 3.5.1.1).[12,96] A variety of related bis-acetylene complexes of the group V and VI metals are known which may also be precursors to related cyclopentadienone derivatives,[11,74,95,145] and free tetracyclone is in fact produced from the photodecomposition of one such species, $CpVCO(PhC \equiv CPh)_2$.[11,95]

Free cyclopentadienones are also obtained on thermal degradation of $Co_2(CO)_6(RC \equiv CC_6F_5)$ (R = Ph or C_6F_5) although higher yields are obtained if acetylene complexes of this type are reacted with excess alkyne.[146] The thermolysis of $Co_2(CO)_6(PhC \equiv CC_6F_5)$ produces, in addition to free cyclopentadienones, an organocobalt species assigned structure **CVIII**. Reactions of terminal acetylenes $RC \equiv CH$, (R = Bu^t, H) with $Co_2(CO)_6(RC \equiv CH)$ lead to a different class of cyclopentadienone complex, **CIX**, but not surprisingly flyover complexes **LXXXII** are formed in a competing reaction.[147]

CVIII (R = Ph or C_6F_5)   CIX

The tetrafluorobenzyne complex $Co_4(CO)_{10}C_6F_4$ has been reported to undergo thermolysis in vacuo giving octafluorenone[148] whereas "normal" acetylene complexes of this type decompose on heating to give dinuclear derivatives $Co_2(CO)_6RC\equiv CR$.[149] Interestingly a benzyne intermediate was proposed to account for the formation of 2,3-diphenylindenone in 67% yield from the thermolysis of the diphenylacetylene complex $CpFe(SnPh_3)CO(PhC\equiv CPh)$.[150] However a more realistic suggestion may be a metallacyclic intermediate related to **XX**, which is formed on thermolysis of $CpMn(CO)_2(Ph_3SnC\equiv CPh)$[36] (Section 3.2).

Although the carbonylation of acetylenes to produce cyclopentadienones in the presence of transition metal derivatives is a reasonably well understood process the same cannot be said for reactions leading to quinone formation. The only reactions in which acetylene complexes are known to be precursors to quinone derivatives are those illustrated in Fig. 32.[1] Little is known about the mechanism of quinone ring formation although maleoyl complexes **CX** have been proposed as intermediates.[151] One such complex has been isolated in low yield from reactions of iron carbonyls with acetylenes[1] but since complexes related to **CXI** have been shown to act as precursors to free quinones the situation remains unresolved at present, the two routes being equally plausible.

3.5.4 Reactions with isonitriles

The simplest reaction observed between a coordinated acetylene and isonitriles involves metallacyclization as illustrated in eq 67.[71,152] More complex reactions are observed with the cobalt complex **LXIV** (Fig. 33), leading ultimately to both 1:2 and 1:3 co-telomerization products **CXII** and **CXIII** via isolable organometallic intermediates.[153] Iminocyclobutenes **CXII**

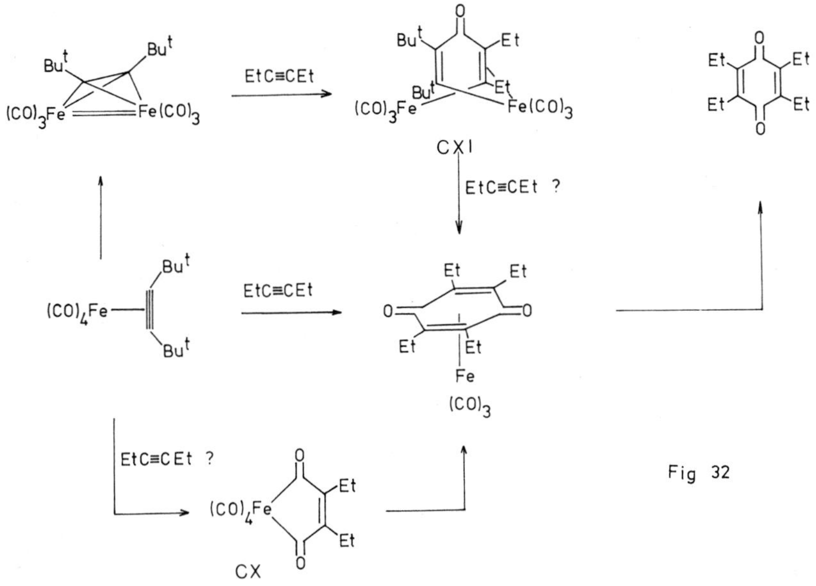

Fig 32

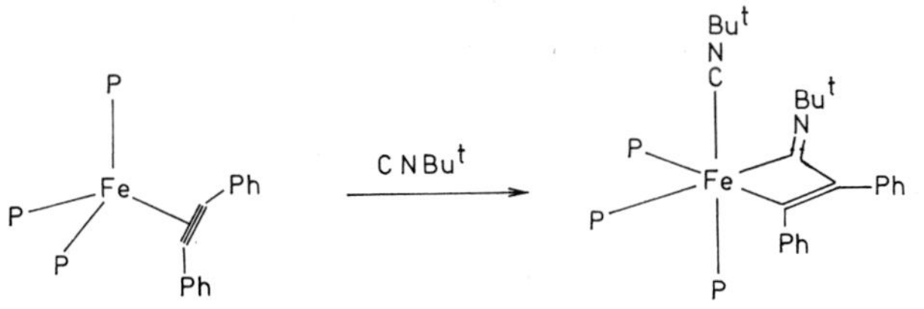

$$P = P(OMe)_3$$

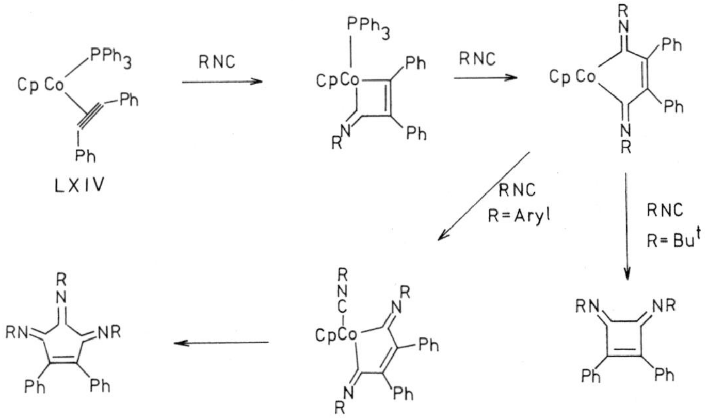

(67)

LXIV

Fig 33 (R = Bu^t, Aryl)

CXIII

CXII

are also produced, albeit in low yield, from the thermolysis of $Ni(CNR)_2(PhC\equiv CPh)$ (R = aryl, Bu^t) but the major product is the cyclopentadienimine **CXIV**.[154] In contrast, bis-acetylene complexes $CpMX(CF_3C\equiv CCF_3)_2$ (M = Mo, X = CF_3; M = W, X = Cl) react with t-butylisocyanide to give cyclopentadienimine complexes **CXV** from which it has not yet proved possible to release the diene ligand.[14] An η^2-vinyl complex resulting from nucleophilic attack of Bu^tNC on an acetylenic carbon has also been isolated from this reaction and shown to act as a precursor to **CXV** (eq 1, Section 2.1).[14c]

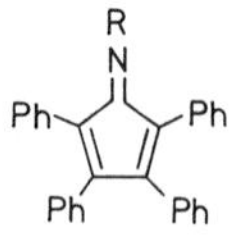

CXIV

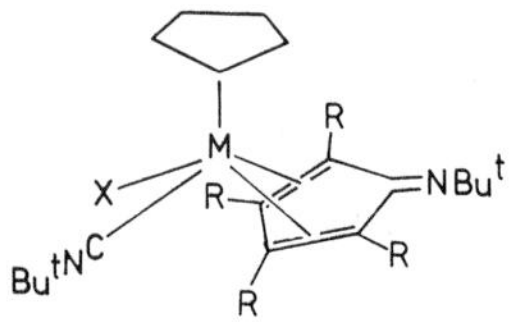

CXV

$M = Mo, X = CF_3$

$M = W, X = Cl$

$R = CF_3$

3.5.5 Reactions with nitriles

Organic nitriles also undergo condensation with acetylenes in the presence of metals to afford pyridines in reactions which are frequently catalytic but which also exhibit low regioselectivity. Cyclopentadienyl cobalt catalysts have been employed in many cases and a mechanism related to the benzene synthesis in Fig. 25, (Section 3.5.1.2.1.), has been proposed in which a metallacyclopentadiene again plays a prominent role, (Fig. 34).[155] Yamazaki has in fact demonstrated that metallacycles **CXVI** (L = PPh_3) undergo stoichiometric condensation reactions with nitriles to produce substituted pyridines in 30-70% yields.[156] As with cyclohexadiene synthesis (Section 3.5.2), catalytic processes exhibit lower yields, probably as a result of the competing acetylene trimerization, although this is less of a problem with α,ω diynes which give annelated pyridines (eq 68) in yields as high as 81%.[157]

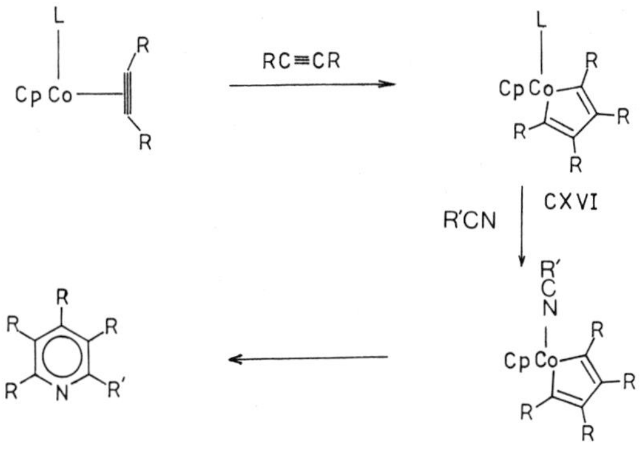

Fig 34

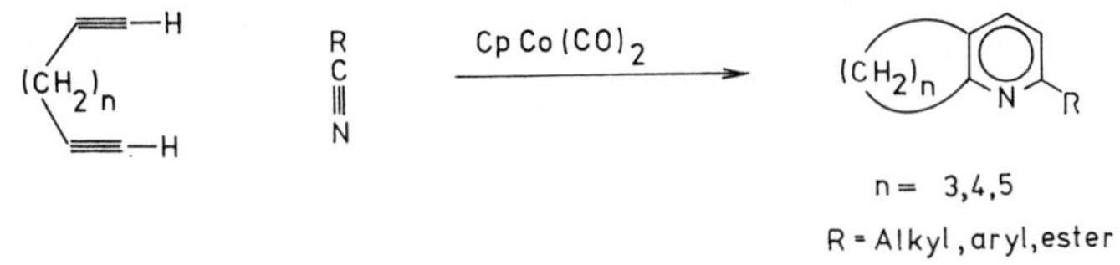

n= 3,4,5
R = Alkyl, aryl, ester

(68)

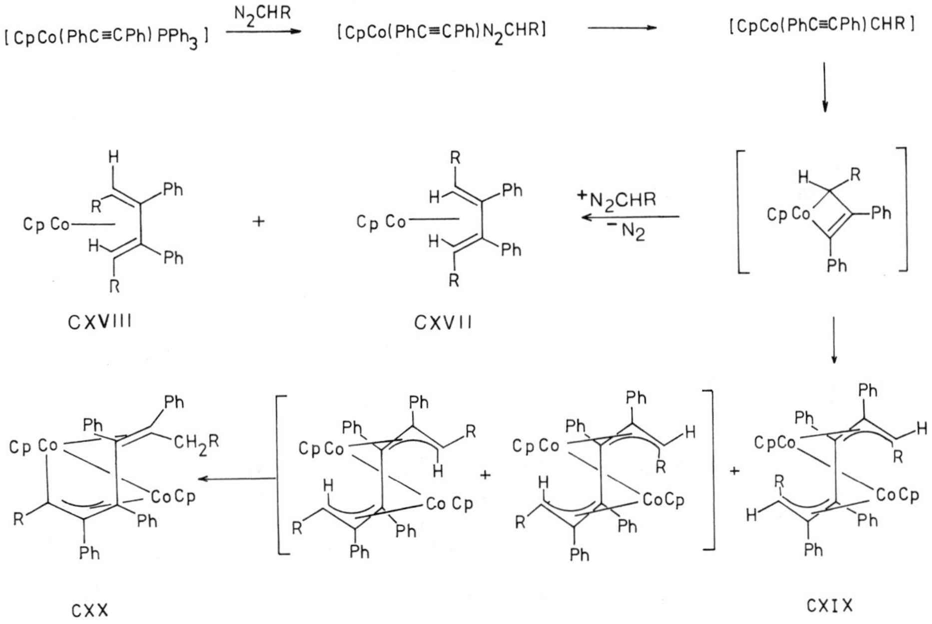

[CpCo(PhC≡CPh)PPh₃] $\xrightarrow{N_2CHR}$ [CpCo(PhC≡CPh)N₂CHR] ⟶ [CpCo(PhC≡CPh)CHR]

CXVIII + CXVII

CXX   CXIX

Fig 35   (R = COOMe)

3.5.6 Reactions with diazo compounds

Yamazaki has shown that alkyldiazoacetates N_2CHCO_2R (R = Me, Et, Bu^t) function as a carbene source in reactions with $CpCoPPh_3(PhC\equiv CPh)$ from which four organometallic products **CXVII-CXX**, have been isolated.[158] The proposed mechanism (Fig. 35) assumes that on coordination the diazo compound loses nitrogen and the resulting carbene ligand inserts into the cobalt acetylene bond to give a metallacycle. This subsequently acts as a precursor to the mono- and dinuclear products isolated.

3.5.7 Reactions with carbonyl sulfides

Sulfur containing molecules frequently exhibit reactivity towards acetylenes particularly those bearing electron withdrawing substituents.[159] This is clearly illustrated by the reactions of CS_2 and COS with coordinated dimethylacetylenedicarboxylate in eq 69, whereas reaction of this acetylene with coordinated CS_2 (eq 70) leads to an isomeric product.[160] The reaction

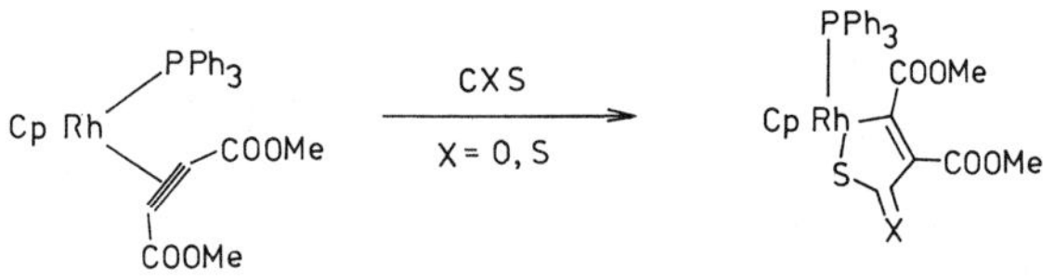

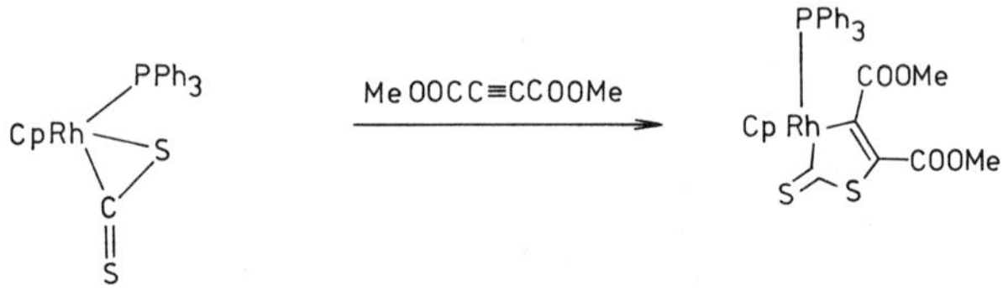

of dimethylacetylenedicarboxylate with $Fe(CO)_2[P(OMe)_3]_2CS_2$ gives **CXXI** via a carbene intermediate **CXXII**.[161] These contrasting results, which have been explained in terms of the ionic mechanisms in Fig. 36, clearly demonstrate that when two potential ligands cyclize in the presence of a metal the product isolated may be determined by which species attaches itself to the metal first.

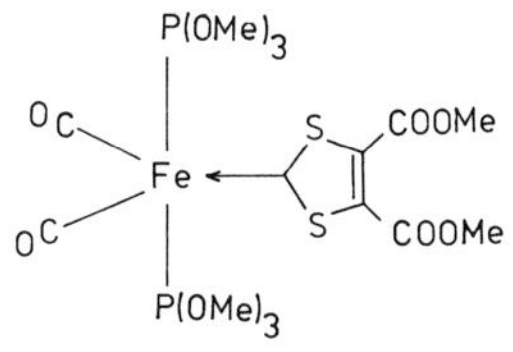

CXXI

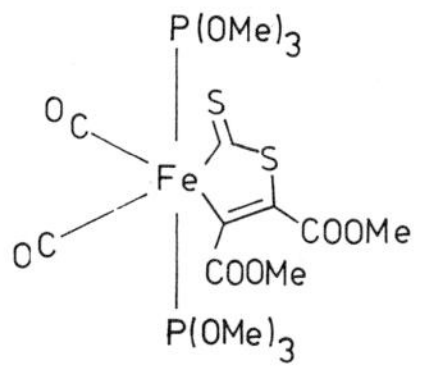

CXXII

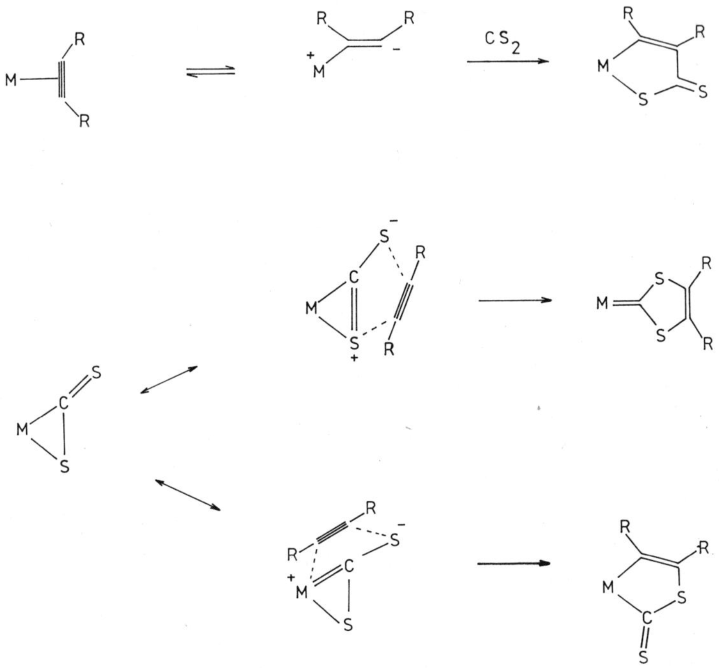

Fig 36 (R = COOMe)

REFERENCES

1. W. Hubel in "Organic Synthesis via Metal Carbonyls", (I. Wender and P. Pino, eds.), Vol. 1, Wiley, New York 1968, p. 273.

2. F.L. Bowden and A.B.P. Lever, Organomet. Chem. Rev. (A) 3:227 (1968).

3. J.H. Nelson and H.B. Jonassen, Coord. Chem. Rev. 6:27 (1971).

4. L.P. Yur'eva, Russ. Chem. Rev. 43:48 (1974).

5. F.R. Hartley, Chem. Rev. 69:799 (1969).

6. P.M. Maitlis, Acc. Chem. Res. 9:93 (1976).

7. K.M. Nicholas, M.O. Nestle and D. Seyferth in "Transition Metal Organometallics in Organic Synthesis", (H. Alper, ed.), Vol. 2, Academic Press, London 1978, p. 1.

8. S.D. Ittel and J.A. Ibers, Adv. Organomet. Chem. 14:33 (1976).

9. (a) S. Otsuka and A. Nakamura, Adv. Organomet. Chem. 14:245 (1976); (b) J.L. Templeton, P.B. Winston and B.C. Ward, J. Am. Chem. Soc. 103:7713 (1981); (c) K. Tatsumi, R. Hoffmann and J.L. Templeton, Inorg. Chem. 21:466 (1982).

10. E.O. Greaves, C.J.H. Lock and P.M. Maitlis, Can. J. Chem 46:3879 (1968).

11. A.N. Nesmeyanov, K.N. Anisimov, W.E. Kolobova and A.A. Pasynskii, Dokl. Akad. Nauk SSSR 182:112 (1968).

12. J.L. Davidson, M. Green, F.G.A. Stone and A.J. Welch, J.C.S. Dalton Trans. 738 (1976).

13. R.B. King, Inorg. Chem 7:1044 (1968).

14. (a) J.L. Davidson, M. Green, J.Z. Nyathi, F.G.A. Stone and A.J. Welch, J.C.S. Dalton Trans. 2246 (1977); (b) J.L. Davidson, unpublished work; (c) J.L. Davidson, W.F. Wilson, Lj. Manojlović-Muir and K.W. Muir, J. Organomet. Chem. 254:C6 (1983).

15. J. Burt, M. Cooke and M. Green, J. Chem. Soc. (A) 1891 (1970).

16. R.S. Dickson and C.N. Pain, J.C.S. Chem. Commun. 277 (1979).

17. J.L. Davidson, M. Green, F.G.A. Stone and A.J. Welch, J.C.S. Dalton Trans. 506 (1979).

18. L.F. Dahl and D.L. Smith, J. Am. Chem. Soc. 84:2450 (1962).

19. P.F. Heveldt, B.F.G. Johnson, J. Lewis, P.B. Raithby and G.M. Sheldrick, J.C.S. Chem. Commun. 340 (1978).

20. E.L. Muetterties, W.R. Pretzer, M.G. Thomas, B.F. Beier, D.L. Thorn, V.W. Day and A.B. Anderson, J. Am. Chem. Soc. 100:2090 (1978).

21. R.S. Dickson and H.P. Kirsch, Aust. J. Chem. 25:1815 (1972).

22. D. Seyferth and D.L. White, J. Organomet. Chem. 32:317 (1971).

23. D. Seyferth, M.O.Nestle and A.I. Wehman, J. Am. Chem. Soc. 97:7417 (1975).

24. S. Aime, L. Milone and D. Osella, J.C.S. Chem. Commun. 704 (1979).

25. P-J. Kim and N. Hagihara, Bull. Chem. Soc. Jpn. 41:1184 (1968).

26. K.M. Nicholas and B. Pettit, Tetrahedron Lett. 3475 (1971).

27. B.E. Connors and K.M. Nicholas, J. Organomet. Chem. 125:C45 (1977).

28. C. Descoins and P. Samain, Tetrahedron Lett. 745 (1976).

29. K.M. Nicholas and R. Pettit, J. Organomet. Chem. 44:C21 (1972).

30. R.F. Lockwood and K.M. Nicholas, Tetrahedron Lett. 4163 (1977).

31. H.D. Hodes and K.M. Nicholas, Tetrahedron Lett. 4349 (1978).

32. J. Burgess, M.E. Howden, R.D.W. Kemmitt and N.S. Sridhara, J.C.S. Dalton Trans. 1577 (1978).

33. V. Raverdino, S. Aime, L. Milone and E. Sappa, Inorg. Chim. Acta. 30:9 (1978).

34. K. Folting, J.C. Huffmann, L.N. Lewis and K.G. Caulton, Inorg. Chem. 18:3483 (1979).

35. A.B. Antonova, N.E. Kolobova, P.V. Petrovskii, B.V. Lokshin and N.S. Obezyuk, J. Organomet. Chem. 137:55 (1977).

36. A.N. Nesmeyanov, N.E. Kolobova, A.B. Antonova and K.N. Anisimov, Dokl. Akad. Nauk S.S.S.R. 220:105 (1975).

37. A. Davison and J.P. Selegue, J. Am. Chem. Soc. 100:7763 (1978).

38. J.M. Bellerby and M.J. Mays, J. Organomet. Chem. 117:C21 (1976).

39. M.I. Bruce and R.C. Wallis, J. Organomet. Chem. 161:C1 (1978).

40. M.H. Chisholm and H.C. Clark, J. Am. Chem. Soc. 94:1532 (1972).

41. M. Franck-Neumann and F. Brion, Angew. Chem. Int. Ed. Engl. 18:688 (1979).

42. (a) R.B. King and C.A. Harman, Inorg. Chem. 15:879 (1976); (b) G.G. Cash, R.C. Pettersen and R.B. King, J.C.S. Chem. Commun. 30 (1977).

43. H. Yamazaki, Y. Wakatsuki and K. Aoki, Chem. Lett. 1041 (1979).

44. T.J. Katz and J. McGinnis, J. Am. Chem. Soc. 97:1592 (1975).

45. J.R. Fritch, K.P.C. Vollhardt, M.R. Thompson and V.W. Day, J. Am. Chem. Soc. 101:2768 (1979); K.P.C. Vollhardt and J.R. Fritch, Angew. Chem. Int. Ed. Engl. 18:409 (1979).

46. (a) R.G. Bergman and J.M. Huggins, J. Am. Chem. Soc. 101:4410 (1979); (b) J.L. Davidson, M. Shiralian, Lj. Manojlovic-Muir and K.W. Muir, J.C.S. Chem. Commun. 30 (1979).

47. (a) M. Green and M. Bottrill, J. Am. Chem. Soc. 99:5795 (1977); (b) M. Green, N.C. Norman and A.G. Orpen, J. Am. Chem. Soc. 103:1267 (1981).

48. (a) D.L. Reger, C.J. Coleman and P.J. McElligott, J. Organomet. Chem. 171:73 (1979); (b) D.L. Reger and C.J. Coleman, Inorg. Chem. 18:3155 (1979).

49. (a) J.L. Davidson, I.E.P. Murray, P.M. Preston, M.V. Russo, Lj. Manojlovic-Muir and K.W. Muir, J.C.S. Chem. Commun. 1059 (1981); (b) J.L. Davidson, M. Green, F.G.A. Stone and A.J. Welch, J.C.S. Dalton Trans. 287 (1977).

50. H. Berke, Angew. Chem. Int. Ed. Engl. 15:624 (1976).

51. S.B. Samuels, S.R. Berryhill and M. Rosenblum, J. Organomet. Chem. 166:C9 (1979).

52. A.J. Carty, M.J. Taylor, W.F. Smith, M.F. Lappert and P.L. Pye, J.C.S. Chem. Commun. 1017 (1978).

53. D.M. Barlex, R.D.W. Kemmitt and G.W. Littlecott, J. Organomet. Chem. 43:225 (1972).

54. P.C. Kong and T. Theophanicles, Inorg. Chim. Acta 7:299 (1973).

55. A.J. Deeming and S. Hasso, J. Organomet. Chem. 112:C39 (1976).

56. Y.S. Wong, H.N. Paik, P.C. Chieh and A.J. Carty, J.C.S. Chem. Commun. 309 (1975); A.J. Carty, N.J. Taylor, H.N. Park, W. Smith and J.E. Yule, J.C.S. Chem. Commun. 41 (1976); A.J. Carty, C.N. Mott, N.J. Taylor and J.E. Yule, J. Am. Chem. Soc. 100:3051 (1978); A.J. Carty, G.N. Mott, N.J. Taylor, G. Ferguson, M.A. Khan and P.J. Roberts, J. Organomet. Chem. 149:345 (1978).

57. M.H. Chisholm and H.C. Clark, Acc. Chem. Res. 6:202 (1973); M.H. Chisholm, Platinum Met. Rev. 19:100 (1975).

58. H.C. Clark and M.H. Chisholm, J.C.S. Chem. Commun. 1484 (1971).

59. T.G. Appleton, M.H. Chisholm, H.C. Clark and K. Yasufuku, J. Am. Chem. Soc. 96:6600 (1974).

60. (a) J.A. Labinger, J. Schwartz and J.M. Townsend, J. Am. Chem. Soc. 96:4009 (1974); (b) J.A. Labinger and J. Schwartz, J. Am. Chem. Soc. 97:1596 (1975).

61. J.L. Thomas, Inorg. Chem. 17:1507 (1978 and references therein.

62. (a) J.W. McDonald, J.L. Corbin and W.E. Newton, J. Am. Chem. Soc. 97:1970 (1975); (b) J.W. McDonald, W.E. Newton, C.T.C. Creedy and W.E. Newton, J. Organomet. Chem. 92:C25 (1975).

63. R.S. Dickson and P.J. Fraser, Adv. Organomet. Chem. 12:323 (1974).

64. B.E. Mann, B.L. Shaw and N.I. Tucker, J. Chem. Soc. (A) 2667 (1971).

65. P.B. Tripathy, B.W. Renoe, K. Adzamli and D.M. Roundhill, J. Am. Chem. Soc. 93:4406 (1971).

66. D.M. Barlex, R.D.W. Kemmitt and G.W. Littlecott, J.C.S. Chem. Commun. 613 (1969).

67. J. Trocha-Grimshaw and H.B. Henbest, J.C.S. Chem. Commun. 757 (1968).

68. M. Cooke and M. Green, J. Chem. Soc. (A) 1200 (1971).

69. R.D.W. Kemmitt, B.Y. Kimura and G.W. Littlecott, J.C.S. Dalton Trans. 636 (1973).

70. G.L. McClure and W.H. Baddely, J. Organomet. Chem. 25:261 (1970).

71. M.A. Bennett and T. Yoshida, J. Am. Chem. Soc. 100:1750 (1978).

72. B. Clarke, M. Green and F.G.A. Stone, J. Chem. Soc. (A) 951 (1970).

73. U. Krüerke and W. Hübel, Chem. Ber. 94:2829 (1961).

74. W. Hubel and R. Merenyi, J. Organomet. Chem. 2:213 (1964).

75. T.G. Appleton, M.A. Bennett, A. Singh and T. Yoshida, J. Organomet. Chem. 154:369 (1978).

76. A. Nakamura and S. Otsuka, J. Mol. Catal. 1:285 (1975).

77. (a) H.C. Clark, C.R. Jablonski and C.S. Wong, Inorg. Chem. 14:1332 (1975); (b) H.C. Clark, P.L. Fiess and C.S. Wong, Can. J. Chem. 55:177 (1977); (c) H.C. Clark and C.S. Wong, J. Organomet. Chem. 92:C31 (1975); (d) H.C. Clark and C.R. Milne, J. Organomet. Chem. 161:51 (1978).

78. D.W. Hart and J. Schwartz, J. Organomet. Chem. 87:C11 (1975).

79. H.C. Clark and K.E. Hine, J. Organomet. Chem. 105:C32 (1976).

80. H.C. Clark and C.S. Wong, J. Am. Chem. Soc. 99:7073 (1977).

81. H. Eshtiagh-Hosseini, J.F. Nixon and J.S. Poland, J. Organomet. Chem. 164:107 (1979) and references therein.

82. (a) T. Blackmore, M.I. Bruce and F.G.A. Stone, J.C.S. Dalton Trans. 106 (1974); (b) L.E. Smart, J.C.S. Dalton Trans. 390 (1976).

83. P.M. Triechel, E. Pitcher and F.G.A. Stone, Inorg. Chem. 1:511 (1962).

84. J.L. Davidson, M. Green, F.G.A. Stone and A.J. Welch, J.C.S. Dalton Trans. 2044 (1976).

85. (a) H.C. Clark and R.J. Puddephatt, Inorg. Chem. 9:2670 (1970); (b) H.C. Clark and R.J. Puddephatt, Inorg. Chem. 10:18 (1971).

86. H.C. Clark and L.E. Manzer, Inorg. Chem. 13:1291 (1974).

87. H.C. Clark and K. von Werner, J. Organomet. Chem. 101:347 (1975).

88. H.G. Alt, J. Organomet. Chem. 127:349 (1977).

89. J.L. Davidson, J.C.S. Chem. Commun. 597 (1979).

90. R.B. Woodward and R. Hoffmann, Angew. Chem. Int. Ed. Engl. 8:781 (1969).

91. A. Efraty, Chem. Rev. 77:692 (1977).

92. W. Reppe, O. Schlichting, K. Klager and T. Toepel, Liebigs Ann. Chem. 560:1 (1948).

93. F.D. Mango, Coord. Chem. Rev. 15:109 (1975).

94. J.F. Helling, S.C. Rennison and A. Merijan, J. Am. Chem. Soc. 89:7140 (1967).

95. A.N. Nesmeyanov, Adv. Organomet. Chem. 10:1 (1972).

96. J.L. Davidson, J.C.S. Chem. Commun. 113 (1980).

97. J.L. Davidson, Lj. Manojlovic-Muir, K. Muir and A.N. Keith, J.C.S. Chem. Commun. 749 (1980).

98. R. Buehler, R. Geist, R. Muendnich and H. Pleninger, Tetrahedron Lett. 1919 (1973).

99. W.S. Lee and H.H. Brintzinger, J. Organomet. Chem. 127:93 (1977).

100. H. Yamazaki and N. Hagihara, J. Organomet. Chem. 21:431 (1970).

101. H. Yamazaki and Y. Wakatsuki, J. Organomet. Chem. 139:157 (1977).

102. M.D. Rausch, G.F. Westover, E. Mintz, G.M. Reisner, I. Bernal, A. Clearfield and J.M. Troup, Inorg. Chem. 18:2605 (1979).

103. K. Yasafuku and H. Yamazaki, J. Organomet. Chem. 121:405 (1976); K. Yasufuku and H. Yamazaki, J. Organomet. Chem. 127:197 (1977).

104. (a) O. Gambino, G. Cetini, E. Sappa and M. Valle, J. Organomet. Chem. 20:195 (1969); (b) M. Tachikawa, J.R. Shapely and C.G. Pierpont, J. Am. Chem. Soc. 97:7173 (1975) and references therein.

105. J.L. Davidson and D.W.A. Sharp, J.C.S. Dalton Trans. 1123 (1976).

106. L.E. Smart, J. Browning, M. Green, A. Laguna, J.L. Spencer and F.G.A. Stone, J.C.S. Dalton Trans. 1777 (1977).

107. C. Hoogzand and W. Hubel in "Organic Synthesis via Metal Carbonyls", (I. Wender and P. Pino eds.), Vol. 1, Wiley, New York (1968), p. 343.

108. G.M. Whitesides and W.J. Ehmann, J. Am. Chem. Soc. 91:3800 (1969).

109. A.N. Nesmayanov, A.I. Gusev, A.A. Pasynskii, K.N. Anisimov, N.E. Kolobova and Y.T. Struchkov, Chem. Commun. 739 (1969).

110. A. Bond, M. Bottrill, M. Green and A.J. Welch, J.C.S. Dalton Trans. 2372 (1977).

111. J.A. Connor and G.A. Hudson, J. Organomet. Chem. 97:C43 (1975).

112. H. Yamazaki and N. Hagihara, J. Organomet. Chem. 7:P22 (1967); E. Muller, Synthesis 761 (1974); K. Mosely and P.M. Maitlis, J.C.S. Dalton Trans. 169 (1974).

113. Y. Wakatsuki, T. Karamitsu and H. Yamazaki, Tetrahedron Lett. 4549 (1974).

114. K.P.C. Vollhardt, Acc. Chem. Res. 10:1 (1977).

115. D.R. McAlister, J.E. Bercaw and R.G. Bergman, J. Am. Chem. Soc. 99:1666 (1977).

116. J.P. Collman, J.W. Kang, W.F. Little and M.F. Sullivan, Inorg. Chem. 7:1298 (1968).

117. J. Browning, M. Green, J.L. Spencer and F.G.A. Stone, J.C.S. Dalton Trans. 97 (1974).

118. H.C. Clark and A. Shaver, Can. J. Chem. 53:3462 (1975).

119. M.A. Bennett and P.B. Donaldson, Inorg. Chem. 17:1995 (1978).

120. (a) R.S. Dickson and H.P. Kirsch, Aust. J. Chem. 25:2535 (1972); (b) S.A. Gardner, P.S. Andrews and M.D. Rausch, Inorg. Chem. 12:2396 (1973).

121. O. Gambino, R.P. Ferrari, M. Chinone and G.A. Vaglio, Inorg. Chim. Acta 12:155 (1975) and references therein.

122. G.N. Schrauzer, P. Glockner and S. Eichler, Angew. Chem. Int. Ed. Engl. 3:185 (1964).

123. H. Hoberg and W. Richter, J. Organomet. Chem. 195:347 (1980).

124. (a) S.A.R. Knox, R.F.D. Stansfield, F.G.A. Stone, M.U. Winter and P. Woodward, J.C.S. Chem. Commun. 221 (1978); (b) M. Green, N.C. Norman and A.G. Orpen, J. Am. Chem. Soc. 103:1269 (1981).

125. P.M. Bailey, B.E. Mann, D.I. Brown and P.M. Maitlis, J.C.S. Chem. Commun. 238 (1976).

126. Y. Wakatsuki, K. Aoki and H. Yamazaki, J. Am. Chem. Soc. 101:1123 (1979).

127. C.E. Dean, R.D.W. Kemmitt, D.R. Russell and M.D. Schilling, J. Organomet. Chem. 187:C1 (1980).

128. Y. Wakatsuki and H. Yamazaki, J. Organomet. Chem. 193:169 (1977).

129. (a) H. Suzuki, K. Itoh, Y. Ishii, K. Simon and J. Ibers, J. Am. Chem. Soc. 98:8494 (1976); (b) L.D. Brown, K. Itoh, H. Suzuki, K. Hirai and J. Ibers, J. Am. Chem. Soc. 100:8232 (1978).

130. P. Caddy, M. Green, E. O'Brien, L.E. Smart and P. Woodward, J.C.S. Dalton Trans. 962 (1980).

131. S.A.R. Knox, R.F.D. Stansfield, F.G.A. Stone, M.J. Winter and P. Woodward, J.C.S. Chem. Commun. 934 (1979).

132. (a) I.V. Khand, G.R. Knox, P.L. Pauson, W.E. Watts and M.I. Forman, J.C.S. Perkin Trans. I 977 (1973); (b) I.U. Khand and P.L. Pauson, J.C.S. Perkin Trans. I 30 (1976).

133. P. Bladon, I.U. Khand and P.L. Pauson, J. Chem. Res. (S) 8 (1977).

134. I.U. Khand and P.L. Pauson, J. Chem. Res. (S) 9 (1977).

135. I.U. Khand and P.L. Pauson, J.C.S. Chem. Commun. 379 (1974).

136. Y. Iwashita, F. Tamura and H. Wakamatsu, Bull. Chem. Soc. Jpn. 43:1520 (1973).

137. H.D. Empsall, B.L. Shaw and A.J. Stringer, J.C.S. Dalton Trans. 185 (1976).

138. G. Varadi, I. Vecsei, I. Otvös, G. Palyi and L. Marko, J. Organomet. Chem. 182:415 (1979), refs. 2-10.

139. D.J.S. Guthrie, I.U. Khand, G.R. Knox, J. Kollmeier, P.L. Pauson and W.E. Watts, J. Organomet. Chem. 90:93 (1975).

140. H.G. Alt, Z. Naturforsch. 32b:144 (1977).

141. H.G. Alt, Z. Naturforsch. 32b:1139 (1977).

142. M. Green, J.Z. Nyathi, C. Scott, F.G.A. Stone, A.J. Welch and P. Woodward, J.C.S. Dalton Trans. 1067 (1978).

143. M. Bottrill and M. Green, J.C.S. Dalton Trans. 820 (1979).

144. E.A. Kelley, G.A. Wright and P.M. Maitlis, J.C.S. Dalton Trans. 178 (1979).

145. A.A. Pasynskii, K.N. Anisimov, N.E. Kolobova and A.N. Nesmeyanov, Dokl. Akad. Nauk. SSSR. 185:610 (1969); (b) A.N. Nesmeyanov, A.I. Gusev, A.A. Pasynskii, K.N. Anisimov, N.E. Kolobova and Yu. T. Struchkov, J.C.S. Chem. Commun. 277 (1969).

146. R.S. Dickson and L.J. Michel, Aust. J. Chem. 28:1957 (1975) and refs. therein.

147. U. Kruerke and W. Hubel, Chem. Ber. 94:2829 (1961).

148. D.M. Roe and A.G. Massey, J. Organomet. Chem. 23:547 (1970).

149. R.S. Dickson and G.R. Tailby, Aust. J. Chem. 23:229 (1970).

150. A.N. Nesmeyanov, N.E. Kolobova, U.V. Skripkin and K.N. Anisimov, Dokl. Akad. Nauk. SSSR 196:606 (1971).

151. R.C. Petterson, J.L. Cihonski, F.R. Young III and R.A. Levenson, J.C.S. Chem. Commun. 370 (1975).

152. T.V. Harris, J.W. Rathke and E.L. Muetterties, J. Am. Chem. Soc. 100:6966 (1978).

153. H. Yamazaki, K. Aoki, Y. Yamamoto and Y. Wakatsuki, J. Am. Chem. Soc. 97:3546 (1975).

154. Y. Suzuki and T. Takizawa, J.C.S. Chem. Commun. 837 (1972).

155. H. Bönnemann, Angew. Chem. Int. Ed. Engl. 17:505 (1978).

156. Y. Wakatsuki and H. Yamazaki, J.C.S. Chem. Commun. 280 (1973).

157. A. Naiman and K.P.C. Vollhardt, Angew. Chem. Int. Ed. Engl. 16:708 (1977).

158. P. Hong, K. Aoki and H. Yamazaki, J. Organomet. Chem. 150:279 (1978).

159. J.L. Davidson and D.W.A. Sharp, J.C.S. Dalton Trans. 2283 (1975); F.Y. Petillon, F. Le Floch-Perennou, J.E. Guerchais and D.W.A. Sharp, J. Organomet. Chem. 173:89 (1979).

160. Y. Wakatsuki, H. Yamazaki and H. Iwasaki, J. Am. Chem. Soc. 95:5781 (1973).

161. H. Le Bozec, A. Gorgues and P. Dixneuf, J.C.S. Chem. Commun. 573 (1978).

NUCLEOPHILIC ATTACK ON UNSATURATED HYDROCARBONS
COORDINATED TO TRANSITION METALS

S.G. Davies, M.L.H. Green and D.M.P. Mingos

Oxford University

1. INTRODUCTION

Nucleophilic addition and substitution reactions do not normally occur with simple unsaturated hydrocarbons such as ethylene, butadiene or benzene. When these unsaturated polyenes are coordinated to electron-withdrawing transition metal centers, however, they readily react with a wide range of nucleophiles such as H^-, alkyl$^-$, aryl$^-$, RO^-, R_3N, R_3P, CN^-, etc.

As might be expected, the less electron-rich a complex the more readily (faster) nucleophilic addition occurs. This is illustrated by the exchange of halogen for methoxide in the complexes $(\eta^6\text{-}C_6H_5X)[M]$, X=F, Cl, where the rate of exchange increases along the series [M] = $[Cr(CO)_3] < [Mo(CO)_3] \ll [Fe(\eta^5\text{-}C_5H_5)]^+ < [Mn(CO)_3]^+$.[1]

Nucleophilic addition to 18-electron organotransition metal cations has been extensively studied. The products are generally stable neutral 18-electron complexes. A series of rules has been introduced that allows the prediction of the regioselectivity of nucleophilic addition reactions to cationic 18-electron transition metal complexes containing unsaturated hydrocarbon ligands.[2] For example, the cation $[Fe(\eta^5\text{-}C_5H_5)(CO)_2(\eta^2\text{-}C_2H_4)]^+$, **I**, undergoes nucleophilic attack regioselectively by hydride addition on the ethylene ligand to give the ethyl compound, **II**.[3]

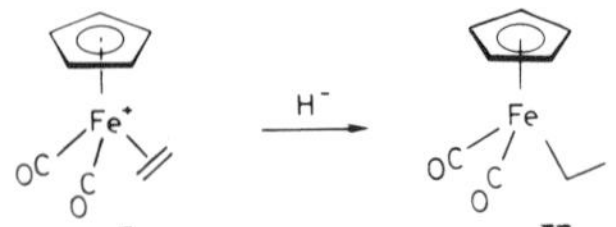

Nucleophilic addition to neutral complexes, for example, $[Cr(\eta^6\text{-arene})(CO)_3]^4$ and $[Fe(\eta^2\text{-olefin})(CO)_4]^5$, also occurs readily although the anionic primary products are normally not isolable. Nucleophilic addition to olefins may also be catalyzed by transition metal species.

In this article we shall be mainly concerned with nucleophilic addition reactions involving reactive cationic organotransition metal species.

2. NUCLEOPHILIC ADDITION TO ORGANOTRANSITION METAL CATIONS

The increased reactivity of unsaturated hydrocarbons coordinated to transition metal cations towards nucleophilic attack results from a net withdrawal of electron density from the unsaturated hydrocarbon to the positively charged metal center. This transfer of electron density can be attributed to the nature of the metal-ligand bonding. The coordination of olefins to a metal cation is, therefore, comparable to the introduction of electron-withdrawing substituents such as $-CO_2R$, $-NO_2$ or to reactive species such as bromonium ions, viz:

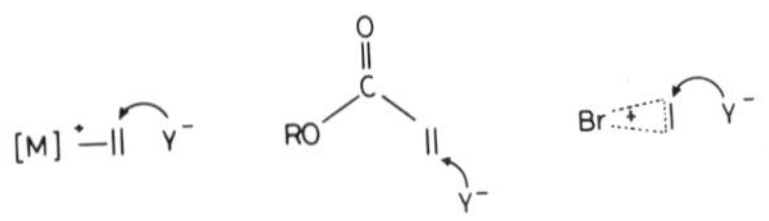

2.1 Stereochemistry of nucleophilic additions to 18-electron cations

Xray crystallographic and spectroscopic studies have shown that nucleophilic attack on unsaturated hydrocarbon ligands coordinated to 18-electron transition metal centers invariably occurs on the <u>exo</u>-face of the ligand, i.e. on the side of the ligand away from the metal.[6-8]

III

IV

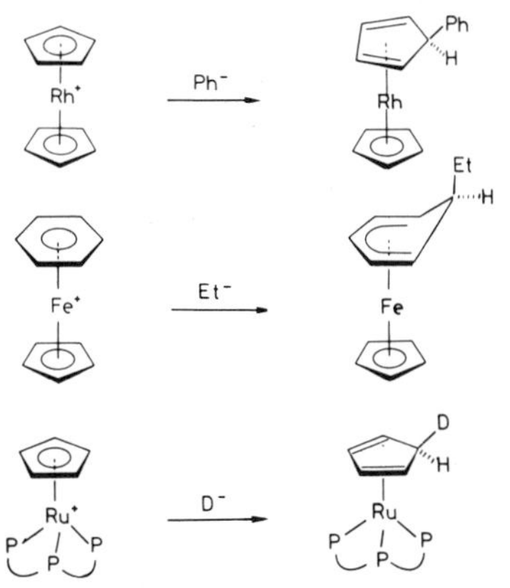

V $P \frown P \frown P = (Ph_2PCH_2CH_2)_2PPh$

VI $P \frown P \frown P = (Ph_2PCH_2)_3CMe$

This may be better described as an S_N2 displacement reaction whereby the incoming nucleophile displaces the M-C bond with inversion at the carbon.

2.2   Regioselectivity rules for nucleophilic additions

Although many organotransition metal cations contain several hydrocarbon ligands, nucleophilic addition reactions are generally very regioselective, occurring only on one of the polyene ligands. We have previously proposed a series of rules which if applied sequentially allow the prediction of the most favorable position of nucleophilic addition on the 18-electron organotransition metal cation for reactions that are <u>kinetically</u> rather than thermodynamically controlled.[2]

Unsaturated hydrocarbon ligands may be classified as <u>even</u> or <u>odd</u> according to the parity of the ligand <u>hapto</u> number (η = number of adjacent carbon atoms bonded to the metal) and as <u>closed</u> if they are cyclically conjugated and <u>open</u> if they are not. All unsaturated hydrocarbon ligands may be defined in terms of <u>even</u> or <u>odd</u> and <u>open</u> or <u>closed</u>. Some examples are given below.

Even η = 2, 4, 6

Odd η = 3, 5, 7

Closed   cyclically conjugated

Open   not cyclically conjugated

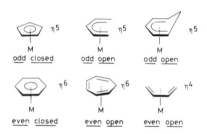

Rule 1. Nucleophilic attack occurs preferentially at <u>even</u> coordinated polyenes which have no unpaired electrons in their h.o.m.o's. (although cyclobutadiene is an <u>even</u> polyene it has unpaired electrons in its h.o.m.o and therefore, according to Rule 1, nucleophilic addition to other <u>even</u> polyenes is preferred. It is, however, attacked in preference to <u>odd</u> polyenyl ligands).

Rule 2. Nucleophilic addition to <u>open</u> coordinated polyenes is preferred to addition to <u>closed</u> polyenes.

Rule 3. For <u>even, open</u> polyenes nucleophilic attack at the terminal carbon atom is always preferred; for <u>odd, open</u> polyenyls attack at C-1 or C-3 occurs only if the metal center is a strongly electron withdrawing group.

Rules 1 and 2 allow the prediction of which ligands will be attacked. Rule 3 is concerned with regioselectivity on the particular ligand already determined by Rules 1 and 2 and concerns only the role of the metal center in determining the regioselectivity. Substituents on the polyenyl may also play a role in determining the position of nucleophilic addition, e.g. by steric restriction.

Rules 1 and 2 may, for the large majority of cases be simply expressed as:

Rule 1.   <u>Even</u> before <u>odd</u>

Rule 2.   <u>Open</u> before <u>closed</u>

2.3   Application of Rule 1

Some illustrations of the application of Rule 1 are given in Table 1. For example, the regioselectivity of nucleophilic attack onto cations **IV** and

Table 1. Application of Rule 1: Regioselectivity of hydride attack on transition metal cations.

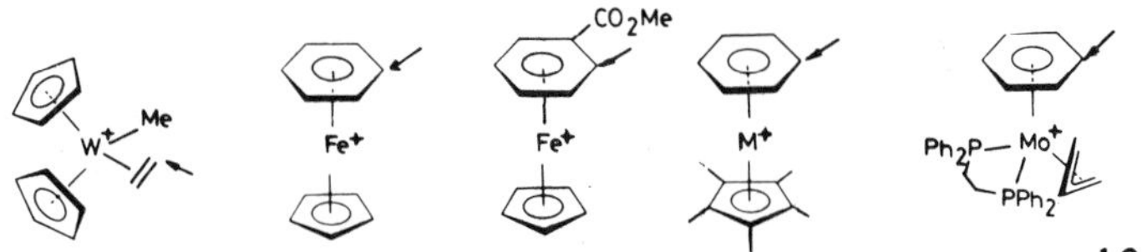

VII^9 $IV^{7,10}$ VIII^{11} IX^{12} M=Rh,Ir X^{13}

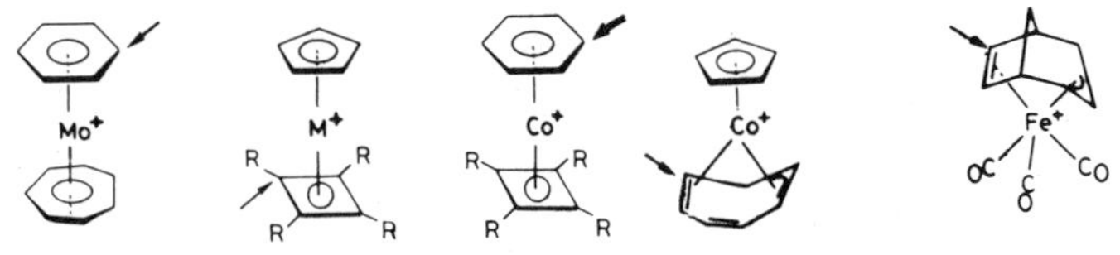

XI^{14} XII^{15} M=Ni,Pd XIII^{16} XIV^{17} XV^{18}

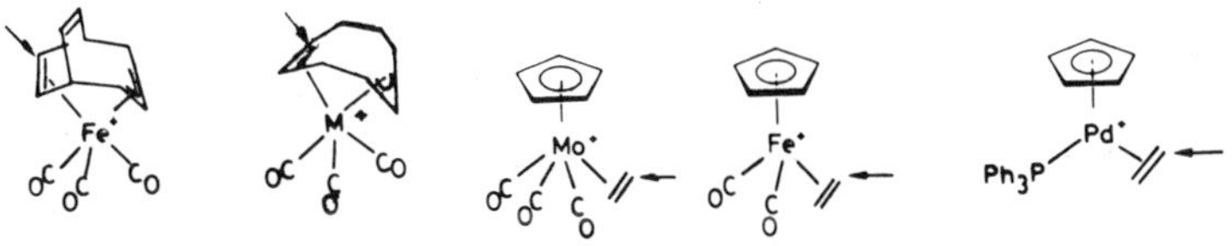

XVI^{19} XVII^{20} M=Fe,Ru,Os XVIII^{21} XIX^{3} XX^{22}

XI follows from Rule 1. In the latter example it is the benzene ligand that is attacked and not the cycloheptatrienyl ligand which one may have intuitively predicted as approximating to the tropylium cation. Cycloheptatriene in $[Mn(\eta^6\text{-}C_7H_8)(CO)_3]^+$ is approximately 10^4 times more reactive towards tertiary phosphine nucleophiles than the benzene ligand in the cation $[Mn(\eta^6\text{-}C_6H_6)(CO)_3]^{+.22b}$

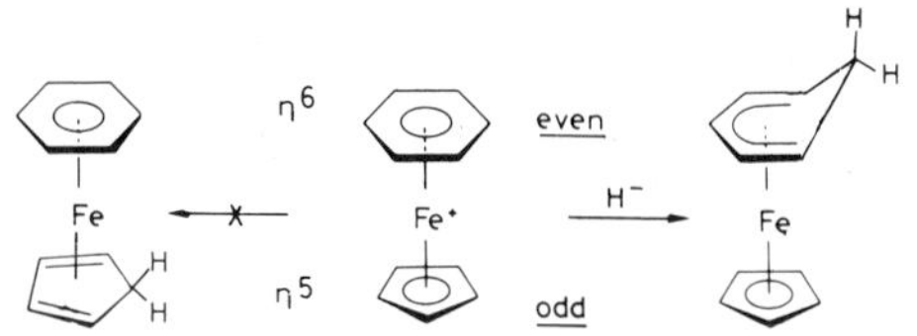

IV

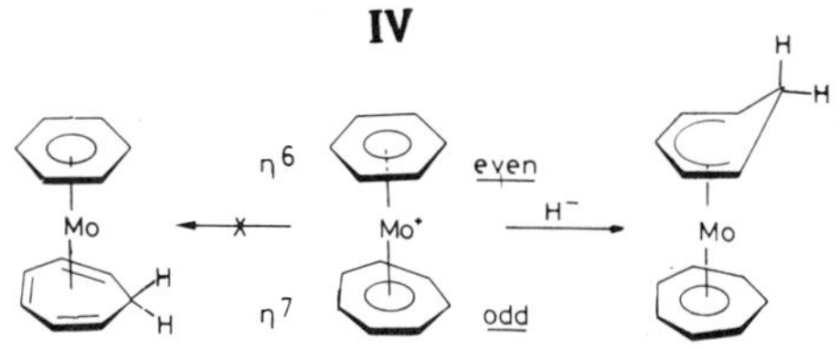

XI

2.4   Application of Rule 2

In many cases after applying Rule 1 there is still a choice between two odd or even ligands.   In these cases Rule 2 should be employed.   Some examples of the application of Rule 2 are given in Table 2.

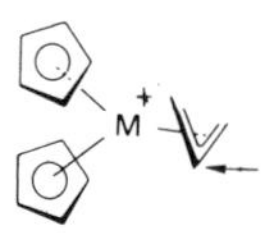

XXI,M=Mo[23]   **XXIII**[24] 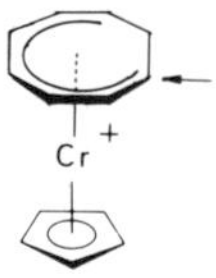   **XXIV**[25]
XXII, M=W

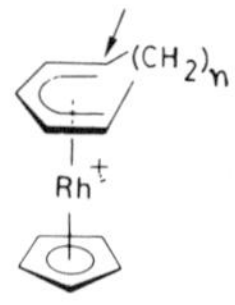

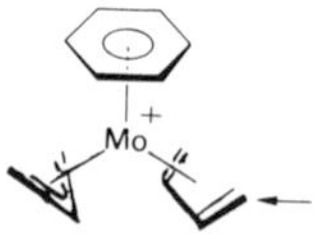

XXV[26] 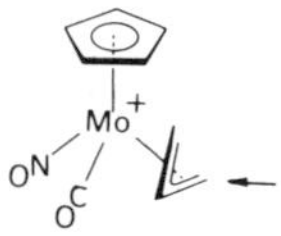   **XXVI**[27]

Table 2. Applications of Rule 2 (open before closed).

Cation **XXV** demonstrates the need to apply the rules sequentially. Rule 1 (even before odd) eliminates attack on the allyl ligand.   Rule 2 (open before closed) indicates nucleophilic attack on the butadiene while terminal attack follows from Rule 3 (see Section 2.5).

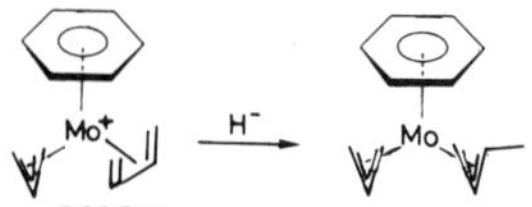

XXV

The Rules provide a ready explanation for the different modes of nucleophilic attack on the dications **XXVII** and **XXVIII** which result in addition to both rings for **XXVII** and the same ring twice for **XXVIII**.[28-9]

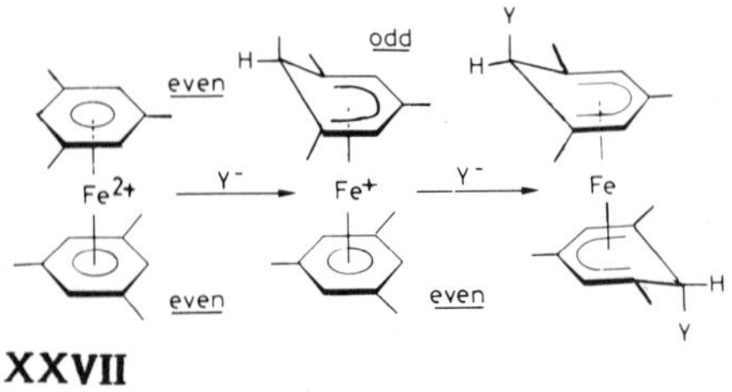

XXVII

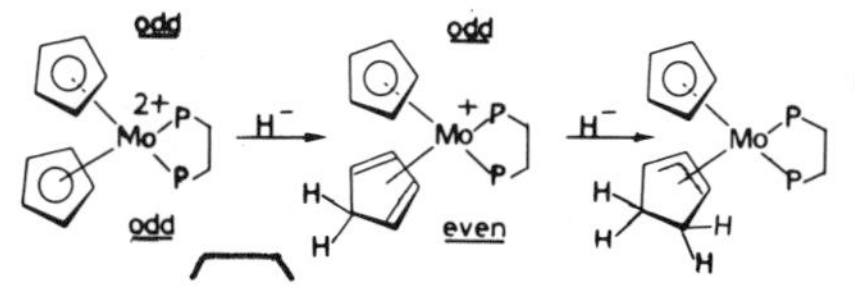

XXVIII (P⌒P = Ph$_2$PCH$_2$CH$_2$PPh$_2$)

Rules 1 and 2 may be used to define a reactivity series of unsaturated hydrocarbon ligands, i.e. preference for attack by nucleophiles of ligands:-

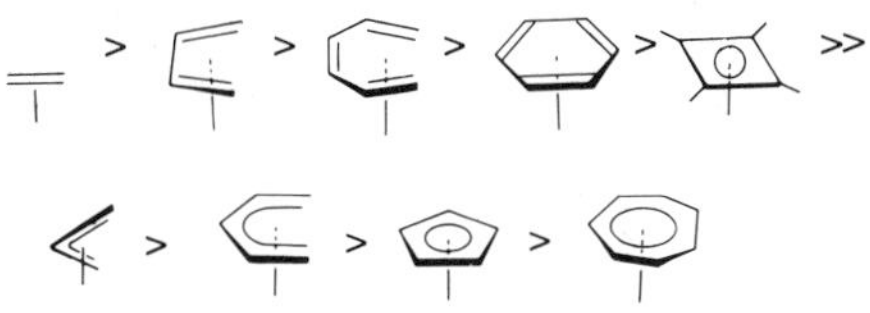

2.5 Application of Rule 3

When the hydrocarbon ligand undergoing nucleophilic attack is <u>open</u> then Rule 3 is applicable. For the η^3-allyl ligand the central carbon or one of the terminal carbon atoms may be attacked. Both modes of addition have been observed, the regioselectivity varying with the electron-richness of the metal.

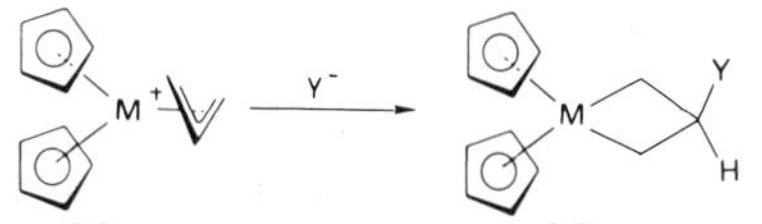

XXI[23] M=Mo   **XXII**[23] M=W

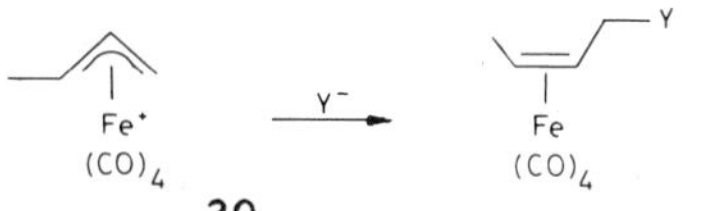

XXIX[30]

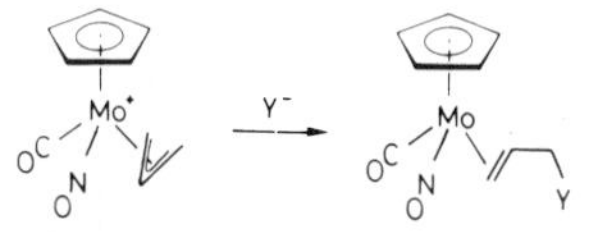

XXVI[27]

When coordinated to an electron-rich metal center the electron density on the allyl ligand is similar to that of the allyl anion, viz:

with the central carbon atom having the least electron density. When coordinated to an electron-poor metal center, however, the electron density on the allyl ligand resembles that on the allyl cation, viz:

with the terminal carbons atoms now having the lowest electron density. Therefore, in so far as additions are charge controlled, attack on allyl ligands coordinated to electron-rich centers (e.g. **XXI** and **XXII**) is expected and observed to occur on the central carbon atom. Attack on allyl ligands coordinated to electron-poor centers (e.g. **XXIX, XXVI**), however, is expected to occur on a terminal carbon atom, as observed.

For <u>even, open</u> ligands the electron density will always tend to be least on the terminal carbon atoms and, as expected, attack always occurs there (see Section 2.6).

2.6   Theoretical basis of the Rules

In Sections 2.2 to 2.5 above it was demonstrated that the classification of ligands according to their <u>hapto</u> number and the presence or absence of cyclic conjugation has important consequences for understanding the regioselectivities of nucleophilic addition reactions of complexes of these ligands.   The simplicity and general applicability of the rules, and in particular Rules 1 and 2, suggest that the observed regioselectivities must have their electronic origins in a general topologically based phenomenon which makes a clear distinction on the basis of the number of carbon atoms coordinated to the metal and the degree of cyclic conjugation.   General topological phenomena of this kind, the most important and dramatic illustration being the Woodward-Hoffmann Rules of organic chemistry, are often effectively understood in terms of perturbation theory arguments rather than detailed and sophisticated molecular orbital calculations on specific molecules.[31]   Therefore, in the following Section, we propose a very straightforward pertubation theory analysis which accounts in general terms for the Regioselectivity Rules.   However, recognising that such arguments may represent either a gross over-simplification or a distortion of the situation pertaining in real molecules we have sought to verify our conclusions by reference to molecular orbital calculations reported by other workers on specific organometallic cations.

The basic assumption underlying the pertubation theory analysis is that electron donation from the highest occupied molecular orbital of the polyene to an orbital of appropriate symmetry on the metal is particularly influential in determining the electron density and frontier orbital characteristics of the coordinated olefin. Such an assumption clearly is not novel, has its origins in second-order perturbation theory, and has been extensively used in organic chemistry.[32] Figure 1 illustrates the relevant interaction diagram between the h.o.m.o. of the polyene, ψ_P^o, and the l.u.m.o. of the metal cationic fragment, ψ_M^o. The energies of these molecular orbitals in the isolated fragments are E_P^o and E_M^o respectively. The energies and wavefunctions of the bonding and antibonding combinations arising from the interaction of ψ_P^o and ψ_M^o can be estimated from pertubation theory to be[33]

$$E_M'' = E_P'' = KS_{MP}^2/(E_M^o - E_P^o) \tag{1}$$

where K = constant and S_{MP} = overlap integral, $\int \psi_M^o \psi_P^o d\tau$, in the complex, and

$$\psi_b = \cos\theta \cdot \psi_M^o + \sin\theta \cdot \psi_P^o \tag{2}$$

$$\psi_a = \sin\theta \cdot \psi_M^o - \cos\theta \cdot \psi_P^o \tag{3}$$

where $\cos\theta = 1/(1+c_{PM}'^2)^{1/2}$ and $\sin\theta = c_{PM}'/(1 + c_{PM}'^2)^{1/2}$

$$c_{PM}' = KS_{PM}/(E_M^o - E_P^o) \tag{4}$$

These expressions are based on the neglect of interfragment overlap and the assumption that the interfragment Hamiltonian matrix element can be approximated by the expression $H_{PM} = KS_{PM}$.

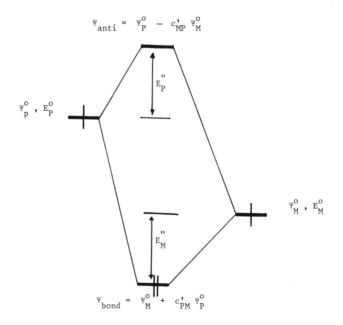

$$\Psi_{anti} = \psi_P^O - c_{MP}' \psi_M^O$$

$$\psi_P^O, E_P^O$$

$$E_P''$$

$$\psi_M^O, E_M^O$$

$$E_M''$$

$$\Psi_{bond} = \psi_M^O + c_{PM}' \psi_P^O$$

Figure 1
Schematic illustration of the interaction between the h.o.m.o. of the polyene and the l.u.m.o. of the metal ligand cationic fragment. Although the interaction diagram has been illustrated in terms of single electron occupancy of the polyene and metal frontier molecular orbitals, the general scheme is also applicable to situations where the polyene h.o.m.o. is doubly occupied and the l.u.m.o. of the metal fragment is empty.

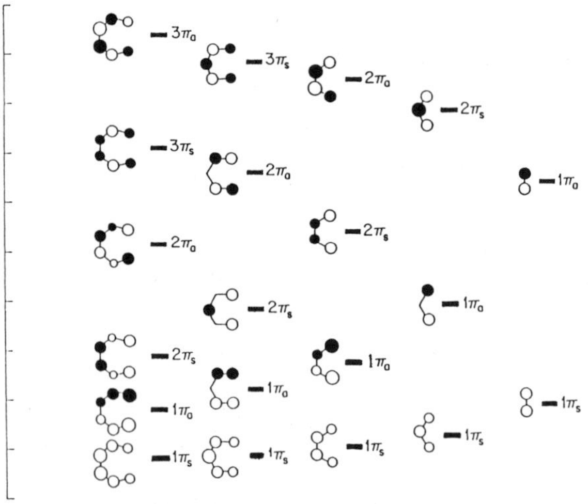

Figure 2
Relative energies and nodal characteristics of the π-molecular orbitals of some linear polyenes.

It is clear from these expressions that the mixing coefficients $\sin\theta$ and $\cos\theta$ will be determined primarily by the overlap integral S_{PM} and the energy difference $E_M^o - E_P^o$. Figure 2 illustrates the relative energies and nodal characteristics of the π-molecular orbitals of some acyclic polyenes. For odd-polyenes the h.o.m.o.s have essentially the same energy (identical within the Hückel approximation) and therefore the mixing coefficient C'_{PM} will only be a function of the overlap integral S_{PM}. For even polyenes, with the exception of ethylene, the h.o.m.o.s have comparable energies and are stabilized relative to the corresponding orbitals in the odd polyenes. If S_{PM} for even and odd polyenes are approximately equal it follows that the mixing coefficient c'_{PM} will be somewhat larger for even polyenes. However, for reasons which will be discussed below, this difference is unlikely to be as important as the different electron occupations of the h.o.m.o.'s in even and odd polyenes.

Figure 3 illustrates three possible situations for even and odd polyenes C_nH_{n+2} coordinated to an ML_n^+ fragment which contains a total of $18 - n$ valence electrons about the metal atom. $\theta = 90^o$ corresponds to a situation where the metal acceptor orbital lies at much higher energies than the polyene h.o.m.o. and consequently c'_{PM} is small and little charge transfer occurs; $\theta = 45^o$ where the metal acceptor orbital and polyene h.o.m.o. are well matched in energy and finally $\theta = 0^o$ where the metal acceptor orbital lies at much lower energies than the polyene h.o.m.o. and extensive charge transfer results. Therefore in general nucleophilic attack at the polyene will be encouraged by electron withdrawing ligands, L, at the metal center which stabilize the metal acceptor orbital ψ_M^o.

The fact that the highest occupied orbital of an even polyene is doubly occupied whereas the highest occupied of an odd polyene is singly occupied gives rise to the important differences in their susceptibility to attack by nucleophiles. In the former case all the electron density in the bonding molecular orbital ψ_b originates from the polyene h.o.m.o., but in the latter requirements of the 18-electron rule dictates that the electron density be contributed by the polyene and the metal. From equation (2) above it is clear that the electron density associated with the metal in ψ_b is equal to

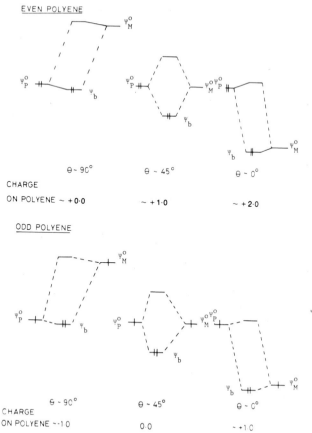

Figure 3

H.o.m.o. - l.u.m.o. interactions in odd and even polyene metal complexes as a function of the relative orbital energies.

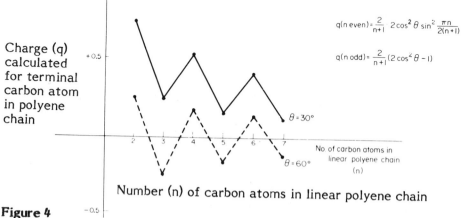

$$q(n\ even) = \frac{2}{n+1}\ 2\cos^2\theta\ \sin^2\frac{\pi n}{2(n+1)}$$

$$q(n\ odd) = \frac{2}{n+1}(2\cos^2\theta - 1)$$

Number (n) of carbon atoms in linear polyene chain

Figure 4

Calculated charge for the terminal carbon atoms in a homologous series of even and odd polyene and polyenyl cationic metal complexes.

$2\cos^2\theta$. The resultant positive charge on the coordinated <u>even</u> polyene in therefore $+2\cos^2\theta$, and for the coordinated <u>odd</u> polyene is $+2\cos^2\theta - 1$, because prior to complex formation the metal orbital ψ_M^0 contained a single electron.

Therefore, as long as the mixing coefficients for <u>odd</u> and <u>even</u> polyenes do not differ greatly, a coordinated <u>even</u> polyene has an extra unit of positive charge associated with the carbon atoms compared to an <u>odd</u> polyene. For the three situations illustrated in Figure 3 this suggests that the charge on an <u>even</u> polyene will vary from 0 to +2.0 and that for an <u>odd</u> polyene from -1.0 to +1.0.

Estimates of the specific charges on particular carbon atoms of the coordinated polyenes have been made by substituting the explicit forms of the polyene h.o.m.o.s into equation (2). The computed charges for the terminal carbon atoms as a function of polyene chain length are plotted in Figure 4 for two values of θ corresponding to ML_n^+ being a strong and weak electron withdrawing group. From Figure 4 it is clear that as long as θ does not vary greatly with n and the other factors discussed above, the terminal carbon atoms of an <u>even</u> polyene are more positively charged; i.e. the charge is dominated by the parity of the polyene. The rates of nucleophilic addition reactions of 18-electron cationic polyene complexes are likely to be charge rather than orbitally controlled, especially if the nucleophile is small and highly charged, and therefore the relative rates of such reactions should directly reflect the charge differences described above.

Support for this simple perturbation theory argument, which leads directly to Rule 1 described in section 2.3, has been gained from the INDO molecular orbital calculations reported by Clack and Kane-Maguire[34] on $[Fe(\eta^5-C_5H_5)(\eta^6-C_6H_5X)]^+$, where X = H, Me, OMe or CO_2Me. For $[Fe(\eta^5-C_5H_5)(\eta^6-C_6H_6]^+$, for example, the total positive charge on the <u>odd</u> cyclopentadienyl ligand of +0.35 is significantly smaller than that on the <u>even</u> benzene ring of +0.89.

Eisenstein and Hoffmann[35] have made a detailed analysis of potential energy surfaces of the reactions between the hydride ion, H^-, and ethylene complexes of the type $[Ni(PH_3)_2(C_2H_4)]$, $[Fe(\eta^5-C_5H_5)(CO)_2(\eta^2-C_2H_4)]^+$ and $[Mo(\eta^5-C_5H_5)_2R(\eta^2-C_2H_4)]^+$. They have argued that these calculations based on the extended Hückel formalism do not support a simple mechanism based on a primary bonding interaction between the incoming H^- and the olefinic carbon atoms which have been polarized by virtue of being adjacent to a positively charged metal ion. This lack of constructive bonding has been attributed to the presence of a significant degree of backbonding from metal to olefin in accordance with the Dewar-Chatt-Duncanson bonding model. They have argued that the major contribution to the activation energy for nucleophilic addition originates from a slipping motion of the ML_n^+ fragment towards an η^1-coordination mode as the nucleophile approaches the coordinated olefin. This suggestion is a most interesting one, but it is difficult to imagine experiments which would verify its occurrence and eliminate the alternative mechanism which is promoted by the presence of small positive charges on the ethylene carbon atoms.

Rule 2 which suggests that nucleophilic attack occurs more readily at open rather than closed polyenes would at first sight appear to be intuitively obvious, but is worthy of consideration as an extension of the perturbation theory analysis developed above. For a cyclic polyene the h.o.m.o. is a degenerate set of occupied or partially occupied molecular orbitals and therefore a comparison of open and closed polyenes requires a consideration of the interaction diagrams illustrated in Figure 5 for pentadienyl and cyclopentadienyl coordinated to an axially symmetric ML_n^+ fragment which is a good electron withdrawing group. Since the second order perturbation theory terms which form the basis of the interactions are additive then it can be readily appreciated that the resultant charges on pentadienyl and cyclopentadienyl are respectively

$$2\cos^2\theta_1 + 2\cos^2\theta_2 - 1 \quad\quad \text{and} \quad\quad 4\cos^2\theta_3 - 1$$

where θ_1, θ_2 and θ_3 represent the mixing coefficients as defined in equations (2), (3) and (4) above for the polyene orbitals ψ_1, ψ_2 and ψ_3 in Figure 5.

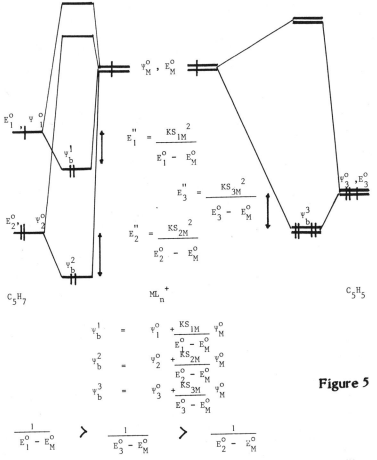

Figure 5

Comparison of the interactions of pentadienyl and cyclo-pentadienyl frontier molecular orbitals.

If the metal polyene overlap intergrals S_{PM} in the expression given in (4) are equal for pentadienyl and cyclopentadienyl then the relative magnitudes of θ_1, θ_2 and θ_3 will be set by the denominator energy term $E_M^O - E_P^O$.

Since $E^O(\psi_1^O) < E^O(\psi_3^O) < E^O(\psi_2^O)$ then θ_3 is likely to take some intermediate value between θ_1 and θ_2. Indeed if

$$\cos^2\theta_3 = (\cos^2\theta_1 + \cos^2\theta_2)/2$$

then the total positive charge on the pentadienyl ligand will be identically equal to that on the cyclopentadienyl ligand. However given that $E^O(\psi_3^O)$ more closely approximates to $E^O(\psi_2^O)$ the total charge on the pentadienyl is likely to be somewhat larger than that on the cyclopentadienyl. This effect, however, is likely to be less significant than the difference in positive charge on specific carbon atoms.   For cyclopentadienyl the degenerate nature of the h.o.m.o.s will result in an equal distribution of positive charge for all carbon atoms on complexation.   However, for pentadienyl the stronger interaction between ψ_1^o, which has a nodal plane passing through atoms 2 and 4 as shown in (1), will result in a larger build up of positive charge on carbon atoms 1, 3 and 5 on coordination.

(1)

This argument can be illustrated by the comparison of the computed charges for the isolated pentadienyl and cyclopentadienyl cations shown in (2) and (3) below.

(2)   (3)

Such an effect would of course be less dramatic for such species coordinated to a metal center because of the additional interaction between ψ_2^o and ψ_M^o. Nonetheless, we consider that the effect will be sufficiently large and general to justify the proposal of Rule 2.

From the perturbation theory expressions given above, it can readily be shown that an <u>even</u> polyene bears a positive charge for all values of $\theta < 90°$, (see Figure 5 for example) whereas an <u>odd</u> polyene only bears a positive charge for values of $\theta < 45°$.   Furthermore, within the approximations of the model the charge on the terminal atom of an <u>even</u> polyene is always more positive than that on the internal carbon atoms.   In contrast the terminal carbon atom of an <u>odd</u> polyene only bears the greatest positive charge when $\theta < 45°$, and therefore we propose in Rule 3 that for such ligands the preferred site for attack should vary as a function of the electron withdrawing capability of the ML_n^+ fragment.   The theoretical justification for Rule 3 from detailed molecular orbital calculations is at the present time incomplete, and there appears to be little correlation between the site of nucleophilic addition and the computed charges. For example, Clack, Monshi and Kane-Maguire[37-8] have reported INDO calculations on $[Fe(\eta^5-C_6H_7)(CO)_3]^+$ where the calculated charges on the coordinated dienes would predict the regioselectivities C(2) > C(3) > C(1), whereas nucleophiles generally add at the C(1) position. These workers suggested that computed free valency indices more correctly predict the site of nucleophilic addition; however, these indices do not always predict the correct regioselectivities in organometallic cations (see for example reference 34). A similar absence of predictability appears to govern attempts to compute the effect of σ- and π-donating or electron withdrawing substituents on the regioselectivities of nucleophilic addition. Clack and Kane-Maguire have studied the systems $[Cr(\eta^6-C_7H_6Z)(CO)_3]^+$, where Z = H, OMe, Cl, and CO_2Me, and concluded that in none of these systems do charge variations alone provide an adequate rationale for the regioselectivity on a single hydrocarbon. In view of earlier observations that the site of nucleophilic attack on the cation $[Fe(\eta^5-C_6H_7)(CO)_3]^+$ correlates well with the dienyl carbon free valencies, C-C bond orders were also calculated. However, the calculated favored site for formation of a bond to a nucleophile did not consistently agree with the experimental observations. These workers also considered the coefficients of the carbon atoms of the l.u.m.o.s in these ions but could find no obvious relationship to the observed regioselectivities. From these results it would appear, either that the current level of calculations is not sufficiently sensitive to predict the charge distribution in these ions, or that the transition state does not depend exclusively on the ability of the nucleophile to bond most effectively with the most positively charged carbon atom. Clearly additional experimental and theoretical work is required to define the applicability and theoretical viability of Rule 3.

Schilling, Hoffmann and Faller[39] have made a detailed experimental analysis of the effect of asymmetry in the ML_n^+ fragment on the regioselectivity of nucleophilic addition to the allyl ligand. It has been established that the addition of nucleophiles to $[Mo(\eta^5-C_5H_5)(\eta^3-C_3H_5)(NO)(CO)]^+$ is highly stereoselective adding <u>cis</u> to the nitrosyl and <u>trans</u> to the nitrosyl in the two epimers illustrated below:

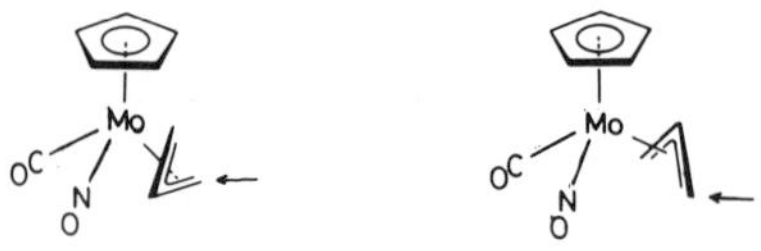

XXXIII (endo)   **XXXIV** (exo)

The net charges calculated on the terminal carbon atoms of the allyl ligand not only differ within the allyl group but change when going from _exo_ to _endo_ as illustrated above. Simple arguments based on these charges thus account for the regioselectivities.

2.7 Evidence for charge control in nucleophilic addition reactions to organometallic cations

Evidence in support of the view that the nucleophile attacks the carbon having the least electron density is provided by the examples given below where the polyenes under attack possess substituents that either donate or withdraw electron density. Nucleophilic addition to cations **XXX** and **XXXI** occurs preferentially onto the rings having the mesomerically electron-withdrawing carbomethoxy group, and in the β-positions.[40] The weakly inductively electron-donating methyl group exerts little, if any, influence.

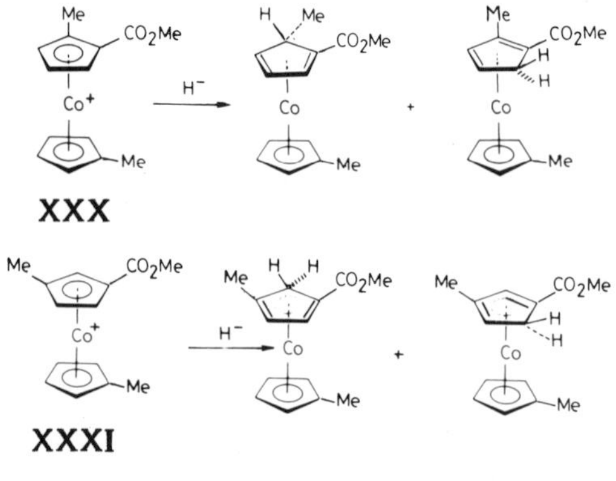

XXX

XXXI

For the cation **XXXII**, nucleophilic attack occurs at the 2-position when X is an electron withdrawing group (e.g. CO_2Me) whereas attack at the 3-position is observed when X is electron releasing (e.g. OMe).[37]

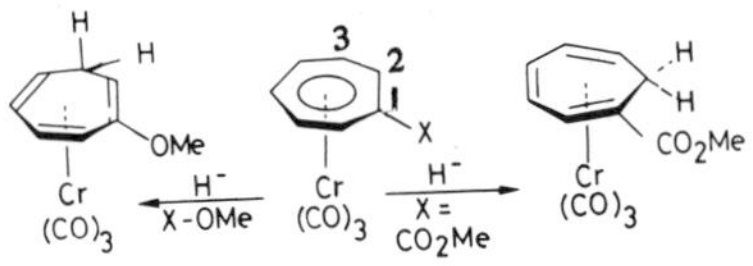

XXXII

Of particular interest are the endo and exo isomers **XXXIII** and **XXXIV** because the terminal positions of the allyl ligand are not equivalent (one end is trans to NO and the other trans to CO). As in other nucleophilic addition reactions, the nucleophile attacks the uncoordinated face of the ligand. Theoretical calculations of charge distribution suggest that in the endo isomer **XXXIII** the carbon trans to NO bears less electron density than the carbon trans to CO; whereas the reverse is found for the exo isomer **XXXIV**[39] (see Section 2.6). Nucleophilic attack occurs trans to NO in isomer **XXXIII** but trans to CO in isomer **XXXIV**.[41] Thus once again regioselectivity appears to be charge controlled.

Some other examples where substituents influence the regioselectivity of nucleophilic addition are given below.

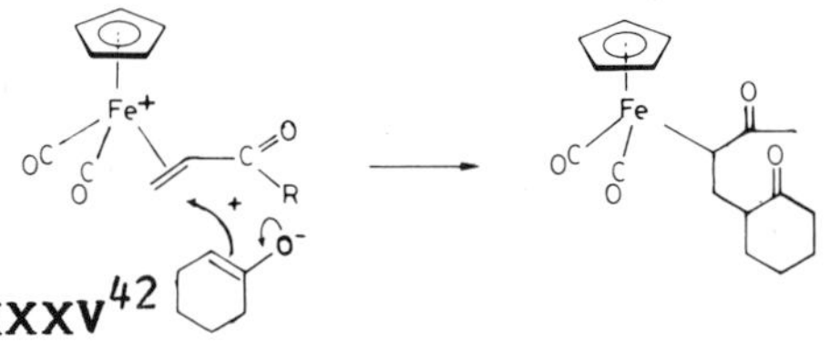

XXXV[42]

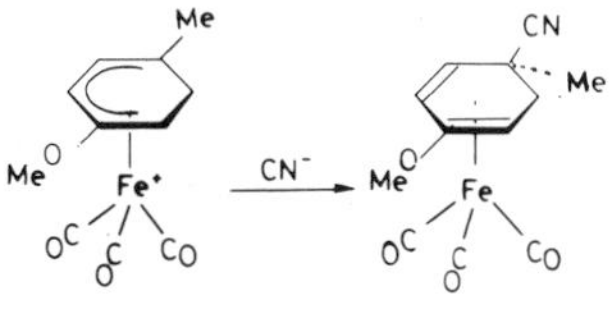

XXXVII[44]

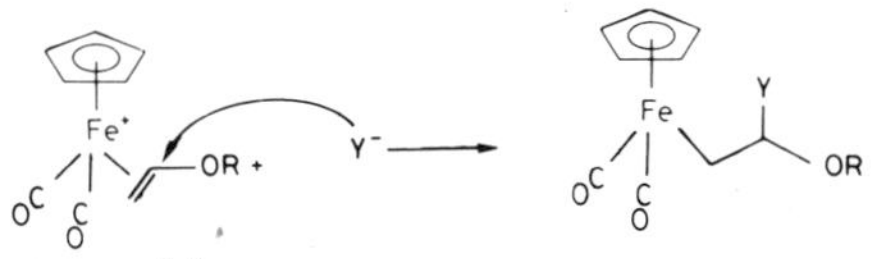

XXXVI[43]

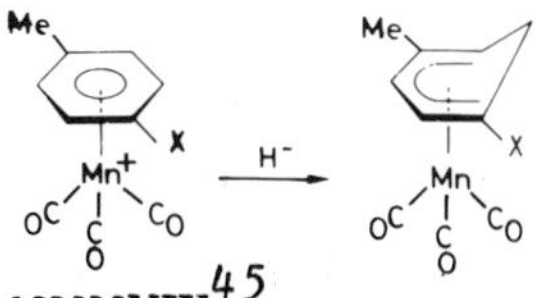

XXXVIII[45]

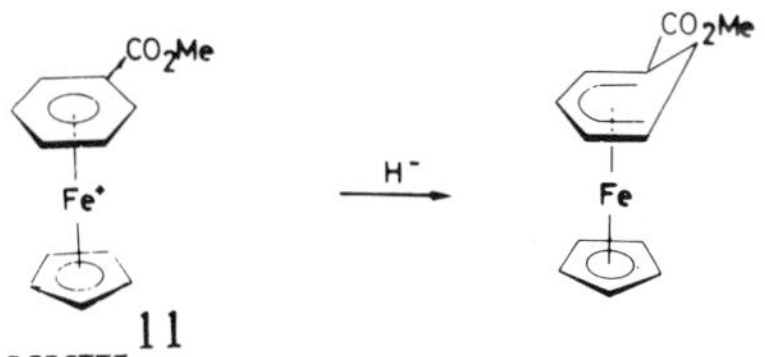

XVIII[11]

X	A:B
= Cl	69:31
= OMe	37:63
= NMe$_2$	3:97

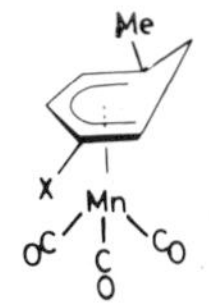

2.8 Nucleophilic addition to cations containing CO and unsaturated hydrocarbon ligands

Cations containing carbon monoxide ligands may be susceptible to nucleophilic attack on the carbon atom of the CO ligand. For atoms containing both CO and unsaturated ligands both types of reaction have been observed. Attack at an unsaturated hydrocarbon ligand is, however, generally preferred except where the choice is between an η^5-C_5H_5 ligand and CO. For example, cation **XXXIX** undergoes attack by hydride at CO to generate the metal formyl complex **XL.**[46]

XXXIX

XL

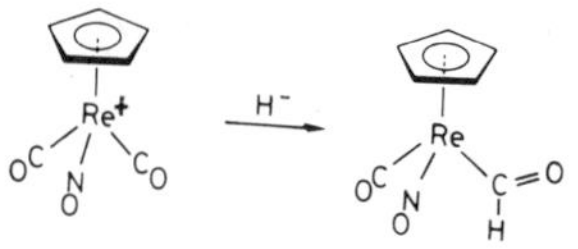

In the hydride reduction of cation **XLI,** attack at CO represents a minor pathway:[47]

XLI

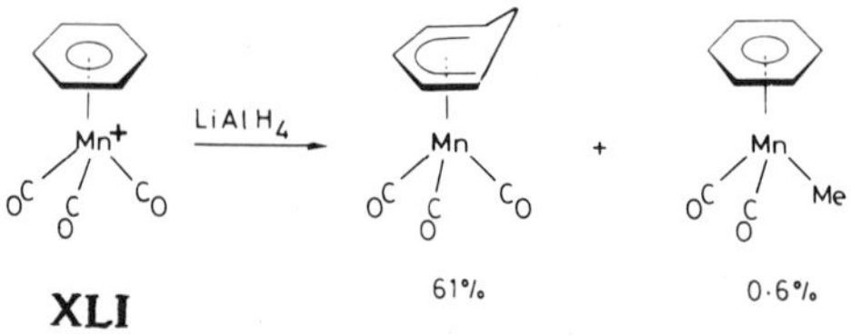

Reduction of cation **XLII** to give the iron hydride **XLIII** is also believed to proceed via initial attack at CO.[48] Methoxide and hydroxide behave similarly[49]

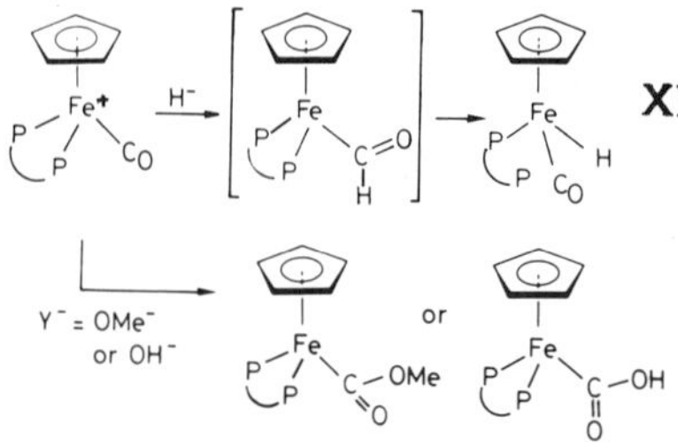

XLII

XLIII

(P P = $Ph_2PCH_2CH_2PPh_2$).

The dication **XLIV** is also readily susceptible to attack on CO:[50]

XLIV

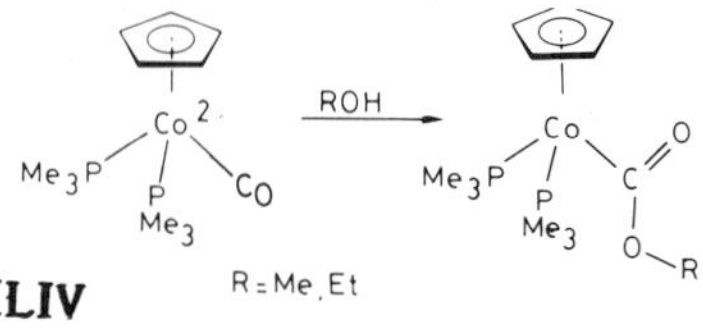

The cyclohexadienyl complex **XLV** undergoes initial attack by methoxide on CO; however, subsequent equilibration produces the exo product **XLVI:**[51]

XLV **XLVI**

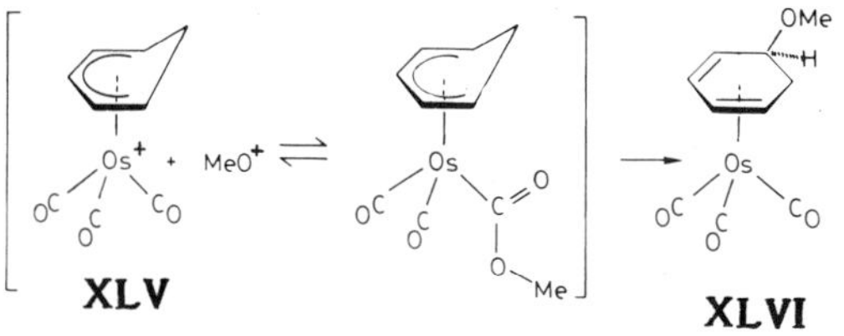

The regioselectivity of nucleophilic addition to hydrocarbon ligands in cations containing two or more unsaturated hydrocarbon ligands and CO ligands can be predicted on the basis of the Rules given above. Some examples are shown in Table 3.

Table 3. Regioselective addition to CO containing cations.

XIX[3,52] **XV**[18] **XVI**[19] **XVIII**[21] **XXVI**[27] **XLVII**[53]

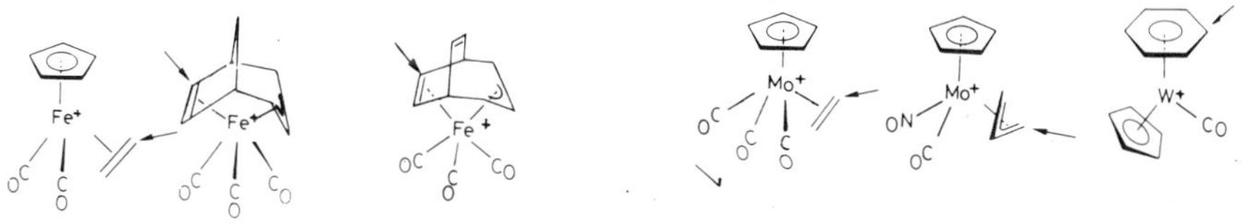

2.9 Direct nucleophilic attack on the metal

Direct nucleophilic addition to the metal occurs when the cationic species under attack has 16 or less electrons. Thus stable 16-electron cations, and 18-electron cations which readily lose a 2-electron ligand to generate a reactive 16-electron intermediate, are attacked at the metal. For example, the 16-electron cation **XLVIII** reacts with a variety of 2-electron ligands L to produce the stable 18-electron cations **XLIX:**[54]

XLVIII **XLIX**

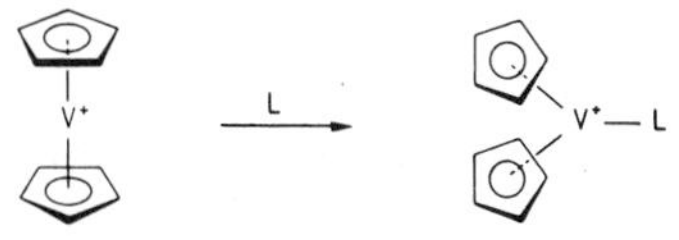

L = H_2O,Me_2CO,pyridine,etc.

The 18-electron cations **L** and **LII** are reduced by hydride to the corresponding iron hydride species by a mechanism involving direct attack on the metal, as demonstrated by deuterium labeling.[55] The mechanism has been shown to involve prior dissociation of a phosphine ligand since cation **LII** gives a mixture of hydrides **LIII** and **LIV**.[56] Hydride attack directly on the metal of **LII** would have given by inversion at iron only the hydride **LIII.**

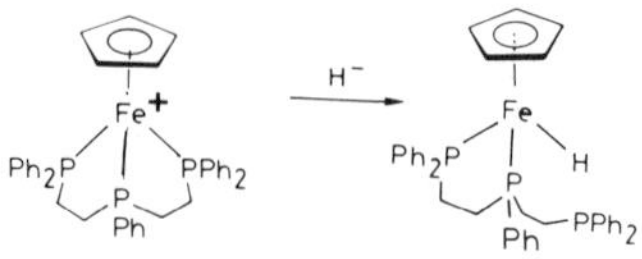

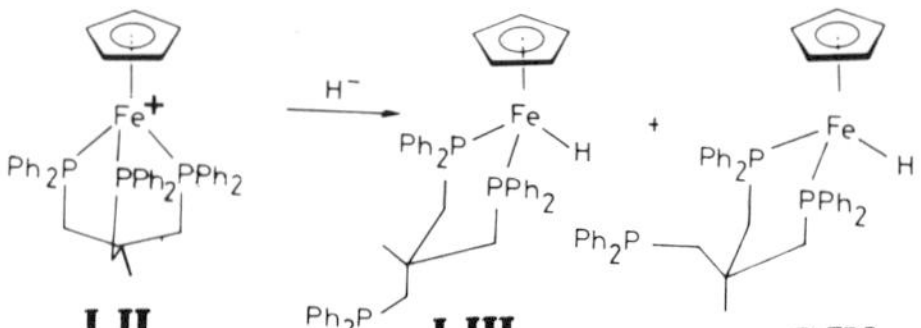

Ionisation of the metal halides **LV** generates the 16-electron intermediates **LVI** which react with 2-electron ligands L' to form the 18-electron cations **LVII**:

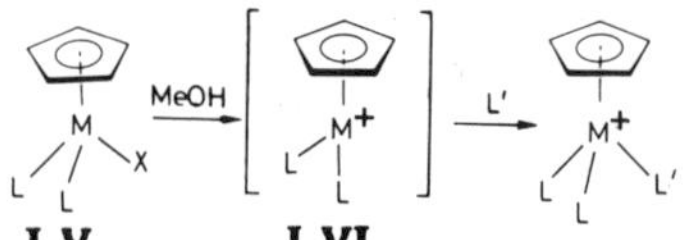

M = Fe, L_2 = $Ph_2PCH_2CH_2PPh_2$, X = Br[57] or M = Ru, L = Me_3P[58], L_2 = $Ph_2PCH_2CH_2PPh_2$[59] and L' = olefin, RCN, pyridine, etc.

2.10 Examples of attack on <u>even</u>, <u>open</u> ligands

<u>Even</u>, <u>open</u> ligands include η^2-olefins, η^4-dienes, and η^6-trienes such as η-cycloheptatriene. It follows from the rules that η^2-olefin ligands bonded to transition metal cations are particularly susceptible to nucleophilic attack. Indeed, as expected, they are found to be attacked in preference to other unsaturated hydrocarbon ligands. Nucleophilic addition to olefins coordinated to $[Fe(\eta^5\text{-}C_5H_5)(CO)_2]^+$ has been extensively studied and the reaction has found some use in organic synthesis:[60]

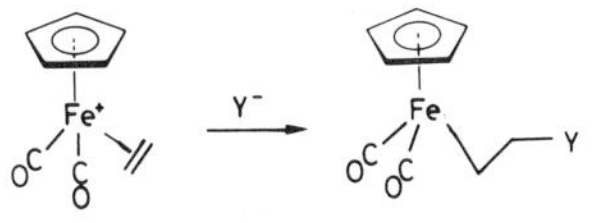

Y^- = stabilised carbanions, enamines, RS^-, RO^-, Me_3N, R_3P, $P(OR)_3$, H^-; $YH = NR_2H$.

Nucleophilic addition to $[Fe(\eta^5-C_5H_5)(CO)_2(\eta^2\text{-olefin})]^+$ cations occurs stereospecifically trans to the metal:[61]

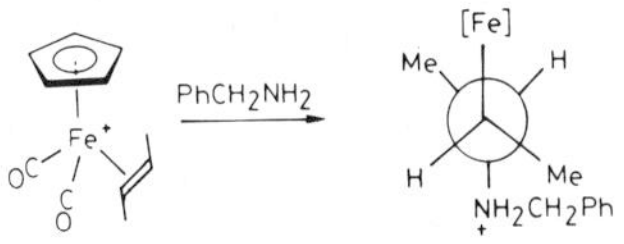

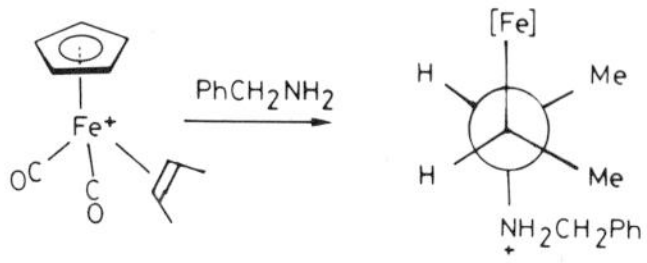

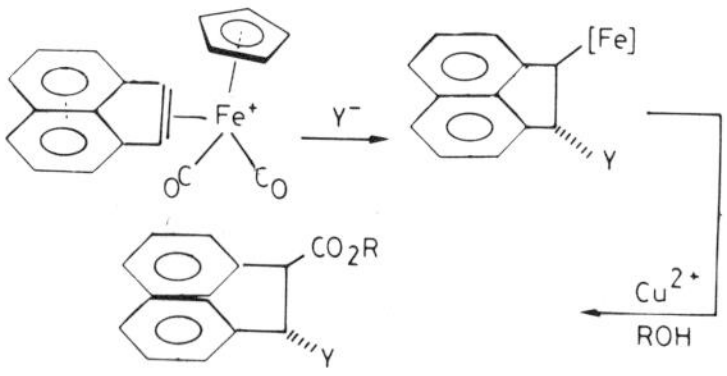

Nucleophilic additions to monosubstituted olefins coordinated to $[Fe(\eta^5-C_5H_5)(CO)_2]^+$ are highly regioselective when there are strongly electron withdrawing or electron donating substituents on the olefin (see Section 2.7).[42-3]

XXXV

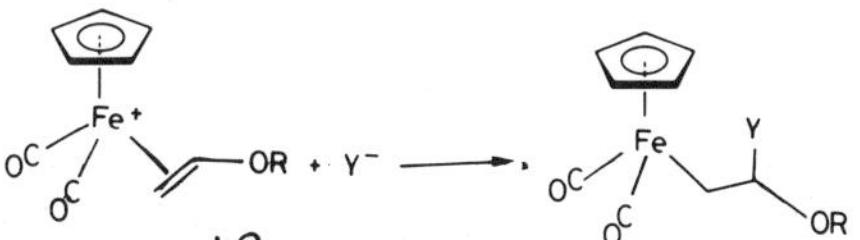

XXXVI [42]

For simple alkyl substituted alkenes, nucleophilic addition to monosubstituted-olefin complexes such as **LVIII** is generally not very regiospecific.[62] For the styrene complex **LIX**, however, only addition to the carbon bearing the phenyl group is observed.[62] This is presumably because the phenyl group stabilizes the carbonium ion in the resonance structure **LX** much better than a simple alkyl group.[35]

$$R = CH(CO_2Me)_2$$

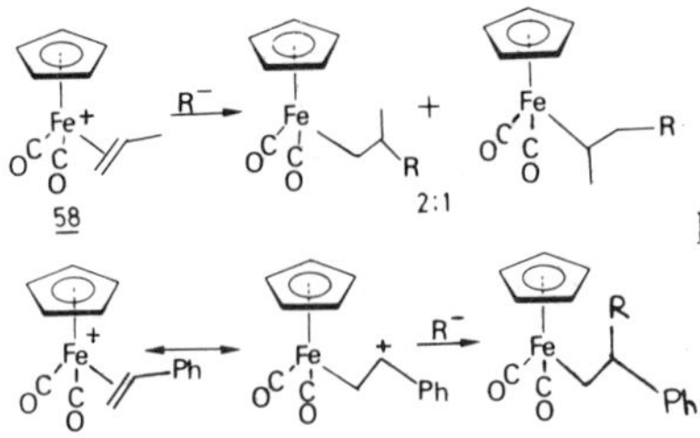

The cations $[Fe(\eta^5\text{-}C_5H_5)(CO)_2(\eta^2\text{-olefin})]^+$ are sufficiently electrophilic to react with $[Fe(\eta^5\text{-}C_5H_5)(CO)_2(\eta^1\text{-allyl})]$ complexes as shown below.[63]

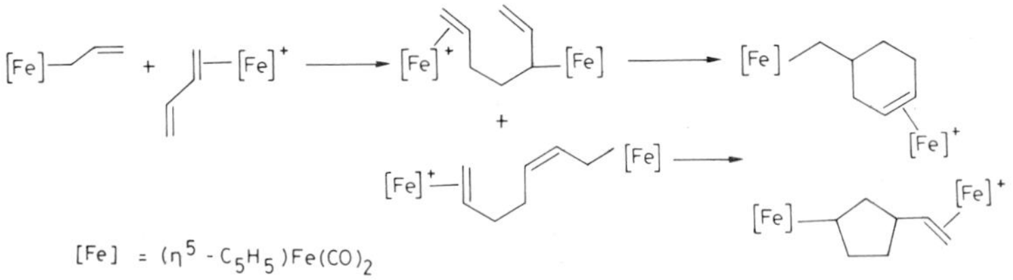

$$[Fe] = (\eta^5\text{-}C_5H_5)Fe(CO)_2$$

Some examples of nucleophilic addition to other η^2-olefin cations are given in Table 4.

The cation **LXI** is reduced by the iron hydride compound **LXII**, attack again occurring on the η^2-olefin ligand.[64]

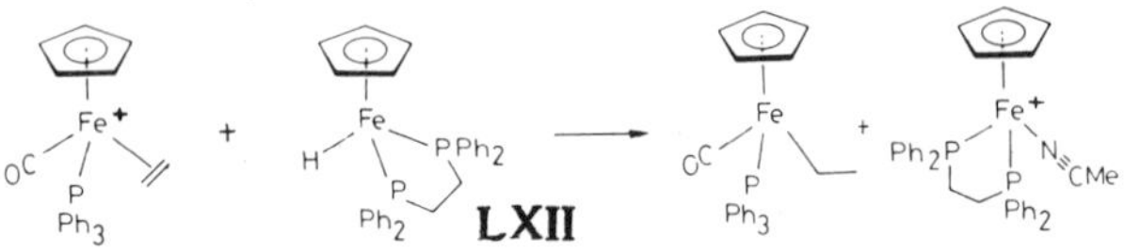

The (η^2-acetylene) ligand in **LXIII** is also readily susceptible to nucleophilic addition reactions. Again nucleophilic addition occurs stereospecifically <u>trans</u> to the metal.[65]

LXIII

$Y^- = HNa, LiMe_2Cu, LiPh_2Cu, C^-H(CO_2Me)_2, {}^-C\equiv CH, CN^-, OEt^-, SBu^-, SPh^-.$

The dication **LXIV** is particularly susceptible to nucleophilic attack, again regioselectively on the olefin.[66]

LXIV

$Y^- = SCN^-, P(OR)_3, PR_3, NR_3.$

The cations $[Mo(\eta^5\text{-indenyl})(CO)_2(\eta^4\text{-diene})]^+$ exist as an equilibrating mixture of <u>exo</u> and <u>endo</u> isomers.[70] Nucleophilic addition occurs on the diene (Rule 1) in a terminal position (Rule 3) onto the face away from the

Table 4. Nucleophilic addition to some η^2-olefin complexes.

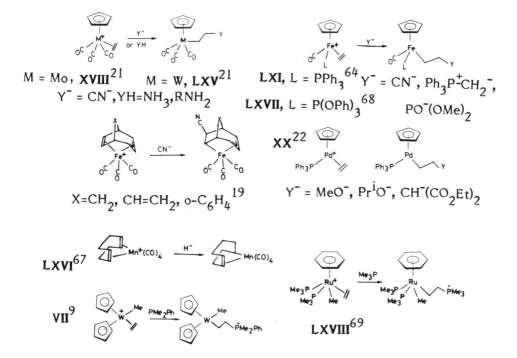

$M = Mo$, **XVIII**[21] $M = W$, **LXV**[21]   **LXI**, $L = PPh_3$[64] $Y^- = CN^-, Ph_3P^+CH_2^-,$
$Y^- = CN^-, YH = NH_3, RNH_2$ **LXVII**, $L = P(OPh)_3$[68] $PO^-(OMe)_2$

XX[22]

$X=CH_2, CH=CH_2, o\text{-}C_6H_4$[19] $Y^- = MeO^-, Pr^iO^-, CH^-(CO_2Et)_2$

LXVI[67]

VII[9]

LXVIII[69]

metal.   For example, hydride attack on **LXIX** and **LXX** generates the thermodynamically less stable <u>anti</u>-isomers **LXXI** and **LXXII** respectively.

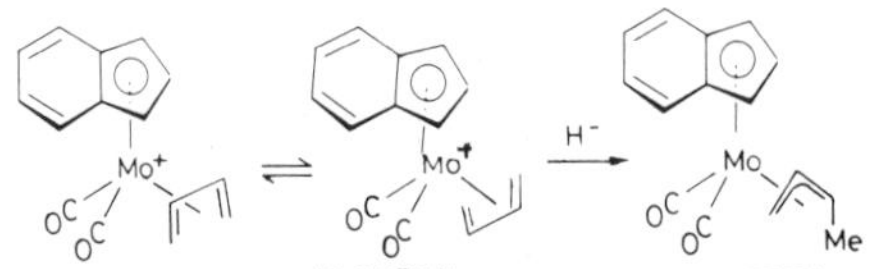

LXIX - endo   **LXIX** - exo   **LXXI**

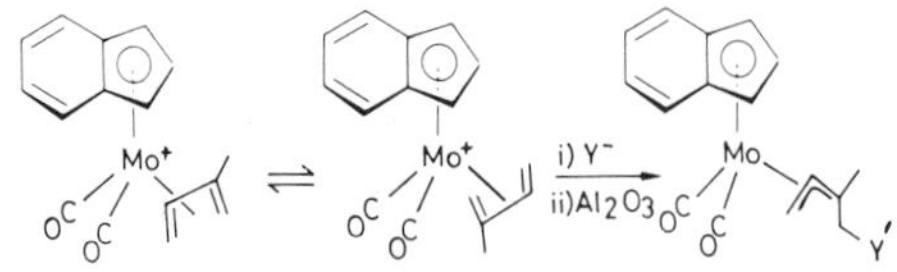

LXX-endo   **LXX** - exo   **LXXII**

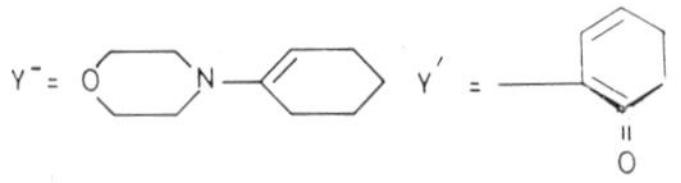

Cation **LXXIII** undergoes hydride addition on the cyclopentadiene ligand in preference to the cyclopentadienyl ring:[71]

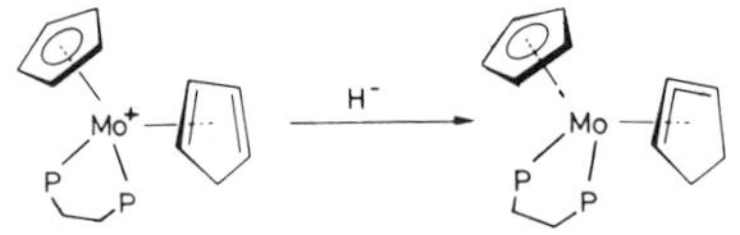

LXXIII   P⌒P = dppe

The cycloheptatriene complex **LXXIV** also undergoes nucleophilic addition on a terminal carbon atom.[74]

LXXIV

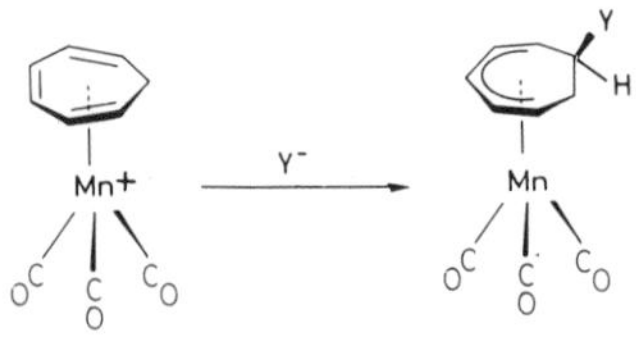

Y⁻ = H⁻, RO⁻, CN⁻, Me⁻, Ph⁻, R₃P.

2.11 Examples of attack onto <u>even closed</u> ligands

The reactions of nucleophiles with a wide variety of η^6-arene transition metal cations have been reported. Some examples are given in Table 5. Cation **LXXXI** is reduced to the cyclohexadiene complex **LXXXII** on treatment with $NaBH_4/H_2O$:[78]

LXXXI   LXXXII

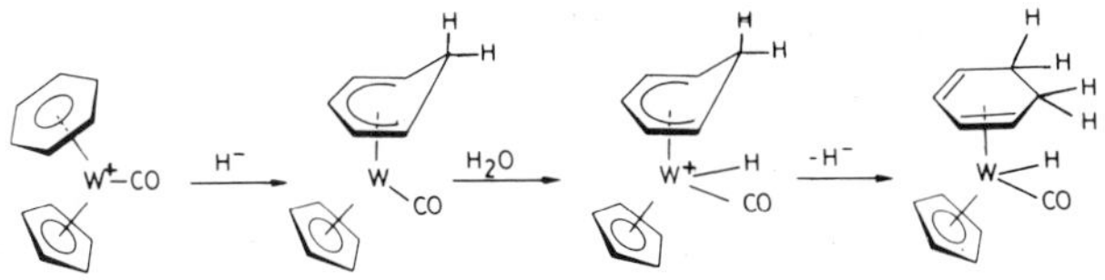

The arene dications shown below are also very susceptible to nucleophilic addition. Products from addition of one or two moles of nucleophile may be observed.

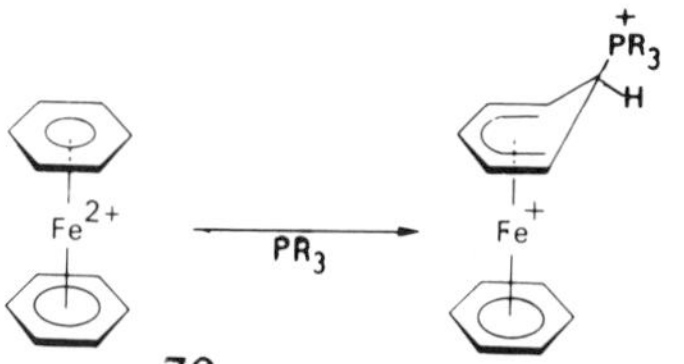

LXXXIII[79]

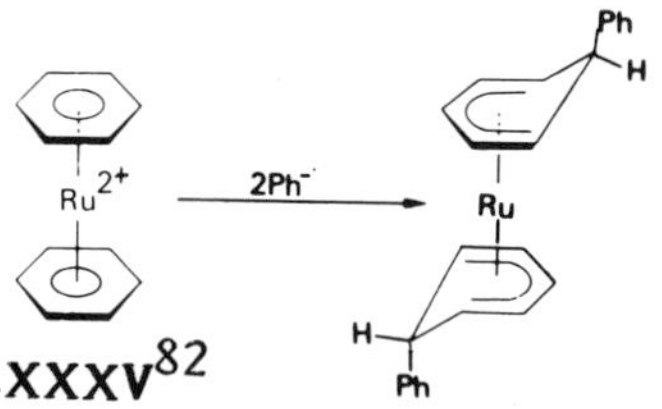

LXXXV[82]

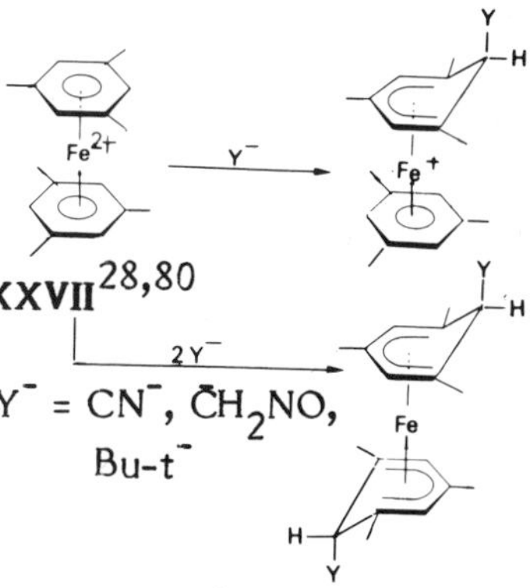

XXVII[28,80]

Y⁻ = CN⁻, ĊH₂NO,
Bu-t⁻

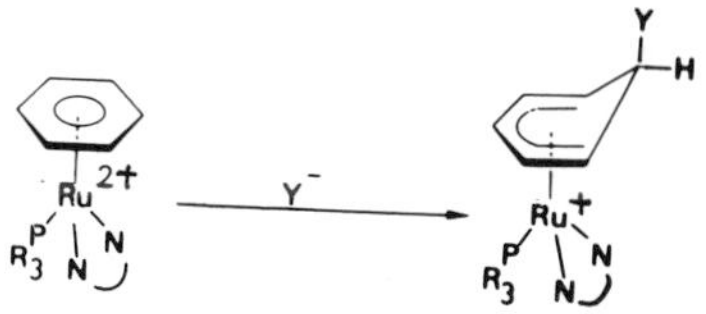

LXXXIV[81]   Y⁻ = H⁻, CN⁻, OH⁻

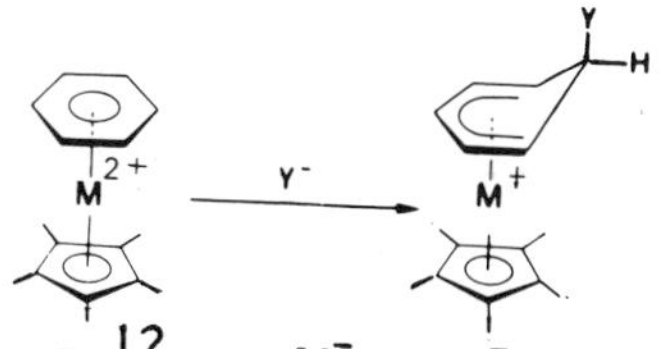

IX[12]   Y⁻ = Me⁻, H⁻

M = Rh, Ir

LXXXVI[83]

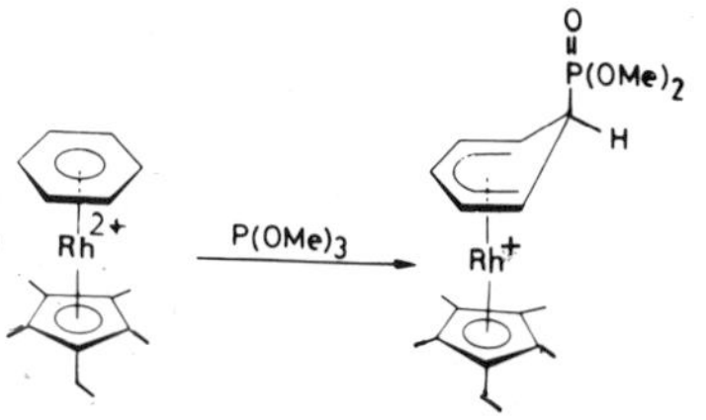

1.12 Examples of attack onto odd, open ligands

The electron poor $[Fe(\eta^3\text{-allyl})(CO)_4]^+$ cations are very electrophilic and readily undergo nucleophilic addition on a terminal carbon atom by a variety of nucleophiles:

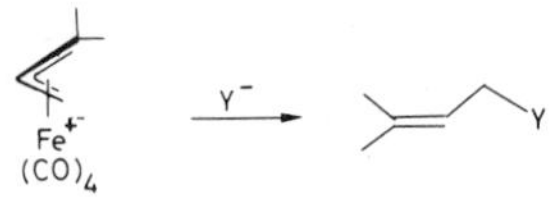

LXXXVII[84] Y⁻= C₅H₅N, Ph₃P, −COCHCOOMe, R₂Cd

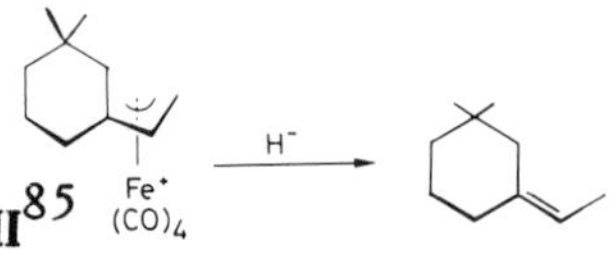

LXXXVIII[85]

Table 5. Nucleophilic addition to some η^6-arene complexes.

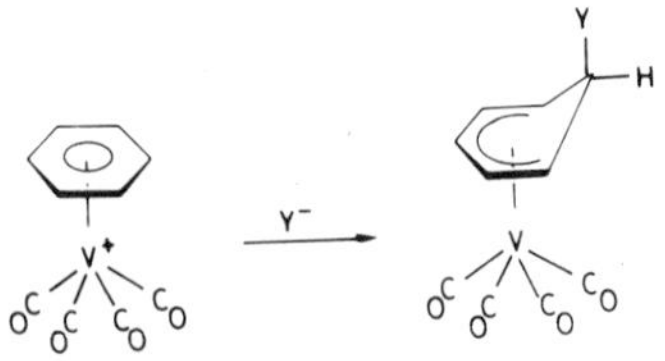

LXXV[73] Y⁻ = H⁻, D⁻

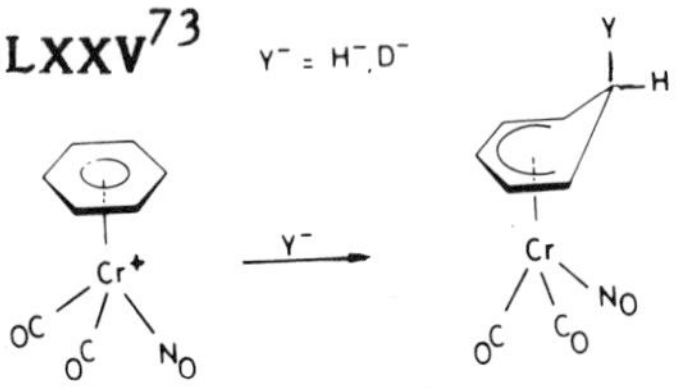

LXXVI[74] Y⁻ = H⁻, Me⁻

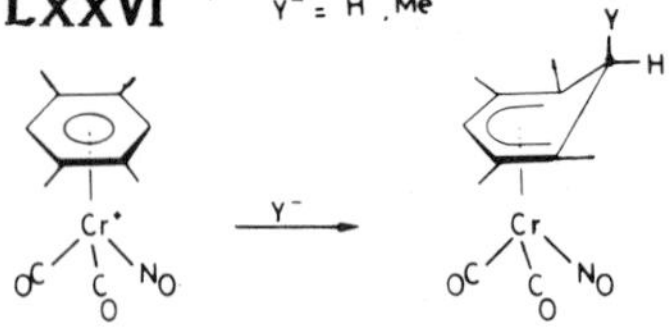

LXXVII[75] Y⁻ = H⁻ Me⁻

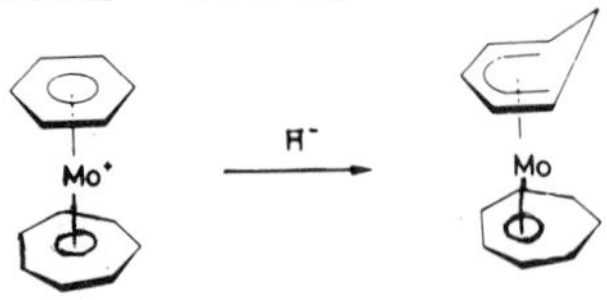

XI

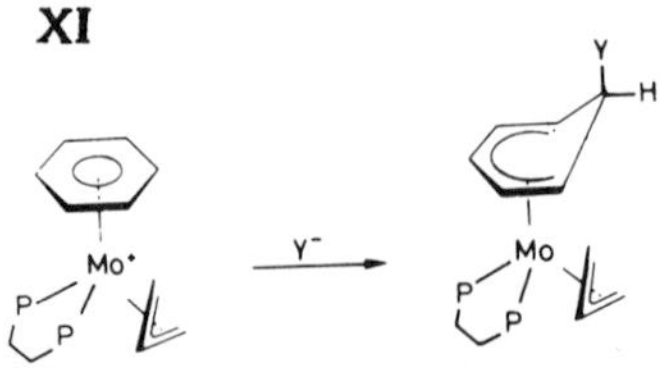

X[13] Y⁻= H⁻, CN⁻, Bu⁻   P–P = Ph₂PCH₂CH₂PPh₂

LXXVIII[76]

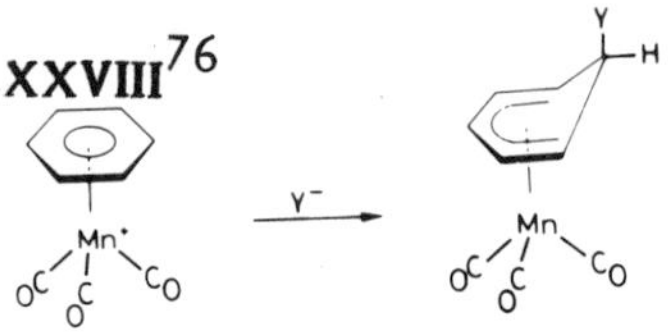

Y⁻= H⁻, CN⁻, CH(CO₂Et)₂, R⁻
N₃⁻, NCS⁻, MeO⁻, PPh₂⁻

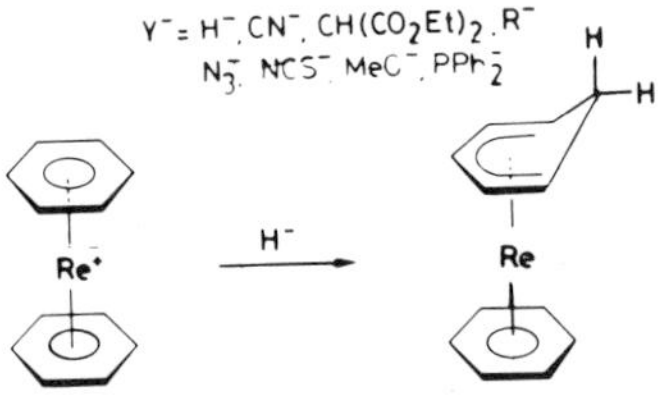

LXXIX

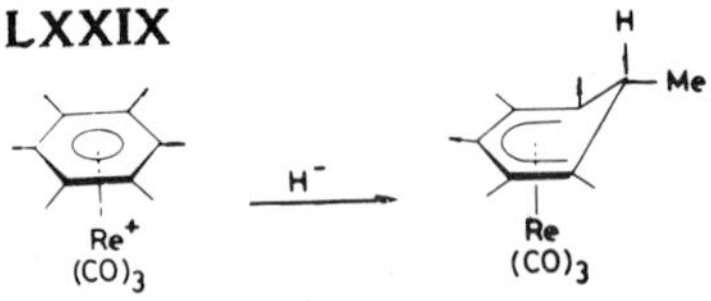

LXXX[77]

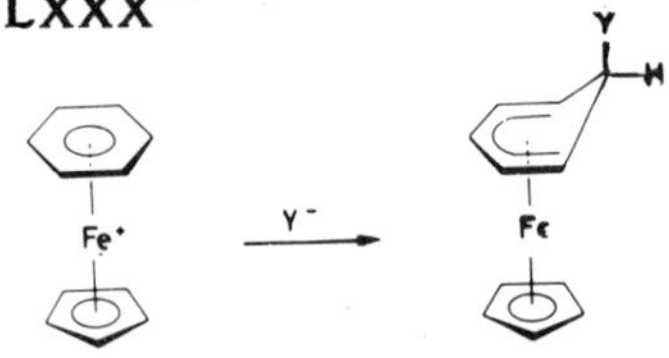

IV[7,10] Y⁻= R⁻, H⁻, Ph⁻, C₅H₅⁻

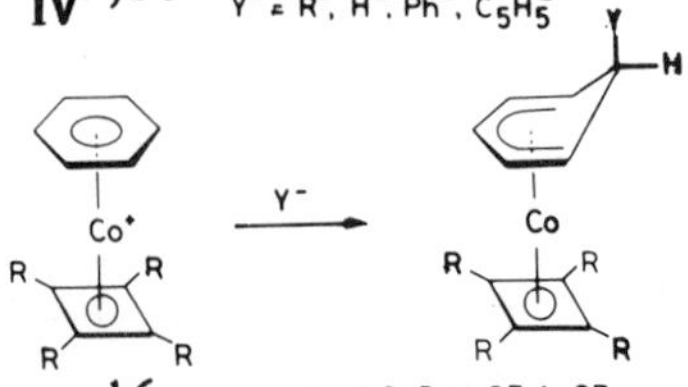

XIII[16] Y⁻= Ph⁻, H⁻, OH⁻, MeO⁻ AcO⁻

The cations $[Mo(\eta^5-C_5H_5)(CO)(NO)(\eta^3-allyl)]^+$ are attacked by a range of nucleophiles (e.g. H^-, $C_5H_5^-$, AcO^-, etc.) on a terminal carbon atom (see Section 1.7).[27,41]

The allyl ligand in the electron rich cations **XXI** and **XXII** is attacked on the central carbon atom, as predicted by Rule 3, to generate metallacyclobutane derivatives.[23,86]

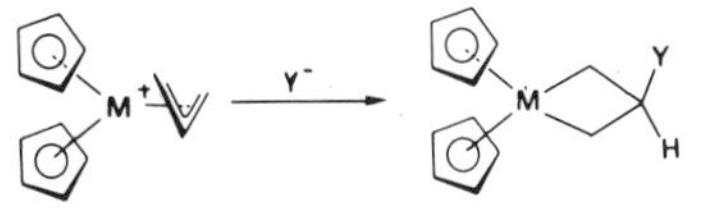

XXI[23] M = Mo; **XXII**[23] M = W

Reaction of $[Pd(\eta^3-allyl)Cl]_2$ with carbanions, e.g. **LXXXIX**, in the presence of excess $(Me_3N)_3PO$ leads to formation of the cyclopropane derivatives **XC**.[87] It is reasonable that under the reaction conditions cation **XCI** is formed which Rule 3 predicts should be attacked by the carbanion in the 2-position generating the metallacyclobutane derivatives **XCII** which would then give **XC**.[87] Attack at C(2) was subsequently confirmed by deuterium labeling studies:

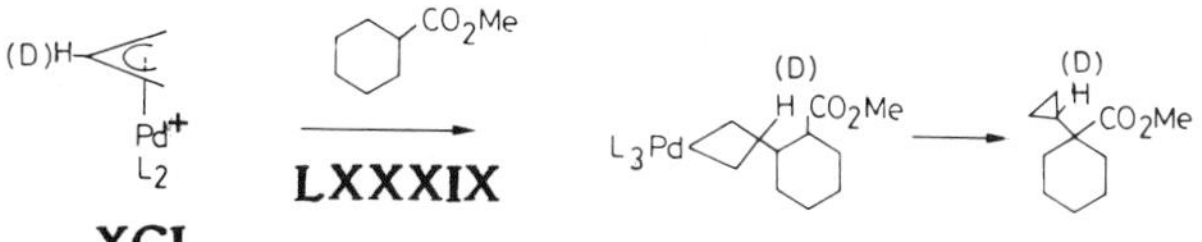

XCI

L = $(Me_3N)_3$ PO

LXXXIX

XC

The electron-poor cations $[Fe(\eta^5-pentadienyl)(CO)_3]^+$ react with nucleophiles on a terminal carbon atom.[88] The stereochemistry of the $Fe(diene)(CO)_3$ product depends on the nature of the nucleophile. Good nucleophiles (e.g. R_2Cd) attack **XCIII** directly and lead to $Fe(\underline{cis}-diene)(CO)_3$ complexes. Weak nucleophiles such as H_2O, however, often lead to $Fe(\underline{trans}-diene)(CO)_3$ products, presumably by the mechanism shown below:

XCIII

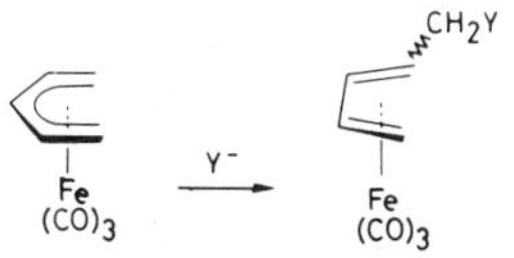

$Y^- = H^-, R_2Cd$; YH = H_2, ROH, RNH_2, etc.

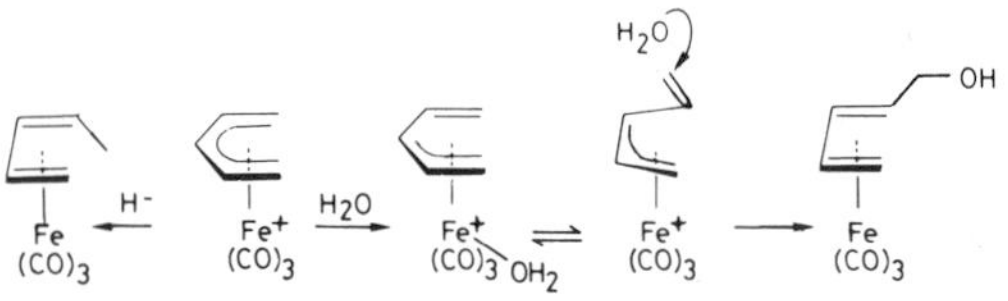

XCIII   **XCIV**

Nucleophilic addition to the substituted $[Fe(\eta^5\text{-dienyl})(CO)_3]^+$ cation
XCV occurs onto the most substituted end of the dienyl ligand. This
regioselectivity may be the result either of the formation of the least
sterically crowded $(\eta^3\text{-allyl})$ product complex analogous to **XCIV** or it may
indicate that the transition state resembles the most stable carbonium ion
i.e. **XCVI:**

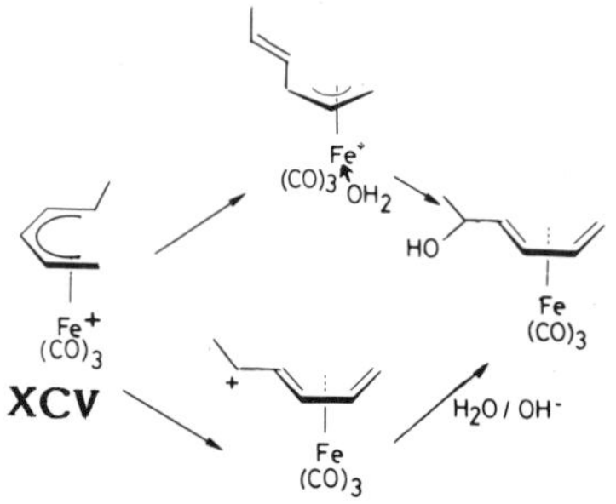

XCVI

Cations $[Fe(\eta^5\text{-pentadienyl})(CO)_3]^+$ undergo nucleophilic addition
exclusively at C(1) even though C(3) of a dienyl ligand bears approximately
the same charge as C(1) and C(5). Attack at C(3) is observed, however, in
the reactions of the rhodium and iridium cations **XCVII** and **XCVIII** with
nucleophiles.[89] This change in regioselectivity can be understood in terms
of the relative product stabilities. The group $Fe(CO)_3$ prefers to form
complexes with conjugated dienes whereas the group $M(\eta^5\text{-}C_5H_5)$ (M=Rh or
Ir) prefers to coordinate non-conjugated dienes. Presumably this difference
in stability is reflected in the transition states for the nucleophilic addition:

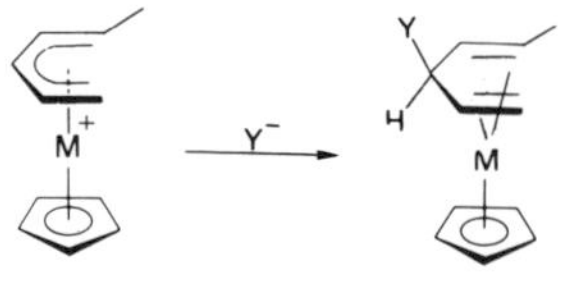

XCVII, M=Rh $Y^- = OMe^-, H^-$
XCVIII, M=Ir

Nucleophilic addition to $[Fe(\eta^5\text{-cyclohexadienyl})(CO)_3]^+$ cations has been extensively studied;[90-2] attack occurs exclusively onto a terminal carbon atom of the η^5-ligand. Yields of products from these reactions seem to be very variable and the appropriate choice of conditions can be critical.[91]

XCIX

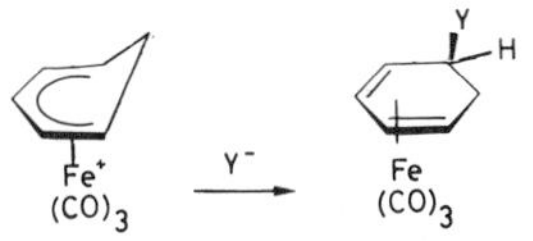

$Y^- = HS^-$, $S_2O_4^{2-}$, adenine, HPO_2^-, F^-, Cl^-, Br^-, CN^-, H^-, R_2Cd, R_2CuLi, enolates, enamines, allylsilanes, amides, thiourea, $CH_2NO_2^-$, etc.

The nucleophile attacks stereospecifically onto the uncoordinated face of the (η^5-cyclohexadienyl) ligand to give the <u>exo</u>-addition products. Under certain conditions the product formed initially may rearrange to a thermodynamically stable isomer. The rearrangement of the <u>exo</u>-isomer C to the <u>endo</u>-isomer CI[93] probably proceeds through a metal hydride species, since a byproduct is the reduced compound CII. The surprising preponderance of CI over C probably reflects the thermodynamic stability of the corresponding protonated species rather than the relative stabilities of C and CI themselves.

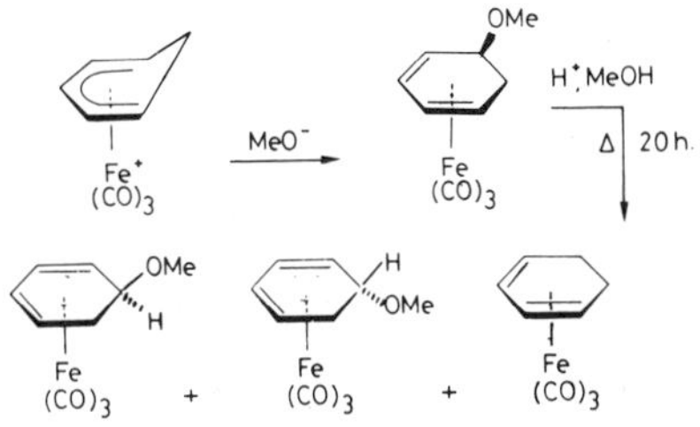

C   CI   CII

The introduction of alkyl substituents onto C(1) of $[Fe(\eta^5\text{-}$ cyclohexadienyl)(CO)$_3]^+$ cations, as in CIII, introduces steric constraints to attack at that position but also increases the positive charge on that carbon relative to C(5) by stabilization of the resonance form CIV:

CIII ⟶ **CIV**

Thus the sterically undemanding nucleophile OH⁻ attacks cation **CV** predominantly at C(1),[94] whereas cation **CVI** undergoes attack exclusively at C(5) by the larger nucleophiles BH_4^-, $(MeO)_3P$, HS^-:[95]

CV   **CVI**

Cations **CVII** and **CVIII** undergo regioselective attack on the terminal carbon remote from the alkyl substituent:[96]

CVII   **CVIII**

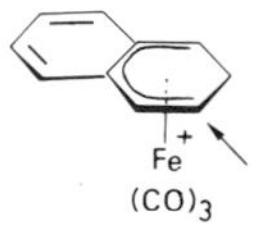

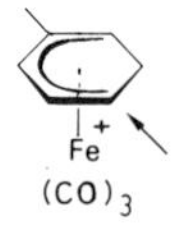

For cyclohexadienyl ligands nucleophilic addition generally occurs exclusively at C(5) if there is a 2-methoxy substituent present, as in **CIX**.[97] This effect is presumably due to the methoxyl group lowering the positive charge on C(1) by mesomeric effects.

CIX

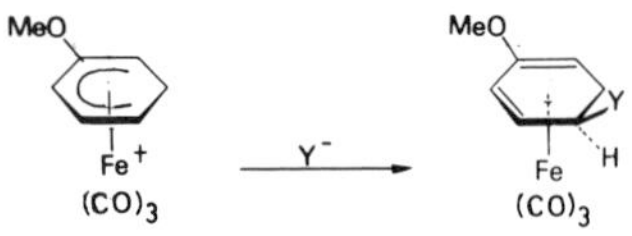

$Y^- = R^-$, CN^-, enamines, RO^-, BH_4^-.

The electronic effects of a 2-methoxy group generally outweigh the steric effect of alkyl substituents, as illustrated by the cation **CX**.[98] However, in extreme cases of steric hindrance (e.g. **CXI**) attack at C(1) is observed.[99]

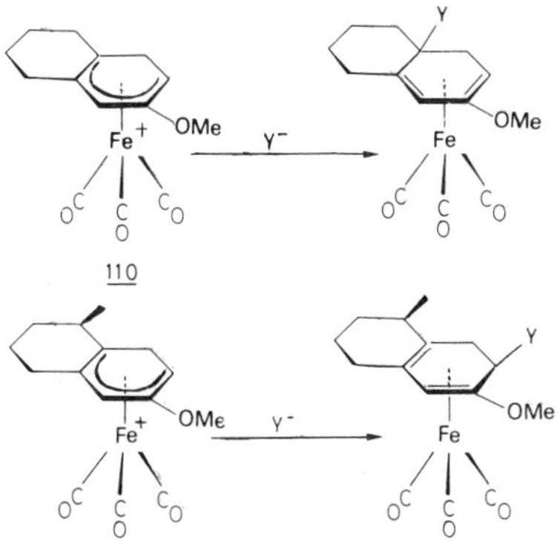

CX

CXI

110

$$Y = CH(CO_2Et)_2$$

For many nucleophiles, addition to $[Os(\eta^5\text{-cyclohexadienyl})CO)_3]^+$, **CXII**, occurs regioselectively at a terminal carbon atom. Hydride and cyanide, however, attack at both C(1) and C(2) [C(1):C(2) = 1:1 and 1:3 respectively].[100]

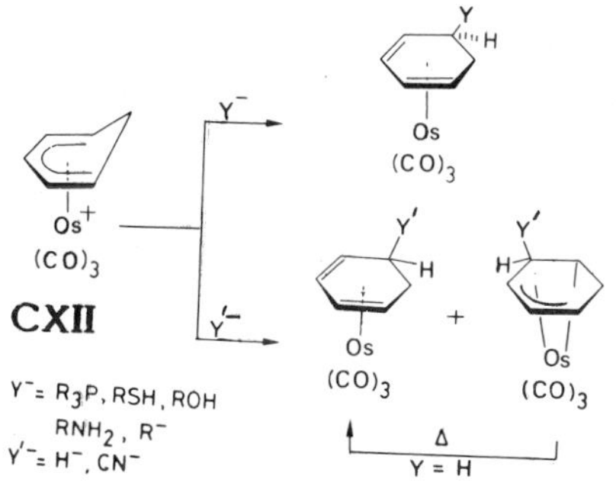

CXII

Y⁻ = R₃P, RSH, ROH
RNH₂, R⁻
Y'⁻ = H⁻, CN⁻

Os (CO)₃

Os (CO)₃   +   Os (CO)₃

Δ
Y = H

The $[Fe(\eta^5\text{-cycloheptadienyl})CO)_3]^+$ cation **CXIII** reacts regio-selectively at C(1) with many nucleophiles. Hydride addition to **CXIII**, however, is less selective and products from C(1) and C(2) attack are observed:[101]

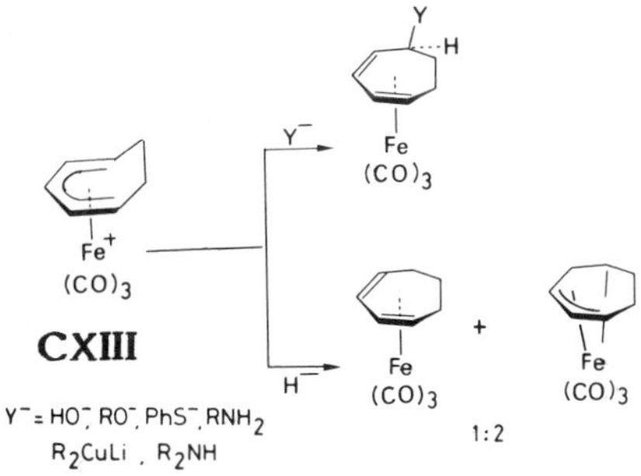

CXIII

Y⁻ = HO⁻, RO⁻, PhS⁻, RNH₂
R₂CuLi , R₂NH

Fe (CO)₃

Fe (CO)₃   +   Fe (CO)₃

1:2

Some other examples of attack onto η^5-open ligands are given below.

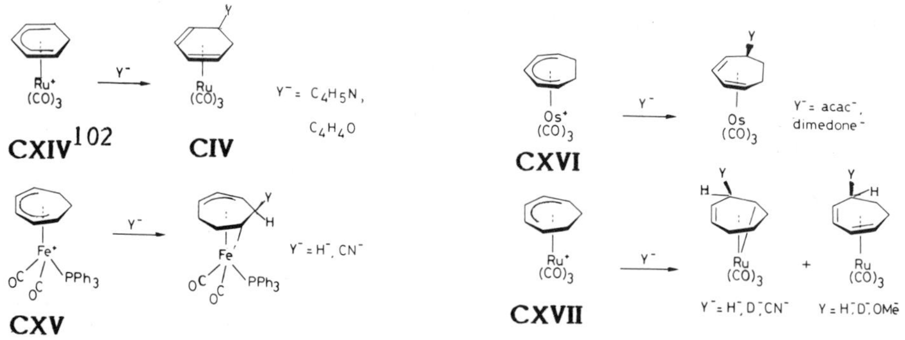
Ru+
(CO)3
Y-
Ru
(CO)3
Y- = C4H5N,
C4H4O
CXIV102
CIV
Fe+
OC
OC
PPh3
Y-
Fe
OC
OC
PPh3
Y- = H-, CN-
CXV
Os+
(CO)3
CXVI
Y-
Os
(CO)3
Y- = acac-,
dimedone-
Ru+
(CO)3
CXVII
Y-
Ru
(CO)3
+
Ru
(CO)3
Y- = H-, D-, CN-
Y = H-, D-, OMe

2.13 **Examples of attack onto odd closed systems**

Odd closed ligands are the least susceptible to attack by nucleophiles and consequently relatively few examples of nucleophilic addition to these ligands have been reported. Attack on such ligands appears to be favorable only when no other unsaturated hydrocarbon ligands are present.

Metallocene cations undergo nucleophilic addition reactions with attack occurring onto the unbound face of the ligand:[106]

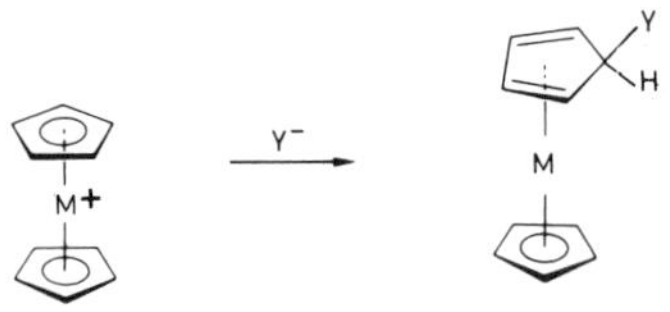
M+
Y-
M

CXVIII M = Co $\quad Y^- = (C_5H_5)^-$, Ph^-, H^-, R^-
CXIX M = Rh

Some further examples of attack onto η^5-C_5H_5 ligands are given in Table 6.

Some examples of nucleophilic addition to the η^7-cycloheptatrienyl ligand are shown below.

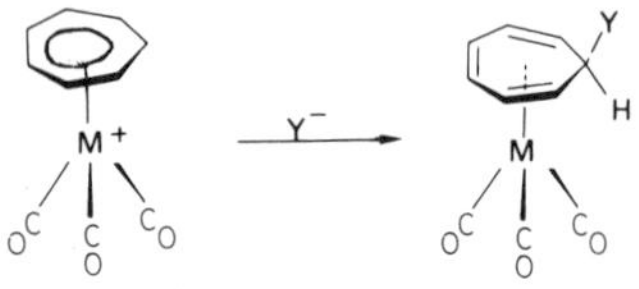
M+
OC
C
O
CO
Y-
M
OC
C
O
CO

CXXV[111] M=Cr; Y^- = R_3P, Me^-, Ph^-, $^-CH_2CO_2Et$, $PhC{\equiv}C^-$, Me_2HN, S_2^{2-}, SH^-, OMe^-, $^-CH(CO_2Et)_2$, CN^-, $acac^-$
CXXVI[112-3] M=Mo; Y^- = MeO^-, $acac^-$, $Bu^{[t]-}$.
CXXVII[112] M=W; Y^- = MeO^-, $acac^-$.

For cation **CXXVIII** if X is an electron releasing substituent (e.g. MeO) attack occurs preferentially at C(3) whereas when X is electron withdrawing attack occurs at C(2):[114]

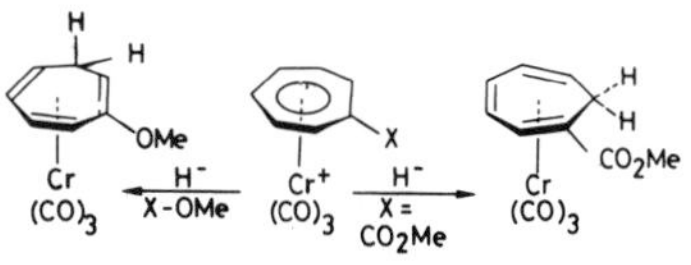

CXXVIII

Table 6. Nucleophilic addition to η^5-cyclopentadienyl ligands.

CXX[56]

CXXIII[109]

CXXI[107]

XXVIII[29]

CXXII[108]

CXXIV[110]

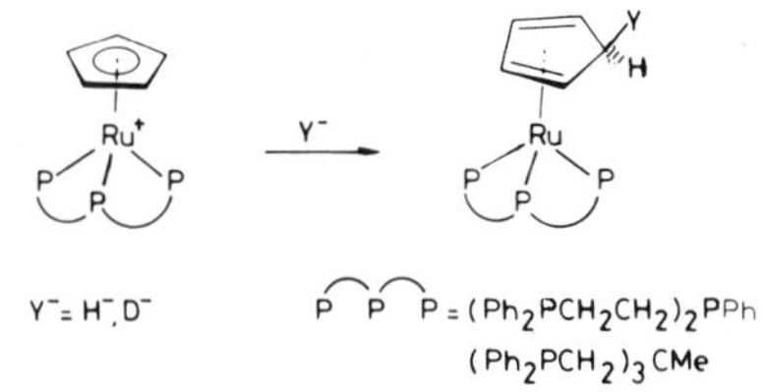

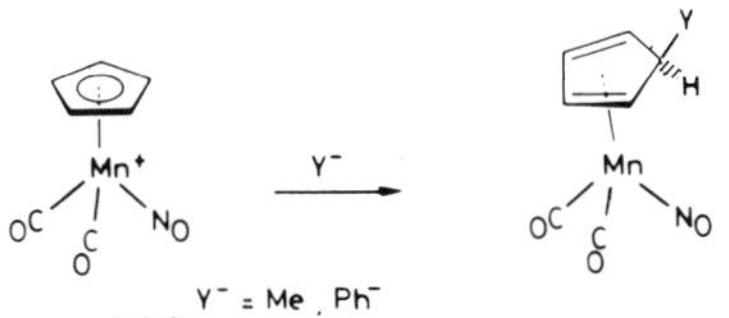

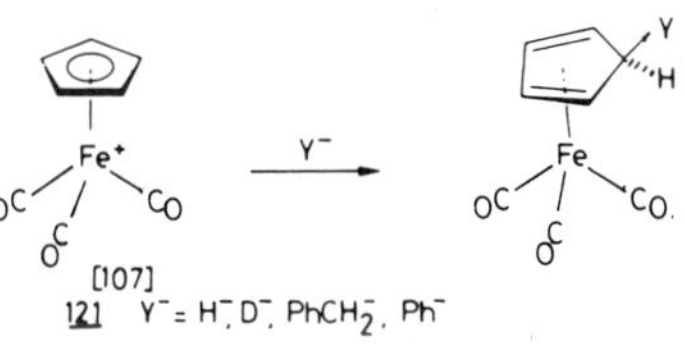

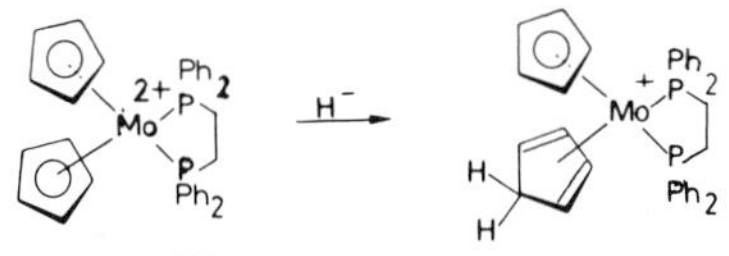

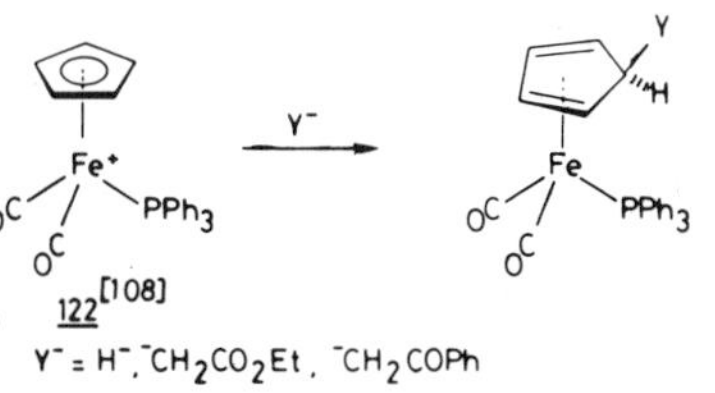

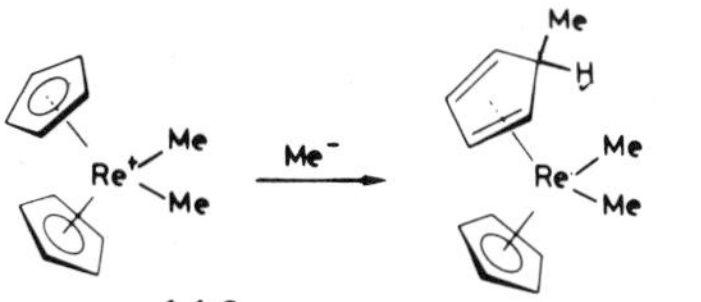

REFERENCES

1. J.F. McGreer and W.E. Watts, J. Organomet. Chem. 110:103 (1976).

2. S.G. Davies, M.L.H. Green and D.M.P. Mingos, Nouveau J. Chim. 1:445 (1977); Tetrahedron Report No. 57, Tetrahedron 34:3047 (1978).

3. M.L.H. Green and P.L.I. Nagy, J. Organomet. Chem. 1:581 (1963).

4. M.F. Semmelhack, Ann. N.Y. Acad. Sci. 36 (1977); G. Jaouen in "Transition Metal Organometallics in Organic Synthesis", Vol. 2, H. Alper ed., Academic Press, London and New York, 1978, p. 105.

5. B.W. Roberts, M. Ross and J. Wong, J.C.S. Chem. Commun. 428 (1980).

6. M.R. Churchill and R. Mason, Proc. Roy. Soc. Ser. A. 279 (1964).

7. A.N. Nesmeyanov and V.A. Vol'kenau, Izvest. Akad. Nauk. S.S.S.R., Ser. Khim. 1151 (1975).

8. S.G. Davies, H. Felkin, T. Fillebeen-Khan, F. Tadj and O. Watts, J.C.S. Chem. Commun. 341 (1981).

9. M.L.H. Green, L.C. Mitchard and W.E. Silverthorn, J.C.S. Dalton Trans. 1952 (1973).

10. I.U. Khand, P.L. Pauson and W.E. Watts, J. Chem. Soc. 2257 (1968); G. Winkhaus, L. Pratt and G. Wilkinson, J. Chem. Soc. 3807 (1971).

11. J.F. McGreer and W.E. Watts, J. Organomet. Chem. 110:103 (1976).

12. C. White and P.M. Maitlis, J. Chem. Soc (A) 3322 (1971).

13. M.L.H. Green, L.C. Mitchard and W.E. Silverthorn, J.C.S. Dalton Trans. 2177 (1973).

14. E.F. Ashworth, M.L.H. Green and J. Knight, Proc. 1st Int. Conf. Chem. Uses of Molybdenum 114 (1973).

15. P.M. Maitlis, A. Efraty and M.L. Games, J. Am. Chem. Soc. 87:719 (1965).

16. A. Efraty and P.M. Maitlis, Tetrahedron Lett. 4025 (1966).

17. J. Evans, B.F.G. Johnson, J. Lewis and D.J. Yarrow, J.C.S. Dalton Trans. 2375 (1974).

18. A. Eisenstadt, J. Organomet. Chem. 60:335 (1973), A. Eisenstadt, J. Organomet. Chem. 113:147 (1976).

19. A. Abramson and A. Eisenstadt, Tetrahedron 36:105 (1980).

20. A.D. Charles, P. Divers, B.F.G. Johnson, K.D. Karlin, J. Lewis, A.D. Rivera and G.M. Sheldrick, J. Organomet. Chem. 128:C31 (1977); F.A. Cotton, A.J. Deeming, P.L. Josty, S.S. Ullah, A.J.P. Domingos, B.F.G. Johnson and J. Lewis, J. Am. Chem. Soc. 93:4624 (1971); F.A. Cotton and M.D. LaPrade, J. Organomet. Chem. 39:345 (1972); A.J. Deeming, S.S. Ullah, A.J.P. Domingos, B.F.G. Johnson and J. Lewis, J.C.S. Dalton Trans. 2093 (1974).

21. W.H. Knoth, Inorg. Chem. 14:1566 (1975).

22. H. Kurosawa and N. Asada, Tetrahedron Lett. 255 (1979); H. Kurosawa, T. Majima and N. Asada, J. Am. Chem. Soc. 102:6996 (1980).

22b. D.M. Birnay, A.M. Crane and D.A. Sweigart, J. Organomet. Chem. 152:87 (1978).

23. F.W. Benfield, B.R. Francis, M.L.H. Green, N-T. Luong Thi, G. Moser, J.S. Poland and D.M. Roe, J. Less Common Metals 36:187 (1974).

24. B.F.G. Johnson, J. Lewis and D. Yarrow, J.C.S. Dalton Trans. 2084 (1972); J. Evans, B.F.G. Johnson and J. Lewis, J.C.S. Chem. Commun. 1252 (1971); J. Evans, B.F.G. Johnson and J. Lewis, J.C.S. Dalton Trans. 2668 (1972).

25. J. Muller and H. Menig, J. Organomet. Chem. 96:83 (1975).

26. M.L.H. Green, L.C. Mitchard and W.E. Silverthorn, J.C.S. Dalton Trans. 1952 (1973).

27. N.A. Bailey, W.G. Kita, J.A. McCleverty, A.J. Murray, B.E. Mann and N.W. Walker, J.C.S. Chem. Commun. 592 (1974).

28. J.F. Helling and D.M. Braitsch, J. Am. Chem. Soc. 92:7207 (1970).

29. T. Aviles, M.L.H. Green, A.R. Dias and C. Romao, J.C.S. Dalton Trans. 1367 (1979).

30. T.H. Whitesides, R.W. Arhart and R.W. Slaven, J. Am. Chem. Soc. 95:5792 (1973); A.J. Birch and I.D. Jenkins in "Transition Metal Organometallics in Organic Synthesis", Vol. 1, H. Alper ed., Academic Press, London and New York, 1973.

31. M.J.S. Dewar, "The Molecular Orbital Theory of Organic Chemistry", McGraw Hill, New York, 1969.

32. K.F. Gukui in "Molecular Orbitals in Chemistry, Physics and Biology", P.O. Lowdin and B. Pullman, ed., Academic Press, New York, 1964, p. 513.

33. L. Libit and R. Hoffmann, J. Am. Chem. Soc. 96:1370 (1974).

34. D.W. Clack and L.A.P. Kane-McGuire, J. Organomet. Chem. 174:199 (1979).

35. O. Eisenstein and R. Hoffmann, J. Am. Chem. Soc. 102:6148 (1980).

36. D.W. Clack and L.A.P. Kane-McGuire, J. Organomet. Chem. 145:201 (1978).

37. D.W. Clack, M. Monshi and L.A.P. Kane-McGuire, J. Organomet. Chem. 107:C40 (1976).

38. D.W. Clack, M. Monshi and L.A.P. Kane-McGuire, J. Organomet. Chem. 120:C25 (1976).

39. B.E.R. Schilling, R. Hoffmann and J.W. Faller, J. Am. Chem. Soc. 101:592 (1979).

40. N. El Murr and E. Laviron, Can. J. Chem. 54:3357 (1976).

41. R.D. Adams, D.F. Ghodosh, J.W. Faller and A.M. Rosan, J. Am. Chem. Soc. 101:2570 (1979).

42. A. Rosan and M. Rosenblum, J. Org. Chem. 40:3621 (1975).

43. C.T. Chang, M. Rosenblum and S.B. Samuels, J. Am. Chem. Soc. 102:5930 (1980).

44. A.J. Pearson, J.C.S. Chem. Commun. 339 (1977).

45. P.L. Pauson and J.A. Segal, J.C.S. Dalton Trans. 1683 (1975).

46. C.P. Casey and S.M. Neumann, J. Am. Chem. Soc. 98:5395 (1976); J.R. Sweet and W.A.G. Graham, J. Organomet. Chem. 173:C9 (1979); C.P. Casey, M.A. Andrews and D.R. McAlister, J. Am. Chem. Soc. 101:3371 (1979); W.K. Wong, W. Tan and J.A. Gladysz, J. Am. Chem. Soc. 101:5440 (1979); W.K. Wong, W. Tan, C.E. Strause and J.A. Gladysz, J.C.S. Chem. Commun. 530 (1979).

47. P.L. Pauson, Israel J. Chem. 15:258 (1977).

48. S.G. Davies, J. Hibberd, S.J. Simpson, S.E. Thomas and O. Watts, J.C.S. Dalton Trans. 701 (1984).

49. N. Grice, S.C. Kao and R. Pettit, J. Am. Chem. Soc. 101:1627 (1979).

50. A. Spencer and H. Werner, J. Organomet. Chem. 177:209 (1979).

51. E.G. Bryan, A.L. Burrows, B.F.G. Johnson, J. Lewis and G.M. Schiavon, J. Organomet. Chem. 129:C19 (1977).

52. A. Rosan, M. Rosenblum and J. Tancrede, J. Am. Chem. Soc. 93:3062 (1973); A. Rosan and M. Rosenblum, J. Organomet. Chem. 80:103 (1974).

53. E.O. Fischer and F.J. Kohl, Chem. Ber. 98:2134 (1965).

54. B. Faschinetti and C. Floriani, J.C.S. Chem. Commun. 578 (1975).

55. S.G. Davies, H. Felkin and O. Watts, J.C.S. Chem. Commun. 159 (1980).

56. S.G. Davies, H. Felkin, T. Fillebeen-Khan, F. Tadj and O. Watts, J.C.S. Chem. Commun. 341 (1981).

57. G.J. Baird and S.G. Davies, J. Organomet. Chem. 262:215 (1984).

58. P.M. Treichel and D.A. Komar, Inorg. Chim Acta 42:277 (1980).

59. S.G. Davies and F. Scott, J. Organomet. Chem. 188:C41 (1980); S.G. Davies and T. Fillebeen-Khan, unpublished observations.

60. L. Busetto, A. Palazzi, R. Ros and U. Belluco, J. Organomet. Chem. 25:207 (1970); P. Lennon, M. Madhavaras, A. Rosan and M. Rosenblum, J. Organomet. Chem. 108:93 (1976).

61. P. Lennon, A.M. Rosen and M. Rosenblum, J. Am. Chem. Soc. 99:8426 (1977) and references therein; K.M. Nicholas and A.M. Rosan, J. Organomet. Chem. 84:351 (1975); A. Sanders, C.V. Magatti and W.P. Giering, J. Am. Chem. Soc. 96:1610 (1974).

62. M. Rosenblum, Acc. Chem. Res. 7:122 (1974) and references therein.

63. A. Rosan, M. Rosenblum and J. Tancrede, J. Am. Chem. Soc. 95:3062 (1973).

64. T. Bodnar, S.J. La Croce and A.R. Cutler, J. Am. Chem. Soc. 102:3292 (1980).

65. D.L. Reger and P.J. McElligott, J. Am. Chem. Soc. 102:5923 (1980); D.L. Reger, C.J. Coleman and P.J. McElligott, J. Organomet. Chem. 171:73 (1979).

66. H. Weiner, R. Feser and R. Weiner, J. Organomet. Chem. 181:C7 (1979).

67. P.J. Harris, S.A.R. Knox and F.G.A. Stone, J. Organomet. Chem. 148:327 (1978).

68. D.L. Reger and C.J. Coleman, Inorg. Chem. 18:3155 (1979).

69. H. Werner and R. Werner, J. Organomet. Chem. 174:C63 (1979).

70. J.W. Faller and A.M. Rosan, J. Am. Chem. Soc. 99:4858 (1977); M. Bottrill and M. Green, J.C.S. Dalton Trans. 2365 (1977).

71. T. Aviles, M.L.H. Green, A.R. Dias and C. Romao, J.C.S. Dalton Trans. 1367 (1979).

72. F. Hague, J. Miller, P.L. Pauson and J.B. Tripathi, J. Chem. Soc. (C) 743 (1971); P.J.C. Walker and R.J. Mawby, Inorg. Chim. Acta 7:621 (1973); P.J.C. Walker and R.J. Mawby, J. Organomet. Chem. 55:C39 (1973); J.C. Burt, S.A.R. Knox, R.J. McKinney and F.G.A. Stone, J.C.S. Dalton Trans. 1 (1977); D.M. Birney, A.M. Crane and D.A. Sweigart, J. Organomet. Chem. 152:187 (1978).

73. F. Calderazzo, Inorg Chem. 5:429 (1966).

74. N.G. Connelly and R.L. Kelly, J.C.S. Dalton Trans. 2334 (1974).

75. D.E. Ball and N.G. Connelly, J. Organomet. Chem. 55:C24 (1973).

76. P.J.C. Walker and R.J. Mawby, Inorg. Chim. Acta 7:691 (1973); R.J. Angelici and L.J. Blacik, Inorg. Chem. 11:1754 (1974); P.J.C. Walker and R.J. Mawby, J.C.S. Dalton Trans. 622 (1973); P.J.C. Walker and R.J. Mawby, J.C.S. Chem. Commun. 330 (1972); T.H. Coffield and R.D. Closson, U.S. Pat. 3,042,693 (1962), CA 59:11558d (1963); A. Mawby, P.J.C. Walker and R.J. Mawby, J. Organomet. Chem. 55:C39 (1973); G. Winkhaus, Z. Anorg. Allg. Chem. 319:404 (1963); P.L. Pauson and J.A. Segal, J.C.S. Dalton Trans. 1677 (1975); G. Winkhaus, L. Pratt and G. Wilkinson, J.C.S. Dalton Trans. 3807 (1965); G. Winkhaus and G. Wilkinson, Proc. Chem. Soc. 311 (1960); D. Jones, L. Pratt and G. Wilkinson, J. Chem. Soc. 4458 (1962); D. Jones and G. Wilkinson, Chem. Ind. (London) 1408 (1965); D. Jones and G. Wilkinson, J. Chem. Soc. 2479 (1964).

77. P.H. Bird and M.R. Churchill, J.C.S. Chem. Commun. 777 (1967); G. Winkhaus and H. Singer, Z. Naturforsch. 18b:418 (1963).

78. E.O. Fischer and F.J. Kohl, Chem. Ber. 98:2134 (1965).

79. S.G. Davis, L.S. Gelfand and D.A. Sweigart, J.C.S. Chem. Commun. 762 (1979).

80. J.F. Helling and G.G. Cash, J. Organomet. Chem. 73:C10 (1974).

81. D.R. Robertson, I.W. Robertson and T.A Stephenson, J. Organomet. Chem. 202:309 (1980).

82. D. Jones and G. Wilkinson, J. Chem. Soc. 2479 (1964).

83. G. Fairhurst and C. White, J. Organomet. Chem. 166:C23 (1979).

84. T.H. Whitesides, R.W. Arhart and R.W Slaven, J. Am. Chem. Soc. 95:5792 (1973); A.J. Pearson, Tetrahedron Lett. 3617 (1975).

85. A.J. Pearson, Aust. J. Chem. 29:1841 (1976).

86. G.S.B. Adam, S.G. Davies, M. Ephritikhine, K.A. Ford, M.L.H. Green and P.F. Todd, J. Mol. Catal. 8:15 (1980).

87. L.S. Hegedus, W.H. Darlington and C.E. Russell, J. Org. Chem. 45:5793 (1980).

88. R.S. Bayoud, E.R Biehl and P.C. Reeves, J. Organomet. Chem. 174:297 (1979) and references therein; J.E. Mahler and R. Pettit, J. Am. Chem. Soc. 84:1511 (1962); J.E. Mahler and R. Pettit, J. Am. Chem. Soc. 85:3955 (1963); J.E. Mahler, D.M. Gibson and R. Pettit, J. Am. Chem. Soc. 85:3959 (1963); P. McArdle and M. Sherlock, J. Organomet. Chem. 52:C29 (1973); M. Anderson, A.D.M. Claque, L.P. Blauw and P.A. Corperus, J. Organomet. Chem. 56:307 (1973); B.F.G. Johnson, J. Lewis, D.G. Parker and S.R. Postle, J.C.S. Dalton Trans. 794 (1977); G. Maglio, A. Musco, R. Palumbo and A. Sirigu, J.C.S. Chem. Commun. 100 (1971); G. Maglio, A. Musco, R. Palumbo and A. Sirigu, J. Organomet. Chem. 32:127 (1971); T.G. Bonner, K.A. Holder, P. Powell and E. Styles, J. Organomet. Chem. 131:105 (1977).

89. P. Powell, J. Organomet. Chem. 165:C43 (1979).

90. A.J. Birch, Ann. N.Y. Acad. Sci. 333:107 (1980) and references therein; A.J. Pearson, Acc. Chem. Res. 13:463 (1980) and references therein; A.J. Birch, B.M.R. Bandasa, K. Chamberlain, B. Chauncy, P. Dahler, A.I. Day, I.D. Jenkins, L.F. Kelly, T.C. Khar, G. Kretschmer, A.J. Lepa, A.S. Narula, W.D. Raverty, E. Rizzardo, C. Sell, G.R. Stephenson, D.J. Thompson and D.H. Williamson, Tetrahedron 37:289 (1981).

91. B.R. Reddy, V. Vaughan and J.S. McKennis, Tetrahedron Lett. 3639 (1980); B.R. Reddy and J.S. McKennis, J. Organomet. Chem. 182:C61 (1979).

92. A.J. Birch, P.E. Cross, L. Lewis, D.A. White and S.B. Wild, J. Chem. Soc. (A) 3183 (1968); Y. Becker, A. Eisenstadt and Y. Shvo, J. Organomet. Chem. 202:175 (1980); A.J. Pearson and D.C. Rees, Tetrahedron Lett. 3937 (1980).

93. A.L. Burrows, K. Hine, B.F.G. Johnson, J. Lewis, D.G. Parker, A. Poe and E.J.S. Vicki, J.C.S. Dalton Trans. 1136 (1980).

94. A.J. Birch, I.D. Jenkins and A.J. Liepa, Tetrahedron Lett. 1723 (1975).

95. A.J. Birch, K.B. Chamberlain, M.A. Haas and D.J. Thompson, J.C.S. Perkin I 1882 (1973).

96. A.J. Birch, A.S. Narula, P. Dahler, G.R. Stephenson and L.F. Kelly, Tetrahedron Lett. 979 (1980); L.F. Kelly, A.S. Narula and A.J. Birch, Tetrahedron Lett. 871, 2455 (1980).

97. R.E. Ireland, G.G. Brown, R.H. Sandford and T.C. McKenzie, J. Org. Chem. 39:51 (1974); A.J. Birch, P.W. Westerman and A.J. Pearson, Aust. J. Chem. 29:1671 (1976).

98. A.J. Pearson, J.C.S. Perkin I 2069 (1979), 400 (1980); A.J. Pearson and P.R. Raithlet, J.C.S. Perkin I 395 (1980).

99. A.J. Pearson, J.C.S. Perkin I 495 (1978).

100. A.L. Burns, B.F.G. Johnson, J. Lewis and D.G. Parker, J. Organomet. Chem. 194:C11 (1980).

101. R. Aumann and J. Knecht, Chem. Ber. 109:175 (1976).

102. G.R. John, C.A. Mansfield and L.A.P. Kane-McGuire, J.C.S. Dalton Trans. 574 (1977).

103. R. Edwards, J.A.S. Howell, B.F.G. Johnson and J. Lewis, J.C.S. Dalton Trans. 2105 (1974).

104. L.A.P. Kane-McGuire, J. Chem. Soc. (A) 1602 (1979).

105. B.A. Osinsky, S.A.R. Knox and F.G.A. Stone, J.C.S. Dalton Trans. 1633 (1975); A.J.P. Domingos, B.F.G. Johnson and J. Lewis, J.C.S. Dalton Trans. 2288 (1975).

106. M.R. Churchill and R. Mason, Proc. Roy. Ser. A. 279:191 (1964); E.O. Fischer and G.E. Herberich, Chem. Ber. 94:1577 (1961); E.O. Fischer and G.E. Herberich, Chem. Ber. 95:2254 (1962); R.J. Angelici and E.O. Fischer, J. Am. Chem. Soc. 85:3733 (1963); M.L.H. Green, L. Pratt and G. Wilkinson, J. Chem. Soc. 3753 (1959).

107. J. Ellerman, H. Behrens and H. Krohberger, J. Organomet. Chem. 46:119 (1972); R.J. Angelici and L. Busetto, J. Am. Chem. Soc. 91:3197 (1969); A.E. Kreuse and R.J. Angelici, J. Organomet. Chem. 24:231 (1970); T.M. Whitesides and J. Selly, J. Organomet. Chem. 92:215 (1975).

108. A. Davison, M.L.H. Green and G. Wilkinson, J. Chem. Soc. 3172 (1961); S.G. Davies, O. Watts, N. Aktogu and H. Felkin, J. Organomet. Chem. 245:C51 (1983).

109. H. Brunner and M. Langer, J. Organomet. Chem. 54:221 (1973).

110. D. Baudry and M. Ephritikhine, J.C.S. Chem. Commun. 895 (1979).

111. P. Hackett and G. Jaouen, Inorg. Chim Acta 12:119 (1975); P.E. Baikie, O.S. Mills, P.L. Pauson, G.H. Smith and J. Valentine, J.C.S. Chem. Commun. 425 (1965); J.D. Munro and P.L. Pauson, J. Chem. Soc. 3475, 3479, 3484 (1961); J.D. Munro and P.L. Pauson, Proc. Chem. Soc. 267 (1959); P.L.Pauson, G.M. Smith and J.H. Valentine, J. Chem. Soc. (C) 1057, 1061 (1967); C.A. Mansfield, K.M. Al-Kathumi and L.A.P. Kane-McGuire, J. Organomet. Chem. 71:C11 (1974).

112. K.M. Al-Kathumi and L.A.P. Kane-McGuire, J. Organomet. Chem. 102:C4 (1975); M.L. Ziegler, H.E. Sasse and B. Nuber, Z. Naturforsch. 30b:22, 26 (1975).

113. J.W. Faller, Inorg. Chem. 19:2857 (1980).

114. D.W. Clack, M. Monshi and L.A.P. Kane-McGuire, J. Organomet. Chem. 120:C25 (1976).

REACTIONS OF COORDINATED DIENES

J.A.S. Howell

University of Keele

1.   INTRODUCTION

It is the purpose of this chapter to review the reactivity of η^4-conjugated dienes complexed to transition metals. The apparently broad scope of this topic is limited by three main factors:

(a)   The necessity for the metal to be in a low formal oxidation state.

(b)   The absence of any significant number of η^4-diene complexes of the titanium and vanadium groups. The preparation of complexes such as $(\eta^4$-butadiene)M[Me$_2$P(CH$_2)_2$PMe$_2$] (M = Ti,Zr), TaCl(η^4-naphthalene) [Me$_2$P(CH$_2)_2$PMe$_2$], $(\eta^3$-1-methylallyl)M-(η^4-butadiene)$_2$ (M = Nb,Ta) and Cp$_2$M(η^4-butadiene) (M = Zr, Hf) may be noted, while photolysis of Cp$_2$ZrPh$_2$ with   2,3-dimethylbutadiene   yields   (η^4-s-cis-2,3-dimethylbutadiene)ZrCp$_2$ which on photolysis undergoes isomerization to the $(\eta^4$-s-trans-diene) complex. The latter is the first example of a conjugated diene coordinated to a single metal atom in an s-trans fashion.[1-5] The low valent group IV metals are powerful reducing agents, and the structural features of these complexes are consistent with a considerable degree of π-acceptor character by the diene (i.e., a considerable contribution from the metallacyclopentene valence structure). Little is known about the reactivity of the complexed diene, but this promises to be an interesting area of research (see Section 7), and (c) the preference of zero-valent six-coordinate chromium group metals and, particularly, three- and four-coordinate nickel group metals for η^4-chelate coordination of unconjugated 1,4- or 1,5-diene systems.   Indeed, zero-valent complexes of the nickel group act as efficient catalysts for the oligomerization of conjugated dienes.

Thus, with the primary exception of the complexes of cyclobutadiene, a systematic discussion of reactivity is limited somewhat to the extensive chemistry of η^4-conjugated diene and polyolefin complexes based on the Fe(0) or Co(I) triads.   The types of reaction which have been studied in most detail, and which will primarily be discussed here, are electrophilic and nucleophilic addition and substitution and cycloaddition reactions. Complexation is known to stabilize the diene, particularly towards electrophilic substitution, while complexation is also known to affect both the regio- and stereo-specificity of cycloaddition reactions.   Sections are also included on ligand isomerization reactions and on the catalytic transformations of conjugated dienes.   Throughout, only leading references to the literature are given.

2.   ELECTROPHILIC ADDITION AND SUBSTITUTION

Much effort has been devoted to the study of the   protonation of (diene)Fe(CO)$_3$ complexes.[6]   Reaction of (1,3-butadiene)Fe(CO)$_3$ with acids HX where X is coordinating (e.g. Cl) results in addition to yield the neutral syn-(1-methylallyl)Fe(CO)$_3$Cl derivative I.   Where X is non-coordinating (e.g. BF$_4$),   the reaction was thought to yield the 16-electron anti-[(methylallyl)Fe(CO)$_3$]X salt II which is now known, however, to be the anti-[(methylallyl)Fe(CO)$_4$]X complex arising from partial decomposition of the tricarbonyl complex II.   Evidence now favors structure III (Scheme 1) as the species present in acid solution.   While not isolable where L is carbon monoxide, the salt where L is P(OMe)$_3$ is stable, and xray structural data agree with the formulation of a 3c-2e$^-$ Fe-H-C bond with the electron density concentrated in the C-H region.   Deuteration experiments are consistent with Scheme 1; protonation is irreversible as only monodeutero salts are isolated, but deuterium is incorporated into the five possible

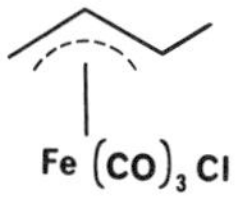

Scheme 1

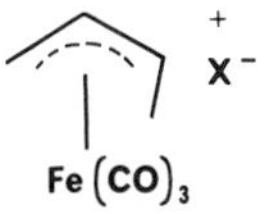

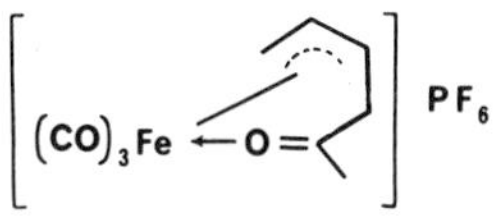

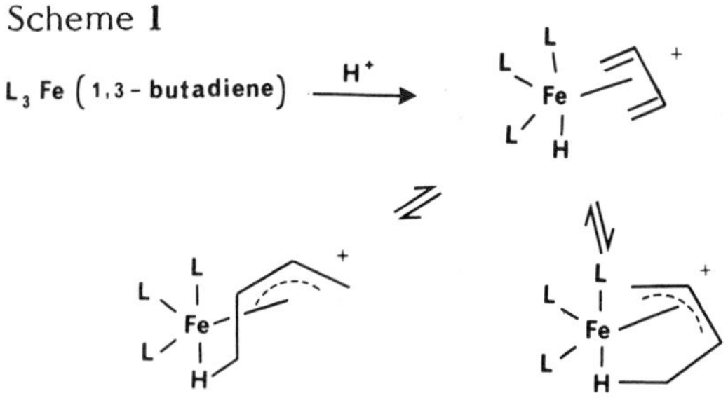

terminal positions, as nmr shows that rotation of the methyl group occurs at room temperature. In the protonation of (1,3-cyclooctadiene)Fe[P(OMe)$_3$]$_3$, where C-C bond rotation is not possible, deuterium is incorporated into the five methylene positions which are <u>endo</u> with respect to the metal. The influence of metal coordination and auxiliary ligand may be noted here. While uncoordinated dienes require strong acids (H$_2$SO$_4$, pK$_a$-10), the electron donating character of the Fe(CO)$_3$ moiety reduces the acid strength required for protonation (HBF$_4$/acetic anhydride, pK$_a$-6), while an increase in the electron donating character of the auxiliary ligands in the Fe[P(OMe)$_3$]$_3$ unit means that weak acids (NH$_4$BF$_4$; pK$_a$ 9) will suffice for protonation of (diene)Fe[P(OMe)$_3$]$_3$ complexes. The nmr spectra of (1,3-butadiene)MCp (M = Rh, Ir) and (1,3-cyclohexadiene)RhCp in strong acid can also be explained in terms of reactions analogous to Scheme 1 provided that equilibration is rapid on the nmr timescale.[7,8] If the protonation is done at low temperature, the metal hydride intermediate may be detected by nmr, and deuteration experiments show that here the protonation is reversible, as, after neutralisation, the (1,3-cyclohexadiene)RhCp recovered has deuterium incorporated into <u>both</u> endo positions of the C$_6$H$_8$ ring. Recently, the isoelectronic $[(\eta^4\text{-cyclohexadiene})Mn(CO)_3]^-$ anion has been prepared and undergoes protonation or methylation to give neutral complexes whose spectroscopic properties indicate them to have 3c-2e⁻ bonds of the type described above.[9]

Acetylation of (butadiene)$Fe(CO)_3$ appears to proceed in a similar fashion. An allylic intermediate of structure **IV** may be isolated in which the 18-electron configuration for iron is maintained by coordination of acyl oxygen.[11] Deprotonation with a bulky, non-coordinating base yields exclusively the <u>anti</u>-acetyl derivative **V** which, however, undergoes thermal and acid and base catalyzed isomerization to the <u>syn</u>-isomer **VI**. The methoxide catalyzed reaction appears to proceed via an initial nucleophilic attack at iron to generate the allyl **VII**, which by C-C bond rotation and elimination of methoxide yields the <u>syn</u>-isomer. The acid catalyzed reaction is thought to proceed via oxygen protonation to yield an intermediate **VIII** of similar structure.[12] Borohydride reduction of the dienone complexes **IX** and **X** has been shown to be highly stereospecific, yielding almost exclusively the diastereoisomers **XI** and **XII** respectively.

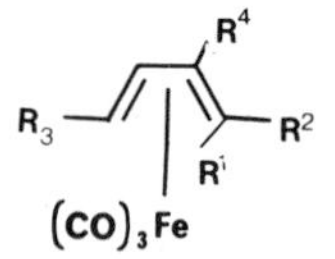

V R^1 = COMe, $R^2\text{-}R^4$ = H

VI R^2 = COMe, R^1, R^3, R^4 = H

IX R^2 = COMe, R^1 = H, R^3, R^4 = Me

X R^1 = COMe, R^2 = H, R^3, R^4 = Me

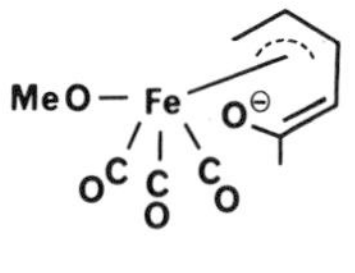

VII

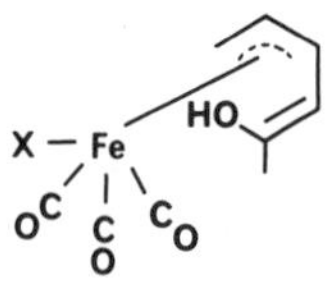

VIII X = HSO_4^-

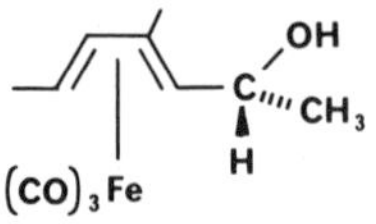

XI

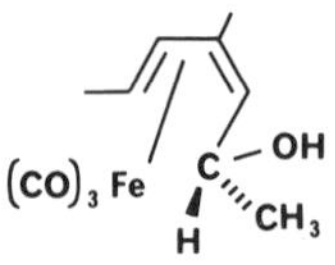

XII

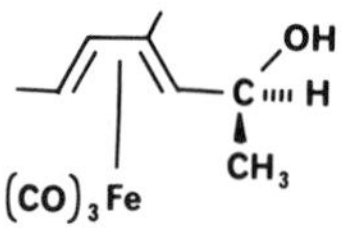

XIII

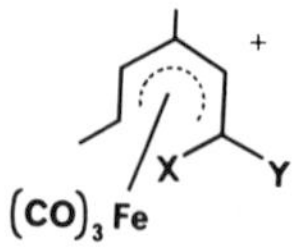

XIV X = Me, Y = H

XV X = H, Y = Me

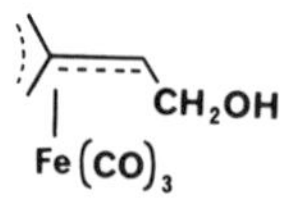

XVI

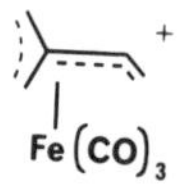

XVII

Reaction of the free dienol of **XI** with $Fe_2(CO)_9$ yields a mixture rich in the other diastereoisomer, **XIII**. Protonation of the syn-alcohols **XI** and **XIII** at low temperature in strong acid leads stereospecifically to the syn, anti- and syn, syn-pentadienyl cations **XIV** and **XV** respectively, as expected if the reaction proceeds via a trans-dienyl intermediate cation which undergoes C(3)-C(4) bond rotation to give the cis-structure. At higher temperature, syn, anti - syn, syn isomerization is observed, and indeed on reaction of **XI** with HBF_4/acetic anhydride, only the syn, syn-salt is isolated. Even at low temperature, **XII** ionizes non-stereospecifically to a 1:3 syn, anti/syn, syn mixture.[14]

(Trimethylenemethane)$Fe(CO)_3$ undergoes protonation to give the [(2-methylallyl)$Fe(CO)_3$]$^+$ cation (although the detailed structure in solution may need to be reinvestigated in the light of the results above) and may be acetylated in low yield using Friedel-Crafts conditions.[15] Recently, protonation of the functionalized trimethylenemethane complex **XVI** has been used to generate the cross-conjugated dienyl cation **XVII** in solution. Nmr clearly shows a substantial barrier to C(2)-C(3) bond rotation, thus eliminating a simple [(allyl)$Fe(CO)_3$]$^+$ structure for this cation.[16]

Protonation in acid solution of (cyclobutadiene)$Fe(CO)_3$ yields a cyclobutenyl cation whose structure is analogous to **III**[17], and the complex also undergoes a wide variety of electrophilic substitution reactions such as acetylation, formylation, chloromethylation, aminomethylation, acetoxymercuration and sulfonation. An extensive organic chemistry of these complexes which is beyond the scope of this article has been described, together with the use of these complexes as in situ sources of cyclobutadienes via oxidative cleavage.[18]

XVIII X=COMe, Y=H   **XX**   **XXI**
XIX X=H,   Y=COMe

Initial electrophilic attack at the metal atom thus seems a reasonable assumption and the similarity to the mechanism proposed[19] for substitution of ferrocene may be noted.   Results on substitution and addition reactions of some cyclic (diene)Fe(CO)$_3$ complexes indicate, however, that direct exo-electrophilic attack may be a competitive pathway.   Thus acetylation of (1,3-cyclohexadiene)Fe(CO)$_3$ by (CH$_3$COCl/AlCl$_3$) yields a 4:1 mixture of the 5-exo and 5-endo isomers **XVIII** and **XIX**,[20] in which the exo-isomer is thought to arise from direct electrophilic attack on the diene face opposite to the Fe(CO)$_3$ moiety.   It is interesting to note that (1,3-cyclohexadiene)Fe(CO)$_3$ reacts with AlCl$_3$ alone to give **XX**, whose yield may be greatly increased if the reaction is performed under an atmosphere of CO. The mechanism of the reaction is not clear; while it is known that reversible adduct formation at the iron atom occurs initially, at least two moles of the Lewis acid are required for conversion to **XX**.[21] Reaction of (1,3-cyclohexadiene)Fe(CO)$_3$ with Tl(CF$_3$COO)$_3$ in ethanol to yield (5-exo-ethoxy-1,3-cyclohexadiene)Fe(CO)$_3$ is also thought to proceed via initial electrophilic attack by Tl(III), but the stereochemistry of the addition is not clear.[22]   Reaction of cyclic (η^4-diene)Fe(CO)$_3$ complexes with [Ph$_3$C]X salts yields stable cyclic dienyl salts such as **XXI**.   In this case, hydride abstraction definitely occurs at the exo-face of the ligand.[8] although isolated examples of endo-abstraction are known.[10]

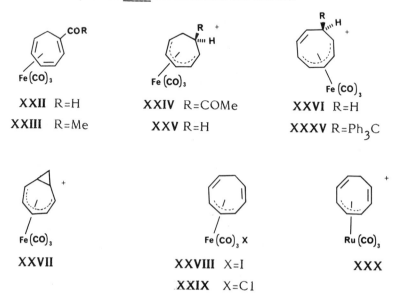

Fe(CO)$_3$	Fe(CO)$_3$	Fe(CO)$_3$
XXII R=H	**XXIV** R=COMe	**XXVI** R=H
XXIII R=Me	**XXV** R=H	**XXXV** R=Ph$_3$C

Fe(CO)$_3$	Fe(CO)$_2$X	Ru(CO)$_3$
XXVII	**XXVIII** X=I	**XXX**
	XXIX X=Cl	

Where the η^4-diene system forms part of a higher conjugated polyene system, more convincing evidence for direct exo electrophilic attack is found. Friedel-Crafts acetylation and Vilsmeier formylation of $(\eta^4$-1,3,5-cycloheptatriene)Fe(CO)$_3$ may be accomplished to yield the substituted complexes **XXII** and **XXIII**. In the acetylation, an ionic product **XXIV** is also isolated which may not be deprotonated to **XXIII**, and is thus not an intermediate in the substitution. Proton addition to yield **XXV** is known to be exo, and to occur at the terminal carbon of the uncoordinated double bond.[23-4]

The reactions towards electrophiles of (cyclooctatetraene)M(CO)$_3$ (M = Fe, Ru) and CpM(cyclooctatetraene) (M = Co, Rh, Ir) have been studied extensively. In non-coordinating acid, (cot)Fe(CO)$_3$ undergoes protonation at -120°C to give the dienyl cation **XXVI**, which on warming undergoes ring closure to the bicyclo[5.1.0] structure **XXVII**. Surprisingly, the reaction is reversible, in that **XXVII** reacts with LiI to give **XXVIII**, and the analogous chloride **XXIX** may be prepared by direct treatment of (cot)Fe(CO)$_3$ with HCl. Reaction of (cot)Ru(CO)$_3$ with HCl yields the ruthenium analogue of

XXXI	**XXXII** R=H	**XXXIV**
	XXXIII R=Me	

XXIX, and a cation of possibly similar structure **XXX** is thought to be the initial product of protonation at low temperature in strong non-coordinating acid. On warming, **XXX** undergoes isomerization to **XXXI** rather than ring closure.[25] Initial protonation is known to be exo, and most probably occurs at uncoordinated rather than coordinated carbon.[26] (cot)Fe(CO)$_3$ May also be formylated and acetylated to yield **XXXII** and **XXXIII**. An ionic bicyclo[3.2.1] product, **XXXIV**, is also isolated from the acetylation reaction. Addition is again exo, and **XXXIV** is thought to arise from a 1,4-suprafacial shift of an initially formed cation of structure **XXVII**. The reaction is not reversible, in the sense that nucleophiles add either to the allyl or olefinic bond of **XXXIV** with retention of the bicyclo[3.2.1] system.[27] The influence of electrophile on the mode of ring closure is not fully understood, and indeed reaction of $(cot)Fe(CO)_3$ with $[Ph_3C]X$ results in addition to yield **XXXV**, which shows no tendency towards ring closure.

Cyclooctatetraene complexes of the CpM series show a reactivity which is dependent on the mode of coordination. Thus, protonation of either $(1\text{-}4\text{-}\eta\text{-}C_8H_8)RhMe_5Cp$ or $(1,2:5,6\text{-}\eta\text{-}C_8H_8)RhMe_5Cp$ leads to the same mixture of bicyclic and monocyclic cations **XXXVI** and **XXXVII**. It is clear, however, that protonation of the 1,3-bonded isomer proceeds by <u>exo</u>-electrophilic attack [cf. $(1\text{-}4\text{-}\eta\text{-}C_8H_8)Fe(CO)_3$], while protonation of the 1,5-isomer is <u>endo</u> [cf. $(1.2:5,6\text{-}\eta\text{-}C_8H_8)Mo(CO)_4$] and proceeds via a metal hydride intermediate, which may actually be detected at low temperature in the protonation of $(1,2:5,6\text{-}\eta\text{-}C_8H_8)IrCp$. Where M is cobalt or rhodium, cation **XXXVI** isomerizes irreversibly to **XXXVII**, but where M is iridium, no isomerization is observed, implying that two routes exist to the monocyclic cation.[28-9] Protonation of $(cot)Ru(C_6Me_6)$ proceeds initially to give a cation in which the C_8H_9 ligand is bound as in **XXVI**, but on warming, this isomerizes to a cation of structure **XXXI**, the eventual product of the protonation of $(cot)Ru(CO)_3$. Protonation of $(C_8Me_6)Ru(1,3\text{-cyclohexadiene})$ proceeds in a fashion similar to that described for the CpRh(1,3-cyclohexadiene) analogue.[30-1]

The functionalized complexes **XXII** and **XXIII** undergo reduction with borohydride or reaction with Grignard reagents to give, after dehydration, heptafulvene complexes **XXXVIII**, which in contrast to free heptafulvenes may also be formylated at the ring. Reaction of **XXII** with trimethyl-orthoformate leads directly to the heptafulvene **XXXIX**.[32]

MCp

XXXVI

MCp

XXXVII

Fe(CO)₃

XXXVIII R=R¹=H,alkyl,aryl

XXXIX R=H,R¹=OMe

3. NUCLEOPHILIC ADDITION REACTIONS

Although neutral η^4-diene complexes based on the iron(0) and cobalt(I) triads in general show little reactivity towards nucleophiles, (but see Section 7) those based on the Ni(II) triad do show such reactivity, in keeping with the increased formal oxidation state. Thus reaction of **XL** with alkoxide and methyl or ethyl malonate yields the cyclobutenyl complexes **XLI**. An xray structure of **XLII**, obtained from reaction of [(tetramethylcyclobuta-diene)NiCl$_2$]$_2$ with NaCp, confirms the exo-addition of the nucleophile.[18]

The reactivity of cationic dienyls such as **XXI** with nucleophiles has, however, been a subject of great mechanistic and synthetic interest. In general, products resulting from attack at the ring, at the metal atom, or at a coordinated carbonyl group have been observed, although the most recent results show that an inference of mechanism based on product structure must be treated with caution. Thus reaction of [(1-5-η-cyclohexadienyl)M(CO)$_3$]$^+$ (M = Fe, Os) with methoxide proceeds to yield initially the carbomethoxy complexes **XLIII**, isolated where M = Os, but only detected by infrared where M = Fe. On heating, these complexes undergo isomerization to the exo-complex **XLIV** rather than the endo-complex **XLV**. The analogous [(1-5-η-cycloheptadienyl)Fe(CO)$_2$(CO$_2$Me)] complex may also be isolated from reaction of [(1-5-η-cycloheptadienyl)Fe(CO)$_3$]$^+$ with methoxide, and competition experiments show that its conversion to (5-exo-methoxy-1,3-cycloheptadiene)Fe(CO)$_3$ is not intramolecular and involves a reversible dissociation of OMe$^-$ from the carbomethoxy complex followed by exo-addition to the ring[33]. If methanol is used in place of methoxide in the reaction of [(1-5-η-cyclohexadienyl)M(CO)$_3$]$^+$, then the endo-isomer **XLV** is obtained as well. The exo-complex **XLIV** is still the initial kinetically controlled product, but the mechanism of its transformation to the endo-isomer is not established.[34]

Y = OR, CH(COOMe)$_2$, CH(COOEt)$_2$

| XL | XLI | XLII |

XLIII M=Fe,Os

XLIV X=OMe,Y=H
XLV X=H,Y=OMe
M=Fe,Os

XLVI

A stopped flow nmr study of the reaction of the [(tropylium)-Mo(CO)$_3$]$^+$ cation with I$^-$ shows that this proceeds via initial attack at the metal, followed by transfer of I$^-$ to yield an _exo_-substituted intermediate of structure **XLVI**. This intermediate reacts further via a solvent-assisted carbonyl displacement to yield as the final product the (tropylium)Mo(CO)$_2$I complex.[35]   Cyclic [(dienyl)Fe(CO)$_3$]$^+$ salts also react with I$^-$ to give (dienyl)Fe(CO)$_2$I complexes and there is some evidence which suggests initial attack at the metal.[36] Kinetic studies of the _exo_-nucleophilic addition of phosphines and phosphites to [(dienyl)M(CO)$_3$]$^+$ systems (M = Fe, Ru, Os) have been made, and based on ring size and substituent effects, a direct nucleophilic attack at the dienyl ligand was postulated, although a mechanism involving initial metal attack could not be ruled out[37] (see Section 7). Calculations of the frontier electron density of the LUMO in the series [(C$_5$H$_5$)Fe(CO)$_3$]$^+$, [(C$_6$H$_7$)Fe(CO)$_3$]$^+$ and [(C$_7$H$_9$)Fe(CO)$_3$]$^+$ are consistent with initial metal attack.[38]

The regiospecificity of nucleophilic attack on these systems has been shown to be dependent on the metal, the ring size, and the auxiliary ligands present.   Nucleophilic attack on [(cyclohexadienyl)Fe(CO)$_3$]$^+$ invariably generates the 5-_exo_-1,3-cyclohexadiene complex, implying attack at the 1-position of the dienyl system.   Complex **XLVII**, the product of hydride

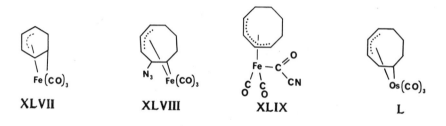

XLVII   **XLVIII**   **XLIX**   **L**

attack at the 2-position, has been prepared by other means, and is known to isomerize to the η^4-1,3-diene system only at $120°C$, while $(\eta^4$-1,4-cyclohexadiene)$Fe(CO)_3$ complexes, the product of attack at the 3-position, are unknown. Reduction of $[(1$-5-η-cyclohexadienyl)$Os(CO)_3]^+$ does, however, generate a mixture of $(\eta^4$-1,3-cyclohexadiene)$Os(CO)_3$, and the osmium analogue of **XLVII**.[39] Borohydride reduction of $[(cyclooctadienyl)Fe(CO)_3]^+$ yields only the η^4-1,3-diene complex in poor yield, but reaction with azide yields **XLVIII**,[40] while reaction with cyanide yields **XLIX**, the product of attack at the carbonyl group.[41] Borohydride reduction of the osmium analogue yields only **L**.[42]

The reactions of the seven-membered $[(cycloheptadienyl)M(CO)_3]^+$ series provide the most interesting comparisons. Borohydride reduction of the iron and ruthenium complexes both yield a mixture of the 1,3-diene and σ,π-allyl structures **LI** and **LII**.[41,43] Again, the σ,π-allyl (M = Fe) has been prepared independently and converts thermally to **LII** only at temperatures greater than $100°C$. In addition, the relative proportions of **LI** and **LII** vary according to the reaction conditions, implying a kinetic rather than thermodynamic control of the reaction. The position of attack is quite sensitive to auxiliary ligand. Thus reduction of $[(cycloheptadienyl)Fe(CO)_2PPh_3]^+$ yields only the σ,π-allyl complex related to **LI**, while reduction of $[(cycloheptadienyl)Fe(CO)(1,3\text{-cyclohexadiene})]^+$ yields only complex **LIII**, resulting from attack at the 3-position.[44] In general, the addition of hard nucleophiles to 18-electron cations to give products of kinetic control appears to be charge, rather than orbital, controlled. From simple molecular orbital arguments, some basic rules have been proposed[45] for the direction of attack, and these are expounded further in Chapter 14 above.

The potential in synthesis of the cyclohexadienyl salts holds some

| LI | LII | LIII | LIV |

promise,[46-7] as do the isoelectronic $[(\eta^5$-cyclohexadienyl)Cr(CO)$_3]^-$ salts produced by nucleophilic attack on (arene)Cr(CO)$_3$ complexes.[48] Recently, the potential for asymmetric synthesis has been investigated. Thus, asymmetric complexation followed by hydride abstraction yields the optically active salt (LIV) which after alkylation, oxidative cleavage, and hydrolysis yields R(-)-cryptone LV.[49] The enantiomeric enrichment is low, but the potential of such reactions is clear (see Section 7).

4.   CYCLOADDITION REACTIONS

Conjugated dienes react with monoenes via 1,4-addition, although with electrophilic olefins such as CF$_2$=CF$_2$, 1,2-addition may be found. As neither reaction would preserve a 4-electron donor ligand in the case of (η^4-diene)M complexes, a different pathway involving formal insertion into a metal-carbon bond is observed. Thus, (1,3-butadiene)Fe(CO)$_3$ reacts photochemically with CF$_2$=CF$_2$ to give LVI, and similar reactions are observed with (tetramethylcyclobutadiene)Fe(CO)$_3$ and (1,3-cyclohexadiene) Fe(CO)$_3$. Where other products are observed, such as LVII isolated from the reaction of (1,3-cyclooctadiene)Fe(CO)$_3$ with CF$_2$=CF$_2$, they may be viewed as being derived from an initial insertion product of structure type LVI.[50]

Reaction of CF$_3$C≡CCF$_3$ with (1,3-butadiene)Fe(CO)$_3$ proceeds in a similar way to yield LVIII, while reaction with (1,3-cyclohexadiene)Fe(CO)$_3$ yields a 2:1 adduct, LIX. On heating, LVIII undergoes rearrangement to give LX, a reaction which is thought to proceed via a π-allyl intermediate which undergoes reductive elimination, to give a (1,4-cyclohexadiene)Fe(CO)$_3$ complex which then isomerizes to the 1,3-isomer.[51] Reaction of (tetramethylcyclobutadiene)Fe(CO)$_3$ with CF$_3$C≡CCF$_3$ yields directly the (arene)Fe(CO)$_3$ complex LXI. The reaction is thought to proceed via formation of an insertion product of structure LVIII which undergoes reductive elimination to give a (Dewar benzene)Fe(CO)$_3$ complex which isomerizes to LXI.[52]

LV   LVI   LVII

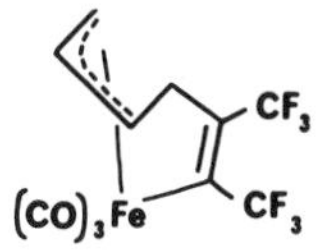

LVIII 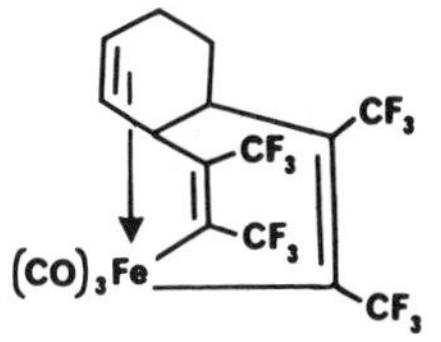 LIX 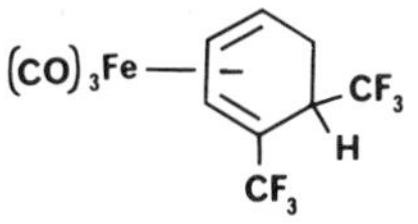 LX

Where the diene forms part of a larger polyene system, differences in reaction are found. Thus, (1,3,5-cycloheptatriene)Fe(CO)$_3$ reacts photochemically with dimethylmaleate to yield an insertion product of structure **LXII**, but with $CO_2MeC\equiv CCO_2Me$ to give **LXIII**, the product of formal 1,6-addition.[53] All of the above reactions are photochemically initiated and the <u>endo</u>-insertion or addition implies an initial metal coordination of the dienophile, either by a reversible CO dissociation, or by a dechelation to yield an η^2-diene complex. Matrix photolysis of (butadiene)Fe(CO)$_3$ is known to yield both (η^4-butadiene)Fe(CO)$_2$ and (η^2-butadiene)Fe(CO)$_3$ fragments.[54] The exact mechanism of insertion or addition is not established.

η^4-triene or tetraene systems also undergo thermal <u>exo</u>-addition of electrophilic olefins such as $(CN_2)C=C(CN)_2$. A mechanism involving a dipolar intermediate (Scheme **2**) has been proposed, but present evidence suggests a concerted addition at the <u>exo</u>-face of the uncoordinated double bond.[56] Diphenylketene yields the product of 1,2-addition,[55] $(CN)_2C=C(CN)_2$ yields the product of 1,3-addition,[56] while N-methyltriazolinedione yields the product of 1,5-addition.[58] (η^4-N-methoxycarbonyl-1H-azepine)Fe(CO)$_3$ reacts with $(CN)_2C=C(CN)_2$ to yield products of both 1,3- and 1,6-addition, although the latter is thought to arise from isomerization of the initial 1,3-adduct.[57]

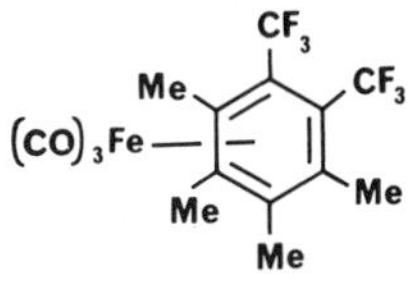

LXI 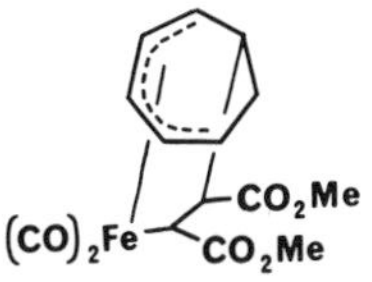 LXII 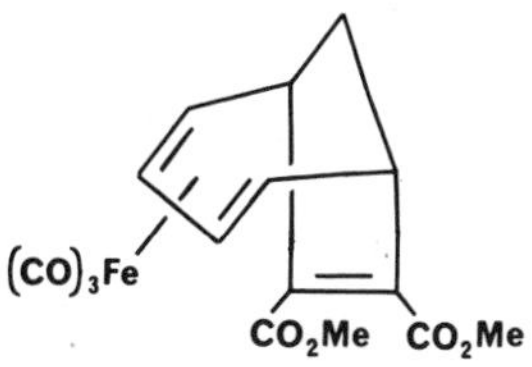 LXIII

(Cyclooctatetraene)Fe(CO)$_3$ reacts with (CN)$_2$C=C(CN)$_2$ to yield a product resulting from 1,3-addition. [59]   It may be noted that free cyclooctatetraene and cycloheptatriene both react with (CN)$_2$C=C(CN)$_2$ in a normal Diels-Alder manner via their bicyclic valence tautomers. [60-f]

Scheme 2

1,6- addition

1,2-addition

1,3- addition

1,5- addition

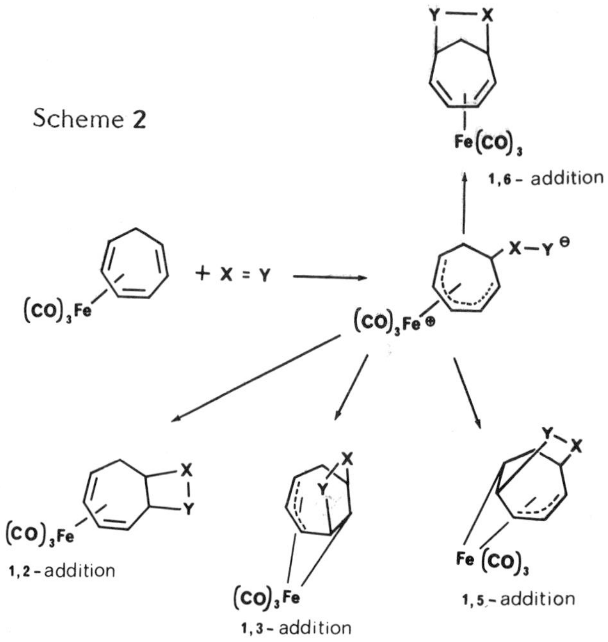

5.   LIGAND ISOMERIZATION REACTIONS

The preference of the Fe(CO)$_3$ moiety in particular for bonding to a 1,3-diene fragment is well established.   Thus, in many cases, reaction of unconjugated dienes with iron carbonyl reagents yields only 1,3-diene complexes, and analysis of the uncomplexed diene recovered frequently indicates that the isomerization is catalytic; the ability of (η^4-diene) Fe(CO)$_3$ complexes to undergo ligand exchange is well documented. [65] While (η^4-1,5-cyclooctadiene)Fe(CO)$_3$, the 1,4-complex **LIII** and the 1-3,5-complex **LII** are known, all undergo clean thermal isomerizations to the η^4-1,3-diene isomer.   Such reactions are thought to proceed by an (allyl)MH inter-mediate, depicted as **LXIV** for the isomerization of **LIII**.   It has been shown that in the Ru$_3$(CO)$_{12}$ catalyzed isomerization of 1,5-cyclooctadiene, free 1,4-cyclooctadiene is produced, consistent with a stepwise mechanism. [63]

The isomerization mechanism implies an <u>endo</u>-transfer of hydrogen from the ligand to the metal, and it has been demonstrated that the complex $(\eta^4$-1,3-cyclohexadiene)Ir(1,5-cyclooctadiene)H exists in a dynamic equilibrium in solution with the allyl derivative **LXV**, with the addition-elimination equilibria occurring via <u>endo</u>-hydrogen transfer.[64] Similarly, while (5-<u>exo</u>-methoxy-1,3-cyclohexadiene)Fe(CO)$_3$ undergoes thermal isomerization to a mixture of the 1- and 2-isomers, the 5- <u>endo</u> isomer does not.[65] Isomerization and hydrogen scrambling in acyclic $(\eta^4$-diene)Fe(CO)$_3$ complexes proceed by a similar mechanism.[66]

A reaction which is irreversible in this sense is the thermolysis of $(\eta^4$-1,3-cyclopentadiene)Fe(CO)$_3$ to yield the dimer Cp$_2$Fe$_2$(CO)$_4$. Although somewhat complicated, it is thought that the reaction proceeds initially by hydrogen transfer to yield **LXVI**, which rapidly undergoes loss of CO and hydrogen to give the dimer.[67] (5,5-dialkylcyclopenta-1,3-diene)Fe(CO)$_3$ complexes also undergo a similar isomerization to yield the stable $(\eta^5$-C$_5$H$_4$R)Fe(CO)$_2$R^1 complex by transfer of the <u>endo</u>-alkyl group.[68]

An intramolecular hydrogen transfer which may proceed by a similar mechanism has been demonstrated in the thermal isomerization of $(\eta^4$-cycloocta-1,5-diene)Ru(η^6-cycloocta-1,3,5-triene) to (C$_8$H$_{11}$)$_2$Ru (**LXVII**).[69] Such a mechanism may be important in metal atom syntheses. Reaction of 1,3,5-cycloheptatriene and 1,3,5-cyclooctatriene with iron atoms yields (C$_7$H$_9$)(C$_7$H$_7$)Fe (**LXVIII**) and (C$_8$H$_{11}$)(C$_8$H$_9$)Fe (**LXIX**) respectively.[70-1] The products obtained on reaction of 1,3-cyclohexadiene with iron atoms, which include the disproportionation products benzene and cyclohexene, can also be explained in terms of initial formation of $(\eta^4$-1,3-cyclohexadiene)$_2$Fe followed by intramolecular hydrogen transfer, and from co-condensation with P(OMe)$_3$, complex **LXX** may be isolated. Reaction of 1,3,5-cycloheptatriene with molybdenum atoms similarly yields **LXXI**, while ligand to metal hydrogen transfer is observed on reaction of tungsten atoms with cyclopentadiene to give Cp$_2$WH$_2$.[72]

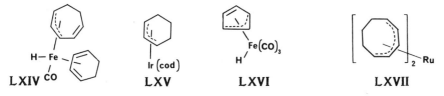

LXIV   LXV   LXVI   LXVII

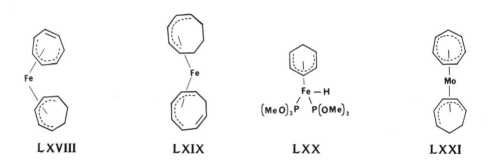

| LXVIII | LXIX | LXX | LXXI |

Reaction of binary metal carbonyl clusters with diene and polyene ligands also frequently yields metal cluster complexes in which the organic ligand is derived formally from diene isomerization, hydrogenation or dehydrogenation, or transfer of hydrogen from ligand to metal. Taking $Ru_3(CO)_{12}$ as an example, reaction with 2,3-dimethyl-1,3-butadiene yields LXXII,[73] while reaction with 1,3-cyclohexadiene yields LXXIII and LXXIV.[74] 1,3,5-cyloheptatriene reacts to give $[Ru_3(CO)_6(C_7H_7)(C_7H_9)]$ (LXXV) and $[Ru_2(CO)_6(C_7H_8)]$ (LXXVI).[75] Cyclooctatetraene undergoes both formal dehydrogenative and hydrogenative ring closure to yield LXXVII and LXXVIII.[76] The low yields and concomitant formation of monometallic complexes in many of these reactions may be noted.

As several cluster complexes containing a 1,3-diene coordinated in the normal η^4-manner are known,[77] initial diene coordination to the cluster seems likely in these reactions. Little is known about the process of hydrogen transfer, and the rather unique ability of clusters to activate C-H and C-C bonds in other saturated and unsaturated organic ligands may indicate a significant difference in mechanism from monometallic complexes.

LXXII

LXXIII L=CO
LXXIV $L_3=\eta^6-C_6H_6$

LXXV

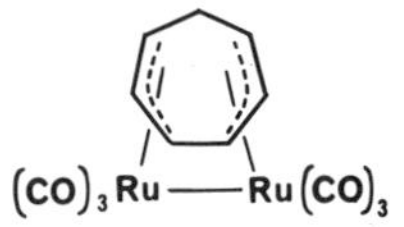

LXXVI 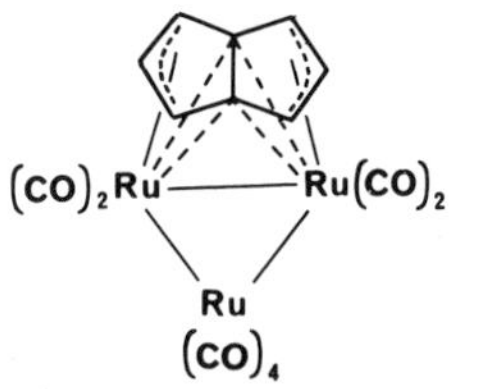   LXXVII 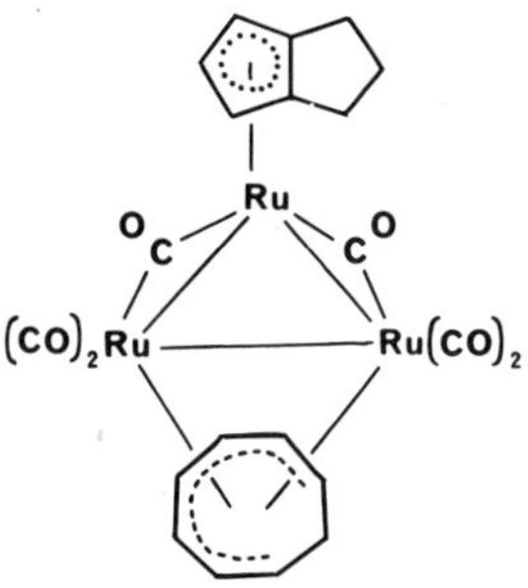   LXXVIII

6.   CATALYTIC TRANSFORMATIONS OF CONJUGATED DIENES

Homogeneous and heterogeneous transition metal catalysts are known for the dimerization and cyclodimerization, co-dimerization, cyclotrimerization and polymerization of conjugated dienes. While an initial η^2-or η^4-coordination to the metal centre is presumed, the active intermediates are almost always π-allyl species derived from insertion into M-H or M-R bonds formed in situ or by electronic rearrangement of two 1,3-diene molecules coordinated to a low valent metal centre. In the latter case, the metal undergoes an increase of two in its formal oxidation state.

6.1   Cyclodimerization and Trimerization

Several Ni(0) complexes such as $Ni(cod)_2$ or (1,5,9-cyclo-dodecatriene)Ni function as cyclodimerization catalysts in the presence of added PR_3 ligands, yielding cis,cis-1,5-cyclooctadiene, vinylcyclohexene and cis-1,2-divinylcyclobutane. Although some details are still in dispute, a mechanism is shown in Scheme 3. Intermediate LXXIX has been detected, and nmr results are consistent with the cis,trans formulation (C_2-C_3; C_6-C_7) as the minimum energy conformation;[78-9] LXXX has been characterized by crystallography [79] and shown to react with CO to release vinyl-cyclohexene.[81] Complex LXXXVI reacts with 2,3-dimethylbutadiene to give (η^4-2,3-dimethylbutadiene)$_2$Ni which on treatment with phosphine yields directly the complex of structure LXXX.[82] The intermediacy of LXXXI is more problematical. While it provides an isomerization route to

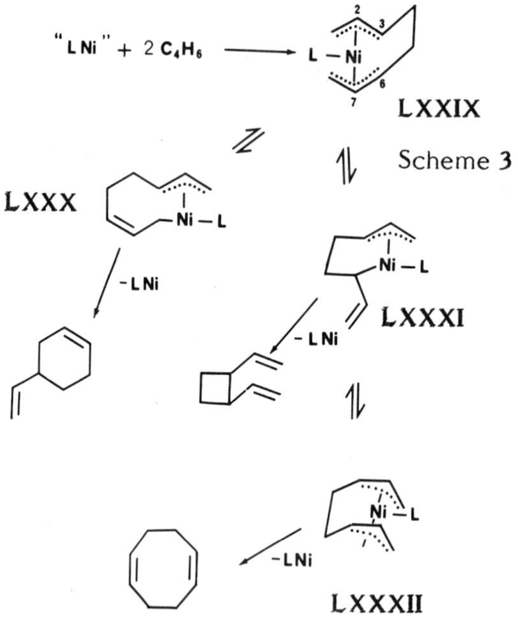

LXXIX

Scheme 3

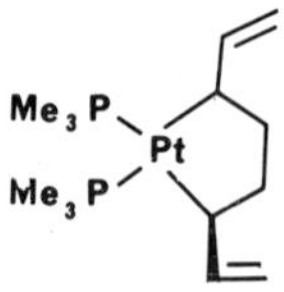

LXXX

LXXXI

LXXXIII

LXXXII

the cis,cis-derivative **LXXXII**, which is sterically suited to elimination to give 1,5-cyclooctadiene, an alternative mechanism involving metallacycle formation has been proposed for formation of divinylcyclobutane. Recently, it has been shown that a platinum analogue of **LXXX** undergoes rearrangement in the presence of phosphine to yield the platinacyclobutane **LXXXIII**, although the stereochemistry of the vinyl groups here is trans.[83] Initial coordination is thought to proceed via an $LNi(\eta^2\text{-butadiene})_2$ complex.

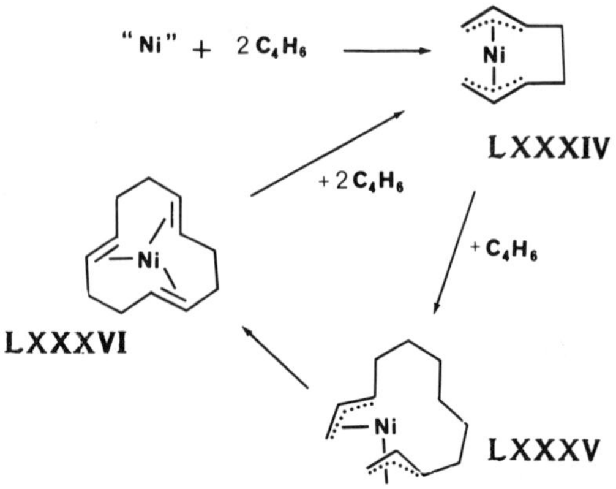

LXXXIV

LXXXVI

LXXXV

Scheme 4

In the absence of phosphine, the intermediate of structure **LXXXIV** reacts further with butadiene to yield **LXXXV**, which may be isolated at low temperature.[84]   Ring closure to **LXXXVI**, followed by displacement by butadiene, completes the catalytic cycle (Scheme 4).

Several Pd(0) complexes also function as dimerization catalysts in the presence of PR_3, but only linear dimers are obtained.[85]

6.2   Linear Dimerization and Polymerization

Several systems such as $CoCl_2/NaBH_4$ or $Co_2(CO)_8/AlEt_3$ catalyze the dimerization of butadiene to yield 3-methyl-1,3,6-heptatriene with some 1,3,6-octatriene.   The mechanism is thought to proceed via a metal hydride intermediate (Scheme 5).   Intermediate **LXXXIX** has been characterized crystallographically; in particular, the proximity to the metal of one of the hydrogens α to the allyl group may be noted with respect to the reaction following involving elimination of triene.[86]   Formation of the more sterically hindered σ-allyl **LXXXVIII** may seem surprising, and an alternative electrocyclic process involving C-C bond formation outside the coordination sphere via the $[Co-CH_2CH = CHCH_3]$ species has been proposed.[87]

Both 1,2- and 1,4-addition have been observed in the polymerization of butadiene.   The 1,2-addition may, in principle, be viewed in terms of

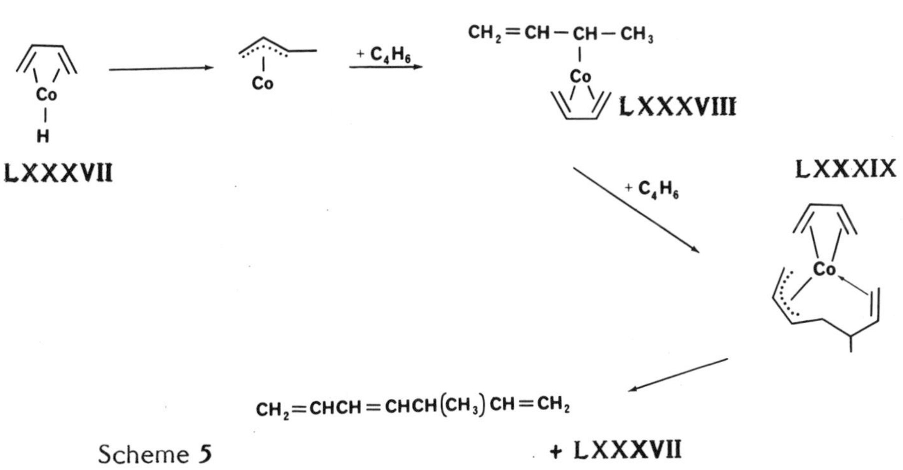

Scheme 5

monoene polymerization, while the 1,4-addition is diene specific and is catalyzed by Zeigler type systems based on titanium or nickel group metals, as well as monometallic nickel group catalysts (Scheme 6). Alkylation of the metal is presumed, followed by diene coordination (**XC**) and insertion to give the allyl **XCI**.   Rearrangement to the σ-allyl in the presence of diene, followed by insertion, provides the growing chain. Recently the insertion of monoene into an M-R bond has been questioned as a step in the Zeigler-Natta polymerization of monoenes.[88]   An alternative mechanism involving metal carbene species was postulated, and a similar mechanism may be written to account for the conversion of **XC** to **XCI** (Scheme 6) and for the previously described conversion of **LXXXVIII** into **LXXXIX**. The mechanism of Scheme 6 also accounts for the ability of complexes such as $Ni(\eta^3\text{-allyl})_2$ to catalyze the reaction homogeneously.[89]

More recently, it has been shown that metallacycle formation may easily be induced in the reaction of $(butadiene)_3Mo$ with $Li(tetramethyl-ethylenediamine)$ to give **XCII**.[90]

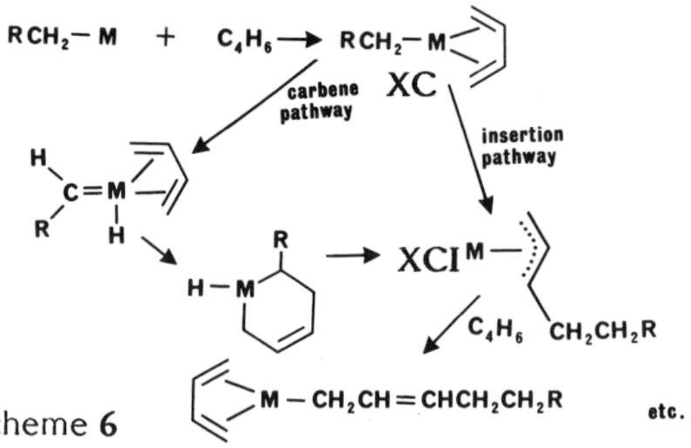

Scheme 6 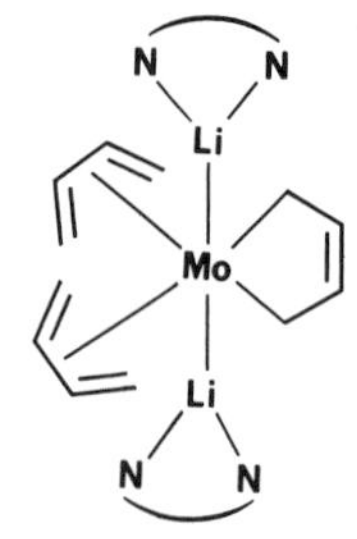   XCII

7.   MORE RECENT DEVELOPMENTS

There has been significant recent activity in the preparation and reactions of Zr-diene complexes.[91]

There is now a substantial chemistry based on nucleophilic attack on neutral $(diene)Fe(CO)_3$ complexes.[92]

For evidence of phosphine attack at the metal of $[(dienyl)Fe(CO)_3]X$, see Ref. 93, and for more recent work on the use of $(diene)Fe(CO)_3$ and $[(dienyl)Fe(CO)_3]X$ complexes in enantioselective synthesis, see Refs. 94-5. Note also the use of $[(diene)Mn(CO)_3]^-$ anions in synthesis.[96]

REFERENCES

1. P.R. Brown, F.G.N. Cloke and M.L.H. Green, J.C.S. Chem. Commun. 1126 (1980).

2. S. Datta, M.B. Fischer and S.S. Wreford, J. Organomet. Chem. 188:353 (1980).

3. R.P. Beatty, S. Datta and S.S. Wreford, Inorg. Chem. 18:3139 (1979).

4. H. Yasuda, Y. Kagihara, K. Mashima, K. Lee and A. Nakamura, Chem. Lett. 519 (1981).

5. G. Erker, J. Wicher, K. Engel, F. Rosenfeldt, W. Dietrich and C. Kruger, J. Am. Chem. Soc. 102:6344 (1980).

6. S.D. Ittel, F.A. Van Catledge and J.P. Jesson, J. Am. Chem. Soc. 101:6905 (1979).

7. L.A. Oro, Inorg. Chim. Acta. 21:L7 (1977).

8. B.F.G. Johnson, D.J. Yarrow and J. Lewis, J.C.S. Dalton Trans. 2084 (1972).

9. W.H. Amma and M. Brookhart, J. Am. Chem. Soc. 103:989 (1981).

10. N. El Murr, J. Organomet. Chem. 208:C9 (1981).

11. G.A. Sim and D.I. Woodhouse, Acta. Cryst. 35B:477 (1979).

12. J.S. Fredericksen, R.E. Graff, D.G. Gresham and C.P. Lillya, J. Am. Chem. Soc. 101:3863 (1979).

13. N.A. Clinton and C.P. Lillya, J. Am. Chem. Soc. 92:3058 (1970).

14. D.G. Gresham, D.J. Kowalski and C.P. Lillya, J. Organomet. Chem. 144:71 (1978).

15. K. Ehrlich and G.F. Emerson, J. Am. Chem. Soc. 94:2464 (1972).

16. P.A. Dobosh, C.P. Lillya, E.S. Magyar and G. Scholes, Inorg. Chem. 19:228 (1980).

17. G.A. Olah and G. Liang, J. Org. Chem. 41:2659 (1976).

18. A. Efraty, Chem. Rev. 77:691 (1977).

19. F.A. Cotton and G. Wilkinson, "Advanced Inorganic Chemistry", 4th Edn., Wiley-Interscience, New York (1980) p.1166.

20. B.F.G. Johnson, J. Lewis and D.G. Parker, J. Organomet. Chem. 141:319 (1977).

21. B.F.G. Johnson, J. Lewis and K.D. Karlin, J. Organomet. Chem. 145:C23 (1978).

22. B.F.G. Johnson, J. Lewis and D.G. Parker, J. Organomet. Chem. 127:C37 (1977).

23. B.F.G. Johnson, J. Lewis, P. McArdle and G.L.P. Randall, J.C.S. Dalton Trans. 456 (1972).

24. M. Brookhart, K.J. Karel and L.E. Nance, J. Organomet. Chem. 140:203 (1977).

25. A.D. Charles, P. Diversi, B.F.G. Johnson and J. Lewis, J. Organomet. Chem. 116:C25 (1976).

26. M. Brookhart, E.R. Davis and D.L. Harris, J. Am. Chem. Soc. 94:7853 (1972).

27. A.D. Charles, P. Diversi, B.F.G. Johnson and J. Lewis, J.C.S. Dalton Trans. 1906 (1981).

28. A.K. Smith and P.M. Maitlis, J.C.S. Dalton Trans. 1773 (1976).

29. J. Evans, B.F.G. Johnson, J. Lewis and D.J. Yarrow, J.C.S. Dalton Trans. 2375 (1974).

30. M.A. Bennett, T.W. Matheson, G.B. Robertson, A.K. Smith and P.A. Tucker, Inorg. Chem. 19:1014 (1980).

31. M.A. Bennett, T.W. Matheson, G.B. Robertson, A.K. Smith and P.A. Tucker, Inorg. Chem. 20:2353 (1981).

32. Z. Goldschmidt and Y. Bakal, J. Organomet. Chem. 168:215 (1979).

33. D.A. Brown, W.K. Glass and F.M. Hussein, J. Organomet. Chem. 186:C58 (1980).

34. A.L. Burrows, K. Hine, B.F.G. Johnson, J. Lewis, D.G. Parker, A. Poe and E.J.S. Vichi, J.C.S. Dalton Trans. 1135 (1980).

35. P. Powell, L.J. Russell and E. Styles, J. Organomet. Chem. 149:C1 (1978).

36. D.A. Brown, S.K. Chawla and W.K. Glass, Inorg. Chim. Acta. 19:L31 (1976).

37. G.R. John and L.A.P. Kane-Maguire, J.C.S. Dalton Trans. 873 (1979).

38. D.A. Brown, J.P. Chester and N.J. Fitzpatrick, J. Organomet. Chem. 155:C21 (1978).

39. A.L. Burrows, B.F.G. Johnson, J. Lewis and D.G. Parker, J. Organomet. Chem. 194:C11 (1980).

40. G. Schiavon, C. Paradisi and C. Boanini, Inorg. Chim. Acta. 14:L5 (1975).

41. G. Schiavon and C. Paradisi, J. Organomet. Chem. 210:247 (1981).

42. A.J. Deeming, S.S. Ullah, A.J.P. Domingos, B.F.G. Johnson and J. Lewis, J.C.S. Dalton Trans. 2093 (1974).

43. B.A. Sosinsky, S.A.R. Knox and F.G.A. Stone, J.C.S. Dalton Trans. 1633 (1975).

44. B.F.G. Johnson, J. Lewis, I.E. Ryder and M.V. Twigg, J.C.S. Dalton Trans. 421 (1976).

45. S.G. Davies, M.L.H. Green and D.M.P. Mingos, Tetrahedron 34:3047 (1978).

46.   A.J. Birch, Ann. N.Y. Acad. Sci. 333:107 (1980).

47.   A.J. Birch, B.M.R. Bandara, K. Chamberlain, P. Dahler, A.I. Day, I.D. Jenkins, L.F. Kelly, T.C. Khor, G. Kretschmer, A.J. Liepa, A.S. Narula, W.D. Raverty, E. Rizzardo, C. Sell, G.R. Stephenson, D.J. Thompson and D.H. Williamson, Tetrahedron 37 (suppl. 1) 289 (1981).

48.   M.F. Semmelhack, G.R. Clark, J.L. Garcia, J.J. Harrison, Y. Thebtaramoth, W. Wulpp and A. Yamashita, Tetrahedron 37:3957 (1981).

49.   A.J. Birch, R.D. Raverty and G.R. Stephenson, Tetrahedron Lett. 21:197 (1980).

50.   A. Bond, B. Lewis, and M. Green, J.C.S. Dalton Trans. 1109 (1975).

51.   M. Bottrill, R. Davies, R. Goddard, M. Green, R.P. Hughes, B. Lewis and P. Woodward, J.C.S. Dalton Trans. 1252 (1977).

52.   A. Bond, M. Bottrill, M. Green and A.J. Welch, J.C.S. Dalton Trans. 2372 (1977).

53.   R.E. Davis, T.A. Dodds, T.H. Hseu, J.C. Wagnon, T. Devon, J. Tancrede, J.S. McKennis and R. Pettit, J. Am. Chem. Soc. 96:7563 (1974).

54.   G. Ellerhorst, W. Gerhartz and F.W. Grevels, Inorg. Chem. 19:67 (1980).

55.   Z. Goldschmidt and S. Antebi, Tetrahedron Lett. 271 (1978).

56.   S.K. Chopra, M.J. Hynes and P. McArdle, J.C.S. Dalton Trans. 587 (1981).

57.   S.K. Chopra, G. Moran and P. McArdle, J. Organomet. Chem. 214:C36 (1981).

58.   G.D. Andreeti, G. Bocelli and P. Sgarabotto, J. Organomet. Chem. 150:85 (1978).

59.   L.A. Paquette, S.V. Ley, S. Maiorana, D.F. Schneider, M.J. Broadhurst and R.A. Boggs, J. Am. Chem. Soc. 97:4658 (1975).

60.   L.A. Paquette, D.R. James, and G.H. Birnberg, J. Am. Chem. Soc. 96:7454 (1974).

61.   G.W. Wahl, J. Org. Chem. 33:2158 (1968).

62.   J.A.S. Howell, P.M. Burkinshaw and D.T. Dixon, J.C.S. Dalton Trans. 999 (1980).

63.   A.J. Deeming, S.S. Ullah, A.J.P. Domingos, B.F.G. Johnson and J. Lewis, J.C.S. Dalton Trans. 2093 (1974).

64.   J. Muller, H. Menig and P.V. Rinze, J. Organomet. Chem. 181:387 (1979).

65.   K.E. Hine, B.F.G. Johnson and J. Lewis, J.C.S. Dalton Trans. 2093 (1974).

66.   T.H. Whitesides, and J.P. Neilan, J. Am. Chem. Soc. 98:63 (1976).

67.   T.H. Whitesides, and J. Shelly, J. Organomet. Chem. 92:215 (1975).

68.   P. Eilbracht and P. Dahler, J. Organomet. Chem. 135:C23 (1977).

69.   P. Pertici, G. Simonelli, G. Vitulli, G. Deganello, P. Sandrini and A. Mantovani, J.C.S. Chem. Commun. 132 (1977).

70.   J.R. Blackborrow, K. Hildenbrand, E.A. Koerner von Gustorf, A. Scrivanti, C.R. Eady and D. Ehntholt, J.C.S. Chem. Commun. 16 (1976).

71.   S.D. Ittel, F.A. van Catledge and J.P. Jesson, J. Am. Chem. Soc. 101:3874 (1979).

72.   E.M. van Dam, W.N. Brent, M.P. Silvon, and P.S. Skell, J. Am. Chem. Soc. 97:465 (1975).

73.   R.P. Ferrari and G.A. Vaglio, J. Organomet. Chem. 182:245 (1979).

74.   S. Aime, L. Milone, D. Osella, G.A. Vaglio, M. Valle, A. Tiripicchio and M.T. Camellini, Inorg. Chim. Acta. 34:49 (1979).

75.   J.C. Burt, S.A.R. Knox and F.G.A. Stone, J.C.S. Dalton Trans. 731 (1975).

76.   R. Bau, B. Chou, S.A.R. Knox, V. Riera and F.G.A. Stone, J. Organomet. Chem. 82:C43 (1974).

77.   C.G. Pierpont, Inorg. Chem. 17:1976 (1978).

78.   C.R. Graham and L.M. Stephenson, J. Am. Chem. Soc. 99:7099 (1977).

79.   W. Brenner, P. Heimbach, H. Hey, E.W. Muller and G. Wilke, Liebigs Ann. Chem. 727:161 (1969).

80.   B. Barret, B. Bussmeirer, P. Heimbach, P.W. Jolly, C. Kruger, I. Tkatchenko and G. Wilke, Tetrahedron Lett. 1457 (1972).

81.   P.W. Jolly, I. Tkatchenko and G. Wilke, Angew. Chem. Int. Ed. Engl. 10:329 (1971).

82.   P. Jolly, R. Mynott and R. Salz, J. Organomet. Chem. 184:C49 (1980).

83.   G.K. Barker, M. Green, J.A.K. Howard, J.L. Spencer, and F.G.A. Stone, J. Am. Chem. Soc. 98:3373 (1976).

84.   B. Bogdanovic, P. Heimbach, M. Kroner, G. Wilke, E.G. Hoffmann and J. Brandt, Liebigs Ann. Chem. 727:143 (1969).

85.   E.J. Smutney, J. Am. Chem. Soc. 89:6794 (1967).

86.   G. Allegra, F.L. Giudice, G. Natta, U. Giannini, G. Fagherazzi and P. Pino, J.C.S. Chem. Commun. 1263 (1967).

87.   R.P. Hughes and J. Powell, J. Am. Chem. Soc. 94:7723 (1972).

88.   K.J. Ivin, J.J. Rooney, C.D. Stewart, M.L.H. Green and R. Mahtab, J.C.S. Chem. Commun. 604 (1978).

89. R. Warin, M. Julemont and P. Teyssie, J. Organomet. Chem. 185:413 (1980).

90. W. Gausing and G. Wilke, Angew. Chem. Int. Ed. Engl. 20:186 (1981).

91. G. Erker, H.J. de Liefde Meijer and J.H. Teuben, Organometallics 2:1473 (1983); G. Erker, K. Engel, J.L. Atwood and W.E. Hunter, Angew. Chem. Int. Ed. Engl. 22:494 (1983).

92. M.F. Semmelhack, J.W. Herndon and J.P. Springer, J. Am. Chem. Soc. 105:2497 (1983); M.F. Semmelhack and J.W. Herndon, Organometallics 2:363 (1983).

93. D.A. Brown, S.K. Shawla, W.K. Glass and F.M. Hussein, Inorg. Chem. 21:2726 (1982).

94. J.A.S. Howell and M.J. Thomas, J.C.S. Dalton Trans. 1401 (1983).

95. A.J. Birch, W.D. Raverty and G.R. Stephenson, J. Org. Chem. 46:5166 (1981).

96. M. Brookhart, W. Lamanna and A.R. Pinhas, Organometallics 2:638 (1983).

REACTIONS OF 5-CARBON AND LARGER LIGANDS

P. Powell

Royal Holloway College, London

1.   SCOPE OF THE CHAPTER

This Chapter is concerned with the reactions of pentahapto or larger delocalized systems bonded to transition elements, with the exception of cyclopentadienyl ligands. Emphasis is laid on the attack of nucleophiles and electrophiles on such ligands and on the mechanisms of such processes. A comprehensive survey is not possible in the available space; reference is made where appropriate to relevant reviews and textbooks. Rearrangements of coordinated cyclic polyolefins are not covered, in view of the recent publication of a detailed text on the chemistry of these compounds.[1]

2.   NUCLEOPHILIC ATTACK ON COORDINATED LIGANDS

2.1   Introduction

The reactivity of unsaturated hydrocarbons such as butadiene or benzene towards nucleophilic addition is often greatly enhanced by coordination to a transition metal, particularly when the complex is cationic.   Moreover many otherwise reactive species such as the cyclohexadienyl cation, 'protonated benzene', can be stabilized by such coordination which permits reactions to be performed at ligand sites.

Nucleophilic addition to coordinated ligands is normally highly stereospecific, taking place on the exo face of the ligand away from the metal. Consequently, synthetic routes which involve coordination of an unsaturated organic compound, followed by stereospecific reaction at the ligand and finally removal of the protecting metal grouping are finding application in organic chemistry.[2-4]

A recent discussion of nucleophilic addition to organotransition metal cations, including rules which allow prediction of the most favorable site of attack in complexes which contain more than one hydrocarbon group,[5] is further developed in Chapter 14. Addition can also take place at the carbon atom of a carbon monoxide ligand, where present, or directly at the metal center. It should also be borne in mind that more than one of these sites could be susceptible to attack in any particular case, and that addition can be either reversible or irreversible. Early work on reactions of coordinated ligands has been summarized by White.[6]

2.2 Nucleophilic attack on η^5-complexes

2.2.1 Tricarbonylcyclohexadienyliron salts[4]

The tricarbonylcyclohexadienyliron cation I was first prepared by Fischer and Fischer by hydride abstraction from tricarbonylcyclohexadieneiron with triphenylmethyltetrafluoroborate.[7] The tetrafluoroborate or hexafluorophosphate salts are air stable, yellow crystalline materials which are soluble in polar organic solvents such as acetone, methyl cyanide and nitromethane. They can even be recrystallized successfully from water.[8] The fact that such moderately basic solvents can be used shows that a very considerable stabilization of the cyclohexadienyl ligand occurs on coordination to $Fe(CO)_3$, compared with the free ion $C_6H_7^+$ which is observed only in superacid media.[9]

There are three sites in the molecule at which nucleophilic attack has been observed. Most commonly direct addition to the organic ligand occurs. This always takes place at a terminal position, and the incoming groups enter exo to the $Fe(CO)_3$ group:[8]

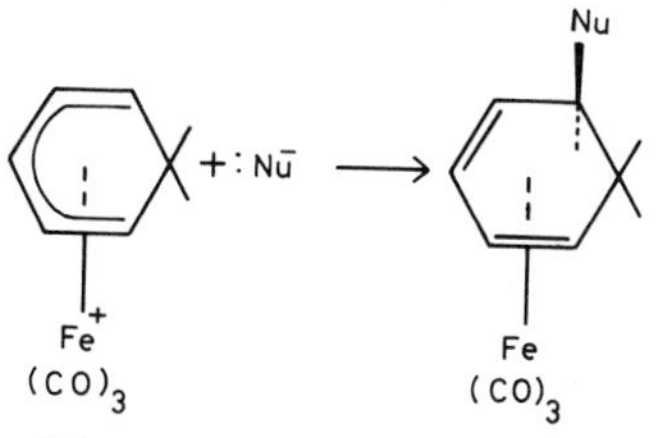

Reversible nucleophilic addition to a carbonyl ligand in **III** takes place with amines and other nitrogen bases:[10]

$$[Fe(C_5H_5)(CO)_3]^+ + 2RNH_2 \rightleftharpoons [Fe(C_5H_5)(CO)_2CONHR] + RNH_3^+$$

III

The 'closed' cyclopentadienyl ligand is however more resistant to attack than the 'open' cyclohexadienyl group,[5] so that reaction at the organic moiety is more likely to compete successfully with addition to CO in the latter case. Attack at the metal with displacement of carbon monoxide is a third possibility, observed with iodide ion.[11-12] These alternative reaction paths are summarized in Scheme 1. A scheme of this type accounts in essence for most observations of nucleophilic attack on $[M(hydrocarbon)(CO)_3]^+$ complexes, explaining many apparently conflicting results.

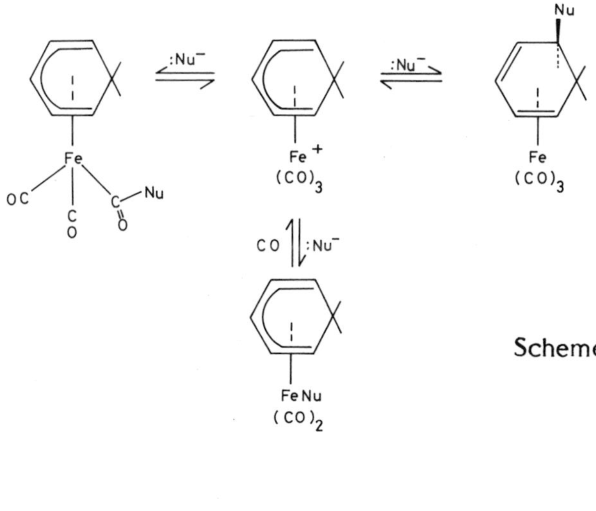

Scheme 1

2.2.1 Addition to the dienyl ligand

A very wide range of nucleophiles add to the terminal carbon atom of the dienyl system.[8] These include hydride, OMe^-, OAr^-,[13] CN^-,[14] water in the presence of sodium hydrogen carbonate, amines, phosphines,[15] sodium diethyl malonate, methyl lithium, β-diketones,[16-17] and activated aromatic hydrocarbons.[18-21] In general the addition occurs stereospecifically exo to the $Fe(CO)_3$ group. This stereochemistry also applies to attack by methoxide ion but when a salt of I is refluxed in methanol for several hours, exo addition is followed by the formation of the endo-methoxy derivative, together with some tricarbonylcyclohexadieneiron.[22-4] It was first suggested that the endo isomer arose via initial formation of a carbomethoxy complex,[22] followed by intramolecular transfer of the OMe group to the underside of the ring. In the case of the ruthenium and osmium cations, the formation of carbomethoxy species has indeed been observed early in the reaction, and they have been isolated.[25] Rearrangement of $Os(C_6H_7)(CO)_2COOMe$, however, yields the exo-methoxy product rather than the endo, showing that it proceeds by dissociation of methoxide ion followed by normal attack at the exo position.[26] The following mechanism has now been proposed (Scheme 2).[24] While methoxide reacts rapidly and irreversibly with $[Fe(C_6H_7)(CO)_3]^+$ to give only the exo isomer, in methanol alone acid is produced which makes the reaction reversible. The exo-isomer isomerizes to a pseudo-equilibrium mixture of endo and exo, perturbed by the formation of $Fe(C_6H_8)(CO)_3$ in a side reaction. Although endo attack of methanol on $[Fe(C_6H_7)(CO)_3]^+$ is very much slower than exo-attack, once formed the endo derivative is much more kinetically inert to protonation and loss of methanol than the exo complex.

exo-$Fe(C_6H_7OMe)(CO)_3$ + HA $\rightleftharpoons$ exo$[Fe(CO)_3(C_6H_7OMeH)]^+A^-$

$$OMe^-$$

$Fe(CO)_2(CO_2Me)(C_6H_7)$ $\rightleftharpoons$ $[Fe(CO)_3(C_6H_7)]^+A^-$ + MeOH

$$OMe^-$$

endo-$Fe(CO)_3(C_6H_7OMe)$ + HA $\rightleftharpoons$ endo$[Fe(CO)_3(C_6H_7OMeH^+)]A^-$

Scheme 2

A reversible addition of OMe⁻ to a carbonyl ligand, important for Ru and Os, but not significant for iron, completes the picture of this rather complicated set of reactions.

I yields **IV** on reaction with potassium cyanide.[14] The _exo_ configuration has been proved by xray crystallography.[27] Methylmagnesium iodide converts **IV** into the _exo_-acetyl complex **V**, which undergoes a Wittig reaction with $Ph_3P=CH_2$. The resulting isopropenyl complex **VI** rearranges on protonation to the tricarbonyl 1-isopropylcyclohexadienyliron cation **VII** (Scheme 3):[28]

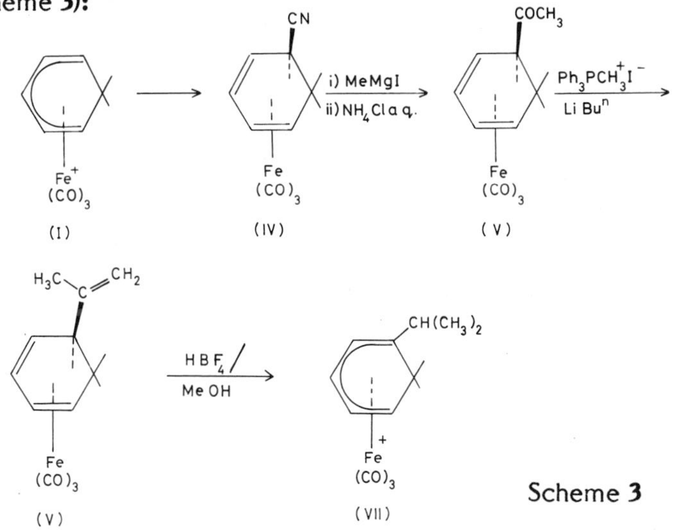

Scheme **3**

In water, the fluoride ion is too poor a nucleophile to react with **I**, but its nucleophilicity is improved by use of KF and 18-crown-6 in methyl cyanide ('naked fluoride').[12] Under these conditions the _exo_ fluoroderivative is obtained in 70% yield. Potassium chloride and bromide in CH_3CN also attack the ring but the products are unstable. Iodide however gives a mixture of an unstable ring adduct and $Fe(C_6H_7)(CO)_2I$ formed by displacement of one CO ligand from the metal.

Alkylations of tricarbonylcyclohexadienyliron cations have been carried out using lithium alkyls,[8] but the milder organozinc or organocadmium derivatives give better yields. In all cases _exo_ attack at the terminal carbon atom occurs.[29-30] Birch, Pearson and their coworkers have studied the effect of these reagents on 2-substituted cyclohexadienyl cations:[31]

(VIII)

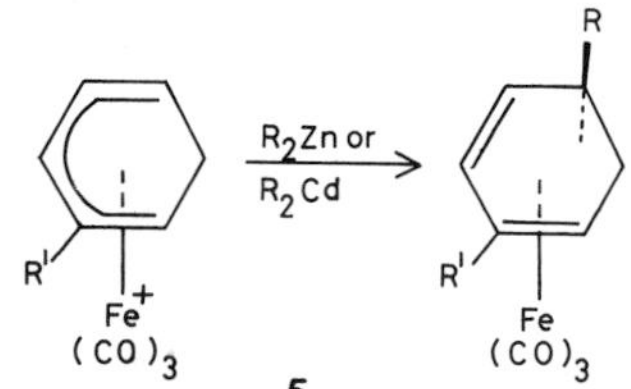

(R' = Me, CO_2Me, OMe)
(R = Ph, Pr^i, $CH_2CH=CH_2$)

Steric effects favor attack at C^5 in all cases. A similar result is obtained with alkyl copper reagents.[32-3] Borohydride reduction, however, seems to be subject mainly to electronic control. When R' = OMe, the hydride attacks selectively at C^5, whereas with R' = Me or CO_2Me, the reduction occurs relatively unselectively at C^1 and at C^5. The directing effect of the 2-methoxy substituent has been used by Pearson as a means of introducing angular substituents and forming quaternary centres via dienetricarbonyliron complexes. Complex **IX** adds nucleophiles (morpholine, hydroxide or cyanide) preferentially at C^5 as the methoxy substituent makes the adjacent C^1 position electron rich on account of its resonance effect, and hence deactivated towards addition.[35] This conclusion is supported by ^{13}C nmr studies.[36]

(IX)

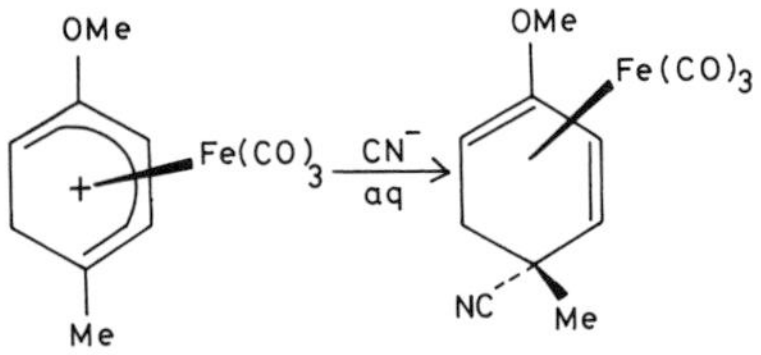

The major product obtained from attack of dimethylmalonate ion on **X** was converted into the spirocyclic compound **XI** as shown in Scheme 4.[37-8]

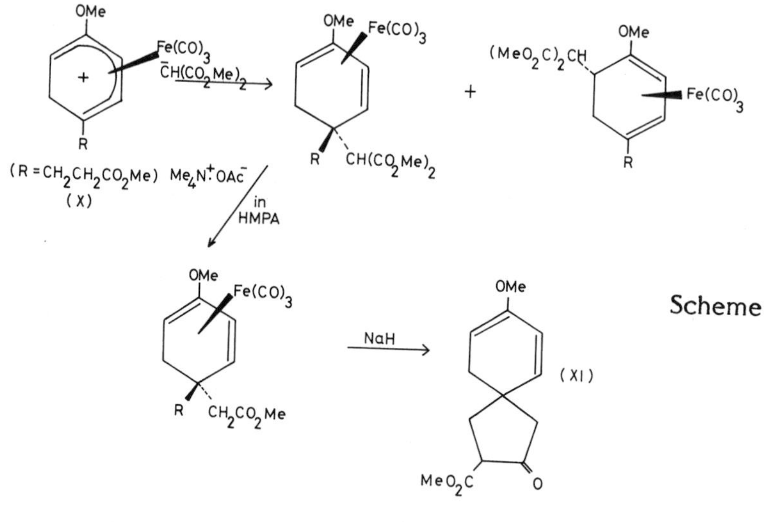

Scheme 4

Similarly, addition of **XIII** to **XII** provides a method for making a direct ring connection between two highly substituted centers.[39] This reaction suggests a potential approach to the synthesis of trichothecane.

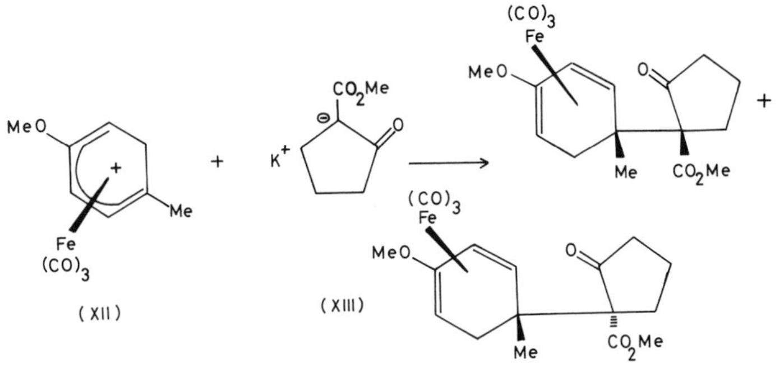

Similar reactions using tricarbonyliron derivatives of bicyclo[4,4,0]-decadienes have been employed to introduce angular substituents (Scheme 5):[36,40]

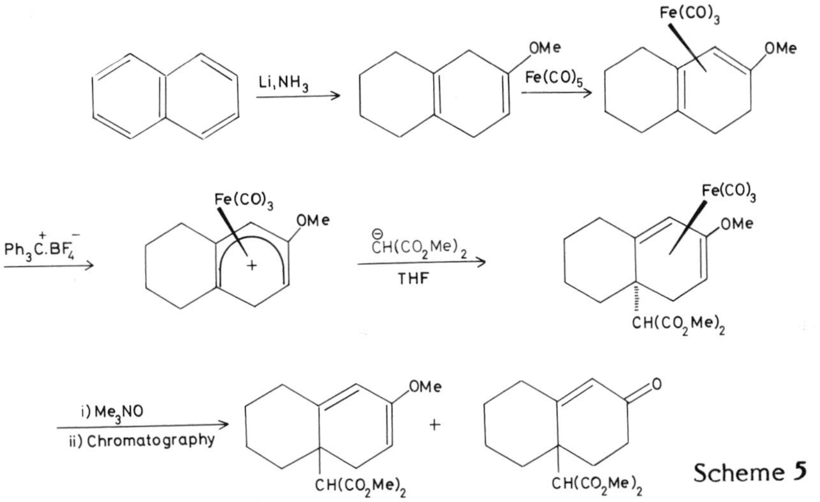

Scheme 5

1-Methoxycyclohexa-1,4-diene, obtained by the Birch reduction of anisole, yields a mixture of two isomeric complexes, **XIV** and **XV**, with $Fe(CO)_5$. Hydride abstraction with $Ph_3C^+BF_4^-$ followed by treatment with warm water gives the 2-methoxycyclohexadienyl salt **XVI** together with η^4-cyclohexadienonetricarbonyliron (**XVIII**) which arises by hydrolysis of **XVII**.[8,41-4]

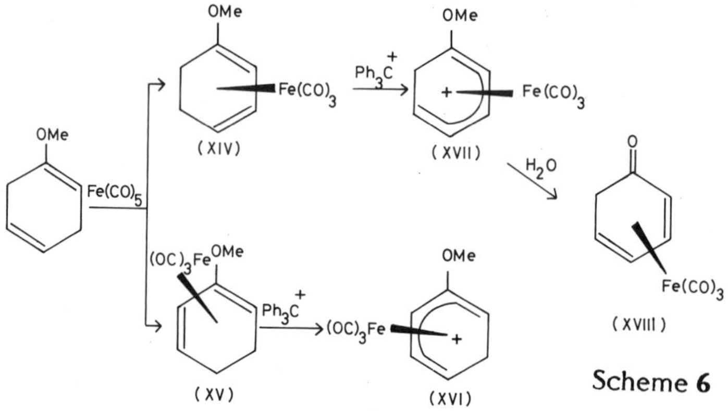

OMe

OMe

Ph$_3$C$^+$

OMe

Fe(CO)$_5$

Fe(CO)$_3$

(XIV)

Fe(CO)$_3$

(XVII)

H$_2$O

Fe(CO)$_3$

(XVIII)

(OC)$_3$Fe^{OMe}

Ph$_3$C$^+$

OMe

(XV)

(OC)$_3$Fe

(XVI)

Scheme 6

Trialkyl and triarylphosphines form stable phosphonium adducts with cyclohexadienyltricarbonyliron salts.[15,45] The <u>exo</u> configuration of the triphenylphosphine adduct has been proved by xray crystallography.[46] Trialkyl phosphites were found to add similarly.[47] Normally phosphonium adducts of phosphites undergo spontaneous Arbusov reactions yielding phosphonates, but this does not occur with tributyl phosphite in acetone.[48] On the other hand, rearrangement does occur when I is treated with trimethyl phosphite in methanol:[49]

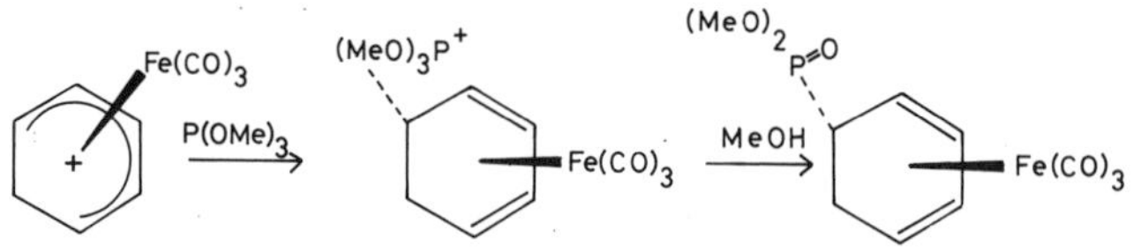

Fe(CO)$_3$

(MeO)$_3$P$^+$

(MeO)$_2$P≈O

P(OMe)$_3$

Fe(CO)$_3$

MeOH

Fe(CO)$_3$

The Fe(CO)$_3$ function can be removed by mild oxidation, for example with iron(III) chloride, cerium(IV), or copper(II). The sequence of reactions thus provides a route to cyclohexadienyl phosphonate esters.

The mechanisms of these additions have been studied kinetically in acetone and in nitromethane.[45] The reactions follow the rate law, Rate = k_2[complex][PR$_3$]. 2-Methoxycyclohexadienyltricarbonyliron tetrafluoroborate reacts 4 to 10 times more slowly than the parent compound. The results are interpreted in terms of direct addition to the organic ligand. It had been suggested that addition of tributylphosphine to [Fe(C$_7$H$_9$)(CO)$_3$]$^+$ proceeds via initial attack at the metal, followed by rearrangement,[50] but this is not borne out by Kane-Maguire's more detailed study. The rates of addition decrease in the order PBu$_3$>P(C$_6$H$_4$Me-p)$_3$>PPh$_3$>P(OBu)$_3$> P(OPh)$_3$, and correlate well with the basicities of the phosphorus nucleophiles as measured by their half neutralisation potentials, with the exception of PPh_3 which reacts anomalously rapidly. The strong rate discrimination by the dienyl cations between the phosphorus nucleophiles [a factor of ca 10^6 between PBu_3 and $P(OPh)_3$] suggests significant bond formation in the transition states. This is supported by a wide variation in $\Delta H^{\dagger}$ values, fairly good isokinetic plots and negative $\Delta S^{\dagger}$ values.

The addition of trialkyl phosphines to the neutral iodo complex **XIX** has also been studied kinetically:[51]

Fe(CO)$_2$I   Fe(CO)$_2$I   Fe(CO)$_2$ + I$^-$ Fe PR$_3$ (CO)$_2$

(XIX)   Step 1   Step 2

For the first step, second order kinetics (Rate = $k[$complex$][PBu_3^n]$) were observed, with $\Delta H^{\dagger} = 32.2$ kJ mol^{-1} and $\Delta S^{\dagger} = 130$ J K^{-1} mol^{-1}. The rate constant, $k_1 = 1.76$ dm^3 mol^{-1}s^{-1} (20°C), is ca 10^5 times slower than that for the addition of PBu_3^n to the parent cation $[Fe(C_6H_7)(CO)_3]^+$ ($k = 1.3 \times 10^5$ dm^3 mol^{-1}s^{-1}). The C_7 complex $Fe(C_7H_9)(CO)_2I$ reacts about 80 times more slowly. This difference is attributed to greater steric hindrance to direct ring addition, whereas the steric requirements for attack at the metal should be comparable for the C_6 and C_7 compounds.

The second step, which involves replacement of I^- at the metal by PBu_3^n, is retarded by iodide ion and shows a non-linear dependence on $[PBu_3^n]$. The results are interpreted in terms of a dissociation of iodide followed by phosphine addition to the 16-electron intermediate.

Triphenylphosphine and triphenyl phosphite react with $Fe(dienyl)(CO)_2I$ (dienyl = C_6H_7 or C_7H_9) at 44-72°C to yield carbonyl substituted products $Fe(dienyl)(CO)L(I)$ without ring addition.[52] This is probably another example where a slow irreversible process competes with the reversible ring addition, for which the equilibrium will be relatively unfavorable with these less nucleophilic phosphorus donors.

The reactions of various nitrogen donors such as pyridines,[53] anilines,[54-5] nucleoside and nucleotide bases,[56] and enamines[57] with $[Fe(C_6H_7)(CO)_3]^+$ and $[Fe(2\text{-}MeOC_6H_6)(CO)_3)]^+$ have been reported. At room temperature in acetonitrile the anilines (R' = H, Me, OMe) react through the nitrogen atom. At reflux, however, <u>ortho</u> and <u>para</u> C-alkylation of the amine by the dienyl cation was observed:[54]

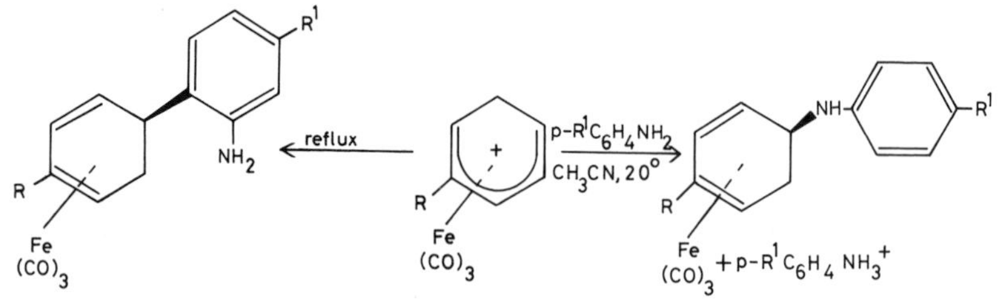

Some preliminary kinetic studies of the first of these reactions and also the quaternisation of various substituted pyridines have been reported.[58]   In general, under pseudo first order conditions, the rate expression k_{obs} = k[pyridine] holds, except for 3-cyanopyridine, where k_{obs} = k[pyridine] + k_{-1}. Both classes of reaction obey the Brønsted relationship, log k = α pKa + constant, with α ~1.0. This high value of α should be compared with α ~0.05 for attack of pyridines on soft Pt(II) centers, α~0.2 on moderately soft carbon centers such as alkyl halides, and 0.5 on free carbonium ions. This seems to indicate that the dienyl groups are 'hard' electrophiles.

Dienyltricarbonyliron cations and their ruthenium and osmium analogues also act as electrophiles towards a wide variety of activated aromatic compounds, including pyrroles, indoles, furan, thiophene, 1,3-dimethoxybenzene, 1,3,5-trimethoxybenzene and N,N-dimethylaniline.[18-21] Many of these reactions have been studied kinetically and the general reaction scheme proposed is illustrated in Scheme 7 for dimethylaniline:[55]

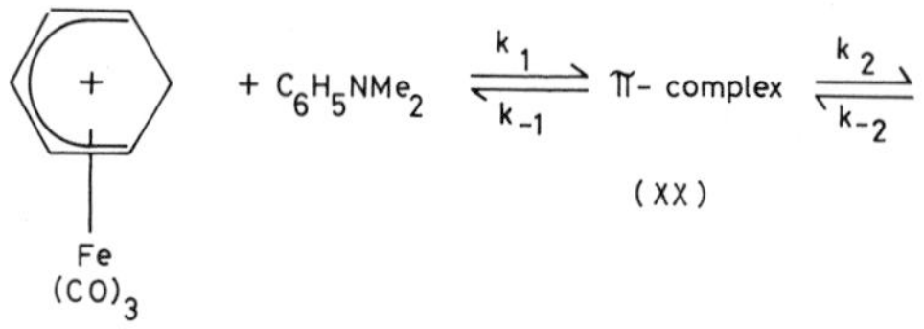

$$ + \; C_6H_5NMe_2 \; \underset{k_{-1}}{\overset{k_1}{\rightleftarrows}} \; \pi\text{- complex} \; \underset{k_{-2}}{\overset{k_2}{\rightleftarrows}} $$

(XX)

Scheme 7

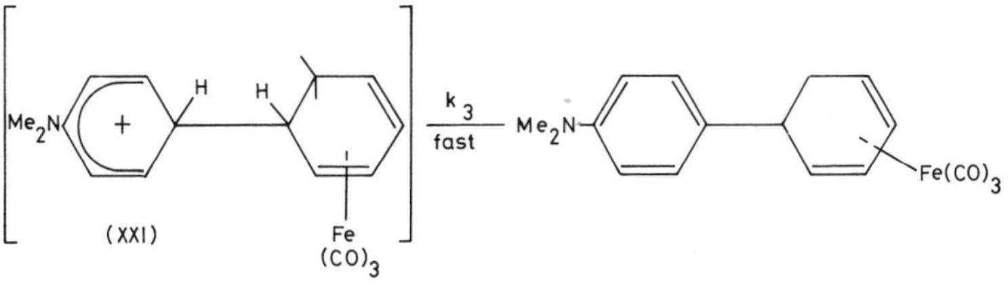

Scheme **7** (cont.)

The reactions follow the rate law $k_{obs} = k[Fe][C_6H_5NMe_2]$ in a wide variety of solvents. It was not possible to distinguish between rate-determining formation of a π-complex **XX** and rate-determining formation of the Wheland intermediate **XXI** with no significant contribution from **XX**. Comparison between species $[M(C_6H_7)(CO)_3]^+$ gives the order of relative rates as Fe >Os >Ru and this also holds for other reactions. Such comparisons however assume a common mechanism down the triad.

The reactions of the iron and ruthenium complexes with indole and pyrrole show similar kinetic behavior, except that for $[Ru(C_6H_7)(CO)_3]^+$ and indole there is evidence of a preequilibrium or a reversible side reaction, the nature of which is not clearly established.[20]

Aryltrimethylsilanes and -stannanes are much more reactive towards electrophiles than the parent arene. Relative rates of electrophilic cleavage of several such derivatives $ArMMe_3$ (M = Si, Sn) with $[Fe(C_6H_7)(CO)_3]^+$ have been reported:[58]

$$[C_6H_7Fe(CO)_3]^+BF_4^- + ArMMe_3 \longrightarrow C_6H_7ArFe(CO)_3 + Me_3M.BF_4.$$

The addition of carbanions to **I** has already been mentioned.[16-17] Neutral β-diketones such as pentane-2,4-dione and dimedone[59] also react, and kinetic studies show an inverse dependence on hydrogen ion concentration. This argues for a mechanism involving rapid predissociation of the dione to the reactive carbanion, which then adds directly to the dienyl ring in a rapid but rate determining step.[17]

An unusual reaction occurs when $[Fe(C_6H_7)(CO)_3]^+BPh_4^-$ is heated to $90°$ in water.[60] A possible mechanism is shown in Scheme **8**.

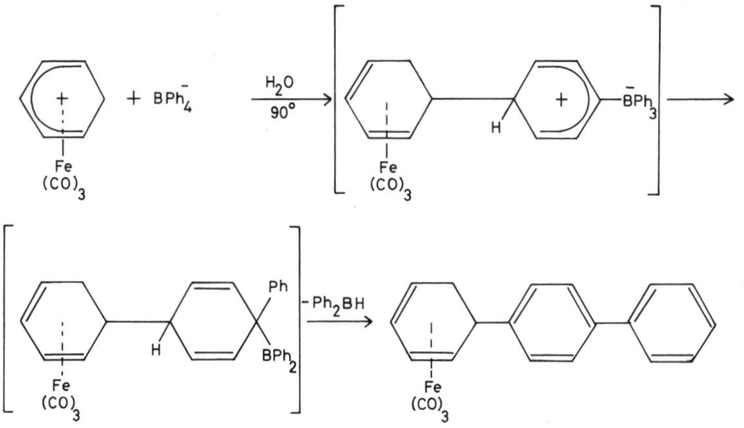

Scheme **8**

INDO molecular orbital calculations on the $[Fe(C_6H_7)(CO)_3]^+$ ion show a higher positive charge on C^2 than on C^1 or C^3, which would predict nucleophilic attack at $C^2 > C^3 > C^1$. As we have seen, attack occurs at C^1. The bond index values, however, lie in the order $C^1 > C^3 > C^2$ for both σ and π contributions in agreement with preferential attack at C^1.[61]

2.2.2 Tricarbonylcycloheptadienyliron salts

The cycloheptadienyltricarbonyliron cation **XXII** can be prepared either by protonation of tricarbonylcycloheptatrieneiron or by hydride abstraction from tricarbonylcycloheptadieneiron.[62-3] The former method is the more convenient because cycloheptatriene is readily available. The tetrafluoroborate is an air stable yellow solid which is even more resistant to mildly nucleophilic solvents than its cyclohexadienyl analogue. For instance, it can be recrystallized from water and is stable in boiling ethanol.

Most of its reported reactions with nucleophiles occur by addition at the organic ring. An exception is the attack of iodide which proceeds by displacement of CO to give $Fe(\eta^5\text{-}C_7H_9)(CO)_2I$. Alkoxide ions in the cold add to a carbonyl group, and the carboethoxy derivative $Fe(\eta^5\text{-}C_7H_9)(CO)_2COOEt$ has been isolated.[64] Such derivatives rearrange either in the solid state or in solution by a first order reaction, presumably by dissociation of OR^- and subsequent exo ring addition. Azide ion behaves similarly, although here the product initially formed in a side equilibrium may involve a $Fe^+N_3^-$ ion pair rather than a carbonyl azide $Fe-CON_3$.[65] In general, however, <u>exo</u> addition at the organic ligand occurs. Where Nu = OMe, OEt, $CH(COOEt)_2$, OH, OPh, SPh, Me_2N, ^tBuNH and with neutral molecules Ph_3P and pyridine, the nucleophile adds to the terminal 1-position of the dienyl system.[11,15,64,66-7] With borohydride, however, both 1- and 2-addition occurs affording a mixture of **XXIII** and **XXIV** (Nu = H, D).[68-9]

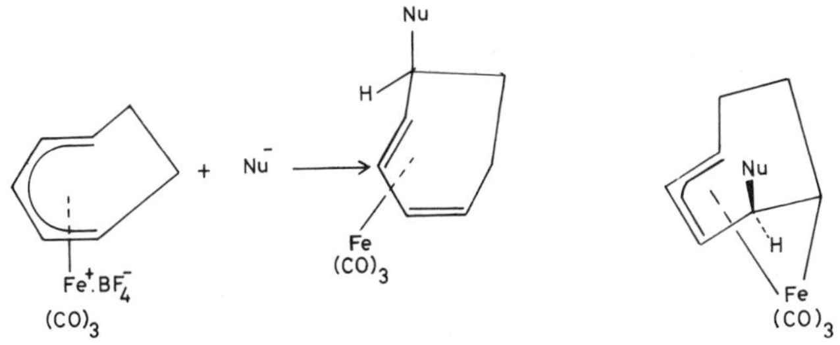

Phosphine substituted complexes **XXV** exhibit a greater tendency towards 2-addition then the unsubstituted derivatives.[80] Thus with L = PPh_3 or $AsPh_3$, reaction with borohydride or cyanide gave exclusively the σ-η-allyl complexes **XXVI**. Similar behavior was observed for the ruthenium complexes **XXV**, L = PPh_3, and L = CO.

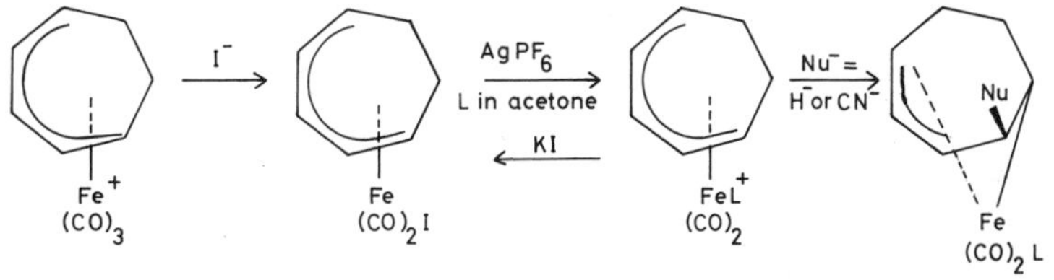

The tendency to form σ-η-allyl complexes increases down the triad and is especially pronounced for osmium. Thus $[Os(C_7H_9)(CO)_3]^+$ is reduced to the σ-η-allyl complex only. While $[M(\eta^5-C_6H_7)(CO)_3]^+$, M = Fe, Ru, show no tendency at all to react at the 2-position, the osmium complex gives a mixture of 1- and 2-adducts with H^- and CN^-.[72] Other nucleophiles, however, react normally at the terminal position even in this case.

Cycloheptadienyl-1-onetricarbonyliron (**XXVIII**) obtained by the protonation of troponetricarbonyliron **XXVII**,[73] also undergoes 2-addition with H^- and CN^- but 1-addition with MeO^-, RNH_2 and azide.[74]

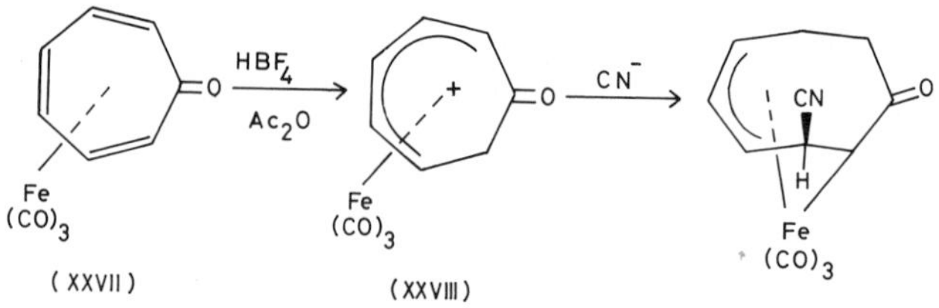

(XXVII)   (XXVIII)

The tropyliumtricarbonyliron cation **XXIX** cannot be obtained by hydride abstraction from cycloheptatrienetricarbonyliron, as addition of Ph_3C^+ occurs yielding $[Fe(Ph_3C.C_7H_9)(CO)_3]^+$. It was made via 7-methoxycycloheptatriene as follows:[75]

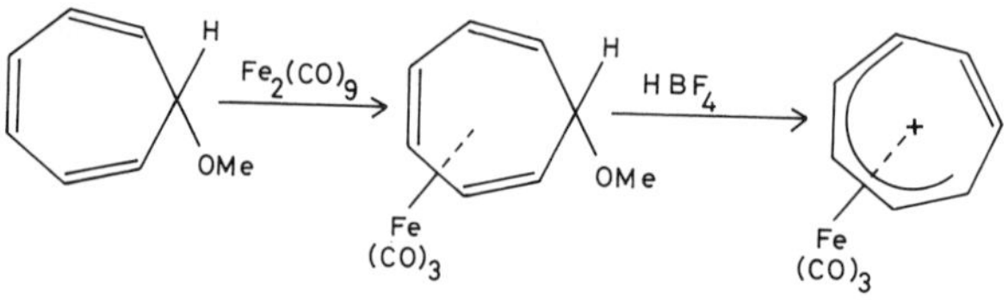

(XXIX)

It has been employed by Rosenblum in a synthesis of hydroazulenes using cyclopentadienyldicarbonyliron reagents.[76]

The addition of methoxide to the terminal carbon atom of a dienyl system is involved in the synthesis of heptafulvene complexes of iron.[77]

In the reactions of $[Fe(C_6H_7)(CO)_3]^+$ and $[Fe(C_7H_9)(CO)_3]^+$ with certain bases, the by-products **XXX** and **XXXI** are sometimes obtained. If **XXII** is treated with sodium hydroxide in a heterogeneous aqueous/organic solvent mixture the yield of the C-C linked dimer **XXXI** rises to 60%. Significant yields (n = 1 and 2) are also obtained merely on heating the salts under reflux in acetonitrile. A free radical mechanism may be involved.[78] Similar reductions are observed with $[Co(C_7H_7)(CO)_3]^+$ and with open chain pentadienyltricarbonyliron cations.[79-80]

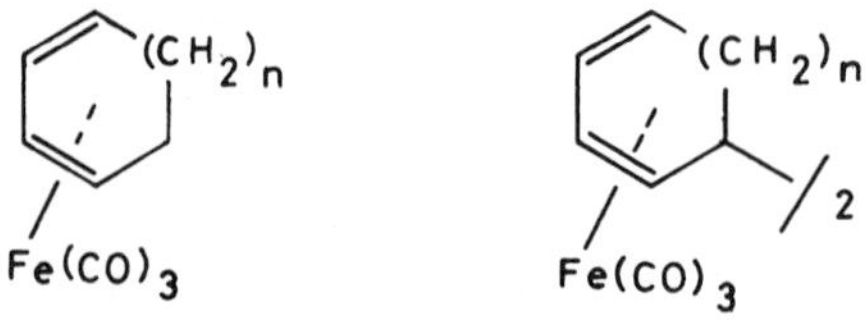

(XXX)   (XXXI)

2.2.3 η^5-cyclooctadienyl and related complexes

Hydride abstraction from tricarbonylcycloocta-1,3-diene complexes of iron, ruthenium and osmium yields the corresponding 1-5-η-cyclooctadienyl cations **XXXII**.[81-2] While the iron complex is reduced by borohydride in low yield (by attack at the 1-position), the osmium complex reacts at the 2-position affording a σ-η-allyl complex.[81] With methoxide and the iron salt attack at the ring giving **XXXIII**, and also the formation of a carbomethoxy-derivative **XXXIV** (which disappeared as the reaction progressed) were observed. It now seems likely that **XXXIV** is formed in a dead-end equilibrium rather than en route to the endo-methoxy complex as originally claimed.[82]

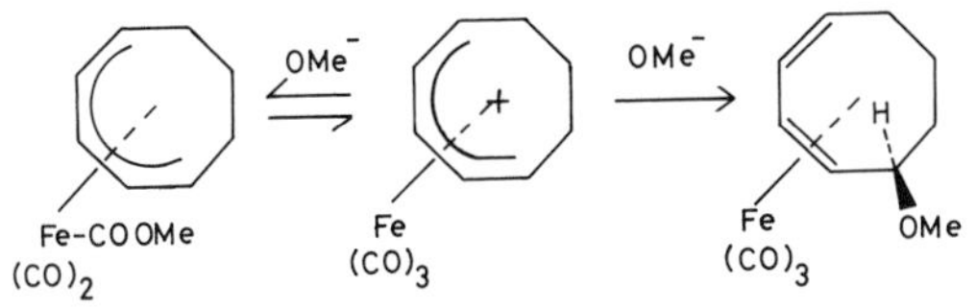

| (XXXIV) | (XXXII) | (XXXIII) |

The cycloocta-1,5-diene complexes **XXXV** react with triphenylmethyl-tetrafluoroborate giving salts **XXXVI**:[81,83-4]

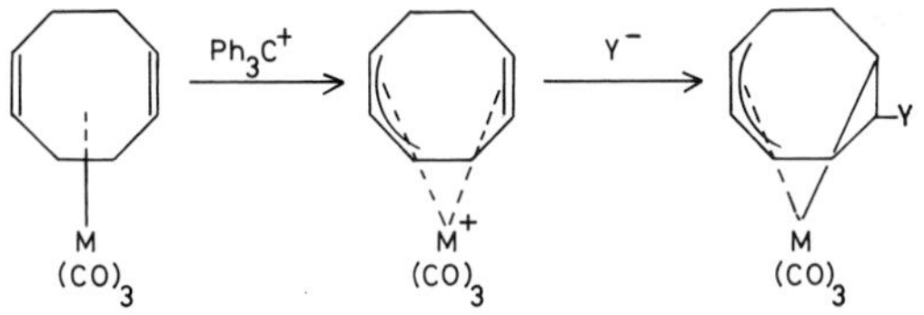

| (XXXV) | (XXXVI) | (XXXVII) |

As predicted by Green's rules, nucleophiles [$Y^- = H^-$, OMe^-, N_3^-] yield **XXXVII** as the major and usually the sole products by attack at the 'even' olefin group rather than the 'odd' η^3-allyl system.[5,81-2] This contrasts with the behavior of similar cobalt complexes, **XXXVIII**, which add nucleophiles at the allyl group reforming cycloocta-1,5-diene complexes **XXXIX**, in apparent contradiction of the rules.[85]

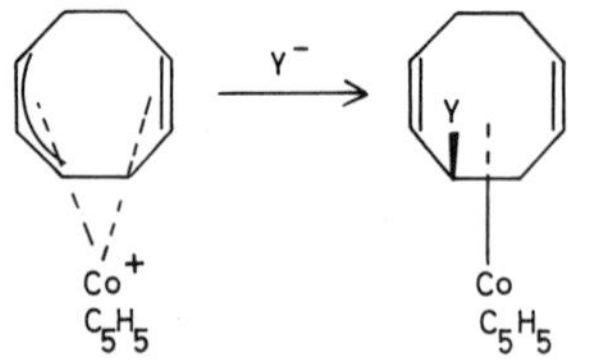

(XXXVIII)   (XXXIX)

$(Y = OMe, CN, Me_2N, acac, etc.)$

The kinetics of the reversible addition of triphenylphosphine to the cobalt complex **XXXVIII** have been measured.[86]

Protonation of tricarbonylcyclooctatetraeneiron (**XL**) yields the bicyclo[5.1.0]octadienyl complex **XLII**, via an η^5-bonded intermediate **XLI**:[87]

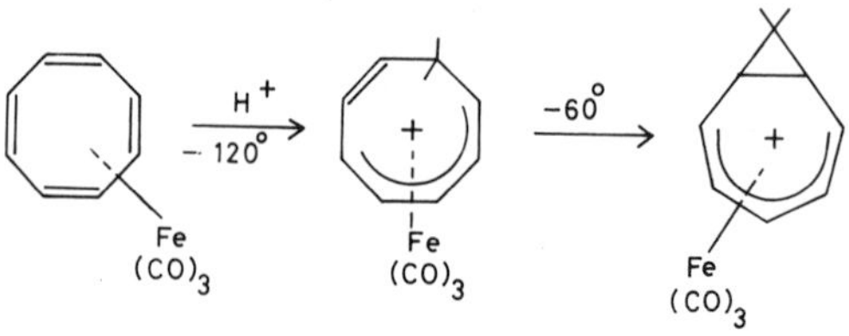

XL   XLI   XLII

Borohydride reduction of **XLII** gives a mixture of **XLIII** and **XLIV** arising from both 1- and 2-addition:[69]

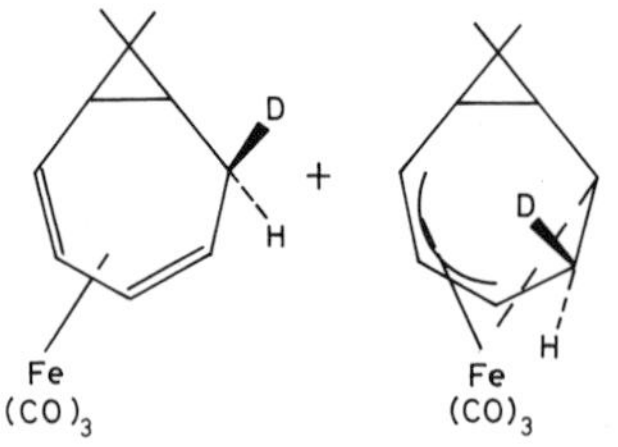

(XLIII)   (XLIV)

Pyridine, however, opens the C_3 ring of **XLII**, regenerating tricarbonylcyclooctatetraeneiron.[88]

Friedel-Crafts acetylation of **XL** gives a mixture of substituted products and an acyl cation which has a bicyclic structure **XLV**.[89] Nucleophiles add to the η^2-bonded olefinic grouping, (except iodide, which as usual attacks the metal with displacement of carbon monoxide):

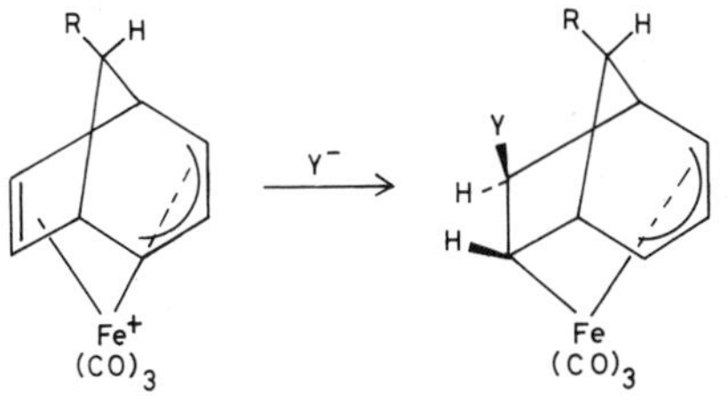

R = H, CH$_3$CO, PhCO

Y = H, D, CN

Some similar reactions have also been reported by Eisenstadt.[90-1]

2.2.4 Open chain η^5-pentadienyl complexes

Open chain pentadienyltricarbonyliron salts **XLVII** were discovered by Mahler and Pettit, who prepared them by protonation of η^4-dienol complexes **XLVI**:[79,82]

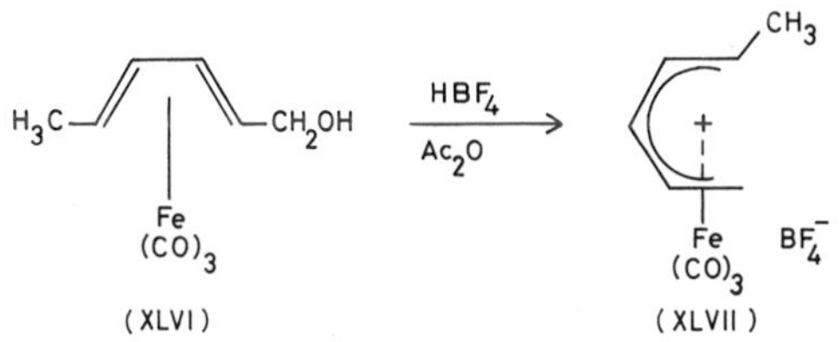

The cis structure of the ion was assigned on the basis of nmr studies and from the observation that the parent ion $[Fe(\eta^5\text{-}C_5H_7)(CO)_3]^+$ could be prepared by hydride abstraction from cis-pentadienetricarbonyliron, using Ph$_3$C$^+$, but not from the trans-pentadiene complex. Nucleophiles attack the open chain pentadienyl cations at the terminal carbon atom. Water, for example, reacts with **XLVIII** to give only one diastereoisomer of the alcohol **XLIX** (Scheme 9). This isomer has been shown by xray diffraction to be a racemate of enantiomers having the respective configurations RR and SS about the two chiral centers.[93] In the favored conformation [H in the most crowded position near the Fe(CO)$_3$ group] the -OH substituent will be in a ψ-exo position with respect to the metal, with the CH$_3$ group in a ψ-endo stereochemistry. It was suggested that the hydrolysis of the pentadienyltricarbonyl complexes proceeds via a trans ion, which is attacked by a nucleophile stereospecifically exo to the metal:

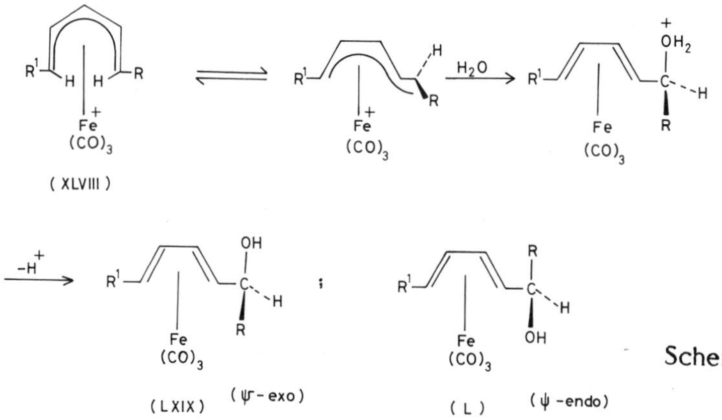

(XLVIII)

(LXIX)   (ψ- exo)   (L)   (ψ -endo)   Scheme 9

Evidence for <u>trans</u> pentadienyl ions has been obtained from kinetic and stereochemical studies on the solvolysis of dinitrobenzoate esters of diastereoisomeric alcohols of the type **XLIX** and **L**. These solvolyses proceed by an S_N1 mechanism essentially with retention of configuration at C^1. Moreover the ψ-<u>endo</u> (RS or SR) esters react appreciably more slowly than the ψ-<u>exo</u> isomers.[94-7]

It has been possible in suitable cases to observe <u>trans</u> ions directly by nmr spectroscopy on low temperature protonation of alcohols, e.g. **XLIX** and **L**. Normally these ions rearrange rapidly even at -78°C to the <u>cis</u> structures. This rearrangement could be inhibited to some extent by destabilising the <u>cis</u> cation by introduction of an anti substituent (**LI**),[98] or by constraining the <u>trans</u> structure partly in a ring system (**LII**):[99]

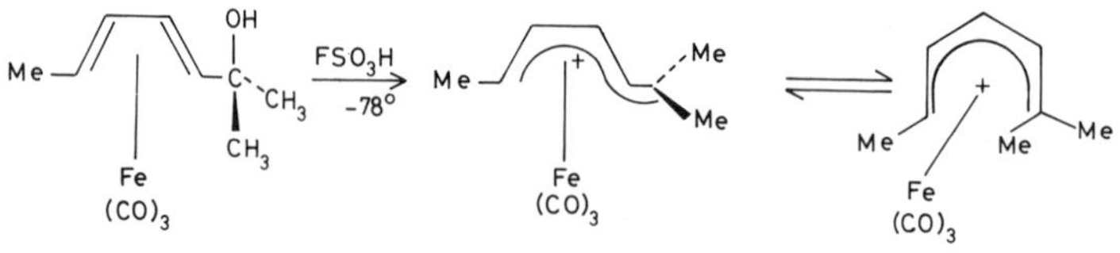

(LI)

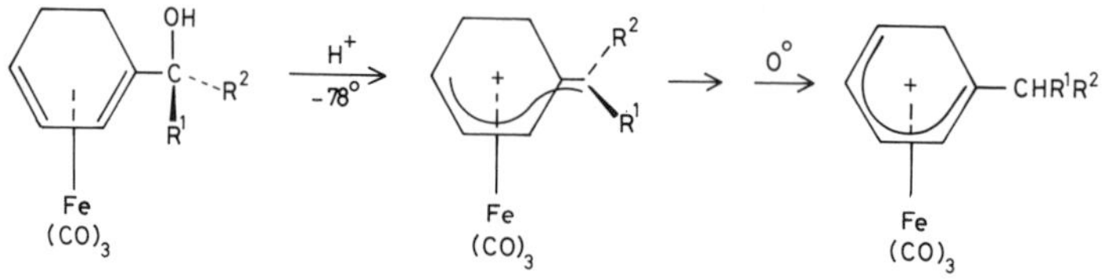

(LII)

Cations having the cis, syn, syn and cis, syn anti structures can be generated stereospecifically by protonation of ψ-exo and ψ-endo alcohol precursors at low temperature. If these are quenched with methanol at $-78°$, the methoxy derivatives are formed with retention of configuration. The cis, syn, anti cations rearrange on warming to room temperature to the thermodynamically more stable cis, syn, syn isomer (Scheme 10):[100]

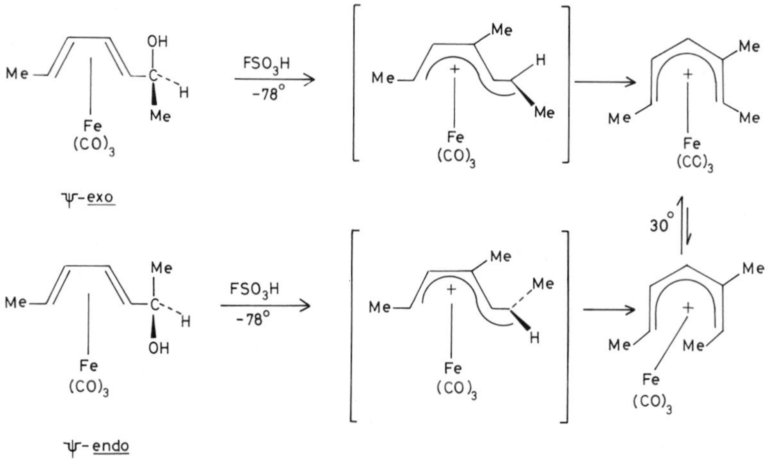

Scheme 10

Maglio, Musco and coworkers found that cis pentadienyltricarbonyl cations react with weakly basic amines such as p-nitroaniline to yield products with a trans,trans structure whereas strongly basic amines such as dimethylamine afford cis,trans diene complexes.[101-3] They proposed that while the weakly basic amines react only with the trans ion LIIIa, the more strongly basic amines can also react with the cis cation LIIIb directly:

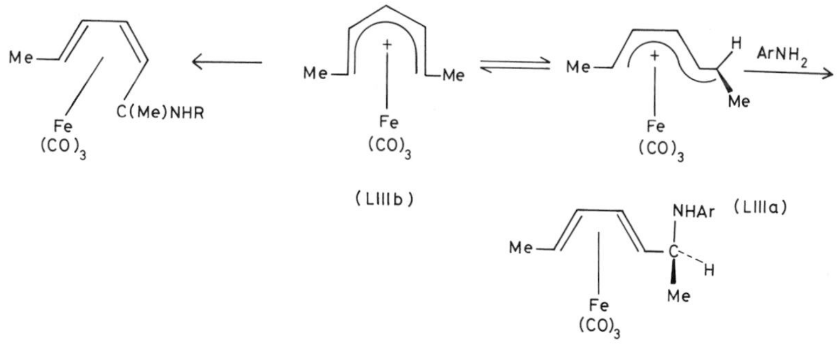

Amines of intermediate basicity give a mixture of both products.

Kinetic studies on the electrophilic substitution of di- and tri-methoxybenzenes by pentadienyltricarbonyliron cations also indicate a pre-equilibrium step, cis ion $\rightleftharpoons$ trans ion, followed by attack of the latter on the aromatic substrate. A common intermediate in the reactions of several different weak nucleophiles is also suggested by this work.[104-5]

In reactions of 1-alkylpentadienyltricarbonyliron salts with water, methoxide etc. the nucleophile normally adds to the carbon atom bearing the substituent, even when this is bulky.[106] The usual explanation of this is that the alkyl substituent stabilizes the trans ion (or similar η^3-allyl intermediate) by virtue of its inductive effect.[79] 1,5-Disubstituted pentadienyl cations are normally attacked preferentially by water at the carbon atom bearing the less bulky substituent (Scheme 11):

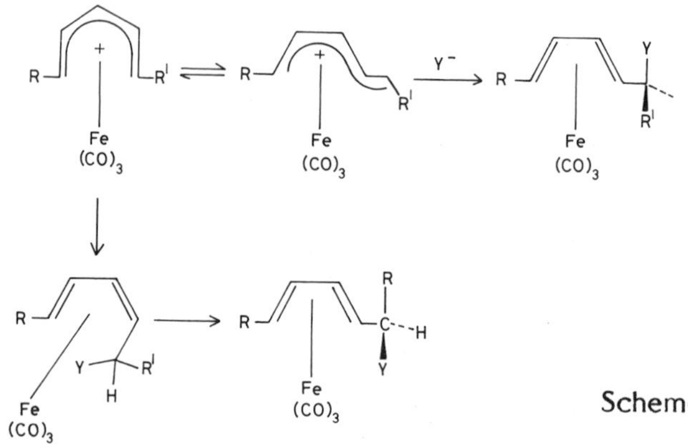

Fe(CO)$_3$ R—[+]—R' $\rightleftharpoons$ R—[+]—R' $\xrightarrow{Y^-}$ R—[Y]—R'   Fe(CO)$_3$

$\downarrow$

R—[Y, H, R']—Fe(CO)$_3$ $\longrightarrow$ R—[C–H, R, Y]—Fe(CO)$_3$

Scheme 11

In contrast to water and primary and secondary amines, tertiary amines PPh$_3$ and AsPh$_3$ add to the unsubstituted end of the methylpentadienyltricarbonyliron cation, showing that steric effects are significant with these very bulky nucleophiles.[107-8]

The reduction of 1-alkylpentadienyltricarbonyliron salts with various hydride donors may not conform completely to the pattern described above. While NaBH$_4$ and NaBH$_3$CN yield complexes **LIV** which could arise, as expected, from direct addition to the cis ion, the even stronger nucleophile NaBEt$_3$H gives mainly products with inverted configuration (**LV** and **LVI**).[109] The authors postulate hydride transfer via a formyl derivative, but this seems unlikely, as the formyl, if formed, would probably form part of a dead-end equilibrium. Nucleophilic attack at iron generating an η^3-allyl iron hydride, followed by endo transfer of H to the ligand, would seem to be a possible alternative:

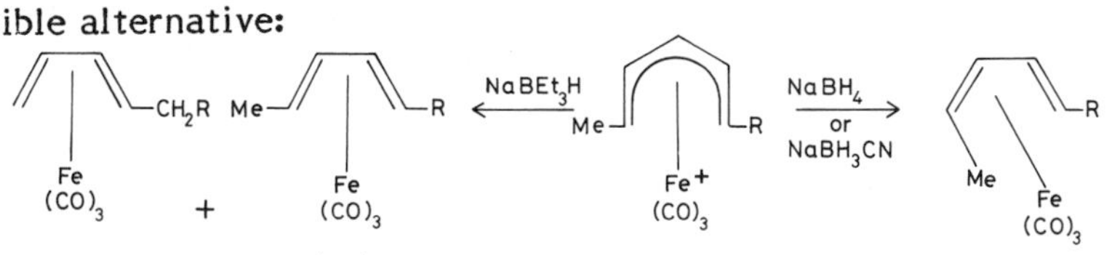

Trimethylamine or triethylamine adducts of pentadienyltricarbonyliron cations decompose in vacuo to afford tricarbonylhexa-1,3,5-triene complexes **LVIII**:

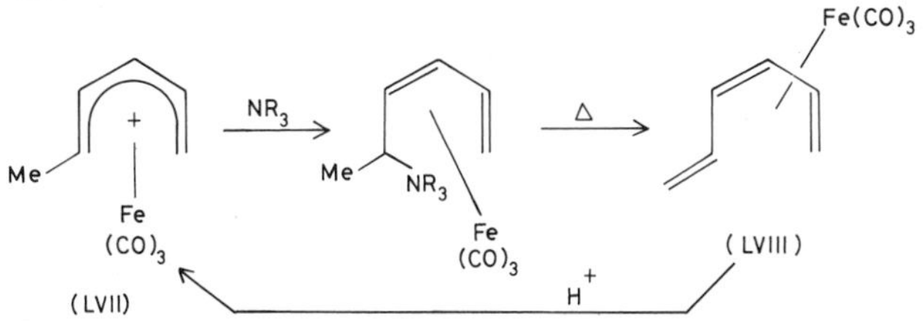

The latter can be protonated by HBF_4 to reform the salts.[108,110-11]

Reduction of **LVII** with zinc dust or reaction with alkali leads to coupling of two pentadienyl groups. This reaction has been exploited in a synthesis of cycloalkanes from diacyl chlorides.

Molecular orbital calculations on the pentadienyltricarbonyliron cation by an extended Hückel method predict C^1, C^3 and C^5 to be electron rich, with the greatest negative charge at C^1.[113-14] These results correlate with ^{13}C chemical shifts.[115] They seem at variance, however, with the observation that nucleophilic attack occurs at C^1 rather than at C^2 in these complexes.[61]

Pentadienyltricarbonyliron cations can also be prepared by protonation of η^4-hexatriene complexes.[116] Recently some similar derivatives $\{M(\eta^5\text{-}C_5H_5)[\eta^5\text{-}R(CH)_5R]\}^+$ (M = Rh and Ir) have been described.[117-19] Cross-conjugated dienyltricarbonyliron cations are also known.[120]

2.3 Nucleophilic attack on η^6-complexes

The chemistry of η^6-arene complexes has been reviewed.[121] Pauson has given an account of the contributions of his research group to the subject of nucleophilic addition to transition metal complexes.[122]

2.3.1 η^6-arenetricarbonyl manganese salts

Arenetricarbonylmanganese salts can react with nucleophiles in four different ways, depending on the attacking group and on the conditions of the experiment. With the benzene complex all these pathways have been observed, viz addition to the arene to yield cyclohexadienyl complexes, addition to a carbonyl ligand, attack at the metal with displacement of carbon monoxide, and attack at the metal with loss of arene.[123]

Arenetricarbonylmanganese salts are usually prepared by the action of an arene on chloropentacarbonylmanganese in the presence of aluminum trichloride.[124-6] They can also be obtained from arenes and $Mn(C_5H_4CH_3)(CO)_3$, which is available commercially at a modest cost, using aluminum tribromide as the Lewis acid.[127] Some redistribution of alkyl substituents occurs when substituted arenes are used, but the method using methylcymantrene gives good yields of the pure benzene complex **LIX**.

Benzenetricarbonylmanganese salts are converted by hydride ion donors or alkyllithium reagents into air stable cyclohexadienyl derivatives **LX** and **LXI**:[125]

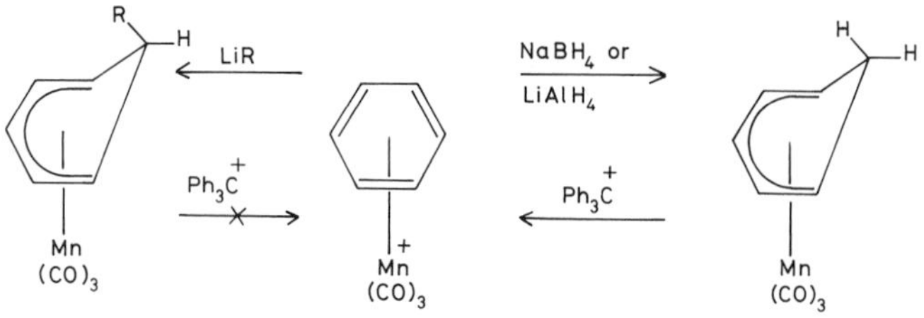

(LXI)   (LIX)   (LX)

It was originally assumed that the added alkyl group R was <u>endo</u> with respect to the $Mn(CO)_3$ group, but spectroscopic correlations, in conjunction with crystallographic studies, soon showed that it adopts an <u>exo</u> configuration.[126] The compounds **LXI** are unaffected by triphenylmethyl tetrafluoroborate. This bulky reagent is able to abstract <u>exo</u>-hydrogen substituents, but is unable to remove an <u>endo</u>-hydrogen on account of steric hindrance.

N-bromosuccinimide can sometimes be used to abstract endo H-substituents, but it cannot be employed here, as it destroys the manganese complexes. Munro and Pauson, however, have developed a synthesis of alkylbenzenetricarbonylmanganese salts from the parent compound which makes use of a thermal shift of the endo hydrogen in **LXI** which proceeds at about 135°C.[127-8] An η^4-arenemetal hydride is a probable intermediate:

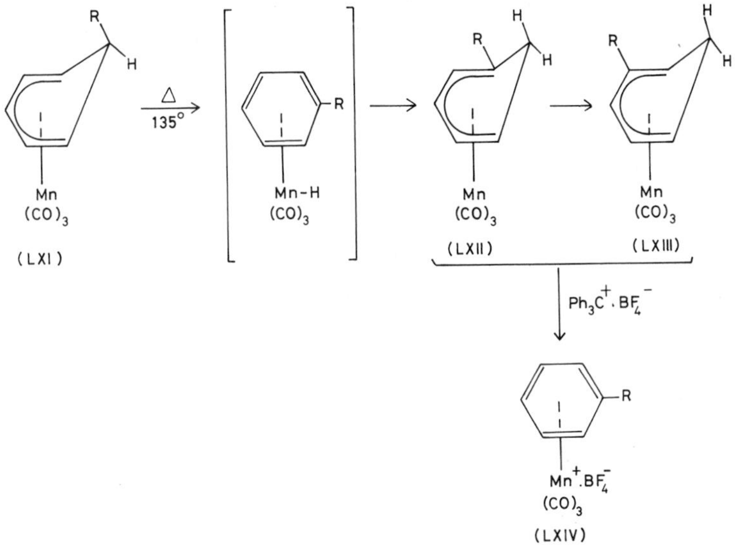

A mixture of 1- and 2-alkylcyclohexadienyl derivatives **LXI** and **LXII** is obtained, which on treatment with $Ph_3C^+.BF_4^-$ yields **LXIV**. Similar thermal shifts have been observed in the species $Cr(\eta^6-C_7H_7R)(CO)_3$ and $Co(\eta^4-C_5H_5R)(C_5H_5)$.[129]

While ring addition is the major reaction pathway with lithium aluminum hydride, **LIX** yields about 0.6% $Mn(C_6H_6)(CO)_2CH_3$ through reduction of a carbonyl ligand, in addition to **LX** (61%).[132] Alkyl lithium also attacks carbonyl groups as well as the ring; with polyalkylbenzene complexes fair yields of acyl derivatives result, which can be decarbonylated photochemically or converted into cationic carbene complexes using $Et_3O^+BF_4^-$ (see Section 2.3.2).

Walker and Mawby made a detailed study of the reaction of cyanide ion with arenetricarbonyl manganese cations.[131-2] In general the nucleophile initially yields an exo-cyanohexadienyl complex **LXV** in which

the cyano-group is attached to an unsubstituted carbon atom. In chloroform
solution however the cyanide slowly migrates to a substituted position (**LXVI**
for arene = mesitylene):

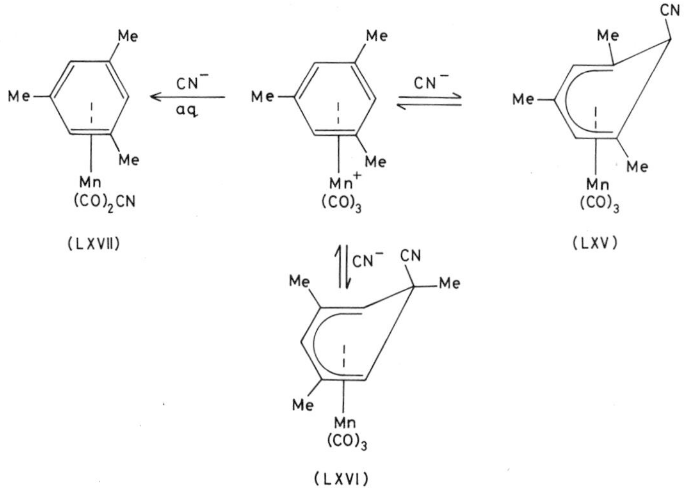

Mild oxidation of the cyanocyclohexadienyl complexes, with cerium
(IV) for example, yields cyanobenzenes. On heating in water or methanol,
complexes **LXV** decompose with evolution of carbon monoxide to form
LXVII. The most probable mechanism is as indicated.

When **LIX** is treated with trialkyl or triarylphosphines in the light
under a nitrogen atmosphere, substitution of one carbonyl ligand occurs
giving **LXVIII**.[123] In the dark, or in the presence of oxygen, an equilibrium
involving _exo_ addition to the ring is set up. Light presumably aids loss of CO
by a dissociative mechanism:[133]

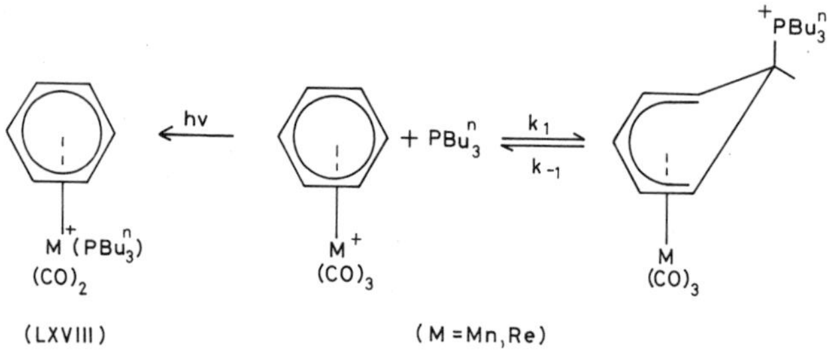

Equilibrium constants for addition of tributylphosphine in acetone in the dark decrease with increasing alkyl substitution of the arene ($K = k_1/k_{-1}$ 1 mol^{-1} at 25°C; M = Mn; arene = benzene, 260; toluene, 60; p-xylene, 4; mesitylene, 1).[134] The corresponding rhenium adducts are somewhat more stable. The additions follow the rate equation $k_{obs} = k_1[PBu_3] + k_{-1}$, as expected for the proposed mechanism.

In contrast to the manganese complexes, arenetricarbonylrhenium cations are rather susceptible to displacement of arene by donor solvents such as water, acetonitrile or acetone. Acetonitrile, for example, displaces arene about 10^4 times faster from rhenium than from manganese.

With weak, hard nucleophiles such as amines and methoxide ion, reversible attack at a carbonyl ligand is observed:

$$[Mn(arene)(CO)_3]^+ + 2RNH_2 \rightleftharpoons [Mn(arene)(CO)_2CONHR] + RNH_3^+$$

Hydrazine yields an isocyanate, presumably via a similar intermediate. The reactivity of ligands towards nucleophiles has been correlated with the carbon-oxygen stretching force constant, assessed by the Cotton-Kraihanzel method.[137]

When a nucleophile (e.g. RLi, LiAlH$_4$) irreversibly adds to the arene ring in the tricarbonylchlorobenzenemanganese cation **LXIX**, cyclohexadienyl complexes result, as discussed above. Thus lithium aluminum hydride affords a 2:1 mixture of the 1- and 2-chlorocyclohexadienyl products, **LXX** and **LXXI**.[138] Where the addition is reversible, as with OMe$^-$, OPh$^-$, SPh$^-$, N$_3^-$ or amines RNH$_2$ or R$_2$NH, the initially formed adducts will be in equilibrium with the starting material, and also with a small amount of ipso adduct **LXII**:[139]

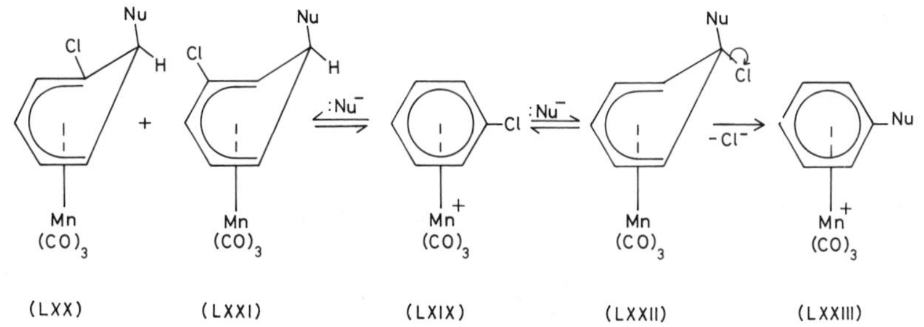

Scheme **12**

This eliminates chloride ion to yield a substituted arenetricarbonylmanganese cation **LXXIII.** The resulting arene is readily released from the complex by heating in acetonitrile. The reaction sequence

$$PhCl \longrightarrow [Mn(PhCl)(CO)_3]^+ \xrightarrow{\quad :Nu^- \quad} [Mn(PhNu)(CO)_3]^+ \longrightarrow PhNu$$

thus provides a mild route for nucleophilic substitution of chlorobenzene. A similar approach using chromium complexes has also been suggested[140-1] (see Section 2.3.2).

$[Mn(\eta^6\text{-}C_6H_5N_3)(CO)_3]^+$, on heating to 130°, loses nitrogen and forms a cyanocyclopentadienyl complex through ring contraction. Electronic effects in arenetricarbonylmanganese cations have been discussed in the light of ^{55}Mn nqr and ^{13}C nmr spectra.

2.3.2 η^6-Arenetricarbonylchromium complexes

The neutral arenetricarbonylchromium complexes are also susceptible to reactions with nucleophiles. Nicholls and Whiting observed that the chlorobenzene complex undergoes ready nucleophilic substitution by methoxide at 65°C in methanol.[144] This reactivity resembles that of p-chloronitrobenzene, and it was suggested that the electron withdrawing effects of the $Cr(CO)_3$ and the para nitro groups are approximately equal (see Section 5). These reactions were extended to various carbanions such as diethylmalonate.[195] In the early stages of the reaction, 1- and 2-ring adducts were observed, but these were gradually replaced by the substitution product.[146] A similar mechanism to that illustrated above for chlorobenzenetricarbonylmanganese cation is proposed. The reactivity towards nucleophiles decreases in the order X = F > Cl > Br > I in $Cr(PhX)(CO)_3$.[147]

These results led Semmelhack and his colleagues to study the addition of carbanions to benzenetricarbonylchromium itself.[3] Carbanions derived from species R-H with $pK_a \geqslant 25$ in tetrahydrofuran form cyclohexadienyl adducts which are stable below 0°C in the absence of air.[146,148-9] Oxidation with iodine at 25°C affords substituted benzenes:

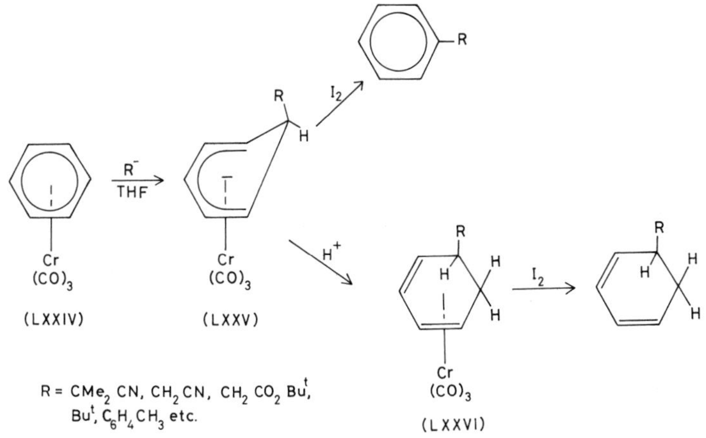

R = CMe$_2$CN, CH$_2$CN, CH$_2$CO$_2$Bu^t,
Bu^t, C$_6$H$_4$CH$_3$ etc.

The lithium enolates of ketones and esters require the addition of HMPA to react. The sequence provides a method of nucleophilic substitution of benzene itself. Protonation of the anionic species **LXXV** with a strong acid yields a cyclohexadiene complex **LXXVI** from which the substituted cyclohexadiene ligand can be released by iodine oxidation. Weak acids such as water, however, merely reverse the addition, regenerating **LXXIV**.[150] Some applications of these reactions in organic synthesis have been reviewed.[151] The preparation of fused ring compounds, for example, can be achieved by intramolecular carbanion attack on arene chromium complexes:[152]

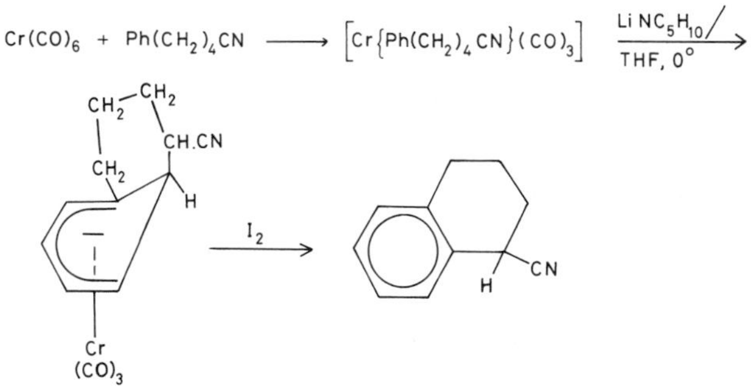

These routes have recently become more attractive as a simple procedure for making arenetricarbonylchromium complexes using a di-n-butyl ether/THF solvent mixture has been reported.[153-4]

Very strong nucleophiles, however, such as phenyllithium, add irreversibly to a carbonyl ligand to afford carbene complexes.[155] A carbyne cation can be obtained using boron trichloride in the usual way:[156]

$$Cr(arene)(CO)_3 \xrightarrow[\;\; ii) \quad Me_3O^+BF_4^-\;\;]{i) \quad LiPh/Et_2O} Cr(arene)(CO)_2 \longleftarrow C \Big\langle {}^{Ph}_{OMe}$$

$$\xrightarrow{2BCl_3} [Cr(arene)(CO)_2(CPh)]^+ BCl_4^-$$

The action of heat on $Cr(C_6H_6)(CO)_2(CH_3CN)$ yields a complex which contains a Cr $\equiv$ Cr triple bond, $(\eta^6\text{-}C_6H_6)Cr(\mu\text{-}CO)_3Cr(\eta^6\text{-}C_6H_6)$.[157]

Arenes react with vanadium hexacarbonyl to form salts **LXXVII**, which are reduced by sodium borohydride to cyclohexadienyl derivatives **LXXVIII**.[158] It does not seem that their chemistry has been further pursued.

$$\text{Arene} + V(CO)_6 \longrightarrow \underset{\textbf{LXXVII}}{[V(CO)_4(arene)]^+[V(CO)_6]^-} \xrightarrow{NaBH_4} \underset{\textbf{LXXVIII}}{[V(CO)_4(\eta^5\text{-}C_6H_7)]}$$

2.3.3 η^6-Arene-η^5-cyclopentadienyliron salts

Arenecyclopentadienyliron salts **LXXIX** are available from the reaction of ferrocene with arenes in the presence of aluminum and aluminum trichloride. Following hydrolysis of the reaction mixture, the cations are precipitated as hexafluorophosphates or tetrafluoroborates.[159-60] In addition to the desired complexes, various by-products, the 'ferrocenophanes', are formed, but they are readily separated as they remain in the organic phase during the hydrolysis step.[162-5]

$$Fe(C_5H_5)_2 + \text{Arene} \xrightarrow[\substack{Al \\ reflux}]{AlCl_3/} \underset{\textbf{LXXIX}}{[Fe(C_5H_5)(Arene)]^+} + \text{ferrocenophanes}$$

The attack of nucleophiles on the cations has been studied in some detail by Pauson, Watts and their groups.[122,166-72] Hydride or methyllithium attack exclusively at the arene ring, preferentially at an unsubstituted carbon atom, in agreement with Green's rules.[5] The only exception is the reaction of methyllithium with the complex $[Fe(C_5H_5)(C_6Me_6)]^+ BF_4^-$ where addition to the cyclopentadienyl ligand is observed, presumably on account of steric effects. These papers describe some useful spectroscopic correlations which have been widely used to identify <u>exo</u> and <u>endo</u> H substituents. Recent work, however, suggests such assignments must be made with caution, particularly in the infrared (see Section 2.4.1).

The orientation of hydride addition to a number of salts $[Fe(C_5H_5)(C_6H_5X)]BF_4^-$ was also considered. Substituents X = Cl, Br, CO_2Me are found to be essentially ortho directing, although significant displacement of bromide ion by hydride occurs for X = Br. A methoxy group is largely meta directing, while methyl groups do not discriminate effectively between ortho, meta or para additions. In the arenetricarbonylchromium series, somewhat similar substituent effects are found.[173]

Table 1. Directing effects of substituents in η^6-arene complexes.

X	$[Fe(C_5H_5)(C_6H_5X)]^+$	$[Cr(CO)_3(C_6H_5X)]$
Cl	o + m	o + m
MeO	m	m
MeO_2C	o	-
Me	o + m + p	m + o
Me_3Si	-	p
CF_3	-	p

An attempt has been made to correlate these directive effects with orbital coefficients in the arene LUMO[174] and by means of INDO molecular orbital calculations,[175] but it would seem that further information is desirable. Perhaps the observed distribution of products is not kinetically controlled in some cases. The strong meta directing characteristic of the methoxy group has, however, been exploited in the preparation of vicinally trisubstituted arenes.[3]

While ethyllithium yields an <u>exo</u> addition product with $[Fe(C_5H_5)(C_6H_6)]^+$ at $0°C$, at $-25°$ reduction to the 19-electron complex $[Fe(\eta^5-C_5H_5)(\eta^6-C_6H_6)]$ occurs (see Section 6).[176,179] At $25°C$ decomposition to ferrocene and benzene is observed.

When nucleophilic addition to the arene is reversible, as with OMe^-, OH^-, SMe^-, SH^- or amines, $[Fe(C_5H_5)(C_6H_5Cl)]^+$ reacts by <u>ipso</u> substitution, as noted above for the tricarbonyl chromium complex.[178-9] Base abstracts a proton from the resulting phenol or thiophenol complexes to yield **LXXX:**

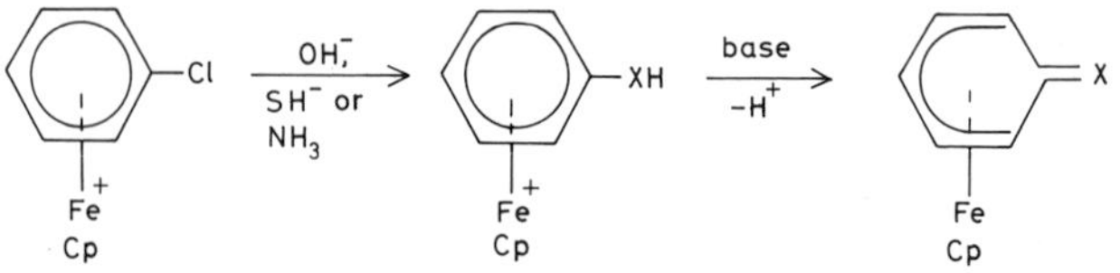

(LXXX)

$$[X = O, S, NH]$$

Similar deprotonations occur with $[Fe(C_5H_5)(C_6H_5X)]^+$ where X = -NHPh, $-CHPh_2$ etc. (see also Section 3.1.1).[180]

2.3.4 Bis-η^6-arene salts

Bis(arene)iron cations are prepared by a remarkably simple method, the reaction of iron (II) chloride with an arene in the presence of aluminum trichloride.[181-3] Nucleophiles including cyanide and various carbanions, add <u>exo</u> to one arene ring to give **LXXXI:**[184]

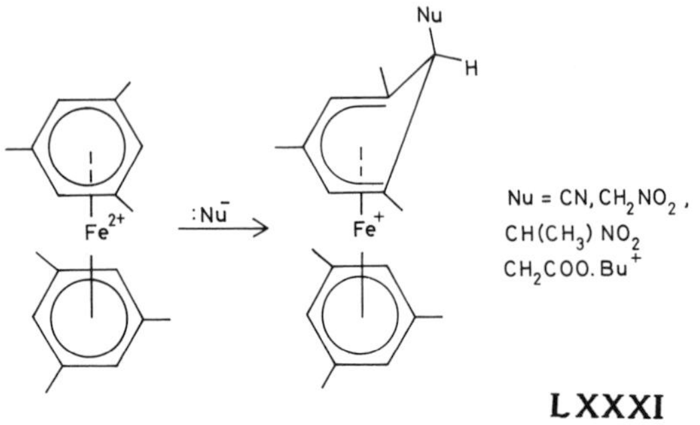

$Nu = CN, CH_2NO_2$, $CH(CH_3)NO_2$ $CH_2COO.Bu^+$

LXXXI

Alkyl- and aryllithium reagents can add either to one ring only (1:1 reactant ratio) or to both rings (2:1 ratio). The addition of aryllithium to bis(arene)iron (II) salts followed by oxidation of the diadduct provides a method of making unsymmetrical biphenyls.[179-81]

Tertiary phosphines add rapidly to bis(benzene) salts of Fe, Ru and Os to form air stable mono-adducts, **LXXXII**.

$$[M(C_6H_6)_2]^{2+} + PR_3 \underset{k_{-1}}{\overset{k_1}{\rightleftharpoons}} [M(\eta^5\text{-}C_6H_6PR_3)(\eta^6\text{-}C_6H_6)]^{2+}$$

LXXXII

The rate and equilibrium constants for triphenylphosphine addition have been measured using an nmr technique.[185-6] The forward reactions is strongly metal dependent (Fe >Ru >Os in ratio 390:7:1). This contrasts with the behaviour of $[M(C_7H_7)(CO)_3]^+$ (M = Cr, Mo, W) where the rate varies very little down the triad. The most likely mechanism involves direct bimolecular attack on the arene ligand. The reasons for these differences are not really clear. A reactivity order for phosphine addition to a number of cationic organotransition metal complexes has been compiled as follows:-

$[Mn(C_7H_8)(CO)_3]^+ \geqslant [Fe(C_6H_6)_2]^{2+} > [Fe(C_4H_4)(CO)_2NO]^+ >$
$[Fe(C_6H_7)(CO)_3]^+ > [Ru(C_6H_6)_2]^{2+} > [Cr(C_7H_7)(CO)_3]^+ >$
$[Fe(C_6H_6OMe)(CO)_3]^+ \sim [Mo(C_7H_7)(CO)_3]^+ \sim [W(C_7H_7)(CO)_3]^+ \sim$
$[Os(C_6H_6)_2]^{2+} > [Fe(C_7H_9)(CO)_3]^+ > [Mn(C_6H_6)(CO)_3]^+$

A specific synthesis of bis(arene)ruthenium cations which contain two different arenes has recently been developed.[187] The chlorobenzene complex, **LXXXIII**, is very susceptible to nucleophilic substitution, forming, for example, the anisole derivative in methanol. Deprotonation of the phenol complex affords an η^5-cyclohexadienone complex, as was noted above for iron.

$$[RuCl_2(arene^1)]_2 + AgPF_6 \xrightarrow{\text{acetone}} [Ru(arene^1)(acetone)_3]^{2+}$$

$$\xrightarrow[\text{acid}]{\text{arene}^2} [Ru(arene^1)(arene^2)]^{2+}$$

$$[Ru(C_6H_3Me_3)(C_6H_5Cl)]^{2+} \xrightarrow{\text{MeOH}} [Ru(C_6H_3Me_3)(C_6H_5OMe)]^{2+}$$

LXXXIII

The isoelectronic cobalt, rhodium and iridium salts $[M(C_5Me_5)(arene)]^{2+}(PF_6^-)_2$ undergo nucleophilic addition, by hydride or methyllithium, exclusively at the arene ring.[188-9] The rhodium salts catalyze intramolecular nucleophilic substitution of o-$FC_6H_4(CH_2)_3OH$ forming chromans.[190] The high reactivity of the arene in these complexes is also indicated by the addition of trimethyl phosphite with Arbusov rearrangement, to yield $\{Rh(\eta^5\text{-}C_5Me_5)[\eta^6\text{-}C_6H_6PO(OMe)_2]\}^+$.[191] Tributylphosphine gives a stable _exo_ adduct, but triphenylphosphine does not react.

Bis(arene) cations of manganese[192] and rhenium[193-5] are also attacked by hydride ion to yield η^5-cyclohexadienyl derivatives e.g.

$$[Re(\eta^6\text{-}C_6H_6)_2]^+ + H^- \longrightarrow Re(\eta^5\text{-}C_6H_7)(\eta^6\text{-}C_6H_6)$$

$[Re(\eta^6\text{-}C_6Me_6)_2]PF_6$ on reduction with sodium in liquid ammonia yields air-stable diamagnetic $Re(\eta^6\text{-}C_6Me_6)(\eta^5\text{-}C_6Me_6H)$, but reduction with molten lithium at 200 - 230°, followed by immediate quenching of the volatile product on a cold finger cooled in liquid nitrogen, affords black $Re(C_6Me_6)_2$. This very reactive, paramagnetic material dimerizes on warming to room temperature. A likely structure for the dimer is **LXXXIV**, which resembles those of rhodocene and technetocene. The cyclobutadiene cobalt complex $[Co(\eta^4\text{-}C_4Ph_4)(\eta^6\text{-}C_6H_6)]^+$, **LXXXV**, is very inert to nucleophilic attack, although the corresponding η^6-cycloheptatriene compound is quite reactive. **LXXXV** is resistant to methoxide and even to methylmagnesium bromide. Borohydride, however, gives a 20% yield of $Co(C_4Ph_4)(\eta^5\text{-}C_6H_7)$.

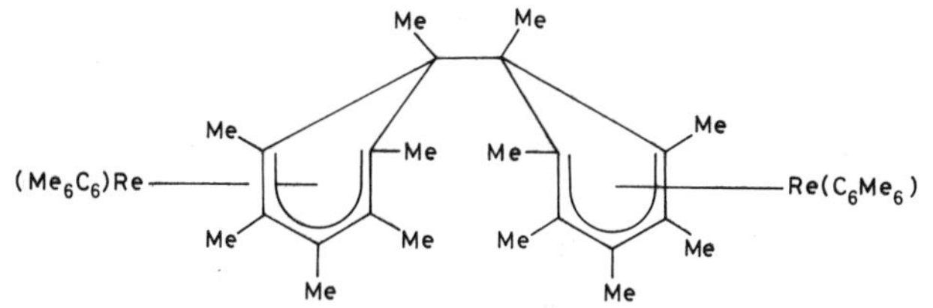

LXXXIV

2.3.5 Other η^6-arene complexes

The reactions of cyclohexa-1,3- or 1,4-dienes with ruthenium trichloride in ethanol afford η^6-arene complexes $[Ru(\eta^6$-arene$)Cl_2]_2$.[197-8] Many nucleophiles attack these complexes at the metal. For example hydroxide and alkoxide yield triply bridged derivatives;[199-203] the corresponding chloride is obtained by heating **LXXXVI** with water.[198]

$$[Ru(\eta^6\text{-}C_6H_6)Cl_2]_2 \longrightarrow [(C_6H_6)Ru(X)_3Ru(C_6H_6)]^+ \ (X = OH, OMe, Cl)$$
LXXXVI

Phosphines and phosphites merely cleave the chloride bridges. While cyanide ion does form an unstable cyano-cyclohexadienyl adduct,[197,204] nucleophilic attack on the arene ligand is better observed in cationic derivatives in which the chloride bridges have been opened by suitable donors:[205]

$$[Ru(arene)Cl_2]_2 \xrightarrow[\substack{\text{i) N-N = bipy} \\ \text{or o-phen} \\ \text{ii) } NH_4PF_6}]{} [Ru(arene)Cl(N\text{-}N)]^+ PF_6^- \xrightarrow{R_3P}$$

$$[Ru(arene)(PR_3)(N\text{-}N)]^{2+} (PF_6^-)_2 \xrightarrow{Y^-} [Ru(\eta^5\text{-}C_6H_6Y)(PR_3)(N\text{-}N)]^+PF_6^-$$
$$(Y = H, CN, OH)$$

2.4 Nucleophilic attack on η^7-complexes

2.4.1 η^7-Cycloheptatrienyl tricarbonyl salts of Group VIA

Tricarbonyltropylium salts of the Group VIA metals are readily obtained from the hexacarbonyls by a two stage synthesis.[206-8]

$$\text{M(CO)}_6 \xrightarrow{\text{C}_7\text{H}_8} [\text{M}(\eta^6\text{-C}_7\text{H}_8)(\text{CO})_3] \xrightarrow{\text{Ph}_3\text{C}^+\text{BF}_4^-} [\text{M}(\eta^7\text{-C}_7\text{H}_7)(\text{CO})_3]^+\text{BF}_6^-$$

In some cases, such as the preparation of derivatives of substituted cycloheptatrienes, it may be advantageous first to convert the hexacarbonyl to the acetonitrile complex $[\text{M(CH}_3\text{CN})_3(\text{CO})_3]$.[209] The groundwork in this field, particularly on the chromium complexes, was done by Pauson and his coworkers.[122] The ion $[\text{Cr}(\text{C}_7\text{H}_7)(\text{CO})_3]^+$ adds Y^- = OMe$^-$, SH$^-$, CH(CO$_2$Et)$_2^-$, H$^-$ and Ph$^-$ to form cycloheptatrienyl complexes Cr(η^6-C$_7$H$_7$Y)(CO)$_3$.[206] The exo configuration of the phenyl derivative was proved by xray diffraction[210] and spectroscopic correlations indicated that this stereospecific addition is quite general.[80, 210-11]

More recently, however, Faller has found that hydride addition to $[\text{Mo}(\text{C}_7\text{H}_7)(\text{CO})_3]^+$ is far from stereospecific, contrary to this generally accepted view. Thus BD$_4^-$ gives 72% exo and 28% endo $[\text{Mo}(\text{C}_7\text{H}_7\text{D})(\text{CO})_3]$, the latter possibly via MoD(η^5-C$_7$H$_7$)(CO)$_3$. It seems that identification of isomers on the basis of ^{1}H nmr and especially infrared spectroscopy, where a band at ca. 2750 cm^{-1} has been taken to be characteristic of an exo-C-H group, must be carried out with caution.[212]

Weakly nucleophilic bases such as sodium hydrogen carbonate, sodium acetate and somewhat surprisingly sodium cyanide in aqueous solution cause reductive coupling (as noted in Section 2.2.2 for some tricarbonyliron complexes) to form mono- and bis-tricarbonylchromium complexes of ditropyl, **LXXXVII** and **LXXXVIII**:[80]

(LXXXVII)   (LXXXVIII)

Cyanide in acetone, however, yields the cyano adduct Cr(C$_7$H$_7$CN)(CO)$_3$ in 92% yield.[209]

7- <u>Endo</u> substituted cycloheptatrienetricarbonylchromium derivatives can be prepared stereospecifically by direct reaction between $Cr(CH_3CN)_3(CO)_3$ and a 7-substituted cycloheptatriene.[212] This allowed direct comparison between corresponding <u>endo</u> and <u>exo</u> isomers. As triphenylmethyl tetrafluoroborate abstracts <u>exo</u> but not <u>endo</u> hydrogen atoms, 7-alkoxytropylium salts could be obtained by the following route (Scheme 13):-

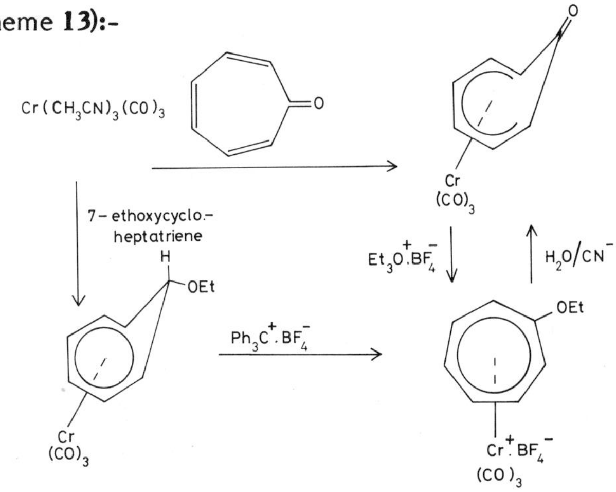

Scheme 13

The 7-<u>exo</u>-methoxy derivative, however, loses its methoxy substituent on treatment with $Ph_3C^+BF_4^-$.

The reactions of $[M(C_7H_7)(CO)_3]^+ BF_4^-$ with alkyl and aryl phosphines give different products under different conditions, and this led to some confusion in the literature. The position now seems to have been clarified by the work of Salzer[213] and of Kane-Maguire.[214-5] The results can be interpreted in terms of the general scheme discussed above:

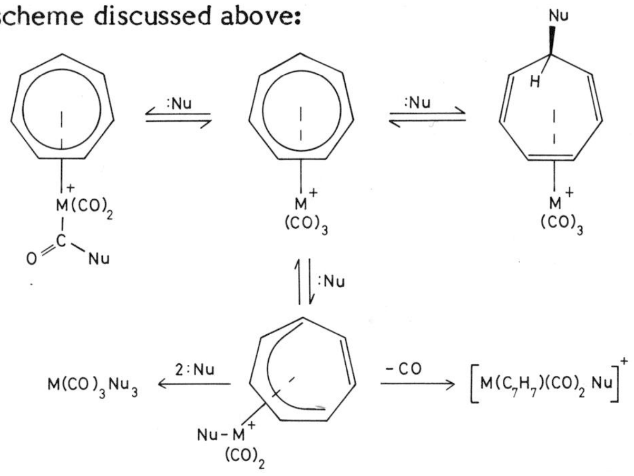

Scheme 14

The particular case of a neutral monodentate nucleophile is illustrated, but Scheme 14 applies equally well in essence for anionic nucleophiles such as OMe$^-$ or I$^-$.

Early observations showed that phosphines (L = PBu$_3^n$, PEt$_2$Ph) add to the ring in [Cr(C$_7$H$_7$)(CO)$_3$]$^+$ to yield phosphonium salts.[216] The molybdenum complex, however, afforded [Mo(C$_7$H$_7$)(CO)$_2$L]$^+$ or Mo(CO)$_3$L$_3$ depending on the reaction conditions.[217] The bidentate ligand Ph$_2$PCH$_2$PPh$_2$ (dppm) gave [Mo(C$_7$H$_7$)(CO)(dppm)]$^+$ on reflux in ethanol, while Ph$_2$PCH$_2$CH$_2$PPh$_2$ (dppe) yielded a mixture of [Mo(C$_7$H$_7$)(CO)(dppe)]$^+$ and [Mo(C$_7$H$_7$)(CO)$_2$(dppe)]$^+$.[218] Salzer found, however, that the initial product is the phosphonium salt, formed in equilibrium with the starting complex.[213] Thus on addition of tri-isopropylphosphine to LXXXIX the intense violet color of the phosphonium salt appears, but disappears within 1 minute at room temperature with formation of M(CO)$_3$L$_3$. The adduct can be isolated at -40°C. The kinetics of ring addition of tri-n-butylphosphine to [M(C$_7$H$_7$)(CO)$_3$]BF$_4$ (M = Cr, Mo, W) have been studied at 0°C in acetone.[214] The reactions exhibit overall second order kinetics (Rate = k$_2$[complex] [PBu$_3^n$]) [k$_2$/dm^{3}mol^{-1}s^{-1} = 17400 (Cr), 7500 (Mo), 8400 (W)]. If initial attack were at the metal, large rate differences might be expected. Thus for ring displacement by P(OMe)$_3$ from M(C$_6$H$_6$)(CO)$_3$ relative rates are Cr : Mo: W = 1 : 2200 : 350. Similarly the rate of ring addition of acetylacetone in 1,2-dichloroethane is almost independent of the metal.[219] In refluxing tetrahydrofuran, however, [Mo(C$_7$H$_7$)(CO)$_3$]$^+$ reacts with acetylacetone to form [Mo(C$_7$H$_7$)(acac)(H$_2$O)]$^+$, XC, from which the aquo ligand is readily displaced by Group V donors, halides or pseudohalides.[221-2] Ketones such as acetone react with LXXXIX (M = Cr) to give low yields of ring addition products such as Cr(CH$_3$COCH$_2$.C$_7$H$_7$)(CO)$_3$ in the presence of base,[223] or on electrolysis of solutions in the presence of 48% HBr.[204]

Several years ago King noticed that on addition of iodide to [Mo(C$_7$H$_7$)(CO)$_3$]$^+$ a brown color develops, which is gradually replaced by the dark green color of the final product [Mo(C$_7$H$_7$)(CO)$_2$I].[225] Chloride, bromide and some pseudohalides react similarly.[226] A kinetic and nmr study of the reactions of LXXXIX (M = Mo, W) with iodide has indicated that the initially formed brown compound is probably a ring adduct $[M(C_7H_7I)(CO)_3]$.[227] The interpretation of the results should be modified, however, in the light of work of Kane-Maguire and Salzer. A more likely scheme is as follows, which seems to be in agreement with the experimental observations:

$$[M(C_7H_7)(CO)_3]^+I^- \overset{I^-}{\rightleftharpoons} [M(C_7H_7)(CO)_3]^+ \overset{I^-}{\rightleftharpoons} M(C_7H_7I)(CO)_3$$

or

$M(C_7H_7)(CO)_2COI$

or

$M(\eta^5\text{-}C_7H_7)(CO)_3I$

$\Big\Updownarrow$ acetone (−CO)

$[M(C_7H_7)(CO)_2\text{acetone}]^+ \overset{I^-}{\longrightarrow} M(C_7H_7)(CO)_2I$

The kinetics of methoxide attack on **LXXXIX** in methanol also show a limiting rate at high methoxide concentrations. This is consistent with a 'dead-end' equilibrium forming a carbomethoxy derivative which is in competition with direct ring addition:[228]

$$M(C_7H_7)(CO)_2(CO_2Me) \overset{OMe^-}{\rightleftharpoons} [M(C_7H_7)(CO)_3]^+ \overset{OMe}{\longrightarrow} M(C_7H_7OMe)(CO)_3$$

Other modes of reaction of **LXXXIX** include displacement of the organic ligand[229] and displacement of carbonyl groups. Of particular interest is the replacement of three carbonyl ligands by arenes under reflux to give the cationic species $[Mo(C_6H_6)(C_7H_7)]^+$.[220-21] The arene ligands are displaced by nitriles affording $[Mo(C_7H_7)(CH_3CN)_n]^+$ (n = 2 or 3), by acetylacetone to give **XC**, and by dppe.[230]

Nucleophilic attack on $[Ru(\eta^4\text{-}1\text{-}2,5\text{-}6\text{-}C_8H_{12})(\eta^7\text{-}C_7H_7)]^+BF_4^-$ takes place at the C_7 ring.[231]

3. ELECTROPHILIC ATTACK ON $\eta^5\text{-},\eta^6\text{-}$ AND η^7-COMPLEXES

3.1 Protonation

Protonation takes place at the metal. The hydride initially formed may then transfer to the <u>endo</u> face of a coordinated ligand, or if this pathway is not available, remain in equilibrium with the unprotonated complex. Arene complexes usually form metal hydrides in strong acids; no transfer of hydrogen occurs, probably on account of the loss of aromaticity of the ligand which would result. (Arene)tricarbonyl chromium complexes, for example, form hydrides $[CrH(arene)(CO)_3]^+$ in FSO_3H/SO_2 and similar acid media, which give rise to signals at $\tau \sim 14$ in the ^1H nmr spectra.[232-4] Substitution of CO by PPh_3 increases the basicity, so that $Cr(arene)(CO)_2PPh_3$ is protonated by CF_3COOH alone. Bis(benzene) tungsten is the only member of the bis(arene) complexes of Group VIA known to form a stable 18-electron hydride, which can be isolated as the salt $[WH(C_6H_6)_2]PF_6$.[236] The chromium and molybdenum compounds are oxidized to the cations $[M(C_6H_6)_2]^+$ with evolution of hydrogen. The transfer of a hydrogen from metal to ligand has been very extensively studied with η^4-diene complexes, which yield η^3-allyl derivatives.[237] There are rather few examples as yet of the protonation of η^6-triene complexes. While $M(\eta^6\text{-}C_7H_8)(CO)_3$, **XCI** (M = Cr), is oxidized to $[Cr(C_7H_7)(CO)_3]^+$ by strong acids, the molybdenum derivative forms the 16-electron cation $[Mo(\eta^5\text{-}C_7H_9)(CO)_3]^+$.[238] No similar complex could be isolated when M = W, but in the presence of a stoichiometric quantity of a Group VB donor ligand, salts $[W(\eta^5\text{-}C_7H_9)(CO)_3L]^+BF_4^-$ [L = PPh_3, $AsPh_3$, $SbPh_3$, $P(OMe)_3$ etc.] were obtained. Hydrogen chloride gives covalent derivatives $M(\eta^5\text{-}C_7H_9)(CO)_3Cl$, (M = Mo, W).[207] $Mo(\eta^6\text{-}C_8H_8)(CO)_3$ (**XCII**) is also readily protonated but in this case an η^7-ligand results on account of the presence of an uncomplexed double bond in the starting material.[239] The corresponding chromium complex cannot be obtained by this method, but it is available by hydride abstraction from $Cr(CO)_3(\eta^6\text{-}C_8H_{10})$.[240] $Cr(\eta^6\text{-}C_8H_8)(\eta^5\text{-}C_5H_5)$ does, however, give $Cr(\eta^7\text{-}C_8H_9)(\eta^5\text{-}C_5H_5)^+$ on treatment with HPF_6.[241]

Bicyclo[6.1.0]nona-2,4,6-trienetricarbonylmolybdenum undergoes proton addition at the metal at $-120°C$. This addition is accompanied by a change in the coordination of the ligand.[242]

η^6-Phenol complexes of cobalt, rhodium and iridium are deprotonated by base to afford η^5-oxocyclohexadienyl derivatives (**XCIII**). The reactions are reversed by acid.[243-4]

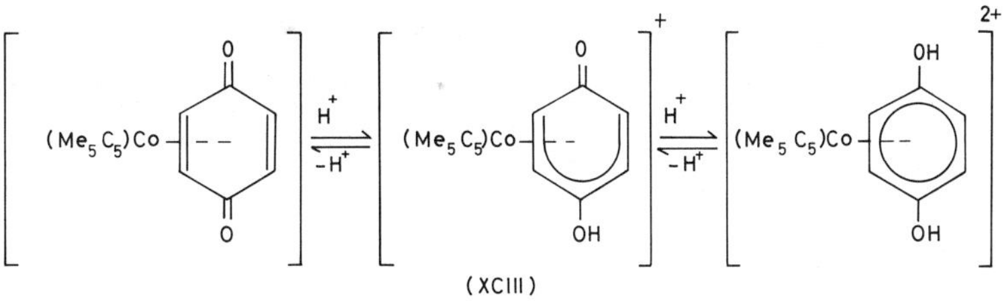

(XCIII)

3.1.1 Metal migration in indene and fluorene complexes

Some interesting migrations of metals between the five and six membered rings of indene and fluorene occur on protonation or deprotonation. While ferrocene is protonated at the metal forming $[FeH(C_5H_5)_2]^+$, bis(indenyl)iron or $[Fe(C_5H_5)(C_9H_7)]$, **XCIV**, on treatment with HCl followed by PF_6^- yield salts in which the metal has migrated to a six-membered ring.[246-7] Stereospecific endo protonation is likely:

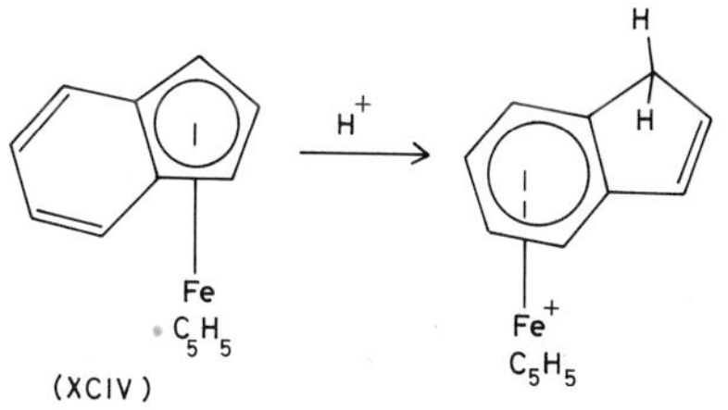

Similar migrations from the 5- to the 6-membered ring occur on protonation of $[M(\eta^5-C_5Me_5)(\eta^5-C_9H_7)]^+$, (M = Rh, Ir), which is reversed by addition of base.[248]

In the fluorenyl complexes $M(C_{13}H_{10})$ [M = $Fe(C_5H_5)$ or $Mn(CO)_3$] the metal is bonded to one of the six-membered rings. In the former case, deprotonation leads to **XCV**, in which the metal remains η^6-bonded. The product possesses appreciable zwitterionic character:[249-52]

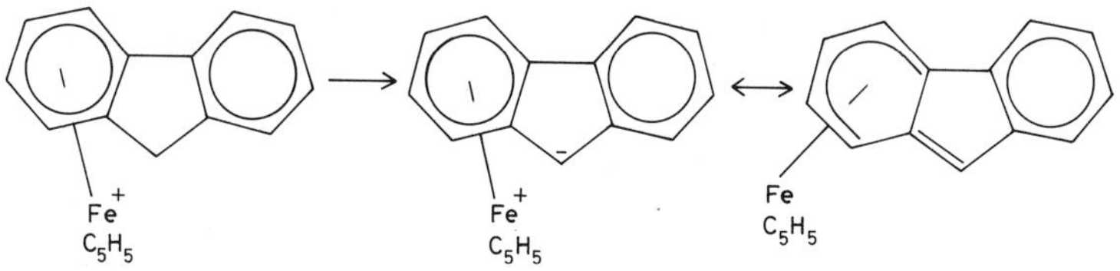

(XCV)

The action of base on $[Mn(\eta^6\text{-}C_{13}H_{10})(CO)_3]^+PF_6^-$ leads initially to **XCVI** but this is followed by slow rearrangement to **XCVII** on heating:[250]

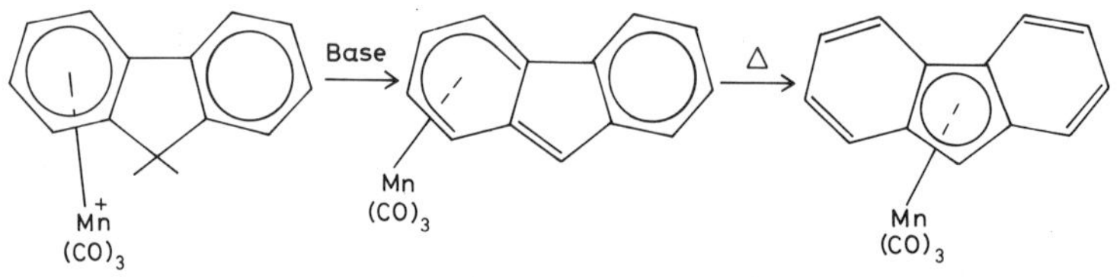

(XCVI)   (XCVII)

The isoelectronic chromium complexes $Cr(\eta^6\text{-}C_{13}H_9)(CO)_3$ behave similarly. Base initially yields η^6-bonded anions, which react with alkyl halides to give exo-9-substituted derivatives, indicating a high charge density at C^9. These η^6-anions rearrange to the isomeric $[Cr(\eta^5\text{-}C_{13}H_9)(CO)_3]^-$, which yield 9-endo alkyl complexes with alkyl halides, probably through initial attack at the metal.[253-4]

3.2 Hydride abstraction

The triphenylmethyl cation abstracts hydride ion from coordinated hydrocarbons where the resulting cation is more stable. Thus the complexes $M(\eta^6\text{-}C_7H_8)(CO)_3$ (M = Cr, Mo, W) yield the cycloheptatrienyl salts $[M(\eta^7\text{-}C_7H_7)(CO)_3]^+BF_4^-$ on treatment with $Ph_3C^+.BF_4^-$.[106-208,255] Triethyl-oxonium tetrafluoroborate has also been employed to abstract hydride.[256] An interesting variation has yielded heptafulvene complexes **XCIX**. They are attacked by electrophiles exclusively at the exocyclic double bond; this treatment with acids regenerates the cationic precursors **XCVIII**:[257]

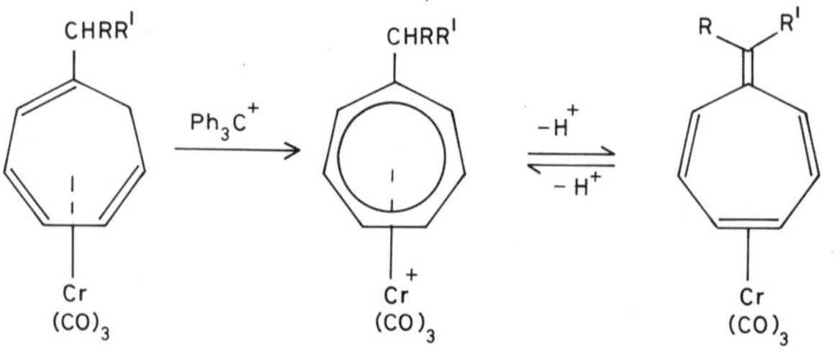

(XCVIII)   (XCIX)

Hydride abstraction by Ph_3C^+ is stereospecific, only exo-hydrogens being removed; **C** is converted into the cation **CI** whereas its exo-phenyl isomer is inert.[258]

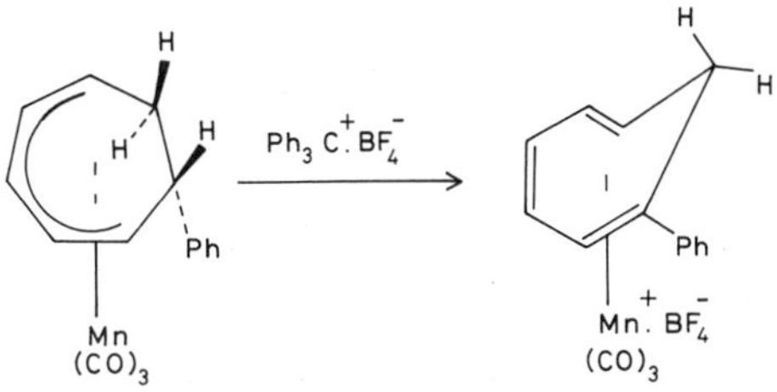

Other η^6-cycloheptatriene complexes which are converted into the corresponding η^7-cycloheptatrienyl salts with $Ph_3C^+BF_4^-$ include $Ru(\eta^6\text{-}C_7H_8)_2$,[231] $Ru(\eta\text{-}1\text{-}2,5\text{-}6\text{-}C_8H_{12})(\eta^6\text{-}C_7H_8)$,[231] and $Mn(\eta^5\text{-}C_5H_5)(\eta^6\text{-}C_7H_8)$.[261] $V(C_7H_8)_2$[259-60] forms the dication CII in a two stage process:

$$V(\eta^6\text{-}C_7H_8)_2 \longrightarrow [V(\eta^6\text{-}C_7H_8)(\eta^7\text{-}C_7H_7)]^+ \longrightarrow [V(\eta^7\text{-}C_7H_7)_2]^{2+}$$
$$CII$$

The method has also been applied to the preparation of η^7-cyclooctatrienyl cations $[M(\eta^7\text{-}C_8H_9)(CO)_3]^+$, $(M = Cr, Mo, W)$ from the neutral η^6-cyclooctatriene complexes.[262-3]

For an excellent survey of the chemistry of transition metal complexes of cyclic polyolefins having more than 6-carbon atoms in any one ring, the book by Deganello should be consulted.[1]

3.3 Friedel-Crafts acylation

While Friedel-Crafts acylation of ferrocenes, η^4-dienetricarbonyliron complexes, η^5-cyclopentadienyltricarbonylmanganese and related compounds has been extensively studied, comparatively little work has been reported on the behavior of materials within the remit of this Chapter. Attempts to acetylate arenetricarbonylchromium complexes under normal conditions (excess $AlCl_3, CH_2Cl_2, 0°C$) lead to extensive decomposition. The Perrier method, which employs preformed $CH_3CO^+AlCl_4^-$ to avoid excess $AlCl_3$, can be applied with success.[264] Benzenetricarbonyl-chromium is deactivated to electrophilic substitution compared with benzene itself.[265] In contrast to the free aromatics, the reactions are characterized by a lack of activation by electron donating substituents such

as methyl or methoxy, especially at the para position. While toluene gives
o:m:p ratios of 1.2:2:96.8, its $Cr(CO)_3$ complex gives 43:17:40. Again
whereas toluene reacts 128 times faster than benzene, $Cr(toluene)(CO)_3$
reacts only 1.1 times as fast as $Cr(benzene)(CO)_3$ itself.[266]

Acylation of $M(C_5H_5)(C_6H_6)$ (M = Cr, Mn) proceeds with expansion of
the arene ring affording a methyltropylium complex **CIII**:[267]

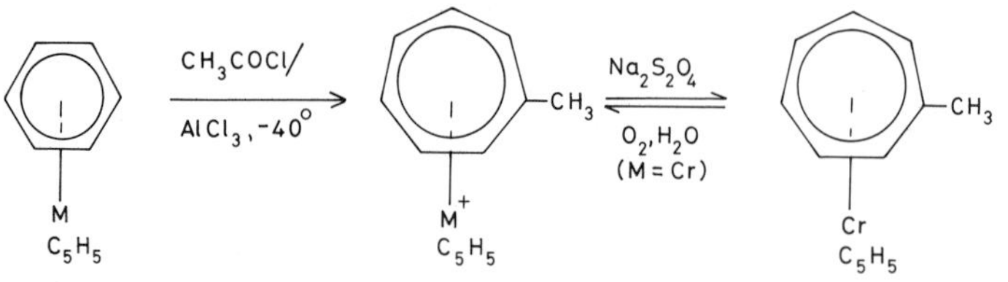

Reduction of the chromium cation **CIII** (μ_{eff} = 2.1 B.M.) yields the
diamagnetic neutral complex, **CIV**. These ring expansions may be related to
similar reactions of borabenzene complexes discussed in Section 7.
Stereospecific ring expansion of arene also occurs when $Cr(arene)(CO)_3$ is
treated at $-40°$ with benzyl chloride and base.[268] The reaction presumably
involves insertion of the weakly electrophilic carbene 'PhCH'. The endo
configuration of the products is confirmed by determination of the
molecular structure of one such compound by xray diffraction.[269]

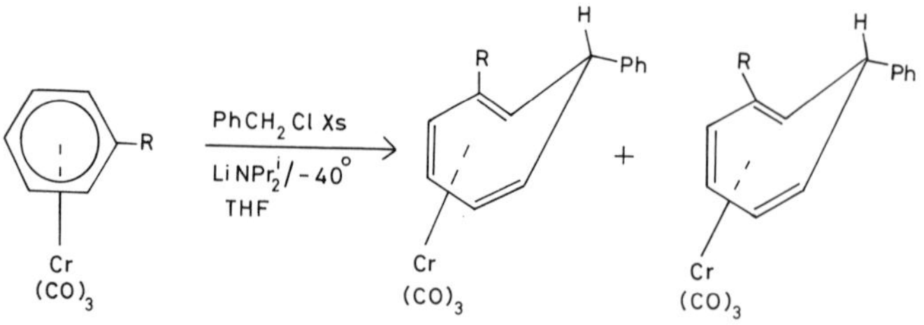

4.   METALATION REACTIONS

Metalation using, for example, lithium alkyls, provides an important
synthetic route to ferrocene derivatives. By analogy, it might be expected
that bis(benzene)chromium **(CV)** would also be susceptible to metalation.
Fischer and Brunner found that **CV** reacts with n-pentylsodium in boiling
hexane to give a mixture of ring metalated products, which on treatment
with iodomethane afford a mixture of methyl derivatives.[270] They were

also able to convert the sodium derivative of **CV** into the ester $Cr(C_6H_6)(C_6H_5COOMe)$ using carbon dioxide followed by dimethyl sulfate, and into various alcohol derivatives using standard procedures.[270] n-Butyllithium alone does not metallate **CV**, but its TMEDA complex in cyclohexane does.[272-5] In competition with benzene, **CV** is more easily metalated, that is, the kinetic acidity of the C-H bonds is enhanced by bonding to chromium.

Fluorobenzenes are more acidic than benzene, and as expected, $Cr(C_6H_6)(C_6F_5H)$, prepared by the metal atom technique, is readily metalated by tert-butyllithium; activation by TMEDA is unnecessary. The product $[Cr(C_6H_6)(C_6F_5Li)]$ has been used to synthesize a wide range of derivatives.[176,277] It was not possible, however, to obtain the acid $[Cr(C_6H_6)(C_6F_5COOH)]$ by carbonation, although its ethyl ester was prepared by reaction with ethyl chloroformate.[278]

The bis(benzene)chromium(I) cation reacts with strong bases, such as potassium in dimethoxyethane, mainly by reduction to $Cr(C_6H_6)_2$. A significant side reaction, however, is deprotonation to give the paramagnetic species $Cr(C_6H_6)(C_6H_5)$ which has been characterized by its esr spectrum.[279]

Bis(chlorobenzene)molybdenum (**CVI**) is metalated by tert-butyllithium at $-78°$ giving a dilithio derivative which reacts with trimethylchlorosilane. n-Butyllithium however does not metalate **CVI** but effects nucleophilic substitution of chloride by n-butyl:[280]

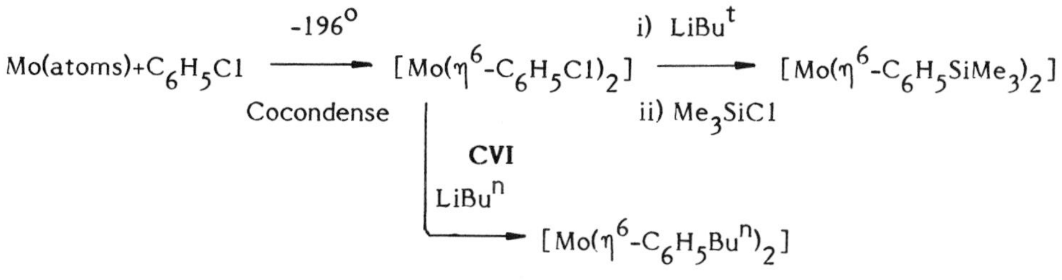

$$Mo(atoms) + C_6H_5Cl \xrightarrow[\text{Cocondense}]{-196°} [Mo(\eta^6-C_6H_5Cl)_2] \xrightarrow[\text{ii) } Me_3SiCl]{\text{i) } LiBu^t} [Mo(\eta^6-C_6H_5SiMe_3)_2]$$

CVI $\xrightarrow{LiBu^n} [Mo(\eta^6-C_6H_5Bu^n)_2]$

The metalation of arenetricarbonylchromium complexes by n-butyllithium in the presence of TMEDA is discussed in Section 5.

The site of metalation in the complexes $M(\eta^5-C_5H_5)(\eta^7-C_7H_7)$ (M = Ti, V, Cr) has been correlated with the charge densities on the rings. While the titanium complex is metalated readily and specifically in the C_7 ring, the chromium derivative reacts with difficulty in the C_5 ring. The vanadium compound is intermediate, metalation occurring in both rings, with a preference for C_5.[281-3] This suggests that in the C_7H_7 ligand the negative charge decreases in the order Ti >V >Cr, while in the C_5H_5 ligand the charge varies only slightly. This conclusion is supported by nmr,[284] ESCA[285] and UVPES measurements and by ab initio molecular orbital calculations.[286] The e_2 m.o.s of C_7H_7 form a bonding combination with the metal 3dδ orbitals (xy, x^2-y^2). The energies of the metal orbitals decrease across the series Ti >V >Cr, so that the e_2 bonding m.o. has a greater ligand character in the titanium complex than in that of chromium. Metalation of $Ti(\eta^5-C_5H_5)(\eta^8-C_8H_8)$ takes place, with difficulty, in the C_5 ring.[287-8]

5. SIDE CHAIN AND OTHER ORGANIC REACTIONS OF η^6-ARENETRICARBONYLCHROMIUM COMPLEXES

The organic reactions of arenetricarbonylchromium complexes and their applications in organic synthesis have been reviewed. The effects on arene reactivity arising from coordination to the $Cr(CO)_3$ group can be summarized as follows:[3-4]

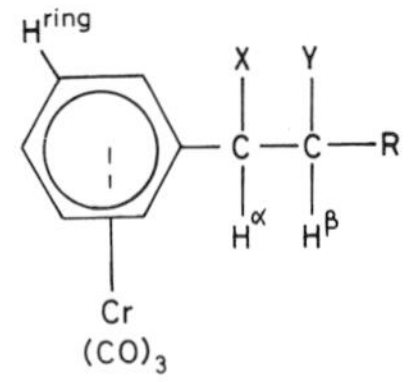

a)   Enhanced acidity of arene ring hydrogens (H^{ring}) (see Section 4).

b)   Reduced electron density on ring; enhanced nucleophilic aromatic addition and substitution (see Section 2.3.2).

c)   Enhanced acidity of side chain hydrogens (H^α, H^β); stabilization of side chain anion sites.

d)   Stabilization of side chain cation sites (benzyl, phenethyl); enhanced rate of solvolysis of X and Y.

e)   Steric effect of $Cr(CO)_3$ group.

a)   Benzenetricarbonylchromium is metalated by n-butyllithium in THF at $-40°$,[281] or in better yields using $LiBu^n$ + TMEDA in THF at $-78°$ [290] or $LiBu^n$ alone in THF/ether.[291] Benzene itself is not metalated under similar conditions.

b)   The electron withdrawing effect of the $Cr(CO)_3$ group is illustrated by the dissociation constants of benzoic acid complexes.[144,292-4] The pK_a* value of $Cr(PhCO_2H)(CO)_3$ in 50% aqueous ethanol at $25°C$, for example, is 4.52 compared with 5.48 for the free acid.

c)   The stabilization of side chain anions sites by coordination to $Cr(CO)_3$ has been exploited synthetically,[295-6] in particular by Jaouen.[297] Thus **CVII** is converted into **CVIII** or **CIX** depending on the base used:

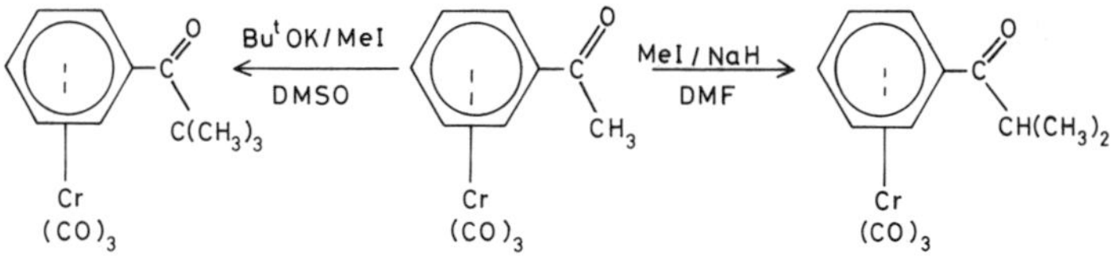

While $PhCH_2CO_2Me$ is inert to NaH/MeI in dimethylformamide, its complex affords $Cr(CO)_3(PhCMe_2COOMe)$ in 97% yield. Some further applications are quoted in a recent review.[2] The anionic species **CX** has been observed by ^{13}C nmr and converted into various derivatives with electrophiles.[298]

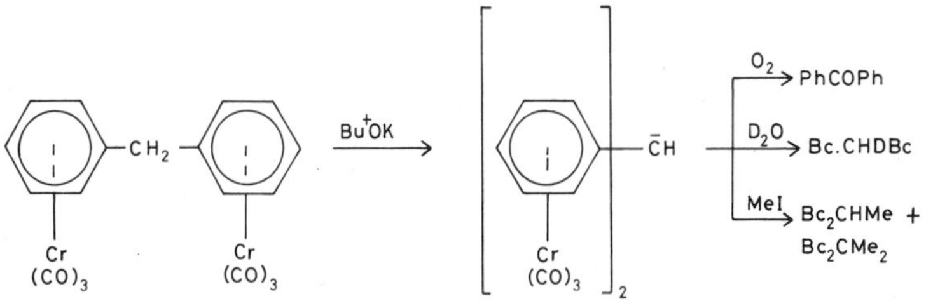

d)   The stabilization of carbonium ions adjacent to organotransition metal units such as ferrocenyl, cymantrenyl and benchrotrenyl has been very widely investigated.[299-301] Compared with the α-ferrocenyl group, the stabilization arising from an α-$C_6H_5Cr(CO)_3$ group is relatively small, but is sufficient to produce noticeable changes in reactivity.[302-4] Thus the pK_{R+} values for $PhCH_2^+$, $PhCH_2Cr(CO)_3^+$ and $PhCH_2Fc^+$ are -17.3, -11.8 and 0.4 respectively. Seyferth and Eschbach isolated the first crystalline salt of this type, $[(OC)_3Cr(C_6H_5CHC_6H_5)Cr(CO)_3]^+PF_6$, which reacts with nucleophiles at the $-CH^+-$ center.[305-6] Other ions have been studied directly using ^{13}C nmr spectroscopy.[307-8] There is much kinetic evidence for their involvement in ionization reactions; the rate of solvolysis of benzyl chloride for example is enhanced by a factor of 10^5 on complexing with $Cr(CO)_3$.[309] Several detailed mechanistic studies include attempts to assign Hammett and related constants to the $Cr(CO)_3$ group and to the reactions for which carbonium ion stabilization is postulated.[310-19] Further discussion of this topic, however, is outside the scope of this article.

Some use is now being made of these effects in organic synthesis. The Ritter reaction[320] which involves electrophilic addition to a nitrile, is greatly accelerated in the case of primary alcohols. While the uncomplexed reagent needs 20-40 hr, the complex reacts within a few minutes:[321]

$$Cr(\eta^6\text{-}R^1C_6H_4CR^2R^3OH)(CO)_3 \xrightarrow[-15^0]{H_2SO_4} [Cr(\eta^6\text{-}R^1C_6H_4CR^2R^3)(CO)_3]^+ \xrightarrow{R^4CN}$$

$$Cr(\eta^6\text{-}R^1C_6H_4CR^2R^3NHCOR^4)(CO)_3 \xrightarrow[O_2]{h\nu} R^1C_6H_4CR^2R^3NHCOR^4$$

Complexed benzyl alcohols can also be readily converted into amines or ethers by reaction with HPF_6 etherate followed by an amine or alcohol.[322]

e)   The steric effect of the $Cr(CO)_3$ group induces stereospecific exo attack of nucleophiles on coordinated ligands. Jaouen and his coworkers have used this principle in some elegant stereoselective syntheses, for example of indanols, tetralols, and substituted benzyl alcohols. One illustration of such an asymmetric synthesis is given in Scheme 15.[323] Further examples are described in a review.[2]

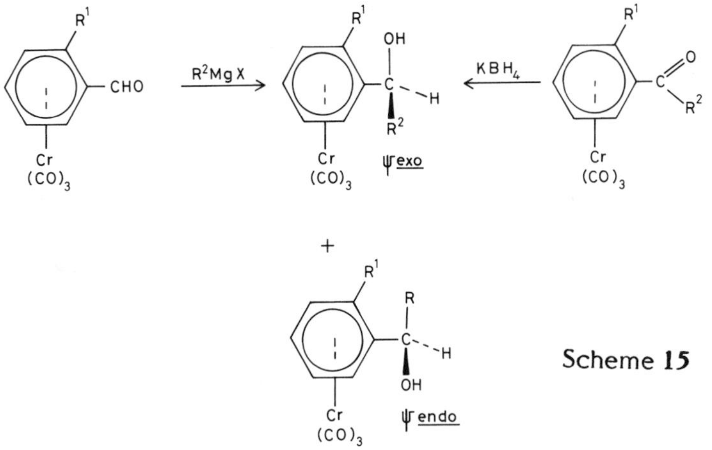

Scheme 15

Attack of the Grignard reagent or borohydride from the side opposite to the $Cr(CO)_3$ group favors formation of the ψ-endo isomer. The two diastereoisomers can be separated by chromatography. Decomplexation gives optically active alcohols in 100% optical purity, when a pure enantiomer of the starting aldehyde or ketone complex is used.

6.   REDOX REACTIONS

When considering the reactions of electrophiles or nucleophiles the possibility should be borne in mind that, respectively, oxidation or reduction of the complex may occur, rather than attack at the ligand. This is well known, of course, in ferrocene chemistry where attempts to effect direct electrophilic halogenation or nitration lead to formation of the ferricinium cation. The electron rich complexes $M(C_6H_6)_2$,[280] $M(C_5H_5)(C_7H_7)$, (M = Cr,Mo)[324] and $Ti(C_5H_5)(C_8H_8)$,[325] for example, are very susceptible to oxidation to the corresponding unipositive cations by mild electrophiles. This also applies to the 19-electron complex cobaltocene[326] and to the isoelectronic sandwich compounds $Fe(C_5H_5)(arene)$.[327-32] Nucleophilic addition to the arene ligand in $[Fe(C_5H_5)(arene)]^+$ was noted in Section 2.3.3. This is usually the dominant pathway with hydride ion donors such as borohydride, although reduction rather than addition is observed with the naphthalene complex.[331] Reduction to $Fe(C_5H_5)(arene)$, however, generally occurs with sodium amalgam or electrochemically. A further irreversible reduction has been observed.[329,333-4] The neutral species are very readily oxidized. Even in the absence of air or other oxidizing agents dimerization occurs at room temperature with the formation of a carbon-carbon bond:[330-34]

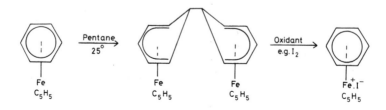

The 19-electron species are best stabilized against such dimerization by peralkylation of the benzene ring; permethylation of the C_5 ring proved destabilizing.[335]

7.   BORABENZENE COMPLEXES

The η^6-borabenzene ligand, which formally contributes five electrons, is in many ways similar to the cyclopentadienyl and cyclohexadienyl groups and it therefore seems appropriate to include some account of its chemistry here. The coverage is not comprehensive as a review of boron heterocycles as ligands in transition metal chemistry has appeared.[336]

Borabenzene complexes were discovered by Herberich and coworkers, who treated cobaltocene with methyl- or phenyl-dichloroborane. Addition of the boron halide to the cyclopentadienyl ring, followed by insertion, leads to mono- or bis-borabenzene complexes, depending on the ratio of reactants taken:[337]

$$3Co(C_5H_5)_2 + RBX_2 \longrightarrow 2[Co(C_5H_5)_2]^+X^- + [Co(C_5H_5)(C_5H_5BR)]$$
$$5Co(C_5H_5)_2 + 2RBX_2 \longrightarrow 4[Co(C_5H_5)_2]^+ X^- + [Co(C_5H_5BR)_2]$$

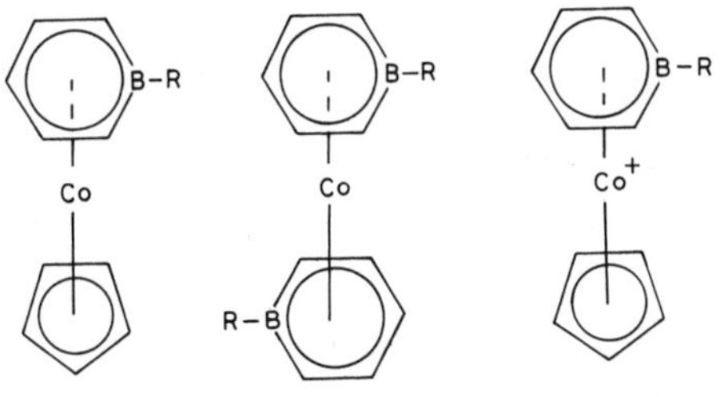

(CXI) (CXII) (CXIII)

Under slightly different reaction conditions, the cation **CXIII** can be obtained. It is probably the initial product, being subsequently reduced by cobaltocene to **CXI**. These ring insertions are related to the formation of the cyclopentadienyl cyclohexadienyl cobalt from cobaltocene and dichloro- or dibromomethane:[338]

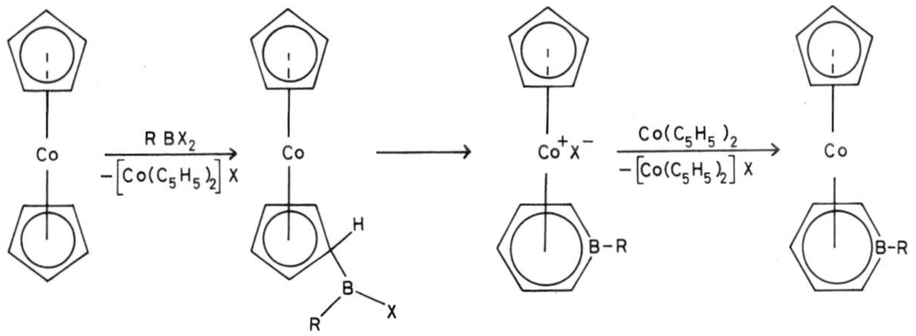

In contrast to the η^5-cyclohexadienyl ligand, however, the borabenzene group is planar, showing that the boron p_z-orbital is involved in the bonding.[339]

The complexes, like cobaltocene, are 19-electron compounds, as indicated by their magnetic moments, μ_{eff} 1.8 B.M., corresponding to one unpaired electron.[340] Cyclic voltammetry has shown that they can be reversibly oxidized to their respective cation and also reversibly reduced to the corresponding anion at the following potentials:-[341]

	Ē/V (cation ⇌ neutral)	Ē/V (neutral ⇌ anion)
$Co(C_5H_5)_2$	-0.945	-1.88 (irreversible)
CXI	-0.435	-1.46
CXII	+0.45	-1.10

Thus **CXII** is reduced by sodium amalgam to the 20-electron anion $[Co(C_6H_5BPh)_2]^-$. **CXI** is rather air-sensitive, and can readily be oxidized to **CXIII**.   During the oxidation of **CXI** or **CXII** in the presence of nucleophiles such as water or methanol, a ring contraction takes place, which is the reverse of the ring expansion which occurs under reducing conditions in the preparation of **CXII**:[342-3]

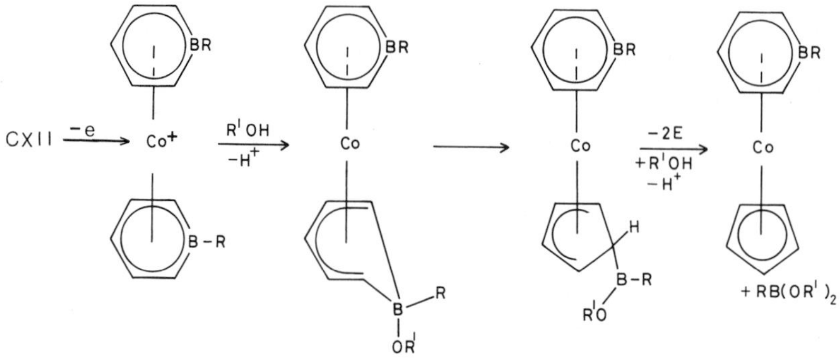

The rhodium cation $[Rh(C_5Me_5)(C_5H_5BPh)]^+$ adds hard nucleophiles such as cyanide and ammonia at the boron atoms. Hydroxide ion similarly gives a reactive adduct initially, which then decomposes by ring contraction. The corresponding iridium derivative gives a stable B-adduct with cyanide; ring contraction does not occur even in the presence of the oxidant hexacyanoferrate(III).   Hydride ($NaBH_4$ in acetonitrile) adds to both the cyclopentadienyl and the borabenzene rings when it reacts with $[Co(C_5H_5)(C_5H_5BPh)]^+$ or with $[Rh(C_5Me_5)(C_5H_5BPh)]^{+}$.[343]

Ring contraction also occurs on oxidation of $[Fe(C_5H_5)(C_5H_5BR)]$ (**CXIV**) and $[Fe(C_5H_5BR)_2]$ (**CXV**) with cerium(IV). In contrast to the cobalt complexes, the group R is transferred from boron to a carbon atom, yielding $[Fe(C_5H_5)(C_5H_4R)]^+$ as the sole product from **CXIV**, while a mixture of $[Fe(C_5H_4R)_2]^+$, $[Fe(C_5H_5)(C_5H_4R)]^+$ and $[Fe(C_5H_5)_2]^+$ results from **CXV**.[344]   Similar ring contractions are also observed in Friedel-Crafts acetylation of borabenzene complexes.[345]   At 0°C **CXV** gives 20% of the 2-acetyl derivative,[346] but when excess acetylating mixture is used $[Fe(\eta^6\text{-}C_5H_5BMe)(\eta^6\text{-}C_6H_5CH_3)]^+$ results. $[Mn(C_5H_5BMe)(CO)_3]$ gives **CXVI** selectively in nitromethane, while **CXVII** is formed as a by-product in dichloromethane.[345]

i) $CH_3COCl/AlCl_3$
ii) H_2O, NH_4PF_6

B—Me
Mn
(CO)$_3$

B—Me
COCH$_3$
Mn
(CO)$_3$

(CXVI)

+

CH$_3$
Mn$^+$.PF$_6^-$
(CO)$_3$

(CXVII)

8. MORE RECENT DEVELOPMENTS

8.1 η^5-dienyl complexes

Since the main manuscript of the Chapter was completed important work on applications of tricarbonylcyclohexadienyliron complexes in organic synthesis has been published. Pearson has reviewed the contributions from his group[347-8] and has demonstrated the potential of the iron reagents in the synthesis of, for example, steroid derivatives, trichothacene analogues and spirocyclic compounds, including some azaspirocycles which could lead to various alkaloids.[349-56] Very significant advances have also been described by Birch and his coworkers.[357-8] Optically active tricarbonylcyclohexadienyliron salts have been prepared and converted into the terpenes R(-)cryptone and S(+)-α-phellandrene. Stereospecific reactions of these salts can direct the formation of new chiral centers at carbon.[359-61] The range of nucleophiles which can usefully be employed with the iron salts has been extended to O-silylated enolates,[362-3] allyl silanes,[364] organolithium compounds (at low temperatures)[365] and the carbanion from nitromethane.[366-7] Kinetic studies have been made on the reactions of tricarbonyl(dienyl)iron salts with di- and trimethoxybenzenes,[368] pyridines,[369] triarylphosphines[370] and acetylacetone.[371] It has been claimed that trialkylphosphines add to tricarbonylcycloheptadienyliron salts in acetonitrile to give, unusually, the endo adducts.[372] The tricarbonylcyclohexadienemanganese anion has been obtained by reduction of either $[Mn(C_6H_6)(CO)_3]^+$ or $[Mn(\eta^5\text{-}C_6H_7)(CO)_3]$.[373-4]

8.2 η^6-Arene complexes

Kinetics of reactions of $[Fe(C_5H_5)(C_6H_5Cl)]^+$ with piperidine[375] and of $[Fe(C_5H_5)(arene)]^+$ with methoxide have been reported.[376-7] Bis(arene)chromium complexes, prepared by cocondensation of arenes with chromium atoms, have been studied by cyclic voltammetry. The rate of base catalyzed transesterification of methyl benzoate coordinated to the electron-withdrawing group $[Cr(C_6H_6)]^+$ is enhanced by a factor of several powers of ten compared with that for the free ester.[378]

Other papers concern the use of arenetricarbonylchromium complexes in organic synthesis,[379-80] nucleophilic attack on arene complexes of cobalt,[381] ruthenium[382] and iridium,[383] and reactions of $[Fe(C_5H_5)(arene)]^+$ salts.[384-7] Of particular interest is the preparation of optically active $[Fe(C_5H_5)(arene)]^+$ complexes.[388]

REFERENCES

1. G. Deganello, "Transition Metal Complexes of Cyclic Polyolefins", Academic Press, New York (1979).

2. G. Jaouen, Ann. N.Y. Acad. Sci. 295:59 (1977).

3. M.F. Semmelhack, Ann. N.Y. Acad. Sci. 295:36 (1977).

4. A.J. Birch and I.D. Jenkins, in "Transition Metal Organometallics in Organic Synthesis", (H. Alper, ed.), Vol. 1, Academic Press, New York (1976).

5. S.G. Davies, M.L.H. Green and D.M.P. Mingos, Tetrahedron 34:3047 (1978).

6. D.A. White, Organomet. Chem. Rev. Sect. A. 4:497 (1968).

7. E.O. Fischer and R.D. Fischer, Angew. Chem. 72:919 (1960).

8. A.J. Birch, P.E. Cross, J. Lewis, D.A. White and S.B. Wild, J. Chem. Soc. (A) 332 (1968).

9. G.A. Olah, H.C. Lin and Y.K. Mo, J. Am. Chem. Soc. 94:3667 (1972).

10. L. Busetto and R.J. Angelici, Inorg. Chim. Acta 2:391 (1968).

11. M.A. Hashmi, J.D. Munro, P.L. Pauson and J.M. Williamson, J. Chem. Soc. (A) 240 (1967).

12. B.F.G. Johnson, K.D. Karlin, J. Lewis and D.G. Parker, J. Organomet. Chem. 157:C67 (1978).

13. N.S. Nametkin, S.P. Gubin, V.I. Ivanov and V.D. Tyurin, Izv. Akad. Nauk SSSR, Ser. Chim. 486 (1974).

14. B.F.G. Johnson, J. Lewis and D.G. Parker, J. Organomet. Chem. 141:319 (1977).

15. J. Evans, D.V. Howe, B.F.G. Johnson and J. Lewis, J. Organomet. Chem. 61:C48 (1973).

16. L.A.P. Kane-Maguire, J. Chem. Soc. (A) 1602 (1971).

17. C.A. Mansfield and L.A.P. Kane-Maguire, J.C.S. Dalton Trans. 2187 (1976).

18. L.A.P. Kane-Maguire and C.A. Mansfield, J.C.S. Chem. Commun. 540 (1973).

19. L.A.P. Kane-Maguire, K.M. Al-Kathumi and C.A. Mansfield, J. Organomet. Chem. 71:C11 (1974).

20. L.A.P. Kane-Maguire and C.A. Mansfield, J.C.S. Dalton Trans. 2192 (1976).

21. G.R. John, C.A. Mansfield and L.A.P. Kane-Maguire, J.C.S. Dalton Trans. 574 (1977).

22. K.E. Hine, B.F.G. Johnson and J. Lewis, J.C.S. Chem. Commun. 81 (1975).

23. K. Hine, B.F.G. Johnson and J. Lewis, J.C.S. Dalton Trans. 1702 (1976).

24. A.L. Burrows, K. Hine, B.F.G. Johnson, J. Lewis, D.G. Parker, A. Poe and E.J.S. Vichi, J.C.S. Chem.Trans. 1135 (1980).

25. R.J.H. Cowles, B.F.G. Johnson, P.L. Josty and J. Lewis, J.C.S. Chem. Commun. 392 (1969).

26. E.G. Bryan, A.L. Burrows, B.F.G. Johnson, J. Lewis and E.M. Schiavon, J. Organomet. Chem. 129:C19 (1977).

27. B.F.G. Johnson, J. Lewis, D.G. Parker, P.R. Raithby and G.M. Sheldrick, J. Organomet. Chem. 150:115 (1978).

28. B.F.G. Johnson, J. Lewis, D.G. Parker and G.R. Stephenson, J. Organomet. Chem. 194:C14 (1980).

29. B.F.G. Johnson, J. Lewis, D.G. Parker and G.R. Stephenson, J. Organomet. Chem. 194:C14.

30. A.J. Birch, P.W. Westerman and A.J. Pearson, Aust. J. Chem. 29:1671 (1976).

31. A.J. Birch and A.J. Pearson, J.C.S. Perkin Trans. 1 954 (1976).

32. A.J. Pearson, Aust. J. Chem. 29:1101 (1976).

33. A.J. Pearson, Aust. J. Chem. 30:345 (1977).

34. A.J. Birch, K.B. Chamberlain, M.A. Haas and D.J. Thompson, J.C.S. Perkin Trans. 1 1882 (1973).

35. A.J. Pearson, J.C.S. Chem. Commun. 339 (1977).

36. A.J. Pearson, J.C.S. Perkin Trans. 1 2069 (1977).

37. A.J. Pearson, J.C.S. Perkin Trans. 1 1255 (1979).

38. A.J. Pearson, J.C.S. Perkin Trans. 1 400 (1980).

39. A.J. Pearson and P.R. Raithby, J.C.S. Perkin Trans. 1 395 (1980).

40. A.J. Pearson, J.C.S. Perkin Trans. 1 495 (1978).

41. A.J. Birch and D.H. Williamson, J.C.S. Perkin Trans. 1 1892 (1973).

42. A.J. Birch and A.J. Pearson, J.C.S. Perkin Trans. 1 638 (1978).

43. A.J. Birch and K.B. Chamberlain, Org. Syn. 57:107 (1977).

44. A.J. Birch and I.D. Jenkins, Tetrahedron Lett. 119 (1975).

45. G.R. John and L.A.P. Kane-Maguire, J.C.S. Dalton Trans. 873 (1979).

46. J.J. Guy and B.E. Reichert, Acta Cryst. B32:2504 (1976).

47. Y.S. Wong, H.N. Paik, P.C. Chieh and A.J. Carty, J.C.S. Chem. Commun. 309 (1975).

48. G.R. John and L.A.P. Kane-Maguire, J. Organomet. Chem. 120:C45 (1976).

49. A.J. Birch, I.D. Jenkins and A.J. Liepa, Tetrahedron Lett. 1723 (1975).

50. D.A. Brown, S.K. Chawla and W.K. Glass, Inorg. Chim. Acta 19:231 (1976).

51. M. Gower, G.R. John, L.A.P. Kane-Maguire, T.I. Odiaka and A. Salzer, J.C.S. Dalton Trans. 2003 (1979).

52. D.A. Brown and S.K. Chawla, Inorg. Chim. Acta 24:71 (1977).

53. L.A.P. Kane-Maguire, T.I. Odiaka, S. Turgoose and P.A. Williams, J. Organomet.. Chem. 188:C5 (1980).

54. A.J. Birch, A.J. Liepa and G.R. Stephenson, Tetrahedron Lett. 3565 (1979).

55. T.I. Odiaka and L.A.P. Kane-Maguire, Inorg. Chim. Acta 37:85 (1979).

56. F. Franke and I.D. Jenkins, Aust. J. Chem. 31:595 (1978).

57. R.E. Ireland, G.G. Brown, R.H. Stanford Jr. and T.C. McKenzie, J. Org. Chem. 39:51 (1974).

58. G.R. John, L.A.P. Kane-Maguire and C. Eaborn, J.C.S. Chem. Commun. 481 (1975).

59. A.J. Birch and K.B. Chamberlain, Org. Syn. 57:16 (1977).

60. B.R. Reddy and J.S. McKennis, J. Am. Chem. Soc. 101:7730 (1979).

61. D.W. Clack, M. Monshi and L.A.P. Kane-Maguire, J. Organomet. Chem. 107:C40 (1976).

62. H.J. Dauben, Jr. and D.J. Bertelli, J. Am. Chem. Soc. 83:497 (1961).

63. R. Burton, L. Pratt and G. Wilkinson, J. Chem. Soc. 594 (1961).

64. D.A. Brown, W.K. Glass and F.M. Hussein, J. Organomet. Chem. 186:C58 (1980).

65. D.A. Brown, S.K. Chawla and W.K. Glass, Inorg. Chim. Acta 19:L31 (1976).

66. R.J.H. Cowles, B.F.G. Johnson, J. Lewis and A.W. Parkins, J.C.S. Dalton Trans. 1768 (1972).

67. B.Y. Shu, E.R. Biehl and P.C. Reeves, Synth. Commun. 8:523 (1978).

68. R. Aumann, J. Organomet. Chem. 47:C29 (1973).

69. R. Aumann and J. Knecht, Chem. Ber. 109:174 (1976).

70. R. Edwards, J.A.S. Howell, B.F.G. Johnson and J. Lewis, J.C.S. Dalton Trans. 2105 (1974).

71. B.A. Sosinsky, S.A.R. Knox and F.G.A. Stone, J.C.S. Dalton Trans. 1633 (1975).

72. A.L. Burrows, B.F.G. Johnson, J. Lewis and D.G. Parker, J. Organomet. Chem. 194:C11 (1980).

73. A. Eisenstadt and S. Winstein, Tetrahedron Lett. 613 (1971).

74. A. Eisenstadt, J. Organomet. Chem. 113:147 (1976).

75. J.E. Mahler, D.A.K. Jones and R. Pettit, J. Am. Chem. Soc. 86:3589 (1964).

76. N. Genco, D. Marten, S. Raghu and M. Rosenblum, J. Am. Chem. Soc. 98:848 (1976).

77. B.F.G. Johnson, J. Lewis, P. McArdle and G.L.P. Randall, J.C.S. Dalton Trans. 456 (1972).

78. B.R. Reddy and J.S. McKennis, J. Organomet. Chem. 182:C61 (1979).

79. J.E. Mahler, D.H. Gibson and R. Pettit, J. Am. Chem. Soc. 85:3959 (1963).

80. J.D. Munro and P.L. Pauson, J. Chem. Soc. 3484 (1961).

81. A.J. Deeming, S.S. Ullah, A.J.P. Domingos, B.F.G. Johnson and J. Lewis, J.C.S. Dalton Trans. 2093 (1974).

82. G. Schiavon, C. Paradisi and C. Boanini, Inorg. Chim. Acta 14:L5 (1975); G. Schiavon and C. Paradisi, J. Organomet. Chem. 210:247 (1981).

83. F.A. Cotton, A.J. Deeming, P.L. Josty, S.S. Ullah and A.J.P. Domingos, J. Am. Chem. Soc. 93:4624 (1971).

84. F.A. Cotton, M.D. La Prade, B.F.G. Johnson and J. Lewis, J. Am. Chem. Soc. 93:4626 (1971).

85. J. Lewis and A.W. Parkins, J. Chem. Soc. (A) 1150 (1967).

86. L.A.P. Kane-Maguire, P.D. Mouncher and A. Salzer, J. Organomet. Chem. 168:C42 (1979).

87. A.D. Charles, P. Diversi, B.F.G. Johnson and J. Lewis, J. Organomet. Chem. 116:C25 (1976).

88. R. Aumann, J. Organomet. Chem. 78:C31 (1974).

89. A.D. Charles, P. Diversi, B.F.G. Johnson, K.D. Karlin, J. Lewis, A.V. Rivera and G.M. Sheldrick, J. Organomet. Chem. 128:C31 (1977).

90. A. Eisenstadt, J. Organomet. Chem. 60:335 (1973).

91. A. Eisenstadt, J. Organomet. Chem. 113:147 (1976).

92. J.E. Mahler and R. Pettit, J. Am. Chem. Soc. 85:3955 (1963).

93. P.E. Riley and R.E. Davis, Acta Cryst. B32:381 (1976).

94. N.A. Clinton and C.P. Lillya, J. Am. Chem. Soc. 92:3058 (1970).

95. N.A. Clinton and C.P. Lillya, J. Am. Chem. Soc. 92:3065 (1970).

96. D.E. Kuhn and C.P. Lillya, J. Am. Chem. Soc. 94:1682 (1972).

97. J.W. Burrill, B.R. Bonazza, D.W. Garrett and C.P. Lillya, J. Organomet. Chem. 104:C37 (1976).

98. T.S. Sorensen and C.R. Jablonski, J. Organomet. Chem. 25:C62 (1970).

99. C.R. Jablonski and T.S. Sorensen, Can. J. Chem. 52:2085 (1974).

100. D.G. Gresham, D.J. Kowalski and C.P. Lillya, J. Organomet. Chem. 144:71 (1978).

101. G. Maglio, A. Musco, R. Pulcino and A. Striga, J.C.S. Chem. Commun. 100 (1971).

102. G. Maglio, A. Musco and R. Palumbo, J. Organomet. Chem. 32:127 (1971).

103. G. Maglio and R. Palumbo, J. Organomet. Chem. 76:367 (1974).

104. T.G. Bonner, K.A. Holder and P. Powell, J. Organomet. Chem. 77:C37 (1974).

105. T.G. Bonner, K.A. Holder, P. Powell and E. Styles, J. Organomet. Chem. 131:105 (1977).

106. R.S. Bayoud, E.R. Biehl and P.C. Reeves, J. Organomet. Chem. 150:75 (1978).

107. P. McArdle and H. Sherlock, J.C.S. Chem. Commun. 537 (1976).

108. P. McArdle and H. Sherlock, J.C.S. Dalton Trans. 1678 (1978).

109. R.S. Bayoud, E.R. Biehl and P.C. Reeves, J. Organomet. Chem. 174:297 (1979).

110. P. McArdle and H. Sherlock, J. Organomet. Chem. 52:C29 (1973).

111. M. Anderson, A.D.H. Clague, L.P. Blaauw and P.A. Couperus, J. Organomet. Chem. 56:307 (1973).

112. R.E. Sapienza, P.E. Riley, R.E. Davis and R. Pettit, J. Organomet. Chem. 121:C35 (1976).

113. R. Hoffmann and P. Hofmann, J. Am. Chem. Soc. 98:598 (1976).

114. P.A. Dobosh, D.G. Gresham, C.P. Lillya and E.S. Magyar, Inorg. Chem. 15:2311 (1976).

115. G.A. Olah, S.H. Yu and G. Liang, J. Org. Chem. 41:2383 (1976).

116. B.F.G. Johnson, J. Lewis, D.G. Parker and S.R. Postle, J.C.S. Dalton Trans. 794 (1977).

117. P. Powell and L.J. Russell, J. Chem. Res. S283 (1978).

118. P. Powell, J. Organometal. Chem. 165:C43 (1979).

119. P. Powell, J. Organomet. Chem 206:239 (1981).

120. B.R. Bonazza, C.P. Lillya, E.S. Magyar and G. Scholes, J. Am. Chem. Soc. 101:4100 (1979).

121. W.E. Silverthorn, Advan. Organomet. Chem. 13:47 (1975).

122. P.L. Pauson, J. Organomet. Chem. 200:207 (1980).

123. P.J.C. Walker and R.J. Mawby, Inorg. Chim. Acta 7:621 (1973).

124. T.H. Coffield, V. Sandel and R.D. Closson, J. Am. Chem. Soc. 79:5826 (1957).

125. G. Winkaus, L. Pratt and G. Wilkinson, J. Chem. Soc. 3807 (1961).

126. D. Jones and G. Wilkinson, J. Chem. Soc. 2479 (1964).

127. G.A.M. Munro and P.L. Pauson, J.C.S. Chem. Commun. 134 (1976).

128. G.A.M. Munro and P.L. Pauson, Z. Anorg. Allg. Chem. 458:211 (1979).

129. M.I. Foreman, G.R. Knox, P.L. Pauson, K.H. Todd and W.E. Watts, J.C.S. Perkin Trans. 2 1141 (1972).

130. G.A.M. Munro and P.L. Pauson, Isr. J. Chem. 15:258 (1977).

131. P.J.C. Walker and R.J. Mawby, Inorg. Chem. 10:404 (1971).

132. P.J.C. Walker and R.J. Mawby, J.C.S. Dalton Trans. 622 (1973).

133. D.A. Sweigart and L.A.P. Kane-Maguire, J.C.S. Chem. Commun. 13 (1976).

134. L.A.P. Kane-Maguire and D.A Sweigart, Inorg. Chem. 18:700 (1979).

135. R.J. Angelici and L.J. Blacik, Inorg. Chem. 11:1754 (1972).

136. D.J. Darensbourg and M.Y. Darensbourg, Inorg. Chem. 9:1691 (1970).

137. F.A. Cotton and C.S. Kraihanzel, J. Am. Chem. Soc. 84:4432 (1962).

138. P.L. Pauson and J.A. Segal, J.C.S. Dalton Trans. 1683 (1975).

139. P.L. Pauson and J.A. Segal, J.C.S. Dalton Trans. 1677 (1975).

140. G. Jaouen, L. Tchissambou and R. Dabard, C.R. Hebd. Acad. Sci. Ser. C 274:654 (1972).

141. L. Tchissambou, G. Jaouen and R. Dabard, C.R. Hebd. Acad. Sci. Ser. C 274:806 (1972).

142. G.A.M. Munro and P.L. Pauson, J. Organomet. Chem. 160:177 (1978).

143. T.B. Brill and A.J. Kotlar, Inorg. Chem. 13:470 (1974).

144. B. Nicholls and M.C. Whiting, J. Chem. Soc. 551 (1959).

145. M.F. Semmelhack and H.T. Hall, Jr., J. Am. Chem. Soc. 96:7091 (1974).

146. M.F. Semmelhack and H.T. Hall, Jr., J. Am. Chem. Soc. 96:7092 (1974).

147. J.F. Bunnett and H. Herrmann, J. Org. Chem. 36:4081 (1971).

148. M.F. Semmelhack, H.T. Hall, Jr., M. Yoshifuji and G. Clark, J. Am. Chem. Soc. 97:1247 (1975).

149. M.F. Semmelhack, H.T. Hall, Jr., and M. Yoshifuji, J. Am. Chem. Soc. 98:6387 (1976).

150. M.F. Semmelhack, H.T. Hall, Jr., F. Farina, M. Yoshifuji, G. Clark, T. Barger, K. Hirotsu and J. Clardy, J. Am. Chem. Soc. 101:3535 (1979).

151. M.F. Semmelhack in "New Applications of Organometallic Reagents in Synthesis" (D. Seyferth, ed.), American Elsevier, New York (1976), p. 36.

152. M.F. Semmelhack, L. Keller and Y. Thebtaranonth, J. Am. Chem. Soc. 99:959 (1977).

153. C.A.L. Mahaffy and P.L. Pauson, Inorg. Syn. 19:154 (1979).

154. S. Top and G. Jaouen, J. Organomet. Chem. 182:381 (1979).

155. H.J. Beck, E.O. Fischer and C.G. Kreiter, J. Organomet. Chem. 26:C41 (1971).

156. E.O. Fischer, P. Stuckler, H-J. Beck and F.R. Kreissl, Chem. Ber. 109:3089 (1976).

157. L. Knoll, K. Reiss, J. Schafer and P. Klufers, J. Organomet. Chem. 193:C40 (1980).

158. F. Calderazzo, Inorg. Chem. 5:429 (1966).

159. A.N. Nesmeyanov, N.A. Vol'kenau and L.S. Shilotseva, Dokl. Akad. Nauk. SSSR 160:1327 (1965).

160. A.N. Nesmeyanov, N.A. Vol'kenau and I.N. Bolesova, Dokl. Akad. Nauk. SSSR 149:615 (1963).

161. A.N. Nesmeyanov, N.A. Vol'kenau and I.N. Bolesova, Tetrahedron Lett. 25:1725 (1963).

162. D. Astruc and R. Dabard, Bull. Soc. Chim. Fr. 2571 (1975).

163. D. Astruc and R. Dabard, Tetrahedron 32:245 (1976).

164. D. Astruc and R. Dabard, J. Organomet. Chem. 111:339 (1976).

165. D. Astruc and R. Dabard, C.R. Hebd. Acad. Sci. Ser. C 273:1248 (1971).

166. I.U. Khand, P.L. Pauson and W.E. Watts, J. Chem. Soc. (C) 2257 (1968).

167. I.U. Khand, P.L. Pauson and W.E. Watts, J. Chem. Soc. (C) 2261 (1968).

168. I.U. Khand, P.L. Pauson and W.E. Watts, J. Chem. Soc. (C) 116 (1969).

169. I.U. Khand, P.L. Pauson and W.E. Watts, J. Chem. Soc. (C) 2024 (1969).

170. M.M. Khan and W.E. Watts, J. Organomet. Chem. 108:C11 (1976).

171. J.F. McGreer and W.E. Watts, J. Organomet. Chem. 110:103 (1976).

172. M.M. Khan and W.E. Watts, J.C.S. Perkin Trans. 1 1879 (1979).

173. M.F. Semmelhack and G. Clark, J. Am. Chem. Soc. 99:1675 (1977).

174. M.F. Semmelhack, G.R. Clark, R. Farina and M. Saeman, J. Am. Chem. Soc. 101:217 (1979).

175. D.W. Clack and L.A.P. Kane-Maguire, J. Organomet. Chem. 174:199 (1979).

176. A.N. Nesmeyanov, N.A. Vol'kenau, L.S. Shilovtseva and V.A. Petrakova, Izv. Akad. Nauk. SSSR, Ser. Khim. 1151 (1975).

177. A.N. Nesmeyanov, N.A. Vol'kenau, L.S. Shilovtseva and V.A. Petrakova, J. Organomet. Chem. 85:365 (1975).

178. A.N. Nesmeyanov, N.A. Vol'kenau and I. Bolesova, Dokl. Akad. Nauk. SSSR 175:606 (1967).

179. J.F. Helling and W.A. Hendrickson, J. Organomet. Chem. 168:87 (1979).

180. J.F. Helling and W.A. Hendrickson, J. Organomet. Chem. 141:99 (1977).

181. J.F. Helling and D.M. Braitsch, J. Am. Chem. Soc. 92:7207 (1970).

182. J.F. Helling and D.M. Braitsch, J. Am. Chem. Soc. 92:7209 (1970).

183. J.F. Helling, S.L. Rice and D.M. Braitsch, J.C.S. Chem. Commun. 930 (1971).

184. J.F. Helling and G.G. Cash, J. Organomet. Chem. 73:C10 (1974).

185. S.G. Davis, L.S. Gelfand and D.A. Sweigart, J.C.S. Chem. Commun. 762 (1979).

186. P.J. Domaille, S.D. Ittel, J.P. Jesson and D.A. Sweigart, J. Organomet. Chem. 202:191 (1980).

187. M.A. Bennett and T.W. Matheson, J. Organomet. Chem. 175:87 (1979).

188. C. White and P.M. Maitlis, J. Chem. Soc. (A) 3322 (1971).

189. N.A. Bailey, E.H. Blunt, G. Fairhurst and C. White, J.C.S. Dalton Trans. 829 (1980).

190. R.P. Houghton, M. Voyle and R. Price, J.C.S. Chem. Commun. 884 (1980).

191. G. Fairhurst and C. White, J. Organomet. Chem. 166:C23 (1979).

192. E.O. Fischer and M.W. Schmidt, Chem. Ber. 100:3782 (1967).

193. E.O. Fischer and M.W. Schmidt, Chem. Ber. 99:2206 (1966).

194. D. Jones, L. Pratt and G. Wilkinson, J. Chem. Soc. 4458 (1962).

195. G. Winkaus and H. Singer, Z. Naturforsch. Teil B 418 (1963).

196. A. Efraty and P.M. Maitlis, J. Am. Chem. Soc. 89:3744 (1967).

197. R.A. Zelonka and M.C. Baird, J. Organomet. Chem. 35:C43 (1972).

198. M.A. Bennett and A.K. Smith, J.C.S. Dalton Trans. 233 (1974).

199. D.R. Robertson and T.A. Stephenson, J. Organomet. Chem. 116:C29 (1976).

200. R.O. Gould, C.L. Jones, D.R. Robertson and T.A. Stephenson, J.C.S. Chem. Commun. 222 (1977).

201. D.R. Robertson and T.A. Stephenson, J. Organomet. Chem. 157:C47 (1978).

202. D.R. Robertson, T.A. Stephenson and T. Arthur, J. Organomet. Chem. 162:121 (1978).

203. T. Arthur and T.A. Stephenson, J. Organomet. Chem. 168:C39 (1979).

204. R.A. Zelonka and M.C. Baird, Can. J. Chem. 50:3063 (1972).

205. D.R. Robertson and T.A. Stephenson, J. Organomet. Chem. 202:309 (1980).

206. J.D. Munro and P.L. Pauson, J. Chem. Soc. 3475 (1961).

207. D.J. Bartelli, Diss. Abs. 22:3850 (1962) cited by R.B. King and M.B. Bisnette, Inorg. Chem. 3:785 (1964).

208. A. Salzer, Z. Anorg. Allg. Chem. 448:88 (1975).

209. P.L. Pauson, G.H. Smith and J.H. Valentine, J. Chem. Soc. (C) 1057 (1967).

210. J.D. Munro and P.L. Pauson, J. Chem. Soc. 3479 (1961).

211. P.L. Pauson and K.H. Todd, J. Chem. Soc. (C) 2315 (1970).

212. J.W. Faller, Inorg. Chem. 19:2857 (1980).

213. A. Salzer, Inorg. Chim. Acta 17:221 (1976).

214. D.A. Sweigart, M. Gower and L.A.P. Kane-Maguire, J. Organomet. Chem. 108:C15 (1976).

215. G.R. John, L.A.P. Kane-Maguire and D.A. Sweigart, J. Organomet. Chem. 120:C47 (1976); J.G. Atton, L.A. Hassan and L.A.P. Kane-Maguire, Inorg. Chim. Acta 41:245 (1980).

216. P. Hackett and G. Jaouen, Inorg. Chim Acta 12:L19 (1975).

217. E.E. Isaacs and W.A.G. Graham, J. Organomet. Chem. 90:319 (1975).

218. E.E. Isaacs and W.A.G. Graham, J. Organomet. Chem. 120:407 (1976).

219. K.M. Al-Kathumi and L.A.P. Kane-Maguire, J. Organomet. Chem. 102:C4 (1975).

220. M. Bochmann, M. Cooke, M. Green, H.P. Kirsch, F.G.A. Stone and A.J. Welch, J.C.S. Chem. Commun. 381 (1976).

221. M. Green, H.P. Kirsch, F.G.A. Stone and A.J. Welch, J.C.S. Dalton Trans. 1755 (1977).

222. M. Green, H.P. Kirsch, F.G.A. Stone and A.J. Welch, Inorg. Chim. Acta 29:101 (1978).

223. F. Zahedi and M.L. Ziegler, Z. Naturforsch. Teil B 34:918 (1979).

224. M. Sommer, K. Weidenhammer, H. Wienand and M.L. Ziegler, Z. Naturforsch. Teil B 33:361 (1978).

225. R.B. King and A. Fronzaglia, Inorg. Chem. 5:1837 (1966).

226. G. Hoch and R. Panter, Z. Naturforsch. Teil B 31:294 (1976).

227. P. Powell, L.J. Russell, E. Styles, A.J. Brown, O.W. Howarth and P. Moore, J. Organomet. Chem. 149:C1 (1978).

228. P. Powell, M. Stephens and K.H. Yassin, unpublished work.

229. K.M. Al-Kathumi and L.A.P. Kane-Maguire, J.C.S. Dalton Trans. 1683 (1973).

230. M.L.H. Green and R.B.A. Pardy, J. Organomet. Chem. 117:C13 (1976).

231. J. Muller and S. Schmitt, J. Organomet. Chem. 97:275 (1975).

232. A. Davison, W. McFarlane, L. Pratt and G. Wilkinson, J. Chem. Soc. 3653 (1962).

233. C.P. Lillya and R.A. Sahatjian, Inorg. Chem. 11:889 (1972).

234. G.A. Olah and S.H. Yu, J. Org. Chem. 41:717 (1976).

235. D.N. Kursanov, V.N. Settina, P.V. Petrovskii, V.I. Zdanovich and N.K. Baranetskaya, J. Organomet. Chem. 37:339 (1972).

236. F.G.N. Cloke, M.L.H. Green and G.E. Morris, J.C.S. Chem. Commun. 72 (1978).

237. M.I. Lobach and V.A. Korner, Russ. Chem. Rev. 48:758 (1979).

238. A. Salzer and H. Werner, J. Organomet. Chem. 87:101 (1975).

239. S. Winstein, H.D. Kaesz, C.G. Kreiter and E.C. Friedrich, J. Am. Chem. Soc. 87:3267 (1965).

240. S. Winstein, C.G. Kreiter and J.J. Brauman, J. Am. Chem. Soc. 88:2047 (1966).

241. J. Muller and H. Menig, J. Organomet. Chem. 96:83 (1975).

242. M. Brookhart and D.L. Harris, Inorg. Chem. 13:1541 (1974).

243. G. Fairhurst and C. White, J.C.S. Dalton Trans. 1531 (1979).

244. C. White, C.J. Thompson and P.M. Maitlis, J. Organomet. Chem. 127:415 (1977).

245. P.M. Treichel and J.W. Johnson, J. Organomet. Chem. 88:207 (1975).

246. P.M. Treichel and J.W. Johnson, J. Organomet. Chem. 88:215 (1975).

247. P.M. Treichel and J.W. Johnson, J. Organomet. Chem. 88:227 (1975).

248. C. White, C.J. Thompson and P.M. Maitlis, J.C.S. Dalton Trans. 1654 (1977).

249. J.W. Johnson and P.M. Treichel, J. Am. Chem. Soc. 90:1427 (1977).

250. P.M. Treichel and J.W. Johnson, Inorg. Chem. 16:749 (1977).

251. C.C. Lee, B.R. Steele, K.J. Demchuk and R.G. Sutherland, Can. J. Chem. 57:946 (1979).

252. R.G. Sutherland, B.R. Steele, K.J. Demchuk and C.C. Lee, J. Organomet. Chem. 181:411 (1979).

253. A.N. Nesmeyanov, N.A. Ustynyuk, L.G. Makarova, S. Andre, Yu. A. Ustynyuk, L.N. Novikova and Yu. N. Luzikov, J. Organomet. Chem. 154:45 (1978).

254. A.N. Nesmeyanov, N.A. Ustynyuk, L.N. Novikeva, T.N. Rybina, Yu. A. Ustynyuk, Yu. F. Oprunenko and O.I. Trifonova, J. Organomet. Chem. 184:63 (1980); A. Ceccon, A. Gambara, G. Agostini and A. Venzo, J. Organomet. Chem. 217:79 (1981).

255. H.J. Dauben, Jr. and L.R. Honnen, J. Am. Chem. Soc. 80:5570 (1958).

256. J.A. Connor and E.J. Rasburn, J. Organomet. Chem. 24:441 (1970).

257. J.A.S. Howell, B.F.G. Johnson and J. Lewis, J.C.S. Dalton Trans. 293 (1974).

258. J.C. Burt, S.A.R. Knox, R.J. McKinney and F.G.A. Stone, J.C.S. Dalton Trans. 1 (1977).

259. P.L. Pauson and J.A. Segal, J. Organomet. Chem. 63:C13 (1973).

260. P.L. Pauson and J.A.Segal, J.C.S. Dalton Trans. 2387 (1975).

261. J. Müller and B. Mertschenk, Chem. Ber. 105:3346 (1972).

262. R. Aumann and S. Winstein, Tetrahedron Lett. 903 (1970).

263. A. Salzer, Inorg. Chim. Acta 18:L31 (1976).

264. G.E. Herberich and E.O. Fischer, Chem. Ber. 95:2803 (1962).

265. D.R. Brown and J.R. Raju, J. Chem. Soc. (A) 40 (1966).

266. W.R. Jackson and W.B. Jennings, J. Chem. Soc. (B) 1221 (1969).

267. E.O. Fischer and S. Breitschaft, Chem. Ber. 99:2213 (1966).

268. G. Simonneaux, G. Jaouen, R. Dabard and M. Louër, Nouveau J. Chim. 2:203 (1978); G. Simmoneaux, G. Jaouen and R. Dabard, Tetrahedron 36:893 (1980).

269. M. Louër, G. Simonneaux and G. Jaouen, J. Organomet. Chem. 16:235 (1979).

270. E.O. Fischer and H. Brunner, Chem. Ber. 95:1999 (1962).

271. E.O. Fischer and H. Brunner, Chem. Ber. 98:175 (1965).

272. C. Elsenbroich, J. Organomet. Chem. 14:157 (1968).

273. C. Elsenbroich, J. Organomet. Chem. 12:677 (1970).

274. C. Elsenbroich and F. Stohler, J. Organomet. Chem. 67:C51 (1974).

275. A.N. Nesmeyanov, L.P. Yur'eva and S.N. Levchenko, Dokl. Akad. Nauk. SSSR 190:118 (1970).

276. M.J. McGlinchey and T.S. Tan, J. Am. Chem. Soc. 98:2271 (1976).

277. A. Agarwal, M.J. McGlinchey and T.S. Tan, J. Organomet. Chem. 141:85 (1977).

278. N. Hao and M.J. McGlinchey, J. Organomet. Chem. 165:225 (1979).

279. C. Elsenbroich, F. Gerson and J. Heinzer, Z. Naturforsch. Teil B 27:312 (1972).

280. M.L.H. Green, J. Organomet. Chem. 200:119 (1980).

281. C.J. Groenenboom, H.J. de Liefde Meijer and F. Jellinek, Recl. Trav. Chim. Pays-Bas 93:6 (1974).

282. C.J. Groenenboom, H.J. de Liefde Meijer and F. Jellinek, J. Organomet. Chem. 69:235 (1974).

283. H. Verkouw, M.E.E. Veldman, C.J. Groenenboom, H.O. van Oven and H.J. de Liefde Meijer, J. Organomet. Chem. 102:49 (1975).

284. C.J. Groenenboom and F. Jellinek, J. Organomet. Chem. 80:229 (1974).

285. C.J. Groenenboom, G. Sawatsky, H.J. de Liefde Meijer and F. Jellinek, J. Organomet. Chem. 76:C4 (1974).

286. J.D. Zeinstra and W.C. Nieuwpoort, Inorg. Chim. Acta 30:103 (1978).

287. M.E.E. Veldman and H.O. van Oven, J. Organomet. Chem. 84:247 (1975).

288. M. Vliek, C.J. Groenenboom, H.J. de Liefde Meijer and F. Jellinek, J. Organomet. Chem. 97:67 (1975).

289. A.N. Nesmeyanov, N.E. Kolobova, K.N. Anisimov and Yu U. Makarov, Izv. Akad. Nauk. SSSR, Ser. Khim. 2665 (1968).

290. M.F. Semmelhack, J. Bisaha and M. Czarny, J. Am. Chem. Soc. 101:768 (1979).

291. M.D. Rausch, G.A. Moser and W.A. Lee, Synth. React. Inorg. Met. Org. Chem. 9:357 (1979).

292. F. van Meurs, J. Organomet. Chem. 142:299 (1977).

293. G. Jaouen and R. Dabard, J. Organomet. Chem. 61:C36 (1973).

294. E.O. Fischer, K. Ofele, H. Essler, W. Fröhlich, J.P. Mottensen and W. Semmlinger, Chem. Ber. 91:2763 (1958).

295. W.S. Trahanovsky and R.J. Card, J. Am. Chem. Soc. 94:2897 (1972).

296. R.J. Card and W.S. Trahanovsky, Tetrahedron Lett. 3823 (1973).

297. G. Jaouen, A. Meyer and G. Simonneaux, J.C.S. Chem. Commun. 813 (1975).

298. G. Jaouen, S. Top and M.J. McGlinchey, J. Organomet. Chem. 195:C5 (1980).

299. L. Haynes and R. Pettit in "Carbonium Ions", (G.A. Olah and P.V.R. Schleyer, eds.), Vol. 5, Wiley, New York (1975).

300. L.M. Loim, L.A. Matutschenko, Z.N. Parnes and D.N. Kursanov, J. Organomet. Chem. 108:363 (1976).

301. R.E. Davis, H.D. Simpson, N. Grice and R. Pettit, J. Am. Chem. Soc. 93:6688 (1971).

302. W.S. Trahanovsky and D.K. Wells, J. Am. Chem. Soc. 91:5870 (1969).

303. A. Ceccon, A. Gobbo and A. Venzo, J. Organomet. Chem. 162:211 (1978).

304. D.K. Wells and W.S. Trahanovsky, J. Am. Chem. Soc. 91:5871 (1969).

305. D. Seyferth and C.S. Eschbach, J. Organomet. Chem. 94:C5 (1975).

306. D. Seyferth, J.S. Merola and C.S. Eschbach, J. Am. Chem. Soc. 100:4124 (1978).

307. G.A. Olah and S.H. Yu, J. Org. Chem. 41:1694 (1976).

308. M. Acampora, A. Ceccon, M. Dal Farra, G. Giacometti and G. Rigatti, J.C.S. Perkin Trans. 2 483 (1977).

309. J.D. Holmes, D.A.K. Jones and R. Pettit, J. Organomet. Chem. 4:324 (1965).

310. G. Klopman and F. Calderazzo, Inorg. Chem. 6:977 (1967).

311. V.S. Khandkarova, S.P. Gubin and B.A. Kvasov, J. Organomet. Chem. 23:509 (1970).

312. S.P. Gubin, V.S. Khandkarova and A.Z. Kreindlein, J. Organomet. Chem. 64:229 (1974).

313. R.S. Bly and R.L. Veazey, J. Am. Chem. Soc. 91:4221 (1969).

314. R.S. Bly, R.C. Strickland, R.T. Swindell and R.L. Veazey, J. Am. Chem. Soc. 92:3722 (1970).

315. R.S. Bly and R.C. Strickland, J. Am. Chem. Soc. 92:7495 (1970).

316. R.S. Bly, R.A. Mateer, K.K. Tse and R.L. Veazey, J. Org. Chem. 38:1518 (1973).

317. R.S. Bly, K.K. Tse and R.K. Bly, J. Organomet. Chem. 117:35 (1976).

318. A. Ceccon and G.S. Biserni, J. Organomet. Chem. 39:313 (1972).

319. A. Ceccon and S. Sartori, J. Organomet. Chem. 50:161 (1973).

320. L.J. Krimen and D.J. Cota, Org. Reactions 17:213 (1969).

321. S. Top and G. Jaouen, J.C.S. Chem. Commun. 224 (1979); idem, J. Org. Chem. 46:78 (1981).

322. S. Top and G. Jaouen, J. Organomet. Chem. 197:199 (1980).

323. A. Meyer, Ann. Chim. (Paris) 8:397 (1973).

324. H. Werner, E.O. Fischer and J. Muller, Chem. Ber. 103:2258 (1970).

325. J. Knol, A. Westerhof, H.D. van Ofen and H.J. de Liefde Meijer, J. Organomet. Chem. 96:257 (1975).

326. N. El Murr, Y. Dusausoy, J.E. Sheats and M. Agnew, J.C.S. Dalton Trans. 901 (1979).

327. A.N. Nesmeyanov, N.A. Vol'kenau, L.S. Shilovtseva and V.A. Petrakova, J. Organomet. Chem. 61:329 (1973).

328. A.N. Nesmeyanov, N.A. Vol'kenau and V.A. Petrakova, Izv. Akad. Nauk. SSSR, Ser. Khim. 2159 (1974).

329. A.N. Nesmeyanov, N.A. Vol'kenau, V.A. Petrakova, L.S. Kotova and L.I. Denisovich, Dokl. Akad. Nauk. SSSR 217:104 (1974).

330. A.N. Nesmeyanov, N.A. Vol'kenau and V.A. Petrakova, J. Organomet. Chem. 136:363 (1977).

331. A.N. Nesmeyanov, S.P. Solodovnikov, N.A. Vol'kenau, L.S. Kotova and N.A. Sinitsyna, J. Organomet. Chem. 148:C5 (1978).

332. S.P. Solodovnikov, A.N. Nesmeyanov, N.A. Vol'kenau and L.S. Kotova, J. Organomet. Chem. 201:C45 (1980).

333. D. Astruc and R. Dabard, Bull. Soc. Chim. Fr. 228 (1976).

334. C. Moinet, E. Roman and D. Astruc, J. Organomet. Chem. 128:C45 (1977).

335. D. Astruc, J.R. Hamon, G. Althoff, E. Roman, P. Batail, P. Michaud, J.P. Mariot, F. Varret and D. Cozak, J. Am. Chem. Soc. 101:5445 (1979).

336. W. Siebert, Advan. Organomet. Chem. 18:301 (1980).

337. G.E. Herberich and G. Greiss, Chem. Ber. 105:3413 (1972).

338. G.E. Herberich and J. Schwarzer, Chem. Ber. 103:2016 (1970).

339. G. Huttner, B. Krieg and W. Gurtzke, Chem. Ber. 105:3424 (1972).

340. G.E. Herberich, T. Lund and J.B. Raynor, J.C.S. Dalton Trans. 985 (1975).

341. U. Kölle, J. Organomet. Chem. 152:225 (1978).

342. G.E. Herberich and W. Pahlmann, J. Organomet. Chem. 97:C51 (1975).

343. G.E. Herberich, E. Cordula and W. Pahlmann, Chem. Ber. 112:607 (1979).

344. G.E. Herberich and K. Carsten, J. Organomet. Chem. 144:C1 (1978).

345. G.E. Herberich, B. Hessner and T.T. Kho, J. Organomet. Chem. 197:1 (1980).

346. A.J. Ashe, E. Meyers, P. Shu, T. von Lehmann and J. Bustide, J. Am. Chem. Soc. 97:6865 (1975).

347. A.J. Pearson, Acc. Chem. Res. 13:463 (1980).

348. A.J. Pearson, Transition Met. Chem. (Weinheim) 6:67 (1981).

349. A.J. Pearson and G.C. Heywood, Tetrahedron Lett. 22:1645 (1981); 22:2929 (1981).

350. A.J. Pearson and C.W. Ong, J. Am. Chem. Soc. 103:6686 (1981).

351. A.J. Pearson, P. Ham and D.C. Rees, Tetrahedron Lett. 21:4637 (1980).

352. A.J. Pearson and D.C. Rees, Tetrahedron Lett. 21:3937 (1980); 22:4033 (1981).

353. A.J. Pearson and C.W. Ong, J.C.S. Perkin Trans. 1 1614 (1981).

354. A.J. Pearson and M. Chandler, J.C.S. Perkin Trans. 1 2238 (1980).

355. A.J. Pearson, E. Mincione, M. Chandler and P.R. Raithby, J.C.S. Perkin Trans. 1 2774 (1980).

356. E. Mincione, A.J. Pearson, P. Bovicelli, M. Chandler and G.C. Heywood, Tetrahedron Lett. 22:2929 (1981).

357. A.J. Birch, Ann. N.Y. Acad. Sci. 333:107 (1980).

358. A.J. Birch, B.M.R. Bandara, K. Chamberlain, B. Chauncy, P. Dahler, A.I. Day, I.D. Jenkins, L.F. Kelly, T-C. Khor, G. Kretschmer, A.J. Liepa, A.S. Narula, W.D. Raverty, E. Rizzardo, C. Sell, G.R. Stephenson, D.J. Thompson and D.H. Williamson, Tetrahedron Suppl. 289 (1981).

359. A.J. Birch, W.D. Raverty and G.R. Stephenson, Tetrahedron Lett. 21:197 (1980); J.C.S. Chem. Commun. 857 (1980); J. Org. Chem. 46:5166 (1981).

360. A.J. Birch and G.R. Stephenson, Tetrahedron Lett. 22:779 (1981).

361. A.J. Birch and B.M.R. Bandara, Tetrahedron Lett. 21:2981 (1980).

362. L.F. Kelly, A.S. Narula and A.J. Birch, Tetrahedron Lett. 21:2455 (1980).

363. F. Effenberger and M. Keil, Tetrahedron Lett. 22:2151 (1981).

364. L.F. Kelly, P. Dahler, A.S. Narula and A.J. Birch, Tetrahedron Lett. 22:1433 (1981).

365. B.M.R. Bandara, A.J. Birch and T-C. Khor, Tetrahedron Lett. 21:3625 (1980).

366. B.F.G. Johnson, J. Lewis, D.G. Parker and G.R. Stephenson, J. Organomet. Chem. 204:221 (1981).

367. A.J. Pearson and M. Chandler, J. Organomet. Chem. 202:175 (1980).

368. G.R. John and L.A.P. Kane-Maguire, Inorg. Chim. Acta 48:179 (1981).

369. T.I. Odiaka and L.A.P. Kane-Maguire, J.C.S. Dalton Trans. 1162 (1981).

370. J.G. Atton and L.A.P. Kane-Maguire, J. Organomet. Chem. 226:C43 (1982).

371. A.J. Birch, D. Bogsanyi and L.F. Kelly, J. Organomet. Chem. 214:C39 (1981).

372. D.A. Brown, W.K. Glass and F.M. Hussein, J. Organomet. Chem. 218:C15 (1981).

373. W. Lamanna and M. Brookhart, J. Am. Chem. Soc. 103:989 (1981).

374. P. Bladon, G.A.M. Munro, P.L. Pauson and C.A.L. Mahaffy, J. Organomet. Chem. 221:79 (1981).

375. V.V. Litvak, L.S. Filatova, N.U. Khalikova and V.D. Shteingarts, Zh. Org. Khim. 16:336 (1980).

376. V.V. Litvak, L.S. Filatova, N.U. Khalikova and V.D. Shteingarts, Zh. Org. Khim. 16:342 (1980).

377. V.V. Litvak, P.O. Kun and V.D. Shteingarts, Zh. Org. Khim. 16:1009 (1980).

378. H. Brunner and H. Koch, Chem. Ber. 115:65 (1982).

379. B. Caro and G. Jaouen, J. Organomet. Chem. 220:309 (1981).

380. J.C. Boutonnet, L. Mordenti, E. Rose, O.C. Martret and G. Precigoux, J. Organomet. Chem. 221:147 (1981).

381. Y-H. Lai, W. Tam and K.P.C. Vollhardt, J. Organomet. Chem. 216:97 (1981).

382. R. Werner and H. Werner, J. Organomet. Chem. 210:C11 (1981).

383. A.C. Sievert and E.L. Muetterties, Inorg. Chem. 20:2276 (1981).

384. C.C. Lee, U.S. Gill and R.G. Sutherland, J. Organomet. Chem. 206:89 (1981).

385. C.C. Lee, C.I. Azogu, P.C. Chang and R.G. Sutherland, J. Organomet. Chem. 220:181 (1981).

386. B.R. Steele, R.G. Sutherland and C.C. Lee, J.C.S. Dalton Trans. 529 (1981).

387. E. Roman, D. Astruc and A. Darchen, J. Organomet. Chem. 219:221 (1981).

388. E. Roman, D. Astruc and H. des Abbayes, J. Organomet. Chem. 219:211 (1981).

ABBREVIATIONS

[Some specialized abbreviations are defined in the relevant Chapters and not here.]

acac	acetylacetonate (2,4-pentanedionate)
ACACEN	see Chapter 2
ad	1-adamantyl
adme	1-adamantylmethyl
aetpor	aetioporphyrin (see Chapter 2)
aq	aqueous; water
Ar	aryl
BAE	= ACACEN
Bc	benchrocenyl, $(OC)_3Cr(\eta^6\text{-}C_5H_5\text{-} ...)$
bipy	2,2'-bipyridine
Bz	benzyl [not, in this book, benzoyl]
chel	chelating ligand or set of ligands occupying four equatorial positions
chgH$_2$	cyclohexane-1,2-dione dioxime
cod	cycloocta-1,4-diene
cot	cyclooctatetraene
Cp	$(\eta^5\text{-}C_5H_5)$
Cp*	$(\eta^5\text{-}C_5Me_5)$
Cy	cyclohexyl
cymantrenyl	$(OC)_3Mn(\eta^5\text{-}C_5H_4\text{-} ...)$
DCCD	dicyclohexylcarbodiimide, CyN=C=NCy

depe	diethylphosphinoethane, $Et_2PCH_2CH_2PEt_2$
DibAH	$^iBu_2AlH$
dien	diethylenetriamine, $H_2NCH_2CH_2NHCH_2CH_2NH_2$
diox	(1,4)-dioxane
diphos	= dppe
DME	1,2-dimethoxyethane
DMF	N,N-dimethylformamide
$dmgH_2$	dimethylglyoxime
dmpe	$Me_2PCH_2CH_2PMe_2$
dpa	di(2-pyridylmethyl)amine, $HN(CH_2-2-C_5H_4N)_2$
dpe	= dppe
dpm, dppm	$Ph_2PCH_2PPh_2$
dppe	$Ph_2PCH_2CH_2PPh_2$
en	$H_2NCH_2CH_2NH_2$
Fp	$(\eta^5-C_5H_5)(OC)_2Fe$
Fp'	$(\eta^5-C_5H_5)(OC)(Ph_3P)Fe$
HMPA	hexamethylphosphoramide
M	molar, mol dm^{-3}
Mecp	$(\eta^5-C_5H_4Me)$
Mes	mesityl, 2,4,6-trimethylphenyl
NORPHOS	5-norbornenebis(diphenylphosphine); (-) is 1S(2-endo,3-exo) isomer
Np	neopentyl
Nph	neophyl
Octetpor	octaethylporphyrin
Pc	phthalocyanine
P-C	$Ph_2P-C_6H_4-\underline{o}-$
pdma	$\underline{o}$-phenylenebis(dimethylarsine)
phen	9,10-phenanthroline

PPN$^+$	$Ph_3P = N^+ = PPh_3$
py	pyridine
(8-)quin	quinolin(-8-)olate
Red-Al	$Na[(CH_3OCH_2CH_2O)_2AlH_2]$
R_2P-C	R_2PCH_2-C_6H_4-$\underline{o}$-
TBP	trigonal bipyramid(al)
TCNE	tetracyanoethylene
terpy	2,2'; 6',2''-terpyridyl
THF or thf	tetrahydrofuran
TMEDA	$Me_2NCH_2CH_2NMe_2$
tmen	= TMEDA
TMS	tetramethylsilane
tn	trimethylenediamine, $H_2N(CH_2)_3NH_2$
tol	tolyl
TPP	tetraphenylporphin
TRANSPHOS	2,11-bis(diphenylphosphinomethyl)benzo[c]phenanthrene
tren	$N(CH_2CH_2NH_2)_3$
triphos	$PhP(CH_2CH_2PPh_2)$
Ts	tosyl
ttp	$PhP(CH_2CH_2CH_2PPh_2)_2$

INDEX